D0164446

FOUNDATIONS OF ELECTRONICS

Dedication

Again, this work is dedicated to my wonderful wife, Elizabeth (Betty) Meade, for her patient love, support, and encouragement during the effort required to produce yet another edition. Additionally, I once again acknowledge the blessing of being able to dedicate this work, as well, to my four children, David, Stephen, Rebecca, and Timothy.

5th Edition

FOUNDATIONS OF ELECTRONICS

Russell L. Meade

Australia • Brazil • Japan • Korea • Mexico • Singapore • Spain • United Kingdom • United States

Foundations of Electronics, Fifth Edition
Russell L. Meade

Vice President, Technology & Trades ABU: David Garza

Director of Learning Solutions: Sandy Clark

Senior Acquisitions Editor: Stephen Helba

Senior Product Manager: Michelle Ruelos Cannistraci

Marketing Director: Deborah Yarnell

Senior Channel Manager: Dennis Williams

Marketing Coordinator: Stacey Wiktorek

Production Director: Mary Ellen Black

Senior Production Manager: Larry Main

Production Coordinator: Benjamin Gleeksman

Senior Project Editor: Christopher Chien

Art/Design Coordinator: Francis Hogan

Technology Project Manager: Kevin Smith

Technology Project Specialist: Linda Verde

Senior Editorial Assistant: Dawn Daugherty

For product information and technology assistance, contact us at
Cengage Learning Customer & Sales Support, 1-800-354-9706
For permission to use material from this text or product,
submit all requests online at **www.cengage.com/permissions**
Further permissions questions can be emailed to
permissionrequest@cengage.com

Library of Congress Control Number: 2005046633

ISBN-13: 978-1-4180-0538-2

ISBN-10: 1-4180-0538-X

Delmar
Executive Woods
5 Maxwell Drive
Clifton Park, NY 12065
USA

Cengage Learning is a leading provider of customized learning solutions with office locations around the globe, including Singapore, the United Kingdom, Australia, Mexico, Brazil, and Japan. Locate your local office at **international.cengage.com/region**

Cengage Learning products are represented in Canada by Nelson Education, Ltd.

For your lifelong learning solutions, visit **www.cengage.com/delmar**

Visit our corporate website at **www.cengage.com**

Printed in China by China Translation & Printing Services Limited
3 4 5 6 7 8 9 14 13 12 11

Contents

PART II Basic Circuit Analysis

PART III Producing and Measuring Electrical Quantities

PART IV Reactive Components

Preface

Aim of the Book

The purpose of *Foundations of Electronics, 5e* is to provide comprehensive training in the fundamentals of electricity and electronics in a technically sound, "student-friendly," and easy-to-understand style. A solid foundation in the fundamentals is critical to success in any area of the electrical or electronics profession and *Foundations of Electronics* is the first step towards a successful career in electronics.

This text provides the vital facts, concepts, and principles related to the understanding of passive and active components, devices, and circuits that you will need for further progress in this field. Not only does this book provide this critical training, it also promotes your ability to apply this knowledge to real-life situations. The instructional design of the book strongly stresses the development of logical thinking patterns and knowledge of practical applications. Ideally, those who study this book will have had some training in basic algebra and basic trigonometry; however, enough data are provided to enable even novices in these areas to successfully learn the material presented.

Approach of the Book

Key instructional design strategies used in the book's presentation include factors such as the following:

- Helping you understand, rather than just memorize, important formulas and concepts
- Giving meaningful previews and overviews to provide a context for content details
- Furnishing important learning objectives and clear targets for learning
- Offering clear and frequent examples
- Providing immediate application of knowledge gained
- Using frequent practice problems to reinforce concepts
- Supplying strategically located checkpoints
- Making liberal use of realistic pictorial and diagrammatic information
- Using color to enhance the instructional value of illustrations and diagrams
- Moving from the known to the unknown in a logical manner
- Progressing from the simple to the more complex
- Including effective summaries that provide the essence of what should be learned
- Presenting many troubleshooting, safety, and other practical hints
- Providing a useful troubleshooting technique called the SIMPLER method to sharpen critical thinking skills and teach both "systems-level" and "component-level" troubleshooting
- Furnishing Troubleshooting Challenge circuit problems for practicing the SIMPLER method of troubleshooting
- Providing practice in using a computer spreadsheet program, Excel, for solving electronics problems
- Offering practice in using a circuit simulation program, MultiSIM, for analyzing and verifying circuit operation
- Formatting that allows for either group or individualized instruction delivery systems

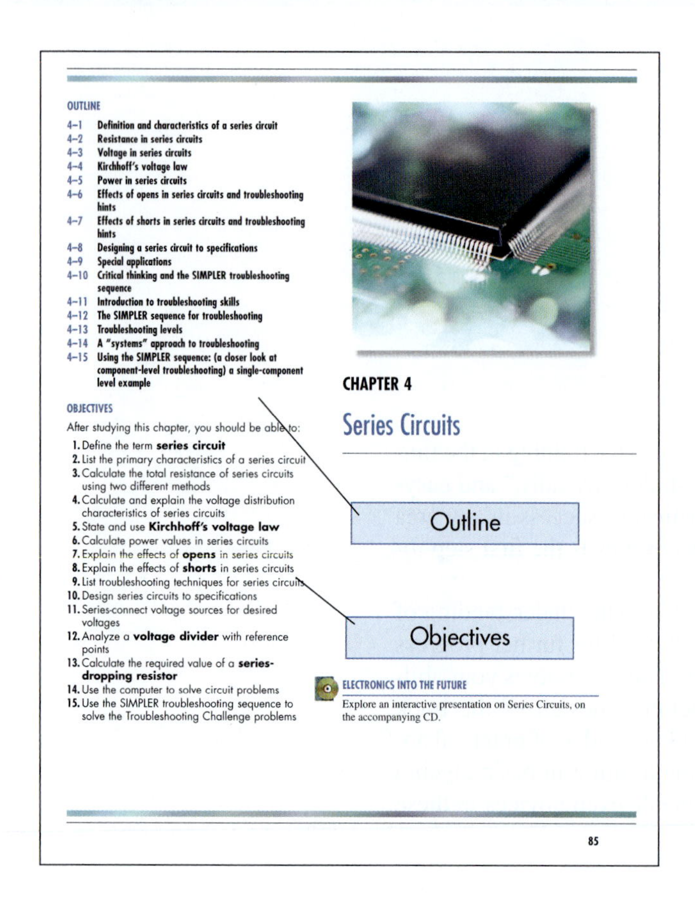

OUTLINE

4–1 Definition and characteristics of a series circuit
4–2 Resistance in series circuits
4–3 Voltage in series circuits
4–4 Kirchhoff's voltage law
4–5 Power in series circuits
4–6 Effects of opens in series circuits and troubleshooting hints
4–7 Effects of shorts in series circuits and troubleshooting hints
4–8 Designing a series circuit to specifications
4–9 Special applications
4–10 Critical thinking and the SIMPLER troubleshooting sequence
4–11 Introduction to troubleshooting skills
4–12 The SIMPLER sequence for troubleshooting
4–13 Troubleshooting levels
4–14 A "systems" approach to troubleshooting
4–15 Using the SIMPLER sequence: (a closer look at component-level troubleshooting) a single-component level example

OBJECTIVES

After studying this chapter, you should be able to:

1. Define the term **series circuit**
2. List the primary characteristics of a series circuit
3. Calculate the total resistance of series circuits using two different methods
4. Calculate and explain the voltage distribution characteristics of series circuits
5. State and use **Kirchhoff's voltage law**
6. Calculate power values in series circuits
7. Explain the effects of **opens** in series circuits
8. Explain the effects of **shorts** in series circuits
9. List troubleshooting techniques for series circuits
10. Design series circuits to specifications
11. Series-connect voltage sources for desired voltages
12. Analyze a **voltage divider** with reference points
13. Calculate the required value of a **series-dropping resistor**
14. Use the computer to solve circuit problems
15. Use the SIMPLER troubleshooting sequence to solve the Troubleshooting Challenge problems

CHAPTER 4

Series Circuits

ELECTRONICS INTO THE FUTURE

Explore an interactive presentation on Series Circuits, on the accompanying CD.

85

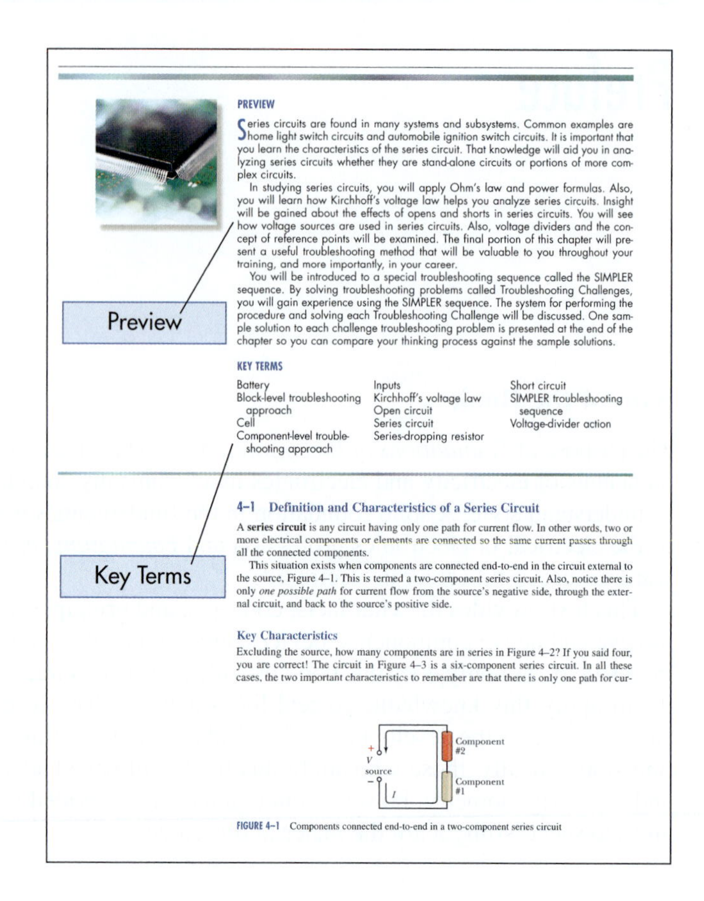

PREVIEW

Series circuits are found in many systems and subsystems. Common examples are home light switch circuits and automobile ignition switch circuits. It is important that you learn the characteristics of the series circuit. That knowledge will aid you in analyzing series circuits whether they are stand-alone circuits or portions of more complex circuits.

In studying series circuits, you will apply Ohm's law and power formulas. Also, you will learn how Kirchhoff's voltage law helps you analyze series circuits. Insight will be gained about the effects of opens and shorts in series circuits. You will see how voltage sources are used in series circuits. Also, voltage dividers and the concept of reference points will be examined. The final portion of this chapter will present a useful troubleshooting method that will be valuable to you throughout your training, and more importantly, in your career.

You will be introduced to a special troubleshooting sequence called the SIMPLER sequence. By solving troubleshooting problems called Troubleshooting Challenges, you will gain experience using the SIMPLER sequence. The system for performing the procedure and solving each Troubleshooting Challenge will be discussed. One sample solution to each challenge troubleshooting problem is presented at the end of the chapter so you can compare your thinking process against the sample solutions.

KEY TERMS

Battery
Block-level troubleshooting approach
Cell
Component-level troubleshooting approach
Inputs
Kirchhoff's voltage law
Open circuit
Series circuit
Series-dropping resistor
Short circuit
SIMPLER troubleshooting sequence
Voltage-divider action

4–1 Definition and Characteristics of a Series Circuit

A **series circuit** is any circuit having only one path for current flow. In other words, two or more electrical components or elements are connected so the same current passes through all the connected components.

This situation exists when components are connected end-to-end in the circuit external to the source, Figure 4–1. This is termed a two-component series circuit. Also, notice there is only *one possible path* for current flow from the source's negative side, through the external circuit, and back to the source's positive side.

Key Characteristics

Excluding the source, how many components are in series in Figure 4–2? If you said four, you are correct! The circuit in Figure 4–3 is a six-component series circuit. In all these cases, the two important characteristics to remember are that there is only one path for cur-

FIGURE 4–1 Components connected end-to-end in a two-component series circuit

Arrangement of the Book

The text begins with a brief overview of the wonders of today's electronic technology and the great opportunities afforded in this field. Next, crucial terms and elemental concepts are presented as foundations for the topics addressed throughout the remainder of the book. Concepts such as the electron theory, basic electrical units and symbols, and how electronic circuitry can be illustrated with diagrams provide the stepping stones to what follows.

Essential circuit fundamentals are then presented so that students can develop the ability to analyze various circuit configurations. In the process of learning these fundamentals, several passive devices are studied, including resistors, inductors, and capacitors. Also, important test instruments and their uses are discussed. The book then applies the fundamental concepts, principles, and theories, first to dc circuits, and then to ac circuits.

It is our hope that as you move from the introductory concepts of electronics through the principles of circuit analysis to practical circuit applications, you will find the book interesting, informative, and very useful in your training.

Features

Key Terms, Chapter Outline, Preview, and Objectives start off each chapter. A list of key terms presents important terminology that will be defined in the chapter. Main topics are listed in an outline. A chapter preview sets the stage and presents the importance of material in the chapter. Finally, critical learning objectives describe the competencies students should achieve upon understanding the chapter material.

In-Process Learning Checks provide an opportunity for you to check your understanding of the subject each step of the way. Answers are provided in Appendix B.

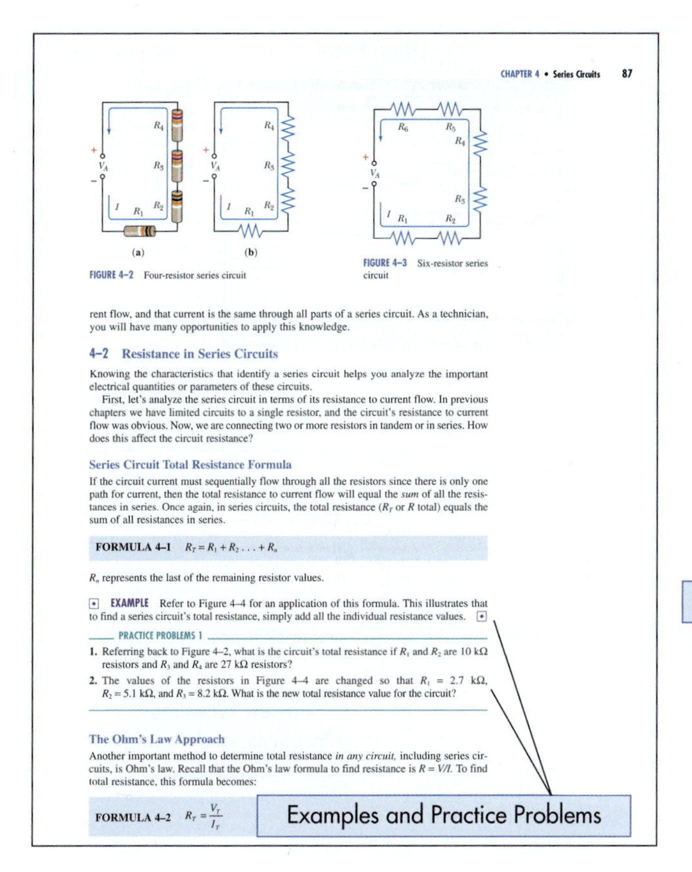

CHAPTER 4 • Series Circuits 87

FIGURE 4–2 Four-resistor series circuit

FIGURE 4–3 Six-resistor series circuit

rent flow, and that current is the same through all parts of a series circuit. As a technician, you will have many opportunities to apply this knowledge.

4–2 Resistance in Series Circuits

Knowing the characteristics that identify a series circuit helps you analyze the important electrical quantities or parameters of these circuits.

First, let's analyze the series circuit in terms of its resistance to current flow. In previous chapters we have limited circuits to a single resistor, and the circuit's resistance to current flow was obvious. Now, we are connecting two or more resistors in tandem or in series. How does this affect the circuit resistance?

Series Circuit Total Resistance Formula

If the circuit current must sequentially flow through all the resistors since there is only one path for current, then the total resistance to current flow will equal the *sum* of all the resistances in series. Once again, in series circuits, the total resistance (R_T or R total) equals the sum of all resistances in series.

FORMULA 4–1 $R_T = R_1 + R_2 \ldots + R_n$

R_n represents the last of the remaining resistor values.

EXAMPLE Refer to Figure 4–4 for an application of this formula. This illustrates that to find a series circuit's total resistance, simply add all the individual resistance values.

PRACTICE PROBLEMS 1

1. Referring back to Figure 4–2, what is the circuit's total resistance if R_1 and R_2 are 10 kΩ resistors and R_3 and R_4 are 27 kΩ resistors?
2. The values of the resistors in Figure 4–4 are changed so that R_1 = 2.7 kΩ, R_2 = 5.1 kΩ, and R_3 = 8.2 kΩ. What is the new total resistance value for the circuit?

The Ohm's Law Approach

Another important method to determine total resistance *in any circuit,* including series circuits, is Ohm's law. Recall that the Ohm's law formula to find resistance is $R = V/I$. To find total resistance, this formula becomes:

FORMULA 4–2 $R_T = \frac{V_T}{I_T}$

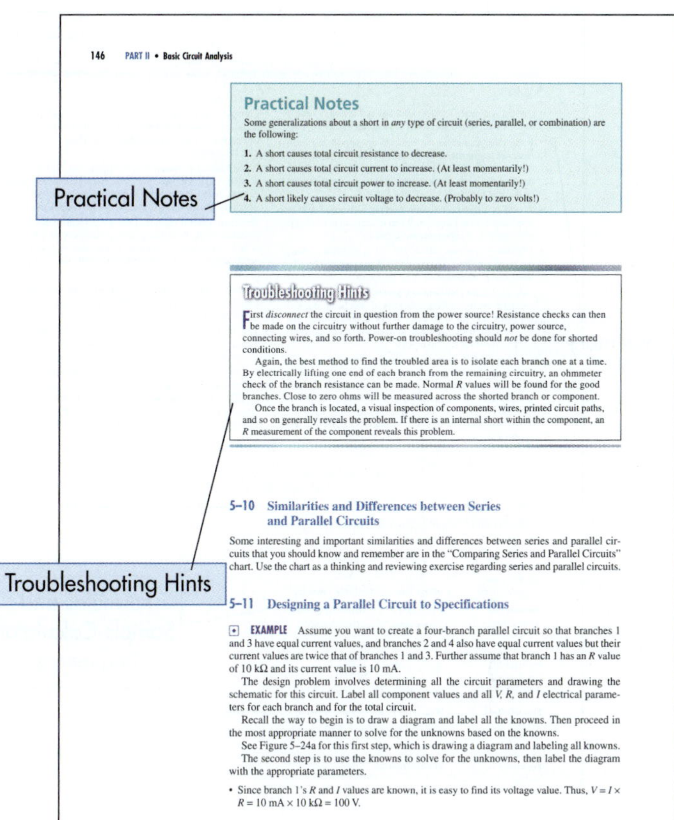

146 PART II • Basic Circuit Analysis

Practical Notes

Some generalizations about a short in *any* type of circuit (series, parallel, or combination) are the following:

1. A short causes total circuit resistance to decrease.
2. A short causes total circuit current to increase. (At least momentarily!)
3. A short causes total circuit power to increase. (At least momentarily!)
4. A short likely causes circuit voltage to decrease. (Probably to zero volts!)

Troubleshooting Hints

First *disconnect* the circuit in question from the power source! Resistance checks can then be made on the circuitry without further damage to the circuitry, power source, connecting wires, and so forth. Power-on troubleshooting should *not* be done for shorted conditions.

Again, the best method to find the troubled area is to isolate each branch one at a time. By electrically lifting one end of each branch from the remaining circuitry, an ohmmeter check of the branch resistance can be made. Normal R values will be found for the good branches. Close to zero ohms will be measured across the shorted branch or component.

Once the branch is located, a visual inspection of components, wires, printed circuit paths, and so on generally reveals the problem. If there is an internal short within the component, an R measurement of the component reveals this problem.

5–10 Similarities and Differences between Series and Parallel Circuits

Some interesting and important similarities and differences between series and parallel circuits that you should know and remember are in the "Comparing Series and Parallel Circuits" chart. Use the chart as a thinking and reviewing exercise regarding series and parallel circuits.

5–11 Designing a Parallel Circuit to Specifications

EXAMPLE Assume you want to create a four-branch parallel circuit so that branches 1 and 3 have equal current values, and branches 2 and 4 also have equal current values but their current values are twice that of branches 1 and 3. Further assume that branch 1 has an R value of 10 kΩ and its current value is 10 mA.

The design problem involves determining all the circuit parameters and drawing the schematic for this circuit. Label all component values and all V, R, and I electrical parameters for each branch and for the total circuit.

Recall the way to begin is to draw a diagram and label all the knowns. Then proceed in the most appropriate manner to solve for the unknowns based on the knowns.

See Figure 5–24a for this first step, which is drawing a diagram and labeling all knowns.

The second step is to use the knowns to solve for the unknowns, then label the diagram with the appropriate parameters.

- Since branch 1's R and I values are known, it is easy to find its voltage value. Thus, $V = I \times R = 10\ \text{mA} \times 10\ \text{k}\Omega = 100\ \text{V}$.

Practical Notes provide special hints, practical techniques, and information to the reader.

Safety Hints highlight requirements for safety, both while learning and when applying electronic principles in "hands-on" situations.

Troubleshooting Hints offer experience-based information about circuit problems, such as what to look for and how to pinpoint them.

Good Idea boxes reinforce key practical ideas.

Examples and Practice Problems show how to apply theory step-by-step and allow immediate practice after learning concepts.

Formulas and Sample Calculator Sequences provide a quick reference of the chapter's most important formulas, followed by an example of how to solve them using a calculator.

Using Excel is a feature that encourages problem solving using spreadsheet exercises and Excel templates available on the accompanying CD. Filenames are placed beside Excel screens for easy reference to the CD.

MultiSIM Exercises provide a feature that explores computer simulation techniques and confirms operational characteristics through the use of pre-created MultiSIM circuit files.

Performance Projects Correlation Charts cross-reference specific projects from the Laboratory Manual that relate to specific topics in the chapter(s).

multiSIM

CHAPTER 3 • Ohm's Law 79

Summary

- Current is directly related to voltage and inversely related to the circuit resistance. Ohm's law states this as $I = V/R$.
- If the other factors in the equation are known, the techniques of transposition allow rearrangement of the equation to solve for an unknown factor. For example, by rearranging the Ohm's law formula, two other useful equations can be derived, namely $V = I \times R$ and $R = V/I$.
- Powers of 10 are useful when working with either large or small numbers in formulas. When using powers of 10, the rules relating to exponents are applied.
- Scientific calculators are very useful to simplify working electrical problems. The [EE] or [EXP] keys are helpful aids when using calculators to solve problems.
- Two good habits to form when solving electrical circuit problems are to (1) draw a diagram of the circuit, if one is not given, and (2) write down any pertinent formulas that be used; then, substitute knowns into them appropri- y to solve for the unknowns.
- direction of electron current flow through a circuit, external to the voltage source, is from the source's negative side, through the circuit, and back to the source's positive side. Arrows are frequently drawn on schematic diagrams to illustrate this. The direction electrons would move through a circuit is the standard being used in this textbook.
- Conventional current, as used in some engineering textbooks, is considered to flow in the direction that a positive charge would move through the circuit. That is, from positive to negative.
- When labeling the polarity of voltage across a given component in a circuit, the component end where the electron current enters is labeled the negative end, and the end where the electron current exits is labeled the positive end.
- Electrical power is the rate of using electrical energy to do electrical work. Power is generally dissipated in a circuit or in a component in the form of heat. One watt of power is the performance of electrical work at the rate of 1 joule per second.
- A joule is the energy used in moving one coulomb of charge between two points that have a difference of potential of one volt between them.
- The formulas commonly used to calculate electrical power, or power dissipation are: $P = V \times I$; $P = I^2R$, and $P = V^2/R$
- Electrical energy usage is computed by multiplying the power used times the amount of time it was used ($W = P \times T$). Units of measure for usage of electrical energy are: wattsecond (Ws); watthour (Wh); and for large amounts of energy usage, kilowatthour (kWh).

Summary

Formulas and Sample Calculator Sequences

Formula 3–1 *(To find current using Ohm's law)*
$I = \frac{V}{R}$; $\left(\text{Amperes} = \frac{\text{Volts}}{\text{Ohms}}\right)$
voltage value, [÷], resistance value, [=]

Formula 3–2 *(To find voltage using Ohm's law)*
$V = I \times R$; (Volts = Amperes × Ohms)
current value, [×], resistance value, [=]

Formula 3–3 *(To find resistance using Ohm's law)*
$R = \frac{V}{I}$; $\left(\text{Ohms} = \frac{\text{Volts}}{\text{Amperes}}\right)$
voltage value, [÷], current value, [=]

Formula 3–4 *(To find foot pounds of work)*
Work (Foot pounds) = Force (Pounds) × Distance (Feet)
force value, [×], distance value, [=]

Formula 3–5 *(To find power in watts)*
$\text{Power (Watts)} = \frac{\text{Energy (Joules)}}{\text{Time (Seconds)}}$
energy value, [÷], time value, [=]

Formulas and Sample Calculator Sequences

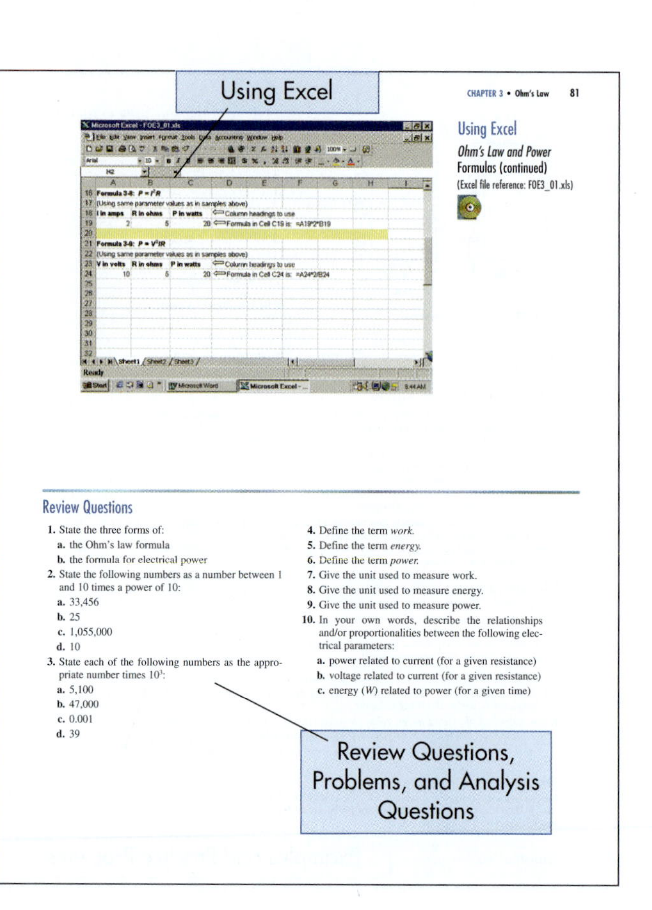

Using Excel
Microsoft Excel - FOE3_01.xls
Formula 3-8: P = I²R
(Using same parameter values as in samples above)
I in amps R in ohms P in watts Column headings to use
2 5 20 Formula in Cell C19 is: =A19^2*B19
Formula 3-9: P = V²/R
(Using same parameter values as in samples above)
V in volts R in ohms P in watts Column headings to use
10 5 20 Formula in Cell C24 is: =A24^2/B24
Sheet1 Sheet2 Sheet3
Ready

CHAPTER 3 • Ohm's Law 81

Using Excel

Ohm's Law and Power Formulas (continued)
(Excel file reference: FOE3_01.xls)

Review Questions

1. State the three forms of:
   a. the Ohm's law formula
   b. the formula for electrical power
2. State the following numbers as a number between 1 and 10 times a power of 10:
   a. 33,456
   b. 25
   c. 1,055,000
   d. 10
3. State each of the following numbers as the appropriate number times 10^3:
   a. 5,100
   b. 47,000
   c. 0.001
   d. 39
4. Define the term *work.*
5. Define the term *energy.*
6. Define the term *power.*
7. Give the unit used to measure work.
8. Give the unit used to measure energy.
9. Give the unit used to measure power.
10. In your own words, describe the relationships and/or proportionalities between the following electrical parameters:
    a. power related to current (for a given resistance)
    b. voltage related to current (for a given resistance)
    c. energy (W) related to power (for a given time)

Review Questions, Problems, and Analysis Questions

Summary, Review Questions, Problems, and Analysis Questions end each chapter with topic highlights, and a variety of questions that progress in difficulty from basic review questions to more advanced critical thinking problems.

SIMPLER Sequences and Troubleshooting Challenges use a unique troubleshooting approach that integrates critical thinking and encourages you to solve both systems-level and component-level troubleshooting problems logically and efficiently.

Practical Use of Color. Another important feature used throughout is the systematic use of color to clarify the illustrations. For example, each resistor pictorial displays an actual color code, adding a practical, real-world dimension to these illustrations. Voltage sources are highlighted with yellow, resistance values are highlighted with brown, capacitive elements with blue, and so forth. The practical use of color is designed to tangibly improve the value of the illustrations for students.

Appendixes provide references tools including color codes, answers, schematic symbols, how to use the "Excel Automated Formulas" listing on the CD, a comprehensive glossary, and more.

Troubleshooting with the SIMPLER Sequence

To become logical troubleshooters, technicians must develop good critical thinking skills. Chapter 4 first introduces a useful troubleshooting approach called the SIMPLER Sequence. You can practice following the SIMPLER Sequence by solving the Troubleshooting Challenges placed at the end of most chapters.

Troubleshooting Challenge

As you move through this book, you will encounter a series of troubleshooting problems called "Troubleshooting Challenges." Each Troubleshooting Challenge includes a challenge

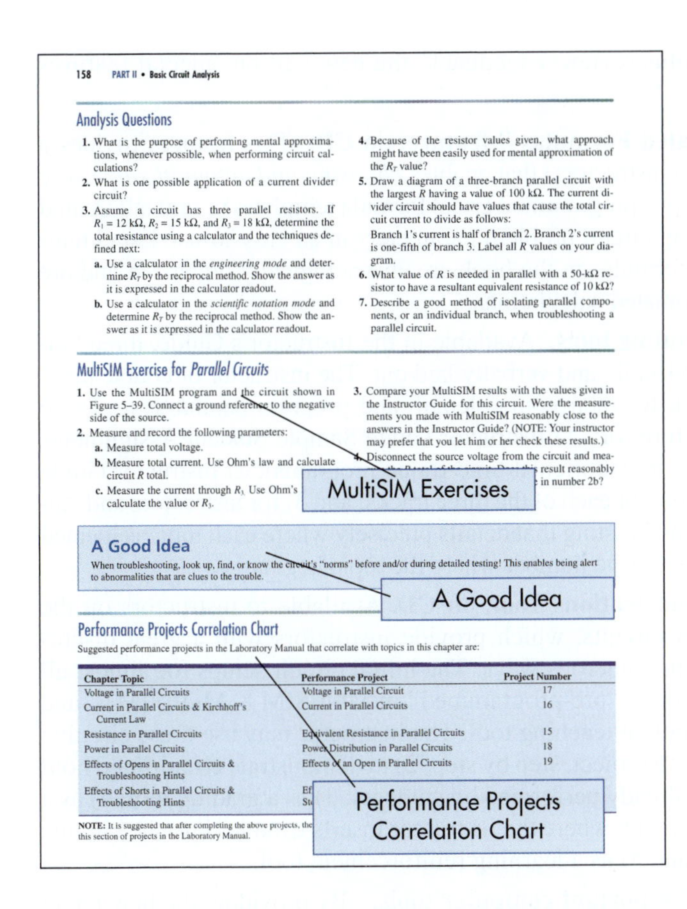

158 PART II • Basic Circuit Analysis

Analysis Questions

1. What is the purpose of performing mental approximations, whenever possible, when performing circuit calculations?
2. What is one possible application of a current divider circuit?
3. Assume a circuit has three parallel resistors. If $R_1 = 12$ kΩ, $R_2 = 15$ kΩ, and $R_3 = 18$ kΩ, determine the total resistance using a calculator and the techniques defined next:
   a. Use a calculator in the *engineering mode* and determine R_T by the reciprocal method. Show the answer as it is expressed in the calculator readout.
   b. Use a calculator in the *scientific notation mode* and determine R_T by the reciprocal method. Show the answer as it is expressed in the calculator readout.
4. Because of the resistor values given, what approach might have been easily used for mental approximation of the R_T value?
5. Draw a diagram of a three-branch parallel circuit with the largest R having a value of 100 kΩ. The current divider circuit should have values that cause the total circuit current to divide as follows:
   Branch 1's current is half of branch 2. Branch 2's current is one-fifth of branch 3. Label all R values on your diagram.
6. What value of R is needed in parallel with a 50-kΩ resistor to have a resultant equivalent resistance of 10 kΩ?
7. Describe a good method of isolating parallel components, or an individual branch, when troubleshooting a parallel circuit.

MultiSIM Exercise for *Parallel Circuits*

1. Use the MultiSIM program and the circuit shown in Figure 5–39. Connect a ground reference to the negative side of the source.
2. Measure and record the following parameters:
   a. Measure total voltage.
   b. Measure total current. Use Ohm's law and calculate circuit R total.
   c. Measure the current through R_3. Use Ohm's … calculate the value or R_3.
3. Compare your MultiSIM results with the values given in the Instructor Guide for this circuit. Were the measurements you made with MultiSIM reasonably close to the answers in the Instructor Guide? (NOTE: Your instructor may prefer that you let him or her check these results.)
4. Disconnect the source voltage from the circuit and mea- … result reasonably … in number 2b?

A Good Idea

When troubleshooting, look up, find, or know the circuit's "norms" before and/or during detailed testing! This enables being alert to abnormalities that are clues to the trouble.

Performance Projects Correlation Chart

Suggested performance projects in the Laboratory Manual that correlate with topics in this chapter are:

Chapter Topic	Performance Project	Project Number
Voltage in Parallel Circuits	Voltage in Parallel Circuit	17
Current in Parallel Circuits & Kirchhoff's Current Law	Current in Parallel Circuits	16
Resistance in Parallel Circuits	Equivalent Resistance in Parallel Circuits	15
Power in Parallel Circuits	Power Distribution in Parallel Circuits	18
Effects of Opens in Parallel Circuits & Troubleshooting Hints	Effects of an Open in Parallel Circuits	19
Effects of Shorts in Parallel Circuits & Troubleshooting Hints	Ef… St…	

NOTE: It is suggested that after completing the above projects, the … this section of projects in the Laboratory Manual.

circuit schematic or block diagram and starting point symptoms information. Using the SIMPLER sequence, you will step through and solve the simulated troubleshooting problems.

Using the SIMPLER troubleshooting method, you will be able to guide yourself through each Troubleshooting Challenge circuit and think critically about the problems you need to solve. One possible approach to solving each Troubleshooting Challenge is provided in step-by-step illustrations following each challenge.

The illustrations on pages xxiv–xxv show a sample Troubleshooting Challenge and illustrate how to complete the challenge. We hope that you and your instructor will have great success with the SIMPLER sequence technique and that these Troubleshooting Challenges will make learning and teaching troubleshooting skills easier and more enjoyable.

Selecting Tests and Finding the Results

As illustrated in Steps 4–6 on pages xxiv–xxv, to simulate making a parameter test on a particular portion of the challenge circuit, you may select a test from the test listing provided with the Troubleshooting Challenge. You may then look up the results of the test you selected by looking in Appendix C. Look in the appendix for the identifier number assigned to the test on the Troubleshooting Challenge page. Next to that number, you will see the parameter value or condition that would be present if you made that same test on the actual circuit simulated by the Troubleshooting Challenge. *Please note:* The tolerances of components and of test instruments will create differences in the theoretical values and those actually shown in the answers. The theoretical values would be achieved only if all components were precisely rated and the test instruments had zero percent error.

New to This Edition

The development of this text began with an extensive review of the previous edition. Reviewers made suggestions for improving content, instructional quality, and art presentations, which formed the core of our plan for developing the new edition.

Based on our own review, plus reviewer feedback, the new edition' special features include:

- **New student "Excel Automated Formulas " feature, on CD.** This feature provides a powerful tool for students (and instructors) that enables *choosing and automatically using* any formula in the text. These pre-programmed Excel formula templates permit automated calculations by simply inserting circuit problem parameters in an easy-to-use Excel template layout. Virtually all the formulas in the book are chronologically organized and are at the user's fingertips for automated calculations.
- **New instructor course scheduling tools.** Available in the Instructor's Guide, three "instructional tracks" have been visually and verbally laid out. The instructor or course manager can choose the best track for their course length and program parameters. These scheduling tools come in the form of: 1) Instructor Guide Sample Student Assignments Matrix for Track #2 and Track #3 (shortened time-frame programs); 2) Instructor Guide short verbal overview descriptors for each of the three tracks; and 3) for instructors and students, a Key Topics "Quick-Find" Listing that details precisely where each topic is located in the text—chapter, section, and page number where the topic begins.
- **New Instructor's Lab Demonstrations Tool, on CD.** Available to instructors on the e.resource CD, are MultiSIM circuits, which provide instructors with a great discretionary tool that can be used in numerous ways. The initial circuit setups for almost all the projects in the Lab Manual are "pre-programmed" in MultiSIM 8, MultiSIM 7, and MultiSIM 2001 formats as a potent teaching tool. The instructor may use these circuits: 1) to demonstrate a complete lab project, step by step; 2) to demonstrate circuits and hold class discussions on projects already performed by students; 3) as a grading tool; 4) as a special troubleshooting assignment, where the instructor inserts troubles into the circuits for students to troubleshoot; and 5) as a learning reinforcement tool.
- **Promotion of student use of important computer tools.** By providing the new Excel and MultiSIM tools and continuing to promote computer use throughout the text, the new edition continues to motivate students to learn and to use the computer. Students learn to use the computer as a strong tool for: 1) calculations and problem solving; 2) circuit simulations and design; and 3) learning circuit analyses and troubleshooting skills.
- **Inclusion of systems-level troubleshooting approaches.** As was done in the last edition, the new edition also discusses both component-level and systems-level approaches to troubleshooting. This concept continues to flow throughout the book in the Troubleshooting Challenges.
- **Use of CD multimedia presentations and simulations.** The CD presentations and simulations used in the fourth edition were well received; and consequently, they are also provided with the fifth edition.

What Is on the CD?

Electronics into the Future

- This interactive CD-ROM is a great leaning tool that can be used to illustrate and visualize difficult concepts. User-friendly and easy to navigate, the CD offers a range of multimedia presentations and interactive simulations designed to develop and expand major concepts in electronics. *Electronics into the Future* gives the user power to learn theory and troubleshooting from interactive applications.
- Contains six interactive modules that cover the following topics:
  - Ohm's Law, Amps, Volts, Ohms, & Power
  - Series Circuits
  - Parallel Circuits
  - Series-Parallel Circuits
  - Network Theorems (Kirchhoff, Superposition, Thevenin)
  - Magnetism and Electromagnetism

- Each Module presents:
  - Presentation video
  - Circuit modeling
  - Interactive conceptualization
  - Troubleshooting
  - Mathematics for electronics

Pre-built MultiSIM Circuit Files

- The CD includes pre-built sample circuits tied directly to the textbook. Students can simulate circuits and transform schematics into live, interactive circuits. A special icon is placed beside selected schematics to direct students to the CD. Students using the MultiSIM circuit files must have MultiSIM installed on their computers to be able to simulate the circuits. To find out more or to purchase MultiSIM, please go to *www.electronicsworkbench.com*, or call Electronics Workbench at (800) 263-5552 or (416) 977-5550.

Excel Tables

The CD includes pre-built formulas templates related to the "Using Excel" features placed at the end of many of the chapters. In addition, the CD contains a special file that provides the complete new "Excel Automated Formulas" feature, containing virtually every formula in the text. This valuable automated formula templates file allows students to select and use formulas in an efficient manner to solve electronics problems and teaches students to use the computer as a powerful aid, during their training and beyond.

The Learning Package

The complete ancillary package was developed to achieve two goals:

1. To assist students in learning the essential information needed to prepare for the exciting field of electronics.
2. To assist instructors in planning and implementing their instructional programs for the most efficient use of time and other resources.

Laboratory Projects

Foundations of Electronics Laboratory Projects is a well-correlated lab manual that can multiply the effectiveness of the learning process. The manual includes 81 easy-to-use projects that provide hands-on experience with the concepts and principles you studied in the text. Projects follow two different formats and can be performed depending on the user's preference. One style assures students will "paint the picture as they go" and will be asked to create conclusions throughout the project while collecting data and observations. "Story Behind the Numbers" are projects that focus on student's analytical technical writing, and communication skills. Projects are grouped together by topic and can be performed one project at a time, or as a topic-related group of projects, as appropriate to your training schedule.

ISBN: 1-4180-4183-1

Instructor's Guide

This comprehensive Instructor's Guide provides all the answers for the text and laboratory manual problems and projects, along with detailed instructional strategies and sample course schedules.

ISBN: 1-4180-0540-1

e.resource

The e.resource is an educational resource that creates a truly electronic classroom. It includes tools and instructional resources that will enrich your classroom and shorten your preparation time.

ISBN: 1-4180-5070-9

e.resource includes:

PowerPoint Presentation Slides

- These slides provide the basis for a lecture outline. Key points and concepts can be graphically highlighted for student retention.

ExamView

- This computerized testbank includes 1,500 true/false, multiple-choice, completion, and short-answer questions to assess student comprehension.

Image Library

- Images from the textbook allow customization of PowerPoint presentation slides, assignments, or creation of transparency masters. The Image Library comes with the ability to browse and search using key words for quick and easy use.

Online Companion

- A Web link brings you directly to the text's Online Companion at www.electronictech.delmar.cengage.com for additional resources and text updates.

WebTutor

This new WebTutor course is created as a student study guide and interactive supplement. The WebTutor offers notes, flashcards, Web links, quizzes, and discussion board topics. The course content will provide supplementary learning material to the learner.

WebTutor is available using the following platforms:

- WebTutor on Blackboard. ISBN: 1-4180-5070-9

Scheduling and Course Planning Information

By using the *Foundations of Electronics* or the *Foundations of Electronics: Circuits and Devices* texts, you have the luxury of selectively designing your in-class program to fit your institution's scheduling needs. At the same time, your students are provided with texts that are comprehensive and can be kept as valuable reference-shelf resources for the rest of their careers.

Ideally, your program can use all the material provided in the text and content-correlated laboratory projects to maximize learning and practice opportunities.

Some programs are becoming more limited in time due to the addition of new courses. This means that instructors and program managers are having to prudently, and carefully be selective about depth, breadth, and number of topics they can cover in their revised programs.

To provide help to instructors and course managers along this line, the author has carefully structured three different training plans that allow course administrators to choose the one training track that will best fit the program's needs and schedule. Obviously, variations can be made for any of these tracks, as may be appropriate at your institution.

Instructors and program managers should carefully look at the details of these training tracks as described in the *Instructor Guide.*

The three tracks detailed in the Instructor Guide are:

Track 1: Most Comprehensive Fundamentals Track (Full-length program)

Track 2: Mid-Level Fundamentals Track (Slightly-shortened program)

Track 3: Fast-Track Fundamentals (Shortest program)

The SIMPLER Sequence for Troubleshooting

Component-Level & Block-Level Troubleshooting Approaches

Follow the seven-step SIMPLER troubleshooting sequence, as outlined below, and see if you can find the problem with this circuit. As you follow the sequence, record your circuit test results for each testing step on a separate sheet of paper. This will aid you and your instructor to see where your thinking is on track or where it might deviate from the best procedure.

STEP 1 **Symptoms**—(Gather, verify, and analyze symptom information) To begin this process, read the data under the "Starting Point Information" heading. Particularly, look for circuit parameters that are not normal for the circuit configuration and component values given in the schematic diagram. For example, look for currents that are too high or too low, voltages that are too high or too low, resistances that are too high or too low, etc. This analysis should give you a first clue (or symptom information) that will aid you in determining possible areas or components in the circuit that might be causing this problem.

STEP 2 **Identify**—(Identify and bracket the initial suspect area) To perform this first "bracketing" step, analyze the clue or symptom information from the "Symptoms" step. Then either on paper or in your mind, bracket, circle, or put parentheses around all of the circuit area that contains any components, wires, etc. that might cause the abnormality that your symptom information has pointed out. NOTE: Don't bracket parts of the circuit that contain items that could not cause the symptom!

STEP 3 **Make**—(Make a decision about the first test you should make: what type and where) Look under the TEST column and determine which available tests shown in that column you think would give you the most meaningful information about the suspect area or components in light of the information you have so far.

STEP 4 **Perform**—(Perform the first test you have selected from the TEST column list) To simulate performing the chosen test (as if you were getting test results in an actual circuit of this type), follow the dotted line to the right from your selected test to the number in parentheses. The number in parentheses in the "Results in Appendix C" column tells you what number to look up in Appendix C to see what the test result is for the test you are simulating.

NOTE: For block-level troubleshooting, several parameters can be tested at each test point: 1) voltage, with respect to ground reference; 2) current level at that test point; 3) resistance at the test point, with respect to ground references *with all power sources disconnected*; and 4) signal conditions at the test point, if appropriate to the system being tested. Be aware that the test results are shown in terms of "high," "low," "normal," "abnormal," or "not present," as appropriate to the system.

STEP 5 **Locate**—(Locate and define a new "narrower" area of uncertainty) With the information you have gained from the test, you should be able to eliminate some of the circuit area or circuit components as still being suspect. In essence, you should be able to move one bracket of the bracketed area so there is a new, smaller area of uncertainty.

STEP 6 **Examine**—(Examine available information and determine the next test: what type and where) Use the information you have obtained thus far to determine what your next test should be in the new, narrower area of uncertainty. Determine the type of test and where, and then proceed, using the TEST listing to select that test, and the numbers in parentheses and Appendix C data to find the result.

STEP 7 **Repeat**—(Repeat the analysis and testing steps until you find the trouble) When you have determined what you would change or do to restore the circuit to normal operation, you have arrived at your solution to the problem. You can check your final result against ours by observing the pictorial step-by-step sample solution on the pages immediately following the "Chapter Troubleshooting Challenge."

STEP 8 **Verify**—(Check system operation) This extra eighth step, beyond the basic SIMPLER steps, is useful for verifying that the system has been restored to normal. This is the final proof that you have done a good job of troubleshooting.

TROUBLESHOOTING WITH THE SIMPLER SEQUENCE

STEP 1

Review the steps in the SIMPLER sequence.

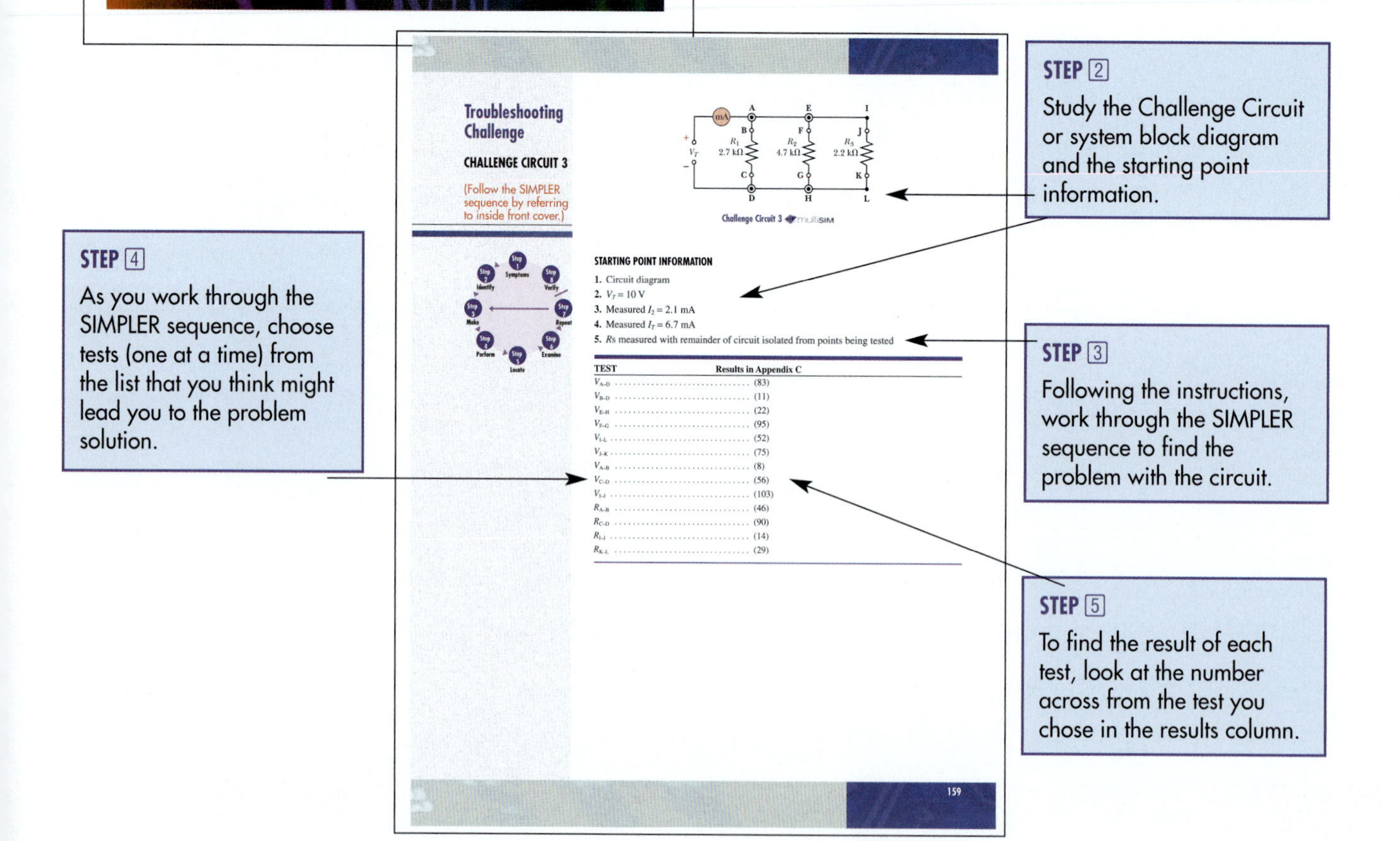

STEP 2
Study the Challenge Circuit or system block diagram and the starting point information.
STEP 3
Following the instructions, work through the SIMPLER sequence to find the problem with the circuit.
STEP 4
As you work through the SIMPLER sequence, choose tests (one at a time) from the list that you think might lead you to the problem solution.
STEP 5
To find the result of each test, look at the number across from the test you chose in the results column.
Step 1
Symptoms
Step 2
Identify
Step 3
Make
Step 4
Perform
Step 5
Locate
Step 6
Examine
Step 7
Repeat
Step 8
Verify
mA
A
B
C
D
E
F
G
H
I
J
K
L
V_T
R_1 2.7 kΩ
R_2 4.7 kΩ
R_3 2.2 kΩ

Troubleshooting Challenge

CHALLENGE CIRCUIT 3

(Follow the SIMPLER sequence by referring to inside front cover.)

Challenge Circuit 3 multiSIM

STARTING POINT INFORMATION

1. Circuit diagram
2. $V_T = 10$ V
3. Measured $I_2 = 2.1$ mA
4. Measured $I_T = 6.7$ mA
5. Rs measured with remainder of circuit isolated from points being tested

TEST	Results in Appendix C
V_{A-D}	(83)
V_{B-D}	(11)
V_{E-H}	(22)
V_{F-G}	(95)
V_{I-L}	(52)
V_{J-K}	(75)
V_{A-B}	(8)
V_{C-D}	(56)
V_{I-J}	(103)
R_{A-B}	(46)
R_{C-D}	(90)
R_{I-J}	(14)
R_{K-L}	(29)

159

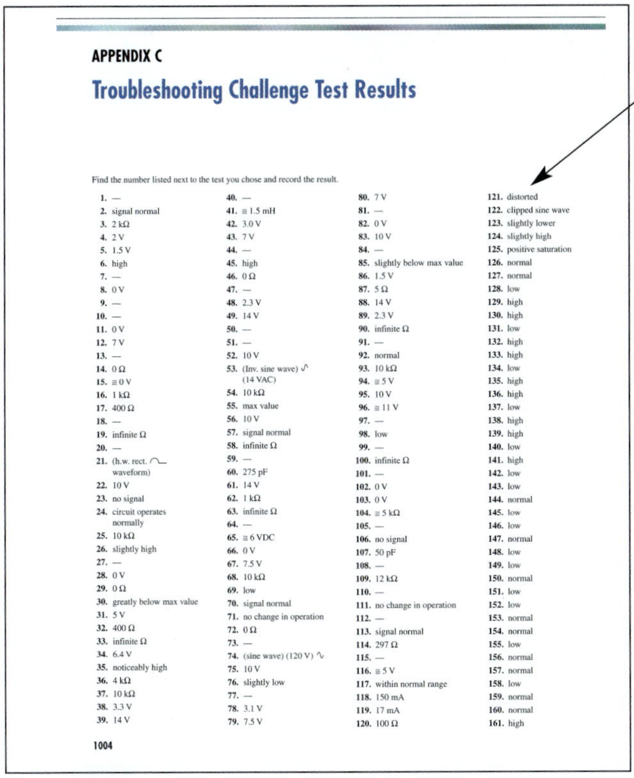

APPENDIX C

Troubleshooting Challenge Test Results

Find the number listed next to the test you chose and record the result.

1. —
2. signal normal
3. 2 kΩ
4. 2 V
5. 1.5 V
6. high
7. —
8. 0 V
9. —
10. —
11. 0 V
12. 7 V
13. —
14. 0 Ω
15. ≅ 0 V
16. 1 kΩ
17. 400 Ω
18. —
19. infinite Ω
20. —
21. (h.w. rect. waveform)
22. 10 V
23. no signal
24. circuit operates normally
25. 10 kΩ
26. slightly high
27. —
28. 0 V
29. 0 Ω
30. greatly below max value
31. 5 V
32. 400 Ω
33. infinite Ω
34. 6.4 V
35. noticeably high
36. 4 kΩ
37. 10 kΩ
38. 3.3 V
39. 14 V
40. —
41. ≅ 1.5 mH
42. 3.0 V
43. 7 V
44. —
45. high
46. 0 Ω
47. —
48. 2.3 V
49. 14 V
50. —
51. —
52. 10 V
53. (Inv. sine wave) (14 VAC)
54. 10 kΩ
55. max value
56. 10 V
57. signal normal
58. infinite Ω
59. —
60. 275 pF
61. 14 V
62. 1 kΩ
63. infinite Ω
64. —
65. ≅ 6 VDC
66. 0 V
67. 7.5 V
68. 10 kΩ
69. low
70. signal normal
71. no change in operation
72. 0 Ω
73. —
74. (sine wave) (120 V)
75. 10 V
76. slightly low
77. —
78. 3.1 V
79. 7.5 V
80. 7 V
81. —
82. 0 V
83. 10 V
84. —
85. slightly below max value
86. 1.5 V
87. 5 Ω
88. 14 V
89. 2.3 V
90. infinite Ω
91. —
92. normal
93. 10 kΩ
94. ≅ 5 V
95. 10 V
96. ≅ 11 V
97. —
98. low
99. —
100. infinite Ω
101. —
102. 0 V
103. 0 V
104. ≅ 5 kΩ
105. —
106. no signal
107. 50 pF
108. —
109. 12 kΩ
110. —
111. no change in operation
112. —
113. signal normal
114. 297 Ω
115. —
116. ≅ 5 V
117. within normal range
118. 150 mA
119. 17 mA
120. 100 Ω
121. distorted
122. clipped sine wave
123. slightly lower
124. slightly high
125. positive saturation
126. normal
127. normal
128. low
129. high
130. high
131. low
132. high
133. high
134. low
135. high
136. high
137. low
138. high
139. high
140. low
141. high
142. low
143. low
144. normal
145. low
146. low
147. normal
148. low
149. low
150. normal
151. low
152. low
153. normal
154. normal
155. low
156. normal
157. normal
158. low
159. normal
160. normal
161. high

1004

STEP 6

Turn to Appendix C and find that number from the results column. Next to the number in the Appendix you will find the result this test would yield. As you locate the number of the test you chose, record the result listed next to it on a sheet of paper.

STEP 7

When you have completed the challenge and pinpointed the system or circuit problem, check your work against the data in the answer pages. This answer offers one possible sequence of steps that might have been used to get to the solution, along with step-by-step color illustrations.

CHALLENGE CIRCUIT 1

STEP 1

SYMPTOMS The total current is too low for voltage applied. This implies that the total resistance has increased, meaning that one or more resistor(s) has changed value.

STEP 2

IDENTIFY initial suspect area: R_1, R_2, and R_3 (i.e., total circuit).

STEP 3

MAKE test decision: Check voltage across R_2 (middle of circuit).

STEP 4

PERFORM **1st Test:** Look up the test result. $V_{B\text{-}C}$ is 1.5 V.

STEP 5

LOCATE new suspect area: R_1 and R_3. NOTE: $V_{B\text{-}C}$ would be greater than 2 V if R_2 had increased and the other resistors had not.

STEP 6

EXAMINE available data.

STEP 7

REPEAT analysis and testing:
2nd Test: Check voltage across R_1. $V_{A\text{-}B}$ is 1.5 V.
3rd Test: Check voltage across R_3. $V_{C\text{-}D}$ is 3 V.
4th Test: Disconnect the V source and check resistance of R_3. The result is that R_3 is 2 kΩ, which is abnormal.

STEP 8

VERIFY **5th Test:** Replace R_3 with a good 1-kΩ resistor and note the current. When this is done, the circuit checks out normal. Each resistor drops 2 V and the circuit current measures 2 mA.

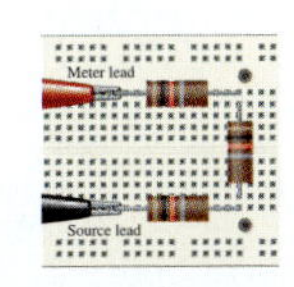

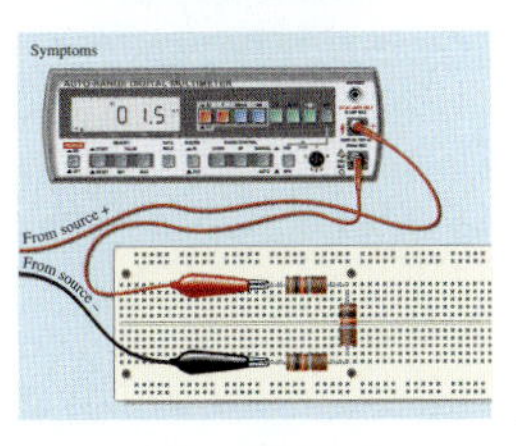

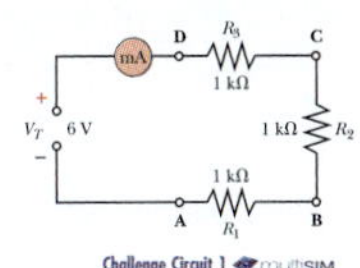

Challenge Circuit 1 multisim

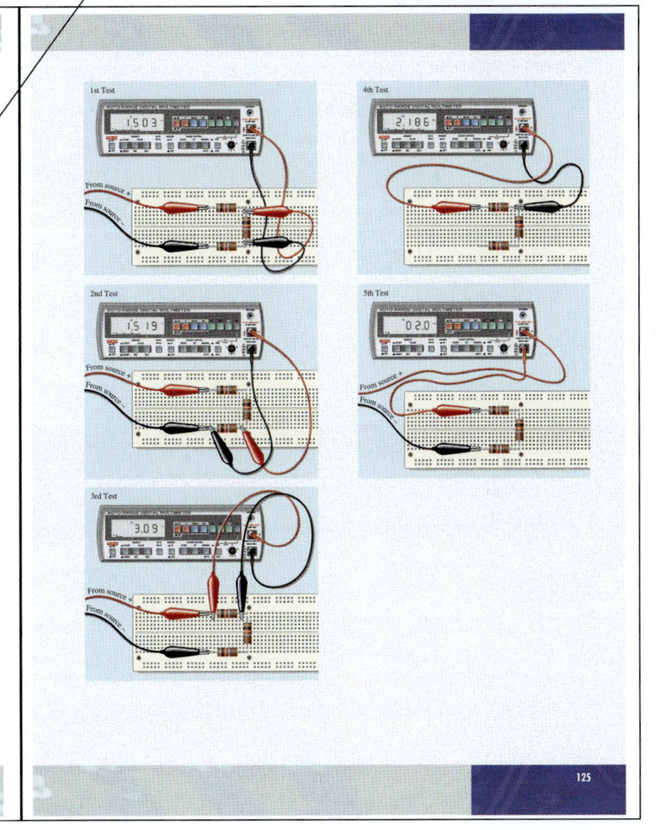

124

125

Acknowledgments

The Author and Delmar, Cengage Learning would like to thank the following reviewers for their feedback:

Russell Bonine
Southwestern College
Chula Vista, CA

Robert Peeler
Lamar State College
Port Arthur, TX

Amy Stephenson
Pittsburgh Community College
Greenville, NC

Cree Stout
York Technical College
Rock Hill, SC

About the Author

Russell Meade has more than 35 years of experience in the areas of technical instruction (USAF, DeVry, and MBI), experience in high-tech industries (Scientific-Atlanta, and Electromagnetic Sciences, among others), state-level technical education administration (Georgia), is a contractor/consultant in instructional design and development of electronics curricula, and is author of several books (Tandy/Radio Shack, Delmar). He holds a B.S. degree, a patent on an electronic training device (patent # 3,340,620), an FCC radio engineering license (with radar endorsement), and an FCC advanced class amateur radio license (W4HIB).

Introduction

The Magnificent Growth of Electronics

Electronics has grown until it permeates almost everything we do! Whether we are at home, at work, or in our automobile, electronic systems, devices, and controls are all around us. The inventors of the telegraph, the telephone, and radio transmitters and receivers would not believe the communications systems used today. Satellite communications, cable TV, the Internet, cell phones, computerized data communications, and Global Positioning Systems would astound these inventors, probably even more than they astound us.

In the *home*:

- Electronic/digital thermostats and controls that precisely control home temperatures, humidity, and air filtering by automated control of the furnace and air-conditioning systems
- Refrigerators that can be used in food inventory control and electronic ordering of groceries
- Cooking ranges and microwave ovens that will do everything but read the recipes and put the necessary ingredients together for the dish you want
- Amazing home security systems that can include anything from simple alarm systems to closed-circuit TV systems and automated alerting of police entities
- Audio sound systems that were "unheard of" in years past. Surround-sound systems with spatial effects that make you feel like you are at a symphony concert (or whatever genre of music you like). For further convenience, wireless multiroom remote controls have been added to a number of these systems. Whether the source is a TV, a DVD, a CD, or an AM or FM radio station, the sound systems available today are several orders of magnitude better than they were just a few years ago. All because of electronics.
- TV systems using satellite dishes, cable, and digital cable for receiving their signals. New TV systems that can provide much improved definition of picture due to HDTV (High-Definition TV), when broadcasts are available in that format

Today our *automobiles* are "loaded" with electronics:

- Electronic control units that automatically control all critical parameters that are involved in engine operations, transmissions, and emissions control
- Keyless entry systems
- Memory systems that automatically set seats, mirrors, and environmental settings based on individual preferences
- Computer-controlled suspension systems that automatically adjust to road conditions
- Electronic security systems
- Digital trip computers that provide an electronic message center with readouts to indicate remaining fuel and fuel consumption projections, current miles-per-gallon fuel use, compass direction, and so forth
- Satellite-linked navigation systems, such as Global Positioning Systems (GPS) that let you know "where on Earth" you are . . . and also may provide specific details on how to get where you want to go
- Sound systems
- Other (such as "variable-assist" steering control)

Manufacturing uses a myriad of industrial controls, computer-aided drafting and design (CADD), and computer-aided manufacturing systems. Automation in industry includes the following:

- Robots for welding, positioning, fastening, and so forth
- Automated drilling, punching, and milling
- Automated shaping and bending
- Counting and sorting
- Process control

Business transactions involve everything from point-of-sale computers that take care of receipting, inventory, and change-making to online computerized transfers of money, documentation, and endless other transactions.

Our *homes* are filled with electronic wizardry that helps us take care of heating and cooling our homes, cooking, washing, cleaning, and providing security. The entertainment systems in our homes today could not have even been dreamed of just a few short years ago.

It is impossible to imagine what tomorrow will bring. It is obvious you have chosen a good field!

Electronics Fundamentals Are Critical to Your Training

To become a competent technician in any specific area within the broad panorama of possibilities offered by electronics, you must get a thorough grasp of the fundamentals. In fact, within any professional field of endeavor, the nitty-gritty fundamentals of that field must be studied, learned, and applied, if you are to succeed. It is necessary to learn the vocabulary, symbols, and basic principles used in that field in order to move through the hierarchy of training needed to become a trained professional in that field. These fundamentals, then, become the foundation upon which you build your critical understanding and all that is required for your career.

Try to establish the attitude that the more diligently you work at learning these fundamentals, the easier the advanced training and work will become. This is true!

If you master the terms, symbols, formulas, and basic principles presented in this book, you will have established a wonderful "launching pad" into this stimulating field of electronics.

It is my wish that each of you will travel as far as your expectations and dreams might carry you, and beyond. I wish you great success as you travel through your learning experiences, and on to your career!

Russ Meade

PART I

Foundational Concepts

OUTLINE

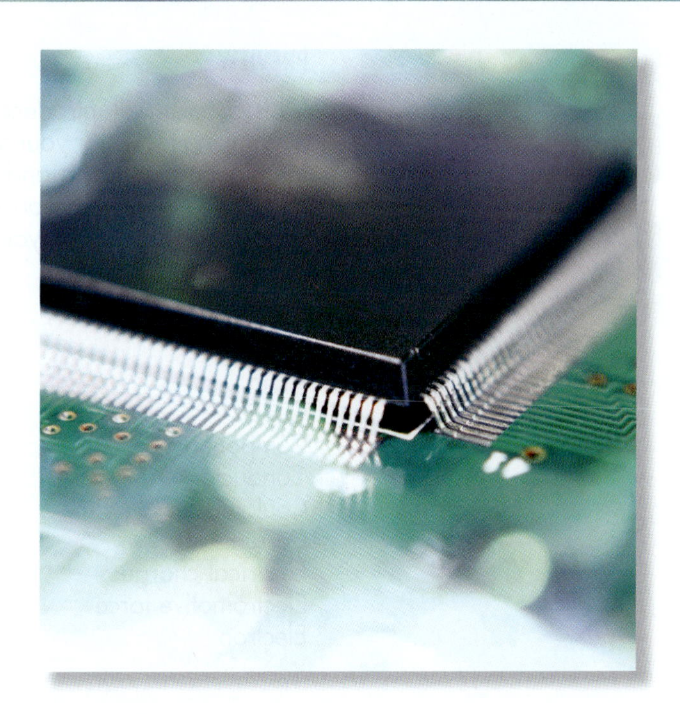

CHAPTER 1

Basic Concepts of Electricity

OBJECTIVES

After studying this chapter, you should be able to:

1. Describe the types of tasks performed by electronics technicians, technologists, and engineers
2. Define the term **matter** and list its physical and chemical states
3. Describe the difference between **elements** and **compounds**
4. Discuss the characteristics and structure of an **atom, molecule,** and **ion**
5. Define the electrical characteristics of an **electron, proton,** and **neutron**
6. Explain the terms **valence electrons** and **free electrons**
7. List the methods used to create electrical imbalances
8. Describe the characteristics of **conductors, semiconductors,** and **insulators**
9. State the law of electrical charges
10. Discuss the terms **polarity** and **reference points**
11. Define charge and its unit of measure, the **coulomb**
12. Define **potential** (emf) and give its unit of measure
13. Define **current** and explain its unit of measure
14. Calculate current when magnitude and rate of charge motion is known
15. Define **resistance** and give its unit of measure
16. List the typical elements of an electrical circuit
17. Describe the difference between closed and open circuits

ELECTRONICS INTO THE FUTURE

Explore an interactive presentation on Ohm's Law, Amps, Volts, Ohms & Power on the accompanying CD.

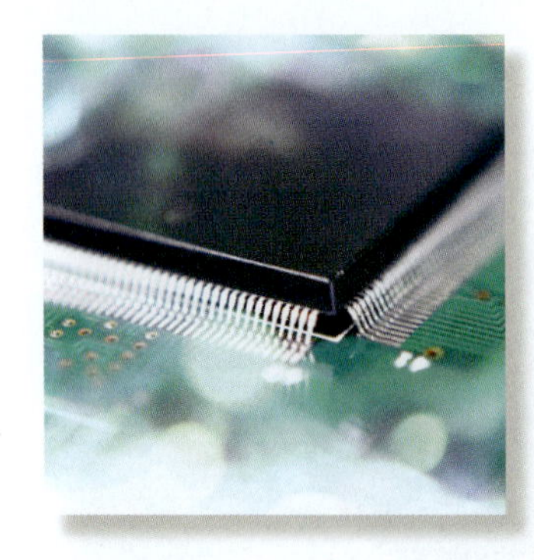

PREVIEW

This chapter will introduce you to a variety of concepts and theories proven useful as "thinking tools" for your study of electricity and electronics. These seemingly divergent foundational concepts and terms will come together in a meaningful way as you proceed in your studies. Consider each item as an important "knowledge capsule" upon which you can build in your continuing learning process.

KEY TERMS

Ampere
Atom
Circuit
Compound
Conductor
Coulomb
Current
Electrical charge
Electromotive force
Electron
Element
Energy
Force
Free electrons
Insulator materials
Ion
Load
Matter
Mixture
Molecule
Neutron
Ohm
Polarity
Potential
Proton
Resistance
Semiconductor
Source
Valence electrons
Volt

1–1 General Information Regarding the Field of Electronics

You are about to start your preparation in a field of work that is both interesting and almost limitless in opportunity. There are three broad categories of work in this field:

1. *Technicians* install, adjust, troubleshoot, repair, and maintain a wide variety of electronic equipment and systems, Figure 1–1.

FIGURE 1–1 Electronics technician making tests on equipment

2. *Technologists* assist in the design, development, and testing of electronic equipment and systems, Figure 1–2.
3. *Engineers* design electronic equipment and systems. Engineers may also be involved as "customer engineers." These engineers serve in a technical service role for industry, providing expertise in making correct choices of equipment or systems for specific applications, Figure 1–3.

Within the field of electronics, you can specialize in many different areas. These include:

- computers (e.g., CPUs, monitors, peripherals)
- communications (e.g., radio, television, telecommunications)

FIGURE 1–2 Electronics technologist performing computer calculations related to equipment design

FIGURE 1–3 Electronics engineer working with industrial customer to determine customer's system needs

FIGURE 1–4 Scenes full of matter

- consumer electronics (e.g., audio, video, alarm systems)
- medical electronics (e.g., x-ray, magnetic imaging, monitoring devices)
- aerospace electronics (e.g., computing, navigation, satellite communications)
- marine electronics (e.g., radar, sonar, navigation, radio)
- automotive electronics (e.g., systems control, user conveniences)
- industrial electronics (e.g., process control, system monitoring, quality control)

As you can see, you have chosen a great field of study!

1–2 Definition, Physical and Chemical States of Matter

Definition of Matter

Everything that we see, touch, or smell represents some form of matter. In Figure 1–4 the plants, water, clouds, and air represent **matter** in various forms. Although many definitions have been used, matter may be defined as anything that has weight and occupies space. We can also say that matter is what all things are made of and what our senses can perceive.

The basic building block of all matter is the atom. Later in this chapter, atoms will be thoroughly discussed because of their importance to the understanding of electronics.

Physical States of Matter

Matter exists in three *physical* states. Matter is either a **solid,** such as the chair you are sitting on; a **liquid,** such as water; or a **gas,** such as oxygen.

When matter exists in the liquid or gaseous state, its dimensions are determined by the container.

Chemical States of Matter

FIGURE 1–5 Some familiar metallic elements *(Photo by Michael A. Gallitelli)*

The chemical states of matter are **elements, compounds,** and **mixtures.** An element, Figure 1–5, is a substance that cannot be chemically broken into simpler substances. In fact, an element has only one kind of atom. Examples of chemical elements are gold, iron, copper, silicon, oxygen, and hydrogen.

A compound is formed by a chemical combination of two or more elements. In other words, compounds are built from two or more atoms that combine into molecules. Also, a compound has a definite structure and characteristic (i.e., same weight and atomic structure). Figure 1–6 shows two examples of compounds. Examples of compounds are water,

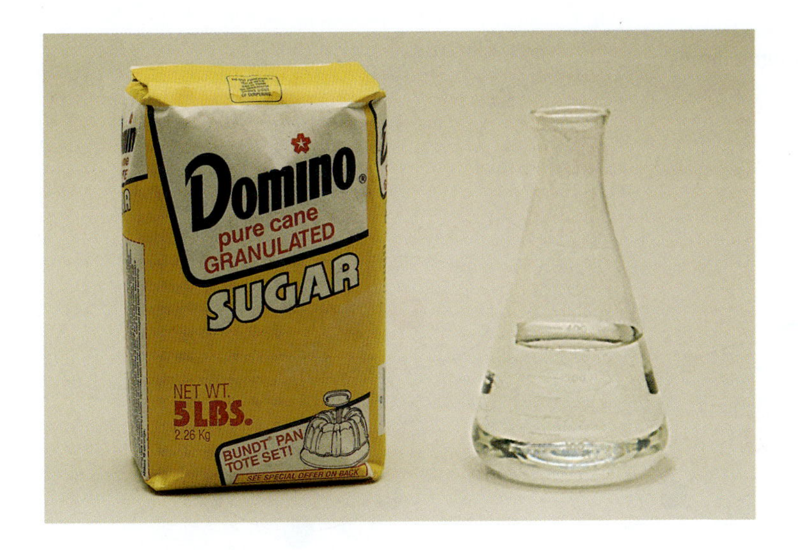

FIGURE 1–6 Some commonly used compounds *(Photo by Michael A. Gallitelli)*

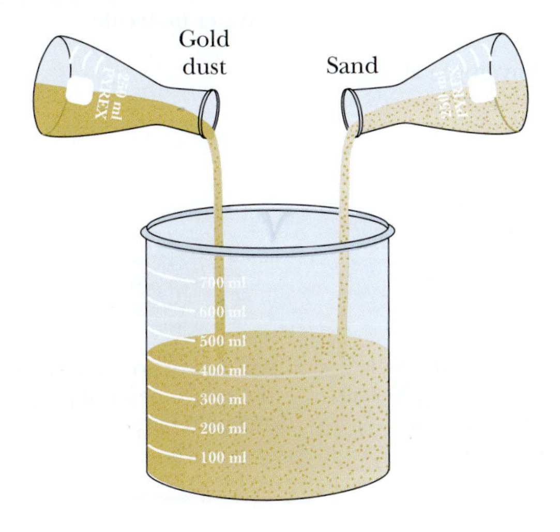

FIGURE 1–7 Mixtures are combinations of substances in which the individual elements possess the same properties as when they are alone. A mixture is different from a compound.

which is the chemical union of hydrogen and oxygen, and sugar, which contains carbon (a black, tasteless solid) plus hydrogen and oxygen (two gases).

Mixtures are a combination of substances where the individual elements possess the same properties as when they are alone. No chemical change results from the combination as it does with compounds. In Figure 1–7, mixing gold dust and sand would not yield a new, chemically different entity or compound. The substances would simply be mixed or combined.

1–3 Composition of Matter

The building blocks of all matter are atoms. Atoms combine chemically to form molecules, and the new compound (matter) created is different from each element that went into the compound's makeup, Figure 1–8. Also, a specific substance's molecules are different from the molecules of other types of matter.

In summary, *all matter is composed of atoms and molecules.* The smallest particle into which a compound can be divided but retain its physical properties is the **molecule.** The smallest particle into which an element can be divided but retain its physical properties is the **atom,** Figure 1–9.

1–4 Structure of the Atom

The Particles

The particles of the atom that interest electronics students are the **electron, proton,** and **neutron.** Although science has identified other particles (mesons, positrons, neutrinos), studying these particles is not necessary to understand the electron theory.

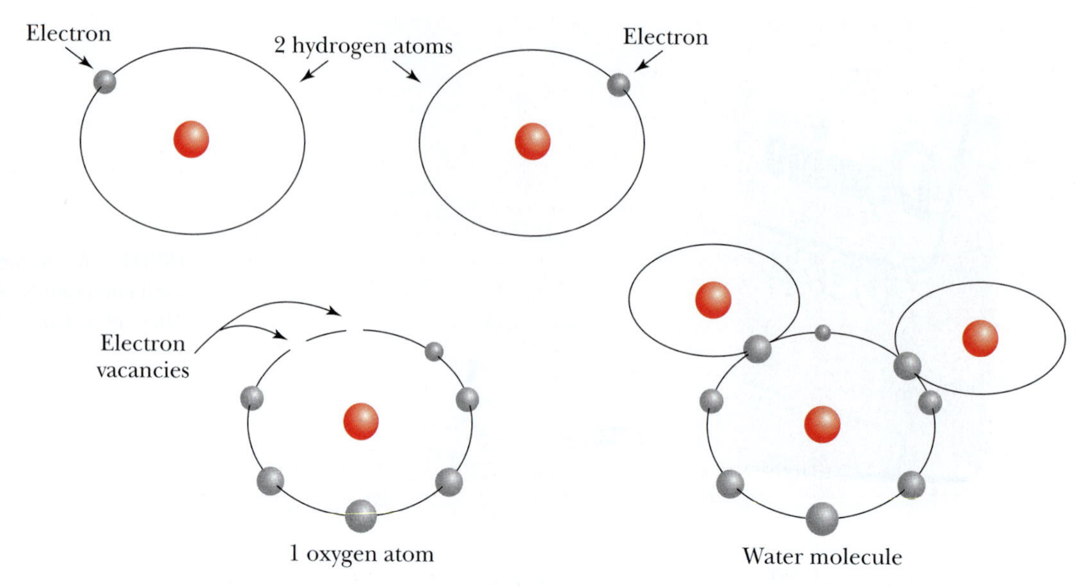

FIGURE 1–8 A molecule of water is formed from oxygen and hydrogen atoms.

FIGURE 1–9 The smallest particle of a compound that has the same characteristics as the compound is a molecule. The smallest part of an element that will show the same characteristics as the element is an atom.

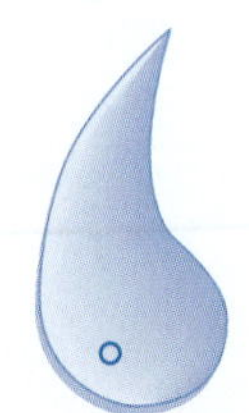

A *molecule* of water will exhibit the same physical characteristics as a drop of water.

An *atom* of gold will show the same chemical properties as a gold bar.

The Model

A Danish scientist, Niels Bohr, developed a model of atomic structure that explains the electron theory. In his model, the atom consists of a nucleus in the center of the atom with electrons orbiting the nucleus. The nucleus has two types of particles: protons, which are electrically positive in charge, and neutrons, which are neutral in electrical charge. The orbiting electrons are negatively charged particles, Figure 1–10. The common analogy for this model is our solar system: Planets orbit the sun like electrons orbit the nucleus of the atom.

The net charge of the atom is neutral because the orbiting electrons' total negative charge strength equals the total positive charge strength of the protons in the nucleus. Also, the number of protons and electrons in an electrically balanced atom are equal, Figure 1–11.

Some interesting characteristics of these three atomic particles are listed in Figure 1–12.

1. *The electron:*
   - has a negative electrical charge;
   - has small mass or weight (9×10^{-28} grams);
   - travels in orbits outside the nucleus;
   - travels around the nucleus at unbelievable speed (trillions of times a second); and
   - helps determine the atom's chemical characteristics.
2. *The proton:*
   - has a positive electrical charge;
   - is approximately 1,800 times heavier than an electron;

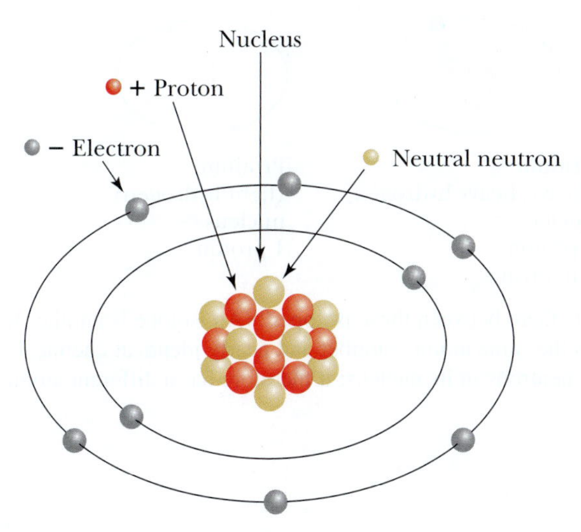

FIGURE 1–10 Bohr's model of the atom. The nucleus (center) contains protons and neutrons. Electrons orbit the nucleus.

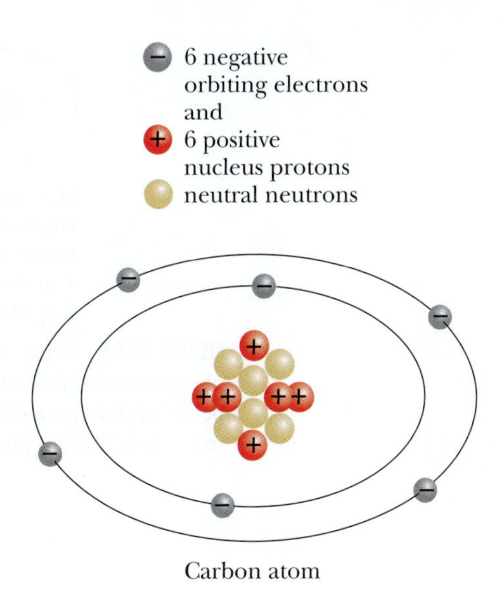

FIGURE 1–11 Typically, atoms are electrically balanced.

ELECTRONS	PROTONS	NEUTRONS
Negative charge Small mass/weight (about 9×10^{-28} grams) Travel in orbits outside of nucleus Rates of speed (trillions of orbits per second)	Positive charge 1,836 times heavier than electron Located in the nucleus Equal in number to atom's electrons	No charge Similar in mass/weight to proton Located in the nucleus

FIGURE 1–12 Some facts about atomic particles

- is located in the nucleus of the atom; and
- is equal in number to the atom's electrons.

3. *The neutron:*
- has no electrical charge;
- is about the same mass or weight as a proton;
- is located in the nucleus of the atom; and
- may vary in number for a given element to form different "isotopes" of the same element; for example, hydrogen has three isotopes: protium, deuterium, and tritium, Figure 1–13.

Atomic Number and Weight

Although it is not necessary to thoroughly understand atomic numbers and atomic weights, it is appropriate to mention them.

The atomic number of an element is determined by the number of protons in each of its atoms. For example, the atomic number for copper is 29 and the atomic number for carbon is 6, Figure 1–14.

The atomic weight of an element is considered to be the average mass of the atom of an element as compared with an atom of carbon-12. (NOTE: Carbon-12 is designated as having 12 atomic mass units.) For example, hydrogen in the simplest form has an atomic weight of about 1.007 "atomic mass units," and copper's atomic weight is 63.54.

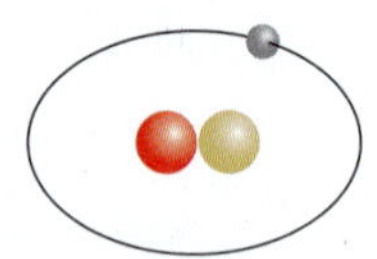
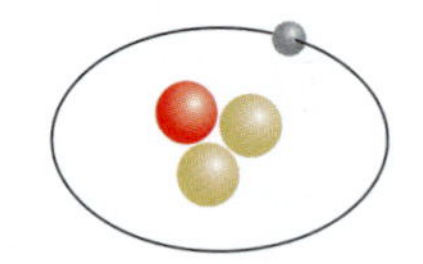
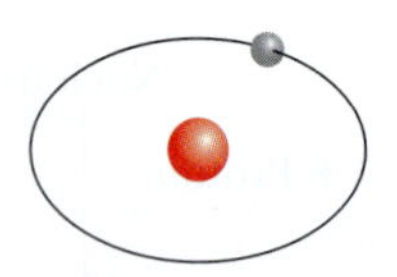

Deuterium	Tritium	Protium
(heavy hydrogen)	(heavy, heavy hydrogen)	(light hydrogen)
nucleus =	nucleus =	nucleus =
1 proton	1 proton	1 proton
1 neutron	2 neutrons	

FIGURE 1–13 What atomic difference is there between these isotopes? (An isotope is another form of a given chemical element, which has the same atomic number and nearly identical chemical behavior, but has a different number of neutrons in its nucleus; it therefore has a different atomic mass and different physical properties.)

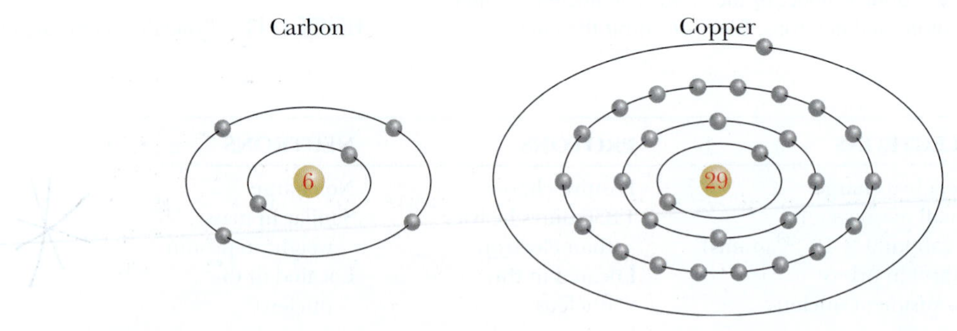

FIGURE 1–14 Carbon and copper atoms

■ IN-PROCESS LEARNING CHECK 1

It will be a good idea at this point to initiate you into a technique that will be used from time to time throughout the book: the "In-Process Learning Check." Rather than wait until the end of the chapter to find out whether you're missing any of the key points, these special "check-ups" will provide you with a chance to assure that you're learning the key points. Fill in the blanks for the following statements. If you have trouble with any of them, simply move back to that topic and refresh your memory. Answers are in Appendix B.

1. Matter is anything that has Weight and occupies Space.
2. Three physical states of matter are Solid, liquid, and gas.
3. Three chemical states of matter are element, compound and Mixtures
4. The smallest particle that a compound can be divided into but retain its physical properties is the Molecule
5. The smallest particle that an element can be divided into but retain its physical properties is the atom.
6. The three parts of an atom that interest electronics students are the electron, proton, and neutron
7. The atomic particle having a negative charge is the electron
8. The atomic particle having a positive charge is the Proton
9. The atomic particle having a neutral charge is the Neutron
10. The Proton and Neutron are found in the atom's nucleus.
11. The particle that orbits the nucleus of the atom is the Electron

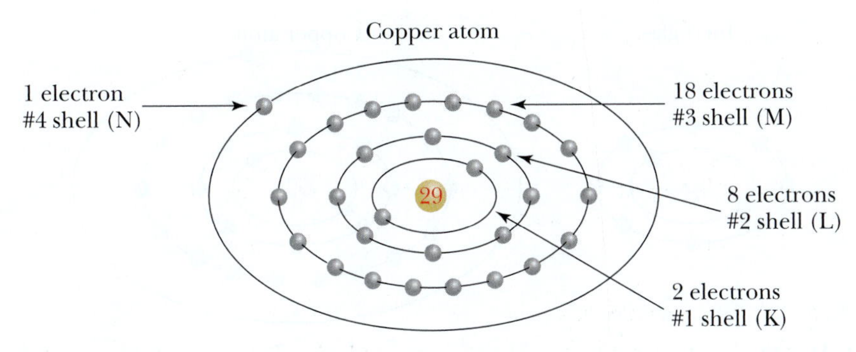

FIGURE 1–15 Electrons align at different distances from the nucleus.

Concept of Atomic Shells

As our illustrations indicate, all electrons traveling around the nucleus of the atom do not travel in the same path or at the same distance from the nucleus. Electrons align themselves in a structured manner, Figure 1–15.

Each ring, or shell, of orbiting electrons has a maximum number of electrons that can locate themselves within that shell if the atom is stable. The formula to use is $2n^2$ where n equals the number of the shell. For example, the innermost shell (closest to the nucleus) may have a maximum of 2 electrons; the second shell, 8 electrons; the third shell, 18 electrons; and the fourth shell, a maximum of 32 electrons. The outermost shell, whichever number shell that is for the given atom, can never contain more than 8 electrons.

1–5 Concept of the Electron Theory

Purpose

The electron theory helps visualize atoms and electrons as they relate to electrical and electronic phenomena, and the previous descriptions of atomic structure are part of this electron theory. The next discussion of valence and free electrons illustrates some practical aspects of the electron theory.

Valence Electrons

Valence electrons are those electrons in the outermost shell of the atom. The number of valence electrons in the atom determines its stability or instability, both electrically and chemically, Figure 1–16.

For all atoms, this outermost shell is *full* when it has eight *electrons.* If there are eight outermost ring electrons, then the material is stable and does not easily combine chemically with other atoms to form molecules. Also, these electrons are not easy to move from the atom. Examples of stable atoms are inert gases, such as neon and argon.

If the material's atoms have fewer than eight outermost ring electrons, the material is chemically and electrically active. Often the material will chemically combine with other atoms to gain stability and to form molecules and/or atomic bonds. Electrically, these valence electrons can be easily moved from their original home atom and are sometimes referred to as free electrons. Examples of such materials are copper, gold, and silver.

Also, there are materials that have four outermost ring electrons. They are halfway between being stable and very unstable. Germanium and silicon are two examples. These materials are used in most of today's solid-state, "semiconductor" devices, such as transistors and integrated circuits.

In summary, the concepts of the electron theory are based on atomic structure and help explain the various electrical phenomena discussed in this book.

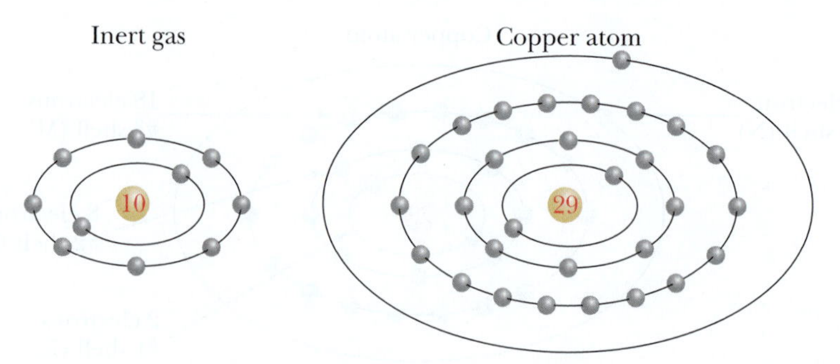

FIGURE 1–16 The number of valence electrons affects the chemical and/or electrical stability of an atom.

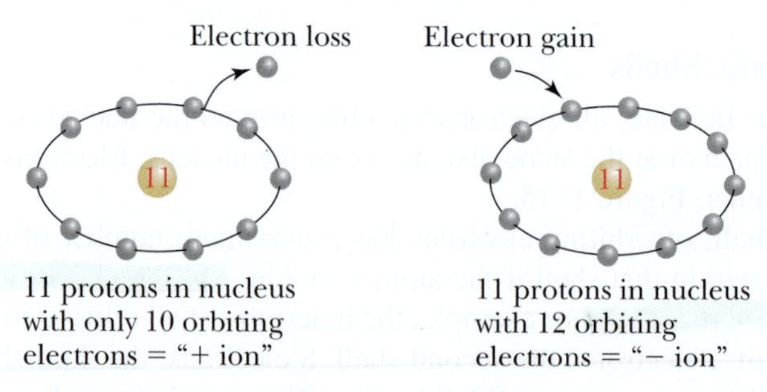

FIGURE 1–17 Ions are atoms that have lost or gained electrons.

1–6 Ions

When an electron leaves its original home atom because of chemical, light, heat, or other types of energy, it leaves behind an atom that is no longer electrically neutral. An **ion** is any atom that is not electrically balanced (or neutral) and that has gained or lost electrons. A positive ion is an atom having fewer electrons than protons (a deficiency of electrons). A negative ion is an atom having more electrons than protons (an excess of electrons), Figure 1–17.

When an electron is torn from a neutral atom, leaving a positive ion, or when an electron is added to a neutral atom, producing a negative ion, the process is called ionization. Later in your study of electronics, you will see how this process can be useful in various electronic devices.

1–7 Energies That Change Electrical Balance

Why should anyone want to change the electrical balance of atoms, or control electron movement? As your study of electronics continues, you will see that the ability to control the movement of electrons, or electron flow, is the basis of electronics.

Figure 1–18 displays some common sources of energy causing electron movement and/or separation of charges. These sources are:

- friction (static electricity);
- chemical energy (batteries);
- mechanical energy (a generator or alternator);
- magnetic energy;
- light energy; and
- heat energy.

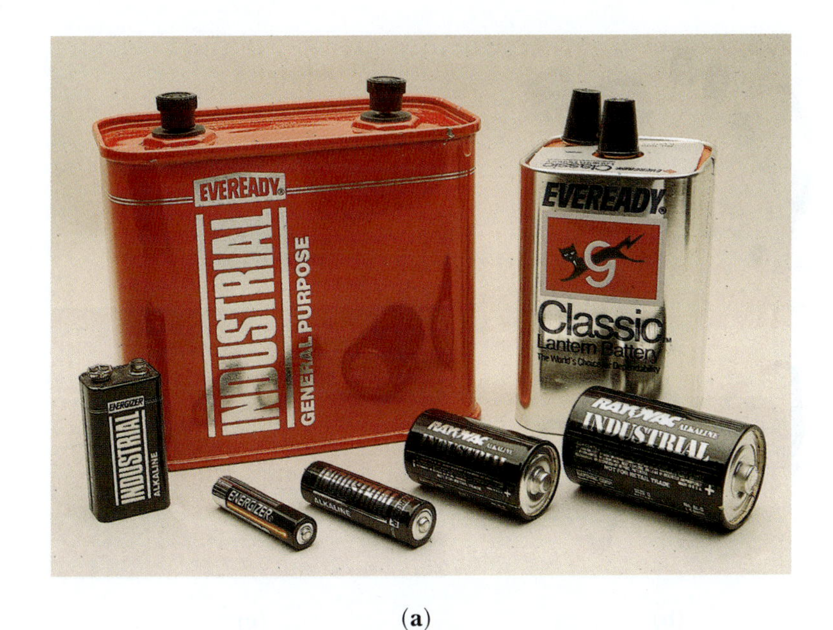

(a)

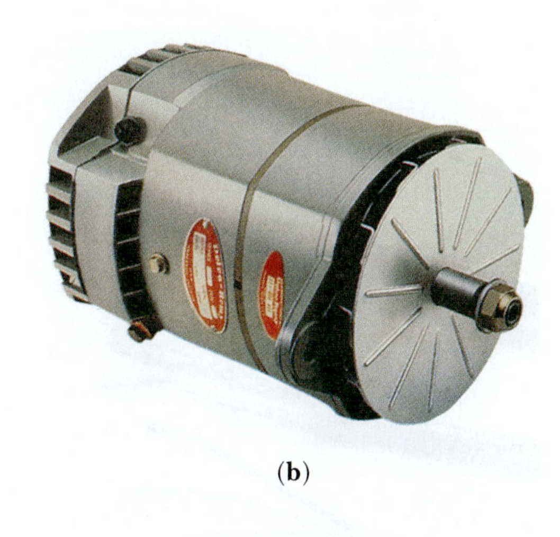

(b)

(c)

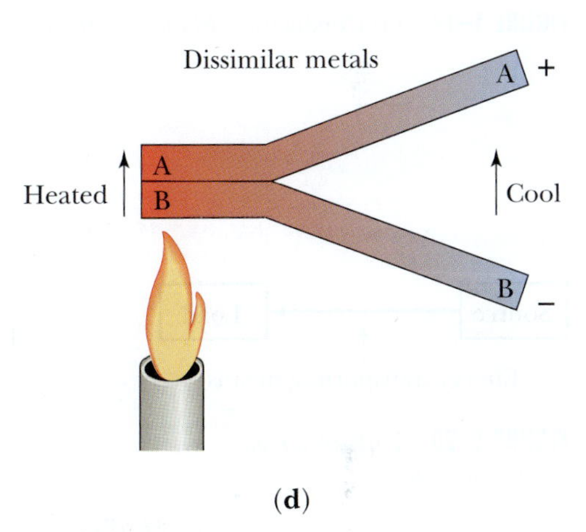

(d)

FIGURE 1–18 Some energy sources used to create useful energy: (a) A battery uses the chemical energy in its cells to create voltage. (b) A generator converts mechanical energy into electricity *(Courtesy of AC-Delco).* (c) Solar cells convert light to electrical energy. (d) A thermocouple uses heat energy to produce an electric current.

By appropriately using one or more of these energies, sources of electrical **energy** may be developed. These sources develop electrical potential, or the potential to move electrons through whatever circuits are electrically connected to these sources. Electrical **circuits** are closed paths designed to carry, manipulate, or control electron flow for some purpose. Later in this chapter, electrical potential and movement of electrons are discussed to show how the concepts presented thus far are foundations for the study of electronics.

1–8 Conductors, Semiconductors, and Insulators

Conductors, such as gold, silver, and copper, have many **free electrons.** These materials conduct electron movement easily because their outermost ring electrons are loosely bound to the nucleus. In other words, their outermost ring contains one, two, or three electrons rather than the eight electrons needed for atomic stability, Figure 1–19a.

Semiconductors, Figure 1–19b, are sample devices that use materials that are halfway between the conductors' characteristic of few outermost shell electrons and the stable, inert

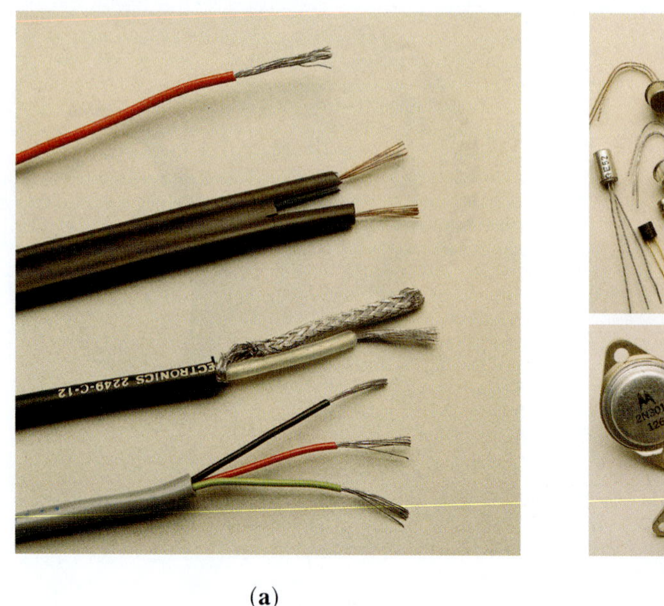

(a)

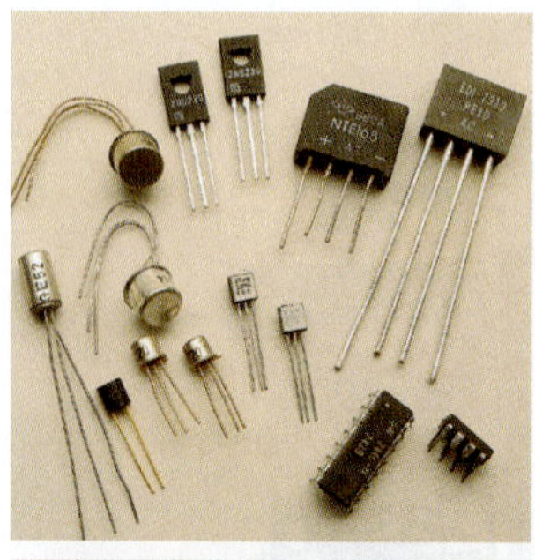

(b)

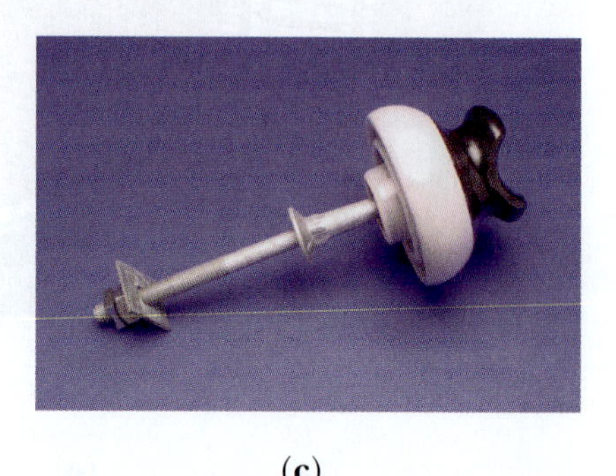

(c)

FIGURE 1–19 (a) Conductors *(Photo by Michael A. Gallitelli);* (b) semiconductors *(Photos by Michael A. Gallitelli);* and (c) insulators

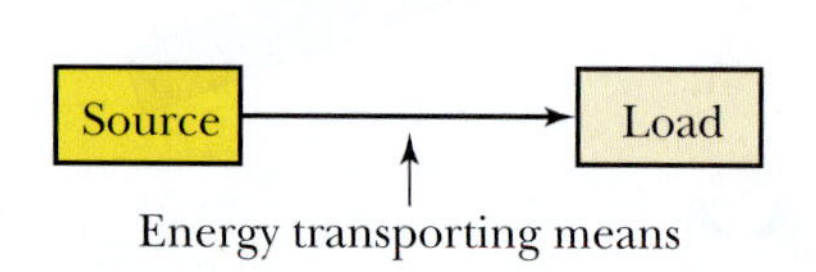

FIGURE 1–20 Parts of a basic electrical system

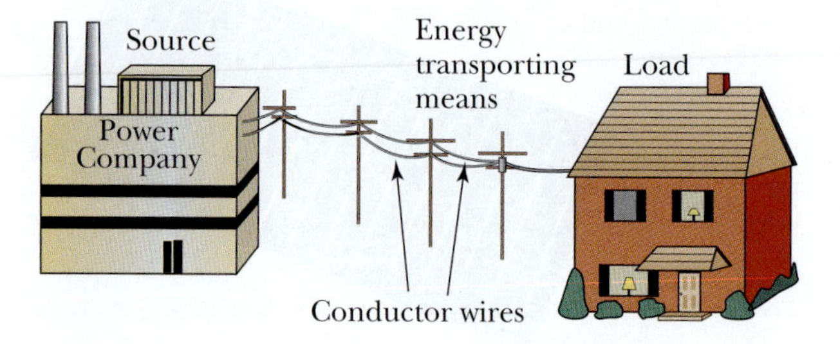

FIGURE 1–21 Components of a commonly used electrical system

materials with eight valence electrons. Semiconductor materials have four outermost ring electrons. Germanium and silicon, as used in the pictured devices, are examples of semiconductor materials.

Insulator materials, Figure 1–19c, do not easily allow electron movement because their five to eight outermost shell electrons are tightly bound to the atom. Insulators have few free electrons. Examples of insulator materials are glass and ceramic.

Electrical and electronic circuits are made up of a variety of components and interconnections consisting of conductors, insulators, and semiconductors.

1–9 Sample of an Electrical System

An electrical system typically has a source of electrical energy, a way to transport that electrical energy from one point to another, and an electrical **load,** Figure 1–20.

The source supplies energy through the transporting means to the load. The load may convert that electrical energy into another form of electrical energy or into another type of energy, such as heat, light, or motion.

Figure 1–21 illustrates an electrical lighting circuit. Observe that the method to transport the electrical energy to the light bulb (the load) is the conductor wires. The load then converts the electrical energy into another useful form, light.

The rest of this chapter will discuss basic electrical laws and principles to help you in understanding what is happening in this basic light circuit. You should seek answers to the following questions: How does the source cause electrical energy or electricity to move through the conductors to the load? In what manner and/or form does the electricity move from one point to another? What determines the quantity of electricity moved from the source to the load?

1–10 Basic Principles of Static Electricity

Before studying the movement of electrons in electrical circuits, it will be helpful to learn some basic principles about charges and static electricity as a frame of reference.

What Is Static Electricity?

Everyone has been electrically shocked after walking on a thick pile rug and touching a door knob or after sliding across an automobile seat and touching the door handle. Where did that electricity come from? Sometimes this electricity is called static electricity. This term is somewhat misleading because static implies stationary and electrons are in constant motion around the nucleus of the atoms. Think of this type of electricity as being associated with nonconductors or insulator materials.

The Basic Law of Electrical Charges

In many science courses, the experiment of rubbing a rubber rod with fur or a glass rod with silk to develop charges is frequently performed, Figure 1–22. After doing this experiment, paper or other light materials are sometimes attracted to the charged body or object. What is meant by a charged body? This means that the object has more or less than its normal number of electrons. In the case of the rubber rod, it gained some electrons from the fur, and the fur lost those electrons to the rod. Thus, the net charge of the rubber rod is now negative, since it has an excess of electrons. Conversely, the fur is positively charged since it lost some of its electrons but retains the same number of protons in its nuclei.

With the silk and glass rod experiment, the glass rod loses electrons and becomes positively charged. Furthermore, another experiment would show there is some attraction between the charged rubber rod (negative) and the charged glass rod (positive). This leads to a conclusion that has been accepted as the basic electrical law: *Unlike charges attract each other, and like charges repel each other.* Remember this important electrical law, Figure 1–23.

Polarity and Reference Points

You might have noticed that a common way of showing the difference between charges is by identifying them with a minus sign for negative or a plus sign for positive. Using positive (+) or negative (–) signs frequently indicates electrical **polarity.** Polarity denotes the relative **electrical charge** of one point in an electrical circuit "with reference to" another point, Figure 1–24. The notion of something "with reference to" something else is not really mysterious. Everyone has heard statements such as "John is taller than Bill," or "Bill is shorter than

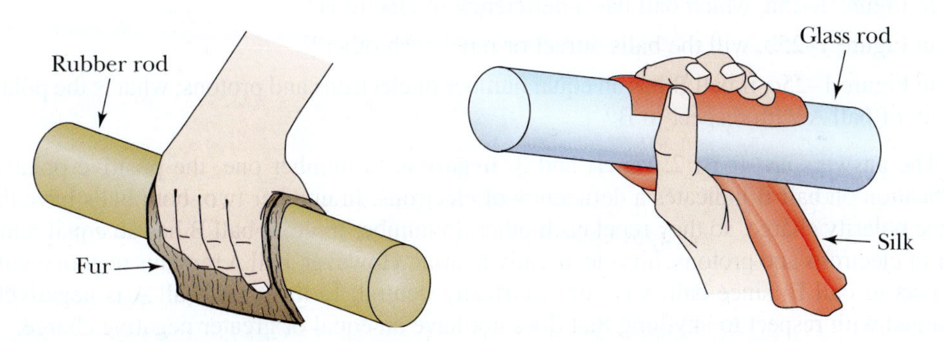

FIGURE 1–22 Static electricity is usually associated with nonconductive materials.

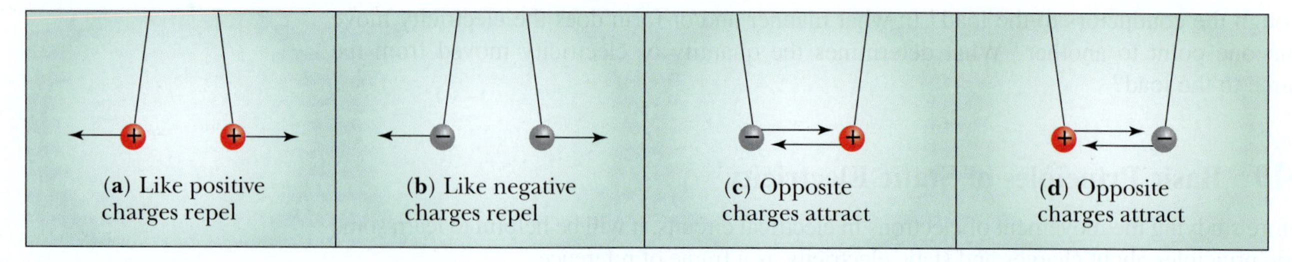

FIGURE 1–23 The basic law of charges is that like charges repel and unlike charges attract.

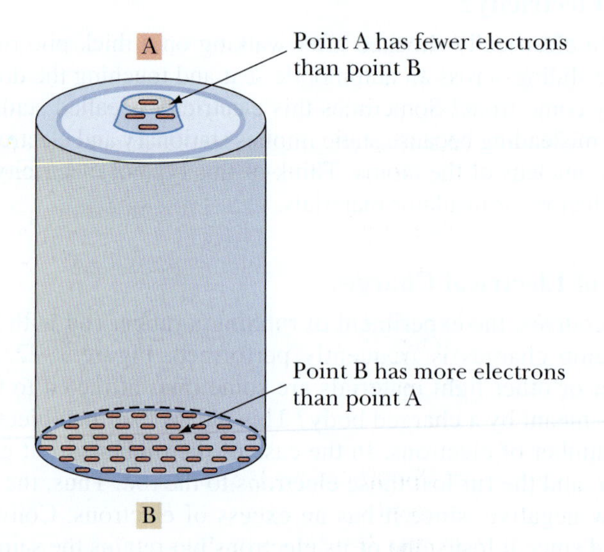

FIGURE 1–24 In a flashlight cell, point A is positive with respect to point B, and point B is negative with respect to point A.

John." In the first statement, Bill is the reference point. In the second statement, the reference point is John.

Another familiar example of polarity and reference points is that the upper tip of the earth's axis is called the North Pole and the lower tip, the South Pole. These terms refer to their geographic locations.

Review some key facts:

1. Electrons are negatively charged particles, and protons are positively charged particles.
2. Any substance or body that has excess electrons is negatively charged.
3. Any substance or body that has a deficiency of electrons is positively charged.
4. Unlike charges attract, and like charges repel.

Refer to Figure 1–25a, b, and c and answer the following questions:

1. In Figure 1–25a, which ball has a deficiency of electrons?
2. In Figure 1–25b, will the balls attract or repel each other?
3. In Figure 1–25c, if ball B has an equal number of electrons and protons, what is the polarity of ball A with respect to B?

The answers are 1. **B;** 2. **repel;** and 3. **negative.** In number one, the positive polarity indication on ball B indicates a deficiency of electrons. In number two, both balls have the same polarity charge, so they repel each other. In number three, if ball B has an equal number of electrons and protons, it is electrically neutral. However, ball A remains negative with respect to ball B, since ball A is not electrically neutral. In this case, ball A is negatively charged with respect to anything that does not have an equal or greater negative charge.

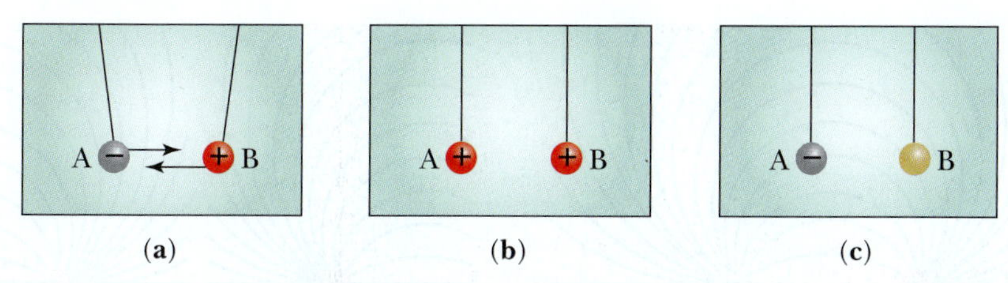

FIGURE 1–25 Some arrangements of electrical charges

Coulomb's Formula Relating to Electrical Charges

Examining the forces of attraction or repulsion (repelling force) between charged bodies, Charles Coulomb, a French physicist, determined that the amount of attraction or repulsion between two charged bodies depended on the amount of charge of each body and the distance between the charged bodies. He expressed this relationship with the formula:

$$\text{Force} = \frac{\text{Charge on body \#1} \times \text{Charge on body \#2}}{\text{Distance between them squared}}$$

FORMULA 1–1 $F = k\dfrac{Q_1 \times Q_2}{d^2}$

where F = force (newtons)
k = the constant 9×10^9 (for air or vacuum)
Q = charge in units of charge (coulombs)
d = distance between charged bodies (meters)

Practical Notes

Wherever you see formulas throughout the book, be aware that "sample calculator sequences" are shown at the end of the chapter for each formula in the chapter. You will find these in the end-of-chapter feature called: "FORMULAS AND SAMPLE CALCULATOR SEQUENCES."

Coulomb's law of charges states that the **force** of attraction or repulsion between two charged bodies is directly related to the product of their charges and inversely related to the square of the distance between them. NOTE: Directly proportional quantities increase or decrease together, with a constant ratio. If one quantity doubles, the other doubles. If one quantity halves, the other quantity halves, and so forth.

Inversely proportional quantities increase and decrease in opposite directions. If one quantity increases in value (for example, doubles), the other inversely proportional quantity decreases in value (for example, halves). If one quantity decreases in value (for example, halves), the other inversely proportional quantity increases in value (for example, doubles).

Unit of Charge

In honor of Charles Coulomb, the unit of charge is designated the **coulomb.** Many electrical units that define quantity or magnitude of electrical parameters are named after famous scientists who performed related experiments.

Electrical units of measure are powerful reference values used every day in electronics. Even as someone had to determine liquid quantities as a gallon, quart, pint, or cup, scientists define units of measure for electrical parameters so that each has its own "reference point" with respect to quantity or value.

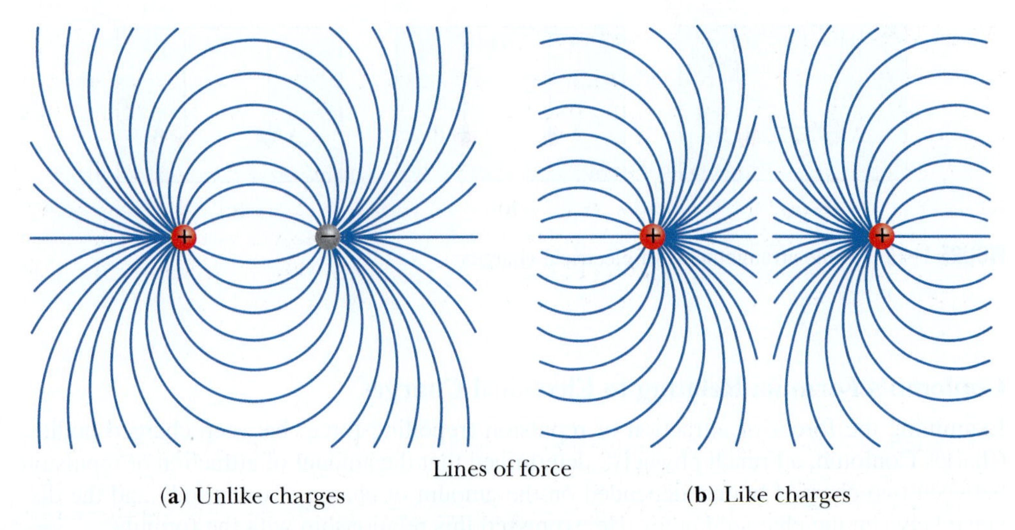

FIGURE 1–26 Electrostatic fields represented by lines of force

In the case of electrical charge, the unit is the coulomb. This is the amount of electrical charge represented by 6.25×10^{18} electrons, or 6,250,000,000,000,000,000 electrons. Now you know the unit for charge used for the "*Q*s" in Formula 1–1.

Fields of Force

You are familiar with several types of "fields," for example, gravitational fields and magnetic fields. Fields of force are represented by imaginary lines, which represent the field of influence for the force involved. Figure 1–26 shows electrostatic fields between unlike charges, Figure 1–26a, and between like charges, Figure 1–26b. This only shows a visualization of the nature of these fields, as represented by the lines. It does not attempt to indicate the strength of the charges, the strength of the field of force, or the distance between the charges. This representation helps us visualize the tangible forces between electrical charges.

1–11 Electrical Potential

What Is Electrical Potential?

The implication drawn from these tangible forces is that they can be harnessed to do work. In this case, the difference of charge levels at two points has the potential to move electrons from a point of excess electrons to a point of electron deficiency, if a suitable path is provided. (Remember, the protons are in the nucleus and are not free to move.) This difference between two points having different charge levels is often termed a **potential** *difference.* This force, which moves electrons from one place to another, is called an **electromotive force** (emf). Remember these important terms because you will use them often.

The Unit of Electrical Potential or Electromotive Force

Scientists have established a measure for the unit of charge (the coulomb), and the unit of measure for this difference of potential. They have named this unit of potential difference the **volt** in honor of Alessandro Volta who invented the electric battery and the electric capacitor. Often, this potential difference between two points, which is measured in volts, is called voltage. An in-depth discussion will come later. For now mentally picture that charges and/or different points are positive or negative relative to each other (that is, have polarity); that there are different quantities of charge depending on the number of excess electrons or electron deficiency; and that there is a potential difference between such points, which provides an electromotive force capable of causing electron movement from one point to another point.

What Means Can Produce Electrical Potential?

Since electrical potential difference (electromotive force), or voltage, performs the electrical work of moving electrons from place to place, what establishes and maintains this voltage?

Earlier in this chapter, we discussed various types of energy that can cause "electrical imbalance," which you now know as electromotive force, or voltage. Static electricity, chemical energy, mechanical energy, magnetic energy, light energy, and heat energy were all mentioned. Look again at Figure 1–18 to review methods of using these various energies to establish and maintain differences of potential (voltage) between two points.

1–12 Charges in Motion

What happens if the two balls in Figure 1–27 touch each other? If you said that the excess electrons on ball A move to ball B and attempt to overcome the deficiency of electrons on ball B, you are right. In fact, electrons would continue to move until ball A and ball B had equal charges, or were neutral with respect to each other.

If the balls do not touch each other but are connected with a copper conductor wire, Figure 1–28, what happens? If you surmise that some electrons move through the conductor wire until the charges on ball A and ball B are equal, you are right.

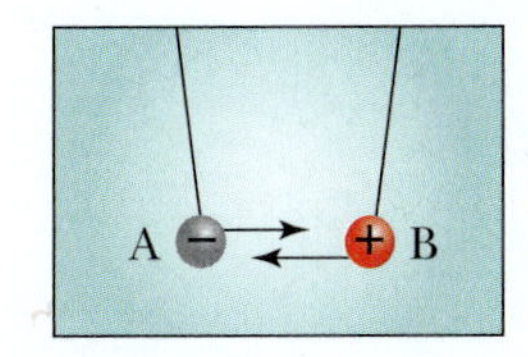

FIGURE 1–27 Two different charges

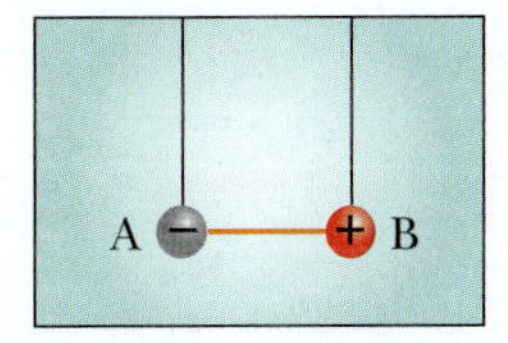

FIGURE 1–28 If a conductor wire is connected to ball A and ball B, what happens?

Current

Are the electrons that move to ball B through the conductor the same electrons that left ball A? Probably not. As you have already studied, the conductor wire has many free electrons moving within it. Having a positive charge at one end of the conductor and a negative charge at the other end (a potential difference between its ends) causes the electrons to move from the negative end to the positive end. This movement of electrons is known as electrical **current flow,** or **current.**

An Analogy of Current Flow

Figure 1–29 illustrates the concept of the movement of electrons (current flow) from atom to atom within the conductor (from negative to positive ends). Bin A has a large quantity of rubber balls (representing electrons) and bin B has a few. The row of people between the two bins represent the conductor. They pass the balls from person to person, similar to the conductor material passing electrons from atom to atom when a current flows through the conductor. The progressive movement of balls from bin A to bin B simulates the movement of

FIGURE 1–29 An analogy relating to electron movement

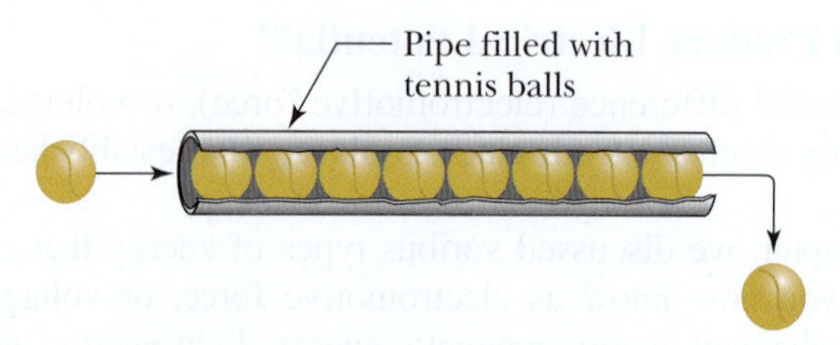

FIGURE 1–30 Another analogy relating to electron movement

electrons from bin A through the conductor to bin B. In the conductor, as a free electron leaves its original atom and moves to an adjacent atom, it is replaced by another electron from another atom. This process is multiplied millions of times when there is current through a conductor. In Figure 1–28, as an electron leaves one end of the conductor (attracted by ball B's positive charge), it is replaced by an electron entering the conductor from ball A. Figure 1–30 also illustrates current flow.

Obviously, these analogies are oversimplifications of the physics involved in current flow, but they will aid your understanding as we discuss current flow and the specific electrical phenomena involved.

The Unit of Current

Recall that the unit of charge, the coulomb, was established in honor of Charles Coulomb, and the unit of potential difference, the volt, was named after Alessandro Volta. In like manner, the unit of measure for current was named in honor of the French mathematician and physicist, Andre Ampere.

An **ampere** of current is the quantity of electron movement represented by a flow rate of one coulomb of charge per second. Restated, *a flow of one coulomb per second = one ampere.*

A Formula That Relates Current to Charge Movement and Time

A common formula that relates current (in amperes) to coulombs of charge and time (in seconds) is:

FORMULA 1–2 $I = \frac{Q}{T}$

where I = current in amperes
Q = charge in coulombs
T = time in seconds

This formula allows you to calculate the current in amperes, if you know the charge in coulombs and the time in seconds. For example, if 10 coulombs of charge move from one point to another point in an electrical circuit over 2 seconds, current equals 5 amperes ($I = 10/2$). Also, this formula can be transposed to calculate the charge if the current and time are known. This transposition yields $Q = I \times T$. If at this point you do not understand how the formula was transposed, be encouraged by knowing that some techniques of transposing will be discussed again later in the book. Look at this formula and the examples, and use your knowledge about direct and inverse relationships to understand the formula. Is current *directly* or *inversely* related to the quantity of charge moved? Is current *directly* or *inversely* related to the duration of time taken to move the given charge? Your answers should have indicated a *direct* relationship between current and quantity of charge moved and an *inverse* relationship between current in amperes and the time it took to move the given charge. Recall that the direct relationship between quantities means that as one quantity increases, the other quantity increases. The inverse relationship indicates that as one quantity increases, the other quantity decreases, or vice versa.

IN-PROCESS LEARNING CHECK 2 Fill in the blanks as appropriate.

1. An electrical system (circuit) consists of a source, a way to transport the electrical energy, and a _____.
2. Static electricity is usually associated with _____ -type materials.
3. The basic electrical law is that _____ charges attract each other and _____ charges repel each other.
4. A positive sign or a negative sign often shows electrical _____.
5. If two quantities are *directly* related, as one increases the other will _____.
6. If two quantities are *inversely* related, as one increases the other will _____.
7. The unit of charge is the _____.
8. The unit of current is the _____.
9. An ampere is an electron flow of one _____ per second.
10. If two points have different electrical charge levels, there is a difference of _____ between them.
11. The volt is the unit of _____ force, or _____ difference.

1–13 Three Important Electrical Quantities

This chapter has discussed the magnitude of electrical charges (amount or value of charge); polarity (either negative or positive); and the difference of charge between two points that creates a potential difference (electromotive force or voltage) that can move electrons from one point to another (a current flow).

Also, you have learned that scientists have defined units of measure for the quantities of charge, potential difference, and current. Recall that the unit of charge is the coulomb (6.25×10^{18} electrons); the unit of electromotive force, or potential difference, is the volt; and the unit of current flow, or current, is the ampere. Of the electrical parameters described thus far, you will most frequently use the units for current and electromotive force (i.e., the ampere and the volt).

A third electrical quantity you will often use is **resistance.** If you have tried to sandpaper a piece of wood, you have experienced *physical resistance.*

Current flow through a conductor encounters molecular resistance to its flow. The amount of resistance to current flow depends on a number of factors. One factor is the type of material through which the current is flowing. Other factors are the dimensions of the conductor wire, including its cross-sectional area and its length, and temperature. An electrical pump (emf or voltage) causes current to flow through the resistance in its path.

As you have probably reasoned, electrical resistance is the opposition shown to current flow. Again, an electrical unit has been named after a scientist, Georg Simon Ohm; thus, the unit of resistance is the **ohm.**

The ohm of resistance has been defined in several ways. One definition states an ohm is the resistance of a column of pure mercury with a cross-sectional dimension of 1 millimeter squared and a length of 106.3 centimeters at a temperature of zero degrees centigrade. Another definition of an ohm is the amount of resistance that develops 0.24 calories of heat when one ampere of current flows through it. *The definition of ohm that is most useful is:*

An ohm of resistance is the amount of electrical resistance that limits the current to one ampere when one volt of electromotive force is applied.

1–14 Basic Electrical Circuit

Generalized Description

Earlier in the chapter, a simple description was given of an electrical circuit. Recall that an electrical circuit has three basic ingredients: the **source,** the means to carry electricity from one point to another point, and a load. Before thoroughly examining this basic circuit, we will discuss the difference between a closed and an open circuit.

The Closed Circuit

You have learned that if a current path is provided when a potential difference exists between two points, electrons move from the point of excess electrons (negative polarity point) to the point of electron deficiency (positive polarity point). One method of providing this path is to connect a conductor wire between the points, thus providing a route through which electrons move and establish current flow. A closed circuit is a complete, unbroken path through which electrical current flows whenever voltage is applied to that circuit, Figure 1–31.

The Open Circuit

You probably already understand that if a closed circuit is an unbroken path for electron flow, then an open circuit has a break in the path for current flow, Figure 1–32. This break may be either a desired (designed) break or an undesired (unplanned) break. The most common method for purposely opening a circuit is an electrical switch, such as the one you use to turn the lights on or off. The switch is a fourth element of the basic electrical circuit, and many circuits contain such a control component, or circuit. In both Figures 1–31 and 1–32 the switch is the control component.

The Basic Electrical Circuit Summarized

The basic electrical circuit has (1) a source, (2) a means of conducting electron movement (current), and (3) a load, and may have (4) a control element, such as a switch.

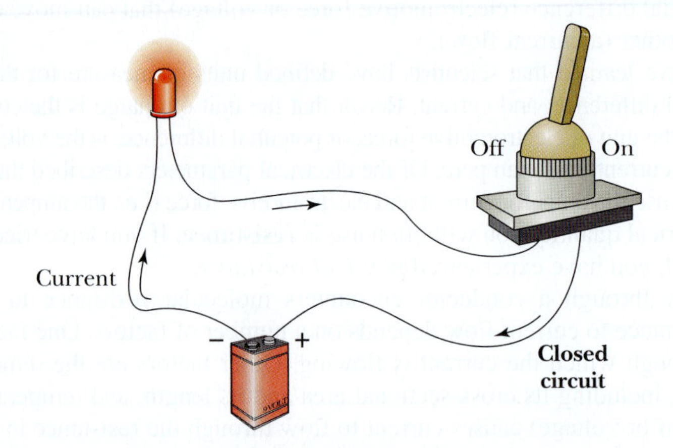

FIGURE 1–31 A closed circuit provides an unbroken path for current flow.

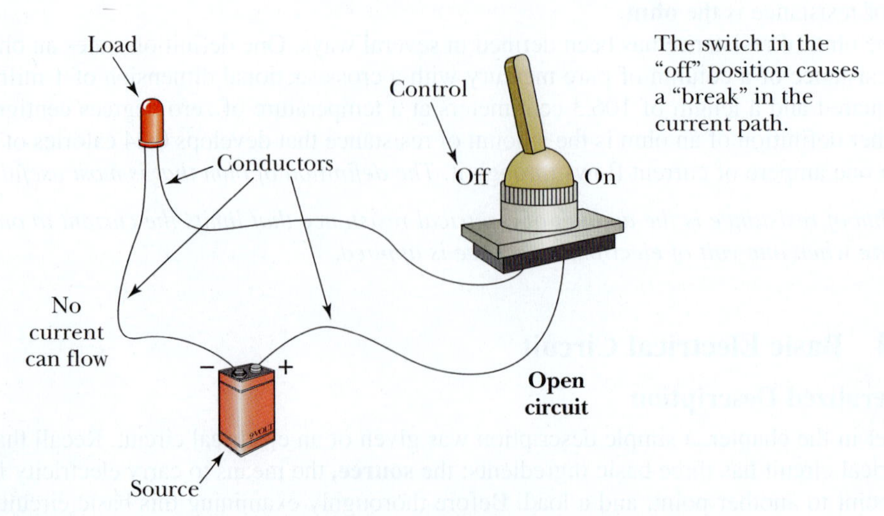

FIGURE 1–32 An open circuit has a break somewhere in the current path.

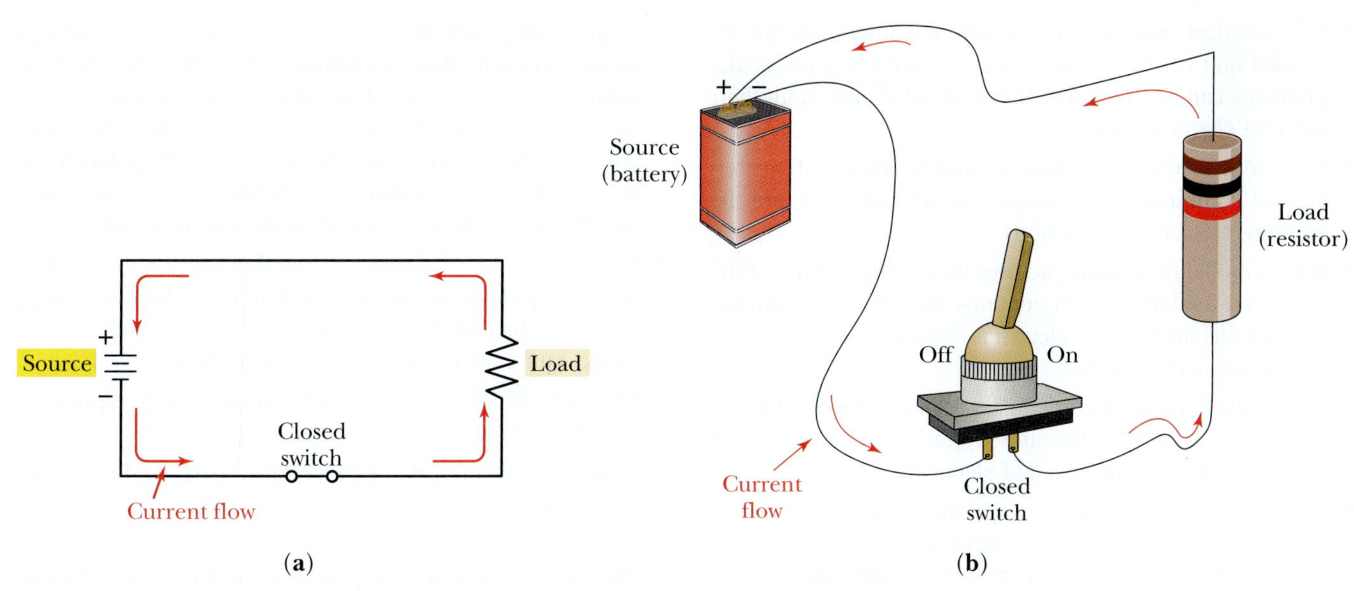

FIGURE 1–33 Direction of current flow through a circuit: (a) schematic; (b) pictorial multiSIM

PARTS OF A CIRCUIT	EXAMPLE OF EACH	IMPORTANT ELECTRICAL QUANTITIES	
1) Source	Power company generators	EMF	(Volts)
2) Transporting means	Conductor wires	Resistance	(Ohms)
3) Load	Light bulb(s)	Current	(Amperes)
4) (Control)	(Switch)		

FIGURE 1–34 Important facts about a basic electrical circuit: as emf increases, current increases; as resistance increases, current decreases.

For current to flow (electron movement) through the circuit, there must be a voltage source and a closed circuit or a complete current-conducting path. If the conductor wire(s) were accidentally connected across the source with no other load component, the wire (with its small resistance) would become the load. Depending on the resistance of the wire and the amount of source voltage, this undesired low resistance condition (sometimes called a short circuit) would probably result in the wire overheating and melting, thus breaking the circuit.

Applying a voltage (electromotive force) to a closed circuit causes current to flow from the source's negative side, through the circuit conductors and load back to the source's positive side, Figure 1–33. The amount of current (amperes) that flows depends on the value of voltage (volts) applied to the circuit and how much resistance (ohms) the circuit offers to current flow, Figure 1–34. Remember these points because they will become more important as your studies proceed.

Summary

- Matter occupies space, has weight, and sometimes can be tangible to one or more of our five senses.
- Matter's physical states include solids, liquids, and gases.
- Matter's chemical states include elements, compounds, and mixtures.
- Elements are comprised of only one type of atom, but compounds are unique combinations of different elements' atoms that result in a new matter or substance.
- A mixture does not cause chemical or physical changes in the elements being mixed.

- The smallest particle into which a compound can be divided and retain its basic characteristics is a molecule. Elements can be divided to the atom level and retain their inherent characteristics.
- The atom is the fundamental building block of matter. Special combinations of atoms, called molecules, are the building blocks of compounds.
- Bohr's model of an atom pictures the central portion of the atom, the nucleus, which contains protons and neutrons. Around this nucleus are electrons, which orbit the nucleus like planets orbit the sun.
- The electron theory helps illustrate electricity and behavior of electrons in matter and relates to the atom model with its protons, neutrons, and electrons.
- The atom's protons have a positive electrical charge, while the orbiting electrons have a negative electrical charge. The neutrons in the center of the atom add mass to the atom but are electrically neutral.
- Protons and neutrons are approximately 1,800 times heavier than electrons.
- The net charge of an atom is typically neutral, since there are equal numbers of electrons and protons and their electrical charges offset each other.
- The atomic number of an element is the number of protons in the nucleus of each of its atoms.
- The atomic weight of an element is a comparison of its weight to that of carbon-12.
- Electrons travel around the nucleus of an atom in shells that are at different distances from the nucleus. The first ring (shell) carries a maximum of 2 electrons; the second shell outward a maximum of 8; the next shell, 18; and the fourth shell, 32. The outermost electron ring containing the valence electrons determines if the atom is electrically and chemically active or stable and is full when it has 8 electrons.
- If the outer shell of an atom contains eight electrons, it is stable. If the outer ring has less than four electrons, it is usually a conductor. If the number of valence electrons is four, it is typically a semiconductor material.
- Materials, such as copper, that have only one outermost ring electron per atom are good conductors. These electrons, sometimes called free electrons, are easily moved within the material from atom to atom.
- An atom that has lost or gained electrons is called an ion. If it has lost electron(s), it is a positive ion. If it has gained electron(s), it is a negative ion.
- Ions and/or movement of electrons within a material are caused by various external and internal energies or forces. Several examples of energy that can electrically unbalance atoms or move electrons are friction, chemical, heat, and light.
- The law of electrical charges is that unlike charges attract and like charges repel each other.
- Static electricity is usually associated with nonconductor materials and is often caused by friction between them.
- Polarity designates differences between two points or objects relative to each other, or opposites. For example, electrical polarities of "–" or "+" show which point has excess electrons and electron deficiency relative to each other. Magnetic polarity indicates differences between magnetic poles (North and South). Also, electrical polarity shows direction of current flow through an electrical component or circuit.
- Coulomb's law of charges states that the force of attraction or repulsion between two charges is directly related to the product of the two charges and is inversely related to the square of the distance between them.
- The unit of charge is the coulomb, which represents 6.25×10^{18} electrons of charge.
- Electrical potential is the potential to perform the electrical work of moving electrons. It is sometimes termed electromotive force.
- The unit of measure for potential difference (electromotive force) is the volt.
- Several types of energy that can establish and maintain potential differences between two points (i.e., create a voltage source) are chemical, heat, light, magnetic, and mechanical energies.
- Organized electron movement, or electron flow, is known as current flow or current.
- Current flow caused by electron movement is from the negative side of the voltage source to the positive side of the voltage source through the circuit connected to the source, if the circuit is a closed circuit.
- The unit of current is the ampere, which represents a rate of electron flow of one coulomb per second.
- A formula that relates the amount of current flow (in amperes) to the amount of charge (in coulombs) moved in a given amount of time (in seconds) is $I = Q/T$. The formula may also be transposed to read $Q = I \times T$.
- Three important electrical quantities of measure are:
  a. volt, the unit of potential difference, or amount of potential difference causing one ampere of current to flow through a resistance of one ohm;
  b. ampere, the unit of current flow, or amount of current flow represented by a rate of charge flow of one coulomb per second; and
  c. ohm, the unit of resistance, or amount of electrical resistance limiting the current to one ampere with one volt applied.
- A basic electrical circuit has an electrical source (emf), a way to transport electrons from one point to another point, and a load, which uses the electrical energy. Another part of many basic circuits is a control device, such as a switch.
- A closed circuit has an unbroken path for current flow: from one side of the source, through the circuitry and back to the other side of the source.
- An open circuit does not provide a continuous path for current flow. The path is broken, either by design (using a

switch) or by accident. The open circuit presents an infinite resistance path for current.

- A short circuit is an undesired, low resistance path for current flow. If the short is across both terminals of a voltage source, it frequently causes undesired effects, such as melted wires, fire, and other damage to the voltage source.
- In the basic electrical circuit, the voltage source provides the electromotive force that moves electrons and causes current flow. The circuit conductors and components provide resistance to current, thereby limiting the current. The various techniques of controlling electrons and electron flow to perform desired effects form the basis of electronic science.

Formulas and Sample Calculator Sequences

NOTE: The calculator sequences shown throughout the book are based on a popular "Algebraic Notation" type of calculator. That is, the sequencing of calculations performed by the device are closely allied to the hierarchy system of operations used in Algebra. For example, operations such as reciprocal, square, and square root are higher priority than powers and roots. Powers and roots are higher priority than multiplication and division. Multiplication and division are higher priority than addition and subtraction, and the equal sign becomes the lowest priority. With algebraic operating systems such as this, lower-priority operations are delayed until higher-priority operations are complete.

Some training programs may have students use calculators that utilize what is known as "Reverse Polish Notation." These calculators use a system that evolved from an earlier system of simplifying arithmetic expressions, as conceived by a Polish logician and scientist named Lukasiewicz. The evolved system somewhat reverses the Polish logician's notation technique, hence the term *Reverse Polish Notation,* or RPN. In some cases the RPN-type calculators can accomplish a given calculation with slightly fewer keystrokes. For this reason, some instructors prefer having their students use this type of device.

No matter which type of calculator your program requires, let us encourage you to begin learning to use the calculator to your advantage at the earliest possible time. Calculators are wonderful tools for the technician and engineer alike! ■

FORMULA 1–1
(To find force)

$$F = k\frac{Q_1 \times Q_2}{d^2}$$

k value, [×], Q_1 value, [×], Q_2 value, [=], [÷], distance value, [x^2], [=]

FORMULA 1–2
(To find current)

$$I = \frac{Q}{T}$$

charge in coulombs, [÷], time in seconds, [=]

EXCEL AUTOMATED FORMULAS

Helpful problem-solving tools, such as the Excel automated formulas available on the CD, allow automatic calculations of formulas.

Review Questions

1. Define the term *matter.*
2. List the three physical states of matter and give examples of each.
3. List the three chemical states of matter and give examples of each.
4. Define the term *element* and give two examples of elements.
5. Define the term *compound* and give two examples of compounds.
6. Sketch a hydrogen atom. Identify each particle.
7. On your hydrogen-atom sketch, designate the electrical charge of each particle.
8. **a.** Which has more weight or mass, the electron or the proton?
   b. Approximately how many times heavier is the one compared to the other?
9. Define the term *free electron.*
10. **a.** Define the term *ion.*
    b. What causes an ion to be classified as a positive ion?
11. Name four energy types that can be used to purposely create electrical imbalances in atoms of materials.

12. How many valence electrons do each of the following materials have?
    a. A conductor.
    b. A semiconductor.
    c. An insulator.
13. Describe the basic law of electrical charges.
14. Describe Coulomb's law of electrical charges.
15. If the distance between two charges is tripled and the charge strength of each charge remains the same, what happens to the force of attraction or repulsion between them?
16. In an electrical circuit, what is potential difference?
17. What unit of measure expresses the amount of electromotive force present between two identified points in a circuit?
18. In an electrical circuit, what is current flow?
19. a. What basic unit of measure expresses the value of current in a circuit?
    b. What is the value of current flow if 20 coulombs of charge move past a given point in 5 seconds?
20. Describe the term *resistance* in terms of voltage and current, and give the unit of measure.
21. Describe the difference between a closed circuit and an open circuit.
22. List three or more elements of a basic electrical circuit.
23. What sign or symbol is used to denote a negative polarity point in an electrical diagram?
24. In a "closed" electrical circuit, electron movement (or electron current) is from the more (negative? positive?) __________ point in the circuit toward the more (negative? positive?) __________ point in the circuit.
25. In the circuit shown in Figure 1–35, is Point B representing the negative or the positive side of the source?

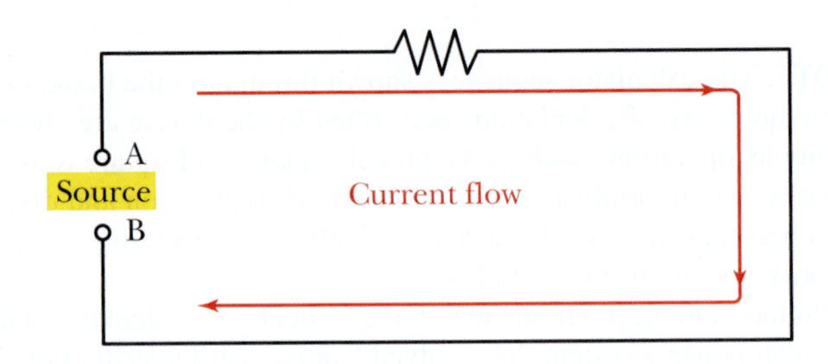

FIGURE 1–35

Analysis Questions

1. Draw a diagram illustrating an electrically balanced atom that has 30 protons in its nucleus and whose first three shells are full, with the fourth (last) shell having two electrons.
2. Label all electrical charges in your diagram.
3. What kind of material is pictured in your diagram?
4. Is this material a conductor, semiconductor, or insulator?
5. What is its atomic number?
6. Define the term *polarity* as used in electrical circuits.
7. Draw an electrical circuit showing voltage source, conductor wires, and load. Show the polarity of that source, and, by using an arrow, show the direction of current flow through the circuit. (Assume a closed circuit.)
8. Referring to Figure 1–33b, list three ways this circuit might be made to become an open circuit.
9. In your own words, define the term *short circuit.*
10. What are some possible consequences of an undesired short circuit?

OUTLINE

OBJECTIVES

After studying this chapter, you should be able to:

1. List the units of measure for charge, potential (emf), **current, resistance,** and **conductance** and give the appropriate abbreviations and symbols for each
2. Use metric system terms and abbreviations to express subunits or multiple units of the primary electrical units
3. List the factors that affect the resistance of a conductor
4. Recognize common types of conductors
5. Use a wire table to find conductor resistance for given lengths
6. Recognize and/or draw the diagrammatic representations for conductors that cross and electrically connect, and that cross and do not connect
7. Define the term **superconductivity**
8. Give the characteristics of several common types of **resistors**
9. Explain the characteristics of surface mount "chip" resistors
10. Use the **resistor color code**
11. Use other special resistor coding systems
12. Explain how to connect meters to measure **voltage,** current, and resistance
13. Recognize and/or draw the diagrammatic symbols for elemental electronic components or devices
14. Interpret basic facts from **block** and **schematic diagrams**
15. List key safety habits to be used in laboratory work

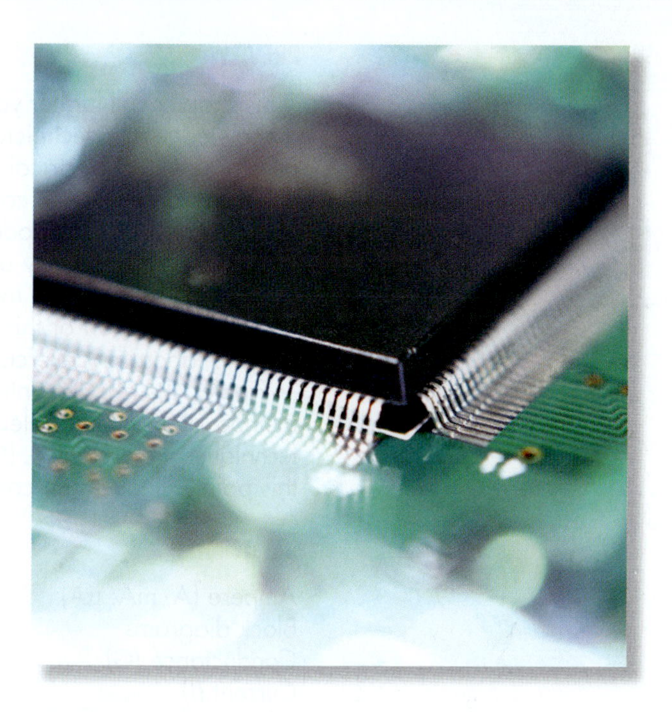

CHAPTER 2

Electrical Quantities and Components

ELECTRONICS INTO THE FUTURE

Explore an interactive presentation on Ohm's Law, Amps, Volts, Ohms & Power on the accompanying CD.

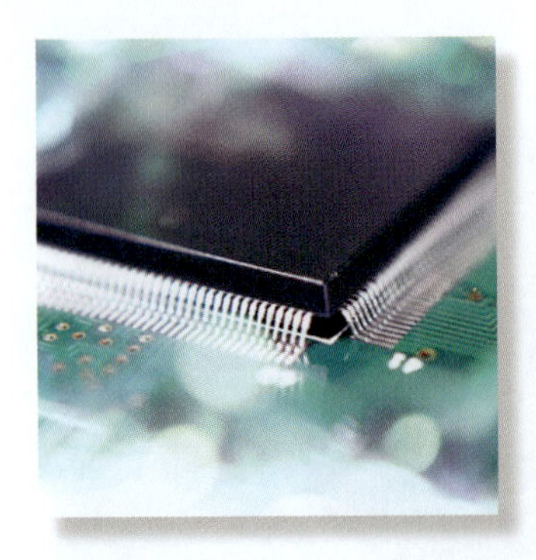

PREVIEW

This chapter will further your study of the basic electrical quantities that are vital to your knowledge as a technician. For example, commonly used metric prefixes and powers of ten related to electrical quantities are presented. Information regarding basic characteristics and types of conductors and resistors is presented. You will also study and learn the resistor color code. (NOTE: Interpreting the color code is one of the most practical and most frequently used skills ever acquired by technicians.)

How to use meters to measure voltage, current, and resistance is also presented. Of course, application of this knowledge, and practice at doing this, is acquired by performing laboratory exercises that require making actual meter measurements.

Finally, the chapter introduces you to the technicians' and engineers' "shorthand" method of presenting electrical circuits. Both block diagram and schematic diagram techniques are presented. (NOTE: These "knowledge capsules" that may seem unrelated at this point will all begin coming together for you in the next chapter.)

KEY TERMS

Ampere (A; mA; μA)
Block diagrams
Conductance *(G)*
Current *(I)*
Ohm (Ω; kΩ; MΩ)
Resistance *(R)*
Resistor (R)
Resistor color code
Schematic diagrams
Siemens (S; mS)
Switch (S)
Volt (V; mV; μV)
Voltage *(V)*

2–1 Basic Electrical Units and Abbreviations

Charge

Recall that the basic unit of charge is the coulomb and that the letter "*Q*" (or *q*) represents charge. However, you need to know that the abbreviation for the coulomb is **C.** (Remember that one C equals 6.25×10^{18} electrons.) Notice that this is a capital "C" and not a lowercase "c." If the unit abbreviation or symbol for electronic measurement is named after an individual, such as coulomb, volt, ampere, or ohm, the abbreviation is a capital letter and not a lowercase letter. Keep this in mind and get in the habit of correctly writing unit abbreviations or symbols.

FIGURE 2–1 A digital multimeter (DMM) used to measure voltage, current, and resistance *(Courtesy of John Fluke Mfg. Co., Inc.)*

Potential

The unit for potential difference, or electromotive force, is the **volt,** named after Alessandro Volta. The abbreviation, or symbol, for this unit is **V. Voltage** is expressed in volts. Recall that one volt equals the amount of electromotive force (emf) that moves a current of one ampere through a resistance of one ohm.

Current

The unit of measure for current flow is the **ampere.** The abbreviation, or symbol, for this basic unit of measure is **A.** Do not confuse the letter *I* (thought of as intensity of electron movement) that represents the general term **current** with the *A* used for the unit of current. Remember that one ampere equals an electron flow of one coulomb per second past a given point.

Resistance

Resistance is another electrical parameter that uses two letters: "***R***" represents the general term resistance and the Greek letter omega (Ω) represents the unit of resistance, the **ohm.** Remember that one ohm equals the resistance that limits the current to one ampere with one volt applied. (NOTE: Figure 2–1 shows a multimeter used to measure V, A, and Ω, the units of measure we have just discussed.)

Conductance

Another electrical parameter is **conductance.** It sometimes is defined as the comparative ease with which current flows through a component or circuit. Conductance is the opposite of resistance. The unit of conductance is the **siemens (S)** named after the scientist Ernst von Siemens. The abbreviation for the general term *conductance* is **G,** and the unit of measure is the siemens (S). Conductance, in siemens, is defined as the reciprocal of resistance in ohms.

$$G \text{ (in siemens)} = \frac{1}{R(\Omega)} \text{ or } G = \frac{1}{R(\Omega)} \text{(S)}$$

These units of measure and abbreviations, or symbols, are important, so a chart has been provided, Figure 2–2. Study this chart until you have memorized all the units and their abbreviations.

BRIEF DEFINITION OF QUANTITY	ELECTRICAL QUANTITY OR PARAMETER	BASIC UNIT OF MEASURE	ABBREVIATION OR SYMBOL FOR UNIT
Excess or deficiency of electrons	Charge (Q)	Coulomb	C (6.25 × 10^{18} electrons)
Force able to move electrons	Potential difference (emf)	Volt (Force that moves one coulomb of charge per second through one ohm of resistance)	V
Progressive flow of electrons	Current (I)	Ampere (An electron flow rate of one coulomb per second)	A
Opposition to current flow	Resistance (R)	Ohm (A resistance that limits current to a value of one ampere with one volt applied)	Ω
Ease with which current can flow through a component or circuit	Conductance (G)	Siemens (The reciprocal of resistance, or, $\frac{1}{R}$)	S

FIGURE 2–2 Important electrical units' abbreviations and symbols

2–2 Using the Metric System to Help

Some Familiar Metrics

Many people have been exposed to the metric system. The meter (approximately 39 inches), centimeter (one one-hundredth of a meter), millimeter (one one-thousandth of a meter), and kilometer (1,000 meters) are not uncommon terms. As you can see, the prefix to "meter" identifies a multiple or subdivision of the basic unit. In other words, the meter is the basic unit and the prefix centi equals 1/100, milli equals 1/1,000, and kilo equals 1,000 times the basic unit.

This technique of using metric prefixes is also useful in identifying multiples or subdivisions of basic electrical units. For example, in electronic circuits it is common to use thousandths or millionths of an ampere, or thousands of ohms of resistance. Although the subunits and multiple units in Figure 2–3 do not represent all prefixes encountered in electronics, they do represent those you will be immediately applying. Study Figure 2–3 until you learn each of the metric terms and their abbreviations or symbols.

There is one other bit of related knowledge that can be of great help to you in practical work. That is, mathematical ways to represent each of these metric prefixes. These mathematical representations are helpful in studying electronic circuits, because simple calculations can be used to analyze, verify, or predict circuit behavior.

Because the subunit and multiple-unit prefixes in Figure 2–3 are based on a decimal system (multiples or submultiples of 10), it is convenient to express these prefixes in powers of ten. Note that:

$$10^0 = \text{(ten to the zero power)} = 1$$
$$10^1 = \text{(ten to the 1st power)} = 10$$
$$10^2 = \text{(ten to the 2nd power)} = 10 \times 10 = 100$$
$$10^3 = \text{(ten to the 3rd power)} = 10 \times 10 \times 10 = 1{,}000$$
$$10^{-1} = \text{(ten to the negative 1st power)} = 1/10 = 0.1$$
$$10^{-2} = \text{(ten to the negative 2nd power)} = 1/100 = 0.01$$
$$10^{-3} = \text{(ten to the negative 3rd power)} = 1/1{,}000 = 0.001$$

Figure 2–4 shows how powers of 10 relate to the electrical units and prefixes discussed earlier. **Learn these because you will use them frequently.**

Study Figure 2–4. You can surmise that powers of ten can simplify writing very large or very small numbers. For example, instead of writing 3.5 microamperes (μA) as 0.0000035 amperes, you can state it as 3.5×10^{-6} amperes. The power of 10 tells us how many decimal places and which direction the decimal moves to state the number as a unit. Thus, 3.5×10^{-6} indicates the

METRIC TERM	SYMBOL	MEANING	TYPICAL USE WITH ELECTRONIC UNITS
pico	p	One millionth of one millionth of the unit	Picoampere (pA)
nano	n	One thousandth of one millionth of the unit	Nanoampere (nA) Nanosecond (ns)
micro	μ	One millionth of the unit	Microampere (μA) Microvolt (μV)
milli	m	One thousandth of the unit	Milliampere (mA) Millivolt (mV)
kilo	k	One thousand times the unit	Kilohms (kΩ) Kilovolts (kV)
mega	M	One million times the unit	Megohm(s) (MΩ)

FIGURE 2–3 Some frequently used metric units

decimal moves six places to the left when stated as the basic unit (i.e., 0.0000035 amperes). This is three and one-half millionths of an ampere.

Another example is 5.5 megohms (MΩ) of resistance simplified as 5.5×10^6 ohms. How would this be stated as a number representing the basic unit? If you said the power of 10 indicates moving the decimal six places to the right, you are correct. This yields 5,500,000 ohms.

Obviously, it is easier to manipulate numbers using powers of 10 than using the bulky multiple decimal places involved in the small and large numbers typically associated with electrical/electronic units. You will get more practice using powers of 10 in the next chapter.

NUMBER	POWER OF 10	TERM	SAMPLE ELECTRONIC TERM
0.000000000001	10^{-12}	pico	pA (1×10^{-12} ampere)
0.000000001	10^{-9}	nano	nA (1×10^{-9} ampere)
0.000001	10^{-6}	micro	μA (1×10^{-6} ampere)
0.001	10^{-3}	milli	mA (1×10^{-3} ampere)
1,000	10^{3}	kilo	kΩ (1×10^{3} ohms)
1,000,000	10^{6}	mega	MΩ (1×10^{6} ohms)
1,000,000,000	10^{9}	giga	GHz (1×10^{9} Hertz)
1,000,000,000,000	10^{12}	tera	THz (1×10^{12} Hertz)

FIGURE 2–4 Powers of 10 related to metric and electronic terms

■ **IN-PROCESS LEARNING CHECK 1** Fill in the blanks as appropriate.

1. Charge is represented by the letter _____. The unit of measure is the _____, and the abbreviation is _____.
2. The unit of potential difference is the _____. The symbol is _____.
3. The abbreviation for current is _____. The unit of measure for current is the _____, and the symbol for this unit is _____.
4. The abbreviation for resistance is _____. The unit of measure for resistance is the _____, and the symbol for this unit is _____.
5. Conductance is the _____ with which current can flow through a component or circuit. The abbreviation is _____. The unit of measure for conductance is the _____, and the symbol for this unit is _____.
6. How many microamperes does 0.0000022 amperes represent? _____ How is this expressed as a whole number times a power of 10? _____
7. What metric prefix represents one-thousandth of a unit? _____ How is this expressed as a power of 10? _____
8. Using a metric prefix, how would you express 10,000 ohms? _____ How is this expressed as a whole number times a power of 10? _____ ■

2–3 Conductors and Their Characteristics

Functions of Conductors

You have already learned that the function of conductors is to carry electrical energy from one point to another. More explicitly, conductors provide a path for the flow of electrical current between components or circuits. As you can observe at home, at work, or at school, there are numerous types of conductors. One type is used to transport power into your home. Another type brings the TV signal from the antenna to your TV set. Still another style of conductor delivers telephone signals. The heavy battery cables in your automobile represent yet another type of conductor.

Types of Conductors

The type of conductor required for a given job is determined by the nature of the application and the amount of electrical energy it must carry. Some ways of classifying conductors include the:

- type of metal used (silver, copper, gold, aluminum);
- physical dimensions of the wire itself (its "size");
- form of the metal conductor(s) (solid, stranded, or braided);
- number of conductors used or packaged together; and
- insulation characteristics (uninsulated, insulated, and type of insulation used).

Several examples of conductor types are shown in Figure 2–5.

Resistance of Solid Conductors

The ideal conductor would have zero resistance, and thus would not dissipate any of the electrical energy being transported. In reality, however, all conductors do have some resistance to the flow of electrons at normal operating temperatures.

Superconductivity

Typically, the resistance of most metal conductors increases as their operating temperature is increased. In 1911, it was found that cooling the temperature of certain metal conductors to almost absolute zero (0° Kelvin, or –273°C) causes virtually all their resistance to current flow to disappear. In other words, their resistance approaches zero ohms. This phenomenon only occurs below some critical, very cold temperature. The production and study of these super cold temperatures is often referred to as "cryogenics." These very low temperature requirements make it difficult to take advantage of this superconductivity characteristic. Recently, scientists have begun experimenting with unique ceramic materials that can have very low resistance at temperatures considerably higher than absolute zero. If, in the future, scientists can find practical materials and methods to move that useable temperature range closer to normal room temperatures, the sky is the limit in terms of applications. Everything from electromagnetically operated devices (motors, generators, etc.) to computers would be greatly impacted.

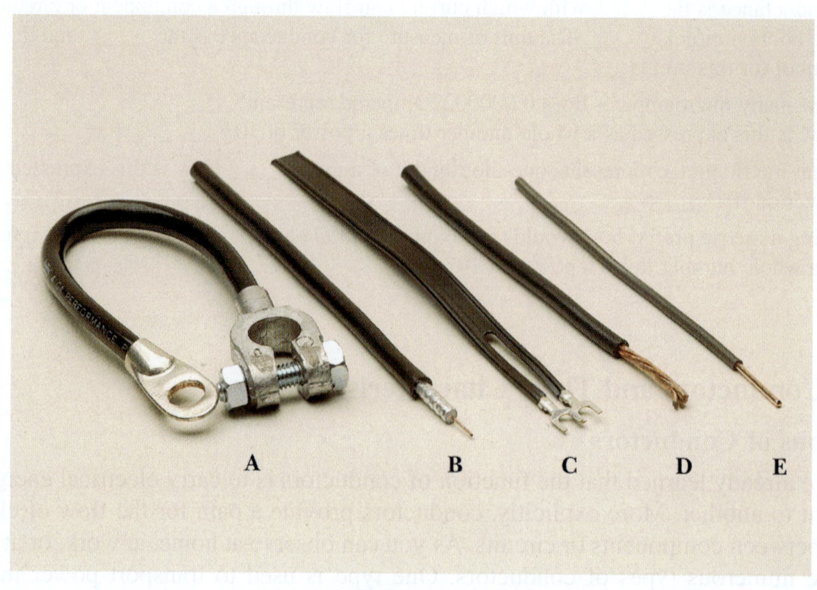

FIGURE 2–5 Examples of several types of conductors: (a) auto battery cable, (b) coaxial cables, (c) TV twin lead, (d) stranded, (e) solid *(Photo by Michael A. Gallitelli)*

Physical Factors Influencing Conductor Resistance

The resistance of a given conductor is related to the following factors:

1. The type of material making up the conductor: The more conductive the material, the lower its resistance per given physical dimensions.
2. The length of the conductor: The longer it is, the higher its total resistance from one end to the other.
3. The cross-sectional area of the conductor: The greater the cross-sectional area, the lower its resistance per given length.
4. The temperature of the conductor: Typically, for metal conductors, as temperature increases, the resistance increases.

Resistivity

The "resistivity" (or "specific resistance") of a given material refers to its characteristic resistance (in ohms) per standard length and cross-sectional area, at a specified temperature. In SI units (International System of Units), resistivity is expressed in terms of meters for length, and square meters for cross-sectional area. The International System of Units (referred to as SI, after the initials of Système International) is basically the metric system. Seven basic categories of units are described in the International System: **length**—meter; **mass**—kilogram; **time**—second; **electric current**—ampere; **thermodynamic temperature**—Kelvin; **amount of substance**—mole; and **luminous intensity**—candela. Different materials each have their own resistivity characteristic. For example, in SI units, silver has a resistivity of 16 nano-ohms per meter; copper, a resistivity of 17 nano-ohms per meter; gold, a resistivity of 24 nano-ohms per meter; and aluminum, a resistivity of 27 nano-ohms per meter. These are expressed in nano-ohms per meter for simplicity of reading, rather than having to state that silver has a resistivity of 0.000000016 ohms/meter, and so on. The symbol used for resistivity is ρ (the Greek letter rho).

In everyday technician work, you are more likely to be dealing with conductors in terms of diameter in mils (thousandths of an inch), area in circular mils (CM), and length in feet. A circular mil describes the cross-sectional area of a round conductor having a diameter of one-thousandth of an inch (1 mil diameter). The cross-sectional area in circular mils is thus found as:

FORMULA 2–1 $A = d^2$

where A = cross-sectional area in circular mils
d = diameter of wire in mils

See Figure 2–6.

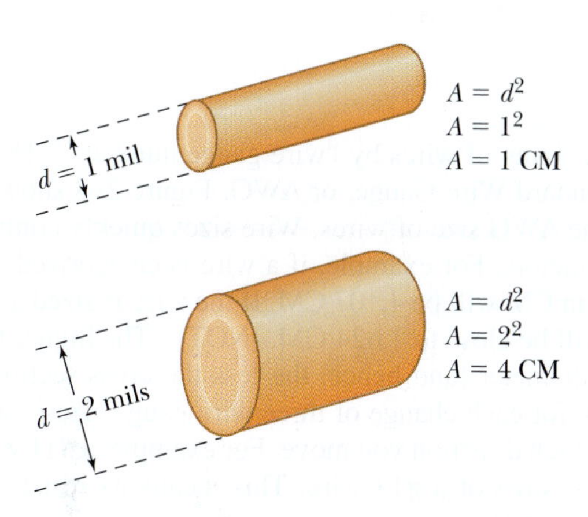

FIGURE 2–6 Cross-sectional area of conductors in circular mils (CM)

EXAMPLE If a wire has a diameter of 4 mils (0.004″), what is its cross-sectional area in circular mils?

Answer: $A = d^2$. $A = 4^2$. $A = 16$ CM

PRACTICE PROBLEMS 1

What is the circular-mil cross-sectional area of a conductor whose diameter is 2.5 mils?

Because resistivity (ρ) of a material is a measure of a material's resistance per standard length and cross-sectional area (at a given temperature), ρ can be expressed in terms of circular-mil ohms per foot. That is, CM-Ω/ft.

Calculating the Resistance of a Conductor

A general formula for finding the resistance of a wire (assuming a given temperature) is:

FORMULA 2–2 $R = \frac{\rho l}{A}$

where R = resistance in ohms
ρ = resistivity in CM-Ω/ft (for the type material in conductor)
l = length in feet
A = cross-sectional area of conductor in CM

EXAMPLE What is the resistance of a copper wire that is 250 feet in length and has a cross-sectional area of 39.75 CM? (NOTE: ρ for copper is 10.4 CM-Ω/ft.)

$$\text{Answer: } R = \frac{\rho l}{A} = \frac{(10.4\ \text{CM-}\Omega/\text{ft})(250\ \text{ft})}{39.75} = 65.4\Omega$$

PRACTICE PROBLEMS 2

What is the resistance of a copper wire that is 300 feet in length and has a cross-sectional area of 2,048 CM? (Remember: ρ for copper is 10.4 CM-Ω/ft)

Summary

The resistance a given conductor exhibits to current flow is directly related to its resistivity characteristic and its length, and inversely related to its diameter squared (its cross-sectional area). That is, the larger the wire (of a given type material), the lower its resistance per given length. The longer the wire, the higher its resistance per given cross-sectional area. See Figure 2–7. These are important facts to remember as you deal with conductors and conductive paths as a technician.

Wire Sizes

There is a standardized sizing of wires by "wire gauge numbers." This standard is referred to as the American Standard Wire Gauge, or AWG. Figure 2–8 shows a tool some technicians use to measure the AWG size of wires. Wire sizes quickly communicate relative size information about conductors. For example, if a wire is categorized as an AWG #14 wire, its cross-sectional area in CM will be 4,107 CM. If the wire is sized as a #18 wire, its cross-sectional area in CM will be close to 1,624 CM. (NOTE: The higher the AWG number, the smaller the wire is in diameter [and hence, the less the cross-sectional area it has].) It is interesting to know that for each change of three wire-gauge sizes, the CM area doubles or halves, depending on which direction you move. For example, a #11 wire has approximately twice the cross-sectional area of a #14 wire. This means its resistance per unit length is approximately half that of #14 wire.

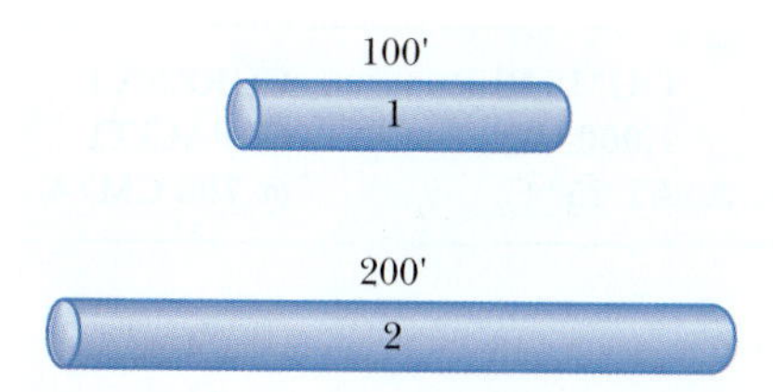

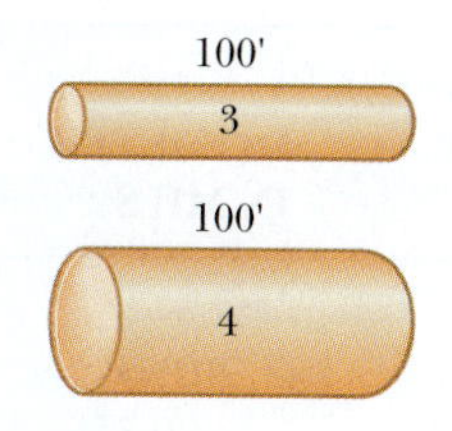

- Same type wire
- Same diameter for 1 and 2
- Conductor 2 is twice the length of conductor 1
- Conductor 2's resistance is twice that of Conductor 1, from one end to the other

- Same type wire
- Conductors 3 and 4 are the same length
- Conductor 4 has twice the diameter of 3 or ≈4 times the CM area
- Conductor 4's resistance is one-fourth that of conductor 3 from end to end

FIGURE 2–7 Relationship of wire resistance to conductor diameter and length

FIGURE 2–8 A wire-gauge tool *(Photo by Michael A. Gallitelli)*

Wire Table

A very useful source of information regarding conductor sizes and their resistances is a wire table, Figure 2–9. The wire table displays data regarding AWG wire numbers, their cross-sectional areas and their resistance in ohms per given length into one concise list or matrix. Most wire tables show the standard resistance per given length in terms of ohms per thousand feet.

PRACTICE PROBLEMS 3

1. Refer to Figure 2–9 and determine the resistance of 500 feet of #12 wire.
2. What gauge wire has a cross-sectional area of about 10,380 CM?
3. What is the resistance in ohms per thousand feet of the wire in question 2?

AWG WIRE SIZE	DIAMETER IN MILS	CIRCULAR-MIL AREA	OHMS PER 1,000 FT AT 25° C	CURRENT CAPACITY @ 700 CM/A
—	—	—	—	—
10	101.9	10,380	1.018	14.8
11	90.7	8,234	1.284	11.8
12	80.8	6,530	1.619	9.33
13	72.0	5,178	2.042	7.4
14	64.1	4,107	2.575	5.87
15	57.1	3,257	3.247	4.65
16	50.8	2,583	4.094	3.69
17	45.3	2,048	5.163	2.93
18	40.3	1,624	6.510	2.32
19	35.9	1,288	8.210	1.84
20	32.0	1,022	10.35	1.46
21	28.5	810	13.05	1.16
22	25.3	642	16.46	0.918
23	22.6	510	20.76	0.728
24	20.1	404	26.17	0.577
25	17.9	320	33.00	0.458
26	15.9	254	41.62	0.363
27	14.2	202	52.48	0.288
28	12.6	160	66.17	0.228
29	11.3	127	83.44	0.181
30	10.0	101	105.2	0.144
31	8.9	80	132.7	0.114
32	8.0	63	167.3	0.090
etc.	etc.	etc.	etc.	etc.

NOTE: Current-carrying capacity varies greatly, depending on environmental conditions. For example, a single #14 wire in open air can carry up to 32 amperes. In conduit, it can only carry a maximum of 17 amperes. At 700 CM/ampere, it can only carry about 5.87 amperes.

FIGURE 2–9 Sample partial wire table

As you might reason, the size of the conductor also dictates the maximum current it can safely carry without overheating and being destroyed. Some wire tables will also show this maximum current-carrying capacity for stated conditions, as well as providing the resistance-per-given length data. For a given type conductor, the larger the diameter of the wire, the greater current-carrying capacity it has. For example, as stated earlier, the cross-sectional area of the #11 wire is about double that of #14 wire. This means that the #11 wire can safely handle about double the current of the #14 wire. The general rule of thumb is that for every three wire sizes changed, the current-carrying capacity of the wire will double, or halve, depending on which direction you move in the wire sizes.

■ IN-PROCESS LEARNING CHECK 2

1. The greater the resistivity of a given conductor, the (higher, lower) _____ its resistance will be per given length.
2. The circular-mil area of a conductor is equal to the square of its _____ in _____.
3. The smaller the diameter of a conductor, the (higher, lower) _____ its resistance will be per given length.
4. The higher the temperature in which a typical metal conductor must operate, the (higher, lower) _____ will be its resistance.
5. Wire tables show the AWG wire _____, the cross-sectional area of wires in _____, the resistance per given length of copper wire, and the operating _____ at which these parameters hold true.

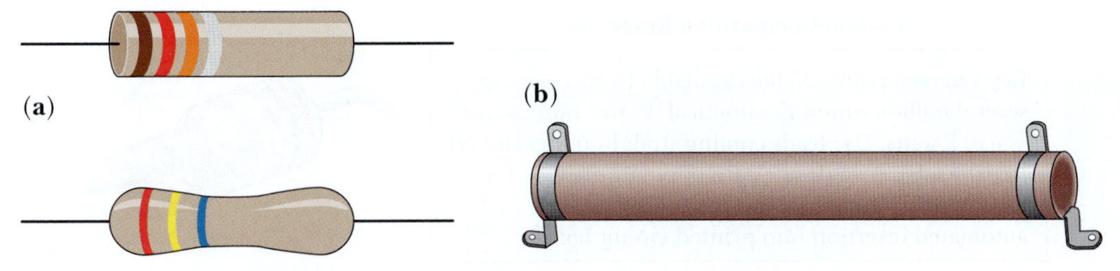

FIGURE 2–10 (a) Carbon resistor physical appearance; (b) wire-wound resistors

2–4 Resistors

Uses

Resistors limit current and help determine current values for a given circuit. This current-controlling function is one of the important roles that resistors perform in electronics. Another function, which you'll study later, is dividing voltage. Both of these purposes are important.

Types

Resistors are designed to perform these functions under a variety of operating conditions. These include numerous voltage, current, and power level requirements. Also, various circuits have different demands related to accuracy of resistance, and stability of resistance value related to temperature variation. Add to these variables the fact that the physical size of a given resistor dictates its physical mounting method and electrical connection method.

Because of this diversity of requirements, several types of resistors have been created to meet these conditions. Two prominent types that you will commonly encounter are the carbon-type resistor, Figure 2–10a, and the higher power-handling wire-wound type, Figure 2–10b. As you might assume, the power-handling capability of any resistor is related to its physical size and ability to dissipate heat. It is likely that you will have many chances to see and use these two common types of resistors during your training.

Look at the chart in Figure 2–11 and study some basic features and characteristics of these, and several other useful types of resistors. More in-depth knowledge of each of these type resistors will be gained as your training proceeds. As a technician or engineer, the considerations you will use in selecting or replacing resistors in circuits and systems will typically focus on power rating, voltage rating, accuracy of value, physical dimensions, and the component connection and mounting method requirements. In some cases, the temperature characteristics of the resistor also may be important.

Some Other Special-Type Resistor Devices

In addition to the devices described in Figure 2–11, there are some special-application components or devices that might be considered in the resistor device family. One of these is called a fusible link, or a fusible resistor. These are resistive elements that are designed to open up the circuit should the current through them exceed a specified amount. (Much like your fuses or circuit breakers do at your home.) Of course, these must be replaced, should they open up. Their purpose is to limit current and also to provide circuit safety should the current get too high.

Because of today's robotics and automated production systems, a "zero-ohm" device has been created. This device allows industry to automate the insertion of a "glorified jumper wire" between two points on a printed circuit board, should the need arise. The very low resistance device (close to zero ohms) is packaged in a way that the automated machinery can insert it into the circuit board, rather than requiring a person to jumper the points with a wire, then solder, and so on.

Carbon-Composition Resistors

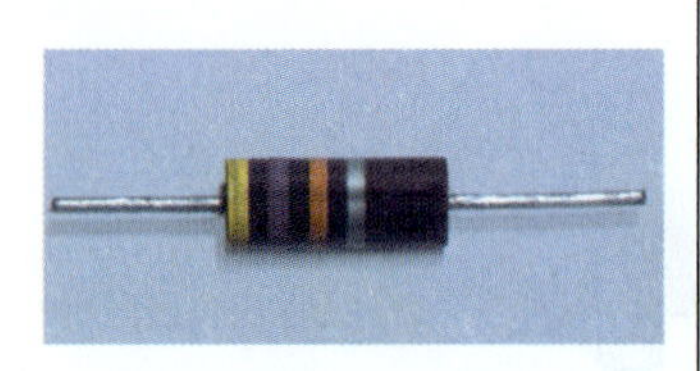

Key Characteristics: Values available from 1 ohm to several million ohms. Economical. Power ratings from 1/10 to 2 watts. Use leads coming straight out each end (axial leads). Two designs are popular: the type for wiring to terminals and the type designed for automated insertion into printed circuit boards.

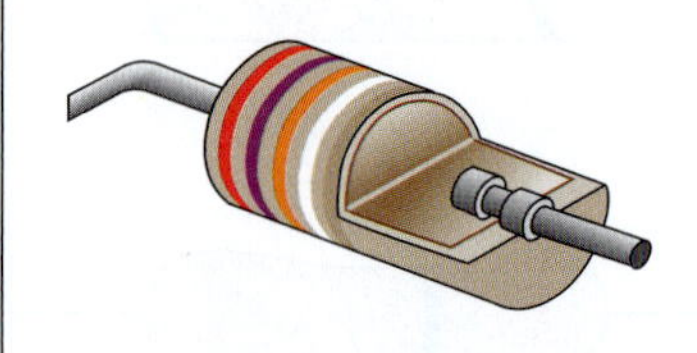

Construction: Resistive element is composed of ground up carbon and insulating material, held together by a resin binder. Resistive element and lead connections to that element are encapsulated in an insulating enclosure.

Precision Film-Type Resistors

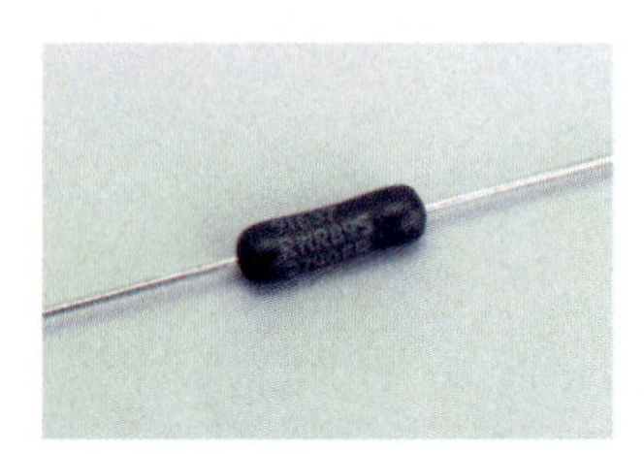

Key Characteristics: Values can be accurately controlled. Quite immune to temperature changes. Have little undesired "noise" generation as well as other good electronic characteristics. Typically are more expensive than standard carbon composition resistors.

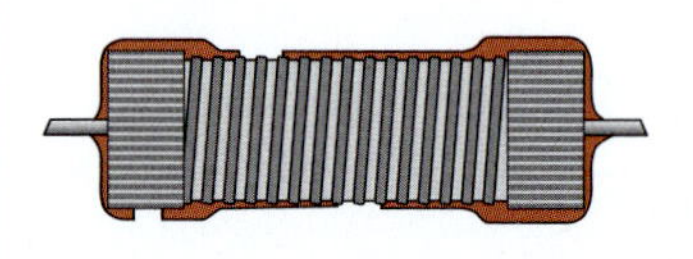

Construction: There are two types of film resistors. One uses a coating of carbon on a cylindrical ceramic substrate; the other uses a thin metal coating sprayed on the substrate. Both types generally have their resistive material cut in a spiral around the tubular form. The amount of resistance depends on the substances used and how much material is left in the spiral resistive element.

Surface Mount "Chip" Resistors

Key Characteristics: Tiny in size, good for compact applications. Have virtually zero lead length, which yields many electronic advantages. Ruggedly built and durable. Have good temperature characteristics. Because of their small size, these devices are typically available only in small power ratings (1/8 to 1/4 watt).

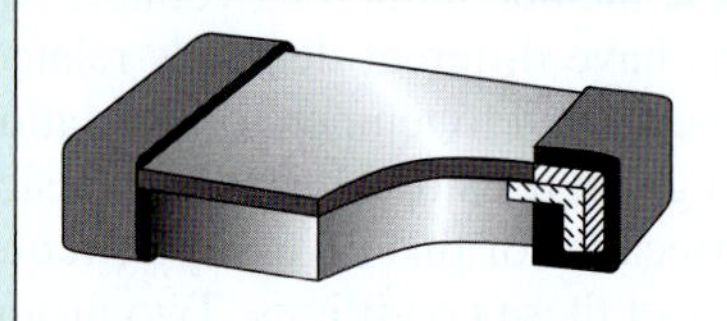

Construction: Carbon layer deposited on a ceramic base or substrate. Contact to the resistive element is via some metal ends or terminals, which create virtually zero lead length. In application, these contacts then are soldered directly to a circuit board's conductive paths usually by automated soldering techniques.

Wire-Wound Resistors

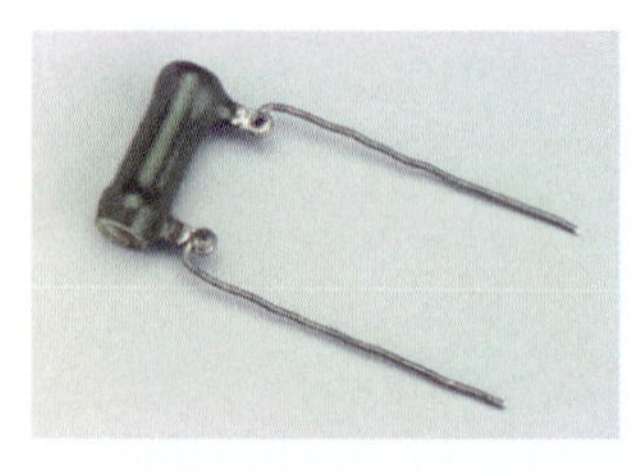

Key Characteristics: In comparison to the other types of resistors, may be relatively large in physical size. Have more power-handling capability because of size and construction (e.g., 5 to 100 W). Can have good accuracy in resistance value. Have good temperature stability. Generally used in high-current applications.

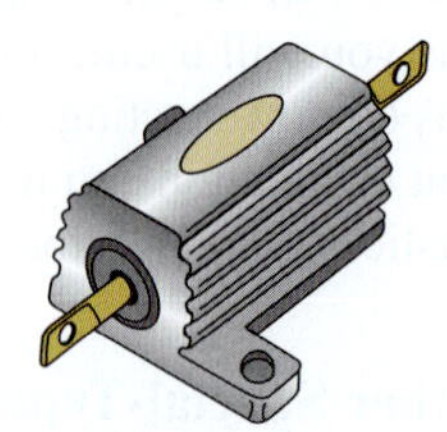

Construction: A special resistance wire is wound around a tubular-shaped ceramic insulator form. The resistance characteristic of the wire and its length dictate the resistor's ohmic value.

FIGURE 2–11 Several common types of "fixed" resistors

Resistor Types and Symbols

You have been looking at some "fixed" resistance devices. Resistors may also come in "semifixed," or "variable" resistance forms. A fixed resistor is one in which the resistance between its terminals or component leads cannot be changed externally. See Figure 2–12a. A semifixed resistor has one contact, or terminal, that can be adjusted to any given spot along the resistance element, and affixed to that position, as shown in Figure 2–12b. Thus

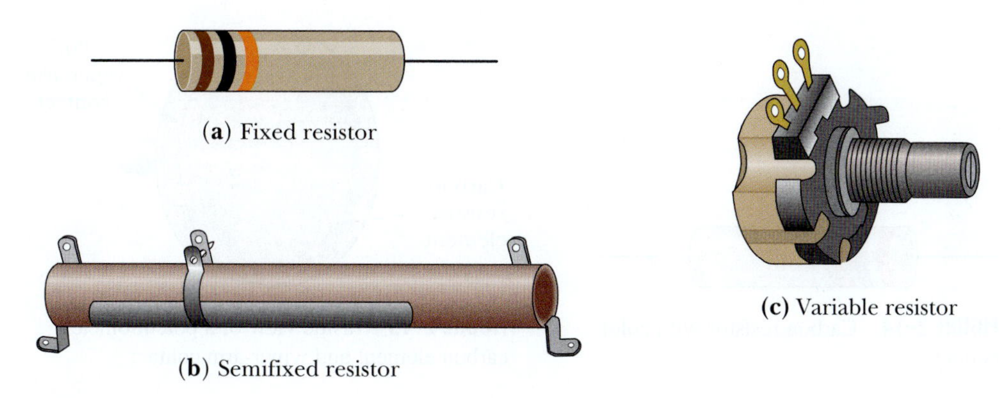

(a) Fixed resistor

(b) Semifixed resistor

(c) Variable resistor

FIGURE 2–12 Fixed, semifixed, and variable resistor constructions

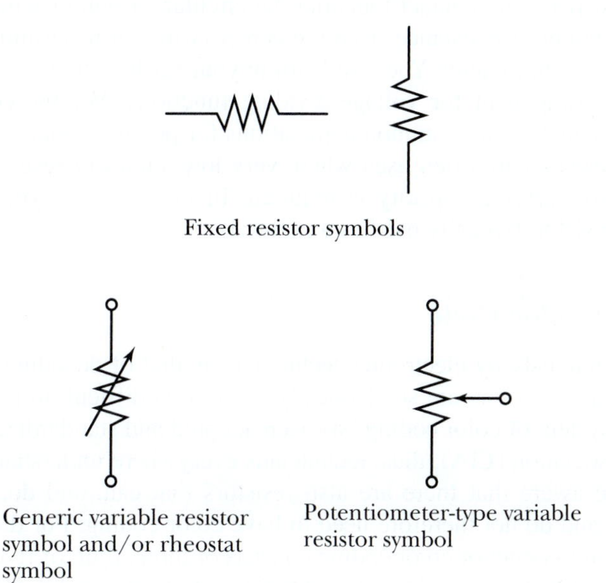

FIGURE 2–13 Schematic symbols for fixed and variable resistors multiSIM

the resistance value can be set between this movable terminal and the end terminals at a desired value. Once adjusted, the movable element is not normally moved again. The variable resistor, however, has a movable contact that can be moved anywhere along the resistance element. In the variable resistor, the movable contact has a shaft attached to purposely enable easy changing of the contact's setting along the resistance element, Figure 2–12c.

Figure 2–13 shows the "schematic symbols" used to represent two of these types. As you will soon learn, schematic symbols are the technician's shorthand for representing electrical components and circuits on paper. You will become very familiar with these, and other electronic diagram symbols as you continue in your studies.

Resistor Construction Comparisons

Recall that cabon-composition resistors use a carbon-based resistive element and have a fixed resistance, depending on the mix of carbon and insulating material. See Figure 2–14. Please note the colored bands. You'll be studying about them later in this chapter.

The wire-wound resistors, on the other hand, use a special resistance wire for the resistive element. Refer again to Figures 2–10b and 2–11. One obvious difference between the "fixed" carbon resistor and the wire-wound types is that wire-wound resistors can be "fixed," "semifixed," or "variable."

Some variable resistors, called "potentiometers," have resistive elements that can be either carbon or wire-wound elements. Look at Figures 2–15 and 2–16 and observe that the

FIGURE 2–14 Carbon resistor with color coding

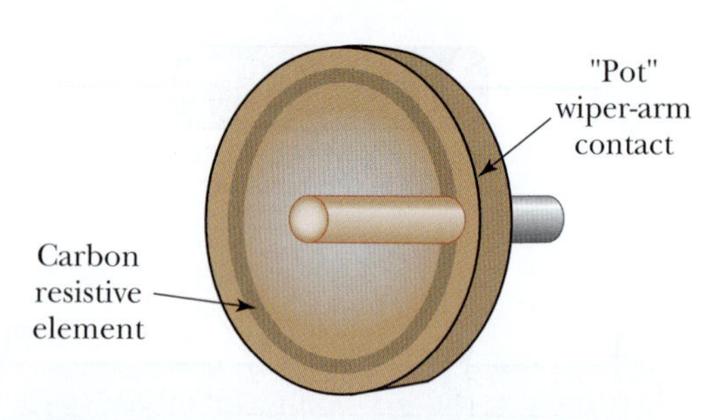

FIGURE 2–15 Inside view of a potentiometer carbon element and wiper-arm contact

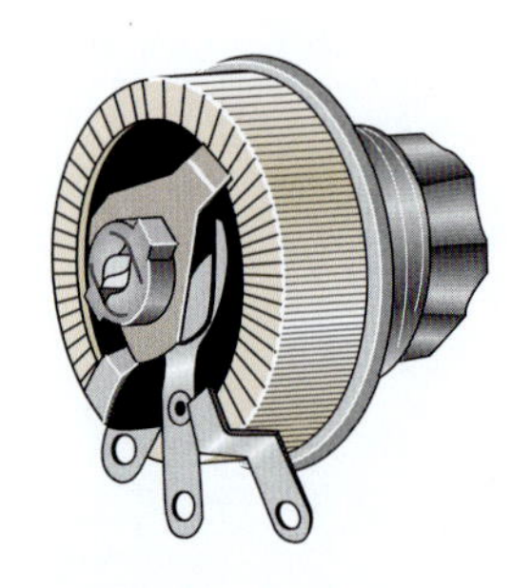

FIGURE 2–16 High current-carrying capacity wire-wound variable resistor. This resistor can be wired as a two-terminal rheostat or as a three-terminal potentiometer

position of the "wiper-arm" contact touching the circular carbon or wire-resistive element determines the amount of resistance from the center (wiper-arm) terminal to the terminals at each end of the potentiometer. You will learn in your further studies that the potentiometer is a valuable component for voltage-dividing functions. Volume controls for radios, audio amplifiers, and TVs are common applications for potentiometers.

Wire-wound resistors are often used where very low values of resistance are needed or where high current-carrying capacity is required. In Figure 2–16, you see a large wire-wound variable resistor, typically used as a rheostat.

2–5 Resistor Color Code

A skill used continuously by electronics technicians is that of decoding the **resistor color code.** Because many resistors are so physically small, it is difficult to print their value in a readable size. A system of color coding has been adopted and standardized by the Electronics Industries Association (EIA); thus, technicians everywhere understand the code.

You should be aware that there are also resistors that can, and do, have their values labeled on them, and do not therefore need to have color coding. For those that do use the color code, the band system of color coding consists of three, four, or five bands (or stripes) of color around the tubular-shaped resistors. On four-band color codes, the first three stripes, when decoded, indicate the resistor's rated resistance value, in ohms. The fourth stripe, when present, indicates the tolerance percentage, or amount of deviation from the rated value that the resistor's actual value may vary in manufacture. If there is not a fourth band, the tolerance is automatically assumed to be ± 20%.

Many precision resistors (resistors with tolerance of ±1%, ±2%, or less) use a five-band color-coding system. See Figure 2–17b. In this case, the first three stripes represent the first three digits of the resistence value. The fourth stripe is the decimal multiplier and the fifth stripe gives the tolerance rating. (See the chart in Figure 2–19 and observe the "Precision Resistor Tolerance Band" information column.) In some cases, extra space is inserted between the tolerance rating stripe and the other stripes; in other cases, all five stripes are equally spaced.

Another variation of the color-coding system involves certain resistors that have a "reliability" rating. On these resistors, the fifth band relates the reliability factor, which is a rating based on the percentage failure rate for 1,000 hours of service. (See the "Reliability Factor. . ." column in Figure 2–19.)

During your training, you will probably be dealing mostly with the four-band carbon-composition resistor-type color codes, and the five-band precision-type resistor color codes.

Significance of Color Band Placement

Look at Figure 2–17 and observe the placement of these color bands. The first color is considered to be the one closest to the end of the resistor. The other color bands progress toward the center of the resistor body. The key to decoding the color code is to learn the significance of each band position and each color.

Four-band General-Purpose Resistor Color Code

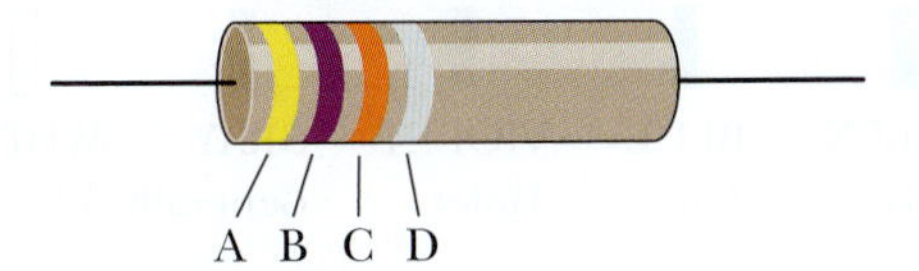

A = First significant figure of resistance value

B = Second significant figure of resistance value

C = Decimal multiplier for resistance value

D = Tolerance rating (in percentage)

(a)

Five-band Precision-Resistor Color Code

A = First significant figure of resistance value

B = Second significant figure of resistance value

C = Third significant figure of resistance value

D = Decimal multiplier for resistance value

E = Tolerance rating (in percentage)

(b)

FIGURE 2–17 Color band significance

Refer again to Figure 2–17a and b and observe that the band positions are interpreted as follows:

For the four-band color code (typical color code for general-purpose resistors)

1. The **first band** (closest to the end of the resistor) is the first significant digit in the resistor's ohmic value rating.
2. The **second band** represents the second significant digit in the resistor's ohmic value rating.
3. The **third band** indicates the multiplier, or number of zeros that should follow the first two numbers to determine the resistor's rated ohmic value. **Note:** For resistors under 10 ohms in value, if the multiplier is gold or silver, the multiplier indicates multiplying the value indicated by the previous digits by 0.1 if the multiplier is gold, or by 0.01 if the multiplier band is silver.
4. The **fourth band** provides the tolerance information in terms of plus or minus percentage (±%) tolerance information. That is how much, in percentage, the resistor can acceptably vary above or below its rated color-coded value and still be within the manufacturer's specification.

For the five-band color code (typical color code for precision resistors)

1. The **first band** (closest to the end of the resistor) is the first significant digit in the resistor's ohmic value rating.
2. The **second band** represents the second significant digit in the resistor's ohmic value rating.
3. The **third band** indicates the third significant digit in the resistor's ohmic value rating.
4. The **fourth band** indicates the multiplier, or number of zeros that shold follow the first two numbers to determine the resistor's rated ohmic value. **Note:** For resistors under 10 ohms in value, if the multiplier band is gold or silver; the multiplier indicates multiplying the value indicated by the previous digits by 0.1 if the multiplier band is gold, or by 0.01 if the multiplier band is silver.
5. The **fifth band** provides the tolerance information in terms of plus or minus percentage (±%) tolerance information. That is how much, in percentage, the resistor can acceptably vary above or below its rated color-coded value and still be within the manufacturer's specification.

NOTE: For resistors using the fifth band as the "reliability" factor rating; the first four bands are interpreted the same as a four-band color code, and the fifth band indicates the reliability factor in terms of percentage failure for 1,000 hours service. (See Figure 2–19, as appropriate to observe colors versus reliability percentage.)

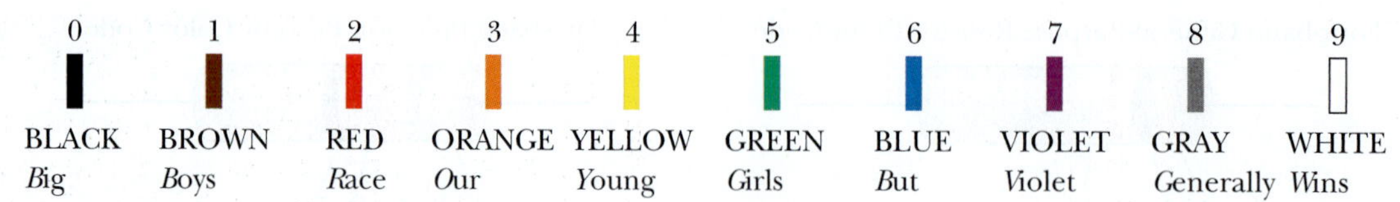

FIGURE 2–18 Sequence of colors memory aid

Significance of Colors

For resistors that are greater than 10 ohms in value, the 10 colors used to represent the values zero through nine (0–9) are used and are fairly easy to remember by means of a memory-aid sentence. As you can see in Figure 2–18, the color code starts with the dark colors at zero, and progresses toward lighter colors as the numbers increase. The figure shows how the sentence "*B*ig *B*oys *R*ace *O*ur *Y*oung *G*irls, *B*ut *V*iolet *G*enerally *W*ins" can be used to remember the color code. These colors represent the sequential numbers zero through nine. The first letter of each word in the sentence is the same as the first letter of each color. Learn these!

Tolerance Band Colors

Gold and silver are frequently used tolerance band colors. Gold indicates ±5% tolerance from rated value. For example, if a resistor is color coded as a 100-ohm resistor with ±5% tolerance rating, its actual value could be anywhere between 95 and 105 ohms, and still be within the manufacturer's rating limits.

Silver represents a tolerance of ±10%. In this case, a resistor color coded 100 ohms could actually have a value between 90 ohms and 110 ohms and still be considered good.

As stated earlier, if there is no fourth band, the tolerance rating is ±20% of rated value.

Use of Gold and Silver Colors as Multipliers

Low-value resistors (below 10 ohms) require a special multiplier value for the third band. This indicates that the numbers decoded from the first color bands are to be multiplied by 0.1 or 0.01 to find the value for the resistor. A gold multiplier band indicates the multiplier is one-tenth (0.1). For example, if the first two color bands are red and violet (27), then a gold third band would indicate a resistor value of 0.1×27, or 2.7 ohms.

A silver third band indicates a multiplier of one-hundredth (0.01). This means if the resistor is color coded brown-black-silver (in that order), the resistor value is 0.1 ohm ($0.01 \times 10 = 0.1$).

A summary chart showing the significance of various colors used in the color code is shown in Figure 2–19.

Examples

Let us see how to apply the combination of color information and band position to decode some resistor values and tolerances. Also, we will look at one case where a reliability factor is given.

Notice in Figure 2–20a that the first band is yellow, representing the number 4. The second band is violet, representing 7. The third band is orange, indicating a multiplier of 1,000, which adds three zeros to the 47. Therefore, the resistor color-coded value is 47,000 ohms (or 47 kΩ). The fourth band is silver, indicating a 10% tolerance rating. This means that the resistor's actual value can be 47,000 ohms, plus or minus 4,700 ohms. That is, its value should be between 47,000 + 4,700 ohms (51,700 ohms) and 47,000 ohms – 4,700 ohms (42,300 ohms).

In Figure 2–20b, the color code is brown, black, orange, gold, in that order. Decoded, this value is 10,000 ohms, ±5%.

In Figure 2–20c, the precision resistor five-band color code is red, red, brown, gold, brown. The value of the resistor is therefore 2, 2, $1 \times 0.1 = 22.1$ ohms. The tolerance is ±11%. (See Figure 2–19.)

Now you try a couple!

COLOR	NUMBER COLOR REPRESENTS (IN 1ST AND 2ND BANDS)	DECIMAL MULTIPLIER COLOR REPRESENTS (IN 3RD BAND)	TOLERANCE COLOR REPRESENTS (IN 4TH BAND)	SPECIAL PRECISION RESISTOR TOLERANCE (IN 5TH BAND)	RELIABILITY FACTOR COLOR REPRESENTS (IN 5TH BAND)
Black	0	1 (no zeros)	—	—	—
Brown	1	10 (1 zero)	—	±1%	1%
Red	2	100 (2 zeros)	—	±2%	0.1%
Orange	3	1,000 (3 zeros)	—	—	0.01%
Yellow	4	10,000 (4 zeros)	—	—	0.001%
Green	5	100,000 (5 zeros)	—	±0.5%	—
Blue	6	1,000,000 (6 zeros)	—	±0.25%	—
Violet	7	10,000,000 (7 zeros)	—	±0.1%	—
Gray	8	100,000,000 (8 zeros)	—	—	—
White	9	1,000,000,000 (9 zeros)	—	—	—
Gold	—	0.1	5%	—	—
Silver	—	0.01	10%	—	—
No band	—	—	20%	—	—

FIGURE 2–19 Color significance summary chart

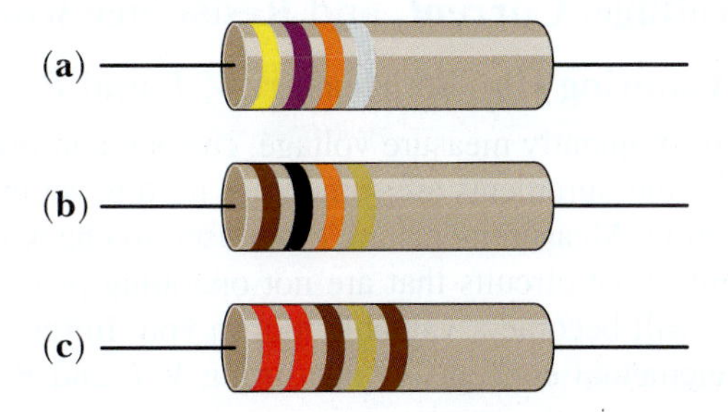

FIGURE 2–20 (a) 47 kΩ, 10% tolerance resistor; (b) 10 kΩ, 5% tolerance resistor; (c) 22.1 Ω 1% tolerance resistor

PRACTICE PROBLEMS 4

1. What is the resistor value and tolerance of the resistor shown?

2. What is the resistor value, and tolerance rating of the precision resistor shown?

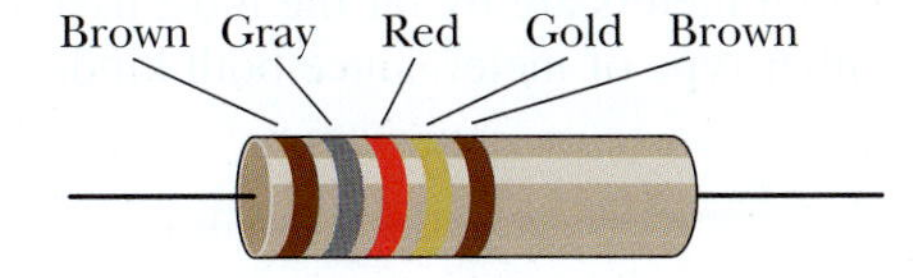

Special Precision Resistors Tolerance Ratings

Although most resistors you will work with use the color coding as presented thus far, there are some precision resistors with tolerances of 2%, or less, that use a special five-band color-code technique. As you have learned, the first three bands are value digits and the fourth band is the multiplier. The fifth band, then, provides the tolerance rating. For these special resistors, the fifth band tolerance rating uses the colors (brown = ±1%, red = ±2%, green = ±0.5%, blue = ±0.25%, and violet = ±0.1%). (See Figure 2–19.)

Other Exceptions

We mentioned earlier that some resistors are labeled rather than color coded. For example, the tiny resistors designed for surface mount technology (or chip resistors), often use a three-digit code to provide the device's resistance information. These three numbers are used in the same way that the first three color bands on carbon composition resistors are used. For example, the first digit on the chip resistor tells you the first number in the resistor's value. The second digit indicates the second number in its value. The third digit is the "multiplier" number, telling you how many zeros to add to the first two numbers to determine the resistor's value. If the chip resistor were coded with the digits 103, its value would be 1, then 0, followed by three zeros, or 10,000 ohms (10 kΩ).

Other types of resistors use a cryptic numbering and lettering system. The letter *R* indicates the decimal point position for resistors under 100 ohms. For example, if a resistor is labeled 2R7, its value is 2.7 ohms. Letters are also used to designate tolerance ratings, where the letter *F* means ±1%, *G* = ±2%, *J* = ±5%, *K* = ±10%, and *M* = ±20%. So if the label shows 2R7K, the resistor is a 2.7-ohm resistor with 10% tolerance.

For resistor values requiring more than three numbers, plus a multiplier, the first three labeled numbers are followed by a fourth number, which is the multiplier, indicating the number of zeros to follow the preceding numbers. An example of this might be a resistor labeled 4702K. In this case, the value is 47,000 ohms and the tolerance is ±10%.

2–6 Measuring Voltage, Current, and Resistance with Meters

The Importance of Knowing How to Measure *V*, *I*, and *R*

As a technician, you will frequently measure voltage, current, and resistance in electrical or electronic circuits. These measurements are sometimes used to set up or adjust circuit conditions for proper operation. Measurements are often used to check that a circuit is operating properly or to troubleshoot circuits that are not operating properly. The skill to make measurements correctly will become a valuable tool to you. In this section, we will introduce you to the most elemental concepts of measuring *V*, *I*, and *R*. Mastery of this measurement skill will only come with actual practice.

Types of Meters

Two general categories of meters you will use in making measurements of *V*, *I*, and *R* are analog meters and digital meters, Figure 2–21. You will learn more about the significance of the terms analog and digital as you progress in your studies. NOTE: In electronics, analog devices are considered those in which voltage or current is represented as continuously varying. Digital devices are those whose electrical parameters vary in discrete levels. (In some cases, such as in digital computers, digital circuitry data varies between two discrete levels, represented by low and high, 0 and 1, or on or off.) For now, you can see that the scale on the digital meter is directly readable, just by reading the numbers in the readout portion of the meter. The analog meter, however, takes some practice in interpreting values in terms of where the pointer rests along the continuous scale. Although the digital meters are by far the most used today, you should strive to become adept at using either type of meter, since both kinds are found in industry.

Symbols for Meters

In Figure 2–22 you can see both the general schematic symbols and the pictorial symbols we will use in this book to represent analog and digital ammeters (ammeters and milliammeters), voltmeters, and the ohmmeter (which measures resistance).

Making a Voltage Measurement with a Voltmeter

To measure the voltage between two points, you simply connect the voltmeter across (or in parallel with) the two points at which you wish to measure the potential difference, Figure 2–23.

(a)

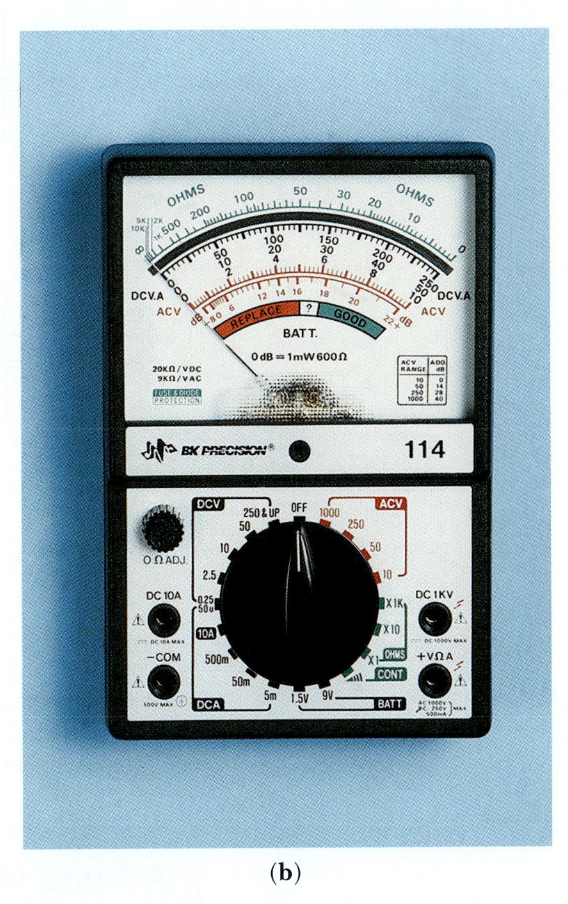

(b)

FIGURE 2–21 (a) Digital meter *(Courtesy of John Fluke Mfg. Co., Inc.)*; (b) analog meter *(Courtesy of B & K Precision)*

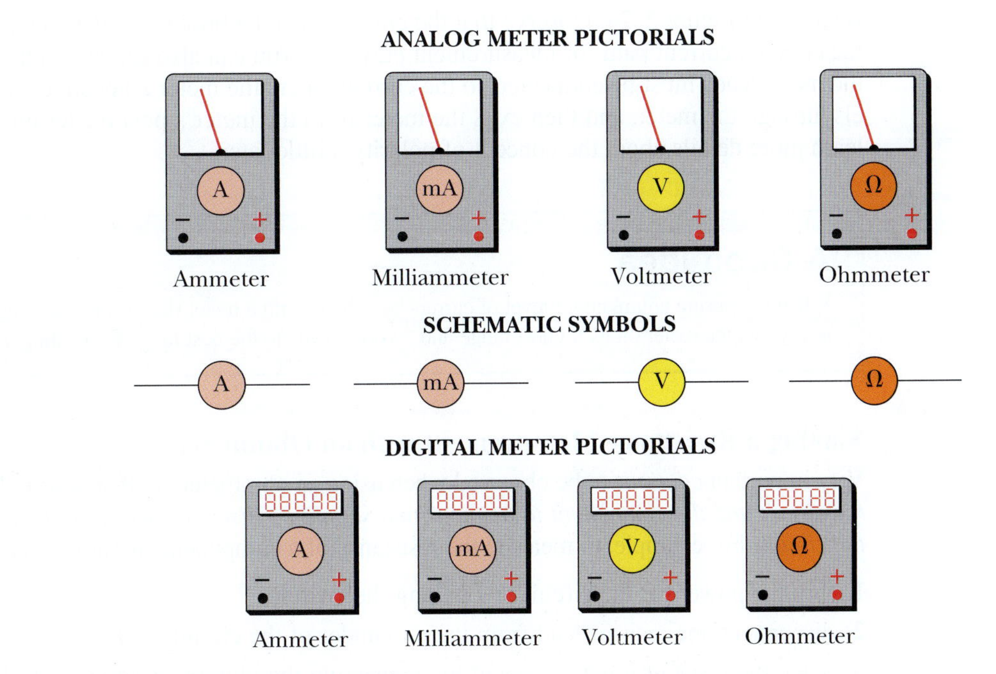

FIGURE 2–22 Pictorial and schematic meter symbols

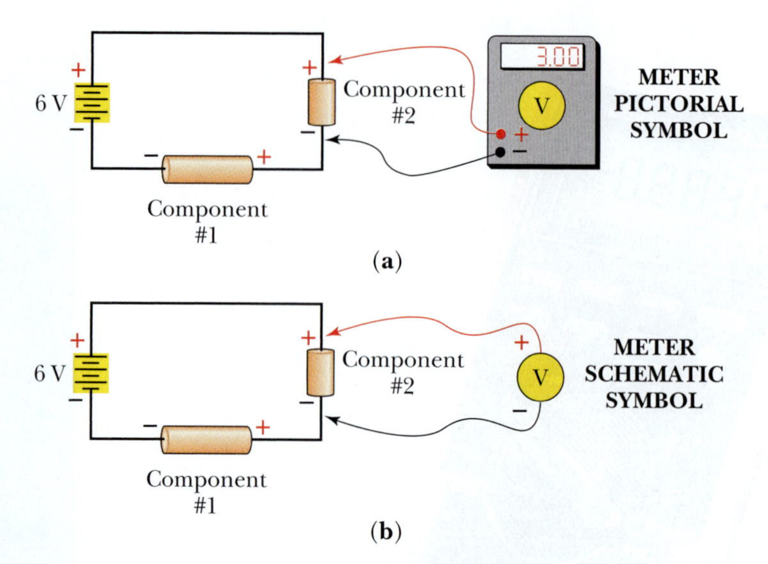

FIGURE 2–23 Measuring voltage "across" a component multiSIM

Notice, when measuring dc voltage, it is necessary to connect the negative test lead from the meter to the more negative point in the circuit and the positive test lead to the more positive point in the circuit. You can see in the figure that the meter is reading the value of voltage across one component in the circuit (in this case, component number 2).

> **A Good Idea**
>
> When preparing to make a circuit test with an electrical meter, turn the power off while connecting the meter's test leads to the circuit (if this is both possible and reasonably practical)!

Making a Current Measurement with an Ammeter

An illustration of how to connect an ammeter or a milliammeter to measure circuit current is shown in Figure 2–24. Observe that the circuit has to be broken for the meter to be put in the circuit's current path for measurement purposes. You can also see that in this dc circuit, the meter leads must be connected so the current enters the meter's negative terminal, travels through the meter, and then exits the meter from the meter's positive terminal. You will learn more details about the concept of polarity a little later.

> **A Good Idea**
>
> When measuring an unknown level of current or voltage with a meter that is not auto-ranging, always set the meter on its highest range and "work down" to the best range for reading the meter.

Making a Resistance Measurement with an Ohmmeter

Very important *cautions* to be observed when using an ohmmeter are that *power should be off the circuit and the component to be measured should be disconnected from the remainder of the circuit!* For example, to measure the resistance of a component, as shown in Figure 2–25:

1. turn the power off the circuit by opening the **switch;**
2. disconnect one resistor lead from the remainder of the circuit; and
3. make the resistance measurement by connecting the ohmmeter across both leads of the resistor.

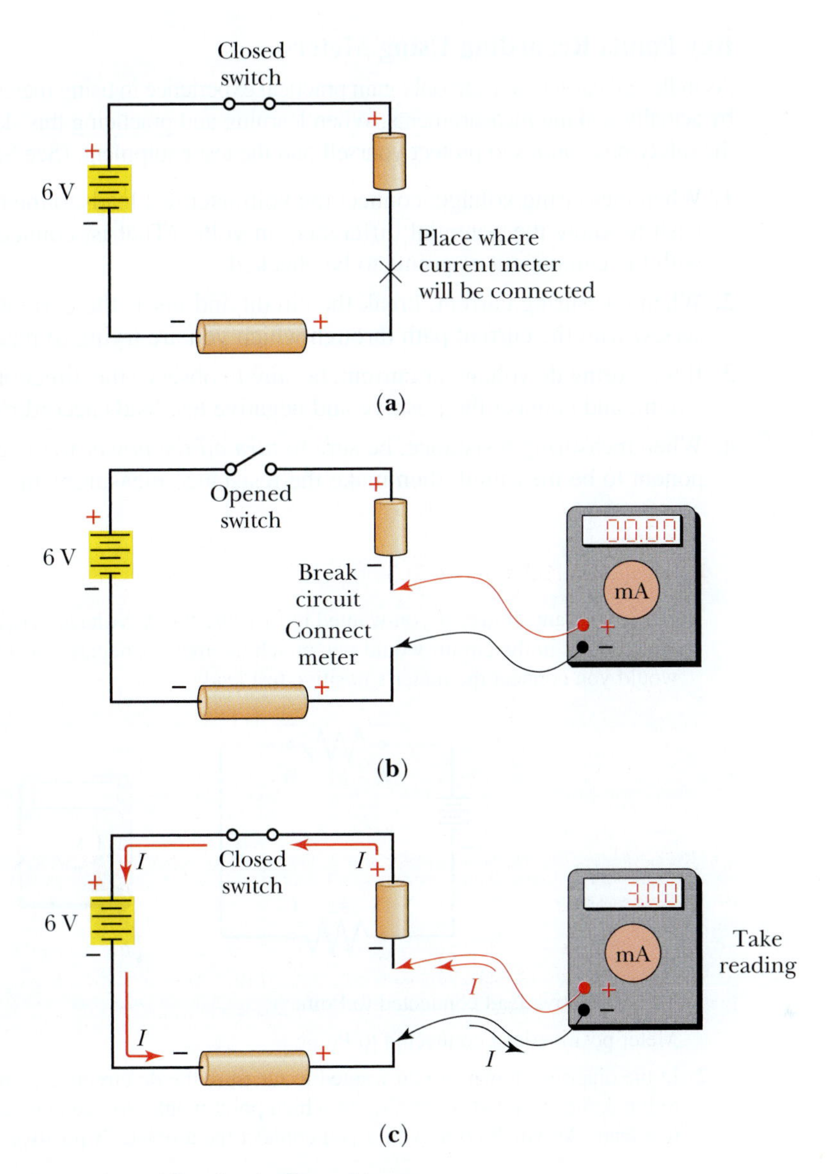

FIGURE 2–24 Measuring current "through" a circuit multiSIM

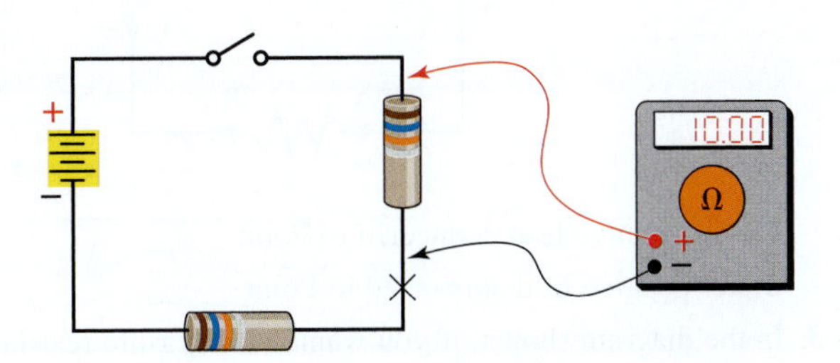

NOTICE: Power is off due to open switch. Component being measured is isolated from the rest of the circuit by disconnecting one of its leads (at "×").

FIGURE 2–25 Ohmmeter measurement

Key Points Regarding Using Meters

As indicated earlier, you can only gain practical experience in using meters to make measurements by actually making measurements. When learning and practicing this skill, be sure to observe all the safety precautions to protect yourself and the test equipment. (See Safety Hint listing.)

1. When measuring voltage, connect the voltmeter test leads to the two points at which you wish to know the potential difference, in volts. (That is, connect across, or in parallel, with the component or points to be checked.)
2. When measuring current, break the circuit and insert the current meter leads in line (in series) with the current path through which you are trying to measure the current.
3. If measuring dc voltage or current, be sure to observe the direction of current through the circuit, and connect the positive and negative test leads accordingly. (Observe polarity.)
4. When measuring resistance, be sure to *turn off the power* to the circuit, isolate the component to be measured, then make the resistance measurement, as appropriate.

IN-PROCESS LEARNING CHECK 3

1. In the diagram shown, if you wanted to measure the dc voltage across component #1, at which point in the circuit would you attach the meter's negative test lead? At which point would you connect the meter's positive test lead?

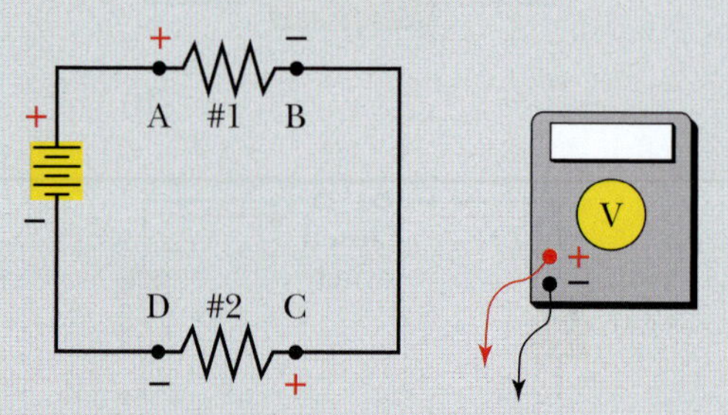

Meter negative lead connected to Point _____.

Meter positive lead connected to Point _____.

2. In the diagram shown, if you wanted to measure the dc circuit current and you were going to break the circuit at point "X," to which point would you connect the ammeter's negative lead? At which point would you connect the ammeter's positive lead?

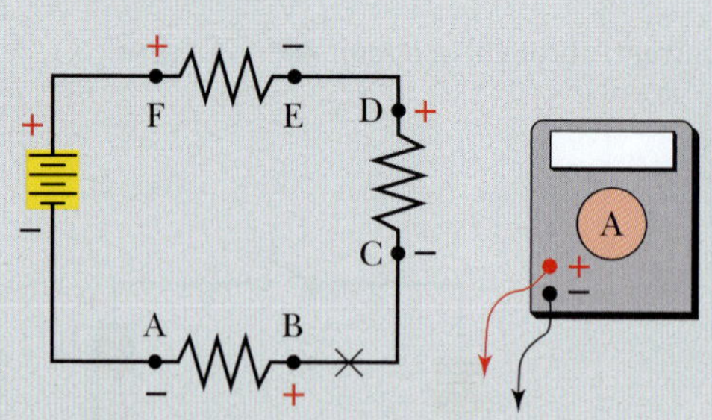

Meter negative lead connected to Point _____.

Meter positive lead connected to Point _____.

3. In the diagram shown, if you wanted to measure resistance, what is the first thing you would do, before connecting the meter across the component to be measured?

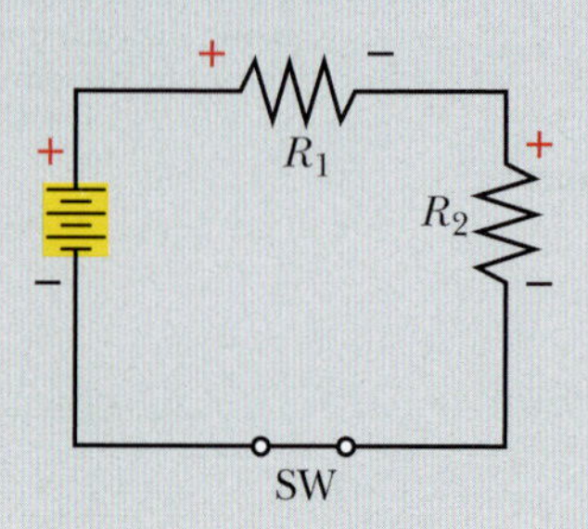

Safety Hint

1. Turn power off in circuit to be tested when connecting meter leads.
2. Appropriately connect meter test leads to circuit. Observe proper polarity if measuring dc voltage or current.
3. Make sure meter is in correct mode (ac, dc, *V, A*, Ω).
4. Make sure range selected on meter is high enough to handle the value that is about to be tested.
5. **a.** If measuring *V* or *I*, turn on power and take reading.
   b. If measuring *R*, *do not turn on power.*
6. Turn off power and remove meter test leads from circuit.
7. If measuring *I*, reconnect circuit, as appropriate.

2–7 Diagrams Used for Electronic Shorthand

Types of Diagrams

During your experiences in electronics you will encounter various types of diagrams that can be helpful. These include **block diagrams, schematic diagrams,** chassis layout diagrams, and others. For now, we'll focus on the block diagram and the schematic diagram since these are the kind you will be using most.

Some Basic Symbols

Electronic circuits are commonly represented by block diagrams, schematic diagrams, or both. In order to interpret and use schematic diagrams, it is necessary to learn some basic electronic symbols used in schematics. Because diagrams can be made to represent an infinite variety of circuits and configurations in a concise manner, schematic diagrams are the electronic technician's or engineer's shorthand.

At this point, we'll introduce you to some of the common schematic symbols you will be using almost immediately. Additional schematic symbols will be introduced throughout the text, as you need to know them. Become familiar with the symbols; the details of the represented components' operation and applications will unfold as your learning continues.

Examine Figure 2–26 and observe a few of the important electronic symbols. Do not be worried about committing these to rote memory at this time. You will automatically learn these by repeated usage as you continue in your studies.

Block Diagrams

Block diagrams are useful because they are simple and they conveniently portray the "functions" relating to each block. Recall that we used a simple block diagram in Chapter 1 when a simple electrical system was discussed.

Refer to Figure 2–27 and see if you can understand the concept of a block diagram. In essence, a block diagram is created by analyzing a system; breaking it into meaningful subsystems, which are drawn as individual blocks; then, interconnecting the subsystem blocks in a way that portrays flow and interaction between the blocks.

Note that the diagram quickly conveys what the system does. That is, this system distributes electrical power from a source to several loads. Also, it graphically shows the direction of flow of the electrical energy between blocks. Additionally, it very simply displays what each part of the system does. The power source supplies power, the distribution components control and allocate portions of the power to the various loads, and the loads use the power delivered to them.

Now, see if you can interpret the block diagram of a system that almost everyone is familiar with.

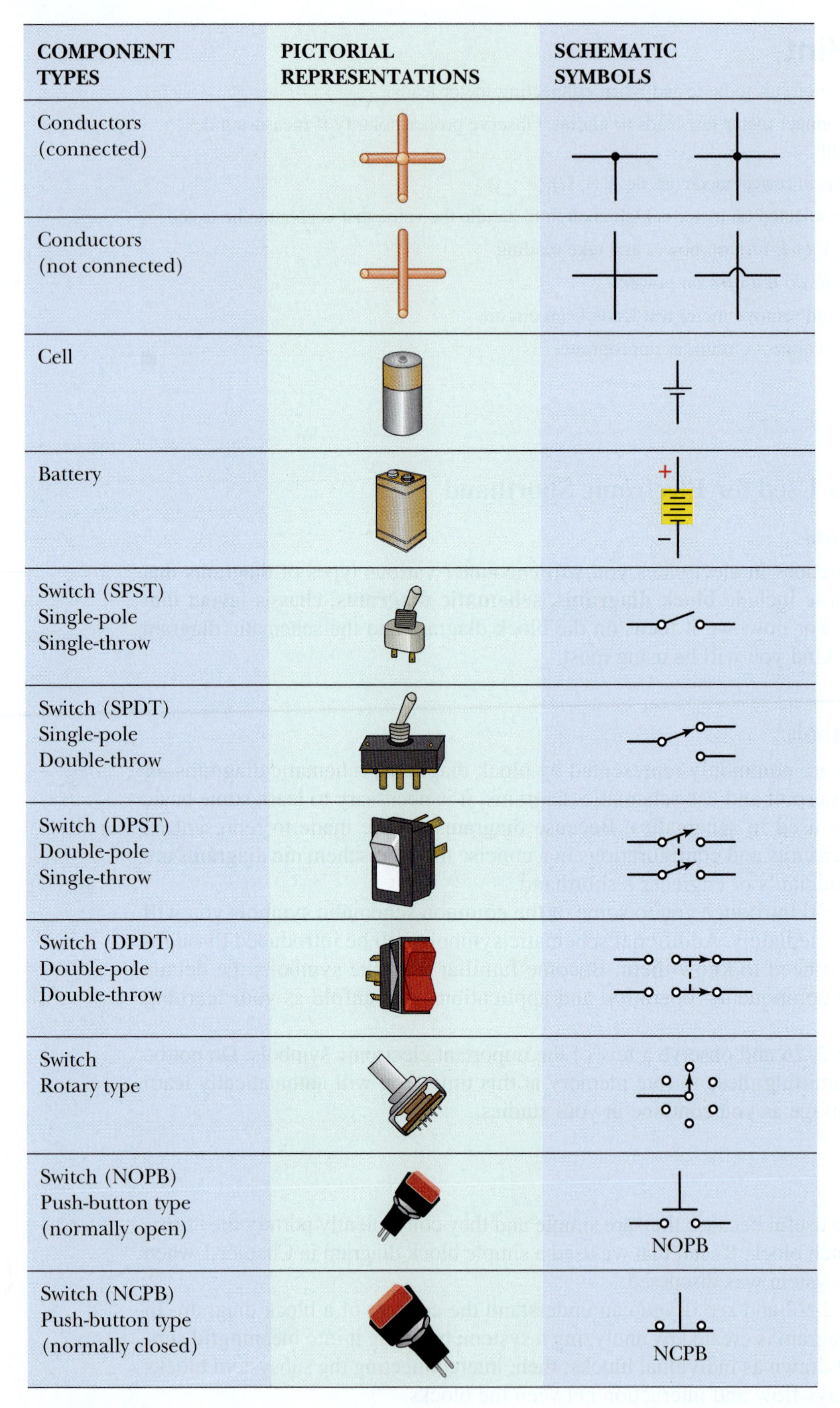

COMPONENT TYPES
PICTORIAL REPRESENTATIONS
SCHEMATIC SYMBOLS
Conductors (connected)
Conductors (not connected)
Cell
Battery
+
−
Switch (SPST) Single-pole Single-throw
Switch (SPDT) Single-pole Double-throw
Switch (DPST) Double-pole Single-throw
Switch (DPDT) Double-pole Double-throw
Switch Rotary type
Switch (NOPB) Push-button type (normally open)
NOPB
Switch (NCPB) Push-button type (normally closed)
NCPB

FIGURE 2–26 Various components and their schematic symbols

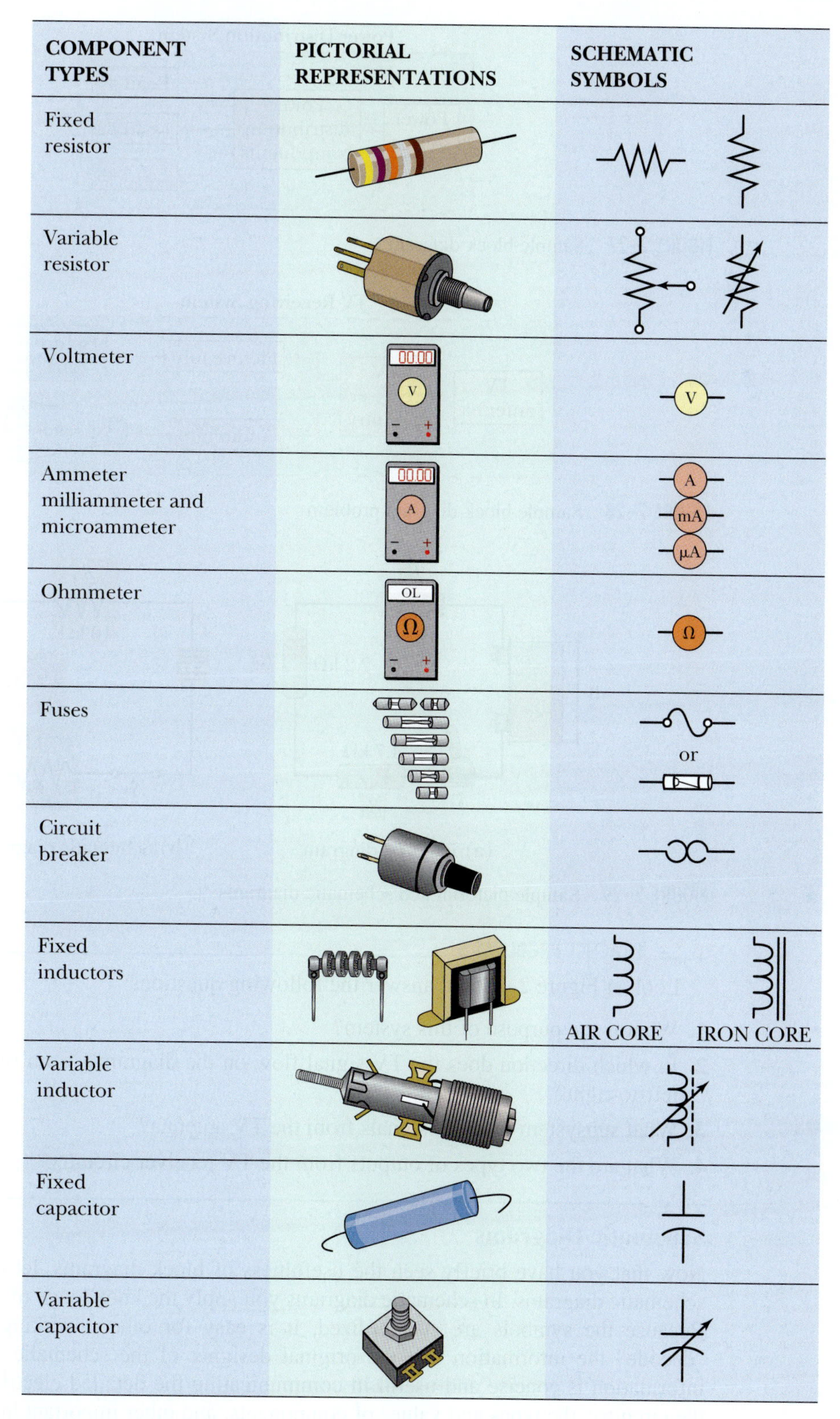

FIGURE 2–26 (cont.) Various components and their schematic symbols

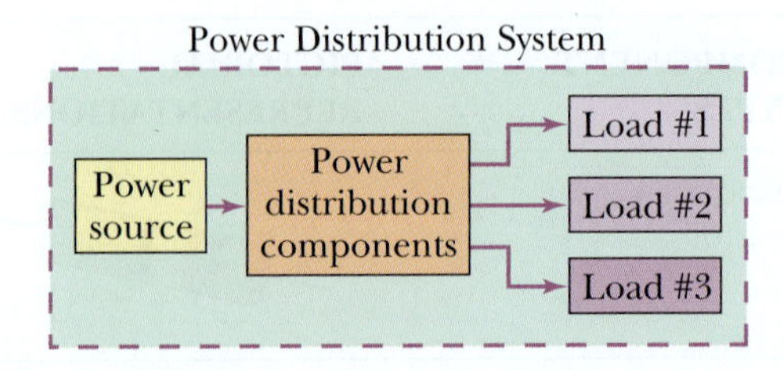

FIGURE 2–27 Sample block diagram

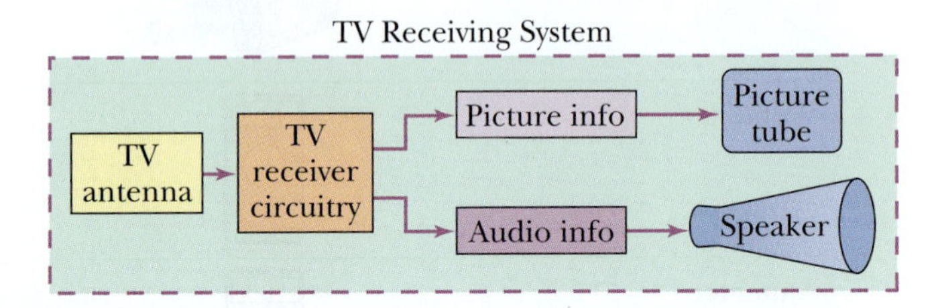

FIGURE 2–28 Sample block diagram problem

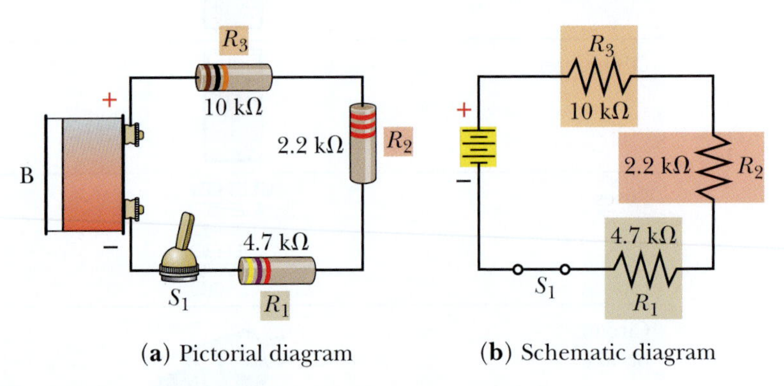

FIGURE 2–29 Sample pictorial and schematic diagrams

PRACTICE PROBLEMS 5

Look at Figure 2–28 and answer the following questions:

1. What is the purpose of this system?
2. In which direction does the TV signal flow on the diagram—from right-to-left or from left-to-right?
3. What subsystem receives signals from the TV antenna?
4. What are the two types of outputs from the TV receiver circuitry?

Schematic Diagrams

Now that you have briefly seen the usefulness of block diagrams, let us move on to the schematic diagrams. In schematic diagrams you apply the knowledge of electrical symbols. Because the symbols are standardized, it is easy for others working in electronics to "decode" the information that the original designer of the schematic is conveying. The information is concise and useful in communicating the detailed electrical connections in the circuitry, the types and values of components, and other important information.

Look at the sample pictorial and schematic diagrams in Figure 2–29 and notice the following facts:

1. Each component has an individualized callout. For example, one resistor is called R_1, another is called R_2, and so on. The switch is labeled S_1. If there were more switches, the series would carry on with S_2, and so on.
2. In the pictorial, you can see that the resistor color-code bands are shown. In this schematic, the resistor values are given. If there were other components with important value data, their values would also be shown on the diagram.

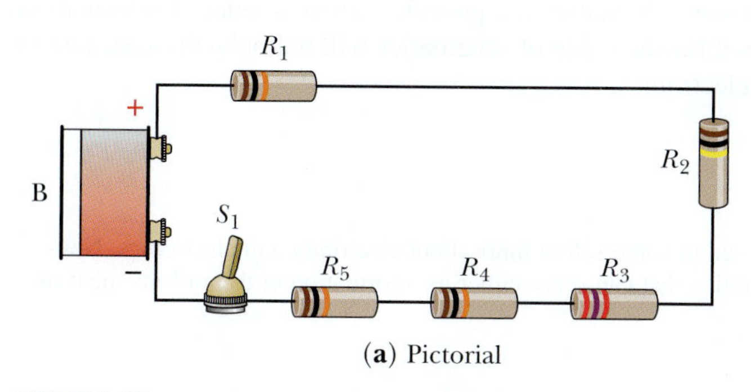

(a) Pictorial

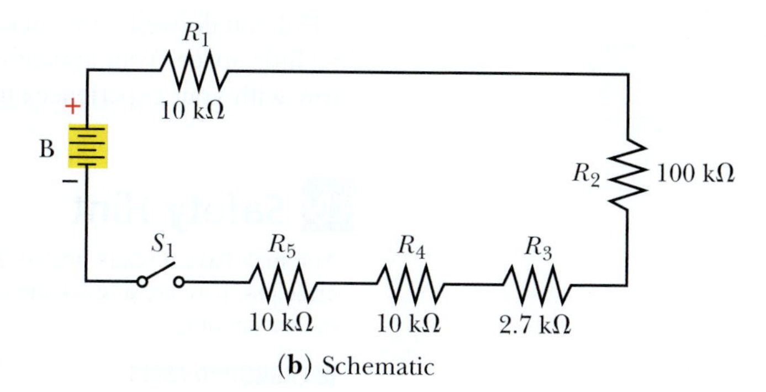

(b) Schematic

FIGURE 2–30 Diagrams for Practice Problems 6

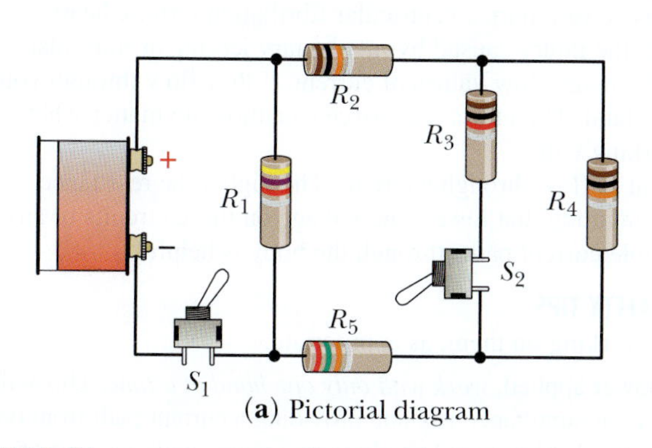

(a) Pictorial diagram

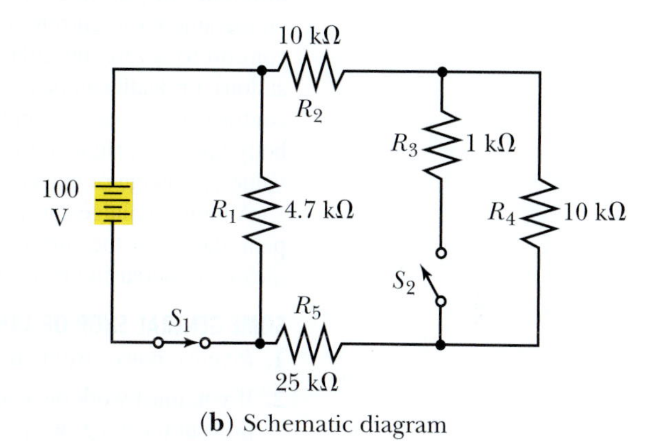

(b) Schematic diagram

FIGURE 2–31 Sample pictorial and schematic diagrams for Practice Problems 7

3. In both the pictorial and the schematic diagrams, the battery terminals are marked with + and –, as appropriate. This marking of polarity can be given at each component; however, since you will study more about polarity in the next chapter, we will wait until then to use these polarity markings throughout the diagram.
4. Finally, an extremely important feature of all schematic, pictorial, or other wiring diagrams is that every circuit interconnection is shown, allowing the reader to understand every component's electrical relationship to every other component in the circuit.

Also, on many schematic diagrams, the values of electrical parameters such as voltage, current, resistance, and so on are called out. These data help immensely when technicians are trying to troubleshoot a circuit.

PRACTICE PROBLEMS 6

Look at Figure 2–30 and answer the following questions:

1. How many different kinds of components are there in the circuit? Name them.
2. What is the highest value resistor?
3. What kind of switch is used, an SPST or an SPDT?
4. Is the voltage source a cell or a battery?

PRACTICE PROBLEMS 7

Look at Figure 2–31 and answer the following questions:

1. How many resistors are in the circuit?
2. Which resistor has the lowest resistance value? What is its value?
3. How many switches are in the circuit?
4. What is the value of the circuit-applied voltage?
5. Which switch is opened and which is closed?

Did you do well? As you can see, schematics can provide a great amount of information in a little space. Your appreciation for the value of schematics will undoubtedly continue to grow with your experiences in electronics.

Safety Hint

You may have already been exposed to some safety hints about electricity and electronics. Nevertheless, here are a few facts and tips that can serve either as information or as reinforcement of earlier learning.

BACKGROUND FACTS

Electrical shock occurs when your body provides a path for current between two points having a difference of potential. Generally, the amount of injury a person receives from a shock depends on the amount of current forced through the path of current. A range of effects from minor discomfort to severe muscular contractions, severe burns, ventricular fibrillation (erratic heart action), or death can occur. Many times the injury caused by involuntary jerking or muscular contractions is worse than the shock. Even very low values of current, if they flow through your body under the right conditions, can be fatal. Therefore, use extreme caution, no matter what value of voltage or current you are working with.

A higher voltage causes more current to flow through a circuit. The higher the resistance path, the lower the current. Therefore, anything that lowers the voltage on those circuits or any means of increasing resistance of possible current paths through the body is helpful.

SOME GENERAL SHOP OR LABORATORY SAFETY TIPS

1. *Remove power* from circuits before working on them, as appropriate.
2. If you must work on a circuit with power applied, *work with only one hand at a time.* This will prevent touching two points in the circuit simultaneously and providing a current path from one hand, through your body, to the other hand. Also, avoid touching two points with any part of your body. For example, your hand touching a point in the circuit, and your arm leaning against the circuit chassis, ground, or another point in the circuit can provide "shocking results."
3. *Insulate your body* from ground so a path is not provided from one point (through your body) to ground.
4. Use only *tools with insulated handles.*
5. *Know where the closest switch or main breaker switch is located* that will turn off power to the system. Make sure everyone working near you also knows this.
6. *Remove metal jewelry.* Dangling jewelry can not only make electrical contact, but, if you are working on "rotating machinery," can entangle you with the machinery.
7. When working on circuits with capacitors that store charge even after the power has been removed, *discharge the capacitors* before working on the circuit.
8. *Ground all equipment that should be grounded,* including equipment chassis, cabinets, and so forth.

SAFETY WHEN WORKING WITH HOT SOLDERING IRONS

1. *Position hot soldering irons properly* to prevent accidentally laying your arm, hands, or other objects on them.
2. *Wear safety glasses* when soldering.
3. *Be careful when wiping hot solder off a soldering iron.* Hot or molten solder splatters easily and causes serious burns.

NOTE: Other safety hints will be provided at strategic times during your training. Don't just read these safety hints; *follow them!* ■

Summary

- Charge (Q or q) is measured in coulombs. The abbreviation for coulomb is C. A typical subunit of the coulomb is the μC, or microcoulomb, meaning one-millionth of a coulomb.
- Potential difference, or voltage, is represented by the letter *V*. The unit of measure is the volt, also abbreviated "V." The submultiples and multiple of voltage commonly found include mV (millivolt), μV (microvolt), and kV (kilovolt).
- Current (*I*) is measured in amperes. The abbreviation for the unit of measure is A. Typical submultiples of this unit are mA (milliampere), μA (microampere), and occasionally pA (picoampere).
- Conductance uses the letter *G* and the unit of measure is the siemens (S). The commonly found submultiples are mS (millisiemens) and μS (microsiemens).
- Resistance uses the letter *R* and the unit of measure is the ohm (Ω). Typical multiples of this unit are the kΩ (kilohm) and the MΩ (megohm).
- Several components found in many circuits are resistors, switches, batteries (or voltage sources), and the conductor wires. Know the standard symbols used for each.
- Commonly used metric subunit prefixes are pico, or micro micro, meaning one millionth of one millionth (one trillionth) of a unit; nano, meaning one thousandth of a millionth (a billionth) of a unit; micro, meaning one millionth of a unit; and milli, meaning one thousandth of a unit.
- Powers of 10 for the above subunit prefixes are 10^{-12} for pico (or micro micro), 10^{-9} for nano, 10^{-6} for micro, and 10^{-3} for milli units.
- Common metric multiple-unit prefixes are kilo (1,000 times the unit) and mega (1,000,000 times the unit).
- Symbols for these metric prefixes are p (pico), n (nano), μ (micro), m (milli), k (kilo), and M (mega).
- Conductors can vary in the materials they are made from, in physical size, in length, and in overall makeup. Solid and stranded wires are two basic types of conductors. Other differences between conductor types may relate to how they are packaged and the type of insulation used.
- There is a standard wire-gauging system called the American Standard Wire Gauge, abbreviated AWG. The lower the gauge number, the larger the wire.
- The cross-sectional area of conductors is measured in circular mils (CM). One mil equals one thousandth of an inch. A one-CM wire has a diameter of 1 mil, or 0.001 inch.
- The larger a conductor's size (i.e., diameter and cross-sectional area) the lower its resistance, and the higher its current-carrying capacity. The greater the length of any given conductor, the greater its resistance from end to end.
- Resistivity (specific resistance) of any given type material is a measure of that material's resistance per standard length and cross-sectional area. The formula for resistivity is $R = \rho l/A$.
- A wire table is a convenient source of information regarding a conductor's resistance per given length and size, as well as its current-carrying capacity for specified conditions.
- Resistors perform several useful functions in electronic circuits: They limit current and divide voltage.
- Some types of resistors are carbon or composition resistors, film resistors, surface mount "chip" resistors, and wire-wound resistors. Another way to categorize resistors is fixed, semifixed, and variable resistors.
- Carbon or composition resistors normally are manufactured with power dissipation ratings from ¼ to 2 watts. Wire-wound resistors generally are rated to handle power levels from 5 watts to over 100 watts.
- Two other types of resistors commonly used are precision resistors (often film-type resistors) and the surface-mount technology "chip" type. Because of their small size, chip resistors come only in very small power ratings from about ⅛ to ¼ watt; however, they can have resistance values ranging from less than an ohm to more than a million ohms.
- Because carbon resistors are typically very small, a color-coding system has been devised to indicate resistor values.
- The sentence *"Big Boys Race Our Young Girls, But Violet Generally Wins"* can help you remember the 10 colors in the code: Black, Brown, Red, Orange, Yellow, Green, Blue, Violet, Gray, and White.
- Block diagrams illustrate the major functions within a circuit or system and convey the flow of signals, electrical power, and so forth. General flow is typically shown from left to right.
- Schematic diagrams provide more information than block diagrams about types of components in the circuit(s), their values, and specific connections and related electrical parameter values.
- Block diagrams provide an overview of the system, whereas schematic diagrams provide the information needed to enable circuit analysis, circuit repairs or maintenance, and/or desired circuit modifications.
- Safety rules relating to electrical circuits, hot soldering irons, and rotating machinery should always be followed.

Formulas and Sample Calculator Sequences

FORMULA 2–1
(To find area of a conductor in circular mils)

$A = d^2$

diameter value (in mils), $\boxed{x^2}$

FORMULA 2–2
(To find resistance of a conductor)

$R = \frac{\rho l}{A}$

Resistivity (CM-Ω/ft), $\boxed{\times}$, length (ft), $\boxed{\div}$, area (CM), $\boxed{=}$

EXCEL AUTOMATED FORMULAS

Helpful problem-solving tools, such as the Excel automated formulas are available on the CD, allow automatic calculations of formulas.

Review Questions

1. Convert 3 mV to an equivalent value in V.
2. If a value of current is 12 μA, this can be stated as a value of 12 × 10 to what power amperes? What is this same value of current when stated as a decimal value of amperes?
3. What submultiple of a milli unit is a micro unit?
4. How many kΩ are there in a MΩ?
5. Draw a schematic diagram illustrating a voltage source (battery), two resistors, and an SPST switch all connected end-to-end. Illustrate the first resistor connected to the source's negative side, the other resistor connected to the source's positive side, and the switch between the resistors.
6. [symbol] is the symbol for a fixed ______.
7. [symbol] is the symbol for a variable ______.
8. [symbol] is the symbol for a ______.
9. [symbol] is the symbol for a ______.
10. Convert the following units:
    - **a.** 15 milliamperes to amperes.
    - **b.** 15 milliamperes to 15 times a power of 10.
    - **c.** 5,000 volts to kilovolts.
    - **d.** 5,000 volts to 5.0 times a power of 10.
    - **e.** 0.5 megohms to ohms.
    - **f.** 100,000 ohms to megohms.
    - **g.** 0.5 ampere to milliamperes.
11. Name two or more types of resistors and discuss their physical construction.
12. Give the resistance value and tolerance rating for the following color-coded resistors: (NOTE: The first color shown is the first band; the second color represents the second band, and so forth.)
    - **a.** Yellow, violet, yellow, and silver.
    - **b.** Red, red, green, and silver.
    - **c.** Orange, orange, black, and gold.
    - **d.** White, brown, brown, and gold.
    - **e.** Brown, red, gold, and gold.
13. Name at least three important parameters, or factors, to consider when selecting a resistor for a given circuit.
14. If a precision resistor has a fifth color band, what does this band designate?
15. What is the value and tolerance rating of a precision film-type resistor having the following sequence of color bands: red, red, red, red, red?
16. What is the tolerance rating of a carbon-composition resistor having only three color-code bands?
17. If you had to replace a resistor having a power rating of 25 watts, what type of resistor would you likely be trying to procure?
18. What is the value of a surface mount "chip" resistor having coding numbers of 443?
19. If a resistor has a "reliability" coding, which band on the color-coded resistor would you use to determine that rating? If that band's color were orange, what percentage failure rate would you expect of that type resistor over 1,000 hours?
20. What is a very basic procedure you should use to prevent electrical shock when you are about to work on an electrical circuit?

21. List at least two physical precautions you can use to minimize the danger of shock if you *must* work on an energized electrical circuit?

22. What added precaution is used when working on circuits containing charged "capacitors"?

23. Name two possible dangers of wearing loose metal jewelry when working with electrical circuits or when using rotating machinery.

24. Name three special precautions you should use when working with hot soldering irons.

25. What is the basic situation that causes someone to get electrically shocked?

Analysis Questions

1. In your own words, describe why a resistor color code is necessary.

2. If the environmental temperature conditions for a certain conductor's operation increase, describe what happens to the following parameters:

a. Its resistance

b. Its current-carrying capacity

3. Explain in your own words what the reliability rating on certain resistors means.

4. Describe in your own words the procedures and precautions for:

a. Measuring dc voltage across a specified component in a multicomponent circuit.

b. Measuring circuit current.

c. Measuring the value of a specified resistor in a multiresistor circuit.

5. Draw a block diagram illustrating a simple power distribution system. Use at least three blocks, and label all blocks and diagram features.

6. Draw a schematic diagram showing two resistors, one meter, two switches, and a battery. Label all components.

7. List the color coding for the following resistors:

a. 1,000 ohms, 10% tolerance.

b. 27 kilohms, 20% tolerance.

c. 1 megohm, 10% tolerance.

d. 10 ohms, 5% tolerance.

8. List at least four safety precautions that should typically be used when connecting *any* type of measurement meter to an existing circuit.

9. Referring to Figure 2–26, if you were wiring a "fixed" resistor into a circuit, how many electrical connection points would you need to connect?

10. Referring to Figure 2–26, if you were wiring a "variable" resistor into a circuit, how many electrical connection points would you need to connect?

Performance Projects Correlation Chart

Suggested performance projects that correlate with topics in this chapter are:

Chapter Topic	Performance Project		Project Number
Making a Voltage Measurement with a Voltmeter	Voltmeters	(Use & Care of Meters series)	1
Making a Current Measurement with an Ammeter	Ammeters	(Use & Care of Meters series)	2
Making a Resistance Measurement with an Ohmmeter	Ohmmeters	(Use & Care of Meters series)	3
Resistor Color Code	Resistor Color Code	(Ohm's law series)	4

NOTE: It is suggested that after completing the above projects, the student should be required to answer the questions in the "Summary" at the end of this section of projects in the Laboratory Manual.

PART II

Basic Circuit Analysis

OUTLINE

- **3–1** Ohm's law and relationships between electrical quantities
- **3–2** Three common arrangements for the Ohm's law equation
- **3–3** Sample application of metric prefixes and powers of 10
- **3–4** Direction of current flow
- **3–5** Polarity and voltage
- **3–6** Comparison of circuit current directions from dc and ac sources
- **3–7** Work, energy, and power
- **3–8** Measuring electrical energy consumption
- **3–9** Commonly used versions of Ohm's law and power formulas

OBJECTIVES

After studying this chapter, you should be able to:

1. Explain the relationships of current, voltage, and resistance
2. Use **Ohm's law** to solve for unknown circuit values
3. Illustrate the direction of current flow and polarity of voltage drops on a schematic diagram
4. Use metric prefixes and powers of 10 to solve Ohm's law problems
5. Use a calculator to solve circuit problems
6. Use a computer spreadsheet program to solve circuit problems
7. Explain **power dissipation**
8. Use appropriate formulas to calculate values of power

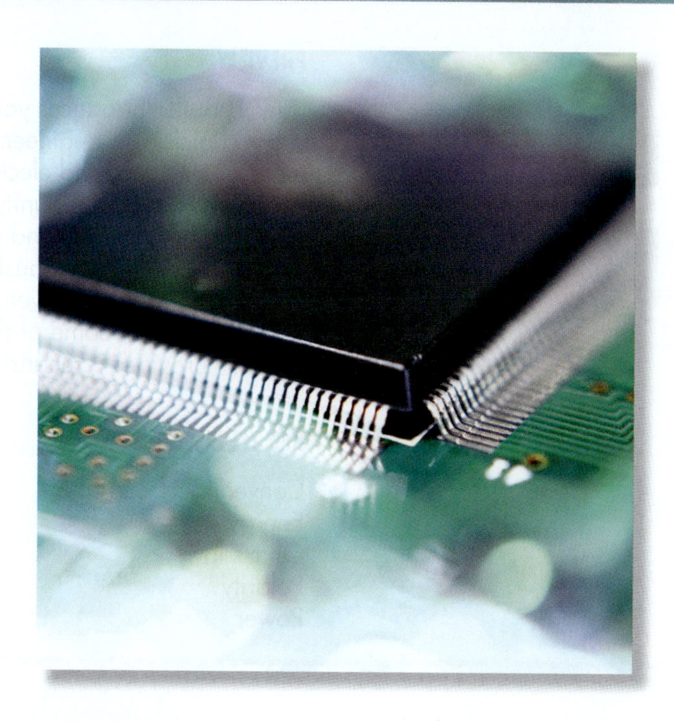

CHAPTER 3

Ohm's Law

ELECTRONICS INTO THE FUTURE

Explore an interactive presentation on Ohm's Law, Amps, Volts, Ohms & Power on the accompanying CD.

PREVIEW

This chapter introduces you to Ohm's law. Ohm's law is fundamental to your career as a technician or engineer. It is the basis for calculations of electrical quantities in all types of electrical and electronic circuits.

You will have opportunity to use the knowledge you have already acquired and combine it with some new and very practical circuit analysis information. For example, you will practice your previously learned knowledge and skills in working with equations, using metrics, using powers of 10, and using a scientific calculator. For those who need a brief review in some of these skills, a file on the CD called "Using the Scientific Calculator" provides a means for doing that.

KEY TERMS

Conventional current
Energy
Ohm's law
Polarity
Power
Power dissipation
Watt (W)
Watthour (Wh)
Wattsecond (Ws)
Work

3–1 Ohm's Law and Relationships between Electrical Quantities

Georg Simon Ohm, after whom the unit of resistance is named, observed and documented important relationships among three fundamental electrical quantities. He verified a relationship that indicated that electrical current in a circuit was *directly related* to the voltage applied and *inversely related* to the circuit's resistance. Recall that if two items are *directly related,* as one item increases, the other item also increases, and it increases proportionately to the first item's increase. If one item goes down, the other item goes down, proportionately. *Inversely related* means that as one item increases, the other item decreases, proportionately, and vice versa. Also recall that in defining some of the electrical quantities, the relationships in **Ohm's law** were implied in the definitions. For example, an ohm is that amount of resistance that limits current to one ampere when one volt is applied.

The Relationship of Current to Voltage with Resistance Constant

These relationships are stated in a formula known as Ohm's law, where:

Current (amperes) = Potential difference (volts) / Resistance (ohms)

FORMULA 3–1 $I = \frac{V}{R}; \left(\text{Amperes} = \frac{\text{Volts}}{\text{Ohms}}\right)$

◆ **EXAMPLE** If a circuit has 10 volts applied voltage and a resistance of 5 ohms, what is the value of current in amperes?

$$I = \frac{V}{R}, \text{ where } I = \frac{10\text{ V}}{5\,\Omega} = 2\text{ A (See Figure 3–1.)}$$ ◆

As you can see from the formula, if V increases and R remains unchanged, then a larger value V divided by the same value R will produce a larger number for I. This means that with a given resistance value, if voltage is increased, then current must increase also, and it will *increase* in direct proportion to the voltage change. Compare the current in Figure 3–1a with the current in Figure 3–1b. As you can see, when voltage increased from 10 V to 15 V and the resistance remained at 5 Ω, the current increased from 2 A to 3 A. That is directly proportional to the voltage change, isn't it? As voltage increased to one and one-half times the original value, current also increased to one and one-half times its original value.

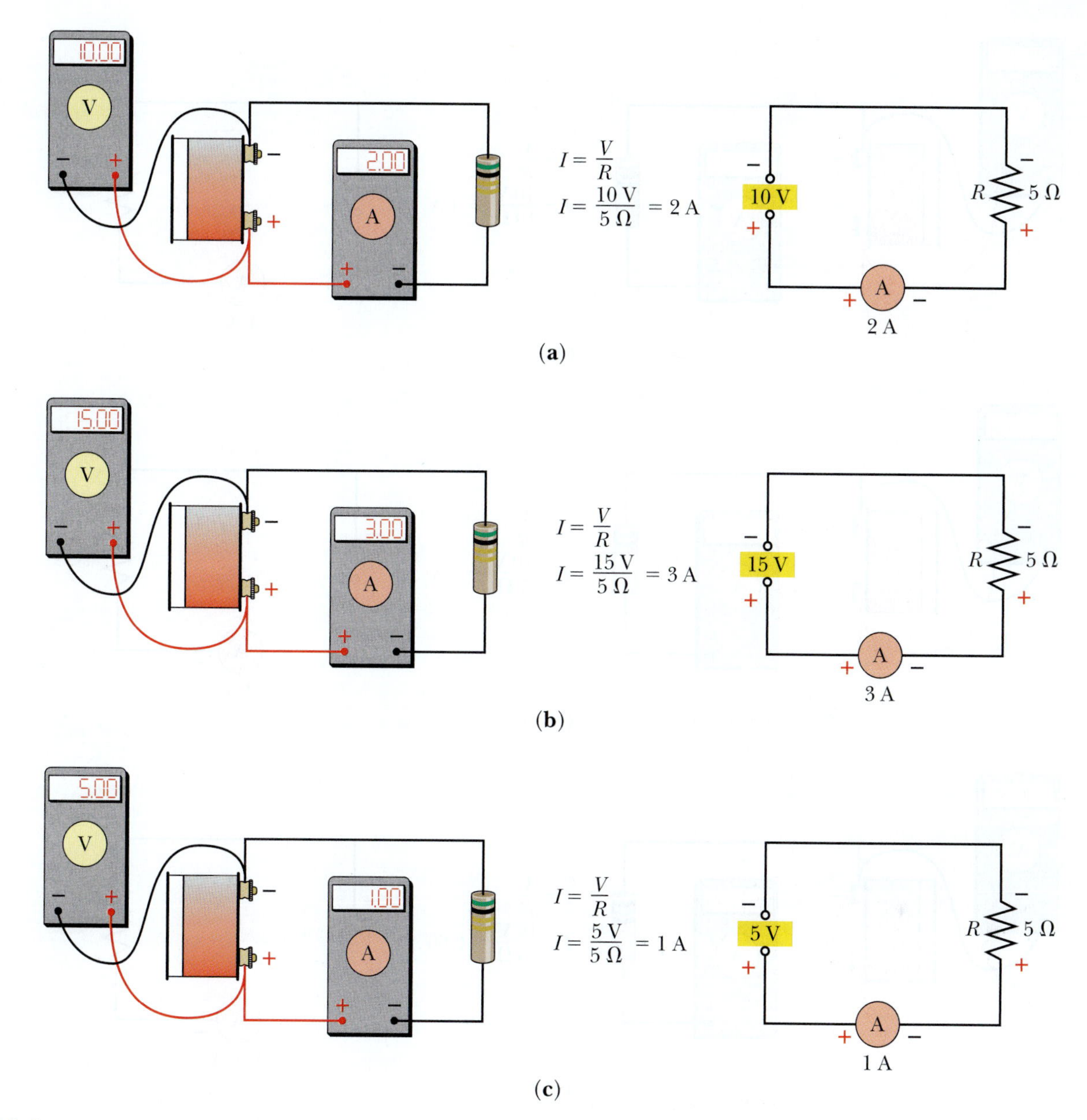

FIGURE 3–1 Ohm's law and the relationship of current to voltage with resistance constant multiSIM

On the other hand, if voltage is decreased while the resistance remains the same, the current will *decrease* proportionately with the voltage change. Compare the current in Figure 3–1c with the current shown for the circuit in Figure 3–1a. Again, you can see a direct proportion. The circuit current in Figure 3–1c is half the value of 2 A, shown in Figure 3–1a. This is due to the fact the applied voltage in the circuit of Figure 3–1c (5 V) is half of the 10-V applied voltage, shown in Figure 3–1a. This type of relationship may be referred to as a "linear proportionality."

We conclude, then, that *current is directly related to voltage* for a given resistance value. If the circuit applied *voltage increases* for a given resistance value, *current increases* in direct proportion to the increase in voltage. If *voltage decreases* for a given resistance value, *current* through the circuit will *decrease* in direct proportion to the voltage change.

The Relationship of Current to Resistance with Voltage Constant

In Figure 3–2a, the voltage is 10 V and the resistance value is 10 Ω. Thus, the current value is 1 A.

Recall, in Figure 3–1a, when the resistance was 5 Ω, and the applied voltage was 10 V, current was 2 A. By *increasing* the resistance to 10 Ω, with the same value of voltage (that

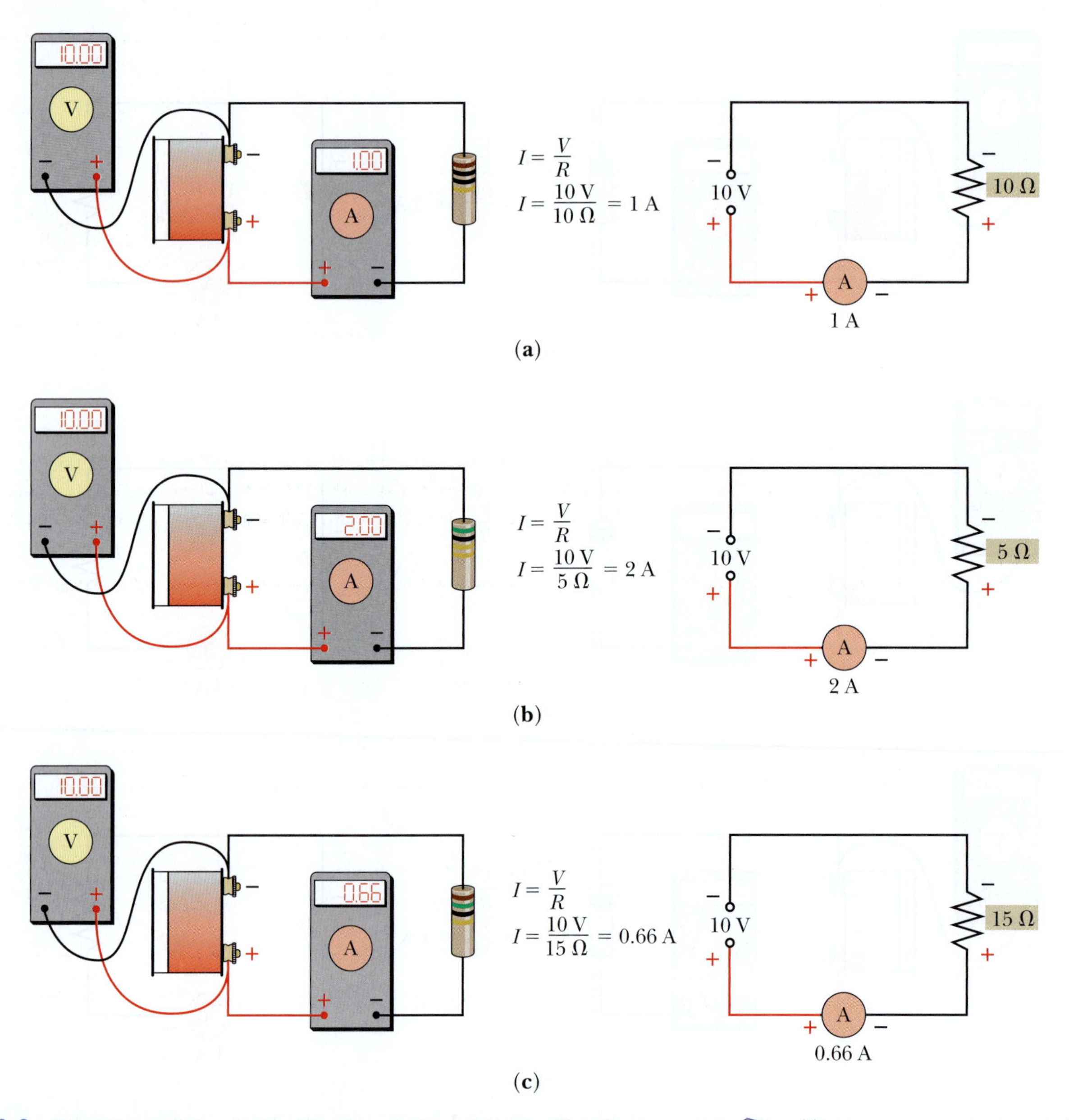

FIGURE 3–2 Ohm's law and the relationship of current to resistance with voltage constant multiSIM

> **Practical Notes**
>
> To solve problems, get in the habit of writing down the appropriate formula, substituting the knowns into the formula, then solving for the unknown. Also, if the problem is stated in words (without a diagram), draw a simple diagram illustrating the circuit in question. It will help you think clearly!

is, 10 V), the circuit current *decreases* to 1 A, as shown in Figure 3–2a. In other words, with *V* remaining the same and *R* increasing, the *I* decreases proportionately, showing an inverse relationship between *I* and *R*. Current is *inversely related to resistance.*

In Figure 3–2b, you can see that if voltage is kept constant at 10 V and the resistance is decreased to 5 Ω, the current will increase proportionately. Likewise, in Figure 3–2c, it is obvious that if the resistance is increased in value, for a given applied voltage, the circuit current will decrease in proportion to the resistance change.

The direct relationship between voltage *(V)* and *(I),* and the inverse relationship between resistance *(R)* and current *(I)* are illustrated with the Ohm's law formula in Figure 3–3.

Now that we have shown you some examples, you can use the knowledge you have gained regarding Ohm's law, and direct and inverse relationships, on the following problems.

If	$\frac{V\uparrow}{R\rightarrow}$	then $I\uparrow$	(I directly related to V. If V doubles, I doubles, if R is not changed)
If	$\frac{V\downarrow}{R\rightarrow}$	then $I\downarrow$	(I directly related to V. If V halves, I halves, if R is not changed)

(a)

If	$\frac{V\rightarrow}{R\uparrow}$	then $I\downarrow$	(I inversely related to R. If R doubles, I halves, if V is not changed)
If	$\frac{V\rightarrow}{R\downarrow}$	then $I\uparrow$	(I inversely related to R. If R halves, I doubles, if V is not changed)

(b)

FIGURE 3–3 Direct and inverse relationships of voltage, current, and resistance: (a) direct; (b) inverse multiSIM

PRACTICE PROBLEMS 1

1. Determine the circuit current for a circuit that has 150 V of applied voltage and a resistance of 10 Ω. (To solve, write down the Ohm's law formula, then substitute values in the formula as appropriate.)
2. If the voltage in problem 1 decreased to 75 V, but the resistance remained at 10 Ω, what would be the new circuit current?
3. Suppose that both voltage and resistance doubled for the circuit in problem 1:
   a. What would happen to the circuit current? Would it increase, decrease, or remain the same?
   b. Use Ohm's law and calculate the actual value resulting from these changes.

3–2 Three Common Arrangements for the Ohm's Law Equation

How can the $I = V/R$ equation be transposed to solve for V or R?

EXAMPLE Use the formula for Ohm's law and solve for V and for R.

$$I = \frac{V}{R}$$

To isolate V, multiply both sides of the equation by R, and cancel factors, as appropriate.

$$\text{Therefore, } I \times R = \frac{V}{R} \times R. \text{ Then } I \times R = \frac{V\cancel{R}}{\cancel{R}}.$$

FORMULA 3–2 $V = I \times R$; (Volts = Amperes × Ohms)

How can we isolate R? If both sides are divided by I, then the I on the right side of the equation will cancel because I divided by $I = 1$.

$$V = IR. \text{ Then, } \frac{V}{I} = \frac{I \times R}{I}. \text{ Then } \frac{V}{I} = \frac{\cancel{I} \times R}{\cancel{I}}$$

FORMULA 3–3 $R = \frac{V}{I}$; $\left(\text{Ohms} = \frac{\text{Volts}}{\text{Amperes}}\right)$

You have now been exposed to all three variations of Ohm's law formula.

$$I = \frac{V}{R}$$
$$V = I \times R$$
$$R = \frac{V}{I}$$

A useful memory aid often helps students remember these three variations of Ohm's law, Figure 3–4.

Restating these three Ohm's law variations:

$$I = \frac{V}{R} \text{ or Amperes} = \frac{\text{Volts}}{\text{Ohms}}$$
$$V = I \times R \text{ or Volts} = \text{Amperes} \times \text{Ohms}$$
$$R = \frac{V}{I} \text{ or Ohms} = \frac{\text{Volts}}{\text{Amperes}}$$

Knowing any two of the three electrical quantities, you can solve for the unknown value by using the appropriate version of the Ohm's law formula. This capability is important!

In the practical world of electronics, you usually will not be dealing with just the basic units of voltage, amperes, and ohms. You will often work with millivolts, milliamperes, microamperes, kilohms, and megohms. By way of review, observe Figure 3–5, which again

V

I × R

Simply look at (or cover up) the desired quantity (I, V, or R) and observe the relationships of the other two quantities. To remember the correct version of the formula:

Look at I, the aid shows $\frac{V}{R}$

$I = \frac{V}{R}$

Look at V, the aid shows $I \times R$

$V = I \times R$

Look at R, the aid shows $\frac{V}{I}$

$R = \frac{V}{I}$

FIGURE 3–4 Ohm's law memory aid

NUMBER	POWER OF 10	TERM	SAMPLE ELECTRONIC TERM
0.000000000001	10^{-12}	pico	pA (1×10^{-12} ampere)
0.000000001	10^{-9}	nano	nA (1×10^{-9} ampere)
0.000001	10^{-6}	micro	µA (1×10^{-6} ampere)
0.001	10^{-3}	milli	mA (1×10^{-3} ampere)
1,000	10^{3}	kilo	kΩ (1×10^{3} ohms)
1,000,000	10^{6}	mega	MΩ (1×10^{6} ohms)
1,000,000,000	10^{9}	giga	GH_z (1×10^{9} hertz)
1,000,000,000,000	10^{12}	tera	TH_z (1×10^{12} hertz)

FIGURE 3–5 Powers of 10 related to frequently used metric and electronic terms

shows the relationships between decimal numbers, powers of 10, metric terms, and electronic terms that are expressed using metric prefixes. Recall that in the preceding chapter, it was indicated that very small and very large electrical quantities can more easily be written and worked with by using powers of 10. In this chapter, we want you to begin to use the powers-of-10 mathematical tool because you will have some actual circuit problems to solve.

EXAMPLE Using powers of 10, the decimal number 250 can be expressed as:

250×10^0, or
25×10^1, or
2.5×10^2, or
0.25×10^3, and so forth.

PRACTICE PROBLEMS 2

Convert the following decimal numbers to a number times a power of 10, as indicated:

1. 25,000 = 25 × 10 to what power?
2. 12.335 = 1.2335 × 10 to what power?
3. 0.100 = 1.0 × 10 to what power?
4. 0.0015 = 1.5 × 10 to what power?

Convert the following numbers to their decimal equivalents:

5. 235×10^2
6. 0.001×10^6
7. 10×10^3
8. 15×10^{-3}

Using the [EE] or [EXP] Key

Now let us try to use the exponent [EE] or [EXP] key on a scientific calculator in conjunction with powers of 10. See Figure 3–6.

1. Input a number with the calculator keypad.
2. Push the [EE] or [EXP] button.
3. Input the power of 10 you want to raise that number to.
4. Press the = sign.

You will see that number expressed as a decimal value.

EXAMPLES Determine the value of (a) 235 × 10 to the 3rd power and (b) 235 × 10 to the minus 2nd power.

1. Input 235, [EE] or [EXP], input 3, [=]
   The readout indicates that 235×10^3 equals 235,000.
   If you want to use a negative power of 10, you simply input the negative sign prior to expressing the number for the power of 10.
2. Input 235, [EE] or [EXP], plus/minus [+/−] or [±], input 2, [=]
   The result indicates that 235×10^{-2} equals 2.35. Now you try the following problems.

PRACTICE PROBLEMS 3

Use the calculator and the [EE] or [EXP] key to find the decimal values for the following:

1. Determine the value of the number 1.456 times 10 to the 6th power.
2. Determine the value of the number 333 times 10 to the minus 4th power.
3. Determine the value of the number 1,500,000 times 10 to the minus 6th power.
4. Determine the value of the number 10 times 10 to the 5th power.

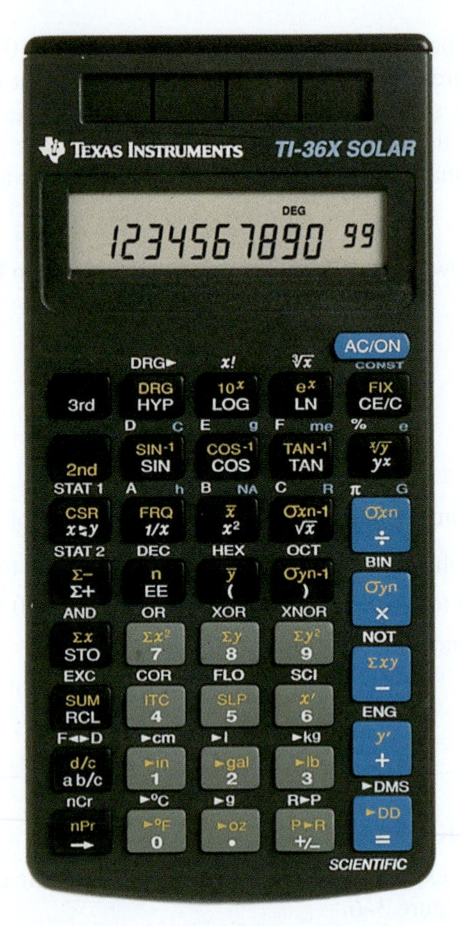

FIGURE 3–6 Photo close-up of a typical scientific calculator *(Courtesy of Texas Instruments)*

3–3 Sample Application of Metric Prefixes and Powers of 10

Suppose that a circuit has 10,000 ohms of resistance with 150 volts of applied voltage. What is the current?

$$I = \frac{V}{R}$$

$$I = \frac{150\text{ V}}{10{,}000\ \Omega}\left(\text{or } \frac{150\text{ V}}{10\text{ k}\Omega}\right)$$

$$I = 0.015 \quad (\text{or } 15\text{ mA})$$

It is cumbersome to divide units by thousands of units (kilounits), isn't it? It turns out that when you divide units (volts) by kilounits (kΩ), the answer is in milliunits (mA). That is a convenient thing to remember, and you will use it often.

However, how would this work if we used powers of 10?

$$I = \frac{V}{R}$$

$$I = \frac{150\text{ V}}{10 \times 10^{3}\ \Omega}$$

$$I = 15 \times 10^{-3}\text{ A (or 15 mA)}$$

Did you notice that when 10^3 is brought from the denominator to the numerator, it becomes 10^{-3}? Suppose we had divided one by one thousand. How many times would one thousand go into one? The answer is one one-thousandth time. Recall that one thousand equals 10^3 and one one-thousandth equals 10^{-3}. Therefore, when 150 volts are divided by 10,000 ohms, the answer is in thousandths of an ampere. Stated mathematically:

$$\frac{1}{1{,}000} = 0.001 \text{ or one thousandth (or } 1 \times 10^{-3}\text{) and}$$

$$\frac{150}{10{,}000} = 0.015 \text{ or 15 thousandths (or } 15 \times 10^{-3}\text{)}$$

See if you can apply Ohm's law, metric prefixes, and powers of 10.

■ IN-PROCESS LEARNING CHECK 1

1. Refer to the diagram below and determine the circuit current. $I =$ _______ amperes, or _______ milliamperes.

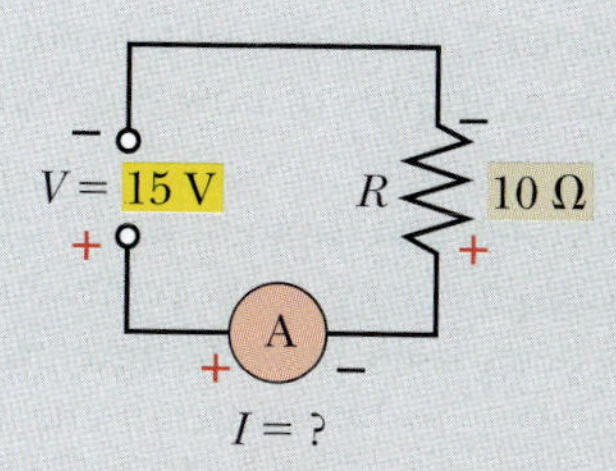

2. Refer to the circuit below and determine the circuit applied voltage. V _______ volts.

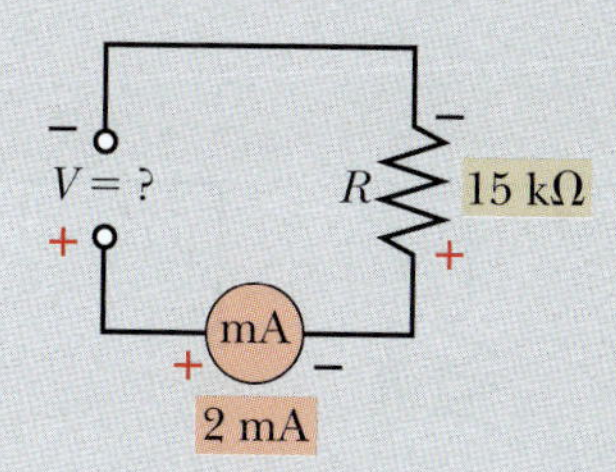

3. Refer to the circuit below and determine the value of circuit resistance. R _______ ohms, or _______ kilohms.

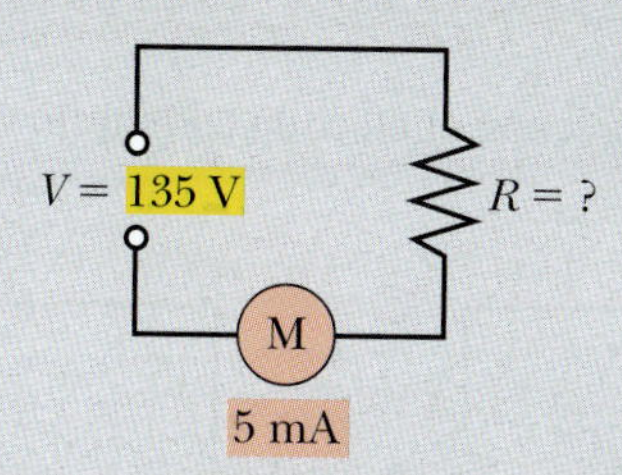

4. a. If R increases and V remains the same in Problem 1, does I increase, decrease, or remain the same?

b. If I increases and R remains the same in Problem 2, would the voltage applied be higher, lower, or the same as stated in the problem?

c. In Problem 3, if I doubles but the V remains the same, what must be true regarding R? ■

3–4 Direction of Current Flow

Electron Flow Approach

In Chapter 1, there was a brief discussion about electrons (negative in charge) moving to a positively charged point in a circuit. When circuit diagrams are shown, it is quite common to show the direction of current flow through the circuit by means of arrows. In the cases shown in Figure 3–7, the arrows are used to show the direction of the electron current flow. *For this text, this is the standard* that will be used for indicating direction of currents through circuits. This is coherent with our discussion of the fact that current flow is composed of the movement of free electrons. Because electrons are negative in charge, they are repelled from negative points in circuits and are attracted toward more positive points; thus, electron current flow moves from negative-to-positive locations in components or circuits external to the voltage source.

Conventional Current Flow Approach

Before the development of the electron theory, from which we derive our electron-flow approach to analyzing electron movement, there was a theory that assumed that current flow was in the direction that a positive charge would drift or move through a circuit. Obviously, this assumption would show current flowing from positive-to-negative points in a circuit. This difference in representing direction of current flow is referred to as **conventional current.** See Figure 3–8 for examples of how diagrams would represent this type of current.

Because many engineering formulas and thought processes were created back when this theory was prevalent, numerous engineering texts represent current flow in this manner. Also, this is the manner that has been used in creating certain electronic component symbols. For these reasons, we want you to be aware of the term *conventional current.* Be assured that regardless of whether electron current flow or conventional current flow is used in circuit analysis, *the answers come out the same!* As indicated earlier, this text will adhere to the electron current flow approach, as indicated in Figure 3–7.

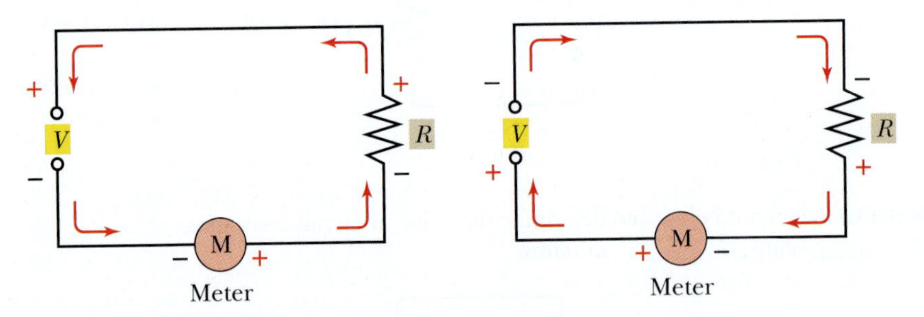

FIGURE 3–7 Method of showing direction of electron current flow with arrows

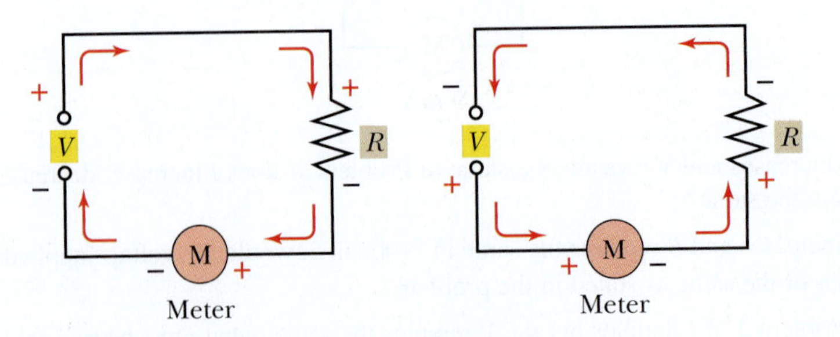

FIGURE 3–8 Arrows showing direction of "conventional current"

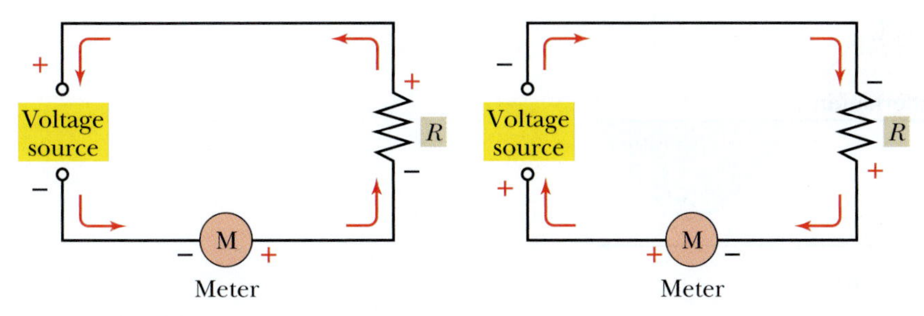

FIGURE 3–9 Illustrating polarity of voltages

3–5 Polarity and Voltage

The concept of **polarity,** or one point in a component or circuit being considered more negative or positive than another point, was discussed earlier. Recall that the more negative point was labeled with a negative sign, and the less negative (or positive point) was labeled with a positive sign. This technique indicates there is a potential difference, or voltage, between the two points.

To label the polarity of voltages across components in a circuit diagram, note the direction of electron current flow, then label the point where the current enters the component as the negative end. The other end of the component (where the current exits) is labeled as the positive point. This process is true for all components in the circuit, except the voltage source. In that case, the negative side of the source is the source of the electron flow, and the positive end of the source is the end where the electrons return after traveling through the external circuit, Figure 3–9.

Practical Notes

You should have observed throughout our discussions that current flows through a component and/or circuit. Voltage does **not** "go" or "flow." It is a difference of potential between points, so think of voltage as being **across,** or a difference of potential **between** two points.

3–6 Comparison of Circuit Current Directions from DC and AC Sources

So far, we have only mentioned dc (direct current) sources, and the resulting direction of electron current flow through circuits connected to them. At this point, we want to introduce you to the fact that there is also another type of voltage and current known as "ac." The ac stands for alternating current. An example of this type of voltage and current is that supplied to your home from the power company. An ac source provides voltage that periodically alternates in direction. The output waveform from such a source is known as a "sine wave." See Figure 3–10a. A complete "cycle" of a sine wave has occurred at the moment the waveform starts repeating itself. For example, in the waveform shown in Figure 3–10a, the voltage starts from zero level, moves to a maximum positive level, and returns back to zero. *Then,* the source's output voltage *reverses* direction, or polarity. The output voltage now moves to a maximum negative point, and then back to zero. At this point, the voltage waveform starts the same sequence of events all over again. This begins the second cycle. See Figure 3–10b.

One-half of the complete cycle is called an "alternation"; therefore, there are two alternations in a cycle of ac, Figure 3–10a. As you can see, there is one positive alternation and one negative alternation in a cycle. You will be studying this type of waveform in more depth in a later chapter. For now, we just want you to see the general differences between dc and ac.

Notice the contrast between dc and ac voltage as shown in Figure 3–11. The dc voltage remains at one polarity. This means the current is always traveling through the circuit in one direction, Figure 3–11a. In the ac circuit, you see that during one alternation of the ac cycle, current travels through the circuit in one direction, Figure 3–11b. (Incidentally, you probably

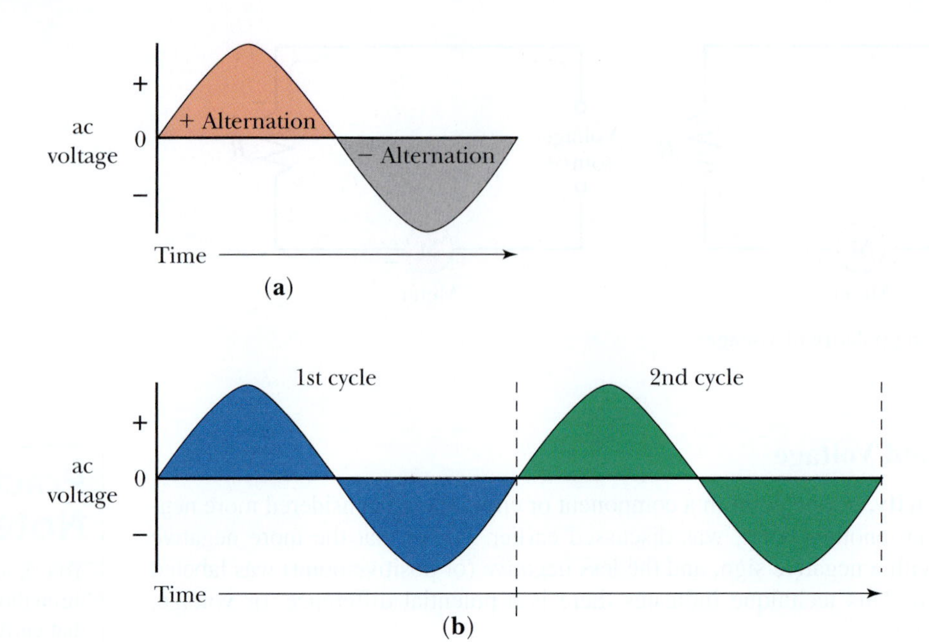

FIGURE 3–10 Representation of ac sine-wave voltages

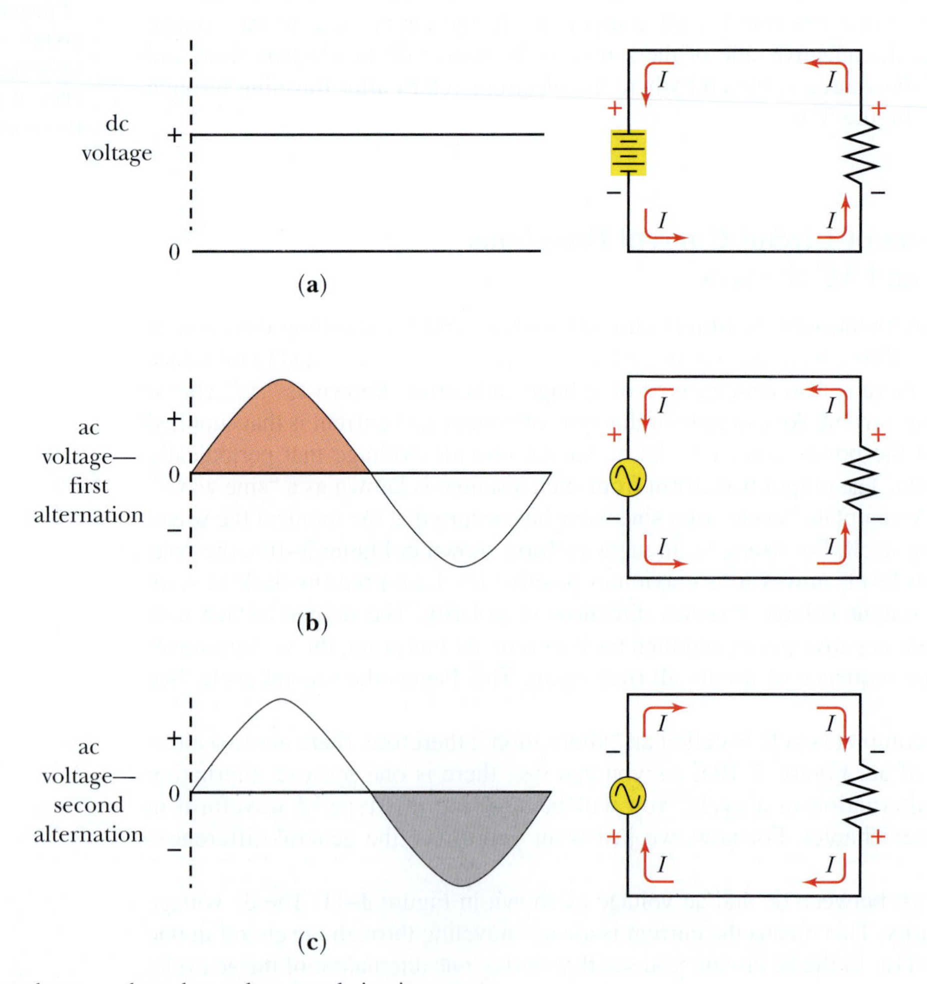

FIGURE 3–11 Contrasts between dc and ac voltages and circuit currents

noticed that the electrons are traveling from the negative side of the source, through the external circuit, back to the positive side of the source, just like they did in the dc circuit.) Obviously, when the source reverses polarity (on the next alternation), the electrons will reverse their direction of travel through the circuit. However, once again, the electrons leave the negative side of the source, travel through the external circuit, and return to the positive side of the source, Figure 3–11c. This indicates that the polarity of the ac source voltage applied to the circuit determines the direction of the electron current flow through the circuit. As the polarity of source voltage reverses, so does the direction of current flow through the circuit, Figure 3–11b and c.

Later in your studies you will learn some more specific facts and terms related to ac voltages and signals. These facts will be presented at the particular time you will be best able to apply them.

3–7 Work, Energy, and Power

Work is the expenditure of energy. **Energy** is the ability to do work, or that which is expended in doing work. **Power** is the rate of using energy. A formula for mechanical work is:

FORMULA 3–4 Work (Foot pounds) = Force (Pounds) × Distance (Feet)

The basic unit of mechanical work is the foot poundal. At sea level, 32.16 foot poundals equals 1 foot pound. Mechanical power is the rate of doing work. For example, the "horsepower" equals a rate of 550 foot pounds per second.

Electrical energy is the ability to do electrical work. The unit of energy used for electrical energy is the joule (J). For electrical applications, the joule is defined as the energy required to move one coulomb of electrical charge between two points having a difference of potential of one volt.

Electrical power is the rate at which electrical energy is used. The unit of electrical power is the **watt,** named after the scientist James Watt. The abbreviation for watt or watts is **W.** *A watt of electrical power is defined as electrical energy expended at the rate of one joule per second.* Since joules can be related to energy, and energy can be related to force multiplied by distance, it can be determined that 746 watts is equivalent to one horsepower of mechanical power. As we already stated, electrical power in watts represents the time rate of expending electrical energy in performing electrical work; for example, the work of causing current flow (moving electrons) through an electrical resistance. Power can be related to electrical energy in the following manner:

FORMULA 3–5 $\text{Power (Watts)} = \dfrac{\text{Energy (Joules)}}{\text{Time (Seconds)}}$

EXAMPLE If 100 joules of energy are used in 10 seconds, determine the power used.

$$P = \frac{100j}{10s} = 10 \text{ W}$$

PRACTICE PROBLEMS 4

1. What is the power, in watts, for a circuit in which 25 joules of energy are being used in 2 seconds?
2. What is the power, in milliwatts, in a circuit where 1 joule is used in 100 seconds?

3–8 Measuring Electrical Energy Consumption

By finding how much power has been used over time, we can determine the total electrical energy that has been consumed. You should be aware that the symbol for both work and energy is *W;* however, the SI unit for both is the joule (J).

(NOTE: Don't get the symbol *W* for work or energy confused with the abbreviation for the unit of power, the watt [W].)

The electric power company charges you for the amount of electrical energy you use for a given period of time. For example, they typically charge you for the total number of "kilowatthours" of electrical energy you have used for that month. When you use one kilowatthour of energy, you are using 1,000 watts for one hour. For example, if you burn ten 100-watt bulbs for 1 hour, you are using one kilowatthour's worth of energy. If you use a 500-watt iron for 2 hours, you are using one kilowatthour of energy, and so on.

FORMULA 3–6 Energy (Watthours) = Power (Watts) × Time (Hours)

The time can be measured in seconds or hours. When the time is expressed in seconds, the unit of energy consumption is called the **wattsecond (Ws).** When the time is expressed in hours, the unit of energy consumption is called the **watthour (Wh),** or, when convenient due to the amount of power involved (as in the case of the power company), the kilowatthour (kWh).

EXAMPLES

1. A power consumption of 200 watts for 3 hours = 600 Wh of electrical energy consumption.

$$W = P \times T; \quad W = 200 \text{ watts} \times 3 \text{ hrs.} = 600 \text{ Wh (or 0.6 kWh)}$$

2. If 450 watts are being used over 9 hours, find the energy consumption.

$$W = P \times T; \quad W = 450 \text{ watts} \times 9 \text{ hrs.} = 4{,}050 \text{ Wh (or 4.05 kWh)}$$

PRACTICE PROBLEMS 5

How many kilowatthours of energy are used by a 250-watt bulb if it is left on for 8 hours?

The Basic Power Formula

Relating electrical power to practical circuits:

- A joule is the energy used to move one coulomb of charge between two points with a one-volt potential difference.
- An ampere is the rate of charge of movement of one coulomb per second.
- A watt is the rate of doing work at one joule per second.

Thus, one watt of power equals one ampere times one volt. The formula is:

Power (in watts) = Voltage (in volts) × Current (in amperes)

FORMULA 3–7 $P = V \times I$ (Watts = Volts × Amperes)

EXAMPLE If a circuit has 50 milliamperes of current flow and an applied voltage of 10 volts, how much power does the circuit dissipate? Using the basic formula for power:

$$P = V \times I$$
$$P = 10 \text{ V} \times 50 \text{ mA}$$
$$P = 10 \times 50 \times 10^{-3} = 500 \times 10^{-3} \text{ W or } 500 \text{ mW}$$

PRACTICE PROBLEMS 6

1. If a circuit has an applied voltage of 150 V, and the circuit resistance limits the current to 10 mA; how much power does the circuit dissipate?
2. If the circuit voltage for Problem 1 remained at 150 V, but the circuit resistance were doubled:
   a. What would happen to circuit current?
   b. What would the circuit power dissipation be?

IN-PROCESS LEARNING CHECK 2

1. Electron current flow is considered to flow from the (negative, positive) _______ side of the source, through the external circuit, back to the (negative, positive) _______ side of the source.
2. "Conventional" current is considered to flow from the (negative, positive) _______ side of the source, through the external circuit, back to the (negative, positive) _______ side of the source.
3. This text uses the (electron, conventional) _______ current flow approach.
4. One of the basic differences between dc and ac is that current in a dc circuit always flows in _______ direction; whereas, current flow in an ac circuit _______ direction for each alternation of the ac voltage.
5. Work is the _______ of energy; energy is the ability to do_______, or that which is expended in doing _______. Power is the rate of using _______.
6. Power in watts is equal to energy in _______ divided by time in _______.
7. Energy = power (in watts) × time. A common measure of energy consumption used by power companies for billing customers is the _______.

Variations of the Basic Power Formula

Just as the Ohm's law formula can be rearranged, so can the power formula. The formula we started with was $P = V \times I$. Two other arrangements are $V = P/I$ and $I = P/V$. Incidentally, the same style of visual aid used for the Ohm's law formula will work here, Figure 3–12.

Other practical variations of the power formula are developed by combining knowledge of Ohm's law and the power formulas. For example, since $V = I \times R$ (Ohm's law) and $P = V \times I$, the formula can be combined to show that $P = I^2 \times R$ or I^2R. You have substituted for V in the Ohm's law formula $P = V \times I$; therefore $P = (I \times R) \times I$.

FORMULA 3–8 $P = I^2R$ (Watts = Amperes squared × Ohms)

Using this same idea, you can derive the formulas that state:

$$P = \frac{V^2}{R}, \quad P = VI, \quad P = V \times \left(\frac{V}{R}\right)$$

FORMULA 3–9 $P = \frac{V^2}{R} \left(\text{Watts} = \frac{\text{Volts squared}}{\text{Ohms}}\right)$

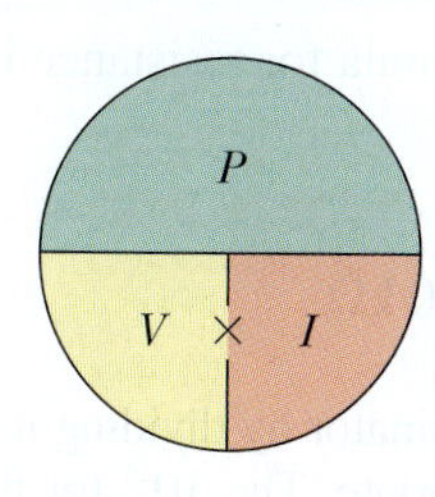

Again, by looking at the desired parameter (or covering it up) and seeing the arrangement of the other two parameters, the correct formula is evident. For example, if you know V and I, to find power take $V \times I$. If you know the power and the current, simply divide P by I to get V, and so forth.

FIGURE 3–12 Power formula memory aid

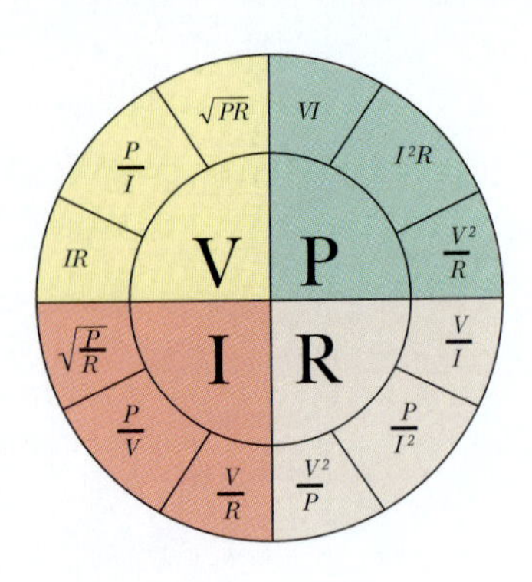

You have simply substituted the Ohm's law formula for I into the $P = V \times I$ power formula. In order to solve for any desired electrical quantity, each of the power formulas can be rearranged using the transposition and cancelling techniques learned earlier in this chapter.

$$P = V \times I, \quad I = \frac{P}{V}, \quad V = \frac{P}{I}$$

$$P = I^2 \times R, \quad I = \sqrt{\frac{P}{R}}, \quad R = \frac{P}{I^2}$$

$$P = \frac{V^2}{R}, \quad V = \sqrt{PR}, \quad R = \frac{V^2}{P}$$

The colored segments of the "wheel" graphic in the margin are aids that illustrate the various formulas that can be used to solve for: V, I, R, and P.

3–9 Commonly Used Versions of Ohm's Law and Power Formulas

The most important equations to remember are:

$$I = \frac{V}{R}, \quad V = IR, \quad \text{and } R = \frac{V}{I} \text{ (Ohm's law formulas)}$$

$$P = VI, \quad P = I^2R, \quad \text{and } P = \frac{V^2}{R} \text{ (power formulas)}$$

By understanding and knowing these Ohm's law and power formulas and by becoming adept at applying them to practical electronic problems, you will be on the way to becoming a good technician.

For a change of pace from simply reading about these formulas, we now want you to see how these formulas can be applied to practical problems. Follow the steps carefully as several sample problems and solutions are shown. You will be doing this sort of thing on your own throughout the remainder of your training.

EXAMPLES Refer to Figure 3–13 for the following.

1. In Diagram 1, find the voltage *(V)*. Ohm's law formula for voltage is $V = I \times R$. Substitute the known values for I and R.
Therefore, $V = 5 \times 10^{-3}$ ampere $\times 10 \times 10^3$ ohms. $V = 50$ volts
Remember: When multiplying powers of 10, add their exponents algebraically: 10^{-3} and 10^3 results in 10^0, which equals 1. Also, milliunits multiplied by kilounits result in units, and, in this case, mA × kΩ result in volts.

2. In Diagram 2, find the circuit current *(I)*. Ohm's law formula for current is $I = V/R$. Substitute the known values for V and R.

$$I = \frac{200 \text{ V}}{100 \text{ k}\Omega} = \frac{200 \times 10^0}{100 \times 10^3} = 2 \times 10^{-3} = 2 \text{ mA}$$

Remember: When dividing powers of 10, subtract the exponents algebraically. Since the voltage did not show a power of 10, show it as 200×10^0 [$10^0 = 1$]. To subtract the denominator 10^3 from the numerator 10^0, change the sign of the denominator power of 10, then add 10^{-3} to 10^0, which equals 10^{-3}. Also, units divided by kilounits result in milliunits. In this case, V/kΩ results in milliamperes (mA).

3. In Diagram 3, find the resistance *(R)*. Ohm's law formula for resistance is $R = V/I$. Substitute the known values for V and I.

$$R = \frac{100 \text{ V}}{5 \text{ mA}} = \frac{100}{5 \times 10^{-3}} = 20 \text{ k}\Omega$$

Remember: When moving a power of 10 in the denominator by dividing it into the numerator, change its sign and add the exponents as appropriate. The 10^{-3} for the milliamperes becomes the 10^3 for kilohms in the answer. Also, in this problem, units divided by milliunits result in kilounits; thus, volts divided by milliamperes result in the answer as kilohms.

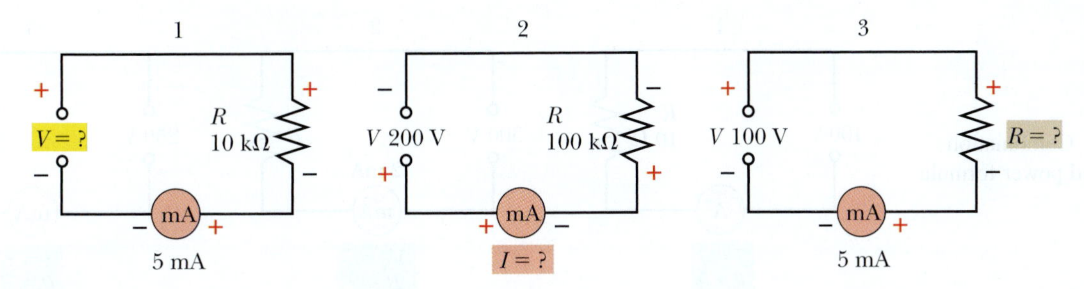

FIGURE 3–13 Ohm's law example problems

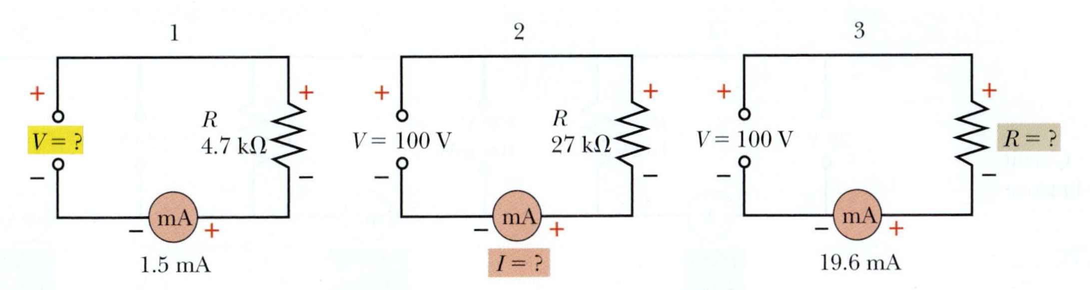

FIGURE 3–14 Ohm's law practice problems

PRACTICE PROBLEMS 7

1. Find the applied voltage in the circuit of Figure 3–14, Diagram 1.
2. Find the circuit current for the circuit of Figure 3–14, Diagram 2.
3. Find the circuit resistance for the circuit of Figure 3–14, Diagram 3.

In summary, when applying Ohm's law and power formulas:

- look at the circuit to determine the knowns and unknown(s);
- select and write the appropriate formula to use;
- substitute the knowns into the formula;
- use metrics and powers of 10 to manage large and small numbers;
- remember that units divided by milliunits equals kilounits and that units divided by kilounits equals milliunits;
- remember that milliunits multiplied by kilounits equals units; and
- recall various expressions for Ohm's law.

Solving Problems Using Both Ohm's Law and Power Formulas

Technicians are often required to combine their knowledge of Ohm's law and the power formulas to find the electrical quantities of interest. The following examples provide an idea about how you need to logically address problems of this nature.

EXAMPLES Refer to Figure 3–15. Find the unknown values shown.

1. In Figure 3–15, Diagram 1, which formula would be easiest to use first? Observe that values for V and R are known.

$$I = \frac{V}{R} = \frac{100\text{ V}}{10\ \Omega} = 10\text{ A}$$

Knowing the current, you can solve the remaining unknown.

$$P = V \times I = 100\text{ V} \times 10\text{ A} = 1{,}000\text{ watts (or 1 kilowatt)}$$

Also, since V, I, and R are known, you can use the other power formulas, as well.

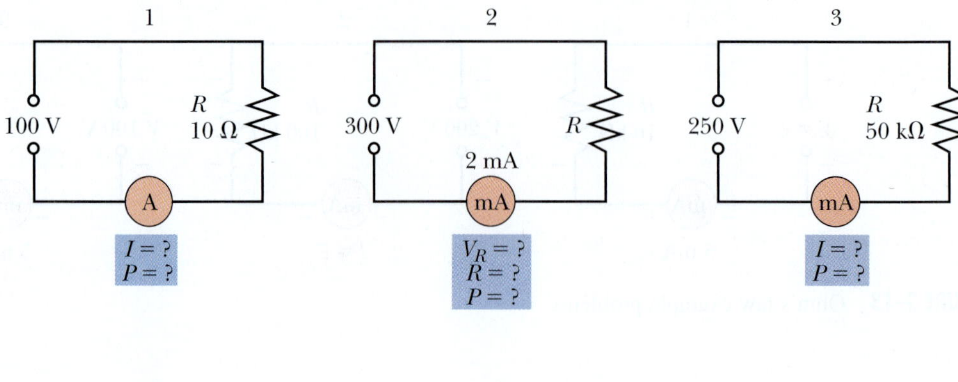

FIGURE 3–15 Combination Ohm's law and power formula problems

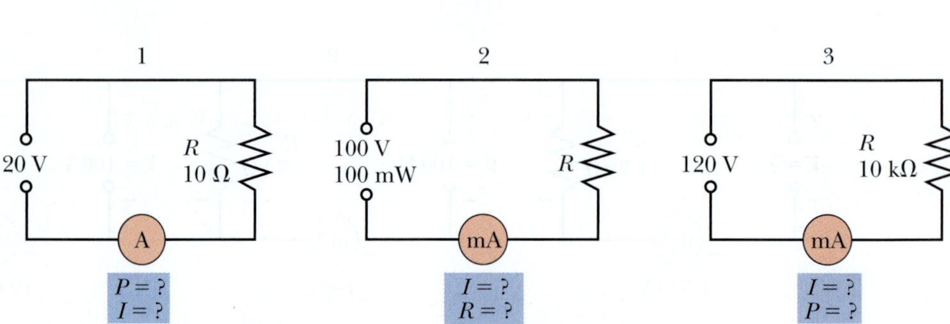

FIGURE 3–16 Circuit problems for Practice Problems

2. In Figure 3–15, Diagram 2, which formula might be used first? In this case, voltage and current are known values, but resistance is unknown. If the current meter does not drop any appreciable voltage, you could assume that V_R equals the applied voltage and first solve for power. In many problems, you can start at more than one place, depending on the known value(s). Let us start with the power formula.

$$P = V \times I = 300 \text{ V} \times 2 \times 10^{-3} \text{ A} = 600 \times 10^{-3} \text{ W (or 600 mW)}$$

Now to solve for the other unknowns, assume 300 V across *R*, where $V_R = 300$ V.

$$R = \frac{V}{I} = \frac{300 \text{ V}}{2 \times 10^{-3} \text{ A}} = 150 \times 10^{3}\ \Omega \text{ (or 150 k}\Omega\text{)}$$

3. In Figure 3–15, Diagram 3, a good starting point is to first solve for *I* since you know the values of *V* and *R*.

$$I = \frac{V}{R} = \frac{250 \text{ V}}{50 \text{ k}\Omega} = \frac{250}{50 \times 10^{3}} = 5 \times 10^{-3} \text{ A (or 5 mA)}$$

Now that voltage and current are known, it is easy to solve for the other unknown.

$$P = V \times I = 250 \text{ V} \times 5 \times 10^{-3} \text{ A} = 1{,}250 \times 10^{-3} \text{ W or 1.25 W}$$ ◆

In summary, now you should know:

- how to find the known value(s) from the given diagram or facts;
- where there are several starting places it is best to look for the formula that might be the easiest to use and start there; and
- how to combine Ohm's law knowledge with power formulas.

PRACTICE PROBLEMS 8

Try the problems in Figure 3–16 to see if you can apply logic and appropriate formulas. Do not cheat yourself by looking ahead.

Now check yourself to see how well you did! You will be getting more practice, both in lab and within this text, as you proceed in your training.

If you have difficulty in solving these, take a few minutes to review appropriate parts of this chapter to find the formulas and the needed guidance or hints.

Summary

- Current is directly related to voltage and inversely related to the circuit resistance. Ohm's law states this as $I = V/R$.
- If the other factors in the equation are known, the techniques of transposition allow rearrangement of the equation to solve for an unknown factor. For example, by rearranging the Ohm's law formula, two other useful equations can be derived, namely $V = I \times R$ and $R = V/I$.
- Powers of 10 are useful when working with either large or small numbers in formulas. When using powers of 10, the rules relating to exponents are applied.
- Scientific calculators are very useful to simplify working electrical problems. The [EE] or [EXP] keys are helpful aids when using calculators to solve problems.
- Two good habits to form when solving electrical circuit problems are to (1) draw a diagram of the circuit, if one is not given, and (2) write down any pertinent formulas that will be used; then, substitute knowns into them appropriately to solve for the unknowns.
- The direction of electron current flow through a circuit, external to the voltage source, is from the source's negative side, through the circuit, and back to the source's positive side. Arrows are frequently drawn on schematic diagrams to illustrate this. The direction electrons would move through a circuit is the standard being used in this textbook.
- Conventional current, as used in some engineering textbooks, is considered to flow in the direction that a positive charge would move through the circuit. That is, from positive to negative.
- When labeling the polarity of voltage across a given component in a circuit, the component end where the electron current enters is labeled the negative end, and the end where the electron current exits is labeled the positive end.
- Electrical power is the rate of using electrical energy to do electrical work. Power is generally dissipated in a circuit or in a component in the form of heat. One watt of power is the performance of electrical work at the rate of 1 joule per second.
- A joule is the energy used in moving one coulomb of charge between two points that have a difference of potential of one volt between them.
- The formulas commonly used to calculate electrical power, or power dissipation are: $P = V \times I$; $P = I^2R$, and $P = V^2/R$
- Electrical energy usage is computed by multiplying the power used times the amount of time it was used ($W = P \times T$). Units of measure for usage of electrical energy are: wattsecond (Ws); watthour (Wh); and for large amounts of energy usage, kilowatthour (kWh).

Formulas and Sample Calculator Sequences

Formula 3–1
(To find current using Ohm's law)

$$I = \frac{V}{R}; \left(\text{Amperes} = \frac{\text{Volts}}{\text{Ohms}}\right)$$

voltage value, [÷], resistance value, [=]

Formula 3–2
(To find voltage using Ohm's law)

$V = I \times R$; (Volts = Amperes × Ohms)

current value, [×], resistance value, [=]

Formula 3–3
(To find resistance using Ohm's law)

$$R = \frac{V}{I}; \left(\text{Ohms} = \frac{\text{Volts}}{\text{Amperes}}\right)$$

voltage value, [÷], current value, [=]

Formula 3–4
(To find foot pounds of work)

Work (Foot pounds) = Force (Pounds) × Distance (Feet)

force value, [×], distance value, [=]

Formula 3–5
(To find power in watts)

$$\text{Power (Watts)} = \frac{\text{Energy (Joules)}}{\text{Time (Seconds)}}$$

energy value, [÷], time value, [=]

Formula 3–6 *(To find watthours of energy used)*

Energy (Watthours) = Power (Watts) × Time (Hours)

power value, [×], time value, [=]

Formula 3–7 *(To find power in watts)*

$P = V \times I$ (Watts = Volts × Amperes)

voltage value, [×], current value, [=]

Formula 3–8 *(To find power in watts)*

$P = I^2 R$ (Watts = Amperes squared × Ohms)

current value, [x^2], [×], resistance value, [=]

Formula 3–9 *(To find power in watts)*

$$P = \frac{V^2}{R} \left(\text{Watts} = \frac{\text{Volts squared}}{\text{Ohms}} \right)$$

voltage value, [x^2], [÷], resistance value, [=]

EXCEL AUTOMATED FORMULAS

Helpful problem-solving tools, such as the Excel automated formulas available on the CD, allow automatic calculations of formulas.

Using Excel

Ohm's Law and Power Formulas

(Excel file reference: FOE3_01.xls)

NOTE: Once the desired formula is placed in the cell under the heading of the parameter you are solving for (the "formula cell"), you do not have to enter it again!

To solve the problem with different parameters, simply type in the new parameter values in the appropriate cells, under each parameter column heading. The new answer will appear in the "formula cell" without you having to change the formula or retype it!

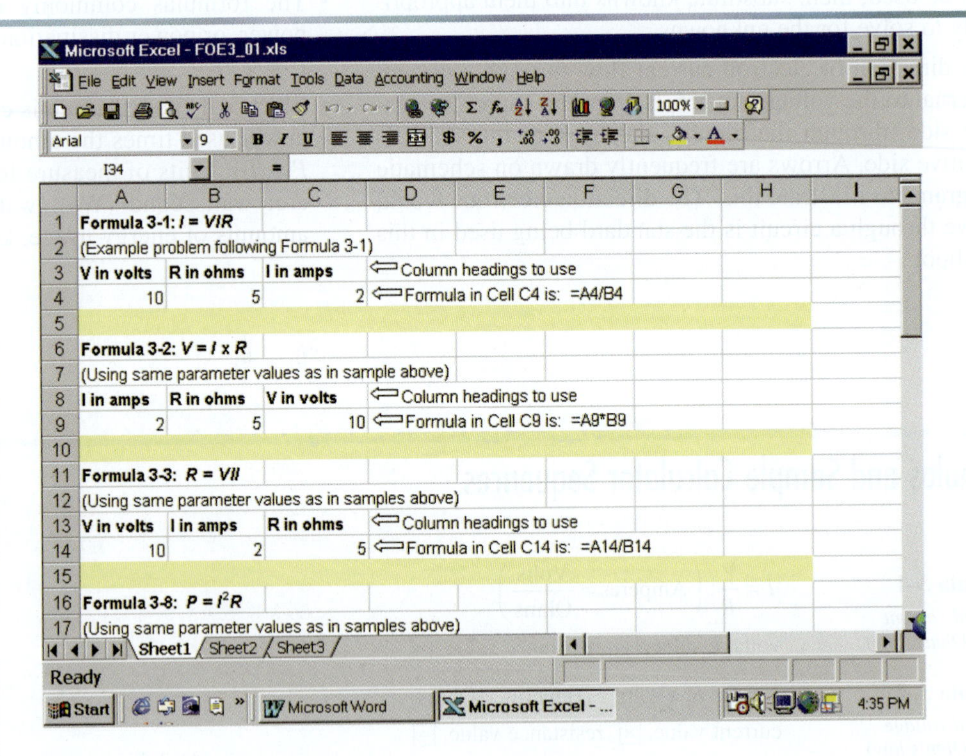

- Refer to Figure 3–1a, b, & c. Use Formula 3–1 spreadsheet sample and solve for I in each case. Compare results to that in figure.
- Refer to Figure 3–16, circuit 1. Use Formula 3–1 spreadsheet sample to solve for I; then, use Formula 3–9 spreadsheet sample to solve for P. Compare results with answers shown in "Answer Appendix" for Practice Problems 8.

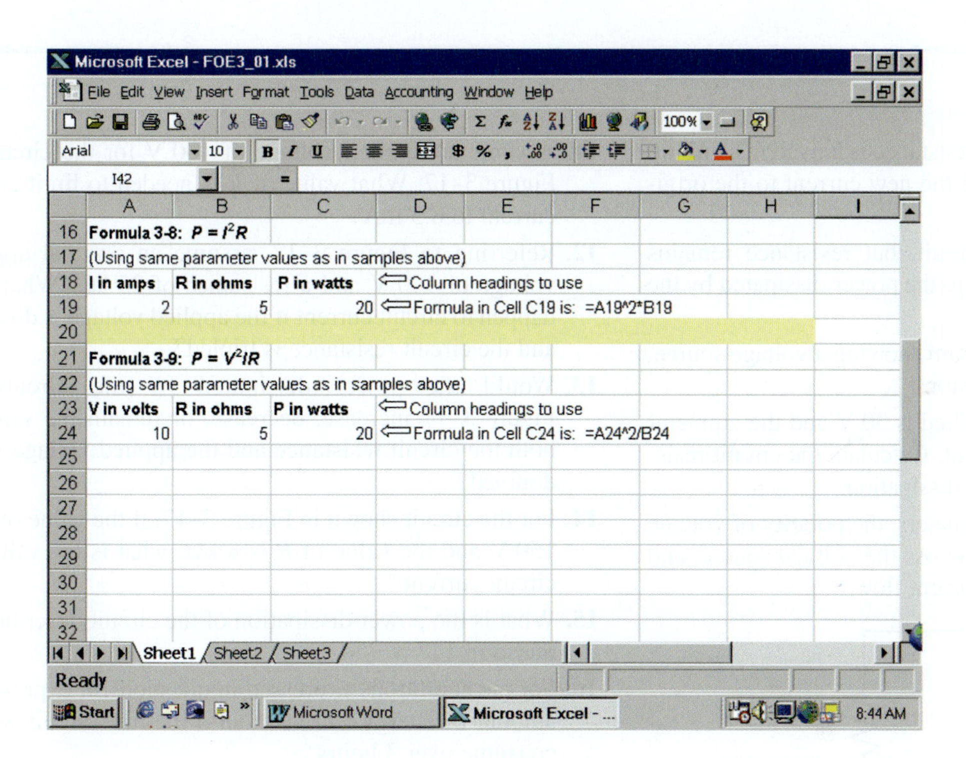

Using Excel

***Ohm's Law and Power* Formulas (continued)**

(Excel file reference: FOE3_01.xls)

Review Questions

1. State the three forms of:
   a. the Ohm's law formula
   b. the formula for electrical power
2. State the following numbers as a number between 1 and 10 times a power of 10:
   a. 33,456
   b. 25
   c. 1,055,000
   d. 10
3. State each of the following numbers as the appropriate number times 10^3:
   a. 5,100
   b. 47,000
   c. 0.001
   d. 39
4. Define the term *work.*
5. Define the term *energy.*
6. Define the term *power.*
7. Give the unit used to measure work.
8. Give the unit used to measure energy.
9. Give the unit used to measure power.
10. In your own words, describe the relationships and/or proportionalities between the following electrical parameters:
    a. power related to current (for a given resistance)
    b. voltage related to current (for a given resistance)
    c. energy (W) related to power (for a given time)

Problems

1. If voltage is doubled and resistance is halved in a circuit, what is the relationship of the new current to the original current?
2. If current triples in a circuit, but resistance remains unchanged, what happens to the power dissipated by the circuit?
3. **a.** Draw a schematic diagram showing a voltage source, an ammeter, and a resistor.
   b. Assume the voltage applied is 30 V and the ammeter indicates 2 mA of current. Calculate the circuit resistance and circuit power dissipation.
   c. Label all electrical parameters, the polarity of voltage across the resistor and across the voltage source, and show the direction of current flow.

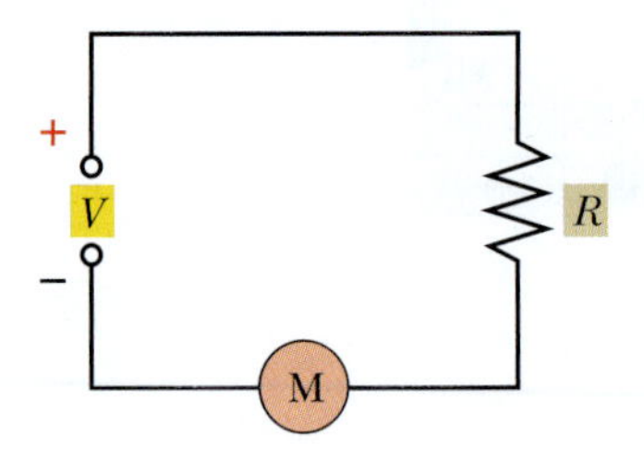

FIGURE 3–17

Refer to Figure 3–17 when answering questions 4 through 21:

4. In Figure 3–17, if the value of V is assumed to be 50 V and the value of R is assumed to be 33 kΩ, what value of current will the meter M read?
5. If the current meter (M) in Figure 3–17 is reading 3 mA, and R has a value of 12 kΩ, what is the value of circuit-applied voltage?
6. Indicate whether the following items would increase, decrease, or remain the same if the value of voltage in question 5 were doubled:
   a. V has ____________________
   b. R will ____________________
   c. I will ____________________
7. If current through the circuit of Figure 3–17 is 50 mA and circuit voltage is 41 V, what is the value of R?
8. In Figure 3–17, what is the voltage applied if the circuit power dissipation is 100 mW and the circuit current is 12.5 mA?
9. For the circuit of Figure 3–17, what is the circuit current if the power dissipated by the 10-Ω resistor is 100 W?
10. If current in the circuit of Figure 3–17 decreases to one-third its original value, and the original power dissipation was 180 W, what is the new power dissipation?
11. Assume an applied voltage of 100 V for the circuit of Figure 3–17. What value of R is needed to limit circuit current to 8.5 mA?
12. Referring to Figure 3–17, assume an original applied voltage of 100 V and a resistance of 25 kΩ. What will happen to circuit current if the applied voltage is doubled and the circuit resistance is tripled?
13. Would the power dissipated by the circuit of Figure 3–17 increase, decrease, or remain the same if both the circuit resistance and the applied voltage were doubled?
14. For the circuit shown in Figure 3–17, if the value of V is 150 V and the value of R is 8 kΩ, what is the value of circuit current?
15. What is the power dissipation of the circuit described in question 14?
16. For the conditions expressed in question 14, express the amount of energy (in watthours) that the circuit would consume over 3 hours.
17. Referring to Figure 3–17, if the value of V is 250 V and the value of R is 12 kΩ, what is the value of circuit current?
18. What is the power dissipation of the circuit described in question 17?
19. What is the energy (in watthours) consumed by the circuit of question 17 over 6.5 hours?
20. For the circuit of Figure 3–17, assume an applied voltage of 100 volts and a power dissipation of 50 mW. What is the value of R?
21. For the circuit of Figure 3–17, assume a power dissipation is 100 mW when the value of R is 10 kΩ. What is the value of applied voltage?
22. Draw a schematic diagram showing a source, a resistor, and an ammeter. Assume the power dissipation is 1,000 watts and the resistance is 10 ohms. Transpose the $P = I^2R$ formula and solve for the value of I.
23. Calculate the applied voltage for the circuit in question 22.
24. Label all the parameters on the circuit diagram you drew for question 22 and indicate the direction of current flow throughout the circuit, and the polarity of voltages across each component.
25. Using powers of 10, change the following as directed:
   a. $\dfrac{10^7 \times 10^{-4}}{10^5 \times 10^2 \times 10^{-2}}$ equals what whole number?
   b. $10^3 \times 10^5 \times 10^6 \times 10^{-2}$ equals 10 to what power?
   c. Add: $(7.53 \times 10^4) + (8.15 \times 10^3) + (225 \times 10^1)$
   d. Subtract: $(6.25 \times 10^3) - (0.836 \times 10^2)$

Analysis Questions

1. Explain what the EE (or EXP) keys on a scientific calculator do when you input a number that you want to express times some power of 10.
2. If someone were to use a calculator and input: 222, EE (or EXP), 4, in that order, what would the calculator indicate as the result?
3. Describe the basic difference between current flow in a circuit having a dc source and that same circuit having an ac source.
4. Describe the difference between a dc voltage and an ac voltage in terms of polarity characteristics.
5. As contrasted with "electron-flow current," "conventional current" is considered to flow in a direction that a (negative, positive) __________ charge would move through the circuit.
6. Determine the electrical power equivalent to 3.5 horsepower.
7. In a given circuit, if power dissipation has increased to 8 times the original power, while resistance has decreased to one-half its original value, what must have happened to the circuit-applied voltage?

A Good Idea

Never set a soldering iron where you might accidentally lay your hand on it or easily touch the hot end of the iron!

A Good Idea

Remove loose jewelry when working on electrical equipment or when using rotating machinery or power tools!

Performance Projects Correlation Chart

Suggested performance projects that correlate with topics in this chapter are:

Chapter Topic	Performance Project	Project Number
The Relationship of Current to Voltage with Resistance Constant	Relationship of I & V with R Constant (Ohm's law series)	5
The Relationship of Current to Resistance with Voltage Constant	Relationship of I & R with V Constant (Ohm's law series)	6
The Basic Power Formula	Relationship of P to V with R Constant (Ohm's law series)	7
Variations of the Basic Power Formula	Relationship of P to I with R Constant (Ohm's law series)	8
	Story Behind the Numbers: Ohm's Law	

NOTE: It is suggested that after completing the above projects, the student should be required to answer the questions in the "Summary" at the end of this section of projects in the Laboratory Manual.

OUTLINE

4–1 Definition and characteristics of a series circuit
4–2 Resistance in series circuits
4–3 Voltage in series circuits
4–4 Kirchhoff's voltage law
4–5 Power in series circuits
4–6 Effects of opens in series circuits and troubleshooting hints
4–7 Effects of shorts in series circuits and troubleshooting hints
4–8 Designing a series circuit to specifications
4–9 Special applications
4–10 Critical thinking and the SIMPLER troubleshooting sequence
4–11 Introduction to troubleshooting skills
4–12 The SIMPLER sequence for troubleshooting
4–13 Troubleshooting levels
4–14 A "systems" approach to troubleshooting
4–15 Using the SIMPLER sequence: (a closer look at component-level troubleshooting) a single-component level example

CHAPTER 4

Series Circuits

OBJECTIVES

After studying this chapter, you should be able to:

1. Define the term **series circuit**
2. List the primary characteristics of a series circuit
3. Calculate the total resistance of series circuits using two different methods
4. Calculate and explain the voltage distribution characteristics of series circuits
5. State and use **Kirchhoff's voltage law**
6. Calculate power values in series circuits
7. Explain the effects of **opens** in series circuits
8. Explain the effects of **shorts** in series circuits
9. List troubleshooting techniques for series circuits
10. Design series circuits to specifications
11. Series-connect voltage sources for desired voltages
12. Analyze a **voltage divider** with reference points
13. Calculate the required value of a **series-dropping resistor**
14. Use the computer to solve circuit problems
15. Use the SIMPLER troubleshooting sequence to solve the Troubleshooting Challenge problems

ELECTRONICS INTO THE FUTURE

Explore an interactive presentation on Series Circuits, on the accompanying CD.

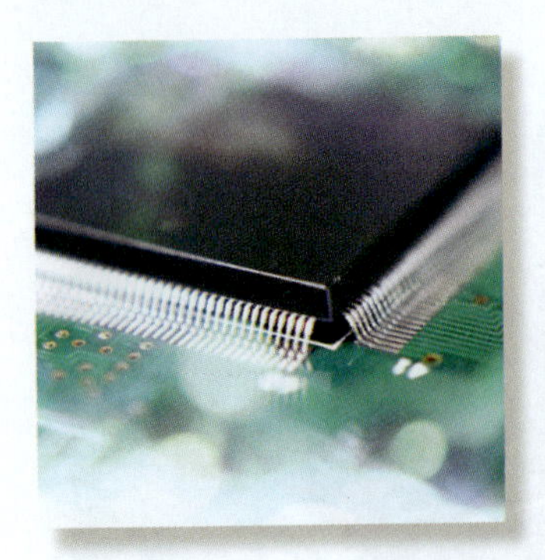

PREVIEW

Series circuits are found in many systems and subsystems. Common examples are home light switch circuits and automobile ignition switch circuits. It is important that you learn the characteristics of the series circuit. That knowledge will aid you in analyzing series circuits whether they are stand-alone circuits or portions of more complex circuits.

In studying series circuits, you will apply Ohm's law and power formulas. Also, you will learn how Kirchhoff's voltage law helps you analyze series circuits. Insight will be gained about the effects of opens and shorts in series circuits. You will see how voltage sources are used in series circuits. Also, voltage dividers and the concept of reference points will be examined. The final portion of this chapter will present a useful troubleshooting method that will be valuable to you throughout your training, and more importantly, in your career.

You will be introduced to a special troubleshooting sequence called the SIMPLER sequence. By solving troubleshooting problems called Troubleshooting Challenges, you will gain experience using the SIMPLER sequence. The system for performing the procedure and solving each Troubleshooting Challenge will be discussed. One sample solution to each challenge troubleshooting problem is presented at the end of the chapter so you can compare your thinking process against the sample solutions.

KEY TERMS

Battery
Block-level troubleshooting approach
Cell
Component-level troubleshooting approach
Inputs
Kirchhoff's voltage law
Open circuit
Series circuit
Series-dropping resistor
Short circuit
SIMPLER troubleshooting sequence
Voltage-divider action

4–1 Definition and Characteristics of a Series Circuit

A **series circuit** is any circuit having only one path for current flow. In other words, two or more electrical components or elements are connected so the same current passes through all the connected components.

This situation exists when components are connected end-to-end in the circuit external to the source, Figure 4–1. This is termed a two-component series circuit. Also, notice there is only *one possible path* for current flow from the source's negative side, through the external circuit, and back to the source's positive side.

Key Characteristics

Excluding the source, how many components are in series in Figure 4–2? If you said four, you are correct! The circuit in Figure 4–3 is a six-component series circuit. In all these cases, the two important characteristics to remember are that there is only one path for cur-

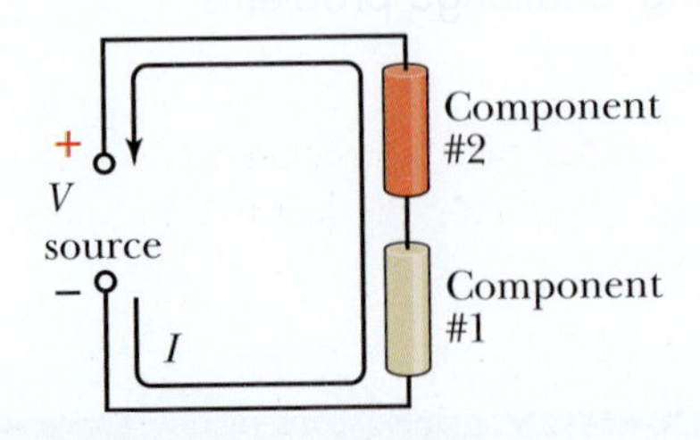

FIGURE 4–1 Components connected end-to-end in a two-component series circuit

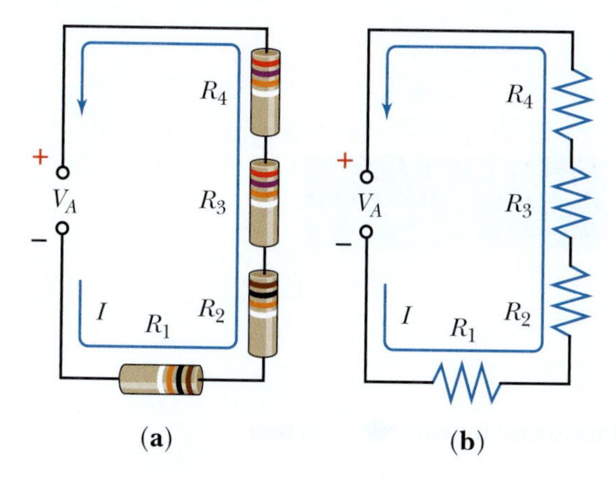

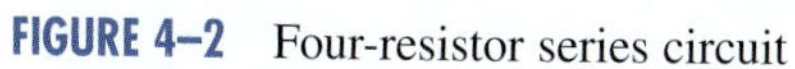
FIGURE 4–2 Four-resistor series circuit

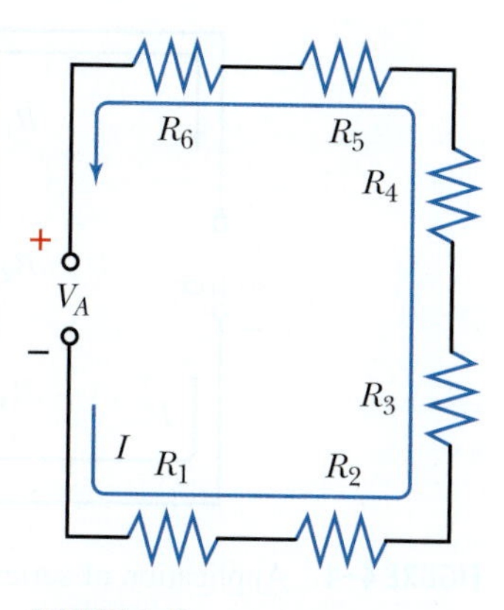

FIGURE 4–3 Six-resistor series circuit

rent flow, and that current is the same through all parts of a series circuit. As a technician, you will have many opportunities to apply this knowledge.

4–2 Resistance in Series Circuits

Knowing the characteristics that identify a series circuit helps you analyze the important electrical quantities or parameters of these circuits.

First, let's analyze the series circuit in terms of its resistance to current flow. In previous chapters we have limited circuits to a single resistor, and the circuit's resistance to current flow was obvious. Now, we are connecting two or more resistors in tandem or in series. How does this affect the circuit resistance?

Series Circuit Total Resistance Formula

If the circuit current must sequentially flow through all the resistors since there is only one path for current, then the total resistance to current flow will equal the *sum* of all the resistances in series. Once again, in series circuits, the total resistance (R_T or R total) equals the sum of all resistances in series.

FORMULA 4–1 $R_T = R_1 + R_2 \ldots + R_n$

R_n represents the last of the remaining resistor values.

EXAMPLE Refer to Figure 4–4 for an application of this formula. This illustrates that to find a series circuit's total resistance, simply add all the individual resistance values.

PRACTICE PROBLEMS 1

1. Referring back to Figure 4–2, what is the circuit's total resistance if R_1 and R_2 are 10 kΩ resistors and R_3 and R_4 are 27 kΩ resistors?
2. The values of the resistors in Figure 4–4 are changed so that R_1 = 2.7 kΩ, R_2 = 5.1 kΩ, and R_3 = 8.2 kΩ. What is the new total resistance value for the circuit?

The Ohm's Law Approach

Another important method to determine total resistance *in any circuit,* including series circuits, is Ohm's law. Recall that the Ohm's law formula to find resistance is $R = V/I$. To find total resistance, this formula becomes:

FORMULA 4–2 $R_T = \dfrac{V_T}{I_T}$

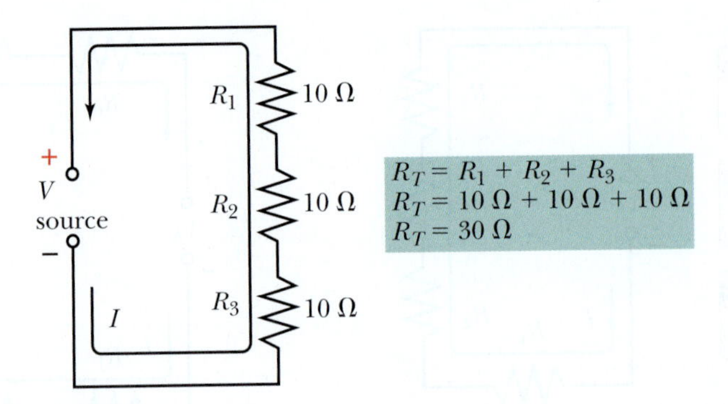

FIGURE 4–4 Application of series-circuit total resistance formula multiSIM

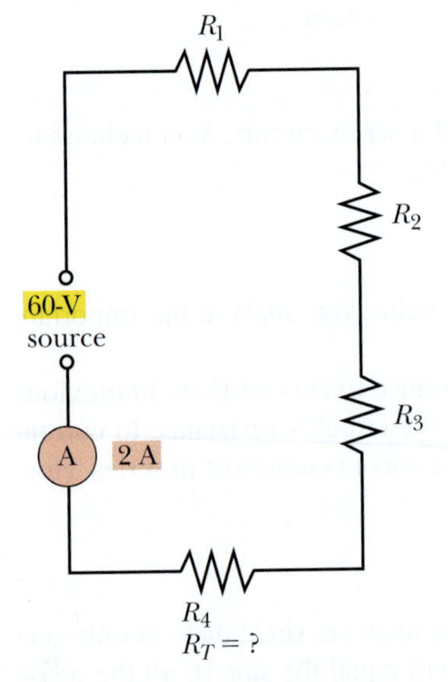

FIGURE 4–5 Finding R_T by Ohm's law

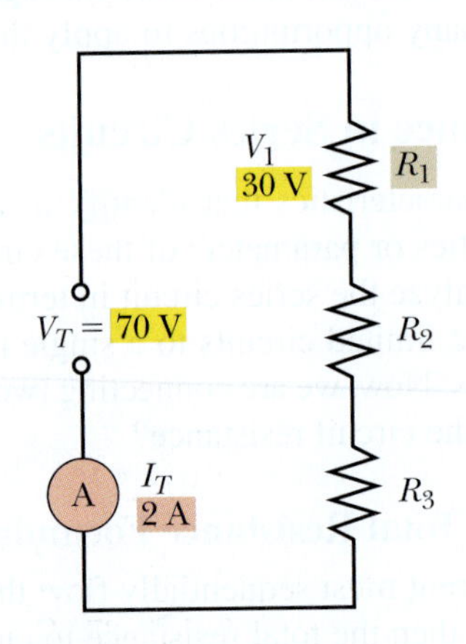

FIGURE 4–6 Finding one resistor's value by Ohm's law

If there is any means of determining the value of total voltage applied to the series circuit *and* the total current (which is the same as the current through any part of the series circuit), Ohm's law can be used to find total resistance by using $R_T = V_T/I_T$.

◆ **EXAMPLE** Look at Figure 4–5 and use Ohm's law to find total resistance. Your answer should be 30 ohms because 60 volts divided by 2 amperes equals 30 ohms. Does this show the value of each of the resistors in the circuit? No, it only reveals the circuit's total resistance. ◆

◆ **EXAMPLE** Now look at Figure 4–6. Again, it is possible to use Ohm's law to determine the circuit's total resistance, where total resistance equals total voltage divided by total circuit current.

$$R_T = \frac{V_T}{I_T} = \frac{70\ V}{2\ A} = 35\ \Omega$$

Because we know the voltage drop across R_1 and know that the current through all parts of a series circuit is the same value, we can also determine the value of R_1. In this case, R_1 equals the voltage across R_1 divided by the 2 amperes of current through R_1. Therefore, R_1 equals 30 volts divided by 2 amperes. R_1 = 15 ohms. ◆

PRACTICE PROBLEMS 2

1. Again refer to Figure 4–6. If $V_T = 100$ V, $I_T = 1$ A, and $V_1 = 47$ V, what is the value of R_1? What is the value of R_T?
2. Assume that the electrical parameters in the circuit of Figure 4–6 are changed so that $V_T = 42$ V and $I_T = 4$ mA. What is the new value of R_T?
3. With the circuit V_T and I_T described in problem 2, if R_2 and R_3 are equal value resistors and $R_1 = 2.7$ kΩ, what is the value of R_2?

We have discussed and illustrated two important facts about series circuits. First, *current* is the same throughout all parts of a series circuit. Second, the total circuit *resistance* equals the sum of all the resistances in series, which indicates total resistance must be greater than any one of the resistances. We will now study voltage, a third important electrical parameter in series circuits.

4–3 Voltage in Series Circuits

To help you understand how voltage is distributed throughout a series circuit, refer to Figure 4–7.

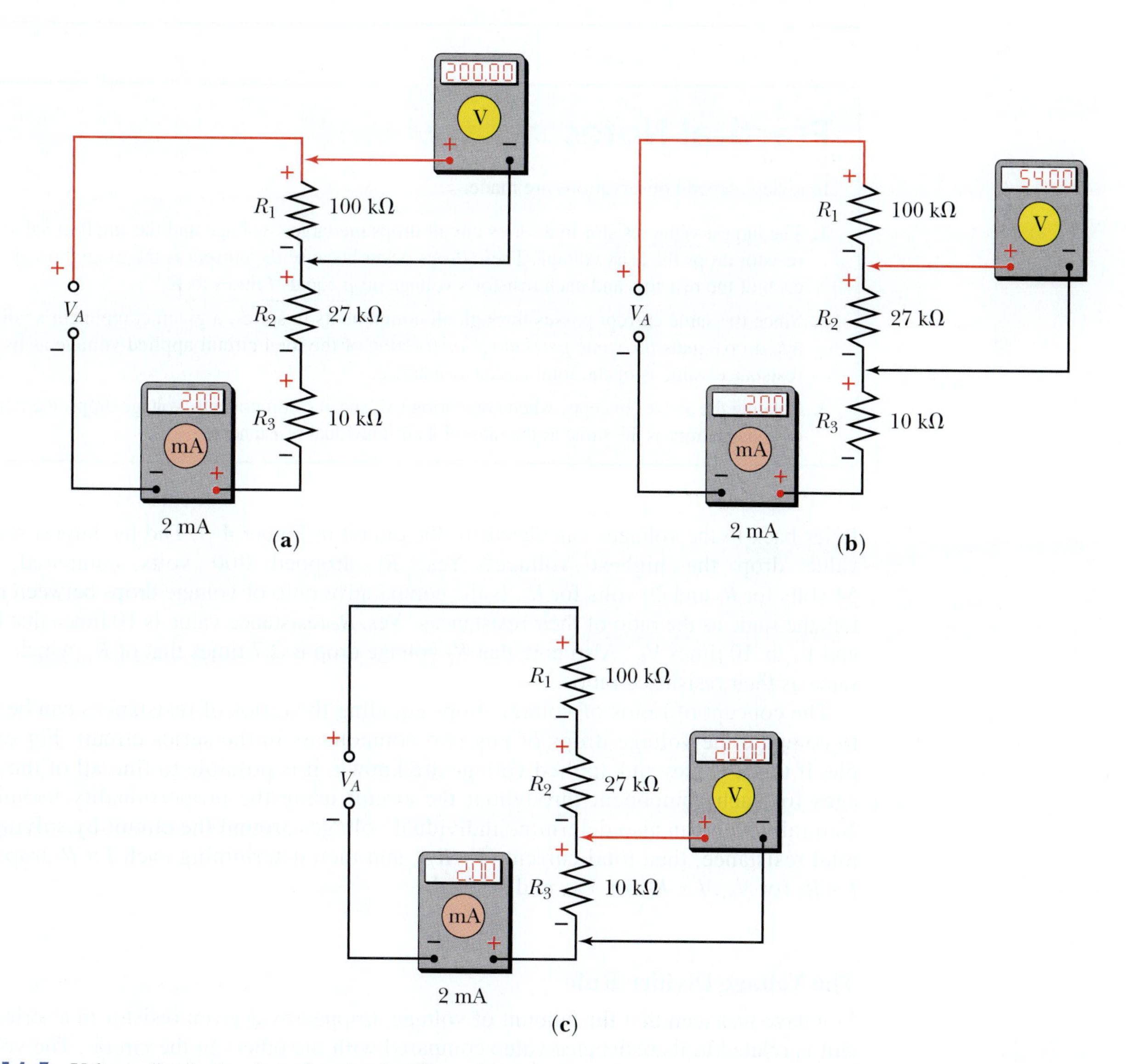

FIGURE 4–7 Voltage distribution in series circuits multiSIM

Individual Component Voltages

Because there is only one path for current, the current *(I)* through each resistor (R_1, R_2, and R_3) must have the same value since it is the same current, Figure 4–7. From the study of Ohm's law, you know the value of voltage dropped by R_1 must equal its I times its R, or

FORMULA 4–3 $V_1 = I_{R_1} \times R_1$

◆ **EXAMPLE** In this case, the current is 2 mA and the resistance of R_1 is 100 kΩ; therefore, V_1 = 200 volts, Figure 4–7a. (Remember that milliunits times kilounits equals units.) Voltage dropped by R_2 must equal 2 mA times 27 kΩ, or 54 volts, Figure 4–7b. The voltmeter, which is measuring the voltage across R_3, would indicate 20 volts because 2 mA × 10 kΩ = 20 volts, Figure 4–7c. ◆

PRACTICE PROBLEMS 3

1. In the circuit shown in Figure 4–7c, if the voltmeter reads 30 V, what are the values of V_A, I_T, V_{R_2}, and V_{R_1}?
2. In the circuit shown in Figure 4–7c, if the voltmeter reads 25 V, what are the values of V_A, I_T, V_{R_2}, and V_{R_1}?

Practical Notes

In review, several observations are made.

1. The largest value resistor in a series circuit drops the largest voltage and the smallest value resistor drops the least voltage. These drops occur because the current is the same through each of the resistors and each resistor's voltage drop equals I times its R.
2. Since the same current passes through all components in series, a given component's voltage drop equals *the same percentage or fraction* of the total circuit applied voltage as its resistance value is of the total circuit resistance.
3. Relating the above concepts, when comparing two specific components' voltage drops, the ratio of voltage drops is the same as the ratio of their individual resistances.

Refer back to the voltages calculated for the circuit in Figure 4–7. Did the largest resistor value drop the highest voltage? Yes, R_1 dropped 200 volts compared with 54 volts for R_2 and 20 volts for R_3. Is the comparative ratio of voltage drops between resistors the same as the ratio of their resistances? Yes, R_1 resistance value is 10 times that of R_3 and V_{R_1} is 10 times V_{R_3}. Also note that R_2 voltage drop is 2.7 times that of R_3, which is the same as their resistance ratios.

The concept of ratios of voltage drops equaling the ratios of resistances can be used to compare the voltage drops of any two components in the series circuit. For example, if the R values and applied voltage are known, it is possible to find all of the voltages for each component throughout the circuit using the proportionality technique. Naturally, you can also determine individual voltages around the circuit by solving for total resistance, then total current (V_T/R_T), and then determining each $I \times R$ drop with $I \times R_1$ for V_{R_1}, $I \times R_2$ for V_{R_2}, and so forth.

The Voltage Divider Rule

You have just seen that the amount of voltage dropped by a given resistor in a series circuit is related to its resistance value compared with the others in the circuit. The voltage divider rule shows how in a series circuit, any given resistor's voltage drop relates to its resistance compared with the circuit's total resistance. Using this rule, you can determine a series resistor's voltage drop, without having to know the circuit current, if you know the applied voltage and circuit total resistance. Here's the simple formula:

FORMULA 4–4 $V_X = \frac{R_X}{R_T} \times V_T$

where: V_X = the voltage drop across the selected resistor
R_X = the resistance value of the selected resistor
R_T = the circuit total resistance
V_T = the circuit applied voltage

Practical Notes

This type of calculation is very simple using a calculator. You should get into the habit of using your calculator for MOST of your calculations (if your program promotes it at this stage of your learning). In some cases, however, the resistor values and the voltage applied value are such that you can almost do this type of calculation in your head. In fact, here's a hint! It is a good idea to learn to approximate answers in your head for any problem that you can, just as a double check on your computations. *Approximating is a powerful skill to learn and practice!* It can often save you from embarrassing results. If you inadvertently make a mathematical error along the way, your approximation will show you that the erroneous results are not logical. Then, you can check back through your procedure, and quickly find the mistake.

EXAMPLE Refer to Figure 4–8 and note how we can find the voltage dropped across R_2 by simply substituting the knowns into the formula. (In this case, R_2 is the value to be used for R_X in the formula.) Calculating R_T for the formula: $R_T = R_1 + R_2 + R_3 = 2.7\text{ k}\Omega + 4.7\text{ k}\Omega + 10\text{ k}\Omega = 17.4\text{ k}\Omega$, therefore,

$$V_{R_2} = \frac{R_2}{R_T} \times V_T; V_{R_2} = \frac{4.7\text{ k}\Omega}{17.4\text{ k}\Omega} \times 50; V_{R_2} = 0.27 \times 50 = 13.5\text{ V}$$

PRACTICE PROBLEMS 4

Now you use the voltage divider formula to find the following answers.

1. For the circuit of Figure 4–8, what is the voltage dropped by R_1?
2. For the circuit of Figure 4–8, what is the voltage dropped by R_2?
3. For the circuit of Figure 4–8, what is the voltage dropped by R_3?
4. Sum the rounded voltages of R_1, R_2, and R_3. Do they add up to close to the applied voltage of 50 volts?

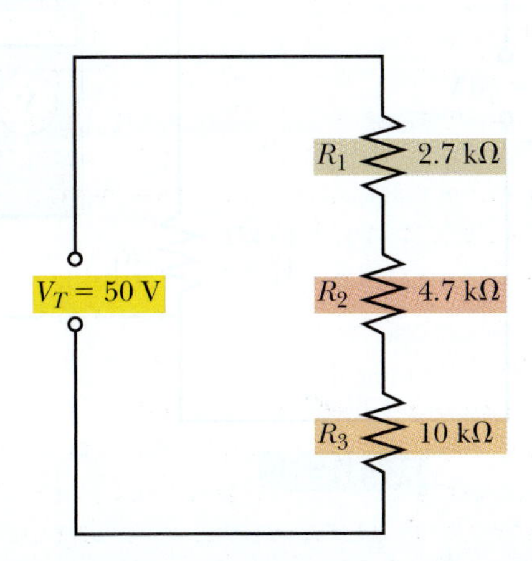

FIGURE 4–8 Voltage-divider rule sample circuit

As you can see, the voltage divider rule can be very handy! You will get more opportunities to use this rule as you proceed in your training.

Finding the Value of Applied Voltage

Again referring to Figure 4–7, let's examine some ways to find the circuit applied voltage. One obvious way is the Ohm's law expression where $V_T = I_T \times R_T$. In this circuit, V total (or V applied) = 2 mA times R_T. As you can easily determine, R_T in this circuit equals 100 kΩ + 27 kΩ + 10 kΩ = 137 kΩ total resistance. Therefore, circuit applied voltage equals 2 mA × 137 kΩ = 274 volts.

Another way to find applied voltage is to add all the individual voltage drops to find total voltage, just as we add all the individual resistances to find total resistance. This yields the same answer where 200 volts + 54 volts + 20 volts = 274 volts. This method suggests an important concept called Kirchhoff's voltage law.

4–4 Kirchhoff's Voltage Law

Kirchhoff's voltage law states the *arithmetic sum* of the voltages around a single circuit loop (any complete closed path from one side of the source to the other) *equals V applied* (V_A). It also says the *algebraic sum* of all the loop voltages, including the source or applied voltage, must *equal zero.* That is, if you observe the polarity and value of voltage drops by the circuit elements and the polarity and value of the voltage source, and add the complete loop's values algebraically, the result is zero. For our purposes, the arithmetic sum approach will be most frequently used. It illustrates how Kirchhoff's voltage law helps determine unknown circuit parameters. Refer to Figure 4–9 as you study the following example.

EXAMPLE If the V_A is 50 volts and V_2 is 20 volts, how can Kirchhoff's voltage law help determine V_1? If the sum of voltages (*not* counting the source) must equal V_A applied, then:

$$V_A = V_1 + V_2$$

Since we know V_A and V_2, transpose to solve for V_1.

Therefore, $V_1 = V_A - V_2 = 50 \text{ V} - 20 \text{ V} = 30 \text{ V}$

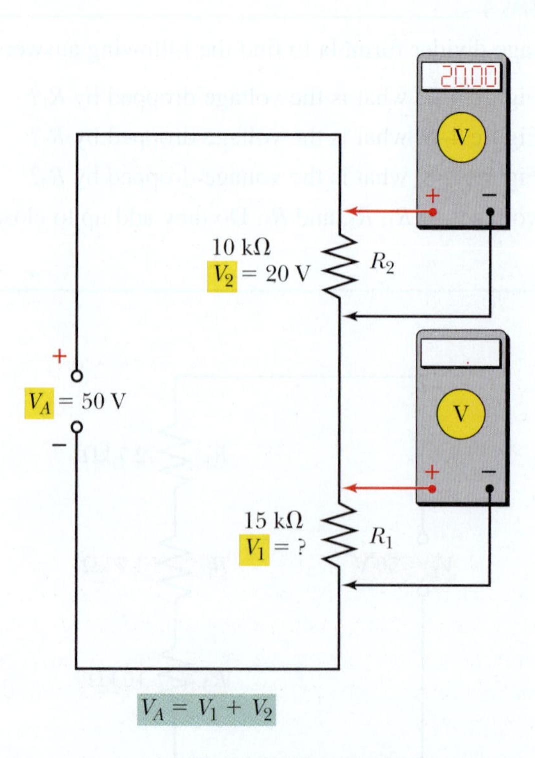

FIGURE 4–9 Using Kirchhoff's voltage law where $V_A = V_1 + V_2$

If individual voltage drops were known but not the applied voltage, find *V* applied by adding the individual voltage drops. In this case, 30 V + 20 V = 50 V. This agrees with Kirchhoff's voltage law that the arithmetic sum equals the applied voltage. Kirchhoff's voltage law can help you find unknown voltages in series circuits through either addition or subtraction, as appropriate. And if there are more than two series components:

FORMULA 4–5 $V_T = V_1 + V_2 \ldots + V_n$

where V_n represents the last of the remaining voltage values.

In Figure 4–10, observe the polarity of voltages around the closed loop. To indicate polarities used in Kirchhoff's voltage law, trace the loop from the source's positive side (Point A), through the resistors and back to the source's negative side (Point B). Consider any voltage a positive voltage whose + point is reached first, and vice versa. In this case, the first voltage reached is +20 V; the next voltage is +30 V, and the source negative terminal is –50 V. Adding these voltages yields:

$$(+20) + (+30) + (-50) = 0; \text{ or } 20 + 30 - 50 = 0$$

Tracing the other direction through the circuit gives:

$$(-30) + (-20) + (+50) = 0; \text{ or } -30 + -20 + 50 = 0$$

In either case, the algebraic sum of the voltage drops *and* the voltage source around the entire closed loop equals zero.

Likewise, the arithmetic sum of the *voltage drops* around a given loop must equal the value of the *applied voltage,* regardless of which direction is used to trace the loop. ◆

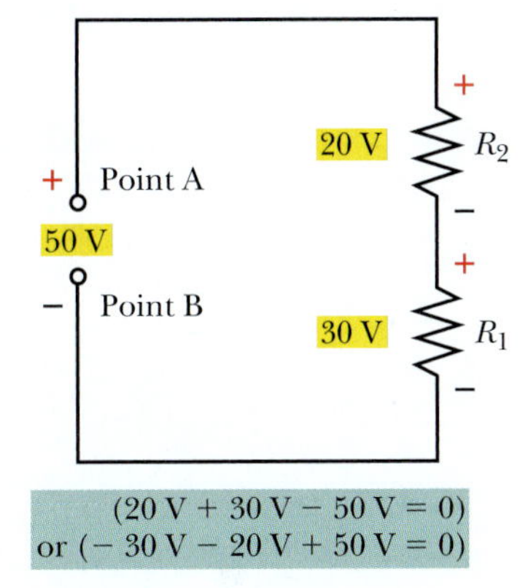

FIGURE 4–10 Kirchhoff's algebraic sum example

PRACTICE PROBLEMS 5

Refer again to Figure 4–10. Using the proportionality of voltage drops concept and Kirchhoff's voltage law, if V_{R_1} is 45 V, what are the values of V_{R_2} and V_A?

IN-PROCESS LEARNING CHECK 1 Fill in the blanks as appropriate.

1. The primary identifying characteristic of a series circuit is that the _____ is the same throughout the circuit.
2. The total resistance in a series circuit must be greater than any one _____ in the circuit.
3. In a series circuit, the highest value voltage is dropped by the _____ value resistance, and the lowest value voltage is dropped by the _____ value resistance.
4. In a two-resistor series circuit, if the applied voltage is 210 volts and one resistor's voltage drop is 110 volts, what must the voltage drop be across the other resistor? _____ V.
5. Answer each part of this question with increase, decrease, or remain the same. In a three-resistor series circuit, if the resistance value of one of the resistors increases, what happens to the circuit total resistance? _____ To the total current? _____ To the adjacent resistor's voltage drop? _____
6. What is the applied voltage in a four-resistor series circuit where the resistor voltage drops are 40 V, 60 V, 20 V, and 10 V, respectively? _____ V.

A Quick Review

Current is the same throughout a series circuit; total resistance equals the sum of all the individual resistances in series, and voltage distribution around a series circuit is directly related to the resistance distribution, since current is the same through all components and each component's voltage drop = current × its resistance. Kirchhoff's voltage law states that the arithmetic sum of voltage drops equals the voltage applied, or the algebraic sum of all the voltage drops and the voltage source equal zero.

Practical Notes

As you continue to work more electronic circuit problems, it is good to remember that:

- Units divided by Kilo units = Milli units. This is often used when finding current: $I = V/R$; where V/kΩ = mA.
- Units divided by Milli units = Kilo units. This is often used when finding resistance: $R = V/I$; where V/mA = kΩ.
- Milli units times Kilo units = Units. This is often used when finding voltage: $V = I \times R$; where mA × kΩ = V.

A Good Idea

Learn to use a calculator's various modes to ease the work involved in circuit calculations!

4–5 Power in Series Circuits

The last major electrical quantity to be discussed in relation to series circuits is power. Recall that power dissipated by a component or circuit (or supplied by a source) is calculated with the formulas $P = V \times I$, $P = I^2R$, or $P = V^2/R$. You have learned that in series circuits the current (I) is the same through all components. Therefore, the I^2 factor in the $P = I^2R$ formula is the same for each component in series and the largest R value dissipates the most power. Conversely, the component with the least R value dissipates the least power (I^2R), Figure 4–11.

Individual Component Power Calculations

Individual component power dissipations can be found if any two of the three electrical parameters for the given component are known. For example, if the component's resistance and the current through it are known, use the $P = I^2R$ formula. If the component's voltage drop and the current are known, use the $P = V \times I$ formula. Knowing the voltage and resistance allows you to use the $P = V^2/R$ version of the power formula.

As implied previously, the individual power dissipations around a series circuit are *directly related* to the resistance of each element, just as the voltage distribution is directly related to each element's resistance. A specific resistor's power dissipation is the same percentage of the circuit's total power dissipation as its resistance value is of the total circuit resistance (R_T). For example, if its resistance is one-tenth (1/10) the total circuit resistance, it dissipates 10% of the total power. If it is half the R_T, it dissipates 50% of the total power, and so forth, Figure 4–11.

Total Circuit Power Calculations

Total power dissipation (or power supplied to the circuit by the power source) is determined by adding all the individual power dissipations, if they are known. See Formula 4–6.

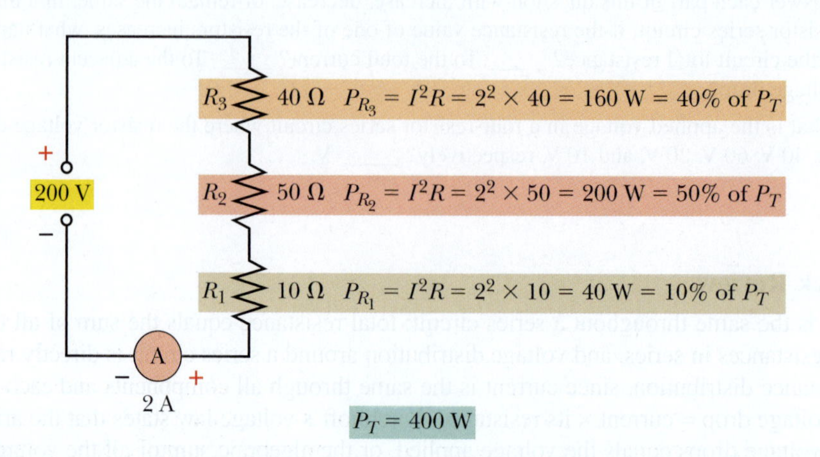

FIGURE 4–11 Largest R dissipates most power, and smallest R dissipates least power.

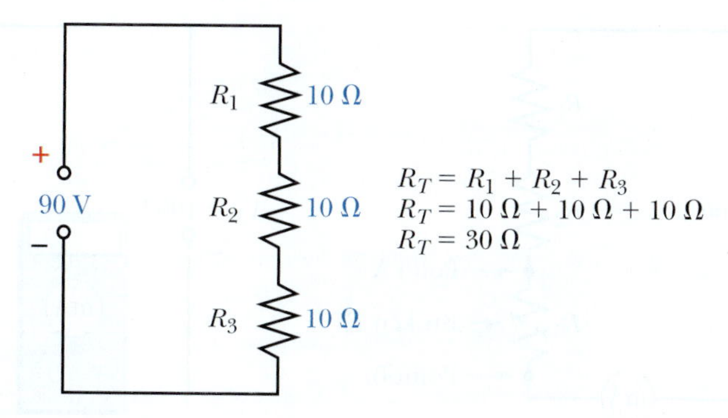

FIGURE 4–12 Diagram for Practice Problems 6

FORMULA 4–6 $P_T = P_1 + P_2 \ldots + P_n$

EXAMPLE Refer to Figure 4–11. Using Formula 4–6:

$$P_T = P_{R_1} + P_{R_2} + P_{R_3} = 40 \text{ W} + 200 \text{ W} + 160 \text{ W} = 400 \text{ W}$$

PRACTICE PROBLEMS 6

Refer to Figure 4–12.
Find the values for:

P_{R_1} = __________ watts V_{R_1} = __________ volts
P_{R_2} = __________ watts V_{R_2} = __________ volts
P_{R_3} = __________ watts V_{R_3} = __________ volts
P_T = __________ watts R_3 = __________ % of R_T
I_T = __________ amperes P_{R_3} = __________ % of P_T

4–6 Effects of Opens in Series Circuits and Troubleshooting Hints

Opens

Thus far, you have studied how to analyze current, resistance, voltage, and power in series circuits. At this point, we will introduce some practical situations that might occur in *abnormal* series circuits. The two extreme conditions are opens and shorts.

First, what are the effects of an open in a series circuit? An **open circuit** occurs anytime a break in the current path occurs somewhere within the series circuit. Because there is only one path for current flow, if a break (or open) occurs *anywhere* throughout the current path, there is no current flow. In effect, the circuit resistance becomes "infinite" and I_T (circuit current) must decrease to zero.

If an open causes circuit resistance to increase to infinity and circuit current to decrease to zero, how does this affect the circuit voltages and power dissipations?

EXAMPLE In Figure 4–13, assume that R_3 is physically broken in half, causing an open or break in the current path. Since there is no complete path for current flow, what would the current meter read? The answer is zero mA.

If there is no current, what is the $I \times R$ drop of R_1? The answer is zero volts because zero times any value of R equals zero. And the answer for V_2 is also zero for the same reason. This means that the difference of potential across each of the good resistors (R_1 and R_2) is zero. That is, R_1 has zero voltage drop and R_2 has zero voltage drop. In effect, the positive side of the source potential is present anywhere from the source itself around the circuit to point A. The negative side of the source potential is present at every point along the circuit on the opposite side of the break all the way to point B.

Troubleshooting Hints

When you suspect an open in a series circuit, you can measure the voltage across each component. The component, or portion of the circuit across which you measure V applied, is the opened component, or circuit portion. *Caution:* To measure these voltages, power must be applied to the circuit. Use all safety precautions possible.

An alternative to this "power-on" approach is to use *"power-off"* resistance measurements. The two points in the circuit at which you measure "infinite ohms" (with the circuit completely disconnected from the voltage/power source) is the opened portion of the circuit. Resistance of the good components should measure their rated values.

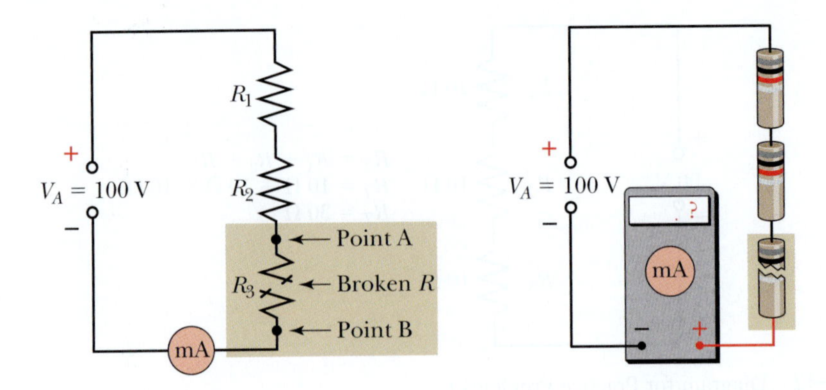

FIGURE 4–13 An open in a series circuit

What is the potential difference across the break, or open? If your answer is 100 volts (or V_A), you are correct! This indicates there will be no $I \times R$ drops across the *good* components in the circuit, and the circuit applied voltage appears across the open portion of the circuit, regardless of where the open appears in the circuit.

An example of a purposely opened circuit is the light switch circuit in your home. When you turn the light on, you *close* an open set of contacts on the switch, which is in series with the light. When you turn the light off, you *open* those switch contacts in series with the light; therefore, you open the path for current flow, causing the applied voltage to appear across the opened switch contacts and zero volts to appear across the light element(s). If you measure the voltage across the switch with the lights *off,* you will measure the source voltage, or V applied. CAUTION: DON'T TRY THIS! The voltage in a 120-volt lighting circuit is DANGEROUS. ◆

PRACTICE PROBLEMS 7

1. In a series circuit having a 100-V source and comprised of four 27-kΩ resistors, indicate whether the following parameters would increase, decrease, or remain the same if one of the resistors opened.
   a. Total resistance would ______________.
   b. Total current would ______________.
   c. Voltage across the unopened resistors would ______________.
   d. Voltage across the opened resistor would ______________.
   e. Total circuit power dissipation would ______________.
2. Would the conditions you have described for question 1 be true no matter which resistor opened?

4–7 Effects of Shorts in Series Circuits and Troubleshooting Hints

Shorts

What are the effects of a short in a series circuit? First, what is a short? A **short circuit** may be defined as an undesired very low resistance path in or around a given circuit. An example is when someone drops a metal object across two wires connected to the output terminals of a power source. In this case you short all the other circuits connected to the power source terminals and provide a *very low* resistance path for current flow through the new metal current path.

A short may occur across one component, several components, or the complete circuit (as in our dropped metal object example). Regardless, consider a short to be a virtually zero resistance path that disrupts the normal operation of the circuit.

◆ **EXAMPLE** Observe Figure 4–14. Notice that the undesired low resistance path is a piece of bare wire with a virtually zero resistance that has fallen across the leads of R_1. *With-*

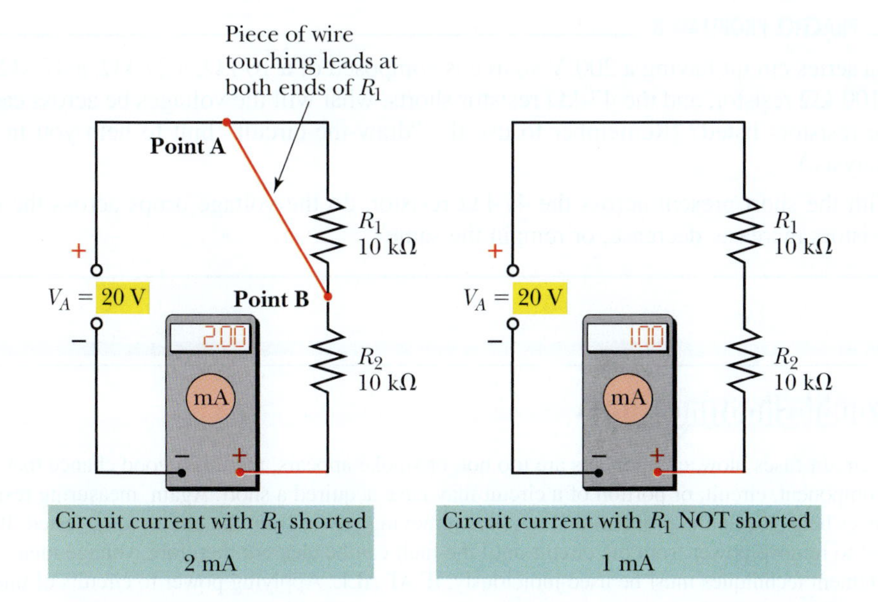

FIGURE 4–14 Effects of a short in a series circuit

out the short, the circuit total resistance would equal 20 kΩ ($R_1 + R_2$). Therefore, with 20 V applied voltage, the current is 1 mA (20 V/20 kΩ). *With* the short, the resistance from point A to point B is close to zero ohms because R_1 is shorted. As far as the voltage source is concerned, "it sees" only 10 kV resistance of R_2 as total circuit resistance. Thus, current is 20 V/10 kΩ, or 2 mA (double what it would be without the short). The short causes R_T to decrease to half its normal value and current to double since the V applied is unchanged.

This circuit analysis points to some important generalizations: (1) If any portion of a series (or any other type circuit) is shorted (or simply decreases in resistance), the total circuit resistance decreases; and (2) the circuit's total current increases, assuming the applied voltage remains unchanged.

What happens to the other circuit parameters if a short (or decreased resistance) occurs in a series circuit? Using our sample circuit in Figure 4–14, let us think through it.

1. R_1 becomes zero ohms.
2. R_T decreases, causing circuit current (I_T) to increase.
3. Because I has increased and R_2 is still the same value, the voltage across R_2 increases. In this case, it will increase to V applied (20 V) because there is no other resistor in the circuit to drop voltage.
4. V_1 (the voltage across R_1) decreases to zero volts because any value of I times zero ohms equals zero volts.
5. The total circuit power dissipation increases because I has doubled and V applied has remained unchanged ($P = V \times I$).
6. Power dissipated by the good resistor (R_2) increases because its R is the same, but the current through it has doubled ($P = I^2R$).
7. Of course, the power dissipated by R_1 decreases to zero because no current flows through it.

What happens if the *total* circuit is shorted instead of part of it? This means both R_1 and R_2 are shorted and the short is across the power supply terminals. The voltage source is looking at zero ohms, and it would try to supply "infinite" current. In reality this cannot happen. Probably the power supply fuse would blow, and/or the power supply and circuit wires/conductors might be damaged. However, the generalizations for the partially shorted circuit would also apply. That is, circuit resistance goes down, total current increases (until a fuse blows or the power supply "dies"), and the voltage across the shorted portion of the circuit decreases to zero. ◆

PRACTICE PROBLEMS 8

1. If a series circuit having a 200-V source is composed of a 10-kΩ, a 27-kΩ, a 47-kΩ, and a 100-kΩ resistor, and the 47-kΩ resistor shorts, what will the voltages be across each of the resistors listed? (Remember to use the "draw-the-circuit" hint to help you in your analysis.)
2. With the short present across the 47-kΩ resistor, do the voltage drops across the other resistors increase, decrease, or remain the same?

Troubleshooting Hints

If circuit fuses blow, components are too hot, or smoke appears, there is a good chance that a component, circuit, or portion of a circuit may have acquired a short. Again, measuring resistances helps locate the problem that requires removing power from the circuit to be tested. It is best to remove power from the circuit until the fault can be cleared; therefore, voltage measurement techniques must be used judiciously, IF AT ALL. Applying power to circuits of this type often causes other components or circuits to be damaged because excessive current may pass through the nonshorted components or circuits.

With power removed from the circuit, it is possible to use an ohmmeter to measure the resistance values of individual components, or of selected circuit portions. If the normal values are known, it is obvious which component(s), or portion(s) of the circuit have the very low resistance value. Often, there will be visual signs of components or wires that have overheated, leaving a trail that indicates where the excessive current flowed as a result of the short.

In special cases, if voltage measurements are used, voltage drops across the good components are higher than normal (due to the higher current), and voltage drops across the shorted component(s) or circuit portion(s) are close to zero.

If there are current meters or light bulbs in the circuit, the meter readings are higher than normal and/or the unshorted light bulbs glow brighter than normal.

A Special Troubleshooting Hint

A simple but useful technique used in troubleshooting circuits that have sequential "in-line" components or subcircuits is called the divide-and-conquer approach, or split technique. This technique can save many steps in series situations where one of many components or subcircuits in the sequential line is the malfunctioning element.

The technique is to make the first test *in the middle* of the circuit. The results tell the technician which half of the circuit has the problem. For example, Figure 4–15 represents a 12-light series Christmas tree circuit with an open bulb. Recall, if one light becomes open, all the lights go out since the only path for current has been interrupted. Note in Figure 4–15, the first test is made from one end of the "string" to the "mid-point," in this case between bulbs 6 and 7. If you make a voltage test with the power on and the meter indicates V applied, then the open bulb must be between bulb 1 and bulb 6. If V measures 0 volts, then the problem is in the other half of the circuit between bulb 7 and bulb 12. With *one* check, half of the circuit has been eliminated as the possible area of trouble.

Incidentally, this technique can be used with the power-off resistance measurement approach. With the circuit *disconnected* from the power source, if the first R measurement shows infinite resistance between bulbs 1 and 6, the problem is in that half of the circuit. If the reading is a low resistance, then the problem is in the other half of the circuit between bulb 7 and bulb 12.

After dividing the abnormally operating circuit's suspect area in half by making the first measurement made at the circuit mid-point, the next step is to divide the remaining suspect area of the circuit in half again by making the second check from one end of the suspect section to its mid-point; thus, narrowing down to a quarter of the circuit the portion that remains in question, Figure 4–16. This splitting technique can continue to be used until only two

A Good Idea

Wear safety glasses when you or someone near you is soldering or using rotating machinery or power tools!

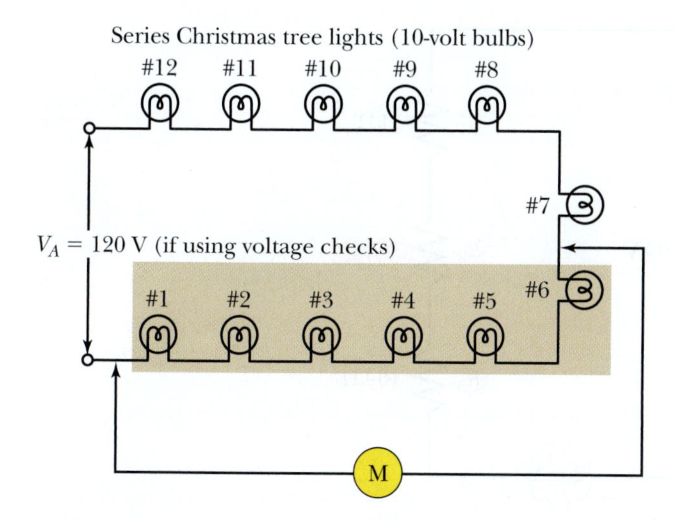

FIGURE 4–15 Divide-and-conquer troubleshooting technique—1st step (NOTE: Meter "M" may be either a voltmeter, if voltage is applied to the circuit under test, or an ohmmeter, if the circuit is disconnected from the power source.)

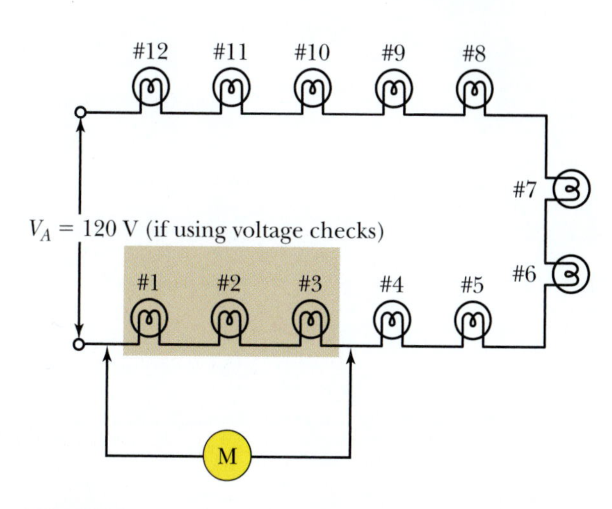

FIGURE 4–16 Divide-and-conquer troubleshooting technique—2nd step (NOTE: Meter "M" may be either a voltmeter, if voltage is applied to the circuit under test, or an ohmmeter, if the circuit is disconnected from the power source.)

components or subcircuits remain. Then each is checked, as appropriate, to find *the* malfunctioning component.

The divide-and-conquer approach can be used in any in-line (linear) system of components or circuits where current flow, power, fluids, or signals must flow sequentially from one component or subcircuit to the next and is useful in troubleshooting electrical, electronic, hydraulic, and many other systems having the in-line condition.

Practical Notes

In review, solve circuit problems with the following steps:

1. Collect all known values of electrical parameters.
2. If no circuit diagram is given, draw a diagram and write known quantities on the diagram.
3. Solve the first part of the problem at a point where you have, or can easily find, sufficient knowns to solve for an unknown. Then proceed by solving the remaining portions of the problem, as appropriate.

4–8 Designing a Series Circuit to Specifications

Let us apply the technique described in the Practical Notes to a sample design problem, then you can try one to see whether you have learned the process.

EXAMPLE Design a three-resistor series circuit where two of the resistors are 10 kΩ, the total circuit current equals 2 mA, and the applied voltage equals 94 V.

The first step is to collect the knowns, which were given to you. The next step is to draw the circuit and label the knowns, Figure 4–17. The third step is to start at a point where sufficient knowns are available to solve for a desired unknown.

To apply the third step, observe that in the circuit we drew, we know both the I and the R values for both R_2 and R_3. This makes it easy to find their voltage drops.

$$V_{R_2} = I \times R_2 = 2 \text{ mA} \times 10 \text{ k}\Omega = 20 \text{ V}$$
$$V_{R_3} = I \times R_3 = 2 \text{ mA} \times 10 \text{ k}\Omega = 20 \text{ V}$$

The unknown is the value of R_1, which completes our design problem.

If the voltage drop of R_1 can be found, it is easy to find its resistance value. This is because we know the current must be 2 mA and its resistance must equal its voltage drop divided by 2 mA. It is convenient to use Kirchhoff's voltage law to find V_{R_1}. According to Kirchhoff, V_A must equal $V_{R_1} + V_{R_2} + V_{R_3}$. Thus, transposing to solve for the unknown V_{R_1}:

$$V_{R_1} = V_A - (V_{R_2} + V_{R_3}) = 94 \text{ V} - (20 \text{ V}+20 \text{ V}) = 54 \text{ V}$$

$$\text{Therefore, } R_1 = \frac{54 \text{ V}}{2 \text{ mA}} = 27 \text{ k}\Omega$$

Another way to solve R_1's value is to find R_T, where $R_T = V_T/I_T$. Thus, R_T = 94 V/2 mA = 47 kΩ. Knowing total resistance equals the sum of all individual resistances in a series circuit and R_2 plus R_3 equals 20 kΩ, R_1 must make up the difference

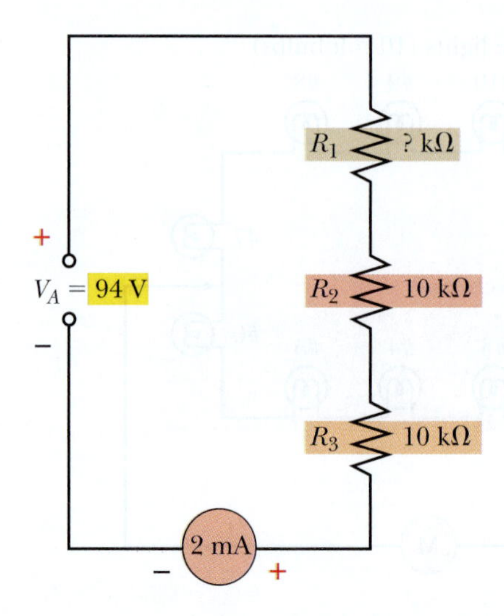

FIGURE 4–17

of 27 kΩ to provide a total resistance of 47 kΩ. The more formal way to illustrate this is to write the R_T formula, then transpose it to solve for R_1. Thus,

$$R_T = R_1 + R_2 + R_3$$
$$R_1 = R_T - (R_2 + R_3) = 47\text{ k}\Omega - (10\text{ k}\Omega + 10\text{ k}\Omega) = 27\text{ k}\Omega$$

Does our design meet the specifications? Yes, it is a three-resistor series circuit where two of the resistors are 10 kΩ, the total circuit current is 2 mA, and the applied voltage equals 94 V. ◆

PRACTICE PROBLEMS 9

Try this next design problem yourself. Complete it on a separate sheet of paper.

Problem: Design a three-resistor series circuit where R_1 is 20 kΩ and drops two-fifths of V applied, R_2 drops 1.5 fifths of V applied, and R_3 has a voltage drop equal to V_{R_2}. Assume V applied is 50 V. Draw the circuit and label all *V, R, P,* and *I* parameter values.

4–9 Special Applications

Voltage Sources in Series

Voltage sources can be connected in series to provide a higher or lower total (resultant) voltage than one of the sources provides alone. The *resultant* voltage of more than one voltage in series depends on the values of each voltage *and* whether they "series-aid" or "series-oppose" each other. Figure 4–18a is a series-aiding type connection. Figure 4–18b is an example of a series-opposing arrangement.

Note that in Figure 4–18, you see the terms **cell** and **battery.** A cell is a single voltaic device that converts chemical energy to electrical energy. The symbols associated with a single cell are shown in each circuit in the figure as the 1.5 V sources. A familiar example of cells are the cells you put into your flashlight. Typically, these are 1.5-V cells.

A battery is simply two or more cells interconnected and put into one package. In the figure, the battery symbol is shown in each circuit related to the 6.0-volt sources. A familiar example of a battery is the 12-volt automobile battery. It is comprised of six 2-volt cells, interconnected and packaged in one battery case.

Note, in Figure 4–18a, that the "resultant" voltage applied to the circuit from the series-aiding sources equals the sum of the two sources. In this case 6.0 V + 1.5 V = 7.5 volts. The resulting current through the resistor is 7.5 amperes,

$$I = \frac{V}{R} = \frac{7.5\text{ V}}{1\ \Omega} = 7.5\text{ A}$$

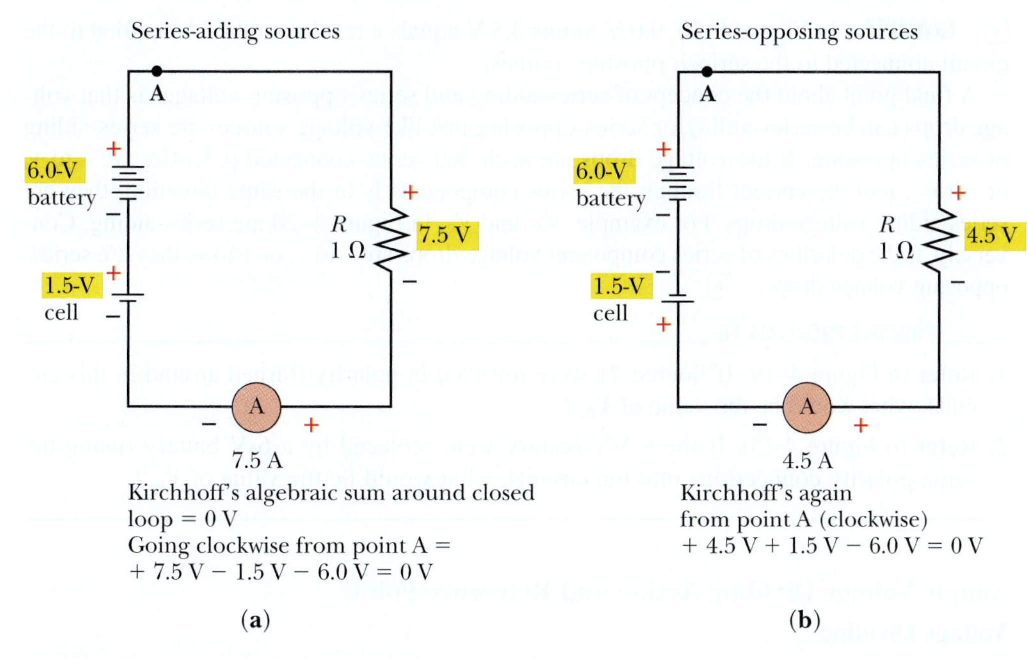

FIGURE 4–18 Series-connected voltage sources

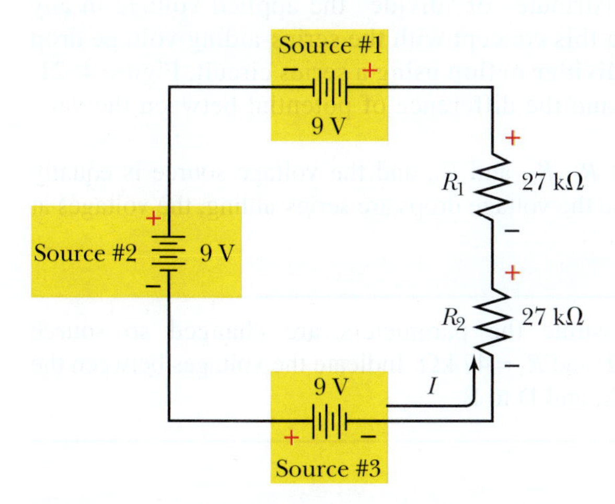

FIGURE 4–19 Series-aiding sources have the following features: (a) Sources are in series. (b) Negative terminal of one source connects to positive terminal of next, and so on. (c) All three sources try to produce current in the same direction through the circuit. (d) They are verifiable using Kirchhoff's law. multiSIM

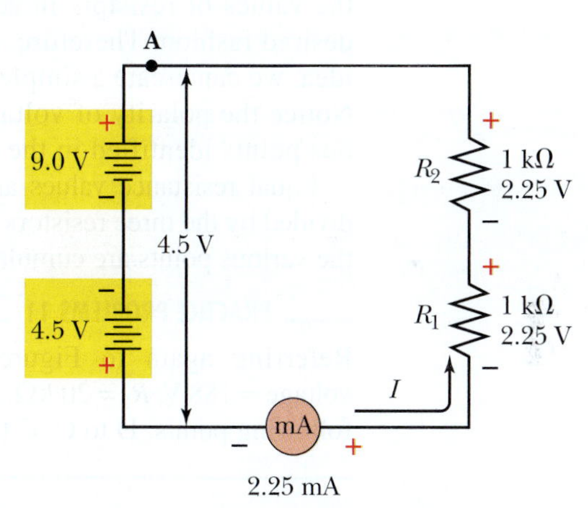

FIGURE 4–20 Series-opposing sources have the following features: (a) Sources are in series. (b) Negative terminal of one source connects to negative terminal of next (or positive connects to positive), and so on. (c) Both sources try to produce current in opposite directions through the circuit. (d) They are verifiable using Kirchhoff's law. multiSIM

Ways to know these sources are series-aiding are (1) the negative terminal of one source is connected to the positive terminal of the next, and (2) both sources try to produce current in the same direction through the circuit.

In Figure 4–19, what is the total voltage applied to the circuit? What is the current value? If you answered 27 V for the resultant voltage and 0.5 mA for the current, you are correct.

As you can see in Figure 4–20, series-opposing sources are connected so that: (1) the negative terminal of one source is connected to the negative terminal of the next, or positive-to-positive; and (2) the sources try to produce current through the circuit in opposite directions.

To determine the resultant or equivalent voltage of series-opposing sources, *subtract* the smaller voltage from the larger voltage.

EXAMPLE In Figure 4–20, 9.0 V minus 4.5 V equals a resultant of 4.5 V applied to the circuit connected to the series-opposing sources.

A final point about the concept of series-aiding and series-opposing voltages is that voltage drops can be series-aiding or series-opposing just like voltage sources are series-aiding or series-opposing. If the voltage drops are such that series-connected polarities are – to +, or + to –, *and* the current through the series components is in the same direction, they are series-aiding voltage drops. For example, V_{R_1} and V_{R_2} in Figure 4–20 are series-aiding. Conversely, if the polarities of series component voltage drops are – to –, or + to +, they are series-opposing voltage drops.

PRACTICE PROBLEMS 10

1. Refer to Figure 4–19. If Source #1 were reversed in polarity (turned around in this circuit), what would be the value of V_{R_1}?
2. Refer to Figure 4–20. If the 4.5-V battery were replaced by a 6-V battery (using the same polarity connections into the circuit), what would be the value of V_{R_2}?

Simple Voltage Dividing Action and Reference Points

Voltage Dividing

As you have learned, voltage drops around a series circuit are proportional to the resistance distribution since the current is the same through all components. You can *select* the values of resistors in series to "distribute" or "divide" the applied voltage in any desired fashion. Therefore, combining this concept with the series-aiding voltage drop idea, we can create a simple **voltage-divider action** using a series circuit, Figure 4–21. Notice the polarity of voltage drops, and the difference of potential between the various points identified in the circuit.

Equal resistance values are used for R_1, R_2, and R_3, and the voltage source is equally divided by the three resistors. Also, since the voltage drops are series-aiding, the voltages at the various points are cumulative.

PRACTICE PROBLEMS 11

Referring again to Figure 4–21, assume the parameters are changed so source voltage = 188 V, R_1 = 20 kΩ, R_2 = 27 kΩ, and R_3 = 47 kΩ. Indicate the voltages between the following points: D to C; C to B; B to A; and D to B.

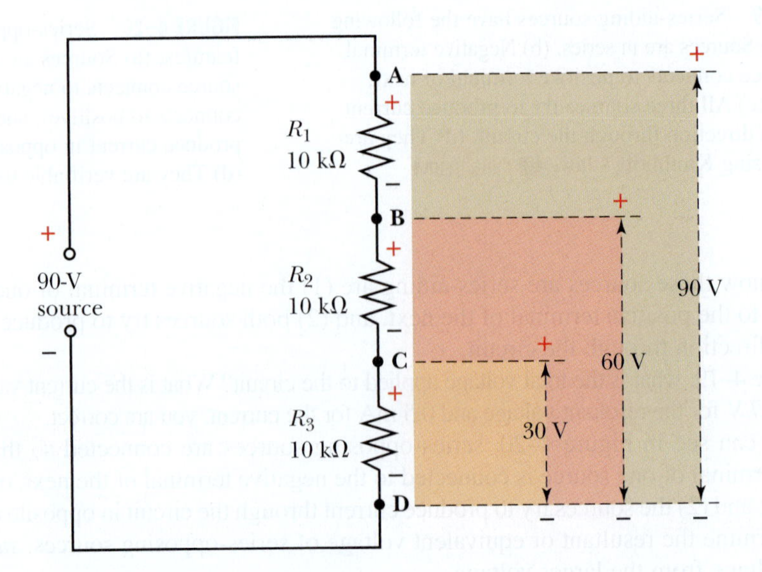

FIGURE 4–21 Example of voltage-divider action

Sample Applications for Voltage Dividers

Volume Controls

You will recall that in an earlier chapter we mentioned that potentiometers were often used as voltage-dividing devices. In Figure 4–22, you can see that if the wiper-arm on the potentiometer is set at point A, the total voltage coming from the preceding stage in an amplifier or radio circuit is applied to the next stage. If the wiper arm is midway on the resistive element, half the voltage is applied to the next stage, and so on. This is the technique used in radio and TV receivers, and in audio amplifiers to control the volume you hear from the system speaker(s). Obviously, the higher the "divided voltage" that is fed to the next (audio amplifier) stage after the potentiometer, the louder the sound you hear.

Transistor Bias Circuit

Later in your studies you will learn appropriate details about transistor amplifier stages. For now, you simply need to see that one very practical and commonly used application for a voltage divider is to supply appropriate voltage levels to various parts of a transistor, Figure 4–23.

Reference Point(s)

Recall in an earlier chapter we discussed the idea of something "with respect to" or "in reference to" something else. The example used was "John is taller than Bill," or "Bill is shorter than John."

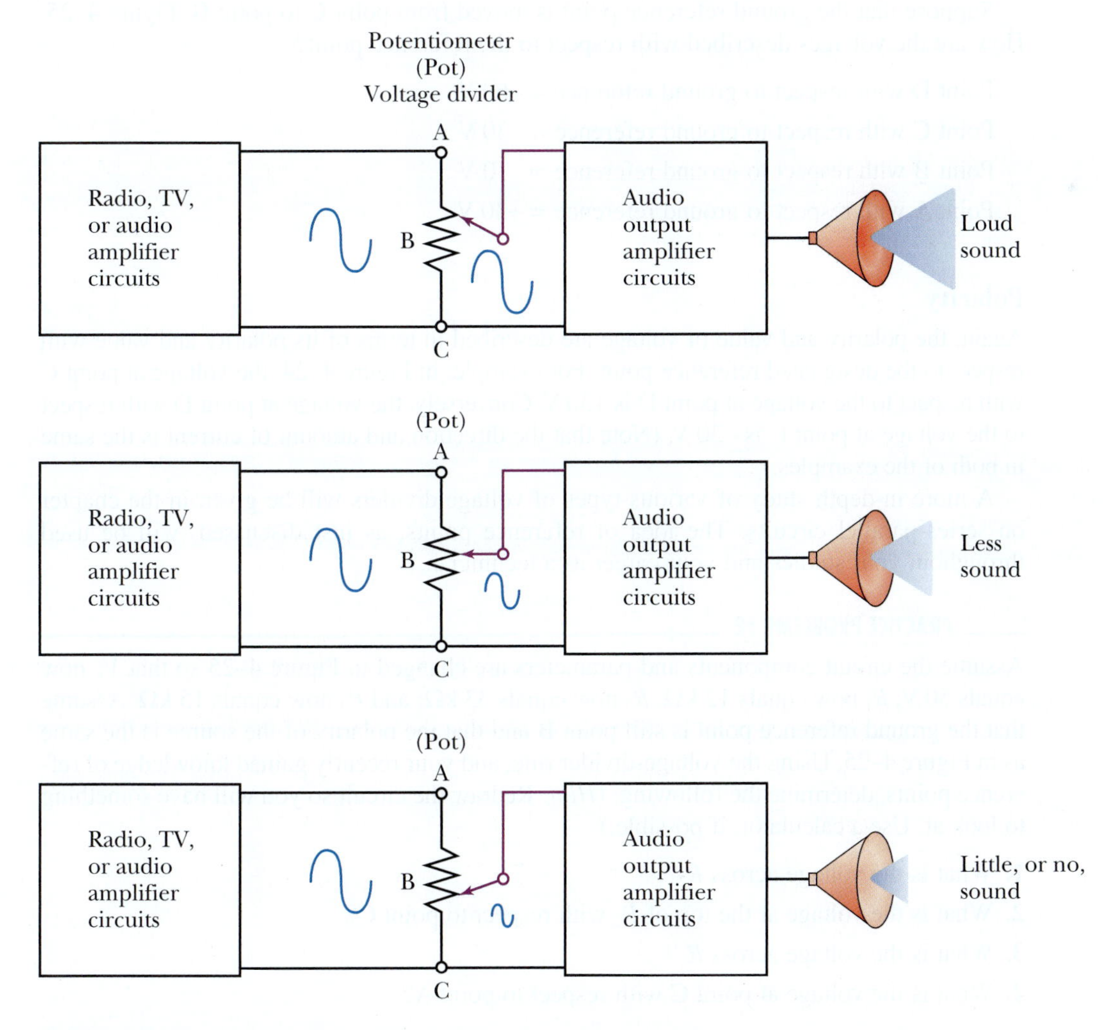

FIGURE 4–22 Potentiometer acting as voltage divider/volume control

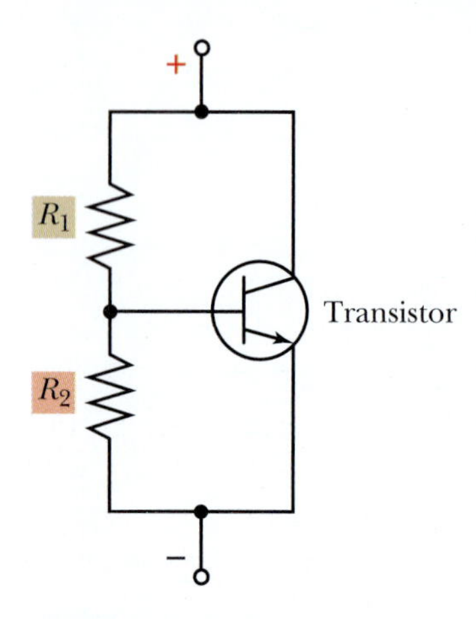

FIGURE 4–23 The voltages applied to the various elements of the transistor depend on the value of the applied voltage and the ratio of the voltage divider resistances R_1 and R_2.

In electronic circuits it is common to relate electrical parameters, such as voltage, to a common reference point in the circuit. Most electronic circuits have some sort of common "return" path or connection point to which many of the circuits in the system are connected. This point becomes the circuit "common" reference point. When the circuits are mounted on a metal chassis, the chassis is often used as this common connection-point path. When the circuits are mounted on an epoxy or glass substrate, such as that used by printed circuit (PC) boards, a copper path around the PC board becomes the common connection point, or return path. This common point on chassis or PC boards is often called "chassis ground." The electrical schematic symbol for "chassis ground" is shown here: ⏊

In many books and diagrams, the chassis ground symbol is not used. The "earth ground" symbol, ⏚, is commonly used to represent the common reference point for circuits. The chassis ground, or common circuit path may or may not be connected to an actual earth ground, as is the power-line system at your home. Throughout this book, we will use the more commonly used ground symbol (⏚) to represent the common reference point in circuits where we do show a common ground reference.

In Figure 4–24, point C is where the ground reference is connected. How is this used as a reference point to describe the voltages along the "divider"? The voltages with respect to this ground reference point can be described as follows:

Point D with respect to ground reference = –30 V
Point C with respect to ground reference = 0 V
Point B with respect to ground reference = +30 V
Point A with respect to ground reference = +60 V

Suppose that the ground reference point is moved from point C to point B, Figure 4–25. How are the voltages described with respect to the reference point?

Point D with respect to ground reference = –60 V
Point C with respect to ground reference = –30 V
Point B with respect to ground reference = 0 V
Point A with respect to ground reference = +30 V

Polarity

Again, the polarity and value of voltage are described in terms of its polarity and value with respect to the designated reference point. For example, in Figure 4–24, the voltage at point C with respect to the voltage at point D is +30 V. Conversely, the voltage at point D with respect to the voltage at point C is –30 V. (Note that the direction and amount of current is the same in both of the examples.)

A more in-depth study of various types of voltage dividers will be given in the chapter on series-parallel circuits. The idea of reference points, as just discussed, will be used throughout your studies and your career as a technician.

PRACTICE PROBLEMS 12

Assume the circuit components and parameters are changed in Figure 4–25 so that V_A now equals 50 V; R_1 now equals 12 kΩ; R_2 now equals 33 kΩ; and R_3 now equals 15 kΩ. Assume that the ground reference point is still point B and that the polarity of the source is the same as in Figure 4–25. Using the voltage-divider rule, and your recently gained knowledge of reference points, determine the following: (*Hint:* Redraw the circuit so you will have something to look at. Use a calculator, if possible.)

1. What is the voltage across R_2?
2. What is the voltage at the top of R_2 with respect to point C?
3. What is the voltage across R_1?
4. What is the voltage at point C with respect to point A?

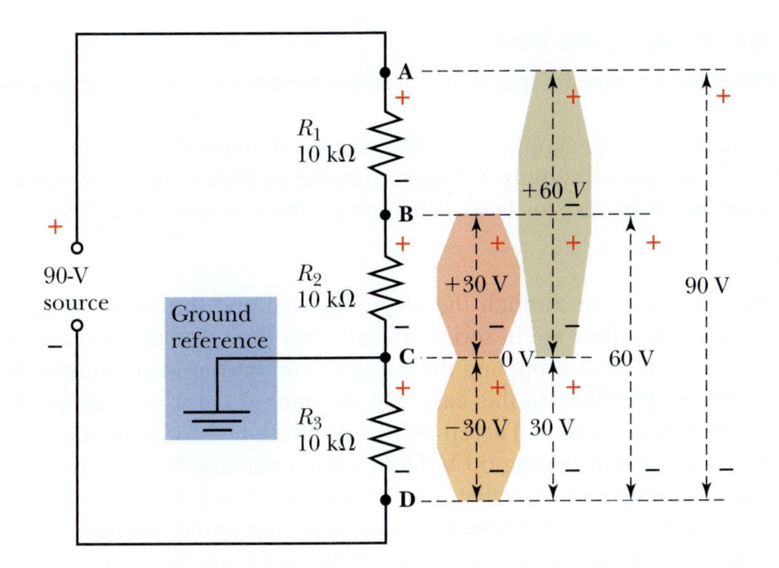

FIGURE 4–24 Ground as a reference point at point C. The 30 V, 60 V, and 90 V indicators show the cumulative voltage drops. The –30 V, +30 V, and +60 V indicators show voltages with respect to point C, where ground reference is connected.

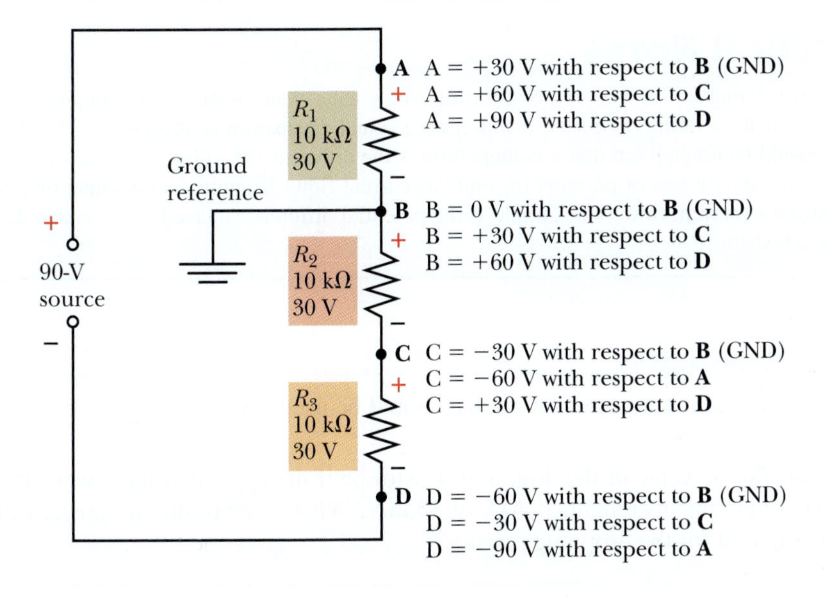

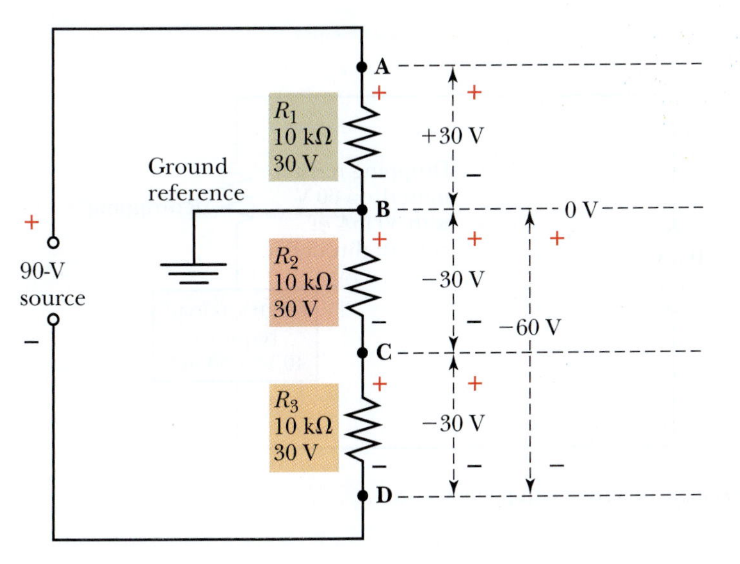

FIGURE 4–25 Ground as a reference point at point B

The Voltage-Dropping Resistor

One final topic relating to some special applications of series circuits is the **series-dropping resistor.**

Often, there is a need to supply a specific voltage at a specific current to an electrical load. If the "fixed" power supply voltage available is higher than the desired voltage value, a series-dropping resistor is used that drops the voltage to the desired level, Figure 4–26.

EXAMPLE The current through the electrical load and the series-dropping resistor must be the same since they are in series. To calculate the value of the dropping resistor, the value of voltage it must drop, and the value of current the load requires, provide the knowns to solve the problem. In this case, the dropping resistor must drop 60 V, and the load requires 30 mA of current. Therefore, the dropping resistor must have 30 mA passing through it *when* it is dropping 60 V. Ohm's law indicates the resistor value must be $R = V/I$, or the dropping resistor value = 60 V/30 mA = 2 kΩ. Naturally, this design calculation technique can be used to select resistor values for an infinite variety of dropping-resistor applications and is another example of the many special applications for series circuits.

Practical Notes

The power rating of the dropping resistor must be greater than the power it is required to dissipate. Usually, a rating *of at least two times* the actual dissipation is chosen.

It should be pointed out that a voltage drop is the difference in voltage between two points caused by a loss of pressure (or emf) as current flows through a component offering opposition to current flow. An example is the potential difference caused by a voltage drop across a resistor.

PRACTICE PROBLEMS 13

1. What is the minimum power rating that should be used for the dropping resistor, Figure 4–26?
2. What should the value of the dropping resistor be if the applied voltage were 150 V and the load requirements remained 40 V at 30 mA? What is the minimum power rating that should be used for the dropping resistor?

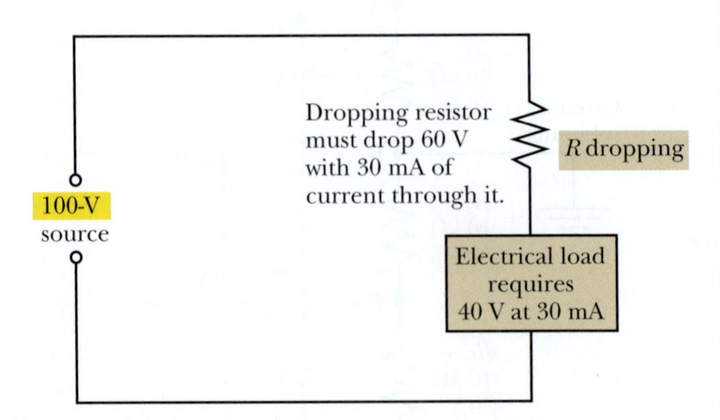

FIGURE 4–26 Example of a series-dropping resistor

4–10 Critical Thinking and the SIMPLER Troubleshooting Sequence

Your most valuable asset as a technician is the ability to think critically and logically. The ability to mentally move from a general principle or general case to a specific situation (deductive reasoning) is valuable. Likewise, moving from a specific case to a probable general case is worthwhile (inductive reasoning).

Throughout the remainder of this text, we will help you develop and enhance your logical reasoning skills. To get started, study the following section called "Introduction to Troubleshooting Skills."

4–11 Introduction to Troubleshooting Skills

All electronic technicians and engineers perform troubleshooting to some degree throughout their careers! Whether designing, installing, testing, or repairing electronic circuits and systems, the skill of locating problem areas in the circuit or system by a *logical* narrowing down process becomes a valuable asset.

To enhance your skills of logical troubleshooting, we will introduce a simple sequence of troubleshooting steps you can easily learn. We'll call it the **"SIMPLER" troubleshooting sequence** (or method).

In many of the remaining chapters, you will have an opportunity to practice this sequence and solve "Troubleshooting Challenge" applications problems. By the time you finish the book, this SIMPLER sequence should be second nature; thus, enhancing your skills and making you an even more valuable technician.

4–12 The SIMPLER Sequence for Troubleshooting

SYMPTOMS Gather, verify, and analyze symptom information.

IDENTIFY Identify initial suspect area for location of trouble.

MAKE Make decisions about: "What type of test to make" and "Where to make it."

PERFORM Perform the test.

LOCATE Locate and define new "narrower" area in which to continue troubleshooting.

EXAMINE Examine available information and again determine "what type test" and "where."

REPEAT Repeat the preceding analysis and testing steps until the trouble is found.

Further Information about Each Step in the Sequence

1. SYMPTOMS may be provided or collected from a number of sources, such as:

   a. What *the user* of the circuit or system tells you.

   b. *Your five senses* frequently can give you strong clues to the trouble. Sight and hearing allow evaluation of the circuit or system output indicators (e.g., normal or abnormal picture, normal or abnormal sound, system meter readings, indicator lights, and signal conditions). Sight also allows you to see obviously overheated/burned components. Smell helps find overheated or burned components. (Once you have smelled certain types of overheated or burned components, you will never forget the smell!) Touch can also be used but should be used *very cautiously!* Bad burns result from touching a component that is running exceedingly hot! (Some components run that way normally, others do not.) Also, you can be dangerously shocked by touching "live" circuits!

   c. By using your evaluation of the *easy-to-check indicators.* For example, if a TV has sound but no picture, or a radio receiver has hum, but you can't tune in radio stations. Most systems have easy-to-get-at switches and controls that aid in narrowing down possible trouble areas before you begin in-depth troubleshooting. For example, TVs

SIMPLER

STEP 1

SYMPTOMS Gather, verify, and analyze symptoms information.

STEP 2

IDENTIFY the initial suspect area.

STEP 3

MAKE a decision about "What type of test to make" and "Where to make it."

STEP 4

PERFORM the test.

STEP 5

LOCATE the new "narrower" suspect area.

STEP 6

EXAMINE the available information and again determine tests to be made.

STEP 7

REPEAT the analysis and testing steps until the trouble is found.

STEP 8

VERIFY Test and verify that the circuit operates normally.

have channel selectors, radios have tuning and volume controls, and electronic circuits frequently have switches and variable resistances.

d. By *comparing actual operation to normal operation and "norms."* "Norms" are normal characteristics of signals, voltages, currents, etc., that are known by experience or determined by referring to diagrams and documentation about the circuit or system being checked. This means you, as a technician, should know how to read and thoroughly understand block and schematic diagrams. (In specific cases, you may have to trace the actual circuit and draw the diagrams yourself.)

2. IDENTIFY the initial suspect area by:

a. analyzing all the symptom data;

b. from the analysis of the symptom data, determining all the possible sections of a system, or components within a circuit, that might cause or contribute to the observed symptoms; and

c. circling, or in some way marking it on a diagram, as appropriate.

3. MAKE a decision about "what kind of test" and "where" to make your first test by:

a. Looking at the initial suspect area, and determining where a test should be made to narrow down the suspect area most efficiently. And, determining what type of test would be most appropriate to make. (NOTE: Many times the "where" will dictate what type of test is to be made. At other times, the easiest type of test to make, which yields useful information, determines both the "where" and "what kind" of test information.)

b. Typically, you begin with general tests, such as looking at obvious indicators, manipulating switches and controls, and similar measures. As you narrow down the suspect area, your tests become more precise, such as voltage, current, or resistance measurements. Finally, substituting a known good component for the suspected bad one confirms your analysis. (In some cases, eradicating the problem is done by soldering, or removing a short, or moving a wire rather than changing a component.)

4. PERFORM the first test based on what your analysis has indicated (i.e., what kind of test and where), and you are provided with new information and insight to use in your quest to narrow down and find the problem.

5. LOCATE and bracket the new narrower questionable area of the circuit or system using information you have. Again, you will identify this new smaller area by circling or bracketing it so you won't make needless checks outside the logical area to be analyzed.

6. EXAMINE the collected symptom information, test results, and other data you have. Now make a decision about the next test that will provide further meaningful information.

7. REPEAT the analysis, testing, and narrowing down process as many times as required, and you will eventually find the fault in the circuit or system. At this point, you will have successfully used the SIMPLER sequence!

4–13 Troubleshooting Levels

There are two basic levels of troubleshooting that technicians may have to perform. One is troubleshooting to the "block" or "module" level. The other is troubleshooting to the single component level.

The **block-level troubleshooting approach** requires knowledge of the normal "inputs" and "outputs" used for each block or module in the total system. These inputs and/or outputs may be audio or video signals, certain voltage levels, certain current values, and so forth.

To troubleshoot to the block level, we'll refer to the "inputs/outputs" theory, where the "inputs" refer to block inputs and the "outputs" refer to the block output(s). If the input quantities (signals, voltages, etc.) check normal for a given block but the output quantities are abnormal, the problem is probably in that block.

On the other hand, if the input quantities are not normal, you trace backward to where the input is supposed to come from until the place of abnormality is found. In some cases, the abnormality can be caused by the input circuitry of the block being tested, rather than from blocks feeding this block. Your job is to isolate the block causing the problem and to replace it. See "A Block-Level Troubleshooting Example," Figure 4–27.

The **component-level troubleshooting approach** requires knowledge of normal parameters throughout the circuitry within each block of the system. By isolating and narrowing down (using the SIMPLER sequence approach), you will eventually narrow the problem to the bad component(s) within a module or block. Replacing the bad component(s) solves the problem. See "A Component-Level Troubleshooting Example," Figure 4–28. Study these two examples; then, proceed in the chapter, as appropriate.

The SIMPLER Sequence and Troubleshooting Levels

The SIMPLER sequence can be applied to both the block and component levels of troubleshooting. You will have a chance to try both troubleshooting levels in the Troubleshooting Challenge problems you will find throughout this book.

After studying the examples in Figures 4–27 and 4–28, and other basic information, you will get to try the first Troubleshooting Challenge, found at the end of this chapter, to begin practicing these concepts.

A Block-Level Troubleshooting Example

Looking at Figure 4–27, if you were told there was no sound from the speaker even though someone was talking into the microphone, the block-level troubleshooting scenario might be as follows:

SYMPTOMS No sound from speaker.

IDENTIFY initial suspect area anywhere in the system for this situation. The trouble could be in the speaker block, amplifier block, or microphone block.

MAKE decisions about test (what kind of test and where). Probably make test about in the middle of the system (at the amplifier input). That way you have cut the area of possible trouble in half with your first test. (Remember the "divide-and-conquer" technique explained with Figure 4–16.)

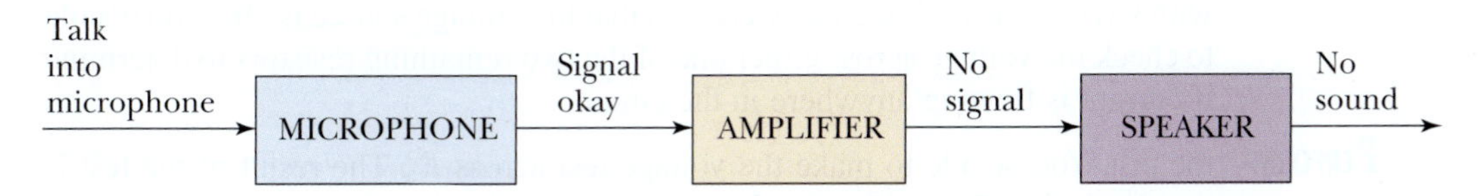

FIGURE 4–27

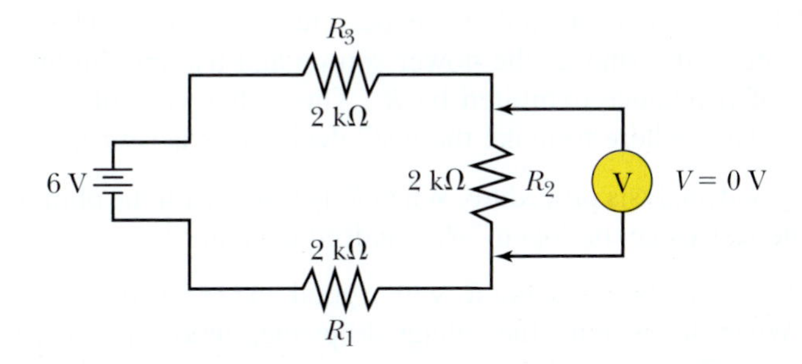

FIGURE 4–28

PERFORM the test. As you can see from the diagram, the signal into the amplifier is OK (the "input" is all right for this block).

LOCATE the new smaller suspect area. Your first test has eliminated the microphone as a possible trouble spot. The remaining possible trouble area is the amplifier block and the speaker block.

EXAMINE available information and make new test decisions. Now see if the amplifier output is normal. The "type" of check is a signal check; the "where" is at the amplifier output.

REPEAT testing and analysis procedure. A check at the amplifier output shows no signal. (The "output" signal is bad.) If the input to the amplifier is good and the output of the amplifier is bad, the chances are the trouble is in the amplifier block. This is troubleshooting to the block level. If we were to go inside the block to find the component in the block causing the trouble, that would be troubleshooting to the component level.

A Good Idea

When performing "block-level troubleshooting," if the output of a block is abnormal, check the condition of all the block's inputs before assuming that the trouble is in this block!

A Component-Level Troubleshooting Example

If you were given the facts in Figure 4-28, and told the voltage source had been checked as being OK, component-level troubleshooting might be done as follows:

SYMPTOMS No voltage across R_2, even though circuit input voltage is normal.

IDENTIFY initial suspect area. The trouble could be with any one of the resistors or with the interconnecting wires in the circuit, since the symptom could be caused by R_2 being shorted or by an open elsewhere in the circuit.

MAKE decisions about test (what type of test and where). Since voltage tests are easy and yield much information for small effort, the "what kind of test" is answered with a voltage test. Since you were told that the voltage source is OK, you decide to check the voltage across either one of the two remaining resistors to determine if current is flowing anywhere in the circuit.

PERFORM the test. You decide to make the voltage test across R_1. The result of the test is that V_{R_1} is found to be 6 volts. This is abnormal for this circuit!

LOCATE the new smaller suspect area. The voltage test across R_1 leads you to believe it may be the suspect! (Recall in a series circuit applied voltage appears across the open portion of the circuit, and zero volts appear across the good components.)

EXAMINE available information and make new test decisions. The logical new test might be to disconnect the power source and use an ohmmeter to check the value of resistance exhibited by R_1. The "what kind of test" is a resistance check. The "where to make the test" decision is across R_1.

REPEAT testing and analysis procedure. When R_1 is tested with the ohmmeter, it measures infinite resistance, or "open." R_1 is indeed the culprit.

VERIFY This is verified by replacing R_1 with a good 2 kΩ resistor and energizing the circuit. When this is done, the voltage drops measured across each of the resistors is 2 volts. This is troubleshooting down to the component level.

4–14 A "Systems" Approach to Troubleshooting

A. The Basic Essence of Any System

Almost any system designed to perform an action, or to produce something, can be simplified to a functional block diagram. The diagram in Figure 4–29 shows the absolute "bare essence" elements in virtually any system. As you can see, a system of any kind requires an input, or inputs. It requires processing element(s), generally illustrated by system and/or sub-systems blocks in the block diagram. These processing elements of the system will process, change, or perform tasks with these input(s). Finally, after the processing has been accomplished, there will be some useful output or outputs, assuming all elements in the system are working properly.

B. Sub-Systems

Of course, our illustration of the "bare essence" system diagram provides a *very* simplified overview of the "systems concept." For a practical block diagram, each of the major systems or sub-systems is usually broken down into more detailed block diagrams. These add further details regarding each of the sub-systems functions, their inputs and outputs, and the system process flow. It is this more detailed type functional block diagram that you will usually use to aid your troubleshooting and thinking process. An example of the simplest type system-flow block diagram for a TV system is shown in Figure 4–30.

Even this type diagram is too simplistic to use for technical troubleshooting. A slightly more detailed block diagram of the TV system is shown in Figure 4–31. But, for a system as complex as this one, the troubleshooter would use a block diagram containing 10, 20, or even 30 blocks. Each block would help show the critical steps taken in processing the inputs (in this case the TV signal), in order to produce the desired outputs. Note that the block diagram does *not* show the hundreds (and in some cases, thousands) of individual components or parts used in the circuits within the blocks. The functional block diagram simply shows the key functions and processing steps used to achieve each desired output from the system.

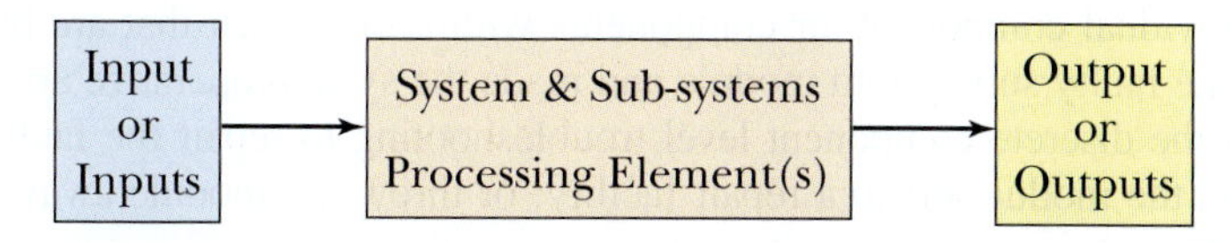

FIGURE 4–29 Simplified system diagram

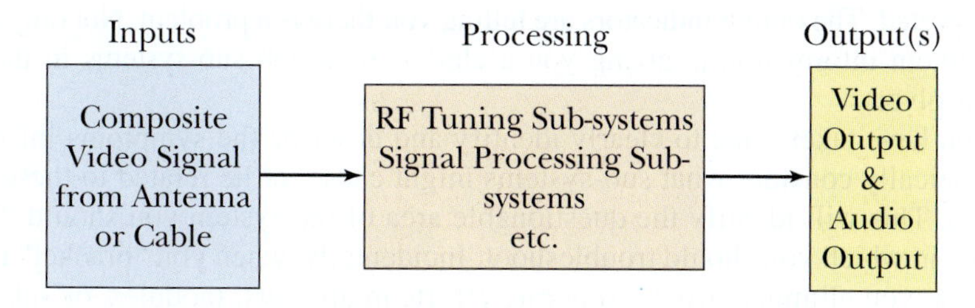

FIGURE 4–30 Simplified television system

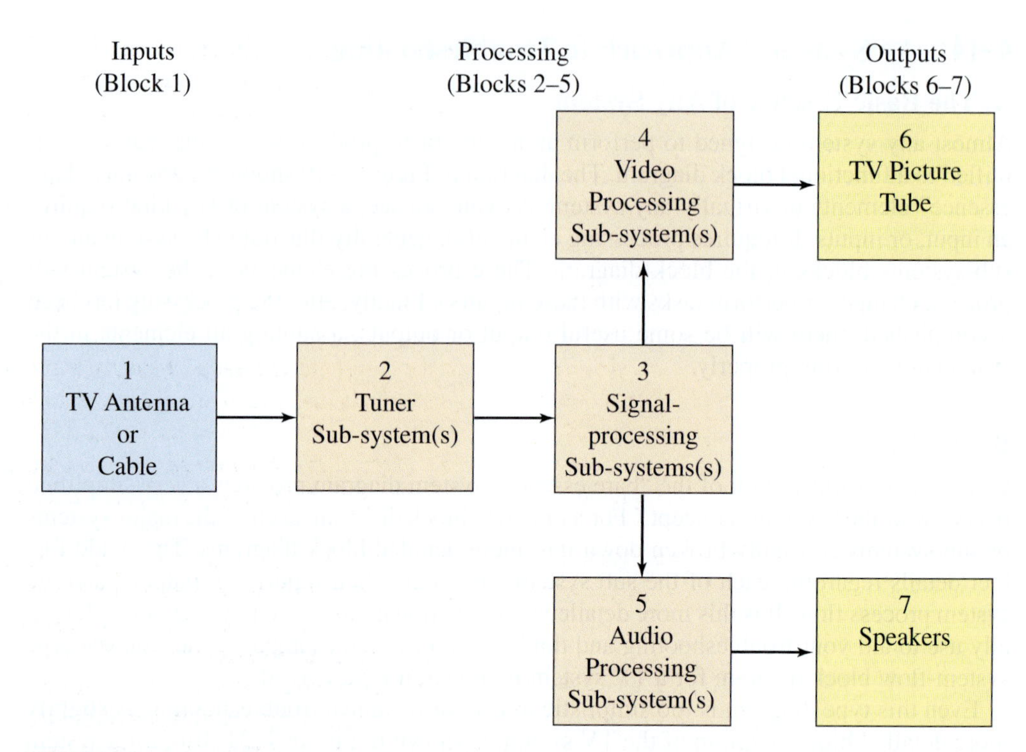

FIGURE 4–31 Simplified TV system block diagram

Conceptually, all of these diagrams convey the importance of the *INPUT-PROCESSING-OUTPUT* flow sequence in any system.

C. System Troubleshooting Concepts

System-style troubleshooting is often performed using the process flow-block diagram approach. This is in contrast with the discrete, item-by-item "component-level" troubleshooting approach. Typically, in system-style troubleshooting, the object is to find the bad "module" or sub-system and replace it, rather than going inside each process block or module and seeking the individual component, or components within that block that are bad. Once you have replaced the faulty sub-system module and made the system operative again, you might opt to perform the discrete component level troubleshooting to repair the faulty module for future use; send the module off to a repair facility; or throw the module away, whichever is most appropriate.

Overview of a Typical Systems Approach Scenario

Initially, symptoms appear that tell you that the system is not working properly. Generally, you are made aware of such a problem because one or more of the "output indicators" for the system shows the output(s) to be abnormal.

For example, look at Figure 4–31 as you think about the following: The TV may have sound but no picture; the TV has picture, but it is abnormal; or, the TV has sound, but the sound is distorted. The output indicators are telling you there is a problem. Not only that, this output symptom information is giving you a clue as to which sub-systems in the system might be involved.

Once you have taken time to clearly identify and describe the symptoms information, you then logically consider what sub-systems might cause, or be related to these specific symptom(s). This will identify the questionable area of the system you should "bracket" and the area in which you should troubleshoot. Incidentally, when you "bracket" the questionable area, you eliminate troubleshooting efforts in all units, modules, or sub-systems that *could not* be related to the symptoms. As in our TV example, if there is good picture but the sound is absent, you would *not* troubleshoot in the video (picture-related) sub-system block(s). Rather, you troubleshoot in the sub-system block(s) related to processing audio.

After identifying the system or sub-system blocks that might cause the trouble symptoms, check inputs and outputs of each of the blocks representing these sub-systems to isolate the specific problem area.

To enable localizing the problem area, there are test points in the actual circuit that relate to the various blocks' inputs and outputs. Measure and observe parameters and their values at these test points to compare them to the "normal" (expected) data for a normally operating system. Then, use your check results to judge how and/or why one segment of the system may be showing abnormal results. At this point, use the functional block diagram's illustration of how the system works to help you find and localize the problem area(s) of the actual circuit. This approach allows a more logical path to troubleshooting than trying to test every part, or even every sub-system in a random fashion. You have logically narrowed down to the probable trouble area.

Once you have identified the malfunctioning "module," either replace the module, or troubleshoot further to find the defective circuit and/or part within the module. As mentioned earlier, you should not only replace the defective module, circuit, or component, you should find the probable cause of its failure as a preventive measure against repeating this failure. You would typically change that module or part to fix the system; however, *not* before you test to see if something else in the system might have caused that unit to go bad. Again, this extra measure prevents ruining the replacement module. When you are convinced that the only problem in this area of the system is the module you have located, then you can replace the module. When this is done, you should follow up by verifying that the system and sub-system is operating properly. NOTE: Often (but not always), when replacing parts or modules, you will have to do some "calibration" work to recalibrate or set the system for desired operation. These recalibrations compensate for subtle differences that may occur when replacement parts or modules are put into the system.

Summarizing This Generalized Systems Approach:

1. Analyze the available *symptoms* of trouble (usually shown by some "output indicator")
2. *Bracket* the system-flow blocks that might relate to causing the symptom(s)
3. *Check inputs and outputs* of each sub-system block within the bracketed "questionable" area
4. Using the "inputs"/"outputs" checking approach, *narrow down* to the likely bad module
5. *Replace* the bad module, after verifying that it is safe to do so
6. *Calibrate* or readjust for minor discrepancies between old and new modules/parts
7. *Verify* proper system operation after replacement and recalibration.

D. Examples of Other Common Systems

Without taking the great amount of space it would take to illustrate each of these examples, let me simply help you think of the types of *inputs, processing,* and *outputs* involved in some systems with which you probably have some familiarity. Be aware that for each item listed, there usually are more sub-systems than noted, each requiring specific inputs and processing to produce desired outputs. For example, in the automobile **inputs** list, the fuel sub-system might contain the gas tank, the fuel pump, the fuel line, the fuel filter, and the fuel injector system.

The Automobile System:

a. **Inputs** are items such as fuel and air, oil, coolant, and electrical energy.
b. **Processing** includes such items as engine, transmission, wheels, and the driver.
c. **Output** is movement in a desired direction, at a desired speed.

SIMPLER

The Multimeter System:

a. **Inputs** are the electrical quantities to be measured (voltage, current, resistance, etc.), brought into the meter system via its test leads.

b. **Processing** may include electromagnetic reaction to, or digital processing of the electrical quantities being measured.

c. **Output** is some readable value via an analog scale, or via digital readout.

Using this quick overview of the "systems approach" to troubleshooting as a background, let's further introduce you to the SIMPLER troubleshooting sequence. You will have opportunity throughout the text to work with challenge problems involving both the systems block-diagram approach, and the more detailed component-level troubleshooting techniques using this SIMPLER sequence.

4–15 Using the SIMPLER Sequence: (A Closer Look at Component-Level Troubleshooting) a Single-Component Level Example

The information given to you is:

A basic circuit is comprised of a voltage source, conducting wires, a switch, and a light bulb (with its socket). It is known that the voltage source and conductor wires have been previously tested and they are good.

The complaint (which reveals symptom information) is that the light bulb does not light, even when the switch is flipped to the "on" position.

1. The **symptom** information gathered is that the light bulb doesn't light! You verify this by connecting the circuit. Analysis of the symptom information can be done by mentally visualizing the circuit, or drawing the circuit diagram. It is then apparent that if the source and the conductor wires are good, the initial suspect areas are the switch, light bulb, and socket.
2. Now you mentally (or on paper) **identify** the portion of the circuit containing the switch, the light bulb, and the socket (by circling or bracketing), Figure 4–32. This becomes the "initial suspect area" for locating the trouble.
3. You now **make** a decision about "what type of test to make" and "where to make it." Some of the possibilities are to (a) make a voltage test with a voltmeter across the light bulb socket; (b) turn the switch to the "off" position and replace the bulb with a known good bulb, then turn the switch to the "on" position to see if the replacement bulb lights; (c) disconnect the power source, unscrew the bulb, and make an ohmmeter test of the suspect bulb to see if it has continuity; (d) measure the voltage across the switch terminals with the switch open and with it closed to see if the voltage changes; and (e) disconnect the power source and use an ohmmeter to check the switch in both the "off" and "on" positions to see if it is operating properly. Because changing the bulb would be reasonably easy to do and bulbs do fail frequently, you decide to make the light bulb test first, Figure 4–33.

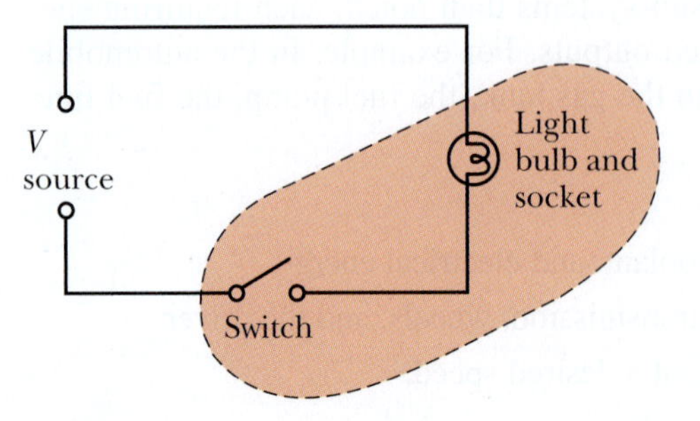

FIGURE 4–32 Initial suspect area

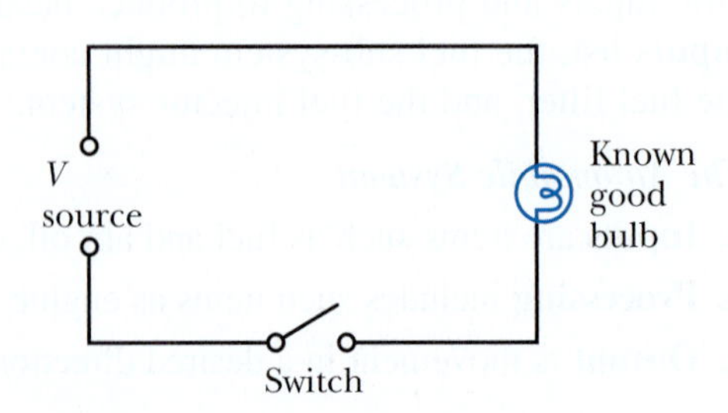

FIGURE 4–33 Checking the light bulb first

4. When you **perform** this test, you find that the "good" replacement bulb does not light, even with the power source connected to the circuit and the switch in the "on" position. Or if you tested the "suspect" bulb using the ohmmeter check approach, you would find the bulb checks good, Figure 4–34.
5. You can now **locate** the new smaller suspect area that includes only the socket and switch, Figure 4–35.
6. When you **examine** all the accumulated information, you decide the most likely suspect between the socket and switch might be the switch since it is somewhat of a mechanical device, which fails more frequently. Three ways you can check into this possibility are to (a) disconnect the power source from the circuit and use an ohmmeter to check the switch continuity in both the "off" and "on" positions; (b) measure the voltage at the light socket to see if voltage is present with the switch in the "on" position; and (c) measure voltage across the switch in both the "off" and "on" positions to see if there is voltage across the switch when in the "off" position and zero volts present when the switch is in the "on" position. After examining the information, your decision about "what type of test" and "where" is to check voltage at the light socket since this is an easy check to make, Figure 4–36.
7. **Repeating** the analysis *and* testing procedures, you make the test at the bulb socket with the switch "on." Voltage is present, Figure 4–37. This indicates the switch is good. With this analysis, your remaining suspect is the light socket. In testing the socket (either by ohmmeter continuity testing with the power off or by replacing with a known good socket), you find the socket is indeed bad. You have found the problem! The solution is obvious: Replace the socket. After replacing the socket, you verify the accuracy of your troubleshooting by testing the circuit. Your reward is that the circuit works properly, Figure 4–38.

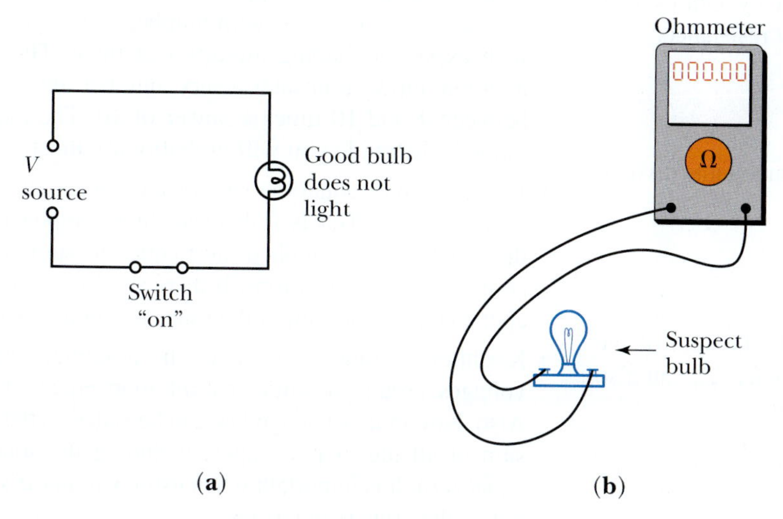

FIGURE 4–34 Results of test: (a) Good bulb does not light. (b) Suspect bulb has continuity.

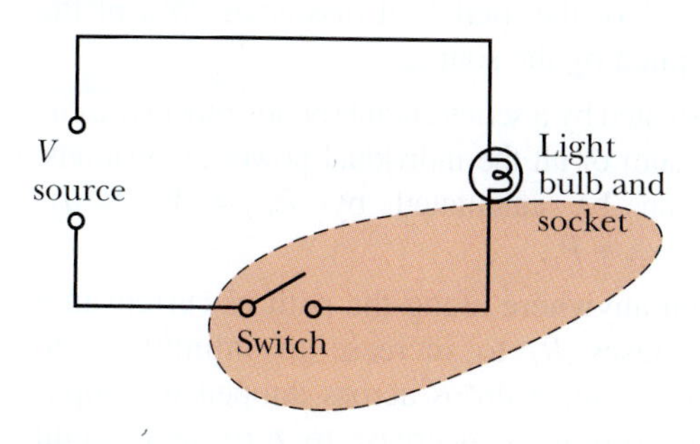

FIGURE 4–35 Narrowed suspect area

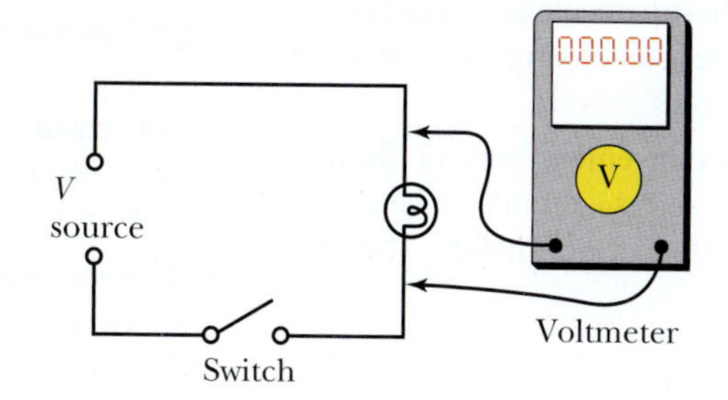

FIGURE 4–36 Voltage check at light socket

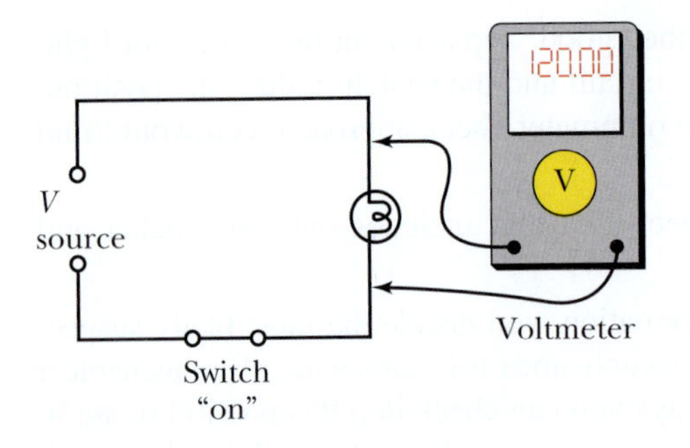

FIGURE 4–37 Voltage present at socket

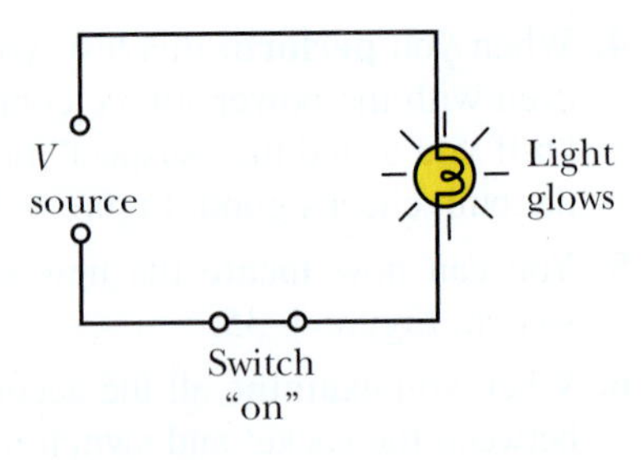

FIGURE 4–38 The circuit works

Although this hypothetical case is simple, it illustrates the concepts of the SIMPLER troubleshooting sequence. Practice using the sequence because it will help you become a logical troubleshooter!

Summary

- The definition and characteristics of a series circuit state that the circuit components are connected so that there is only one current path; therefore, the current is the same through all components in series.
- Current in a series circuit is calculated by Ohm's law, using either total voltage divided by total resistance:

$$I = \frac{V_T}{R_T}$$

or the voltage drop across one of the components divided by its resistance; for example:

$$I = \frac{V_1}{R_1}$$

- Total resistance in a series circuit equals the sum of the individual resistances ($R_T = R_1 + R_2 \ldots + R_n$). R_T can also be calculated if V_T and I are known using:

$$R_T = \frac{V_T}{I_T}$$

- Because I is the same throughout all components in series, each component's voltage drop ($I \times R$) is proportional to its resistance compared with the total resistance. The largest value resistance drops the highest percentage or fraction of V applied; the smallest value resistance drops the smallest portion of the total voltage.
- The voltage-divider rule is a convenient way of finding individual voltage drops in series circuits where the applied voltage and total resistance are known. The formula states:

$$V_X = \frac{R_X}{R_T} \times V_T$$

- Calculators are very useful for most calculations in electronics. Calculators may be used in several different modes of operation. One method is simply entering the literal numbers. Other methods use the advantages of a "scientific calculator" by application of powers of 10. The engineering mode deals with numbers times powers of 10 with exponents having multiples of three. The scientific notation mode translates every number into a number between 1 and 10 times a power of 10. The engineering mode is the mode you will probably use most.
- V applied must equal the sum of the voltage drops around the closed loop. That is, if there are three components around the series string, V applied must equal the sum of the three components' voltage drops. If there are four components, V applied equals the sum of the four individual voltage drops.
- Kirchhoff's voltage law states the arithmetic sum of the voltages around a single circuit loop equals V applied. Also, Kirchhoff's voltage law can be stated as the algebraic sum of all the loop voltages, including the source, must equal zero. It is important when using Kirchhoff's law(s) to notice the voltage polarities.
- The power distribution throughout a series circuit is directly related to the resistance distribution. For example, if a component's R is 25% of the total R, it dissipates 25% of the total power supplied by the source.
- Total power dissipated by a series circuit (or any other type circuit) equals the sum of all the individual power dissipations. Total power can be calculated by $P_T = V_T \times I_T$, or $P_T = P_1 + P_2 \ldots + P_n$.
- A break or open anywhere along the path of current in a series circuit causes R_T to increase to infinity; I_T to decrease to zero; voltage drops across unopened components or circuit portions to decrease to zero; and circuit

applied voltage to be felt across the open portion of the circuit. P_T decreases to zero since I equals zero.

- A short (very low resistance path) across any component, or portion, of a series circuit causes R_T to decrease; I_T to increase; voltage drops across the normal portions of the circuit to increase; and voltage dropped across the shorted portion of the circuit to decrease to virtually zero. A short also causes P_T to increase and the power dissipated by the remaining unshorted components to increase.
- When designing a circuit or solving for unknown quantities in an electrical circuit, it is good to (1) collect all known values; (2) draw a diagram and label known value(s); and (3) solve at any point where enough knowns are available to solve for an unknown. (NOTE: Usually, Ohm's law and/or Kirchhoff's law[s] will be used in the solution.)
- When troubleshooting a series circuit with an open condition, voltage measurements will reveal zero voltage drop across good components (or circuit portions) and V applied across the open points. Take resistance measurements only with the circuit disconnected from the source. Then normal components will reveal normal readings, and the opened component/portion will reveal an infinite ohms reading.
- When troubleshooting a series circuit with a short condition, use *resistance* measurements with the power source disconnected from the circuit. Normal components (or circuit portions) will measure normal values of R, while the shorted component (or circuit portion) will measure very low, or zero R.
- Series circuits are useful for several special applications, such as series-aiding and series-opposing voltage sources, simple voltage-dividing action, and series-dropping an available voltage down to a desired level.

Formulas and Sample Calculator Sequences

FORMULA 4–1
(To find total resistance of series resistors)

$R_T = R_1 + R_2 \ldots + R_n$

R_1 value, [+], R_2 value, [+], . . . , [=]

FORMULA 4–2
(To find total resistance using Ohm's law)

$R_T = \frac{V_T}{I_T}$

total voltage value, [÷], total current value, [=]

FORMULA 4–3
(To find voltage across a selected resistor)

$V_{R_1} = I_{R_1} \times R_1$

current value, [×], resistor #1 value, [=]

FORMULA 4–4
(To find voltage across a selected resistor using the voltage-divider rule)

$V_X = \frac{R_X}{R_T} \times V_T$

R_X value, [÷], R_T value, [×], V_T value, [=]

FORMULA 4–5
(To find total voltage in a series circuit)

$V_T = V_{R_1} + V_{R_2} \ldots + V_n$

V_{R_1} value, [+], V_{R_2} value, [+], . . . , [=]

FORMULA 4–6
(To find total power in a series circuit)

$P_T = P_{R_1} + P_{R_2} \ldots + P_n$

P_{R_1} value, [+], P_{R_2} value, [+], . . . , [=]

EXCEL AUTOMATED FORMULAS

Helpful problem-solving tools, such as the Excel automated formulas available on the CD, allow automatic calculations of formulas.

Using Excel

Series Circuits Formulas

File: FOE4_01.xls

***DON'T FORGET!** It is NOT necessary to retype formulas once they are entered on the worksheet! Just input new parameters data for each new problem using that formula, as needed.*

- Refer to Figure 4–8. Make R_2 be the R_x in the formula. Use the Formula 4–4 spreadsheet sample and solve for V across R_2. Compare with results in the text example.
- Refer to Figure 4–8, again. Make R_1 be the R_x in the formula. Use the Formula 4–4 spreadsheet sample and solve for V across R_1. Compare with answer given in Answer Appendix for Practice Problems 4, question 1.

Microsoft Excel - FOE4_01.xls

	A	B	C	D	E	F	G	H
1	Formula 4-1: $R_T = R_1 + R_2 ... + R_n$							
2	(Sample parameters from Figure 4-7)							
3	R1 in kΩ	R2 in kΩ	R3 in kΩ	RT in kΩ	⇐ Column headings to use			
4	100	27	10	137	⇐ Formula in Cell D4 is: =sum(A4:C4)			
5								
6	Formula 4-4: $V_x = R_x/R_T \times V_T$							
7	(Using select parameters from Figure 4-7 and using R_2's value as R_x)							
8	Rx in kΩ	RT in kΩ	VT in volts	Vx in volts	⇐ Column headings to use			
9	27	137	274	54	⇐ Formula in Cell D9 is: = A9/B9*C9			

Review Questions

1. Explain the most distinguishing electrical characteristic of a "series circuit" and clarify why this characteristic is true for this type circuit.
2. The total resistance of a series circuit is:
   - **a.** more than the largest value resistor in the circuit.
   - **b.** less than the largest value resistor in the circuit.
   - **c.** equal to the largest value resistor in the circuit.
3. Explain why your answer for question 2 is true.
4. In a series resistive circuit, the largest voltage drop appears across:
   - **a.** the smallest R value resistor.
   - **b.** the largest R value resistor.
   - **c.** neither the smallest nor largest R value resistor.
5. In a series resistive circuit, the applied circuit voltage is equal to the:
   - **a.** difference between the smallest and largest voltage drops around the circuit.
   - **b.** sum of all the individual voltage drops around the circuit.
   - **c.** product of the smallest and largest voltage drops around the circuit.
6. Kirchhoff's voltage law indicates that the *mathematical sum* of the voltage drops around a complete circuit loop must equal:
   - **a.** zero.
   - **b.** V_A.
   - **c.** $V_{R_1} + V_{R_2}$.
   - **d.** none of the above.
7. The power distribution throughout a series circuit is:
   - **a.** directly related to the resistance distribution around the circuit.
   - **b.** inversely related to the resistance distribution around the circuit.
   - **c.** has no relationship to the resistance distribution around the circuit.
8. If a break occurs in a series circuit:
   - **a.** R_T decreases to zero ohms.
   - **b.** R_T increases to infinite ohms.
   - **c.** neither of the above is true.
9. If a short is placed across part of a series circuit:
   - **a.** R_T decreases to zero ohms.
   - **b.** R_T does not change.
   - **c.** R_T decreases.
   - **d.** R_T increases.
10. What type of measurement would you make in troubleshooting a series circuit containing a short?
    - **a.** Resistance measurements with the power on
    - **b.** Resistance measurements with the power off
    - **c.** Voltage measurements with the power off
    - **d.** None of the above

Problems

NOTE: While working on these problems, don't forget to draw and label diagrams, as appropriate. It is a good habit to form.

1. If a series string of two equal resistors draws 5 amperes from a 10-volt source, what is the value of each resistor?
2. What is the current through a series circuit consisting of one 10-kΩ, one 20-kΩ, and one 30-kΩ resistor? Assume the circuit applied voltage is 40 V.
3. Draw a diagram of a series circuit containing resistors having values of 50 Ω, 40 Ω, 30 Ω, and 20 Ω, respectively. Label the 50-Ω resistor as R_1, the 40-Ω resistor as R_2, and so forth. Calculate the following and appropriately label your diagram:
   - **a.** What is the value of R_T?
   - **b.** What is V applied if I equals 2 A?
   - **c.** What is the voltage drop across each resistor?
   - **d.** What is the value of P_T?
   - **e.** What value of power is dissipated by R_2? by R_4?
   - **f.** What fractional portion of V_T is dropped by R_4?
   - **g.** If R_3 increases in value while the other resistors remain the same, would the following parameters increase, decrease, or remain the same?
     - **(1)** Total resistance
     - **(2)** Total current
     - **(3)** V_{R_1}, V_{R_2}, and V_{R_4}
     - **(4)** P_T
   - **h.** If R_2 shorted, would the following parameters increase, decrease, or remain the same?
     - **(1)** Total resistance
     - **(2)** Total current
     - **(3)** V_{R_1}, V_{R_3}, and V_{R_4}
4. According to Kirchhoff's voltage law, if a series circuit contains components dropping 10 V, 20 V, 30 V, and 50 V, respectively, what is V applied? What is the algebraic sum of voltages around the complete closed loop, including the source?
5. Draw a diagram showing how you would connect three voltage sources to acquire a circuit applied voltage of 60 V, if the three sources equaled 100 V, 40 V, and 120 V, respectively.
6. **a.** Calculate the P_T and P_{R_1} values for the circuit shown in Figure 4–39.
   - **b.** What is the V_T?
   - **c.** What is the R_T?
   - **d.** How many times greater is P_{R_4} than P_{R_1}?
   - **e.** If R_4 shorted, what would the value of P_T become? of P_{R_1}?
7. **a.** Find the value of V_A for the series circuit shown in Figure 4–40.
   - **b.** Assume that R_3 and R_4 are equal. What is the value of R_3?
   - **c.** What is the value of R_2?

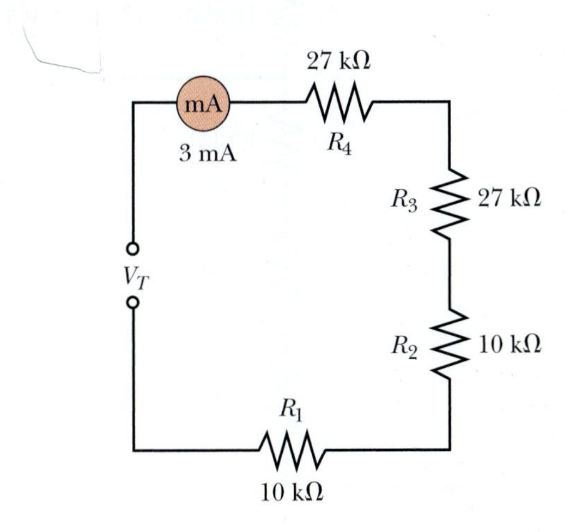

FIGURE 4–39

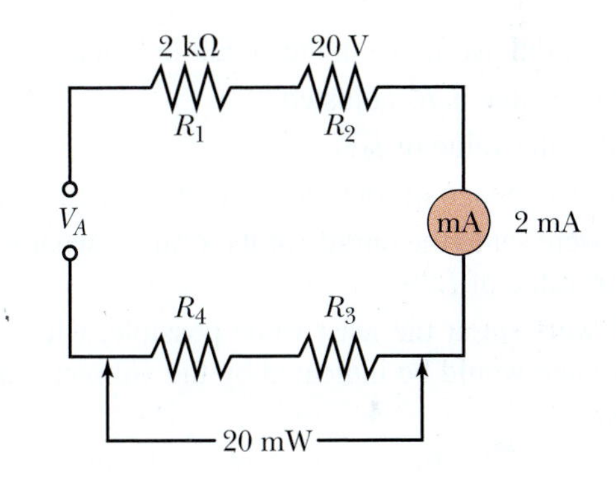

FIGURE 4–40

8. Draw a simple three-resistor series voltage divider that provides equal voltage division of a 180-V source and draws 2 mA of current.

Referring to Figure 4–41, calculate and answer questions 9–16:

9. What value of voltage will be read by the voltmeter?
10. What value of current will be read by the milliammeter?
11. What is the voltage drop across R_2?
12. What is the voltage drop across R_3?
13. What is the voltage drop across R_4?
14. If R_3 were to short out, what would be the voltage reading on the voltmeter?
15. Which resistor in the circuit of Figure 4–41 dissipates the most power? The least power?
16. If V_A doubled for the original circuit of Figure 4–41, would the percentage of applied voltage dropped by R_3 increase, decrease, or remain the same?
17. In a three-resistor series circuit, if V applied is 200 V, V_{R_1} is 40 V, and V_{R_2} is 90 V, what is the value of V_{R_3}?
18. In a three-resistor series circuit, if V_{R_1} is 25 V, V_{R_2} is 50 V, and V_{R_3} is twice V_{R_1}, what is the value of V applied?

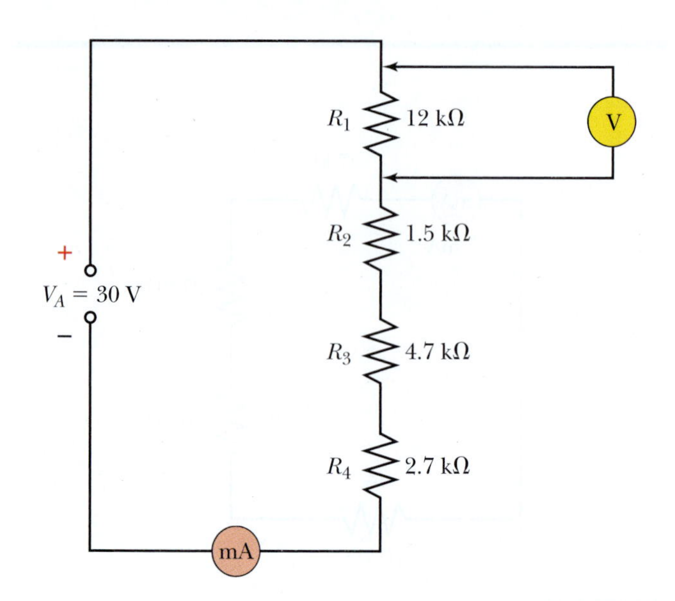

FIGURE 4–41 Review questions problem circuit

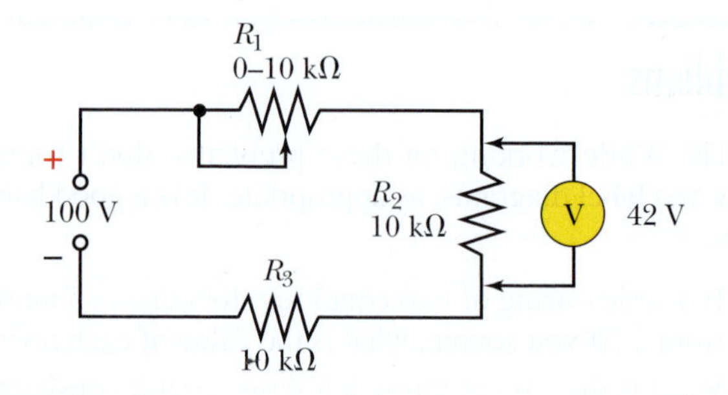

FIGURE 4–42

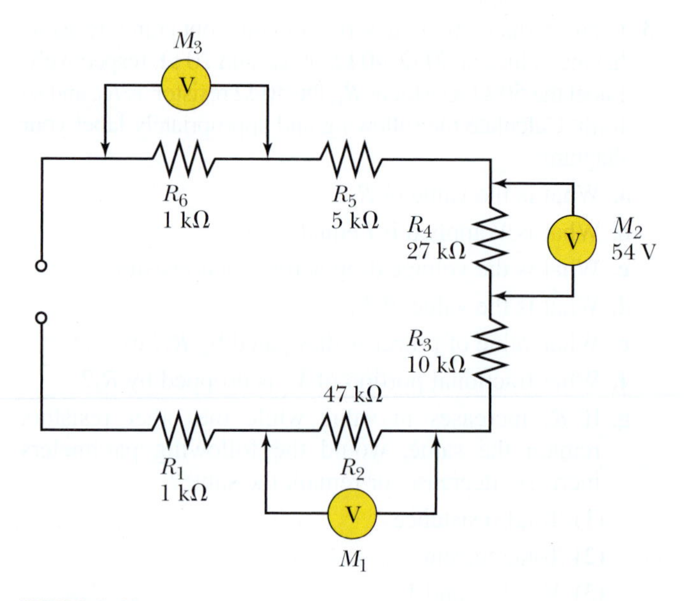

FIGURE 4–43

19. For the conditions shown in the circuit in Figure 4–42:

a. To what value is R_1 adjusted?

b. What is the value of I_T?

c. What percentage of total power is dissipated by R_1?

d. If R_1 were set to the middle of its R range, what would be the value of I_T?

e. If R_1 were set at the least value possible, what value of voltage would be indicated by the voltmeter in the circuit?

20. Find R_T and V_T for the circuit shown in Figure 4–43.

a. What value will meter M_1 indicate?

b. What value will meter M_3 indicate?

c. What is the value of I_T for this circuit?

d. If all R values remain the same, but P_T doubles, what is the new value of V_T? What is the new I_T?

21. Refer to Figure 4–43. If all R values are as shown, but P_{R_5} equals 45 mW:

a. What is the value of P_{R_2}?

b. What is the value of P_T?

c. What is the value of I_T?

22. Draw a circuit diagram of a series circuit where V_{R_3} is three times the value of V_{R_1} and V_{R_2} is twice V_{R_1}. Assume V applied is 60 V, R_T is 120 kΩ, and calculate:

a. value of R_1.

b. values of V_{R_1}, V_{R_2}, and V_{R_3}.

c. values of I and P_T.

23. Determine the value of a series-dropping resistor needed to drop a source voltage of 100 V to the appropriate value for a load rated as 50 V at 5 mA. Draw the circuit and label all components and electrical parameters.

Refer to Figure 4–43 to answer questions 24–28.

24. If R_5 were to be shorted out by a jumper wire, would the voltage drop across R_1 increase, decrease, or remain the same?

25. If R_2 increased in value, because of age, would the meter reading of meter M_3 increase, decrease, or remain the same?

26. If R_4 were shorted, would the circuit applied voltage increase, decrease, or remain the same?

27. With the parameters shown in Figure 4–43, determine the value of circuit applied voltage.

28. All other factors (parameters) remaining the same, if the power dissipated by R_3 quadruples, what must have happened to the value of circuit applied voltage and the circuit current? (Be specific!)

Analysis Questions

1. In your own words, explain the difference between calculator scientific-mode notation and calculator engineering-mode notation.
2. List at least two other possible applications for simple voltage dividers, other than those mentioned in this chapter.
3. In your own words, express the key facts you have learned in this chapter regarding the "divide-and-conquer" rule.
4. In the SIMPLER troubleshooting sequence of steps, the first step relates to gathering and analyzing symptom information. Name at least three tools and techniques that can help you in this effort.
5. In the SIMPLER troubleshooting method, how do you go about identifying the initial suspect area?
6. In the SIMPLER troubleshooting method, what are two decisions you must make before making each test as you proceed through the troubleshooting sequence.
7. In your own words, define what the term *bracketing* means in reference to the SIMPLER troubleshooting system.
8. In what type of circuit or system situation is the divide-and-conquer technique most appropriately used?
9. What type of documentation aid do you think would be most helpful when you are beginning to troubleshoot a circuit of any kind?
10. What precaution should be taken when troubleshooting a series circuit suspected of having a short? Briefly explain why this precaution should be taken.
11. If a series circuit's current suddenly drops to zero, what has happened in the circuit?
12. For the circuit of Figure 4–40, if certain components are overheating, circuit current has increased, and the voltage drops across R_2 and R_3 have drastically decreased, which of the following has occurred?
    a. The total circuit has been shorted.
    b. The total circuit has been opened.
    c. R_1 has shorted.
    d. R_2 and R_3 have shorted out.
    e. None of the above
13. What basic technique is often used in troubleshooting a defective series circuit that has a number of components in series?

MultiSIM Exercise for *Series Circuits*

1. Use the MultiSIM program and the circuit shown in Figure 4–24 for the following exercise.
2. Measure and record the following parameters. Record both the value and the polarity. (NOTE: Connect the multimeter negative lead to the ground reference point in the circuit when making the MultiSIM measurements.)
    a. Measure and record the voltage from point C to point D.
    b. Measure and record the voltage from point C to point A.
    c. Measure and record the voltage from point C to point B.
3. Do the results agree with the parameters shown in Figure 4–24?

Performance Projects Correlation Chart

Suggested performance projects in the Laboratory Manual that correlate with topics in this chapter are:

Chapter Topic	Performance Project	Project Number
Resistance in Series Circuits	Total Resistance in Series Circuits	9
Definition and Characteristics of a Series Circuit	Current in Series Circuits	10
Voltage in Series Circuits	Voltage Distribution in Series Circuits	11
Power in Series Circuits	Power Distribution in Series Circuits	12
Effects of Opens in Series Circuits and Troubleshooting Hints	Effects of an Open in Series Circuits	13
Effects of Shorts in Series Circuits and Troubleshooting Hints	Effects of a Short in Series Circuits Story Behind the Numbers: Series Circuits	14

NOTE: It is suggested that after completing the above projects, the student should be required to answer the questions in the "Summary" at the end of this section of projects in the Laboratory Manual.

Troubleshooting Challenge

CHALLENGE CIRCUIT 1

(Follow the SIMPLER sequence by referring to inside front cover.)

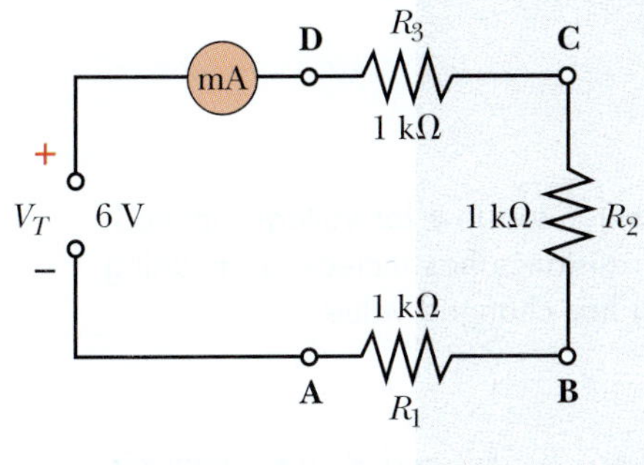

Challenge Circuit 1 multiSIM

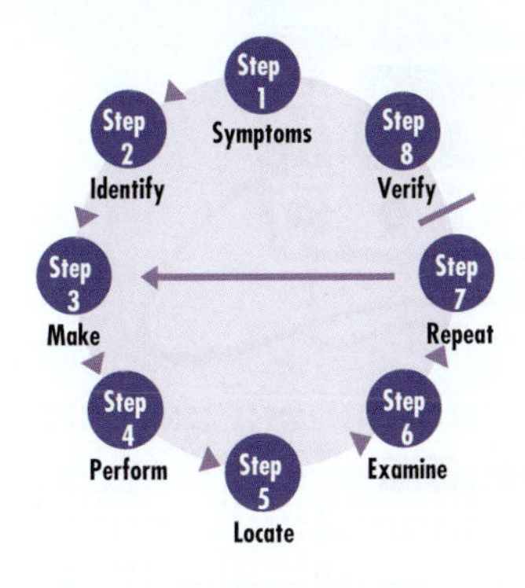

STARTING POINT INFORMATION

1. Circuit diagram
2. $V_T = 6$ V
3. Measured $I_T = 1.5$ mA

TEST	Results in Appendix C
$V_{A\text{–}B}$	(5)
$V_{B\text{–}C}$	(86)
$V_{C\text{–}D}$	(42)
Res. of R_1	(16)
Res. of R_2	(62)
Res. of R_3	(3)
R_T	(36)

CHALLENGE CIRCUIT 1

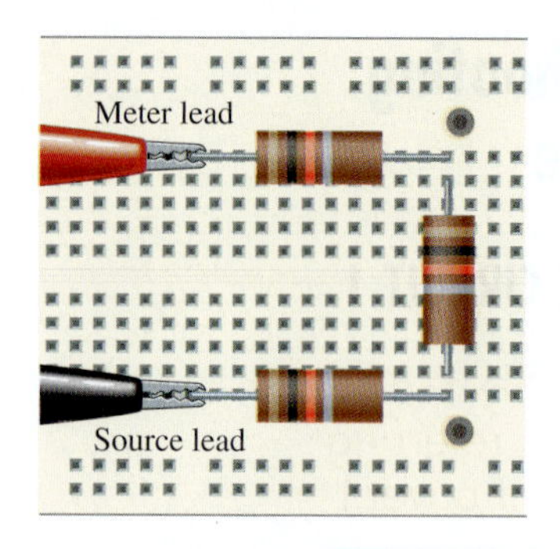

STEP 1

SYMPTOMS The total current is too low for voltage applied. This implies that the total resistance has increased, meaning that one or more resistor(s) has changed value.

STEP 2

IDENTIFY initial suspect area: R_1, R_2, and R_3 (i.e., total circuit).

STEP 3

MAKE test decision: Check voltage across R_2 (middle of circuit).

STEP 4

PERFORM **1st Test:** Look up the test result. $V_{B\text{–}C}$ is 1.5 V.

STEP 5

LOCATE new suspect area: R_1 and R_3. NOTE: $V_{B\text{–}C}$ would be greater than 2 V if R_2 had increased and the other resistors had not.

STEP 6

EXAMINE available data.

STEP 7

REPEAT analysis and testing:
2nd Test: Check voltage across R_1. $V_{A\text{–}B}$ is 1.5 V.
3rd Test: Check voltage across R_3. $V_{C\text{–}D}$ is 3 V.
4th Test: Disconnect the V source and check resistance of R_3. The result is that R_3 is 2 kΩ, which is abnormal.

STEP 8

VERIFY **5th Test:** Replace R_3 with a good 1-kΩ resistor and note the current. When this is done, the circuit checks out normal. Each resistor drops 2 V and the circuit current measures 2 mA.

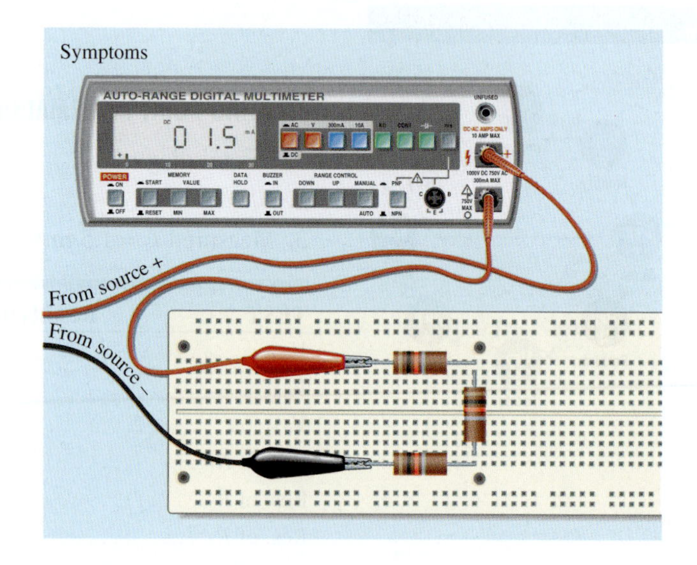

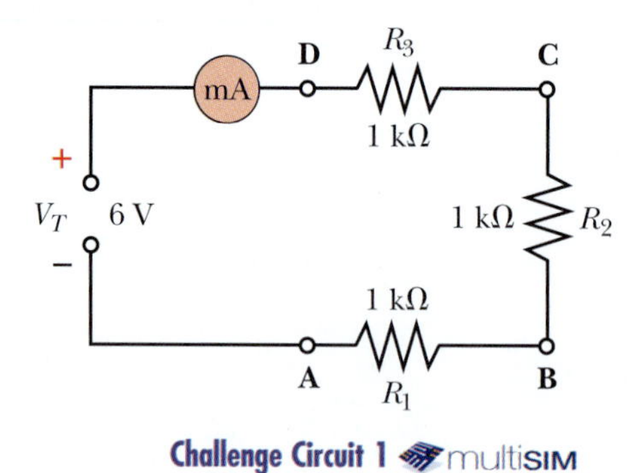

Challenge Circuit 1 multiSIM

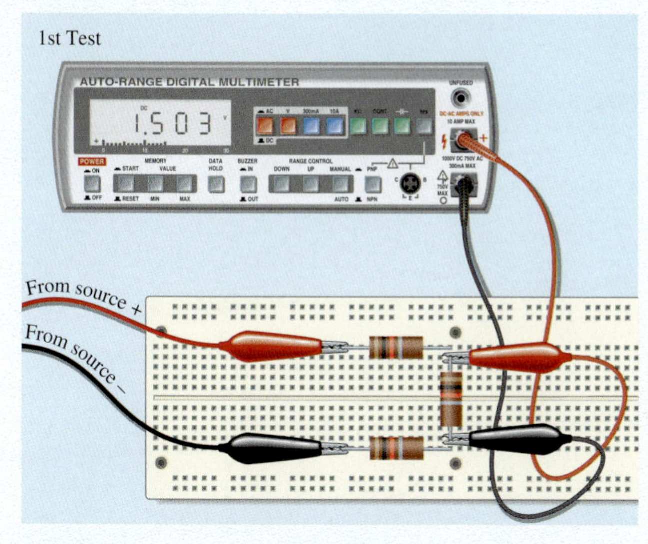
1st Test
AUTO-RANGE DIGITAL MULTIMETER
1.503
From source +
From source –

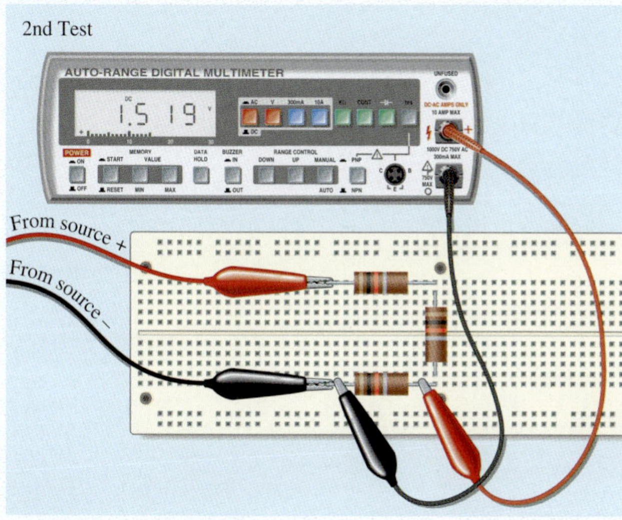
2nd Test
AUTO-RANGE DIGITAL MULTIMETER
1.519
From source +
From source –

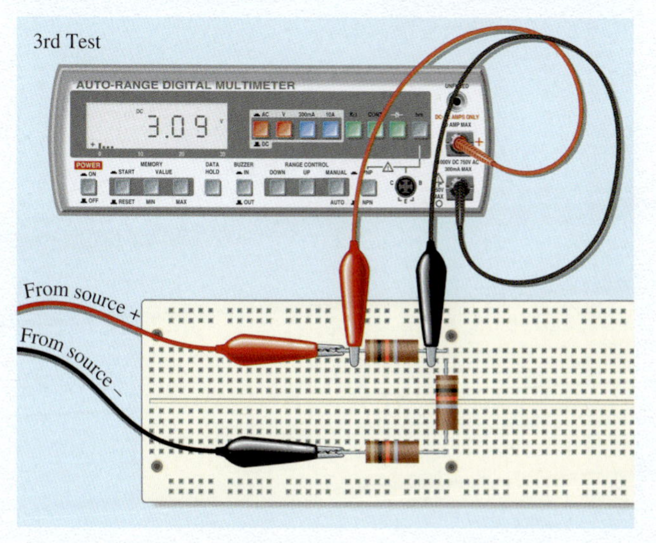
3rd Test
AUTO-RANGE DIGITAL MULTIMETER
3.09
From source +
From source –

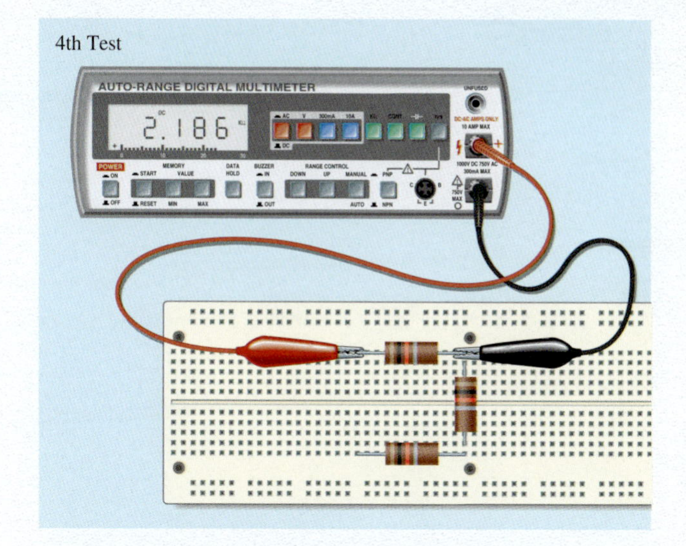
4th Test
AUTO-RANGE DIGITAL MULTIMETER
2.186

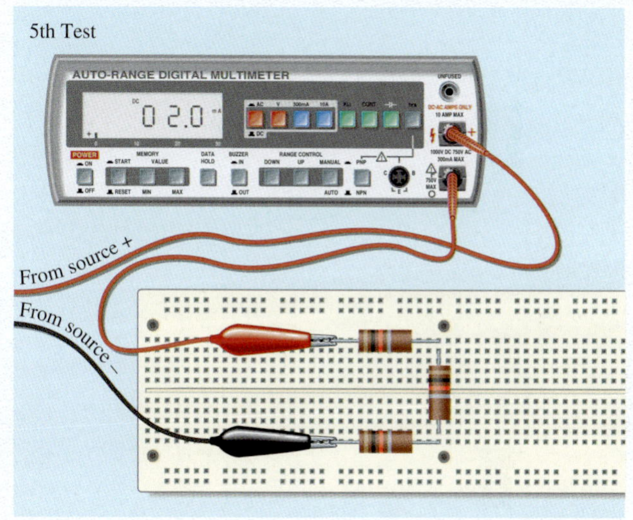
5th Test
AUTO-RANGE DIGITAL MULTIMETER
02.0
From source +
From source –

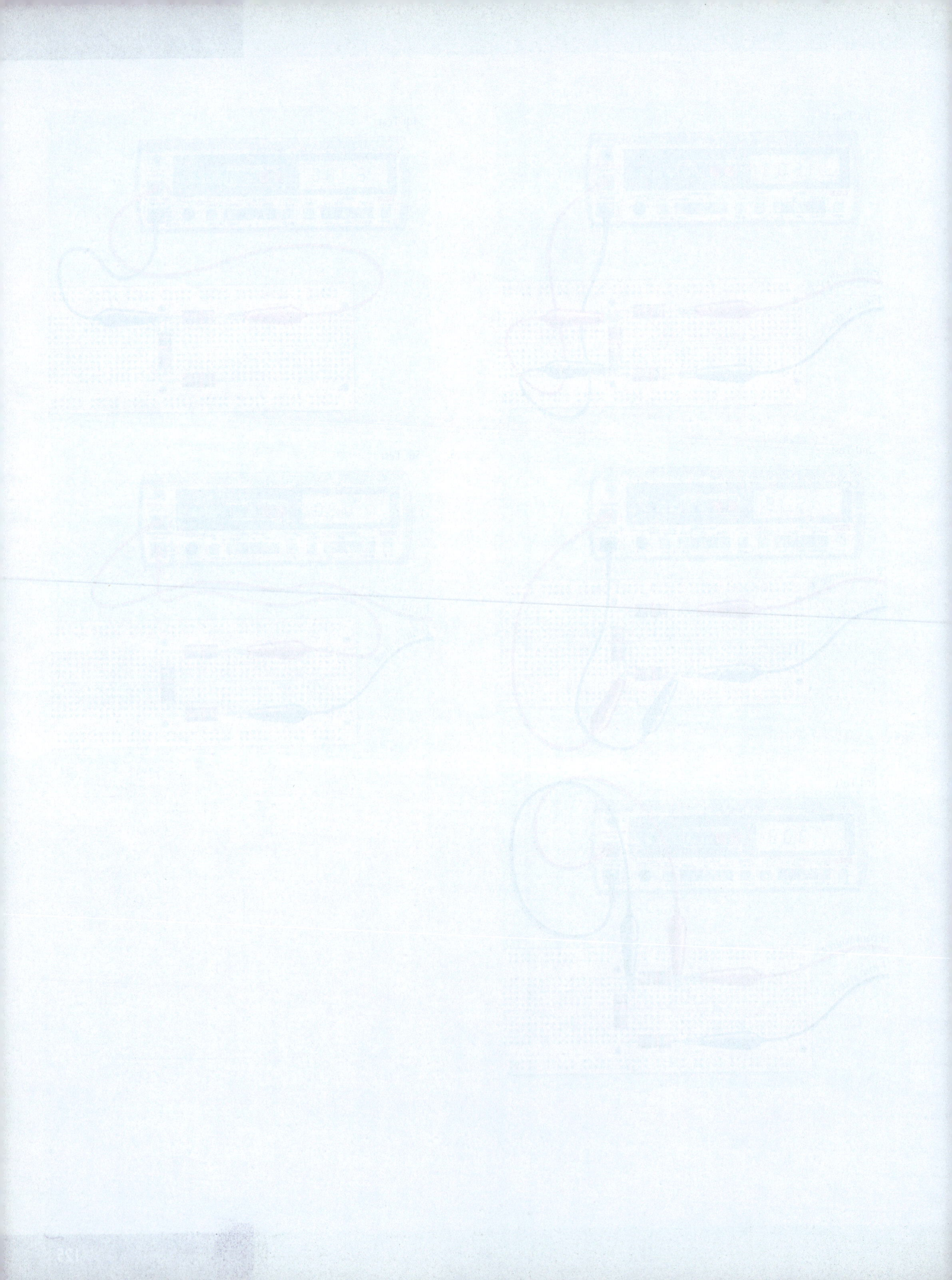

CHALLENGE CIRCUIT 2 (Series Circuit Block Diagram)

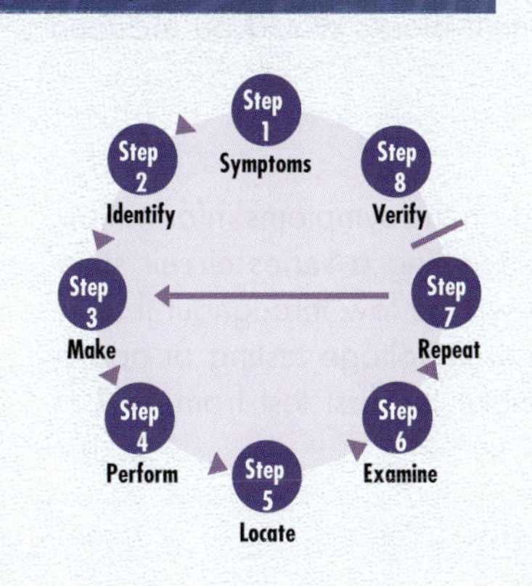

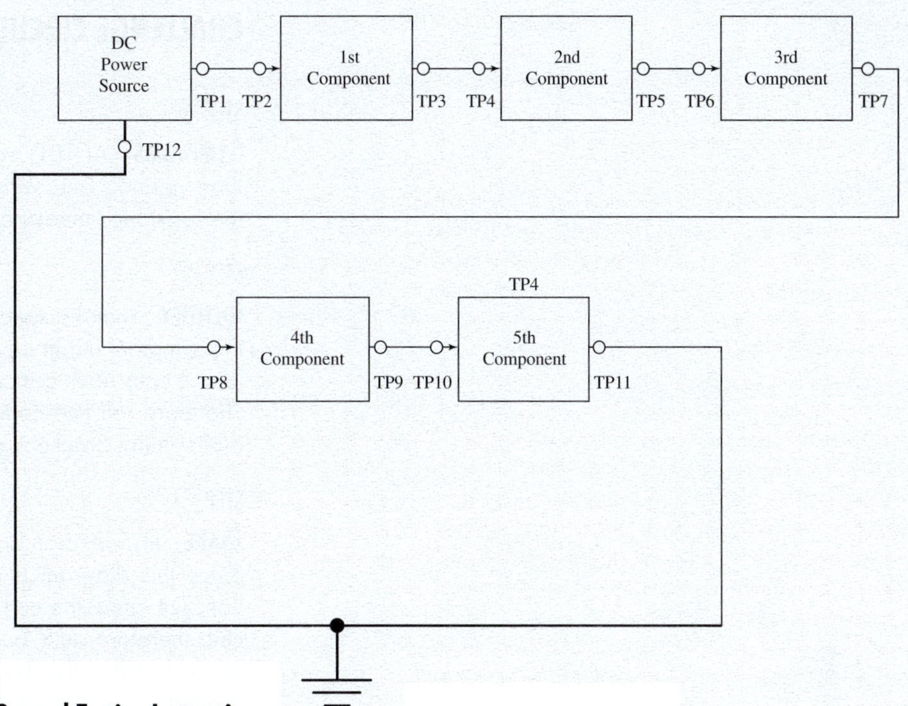

General Testing Instructions:

Measurement Assumptions:

I = at **TP**

V = from **TP** to ground

R = from **TP** to ground (with the power source disconnected from the circuit)

Possible Tests & Results:

Current (high, low, normal)

Voltage (high, low, normal)

Resistance (high, low, normal)

Starting Point Information	Test Points	Test Results in Appendix C *V*	*I*	*R*	*Signal*
At TP1:	TP2	(127	128	129	NA)
	TP3	(130	131	132	NA)
V = normal	TP4	(133	134	135	NA)
I = low	TP5	(136	137	138	NA)
R = high	TP6	(139	140	141	NA)
	TP7	(142	143	144	NA)
	TP8	(145	146	147	NA)
	TP9	(148	149	150	NA)
	TP10	(151	152	153	NA)
	TP11	(154	155	156	NA)
	TP12	(157	158	159	NA)

CHALLENGE CIRCUIT 2: SERIES CIRCUIT

STEP [1]

SYMPTOMS At TP1, voltage is normal, current is lower than normal, and with the power supply disconnected, the resistance measures higher than normal.

STEP [2]

IDENTIFY initial suspect area. Since this is a series circuit, the symptom might be caused by any one of the component blocks having increased in its opposition to current. Therefore, all five component blocks would be included in the initial bracketing.

STEP [3]

MAKE test decision based on the symptoms information. Since this diagram is representing a series circuit situation, we know the current will be low throughout the circuit; therefore, let's begin with voltage testing at appropriate test point(s). Let's make the first test from **TP4** to ground.

STEP [4]

PERFORM

First Test: Voltage from **TP4** to ground is slightly high.

STEP [5]

LOCATE new suspect area. In order for voltage to be high at this point **(TP4),** when we know current is lower than normal suggests that the R between **TP4** and ground must be higher than normal. (But which component is it?) It could be Component 2, 3, 4, or 5.

STEP [6]

EXAMINE available data.

STEP [7]

REPEAT analysis and testing.

Second Test: measure voltage at **TP6.**

Voltage at **TP6** reads slightly high.

Third Test: measure voltage at **TP8.**

Voltage at **TP8** is slightly lower than normal, suggesting that the third Component Block may contain the problem.

STEP [8]

VERIFY

Fourth Test: Replacing the 3rd Component Block with a new module that meets specifications causes the circuit current to increase back to its normal specified value.

OUTLINE

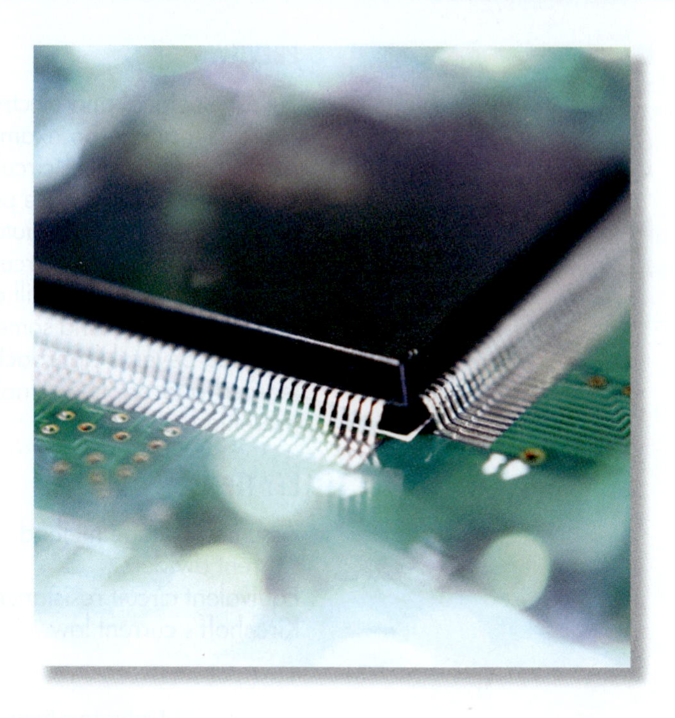

CHAPTER 5

Parallel Circuits

OBJECTIVES

After studying this chapter, you should be able to:

1. Define the term **parallel circuit**
2. List the characteristics of a parallel circuit
3. Determine voltage in parallel circuits
4. Calculate the total current and branch currents in parallel circuits
5. Compute total resistance and branch resistance values in parallel circuits using at least three different methods
6. Determine conductance values in parallel circuits
7. Calculate power values in parallel circuits
8. List the effects of opens in parallel circuits
9. List the effects of shorts in parallel circuits
10. Describe troubleshooting techniques for parallel circuits
11. Use **current divider** formulas
12. Use the computer to solve circuit problems
13. Use the SIMPLER troubleshooting sequence to solve Troubleshooting Challenge problems

ELECTRONICS INTO THE FUTURE

Explore an interactive presentation on Parallel Circuits on the accompanying CD.

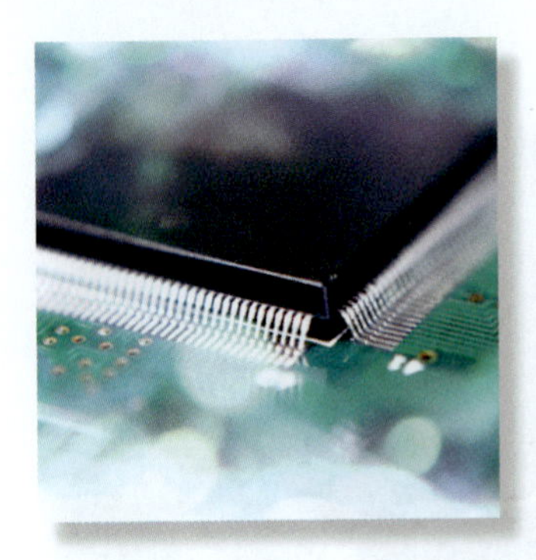

PREVIEW

Most electrical and electronic circuits, no matter how simple or complex, contain portions that can be examined using parallel circuit analysis. Of course, some circuits contain only parallel circuit arrangements. For example, the lights and wall outlets in your home generally use parallel circuitry to connect several lights or outlets on one circuit, Figure 5–1. Many automobile accessories (heaters and radios) are connected to the battery using parallel circuit methods, Figure 5–2.

In this chapter, you will again apply Ohm's law and the power formulas. Also, Kirchhoff's current law and some *vital contrasts* between parallel and series circuits will be examined. Several approaches to solve for total resistance in parallel circuits will be learned and applied. Finally, information on circuit troubleshooting techniques will be presented.

KEY TERMS

Assumed voltage method
Current dividers
Equivalent circuit resistance
Kirchhoff's current law
Parallel branch
Parallel circuit
Product-over-the-sum method

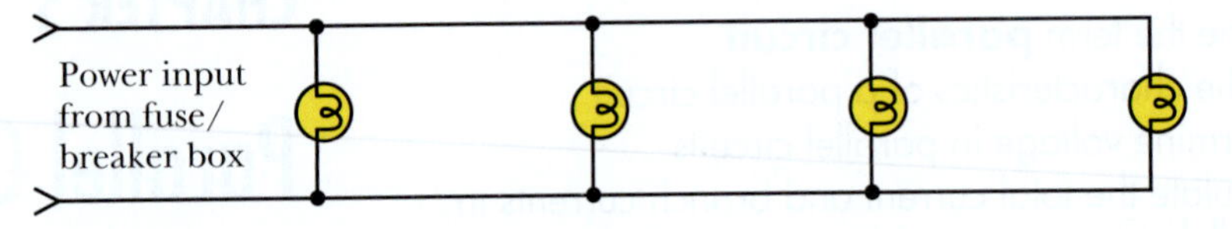

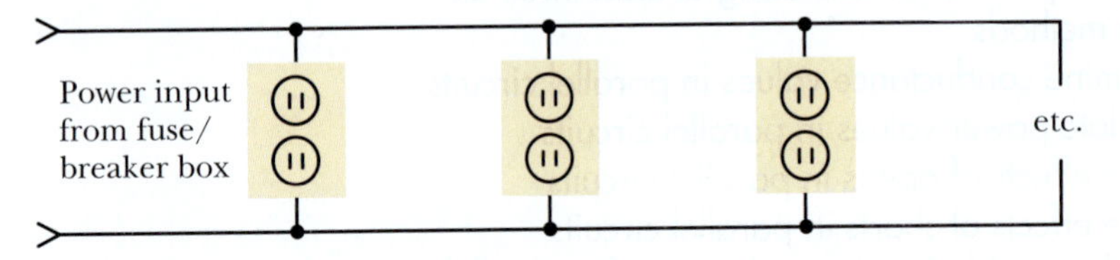

FIGURE 5–1 Parallel circuits are common in a home.

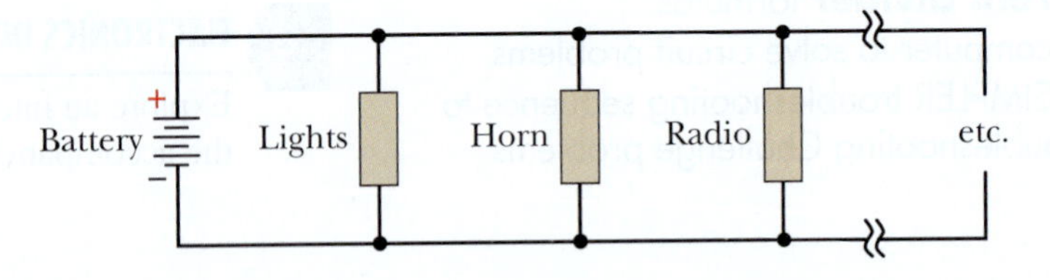

FIGURE 5–2 Automotive accessories are often connected in parallel across the battery. NOTE: Series switches in circuits are not shown.

5–1 Definition and Characteristics of a Parallel Circuit

Recall from the last chapter that the important characteristic defining a series circuit is that only one path for current exists. Therefore, the current through all components is the same. Because of this fact, the voltage distribution throughout the circuit (voltage across each of the components) depends on each component's resistance value (i.e., the highest R value dropping the highest voltage or the lowest R value dropping the lowest voltage).

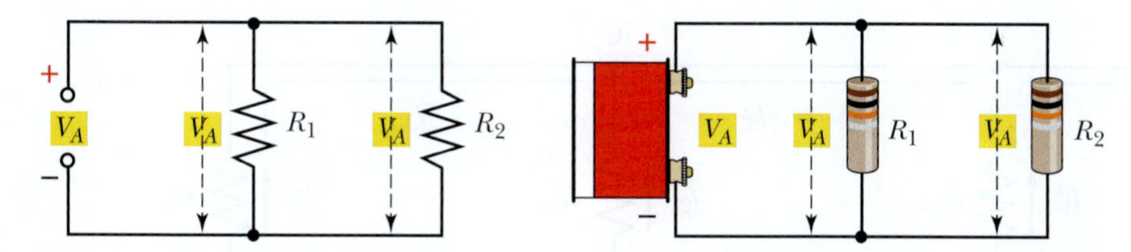

FIGURE 5–3 Voltage in parallel circuits is the same across all components, where $V_{R_1} = V_A$, and $V_{R_2} = V_{R_1} = V_A$. multiSIM

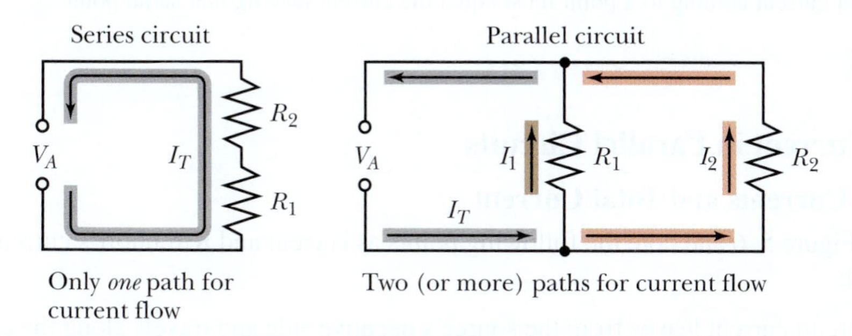

FIGURE 5–4 A basic difference between series and parallel circuits

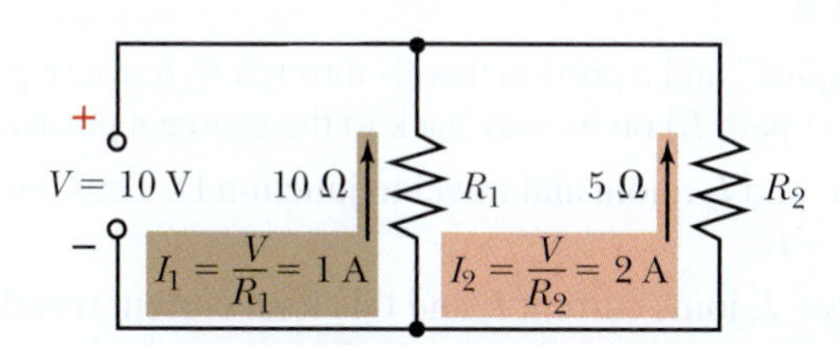

FIGURE 5–5 In parallel circuits, current through each branch is *inversely* proportional to each branch's resistance.

In contrast to the series circuit, the significant features of **parallel circuits** are (1) *the voltage across all parallel components must be the same,* Figure 5–3; and (2) *there are two or more paths (branches) for current flow,* Figure 5–4. Because the voltage is the same across each **parallel branch,** the current distribution throughout the circuit depends on each branch's resistance value. Each branch current is *inverse* to its resistance value, which means the higher the resistance value, the lower the branch current, Figure 5–5.

Thus, repeating the important parallel circuit descriptors, a parallel circuit is defined as one where the voltage across all (parallel branch) components is the same and where there are two or more branches for current flow.

5–2 Voltage in Parallel Circuits

If you know the voltage applied to a parallel circuit or the voltage across any one of the branches, you automatically know the voltage across any one or all of the parallel branches. This voltage is the same across all branches of a parallel circuit, Figure 5–3.

PRACTICE PROBLEMS 1

Again refer to Figure 5–3. If R_2 is 27 kΩ and current through R_2 is 2 mA, what is the value of V_{R_1}? Of V_A?

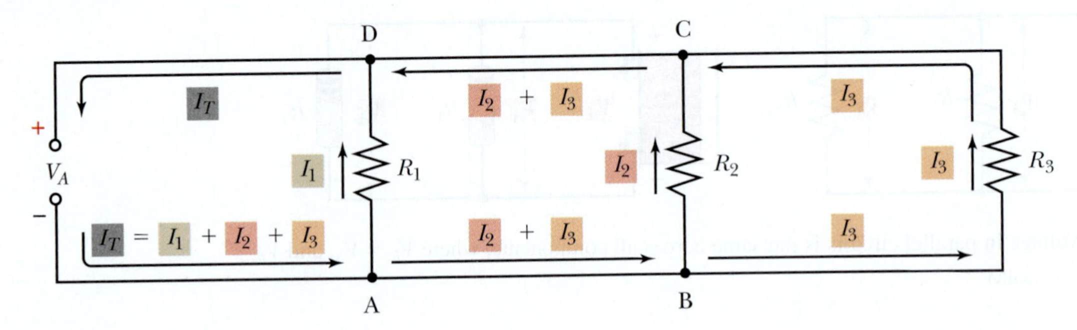

FIGURE 5–6 Kirchhoff's current law states that current coming to a point must equal the current leaving that same point.

5–3 Current in Parallel Circuits

Branch Currents and Total Current

Refer to Figure 5–6 and note the following points as current and Kirchhoff's current law are discussed:

- Total circuit current leaves from the source's negative side and travels along the conductor until it reaches the circuit "junction" at point A.
- At point A, a portion of the total current travels through branch resistor R_1 (current path I_1) on its way back to the source's positive side. The remaining portion travels on to the circuit junction at point B.
- At point B, current "splits" and a portion travels through R_2 (current path I_2) and the remainder through R_3 (current path I_3) on its way back to the source's positive side.
- At point C, currents I_2 and I_3 rejoin and travel to junction D. Thus, current value from point C to D must equal $I_2 + I_3$.
- At point D, current $I_2 + I_3$ joins current I_1 and this total current travels back to the source's positive side. NOTE: *The total current equals the sum of the branch currents.* This is an important fact; remember it! Mathematically:

FORMULA 5–1 $I_T = I_1 + I_2 + I_3 \ldots + I_n$

PRACTICE PROBLEMS 2

Assume a five-branch parallel circuit. What is the formula for I_T? Which resistor passes the most current? Which resistor passes the least current?

5–4 Kirchhoff's Current Law

The previous detailed statements express **Kirchhoff's current law.** Simply stated, this law is *the value of current entering a point must equal the value of current leaving that same point.*

In parallel circuit branch currents, since V across each branch is the same, the current divides through the branches in an inverse relationship to the individual branch resistances. That is, the highest R value branch allows the least current, and the lowest R value branch allows the highest branch current.

EXAMPLE In using Ohm's law to solve for branch currents, this fact becomes apparent, Figure 5–7. It suggests an interesting proportionality approach in thinking about branch currents in parallel circuits, Figure 5–8. You can see that if branch 1 has twice the resistance of branch 2, it will have half the current value through it. Conversely, branch 2 will have twice the current value of branch 1. If branch 1 were to have one-fourth the R value of branch 2, then it would have four times the current of branch 2, and so forth.

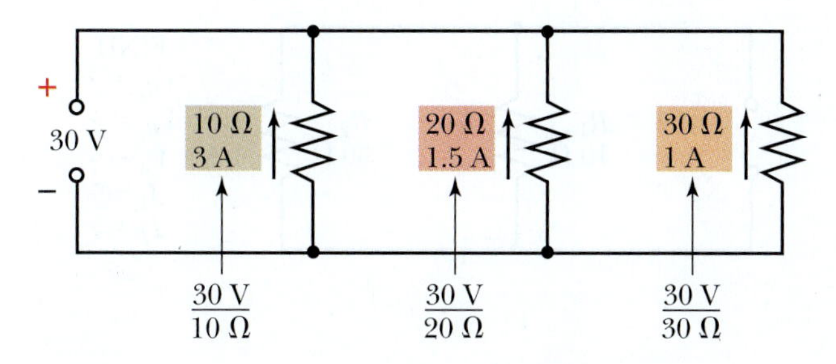

FIGURE 5–7 Branch current is inverse to branch R value. multiSIM

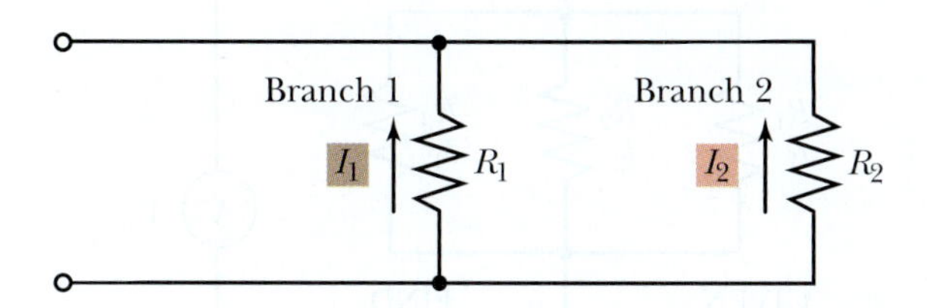

FIGURE 5–8 Looking at branch 1 via the proportionality approach: if R_1 has twice the R of R_2, then I_1 = half of I_2 (or $I_2 = 2 \times I_1$). If R_1 has one-fourth the R of R_2, then $I_1 = 4 \times I_2$ (or I_2 is one-fourth of I_1).

At this point, you have learned that voltage is the same across each parallel branch, current to and from any given circuit point must be the same, branch currents are inverse to their branch resistances, and total current equals the sum of the branch currents. It is time to see if you can apply this knowledge and Ohm's law and Kirchhoff's law to solve some problems. ◆

PRACTICE PROBLEMS 3

Look at Figure 5–9 and, without looking ahead, determine the values for V_{R_2}, V_{R_1}, V_A, I_1, and I_T. Now, read the following explanation.

To find V_2, use Ohm's law, where $V_2 = I_2 \times R_2 = 1\text{ A} \times 50\ \Omega = 50\text{ V}$. Since the voltage must be the same across all parallel branches, then V_1 and V_A must also equal 50 V in this circuit. Since you know V_1 and R_1, then Ohm's law will again enable you to find the answer for I_1. That is, I_1 must equal V_1/R_1, or 5 A. And I_T equals the sum of the branch currents; thus, 1 A + 5 A = 6 A.

Using Ohm's law, what must the value of R_T be? The answer is

$$R_T = \frac{V_T}{I_T} = \frac{50\text{ V}}{6\text{ A}} = 8.33\ \Omega$$

It is interesting to note the total circuit resistance is *less* than either branch's resistance. With respect to the power source, the two branch resistances of 10 Ω and 50 Ω could be replaced by one resistor with a value of 8.33 Ω and it would still supply the same total current and circuit power. Later in the chapter you will study several methods to determine total circuit resistance for parallel circuits. However, at this point, remember that *total resistance of parallel circuits is always less than the smallest value branch resistance.*

PRACTICE PROBLEMS 4

1. Refer to Figure 5–10 and with your knowledge of parallel circuits, solve for the unknowns. Do not cheat yourself by looking up the answers before working the problems.
2. Again referring to Figure 5–10, assume the parameters are changed to: $V_A = 125$ V, $I_T = 10.64$ mA, $I_1 = 4.63$ mA, and $R_3 = 56\text{ k}\Omega$. Find R_1, I_2, I_3, R_2, and the value of R_T.

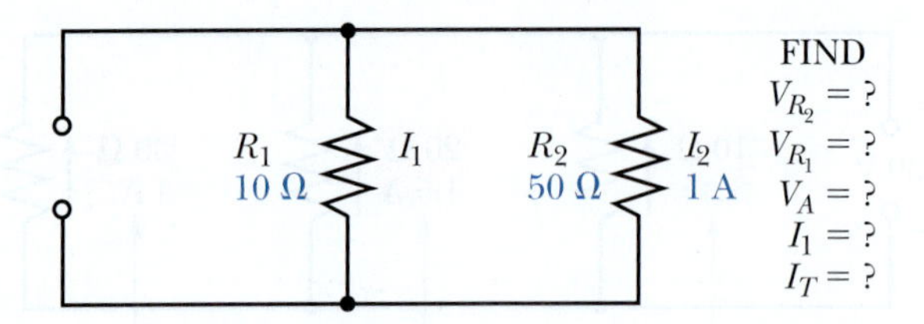

FIGURE 5–9

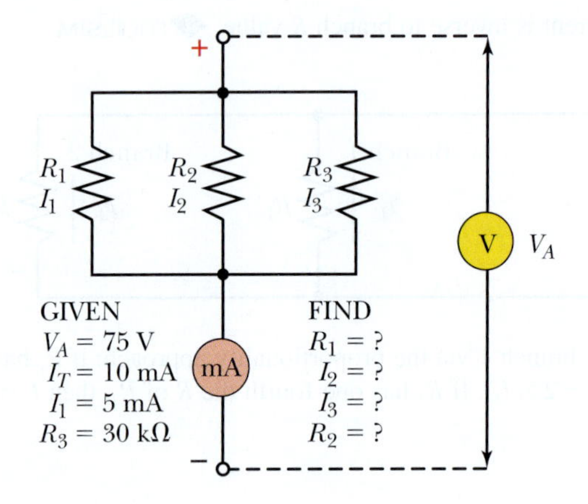

FIGURE 5–10

5–5 Resistance in Parallel Circuits

Again, total resistance of parallel circuits is always *less than the smallest value branch resistance.* Let us examine the reason for this and learn several useful ways to solve for total resistance in parallel circuits.

Since each parallel branch provides another current path where the voltage/power source must supply current, then the source must provide more current for each additional branch in a parallel circuit. Since the V is the same from the source, but the I supplied by the source is higher for every branch added in parallel, then Ohm's law ($R = V/I$) tells us that the total or **equivalent circuit resistance** must decrease every time another branch (or current path) is added to the parallel circuitry. For example, if a circuit comprises only the lowest R value branch, the total resistance equals that value. If *any* value R branch is added in parallel with that branch, another current path has been provided. Therefore, total current supplied by the source increases, source voltage remains constant, and total or equivalent circuit resistance decreases. Moving forward from that thought, let us look at several methods available to determine parallel circuit total resistance.

5–6 Methods to Calculate Total Resistance (R_T)

The Ohm's Law Method

Recall the formula $R_T = V_T/I_T$ to find total resistance.

EXAMPLE Look at Figure 5–11. The total resistance according to Ohm's law is

$$R_T = \frac{100 \text{ V}}{20 \text{ mA}} = 5 \text{ k}\Omega$$

A logical extension of this law for parallel circuits is that each branch resistance can be solved by knowing branch voltage and branch current. This is stated as

$$R \text{ branch} = \frac{V \text{ branch}}{I \text{ branch}}$$

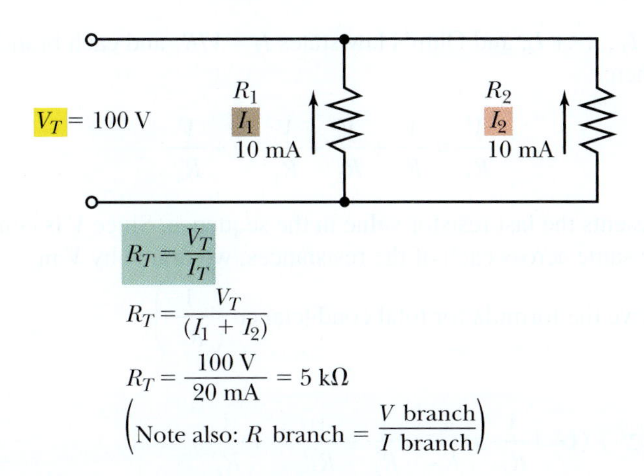

FIGURE 5–11 Ohm's law method to find R_T multiSIM

Again, refer to Figure 5–11. Since we know the voltage is the same across all branches and must equal V applied, then

$$R_1 = \frac{100\text{ V}}{10\text{ mA}} = 10\text{ k}\Omega$$

$$R_2 = \frac{100\text{ V}}{10\text{ mA}} = 10\text{ k}\Omega$$

$$R_T = \frac{100\text{ V}}{20\text{ mA}} = 5\text{ k}\Omega$$

Note this result of $R_T = 5\text{ k}\Omega$ agrees with our earlier statement that parallel circuit total resistance must be less than any one of the branch resistances in parallel.

Assume the following changes in parameters for the circuit of Figure 5–11. $V_T = 50$ V, $I_1 = 1.06$ mA, and $I_2 = 1.85$ mA. Again, total resistance can be found using Ohm's law. (NOTE: Rounding values to two decimal places, the answers will be close, but not absolutely precise in value.)

$$R_T = \frac{V_T}{I_T} = \frac{50\text{ V}}{(1.06\text{ mA} + 1.85\text{ mA})} = \frac{50\text{ V}}{2.91\text{ mA}} = 17.18\text{ k}\Omega$$

$$R_1 = \frac{V_1}{I_1} = \frac{50\text{ V}}{1.06\text{ mA}} = 47\text{ k}\Omega$$

$$R_2 = \frac{V_2}{I_2} = \frac{50\text{ V}}{1.85\text{ mA}} = 27\text{ k}\Omega$$

PRACTICE PROBLEMS 5

1. Again, refer to Figure 5–11 and apply Ohm's law. If $I_1 = 15$ mA and $R_1 = 2\text{ k}\Omega$, what are the values of V_{R_1} and V_T?
2. If $R_2 = 10\text{ k}\Omega$ and R_1 is $2\text{ k}\Omega$, as indicated in question 1, what is the value of R_T?

The Conductance-Related Method

Recall that resistance is the opposition shown to current flow and that conductance, conversely, is the ease with which current passes through a component or circuit. To calculate conductance, we use the reciprocal of resistance. That is $G = 1/R$, where *G is conductance in siemens* (S), and R is resistance in ohms (Ω).

In parallel circuits, total circuit current equals the sum of the branch currents. Since V is the same across all parallel branches, we can derive a method of calculating total circuit conductance (G_T or $1/R_T$) from branch conductances as shown in the following discussion.

Since $I_T = I_1 + I_2 \ldots + I_n$, and Ohm's law states $I_T = V/R_T$ and each branch current equals its V over its R, then:

$$\frac{V}{R_T} = \frac{V}{R_1} + \frac{V}{R_2} + \frac{V}{R_3} \cdots + \frac{V}{R_n}$$

Recall R_n represents the last resistor value in the sequence. Since V is common in parallel circuits and is the same across each of the resistances, we divide by V in each case and derive the formula for total conductance $\left(\frac{1}{R_T}\right)$.

FORMULA 5–2 $(G_T)\frac{1}{R_T} = \frac{1}{R_1} + \frac{1}{R_2} + \frac{1}{R_3} \cdots + \frac{1}{R_n}$

Once total conductance has been determined, it is very easy to determine the circuit's R_T by simply finding the reciprocal of total conductance, using the calculator's reciprocal function.

As shown in the following example, the reciprocal function of the calculator allows very easy determination of total circuit conductance by simply summing the reciprocals from each branch's resistance. Once total conductance is found, it is a simple matter to take the reciprocal of the total conductance in order to find the circuit total resistance. Follow the example to see how this method works.

◆ **EXAMPLE** If a four-resistor parallel circuit consists of resistors having values of 22 Ω, 39 Ω, 56 Ω, and 82 Ω, the total resistance of the circuit may be found, using a calculator, as follows:

Step 1: Clear the calculator by pressing the AC/ON button. The readout should show 0.

Step 2: Enter 22, then press the 1/x button, then the + button. The readout should show 0.04545.

Step 3: Enter 39, then the 1/x button, then the + button. The readout should show 0.07109.

Step 4: Enter 56, then the 1/x button, then the + button. The readout should show 0.0889.

Step 5: Enter 82, then the 1/x button, then the = button. The readout should show 0.1011. This number is the $\frac{1}{R_T}$ value, or G_T (total conductance).

Step 6: Press the 1/x button to find the circuit total resistance. $R_T = 9.886\ \Omega$ (rounded to two places, $R_T = 9.89\ \Omega$, or approximately 10 Ω). ◆

(Is this less than the smallest value branch R? Yes, and it does seem reasonable, using the mental approximating technique.) Again, R equivalent is less than the smallest value branch resistance.

A formula that states the 6 steps you have just performed is shown here.

FORMULA 5–3 $R_T = \dfrac{1}{\frac{1}{R_1} + \frac{1}{R_2} + \frac{1}{R_3} \cdots + \frac{1}{R_n}}$

Practical Notes

As you have seen in the example, the reciprocal of the branch conductances method is very easy with calculators. Practice this method until it is second nature to you. In most cases, it is the easiest method to use.

PRACTICE PROBLEMS 6

1. Assume a parallel circuit has branch resistances of 10 kΩ, 27 kΩ, and 20 kΩ, respectively. Use the conductance-related method and find R_T.
2. Assume a parallel circuit has branch resistances of 1.5 kΩ, 3.9 kΩ, 4.7 kΩ, and 5.6 kΩ, respectively. Again, use the conductance-related method and find the circuit R_T.

FORMULA 5–4 $G_T = G_1 + G_2 + G_3$

In other words, $\frac{1}{R_T}(\text{or } G_T) = \frac{1}{R_1} + \frac{1}{R_2} + \frac{1}{R_3}$

PRACTICE PROBLEMS 7

1. Use the conductance-related method and confirm the calculations shown in Figure 5–12.
2. Refer to Figure 5–13, and use the conductance-related method to find the circuit total resistance.

The Product-Over-the-Sum Method

One popular method to solve for total resistance in parallel circuits is the **product-over-the-sum method.** This method is limited to solving total resistance values for two-resistor circuits, or when circuits have more than two resistors, solving them, two resistors at a time. The mathematical formula is derived from the basic reciprocal resistance (conductance) formulas.

That is:

$$G_T = \frac{1}{R_T} = \frac{1}{R_1} + \frac{1}{R_2}$$

Since resistance is the reciprocal of conductance, the formula can be restated as:

$$R_T = \frac{1}{G_T}, \text{ or } R_T = \frac{1}{\frac{1}{R_1} + \frac{1}{R_2}}$$

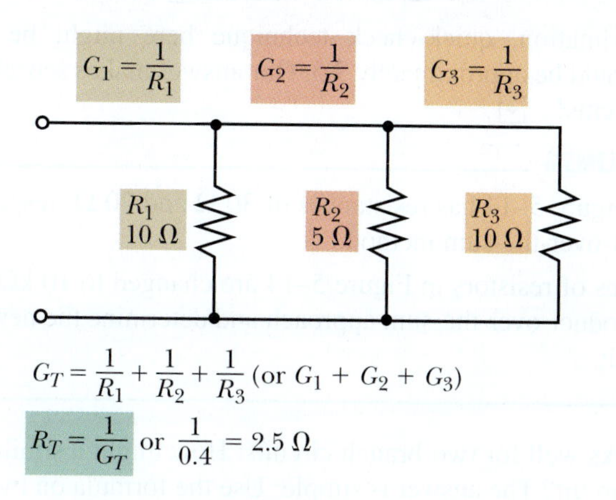

FIGURE 5–12 The conductance method to find R_T multiSIM

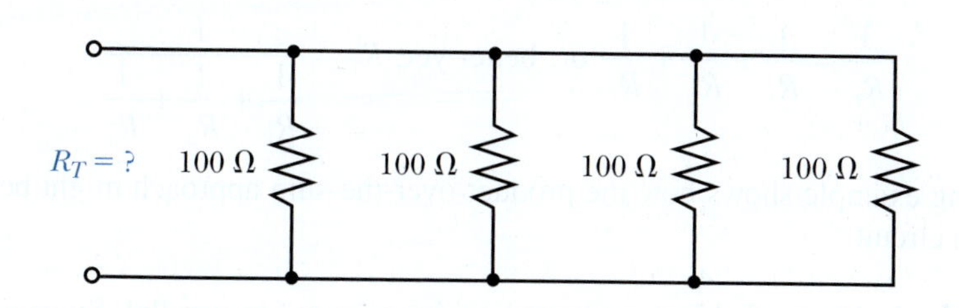

FIGURE 5–13

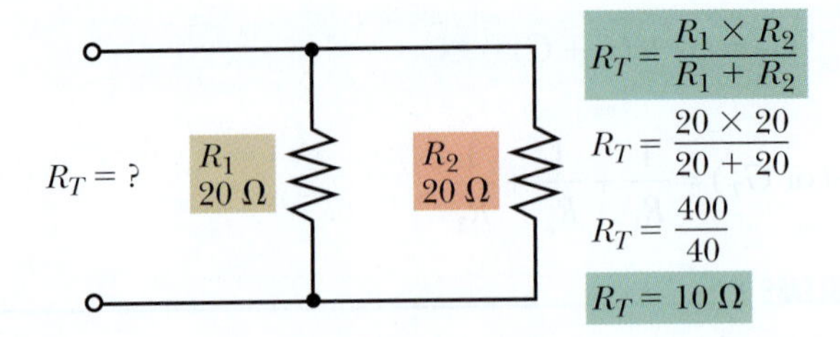

FIGURE 5–14 The product-over-the-sum method to find R_T

From this, the following formula may be derived:

FORMULA 5-5 $R_T = \frac{R_1 \times R_2}{R_1 + R_2}$

EXAMPLE An illustration of this formula is in the two-resistor parallel circuit of Figure 5–14.

What would be the total circuit resistance if $R_1 = 100$ ohms and $R_2 = 50$ ohms? In this case, the product is $100 \times 50 = 5{,}000$, and the sum is $100 + 50 = 150$. Therefore, $R_T = 5{,}000/150$, and total resistance = 33.33 ohms. This product-over-the-sum method works for any two branch values.

EXAMPLE Let us assume the resistor values in Figure 5–14 are changed so that one resistor's value is 3.9 kΩ and the other resistor's value is 6.8 kΩ.

$$R_T = \frac{R_1 \times R_2}{R_1 + R_2} = \frac{(3.9\times10^3)\times(6.8\times10^3)}{(3.9\times10^3)+(6.8\times10^3)} = \frac{26.52\times10^6}{10.7\times10^3}$$

$$R_T = 2.48\times10^3 \text{ or } 2.48\ \text{k}\Omega$$

Our mental approximation, quick-check technique here might be $4 \times 7 = 28$, and $28 \div 4 + 7$ (or 11) would be approximately 2.5. Our answer makes sense! Now you try a couple of practice problems!

PRACTICE PROBLEMS 8

1. If the circuit in Figure 5–14 has resistances of 30 Ω and 20 Ω, respectively, solve for R_T using the product-over-the-sum method.
2. Assume the values of resistors in Figure 5–14 are changed to 10 kΩ and 33 kΩ, respectively. Use the product-over-the-sum approach and determine the new value of R_T (or R_e) for the new circuit.

This method works well for two-branch circuits. How might a similar technique be used for a three-branch circuit? The answer is simple: Use the formula on two branches, *then use the results* from that calculation as one equivalent resistance in parallel with the third resistor, and repeat the process. Of course, you can also use the reciprocal or conductance approach, which was recommended earlier, where

$$\frac{1}{R_T} = \frac{1}{R_1} + \frac{1}{R_2} + \frac{1}{R_3} \text{ or, better yet, } R_T = \frac{1}{\frac{1}{R_1} + \frac{1}{R_2} + \frac{1}{R_3}}$$

The following example shows how the product-over-the-sum approach might be used for a three-branch circuit.

EXAMPLE In Figure 5–15, three branch resistances are in parallel. Since each branch has a resistance of 3 Ω, what is the circuit total resistance?

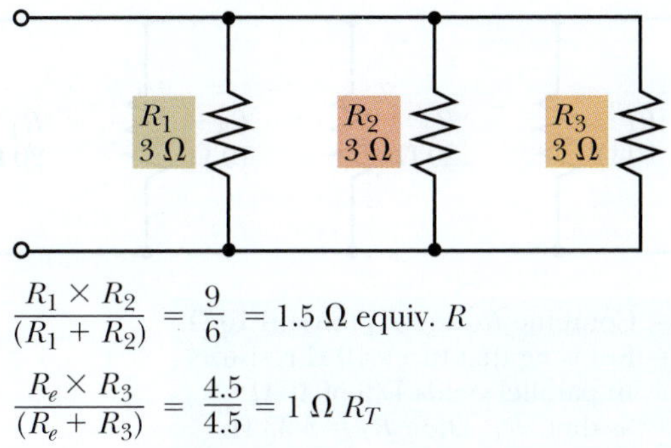

R_1 3 Ω R_2 3 Ω R_3 3 Ω

$$\frac{R_1 \times R_2}{(R_1 + R_2)} = \frac{9}{6} = 1.5\ \Omega \text{ equiv. } R$$

$$\frac{R_e \times R_3}{(R_e + R_3)} = \frac{4.5}{4.5} = 1\ \Omega\ R_T$$

FIGURE 5–15 Using product-over-the-sum method with more than two branches multiSIM

Using the product-over-the-sum formula for the first two branches yields:

$$\frac{R_1 \times R_2}{R_1 + R_2} = \frac{3 \times 3}{3+3} = \frac{9}{6} = 1.5\ \Omega$$

indicating that the equivalent resistance of branches 1 and 2 = 1.5 Ω. However, there is another branch involved! The equivalent resistance of 1.5 Ω is in parallel with branch R_3 (3 Ω). Repeating the product-over-the-sum formula for this situation gives:

$$\frac{R_e \times R_3}{R_e + R_3} = \frac{1.5 \times 3}{1.5+3} = \frac{4.5}{4.5} = 1\ \Omega$$

Note that using the reciprocal approach would also yield:

$$\frac{1}{R_T} = \frac{1}{3} + \frac{1}{3} + \frac{1}{3} = \frac{3}{3} = \frac{1}{1} = 1\text{ S; and } \frac{1}{G_T} = \frac{1}{1} = 1\ \Omega$$

PRACTICE PROBLEMS 9

If the resistances in Figure 5–15 are 47 Ω, 100 Ω, and 180 Ω, respectively, what is the value of R_T? Use the product-over-the-sum approach to calculate.

What would happen if there were four branches? To find the equivalent resistance of any two branches, use the product-over-the-sum formula. Then find the equivalent resistance of the other two branches in the same way, and use the product-over-the-sum formula for the two equivalent resistances to find total circuit resistance.

A Useful Simplifying Technique

You have undoubtedly observed in Figure 5–14, which was described earlier, and had two 20-Ω resistors in parallel, that the total $R = 10\ \Omega$. This number is half the value of one of the equal branch resistances. What do you suppose the total resistance of a three-branch parallel circuit is if the circuit has three equal branches of 30 Ω each? If you said 10 Ω again, you are right. That is, two equal parallel branches are equivalent to half the R value of one of the equal branches, three equal branches are equivalent to one-third the value of one of the equal branches, and so on. The idea is to mentally *combine* equal resistance branches to find their equivalent resistance. Doing this mental process reduces the complexity of the problem.

Look at Figure 5–16. By realizing that R_2 and R_4 can be combined as an equivalent resistance of 10 Ω, then combined with the two remaining 10-Ω branches, you can see the total circuit resistance will be one-third of 10 Ω, or 3.33 Ω.

This same concept can be used whenever branches of a multi-branched parallel circuit have equal values. Combine equal branches in a way to simplify to the fewest remaining branches. Then use the most appropriate method to solve for total circuit resistance, such as the product-over-the-sum method, reciprocal method, Ohm's law, and so on.

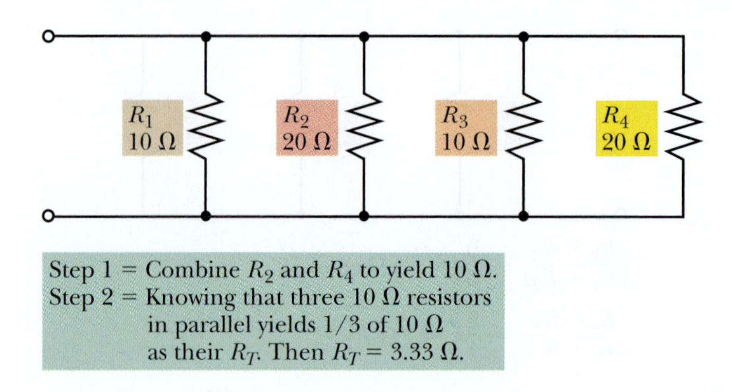
R_1
10 Ω
R_2
20 Ω
R_3
10 Ω
R_4
20 Ω
Step 1 = Combine R_2 and R_4 to yield 10 Ω.
Step 2 = Knowing that three 10 Ω resistors in parallel yields 1/3 of 10 Ω as their R_T. Then R_T = 3.33 Ω.

FIGURE 5–16 A useful simplifying technique
multiSIM

NOTE: The unequal values of resistances used in many circuits do not readily lend themselves to using this simplifying technique. However, whenever you can use it to help, do it! It will simplify the computation efforts.

The Assumed Voltage Method

One technique to solve for total resistance of parallel circuits containing miscellaneous "mismatched" resistance values is the **assumed voltage method.** You'll often find problems of this sort on FCC radio licensing tests.

The technique uses a two-step approach:

1. Assume an uncomplicated or easy circuit applied voltage value to determine each branch current, should the "assumed voltage" actually be applied to the circuit. For example, choose a value that is an even multiple of all the branch resistor values (the "least common denominator" value).
2. Determine the branch currents by dividing the assumed voltage by each branch R value. Then add branch currents to find total current. Now *divide "assumed voltage" by calculated total current to find total circuit resistance.*

◆ **EXAMPLE** Refer to Figure 5–17. Following the preceding steps, the solution for this problem is as follows:

1. Assume 300 V applied voltage since it is easily divided by each of the branch R values.
2. Solve each branch current using Ohm's law. Branch currents are 12 A, 6 A, 3 A, and 2 A, respectively.

   Total current = sum of branch currents = 23 A

   $$\text{Thus, } R_T = \frac{300\text{ V}}{23\text{ A}} = 13.04\ \Omega$$ ◆

In essence, this technique often allows you to mentally perform the first step without paper and pencil to find the total current. All that remains is one division problem, V_T/I_T, to find total circuit resistance.

In Figure 5–17, if the R values were changed to 10, 13, 20, and 26 Ω, respectively, what would be the total circuit resistance? Try using the assumed voltage method to find the answer.

Your answer should be 3.77 Ω. One value of voltage that could have been assumed is 260 V, since it is evenly divisible by all the R values in the circuit. NOTE: It does not matter if you assumed another value of voltage applied, your answer will be the same!

A Useful Parallel Circuit Resistance Design Formula

As a technician, there may be times when you will want to purposely decrease the resistance of an existing circuit by placing a resistor in parallel with it. Or you may need a resistor of some unavailable specific value. In either case, you may have a number of resistors, but

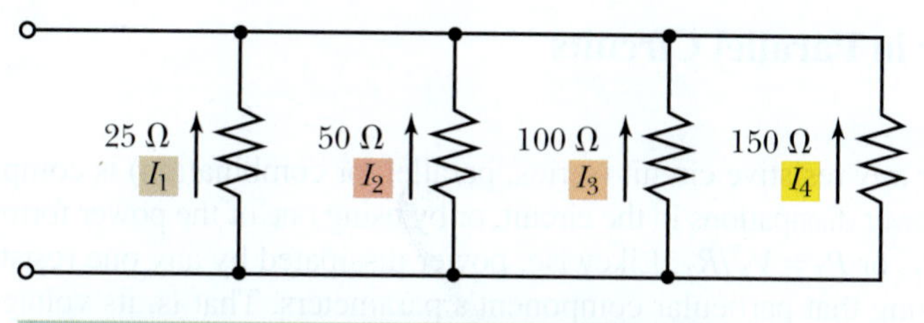

$I_1 = 12$ A $\quad I_2 = 6$ A $\quad I_3 = 3$ A $\quad I_4 = 2$ A

I_T = sum of branch currents = 23 A

$$R_T = \frac{300\text{ V}}{23\text{ A}} = 13.04\ \Omega$$

FIGURE 5–17 The assumed voltage method to find R_T

none are the right value. This happens frequently! How can we determine the unknown resistor value to be paralleled with a given (known) value to achieve the desired resultant resistance? (NOTE: We interchangeably use the terms R_T [R total] and R_e [R equivalent] to indicate the resultant total resistance of resistors in parallel. Be familiar with both terms as you will see them used frequently.)

The answer is a formula derived from the product-over-the-sum method:

FORMULA 5–6 $$R_u = \frac{R_k \times R_e}{R_k - R_e}$$

Where $R_u = R$ unknown

R_e = (desired) equivalent R of resistors in parallel.

R_k = known R value that will be placed in parallel with R unknown achieving the desired R equivalent.

EXAMPLE What value of R is needed in parallel with a 10-ohm resistor to achieve a total resistance of 6 ohms? See the solution in Figure 5–18.

PRACTICE PROBLEMS 10

Use this technique and solve the problem in Figure 5–19.

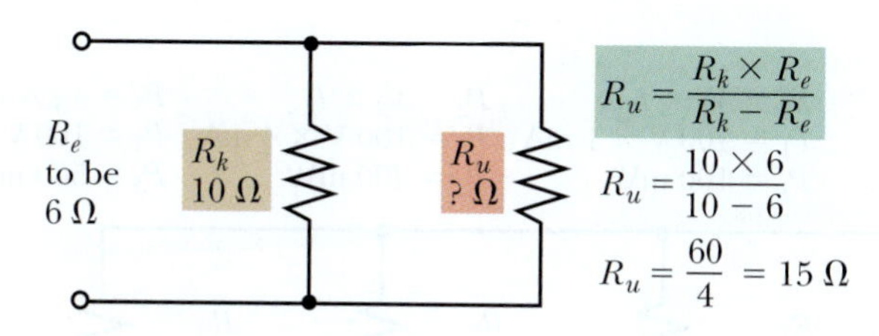

R_e to be 6 Ω
R_k 10 Ω
R_u ? Ω
$R_u = \frac{R_k \times R_e}{R_k - R_e}$
$R_u = \frac{10 \times 6}{10 - 6}$
$R_u = \frac{60}{4} = 15\ \Omega$

FIGURE 5–18 A practical circuit design formula

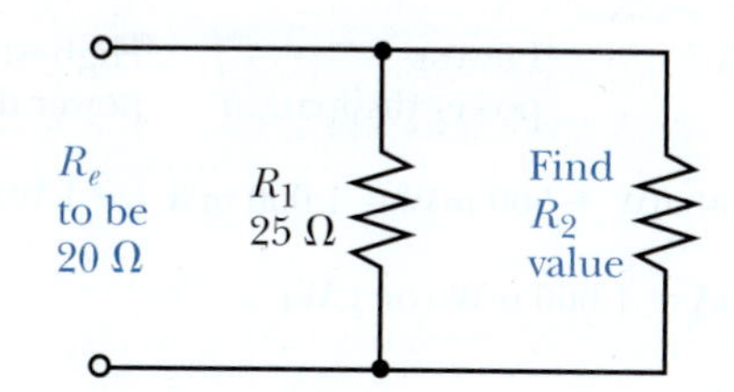

FIGURE 5–19

5–7 Power in Parallel Circuits

Total Power

Total power for *any* resistive circuit (series, parallel, or combination) is computed by adding all individual power dissipations in the circuit, or by using one of the power formulas, $P_T = V_T \times I_T$; $P_T = I_T^2 \times R_T$; or $P_T = V_T^2/R_T$. Likewise, power dissipated by any one resistive component is calculated using that particular component's parameters. That is, its voltage drop, current through it, and its resistance value.

Look at the circuit in Figure 5–20 as you read the next section about power in parallel circuits.

Power Dissipated by Each Branch

To find a branch power dissipation, use the parameters applicable to that branch. In the case of branch 1 (R_1), $P = 100\text{ V} \times I_1$. To find I_1, divide branch V (100 V) by branch R (100 kΩ). The answer is 1 mA. Thus, $P_1 = 100\text{ V} \times 1\text{ mA} = 100\text{ mW}$.

$$\text{Branch 2: Branch } I = \frac{100\text{ V}}{25\text{ k}\Omega} = 4\text{ mA}$$

$$\text{Branch } P = 100\text{ V} \times 4\text{ mA} = 400\text{ mW}$$

$$\text{Branch 3: Branch } I = \frac{100\text{ V}}{20\text{ k}\Omega} = 5\text{ mA}$$

$$\text{Branch } P = 100\text{ V} \times 5\text{ mA} = 500\text{ mW}$$

◆ **EXAMPLE** If we added all the individual power dissipations in the circuit in Figure 5–20, the total power = 100 mW + 400 mW + 500 mW = 1 W. Using the power formula, since V_T equals 100 volts and I_T equals 10 mA, the total power is 100 V × 10 mA = 1,000 mW (or 1 W). In either case, the total power supplied to and dissipated by the circuit is calculated as 1 W. ◆

PRACTICE PROBLEMS 11

What is the total power in the circuit of Figure 5–20 if $V_T = 150\text{ V}$?

Relationship of Power Dissipation to Branch Resistance Value

Refer again to Figure 5–20 and note that the *lowest* branch R value dissipates the *most* power; the *highest* R branch value dissipates the *least* power. This is because V is the same for all

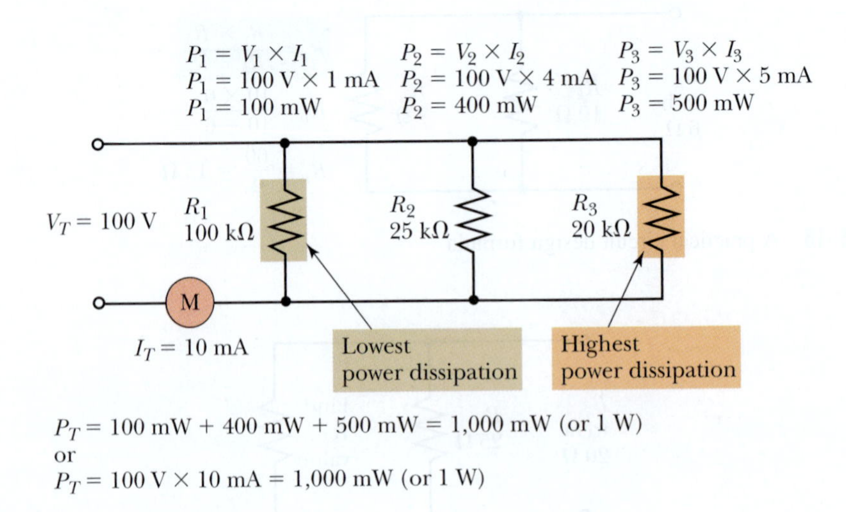

FIGURE 5–20 Power in parallel circuits

branches, and current through each branch is *inverse* to its R value. **This is the opposite of what you learned about series circuits.** In *series circuits,* the largest value R dissipates the most power because its I is the same as all others in the circuit; however, its V is directly proportional to its R value.

Soon we will look at other interesting similarities and differences between series and parallel circuits that you should learn. However, before we do that, let us look at some variations from the norm when parallel circuits develop opens or shorts.

5–8 Effects of Opens in Parallel Circuits and Troubleshooting Hints

Opens

Analysis of one or more opened branches in a parallel circuit is simple. Refer to Figure 5–21.

1. Normal operation for this circuit is: $V_T = 100$ V; $I_T = 40$ mA; $P_T = 4{,}000$ mW (4 W); each branch $I = 10$ mA; and each branch $P = 1{,}000$ mW (1 W).
2. If branch 2 opens (R_2), then total current decreases to 30 mA, since there is no current through branch 2. Under these conditions, the following are true:
   - Total resistance increases.
   - Total current decreases (30 mA rather than 40 mA).
   - Total voltage and branch voltages remain unchanged.
   - Total power decreases (100 V × 30 mA = 3,000 mW, or 3 W).
   - Current through the good (unopened) branches remains unchanged because both their V and R values stay the same.
   - Also, since voltage and current through unopened branches are unaffected, individual branch power dissipations stay the same, *except* for the opened branch or branches.
   - Since current decreases to zero through the opened branch, its power dissipation also drops to zero. The voltage across the opened branch, however, stays the same.

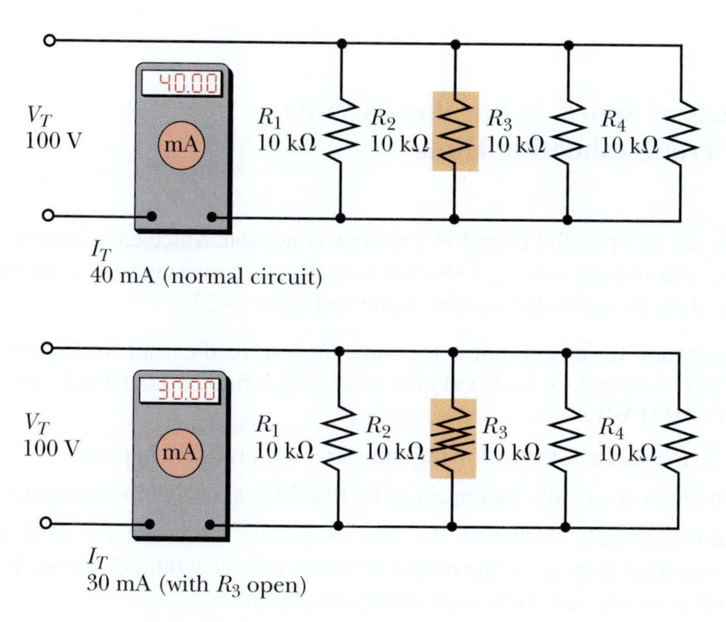

FIGURE 5–21 Effects of an open in a parallel circuit

Troubleshooting Hints

Since an open in a parallel branch causes total resistance to increase, total current to decrease, and current through the opened branch to decrease to zero, troubleshooting hints help isolate and identify these facts.

With power removed from the circuit, you can measure its total resistance with an ohmmeter. If the measured value indicates a higher-than-normal value for the circuit, then there is a good chance an open has occurred or at least some part of the circuit has opened or increased in R value. If it is convenient to electrically "lift" (electrically isolate) one end of each branch one at a time from the remainder of the circuit as you watch the meter, you can spot the troubled branch. For example, if lifting one end of a branch from the circuit causes a change in R (increase), the branch just lifted is probably good. If lifting of the branch causes no change in the R reading, you have probably found the branch with the problem. Because it was already opened, opening it again did not change anything!

If you prefer using "power-on" current readings, the first indication of trouble you will see is a lower-than-normal total circuit current. If you can conveniently connect the meter to measure each branch's current individually, the branch with zero current is the open branch. Other branches should measure normal values.

Practical Notes

Some generalizations worth noting about an open in *any* circuit (series, parallel, or combination) are the following:

1. An open causes total circuit resistance to increase.
2. An open causes total circuit current to decrease.
3. An open causes total circuit power to decrease.

Incidentally, opens can be caused either by *components* opening, or by *conductor paths* opening. On printed circuit boards, a cracked conductor path or poor solder joint can cause an open. On "hard-wired" circuitry, a broken wire (or poor solder joint) can cause an open conductor path, Figure 5–22.

FIGURE 5–22 Open circuits can be caused by broken conductor paths or poor solder joints.

5–9 Effects of Shorts in Parallel Circuits and Troubleshooting Hints

Shorts

A shorted branch in a parallel circuit is a serious condition. Since each branch is connected directly across the voltage source, a shorted branch puts a resistance of zero ohms, or close to zero ohms, directly across the source. Refer to Figure 5–23.

1. Normal operation for this circuit is the same as described earlier for Figure 5–21, where $V_T = 100$ V; $I_T = 40$ mA; $P_T = 4{,}000$ mW (4 W); each branch $I = 10$ mA; and each branch $P = 1{,}000$ mW (1 W).
2. If branch 2 (R_2) shorted, the following things would probably happen:
   - I_T would try to drastically increase until a fuse blows, or the power supply "dies"!
   - V_T would drastically decrease! Because of greatly increased $I \times R$ drop across the source's internal resistance, the output terminal voltage would decrease. It would probably drop to nearly zero volts with most power sources.
   - P_T would drastically increase until the power source quits, or a fuse blows!

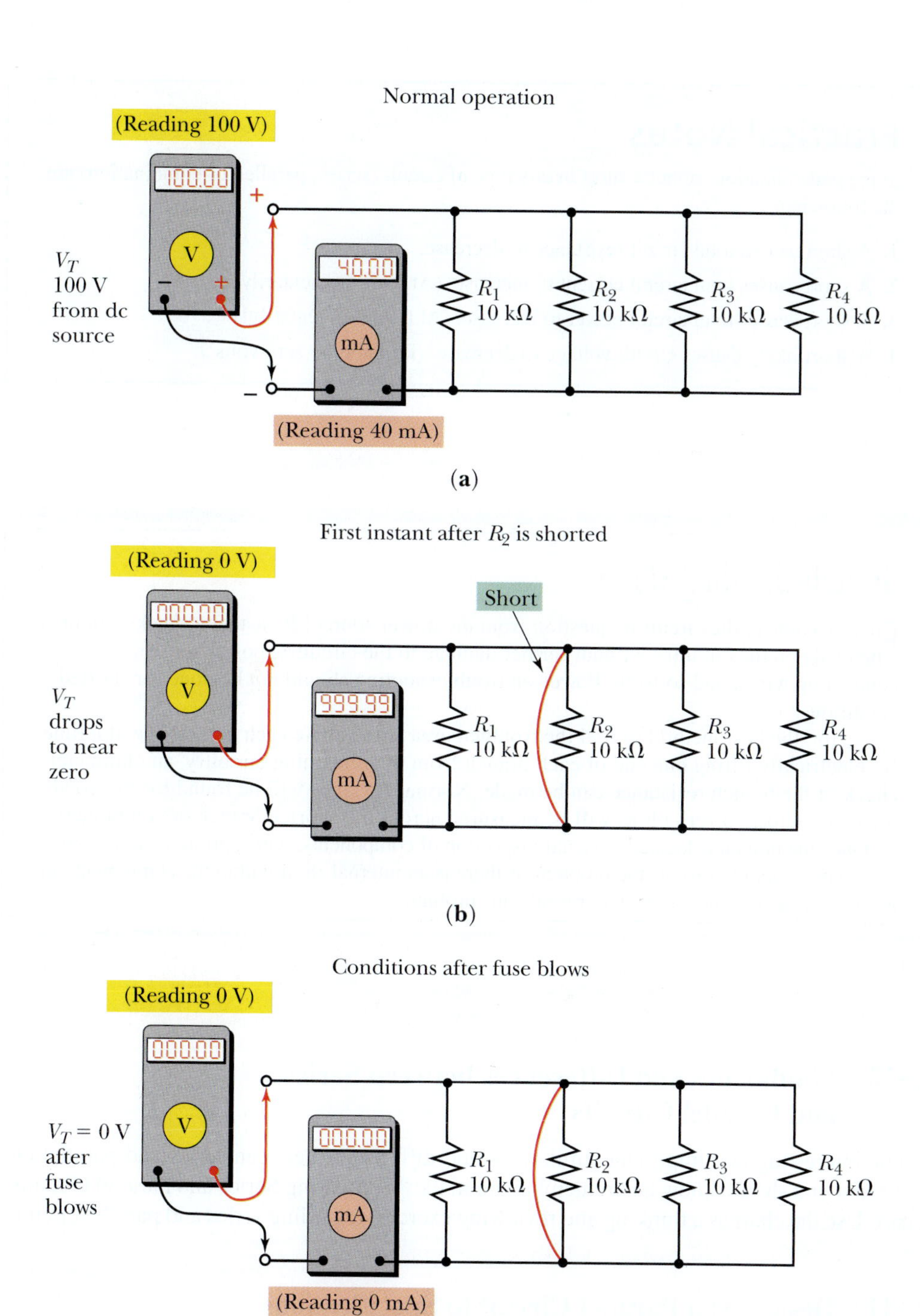

FIGURE 5–23 Effects of a short in a parallel circuit. R_T decreases and I_T increases until the fuse blows or the power source quits.

- Current through the other branches would decrease to zero, or close to it, because the V has dropped to almost zero, or zero.
- Total circuit resistance would drastically decrease! Remember the total resistance will be less than the least branch R.
- If there were any power supply fuses or house circuit fuses associated with the faulty circuit, they would probably blow. Also, there might be some smoke rising from some sections of the circuit or power supply.

Practical Notes

Some generalizations about a short in *any* type of circuit (series, parallel, or combination) are the following:

1. A short causes total circuit resistance to decrease.
2. A short causes total circuit current to increase. (At least momentarily!)
3. A short causes total circuit power to increase. (At least momentarily!)
4. A short likely causes circuit voltage to decrease. (Probably to zero volts!)

Troubleshooting Hints

First *disconnect* the circuit in question from the power source! Resistance checks can then be made on the circuitry without further damage to the circuitry, power source, connecting wires, and so forth. Power-on troubleshooting should *not* be done for shorted conditions.

Again, the best method to find the troubled area is to isolate each branch one at a time. By electrically lifting one end of each branch from the remaining circuitry, an ohmmeter check of the branch resistance can be made. Normal *R* values will be found for the good branches. Close to zero ohms will be measured across the shorted branch or component.

Once the branch is located, a visual inspection of components, wires, printed circuit paths, and so on generally reveals the problem. If there is an internal short within the component, an *R* measurement of the component reveals this problem.

5–10 Similarities and Differences between Series and Parallel Circuits

Some interesting and important similarities and differences between series and parallel circuits that you should know and remember are in the "Comparing Series and Parallel Circuits" chart. Use the chart as a thinking and reviewing exercise regarding series and parallel circuits.

5–11 Designing a Parallel Circuit to Specifications

EXAMPLE Assume you want to create a four-branch parallel circuit so that branches 1 and 3 have equal current values, and branches 2 and 4 also have equal current values but their current values are twice that of branches 1 and 3. Further assume that branch 1 has an *R* value of 10 kΩ and its current value is 10 mA.

The design problem involves determining all the circuit parameters and drawing the schematic for this circuit. Label all component values and all *V, R,* and *I* electrical parameters for each branch and for the total circuit.

Recall the way to begin is to draw a diagram and label all the knowns. Then proceed in the most appropriate manner to solve for the unknowns based on the knowns.

See Figure 5–24a for this first step, which is drawing a diagram and labeling all knowns.

The second step is to use the knowns to solve for the unknowns, then label the diagram with the appropriate parameters.

- Since branch 1's *R* and *I* values are known, it is easy to find its voltage value. Thus, $V = I \times R = 10\text{ mA} \times 10\text{ k}\Omega = 100\text{ V}$.

Comparing Series and Parallel Circuits	
Series Circuit Characteristics	**Parallel Circuit Characteristics**
R_T = Sum of Rs	R_T = Less than lowest R in parallel
$I_T = \frac{V_T}{R_T}$	$I_T = \frac{V_T}{R_T}$
I_T = Same current throughout the circuit	I_T = Sum of branch currents, where branch current $(I_b) = \frac{V_b}{R_b}$ Branch I is inverse to branch R.
$V_T = I_T \times R_T$	$V_T = I_T \times R_T$
V_T = Sum of all IR drops	V_T = Same as across each branch
$V_{R_X} = I_{R_X} \times R_X$; any given R's voltage drop is related to the ratio of its R to R total. $V_{R_X} = \left(\frac{R_X}{R_T}\right) \times V_T$	Voltage across all resistors is same value and equals the circuit applied voltage.
P dissipated by a given R is *directly related* to its R value. Largest R in circuit dissipates most power; smallest R dissipates least power.	P dissipated by a given R is *inverse* to its R value. Largest R value dissipates least power, smallest R value dissipates most power.
$P_T = V_T \times I_T$; or sum of all Ps	$P_T = V_T \times I_T$; or sum of all Ps
Opens cause 0 current; 0 V across good components; and V_T (applied circuit voltage) to appear across open.	*Opens* cause no change in good branches; opened branch has zero current, but V across all branches stays the same.
Shorts cause I to increase; V to decrease across short; V to increase across remaining components; and total circuit power to increase.	*Shorts* cause I_T and P_T to increase (at least momentarily); V to decrease across all branches; and current through good branches to decrease.

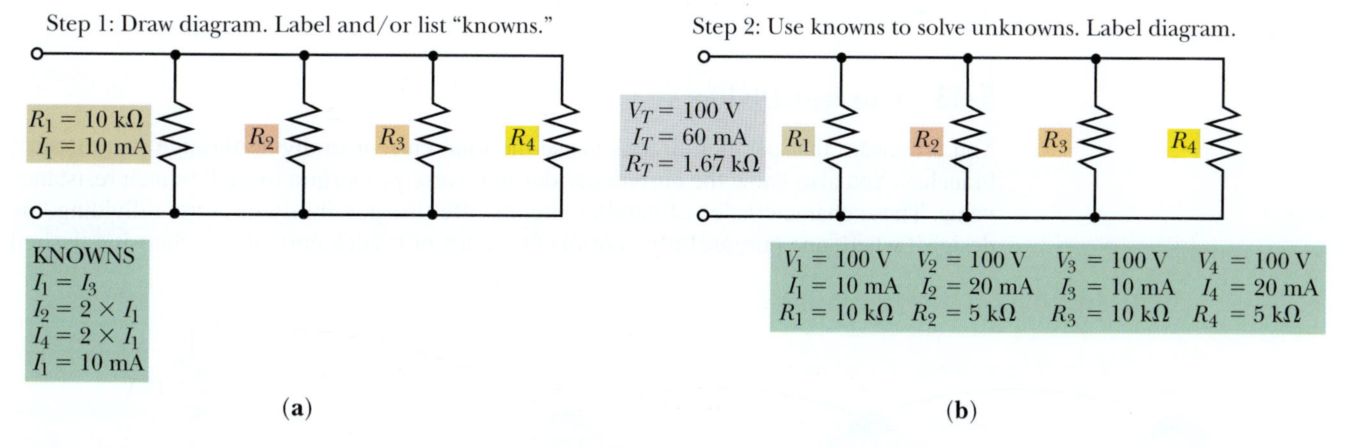

FIGURE 5–24

- Because this is a parallel circuit, 100 V is the voltage across all the branches as well as the V applied value.
- Because branches 3 and 1 have the same current value, and since we know the V applied to both branches is the same, it is obvious branch 3 resistance value must be the same as branch 1. So we now know branch 3 R_3 = 10 kΩ, I_3 = 10 mA, and V_3 = 100 V.

- Because branches 2 and 4 have twice the current value as branches 1 and 3, it is obvious their current values must be 20 mA each. Since the V across each branch is 100 V, each branch's R value is

$$\frac{100\text{ V}}{20\text{ mA}} = 5\text{ k}\Omega$$

- Solving the total circuit current and resistance are simple, and you already know the V applied value. Total current (I_T) equals the sum of the branch currents = 60 mA.

$$R_T = \frac{V_T}{I_T} = \frac{100\text{ V}}{60\text{ mA}} = 1.67\text{ k}\Omega$$

Incidentally, you can use the R values and solve for the total resistance. The two 10-kΩ branches are the equivalent of 5 kΩ. This 5 kΩ in parallel with two other 5-kΩ branches is equivalent to 5 kΩ divided by 3, or 1.67 kΩ. Remember the technique with equal value branches?

- The last task is to label all appropriate parameters on the diagram, Figure 5–24b. ◆

PRACTICE PROBLEMS 12

Try this problem yourself. Design and draw a three-resistor parallel circuit so that:

1. V applied to the circuit is 120 V.
2. The total circuit power dissipation is 1.44 W.
3. Branch 1 has half the current value of branch 2 and one-third the current value of branch 3.

Label your diagram with all V, I, R, and P values for the circuit and for each branch.

5–12 Sources in Parallel

Voltage and power sources may be connected in parallel to provide higher current delivering capacity (and power) to circuits or loads (at a given voltage) than can be delivered by only one source. The sources should be equal in output terminal voltage value when using this technique. A common application of this method is the "paralleling" of batteries or cells, Figure 5–25.

5–13 Current Dividers

You are aware that parallel circuits cause the total current to divide through that circuit's branches. You also know the current divides in inverse proportion to each branch resistance value. These characteristics of parallel circuits offer the possibility of current dividing "by design," where one purposefully controls the ratios of branch currents by choosing desired

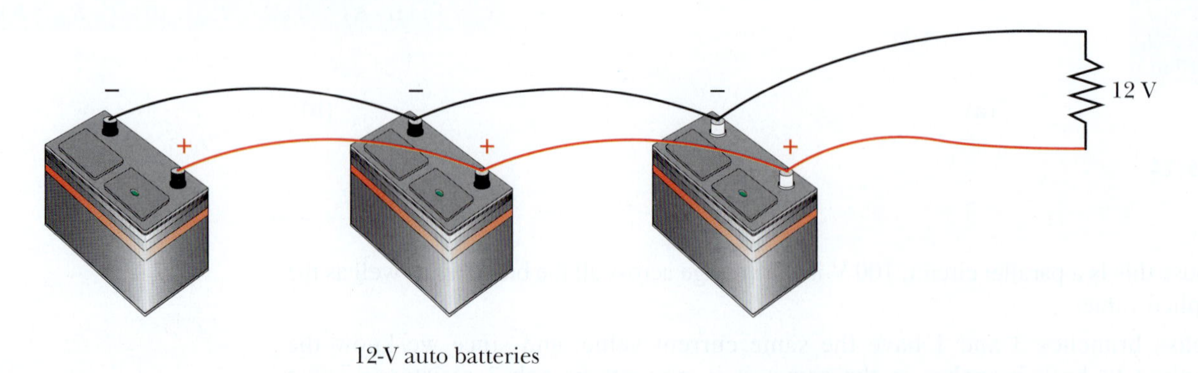

FIGURE 5–25 Equal voltage sources in parallel can deliver more current and power than one source alone.

values of resistances. Also, this current-dividing trait gives another method to find current through branches if only the total resistance, branch resistances, and total current are known.

General Formula for Parallel Circuit with Any Number of Branches

A general formula to find any given branch's current is:

FORMULA 5–7 $I_X = \frac{R_T}{R_X} \times I_T$

where I_X is a specified branch's current value, R_T is the parallel circuit total resistance, and R_X is the value of resistance of the specified branch. Refer to Figure 5–26 for example applications of this formula.

PRACTICE PROBLEMS 13

1. Refer again to Figure 5–26. Assume that $R_1 = 27\ \text{k}\Omega$, $R_2 = 47\ \text{k}\Omega$, $R_3 = 100\ \text{k}\Omega$, $R_T = 14.64\ \text{k}\Omega$, and $I_T = 3.42$ mA. Find each of the branch currents using the divider formula, as appropriate.
2. Assume the circuit of Figure 5–26 is changed so that $I_T = 4$ mA, $R_1 = 10\ \text{k}\Omega$, $R_2 = 5.6\ \text{k}\Omega$, and $R_3 = 2.7\ \text{k}\Omega$.
   a. Find R_T.
   b. Use the general current divider formula to find the current through each branch.

Simple Two-Branch Parallel Circuit Current Divider Formula

If there are only two branches, Figure 5–27, a formula is available to solve for each branch current, knowing I_T and branch R values:

FORMULA 5–8 $I_1 = \frac{R_2}{R_1 + R_2} \times I_T$

$$I_1 = \frac{10\ \text{k}\Omega}{11\ \text{k}\Omega} \times 50\ \text{mA} = 0.909 \times 50\ \text{mA} = 45.45\ \text{mA}$$

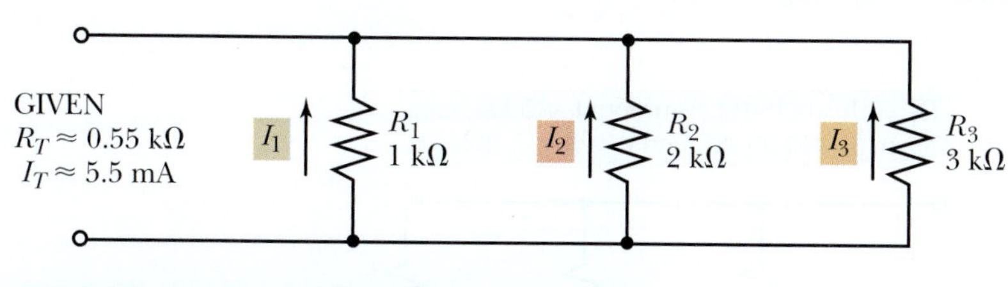

Find approximate values for I_1, I_2, and I_3

$$I_1 \approx \left(\frac{R_T}{R_1}\right) \times I_T = \left(\frac{0.55\ \text{k}\Omega}{1\ \text{k}\Omega}\right) \times 5.5\ \text{mA} \approx 3.0\ \text{mA}$$

$$I_2 \approx \left(\frac{R_T}{R_2}\right) \times I_T = \left(\frac{0.55\ \text{k}\Omega}{2\ \text{k}\Omega}\right) \times 5.5\ \text{mA} \approx 1.5\ \text{mA}$$

$$I_3 \approx \left(\frac{R_T}{R_3}\right) \times I_T = \left(\frac{0.55\ \text{k}\Omega}{3\ \text{k}\Omega}\right) \times 5.5\ \text{mA} \approx 1.0\ \text{mA}$$

FIGURE 5–26 A general current divider example

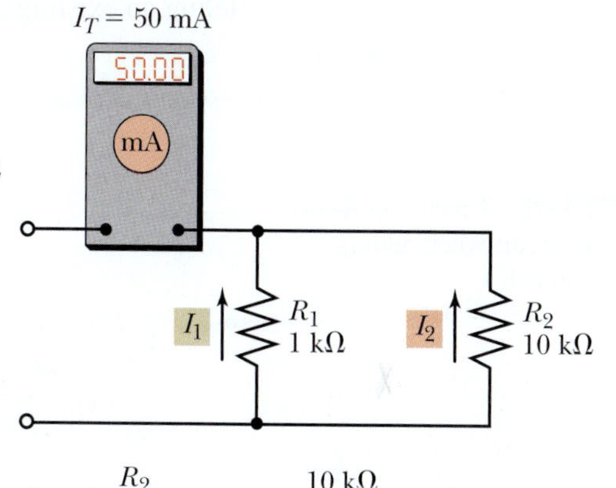

$$I_1 = \frac{R_2}{R_1 + R_2} \times I_T = \frac{10\ \text{k}\Omega}{11\ \text{k}\Omega} \times 50\ \text{mA} = 45.45\ \text{mA}$$

$$I_2 = \frac{R_1}{R_1 + R_2} \times I_T = \frac{1\ \text{k}\Omega}{11\ \text{k}\Omega} \times 50\ \text{mA} = 4.55\ \text{mA}$$

FIGURE 5–27 Two-resistor current-divider example

FORMULA 5–9 $I_2 = \frac{R_1}{R_1 + R_2} \times I_T$

$$I_2 = \frac{1\ \text{k}\Omega}{11\ \text{k}\Omega} \times 50\ \text{mA} = 0.0909 \times 50\ \text{mA} = 4.55\ \text{mA}$$

NOTE: There is an inverse relationship of current to branch resistances. Thus, when solving for I_1, R_2 is in the numerator; when solving for I_2, R_1 is in the numerator.

PRACTICE PROBLEMS 14

1. Refer again to Figure 5–27 and assume the values of R_1 and R_2 change to 100 kΩ and 470 kΩ, respectively. Also, assume the total current is 2 mA. Find the values of current through each branch using the appropriate two-branch divider formula.
2. Assume that the circuit in Figure 5–27 is changed so that total current measures 20 mA, R_1 equals 1.8 kΩ, and R_2 equals 8.2 kΩ. Using the two-resistor current divider formula, determine the currents through each branch.

Practical Notes

You have learned that whenever *any* resistance value is placed in parallel with an existing circuit, it alters the circuit resistance and current because another branch has been added; thus providing another current path.

In a later chapter, we will discuss how this feature impacts the types and uses of measurement devices, such as voltmeters. For now, we only want to alert you to an important observation about parallel circuits.

As already stated, no matter what value of R is paralleled with (or "bridged across") an existing circuit, the circuit R will decrease. But the question is *how much* change will there be?

Look at Figure 5–28. Note the impact (change) that adding a 1 megohm resistor branch has on the circuit resistance. As you can see, adding a *high R* in parallel causes a small change in total circuit R.

Look at Figure 5–29. Note the change in R_T caused by introducing a 1-kΩ resistor in parallel with the same original circuit. Adding a *low R* in parallel with an existing circuit causes a much more significant change in the circuit total resistance (and total current) than paralleling an existing circuit with a high R value.

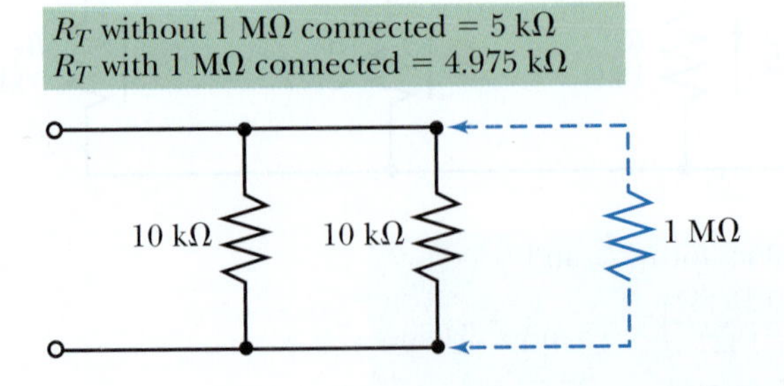

FIGURE 5–28 Change in R_T of a given circuit when adding 1-MΩ branch

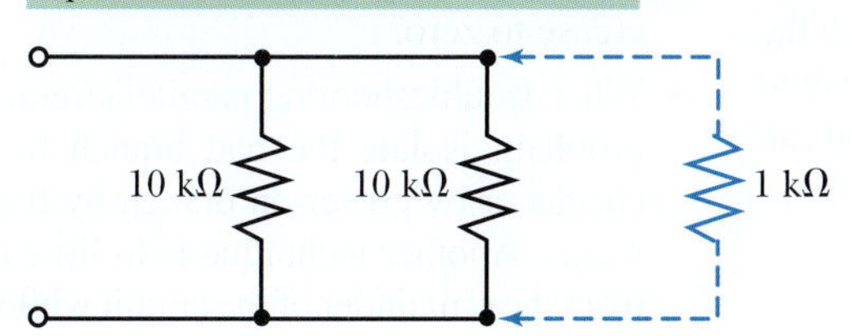

FIGURE 5–29 Change in R_T of a given circuit when adding 1-kΩ branch

The amount of change caused by adding a new branch *R* is related to *both* the *R* value of the "new" *R* branch *and* to the total circuit resistance of the original circuit. If the original circuit's equivalent resistance is low initially, adding a low value *R* in parallel will not have nearly the effect that it does when the original circuit has a high R_T.

Summary

- A parallel circuit is one in which there are two or more paths (or branches) for current flow, and all the branches have the same voltage applied across them.
- Total circuit current divides through the parallel branches in an inverse ratio to the branch resistances.
- Total circuit current in a parallel circuit equals the sum of the branch currents.
- Kirchhoff's current law states the value of current entering a point must equal the value of current leaving that same point in the circuit.
- The total resistance of a parallel circuit must be less than the smallest resistance branch *R* value.
- There are several methods to find the total resistance of parallel circuits:

  1. *Ohm's law*

$$R_T = \frac{V_T}{I_T}$$

  2. *Reciprocal resistance; conductance-related method*

$$\frac{1}{R_T} = \frac{1}{R_1} + \frac{1}{R_2} \cdots + \frac{1}{R_n}; \text{ or } R_T = \frac{1}{\frac{1}{R_1} + \frac{1}{R_2} \cdots + \frac{1}{R_n}}$$

  3. *Conductance method*

$$R_T = \frac{1}{G_T}$$

  4. *Product-over-the-sum method*

$$R_T = \frac{R_1 \times R_2}{R_1 + R_2}$$

  5. *Assumed voltage method*

- One technique to solve for parallel circuit total resistance is to combine equal branch *R* values into their equivalent resistance value, whenever possible. For example, two 10-kΩ resistors in parallel equal 5-kΩ equivalent resistance (or half of one equal branch *R* value). Three 10-kΩ resistors in parallel equal 3.33-kΩ equivalent resistance (or one-third of one equal branch *R* value).
- A useful formula to find the *R* value needed in parallel with a given *R* value to obtain a desired equivalent (total resistance) value is:

$$R_u = \frac{R_k \times R_e}{R_k - R_e}$$

where R_u is the unknown R value being calculated, R_e is the resultant total resistance desired from the "paralleled" resistors, and R_k is the known value resistor across which the unknown value resistor is paralleled to achieve the desired equivalent, or total parallel circuit resistance. Also, you can use:

$$\frac{1}{R_e} - \frac{1}{R_k} = \frac{1}{R_u}$$

- Total power supplied to, or dissipated by, a parallel circuit is found with the power formulas $P_T = V_T \times I_T$, $I_T^2 \times R_T$, or V_T^2/R_T, or by summing the power dissipations of all the branches.
- Power dissipated by any branch of a parallel circuit is computed with the power formulas by substituting the specified branch V, I, and/or R parameters into the formulas, as appropriate.
- Branch power dissipations are such that the highest R value branch dissipates the least power, and the lowest R value branch dissipates the most power.
- An open branch in a parallel circuit causes total circuit current (and power) to decrease, total circuit R to increase, current through the unopened branches to remain the same, and current through the opened branch to decrease to zero. The circuit voltage remains the same.
- A shorted branch in a parallel circuit may cause damage to the circuit wiring and/or to the circuit power supply and/or may cause fuse(s) to blow. Until a fuse blows or the circuit power source fails, the short causes total circuit current to drastically increase; circuit R to decrease to zero ohms; and power demanded from the source to greatly increase. When a fuse blows or the source fails, circuit V, I, and P decrease to zero.
- When troubleshooting parallel circuits with an open branch problem, isolate the bad branch by power-off resistance checks or by power-on branch-by-branch current measurements. Another technique is to lift each branch electrically from the remainder of the circuit while monitoring circuit resistance or circuit current. The branch that causes no change is the bad branch.
- When troubleshooting parallel circuits with a shorted branch problem, *turn the power off and remove the circuit from the power source terminals before performing power-off checks to isolate the problem.* Branch-by-branch resistance measurements usually locate the probem. Also, you can electrically lift and reconnect each branch sequentially while monitoring the total circuit resistance.
- There are several significant *differences* between parallel and series circuits. Current is *common* in series circuits; voltage is *common* in parallel circuits. *Voltage* divides around a series circuit in *direct* relationship to the R values. *Current* divides throughout a parallel circuit in *inverse* relationship to the branch R values. In series circuits, the largest R dissipates the most power; in parallel circuits, the smallest R dissipates the most power. In series circuits, the total resistance equals the sum of all resistances. In parallel circuits, the total circuit resistance is *less* than the smallest branch resistance value.
- Voltage/power sources can be connected in parallel to provide higher current/power at a given voltage level.

Formulas and Sample Calculator Sequences

Formula 5–1
(To find total current in a parallel circuit)

$$I_T = I_1 + I_2 + I_3 \ldots + I_n$$

I_1 value, [+], I_2 value, [+] I_3 value, . . . , [=]

Formula 5–2
(To find 1/RT or total conductance in a parallel circuit)

$$\frac{1}{R_T} = \frac{1}{R_1} + \frac{1}{R_2} + \frac{1}{R_3} \cdots + \frac{1}{R_n}$$

R_1 value, [1/x], [+], R_2 value, [1/x], [+], R_3 value, [1/x], . . ., [=]

Formula 5–3
(To find total resistance in a multiple-branch parallel circuit)

$$R_T = \frac{1}{\frac{1}{R_1} + \frac{1}{R_2} + \frac{1}{R_3} \cdots + \frac{1}{R_n}}$$

R_1 value, [1/x], [+], R_2 value, [1/x], [+], R_3 value, [1/x], . . ., [=], [1/x]

Formula 5–4
(To find total conductance in a parallel circuit)

$$G_T = G_1 + G_2 + G_3$$

G_1 value, [+], G_2 value, [+], G_3 value, [=]

Formula 5–5
(To find total resistance in a two-branch parallel circuit)

$$R_T = \frac{R_1 \times R_2}{R_1 + R_2}$$

R_1 value, [×], R_2 value, [÷], [(], R_1 value, [+], R_2 value, [)], [=]

Formula 5–6
(To find the value of R needed in parallel to get a desired Requiv)

$$R_u = \frac{R_k \times R_e}{R_k - R_e}$$

R known value, [×], R equivalent value, [÷], [(], R known value, [−], R equivalent value, [)], [=]

Formula 5–7
(To find a select branch current when R_T and R_{branch} are known)

$$I_X = \frac{R_T}{R_X} \times I_T$$

R_T value, [÷], R_x value, [×], I_T value, [=]

Formula 5–8
(To find a select branch current using the two-branch, current-divider rule)

$$I_1 = \frac{R_2}{R_1 + R_2} \times I_T$$

R_2 value, [÷], [(], R_1 value, [+], R_2 value, [)], [×], I_T value, [=]

Formula 5–9
(To find a select branch current using the two-branch, current-divider rule)

$$I_2 = \frac{R_1}{R_1 + R_2} \times I_T$$

R_1 value, [÷], [(], R_1 value, [+], R_2 value, [)], [×], I_T value, [=]

EXCEL AUTOMATED FORMULAS

Helpful problem-solving tools, such as the Excel automated formulas available on the CD, allow automatic calculations of formulas.

Using Excel

Parallel Circuits Formulas

(Excel file reference: FOE5_01.xls)

***DON'T FORGET!** It is NOT necessary to retype formulas once they are entered on the worksheet! Just input new parameters data for each new problem using that formula, as needed.*

- Refer to the parameters given in Practice Problems 6, question 2. Use the Formula 5–3 spreadsheet sample and solve for R_T. Compare with results given in the text appendix for this practice problem.
- Refer to Figure 5–27. Assume R_1 is changed to 27 k ohms and R_2 is changed to 12 k ohms. Use the Formula 5–8 spreadsheet sample and solve for I_1.

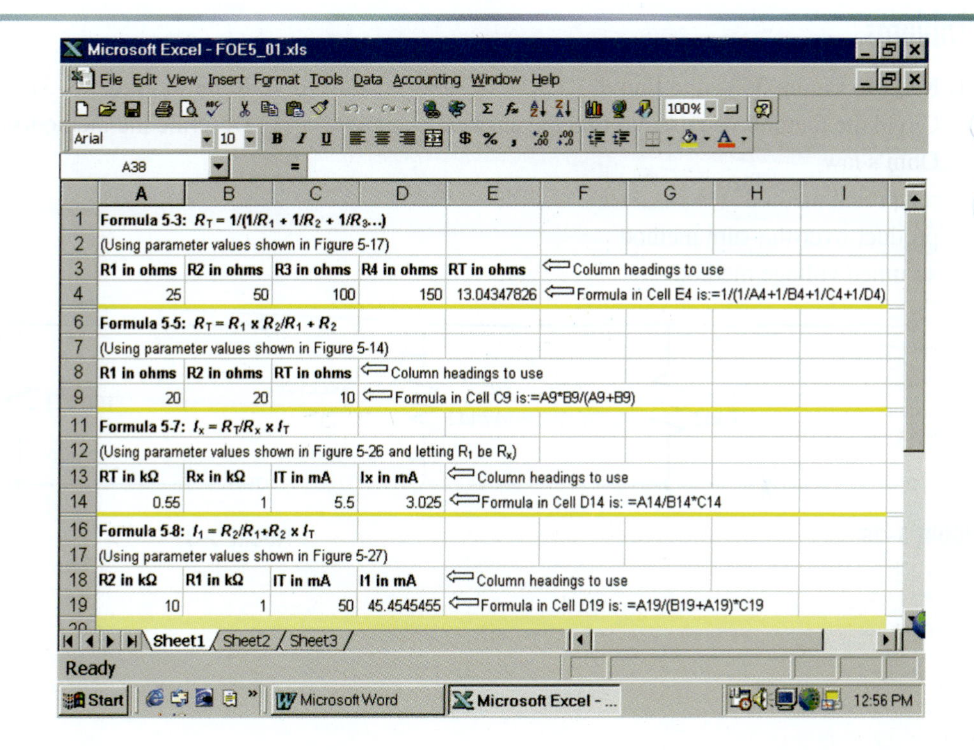

	A	B	C	D	E	F
1	Formula 5-3: $R_T = 1/(1/R_1 + 1/R_2 + 1/R_3...)$					
2	(Using parameter values shown in Figure 5-17)					
3	R1 in ohms	R2 in ohms	R3 in ohms	R4 in ohms	RT in ohms	⇐ Column headings to use
4	25	50	100	150	13.04347826	⇐ Formula in Cell E4 is:=1/(1/A4+1/B4+1/C4+1/D4)
6	Formula 5-5: $R_T = R_1 \times R_2/R_1 + R_2$					
7	(Using parameter values shown in Figure 5-14)					
8	R1 in ohms	R2 in ohms	RT in ohms	⇐ Column headings to use		
9	20	20	10	⇐ Formula in Cell C9 is:=A9*B9/(A9+B9)		
11	Formula 5-7: $I_x = R_T/R_x \times I_T$					
12	(Using parameter values shown in Figure 5-26 and letting R_1 be R_x)					
13	RT in kΩ	Rx in kΩ	IT in mA	Ix in mA	⇐ Column headings to use	
14	0.55	1	5.5	3.025	⇐ Formula in Cell D14 is: =A14/B14*C14	
16	Formula 5-8: $I_1 = R_2/R_1+R_2 \times I_T$					
17	(Using parameter values shown in Figure 5-27)					
18	R2 in kΩ	R1 in kΩ	IT in mA	I1 in mA	⇐ Column headings to use	
19	10	1	50	45.4545455	⇐ Formula in Cell D19 is: =A19/(B19+A19)*C19	

Review Questions

1. What is the minimum number of current paths required for a circuit to be considered a parallel circuit?
   a. 1
   b. 2
   c. 3
   d. 4
2. Total current in a parallel circuit divides itself through various branches:
   a. in direct relationship to each branch resistance value.
   b. in inverse relationship to each branch resistance value.
   c. according to the power rating of each branch resistor.
   d. none of the above.
3. Total current in a parallel circuit is equal to:
   a. the sum of the two highest branch currents.
   b. the product of the lowest branch current times the highest branch current.
   c. the applied voltage divided by the circuit equivalent resistance.
   d. none of the above.
4. Total resistance in a parallel circuit is always:
   a. the sum of the branch resistances.
   b. higher than the highest branch resistance.
   c. lower than the lowest resistance branch.
   d. equal to the lowest resistance branch.
5. Power distribution in a parallel circuit is such that:
   a. the lowest value resistor dissipates the most power.
   b. the highest value resistor dissipates the most power.
   c. since V is the same across all R's, all R's dissipate the same power.
   d. none of the above is true.
6. Give the formula for using the Ohm's law method to find total (or equivalent) resistance of a parallel circuit.
7. Give the formula(s) for using the reciprocal resistance method to find total resistance of a parallel circuit.
8. Give the formula for using the conductance method to find total resistance of a parallel circuit.
9. Give the formula for using the product-over-the-sum approach to find total resistance of a two-resistor parallel circuit.
10. List the steps required to find the total resistance of a parallel circuit using the "assumed voltage" method.
11. Give the formula for finding the value of resistor needed to be placed in parallel with a known value resistor in order to achieve a specified total resistance.
12. When troubleshooting a parallel circuit, the circuit power should always be removed before troubleshooting if the symptom information indicates that:
    a. the circuit resistance is too high and the circuit current is too low.
    b. the circuit resistance is too low and the circuit current is too high.

Problems

1. Refer to Figure 5–30 and find R_T.
   DO Circle the method you used to find the answer:
   Ohm's law
   This reciprocal method
   product-over-the-sum method
   assumed voltage method

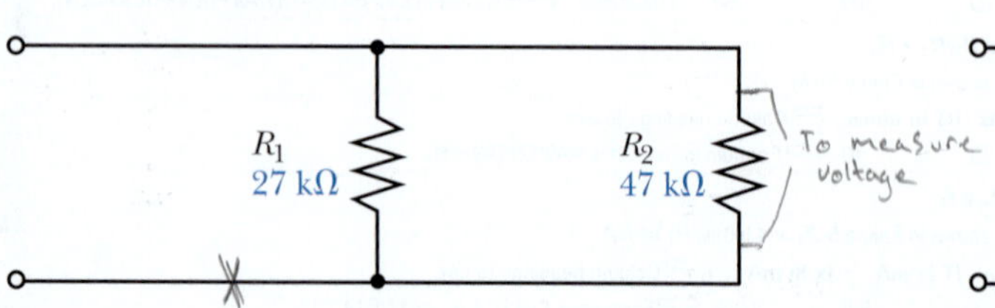

FIGURE 5–30

2. Refer to Figure 5–31 and find R_T.
   Describe the method(s) you used to solve this problem.

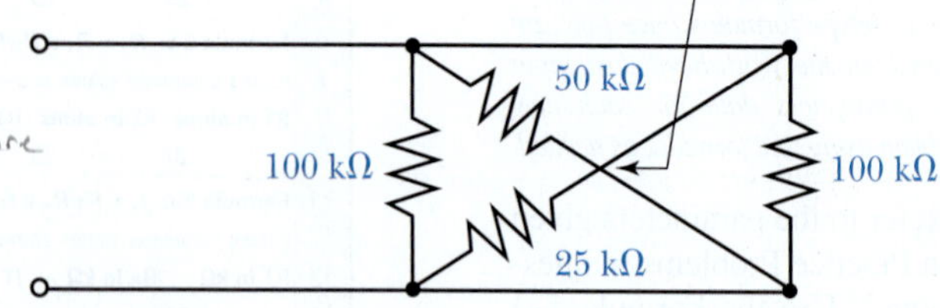

FIGURE 5–31

3. Refer to Figure 5–32 and find I_2.
Describe the method(s) you used to solve this problem.

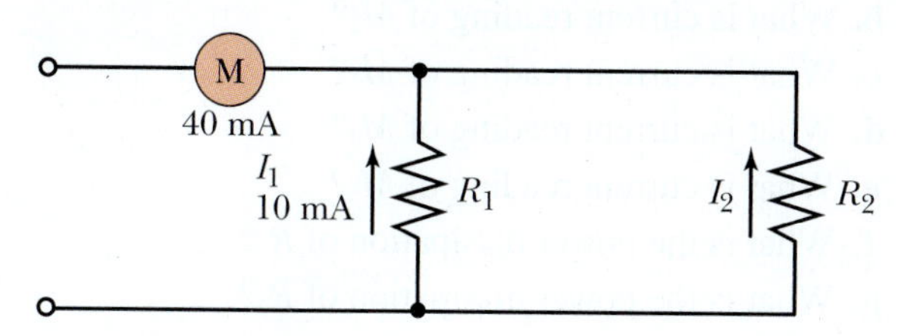

FIGURE 5–32

4. Assume in Figure 5–33 that each of the six resistors is 150 kΩ and determine the following circuit parameters:
R_T, I_T, and P_T
Current reading of meter 1
Current reading of meter 2

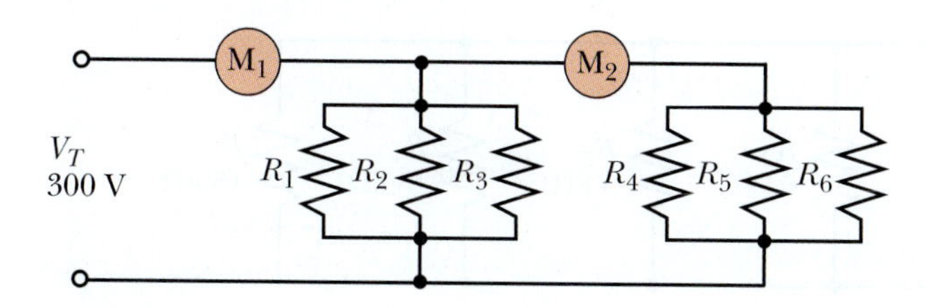

FIGURE 5–33

5. Refer to Figure 5–34 and using the information provided, answer the following with:
"I" for increase
"D" for decrease
"RTS" for remain the same

a. If R_2 opens,

I_1 will _____ P_T will _____
I_2 will _____ V_{R_1} will _____
I_3 will _____ V_T will _____
I_T will _____

b. Assume that R_3 shorts internally. Before fuse or power supply failure,

I_1 will _____ total circuit current
I_2 will _____ will _____
I_3 will _____

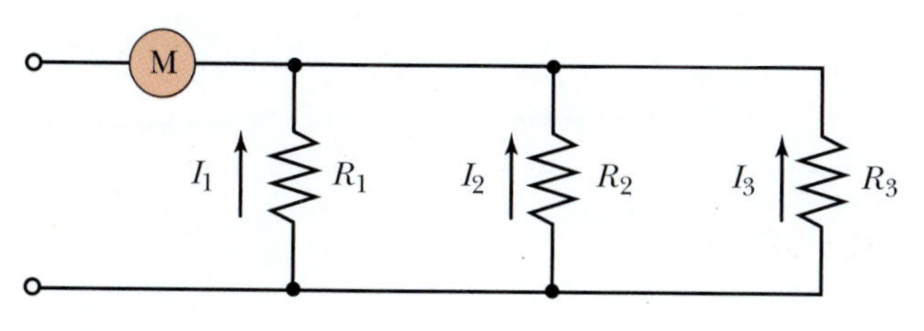

FIGURE 5–34

6. Refer to Figure 5–35 and find the value of R_1.

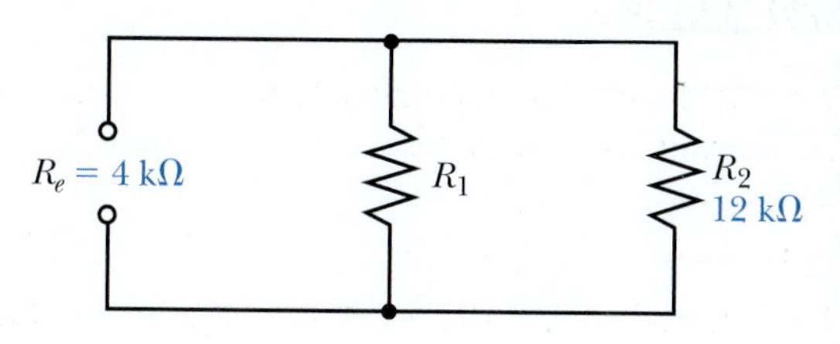

FIGURE 5–35

7. Refer to Figure 5–36. If I_T is 18 mA, what are the values of I_1 and I_2?

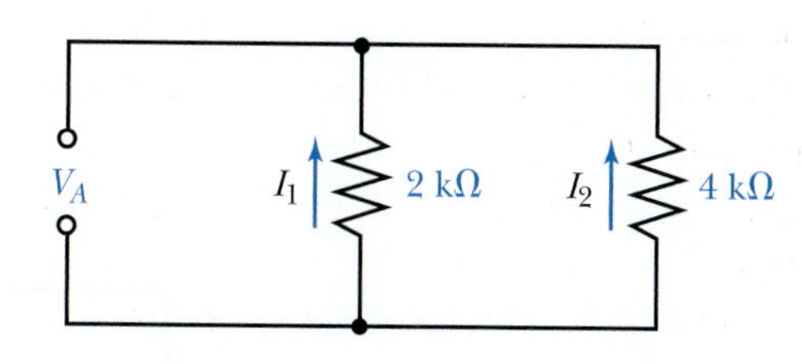

FIGURE 5–36

8. Refer to Figure 5–37. What is the value of R_1?

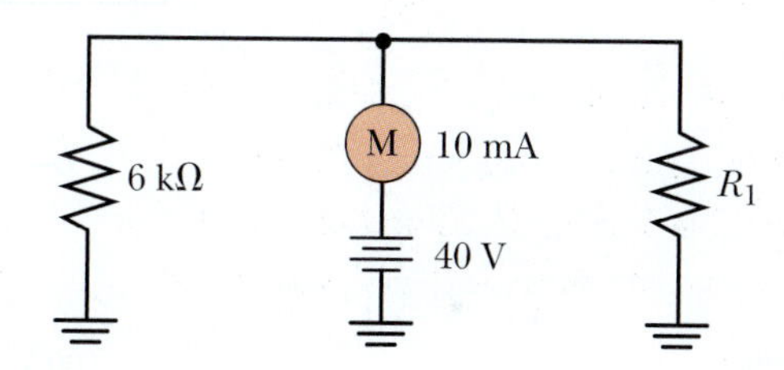

FIGURE 5–37

9. Refer to Figure 5–38. What is the value of P_{R_1}?

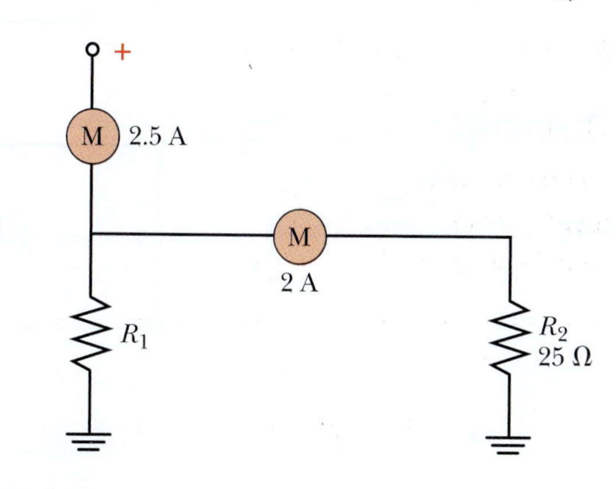

FIGURE 5–38

10. Refer to Figure 5–39 and find the following parameters:

$R_T =$ _____

$I_T =$ _____

$P_T =$ _____

$I_{R_1} =$ _____

$I_{R_2} =$ _____

$I_{R_3} =$ _____

$I_{R_4} =$ _____

$I_{R_5} =$ _____

$P_{R_1} =$ _____

$P_{R_2} =$ _____

$P_{R_3} =$ _____

$P_{R_4} =$ _____

$P_{R_5} =$ _____

11. Refer to Figure 5–40 and answer the following:

a. What is current reading of M_1?

b. What is current reading of M_2?

c. What is current reading of M_3?

d. What is current reading of M_4?

e. What is current reading of M_5?

f. What is the power dissipation of R_1?

g. What is the power dissipation of R_2?

h. What is the power dissipation of R_3?

i. What is total power dissipation of the circuit?

j. If R_2 opens, what is the new value of P_T?

Refer to this circuit, Figure 5–41, when answering questions 12 through 16.

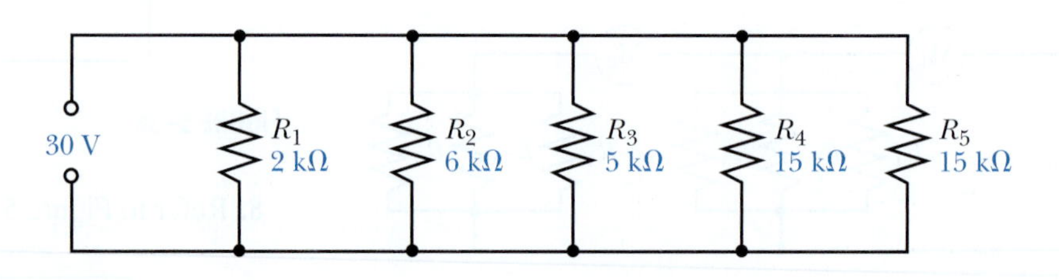

FIGURE 5–39 multiSIM

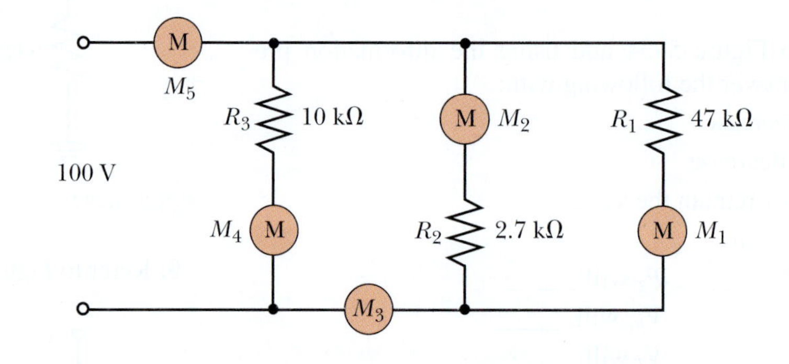

FIGURE 5–40

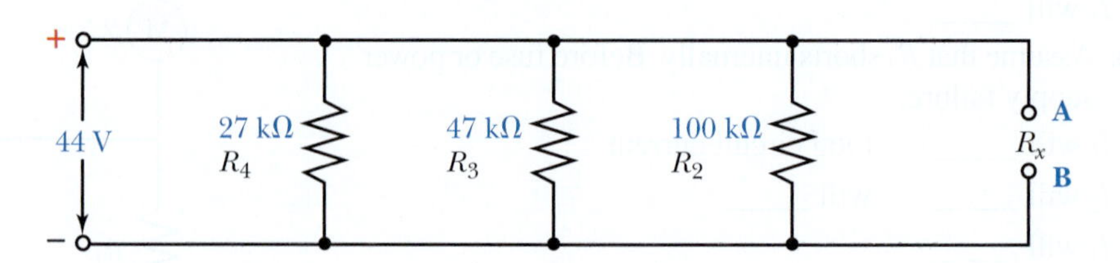

FIGURE 5–41

12. With nothing connected between points A and B, show your "step-by-step" work and solve for R_T using:

a. the reciprocal method.

b. the conductance method.

c. the assumed voltage method.

d. the Ohm's law method.

13. What resistance (R_X) needs to be connected between points A and B for the circuit total resistance to equal 10 kΩ?

14. Assume the R_X of the preceding problem is left connected to the circuit. What happens to the circuit R_T if R_4 is doubled in value and R_2 is halved in value? Would the answer be the same if R_X were removed?

15. With the R_X of question 13 still connected:

a. Which resistor in the circuit dissipates the most power?

b. What is the value of power dissipated by this resistor?

16. With the R_X of question 13 *disconnected:*

a. Which resistor in the circuit dissipates the least power?

b. What is the value of power dissipated by this resistor?

Refer to Figure 5–42 as you answer questions 17 through 25.

17. What is the value of V_T?

18. What is the value of R_1?

19. What is the value of current through R_2?

20. What is the value of R_2?

21. Which resistor dissipates the most power?

22. Which resistor dissipates the least power?

23. What would the total current be if R_2 were changed to a 1.5-kΩ resistor (assuming V_T remained the same)?

24. If R_1 were to increase in value (because of aging and use), what would happen to the total power supplied by the source?

25. For the conditions described in question 24, what would happen to the power dissipated by R_3?

Refer to Figure 5–43 when answering questions 26–30. (NOTE: Round off answers to the nearest whole number, as appropriate.)

26. What value resistors are R_2 and R_4?

27. What is the value of the circuit applied voltage?

28. If R_1 were changed to a resistance value of 27 kΩ, what would the total circuit power dissipation be?

29. What is the total equivalent resistance of the original circuit?

30. What value resistance would need to be added in parallel with the original circuit to have the total circuit resistance equal 4 kΩ?

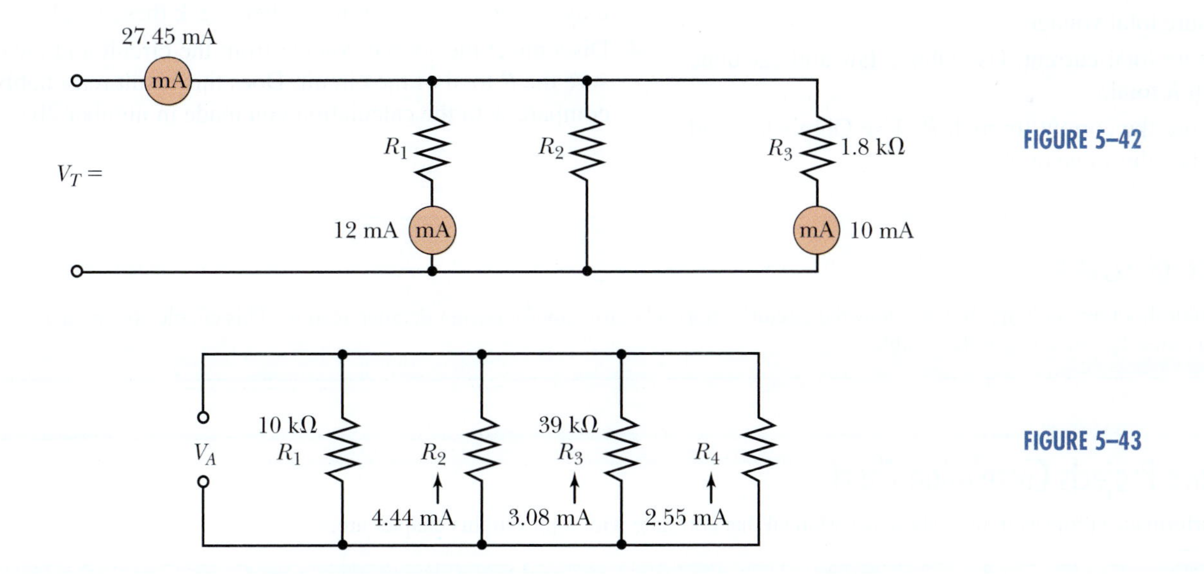

FIGURE 5–42

FIGURE 5–43

Analysis Questions

1. What is the purpose of performing mental approximations, whenever possible, when performing circuit calculations?
2. What is one possible application of a current divider circuit?
3. Assume a circuit has three parallel resistors. If $R_1 = 12\ k\Omega$, $R_2 = 15\ k\Omega$, and $R_3 = 18\ k\Omega$, determine the total resistance using a calculator and the techniques defined next:
   a. Use a calculator in the *engineering mode* and determine R_T by the reciprocal method. Show the answer as it is expressed in the calculator readout.
   b. Use a calculator in the *scientific notation mode* and determine R_T by the reciprocal method. Show the answer as it is expressed in the calculator readout.
4. Because of the resistor values given, what approach might have been easily used for mental approximation of the R_T value?
5. Draw a diagram of a three-branch parallel circuit with the largest R having a value of 100 kΩ. The current divider circuit should have values that cause the total circuit current to divide as follows:

   Branch 1's current is half of branch 2. Branch 2's current is one-fifth of branch 3. Label all R values on your diagram.
6. What value of R is needed in parallel with a 50-kΩ resistor to have a resultant equivalent resistance of 10 kΩ?
7. Describe a good method of isolating parallel components, or an individual branch, when troubleshooting a parallel circuit.

MultiSIM Exercise for *Parallel Circuits*

1. Use the MultiSIM program and the circuit shown in Figure 5–39. Connect a ground reference to the negative side of the source.
2. Measure and record the following parameters:
   a. Measure total voltage.
   b. Measure total current. Use Ohm's law and calculate circuit R total.
   c. Measure the current through R_3. Use Ohm's law and calculate the value or R_3.
3. Compare your MultiSIM results with the values given in the Instructor Guide for this circuit. Were the measurements you made with MultiSIM reasonably close to the answers in the Instructor Guide? (NOTE: Your instructor may prefer that you let him or her check these results.)
4. Disconnect the source voltage from the circuit and measure the R total of the circuit. Does this result reasonably compare with the calculation you made in number 2b?

A Good Idea

When troubleshooting, look up, find, or know the circuit's "norms" before and/or during detailed testing! This enables being alert to abnormalities that are clues to the trouble.

Performance Projects Correlation Chart

Suggested performance projects in the Laboratory Manual that correlate with topics in this chapter are:

Chapter Topic	Performance Project	Project Number
Voltage in Parallel Circuits	Voltage in Parallel Circuit	17
Current in Parallel Circuits & Kirchhoff's Current Law	Current in Parallel Circuits	16
Resistance in Parallel Circuits	Equivalent Resistance in Parallel Circuits	15
Power in Parallel Circuits	Power Distribution in Parallel Circuits	18
Effects of Opens in Parallel Circuits & Troubleshooting Hints	Effects of an Open in Parallel Circuits	19
Effects of Shorts in Parallel Circuits & Troubleshooting Hints	Effects of a Short in Parallel Circuits Story Behind the Numbers: Parallel Circuits	20

NOTE: It is suggested that after completing the above projects, the student should be required to answer the questions in the "Summary" at the end of this section of projects in the Laboratory Manual.

Troubleshooting Challenge

CHALLENGE CIRCUIT 3

(Follow the SIMPLER sequence by referring to inside front cover.)

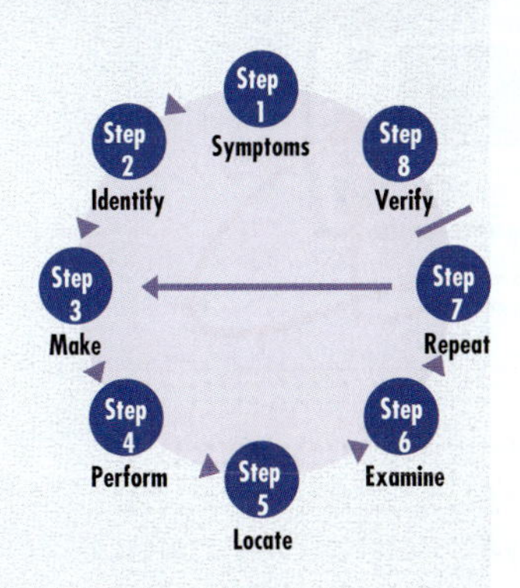

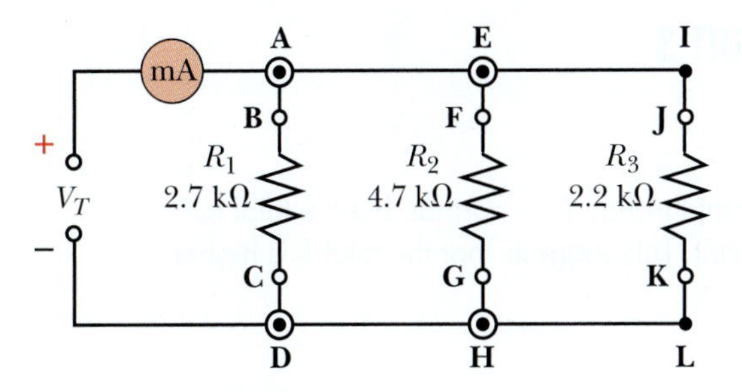

Challenge Circuit 3 multiSIM

STARTING POINT INFORMATION

1. Circuit diagram
2. $V_T = 10$ V
3. Measured $I_2 = 2.1$ mA
4. Measured $I_T = 6.7$ mA
5. *R*s measured with remainder of circuit isolated from points being tested

TEST	Results in Appendix C
$V_{A\text{–}D}$	(83)
$V_{B\text{–}D}$	(11)
$V_{E\text{–}H}$	(22)
$V_{F\text{–}G}$	(95)
$V_{I\text{–}L}$	(52)
$V_{J\text{–}K}$	(75)
$V_{A\text{–}B}$	(8)
$V_{C\text{–}D}$	(56)
$V_{I\text{–}J}$	(103)
$R_{A\text{–}B}$	(46)
$R_{C\text{–}D}$	(90)
$R_{I\text{–}J}$	(14)
$R_{K\text{–}L}$	(29)

CHALLENGE CIRCUIT 3

STEP [1]

SYMPTOMS The current through R_2 is normal, but the total current is low for this circuit. This suggests that the total R is higher than normal.

STEP [2]

IDENTIFY initial suspect area: Branch 1 (points A to D) and branch 3 (points I to L) are suspect areas since branch 2 seems to be operating normally. (Current value is normal.)

STEP [3]

MAKE test decision: Check the voltage between points A and D (branch 1).

STEP [4]

PERFORM **1st Test:** Look up the test result. V_{A-D} is 10 V, which is normal.

STEP [5]

LOCATE new suspect area: Points A to B and C to D in branch 1 and all of branch 3 are the new suspect areas.

STEP [6]

EXAMINE available data.

STEP [7]

REPEAT analysis and testing:

2nd Test: The V_{B-C}. $V_{B-C} = 0$ V, which is abnormal. This could result from a bad connection between points A and B or between points C and D.

3rd Test: Check V_{A-B}. V_{A-B} is 0 V, which is normal.

4th Test: Check V_{C-D}. V_{C-D} is 10 V, which is abnormal.

5th Test: Isolate branch 1 and check the resistance between points C and D. R_{C-D} is infinite ohms, indicating an open. (NOTE: This meter shows a blinking 30.00 to indicate infinite ohms. Other digital multimeters may use different readouts to indicate infinite ohms.)

STEP [8]

VERIFY **6th Test:** Make a solid connection between points C and D and note the total current reading. It should be 10.4 mA if all the branches are operating properly. That is, the result of a good connection is that total current is approximately 10.4 mA. When this is done, the circuit operates normally.

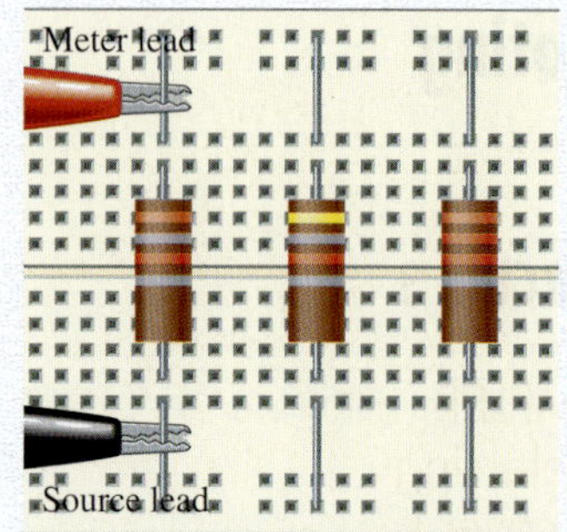

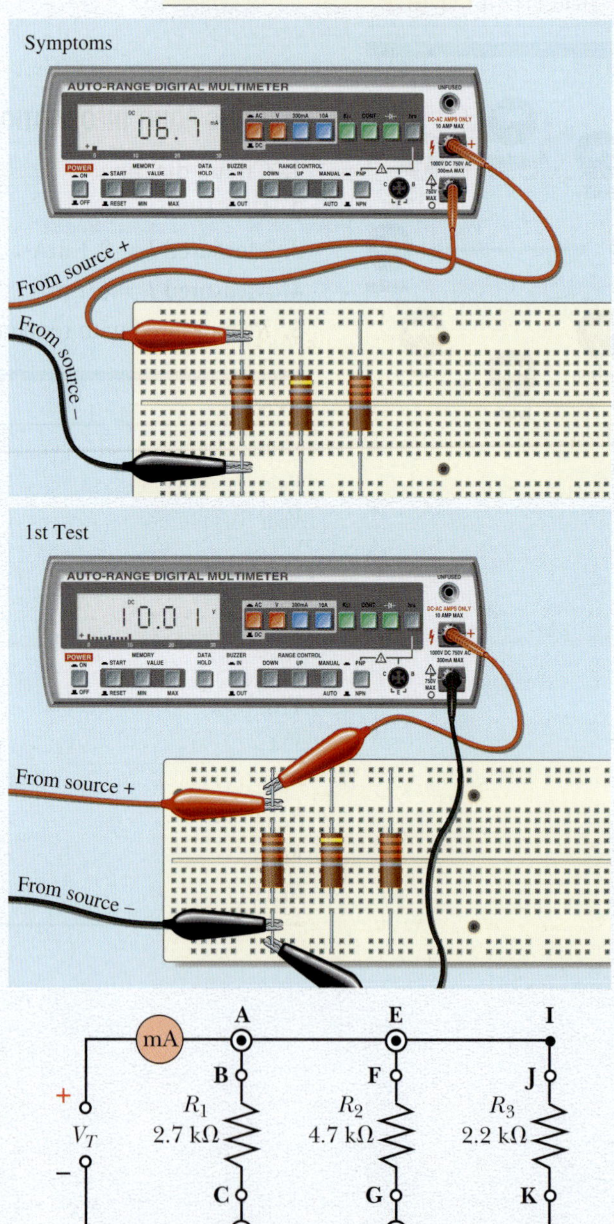

Challenge Circuit 3 multiSIM

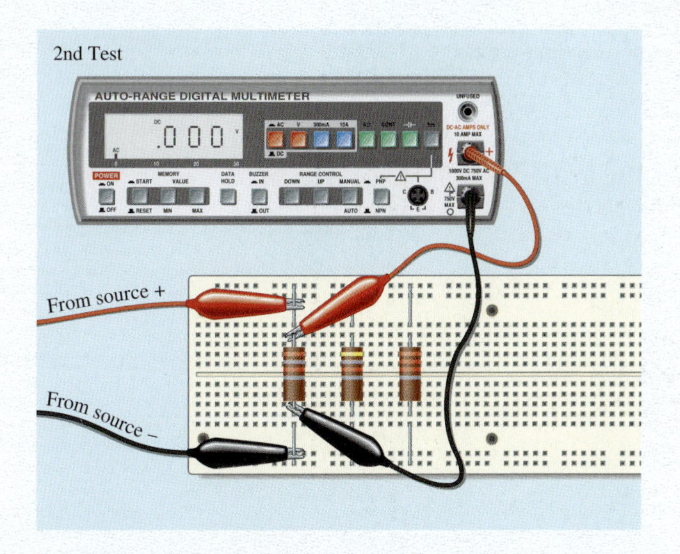

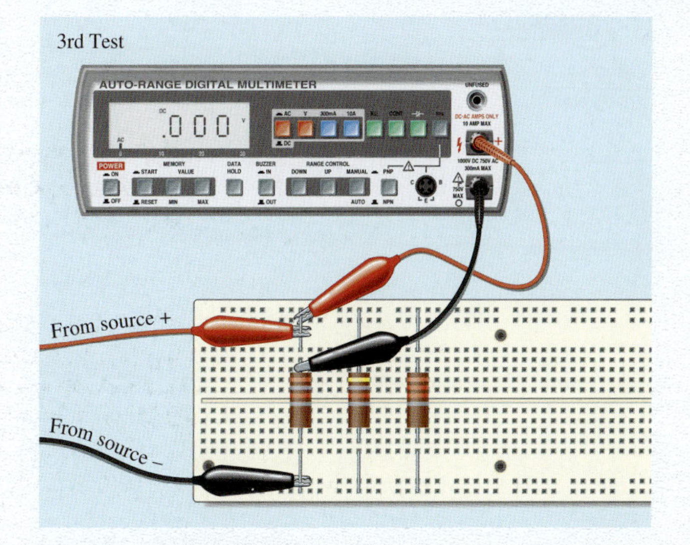

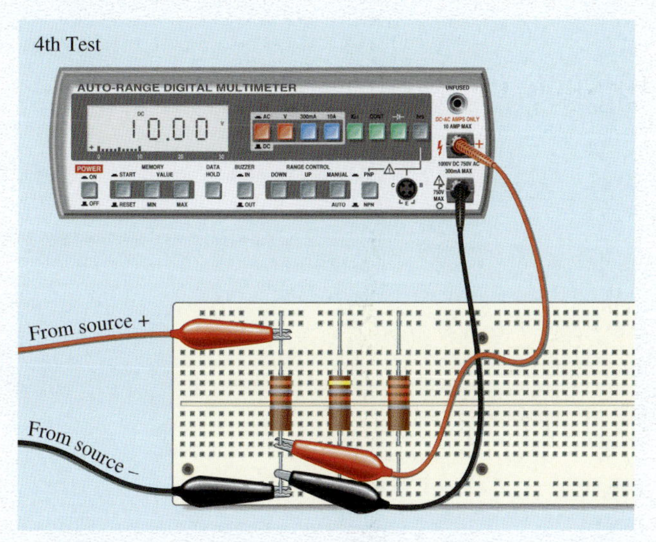

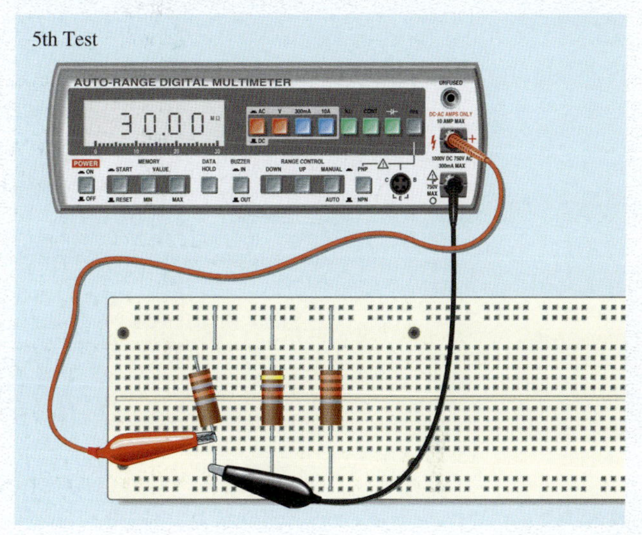

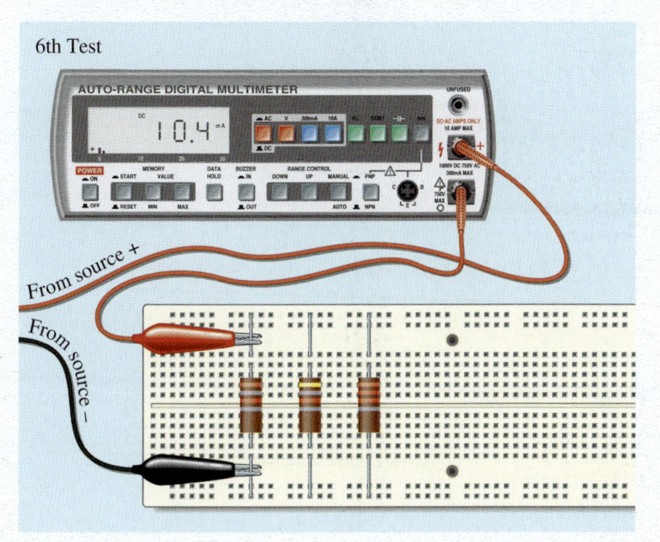

SPECIAL NOTE: You could have found the problem more quickly by calculating what the total current should be, then determining that the amount of current missing was equal to the value branch 1 should have. Then you would have immediately known the problem was with branch 1. You still would have to perform isolation tests within that branch to locate the open.

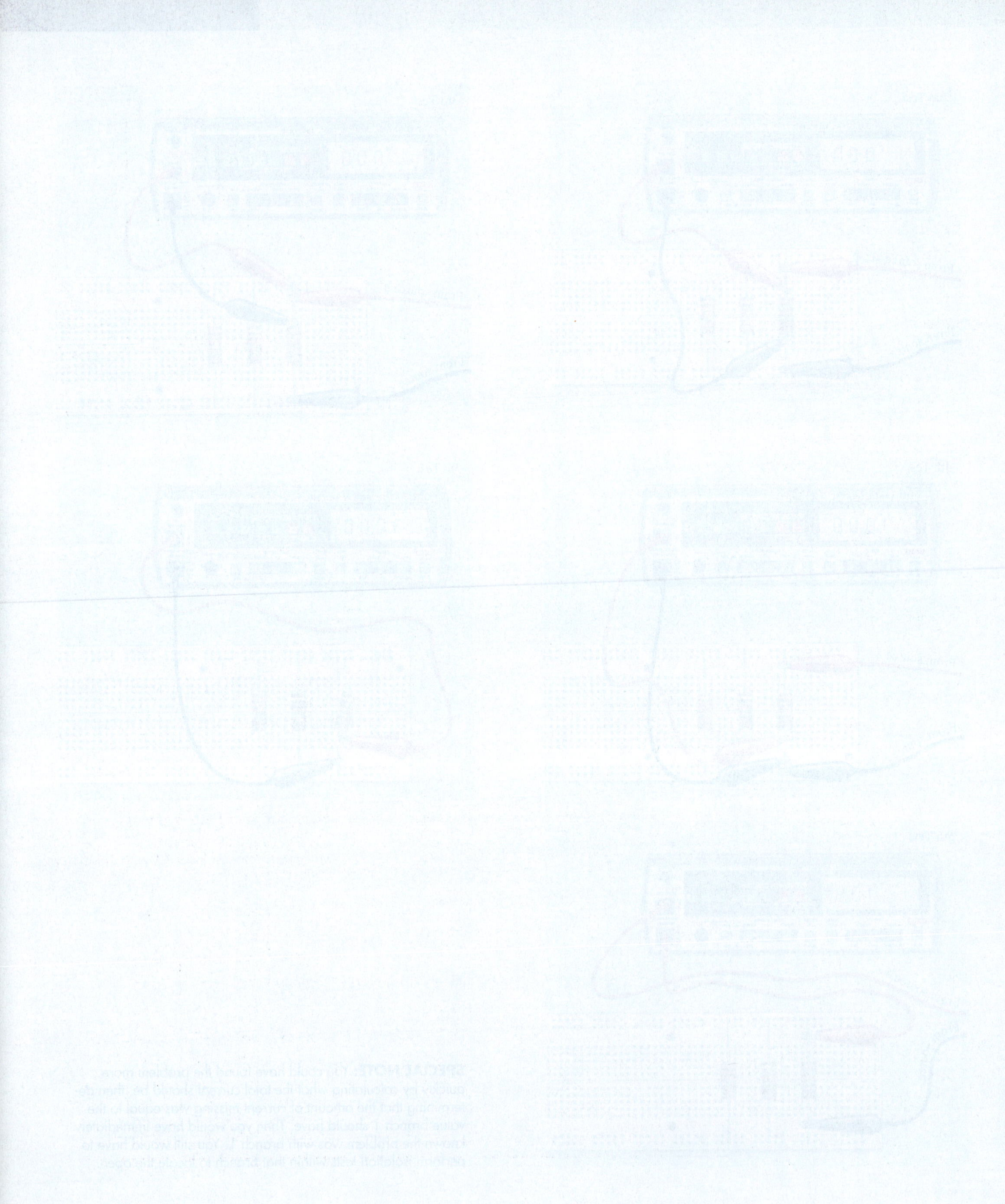

CHALLENGE CIRCUIT 4 (Parallel Circuit Block Diagram)

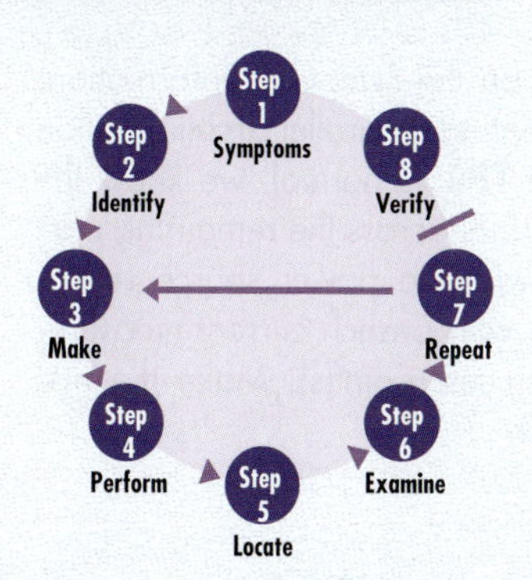

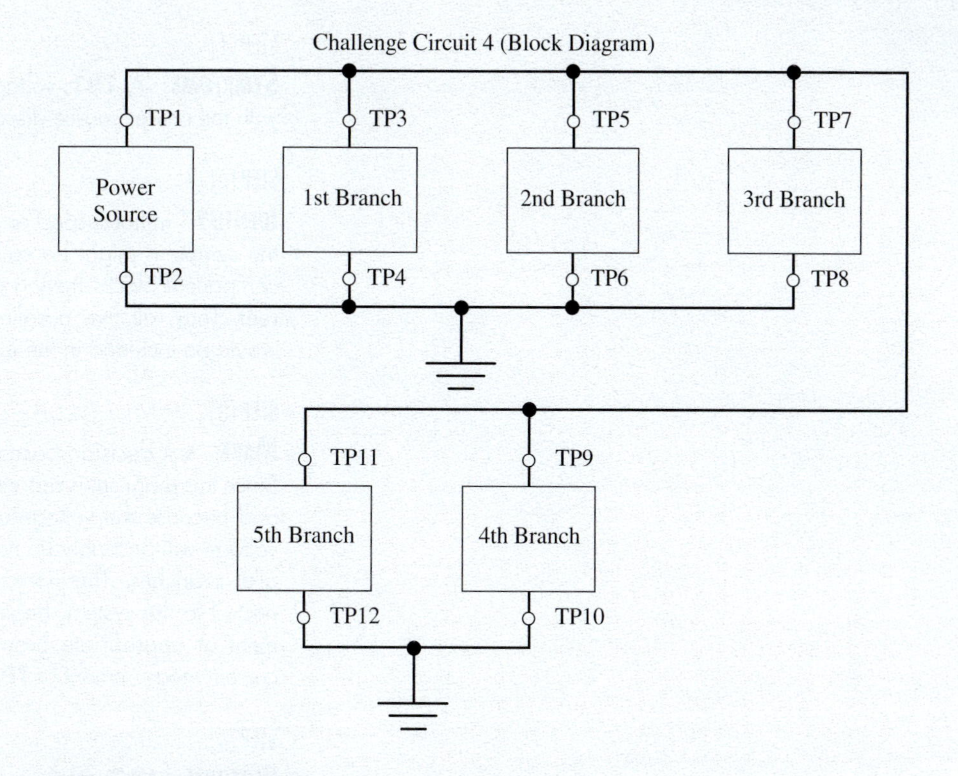

General Testing Instructions:

Measurement Assumptions:
I = at **TP**
V = from **TP** to ground
R = from **TP** to ground (with the power source disconnected from the circuit)

Possible Tests & Results:
Current (high, low, normal)
Voltage (high, low, normal)
Resistance (high, low, normal)

Starting Point Information	Test Points	Test Results in Appendix C *V*	*I*	*R*	*Signal*
At TP1:	TP2	(160	161	162	NA)
	TP3	(163	164	165	NA)
V = normal	TP4	(166	167	168	NA)
I = high	TP5	(169	170	171	NA)
R = low	TP6	(172	173	174	NA)
	TP7	(175	176	177	NA)
	TP8	(178	179	180	NA)
	TP9	(181	182	183	NA)
	TP10	(184	185	186	NA)
	TP11	(187	188	189	NA)
	TP12	(190	191	192	NA)

STEP [1]

SYMPTOMS At **TP1,** voltage is normal, current is high and with the power source disconnected, resistance is low.

STEP [2]

IDENTIFY initial suspect area. Since this is a parallel circuit, the symptom might be caused by any one of the branch component blocks having decreased its opposition to current. Thus, all five parallel branches (component blocks) would be included in the initial bracketing.

STEP [3]

MAKE test decision based on the symptoms information. Since this diagram is representing a parallel circuit situation and because the voltage at **TP1** is normal, we know the voltage will probably be normal across the remaining parallel branches. Therefore, with the power source reconnected to the system, begin with branch current measurement at appropriate branch test point(s). Make the first current measurement at **TP3**.

STEP [4]

PERFORM **1st Test:** Current value at **TP3** is normal.

STEP [5]

LOCATE new suspect area. The remaining 4 branches (branches 2, 3, 4, and 5) are still in question. Since there is not a simpler way to narrow down, at this point, just methodically check the current through these questionable branches.

STEP [6]

EXAMINE available data.

STEP [7]

REPEAT analysis and testing.
2nd Test: measure current at **TP5.**
Current at **TP5** reads normal.
3rd Test: measure current at **TP7.**
Current at **TP7** checks normal.
4th Test: measure the current at **TP9.**
Current at **TP9** measures higher than normal. This suggests that Branch 4 is the branch that is out of specifications.

STEP [8]

VERIFY **5th Test:** When the 4th Branch Block is replaced with a new module that meets specifications, the circuit current comes back into the system specifications.

OUTLINE

OBJECTIVES

After studying this chapter, you should be able to:

1. Define the term **series-parallel circuit**
2. List the primary characteristic(s) of a series-parallel circuit
3. Determine the total resistance in a series-parallel circuit
4. Compute total circuit current and current through any given portion of a series-parallel circuit
5. Calculate voltages throughout a series-parallel circuit
6. Determine power values throughout a series-parallel circuit
7. Analyze the effects of an open in a series-parallel circuit
8. Analyze the effects of a short in a series-parallel circuit
9. Design a simple series-parallel circuit to specifications
10. Explain the **loading effects** on a series-parallel circuit
11. Calculate values relating to a **loaded voltage divider**
12. Make calculations relating to **bridge circuits**
13. Use the computer to solve circuit problems
14. Use the SIMPLER troubleshooting sequence to solve the Troubleshooting Challenge problems

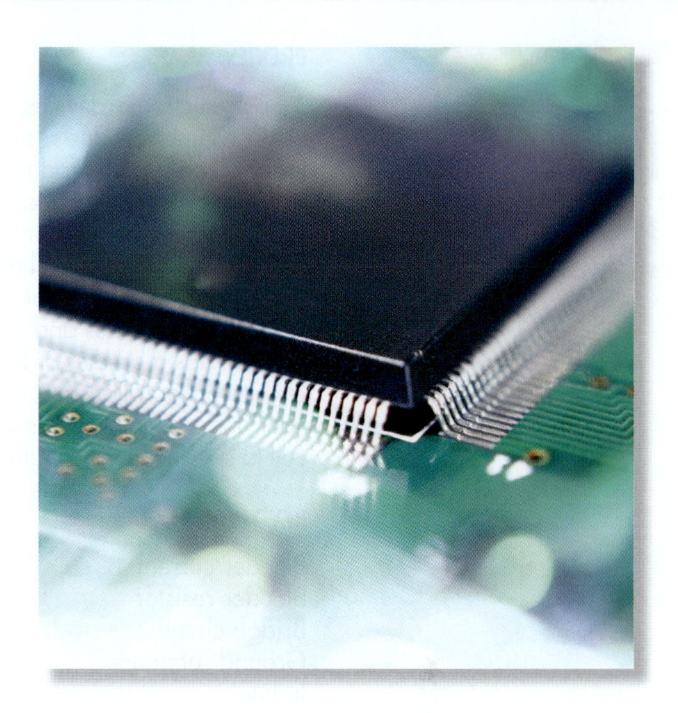

CHAPTER 6

Series-Parallel Circuits

ELECTRONICS INTO THE FUTURE

Explore an interactive presentation on Series-Parallel Circuits on the accompanying CD.

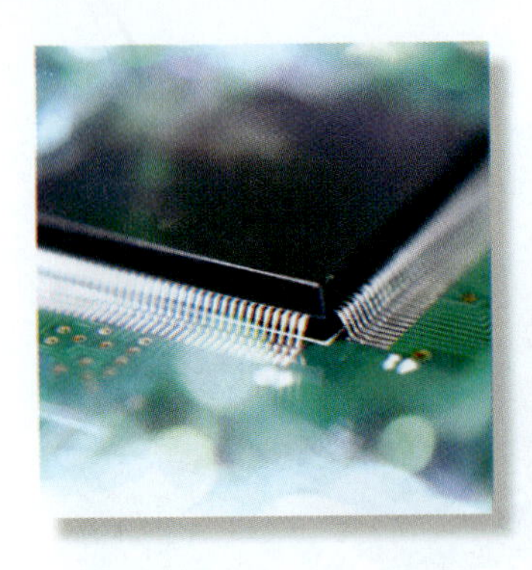

PREVIEW

Most electronic devices, equipment, and systems consist of series-parallel circuitry. Consumer products, such as high fidelity amplifiers, televisions, computers, and virtually all electronic products are filled with series-parallel circuits. A combination of both series and parallel connections form circuits that are coupled, or combined, to perform a desired task or function.

In this chapter, you will examine various configurations of these "combinational" circuits and will become familiar with analyzing them, using the principles you have already learned about series and parallel circuits. Also, you will see how a change in one part of a circuit has a greater or lesser effect on another portion of the circuit, depending on total circuit configuration and component values. Troubleshooting concepts will be examined. Your understanding of these circuits will be used by designing a simple series-parallel circuit to specifications. Finally, special applications of series-parallel circuitry will be studied.

KEY TERMS

Bleeder current
Bleeder resistor
Bridge circuit
Ground reference
In-line
Load
Load current
Loaded voltage divider
Potentiometer
R_L
Series-parallel circuit
Wheatstone bridge circuit

6–1 What Is a Series-Parallel Circuit?

Definition

A **series-parallel circuit** contains a combination of both series-connected and parallel-connected components. There are both **in-line** series current paths and "branching-type" parallel current paths, Figure 6–1.

Characteristics

Series components in a series-parallel circuit may be in series with other *individual* components, or with other *combinations* of components, Figure 6–2.

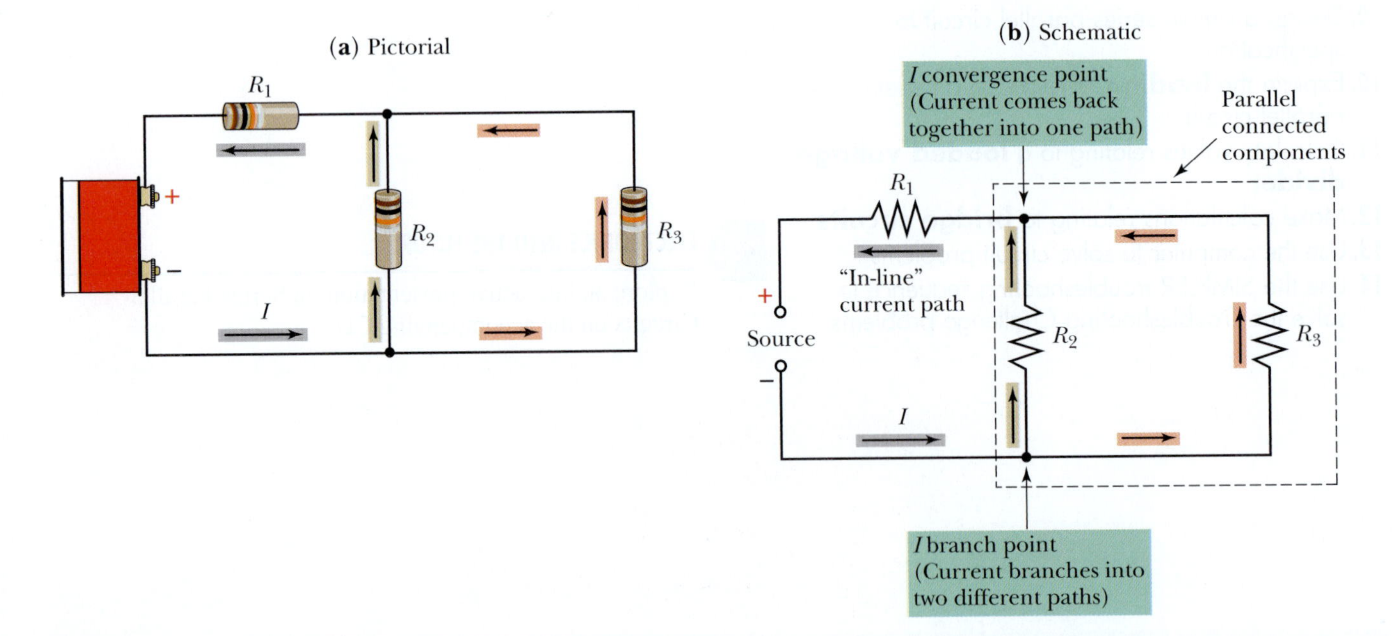

FIGURE 6–1 A basic series-parallel circuit

Parallel components in a series-parallel circuit may be in parallel with other *individual* components, or with other *combinations* of components, Figure 6–3.

6–2 Approaches to Recognize and Analyze Series and Parallel Circuit Portions

1. Start analysis at the portion of the circuit farthest from the source and work back toward the source to identify those components or circuit portions that are in series and in parallel.
2. Trace common current paths to identify components in series, Figure 6–4. Components, or combinations of components with common current are in series with each other.
3. Observe voltages shared (in common) to identify components in parallel, Figure 6–5. Components, or combinations of components with the same voltage connection points (at both ends) are in parallel with each other.
4. Observe current branching and converging points to identify components, or combinations of components that are in parallel with each other, Figure 6–6.
   a. A point where current splits, or branches, is one end of a parallel combination of components, Figure 6–6, points A and B.
   b. The point where those same currents converge (or rejoin) is the other end of that same parallel combination, Figure 6–6, points C and D.

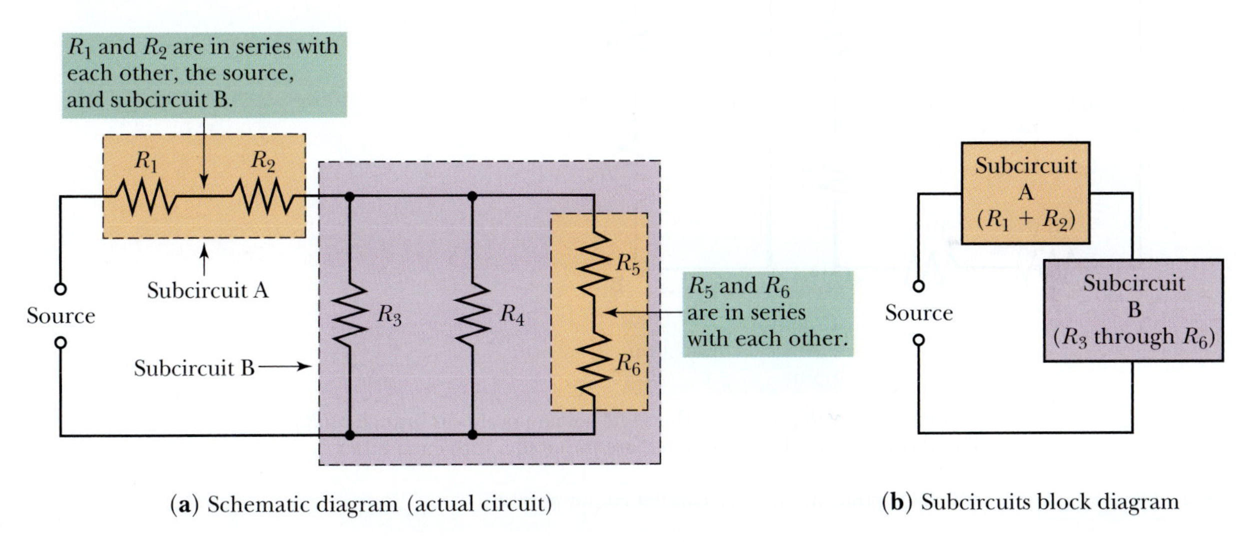

(a) Schematic diagram (actual circuit)

(b) Subcircuits block diagram

FIGURE 6–2 Series components in a series-parallel circuit

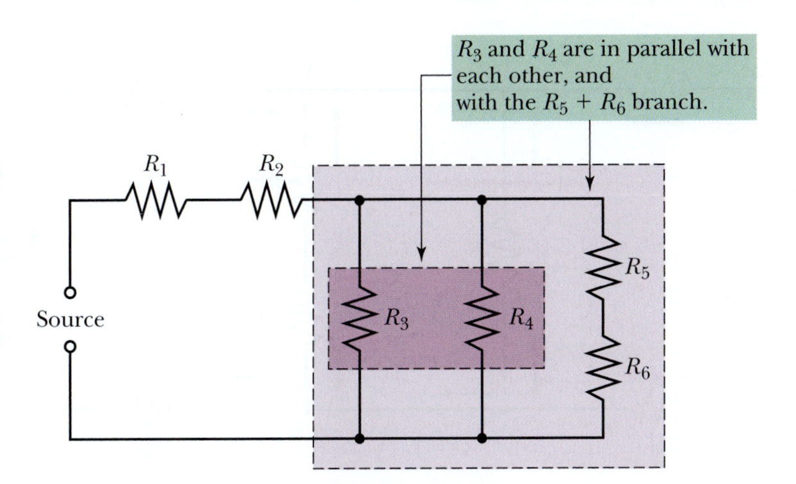

FIGURE 6–3 Parallel components in a series-parallel circuit

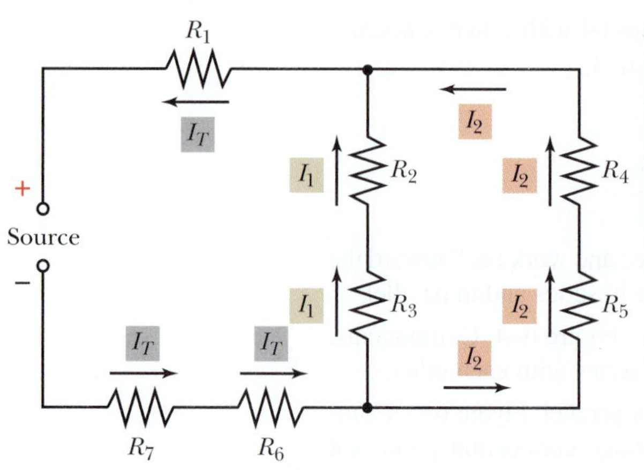

Starting "away from the source" and working back:
1. R_4 and R_5 have common current (I_2) and are in series with each other.
2. R_2 and R_3 have common current (I_1) and are in series with each other.
3. R_1, R_6, and R_7 have common current (I_T) and they are in series with each other, and with the total parallel combination of the $R_2 + R_3$ branch in parallel with the $R_4 + R_5$ branch.

FIGURE 6–4 Tracing common current paths to identify series circuit portions

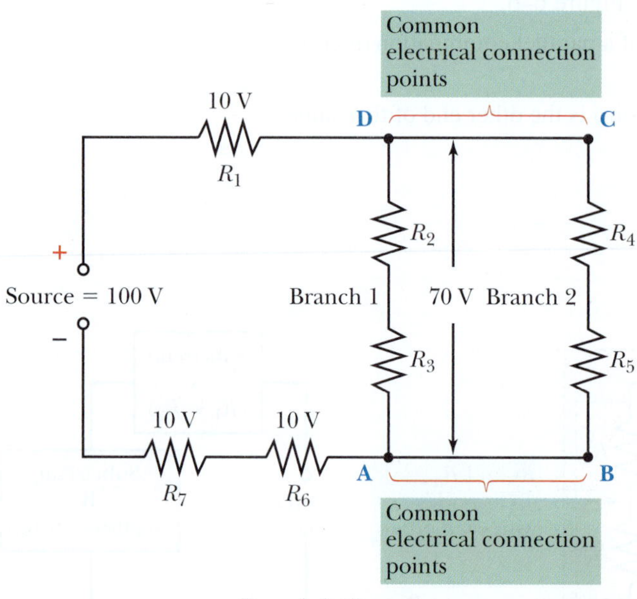

NOTE: Points **A** and **B** are electrically the same point and points **C** and **D** are electrically the same point. Therefore, the voltage from point **B** to point **C** is the same voltage as the voltage from point **A** to point **D**.

Branch 2 ($R_4 + R_5$) is across the same voltage points as branch 1 ($R_2 + R_3$). Therefore, branch 1 and branch 2 are in parallel with each other.

FIGURE 6–5 Observing common voltage points to identify parallel circuit portions

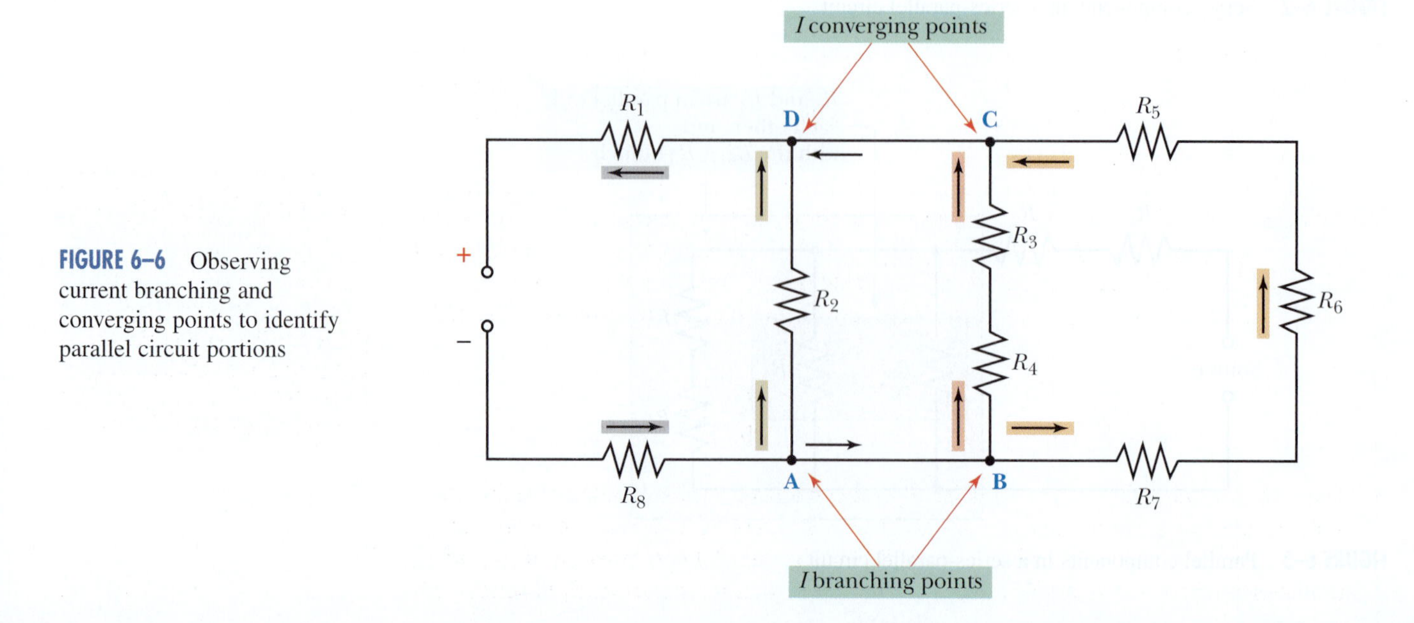

FIGURE 6–6 Observing current branching and converging points to identify parallel circuit portions

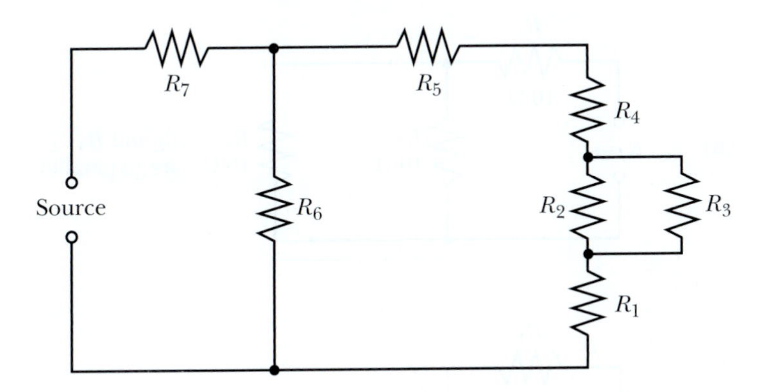

FIGURE 6–7

5. In summary, in the series portions of the circuit, current is common and the voltages (and resistances) are additive. In the parallel portions of the circuit, voltage is common and the branch currents are additive.

PRACTICE PROBLEMS 1

Refer to Figure 6–7 and identify *individual* or single components that:

1. carry total current;
2. are in parallel with each other; that is, have the same voltage across them; and
3. are in series with each other; that is, have the same current through them.

Let's move to some analyses of series-parallel circuit electrical parameters.

6–3 Total Resistance in Series-Parallel Circuits

If total circuit current and circuit applied voltage are known, then solving for total circuit resistance is easy.

Using Ohm's Law

Simply use Ohm's law, where $R_T = V_T/I_T$. If source voltage (V_S) and total current are unknown, then apply the concepts discussed in the preceding section.

Using the "Outside toward the Source" Approach

1. Start at the end of the circuit farthest from the source and work toward the source, identifying and solving the series, the parallel, and the series-parallel combinations, Figure 6–8. You will apply series circuit rules for series portions, and parallel circuit rules for parallel portions in the circuitry.
2. First, identify and solve parallel component segments. The common voltage and the current branching and converging concepts are true for R_2 and R_3; therefore, these two components are in parallel, Figure 6–8a. As you recall, the equivalent resistance of two equal value resistors in parallel is equal to half the value of one equal branch R. Thus, R_e of R_2 and R_3 = 5 Ω. Naturally, you can also solve it with the product-over-the-sum formula. That is:

$$R_e = \frac{R_2 \times R_3}{R_2 + R_3} = \frac{(10\ \Omega) \times (10\ \Omega)}{(10\ \Omega) + (10\ \Omega)} = \frac{100\ \Omega}{20\ \Omega} = 5\ \Omega$$

3. Next, work toward the source to solve the series and/or series-parallel combinations. Total current splits and rejoins as it flows through the parallel combination of R_2 and R_3. Total current also flows through R_1. Thus, R_1 and the parallel combination of R_2 and R_3 are in series with each other, Figure 6–8b. Therefore, total resistance of R_1 in series with

A Good Idea

When given a circuit analysis problem without a diagram, sketch a simple circuit diagram to help you visualize the problem more clearly!

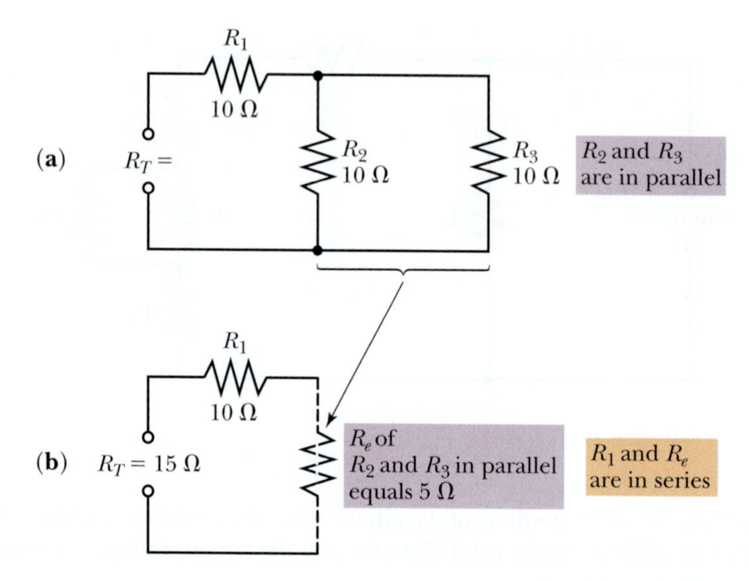

FIGURE 6–8 Analyzing a circuit to find R_T

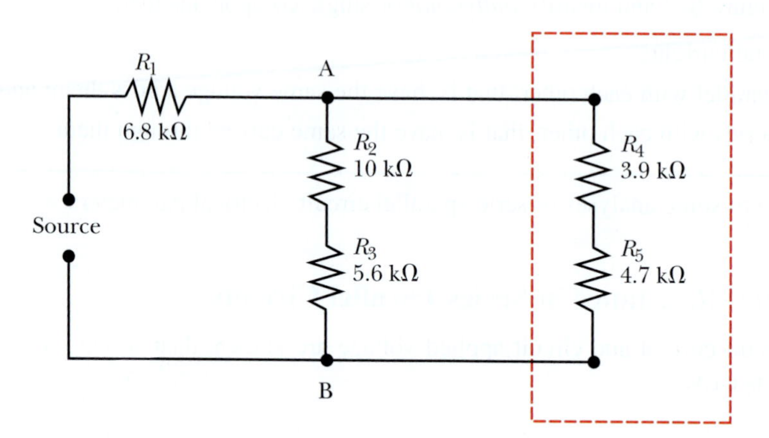

FIGURE 6–9 multiSIM

the parallel combination of R_2 and R_3 equals 10 Ω + 5 Ω = 15 Ω. And the circuit total resistance is 15 Ω.

◆ **EXAMPLE** Referring to Figure 6–9, let's look at another example of using the "outside-in" approach. This time we will use more typical resistor values that aren't quite as easy to analyze as our previous example.

1. Starting with the outside branch, you can see that R_4 and R_5 are in series. Using the calculator, the resistance of this outer branch is: 3.9, [EE], 3, [+], 4.7, [EE], 3, [=] 8,600 ohms, or 8.6 kΩ.
2. Working to the next branch toward the source, we can see that the resistance between points A and B will be equivalent resistance of the outer branch in parallel with the series resistors R_2 and R_3. $R_2 + R_3$ = 10 kΩ + 5.6 kΩ = 15.6 kΩ. The equivalent resistance of this 15.6 kΩ in parallel with the outer branch value of 8.6 kΩ can be quickly calculated with the calculator: 15.6, [EE], 3, [1/x], [+], 8.6, [EE], 3, [1/x], [=], [1/x]. The answer for the value of resistance between points A and B is 5,543 ohms, or 5.54 kΩ.
3. Now, all that remains is to add R_1's value to the value of resistance between points A and B. Thus R_{total} for the circuit is: 6.8, [EE], 3, [+], 5.54, [EE], 3, [=]. R_T for this circuit equals 12.34 kΩ. ◆

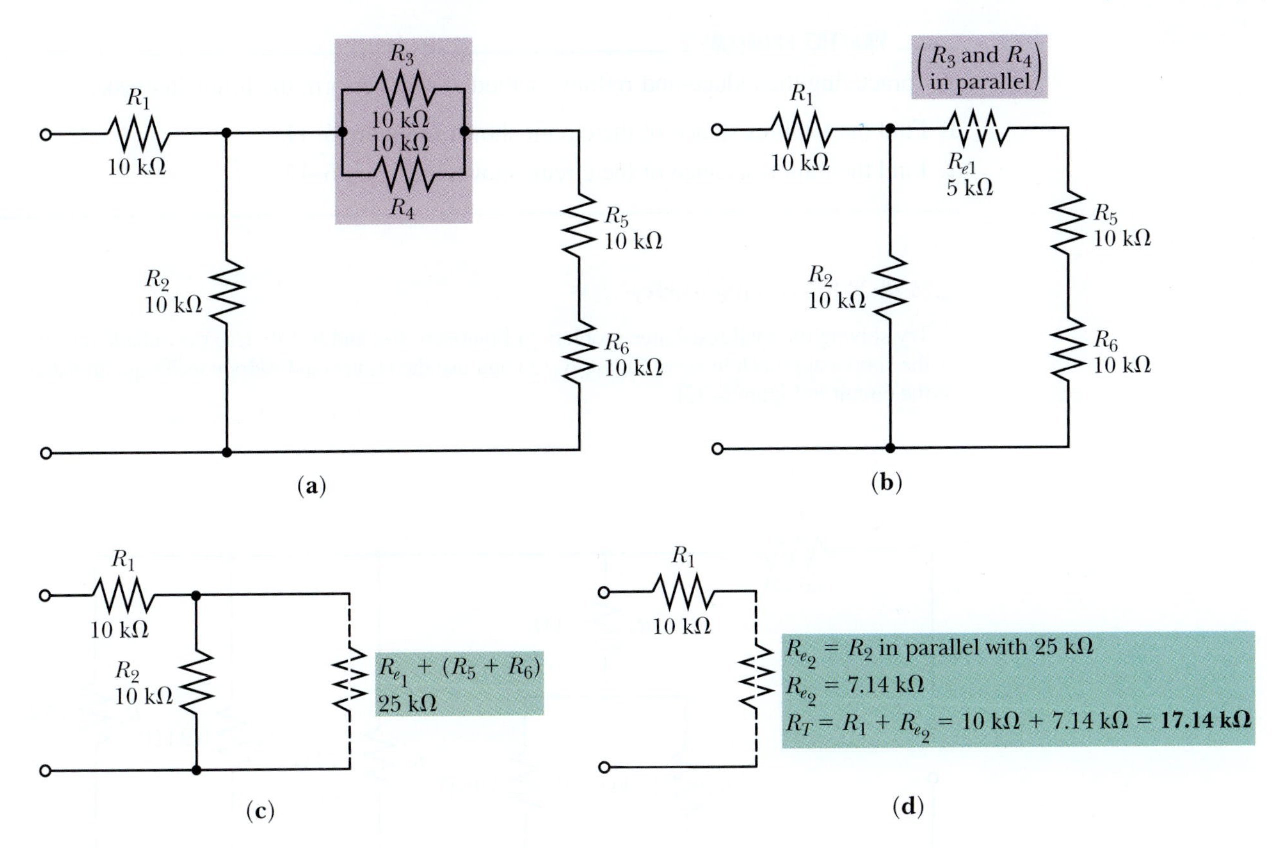

FIGURE 6–10 Simplifying by the reduce-and-redraw technique

The Reduce-and-Redraw Approach

A useful method to analyze series-parallel circuits is the reduce-and-redraw approach. This technique simplifies circuit analysis.

EXAMPLE Look at Figure 6–10 as R_T is discussed for this series-parallel combination circuit.

1. R_3 and R_4 are in parallel with each other and have an equivalent resistance that we will call R_{e_1}. Thus, $R_{e_1} = 5$ kΩ. R_{e_1}, the resistance of R_3 and R_4 in parallel equals:

$$\frac{(10\text{ k}\Omega)(10\text{ k}\Omega)}{(10\text{ k}\Omega)+(10\text{ k}\Omega)} = \frac{100\times10^6}{20\times10^3} = 5\times10^3 = 5\text{ k}\Omega \text{ (Figure 6–10b)}$$

   Of course, a simpler way is to remember that two equal branch resistances in parallel have an equivalent resistance equal to half of one of the branch resistances.

2. R_5 and R_6 are in series with each other and with R_{e_1} of parallel resistors R_3 and R_4. See Figure 6–10c and the calculations that follow.

   The resistance of $R_{e_1} = 5$ kΩ
   The resistance of $R_5 + R_6$ in series = 10 kΩ + 10 kΩ = 20 kΩ
   The resistance of $R_{e_1} + R_5 + R_6 = 25$ kΩ

3. R_2 (10 kΩ) is in parallel with the total 25-kΩ combination of components R_3 through R_6 ($R_{e_1} + R_5 + R_6$). See Figure 6–10c and 6–10d and the following calculations.

$$R_{e2} = \frac{(10\text{ k}\Omega)(25\text{ k}\Omega)}{(10\text{ k}\Omega)+(25\text{ k}\Omega)} = \frac{250\times10^6}{35\times10^3} = 7.14\times10^3 = 7.14\text{ k}\Omega$$

4. The overall combination of components R_2 through R_6 (R_{e_2}) is in series with R_1. See Figure 6–10d and the following calculations.

Total circuit resistance $(R_T) = R_1 + R_{e_2} = 10$ kΩ + 7.14 kΩ = 17.14 kΩ

PRACTICE PROBLEMS 2

Try practicing the reduce-and-redraw method as you perform the following tasks.

1. Find the total resistance of the circuit shown in Figure 6–11.
2. Find the total resistance of the circuit shown in Figure 6–12.

■ IN-PROCESS LEARNING CHECK 1

Try solving the total resistance problem in Figures 6–13a and 6–13b. Use the outside toward the source approach to solve Figure 6–13a and use the reduce-and-redraw technique to solve the circuit in Figure 6–13b.

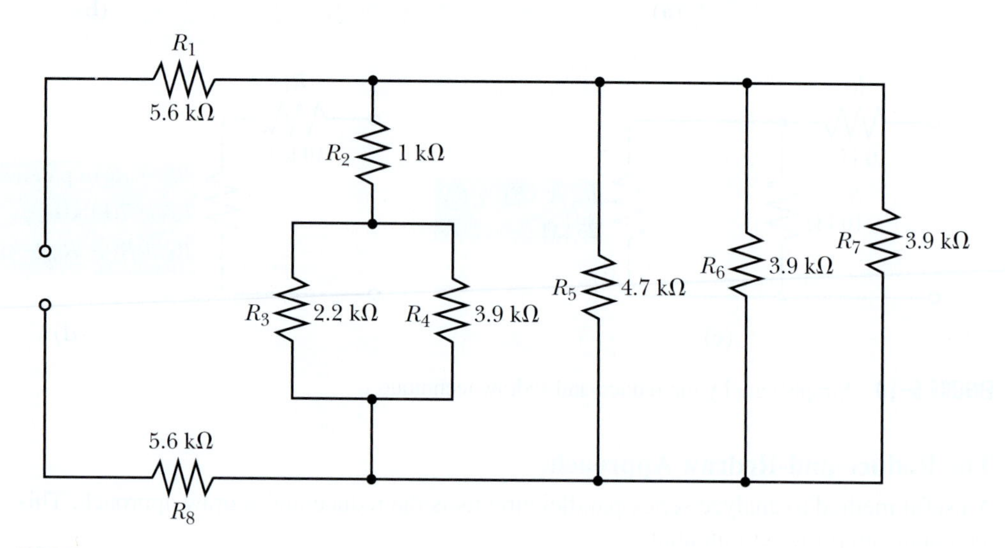

FIGURE 6–11

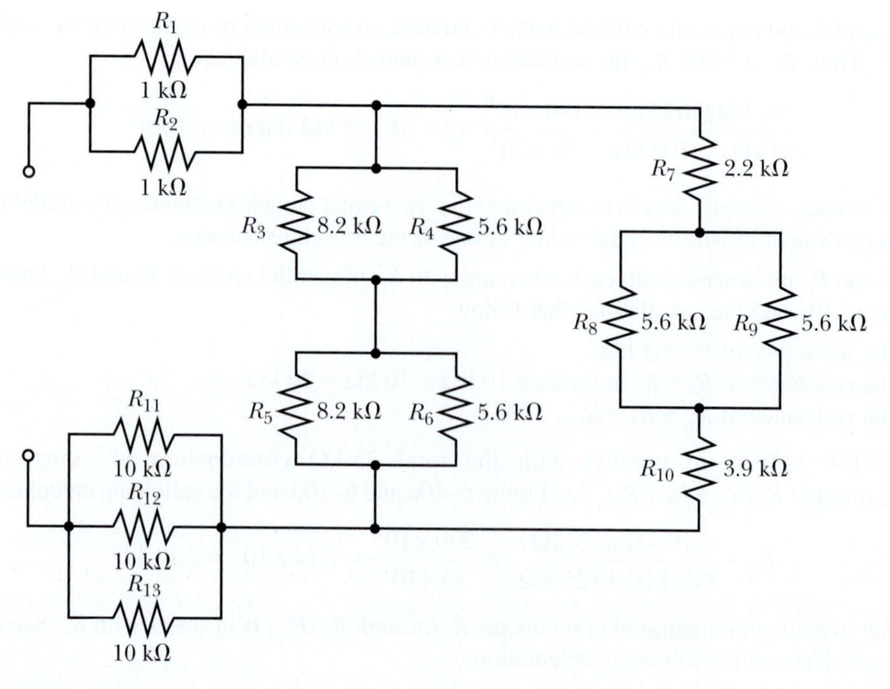

FIGURE 6–12 multiSIM

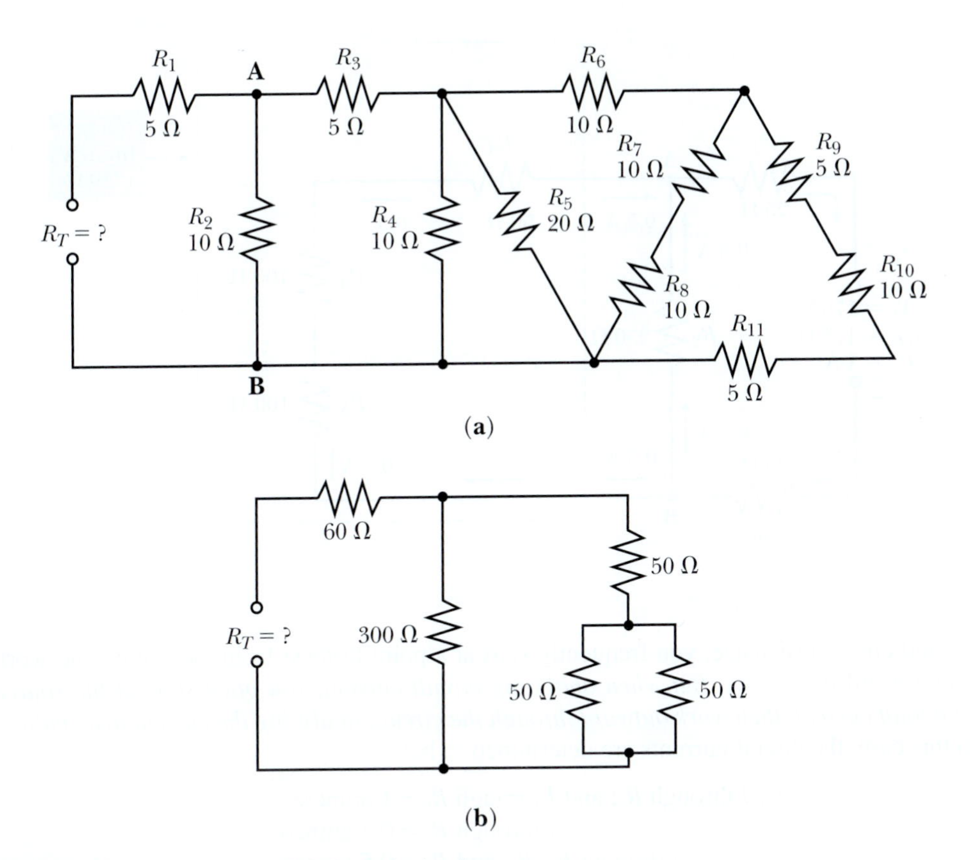

FIGURE 6–13 Solve for R_T values in each circuit.

6–4 Current in Series-Parallel Circuits

Examine Figure 6–14 as we now discuss current in a series-parallel circuit. One can start by finding the total circuit current, then work from that point to determine current distribution throughout the circuit using Ohm's law, Kirchhoff's laws, current divider formulas, and/or other techniques.

It is easier if Ohm's law is used to find total current, where $I_T = V_S/R_T$. In this circuit, the source voltage (V_S) is 175 volts and the total resistance is 175 Ω. For practice, verify the R_T value using the techniques you learned in the preceding section.

Now that R_T and V_S are known, it is easy to solve for I_T.

$$I_T = \frac{V_S}{R_T} = \frac{175 \text{ V}}{175 \ \Omega} = 1 \text{ A}$$

What about the distribution of current throughout the circuit? For this circuit, distribution of current is easy to find because of the values used. One example of a thought process that can be used is as follows:

1. R_1 and R_6 must have 1 ampere through them, since they are in series with the source and the remaining circuitry. In other words, I_T passes through both these resistors, Figure 6–14.
2. Since R_2 (250 Ω) and the outer branch (250 Ω) are equal resistance value branches, the total current equally divides between them. Thus, current through R_2 = 0.5 ampere, and current through the outer branch must equal 0.5 ampere, Figure 6–14.
3. It is apparent in tracing current through the outer branch that all of the 0.5 ampere must pass through R_3, R_4, and R_5, Figure 6–14.

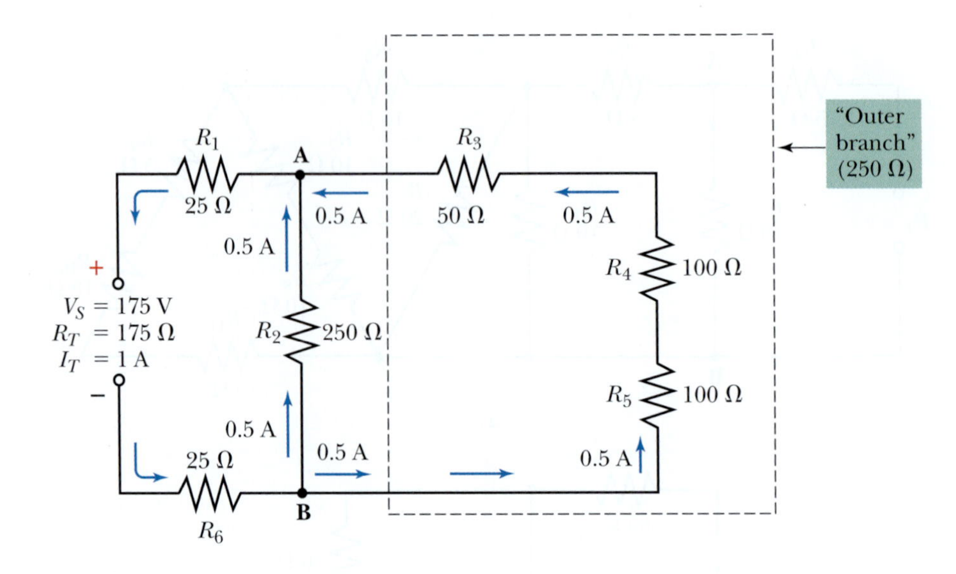

FIGURE 6–14 Solve for current values throughout a series-parallel circuit.

To find circuit resistance, you frequently start at a point farthest from the source and work back toward the source. *But when analyzing circuit current, you often start at the source with total current, then work outward through the circuit, analyzing the current distribution.* In this case, the circuit currents are determined to be:

$$I_T;\ I \text{ through } R_1; \text{ and } I \text{ through } R_6 = 1 \text{ ampere.}$$
$$I \text{ through } R_2 = 0.5 \text{ ampere.}$$
$$I \text{ through } R_3, R_4, \text{ and } R_5 = 0.5 \text{ ampere.}$$

Of course, several other methods, or combinations of methods can be used to provide solutions, such as Ohm's law, Kirchhoff's voltage and current laws, and current divider formulas. In those cases where the values are not as convenient, you will probably have to use these other methods.

EXAMPLE Having solved for total current (1 ampere), you can refer to Figure 6–14 and apply known values to solve for the unknown parameters.

1. I_T passes through both R_1 and R_6. We can use Ohm's law to find the voltage drops. Thus,

$$V_{R_1} = I_T \times R_1 = 1 \text{ A} \times 25\ \Omega = 25 \text{ V}$$
$$V_{R_6} = I_T \times R_6 = 1 \text{ A} \times 25\ \Omega = 25 \text{ V}$$

2. According to Kirchhoff's voltage law, the sum of the voltages around any closed loop must equal V applied. This means the sum of $V_1 + V_{A\text{–}B} + V_6$ must equal 175 V.
   Thus, $25 \text{ V} + V_{A\text{–}B} + 25 \text{ V} = 175 \text{ V}$

$$50 \text{ V} + V_{A\text{–}B} = 175 \text{ V}$$
$$V_{A\text{–}B} = 175 \text{ V} - 50 \text{ V} = 125 \text{ V}$$

3. Kirchhoff's current law indicates that the current leaving point A must equal the current entering point A. Stated another way, the total of the currents through R_2 and the outer branch must equal I_T in this case.
   a. Use Ohm's law to find current through R_2.

$$I \text{ through } R_2 = \frac{V_{R_2}}{R_2} = \frac{125 \text{ V}}{250\ \Omega} = 0.5 \text{ A}$$

   b. Using Kirchhoff's current law:

$$I \text{ through outer branch} = I_T - I \text{ (through } R_2);$$
$$= 1 \text{ A} - 0.5 \text{ A} = 0.5 \text{ A}$$

4. Ohm's law and Kirchhoff's voltage law are now used to solve the voltage drops across the resistors in the outer branch.

Knowing the outer branch current is 0.5 A:

a. $V_{R_5} = 0.5 \text{ A} \times R_5 = 0.5 \text{ A} \times 100\ \Omega = 50 \text{ V}$

b. $V_{R_4} = 0.5 \text{ A} \times R_4 = 0.5 \text{ A} \times 100\ \Omega = 50 \text{ V}$

c. $V_{R_3} = 0.5 \text{ A} \times R_3 = 0.5 \text{ A} \times 50\ \Omega = 25 \text{ V}$ ◆

PRACTICE PROBLEMS 3

1. For practice in solving for current distribution throughout a series-parallel circuit, refer to Figure 6–15. Solve for I_T, I_1, I_2, and I_3.
2. Refer to Figure 6–16 and find the values of I_T, I_1, I_2, and I_3.

6–5 Voltage in Series-Parallel Circuits

Voltage distribution throughout a series-parallel circuit is such that (1) voltage across components or circuit portions that are in series with each other are additive and (2) voltage across components or circuit portions that are in parallel must have equal values.

As you would expect, a given component's voltage drop in a series-parallel circuit is contingent on its circuit location and its resistance value. The location dictates what portion of the circuit's total current passes through it. Consequently, location of any given component within the series-parallel circuit has direct influence on its $I \times R$ drop.

A Good Idea

Use the "approximation" technique to help prevent needless errors when performing calculations!

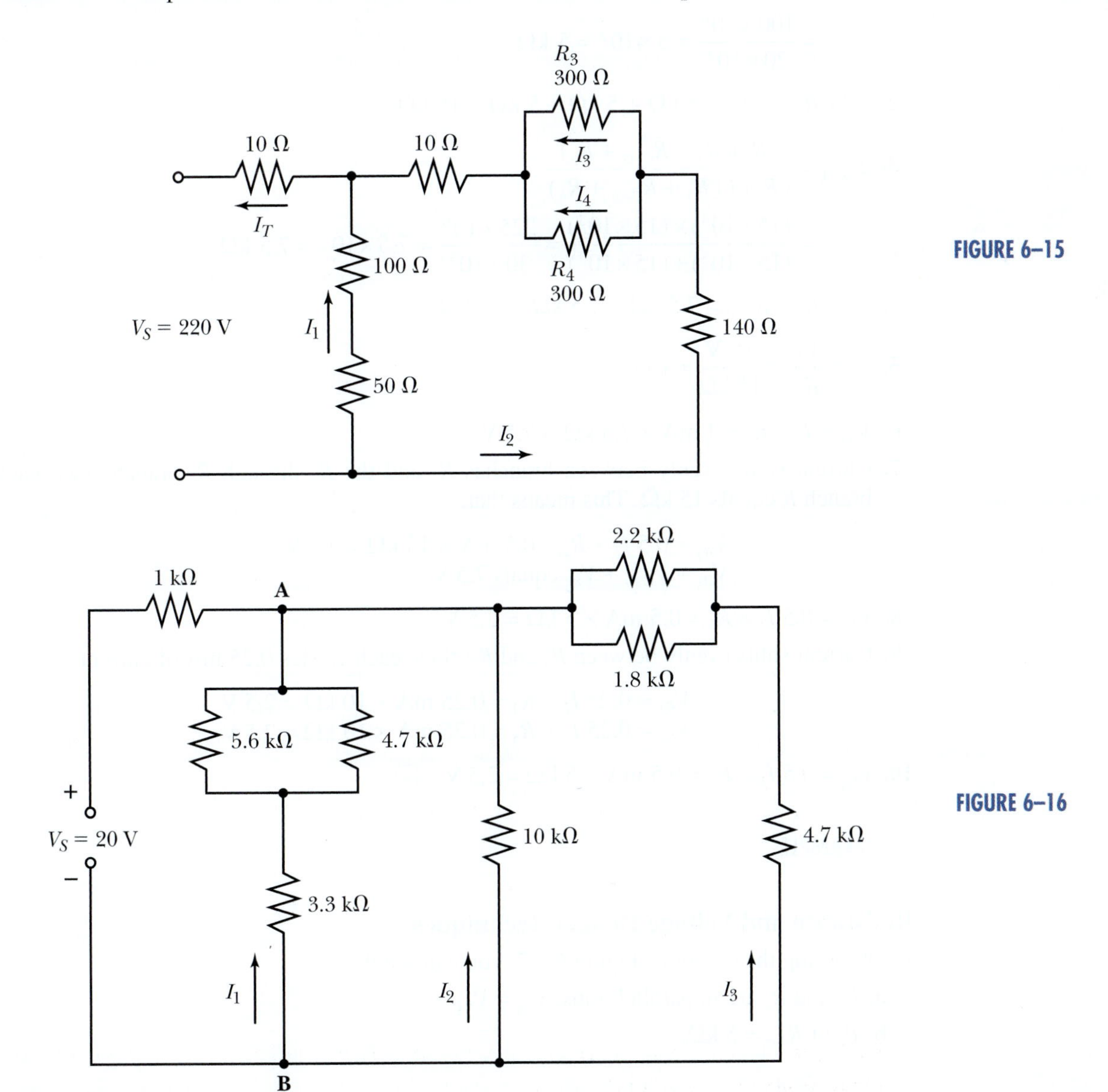

FIGURE 6–15

FIGURE 6–16

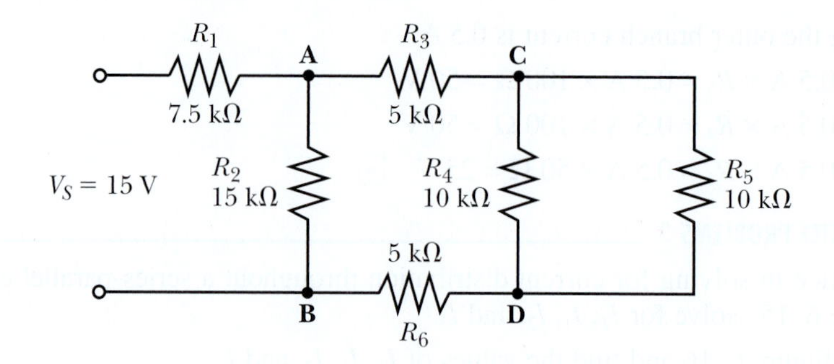

FIGURE 6–17 Analyzing voltages in a series-parallel circuit

Refer to Figure 6–17 as the voltage distribution characteristics are analyzed with two methods: the Ohm's law method and resistance and voltage divider techniques.

EXAMPLE

Ohm's Law Method

Solve circuit R and I values, then use Ohm's law to compute voltages.

1. $R_{C-D} = \dfrac{R_4 \times R_5}{R_4 + R_5} = \dfrac{(10\times10^3)\times(10\times10^3)}{(10\times10^3)+(10\times10^3)}$

   $= \dfrac{100\times10^6}{20\times10^3} = 5\times10^3 = 5\text{ k}\Omega$

2. $R_3 + R_{C-D} + R_6 = 5\text{ k}\Omega + 5\text{ k}\Omega + 5\text{ k}\Omega = 15\text{ k}\Omega$

3. $R_{A-B} = \dfrac{(R_2)(R_3 + R_{C-D} + R_6)}{(R_2)+(R_3 + R_{C-D} + R_6)}$

   $= \dfrac{(15\times10^3)\times(15\times10^3)}{(15\times10^3)+(15\times10^3)} = \dfrac{225\times10^6}{30\times10^3} = 7.5\times10^3 = 7.5\text{ k}\Omega$

4. $R_T = R_1 + R_{A-B} = 7.5\text{ k}\Omega + 7.5\text{ k}\Omega = 15\text{ k}\Omega$

5. $I_T = \dfrac{V_T}{R_T} = \dfrac{15\text{ V}}{15\text{ k}\Omega} = 1\text{ mA}$

6. $V_{R_1} = I_T \times R_1 = 1\text{ mA} \times 7.5\text{ k}\Omega = 7.5\text{ V}$

7. Current splits evenly between branches R_2 and the R_3 through R_6 branch since each branch R equals 15 kΩ. This means that:

   $V_{R_2} = 0.5\ I_T \times R_2 = 0.5\text{ mA} \times 15\text{ k}\Omega = 7.5\text{ V}$
   $V_{R_3} + V_{C-D} + V_{R_6}$ equals 7.5 V

8. $V_{R_3} = 0.5\ I_T \times R_3 = 0.5\text{ mA} \times 5\text{ k}\Omega = 2.5\text{ V}$

9. Current splits evenly between R_4 and R_5; thus, each carries 0.25 mA of current.

   $V_{R_4} = 0.25\ I_T \times R_4 = 0.25\text{ mA} \times 10\text{ k}\Omega = 2.5\text{ V}$
   $V_{R_5} = 0.25\ I_T \times R_5 = 0.25\text{ mA} \times 10\text{ k}\Omega = 2.5\text{ V}$

10. $V_{R_6} = 0.5\ I_T \times R_6 = 0.5\text{ mA} \times 5\text{ k}\Omega = 2.5\text{ V}$

EXAMPLE

Resistance and Voltage Divider Techniques

1. Observing the circuit of Figure 6–17, you can see that:

   a. R_4 and R_5 are in parallel; thus, $V_{R_4} = V_{R_5}$.

   b. R_e of $R_{4-5} = 5\text{ k}\Omega$.

c. R_3, $R_{4\text{–}5}$, and R_6 are in series and together form a 15-kΩ branch of the circuit. Furthermore, each of these drops one-third the voltage present from point A to point B, since each of the three entities represents 5 kΩ.

d. This 15-kΩ branch is in parallel with R_2, another 15-kΩ branch. Thus, R_e from point A to point B equals 7.5 kΩ.

e. R_1 is also 7.5 kΩ and is in series with the equivalent 7.5 kΩ found from point A to point B. Thus V_{R_1} and $V_{\text{A–B}}$ are equal, and each equals half the voltage applied (Kirchhoff's voltage law).

2. Applying the principles of voltage dividers:

a. $V_{\text{A–B}} = \dfrac{R_{\text{A–B}}}{R_T} \times V_T = \dfrac{7.5\ \text{k}\Omega}{15\ \text{k}\Omega} \times 15\ \text{V} = 7.5\ \text{V}$

b. $V_{\text{C–D}} = \dfrac{R_{4\text{–}5}}{R_3 + R_{4\text{–}5} + R_6} \times V_{\text{A–B}} = \dfrac{5\ \text{k}\Omega}{15\ \text{k}\Omega} \times 7.5\ \text{V} = 2.5\ \text{V}$

c. $V_{R_3} = V_{R_4}$ and $V_{R_5} = V_{R_6}$

d. $V_{R_2} = V_{\text{A–B}}$

e. $V_{R_1} = \dfrac{R_1}{R_T} \times V_T = \dfrac{7.5\ \text{k}\Omega}{15\ \text{k}\Omega} \times 15\ \text{V} = 7.5\ \text{V}$ ◆

◆ **EXAMPLE** Let's look at one more example of analyzing a series-parallel circuit, using the Ohm's law method. Solve for circuit R and I values, then compute voltages, as appropriate, for the circuit of Figure 6–18.

1. $R_{\text{C–D}} = R_4 \parallel (R_5 + R_6)$

$$R_{\text{C–D}} = \frac{R_4 \times (R_5 + R_6)}{R_4 + (R_5 + R_6)} = \frac{(3.9\times 10^3)\times(4.9\times 10^3)}{(3.9\times 10^3)+(4.9\times 10^3)}$$

$$R_{\text{C–D}} = \frac{19.11\times 10^6}{8.8\times 10^3} = 2.17\ \text{k}\Omega$$

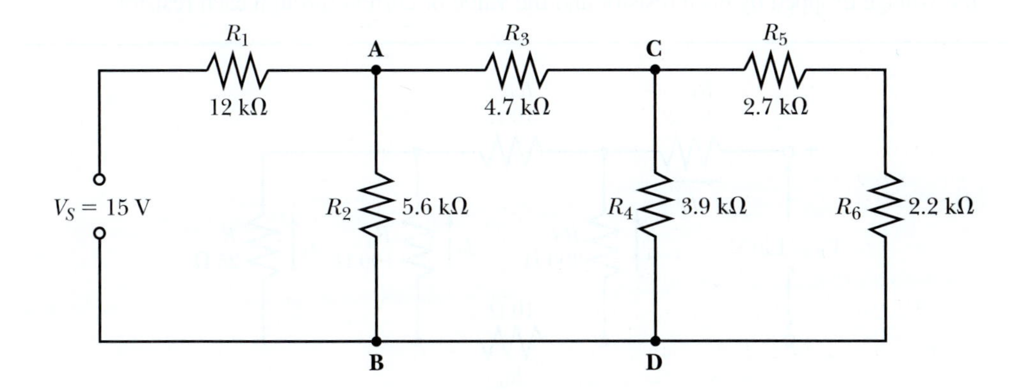

FIGURE 6–18 multiSIM

Practical Notes

An approach often used to simplify notation for indicating that components are in parallel with each other is to use two vertical bars, or short vertical lines between component designators. For example, in the circuit of Figure 6–17 to show that R_4 and R_5 are in parallel, we could state $R_4 \parallel R_5$. If we were going to use this type of notation, we could indicate the resistors and their connections, which make up the resistance at points A–B, as follows:

$$R_{\text{A–B}} = R_2 \parallel [R_3 + (R_4 \parallel R_5) + R_6]$$

2. $R_3 + R_{C\text{–}D} = 4.7\ k\Omega + 2.17\ k\Omega = 6.87\ k\Omega$

3. $R_{A\text{–}B} = R_2 \parallel (R_3 + R_{C\text{–}D}) = \dfrac{(5.6\times10^3)\times(6.87\times10^3)}{(5.6\times10^3)+(6.87\times10^3)} = \dfrac{38.47\times10^6}{12.47\times10^3}$
$= 3.09\ k\Omega$

4. $R_T = R_1 + R_{A\text{–}B} = 12\ k\Omega + 3.09\ k\Omega = 15.09\ k\Omega$

5. $I_T = \dfrac{V_S}{R_T} = \dfrac{15\ V}{15.09\ k\Omega} = 0.994\ mA$

6. $I_{R1} = 0.994\ mA$
$V_{R1} = I_{R1} \times R_1 = 0.994\ mA \times 12\ k\Omega = 11.93\ V$

7. $I_{R2} = \dfrac{V_{R2}}{R_2} = \dfrac{3.07\ V}{5.6\ k\Omega} = 0.5482\ mA$

8. $I_{R_3} = I_T - I_{R_2} = 0.994\ mA - 0.5482\ mA = 0.4458\ mA$
$V_{R_3} = I_{R_3} \times R_3 = 0.4458\ mA \times 4.7\ k\Omega = 2.095\ V$

9. $I_{R4} = \dfrac{V_{R4}}{R_4} = \dfrac{V_S - (V_{R1} + V_{R3})}{R_4} = \dfrac{15\ V - 14.025\ V}{3.9\ k\Omega} = \dfrac{0.975\ V}{3.9\ k\Omega} = 0.25\ mA$

10. $I_{R_5} = I_{R_6} = \dfrac{V_{C\text{–}D}}{R_5 + R_6} = \dfrac{0.975\ V}{4.9\ k\Omega} = 0.198\ mA$
$V_{R_5} = I_{R_5} \times R_5 = 0.198\ mA \times 2.7\ k\Omega = 0.5346\ V$
$V_{R_6} = I_{R_6} \times R_6 = 0.198\ mA \times 2.2\ k\Omega = 0.4356\ V$ ◆

PRACTICE PROBLEMS 4

1. Analyze the voltage and current parameters throughout the circuit of Figure 6–19. List the voltage dropped by each resistor and the current through each resistor.
2. Analyze the voltage and current parameters throughout the circuit of Figure 6–20. Again, list the voltage dropped by each resistor and the value of current through each resistor.

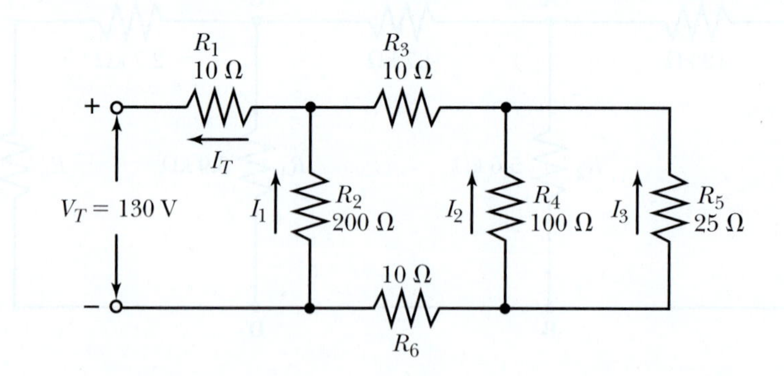

FIGURE 6–19

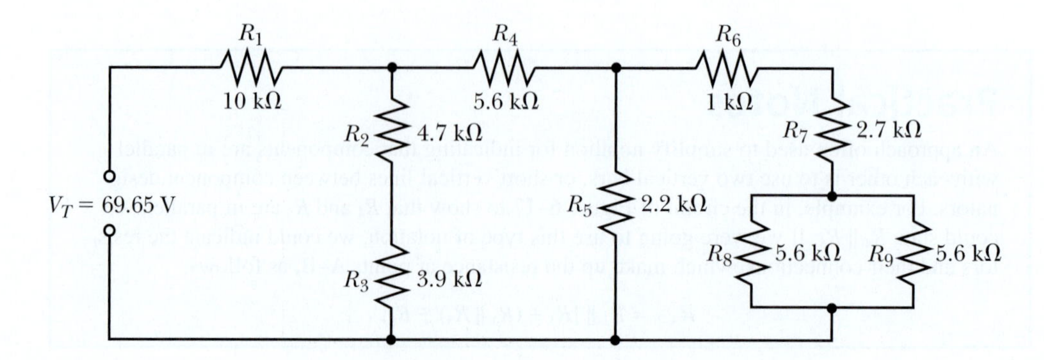

FIGURE 6–20

6–6 Power in Series-Parallel Circuits

Analysis of power dissipations throughout a series-parallel circuit employs the same techniques used in series or parallel circuits. Important points to remember are:

1. total power = $V_T \times I_T$ (or $I_T^2 \times R_T$; or, V_T^2/R_T);
2. total power also equals the sum of all individual power dissipations; and
3. individual component power dissipations are calculated using the individual component's parameters. That is, $V \times I$; I^2R; or V^2/R values.

EXAMPLE Notice the power dissipations throughout the circuit shown in Figure 6–21. If you take time to analyze this circuit completely, you find that the circuit's total resistance is 30 Ω. This means that the circuit's total current is 30 V/30 Ω, or 1 A. Current divides evenly in the two outer branches. Therefore, current through R_4 and $R_5 = 0.5$ A, and current through R_6 and R_7 is also 0.5 A. As you can see from the circuit diagram, the total 1 A current also evenly splits through R_2 and R_3, with each carrying 0.5 A.

With the current through each of the resistors known, you can then find the voltage drops across each component, and the power dissipated by each component, Figure 6–21. Since we know current, resistance, and voltage, any one of the power formulas ($P = V \times I$; $P = I^2R$; or $P = V^2/R$) might be used to compute each of the component's power dissipations. For example: $P_{R_7} = V \times I = 10\text{ V} \times 0.5\text{ A} = 5\text{ W}$; $P_{R_1} = V \times I = 10\text{ V} \times 1\text{ A} = 10\text{ W}$; and so on.

Practical Notes

1. In series circuits, the largest value resistor dissipates the most power, and the lowest value resistor dissipates the least power.
2. In parallel circuits, the smallest resistance branch dissipates the most power, and the largest resistance branch dissipates the least power.
3. In series-parallel circuits, the resistor dissipating the most power *may or may not* be the highest value resistor in the circuit. The resistor dissipating the lowest power *may or may not* be the lowest value resistor in the circuit.

The power dissipated is determined by the position of the given resistor in the total circuit configuration. An illustration of this is in Figure 6–21. Note the following:

- R_7 is twice the value of R_1 but dissipates only half the power.
- R_1 and R_2 are equal value resistors but each dissipates a different value of power.

To make sure you are learning to analyze the key parameters for a series-parallel circuit, perform the following In-Process Learning Check.

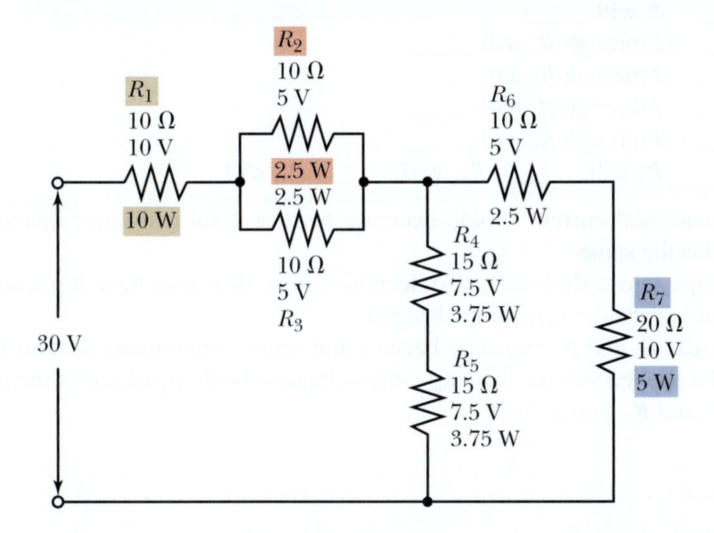

FIGURE 6–21 Effect of location on power, current, and voltage

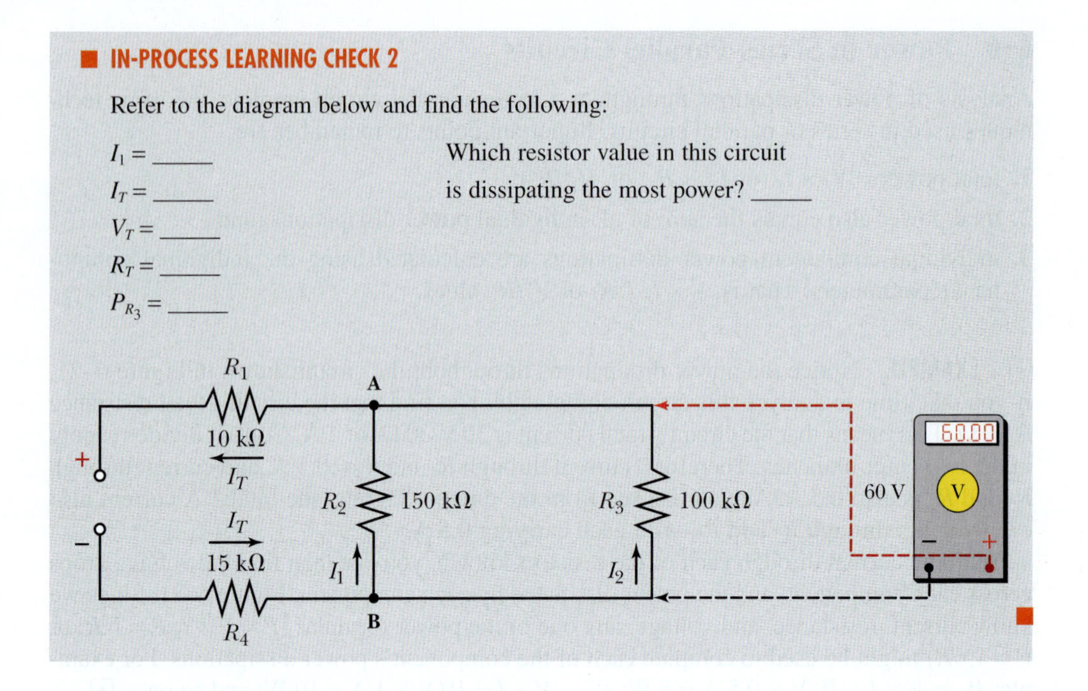

IN-PROCESS LEARNING CHECK 2

Refer to the diagram below and find the following:

I_1 = _____

I_T = _____

V_T = _____

R_T = _____

P_{R_3} = _____

Which resistor value in this circuit is dissipating the most power? _____

6–7 Effects of Opens in Series-Parallel Circuits and Troubleshooting Hints

Effects

As you have seen, the electrical location of a given component in a series-parallel circuit has great influence on all its electrical parameters. The location of an open in the circuit configuration also has a strong influence on all the parameters throughout the circuit.

Looking at the circuit used in In-Process Learning Check 2, you can see that if either R_1 or R_4 opens, then the path for total current is broken. Thus, all other components' voltage drops, currents, and power dissipations decrease to zero!

On the other hand, if R_3 opens, total current decreases, since the total resistance increases from 85 kΩ to 175 kΩ. But current does not decrease to zero. Thus, individual component voltages, currents, and power dissipations change; however V applied remains the same.

Here is a **special thinking exercise** for you. Again refer to the In-Process Learning Check circuit and assume R_2 opens. Will the following parameters increase, decrease, or remain the same?

V_T will _____ I_T will _____

V_{R_1} will _____ I through R_1 will _____

V_{R_2} will _____ I through R_2 will _____

V_{R_3} will _____ I through R_3 will _____

V_{R_4} will _____ I through R_4 will _____

P_T will _____ P_{R_3} will_____ P_{R_4} will _____

Total power and total current should decrease because total resistance increases while V applied remains the same.

Voltage drops across R_1 and R_4 also decrease, since they now have less current through them and their resistances remain unchanged.

Voltage across R_2 and R_3 increases because the series components (R_1 and R_4) are dropping less of the applied voltage, leaving more voltage to be dropped across the parallel combination of R_2 and R_3, points A–B.

Current decreases through R_1 and R_4 (remember I_T decreased); however, current through the unopened parallel branch, R_3, increases since current is no longer splitting between R_2 and R_3. Obviously, current through R_2 decreases to zero, since it is an open path.

Finally, power dissipations change. P_T, P_{R_1}, and P_{R_4} decrease due to the drop in I_T, and P_{R_3} increases because its current increases.

By now you should realize logical thinking is required when analyzing the sequence of events and the effects of an open in a series-parallel circuit. There is not just one simple approach. Rather, think of the interaction between the various portions of the circuit on each other and on the total circuit parameters.

Troubleshooting Hints

Because the component's position in the circuit is important due to the interaction it has on the remaining circuitry, the hints given here are somewhat general.

If an open occurs in *any* component in the circuit, total resistance increases; consequently, total current decreases. If you measure (or monitor) total current and it is lower than normal with normal applied voltage, an open, or at least an increased resistance value, has occurred somewhere in the circuit.

NOTE: Two typical ways to determine total current in a circuit are shown in Figure 6–22.

a. Insert current meter in series with source and the circuitry.
b. Measure voltage across a known resistor value that carries total current, and use Ohm's law (V/R) to determine current.

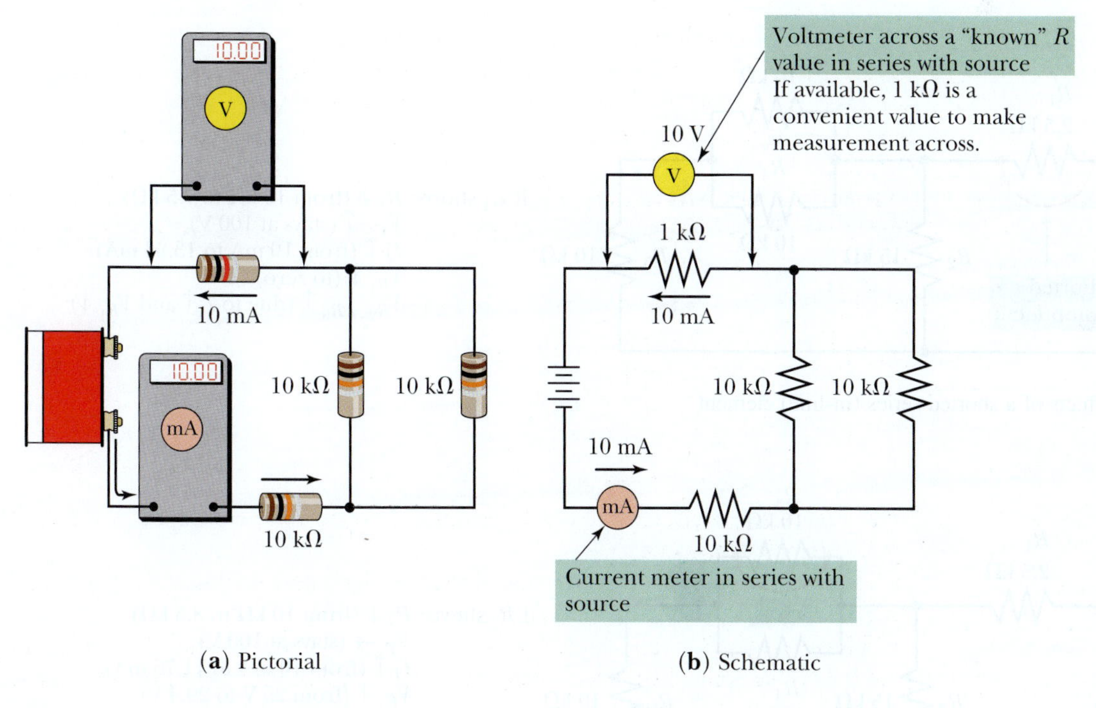

FIGURE 6–22 Two ways of checking total current

Practical Notes

A convenient resistance value for this approach is a 1-kΩ resistor, if available, because its voltage drop equals the number of mA through it. In other words,

$$\frac{10 \text{ V}}{1 \text{ k}\Omega} = 10 \text{ mA}$$

(See Figure 6–22.)

1. If an open occurs in any component in series with the source, total R increases to ∞, and total I decreases to zero. As you learned in series circuits, V applied appears across the opened series component and zero volts are dropped across the good components.
2. If an open occurs in any component other than those in series with the source, current measurements and *common sense* will isolate the problem.
3. If current measurements cannot be easily performed, you can make a number of voltage checks throughout the circuit. Observe where voltages are higher and lower than normal, then logically determine where the open is located.

6–8 Effects of Shorts in Series-Parallel Circuits and Troubleshooting Hints

Effects

Again, the location of the faulty component or circuit portion in the total circuit configuration controls what happens to the parameters of the remaining good components.

Refer to Figure 6–23. If the shorted component is in a line that carries total current, that is, in series with the source, then (1) total current increases, (2) voltage across the shorted component drops to zero, and (3) all other voltages throughout the circuit increase (except V_T).

Refer to Figure 6–24. If the shorted component is elsewhere in the circuit, then (1) total current still increases since total R decreases; however, (2) voltage across the shorted component

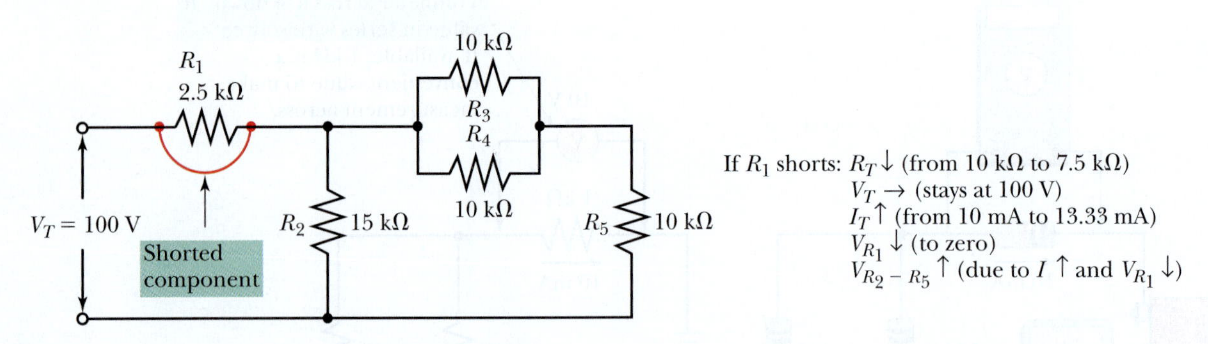

FIGURE 6–23 Effects of a shorted series (in-line) element

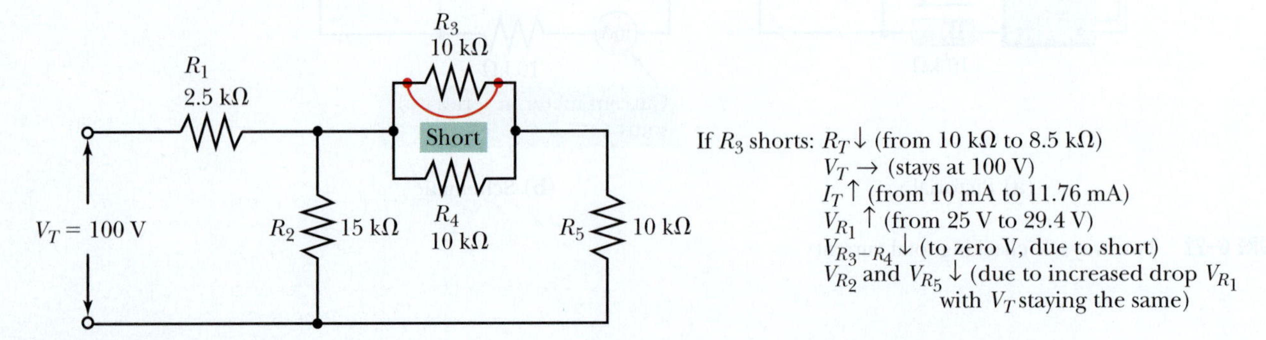

FIGURE 6–24 Effect of shorted element in the combinational portion of a series-parallel circuit multiSIM

and components *directly* in parallel with the shorted component decreases to zero, since the shorted component is shunting the current around components in parallel with it; and (3) other components' voltage drops may increase or decrease, depending on their location.

Troubleshooting Hints

Some general thoughts about troubleshooting shorts in series-parallel circuits are as follows:

1. A short of *any* component in the circuit causes total resistance to decrease and total current to increase.
2. A higher-than-normal total current (or lower than normal R_T) implies that either a short or a lowered R has occurred somewhere in the circuit.
3. Voltage measurements across the circuit components reveal close to zero volts across the shorted component and those directly in parallel with it. The good components may have either higher- or lower-than-normal voltage levels, depending on their individual positions within the circuit. Voltage measurement is one simple way to locate the component or combination of components that are shorted.
4. Once the shorted portion of the circuit is located, resistance measurements will isolate the exact component or circuit portion containing the short. Resistance measurement involves turning the power off during measurement and lifting one end of each branch from the circuit.

Let's see if your analysis skills are developing. Try the troubleshooting problems in the following In-Process Learning Check.

IN-PROCESS LEARNING CHECK 3

1. Refer to Figure 6–25. Determine the defective component and the nature of its defect.
2. Refer to Figure 6–26. Indicate which components or component might be suspect and what the trouble might be.
3. Refer to Figure 6–27. Indicate which voltages will change and in which direction, if R_2 increases in value.

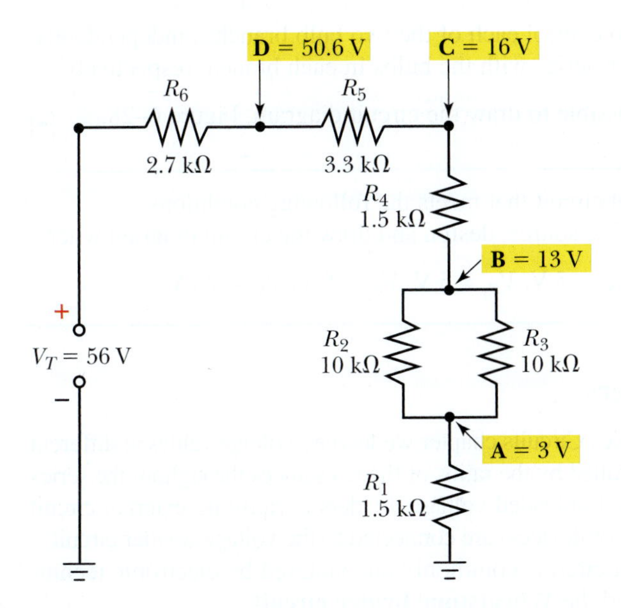

NOTE: Voltages indicated are from specified test points to ground reference.

FIGURE 6–25 multiSIM

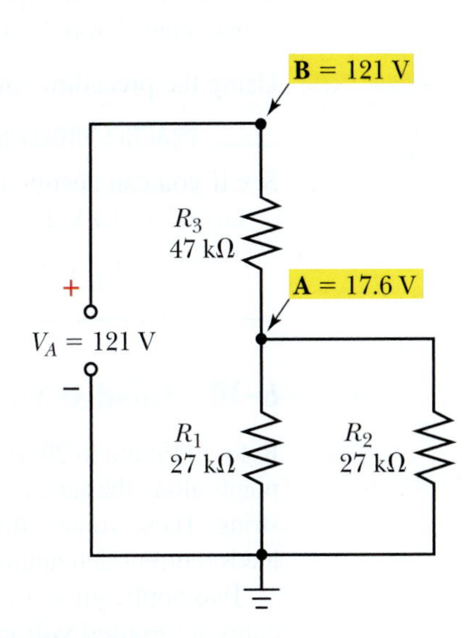

NOTE: Voltages indicated are from specified test points to ground reference.

FIGURE 6–26

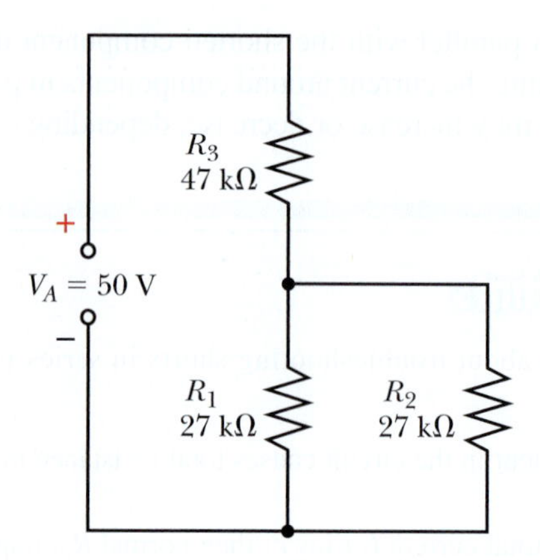

FIGURE 6–27

6–9 Designing a Series-Parallel Circuit to Specifications

For a design problem, let's assume we are designing and drawing the circuit diagram for a simple lighting circuit that contains four light bulbs, each of which normally operates at 10 volts. Using four lights and three SPST switches, configure the circuit so switch 1 acts as an on-off switch for all the lights; switch 2 controls two of the lights' on-off modes (assuming switch 1 is on), and switch 3 controls the remaining lights' on-off conditions (assuming switch 1 is on). The source voltage is 20 volts.

EXAMPLE The thought processes to solve this problem are as follows:

1. Since the source equals 20 V and the bulbs are 10-V bulbs, it is logical to assume two bulbs are in series across the 20-V source.
2. Since there are four lights, it is also logical to assume that there are two, two-bulb branches connected to the 20-V source voltage.
3. Since switch 1 controls the on-off condition of the total circuit connected to the source, it is logical that this switch is in series with the source and connected between the source and the remaining circuitry.
4. Since the circuit specification is to control each of the two bulb branches independently, switches 2 and 3 must be placed in series with the bulbs in each branch, respectively.

Using the preceding knowns, it is possible to draw the circuit diagram, Figure 6–28.

PRACTICE PROBLEMS 5

See if you can design a series-parallel circuit that meets the following conditions:

Given six 10-kΩ resistors and a 45-V source, design and draw the circuit diagram where:

$$V_{R_1} = 20\text{ V};\ V_{R_2} = 5\text{ V};\ V_{R_3} = 5\text{ V};\ V_{R_4} = 5\text{ V};\ V_{R_5} = 5\text{ V};\ V_{R_6} = 20\text{ V}$$

6–10 Loaded Voltage Dividers

Refer to Figure 6–29. Recall in the Series Circuits chapter we learned voltage values at different points along the series circuit are controlled by the ratios of the resistances throughout the series string. These voltage dividers are called unloaded voltage dividers because no external circuit loads (current-demanding components or devices) are connected to the voltage divider circuit.

Two applications of series-parallel circuits commonly encountered by electronic technicians are **loaded voltage dividers** and the **Wheatstone bridge circuit.**

Frequently in electronic circuits, a requirement is to provide different levels of voltage for various portions of the circuit or system. Also, these different portions may demand dif-

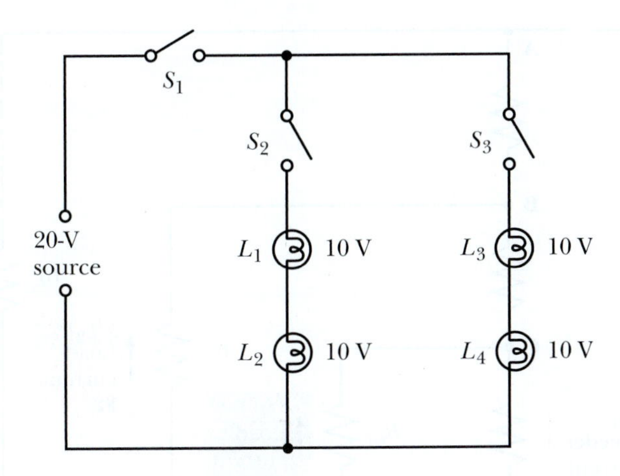

FIGURE 6–28 Sample design problem circuit

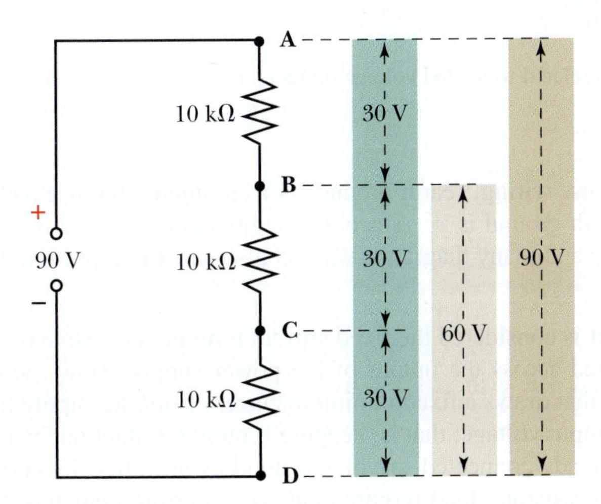

FIGURE 6–29 Unloaded voltage divider

ferent load currents. Voltage dividers are often used with power supplies to provide and distribute these different levels of voltages at the different current values demanded.

To help in our discussion of loaded voltage dividers, it is appropriate to define some important terminology used in our discussion. Refer to Figure 6–30 as you study these definitions.

1. **Load** is a component/device or circuit drawing current from a power source.
2. **Load current** is the current required by the components or circuits that are connected to the power source and/or its output voltage divider.
3. R_L is the resistance value of the load resistor or component.
4. **Ground reference** is an electrical *reference point* used in defining or measuring voltage values in a circuit.

Frequently the electrical ground reference point in a circuit is chassis ground, which is the common metal conducting path through which many components or circuits are electrically connected to one side of the power supply. In printed circuits, it is a common metal conductor *path* where many components are electrically connected, or "commoned." In circuits built on a metal chassis, the metal chassis is often the "return" path, or common conductor path; hence, the term "chassis ground." The symbol for chassis ground is ⏊. In electrical wiring

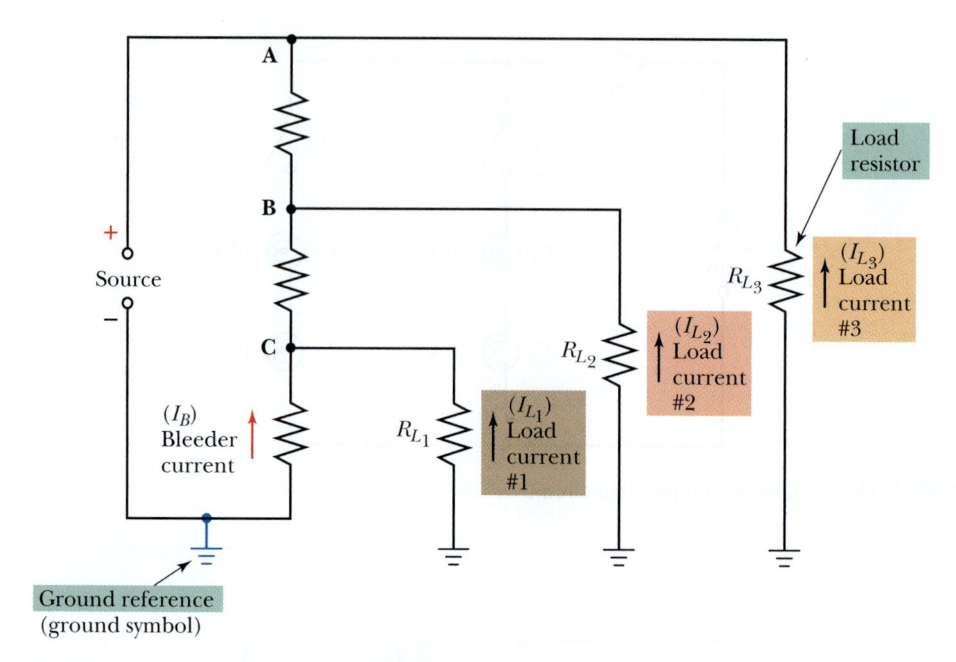

FIGURE 6–30 Terms related to loaded voltage dividers

circuits, such as house wiring, "earth ground" is a common reference and connecting point. The symbol for earth ground is ⏚. Since this symbol is often used to represent the "common" in many texts and many diagrams, we will also use it to represent the circuit common reference point.

5. **Bleeder current** is considered the fixed current through a resistive network, or a bleeder resistor connected across the output of the power supply. Many power supplies use a bleeder resistor that draws a fixed minimum current from the supply to help regulate the power supply output voltage; that is, keeping it more constant under varying load conditions. Because loads connected to voltage dividers are often functional circuits rather than fixed-value resistors, load requirements (i.e., current demands by the loads) often change during operation. For this reason voltage dividers need to be designed so that reasonable variations in load current demands do not unreasonably upset the voltage level at the various divider output points. The higher the bleeder current is, with respect to the load currents drawn, the "stiffer" the voltage supply acts; that is, the less the output voltage will vary with changes in load currents. *Bleeder currents of approximately 10% to 25% of load current demands are quite typical. If higher percentages are used, it is in cases where a stiffer supply is needed.*
6. **Bleeder resistor** is the resistor, or resistor network, connected in parallel with the power supply circuitry, that draws the bleeder current from the source. Another common function of a bleeder resistor is to discharge power supply capacitors after the circuit is turned off, for safety purposes. You will learn more about this in later studies regarding power supply circuits.
7. **Potentiometer** (Figure 6–31) is a three-terminal resistive device used as a voltage-dividing component. One terminal is at one end of the resistive element, a second contact is at the other end of the resistive element, and the third contact is a "wiper arm" that can move to any position along the resistive element. The position of the wiper arm along the resistive element determines the resistance, and thus the voltage from it to either end contact.

With this background information in mind, let's look at a brief analysis of a typical loaded voltage divider circuit.

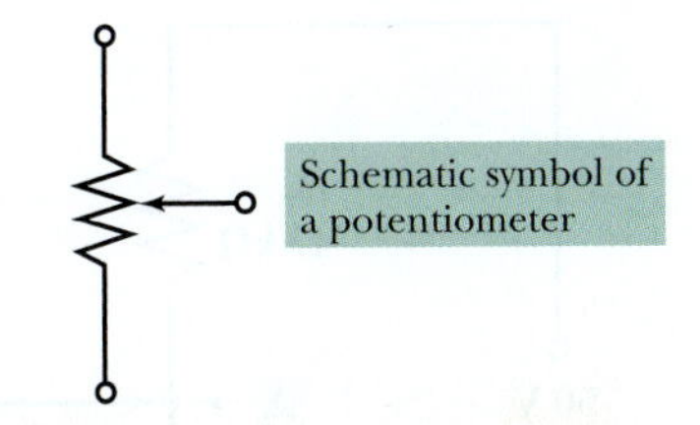

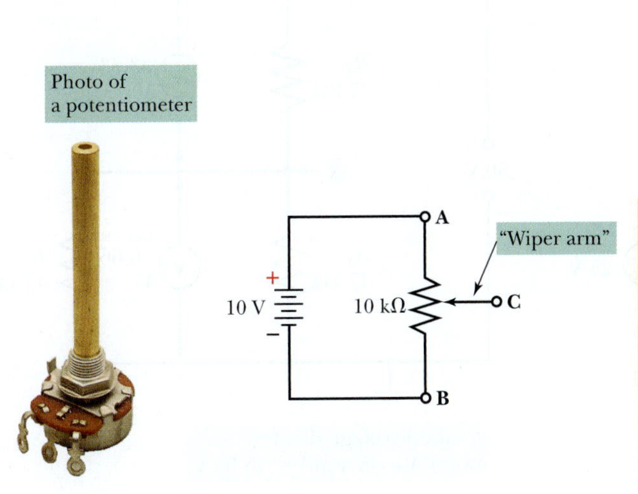

Voltage-dividing action of a potentiometer:

Depending on the position of the wiper arm, point C can be made anywhere in the range of:

0 V to +10 V with respect to **B**
or
0 V to −10 V with respect to **A**

When at its midpoint, point **C** is +5 V with respect to **B** and −5 V with respect to **A**.

Resistance characteristics:

$R_{AC} + R_{CB} = 10\ k\Omega$
$R_{AC} = 0 - 10\ k\Omega$
$R_{CB} = 0 - 10\ k\Omega$

FIGURE 6–31 The potentiometer *(Photo courtesy of Bourns, Inc.)*

Practical Notes

Safety must be observed when dealing with equipment where the equipment chassis is used as the "common" circuit return path. In many (or most) cases, it is advisable to make sure that the chassis is also connected to a good earth ground via some means to prevent inadvertent shock or life-threatening danger caused by differences of potential between equipment chassis and earth ground. Earth ground connections are frequently made via electrical connection made to a water pipe, or a metal rod driven into the ground. The ac electrical power system normally has one of the conductor paths "neutralized" by its being connected to an electrical earth ground.

There are some pieces of equipment that are designed to be operated with a "floating ground," which is not connected to earth ground. That is, the chassis is NOT to be connected to earth ground. In those cases, it is a desirable safety habit to use an "isolation transformer" to isolate the ac power source earth ground from the chassis common circuit return path. You will learn more about transformers in a later chapter. We simply wanted to CAUTION you to be careful when working with circuits and equipment that might shock (or kill) you if you touch both the chassis and an earth ground at the same time. Operate the equipment properly, and be aware of whether the chassis is properly grounded or isolated.

You must be aware of the ground conditions that exist for the equipment you are using, working on, or testing. It is always a good idea to check for differences of potential between equipment and earth grounds using a voltmeter. If the chassis is "hot" with respect to earth or power wiring grounds, you must determine whether you should even work on it at all, if it needs to be repaired before working on it, or whether the power plug needs to be reversed.

Two-Element Divider with Load

Observe Figure 6–32a and 6–32b as you study the following discussion. Let's think through the circuit.

What causes voltage between point A and ground reference to decrease from 25 volts to about 16.7 volts when load resistor R_L is connected? Before the load was connected, the two-series-connected equal value resistors divided the input voltage equally (25 volts each) since they were passing the same current.

When the load resistor was connected, a new branch path for current was created through R_L. The current through R_L must also pass through R_1 to return to the source's positive side. Now R_1 has a higher amount of current through it.

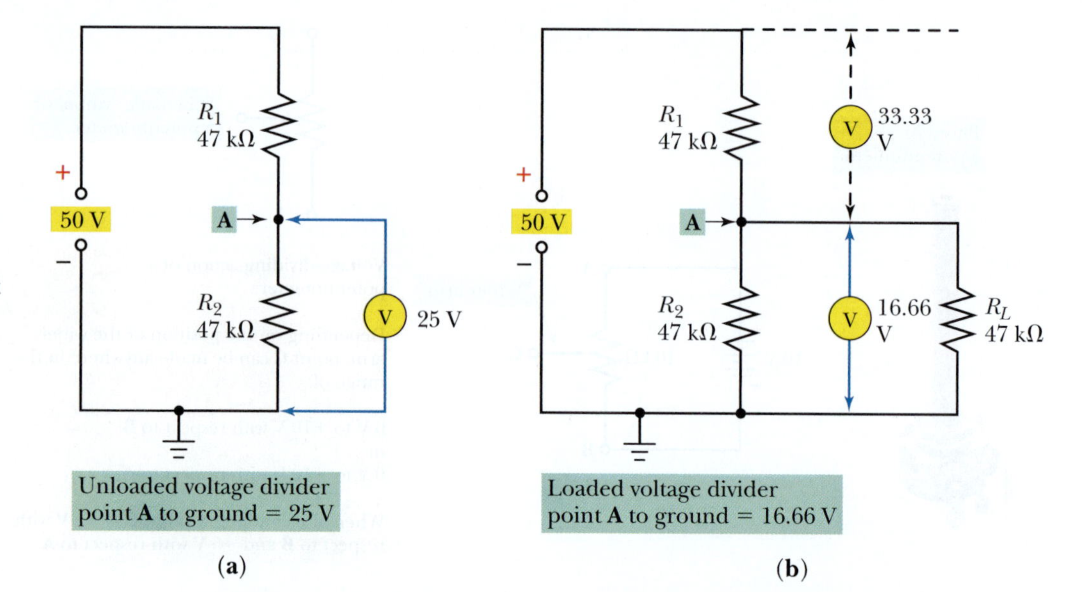

FIGURE 6–32 Two-element divider circuit

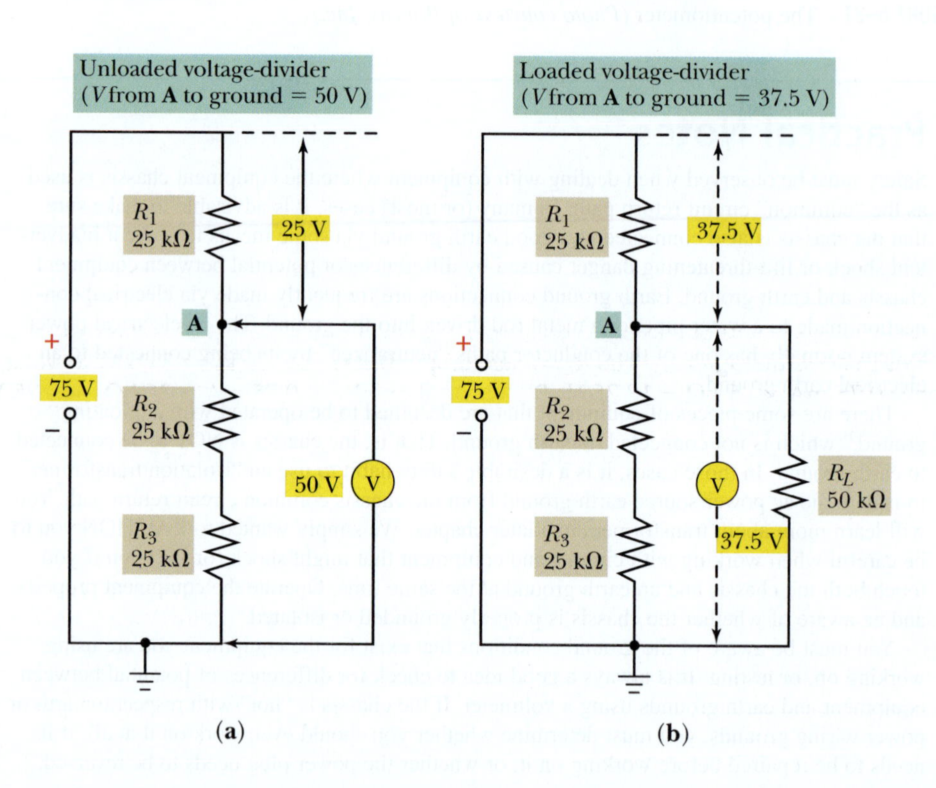

FIGURE 6–33 Comparison of loaded and unloaded voltage dividers (effect of connecting a load to a voltage divider circuit)

If the current through R_1 increases and its R value is unchanged, then its $I \times R$ drop must increase. In this case it increased from 25 volts to about 33.33 volts. Thus, only 16.66 volts remain to be dropped from point A to ground compared with 25 volts previously dropped before the load was connected.

Three-Element Divider with Multiple Loads

Observe in Figure 6–33a and 6–33b that the changes caused by connecting a load to a three-element voltage divider are similar to those changes that occur in a two-element divider.

Look at the three-element loaded voltage divider in Figure 6–33b from the perspective of Ohm's law and Kirchhoff's voltage law.

R_T is solved by using your knowledge of series and parallel circuits. R_L is in parallel with the series combination $R_{2\text{–}3}$. This indicates the R_L value of 50 kΩ is in parallel with 50 kΩ so the equivalent resistance from point A to ground equals 25 kΩ. This 25 kΩ is in series with the R_1 value of 25 kΩ yielding a total circuit resistance of 50 kΩ.

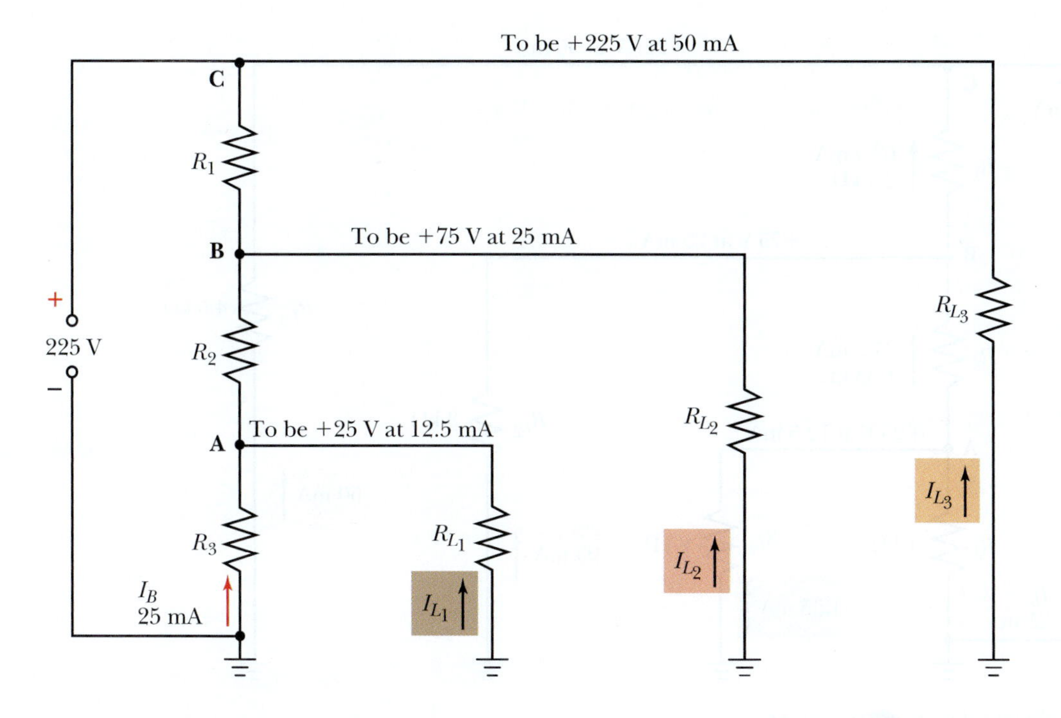

FIGURE 6–34 Analyzing and designing a loaded voltage divider circuit

Solve I_T using Ohm's law, where

$$I_T = \frac{V_T}{R_T} = \frac{75 \text{ V}}{50 \text{ k}\Omega} = 1.5 \text{ mA}$$

Since I_T passes through R_1, $V_1 = I_T \times R_1 = 1.5 \text{ mA} \times 25 \text{ k}\Omega = 37.5$ volts.

Since point A to ground is composed of two equal 50-kΩ branches (R_L and $R_{2\text{-}3}$), total current evenly divides between the branches. This means 0.75 mA passes through R_L and 0.75 mA passes through $R_{2\text{-}3}$. The R_L voltage drop = 0.75 mA × 50 kΩ = 37.5 V; R_2 voltage drop = 0.75 mA × 25 kΩ = 18.75 V; R_3 voltage drop also = 0.75 mA × 25 kΩ = 18.75 V; therefore, $V_{2\text{-}3} = 37.5$ V.

Of course, by Kirchhoff's voltage law (the sum of the voltage drops around a closed loop must equal V applied), you know the voltage from point A to ground has to be 37.5 V. Since the calculated R_1 drop is 37.5 V, the remainder of V applied has to drop from A to ground reference.

Let's move one step further in studying loaded voltage dividers by performing a design and analysis problem to achieve specified voltage levels and load currents.

EXAMPLE Refer to Figure 6–34 for the desired parameters of the circuit to be designed. Note the voltage-dividing system provides voltage and current levels as follows:

Load 1 (R_{L_1}) = 25 V at 12.5 mA

Load 2 (R_{L_2}) = 75 V at 25 mA

Load 3 (R_{L_3}) = 225 V at 50 mA

Bleeder current (I_B) = 25 mA

For additional information, we'll make one task to determine the resistance values of loads 1, 2, and 3. Using the parameters provided, this is not critical to the design tasks but will be of interest.

Since you know the desired voltage across each load and the currents passing through them, simply apply Ohm's law to find load resistance values.

$$R_{L_1} = \frac{25 \text{ V}}{12.5 \text{ mA}} = 2 \text{ k}\Omega$$

$$R_{L_2} = \frac{75 \text{ V}}{25 \text{ mA}} = 3 \text{ k}\Omega$$

$$R_{L_3} = \frac{225 \text{ V}}{50 \text{ mA}} = 4.5 \text{ k}\Omega$$

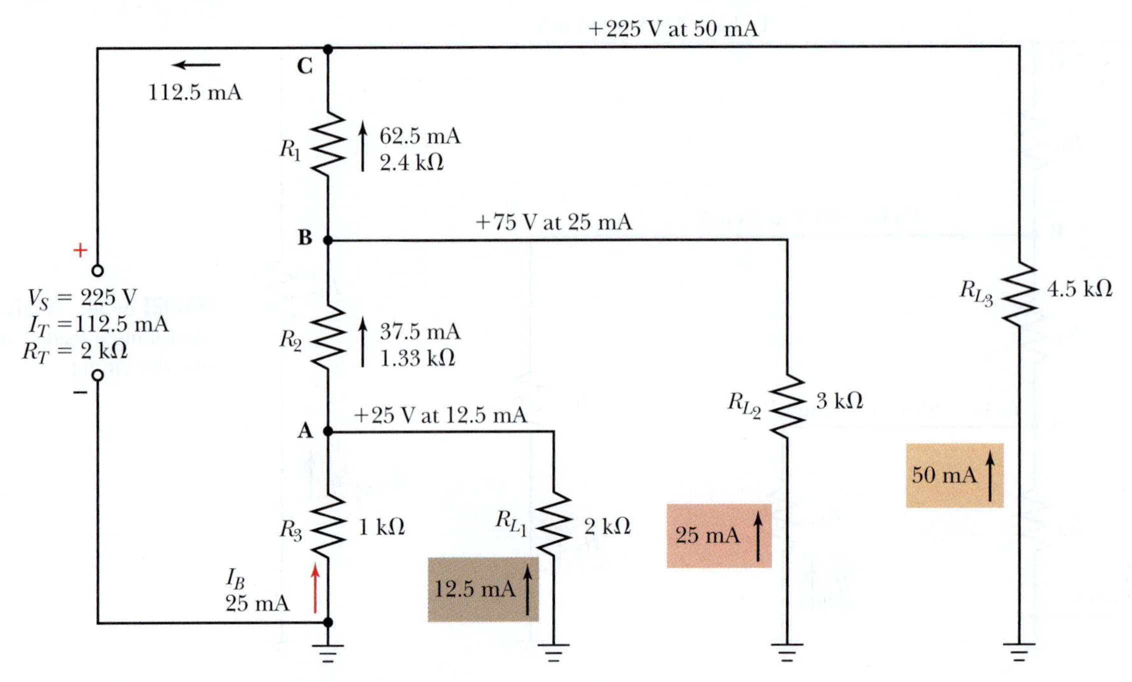

FIGURE 6–35 Voltage divider problem solution multiSIM

The second task, a design task, is to determine the values of voltage divider resistors R_1, R_2, and R_3 that will provide the appropriate voltages to the loads.

Again refer to Figure 6–34 as we examine the design task as follows:

1. Looking at the circuit, you can see that none of the load currents pass through R_3. This means only the given bleeder current (I_B) of 25 mA is passing through R_3. Because its voltage must equal the load 1 voltage of 25 V since they are in parallel, then

$$R_3 = \frac{25\text{ V}}{25\text{ mA}} = 1\text{ k}\Omega$$

2. The R_2 current must equal the 25-mA bleeder current, *plus* the 12.5-mA load 1, or 37.5 mA. The voltage dropped by R_2 must equal 75 V minus the voltage dropped by R_3. Thus, V_{R_2} equals 50 V. Therefore,

$$R_2 = \frac{50\text{ V}}{37.5\text{ mA}} = 1.33\text{ k}\Omega$$

3. The R_1 current must equal the bleeder current plus load 1 current plus load 2 current. Therefore, the current through R_1 equals 25 mA (bleeder current) plus 12.5 mA (I_{L_1}) plus 25 mA (I_{L_2}) equals 62.5 mA. Thus, R_1 voltage drop must equal 225 V minus the drop from point B to ground, or 225 V − 75 V = 150 V. This means

$$R_1 = \frac{150\text{ V}}{62.5\text{ mA}} = 2.4\text{ k}\Omega$$

Look at Figure 6–35 for the completed design. ◆

PRACTICE PROBLEMS 6

1. Draw the circuit of a three-resistor voltage divider with three loads connected as they were in our example. Let the top divider resistor be R_1, the middle one R_2, and the bottom one R_3.

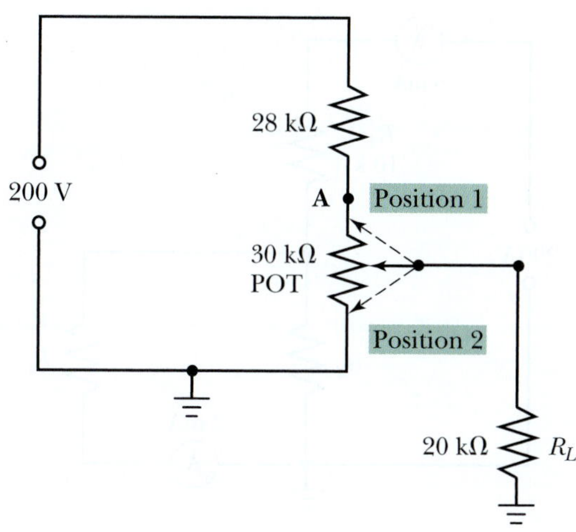

The potentiometer enables varying V across R_L from 0 to 60 V

FIGURE 6–36 Example of potentiometer varying voltage output

2. Determine the value of each of the load resistances and find the values for the divider resistors R_1, R_2, and R_3 assuming the following:
   a. source voltage = 100 V;
   b. bleeder current = 200 mA;
   c. load #1 needs 10 V and demands 10 mA of current;
   d. load #2 needs 30 V and demands 20 mA of current; and
   e. load #3 needs 100 V and demands 40 mA of current.
3. Completely label your drawing showing all resistances, voltages, and currents.

A Practical Voltage Divider Variation

Look at the circuit in Figure 6–36. With the potentiometer's wiper arm at the top (position 1), the 20-kΩ R_L is in parallel with total potentiometer resistance of 30 kΩ, yielding an equivalent resistance of 12 kΩ from point A to ground. The circuit voltage-dividing action is dividing the 200 V across the 40-kΩ total circuit resistance consisting of the 28-kΩ resistor and the 12-kΩ parallel circuit equivalent resistance in series. Therefore, R_L drops twelve-fortieths of 200 V, or 60 V. The 28-kΩ resistor drops the remaining 140 V.

On the other hand, if the wiper arm is at position 2 (or ground), both ends of R_L will connect to ground. Therefore, the voltage across R_L is zero V. This means that with the potentiometer, we can vary the voltage across R_L from 0 to 60 volts. As you will see throughout your electronics experience, this "variable-voltage-divider" capability of the potentiometer is convenient.

PRACTICE PROBLEMS 7

Refer to Figure 6–37 and answer the following:

1. What is the value of voltage across R_L? How much power is dissipated by R_L?
2. What is the value of R_L?
3. What is the value of R_2?
4. If V_{R_L} suddenly decreases due to a change in the R_2 value, does this indicate R_2 has increased or decreased in value?
5. If the decrease in V_{R_L} in question 4 was caused by a change in R_1 value, would it indicate R_1 has increased or decreased in value?
6. If R_L is disconnected from the voltage divider in the circuit, what is the new value of V_{R_2}?
7. What happens to V_{R_L} if R_1 opens? If R_2 shorts?

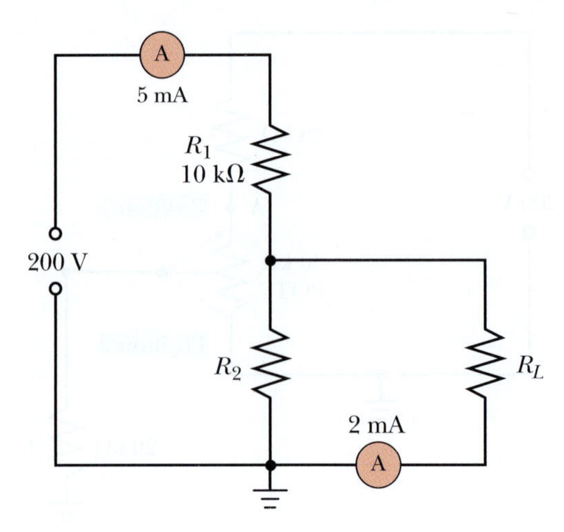

FIGURE 6–37

6–11 The Wheatstone Bridge Circuit

Various types of special series-parallel circuits, called **bridge circuits,** are used to make measurements in electronic circuits, Figure 6–38.

Note that zero volts appear between points A and B when the bridge is balanced. You will see this balance *only* when the ratio of the resistances in the left arm and in the right arm of the bridge are equal.

For example, observe Figure 6–38. In the left arm of the bridge, $R_1 = 10\ k\Omega$ and $R_2 = 20\ k\Omega$. In the right arm of the bridge, $R_3 = 50\ k\Omega$ and $R_4 = 100\ k\Omega$. If V applied is 150 V, then R_1 drops one-third the voltage across the left arm of the bridge, or 50 V (since it is one-third of the left arm's total resistance). R_2 drops the other two-thirds of the 150 volts, or 100 V.

What about the other arm of the bridge? The same 150 V is applied across that arm's 150 kΩ, with R_3 (50 kΩ) dropping one-third the applied voltage, or 50 V, and R_4 dropping two-thirds the applied voltage (R = two-thirds the right arm's total R), or 100 V.

Note that the voltage differential between points A and B is zero. This is because point A is at a potential of + 100 volts with respect to the power supply's negative terminal. Point B is also at + 100 volts with respect to the power supply's negative terminal. Thus, the difference of potential between points A and B is zero volts, since they are at the same potential.

This balanced condition (zero volts between points A and B) exists whenever the resistance ratios of the top resistor to the bottom resistor in the left and right arm of the bridge are the same. That is,

FORMULA 6–1 Bridge balanced when: $\frac{R_1}{R_2} = \frac{R_3}{R_4}$

NOTE: The values of the resistors do not matter. It's the *ratio* of the resistances in each arm that determines the voltage distribution throughout that arm. Recall the series circuit concepts. If both arms have equal voltage distribution, then the midpoint of each arm is at the same potential and the bridge is balanced.

The **Wheatstone bridge circuit,** a special application of the bridge circuit, measures unknown resistance values. When the bridge circuit is balanced, the output terminals (points A and B) have zero potential difference between them. On the other hand, if the bridge circuit is unbalanced, there is a difference of potential between the output terminals. Refer to Figure 6–39 as you read the following discussion.

The following is one method of using the Wheatstone bridge circuit to determine an unknown resistance:

1. Carefully select R_1 and R_2 to be matched (equal) values.
2. Place a sensitive "zero-center-scale" current meter between points A and B, which are the output terminals of the bridge circuit.

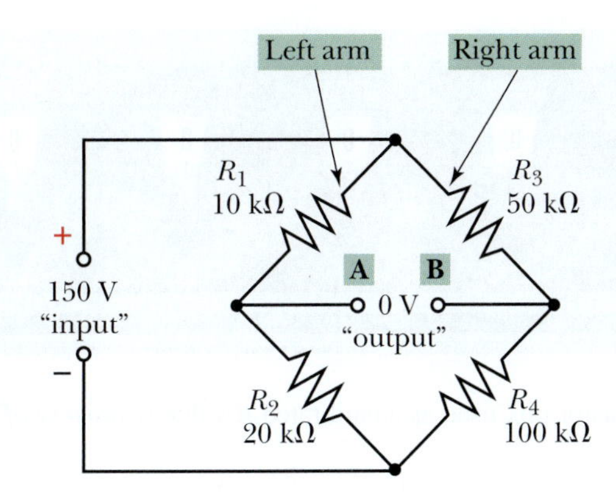

FIGURE 6–38 A balanced bridge circuit exists when $\frac{R_1}{R_2} = \frac{R_3}{R_4}$

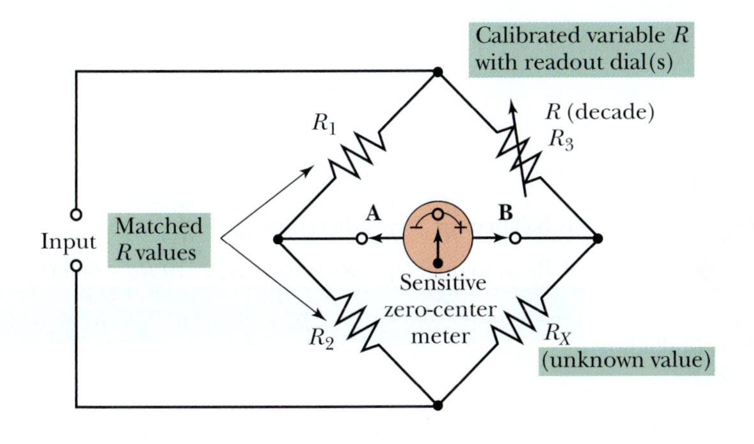

FIGURE 6–39 Concepts of the Wheatstone bridge circuit

3. Use a calibrated variable resistance, such as a "resistance decade box," as R_3. NOTE: This type variable resistance has calibrated dials to enable you to read its resistance setting, Figure 6–40.
4. Place the unknown resistor *(R_X)* in the same arm of the circuit as the calibrated variable resistance.
5. Since R_1 and R_2 are equal value in the left arm, then the bridge is balanced when the calibrated variable resistance and the unknown resistance equal each other in the right arm. In this case, the sensitive current meter's pointer is pointing at "0" in the center of the scale. If the bridge is unbalanced, the pointer points to the left of center, indicating current through the meter in one direction, or to the right of center, indicating current in the other direction. The pointer's direction is dependent on the polarity of the voltage difference between points A and B, as determined by the comparative voltages at points A and B in the bridge network.
6. Once the calibrated variable resistance is adjusted for zero current reading (balanced), the value of the unknown resistor is read directly from the calibrated variable resistance's dial(s).

Another method to determine the unknown resistance value is to use the known resistance values and the concepts of the balanced bridge circuit.

1. Recall that balance occurs when $\frac{R_1}{R_2} = \frac{R_3}{R_4}$
2. With these relationships, it is mathematically shown that:

   $R_1 \times R_4 = R_2 \times R_3$ (cross-multiplying factors in formula)
3. If R_4 is the unknown resistance, *(R_X)*, the formula can be rearranged to solve R_X as follows:

FORMULA 6–2 $R_X = \frac{R_2 \times R_3}{R_1}$

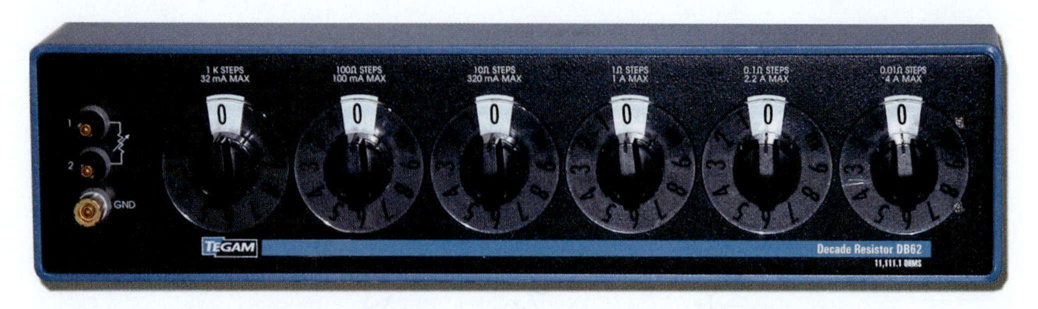

FIGURE 6–40 Decade resistor box used as a calibrated R value *(Courtesy of TEGAM, Inc.)*

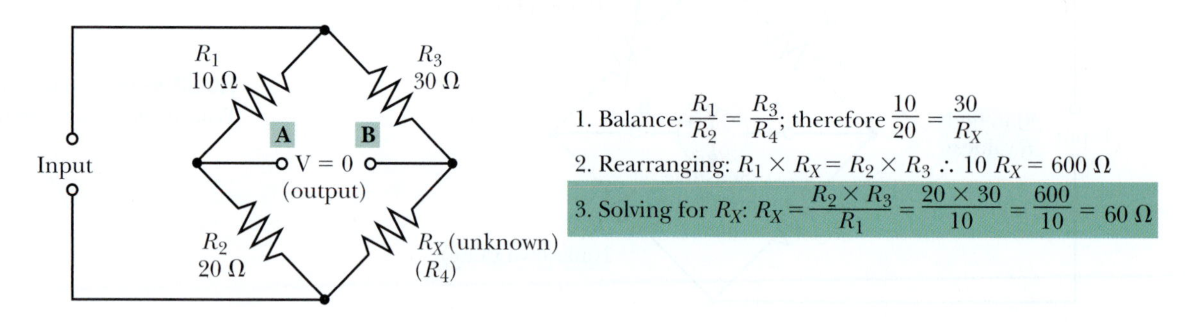

FIGURE 6–41 The calculation approach may be used to solve for the R_X value in a bridge circuit. multiSIM

4. Look at Figure 6–41 with the values shown to see if the steps described earlier will work.
5. Verifying results by substituting our answer into the original balanced bridge formula: $\frac{R_1}{R_2} = \frac{R_3}{R_X} = \frac{10}{20} = \frac{30}{60} = \frac{1}{2}$. (It checks out!)

Some Applications for Bridge Circuits

In industry, there are numerous types of sensors and control circuits used to control the operation of machinery and manufacturing systems. For example, sensors for detecting changes in temperature, flow, pressure, or changes in electrical conditions are all quite common. The output of one of these sensors can be used as one leg of a bridge circuit, with the bridge circuit "zeroed" at the desired reference level for temperature, pressure, flow, and so on, Figure 6–42. Any change in the condition being sensed will then cause the bridge to become unbalanced. This differential at the output terminals of the bridge is then typically used as an input to circuits or devices that eventually translate the change into a readable output indication, such as a digital readout, for monitoring purposes, or a feedback control signal, Figure 6–42. This system condition change can then be managed, either manually, or in many cases automatically, to bring the system back to the desired operational characteristics.

We have already named one common application of the Wheatstone bridge; that is, measuring the value of an unknown resistance. Variations of the bridge circuit can also be used to measure the values of other types of components, such as capacitors, about which you

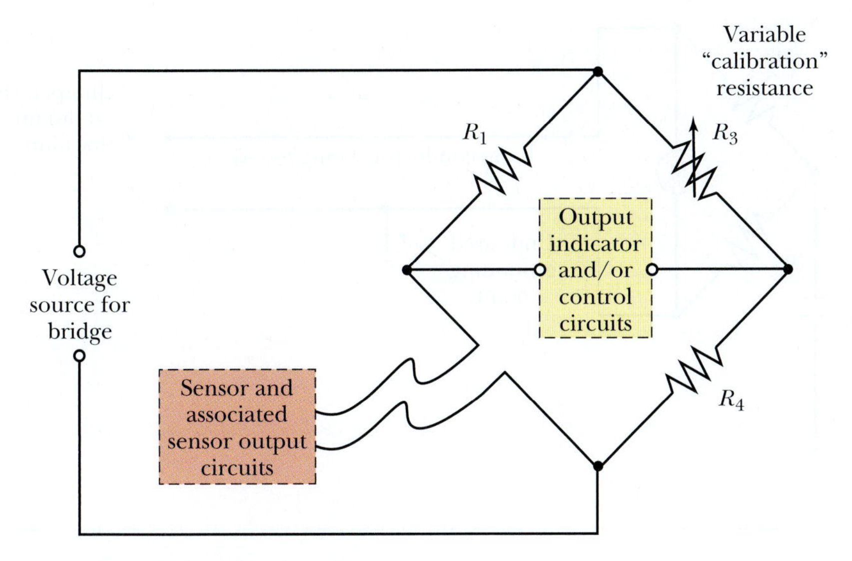

FIGURE 6–42 Bridge circuit used for industrial control

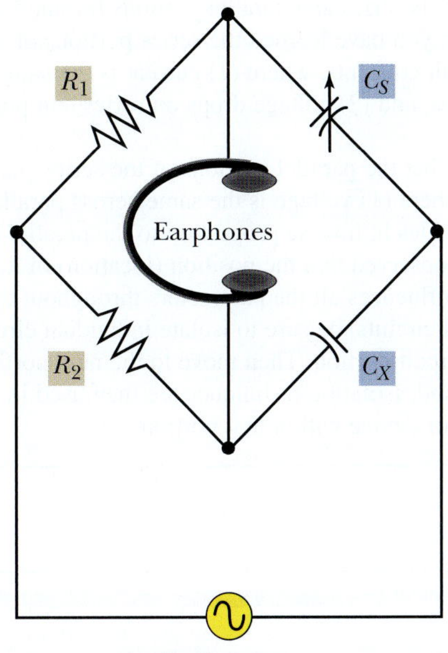

FIGURE 6–43 Bridge circuit used to measure unknown capacitor value

will learn details in a later chapter. Again, the component to be measured is put into one leg of the bridge, Figure 6–43. When the bridge is balanced, by adjusting the "standard capacitor" (C_S) to the same value as the unknown capacitor's value (C_X), the audio tone in the earphones is "nulled" or minimized.

Still another application of a modified bridge circuit, which has been used for years by the power and telephone companies, is the Murray loop. The Murray loop is used to locate where an undesired conductor ground has occurred along the miles of lines used to transport power or telephone messages, Figure 6–44. By knowing the resistance per foot of the conductors, and using the outgoing and return wires in a loop that forms part of the bridge circuit, the precise location of the ground fault can be determined. A service person can then be sent to the location to perform the necessary repair.

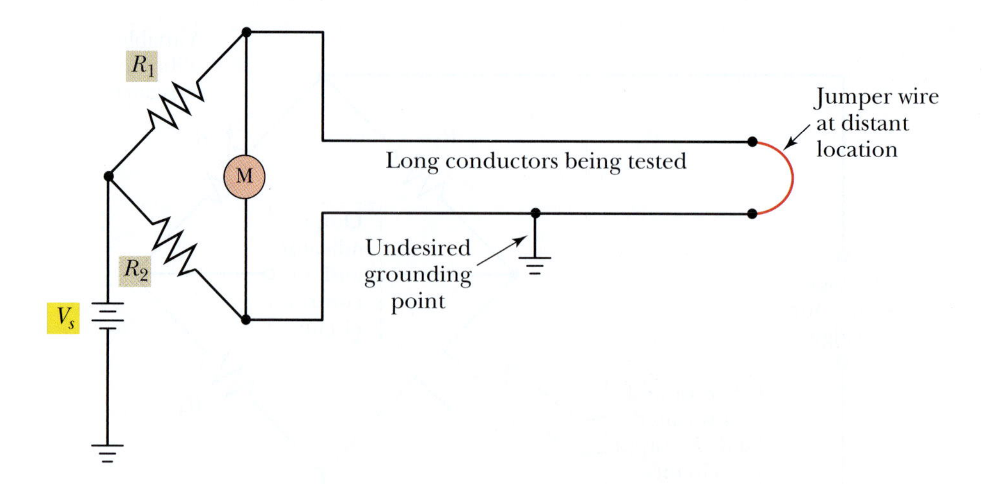

FIGURE 6–44 Example of the Murray loop

Practical Notes

Because series-parallel circuits are combinational circuits that are connected in an infinite variety of configurations, no single analysis technique is applicable. Rather, it is necessary to break the circuit down into its series and parallel portions for analysis.

In studying this chapter, you have learned the series portions of series-parallel circuits are analyzed using series-circuit concepts, where (1) current is the same through series components and/or circuit portions, and (2) voltage drops on series components are proportional to their resistance ratios.

Also, you have learned that the parallel portions of the series-parallel circuits abide by the rules of parallel circuits, where (1) voltage is the same across parallel components or combinations, and (2) current divides in inverse proportion to the parallel branch resistances.

Furthermore, you have observed that the position (location) of a given component or circuit portion in the circuit influences all the parameters throughout the circuit. Therefore, to troubleshoot series-parallel circuits, be sure to isolate individual circuit portions using concepts appropriate to that circuit portion. Then move to the next portion, and so forth, until the faulty circuit portion is found. Isolation techniques are then used in that portion to isolate the specific faulty component or device within that portion.

Summary

- A series-parallel circuit is a combination of series and parallel connected components and/or circuit portions.
- Series components or circuit portions are analyzed using the rules of series circuits.
- Parallel components or circuit portions are analyzed using the rules of parallel circuits.
- Points throughout the circuit where current divides and rejoins help identify parallel portions.
- Points throughout the circuit that share common voltage also help identify components or circuit portions that are in parallel.
- Paths in the circuitry that carry the same current help identify the series circuit elements.
- Analysis of total circuit resistance is approached by starting away from the source, then solving and combining results as one works back toward the source. (NOTE: If V_T and I_T are known, this is unnecessary. Use the $R_T = V_T/I_T$ formula.)
- Analysis of current throughout the circuit is approached by finding the total current value. Then, working outward from the source, analyze currents through various circuit portions.
- Power dissipations throughout the circuit are analyzed by finding the appropriate *I, V,* and *R* parameters. Then use the appropriate power formulas, such as $V \times I$, I^2R, or V^2/R.
- A short, or decreased value of resistance, *anywhere* in the circuit causes R_T to decrease and I_T to increase.
- An open, or increased value of resistance, *anywhere* in the circuit causes R_T to increase and I_T to decrease.
- Two special applications of series-parallel circuits are voltage dividers and bridge circuits used for various purposes.

Formulas and Sample Calculator Sequences

Formula 6–1
(To find a set of balanced bridge circuit R ratios)

Bridge balanced when: $\frac{R_1}{R_2} = \frac{R_3}{R_4}$

R_1 value, [÷], R_2 value, [=], (check for equaling) R_3 value, [÷], R_4 value, [=]

Formula 6–2
(To find an unknown R value using a balanced bridge formula)

$$R_X = \frac{R_2 \times R_3}{R_1}$$

R_2 value, [×], R_3 value, [÷], R_1 value, [=]

EXCEL AUTOMATED FORMULAS

Helpful problem-solving tools, such as the Excel automated formulas available on the CD, allow automatic calculations of formulas.

Using Excel

Series-Parallel Circuits Formulas

(Excel file reference: FOE6_01.xls)

DON'T FORGET! It is NOT necessary to retype formulas once they are entered on the worksheet! Just input new parameters data for each new problem using that formula, as needed.

- Refer to the parameters given in Figure 6–41. Assume that the value of R_3 is changed to 50 ohms. Use the Formula 6–2 spreadsheet sample and solve for R_x.

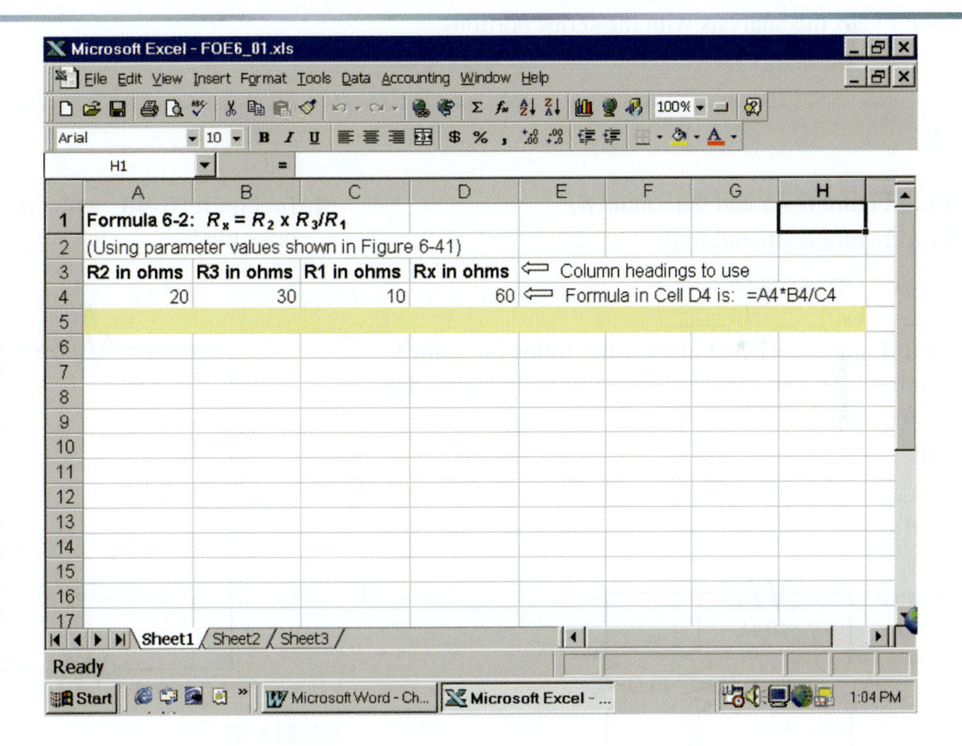

Review Questions

1. In your own words, describe the key characteristics of series-parallel circuits.
2. Name two methods of identifying portions of combination (series-parallel) circuits in which components or branches are in parallel.
3. Name two methods of identifying portions of combination (series-parallel) circuits in which components are in series.
4. Describe a generally useful approach for analyzing a combination (series-parallel) circuit in terms of the circuit's total resistance.
5. Describe a generally useful approach for analyzing a combination (series-parallel) circuit in terms of the circuit's current distribution.

6. In series-parallel circuits, the largest resistor value drops the highest voltage.
   a. True all the time, regardless of the resistor's location in the circuit.
   b. False all the time, regardless of the resistor's location in the circuit.
   c. True sometimes, depending on the resistor's location in the circuit.
   d. Never true, no matter the location of the resistor in the circuit.
7. In series-parallel circuits, total power dissipation always equals the sum of all the individual component power dissipations.
   a. True all the time, no matter how many components are involved.
   b. False all the time, because total power equals $V_T \times I_T$.
   c. True only if all the components have the same value.
   d. Only true for the series portions of the circuit.
8. When troubleshooting a series-parallel circuit:
   a. always start the analysis with the series portions.
   b. always start the analysis with the parallel portions.
   c. it doesn't matter where you begin the analysis.
   d. break the circuit down into series and parallel portions when beginning the analysis.
9. If a series-parallel circuit's total circuit resistance has increased, then:
   a. a circuit portion or a component has definitely opened.
   b. a circuit portion or a component has definitely shorted.
   c. a component has decreased in value or a short has occurred in the circuit.
   d. a component has increased in value or an open has occurred in the circuit.
10. Adding a resistor in parallel with any series portion component, or any parallel portion components in a series-parallel circuit will cause the circuit total resistance to:
    a. increase.
    b. decrease.
    c. remain the same.

Problems

1. Refer to Figure 6–45 and determine R_T.
2. Refer to Figure 6–46 and find:

   I_T _____ V_{R_3} _____ I_2 _____

   V_{R_2} _____ V_{R_5} _____ I_3 _____
3. Refer to Figure 6–47 and answer the following with I, for increase; D, for decrease; or RTS, for remains the same. Assume that if R_5 is shorted:

   R_T will _____ V_{R_3} will _____

   R_4 will _____ P_T will _____

   V_{R_1} will _____
4. In Figure 6–48, what is the maximum total current possible?

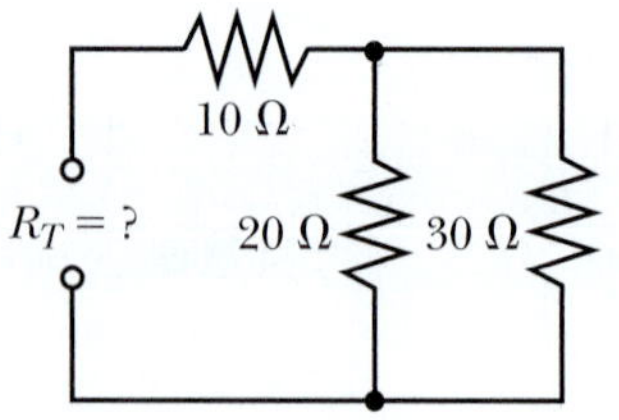

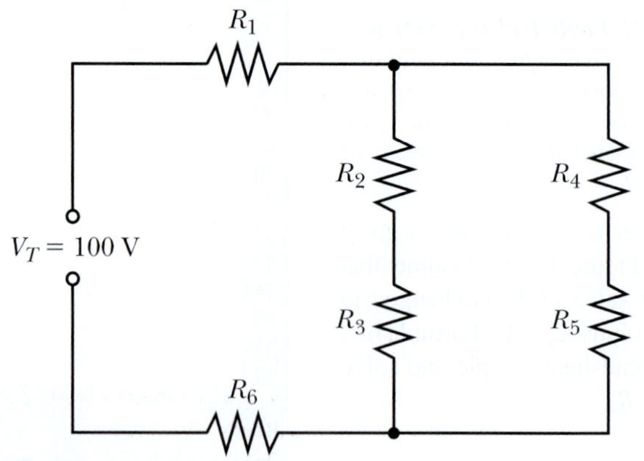

FIGURE 6–45

FIGURE 6–47

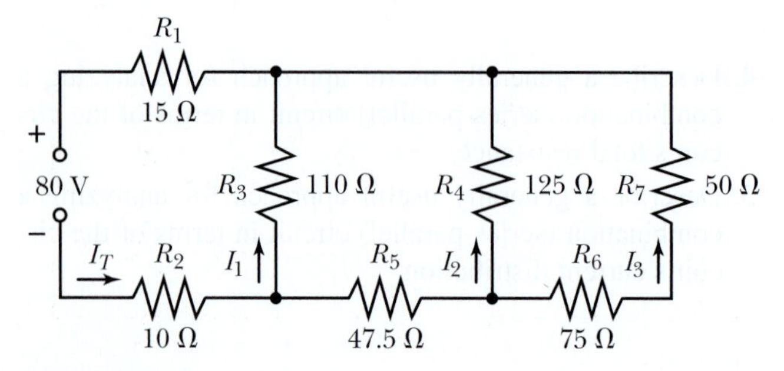

FIGURE 6–46

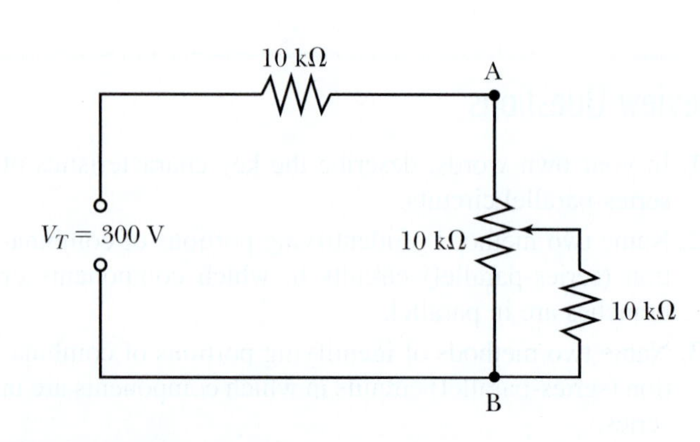

FIGURE 6–48 multiSIM

5. In Figure 6–49, what are the minimum and maximum voltages possible from point A to ground, if the setting of the potentiometer is varied? From _____ V to _____ V
6. In Figure 6–50, I_T is 10 amperes. Find R_1 and P_{R_2}.
7. In Figure 6–51, find the value of R_4 and I_2.

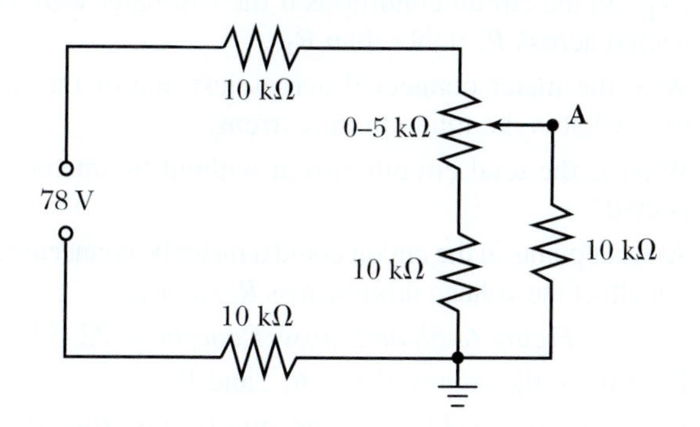

FIGURE 6–49

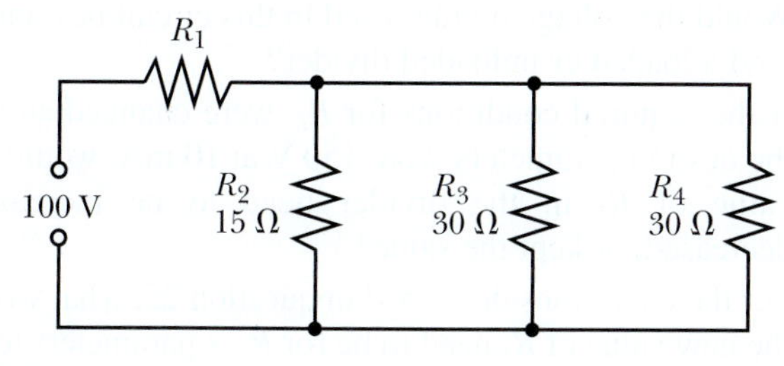

FIGURE 6–50

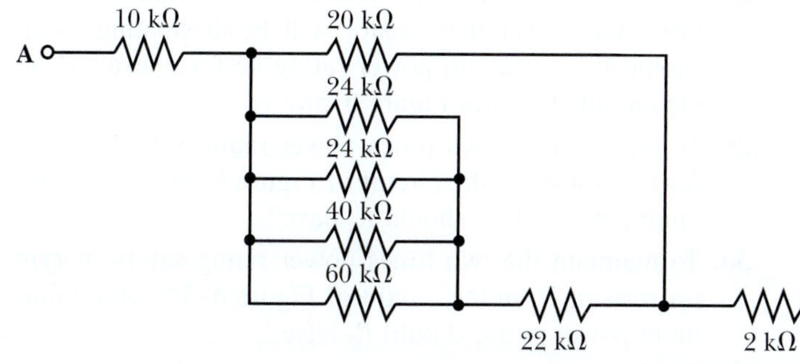

FIGURE 6–52 multiSIM

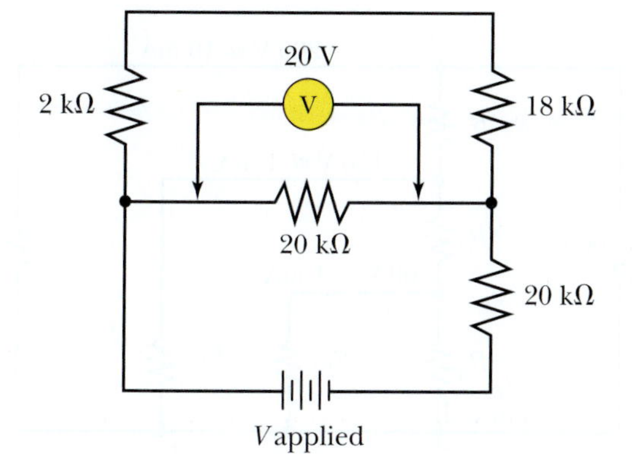

FIGURE 6–54

8. In Figure 6–52, find the resistance between points A and B.
9. In Figure 6–53, will the light get brighter, dimmer, or remain the same after the switch is closed?
10. Find the applied voltage in Figure 6–54.
11. Accurately copy the diagram, Figure 6–55, on a separate sheet of paper. When you are done, draw arrows on the diagram showing the direction of current through each resistor. Using this circuit, assume all resistors have a value of 100 kΩ. Find the following values. (Round answers to whole numbers.)

a. Find R_T
b. Find I_T
c. Find V_{R_1}
d. Find V_{R_2}
e. Find V_{R_3}
f. Find V_{R_4}
g. Find V_{R_5}
h. Find V_{R_6}
i. Find V_{R_7}
j. Find V_{R_8}

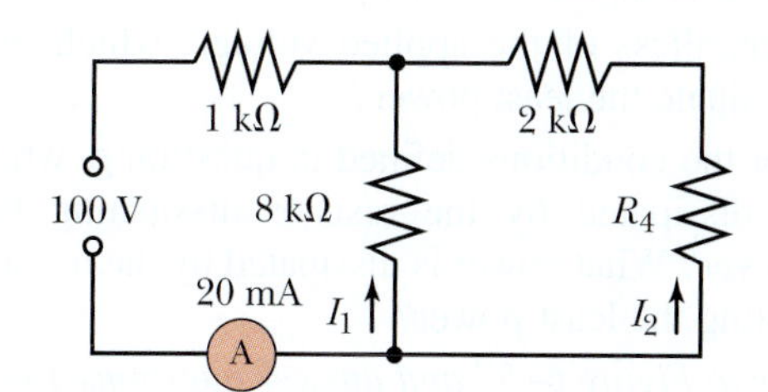

FIGURE 6–51

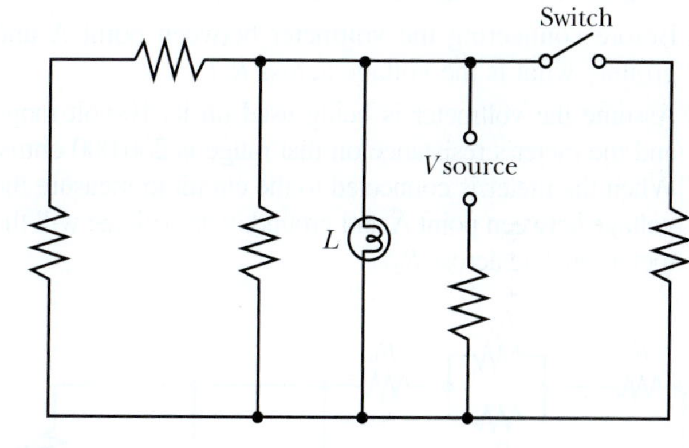

FIGURE 6–53

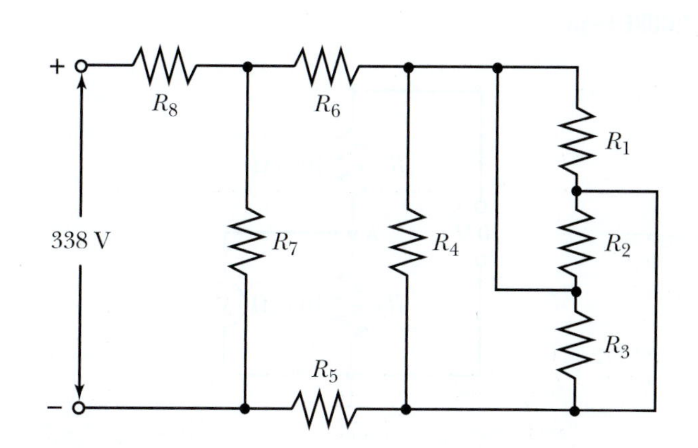

FIGURE 6–55

12. Refer to the circuit in Figure 6–56 and answer the following questions:
    a. If R_1 increases, what happens to the current through R_5?
    b. If R_7 decreases, what happens to the voltage across R_9?
    c. If R_4 shorts, what happens to the voltage across R_2? Across R_8?
    d. If all resistors are 10 kΩ, what is the value of R_T? (All conditions are normal!)
    e. If the circuit applied voltage is 278.5 volts, what is the voltage drop across R_1?
    f. If the circuit applied voltage is reduced to half the value shown in question e, what is the power dissipation of R_9?
    g. If the circuit applied voltage is 557 volts, which resistor(s) dissipate the most power?
    h. Regardless of the applied voltage, which resistor(s) dissipate the least power?
    i. For the conditions defined in question g, what power is dissipated by the resistor dissipating the most power? What power is dissipated by the resistor dissipating the least power?

Refer to Figure 6–57 and answer questions 13–21.

13. Before connecting the voltmeter between point A and ground, what is the voltage across R_1?
14. Before connecting the voltmeter between point A and ground, what is the voltage across R_2?
15. Assume the voltmeter is being used on its 10-volt range and the meter's resistance on that range is 200,000 ohms. When the meter is connected to the circuit to measure the voltage between point A and ground, what voltage will the meter indicate across R_2?

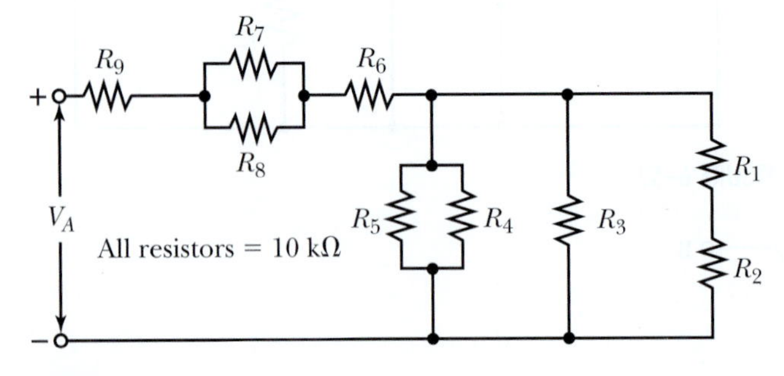

FIGURE 6–56

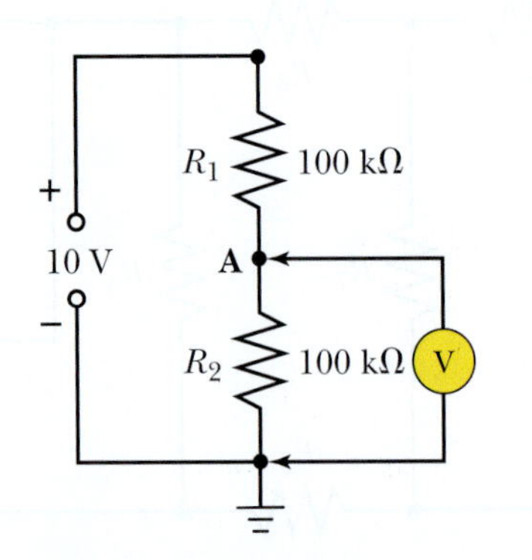

FIGURE 6–57 Circuit for Review Questions 13–21

16. For the conditions described in question 15, and based on Kirchhoff's voltage law, what must be the voltage drop across R_1 at that time?
17. Does connecting the meter to the circuit change the circuit's electrical parameters? Explain.
18. Explain the circuit conditions if the voltmeter were connected across R_1 rather than R_2.
19. With the meter connected across just one of the resistors, what is the total circuit current?
20. What is the total circuit current without the meter connected?
21. At what points in the circuit could a meter be connected and not affect the voltage drops across R_1 and R_2?

Refer to Figure 6–58 and answer questions 22–31.

22. Determine the values of R_{L_1}, R_{L_2}, and R_{L_3}.
23. Determine the values for the divider resistors R_1, R_2, and R_3.
24. Would the voltage divider used in this circuit be considered a loaded or unloaded divider?
25. If the required conditions for R_{L_2} were changed so that the desired parameters were 150 V at 10 mA, would the value of R_1 in the divider need to be increased, decreased, or kept the same?
26. For the conditions described in question 25, what would the new value of R_1 need to be for R_{L_2}'s parameters to be 150 V at 10 mA?
27. If all loads were removed from the circuit and R_1's value is changed to that calculated in question 26, what would the value of divider bleeder current (I_B) become?
28. To maintain a minimum resistor power rating of two times the power the resistor will be dissipating, what should the minimum power rating be for resistor R_3 in the circuit shown in Figure 6–58?
29. To maintain the two times power rating safety margin for resistor R_2 in the circuit of Figure 6–58, what minimum power rating should R_2 have?
30. To maintain the two times power rating safety margin for resistor R_3 in the circuit of Figure 6–58, what minimum power rating should R_3 have?
31. What is the total power being supplied by the power supply in the circuit of Figure 6–58?

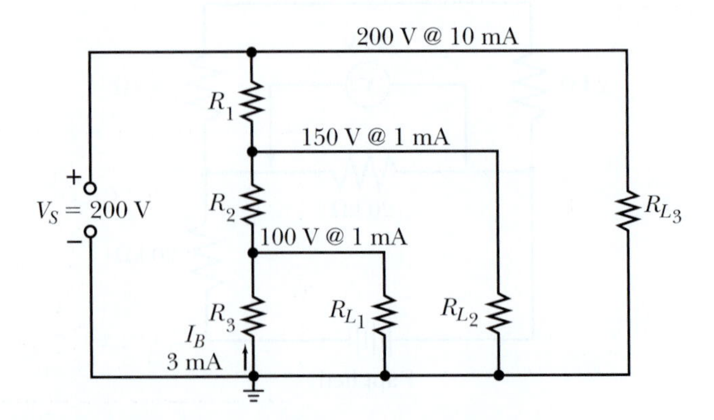

FIGURE 6–58 Circuit for Review Questions 22–31

Analysis Questions

Perform appropriate analysis of the circuit shown in Figure 6–59 and answer the following statements with I for increase, D for decrease, or RTS for remain the same:

1. If R_2 decreased in value by a great amount:
   - a. The voltage across R_3 would _____.
   - b. The voltage across R_1 would _____.
   - c. The current through the load connected in parallel with R_4 would _____.
2. If R_2 "opened":
   - a. The voltage across the load resistor connected to the bottom of R_1 (the "outer loop" load) would _____.
   - b. The voltage across R_1 would _____.
3. Briefly explain how an industrial sensor's output may be used in conjunction with a bridge circuit to help control industrial processes.
4. Determine the values of R_1, R_2, R_3, and R_4 for the circuit, Figure 6–59.
5. Determine the value of R_X in the circuit, Figure 6–60: Assume the bridge is balanced by the variable R set at 100 kΩ.
6. For the circuit of Figure 6–60, if R_1 is 47 kΩ, R_2 is 27 kΩ, and the variable R is set at 100 kΩ when the bridge is balanced, what is the value of R_X?
7. For the circuit of Figure 6–60, if the variable R is set at 1.5 kΩ and the resistor in the R_X position in the circuit is 27 kΩ, what must be the ratio of the resistances in the left arm of the bridge circuit for a balanced bridge?
8. For the circuit of Figure 6–60, assume that the input V is positive toward the top side of the circuit and negative toward the bottom. If the dc meter's negative lead is connected to the top of R_2 and its positive lead is connected to the top of R_X and the meter's pointer is pointing toward the left side of zero, does this indicate that R_2 or R_X has the greater voltage drop?
9. Briefly define the sequence of keystrokes or steps needed to perform calculation of $R_{equivalent}$ of three parallel resistors, using the reciprocal function of a calculator. Assume that you want to use the *engineering mode* of the calculator for this task.
10. Explain how using a voltmeter to measure voltage across part of a series portion of a series-parallel circuit causes circuit conditions to change.

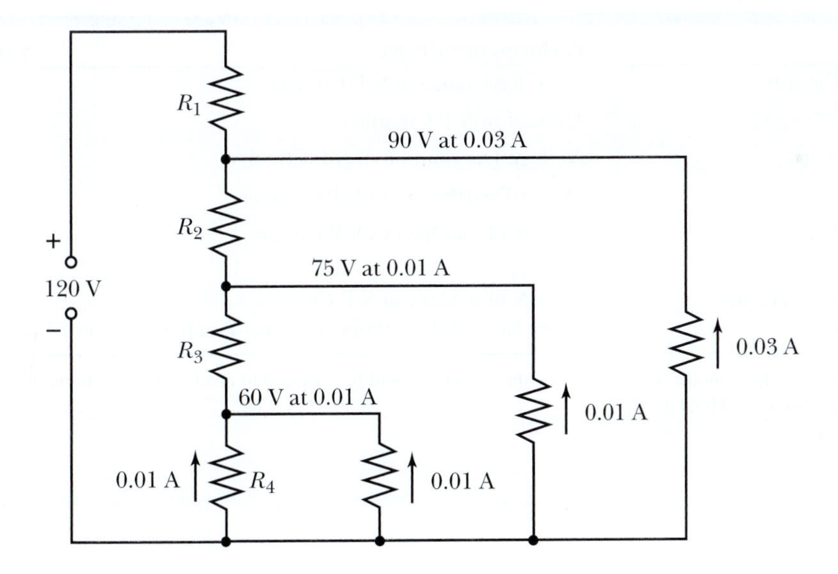

FIGURE 6–59

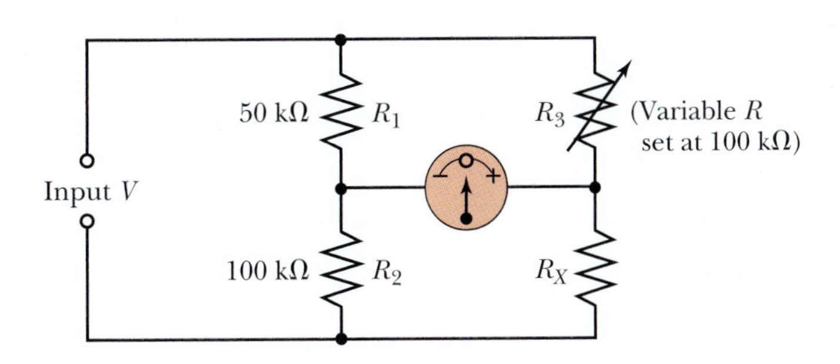

FIGURE 6–60

MultiSIM Exercise 1 for *Series-Parallel Circuits*

1. Use the MultiSIM program and utilize the circuit shown in Figure 6–52. (NOTE: Connect ground at point B.
2. Measure and record the value of resistance from point A to point B.
3. If allowed, check your MultiSIM results against the answer(s) given in the Instructor Guide for Chapter 6—Problems—question 8. Did the MultiSIM result agree with the Instructor Guide answer?

MultiSIM Exercise 2 for *Series-Parallel Circuits*

1. Use the MultiSIM program and utilize the circuit shown in Figure 6–48. (NOTE: Connect ground reference to point B.)
2. Connect the top of R_3 to point B; then, measure and record total circuit current.
3. Remove this connection and move the connection at the top of R_3 so that it is connected to point A; then, measure and record total circuit current again.
4. What was the maximum current possible with this circuit? Was maximum current flowing when the top of R_3 was connected to point A or to Point B? Why was this true?
5. If allowed, check your MultiSIM results against the answer(s) given in the Instructor Guide for Chapter 6—Problems—question 4. Did the answer(s) reasonably compare?

Performance Projects Correlation Chart

Suggested performance projects in the Laboratory Manual that correlate with topics in this chapter are:

Chapter Topic	Performance Project	Project Number
Total Resistance in S-P Circuits	Total Resistance in S-P Circuits	21
Current in S-P Circuits	Current in S-P Circuits	22
Voltage in S-P Circuits	Voltage Distribution in S-P Circuits	23
Power in S-P Circuits	Power Distribution in S-P Circuits	24
Effects of Opens in S-P Circuits and Troubleshooting Hints	Effects of an Open in S-P Circuits	25
Effects of Shorts in S-P Circuits and Troubleshooting Hints	Effects of a Short in S-P Circuits Story Behind the Numbers: Series-Parallel Circuits	26

NOTE: It is suggested that after completing the above projects, the student should be required to answer the questions in the "Summary" at the end of this section of projects in the Laboratory Manual.

Troubleshooting Challenge

CHALLENGE CIRCUIT 5

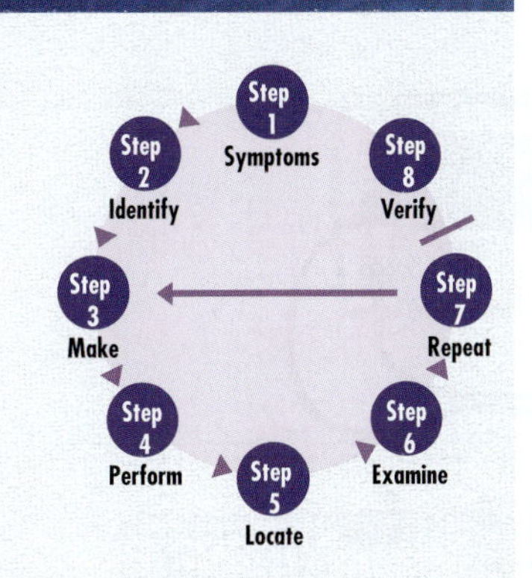

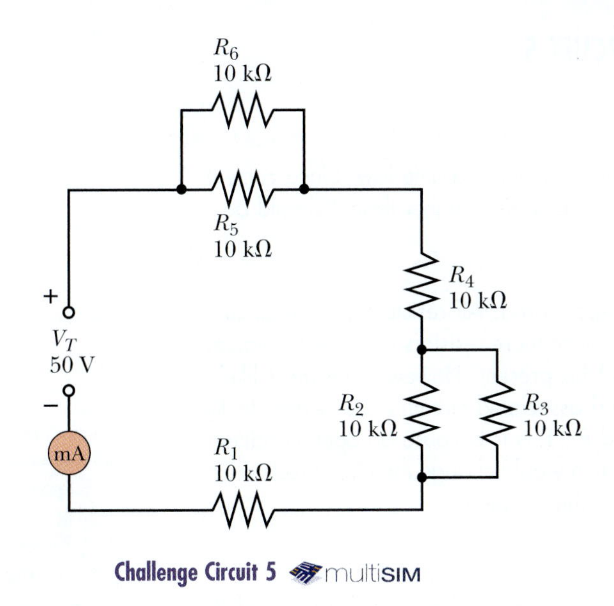

Challenge Circuit 5 multiSIM

STARTING POINT INFORMATION

1. Circuit diagram
2. I_T measures 1.4 mA
3. V_{R_1} measures 14 V
4. Resistances are measured with remainder of circuit being isolated from points being tested.

TEST	Results in Appendix C
V_{R_1}	(49)
V_{R_2}	(80)
V_{R_3}	(12)
V_{R_4}	(39)
V_{R_5}	(61)
V_{R_6}	(88)
Res. of R_1	(25)
Res. of R_2	(54)
Res. of R_3	(93)
Res. of R_4	(37)
Res. of R_5	(68)
Res. of R_6	(33)

CHALLENGE CIRCUIT 5

STEP 1

SYMPTOMS The total current is a little low. Since current should be close to 1.7 mA, R_T is higher than it should be.

STEP 2

IDENTIFY initial suspect area: All resistors are initial suspect areas since if one or more resistors increased in value, the symptom would be present. However, the most likely suspect resistors and associated circuitry are related to R_2 and R_3, plus R_5 and R_6. This is because an open condition for any resistor branch would change the circuit resistance just enough to cause the symptom.

STEP 3

MAKE test decision. **1st Test:** Check the resistance of R_2 (with it isolated).

STEP 4

PERFORM Look up the test result. R_2 is 10 kΩ, which is normal.

STEP 5

LOCATE new suspect area: R_3, R_5, and R_6.

STEP 6

EXAMINE available data.

STEP 7

REPEAT analysis and testing:

2nd Test: Measure the resistance of R_3. R_3 is 10 kΩ, which is normal.

3rd Test: Measure the resistance of R_5. R_5 is 10 kΩ, which is normal.

4th Test: Measure the resistance of R_6. R_6 is infinite ohms. NOTE: This meter shows 30.00 MΩ when measuring an open (or an R value from 30 MΩ to ∞).

STEP 8

VERIFY **5th Test:** Replace R_6 and note the circuit parameters. When this is done, the circuit operates properly.

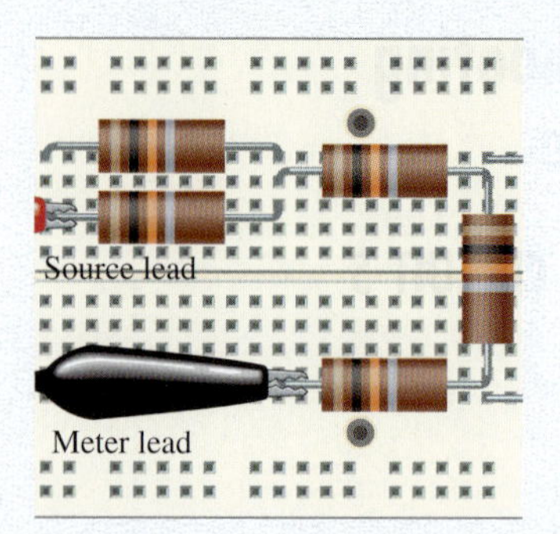

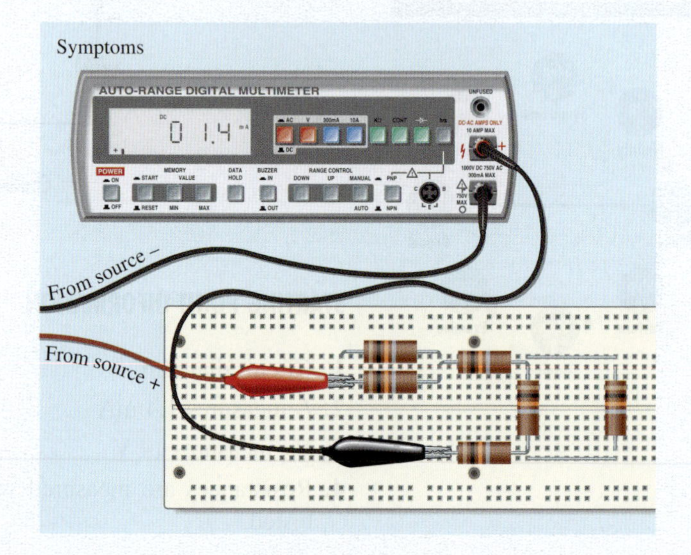

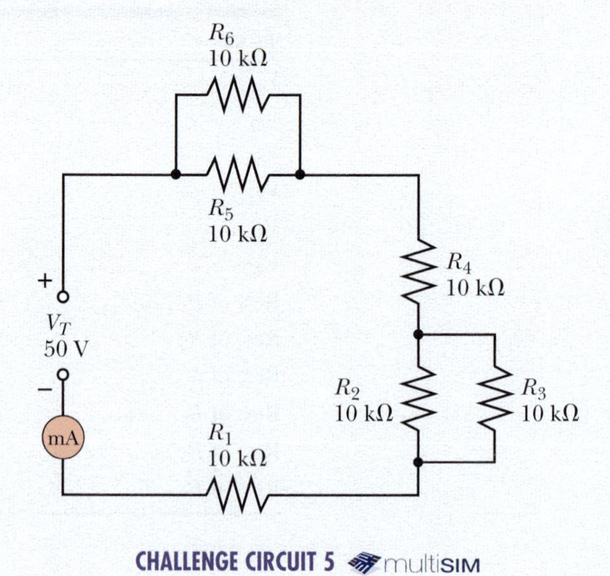

CHALLENGE CIRCUIT 5 multiSIM

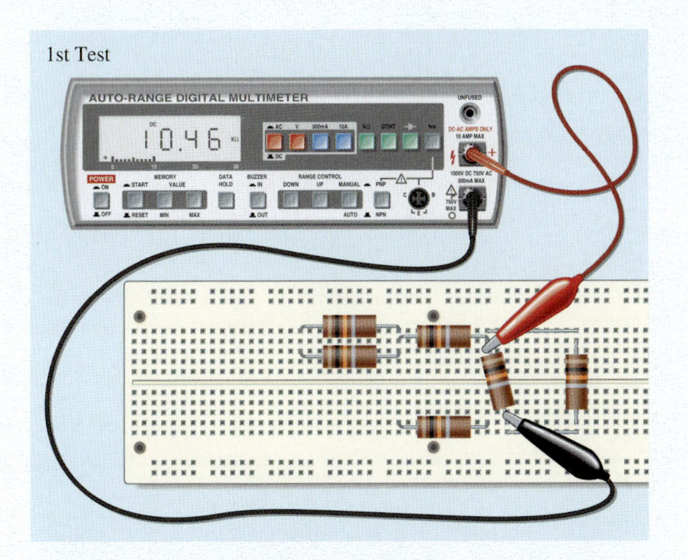

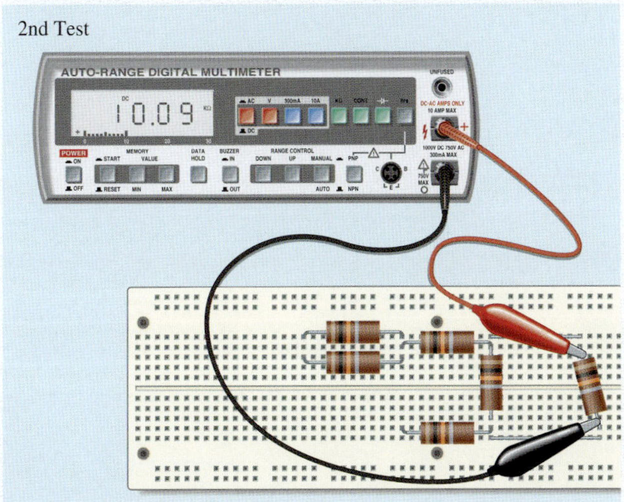

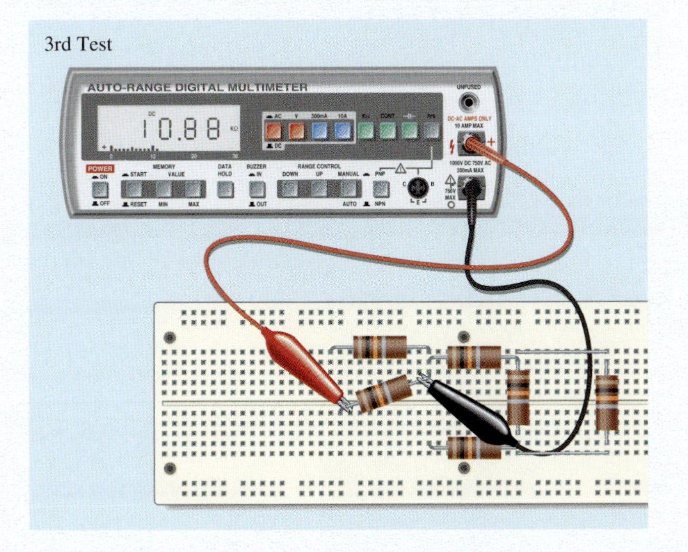

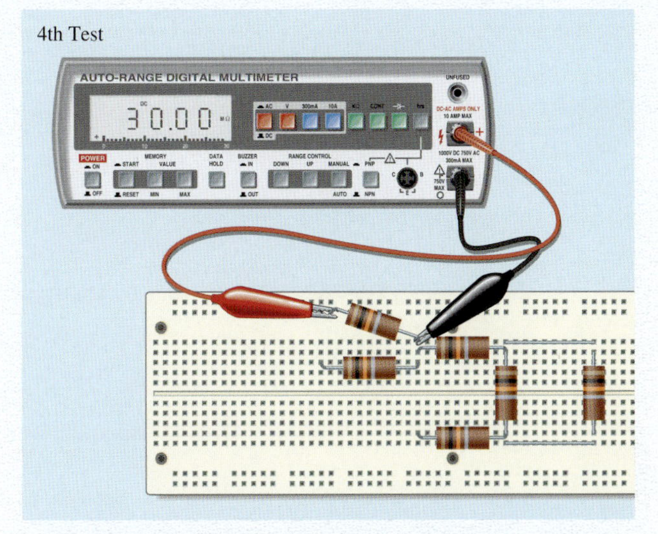

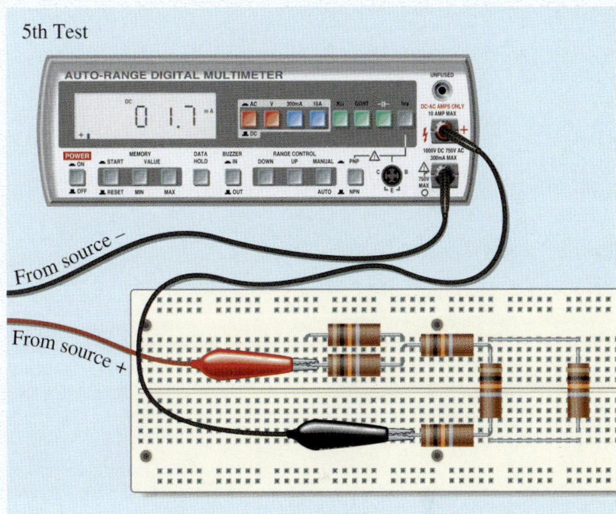

SPECIAL NOTE: Another way to check is to lift one end of each resistor and observe whether total current changed. The resistor you lift without having the total current change is the open one, since opening an open does not change the circuit conditions.

CHALLENGE CIRCUIT 6 (Block Diagram)

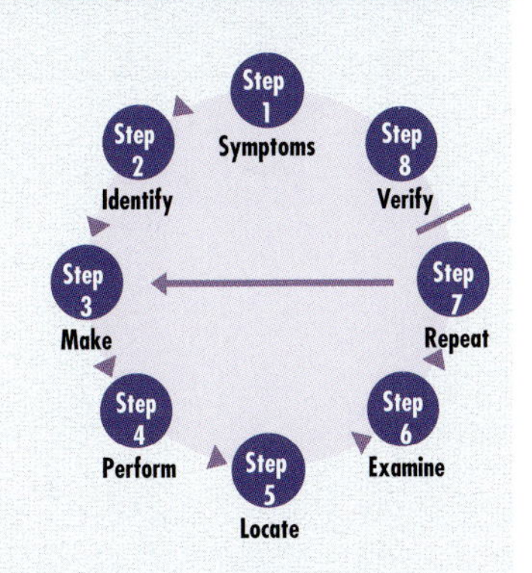

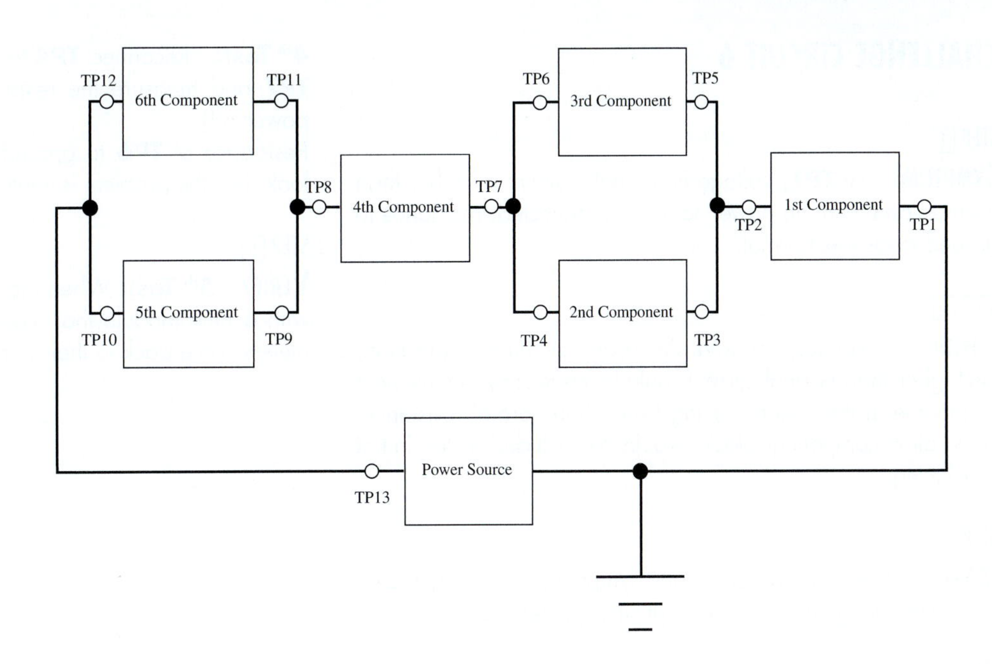

General Testing Instructions:

Measurement Assumptions:

I = at **TP**

V = from **TP** to ground

R = from **TP** to ground

(with the power source disconnected from the circuit)

Possible Tests & Results:

Current (high, low, normal)

Voltage (high, low, normal)

Resistance (high, low, normal)

Starting Point Information	Test Points	Test Results in Appendix C *V*	*I*	*R*	*Signal*
At TP1:	TP2	(193	194	195	NA)
	TP3	(196	197	198	NA)
V = normal	TP4	(199	200	201	NA)
I = high	TP5	(202	203	204	NA)
R = low	TP6	(205	206	207	NA)
	TP7	(208	209	210	NA)
	TP8	(211	212	213	NA)
	TP9	(214	215	216	NA)
	TP10	(217	218	219	NA)
	TP11	(220	221	222	NA)
	TP12	(223	224	225	NA)
	TP13	(226	227	228	NA)

CHALLENGE CIRCUIT 6

STEP [1]

SYMPTOMS At **TP1**, voltage is normal, current is higher than normal, and with the power source disconnected, resistance to ground measures normal.

STEP [2]

IDENTIFY initial suspect area. As in any circuit configuration, the higher-than-normal current could be caused by any element or module in the system having lower-than-normal resistance. Thus, all 6 component blocks would be included in the initial bracketing.

STEP [3]

MAKE test decision based on the symptoms information. Let's start with voltage measurements at strategic test points.

STEP [4]

PERFORM **1st Test:** Voltage value at **TP2.**
Voltage measures higher than normal. This is coherent with higher-than-normal current through a block with normal resistance.

STEP [5]

LOCATE new suspect area. The remaining 5 modules are still suspect.

STEP [6]

EXAMINE available data.

STEP [7]

REPEAT analysis and testing.
2nd Test: Measure the voltage at **TP7.**
Voltage at this point is lower than normal. All right! This indicates that something between TP2 and TP7 has lowered its resistance value. It could be either Component Block #2, or Component Block #3. It will be necessary to isolate these two blocks, one at a time in order to find the one that is out of specifications.
3rd Test: Isolate Component Block #2 from Component Blocks 3–6 by disconnecting **TP4's lead** from **TP6** and **TP7.** Measure resistance to ground at **TP4** (with power off).
Resistance at **TP4** to ground checks normal.
4th Test: Reconnect **TP4** to circuit. Isolate **TP6** from **TP4** and **TP7** and measure the resistance from **TP6** to ground (with power off).
Resistance at **TP6** to ground measures lower than normal. It looks like the problem is contained in the 3rd component block.

STEP [8]

VERIFY **5th Test:** When the 3rd component block is replaced with a new module that meets specifications, all circuit parameters come back to their norms.

OUTLINE

OBJECTIVES

After studying this chapter, you should be able to:

1. State the **maximum power transfer theorem**
2. Determine the R_L value needed for maximum power transfer in a given circuit
3. State the **superposition theorem**
4. Solve circuit parameters for a circuit having more than one source
5. State **Thevenin's theorem**
6. Determine V_L and I_L for various values of R_L connected across specified points in a given circuit or network using Thevenin's theorem
7. State **Norton's theorem**
8. Apply Norton's theorem in solving specified problems
9. Convert between Norton and Thevenin equivalent parameters
10. Use the computer to solve circuit problems

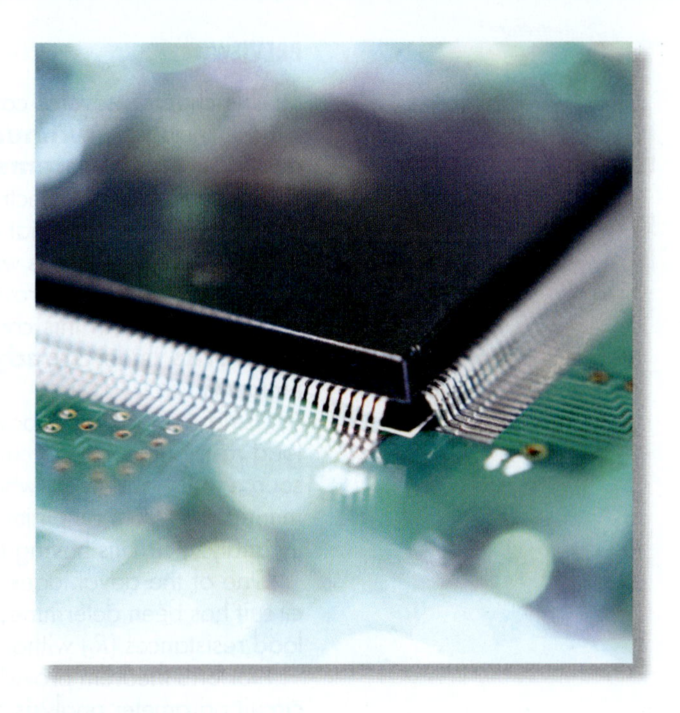

CHAPTER 7

Basic Network Theorems

ELECTRONICS INTO THE FUTURE

Explore an interactive presentation on Network Theorems (Kirchoff, Superposition, Thevenin) on the accompanying CD.

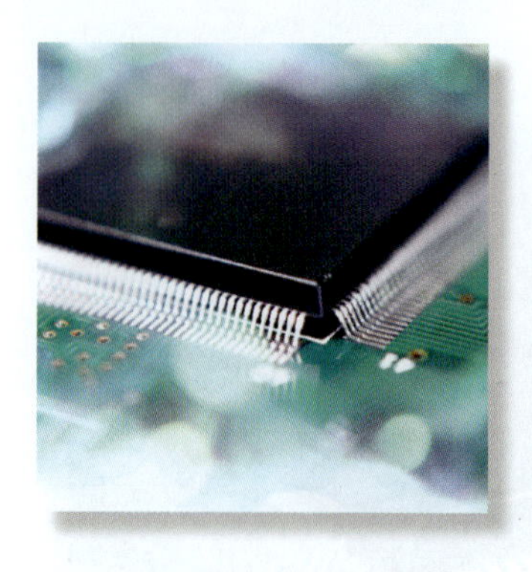

PREVIEW

In this chapter, several commonly used electrical network theorems will be discussed. These include **maximum power transfer, superposition,** and **Thevenin's** and **Norton's theorems.** For our purposes, a network is defined as a combination of components that are electrically connected. Theorems are ideas or statements that are not obvious at first but that we can prove using some more accepted premises. When the theorems are proved this way, they also become laws. The theorems you will learn about in this chapter can be proven in theory and in practice.

The theorems in this chapter deal with **bilateral resistances** that are in **linear networks** under **steady-state conditions,** not during transient switch on or off moments.

Using the maximum power transfer theorem, you can know what value of dc circuit load resistance or ac circuit load impedance allows maximum transfer of power from a source to its load (e.g., what speaker impedance in your stereo allows transfer of maximum power from amplifier to speaker). Using the superposition theorem, you can readily analyze circuits having more than one source.

One of the advantages in using Thevenin's theorem is once the Thevenin equivalent circuit has been determined, you can easily predict electrical parameters for a variety of load resistances (R_L) without having to recalculate the whole network each time.

Norton's theorem provides another approach to simplify complex networks and make circuit parameter analysis easier.

KEY TERMS

Bilateral resistance
Constant current source
Constant voltage source
Efficiency
Impedance
Linear network
Maximum power transfer theorem
Multiple-source circuit
Network
Norton's theorem
Steady-state condition
Superposition theorem
Thevenin's theorem

7–1 Some Important Terms

In the Preview we mentioned several terms that should be defined at this point. We indicated that the theorems you will study in this chapter relate to circuits that are linear networks having bilateral resistances that are being operated under steady-state conditions. These are rather high-sounding words, but their meanings are quite simple. Let's briefly look at definitions for each term:

- **Bilateral resistance**—a resistance having equal resistance in either direction. That is, the resistance is the same for current passing either way through the component. This type of component produces the same voltage drop across it, no matter which direction the current flows through it.
- **Linear network**—a circuit whose electrical behavior does not change with different voltage or current levels or values.
- **Steady-state condition**—the condition where circuit values and conditions are stable or constant. The opposite of transient conditions, such as switch-on or switch-off conditions, when values are changing from one state or level to another.

7–2 Maximum Power Transfer Theorem

In many electronic circuits and systems, it is important to have maximum transfer of power from the source to the load. For example, in radio or TV transmitting systems, it is desired to transfer the maximum power possible from the last stage of the transmitter to the antenna system. In your high fidelity amplifier, you want maximum power transferred from the last amplifier stage to the speaker system. This is usually accomplished by proper "**impedance** matching," which means the load resistance or impedance matches the source resistance or impedance. NOTE: In dc circuits, such as you have been studying, the term for opposition to current flow is *resistance.* In ac circuits, which you will soon study, the term for opposition to current flow is *impedance.*

If it were possible to have a voltage source with zero internal resistance that was able to maintain constant output voltage (even if a load resistance of zero ohms were placed across its output terminals), the source theoretically could supply infinite power to the load, where $P = V \times I$. In this case, V would be some value and I would be infinite. Thus, $P = V$ value $\times$ infinite current = infinite power in watts.

Since these conditions are impossible, the practical condition is that there is an optimum value of R_L where maximum power is delivered to the load. Any higher or lower value of R_L causes less power to be delivered to the load.

The Basic Theorem

The **maximum power transfer theorem** states *maximum power is transferred from the source to the load (R_L) when the resistance of the load equals the resistance of the source.* In other words, it is a series circuit. In practical circuits, the internal resistance (r_{int}) of the source causes some of the source's power to be dissipated within the source, while the remaining power is delivered to the external circuitry, Figure 7–1.

Refer to the circuits in Figure 7–2a, b, and c. Assuming the same source for all three circuits, notice the power delivered to the load in each case.

Observe in the first circuit, Figure 7–2a, the load resistance (R_L) equals the internal resistance of the source (r_{int}). The power delivered to the load is 10 watts.

In the circuit of Figure 7–2b, the load resistance is less than the source's internal resistance. The power delivered to the load is 8.84 watts.

In the third circuit, Figure 7–2c, R_L is greater than r_{int}, and the power delivered to the load is 7.5 watts. You can make calculations for a large number of values to confirm that maximum power is delivered when $R_L = r_{int}$.

Practical Notes

It is important to realize that all power sources have some internal resistance, for which we are using the designation r_{int} in our discussions. We are showing you the r_{int} as an internal source series resistance, which alters the source output voltage delivered to the load, and limits the maximum amount of current that can be delivered by the source, Figure 7–1.

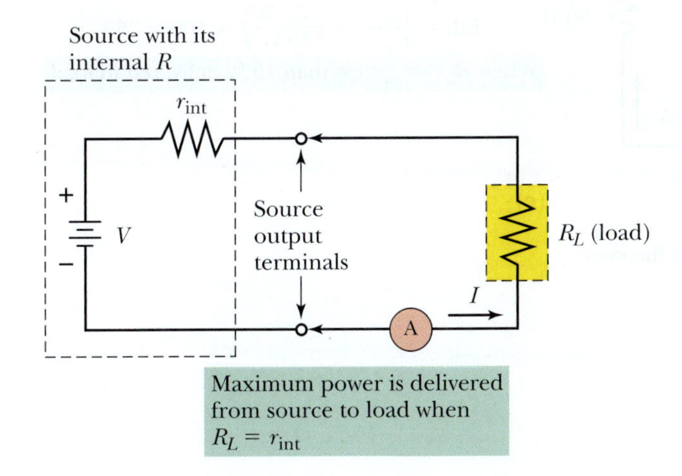

FIGURE 7–1 Maximum power transfer theorem (NOTE: R_T seen by source is the series total of $r_{int} + R_L$).

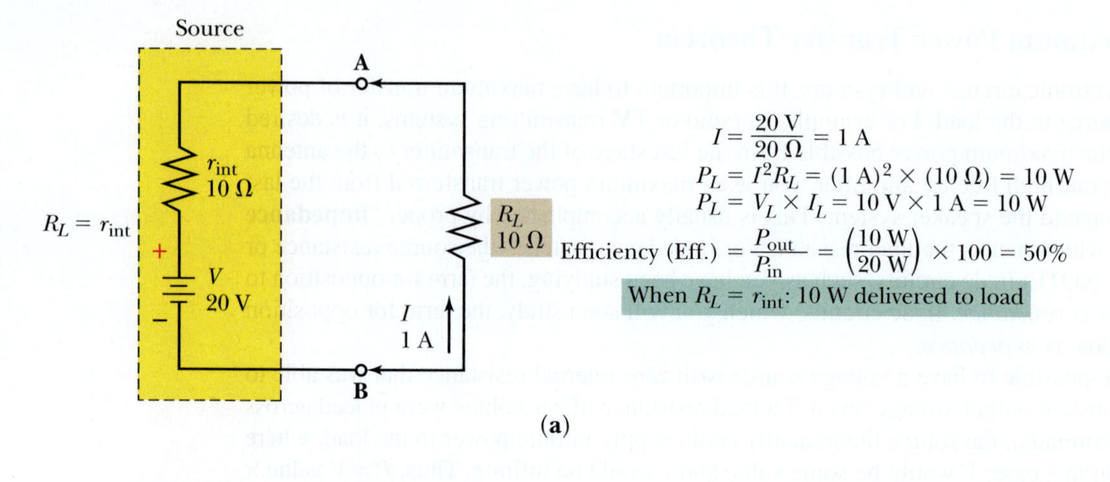

(a)

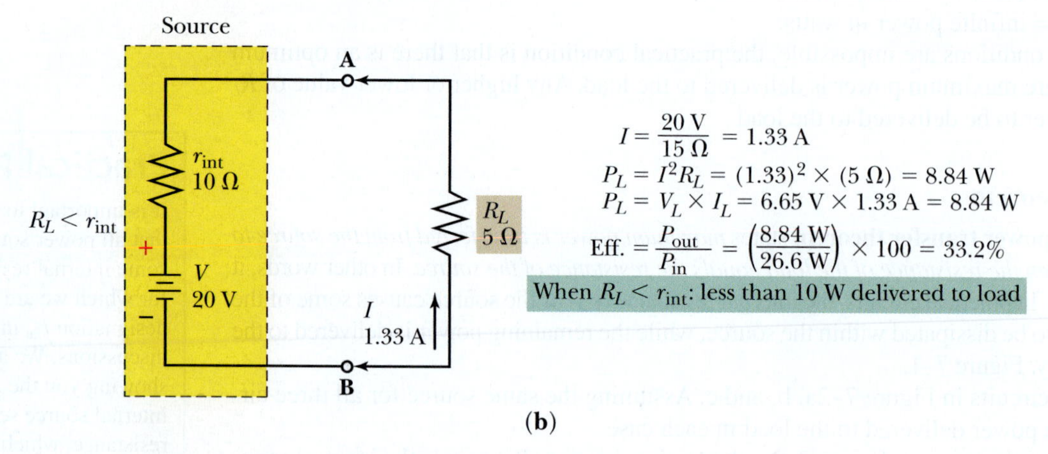

(b)

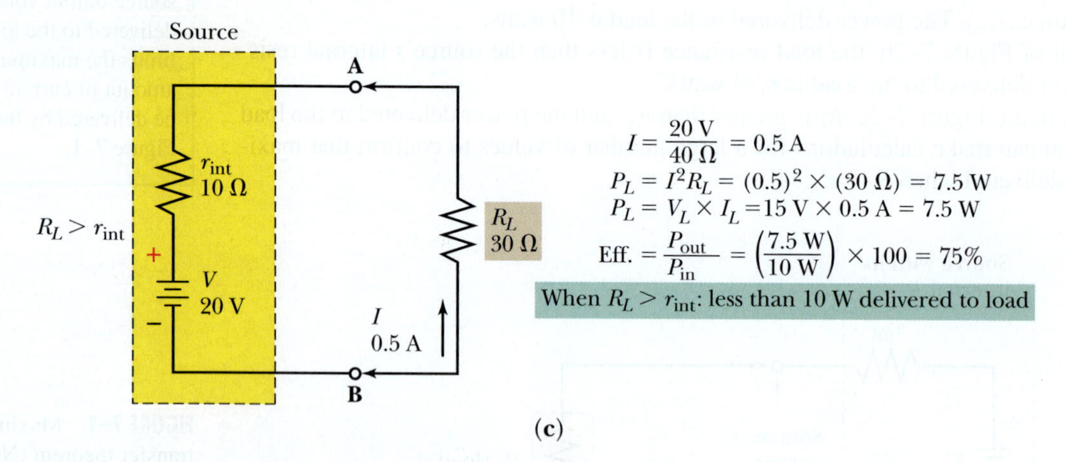

(c)

FIGURE 7–2 Verification of maximum power transfer theorem

Efficiency Factor

FORMULA 7–1 $\text{Efficiency (\%)} = \frac{P_{out}}{P_{in}} \times 100$

If you calculate the **efficiency** factor, based on what percentage of total power generated by the source is delivered to the load, you will find efficiency equals 50% when R_L equals r_{int}.

In Figure 7–2a, $V = 20$ V and $I = 1$ A. Thus, P_T (or P_{in}) = 20 V × 1 A = 20 W. The load is dissipating $(1\text{ A})^2\,(10\ \Omega) = 10$ W, or half of the 20 watts.

In Figure 7–2b, $V = 20$ V and $I = 1.33$ A. Thus, $P_T = 20\text{ V} \times 1.33\text{ A} = 26.6$ W, but only 8.84 W is delivered to the load. Therefore,

$$\text{Efficiency} = \left(\frac{8.84}{26.6}\right) \times 100 = 33.2\%$$

On the other hand, in Figure 7–2c, $P_T = 10$ W and $P_L = 7.5$ W. Thus, 75% of the power generated by the source is delivered to the load. However, the smaller value of total power produced by the source causes P_L to be less than when there was 50% efficiency (R_L matched r_{int}).

Where is the remainder of the power consumed in each case? If you said by the internal resistance of the source, you are right! In other words, when there is maximum power transfer, half the total power of the source is dissipated by its own internal resistance; the other half is dissipated by the external load. When the efficiency is about 33%, Figure 7–2b, 67% of the total power is dissipated within the source. When efficiency is 75%, Figure 7–2c, 25% of the total power is dissipated within the source, while 75% is delivered to the load.

Summary of the Maximum Power Transfer Theorem

- Maximum power transfer occurs when the load resistance equals the source resistance.
- Efficiency at maximum power transfer is 50%.
- When R_L is greater in value than r_{int}, efficiency is greater than 50%.
- When R_L is less than r_{int}, efficiency is less than 50%.

To visually see the effects on load power and efficiency when R_L is varied, using a given source, observe the graphs in Figures 7–3 and 7–4.

FORMULA 7–2 $r_{int} = \frac{V\text{(no-load)} - V\text{(loaded)}}{I\text{(load)}}$

If V (no-load) is 13 V and V (loaded) is 12 V with 100 A (load) current, then:

$$\frac{13\text{ V} - 12\text{ V}}{100\text{ A}} = 0.01\ \Omega$$

Practical Notes

You may or may not have noticed that the amount of voltage at the output terminals of the source varies with load current. This is due to the internal *IR* drop across r_{int}. That is, higher currents result in greater internal *IR* drop and less voltage available at the output terminals of the source. In fact, one way to obtain an idea of a source's internal resistance is to measure its open circuit output terminal voltage, then put a load that demands current from the source across its terminals. The change in output terminal voltage from no-load (zero current drain) to loaded conditions allows calculation of r_{int}.

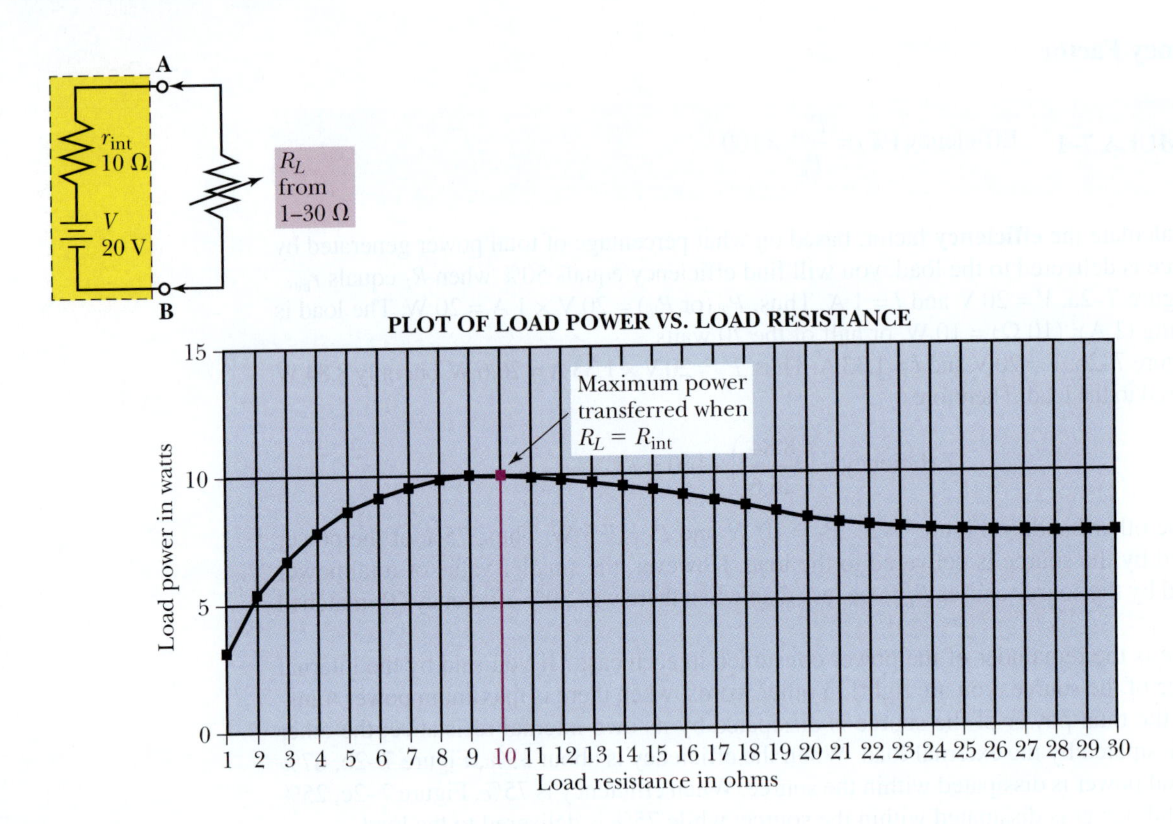

FIGURE 7–3

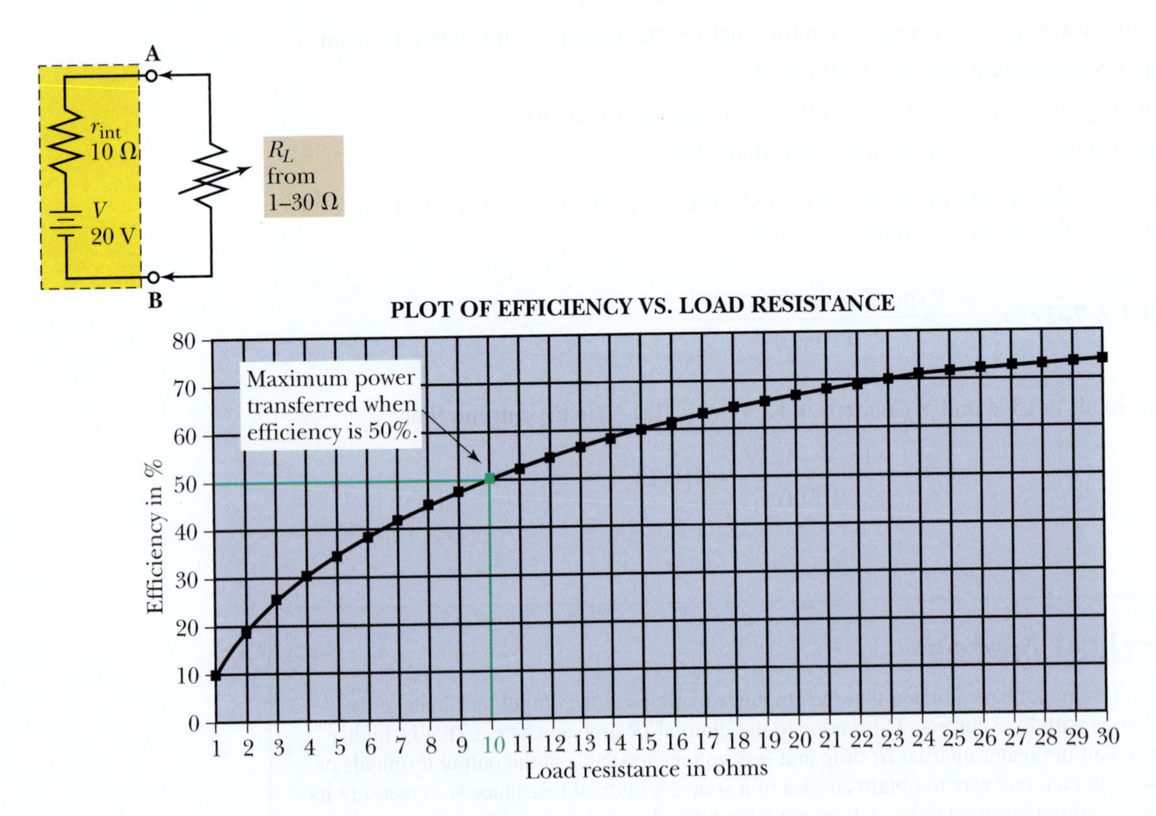

FIGURE 7–4

A practical example of this approach to determine if a source's internal resistance is higher than it should be is the automotive technician's technique of checking a battery's output voltage at no-load, then under heavy load. If the terminal voltage of the battery drops too much under load, its internal resistance is too high, and the battery is either defective or needs charging.

PRACTICE PROBLEMS 1

Try solving the maximum power transfer problems in Figure 7–5.

7-3 Superposition Theorem

Often, you will analyze circuitry having more than one source. The superposition theorem is a useful tool in these cases. The **superposition theorem** states that *in linear circuits having more than one source, the voltage across or current through any given element equals the algebraic sum of voltages or currents produced by each source acting alone with the other sources disabled,* Figure 7–6. NOTE: To disable voltage sources for purposes of calculations, consider them shorted. To disable current sources, consider them opened.

EXAMPLE A circuit containing two voltage sources—a **multiple-source circuit**—is illustrated in Figure 7–6a. In Figure 7–6b through 7–6d, the step-by-step application of the superposition theorem is shown. ***Important:*** **Noting the direction of current flow and polarity of voltage drops is important for each step.** This allows you to find the actual resultant current through a given component, or the resultant voltage drop of that component by adding or subtracting currents or voltages calculated from each source, as appropriate. When currents through a given component from the individual sources pass through that component in the same direction, the resultant current is calculated as the sum of the two individually computed currents. If currents pass through that component in opposite directions, subtract them. When computed voltage drops across a given component from each source are the same polarity, the resultant voltage equals the sum. If the sources cause opposite polarity voltage drops across the component, subtract them to find the resultant voltage drop across that component.

To get proper results from the final algebraic additions, use the same *reference point* in every case to determine polarity of voltage drop across a given component. Then you can combine the individual results properly to get the final resultant value.

In Figure 7–6b, the 60-V source is considered disabled (shorted out). Notice this effectively places R_3 in parallel with R_2. This means the R_1 value of 2 kΩ is in series with the parallel combination of $R_2 - R_3$ across the 30-V source. Thus, total R is 3 kΩ. Total I is 10 mA.

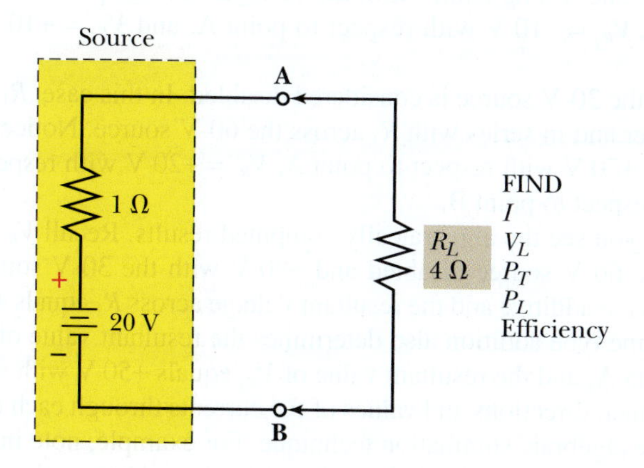

FIGURE 7–5

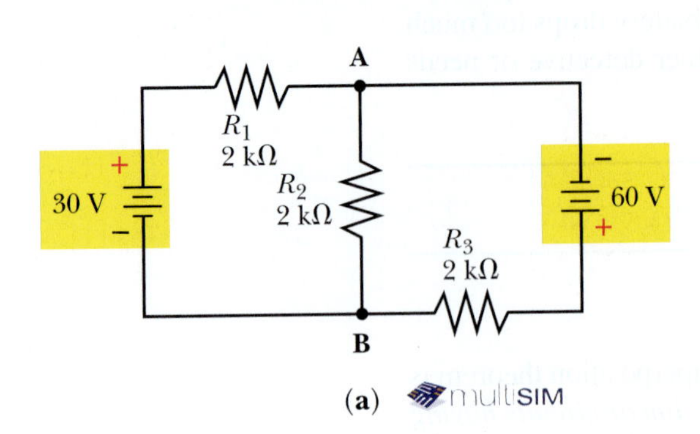

(a)

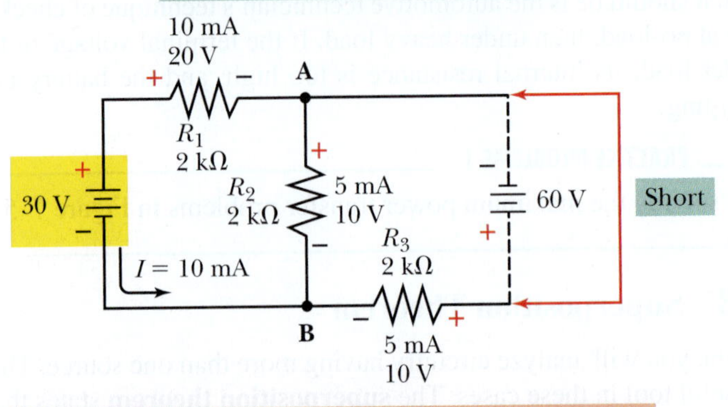

NOTE: If you were going to physically do this, you would *remove* the 60-V source and replace it with a conductor.

Step 1: Assume 60-V source is replaced by a short (0 Ω).

Step 2: Analyze R_T of circuit:

a. $R_2 \| R_3 = \dfrac{2\text{ k}\Omega \times 2\text{ k}\Omega}{2\text{ k}\Omega + 2\text{ k}\Omega} = \dfrac{4 \times 10^6}{4 \times 10^3} = 1\text{ k}\Omega$

b. R_e of $R_2 \| R_3$ is in series with R_1

$R_T = R_1 + R_e = 2\text{ k}\Omega + 1\text{ k}\Omega = 3\text{ k}\Omega$

Step 3: Solve for total current and branch currents:

a. $I_T = \dfrac{30\text{ V}}{3\text{ k}\Omega} = 10\text{ mA}$

b. Since each branch = 2 kΩ, current divides equally:

$I_{R_2} = 5\text{ mA}$; $I_{R_3} = 5\text{ mA}$

Step 4: Use Ohm's law and solve the voltage drops:

a. $V_{R_1} = I_T \times R_1 = 10\text{ mA} \times 2\text{ k}\Omega = +20\text{ V}$ (with respect to point **A**)

b. $V_{R_2} = I_{R_2} \times R_2 = 5\text{ mA} \times 2\text{ k}\Omega = -10\text{ V}$ (with respect to point **A**)

c. $V_{R_3} = I_{R_3} \times R_3 = 5\text{ mA} \times 2\text{ k}\Omega = +10\text{ V}$ (with respect to point **B**)

(b)

FIGURE 7–6 Verification of superposition theorem

R_1 passes the total 10 mA, whereas, the 10 mA is equally divided through R_2 and R_3 with 5 mA through each one. Using Ohm's law, the voltages are computed as: $V_{R_1} = +20$ V with respect to point A, $V_{R_2} = -10$ V with respect to point A, and $V_{R_3} = +10$ V with respect to point B.

In Figure 7–6c, the 30-V source is considered disabled. In this case, R_1 and R_2 are in parallel with each other and in series with R_3 across the 60-V source. Notice the voltage drops are such that $V_{R_1} = +20$ V with respect to point A, $V_{R_2} = +20$ V with respect to point A, and $V_{R_3} = +40$ V with respect to point B.

In Figure 7–6d, you see the algebraically computed results. Recall V_{R_1} is +20 V with respect to A with the 60-V source disabled and +20 V with the 30-V source disabled. This means the voltages are additive and the resultant voltage across R_1 equals +40 V with respect to point A. This same type addition also determines the resultant value of V_{R_2} equals +10 V with respect to point A, and the resultant value of V_{R_3} equals +50 V with respect to point B.

In a similar manner, directions and values of the currents through each component are determined, using the algebraic summation technique. For example, note in Figure 7–6b, current flows from point B toward point A through resistor R_2 and has a value of 5 mA. In Figure 7–6c, current flows from point A toward point B through R_2 and has a value of 10 mA. This means the resultant current through R_2 with both sources active is 5 mA traveling from point A toward point B. You get the idea! Refer to Figure 7–6d to see the other resultant values, as appropriate. ▪

Assuming 30-V source shorted out

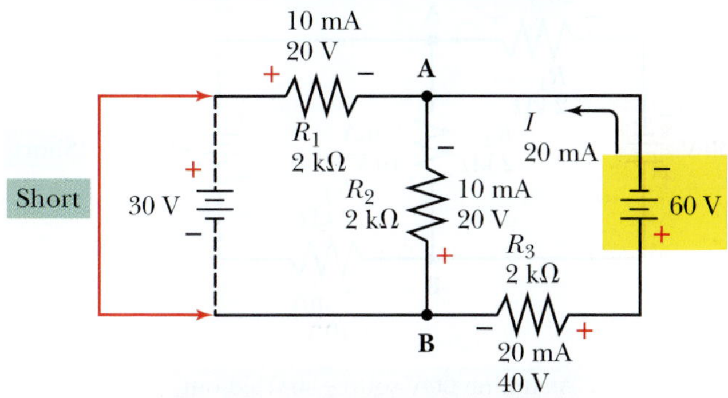

NOTE: To physically perform this step, you would *remove* the 30-V source and replace it with a "short."

Step 5: Assume 30-V source is replaced by a short (0 Ω).

Step 6: Analyze R_T of circuit:

a. $R_1 \| R_2 = \dfrac{2\text{ k}\Omega \times 2\text{ k}\Omega}{2\text{ k}\Omega + 2\text{ k}\Omega} = \dfrac{4 \times 10^6}{4 \times 10^3} = 1\text{ k}\Omega$

b. R_e of $R_1 \| R_2$ is in series with R_3

c. $R_T = R_3 + R_e = 2\text{ k}\Omega + 1\text{ k}\Omega = 3\text{ k}\Omega$

Step 7: Solve for total current and branch currents:

a. $I_T = \dfrac{60\text{ V}}{3\text{ k}\Omega} = 20\text{ mA}$

b. Since each branch = 2 kΩ, current divides equally:
$I_{R_2} = 10\text{ mA}$; $I_{R_3} = 10\text{ mA}$

Step 8: Use Ohm's law and solve the voltage drops:

a. $V_{R_1} = I_{R_1} \times R_1 = 10\text{ mA} \times 2\text{ k}\Omega = +20\text{ V}$ (with respect to point **A**)

b. $V_{R_2} = I_{R_2} \times R_2 = 10\text{ mA} \times 2\text{ k}\Omega = +20\text{ V}$ (with respect to point **A**)

c. $V_{R_3} = I_T \times R_3 = 20\text{ mA} \times 2\text{ k}\Omega = +40\text{ V}$ (with respect to point **B**)

(c)

Resultant circuit *V*s and *I*s with two sources

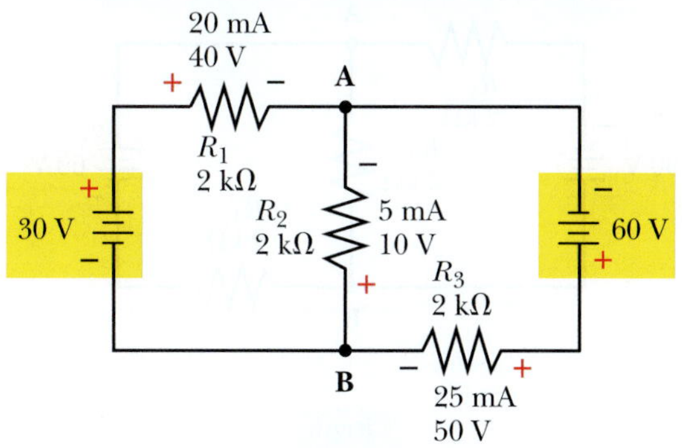

Step 9: Algebraically add voltages and currents from the two sets of computations to find resultant component electrical parameters:

a. $V_{R_1} = +20\text{ V} + 20\text{ V} = +40\text{ V}$ (with respect to point **A**)

b. $V_{R_2} = -10\text{ V} + 20\text{ V} = +10\text{ V}$ (with respect to point **A**)

c. $V_{R_3} = +10\text{ V} + 40\text{ V} = +50\text{ V}$ (with respect to point **B**)

d. I_{R_1} resultant current = 20 mA (away from point **A**)

e. I_{R_2} resultant current = 5 mA (away from point **A**)

f. I_{R_3} resultant current = 25 mA (away from point **B**)

(d)

FIGURE 7–6 *Continued*

EXAMPLE If the polarity of each of the sources in Figure 7–6 is reversed (while the circuit configuration remains the same), the results are like those in Figure 7–7.

The same analysis is used. That is, disable each voltage source, one at a time by viewing them as shorted. Then determine the currents and voltages throughout the circuit, using Ohm's law and so on.

With the 60-V source assumed as shorted, V_{R_1} equals –20 V with respect to point A; V_{R_2} equals +10 V with respect to point A; and V_{R_3} equals –10 V with respect to point B, Figure 7–7b.

With the 30-V source assumed as shorted, V_{R_1} equals –20 V with respect to point A; V_{R_2} equals –20 V with respect to point A; and V_{R_3} equals –40 V with respect to point B, Figure 7–7c.

Combining the results of the two assumed conditions, V_{R_1} equals –40 V with respect to point A; V_{R_2} equals –10 V with respect to point A; and V_{R_3} equals –50 V with respect to point B, Figure 7–7d.

As you can see, when only the polarity of the sources changes, the resultant direction of currents and polarities of the voltage drops reverses but their values remain the same. Compare Figures 7–6d and 7–7d.

Summary of the Superposition Theorem

- Ohm's law is used to analyze the circuit, using one source at a time.
- Then the final results are determined by algebraically superimposing the results of all the sources involved.

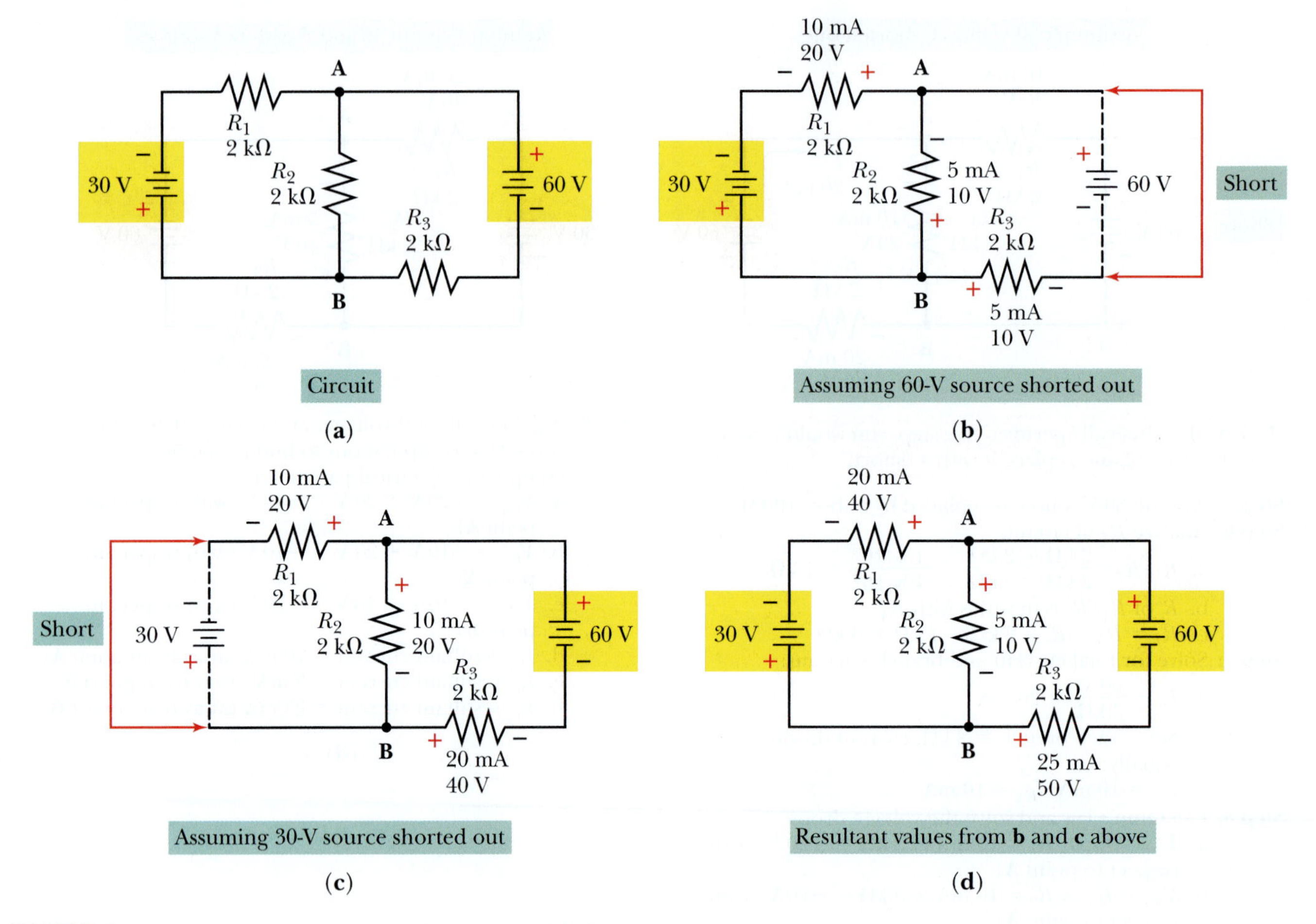

FIGURE 7–7 Superposition theorem example: reversed polarities

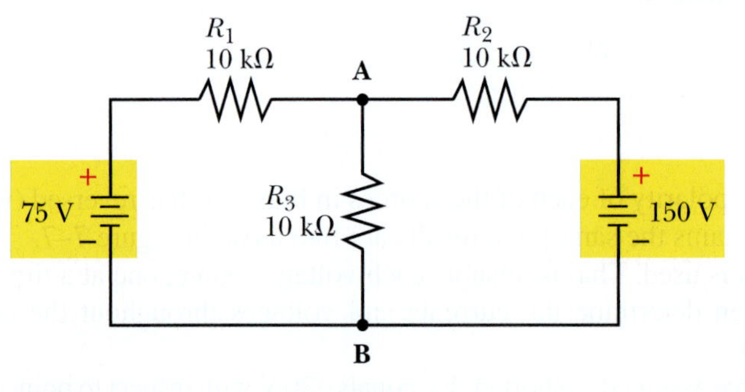

FIND
I through R_1
V across R_1, with respect to Point **A**
I through R_2
V across R_2, with respect to Point **A**
I through R_3
V across R_3, with respect to Point **A**

FIGURE 7–8

PRACTICE PROBLEMS 2

For practice in using the superposition theorem, try the following problems:

1. Solve for the resultant current(s) and voltage(s) for the circuit in Figure 7–8.
2. Assume only the 150-V source is reversed in polarity and solve for the resultant current(s) and voltage(s). **Caution: Only one source is being reversed!**

■ **IN-PROCESS LEARNING CHECK 1** Fill in the blanks as appropriate.

1. For maximum power transfer to occur between source and load, the load resistance should _____ the source resistance.
2. The higher the efficiency of power transfer from source to load, the _____ the percentage of total power is dissipated by the load.
3. Maximum power transfer occurs at _____ % efficiency.
4. If the load resistance is less than the source resistance, efficiency is _____ than the efficiency at maximum power transfer.
5. To analyze a circuit having two sources, the superposition theorem indicates that Ohm's law (can, cannot) _____ be used.
6. What are two key observations needed in using the superposition theorem to analyze a circuit with more than one source? ______________________________

   ____
7. Using the superposition theorem, if the sources are considered "voltage" sources, are these sources considered shorted or opened during the analysis process? _____
8. When using the superposition theorem when determining the final result of your analysis, the calculated parameters are combined, or superimposed, (arithmetically, algebraically) _____. ■

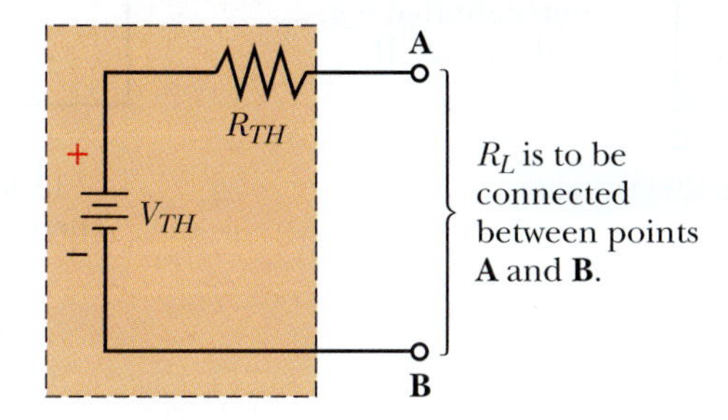

FIGURE 7–9 Thevenin's theorem equivalent circuit

7–4 Thevenin's Theorem

Thevenin's theorem is used to simplify complex **networks** so the calculation of voltage(s) across any two given points in the circuit (and consequently, current[s] flowing between the two selected points) is easily determined.

Once the Thevenin equivalent circuit parameters are known, solving circuit parameters for *any value of* R_L merely involves solving a two-resistor series circuit problem. A **constant voltage source** is assumed. This is a source whose output voltage remains constant under varying load current (or current demand) conditions.

As you recall, voltage distribution was considered when you studied series circuits. Thevenin's theorem deals with a simple equivalent series circuit to simplify voltage distribution analysis. Let's see how this is applied.

The Theorem Defined

One version of **Thevenin's theorem** states that *any* ***linear*** *two-terminal network (of resistances and source[s]) can be replaced by a simplified equivalent circuit consisting of a single voltage source (called* V_{TH}*) and a single series resistance (called* R_{TH}), Figure 7–9.

V_{TH} is the open-terminal voltage felt across the two load terminals (A and B) with the load resistance removed. R_{TH} is the resistance seen when looking back into the circuit from the two load terminals (without R_L connected) with voltage sources replaced by their internal resistance. For simplicity, we assume them shorted.

Refer to Figure 7–10a, b, c, and d. Figure 7–10a shows the original circuit from which we will determine the Thevenin equivalent circuit, a process called "Thevenizing" the circuit.

Figure 7–10b illustrates how the value for V_{TH} is determined. Notice R_L is considered opened (between points A and B). V_{TH} is considered the value of voltage appearing at network

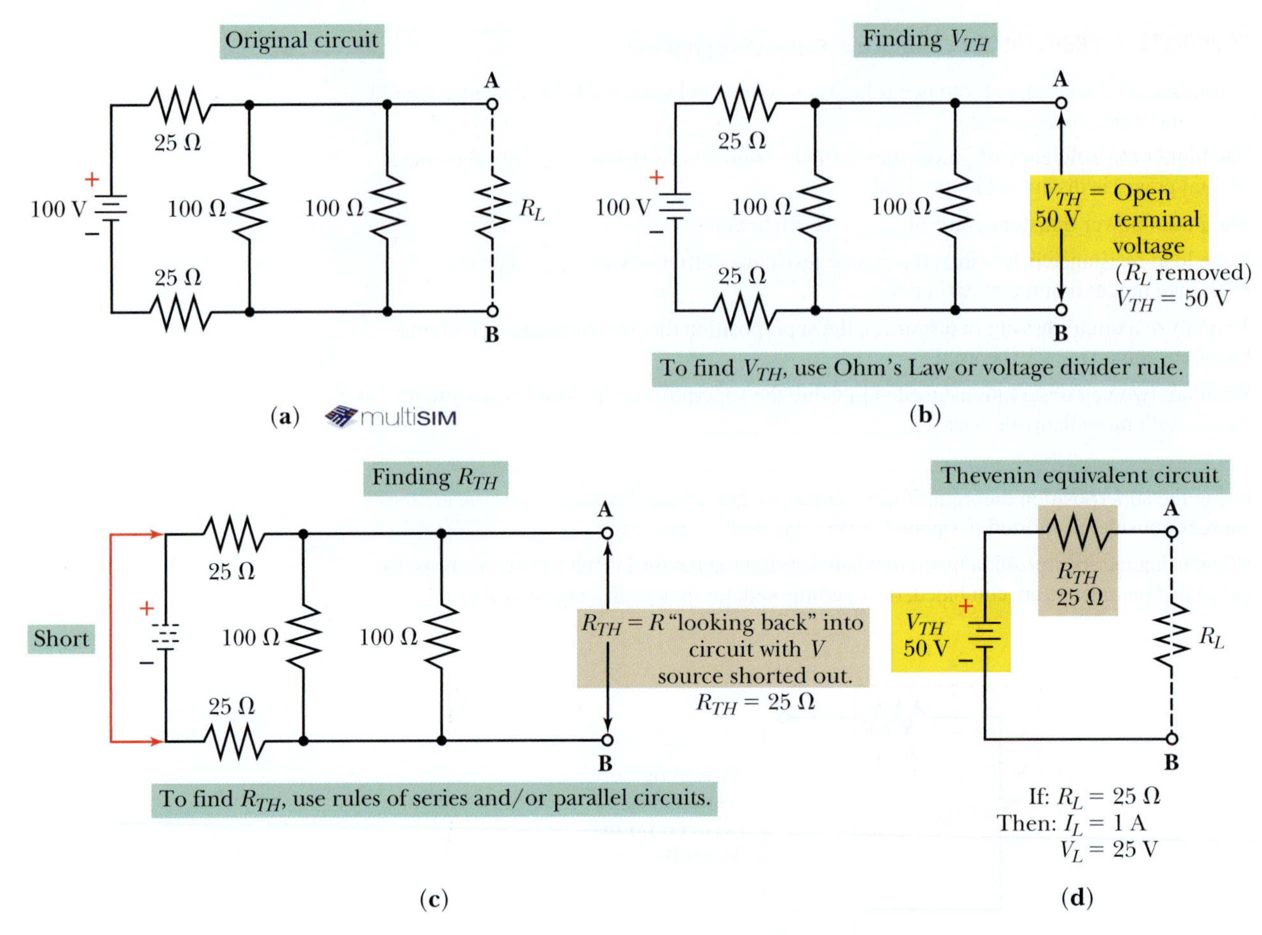

FIGURE 7–10 Thevenin's theorem illustration

terminals A and B (with R_L removed). For example, if Ohm's law is used, R_T is 100 Ω; V_T is 100 V; thus $I_T = 1$ A. Each of the 25-Ω series resistors drops 25 V (1 A × 25 Ω). The 1 A evenly splits between the two 100-Ω parallel resistors. Thus, voltage between points A and B = 0.5 A × 100 Ω, or 50 V. Therefore, $V_{TH} = 50$ V.

Figure 7–10c shows the process to find R_{TH}. Mentally short the source and determine the circuit resistance by looking toward the source from the two network points A and B. With the source shorted, the two 100-Ω parallel resistors are effectively in parallel with 50 Ω (25 Ω + 25 Ω). The equivalent resistance of two 100-Ω resistors is 50 Ω, and that 50 Ω in parallel with the remaining 50-Ω branch yields an R_T of 25 Ω. This is the value of R_{TH}.

Figure 7–10d shows the Thevenin equivalent circuit of a 50-V source (V_{TH}) and the series R_{TH} of 25 Ω. It is thus possible to determine the current through any value of R_L and the voltage across R_L by analyzing a simple two-resistor series circuit.

EXAMPLE If R_L has a value of 25 Ω, what are the values of I_L and of V_L? If you answered 1 A and 25 V, you are correct!

Using formula 7–3, solve this way:

$$I_L = \frac{50\text{ V}}{25\ \Omega + 25\ \Omega} = \frac{50\text{ V}}{50\ \Omega} = 1\text{ A}$$

FORMULA 7–3 $I_L = \dfrac{V_{TH}}{R_{TH} + R_L}$

V_L is calculated by using Ohm's law ($I_L \times R_L$), or the voltage divider rule where:

FORMULA 7–4 $V_L = \frac{R_L}{R_L + R_{TH}} \times V_{TH}$

Using formula 7–4, solve this way:

$$V_L = \frac{25\ \Omega}{25\ \Omega + 25\ \Omega} \times 50\ \text{V} = \frac{25\ \Omega}{50\ \Omega} \times 50\ \text{V} = 0.5 \times 50\ \text{V} = 25\ \text{V}$$ ◆

Summary of Thevenin's Theorem

The Thevenin equivalent circuit consists of a source, V_{TH} (often called the Thevenin generator), and a single resistance, R_{TH} (often called the Thevenin internal resistance), in series with the load (R_L), which is across the specified network terminals. Therefore, this equivalent circuit is useful to simplify calculations of electrical parameters for various values of R_L since it reduces the problem to a simple two-resistor series circuit analysis.

◆ **EXAMPLE** Let's assume that we are to put a load resistance (R_L) across the output terminals (points A and B) of a Wheatstone bridge circuit, Figure 7–11. Further, we would like to determine the *V* and *I* parameters associated with the load resistor. By using Thevenin's theory, we can define the simple Thevenin equivalent circuit for the bridge. Once that is found, simple two-resistor series circuit analysis techniques make it easy to find the *V* and *I* parameters for any given R_L value.

Refer to Figure 7–11 as you study the following steps.

1. Find V_{TH} (the difference of potential between points A and B, with the load resistor removed):

$$V_{TH} = V_A - V_B = \left(\frac{R_2}{R_1 + R_2}\right) \times V_S - \left(\frac{R_4}{R_3 + R_4}\right) \times V_S$$

$$V_{TH} = \left(\frac{220\ \Omega}{320\ \Omega}\right) \times 12\ \text{V} - \left(\frac{470\ \Omega}{800\ \Omega}\right) \times 12\ \text{V}$$

$$V_{TH} = 8.25\ \text{V} - 7.05\ \text{V} = 1.2\ \text{V}$$

2. Find R_{TH} by assuming source to be zero ohms:

$$R_{TH} = R_e \text{ of } (R_1 \parallel R_2) + R_e \text{ of } (R_3 \parallel R_4) \cong 263\ \Omega, \text{ as shown next.}$$

$$R_{TH} = \frac{R_1 \times R_2}{R_1 + R_2} + \frac{R_3 \times R_4}{R_3 + R_4}$$

$$R_{TH} = \frac{100\ \Omega \times 220\ \Omega}{100\ \Omega + 220\ \Omega} + \frac{330\ \Omega \times 470\ \Omega}{330\ \Omega + 470\ \Omega}$$

$$R_{TH} = 68.75\ \Omega + 193.875\ \Omega = 262.625\ \Omega$$

3. Draw the equivalent circuit. See Figure 7–12.

Let assume that an R_L value of 270 Ω is to be connected between points A and B of the bridge circuit. What is the value of V_{R_L} and of I_L?

$$V_{R_L} = \left(\frac{R_L}{R_{TH} + R_L}\right) \times V_{TH} = \left(\frac{270\ \Omega}{263\ \Omega + 270\ \Omega}\right) \times 1.2\ \text{V} \approx 0.61\ \text{V}$$

$$I_L = \frac{V_{TH}}{R_T} = \frac{1.2\ \text{V}}{533\ \Omega} = 2.25\ \text{mA}$$ ◆

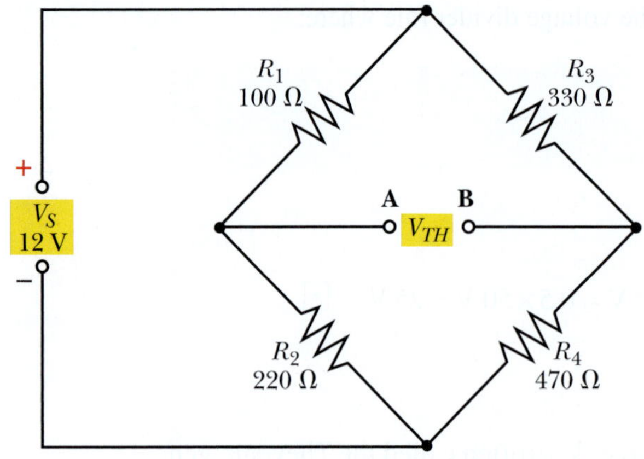

$$V_{TH} = V_A - V_B,$$

where (using the voltage divider method): $V_A = \left(\frac{R_2}{R_1 + R_2}\right) \times V_S$

$$V_B = \left(\frac{R_4}{R_3 + R_4}\right) \times V_S$$

$$V_{TH} = 1.2 \text{ V}$$

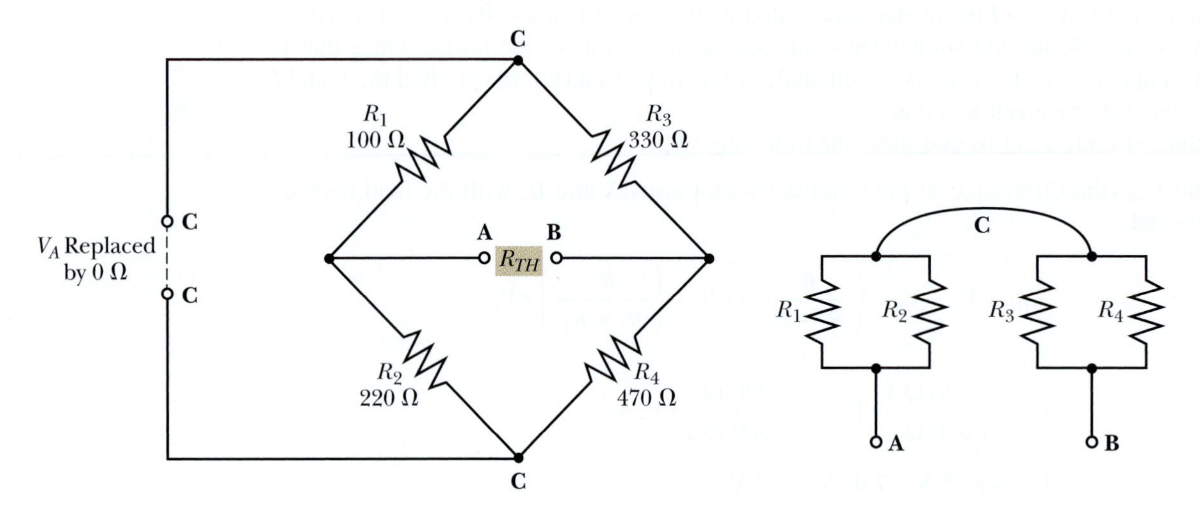

FIGURE 7–11 Analyzing a bridge circuit with Thevenin's theorem

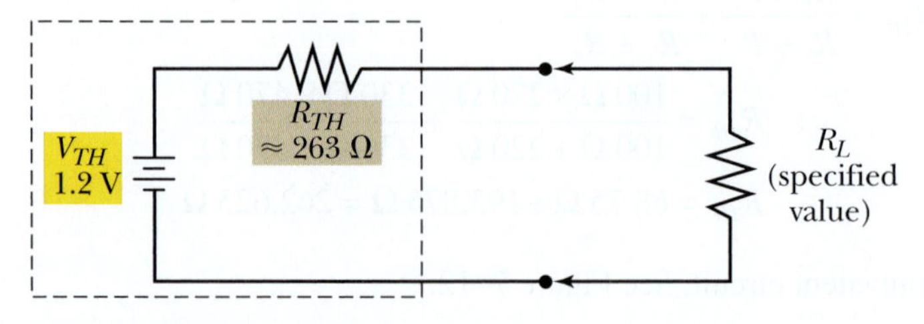

FIGURE 7–12 Thevenin equivalent circuit

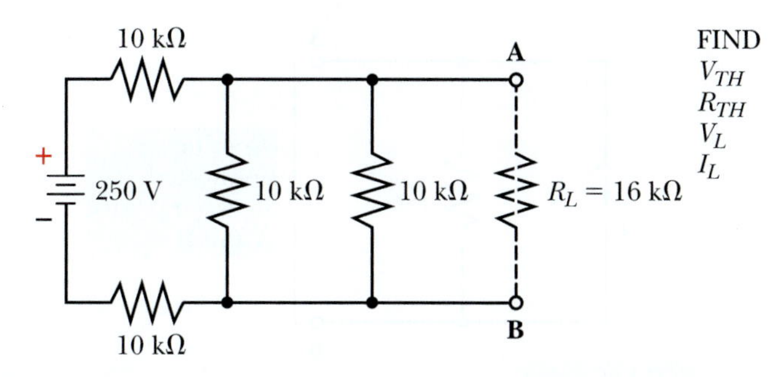

FIGURE 7–13 multiSIM

Practical Notes

In a practical circuit, you can remove the load resistor (R_L) and measure the open-terminal voltage at those two points with a voltmeter to find V_{TH}. Then you can remove the power supply from the circuit and short the points where its output terminals are connected. Next, you can use an ohmmeter to measure the circuit resistance (at the open-terminal points) by looking back toward where the source points are shorted. Thus, the measured value of R is R_{TH}.

As you have seen, one example of a practical application of Thevenin's theorem is in analyzing an unbalanced bridge circuit. Try applying your Thevenin's theorem knowledge to the following practice problems.

PRACTICE PROBLEMS 3

1. Refer again to Figure 7–10. What are the values of I_L and V_L, if R_L is 175 Ω?
2. In Figure 7–10, what value of R_L must be present if I_L is 125 mA?
3. Assume that the parameters in our sample bridge circuit, Figure 7–11, were changed so that $V_A = 18$ V, $R_1 = 220$ Ω, $R_2 = 330$ Ω, $R_3 = 470$ Ω, and $R_4 = 560$ Ω. If R_L is 470 Ω, what are the values of I_L and V_{R_L}?
4. In Figure 7–13, what are the values of V_{TH} and R_{TH}? What are the values of I_L and V_L, if R_L equals 16 kΩ?

7–5 Norton's Theorem

Norton's theorem is another method to simplify complex networks into manageable equivalent circuits for analysis. Again, the idea is to devise the equivalent circuit so parameter analysis at two specified network terminals (where R_L is connected) is easy.

Thevenin's theorem uses a voltage-related approach with a voltage source/series resistance equivalent circuit where voltage divider analysis is applied. On the other hand, Norton's theorem uses a current-related approach with a current source/shunt resistance equivalent circuit where a current-divider analysis approach is applied.

The Theorem Defined

Norton's theorem says *any linear two-terminal network can be replaced by an equivalent circuit consisting of a single* ***(constant) current source*** *(called* I_N*) and a single shunt (parallel) resistance (called* R_N*)*, Figure 7–14.

Refer to Figure 7–15, points A and B. I_N is the current available from the Norton equivalent current source. I_N is determined as the current flowing between the load resistance terminals, if R_L were shorted (i.e., if R from points A to B is zero ohms). R_N is the resistance of the Norton equivalent shunt resistance. This value is computed as the R value seen when looking into the network with the load terminals open (R_L not present) and sources of *voltage* replaced by their r_{int}, Figure 7–16.

FIGURE 7–14 Norton's theorem equivalent circuit

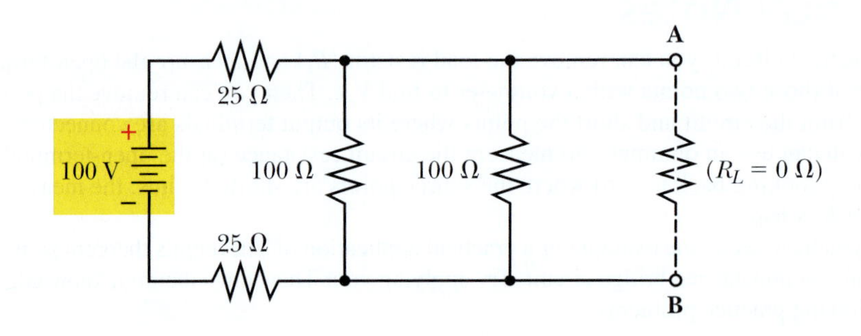

FIGURE 7–15 Norton's I_N = Current from A to B, if R_L shorted (0 Ω). (In this case, I_N would equal 100 V/50 Ω = 2 A.)

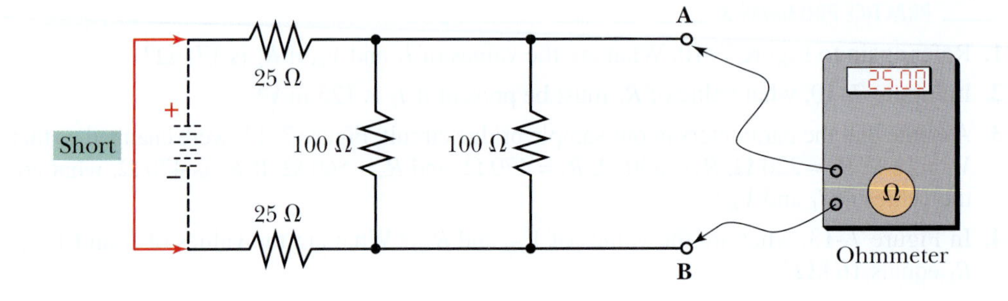

FIGURE 7–16 Norton's $R_N = R_{A\text{–}B}$ with R_L removed and source considered as a short. (In this case, R_N = 25 Ω.)

This method is the same one used to calculate R_{TH} in Thevenin's theorem. In fact, the R_{TH} and R_N values are the same. The only difference is that in Norton's theorem, this resistance value is considered in parallel (or shunt) with the I_N equivalent current source. In Thevenin's theorem, this resistance value is considered in series with the V_{TH} equivalent voltage source.

Observe the similarities and differences between the two types of equivalent circuits in Figure 7–17b and 7–17c. Note these equivalent circuits are derived from the same original circuit used to explain Thevenin's theorem earlier in the chapter. (See Figure 7–10a and see Figure 7–17a.)

Correlating the prior thinking of the Thevenin analysis to the Norton analysis, let's assume the same 25-Ω R_L (load resistance) used in Figure 7–10d, and apply the Norton equivalent circuit approach.

First, let's see how the Norton equivalent circuit parameters, Figure 7–17c are derived. Analysis starts with the original circuit, Figure 7–18a.

I_N equals current between points A and B with R_L shorted. This short also shorts the two 100-Ω parallel resistors, so the source sees a total circuit R of 50 Ω (the two 25-Ω resistors in series, through the short). Thus, current between points A and B is 100 V/50 Ω, or 2 A (I_N = 2 A), Figure 7–18b.

R_N equals R looking back into the network with points A and B open and with *voltage* source(s) shorted. R_N thus equals the equivalent resistance of the two 100-ohm parallel resistors (or 50 Ω) in parallel with the two 25-Ω series resistors' branch (or 50 Ω). This yields a resultant circuit equivalent resistance of 25 Ω (R_N = 25 Ω), again see Figure 7–18b.

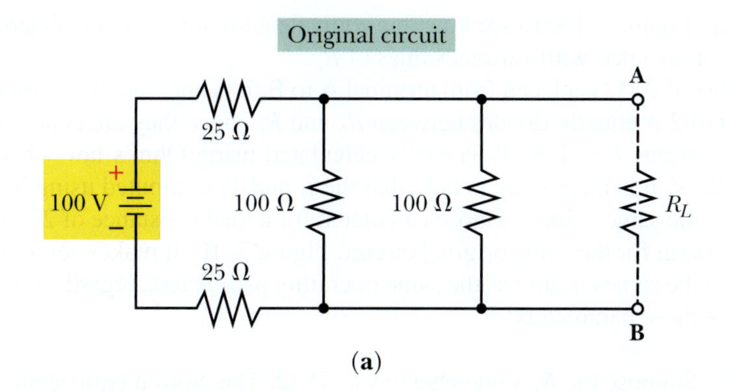

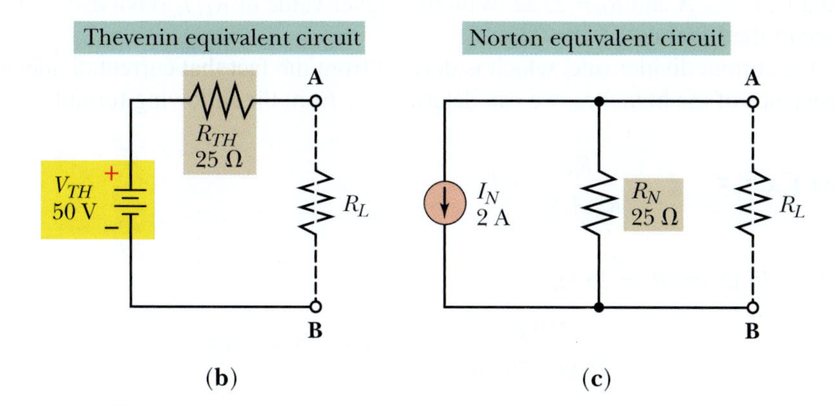

FIGURE 7–17 Comparisons of Thevenin and Norton equivalent circuits

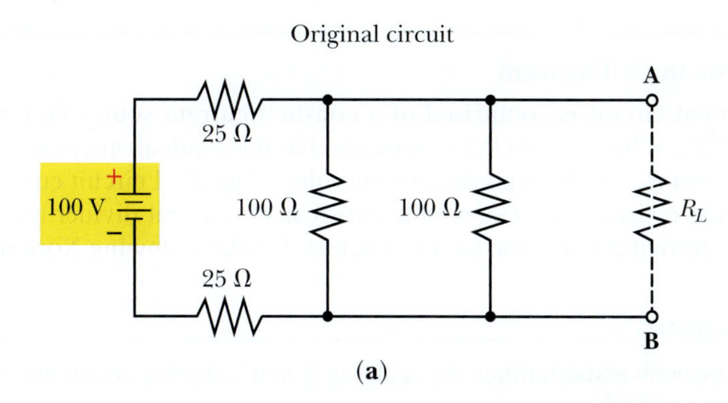

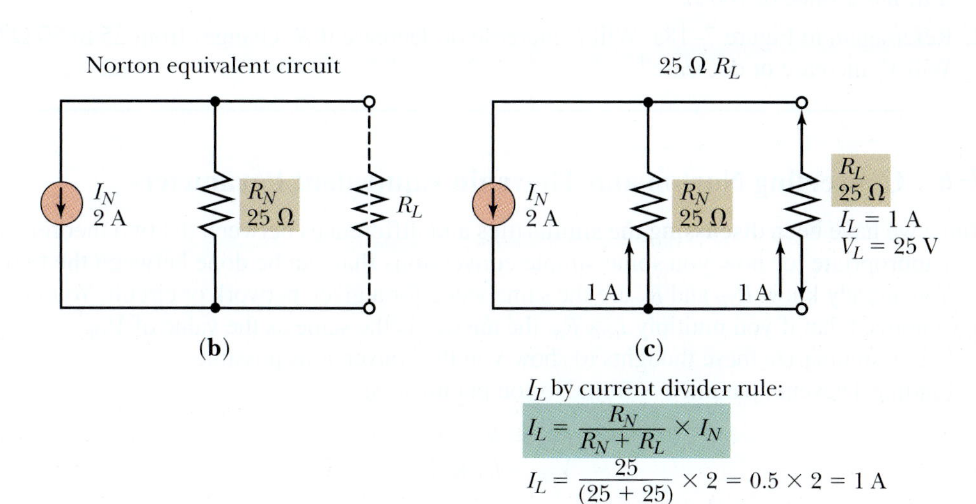

FIGURE 7–18 Application sample of Norton's theorem

Now, refer to Figure 7–18c to see how this equivalent circuit is used to determine electrical parameters associated with various values of R_L.

If an R_L value of 25 Ω is placed from terminal A to B, it is apparent the Norton equivalent circuit current of 2 A equally divides between R_N and R_L, since they are equal value parallel branches. This means $I_L = 1$ A. V_L is easily calculated using Ohm's law where, $I_L \times R_L = 1\text{ A} \times 25\ \Omega = 25$ V. It is interesting to note that the I_L and V_L computed using Norton's theorem are exactly the same values as those calculated for a load resistance of 25 Ω when using Thevenin's theorem for the same original circuit, Figure 7–10. It makes sense that the same R_L connected to the same circuit has the same operating parameters, regardless of the method used to analyze these parameters!

◆ **EXAMPLE** Suppose the R_L value changes to 75 Ω. The Norton equivalent circuit is the same. That is, $I_N = 2$ A and $R_N = 25\ \Omega$. With the higher value of R_L, I_L obviously is a smaller percentage of the I_N value.

Using the current divider rule, which is derived from the fact that current divides inverse to the resistance of the branches, we can determine I_L from the following formula:

FORMULA 7–5 $$I_L = \frac{R_N}{R_N + R_L} \times I_N$$

With $R_L = 75\ \Omega$ and $R_N = 25\ \Omega$,

$$I_L = \frac{25\ \Omega}{25\ \Omega + 75\ \Omega} \times 2\text{ A} = 0.25 \times 2\text{ A} = 0.5\text{ A}$$

To find V_L, again use Ohm's law where $V_L = 0.5\text{ A} \times 75\ \Omega = 37.5$ V ◆

PRACTICE PROBLEMS 4

In Figure 7–18a, assume R_L is 60 Ω. What are the values of I_L and V_L?

Summary of Norton's Theorem

Norton's equivalent circuit is comprised of a constant current source (I_N) and a shunt resistance called R_N. When a load (R_L) is connected to this equivalent circuit, the current (I_N) divides between R_N and R_L according to the rules of parallel circuit current division (i.e., inverse to the resistances). The simple two-resistor current divider rule is applied to determine the current through the load resistance. Try the following Norton's theorem problems.

PRACTICE PROBLEMS 5

1. Use Norton's theorem and determine the values of I_L and V_L for the circuit in Figure 7–18a, if R_L has a value of 100 Ω.
2. Refer again to Figure 7–18a. Will I_L increase or decrease if R_L changes from 25 to 50 Ω? Will V_L increase or decrease?

7–6 Converting Norton and Thevenin Equivalent Parameters

Since we have been discussing the similarities and differences between the two theorems, it is appropriate to show you some simple conversions that can be done between the two.

You already know R_{TH} and R_N are the same value for a given network or circuit. You may have noticed that if you multiply $I_N \times R_N$, the answer is the same as the value of V_{TH}.

Let us summarize these thoughts to show you the conversions possible.

Finding Thevenin parameters from Norton parameters:

$$R_{TH} = R_N$$
$$V_{TH} = I_N \times R_N$$

Finding Norton parameters from Thevenin parameters:

$$R_N = R_{TH}$$

$$I_N = \frac{V_{TH}}{R_{TH}}$$

◆ **EXAMPLE** Let's look briefly at a sample circuit, analyze it with Thevenin's theorem and with Norton's theorem, and then plug values into our simple conversion formulas. This will serve as a quick review of the two theorems, plus illustrate the simplicity of conversion from one set of parameters to the other. See Figure 7–19. ◆

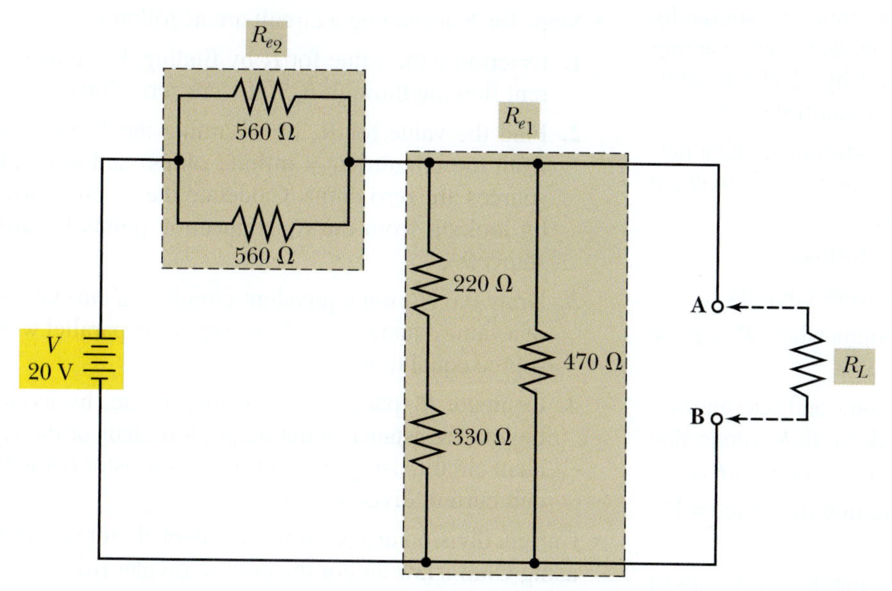

Thevenin's theorem:

$$V_{TH} = \left(\frac{R_{e_1}}{R_{e_1} + R_{e_2}}\right) \times 20\text{ V} = \left(\frac{253\ \Omega}{253\ \Omega + 280\ \Omega}\right) \times 20\text{ V} \cong 9.5\text{ V}$$

$$R_{TH} = \frac{R_{e_1} \times R_{e_2}}{R_{e_1} + R_{e_2}} = \frac{253\ \Omega \times 280\ \Omega}{253\ \Omega + 280\ \Omega} = 133\ \Omega$$

Norton's theorem:

$$I_N = \frac{V}{R_{e_2}} = \frac{20\text{ V}}{280\ \Omega} = 71\text{ mA}$$

$$R_N = \frac{R_{e_1} \times R_{e_2}}{R_{e_1} + R_{e_2}} = 133\ \Omega$$

CONVERSIONS

Thevenin from Norton:

$$R_{TH} = R_N$$
$$133\ \Omega = 133\ \Omega$$
$$V_{TH} = I_N \times R_N$$
$$V_{TH} = 71 \times 10^{-3} \times 133\ \Omega \cong 9.5\text{ V}$$

Norton from Thevenin:

$$R_N = R_{TH}$$
$$133\ \Omega = 133\ \Omega$$
$$I_N = \frac{V_{TH}}{R_{TH}}$$
$$I_N = \frac{9.5\text{ V}}{133\ \Omega} \cong 71\text{ mA}$$

FIGURE 7–19 Example conversions

Summary

- The theorems studied apply to two-terminal linear networks that have bilateral resistances or components.
- The maximum power transfer theorem indicates maximum power is transferred from source to load when the load resistance equals the source resistance. Under these conditions, the efficiency is 50%.
- The superposition theorem says electrical parameters are computed in linear circuits with more than one source by analyzing the parameters produced by each source acting alone, then superimposing each result by algebraic addition to find the circuit parameters with multiple sources.
- Thevenin's theorem simplifies any two-terminal linear network into an equivalent circuit representation of a voltage source (V_{TH}) and series resistance (R_{TH}).
- Steps for Thevenizing a circuit are as follows:
  1. Open (or remove R_L) from the network circuitry.
  2. Determine the open-circuit V at points where R_L is to be connected. This is the value of V_{TH}.
  3. Mentally short the circuit voltage source and determine resistance by looking into the network (from R_L connection points, with R_L disconnected). This is the value of R_{TH}.
  4. Draw the Thevenin equivalent circuit with source = V_{TH} and series resistor of the R_{TH} value.
  5. Make calculations of I_L and V_L for the desired values of R_L (where R_L is assumed to be across the equivalent circuit output terminals) by using two-resistor series circuit voltage analysis.
- Voltage distribution in a Thevenin equivalent circuit is computed using Ohm's law and/or the voltage divider rule.
- Norton's theorem simplifies any two-terminal linear network into an equivalent circuit representation of a current source (I_N) and a parallel or shunt resistance (R_N).
- Steps for Nortonizing a circuit are as follows:
  1. Determine the value for I_N by finding the value of current flowing through R_L if it were zero ohms.
  2. Find the value for R_N by assuming the R_L is removed from the circuit (R_L = infinite ohms) and any voltage sources are zero ohms. Calculate the circuit resistance by looking from the R_L connection points toward the source(s).
  3. Draw the Norton equivalent circuit as a current source of a value equal to I_N and a resistance in parallel with the source equal to R_N.
  4. Compute R_L parameters, as appropriate, by assuming the load resistance at the output terminals of the equivalent circuit, using concepts of two-resistor parallel circuit current division analysis.
- Current division in a Norton equivalent circuit is computed using Ohm's law and/or the current divider rule.

Formulas and Sample Calculator Sequences

Formula 7–1
(To find efficiency of power transfer)

$$\text{Efficiency } (\%) = \frac{P_{\text{out}}}{P_{\text{in}}} \times 100$$

P_{out} value, [÷], P_{in} value, [×], 100, [=]

Formula 7–2
(To find internal resistance of a source)

$$r_{\text{int}} = \frac{V\text{ (no-load)} - V\text{ (loaded)}}{I\text{ (load)}}$$

V (no-load) value, [−], V (loaded) value, [=], [÷], I (load) value, [=]

Formula 7–3
(To find load current using Thevenin equivalent circuit parameters)

$$I_L = \frac{V_{TH}}{R_{TH} + R_L}$$

V_{TH} value, [÷], [(], R_{TH} value, [+], R_L value, [)], [=]

Formula 7–4
(To find load voltage using Thevenin equivalent circuit parameters)

$$V_L = \frac{R_L}{R_L + R_{TH}} \times V_{TH}$$

R_L value [÷], [(], R_L value, [+], R_{TH} value, [)], [=], [×], V_{TH} value, [=]

Formula 7–5
(To find load current using Norton equivalent circuit parameters)

$$I_L = \frac{R_N}{R_N + R_L} \times I_N$$

R_N value, [÷], [(], R_N value, [+], R_L value, [)] , [=], [×], I_N value, [=]

EXCEL AUTOMATED FORMULAS

Helpful problem-solving tools, such as the Excel automated formulas available on the CD, allow automatic calculations of formulas.

Using Excel

Basic Network Theorems Formulas

(Excel file reference: FOE7_01.xls)

***DON'T FORGET!** It is NOT necessary to retype formulas once they are entered on the worksheet! Just input new parameters data for each new problem using that formula, as needed.*

- Refer to the parameters given in Figure 7–10. Assume that the value of R_L is changed to 47 ohms. Use the Formula 7–3 spreadsheet sample and solve for the value of I_L.
- Refer to the parameters given in Figure 7–10. If R_L is 47 ohms, use the Formula 7–4 spreadsheet sample and solve for the value of V_L.

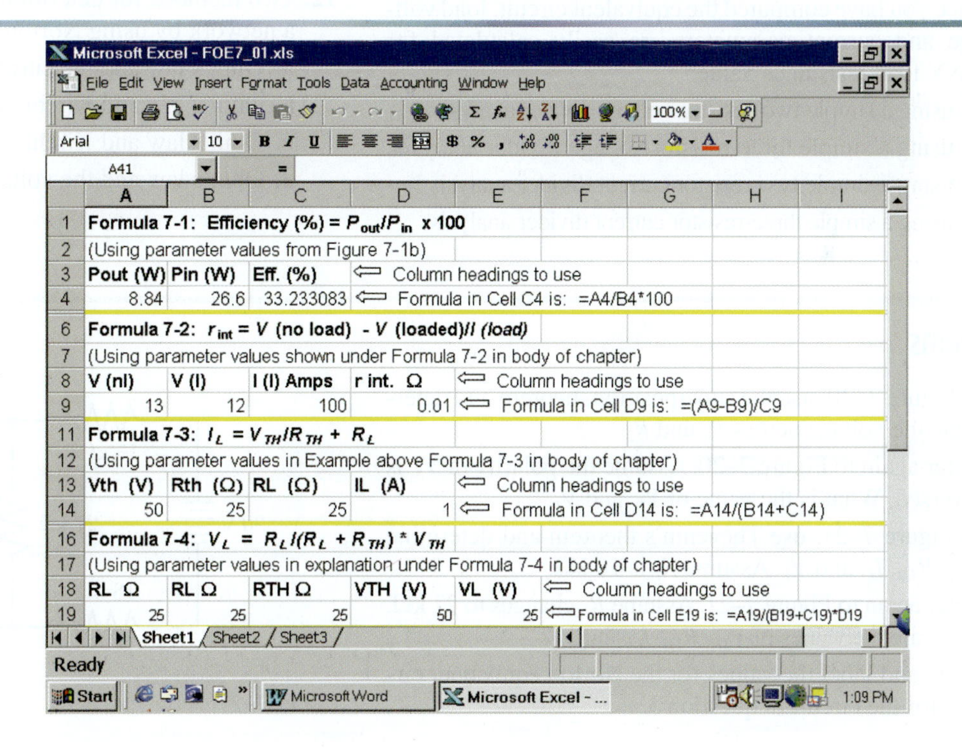

	A	B	C	D	E	F
1	Formula 7-1: Efficiency (%) = P_{out}/P_{in} x 100					
2	(Using parameter values from Figure 7-1b)					
3	Pout (W)	Pin (W)	Eff. (%)	⇐ Column headings to use		
4	8.84	26.6	33.233083	⇐ Formula in Cell C4 is: =A4/B4*100		
6	Formula 7-2: r_{int} = V (no load) - V (loaded)/I (load)					
7	(Using parameter values shown under Formula 7-2 in body of chapter)					
8	V (nl)	V (l)	I (l) Amps	r int. Ω	⇐ Column headings to use	
9	13	12	100	0.01	⇐ Formula in Cell D9 is: =(A9-B9)/C9	
11	Formula 7-3: $I_L = V_{TH}/R_{TH} + R_L$					
12	(Using parameter values in Example above Formula 7-3 in body of chapter)					
13	Vth (V)	Rth (Ω)	RL (Ω)	IL (A)	⇐ Column headings to use	
14	50	25	25	1	⇐ Formula in Cell D14 is: =A14/(B14+C14)	
16	Formula 7-4: $V_L = R_L/(R_L + R_{TH}) * V_{TH}$					
17	(Using parameter values in explanation under Formula 7-4 in body of chapter)					
18	RL Ω	RL Ω	RTH Ω	VTH (V)	VL (V)	⇐ Column headings to use
19	25	25	25	50	25	⇐ Formula in Cell E19 is: =A19/(B19+C19)*D19

Review Questions

1. Name at least two applications for the maximum power transfer theorem.
2. What is the efficiency of a circuit where there is maximum transfer of power?
3. How do you calculate circuit efficiency when evaluating power transfer in a circuit?
4. Define the term *bilateral linear network.*
5. State the sequence of steps used in "Thevenizing" a circuit.
6. State the sequence of steps used in "Nortonizing" a circuit.
7. Describe the most important application of the superposition theorem.
8. One of the benefits of using Thevenin's theorem is that once you have computed the equivalent circuit, load voltage and current parameters are easily calculated for ANY load resistance value:
   a. using a simple two-resistor voltage divider analysis.
   b. using a simple three-resistor voltage divider analysis.
   c. using a simple two-resistor current divider analysis.
   d. using a simple three-resistor current divider analysis.
9. One of the benefits of using Norton's theorem is that once you have computed the equivalent circuit, load voltage and current parameters are easily calculated for ANY load resistance value:
   a. using a simple two-resistor voltage divider analysis.
   b. using a simple three-resistor voltage divider analysis.
   c. using a simple two-resistor current divider analysis.
   d. using a simple three-resistor current divider analysis.
10. When using the superposition theorem to solve a network's parameters that has two voltage sources, the basic technique used is:
    a. analyzing parameters produced by each source acting alone, then superimposing each result by mathematical addition to find the resultant circuit parameters.
    b. analyzing parameters produced by each source acting alone, then superimposing each result by algebraic addition to find the resultant circuit parameters.
    c. removing both sources and considering them zero ohms each, then calculating circuit total resistance, then voltages and currents using Ohm's law, using the sum of or difference between the two supply voltages.
    d. none of the above.
11. Two methods for determining the voltage distribution in a network by using Thevenin's theorem are:
    a. Ohm's law and the current-divider rule.
    b. Ohm's law and Kirchhoff's voltage law.
    c. Ohm's law and Kirchhoff's current law.
    d. Ohm's law and the voltage-divider rule.
12. Two methods for determining the current distribution in a network by using Norton's theorem are:
    a. Ohm's law and the current-divider rule.
    b. Ohm's law and Kirchhoff's voltage law.
    c. Ohm's law and Kirchhoff's current law.
    d. Ohm's law and the voltage-divider rule.

Problems

1. In Figure 7–20, use the superposition theorem and determine the voltage across R_1 and R_2.
2. Refer again to Figure 7–20. Assume the polarity of V_{S_2} is reversed. What is the value of V_{R_1} and V_{R_2}?
3. In Figure 7–21, use Thevenin's theorem and determine V_{TH}, R_{TH}, I_L, and V_L. Assume R_L is 10 kΩ.
4. Refer again to Figure 7–21. Assume R_L changes to 27 kΩ. What are the values of V_{TH}, R_{TH}, I_L, and V_L?
5. Draw and label the values for the Norton equivalent circuit for the circuit in question 3.

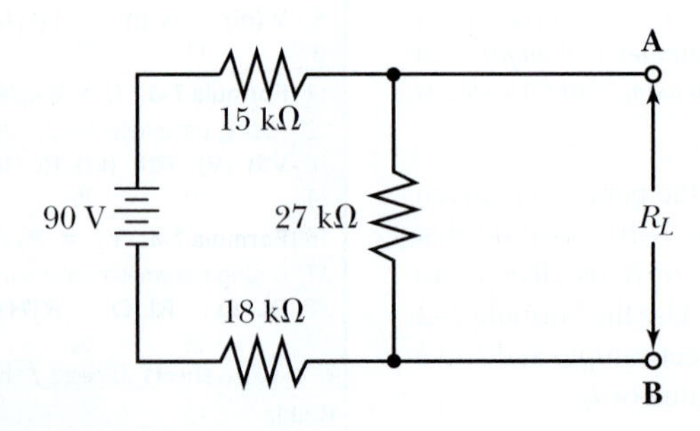

FIGURE 7–21

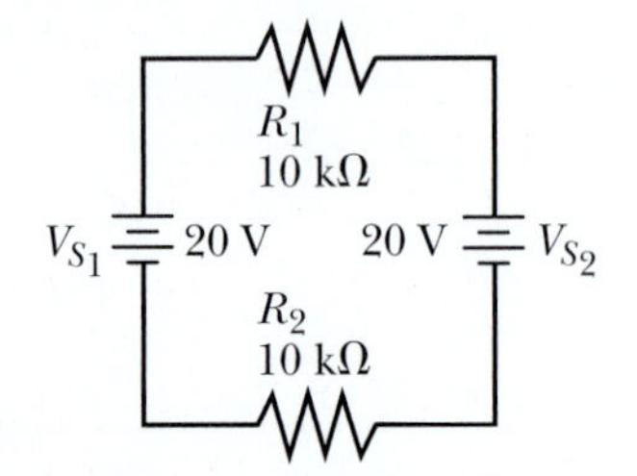

FIGURE 7–20

6. Draw and label the values for the Norton equivalent circuit for the circuit in question 4.
7. What is the maximum power that can be delivered to the load in Figure 7–22?
8. If the voltage at the source terminals of a circuit is 100 V without load and 90 V when 10 A are drawn from the source, what value is the load resistance to afford maximum power transfer?

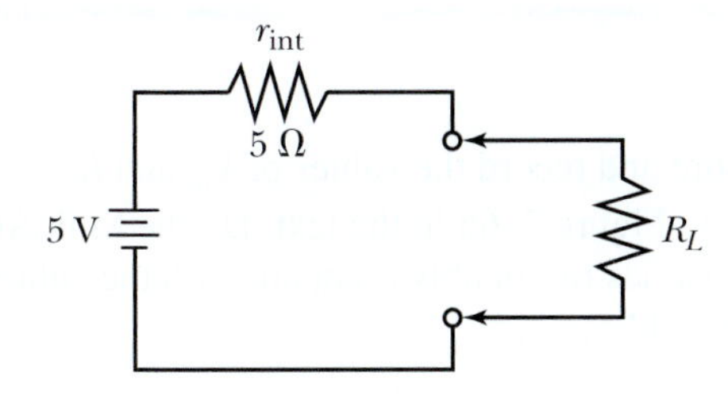

FIGURE 7–22

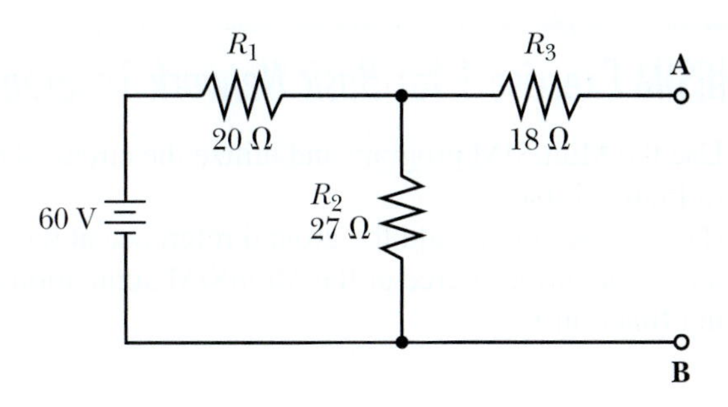

FIGURE 7–23 Find R_{TH} and V_{TH}. multiSIM

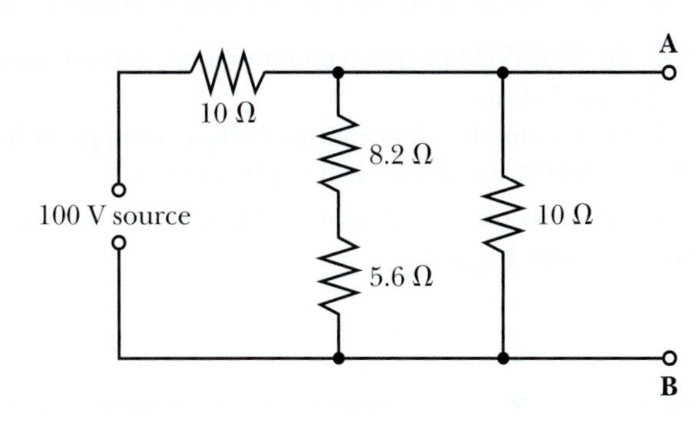

FIGURE 7–24 Find R_N and I_N. multiSIM

9. For the circuit in Figure 7–23, what are the values of R_{TH} and V_{TH}?

10. For the circuit in Figure 7–24, what are the values of R_N and I_N?

11. Convert Norton parameters to Thevenin parameters, given $I_N = 3$ A and $R_N = 10\ \Omega$.

12. Convert Thevenin parameters to Norton parameters, given $R_{TH} = 15\ \Omega$ and $V_{TH} = 25$ V.

13. What is the percentage of efficiency for a circuit whose 100-V source has an internal resistance of 5 Ω and a load resistance of 25 Ω?

14. Would the percentage efficiency be the same for the circuit described in question 13 if the source voltage were 200 V? Does the value of source voltage have any impact on the circuit's efficiency factor?

15. What load resistance value is needed for the circuit described in question 13 to have maximum power transfer?

Analysis Questions

1. Explain why it is important for a voltage source to have as low an internal resistance (r_{int}) as possible.

2. For a given power source, why is it important for the load's opposition to current flow to be equal to that of the source?

3. If the efficiency of power transfer from source to load is 60%, should the load resistance be increased or decreased to enable maximum power transfer?

4. Which theorem relates most closely to voltage divider analysis techniques?

5. Which theorem relates most closely to current divider analysis techniques?

6. a. Give the main reason it is important to observe the direction of current flow through components when using the superposition theorem to analyze a two-source circuit.

b. Describe which component(s) in the circuit are especially impacted by the fact described in part (a) of this question.

7. In your own words, explain the significance of R_{TH} in using Thevenin's theorem, and explain how it is determined.

8. In your own words, explain the significance of V_{TH} in using Thevenin's theorem, and explain how it is found.

9. In your own words, explain the significance of R_N in using Norton's theorem, and explain how it is found.

10. In your own words, explain the significance of I_N in using Norton's theorem, and explain how it is found.

11. a. Perform appropriate research to enable listing at least one type circuit or application in which there may be two or more sources.

b. Which network theorem would you use to help you in analyzing such a circuit?

12. For the circuit of Figure 7–23, explain how you could use a voltmeter and an ohmmeter to determine the Thevenin equivalent circuit parameters of R_{TH} and V_{TH}.

13. For the circuit of Figure 7–23, is it possible to use a meter to measure the Norton equivalent circuit R_N value? If so, what is the procedure to do this?

14. Explain how to determine the internal resistance (r_{int}) of a voltage source.

15. What would be the internal resistance of a battery whose output terminal voltage was 12 V under no-load conditions and 11 V when the load current was 50 A?

MultiSIM Exercise 1 for *Basic Network Theorems*

1. Use the MultiSIM program and utilize the circuit shown in Figure 7–6a.
   (NOTE: Be sure to add the ground reference at the negative side of the source or the MultiSIM simulation will not function.)
2. Measure and record the values of V_{R_1} and I_{R_2}.
3. Refer to Figure 7–6d in the text. Do the MultiSIM measured values reasonably compare with the values shown in the text?

MultiSIM Exercise 2 for *Basic Network Theorems*

1. Use the MultiSIM program and utilize the circuit shown in Figure 7–10a.
   (NOTE: Be sure to add the ground reference at point B or the MultiSIM simulation will not function.)
2. Use the appropriate techniques and measure and record the values of V_{TH} and R_{TH}.
3. Refer to Figure 7–10b and Figure 7–10c and compare your measured values with those shown in the text. Do the MultiSIM measured values reasonably compare with the values shown in the text?
4. Connect an R_L with a value of 270 ohms to points A and B, as appropriate. Measure and record the value of V_L; then, calculate or measure and record I_L.

Performance Projects Correlation Chart

Suggested performance projects in the Laboratory Manual that correlate with topics in this chapter are:

Chapter Topic	Performance Project	Project Number
Thevenin's Theorem	Thevenin's Theorem	27
Norton's Theorem	Norton's Theorem	28
Maximum Power Transfer	Maximum Power Transfer Theorem	29
	Story Behind the Numbers: Basic Network Theorems	

NOTE: It is suggested that after completing the above projects, the student should be required to answer the questions in the "Summary" at the end of this section of projects in the Laboratory Manual.

OUTLINE

OBJECTIVES

After studying this chapter, you should be able to:

1. Define the terms **mesh, loop,** and **node**
2. Analyze a single-source circuit using a loop procedure
3. Use the **assumed mesh currents** approach to find voltage and current parameters for each component in a network having two sources
4. Use the nodal analysis approach to find voltage and current parameters for each component in a network having two sources
5. Convert from delta (Δ) circuit configuration parameters to wye (*Y*) circuit configuration parameters, and vice versa
6. Use the computer to solve circuit problems

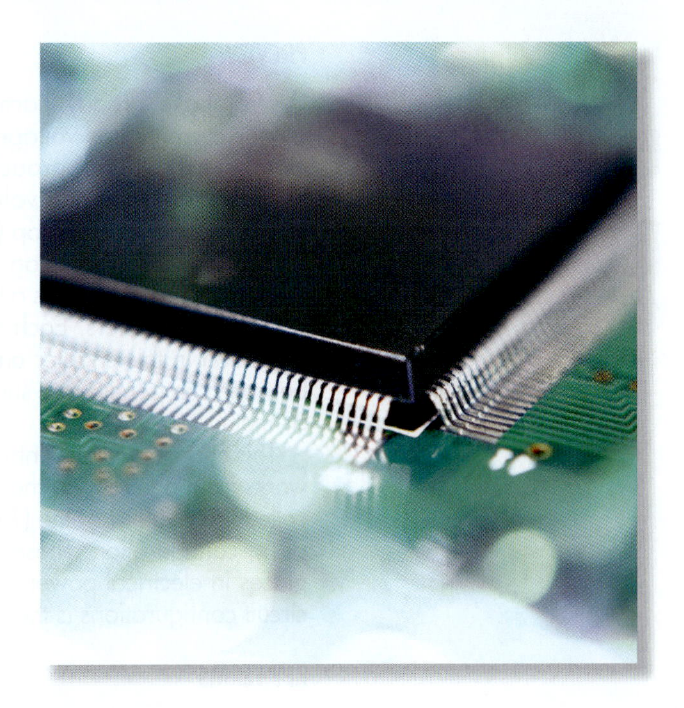

CHAPTER 8

Network Analysis Techniques

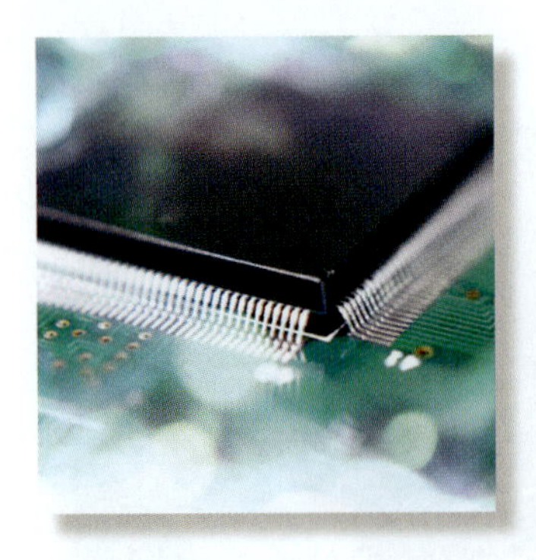

PREVIEW

As you have already learned, circuit and network analysis enable you to find the various currents and voltages throughout the circuit, or network, being analyzed. In this chapter, two other approaches will be introduced for solving network parameters. One method uses Kirchhoff's voltage law as a basis for writing equations for electrical quantities around a closed loop (also sometimes called a mesh). These equations then enable finding current and voltage parameters throughout the network. Another method uses Kirchhoff's current law as a basis for writing "nodal" equations, which also enable solving network parameters. Each method has advantages and disadvantages. The techniques presented in this chapter enable you to analyze complex networks in which connections between components are such that you cannot simply use basic series-parallel circuit analysis methods.

This chapter will also introduce some techniques for converting two specific types of networks from one form to the other when this would be advantageous. One type network is called the wye (Y) or the (T) network configuration. The other is known as the delta (Δ) or pi (π) configuration. These networks, or circuit configurations, are used in a number of places in electrical power systems and in electronic circuitry. Knowledge regarding these circuit configurations is therefore very useful, and has value to you as a technician.

KEY TERMS

Assumed mesh current(s)
Delta-wye conversion
Loop
Mesh
Node
Wye-delta conversion

8–1 Assumed Loop or Mesh Current Analysis Approach

Background Information

A **loop** is considered to be a closed, or complete, path within a circuit. A **mesh** is also considered to be an unbroken, or closed, loop. Observe in Figure 8–1a that two closed loops are shown in the circuit with arbitrary mesh current assignments and identifiers. Figure 8–1b illustrates the commonly accepted single "window-frame" definition for meshes and mesh currents.

For each mesh, or loop, we will use an **assumed mesh current** direction in which the mesh current does NOT branch, as an actual circuit current would. These arbitrarily assigned currents, for analysis purposes, are considered to flow only in their own loop, or mesh. (This nonbranching aspect simplifies matters, and is one advantage of the mesh analysis method.) Kirchhoff's voltage law equations will then be written, using the assumed mesh current designators.

The version of Kirchhoff's voltage law that will be used stipulates that the sum of the voltage drops around a closed loop must equal the voltage source(s). That is: $V_s = V_1 + V_2 \ . \ . \ . + V_n$. From the results of these equations, we will then be able to compute actual circuit voltages and currents throughout the network.

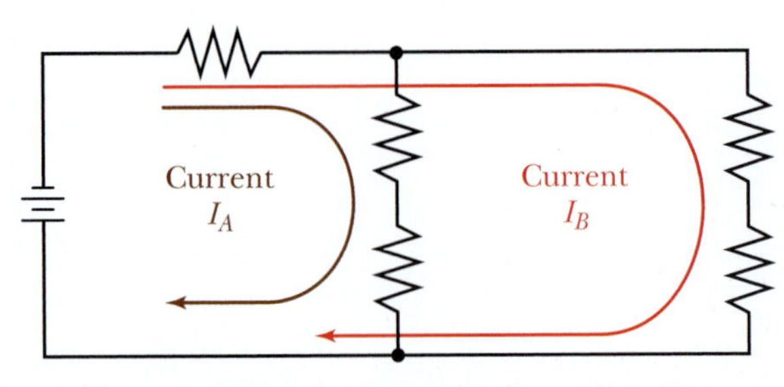

(**a**) Mesh currents—generalized loop definition

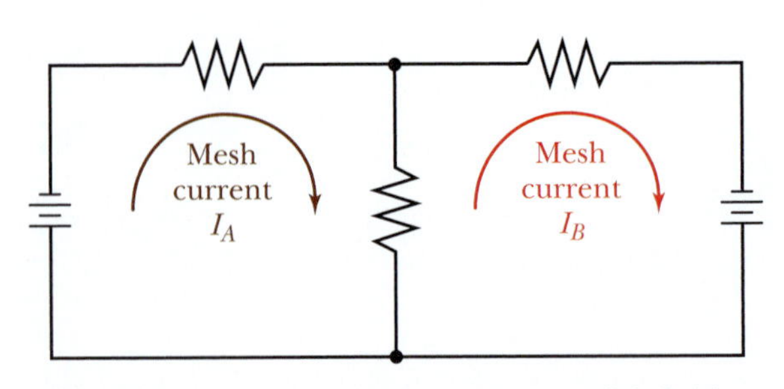

(**b**) Mesh currents—single "window-frame" definition

FIGURE 8–1 Examples of assumed mesh currents

Polarity Assignments

It is important to note directions of currents and polarities of voltages in the analysis steps. The general rules to follow during the analysis are to:

1. Use the arbitrary convention that assumed loop currents will be shown circulating in a clockwise direction through each loop;
2. Assume any voltage drop created across a resistor by its own assumed loop or mesh current to be positive in that mesh's equation. Any voltage drop across a resistor due to another loop or mesh current is considered negative for that mesh's equation; and
3. For source voltage polarity assignment (the algebraic sign to be used in equations), use the arbitrary convention that if the assumed mesh current is returning to the source into its positive terminal, the source voltage will be considered positive. If the assumed mesh current is returning (going into) the negative terminal, it will be considered negative.

Steps to Use the Loop/Mesh Analysis Approach

1. Label the assumed mesh currents on the diagram. (Use the arbitrary clockwise direction convention.)
2. When assigning polarities to sources in the Kirchhoff equations, consider the source voltage positive if the assumed mesh current is entering its positive terminal, and negative if the assumed mesh current enters its negative terminal.
3. Write the Kirchhoff equation for each loop's voltage drops. Assume voltages across resistors in a given loop to be positive when caused by that mesh's assumed current, and negative when caused by an adjacent mesh's assumed current.
4. Solve the resulting equations using simultaneous equation methods.
5. Verify the results using Ohm's law, Kirchhoff's law, and so on, as appropriate.

EXAMPLE Let's try this system out with a single-source bridge circuit, and see how it works. For the given circuit, find the value of voltage between points A and B.

Step 1. Draw assumed loop/mesh currents on the diagram. See Figure 8–2. Use the generalized loop definition approach.

Step 2. Determine polarity of source polarity assignment(s). In this case, both mesh current I_A and mesh current I_B are shown entering the positive terminal of V_S. Thus, the source voltage will have a positive sign in our equations.

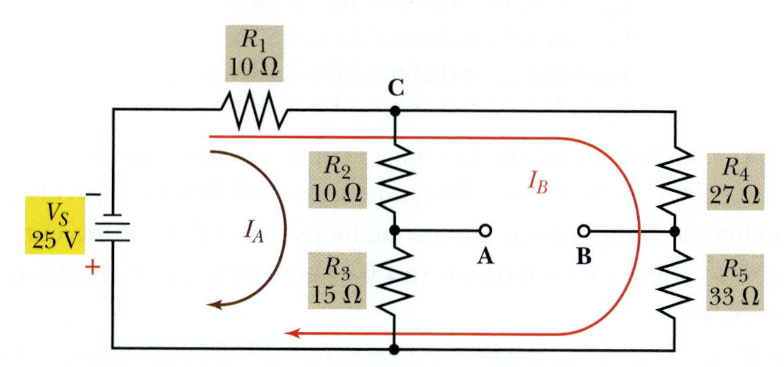

Kirchhoff's voltage law equation 1: $35\ I_A + 10\ I_B = 25$ V
Kirchhoff's voltage law equation 2: $10\ I_A + 70\ I_B = 25$ V
$I_A = 0.638$ A
$I_B = 0.266$ A

FIGURE 8–2 Illustration showing results of several key steps in the single-source "Example" problem

Step 3. Write the Kirchhoff equation for each loop's voltage drops:

a. Loop *A:*

$$(I_A + I_B)\,R_1 + I_AR_2 + I_AR_3 = V_S$$

Switching positions of terms for ease of notation and manipulation:

$$10(I_A + I_B) + 10I_A + 15I_A = 25$$
$$35I_A + 10I_B = 25 \text{ (equation 1)}$$

b. Loop *B:*

$$(I_A + I_B)\,R_1 + I_BR_4 + I_BR_5 = V_S$$

Switching positions of terms for ease of notation and manipulation:

$$10(I_A + I_B) + 27(I_B) + 33(I_B) = 25$$
$$10I_A + 70I_B = 25 \text{ (equation 2)}$$

Step 4.

$$\text{Equation 1: } 35I_A + 10I_B = 25$$
$$\text{Equation 2: } 10I_A + 70I_B = 25$$

Multiplying equation 1 by –7 to make equation 1's "$10I_B$" become "$-70I_B$," and thus cancel the I_B term and solve for I_A.

$$-245I_A - 70I_B = -175$$
$$10I_A + 70I_B = 25$$

Adding to cancel the I_B term:

$$-235I_A = -150$$
$$I_A = 0.638 \text{ A}$$

Substituting 0.638 for I_A into equation 2 to solve for I_B:

$$10(0.638) + 70I_B = 25$$
$$6.38 + 70I_B = 25$$
$$70I_B = 18.62$$
$$I_B = 0.266 \text{ A}$$

Thus:

$$I_A = 0.638 \text{ A}$$
$$I_B = 0.266 \text{ A}$$

Using mesh currents to solve voltage drops:

$$V_{R_1} = (I_A + I_B) \times 10 = 0.904 \times 10 = 9.04 \text{ V}$$
$$V_{R_2} = I_A \times 10 = 0.638 \times 10 = 6.38 \text{ V}$$
$$V_{R_3} = I_A \times 15 = 0.638 \times 15 = 9.57 \text{ V}$$
$$V_{R_4} = I_B \times 27 = 0.266 \times 27 = 7.18 \text{ V}$$
$$V_{R_5} = I_B \times 33 = 0.266 \times 33 = 8.78 \text{ V}$$

Point A is +6.38 V with respect to point C in the circuit. Point B is +7.18 V with respect to point C in the circuit. Thus, point A is – 0.8 V with respect to point B.

Step 5. Checking the voltage drops around the loops to see if it verifies our calculations:

Loop A: $V_{R_1} + V_{R_2} + V_{R_3} = 9.04 + 6.38 + 9.57 = 24.99$ V, which checks out against our 25-V source.

Loop B: $V_{R_1} + V_{R_4} + V_{R_5} = 9.04 + 7.18 + 8.78 = 25$ V, which also checks out against our 25-V source. ◆

PRACTICE PROBLEMS 1

Refer to Figure 8–3, and using the mesh analysis approach, solve the following:

1. I_A
2. I_B
3. V_{R_1}
4. V_{R_2}
5. V_{R_3}
6. V_{R_4}
7. V_{R_5}
8. V_{A-B}

EXAMPLE Now, try the mesh approach with a circuit containing two sources. Refer to Figure 8–4 as you follow the steps for this example.

Step 1. Label the assumed mesh currents on the diagram, using the arbitrary clockwise convention. See Figure 8–4. Use the single window-frame definition.

Step 2. Determine polarity of sources. By observation you can see that the mesh currents are such that current goes into the positive terminal of the 15-V source, thus it will be designated as a positive number. The mesh current goes into the negative terminal of the 12-V source, thus it will be signed as a negative number in equations.

Step 3. Write the Kirchhoff voltage equation for each loop's voltage drops:

$$(3.9I_A + 5.6I_A) - 5.6I_B = 15\text{ V}$$
$$9.5I_A - 5.6I_B = 15\text{ V} \quad \text{(equation 1)}$$
$$-5.6I_A + (5.6I_B + 6.8I_B) = -12\text{ V}$$
$$-5.6I_A + 12.4I_B = -12\text{ V} \quad \text{(equation 2)}$$

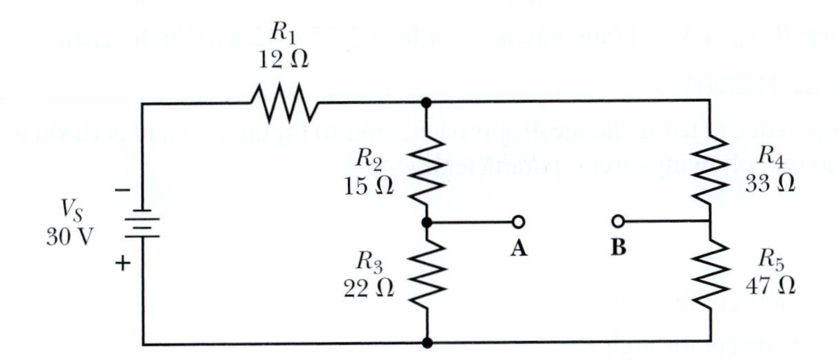

FIGURE 8–3 multiSIM

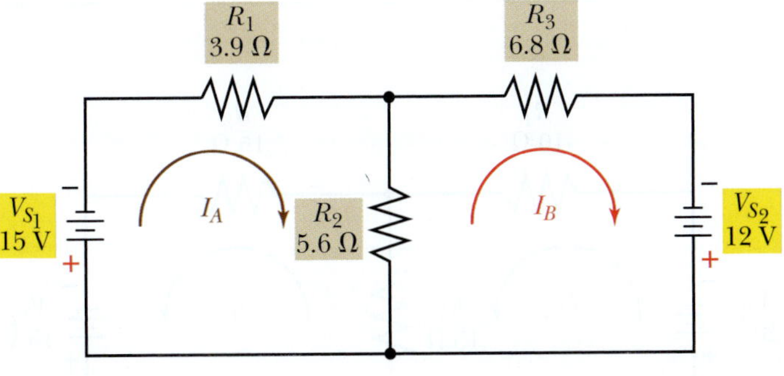

FIGURE 8–4 Illustration showing results of several key steps in the two-source "Example" problem

Step 4. To cancel the I_B terms and solve for I_A, we multiply equation 1 by the factor 2.214 (which is derived from 12.4 being divided by 5.6). This yields:

$$21I_A - 12.4I_B = 33.21$$
$$-5.6I_A + 12.4I_B = -12$$

Adding to cancel the I_B term produces:

$$15.4I_A = 21.21$$
$$I_A = 1.377 \text{ A}$$

To find I_B we substitute 1.377 A for I_A in the second equation, which yields:

$$-5.6(1.377) + 12.4I_B = -12$$
$$-7.71 + 12.4I_B = -12$$
$$12.4I_B = -4.29$$
$$I_B = -0.346 \text{ A}$$

The negative sign of the outcome tells us that the actual current is in the opposite direction from that which we assumed in creating the mesh currents.

Now using the values of 1.377 A for I_A and 0.346 A for I_B, we can calculate circuit parameters quite easily.

$$V_{R_1} = I_A \times 3.9 = 5.37 \text{ V}$$
$$V_{R_2} = (I_A + I_B) \times 5.6 = (1.377 + 0.346) \times 5.6 = 9.65 \text{ V}$$
$$V_{R_3} = I_B \times 6.8 = 0.346 \times 6.8 = 2.35 \text{ V}$$

Step 5. Checking the loop voltages as a verification of our calculations, we find:

Loop A: $V_{R_1} + V_{R_2}$ should equal V_{S_1}: $5.37 + 9.65 = 15.02$ V (which verifies loop *A*)

Loop B: $V_{R_2} + V_{R_3}$ should equal V_{S_2}: $9.65 + 2.35 = 12$ V (which verifies loop *B*) ◆

PRACTICE PROBLEMS 2

For further practice in using the mesh approach, refer to Figure 8–5 and perform the required work to find the following circuit parameters.

1. I_A
2. I_B
3. V_{R_1} and the current through R_1
4. V_{R_2} and the current through R_2
5. V_{R_3} and the current through R_3

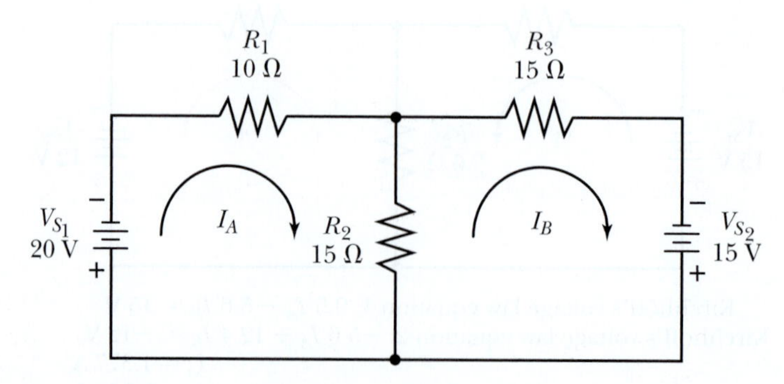

FIGURE 8–5 multiSIM

■ **IN-PROCESS LEARNING CHECK 1**

1. A loop is considered to be a closed, or complete _____ within a circuit.
2. In the loop or mesh approach, it is important to label the assumed _____ currents on the diagram you use for analysis.
3. When assigning polarities to sources in the Kirchhoff equations, a source voltage is considered (negative, positive) _____ if the assumed mesh current enters the positive terminal of the source.
4. When writing Kirchhoff equations for each loop's voltage drops, the component voltage is considered (negative, positive) _____ when caused by its own mesh current, and (negative, positive) _____ when caused by an adjacent mesh's current flow.

8–2 Nodal Analysis Approach

Background Information

A **node** is a current junction point of two or more components in a circuit; that is, a point where currents enter and leave. Kirchhoff's current law can be used to write equations related to these nodes in circuits. Nodes where three or more components are joined are sometimes identified as "major nodes."

For nodal analysis, one major node is specified as the "reference node." Typically, a node that is common to as many parts of the network as possible is chosen as the reference node. Voltages within the circuit are then designated for each of the other nodes in the circuit, relative to the reference node. A Kirchhoff current equation is written for each of the nodes, other than the reference node. For each current term, we use the appropriate V_R/R expression. This is in contrast to our having used the appropriate $I \times R$ expression for voltage terms in the mesh/loop analysis approach.

Steps to Use the Nodal Analysis Approach

1. Find the number of major nodes that are in the circuit, then select the one major node you will use as the reference node.
2. Designate currents into and out of all major nodes, with the exception of the reference point node.
3. Write Kirchhoff Current Law (KCL) equations for currents into and out of each major node, using the V_R/R approach to represent each current. The number of equations required is generally one less than the number of major nodes.
4. Incorporate known values of voltage and resistance in the equations, and solve for the unknown node voltages.
5. Based on the results of step 4, solve for current and voltage values for each component, as appropriate.
6. Verify results by using Kirchhoff's voltage law for each loop in the circuit.

EXAMPLE Refer to the circuit of Figure 8–6 as you observe the following steps in a sample nodal analysis.

Step 1. Find the major nodes and select the reference node. Observe in the circuit that only two major nodes exist. One is at the top of R_2, and the other is at other end of R_2, which happens to be ground reference in this case. Let the ground reference be the reference node. The voltage across R_2 will then be our voltage of interest, against which other parameters will be determined.

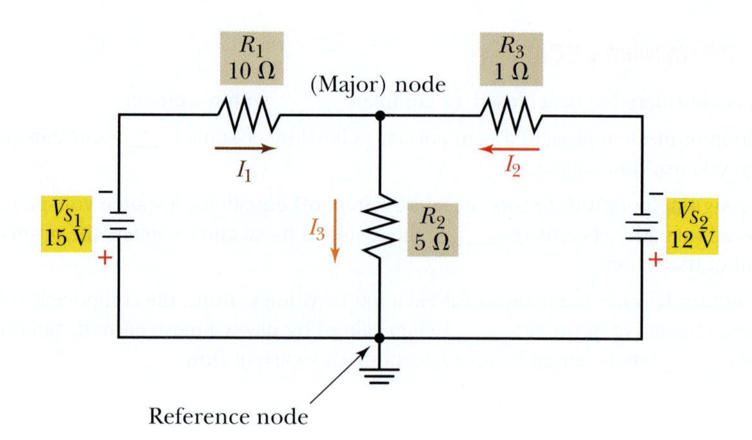

NOTE: I_1 is expressed as V_{R_1}/R_1
I_2 is expressed as V_{R_3}/R_3
I_3 is expressed as V_{R_2}/R_2

FIGURE 8–6 Illustration showing results of several key steps for first nodal analysis "Example" problem multiSIM

Step 2. Designate currents into and out of major nodes. See Figure 8–6 again.

Step 3. Write the Kirchhoff current equations for currents into and out of these major nodes. For the loop with the 15-V source:

$$I_1 + I_2 = I_3$$

(NOTE: The way our circuit components are labeled, current I_2 is the current through R_3, and current I_3 is the current through R_2 and the reference node.)

$$\text{(Term for } I_1\text{)}\ \frac{V_{R1}}{R_1} + \text{(Term for } I_2\text{)}\ \frac{V_{R3}}{R_3}\ \text{(Term for } I_3\text{)}\ \frac{V_{R2}}{R_2}\ \text{(which is the current through } R_2\text{)}$$

(NOTE: V_{R_2} is the voltage of special interest, since it is associated with the reference node; V_{R_2} is the voltage drop of R_2 caused by I_3.)

Step 4. Since we know the V_{S_1} value, we can state that:

$$V_{R_1} + V_{R_2} = 15\text{ V or } V_{R_1} = 15 - V_{R_2}$$

For the loop with the 12-V source, we can state that:

$$V_{R_3} + V_{R_2} = 12\text{ V or } V_{R_3} = 12 - V_{R_2}$$

Now, incorporating these terms into our V/R current statements:

$$\frac{15 - V_{R2}}{10} + \frac{12 - V_{R2}}{1} = \frac{V_{R2}}{5}$$

Multiplying each term by 10 to remove the denominators:

$$\left(\frac{\cancel{10}}{1}\right) \times \frac{15 - V_{R2}}{\cancel{10}} + \left(\frac{10}{1}\right) \times \frac{12 - V_{R2}}{1} = \frac{\cancel{10}}{1} \times \frac{V_{R2}}{5}$$

$$15 - V_{R_2} + 10(12 - V_{R_2}) = 2V_{R_2}$$

$$15 - V_{R_2} + 120 - 10V_{R_2} = 2V_{R_2}$$

$$135 = 13V_{R_2}$$

$$V_{R_2} = \frac{135}{13}$$

$$V_{R_2} = 10.38\text{ V}$$

Step 5. Therefore:

$$I_{R_2} = \frac{10.38}{5} = 2.076 \text{ A}$$

From this, we can continue with:

$$V_{R_3} = 12 - 10.38 = 1.62 \text{ V}$$
$$I_{R_3} = \frac{1.62}{1} = 1.62 \text{ A}$$
$$V_{R_1} = 15 - 10.38 = 4.62 \text{ V}$$
$$I_{R_1} = \frac{4.62}{10} = 0.462 \text{ A}$$

Step 6. Verifying results:

$$V_{R_1} + V_{R_2} = V_{S_1}$$
$$4.62 + 10.38 = 15 \text{ V (It checks!)}$$
$$V_{R_2} + V_{R_3} = V_{S_2}$$
$$10.38 + 1.62 = 12 \text{ V (It checks!)}$$ ◆

◆ **EXAMPLE** The same procedures we have been showing you will work even if the component values are not as convenient to work with as the ones in our first example. Follow along in this example, which does not use a least common denominator to eliminate the denominators, but rather simply uses one of the denominators as the factor to move things from the denominator to the numerator. The steps used in the solution process are essentially the same as those used in the preceding example, so we will not detail the step numbers.

Refer to Figure 8–7 as you follow the procedure shown.

$$I_1 + I_2 = I_3$$

(Again, note that I_2 is R_3's current, and I_3 is R_2's current, where the R_2 parameters are the ones of special interest, since this is the major reference node part of the circuit.)

Currents restated: (current through R_1) $\frac{V_{R1}}{R_1}$ + (current through R_3) $\frac{V_{R3}}{R_3}$ =

(current through R_2) $\frac{V_{R2}}{R_2}$

15-V loop: $V_{R_1} + V_{R_2} = 15$ V or $V_{R_1} = 15 - V_{R_2}$

12-V loop: $V_{R_3} + V_{R_2} = 12$ V or $V_{R_3} = 12 - V_{R_2}$

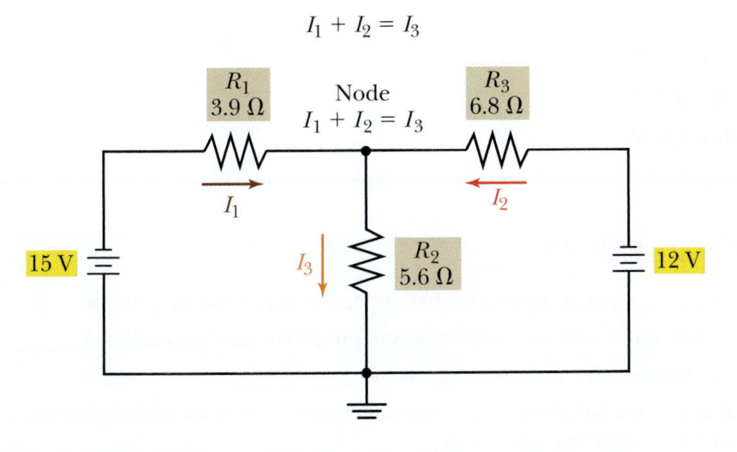

FIGURE 8–7 Reference diagram for nodal analysis second "Example" problem

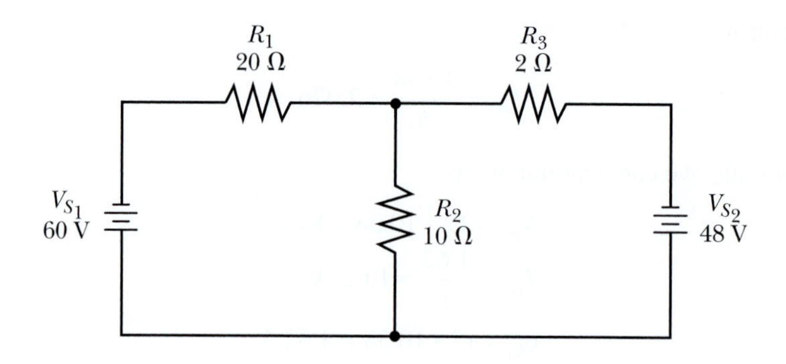

FIGURE 8–8

Substituting these terms into the $I_1 + I_2 = I_3$ equation:

$$\frac{15 - V_{R2}}{3.9} + \frac{12 - V_{R2}}{6.8} = \frac{V_{R2}}{5.6}$$

Multiplying each term by 6.8 to move the denominators:

$$1.74(15 - V_{R2}) + (12 - V_{R2}) = 1.214\ V_{R2}$$
$$(26.1 - 1.74\ V_{R2}) + (12 - V_{R2}) = 1.214\ V_{R2}$$
$$38.1 = 3.95\ V_{R2}$$
$$V_{R_2} = 9.64\ \text{V}$$

Using the reference node voltage, we find the other parameters as follows:

$$V_{R_1} = 15 - 9.64 = 5.36\ \text{V}$$
$$I_{R_1} = \frac{5.36\ \text{V}}{3.9\ \Omega} = 1.37\ \text{A}$$
$$V_{R_3} = 12 - 9.64 = 2.36\ \text{V}$$
$$I_{R_3} = \frac{2.36\ \text{V}}{6.8\ \Omega} = 0.347\ \text{A}$$
$$\text{Current through } R_2 = \frac{9.64\ \text{V}}{5.6\ \Omega} = 1.72\ \text{A or}$$
$$I_1 + I_2 = I_3 : 1.37\ \text{A} + 0.347\ \text{A} = 1.717\ \text{A}$$ ◆

PRACTICE PROBLEMS 3

Refer to Figure 8–8 and find the following circuit parameters, using the nodal analysis method.

1. V_{R_1}
2. V_{R_2}
3. V_{R_3}
4. Current through R_1
5. Current through R_2
6. Current through R_3

■ IN-PROCESS LEARNING CHECK 2

1. A node is a _____ junction point for two or more components in a circuit.
2. A node where three or more components are joined is sometimes called a _____ _____.
3. For nodal analysis, one node is chosen as the _____ _____.
4. In nodal analysis, a Kirchhoff _____ law equation is written for each major node, with the exception of the reference point node.

8–3 Conversions of Delta and Wye Networks

As a technician, you will occasionally deal with two circuit configurations found in numerous applications: the wye or tee configuration, shown in Figure 8–9, and the delta, or pi configuration, shown in Figure 8–10. For example, both types of connections are found in ac power applications. The pi configuration is found in many filter circuits used in electronic systems. As you can see, the names of these configurations are derived from the shapes they form. The delta configuration has the triangular shape of the Greek letter delta (Δ). The wye configuration is often represented in the shape of the letter *Y*.

There are times when it is convenient and helpful to be able to convert from one circuit configuration to the other. One example of simplifying calculations using this technique is the analysis of a bridge circuit, which we will show you a little later. To do this conversion, or transformation, we need to be able to derive equivalent circuits from one configuration to the other. So, let's look a little closer at each of these common network connection configurations, and how they might be converted, or transformed.

For ease of reference, we will label the components and their connection points as shown in Figure 8–11. Notice that we have superimposed one configuration on the other, so you can easily see the cross-referencing between components when transformation formulas are used.

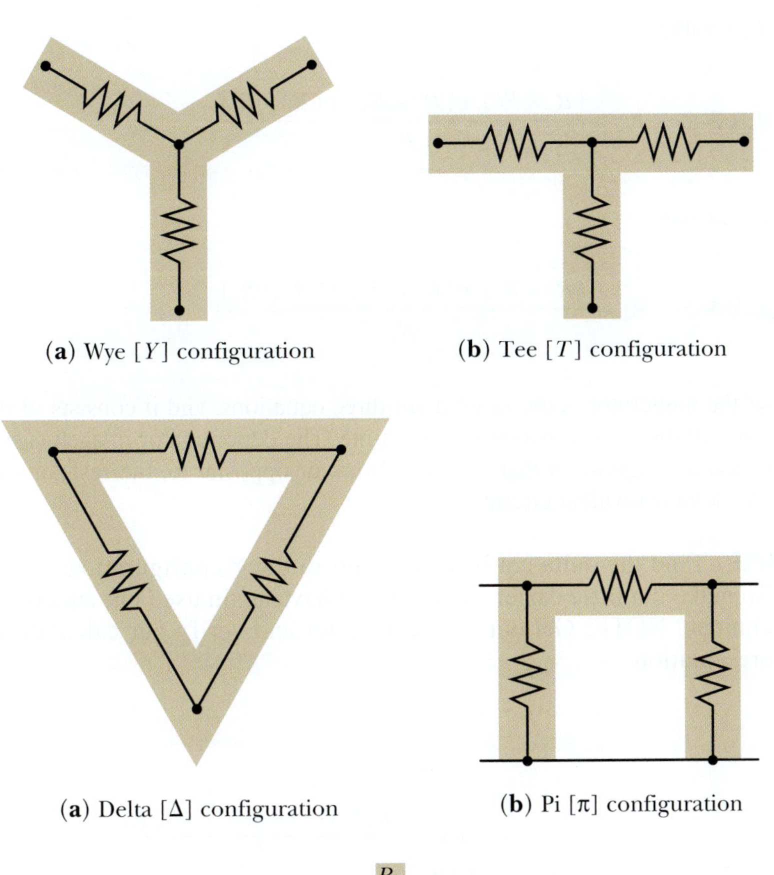

(**a**) Wye [*Y*] configuration (**b**) Tee [*T*] configuration

FIGURE 8–9 Commonly used *Y* and *T* circuit configurations

(**a**) Delta [Δ] configuration (**b**) Pi [π] configuration

FIGURE 8–10 Commonly used Δ and π configurations

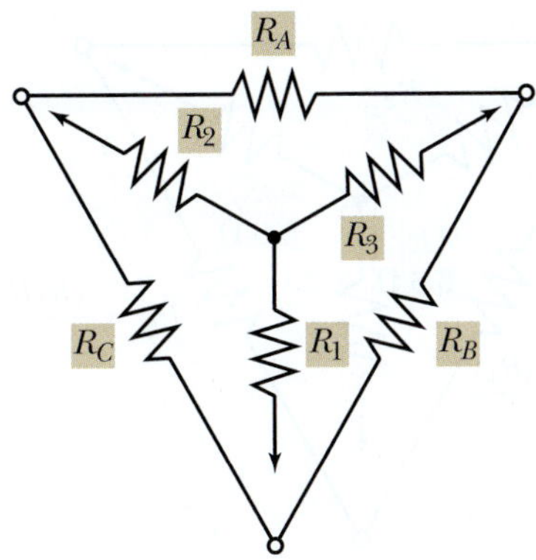

FIGURE 8–11 Superimposed Δ and *Y* configurations with components labeled

The wye components are labeled as R_1, R_2, and R_3; the delta components are labeled as R_A, R_B, and R_C, respectively.

Converting a Wye Network to an Equivalent Delta Network

The object of our conversion is to find out what values the delta components must be in order to provide the same electrical load to the source as the original wye network does. In other words, we are looking for the delta equivalent to the wye circuit. The equivalent circuit will present the same resistance, and draw the same current from the source as the original wye network does. It should be noted, however, that voltages between connection points will differ for each configuration.

The equations used to convert from wye to delta networks **(wye-delta conversion)** are:

1. To find R_A's value:

FORMULA 8–1 $$R_A = \frac{(R_1 \times R_2) + (R_2 \times R_3) + (R_3 \times R_1)}{R_1}$$

2. To find R_B's value:

FORMULA 8–2 $$R_B = \frac{(R_1 \times R_2) + (R_2 \times R_3) + (R_3 \times R_1)}{R_2}$$

3. To find R_C's value:

FORMULA 8–3 $$R_C = \frac{(R_1 \times R_2) + (R_2 \times R_3) + (R_3 \times R_1)}{R_3}$$

Observe that the numerator is the same in all three equations, and it consists of the sum of the products of all the wye resistors, two at a time. The denominator in each case is the resistor in the wye configuration that is across from, or opposite to, the resistor value being sought for the delta equivalent circuit.

EXAMPLE Find the delta equivalent circuit to the *Y* configuration shown in Figure 8–12. To make it a little easier to follow, we have again used our superimposed diagrams technique. NOTE: Get out your calculator and verify our calculations as you track our presentation.

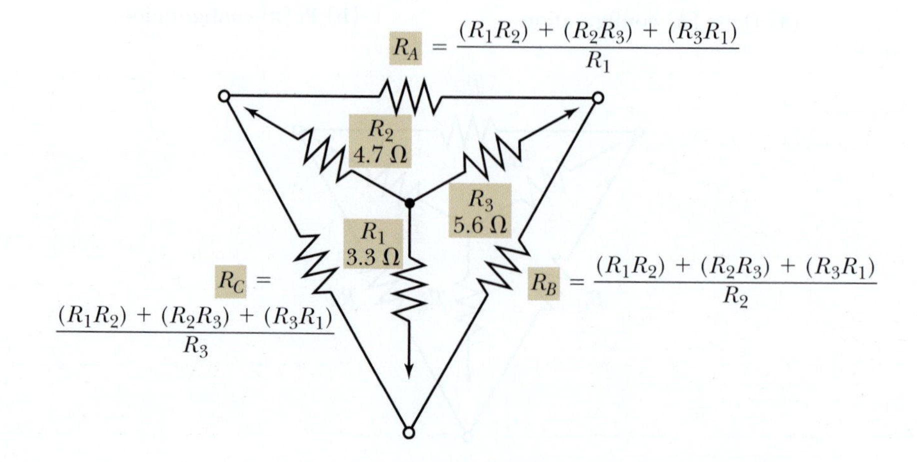

FIGURE 8–12 Reference diagram for wye-delta conversion "Example" problem

Assume that $R_1 = 3.3\ \Omega$, $R_2 = 4.7\ \Omega$, and $R_3 = 5.6\ \Omega$.

$$R_A = \frac{(R_1 R_2) + (R_2 R_3) + (R_3 R_1)}{R_1}$$

$$R_A = \frac{(3.3 \times 4.7) + (4.7 \times 5.6) + (5.6 \times 3.3)}{3.3} = \frac{60.31}{3.3} = 18.28\ \Omega$$

The numerator is the same for all three equations. All we need to do is divide the 60.31 by the appropriate resistor value from the *Y* network to find the resistor of interest in our equivalent delta network. Thus:

$$R_B = \frac{60.31}{4.7} = 12.83\ \Omega$$

$$R_C = \frac{60.31}{5.6} = 10.77\ \Omega$$

Converting a Delta Network to an Equivalent Wye Network

As an interesting way to find if the conversion techniques will work, let's look at the formulas for converting from delta to wye, insert the values we just calculated for the delta network, and see if, after conversion, we arrive back at the same values given in our original wye network. The equations used to convert from delta to wye networks **(delta-wye conversion)** are:

FORMULA 8–4 $$R_1 = \frac{R_B \times R_C}{R_A + R_B + R_C}$$

FORMULA 8–5 $$R_2 = \frac{R_C \times R_A}{R_A + R_B + R_C}$$

FORMULA 8–6 $$R_3 = \frac{R_A \times R_B}{R_A + R_B + R_C}$$

Notice in these equations, the denominator is the same in every case. That is, the denominator equals the sum of the resistances in the delta network. Notice also that the numerator for each equation shows the product of the two delta resistors that connect to one end of the wye resistor of interest in our superimposed networks drawing, or the product of the two delta resistors that are adjacent to the wye network resistor of interest.

Practical Notes

On occasion, you will find that the same value resistors are used for R_1, R_2, and R_3 in the wye configuration. This simplifies conversion calculations. For example, if each of the wye resistors is a 10-Ω resistor, then the resulting numerator in our conversion formula (the sum of the products part) is 300. The denominator is 10. The resulting answer is that the delta resistor must equal 30 Ω, or three times the *Y* resistor. This is true for all three legs. It is also true that the equivalent delta resistor will always be three times its related *Y* resistor value, as long as all the wye network resistors are equal in value.

This fact can also be used in reverse order. When converting from a symmetrical value delta configuration to a wye configuration, the wye resistors will each be one-third the value of their related delta resistors. In other words, if, in our previous example, the delta resistors had been 30 Ω each, the wye equivalent circuit would have had resistors equal to 10 Ω each.

In summary, when converting from wye to delta (where all resistors are equal in value), use three times each wye resistor value for each delta resistor. When converting from delta to wye (where all resistors are equal in value), make each wye resistor one-third the value of each of the delta resistors.

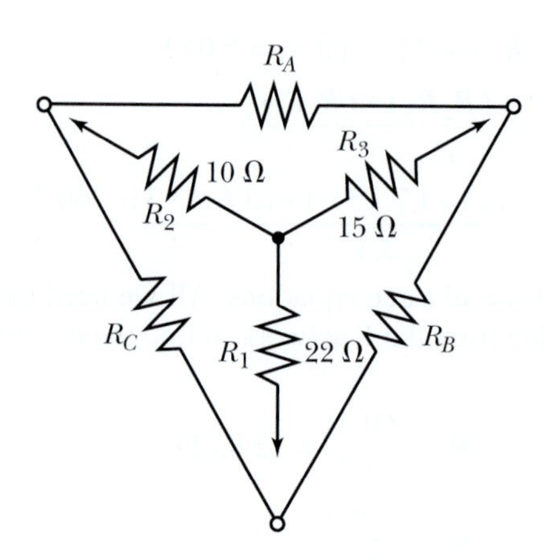

FIGURE 8–13

EXAMPLE Let's insert our computed delta resistor values into these formulas and see if we come close to the original wye circuit values shown in Figure 8–12.

$$R_1 = \frac{R_B \times R_C}{R_A + R_B + R_C} = \frac{12.83 \times 10.77}{18.28 + 12.83 + 10.77} = \frac{138.18}{41.88} = 3.3\ \Omega$$

For these equations, the denominator will be constant, so we only need to determine the numerator in each case, then divide by 41.88. Therefore,

$$R_2 = \frac{R_C \times R_A}{41.88} = \frac{10.77 \times 18.28}{41.88} = \frac{196.88}{41.88} = 4.7\ \Omega$$

$$R_3 = \frac{R_A \times R_B}{41.88} = \frac{18.28 \times 12.83}{41.88} = \frac{234.53}{41.88} = 5.6\ \Omega$$

Observe that our values do check out and verify the techniques!

PRACTICE PROBLEMS 4

1. Perform the necessary conversion from wye to delta configuration for Figure 8–13.
2. Use the values you computed for the delta resistor values in question 1 and determine the values of the wye network resistors, using the delta-wye conversion equations.
3. Did the answers you got in problem 2 appropriately correspond with the original circuit parameters shown in Figure 8–13?

IN-PROCESS LEARNING CHECK 3

1. Because of their visual shapes and electrical connections:
   a. The wye circuit configuration is sometimes also called the _____ configuration.
   b. The delta circuit configuration is sometimes also called the _____configuration.
2. Delta and wye circuit configurations are often found in _____ power applications.
3. The value in knowing how to determine equivalent circuits, or convert from delta-to-wye and wye-to-delta circuits is that the equivalent circuits will present the same resistance and draw the same _____ from the source feeding the circuit.
4. The voltage between connection points in wye and delta configurations (is, is not) _____ the same.

Bridge Circuit Analysis Using Conversion Techniques

Observe the bridge circuit of Figure 8–14. When shown as a bridge circuit, it appears that the network is essentially two delta networks, back-to-back. Simple series-parallel circuit analysis techniques cannot be used here. If we convert the bottom delta portion of this circuit to a wye configuration, it will simplify the analysis of the circuit considerably.

In Figure 8–15, you see that we have superimposed the proposed wye network in the bottom delta of the bridge diagram. By converting this delta circuit portion to an equivalent wye circuit, we can redraw the circuit as shown in Figure 8–16. As stated earlier, this lends itself to much easier analysis.

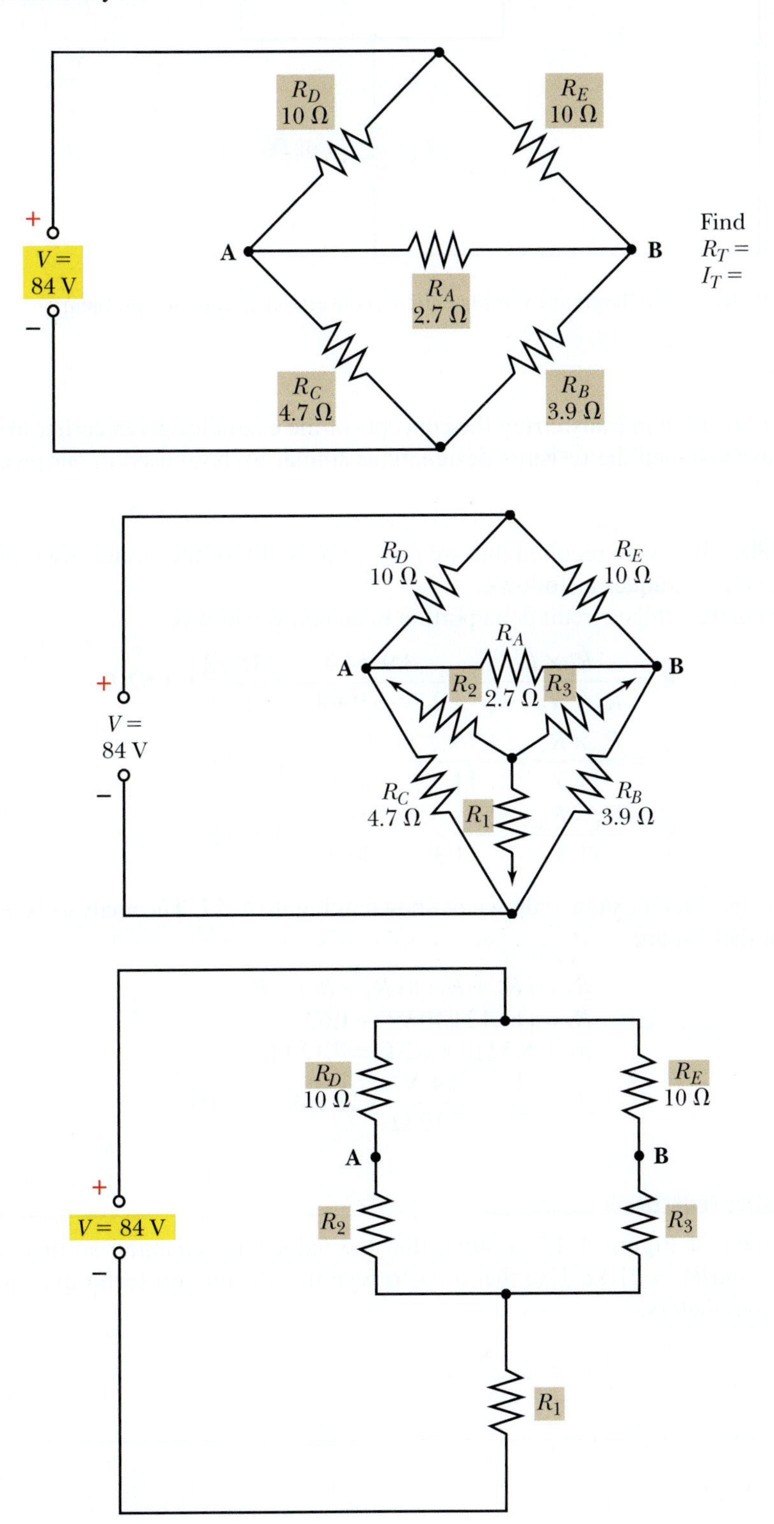

FIGURE 8–14 Sample bridge circuit multiSIM

FIGURE 8–15 Sample bridge circuit analysis

FIGURE 8–16 Redrawn bridge network equivalent with bottom delta transformed to wye

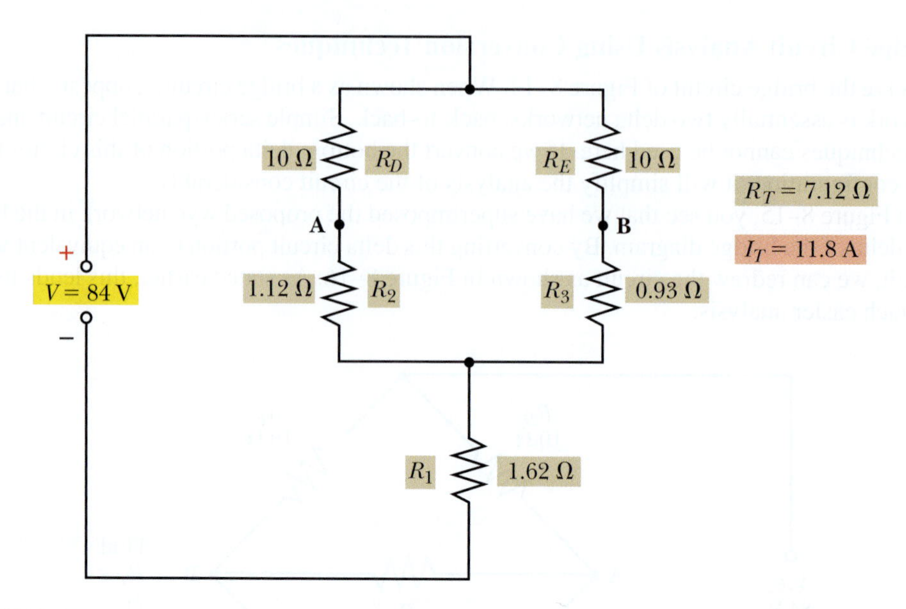

FIGURE 8–17 Reference diagram for bridge circuit conversion "Example" problem

For simplification in transferring the concepts of the examples given earlier to this situation, we have assigned the resistors designations similar to those used in our preceding examples.

EXAMPLE If it were required that we find the R_T and I_T of the circuit, we might use the transformation techniques as follows.

Converting the bridge circuit delta portion to an equivalent wye:

$$R_1 = \frac{R_B \times R_C}{R_A + R_B + R_C} = \frac{3.9 \times 4.7}{2.7 + 3.9 + 4.7} = \frac{18.33}{11.3} = 1.62\ \Omega$$

$$R_2 = \frac{R_C \times R_A}{11.3} = \frac{4.7 \times 2.7}{11.3} = \frac{12.69}{11.3} = 1.12\ \Omega$$

$$R_3 = \frac{R_A \times R_B}{11.3} = \frac{2.7 \times 3.9}{11.3} = \frac{10.53}{11.3} = 0.93\ \Omega$$

Now the bridge circuit can be redrawn as shown in Figure 8–17. The analysis is now that of a series-parallel circuit.

$$R_T = (R_D + R_2) \parallel (R_E + R_3) + R_1$$
$$R_T = (11.12 \parallel 10.93) + 1.62$$
$$R_T = 5.5\ \Omega + 1.62\ \Omega = 7.12\ \Omega$$
$$I_T = \frac{V}{R_T} = \frac{84\ \text{V}}{7.12\ \Omega} = 11.8\ \text{A}$$

PRACTICE PROBLEMS 5

For the circuit of Figure 8–15, assume that the values are changed so that $R_A = 7\ \Omega$, $R_B = 14\ \Omega$, and $R_C = 21\ \Omega$. Use the delta-to-wye transformation techniques to find the following parameters:

1. R_T
2. I_T

Summary

- A loop is a complete path in a circuit, from one side of the source, through a series of circuit components, back to the other side of the source.
- A mesh is an unbroken, closed loop within a network.
- Mesh currents are assumed currents (and current directions) used in conjunction with Kirchhoff's voltage law to analyze networks. Mesh currents are different from actual circuit currents, in that they are assumed not to branch.
- A common convention regarding labeling mesh currents on a diagram is that they are shown in a clockwise direction.
- Voltages across resistors in a network that are caused by their own mesh current are considered positive. Voltage drops due to another mesh current are considered negative.
- An arbitrary convention regarding assignment of source polarities in a network is that if a mesh current returns to the positive side of the source, the source is denoted as a positive term in the equations. If the mesh current returns to the negative terminal of the source, the source is signed as negative in the equations.
- For mesh analysis, voltages are expressed as $I \times R$ terms.
- Simultaneous equation techniques are used to eliminate one unknown from the resulting Kirchhoff equations. This enables solving for the other unknown factor.
- A node is a point where two or more circuit components join and where, when circuit current is flowing, circuit currents may branch. A major node is where three or more components join.
- Nodal analysis uses Kirchhoff's current law to define parameters in the network. Currents to be used in the analysis, however, are expressed in terms of *V/R*.
- Delta and wye circuit configurations are quite common to many electrical and electronic circuits. The ability to analyze these circuits is enhanced by knowing how to convert, or transform, values of one type of network to the other.
- In the conversion formulas for converting from wye to equivalent delta circuit values, the numerator is the common factor used in determining each of the three resistor values.
- In the conversion formulas for converting from delta to equivalent wye circuit values, the denominator is the common factor used in determining each of the three resistor values.
- In symmetrical wye circuits (those where all three legs of the wye have equal value resistors), the delta equivalent resistances will be three times the value of the wye resistors.
- In symmetrical delta circuits (those where all three sides of the delta have equal value resistors), the wye equivalent resistances will be one-third the value of the delta resistors.

Formulas and Sample Calculator Sequences

Formula 8–1
(To find delta resistor R_A's value using the wye-to-delta conversion approach)

$$R_A = \frac{(R_1 \times R_2) + (R_2 \times R_3) + (R_3 \times R_1)}{R_1}$$

[(], R_1 value, [×], R_2 value, [)], [+], [(], R_2 value, [×], R_3 value, [)], [+], [(], R_3 value, [×], R_1 value, [)], [=], [÷], R_1 value, [=]

Formula 8–2
(To find delta resistor R_B's value using the wye-to-delta conversion approach)

$$R_B = \frac{(R_1 \times R_2 + (R_2 \times R_3) + (R_3 \times R_1)}{R_2}$$

[(], R_1 value, [×], R_2 value, [)], [+], [(], R_2 value, [×], R_3 value, [)], [+], [(], R_3 value, [×], R_1 value, [)], [=], [÷], R_2 value, [=]

Formula 8–3
(To find delta resistor R_C's value using the delta-to-wye conversion approach)

$$R_C = \frac{(R_1 \times R_2) + (R_2 \times R_3) + (R_3 \times R_1)}{R_3}$$

[(], R_1 value, [×], R_2 value, [)], [+], [(], R_2 value, [×], R_3 value, [)], [+], [(], R_3 value, [×], R_1 value, [)], [=], [÷], R_3 value, [=]

Formula 8–4
(To find wye resistor R_1's value using the delta-to-wye conversion approach)

$$R_1 = \frac{R_B \times R_C}{R_A + R_B + R_C}$$

R_B value, [×], R_C value, [÷], [(], R_A value, [+], R_B value, [+], R_C value, [)], [=]

Formula 8–5
(To find wye resistor R_2's value using the delta-to-wye conversion approach)

$$R_2 = \frac{R_C \times R_A}{R_A + R_B + R_C}$$

R_C value, [×], R_A value, [÷], [(], R_A value, [+], R_B value, [+], R_C value, [)], [=]

Formula 8–6
(To find wye resitor R_3's value using the delta-to-wye conversion approach)

$$R_3 = \frac{R_A \times R_B}{R_A + R_B + R_C}$$

R_A value, [×], R_B value, [÷], [(], R_A value, [+], R_B value, [+], R_C value, [)], [=]

EXCEL AUTOMATED FORMULAS

Helpful problem-solving tools, such as the Excel automated formulas available on the CD, allow automatic calculations of formulas.

Using Excel

Network Analysis Techniques Formulas

(Excel file reference: FOE8_01.xls)

DON'T FORGET! *It is NOT necessary to retype formulas once they are entered on the worksheet! Just input new parameters data for each new problem using that formula, as needed.*

- Refer to Figure 8–12. Assume that the values of R_1, R_2, and R_3 are changed so that R_1 = 10 ohms, R_2 = 12 ohms, and R_3 = 18 ohms. Use the Formula 8–1 spreadsheet sample and solve for the value of R_A.
- Refer to Figure 8–12, again. This time, assume that values of all three "delta" resistors are 10 ohms. Use the Formula 8–4 spreadsheet sample and solve for the value of R_1 when designing an equivalent Y configuration.

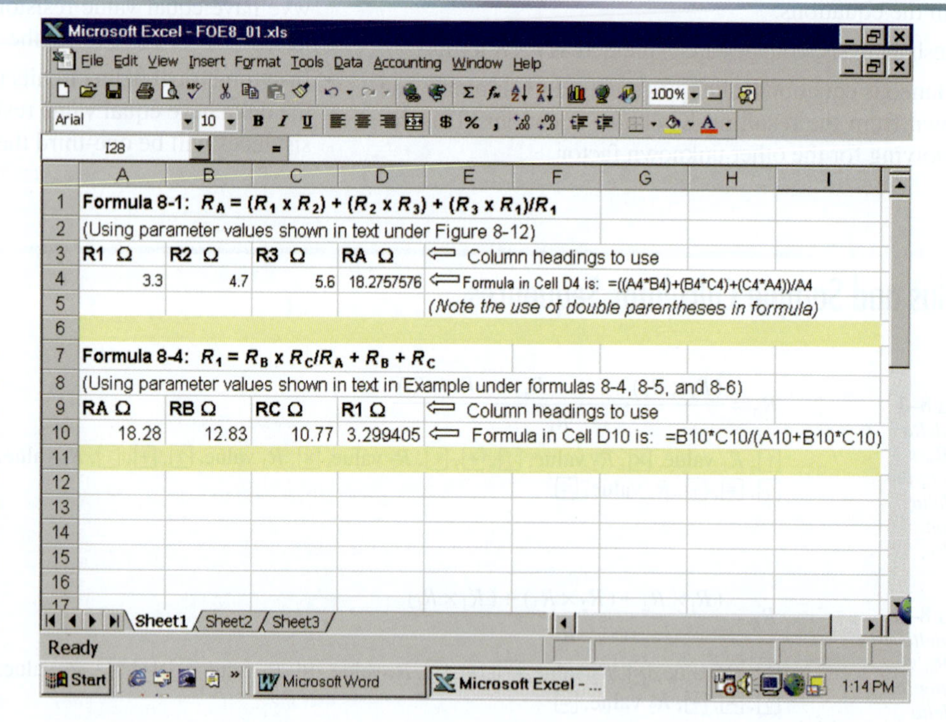

Review Questions

1. Define the term *loop*.
2. Define the term *mesh*.
3. Explain the basic difference between a mesh current and an actual circuit current.
4. A common convention regarding labeling mesh currents on a diagram is that they are shown in a:
   a. counterclockwise direction.
   b. clockwise direction.
   c. different direction for each mesh window.
5. Voltages across resistors in a network that are caused by their own mesh current are considered:
   a. negative.
   b. neutral.
   c. positive.
   d. self-defining.
6. Voltages across resistors in a network that are caused by mesh currents other than their own mesh current are considered:
   a. negative.
   b. neutral.
   c. positive.
   d. self-defining.
7. When assigning source polarities in a network, an arbitrary convention is that if a mesh current returns to the positive side of the source:
   a. the source is indicated as positive in equations.
   b. the source is indicated as negative in equations.
   c. it makes no difference which assignment you use.
   d. sources are not considered in the equations.
8. The purpose of using "simultaneous equations" when using Kirchhoff equations is to:
   a. view two sets of parameters, simultaneously.
   b. eliminate one unknown in order to solve for another.
   c. neither of the above reasons.
9. A "node" is a point in a network where:
   a. two or more circuit components join, and circuit currents branch.
   b. no less than three circuit components join.
   c. where voltages are equal and cancel.
   d. none of the above.
10. A "delta" circuit configuration is the same as a:
    a. wye circuit configuration.
    b. pi circuit configuration.
    c. tee circuit configuration.
    d. none of these configurations.
11. A "wye" circuit configuration is the same as a:
    a. pi circuit configuration.
    b. tee circuit configuration.
    c. delta circuit configuration.
    d. none of these configurations.
12. In ac power circuitry, it is common to find these two different circuit configurations:
    a. Wye and tee
    b. Delta and pi
    c. Wye and delta

Problems

Refer to Figure 8–4. Assume that all three resistors now equal 10 Ω. Assume V_{S_1} is changed to 20 V and V_{S_2} is changed to 16 V. Use the mesh analysis approach to determine the parameters called for in the following questions.

1. I_A
2. I_B
3. V_{R_1}
4. V_{R_2}
5. V_{R_3}
6. I_{R_1}
7. I_{R_2}
8. I_{R_3}

Refer to Figure 8–6. Assume that all three resistors equal 4.7 Ω. Assume V_{S_1} is changed to 21 V and V_{S_2} is changed to 14 V. Use the nodal analysis approach to determine the parameters called for in the following questions.

9. I_1
10. I_2
11. I_3
12. V_{R_1}
13. V_{R_2}
14. V_{R_3}

Refer to Figure 8–11. Assume that $R_1 = 10\ \Omega$, $R_2 = 12\ \Omega$, and $R_3 = 15\ \Omega$. Use the wye-to-delta conversion formulas to find the values of:

15. R_A

16. R_B

17. R_C

18. For the circuit of Figure 8–11, assume that $R_A = 150\ \Omega$, $R_B = 150\ \Omega$, and $R_C = 150\ \Omega$. What would be the values of R_1, R_2, and R_3 in an equivalent wye circuit?

Analysis Questions

1. When would mesh or nodal analysis be used rather than standard series-parallel analysis of a network?
2. How many equations must be written for a three-mesh circuit in order to find the circuit parameters?
3. Perform research and find where you might find delta and wye configurations of electrical circuits in an industrial plant setting.
4. Perform research and find what type(s) of electronic circuitry might commonly contain the pi circuit configuration.

MultiSIM Exercise for *Network Analysis Techniques*

1. Use the MultiSIM program and utilize the circuit shown in Figure 8–6. (NOTE: Connect a ground reference to the bottom of the 5 Ω resistor.)
2. Measure and record the following parameters:
   a. V_{R_1}
   b. V_{R_2}
   c. V_{R_3}
3. Compare your results with those given in the text example for this circuit. Do they reasonably compare?

Performance Projects Correlation Chart

Chapter Topic	Performance Project	Project Number
Assumed Loop or Mesh Current Analysis Approach	Loop/Mesh Analysis	30
Nodal Analysis Approach	Nodal Analysis	31
Conversions of Delta and Wye Networks	Wye-Delta and Delta-Wye Conversions	32
	Story Behind the Numbers: Network Analysis Techniques	

NOTE: It is suggested that after completing the above projects, the student should be required to answer the questions in the "Summary" at the end of this section of projects in the Laboratory Manual.

PART III

Producing and Measuring Electrical Quantities

OUTLINE

OBJECTIVES

After studying this chapter, you should be able to:

1. Define **magnetism, magnetic field, magnetic polarity,** and **flux**
2. Draw representations of magnetic fields related to permanent magnets
3. State the magnetic attraction and repulsion law
4. State at least five generalizations about magnetic lines of force
5. Draw representations of fields related to current-carrying conductors
6. Determine the polarity of electromagnets using the **left-hand rule**
7. List and define at least five magnetic units of measure, terms, and symbols
8. Draw and explain a *B-H* curve and its parameters
9. Draw and explain a hysteresis loop and its parameters
10. Explain motor action and generator action related to magnetic fields
11. List the key factors related to induced emf
12. Briefly explain the relationships of quantities in **Faraday's law**
13. Briefly explain **Lenz's law**
14. Use the computer to solve problems related to magnetic induction

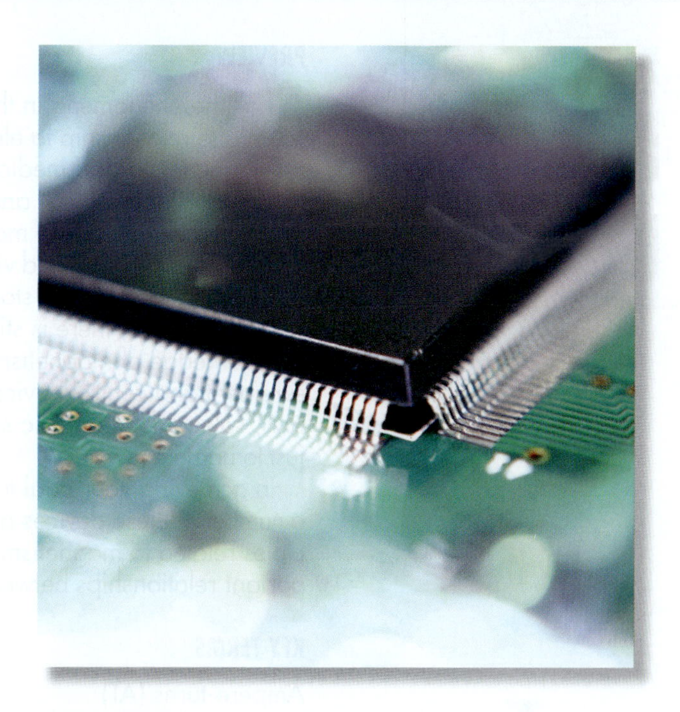

CHAPTER 9

Magnetism and Electromagnetism

ELECTRONICS INTO THE FUTURE

Explore an interactive presentation on Magnetism and Electromagnetism on the accompanying CD.

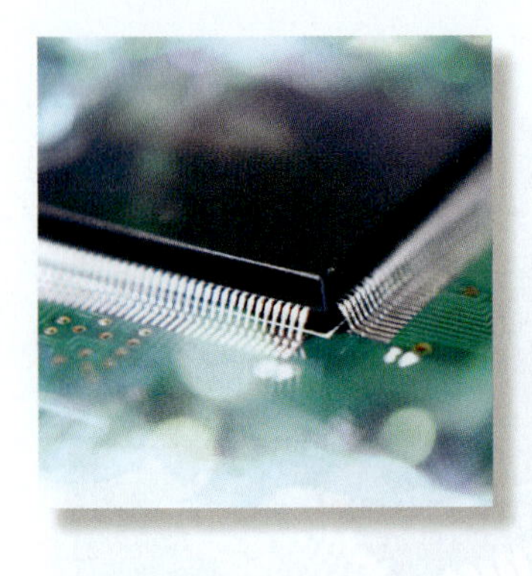

PREVIEW

Of all the phenomena in the universe, one of the most fascinating is magnetism and its unique relationship to electricity.

Magnetic storage media are extremely important in today's world. Computers use magnetic "floppy disks" and "hard disks" to store tremendous amounts of data. Audio and video recorders use magnetic tape to magnetically store magnetic patterns that are transformed to sound and visual intelligence, which we can enjoy at our choosing. (Note: Although optical-digital storage media, such CDs and DVDs are becoming the storage media of choice, there is still a large amount of magnetic storage media in use today.)

Knowledge of magnetism is important for understanding a variety of industrial measurement and control devices and systems. Some examples include electrical meters, relays, solenoids, magnetic switches, inductors, transformers, and motors and generators, just to name a few.

In achieving your goal to become a technician, you will have the opportunity to study a number of these devices and components. In this chapter, you will focus on basic principles related to magnetism, some units of measure used with magnetic circuits, and important relationships between magnetic and electrical phenomena.

KEY TERMS

Ampere-turns (AT)
Faraday's law
Flux
Flux density
Induction
Left-hand rules
Lenz's law
Lines of force (flux lines)
Magnetic field
Magnetic field intensity (H)
Magnetic polarity
Magnetism
Magnetomotive force
Permeability (μ_o)
Relative permeability (μ_r)
Reluctance
Residual magnetism
Saturation
Tesla (T)
Weber

9–1 Background Information

Early History

Early discovery of magnetism happened by people noticing the unique characteristics of certain stones called lodestones. These stones (magnetite material), which were found in the district of Magnesia in Asia Minor, attracted small bits of iron. It was also noted that certain materials stroked by lodestones became magnetized.

Materials or substances that have this unique feature are called magnets, and the phenomenon associated with magnets is called **magnetism.** Materials that are attracted by magnets are called magnetic materials.

As early as the eleventh century, the Chinese used magnets as navigational aids, since magnets that are suspended, or free to move, align approximately in a north-south geographic direction.

Today we have materials, called permanent magnets, that retain their magnetism for long periods. Examples of permanent magnets include hard iron or special iron-nickel alloys.

Magnetic materials that lose their magnetism after the magnetizing force is removed are termed *temporary magnets.*

Later Discoveries

Centuries later the relationship between electricity and magnetism was discovered. In the 1800s several scientists made important observations. Hans Oersted noted that a free-swinging magnet would react to a current-carrying wire, if they were in close proximity. As long as current passed through the wire the reaction occurred. But when there was no current, there was no interaction between the magnet and wire. This indicated that a steady current caused a steady magnetic effect near the current-carrying conductor. This "field of influence" would eventually be called a **magnetic field.**

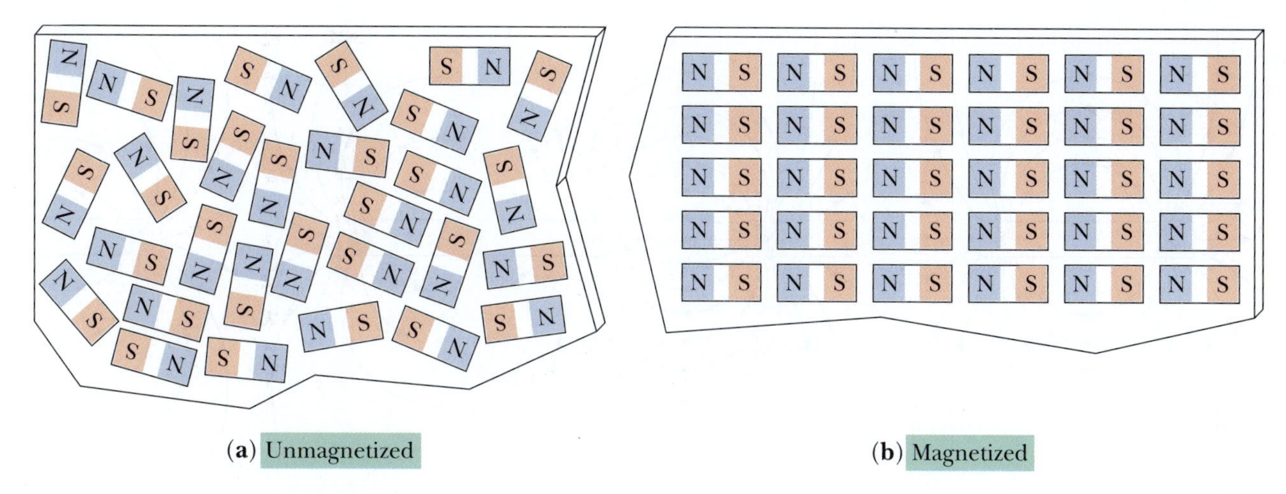

FIGURE 9–1 Magnetized and unmagnetized conditions

Later in the same century, Joseph Henry and Michael Faraday discovered interesting phenomena that occurred when changing current levels or moving magnetic fields were present. Henry determined that when current changed levels in a conductor, current was induced in a nearby conductor (having a closed current path). Faraday determined that a permanent magnet moving in the vicinity of a conductor induced current in the wire (if there existed a complete path for current through the conductor). It should also be stated that even without a closed current path, voltage is induced.

All of these discoveries are very important! You will see these scientists' names again since several units of magnetic measure are named in their honor.

9–2 Fundamental Laws, Rules, and Terms to Describe Magnetism

When a magnet is created, there is an area of magnetic influence near the magnet called a magnetic field. A magnetic field is established either by alignment of internal magnetic forces within a magnetic material, Figure 9–1, or by organized movement of charges (current flow) through conductor materials. This magnetic field is composed of magnetic **lines of force,** sometimes called **flux lines.** Thus, magnetic flux refers to all the magnetic lines of force associated with that magnet. The symbol for flux is the Greek letter phi (ϕ). Each magnetic field line is designated by a unit called the maxwell (Mx). That is, one magnetic field line equals one maxwell, or 50 magnetic field lines equal 50 maxwells.

Another basic unit associated with flux is the **weber.** Whereas the maxwell equals one line of flux, the weber (Wb) equals 10^8 lines of flux, or 100 million flux lines. Thus, a microweber equals 100 lines, or 100 Mx.

Another fundamental term related to flux is **flux density,** which refers to the number of lines per given unit area. We will discuss flux density later in the chapter.

If a nonmagnetic material, like cardboard, is laid on top of a bar magnet (or other magnets), sprinkled with small iron filings, then tapped or vibrated, the iron filings position themselves in a pattern illustrating the effect of the magnetic field around the magnet, Figure 9–2.

A diagram frequently used to show this effect is in Figure 9–3. Notice that at each end of the bar magnet, the lines of force are concentrated, but as they get farther from the ends, they spread. As you can see, the ends of the bar magnets (where the lines of force are concentrated) are labeled "N" and "S" and are the "poles" of the magnet. That is, the N and S indicate North-seeking and South-seeking poles.

To represent directions of the lines of force, it is customary to say the flux lines exit the North pole and reenter the South pole of the magnet. Inside the magnet, of course, the flux lines travel from the South pole to the North pole. Each flux line is an unbroken loop, or ring.

In Figure 9–4 you can see two bar magnets used to illustrate that *like poles repel each other, and unlike poles attract each other.* This is an important law, so remember it!

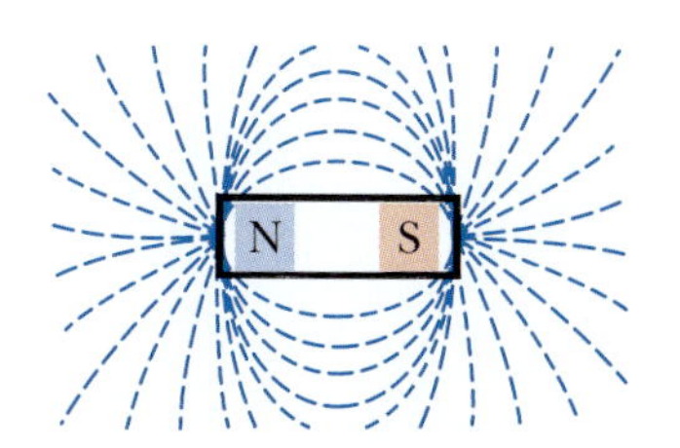

FIGURE 9–2 Iron filings displaying magnetic field pattern of a bar magnet

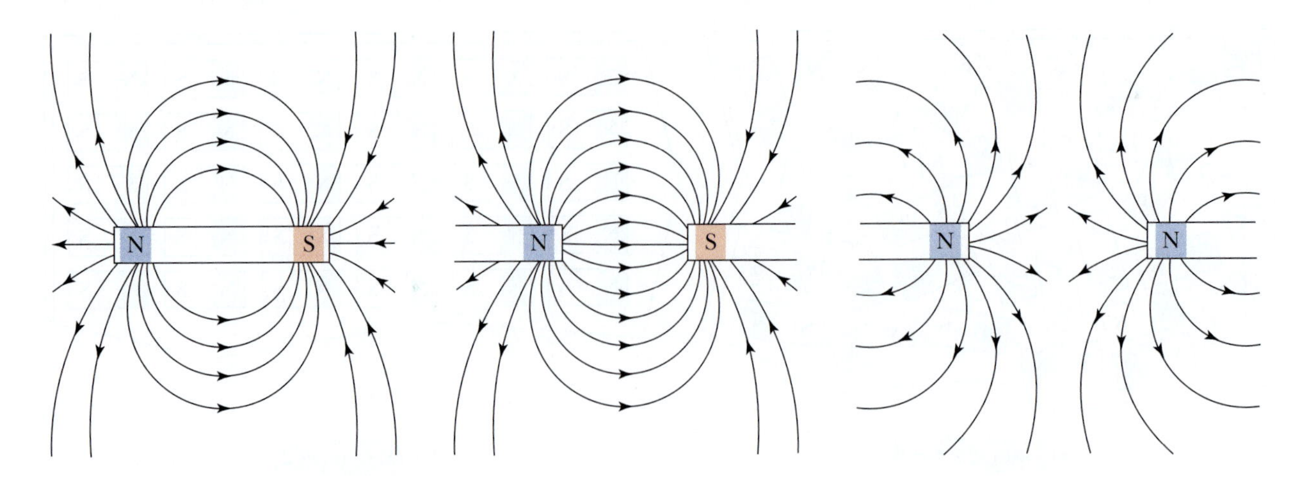

FIGURE 9–3 Lines of force

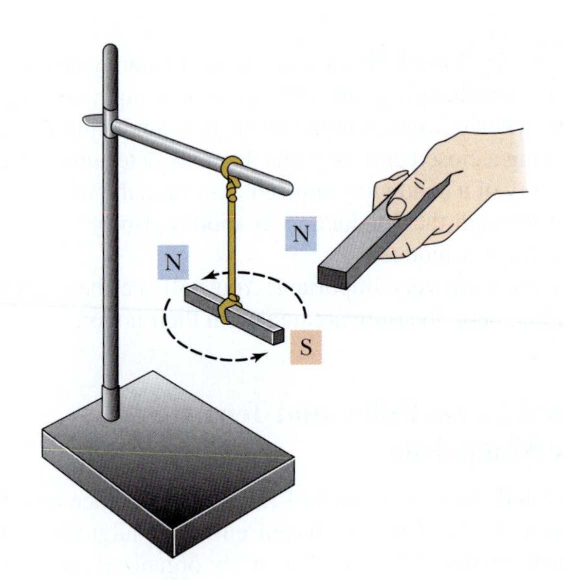

FIGURE 9–4 Like poles repel each other, and unlike poles attract each other

Rules Concerning Lines of Force

Along with the law defined previously, you need to know the following important generalizations related to *lines of force* (flux lines), Figure 9–5.

1. Magnetic lines are continuous, Figure 9–5a.
2. Magnetic lines flow from North to South poles outside the magnet and from South to North inside the magnet, Figure 9–5a.
3. Magnetic lines take the shortest or easiest path, Figure 9–5b.
4. Lines in the same direction repel each other but are additive, strengthening the overall field. Lines in opposite directions attract and cancel each other, weakening the overall field, Figure 9–5c.
5. Magnetic lines penetrate nonmagnetic materials, Figure 9–5d.
6. Lines of force do not cross each other, Figure 9–5e.

Some Practical Points about Permanent Magnets

Before discussing electromagnetism, here are some practical points about permanent magnets.

1. They are not really permanent because they lose their magnetic strength over time, or if heated to a high temperature, or if physically pounded upon, and so on.

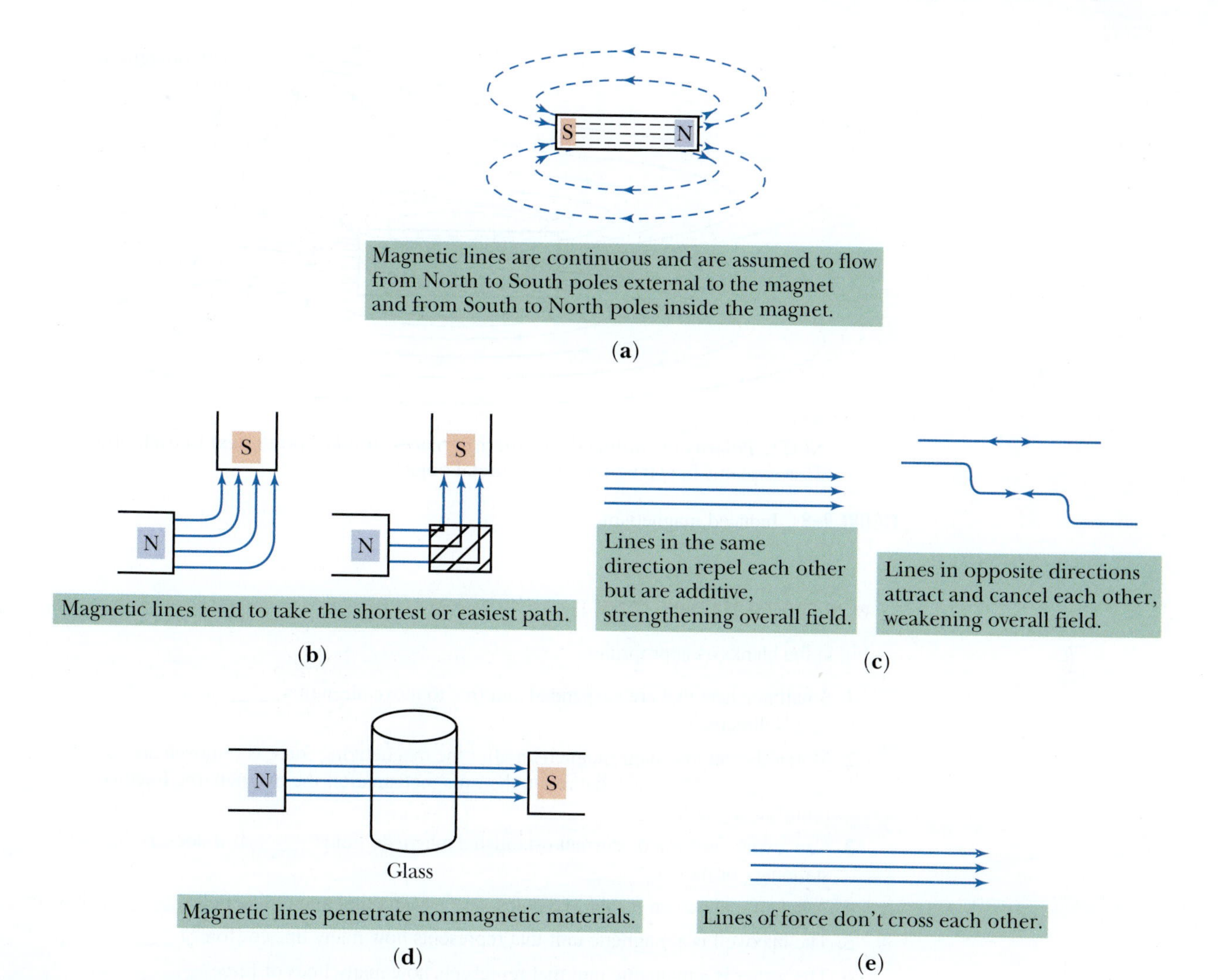
S
N
Magnetic lines are continuous and are assumed to flow from North to South poles external to the magnet and from South to North poles inside the magnet.
(a)
S
N
S
N
Magnetic lines tend to take the shortest or easiest path.
(b)
Lines in the same direction repel each other but are additive, strengthening overall field.
Lines in opposite directions attract and cancel each other, weakening overall field.
(c)
N
S
Glass
Magnetic lines penetrate nonmagnetic materials.
(d)
Lines of force don't cross each other.
(e)

FIGURE 9–5 Rules about lines of force

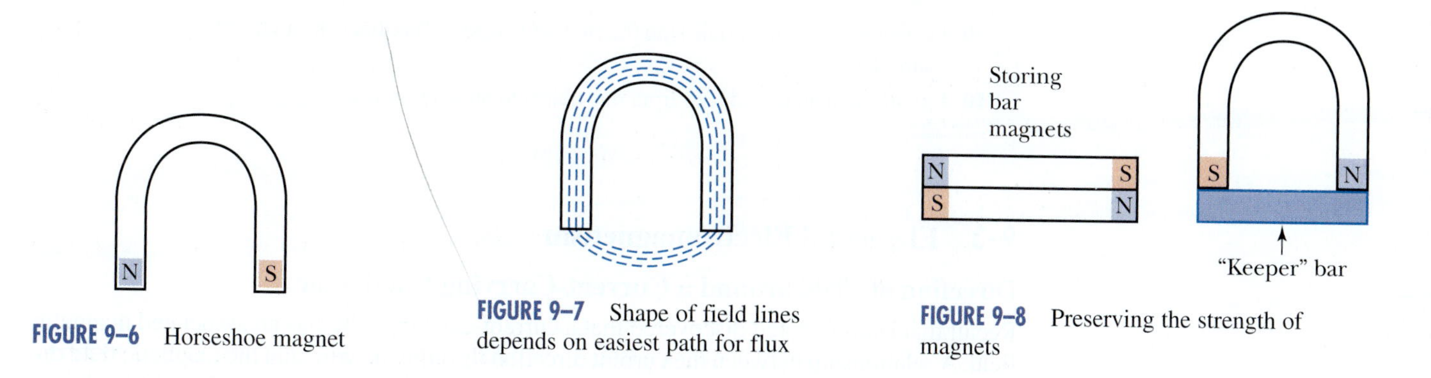
N
S
Storing bar magnets
N
S
S
N
S
N
"Keeper" bar

FIGURE 9–6 Horseshoe magnet

FIGURE 9–7 Shape of field lines depends on easiest path for flux

FIGURE 9–8 Preserving the strength of magnets

2. They can be configured into different shapes in addition to bar magnets. One common shape is termed a *horseshoe magnet* because of its appearance, Figure 9–6.
3. The field lines' shape depends on the easiest path for flux, Figure 9–7.
4. Appropriate methods to store magnets help preserve their field strengths for longer periods. For example, store bar magnets so opposite poles are together, or use a "keeper bar" of magnetic material across the ends of a horseshoe magnet, Figure 9–8.
5. A magnet induces magnetic properties in a nearby object of magnetic material, if the object is in the path of the original magnet's field lines, Figure 9–9.

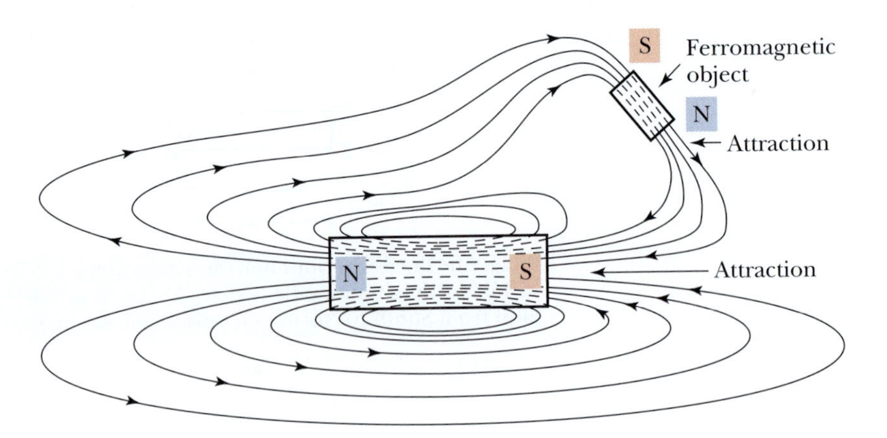

NOTE: Polarity of "induced" magnetism places "unlike" poles next to each other. This explains how magnets attract metal objects.

FIGURE 9–9 Induced magnetism

IN-PROCESS LEARNING CHECK 1

Fill in the blanks as appropriate.

1. Small magnets that are suspended and free to move align in a _____ and _____ direction.
2. Materials that lose their magnetism after the magnetizing force is removed are called _____ _____. Materials that retain their magnetism after the magnetizing force is removed are called _____ _____.
3. Can a wire carrying dc current establish a magnetic field? _____ If it does, is this field stationary or moving? _____
4. A law of magnetism is that like poles _____ each other and unlike poles _____ each other.
5. The maxwell is a magnetic unit that represents how many lines of force? _____
6. The weber is a magnetic unit that represents how many lines of force? _____
7. Are magnetic lines of force continuous or noncontinuous? _____
8. Lines of force related to magnets exit the _____ pole of the magnet and enter the _____ pole.
9. Do nonmagnetic materials stop the flow of magnetic flux lines through themselves? _____
10. Can magnetism be induced from one object to another object? _____

9–3 Elemental Electromagnetism

Direction of Field around a Current-Carrying Conductor

Recall that Hans Oersted discovered that a current-carrying wire has an associated magnetic field. A relationship between the current direction through the wire and the magnetic field direction around the current-carrying conductor is established while moving the compass around the conductor and noting the effect on the compass needle.

Note in Figure 9–10a you are looking at the end of a wire (black dot), and the current is coming toward you. In Figure 9–10b you get an overview of the relationship of current direction with the magnetic field around the current-carrying conductor. Both of these figures illustrate one of the **left-hand rules.**

The Left-Hand Rule for Current-Carrying Conductors

Grasping the wire with your left hand so your extended thumb points in the direction of current flow, the magnetic field around the conductor is in the direction of the fingers that are

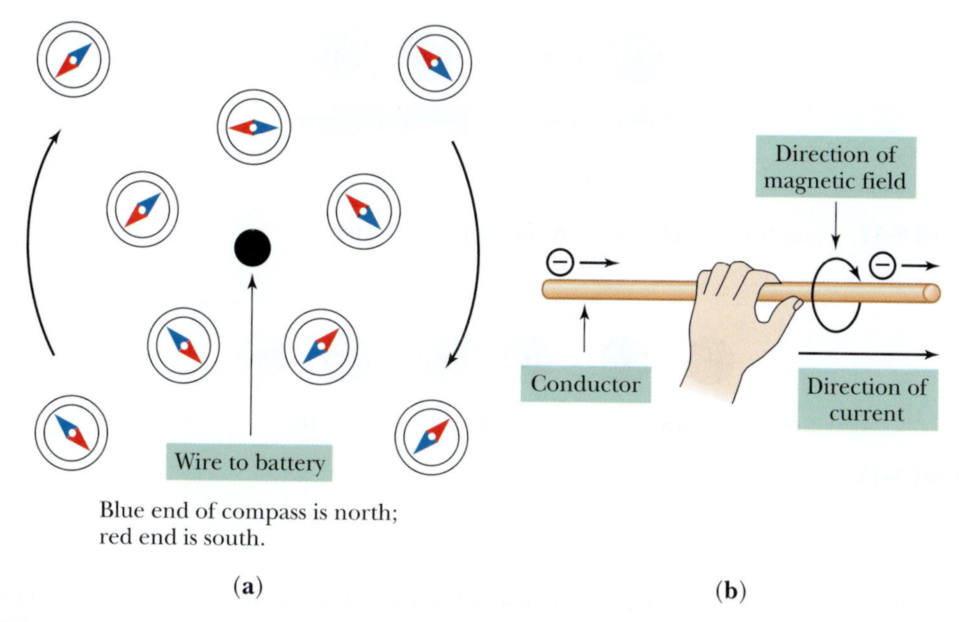

FIGURE 9–10 (a) Field around current-carrying conductor; (b) left-hand rule for current-carrying conductor

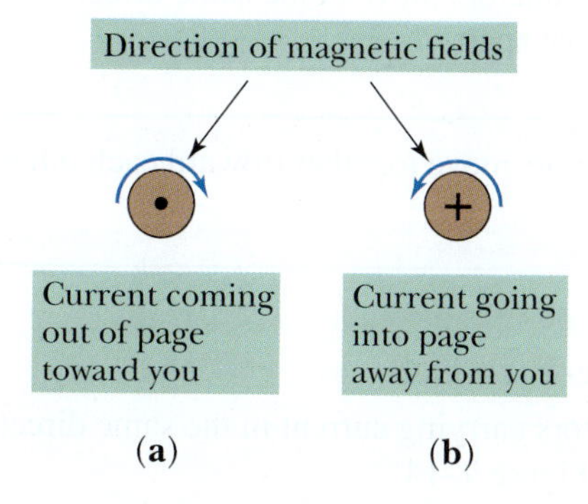

FIGURE 9–11 Magnetic fields around current-carrying conductors

wrapped around the conductor. (NOTE: For those who use "conventional current," where the direction of current flow is considered in the opposite direction, the "right-hand" rule is used instead of this left-hand rule.)

EXAMPLE Another way to illustrate this rule is shown in Figure 9–11. In this illustration, the current (and your thumb) is coming toward you from the conductor with the dot, Figure 9–11a. The dot is like looking at the tip of an arrow coming toward you. In this case, the magnetic field is in a clockwise direction.

In the other picture, Figure 9–11b, the conductor has a plus sign, indicating current is going away from you into the page. This is similar to looking at the tail feathers of an arrow. In this case, the direction of magnetic field is counterclockwise.

PRACTICE PROBLEMS 1

1. Draw a picture of two parallel conductors with the direction of current going away from you in both cases.
2. Using arrows, show the directions of the magnetic fields around each conductor.

Force between Parallel Current-Carrying Conductors

Look at Figure 9–12a. Notice that when two parallel current-carrying wires have current in the same direction, the flux lines between the wires are in opposite directions. This situation causes the flux lines to attract and cancel each other, and the magnetic field pattern is modified

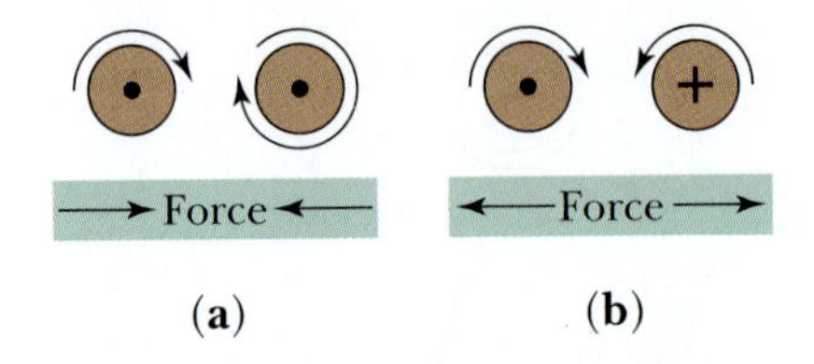

FIGURE 9–12 Force between adjacent current-carrying conductors

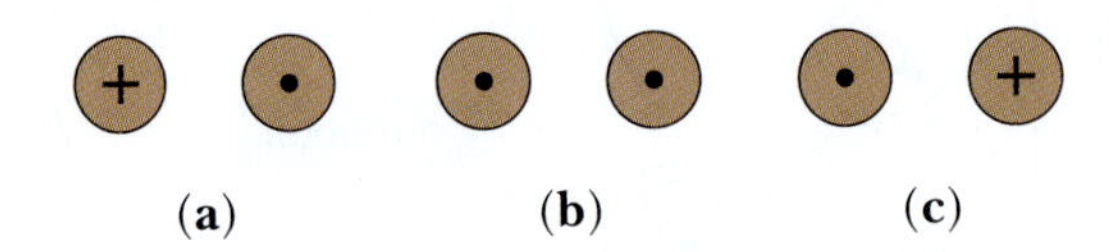

FIGURE 9–13

since lines of force cannot cross. This indicates that the two adjacent wires move toward the weakened portion of the field or toward each other.

On the other hand, if the two wires have current in opposite directions, Figure 9–12b, the lines of force between the two conductors move in the same direction. Hence, the flux lines repel each other and the wires move apart.

PRACTICE PROBLEMS 2

Indicate whether the conductors move together (toward each other) or apart in Figure 9–13a, b, and c.

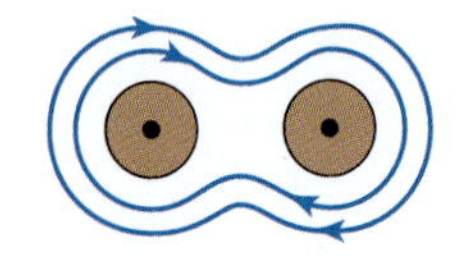

FIGURE 9–14 Stronger "combined" field

Field around a Multiple-Turn Coil

Individual fields of conductors carrying current in the same direction modify to a larger and stronger field, as shown in Figure 9–14.

When a single current-carrying wire forms multiple loops (such as in a coil), there are a number of fields that combine from adjacent turns of the coil, producing a large field through and around the coil, Figure 9–15.

Also notice in Figure 9–15 that the magnet established by this coil has North and South magnetic poles, like a permanent bar magnet. The **magnetic polarity** of this electromagnet is determined by another left-hand rule.

Left-Hand Rule to Determine Polarity of Electromagnets

Grasping the coil so your fingers are pointing in the same direction as the current passing through the coil, the extended thumb points toward the end of the coil that is the North pole, Figure 9–16.

PRACTICE PROBLEMS 3

See if you can determine the magnetic polarity of the coils in Figures 9–17a, b, and c.

Factors Influencing the Field Strength of an Electromagnet

From the previous discussions, you should surmise that the number of turns influences the strength of the electromagnet. That is, the higher the number of turns per unit length of coil, the greater the strength of the magnetic field, Figure 9–18a.

Also, you might have reasoned the amount of current through the conductor directly affects the magnetic strength, since a higher current indicates more moving charge per unit time, Figure 9–18b. These are correct assumptions. Furthermore, the larger the cross-sectional area of the coil, the greater the strength, Figure 9–18c.

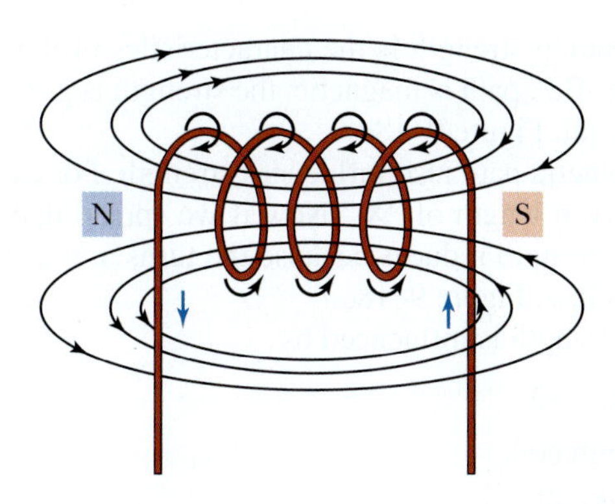

FIGURE 9–15 Strong field of coil and establishment of magnetic polarity

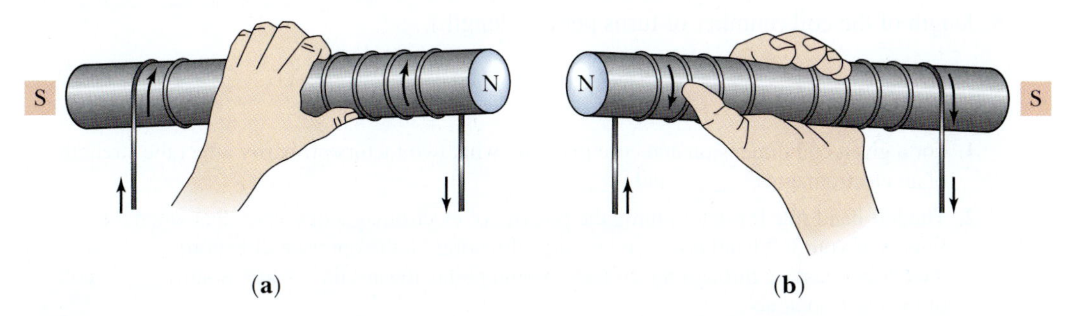

FIGURE 9–16 Left-hand rule to determine magnetic polarity of an electromagnet

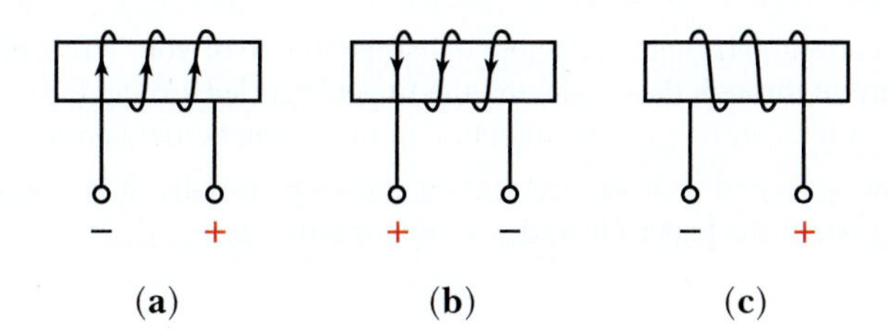

FIGURE 9–17

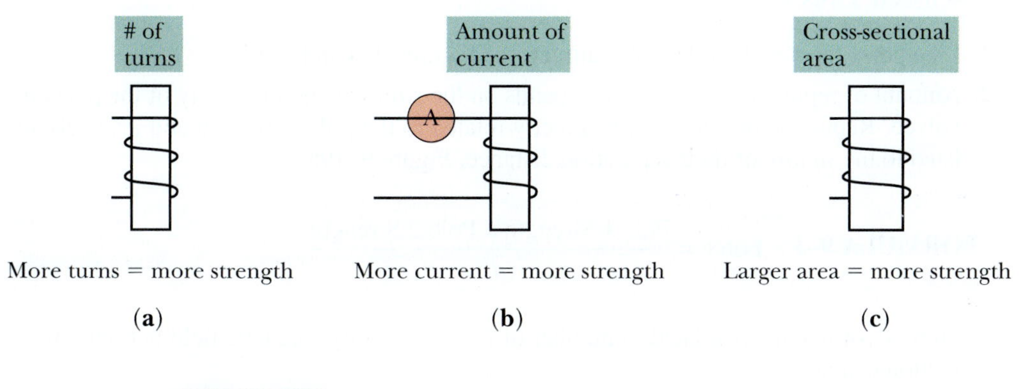

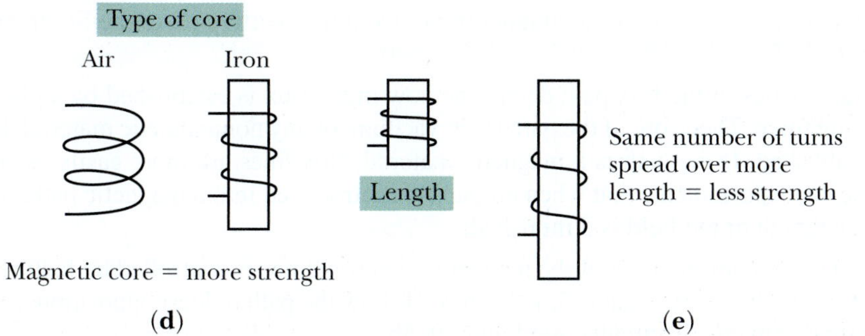

FIGURE 9–18 Some factors that determine strength of an electromagnet

Another factor influencing strength is the characteristics of the material in the flux-line path. If the material in the flux path is magnetic, the strength is greater compared with a path using nonmagnetic material, Figure 9–18d.

The length of the magnetic path also influences strength. For example, suppose we have a 100-turn coil with a given length of "X." Now if we spread that 100 turns over a length twice as great (2X), the strength reduces because the turns are farther apart, thus the effective magnetizing force is less, Figure 9–18e.

To summarize, field strength is influenced by:

1. number of turns on the coil;
2. value of current through coil;
3. cross-sectional area of coil;
4. type of core (magnetic characteristics of material in flux path); and
5. length of the coil (number of turns per unit length).

IN-PROCESS LEARNING CHECK 2

1. For a given coil dimension and core material, what two factors primarily affect the strength of an electromagnet? _____ and _____
2. The left-hand rule for determining the polarity of electromagnetics states that when the fingers of your left hand point (in the same direction, in the opposite direction) _____ as the current passing through the coil, the thumb points toward the (North, South) _____ pole of the electromagnet.
3. Adjacent current-carrying conductors that are carrying current in the same direction tend to (attract each other, repel each other) _____.
4. If you grasp a current-carrying conductor so that the thumb of your left hand is in the direction of current through the conductor, the fingers "curled around the conductor" (will indicate, will not indicate) _____ the direction of the magnetic field around the conductor.
5. When representing an end view of a current conductor pictorially, it is common to show current coming out of the paper via a (dot, cross, or plus sign) _____.

Highlighting Magnetism and Electromagnetism Facts Discussed Thus Far

1. Like poles repel and unlike poles attract, see Figures 9–4 and 9–19a.
2. Amount of repulsion or attraction depends on the strength and proximity of the poles involved. Repulsion or attraction is directly related to the poles' strength and inversely related to the *square* of their separation distance, Figure 9–19a.

FORMULA 9–1 $$\text{Force} = \frac{\text{Pole 1 Strength} \times \text{Pole 2 Strength}}{d^2}$$

3. Strength of magnet is related to number of flux lines in its magnetic field per unit cross-sectional area.
4. Magnetic lines outside of the magnet travel from the North pole to the South pole, see Figures 9–5a and 9–19a.
5. Characteristics of the flux path determine how much flux is established by a given magnetizing force. Therefore, if the path is air, vacuum, or any nonmagnetic material, less flux is established. If the path is a magnetic material, flux lines are more easily established. These concepts indicate that when air gaps are introduced into a magnetic path, the magnetic strength of the field is diminished.
6. The shape or configuration of the magnetic field flux lines is controlled by features of the materials in their path. That is, lines of force follow the path of least opposition, called **reluctance** in magnetic circuits, see Figure 9–5b.

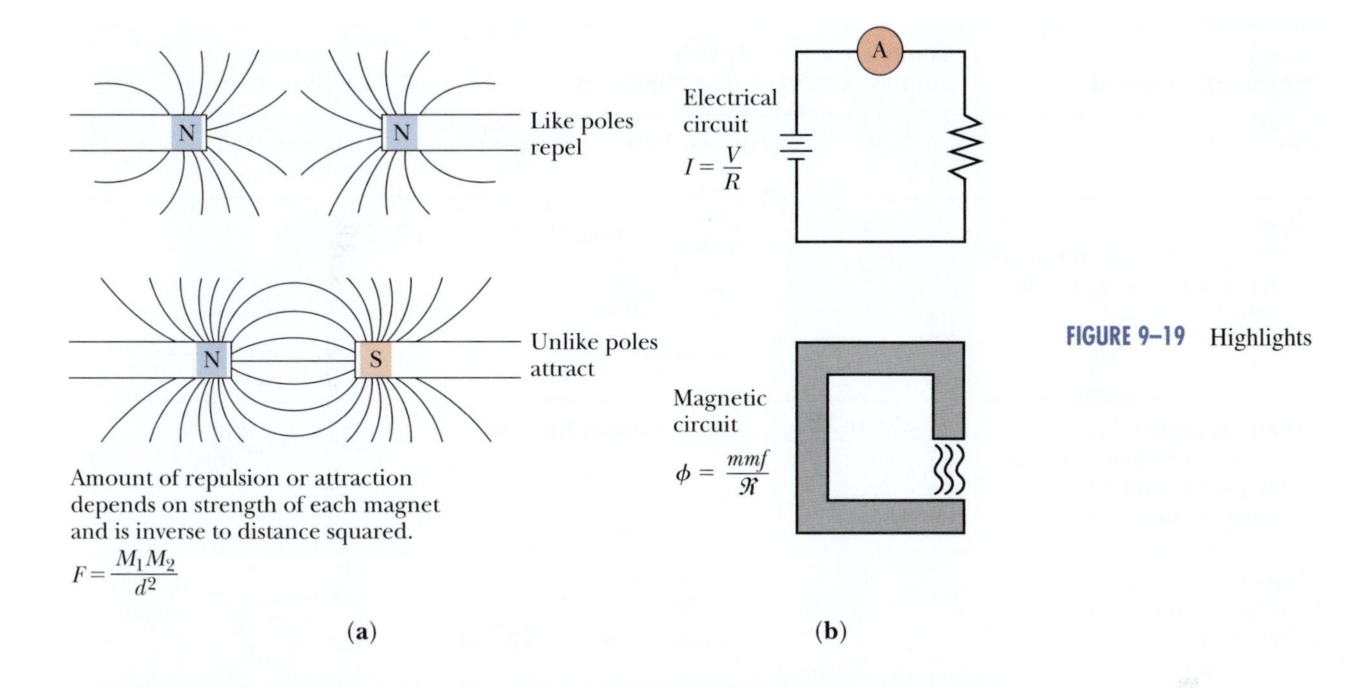
N
N
Like poles repel
N
S
Unlike poles attract
Amount of repulsion or attraction depends on strength of each magnet and is inverse to distance squared.
$F = \frac{M_1 M_2}{d^2}$
(a)
A
Electrical circuit
$I = \frac{V}{R}$
Magnetic circuit
$\phi = \frac{mmf}{\Re}$
(b)

FIGURE 9–19 Highlights

7. An analogy can be drawn between Ohm's law in electrical circuits and the relationships of a magnetic circuit. In Ohm's law, the amount of current through a circuit is directly related to the amount of electromotive force (V) applied to the circuit and inversely related to the circuit's opposition to current flow (R). In a magnetic circuit, the number of flux lines through the magnetic circuit is directly related to the value of the magnetizing force establishing flux and inversely related to the circuit's opposition to those flux lines. This opposition is the reluctance of the path, and the symbol is $\Re$, Figure 9–19b.

9–4 Important Magnetic Units, Terms, Symbols, and Formulas

Now, let's look at some units of measure for the magnetic characteristics discussed earlier. Recall that a single flux line (or magnetic line of force) is a maxwell, and 10^8 flux lines equal a Weber (Wb). Also, although magnetizing force was presented, we did not specifically define it.

Before briefly discussing magnetic units, some background about systems of units would be helpful. Most of us daily use units of measure pertaining to dimensions, volume, weight, speed, and so on. In the field of electronics, it is critical that some meaningful system of units exists.

Before 1960 there were several systems of units related to electricity, electronics, and magnetism. One system was the common unit of the volt, ampere, and watt; another system dealt with electrostatics; and a third system explained magnetics. Both electrostatics and magnetics used a system called the *cgs* system, where *c* was distance in centimeters, *g* was weight in grams, and *s* was time in seconds.

However, since the 1960s, a system has been accepted that uses the *Meter* for distance, the *Kilogram* for weight, the *Second* for time, and the *Ampere* (necessary for electrical systems) for current measure. This system is abbreviated as the *MKSA*. Since this system has been adopted internationally, it is called the International System, abbreviated SI, *Système internationale d'unités.*

The Basic Magnetic Circuit

Because technicians rarely have to deal with the systems of units presented in Figure 9–20, we simply call your attention to the chart as a reference, or source of information. It is good that you look over the chart so that if these terms should arise later, you will have been exposed to them.

TERM OR QUANTITY	SYMBOL OR ABBREVIATION	SI UNIT AND FORMULA	CGS UNIT AND FORMULA
Flux (lines)	ϕ	weber (Wb) $= \frac{\text{number lines}}{10^8}$	Maxwell (Mx) number lines in field
Flux Density (Magnetic flux per unit cross-sectional area at right angles to the flux lines)	B	$\frac{\text{webers}}{\text{sq meter}}$ = **tesla** (T) $B = \frac{\phi\ (\text{mks})}{A\ (\text{mks})}$ or teslas $= \frac{\text{Wb}}{\text{sq mtr}}$	$\frac{\text{lines}}{\text{sq cm}}$ $B = \frac{\phi\ (\text{cgs})}{A\ (\text{cgs})}$ or gauss $= \frac{\text{Mx}}{\text{sq cm}}$
Magnetomotive Force (That which forces magnetic lines of force through a magnetic circuit)	MMF or F_{mm}	Ampere-turns or $AT = NI$	F (cgs) $= 0.4\pi\ NI$ or F (gilberts) $= 1.26 \times N \times I$
Magnetic Field Intensity (Magnetomotive force per unit length)	H	Ampere-turns per meter $\frac{NI}{\text{length}}$ or $\frac{AT}{\text{meter}}$ or $\frac{F_{mm}}{\text{length}}$	H (oersteds) $= \frac{0.4\pi\ NI}{\text{length}}$ or $\frac{1.26\ NI}{1\ \text{cm}}$
Permeability (Ability of a material to pass, conduct, or concentrate magnetic flux; analogous to conductance in electrical circuits), i.e., the ease of establishing magnetic flux through the material.	μ	Webers per ampere-turn per meter $\mu = \frac{l}{\Re A}$ where length (l) is length in meters reluctance ($\Re$) is ampere-turns per weber area ***(A) is cross-sectional area in square meters*** Note: Free space, or vacuum permeability (μ_o) is considered to be: $4\pi \times 10^{-7}$ or 12.57×10^{-7}	Free space: $\mu_o = B/H$. B is in gauss; H is in oersteds; In cgs system, $\mu_o = 1$
Relative Permeability (Not constant because it varies with the degree of magnetization)	μ_r	Relative permeability of a material is a ratio. Thus, $\mu_r = \frac{\text{Flux density with core material}}{\text{Flux density with vacuum core}}$ where, flux density in the core material is: $B = \mu_o\mu_r H$ teslas, and absolute permeability of core materials is: $\mu = B/H = \mu_o\mu_r$ (SI units)	Same concept in both systems
Reluctance Opposition to the establishment of magnetic flux	$\Re$	**Ampere turns** per weber $\Re = \frac{F_{mm}}{\phi}$	Same concept in both systems

FIGURE 9–20 Magnetic units, symbols, and formulas chart

Additionally, you should briefly look at Figure 9–21 to see the analogies that might be drawn between Ohm's law for electrical circuits, and Rowland's law for magnetic circuits.

Relationships among *B, H,* and Permeability Factors

From the charts in our Figures, it is possible to observe that H (magnetic field intensity or mmf per unit length) is the factor producing flux per unit area (or flux density, B) within a given magnetic medium.

OHM'S LAW	ROWLAND'S LAW	EXPLANATION
$R = \frac{V}{I}$, where: Resistance in ohms equals electromotive force in volts divided by current in amperes.	$\mathcal{R} = \frac{F(mmf)}{\phi}$, where: Reluctance (the magnetic circuit's opposition to the establishment of flux) equals the magnetomotive force (F) divided by the flux (ϕ). Also, it can be shown that: $\mathcal{R} = \frac{l}{\mu A}$ where l = length of the path in meters μ = permeability of the path A = cross-sectional area of the path perpendicular to flux, in meters	Reluctance equals magnetomotive force divided by flux. Flux equals magnetomotive force divided by reluctance. $F(mmf) = \phi \times \mathcal{R}$ or flux times reluctance.

FIGURE 9–21 Rowland's law

It is further observed that the amount of B a given H produces is directly related to the permeability of the flux path (and inversely related to the path's reluctance).

EXAMPLES ***Problem 1:*** What is the **magnetomotive force,** if a current of 5 amperes passes through a coil of 100 turns?

Answer:

FORMULA 9–2 $mmf = NI$

Thus, $mmf = 100 \times 5 = 500$ ampere-turns.

Problem 2: If the length of the coil in problem 1 is 0.1 meters, what is the **magnetic field intensity?**

Answer:

FORMULA 9–3 $H = \frac{AT}{\text{meter}}$

Thus, $H = \frac{500}{0.1} = 5{,}000$ ampere-turns per meter.

9–5 Practical Considerations about Core Materials

Since many of the components and devices you will be studying use principles related to electromagnetism, we will now consider the practical aspects of the core materials used in these components. You know that for a given magnetizing force, a coil's magnetic field strength is weaker if its core is air than if its core has a ferromagnetic substance, Figure 9-22. Also, we have indicated that certain magnetic properties of a material are not constant but vary with degrees of magnetization. Therefore, there is not a constant relationship between the magnetizing force field intensity (H) and the flux density (B) produced.

9–6 The *B-H* Curve

Manufacturers of magnetic materials often provide data about their materials to help select materials for a given purpose. One presentation frequently used is a B-H curve, or a magnetization curve, Figure 9–23. Notice in Figure 9–23 that as the magnetizing intensity (H)

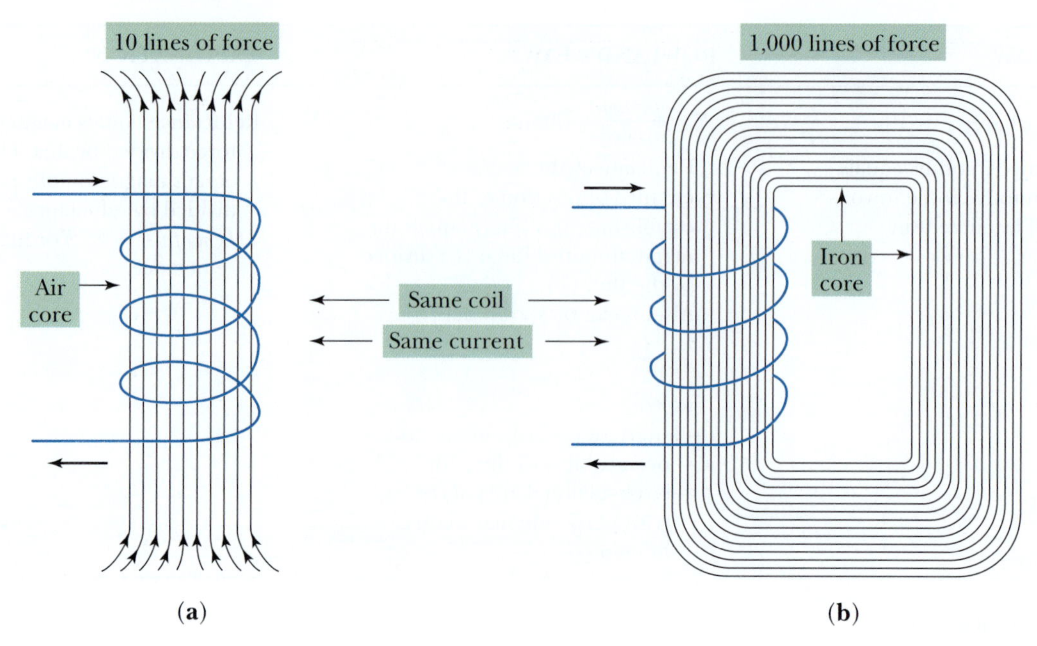

FIGURE 9–22 Effects of core material(s)

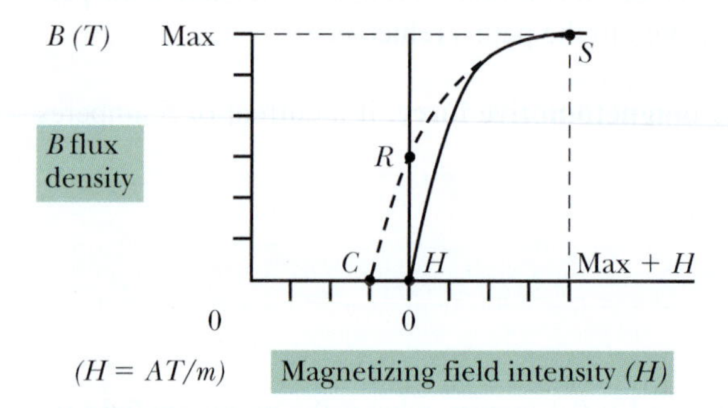

FIGURE 9–23 *B-H* curve

increases, the flux density (*B*) increases but not in a linear fashion. This graph provides several important points.

1. During magnetization (the solid line), a point is reached when increasing *H* (by increasing current through the coil) does not cause a further significant flux density increase. This is due to the core reaching "saturation," represented at point S on the graph. At this point, it cannot contain more flux lines.
2. During demagnetization (with current decreasing to zero through the coil), the dotted line indicates some other interesting characteristics:
   a. When the *H* reaches zero (no current through coil), some flux density (*B*) remains in the core due to **residual** or "remanent" **magnetism.** The amount is indicated by the vertical distance from the graph baseline to point *R*.
   b. To remove the residual magnetism and reduce the magnetism in the core to zero, it takes a magnetizing intensity in the opposite direction equal to that represented horizontally from point 0 to point C. This value indicates the amount of "coercive" force required to remove the core's residual magnetism.

An example of how the *B-H* curve is used to illustrate important magnetic features of different materials is shown in Figure 9–24. For example, notice different materials saturate at different levels of flux density (*B*). **Saturation** occurs when a further increase in *H* does not result in an appreciable increase in *B*.

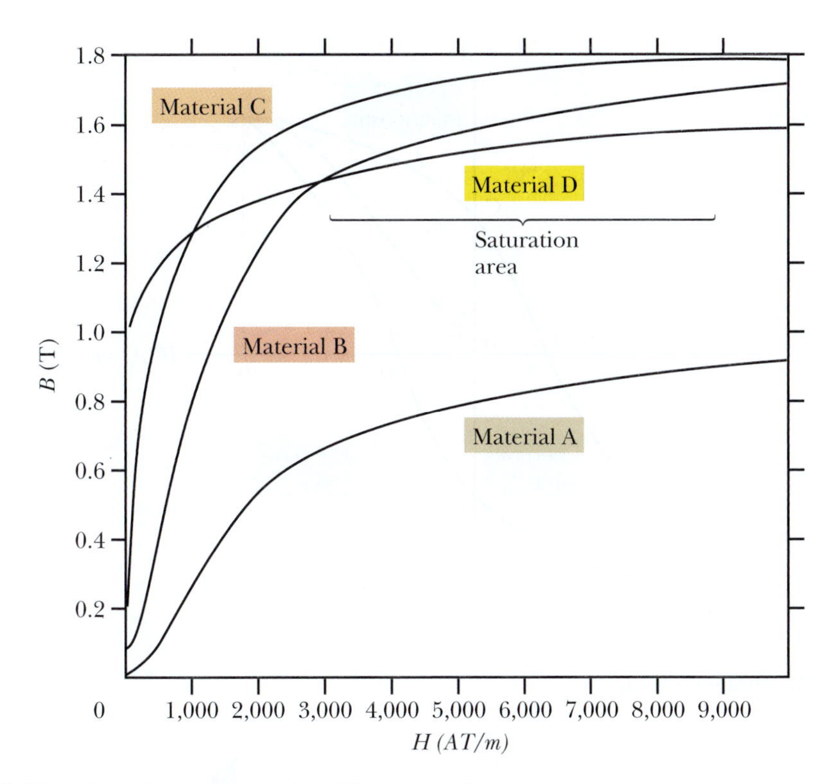

FIGURE 9–24 Magnetization curves for different metals

9–7 The Hysteresis Loop

Another useful graph is the hysteresis loop. It portrays what happens when a core is saturated with current in one direction through its coil, and then the current direction is reversed and increased until the core is saturated in the opposite polarity. If this is continually repeated, a hysteresis loop graph, similar to that in Figure 9–25, can be created that enables meaningful analysis of the material in question. This is typically done by application of an alternating current source to the coil. NOTE: "Hysteresis" means "lagging behind." As this graph demonstrates, the flux buildup and decay lags behind the changes in the magnetizing force.

The following important points are provided by this graph:

1. Points A to B represent the initial magnetization curve from zero *H* until the core reaches saturation.
2. Points B to C represent the flux density decay as *H* changes from its maximum level back to zero. The vertical distance from A to C is the amount of residual magnetism in the core.
3. Points C to D represent the flux density decay back to zero as the opposite polarity *H* is applied. The horizontal distance from A to D represents the coercive force.
4. Points D to E represent the buildup of flux density to core saturation in the opposite polarity from the initial magnetization. Of course, the total horizontal distance from point A to I shows the *H* required for this saturation. The vertical axis of the graph shows the amount of flux density (*B*) involved.
5. Points E to F show decay of flux density as *H* changes from its maximum level of this polarity back to zero. The residual magnetism is represented from points A to F.
6. Points F to G illustrate the decay of flux density back to zero due to the coercive force of the reversed *H* overcoming the residual magnetism. The coercive force is from points A to G.
7. Points G back to B complete the loop, displaying the increase of *B* to saturation in the original direction. For this segment of the loop, the vertical distance from point B to H is the flux density value at saturation.

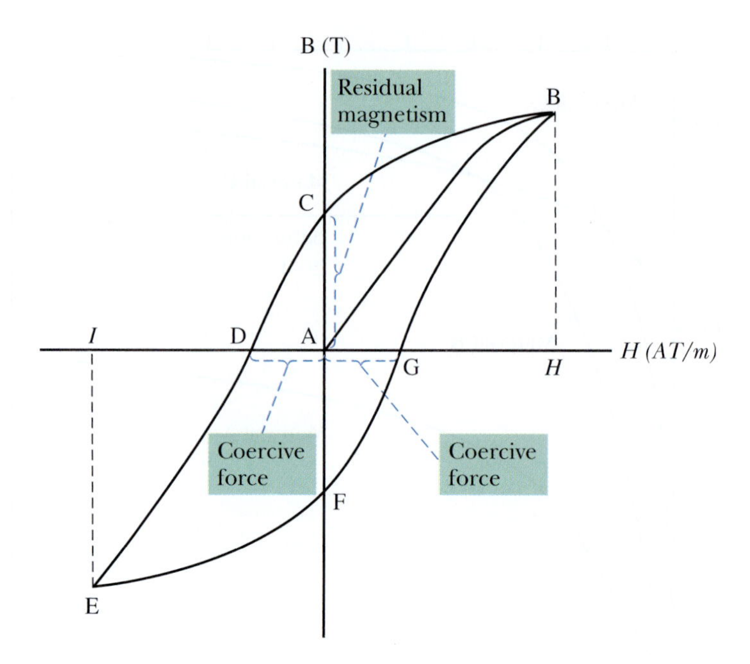

FIGURE 9–25 Hysteresis loop

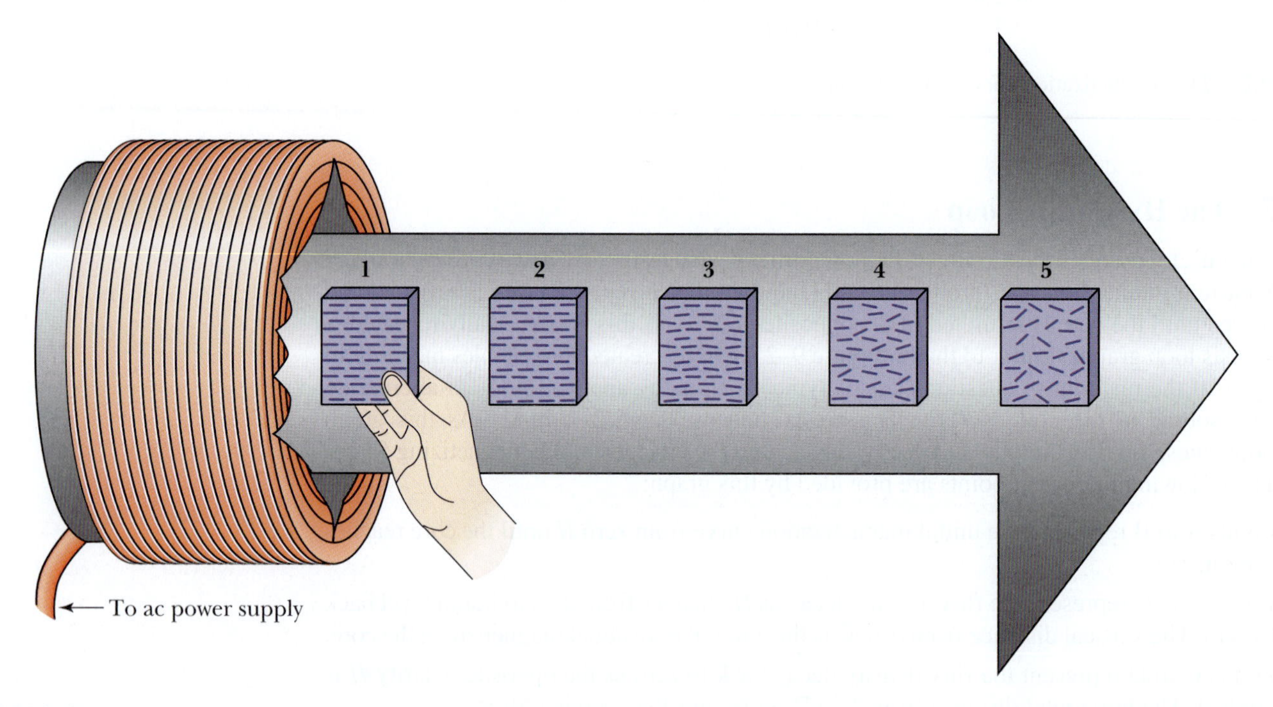

FIGURE 9–26 Example of demagnetizing coil

8. The area inside the hysteresis loop represents the core losses from this cycle of magnetizing, demagnetizing, magnetizing in the opposite polarity, and demagnetizing the core. These losses are evidenced by the heating of the core as a result of the "magnetic domains" continuously having to be realigned with alternating current (ac) applied to the coil. We will discuss core losses in greater detail at a later point in this book.

Incidentally, you may wonder how material that has been magnetized can be demagnetized. One method is to put the material through this hysteresis loop cycle while gradually decreasing the H swing. The result would be shown as a shrinking area under the hysteresis loop. One practical way to accomplish a decrease in H swing is to use a large ac demagnetizing coil so the material is placed within the coil. Gradually pull the material away from the ac energized coil, Figure 9–26. In the past, you may have seen a TV technician use a "degaussing" coil on color picture tubes to demagnetize them. A color picture tube that has become magnetized will have a degraded picture. Demagnetizing it usually helps.

IN-PROCESS LEARNING CHECK 3

1. A *B-H* curve is also known as a _____ curve.
2. *B* stands for flux _____.
3. *H* stands for magnetizing _____.
4. The point where increasing current through a coil causes no further significant increase in flux density is called _____.
5. The larger the area inside a "hysteresis loop" the (smaller, larger) _____ the magnetic losses represented.

9–8 Induction and Related Effects

Motor Action

When talking about motor action, we refer here to converting electrical energy to mechanical energy. You have studied the law that states lines of force in the same direction repel each other, and lines of force in opposite directions attract and cancel each other. One practical application of this phenomenon is motor action. The simplest example of this is a current-carrying conductor placed in a magnetic-field environment.

Look at Figure 9–27 and note the following:

1. The direction of the magnet's flux lines—North to South.
2. The direction of the current-carrying conductor's flux lines. (Use the left-hand rule for current-carrying conductors.)
3. The interaction between the conductor's flux lines and the magnet's lines of force is such that the lines at the top of the conductor are repelled (lines in the same direction repel); the lines at the bottom of the conductor are attracted (lines in opposite directions attract); and the field is weakened at the bottom. Therefore, the conductor will be pushed down. In other words, the lines at the top aid each other and the lines at the bottom cancel each other.
4. The amount of force that pushes the conductor down depends on the strength of both fields. The higher the current through the conductor, the stronger its field. And the higher the flux density of the magnet, the stronger its field.
5. Another method to cause stronger motor action is to make the single conductor into a coil (in the form of an armature) with many turns, thus increasing the ampere-turns and the strength of its field, Figure 9–28.

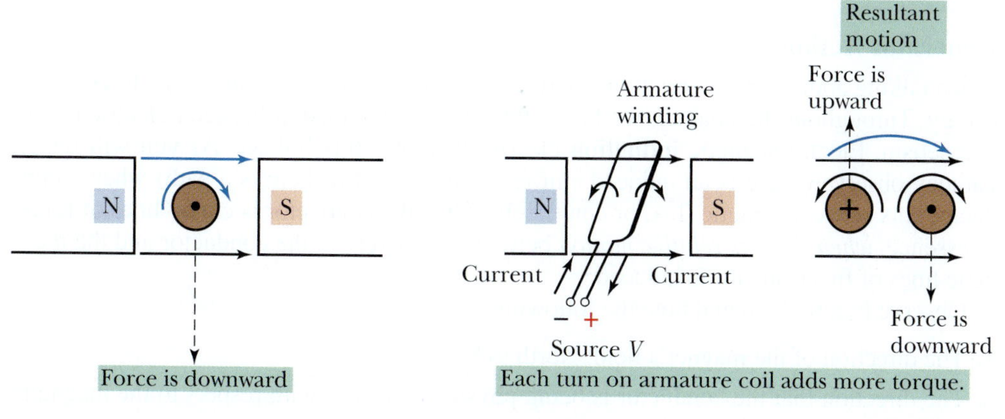

FIGURE 9–27 Motor action on a single conductor

FIGURE 9–28 Concept of stronger motor action with an armature coil

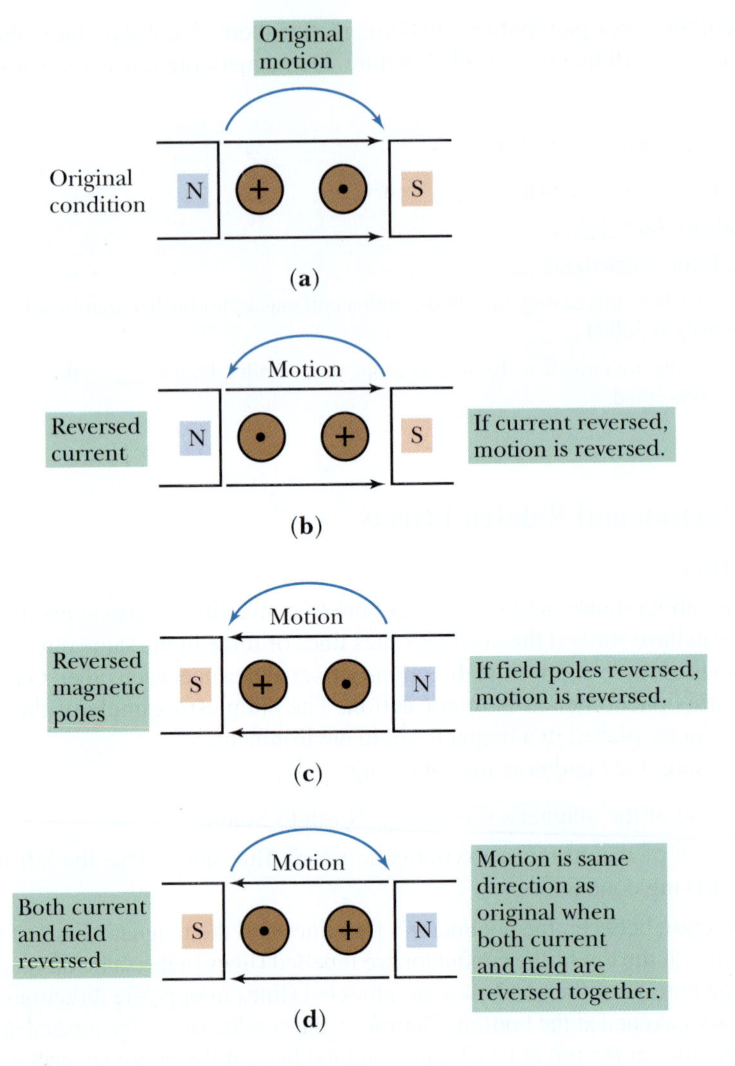

FIGURE 9–29 Effects of changing direction of current or field

6. Refer to Figure 9–29b and c. When the current through the conductor is reversed or the magnetic poles of the magnet are switched, the force (motor action) is in the opposite direction of that in Figure 9–29a. When both current and poles are simultaneously reversed, Figure 9–29d, the force remains in the original direction, prior to being reversed, as that shown in Figure 9–29a.

Generator Action

When talking about generator action, we refer to converting mechanical energy to electrical energy. Throughout the chapter, we have alluded to a relationship between electricity and magnetism. Electromagnetic **induction** clearly illustrates this linkage. As you will recall, Faraday observed there is an induced emf (causing current if there is a path) when a conductor cuts across magnetic flux, or conversely, when the conductor is cut by lines of force. In essence, *when there is relative motion* between the two (i.e., the conductor and the magnetic lines of flux), an emf is induced.

Observe Figure 9–30 and note the following:

1. The direction of the magnet's field: North to South.
2. The direction that the conductor is being physically moved with respect to the magnetic field of the magnet: Upward.
3. The fact that when the conductor cuts the field in this direction, the direction of induced current is toward you, Figure 9–30a. In fact, the left-hand rule for current-carrying con-

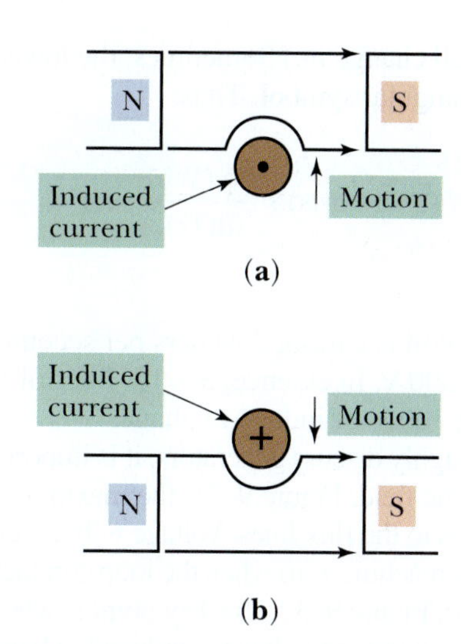

FIGURE 9–30 Converting mechanical energy into electrical energy

ductors (studied earlier) would indicate the direction of current, as shown. For example, if you aim your fingers in the direction of the flux (N to S) and bend your fingers in the same direction the magnet's field is bent, then the induced current is in the same direction as your extended thumb.

4. If the conductor is moved down into the field, rather than up, the induced current is in the opposite direction, Figure 9–30b. (Try the left-hand rule again!)
5. The faster we move the conductor relative to the field, the higher the induced current (and/or emf), since more lines are being cut per unit time.

You may be wondering how cutting flux lines with a conductor causes induced emf, thus producing current flow (assuming a closed current path). Conceptually, it is briefly explained by indicating the magnetic fields involved with each of the electrons in the wire are caused to align, causing electron movement, which is current flow. When the conductor is open (with no complete path for current flow), one end of the conductor will be caused to have an excess of electrons (negatively charged); the other end has a deficiency of electrons (positively charged). Thus, we have an induced emf. If a closed path for current is provided, this induced emf causes current to flow.

What factors affect the amount of induced voltage? The key factors are:

1. amount of flux;
2. number of turns linked by the flux;
3. angle of cutting the flux; and
4. rate of relative motion.

All of these factors relate to the number of flux lines cut per unit time.

9–9 Faraday's Law

Faraday's law states *the amount of induced emf depends on the rate of cutting the flux (ϕ).* The formula is:

FORMULA 9–4 $V_{ind} = \dfrac{\Delta\phi}{\Delta T}$

where, V_{ind} is the induced emf, and $\Delta\phi/\Delta T$ indicates the *rate* of cutting the flux. Using the SI system, if a single conductor cuts one weber of flux (10^8 lines) in one second, the induced voltage is 1 volt, a two-turn coil would have 2 volts induced, and so forth. (NOTE: The symbol Δ is the fourth letter in the Greek alphabet called delta, which, for our purposes, represents the amount of change in, or a small change in.) Sometimes, the lowercase letter "d" is used to indicate delta rather than the triangular symbol. Thus:

FORMULA 9–5 $V_{ind} = N\,(\text{\# of turns}) \times \dfrac{d\phi\ (\text{Wb})}{dt\ (\text{s})}$

EXAMPLE If a 100-turn coil is cutting 3 webers per second ($d\phi/dt$ in formula), then the induced voltage is $100 \times 3 = 300$ V. In essence, 3×10^8 lines of flux are cutting the 100-turn coil each second, resulting in 300 V of induced voltage.

Although we will not thoroughly discuss generators, it is important you know that if a loop of wire is rotated within a magnetic field, Figure 9–31, the maximum voltage is induced when the conductor(s) cuts at right angles to the flux lines. Voltage will incrementally decrease as the loop rotates at less than right angles, reaching zero when the loop conductor moves parallel to the lines of flux (in the vertical position), Figure 9–32. The key point is when the conductor moves parallel to the lines of flux (an angle of 0°), no voltage is induced; when the conductor moves at right angles to the flux lines, maximum voltage is induced; and when the conductor moves at angles between 0° and 90°, some value between zero and the maximum value is induced.

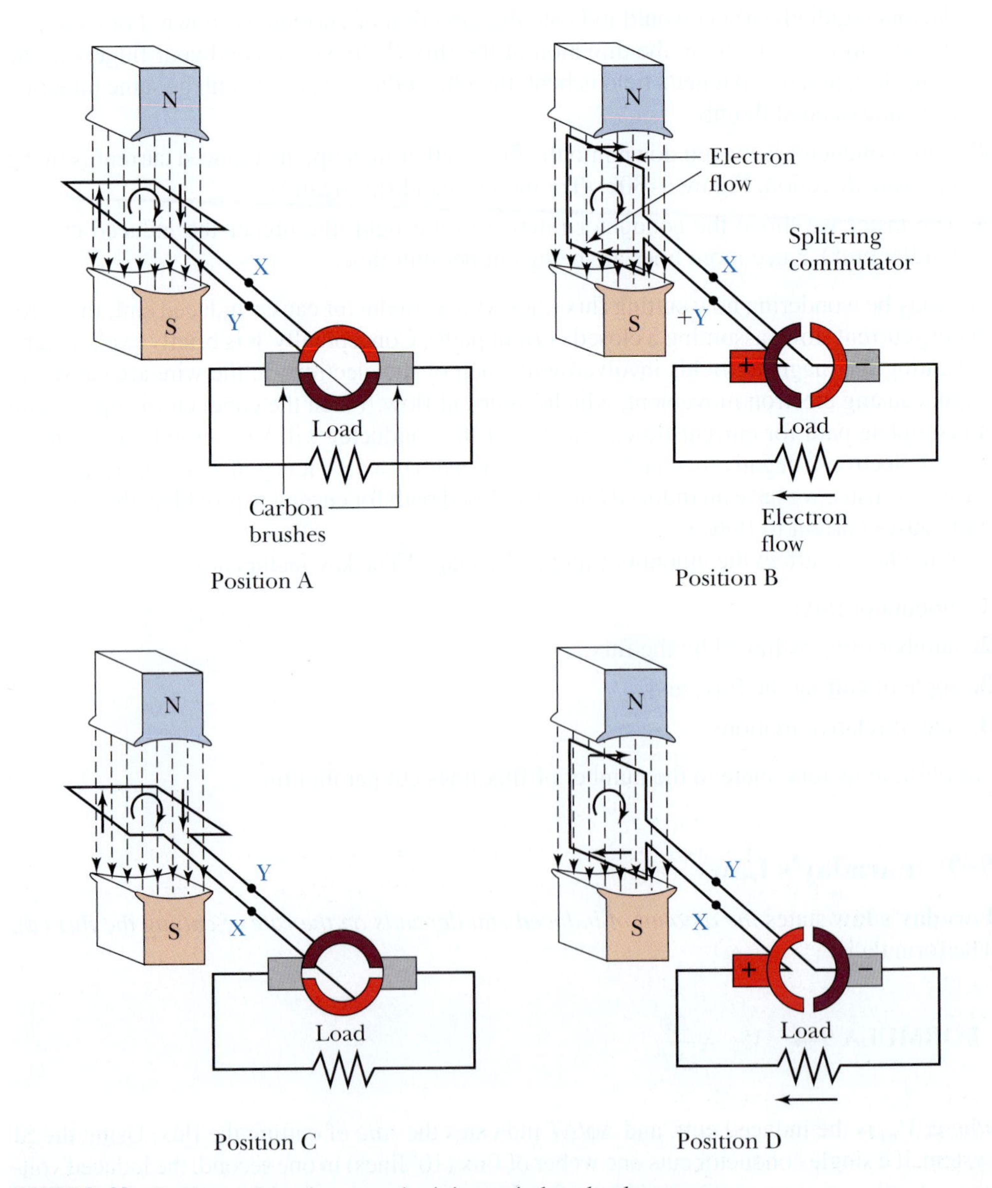

FIGURE 9–31 Regions of maximum and minimum induced voltage

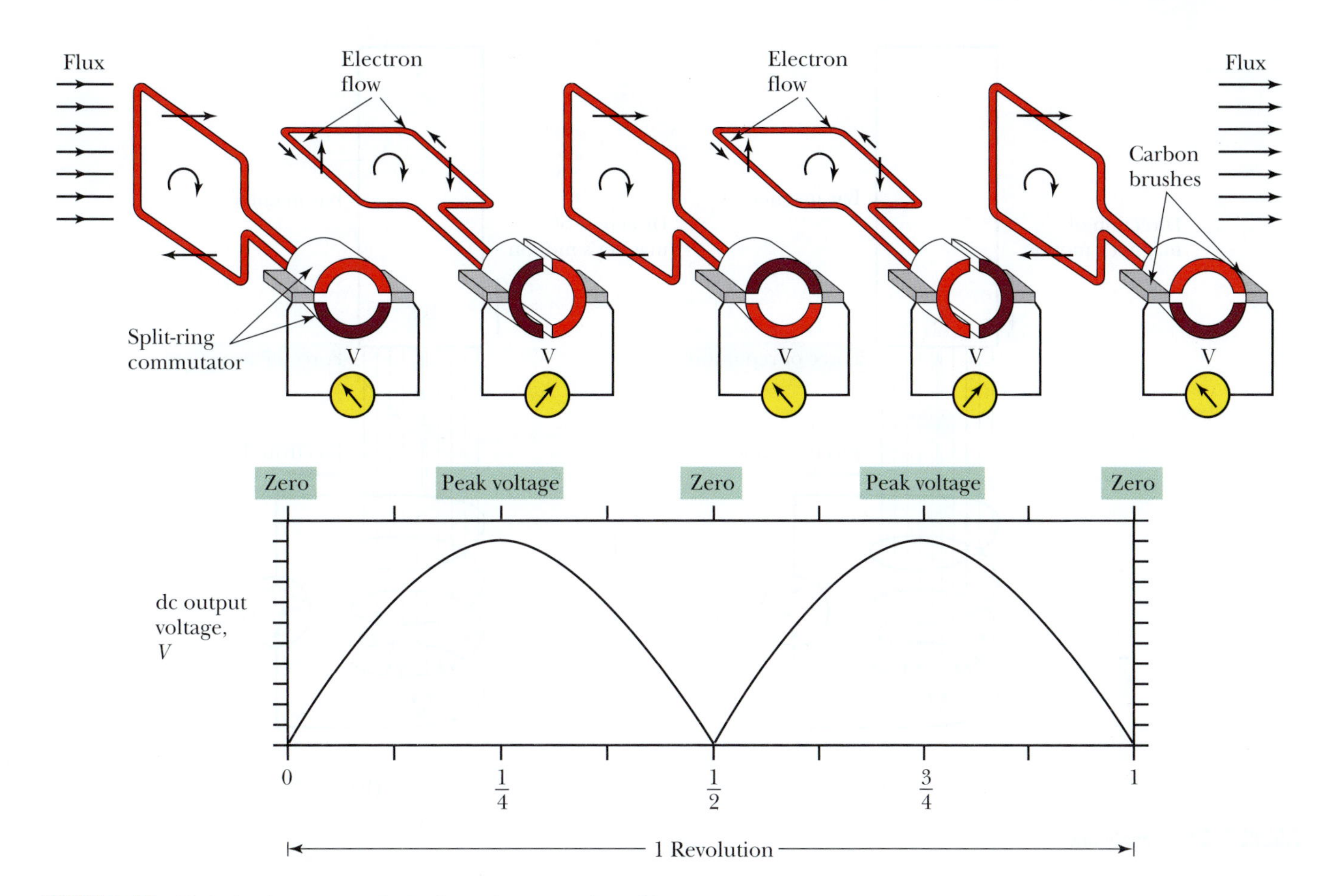

FIGURE 9–32 Variation in output voltage through one rotation of loop

9–10 Lenz's Law and Reciprocal Effects of Motors and Generators

Lenz's Law

As you have already noted, the law of energy conservation is that you do not get something for nothing! In the case of motor action in the generator, to generate electricity with the generator we have to supply mechanical energy that turns the armature shaft and overcomes the force of the opposing motor effect that results from the induced current and its resultant field(s).

Lenz's law states *the direction of an induced voltage, or current, is such that it tends to oppose the change that caused it.* Another way of stating this law is when there is a change in the flux linking a circuit, an emf/current is induced that establishes a field that opposes the change.

Look at Figure 9–33 and use the left-hand rules for current-carrying conductors *and* for electromagnets. Notice when the magnet is moved *down* in the coil, the induced current establishes a North pole at the top end of the coil, facing (and repelling) the North pole of the magnet, whose flux linking the coil caused the original induced current. In other words, as we force the magnet down into the coil, the induced current produces a field to oppose the motion, as stated by Lenz. Conversely, when we pull the magnet out of the coil, the induced current is in the reverse direction, establishing a South pole at the top of the coil and tending to oppose retraction of the magnet. Again, the magnetic polarity can be defined in accordance with Lenz's law.

Motor Effect in a Generator

The generator converts mechanical energy (armature shaft rotation) into electrical energy (generator's output voltage). As you know, when the generator's conductors (armature windings) move through the magnetic field of the generator's field magnets, voltage is induced

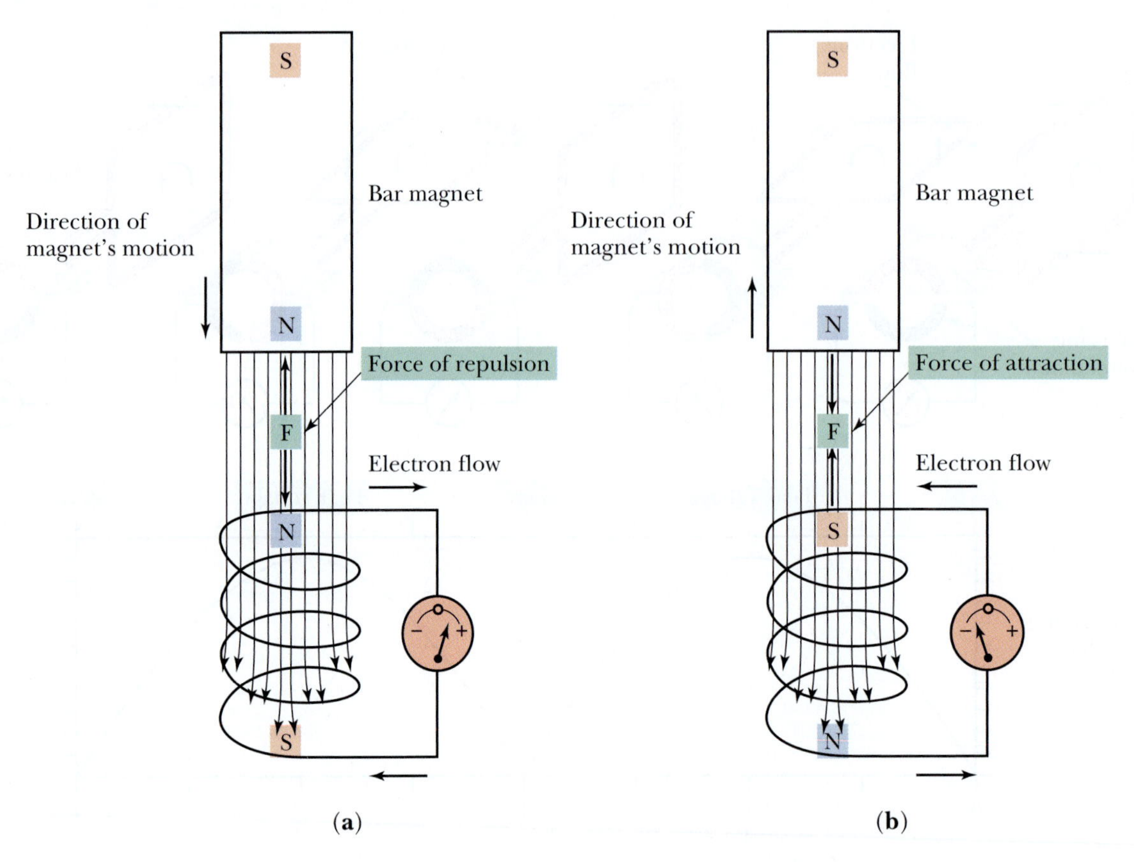

FIGURE 9–33 Lenz's law

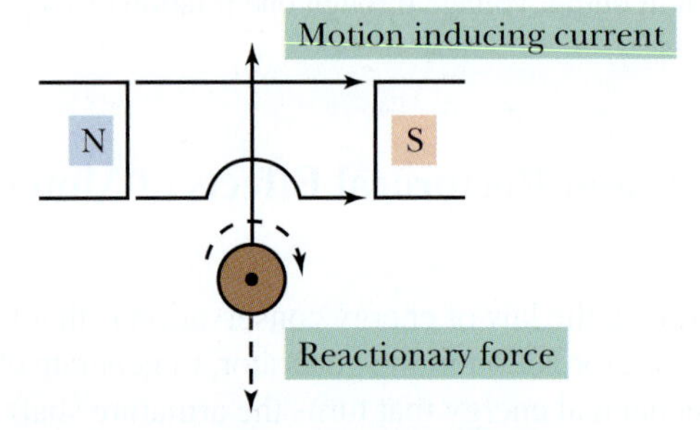

FIGURE 9–34 Motor effect in a generator

according to Faraday's law. This induced voltage causes current through the conductors, assuming a closed path for current. The *direction* of the resulting induced current is such that the magnetic field established around the conductor(s), resulting from the induced current, *opposes* the motion causing it, Figure 9–34.

Generator Effect in a Motor

In Figure 9–35, you can see that the direction of current we are passing through the conductor causes the loop to rotate clockwise because of appropriate motor action. You can also observe that when the conductor rotates clockwise, the wire loop's left side moves up into the

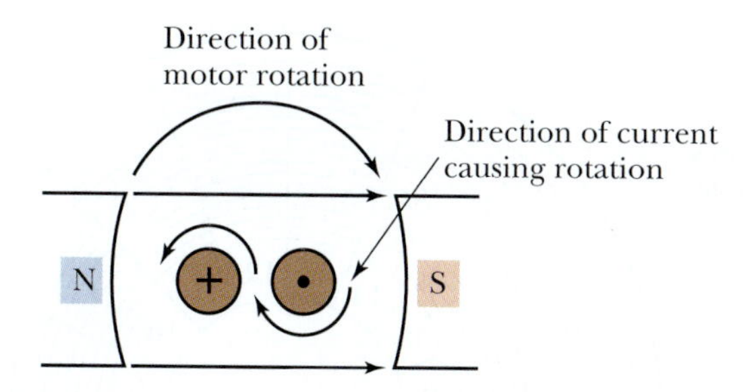

FIGURE 9–35 Generator effect in a motor is such that the induced current due to cutting flux will be in opposite direction of current causing motor to rotate.

FIGURE 9–36 Several magnetic media

flux and the wire loop's right side moves down into the flux. Since we have movement between a conductor and flux lines, there is an induced voltage/current.

This induced current is in the opposite direction from the current that is causing the motor's armature to move clockwise. This phenomenon is often termed *generator effect* in a motor and is related to Lenz's law.

9–11 Some Practical Applications of Magnetism

Magnetic Storage Media

In Figure 9–36 you see several commonly used magnetic storage media elements.

Information on the computer diskettes is stored in structured track and sector formats, Figure 9–37. This allows the computer to selectively access and recover data at any desired location in the "magnetically memorized" data, by virtue of that data's location on the diskette.

On audio cassette tapes, of course, sound waves have been translated into electrical signals at the recording tape head, which in turn causes magnetic patterns to be induced onto the magnetic tape, Figure 9–38a. When the recorded tape is run over the playback head of the recorder, the reverse process happens. That is, the magnetic patterns on the tape induce electrical signals into the pickup head, Figure 9–38b. These signals are then processed, amplified, and eventually become sound waves, which activate your ear drums.

A Good Idea

Don't lay magnetic storage devices, such as computer disks or audio tapes, on or near equipment that may have strong magnetic fields! (Examples of these might be motors, generators, and TVs.)

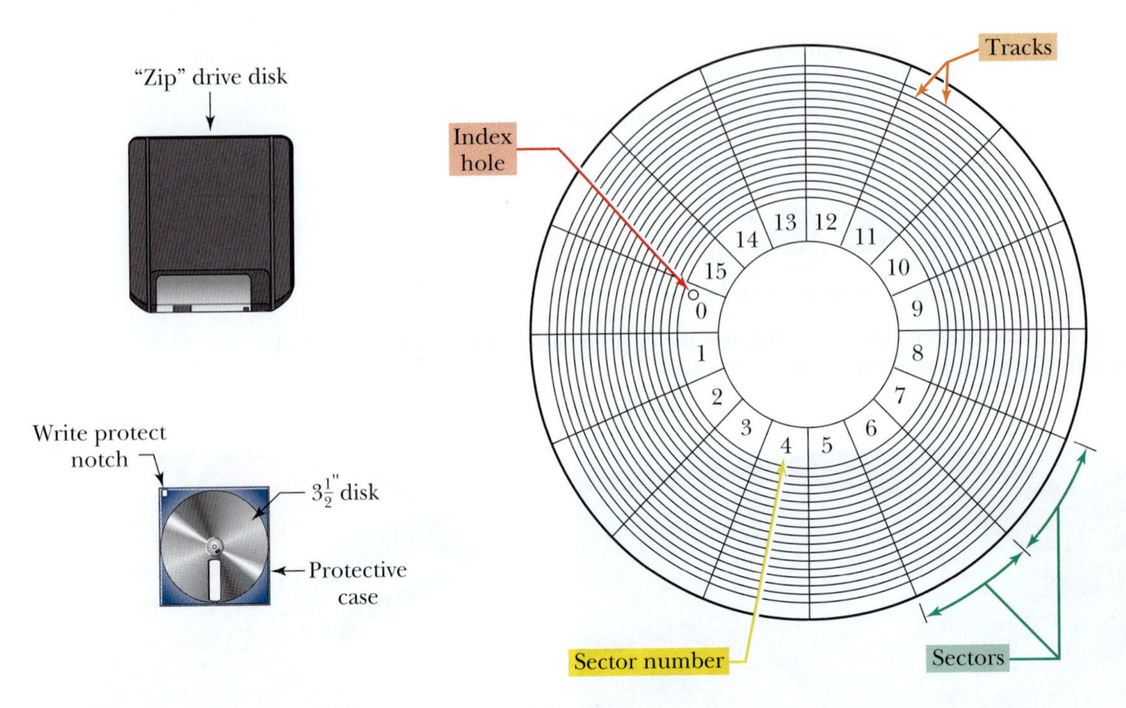

FIGURE 9–37 Computer "floppy" disks typical structure and format

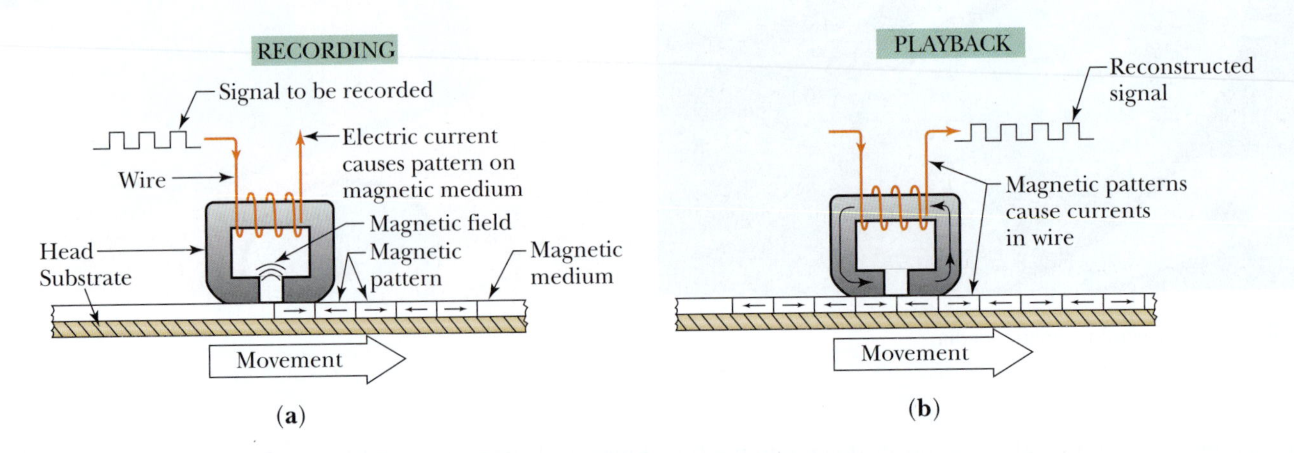

FIGURE 9–38 Using magnetism to record and play back signals (audio, video, or digital data)

Video cassette tape recording systems use a technique similar to, but somewhat more complicated than, the audio recording system. Both audio and video signals are magnetically stored on a videotape's magnetic coating. Once recorded, you can enjoy both seeing and hearing the results by playing them back on your VCR.

Industrial Components and Devices

In industry (and in many home appliances, as well), there are a myriad of applications for relays, solenoids, and magnetic switches, Figure 9–39. Although solid-state switching devices have begun replacing electromechanical relays and solenoids in low-power circuitry, relays and solenoids are still in wide use. The contacts on relays are used to make and break (open and close) circuits as the relay is energized by current flowing through its electromagnetic coil, or is de-energized by current ceasing to flow through its coil, Figure 9–40.

The movable element (plunger, or movable core) in a solenoid is typically used to physically move something. It changes electrical energy into linear, or straight-line, motion,

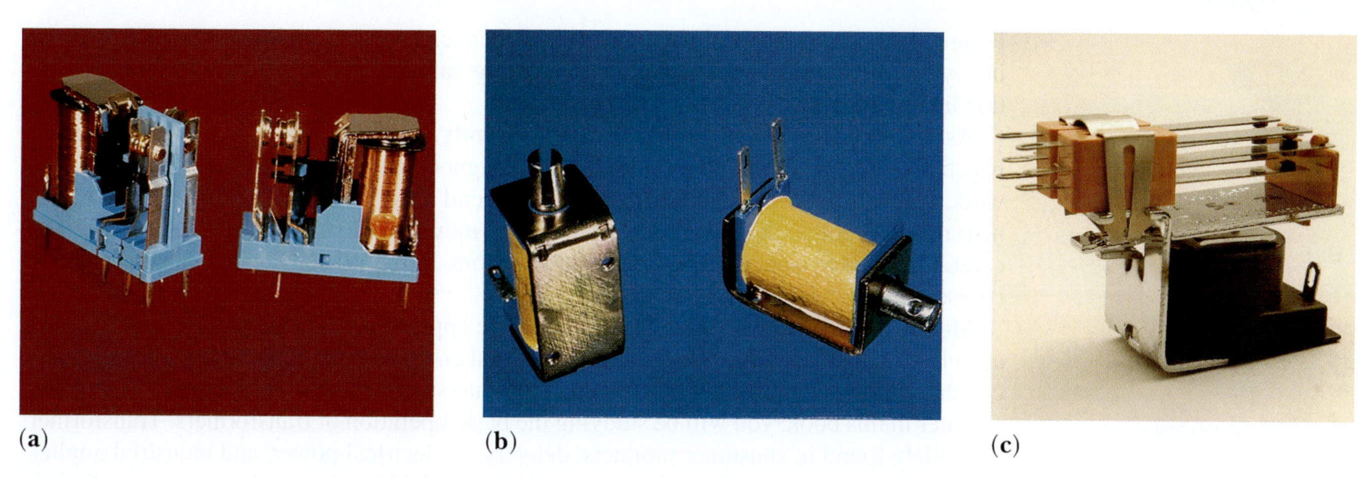

FIGURE 9–39 Common industrial electromagnetic devices *(Photos **a** and **b** courtesy of Guardian Electric Mfg. Co.; photo **c** by Michael A. Gallitelli)*

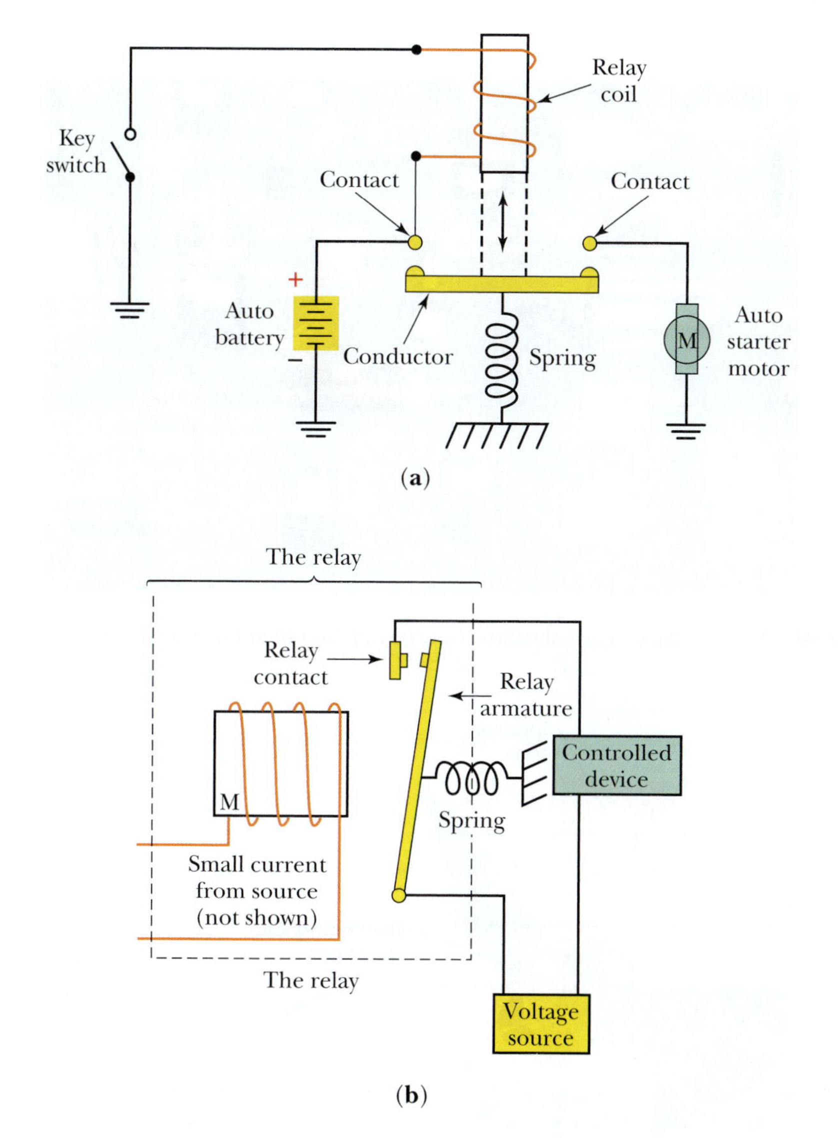

FIGURE 9–40 Simplified diagrams of relays in use: (a) automotive starter system contactor relay; (b) single-pole, single-throw relay operated by low power but controlling higher-power device

Figure 9–41. The travel of this movable core, then, can be used to do everything from shifting cycles in your washing machine, to operating valves, or positioning items on a production line in industry.

As a technician in industry, you will frequently use metering devices or instruments, which also depend on magnetism and electromagnetism for their operation. At home or at work, you will hear speakers. Most speakers depend upon the interaction of a fixed (permanent magnet) and a changing current through a movable electromagnet (the voice coil) to cause the speaker cone to move the air in the room, Figure 9–42. This movement of air, of course, sets up the sound waves you hear.

Motors and generators represent another large application area for the use of magnetism and electromagnetism. A motor changes electrical energy into rotary motion. In contrast, recall that the solenoid changes electrical energy into straight-line motion.

Later in this book, you will be studying the basic operation of transformers. Transformers are widely found in consumer products, delivery of electrical power, and industrial applications. They represent another important application of magnetism and electromagnetism.

In-depth details of all these magnetic devices will be taught at the appropriate times in your studies. For now, we wanted you to see just a few of the applications of permanent and electromagnets and realize how important the knowledge of magnetics is to you as a technician or engineer.

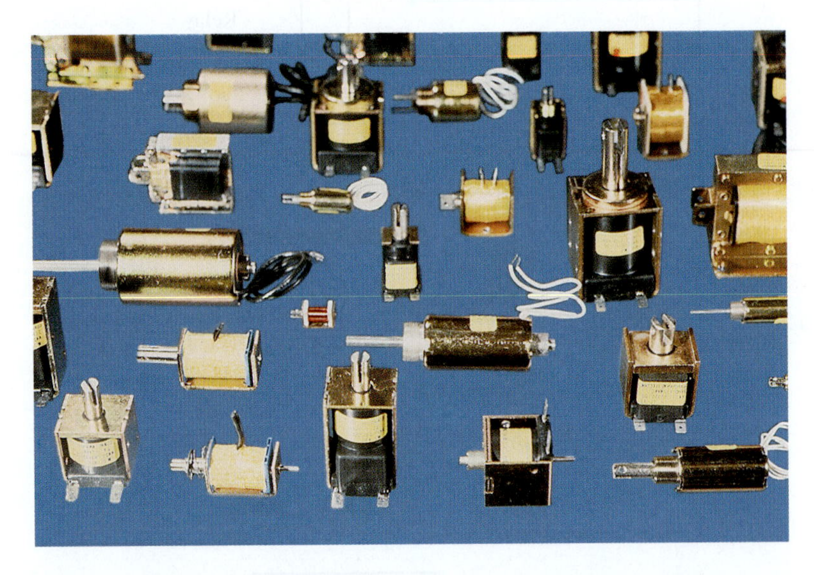

FIGURE 9–41 Several frame-style solenoids *(Courtesy of Guardian Electric Mfg. Co.)*

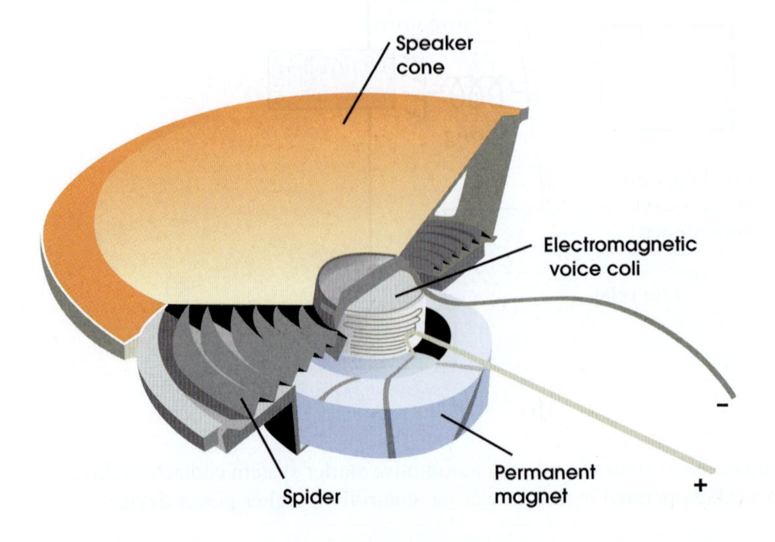

FIGURE 9–42 Permanent magnet dynamic speaker

9–12 Other Topics Related to Magnetism

Several topics not previously discussed are presented next since you now have some general understanding of magnetism. These topics are:

1. the toroidal coil form (closed magnetic path);
2. magnetic shields;
3. shaping fields across a gap via pole pieces;
4. the Hall effect; and
5. special classifications of materials.

Toroidal Coil Form

Notice in Figure 9–43 that the toroidal coil form resembles a doughnut and is made of soft iron, ferromagnetic, or other low-reluctance, high-permeability material. The key feature is this coil effectively confines the lines of flux within itself (because there are no air gaps); therefore, it is a very efficient carrier of flux lines. It allows little "leakage flux" (flux outside the desired path, generally in the surrounding air), thus does not create magnetic effects on nearby objects. At the same time, its internal field is affected very little by other magnetic fields near it.

It is interesting to note that there are no poles, as such, in the toroid configuration. However, if one takes a slice out of the doughnut, creating an air gap, North and South poles are established at opposite sides of the air gap.

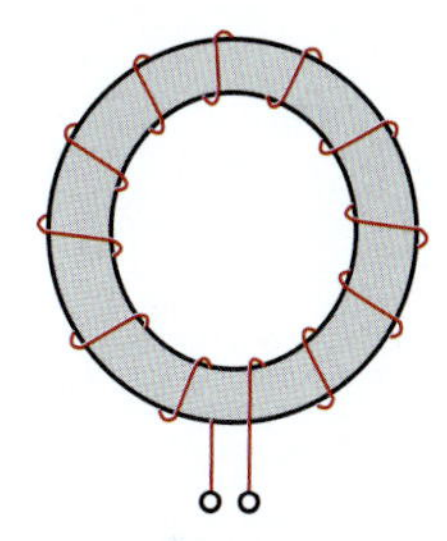

Flux is virtually contained within core with very little "leakage" flux

FIGURE 9–43 Toroid coil

Magnetic Shields

Remember magnetic lines of force penetrate nonmagnetic materials. To protect components, devices, and so forth from stray or nearby magnetic fields, a magnetic shield is used, Figure 9–44. A high-permeability, soft iron enclosure is useful for the shielding function. Actually, it is simply diverting the lines of flux through its low-reluctance path, thus protecting anything contained inside of it. One application for this type shielding is in electrical measuring instruments or meter movements.

Shaping Fields via Pole Pieces

Notice in Figure 9–45 the use of pole pieces. These pieces control the shape of the field across the air gap to afford a more linear field with almost equal flux density across the area of the gap. This control is beneficial for many applications.

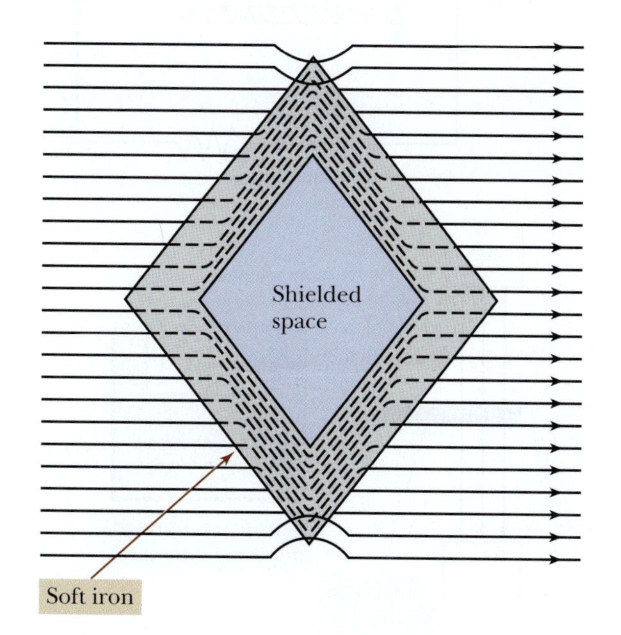

FIGURE 9–44 Concept of the magnetic shield

The Hall Effect

Figure 9–46 illustrates the Hall effect. Named after its discoverer, E. H. Hall, the phenomenon states that when a current-carrying material is in the presence of an external magnetic field, a small voltage develops on two opposite surfaces of the conductor.

As the illustration indicates, this effect is observed when the positional relationships of the current, the field's flux lines, and the surfaces where the difference of potential develops are such that the following are true:

1. Flux is perpendicular to direction of current flow through the conductor. (NOTE: The amount of voltage developed is directly related to the flux density, B.)
2. If current is traveling the length of the conductor, the Hall voltage (v_H) develops between the sides or across the width of the conductor.

Indium arsenide is a semiconductor material used in devices, such as gaussmeters, to measure flux density via the Hall effect. This material develops a relatively high value of v_H per given flux density. Because the amount of Hall-effect voltage developed directly relates to the flux density where the current-carrying conductor is immersed, a meter connected to this probe can be calibrated to indicate the flux density of the probed field, Figure 9–47.

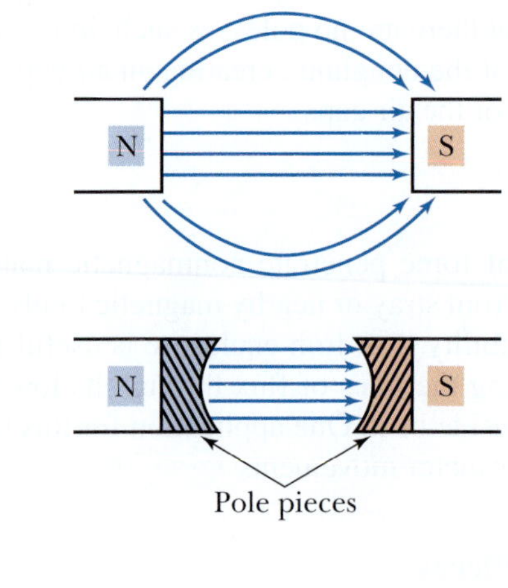

FIGURE 9–45 Concept of pole pieces linearizing a field

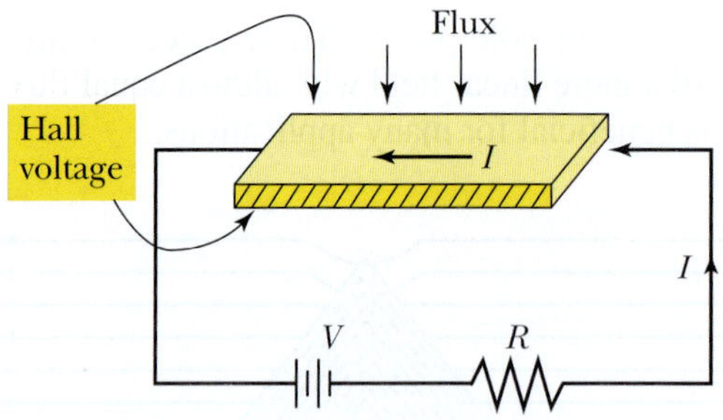

FIGURE 9–46 The Hall effect

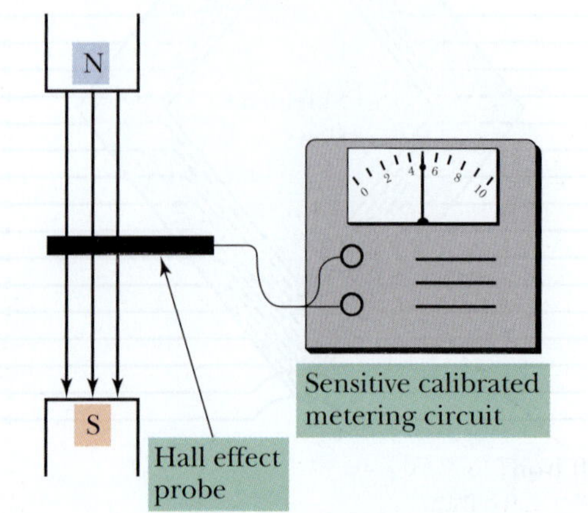

FIGURE 9–47 Using the Hall effect to measure flux density

Special Classifications of Materials

In this chapter, we have primarily been discussing ferromagnetic materials that typically have high permeabilities and can be strongly magnetized. Examples include iron, steel, nickel, and various alloys. Other classifications of materials worth noting are:

1. *paramagnetic* materials that have weak magnetic properties. (Examples are aluminum and chromium.);
2. *diamagnetic* materials that have permeabilities of less than one and align perpendicular to the direction of magnetic fields. (Examples are mercury, bismuth, antimony, copper, and zinc.); and
3. *ferrites* that are powdered and compressed materials with high permeabilities but high electrical resistance that minimizes I^2R/eddy current losses when used in high-frequency applications. (Examples are nickel ferrite, nickel-cobalt ferrite, and yttrium-iron garnet.) Note that eddy currents are those induced into the core material that cause undesired I^2R losses within the core.

Summary

- A magnet is an object that attracts iron, steel, or certain other materials. It can be natural (magnetite or lodestone) or produced in iron, steel, and other materials in a form called a permanent magnet. Also, a magnet can exist in soft iron and other materials as a temporary magnet that requires a current-carrying coil around it.
- Magnetic materials, such as iron and nickel, are attracted by magnets and can be made to have magnetic properties.
- Magnets that are suspended and free to move align in an approximate North/South direction. The end of the magnet seeking the earth's North pole is the North-seeking pole. The other end of the magnet is the South-seeking pole.
- A magnetic field composed of magnetic lines of force called flux surrounds a magnet. This region is where the magnetic forces act.
- Lines of force are continuous; travel outside the magnet from the magnet's North pole to the magnet's South pole; and flow inside the magnet from the South pole to the North pole. They travel the shortest path or the path with least opposition to the flow of flux, and when traveling in the same direction, repel each other. When traveling in opposite directions, they attract each other. Furthermore, lines of force do not cross each other and will penetrate nonmagnetic materials.
- A current-carrying conductor or wire is surrounded by a magnetic field. The strength of the magnetic field is directly related to the value of current passing through it.
- A coil of wire carrying current develops a magnetic field that is stronger than a straight conductor with the same current. The magnetic field becomes stronger if an iron core is inserted in the coil, making it an electromagnet.
- A basic law of magnetism is that like poles repel each other and unlike poles attract each other.
- Permanent magnets come in a variety of shapes; for example, bar-shaped (rectangular), horseshoe-shaped, and disc-shaped. Permanent magnets are not really permanent since they lose their magnetism over time.
- Induction of magnetism occurs when a magnetized material is brought in proximity to or rubbed across some magnetic material.
- The left-hand rule to determine direction of the magnetic field surrounding a current-carrying conductor states that when the wire is grasped in the left hand with the extended thumb in the direction of current flow through the wire, then the fingers are wrapped around the conductor in the same direction as the magnetic field.
- The left-hand rule to determine the polarity of electromagnets (i.e., which end is the North pole and which end is the South pole) states that when the coil is grasped with the fingers pointing in the same direction as the current, the extended thumb points toward the end of the coil that is the North pole.
- Factors influencing the strength of an electromagnet are: the number of turns on the coil; the coil's length; the coil's value of current; the type of core; and the coil's cross-sectional area.
- Important terms and units relating to magnetism include flux, weber, tesla, magnetomotive force, magnetic field intensity, permeability, and relative permeability.
- Comparing magnetic circuit parameters with electrical circuits and Ohm's law can be performed by stating that the amount of flux flowing through a magnetic circuit is directly related to the magnetomotive force and inversely related to the path's reluctance (opposition) to flux.
- Core materials differ in their magnetic and electrical features. For example, some materials saturate at lower levels of magnetization.

- An ideal core material has very high permeability; loses all its magnetism when there is no current flow in the coil; does not easily saturate; and has low I^2R loss due to eddy currents.
- An important relationship exists between flux density produced by a given magnetic field intensity and the permeability of the material involved. This relationship stated for the permeability of a vacuum is:

$$\mu_0 = \frac{B}{H}, \text{ and } B = \mu_0 \times H, \text{ and } H = \frac{B}{\mu_0}$$

where: μ_0 is the absolute permeability of a vacuum
B is flux density and
H is the magnetic field intensity

- When an ac passes through an electromagnet's coil, there is a lag between the magnetizing force and the flux density produced. This is illustrated by the *B-H* curve and the hysteresis loop. These graphs also illustrate residual magnetism and the coercive force needed to overcome it.
- Faraday's law states that the amount of voltage induced when there is relative motion between conductor(s) and magnetic flux lines is such that when 10^8 lines of flux are cut by one conductor (or vice versa) in one second, one volt is induced. In other words, induced voltage relates to the number of lines of force cut per unit time.
- Lenz's law states that when there is an induced voltage (or current), the direction of the induced voltage or current opposes the change causing it.
- A motor converts electrical energy into mechanical energy.
- A generator converts mechanical energy into electrical energy.

Formulas and Sample Calculator Sequences

Formula 9–1
(To find relative force of attraction or repulsion between two magnetic poles)

$$\text{Force} = \frac{\text{Pole 1 Strength} \times \text{Pole 2 Strength}}{d^2}$$

pole 1 value, [×], pole 2 value, [÷], [(], distance value, [x^2], [)], [=]

Formula 9–2
(To find magnetomotive force)

$mmf = NI$

number of turns value, [×], current value, [=]

Formula 9–3
(To find magnetic field intensity)

$$H = \frac{AT}{\text{meter}}$$

value of current (amperes), [×], number of turns, [÷], number of meters, [=]

Formula 9–4
(To find value of induced voltage)

$$V_{ind} = \frac{\Delta\phi}{\Delta T}$$

change in flux value, [÷], amount of time value, [=]

Formula 9–5
(To find value of induced voltage)

$$V_{ind} = N(\#\text{ of turns}) \times \frac{d\phi\,(\text{Wb})}{dt\,(\text{s})}$$

number of turns, [×], [(], number of webers cut, [÷], time to cut, [)], [=]

EXCEL AUTOMATED FORMULAS

Helpful problem-solving tools, such as the Excel automated formulas available on the CD, allow automatic calculations of formulas.

Using Excel

Magnetism and Electromagnetism Formulas

(Excel file reference: FOE9_01.xls)

DON'T FORGET! *It is NOT necessary to retype formulas once they are entered on the worksheet! Just input new parameters data for each new problem using that formula, as needed.*

- Use the Formula 9–5 spreadsheet sample and solve for V_{ind} if the number of turns = 250 and these turns are being cut at a rate of 5 webers in 2 seconds.

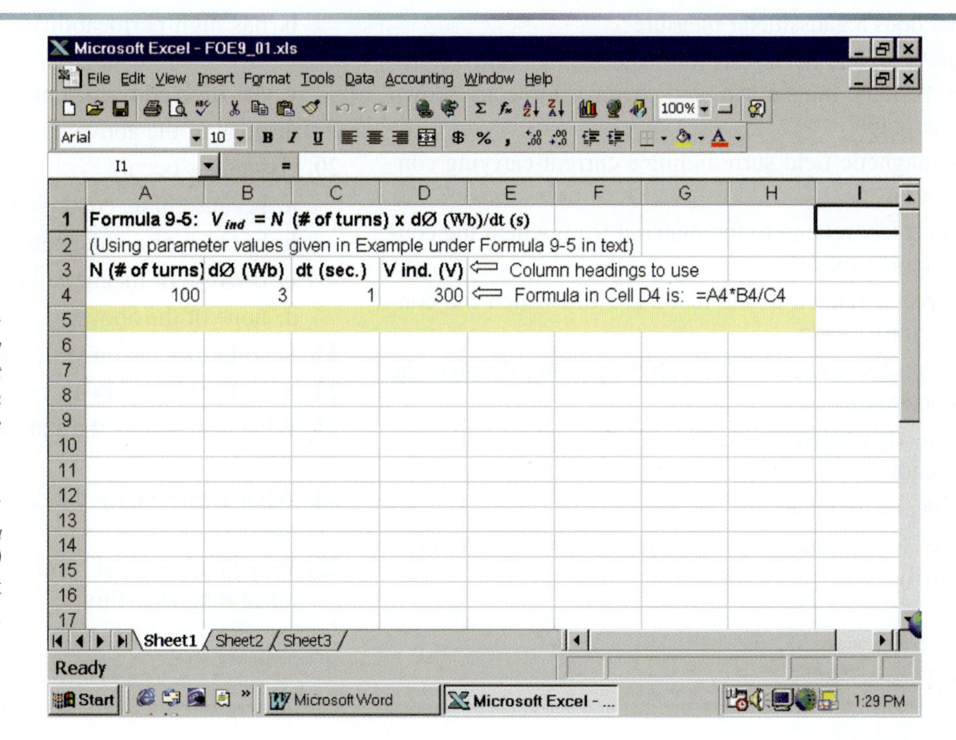

Review Questions

1. Magnetism is:
   - **a.** a material with unique characteristics.
   - **b.** the phenomenon associated with magnetic materials.
   - **c.** the property that attracts metals, woods, and papers.
   - **d.** none of the above.
2. Flux is:
   - **a.** magnetic lines of force related to magnetism.
   - **b.** the space between magnetic poles.
   - **c.** magnetic poles that are identified as N or S.
   - **d.** none of the above.
3. A magnetic field is:
   - **a.** the space at the ends of magnetic poles.
   - **b.** the field of magnetic influence surrounding a magnet.
   - **c.** the space found only within the magnet itself.
   - **d.** none of the above.
4. Magnetic polarity refers to:
   - **a.** a convention that defines the direction of flux relative to a magnet's end poles.
   - **b.** the conventional means of defining a magnet's end faces as either North-seeking or South-seeking.
   - **c.** neither a or b.
   - **d.** both a and b.
5. The law of attraction and repulsion states that:
   - **a.** like poles attract each other.
   - **b.** like poles are neutral to each other.
   - **c.** unlike poles repel each other.
   - **d.** unlike poles attract each other.
6. List five statements about the behavior of magnetic lines of force.
7. What term indicates the ease with which a material passes, conducts, or concentrates magnetic flux?
8. What term indicates the opposition to magnetic lines of force?
9. What is the symbol for flux and what is its SI unit of measure?
10. What is the symbol for flux density and what is its SI unit of measure?
11. What is the symbol for magnetomotive force and what is its SI unit of measure?
12. What is the symbol for magnetic field intensity and what is its SI unit of measure?
13. How many lines of force are represented by a Weber?
14. What is meant by saturation in a magnetic path?
15. How does a magnetic shield give the effect of shielding something from magnetic lines of force?

16. What does a gaussmeter measure?
17. Draw a typical hysteresis loop and explain how saturation, residual magnetism, and coercive force are shown on the graph.
18. The magnetic field surrounding a current-carrying conductor is:
    a. inversely related to the material from which the conductor is made.
    b. directly related to the material from which the conductor is made.
    c. inversely related to the current passing through the conductor.
    d. directly related to the current passing through the conductor.
19. An ideal core material:
    a. has low permeability, low resistance, and saturates easily.
    b. has high permeability, high resistance, and does not easily saturate.
    c. has high eddy currents, and low I^2R losses.
    d. none of the above.
20. Hysteresis occurs:
    a. because of ac coercive force.
    b. because of dc magnetizing force.
    c. because of magnetic lagging effects.
    d. none of the above.
21. State Lenz's law in your own words.
22. State Faraday's law in your own words.
23. What is meant by the statement "motor effect in a generator"?
24. What is meant by the statement "generator effect in a motor"?
25. Why is a "toroidal" coil form such an efficient conductor of magnetic flux?

Problems

NOTE: Redraw illustrations on separate paper to answer questions involving diagrams.

1. Draw a typical field pattern for the magnetic situations in Figure 9–48. Indicate direction(s) of flux lines with arrows.

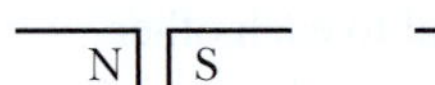

FIGURE 9–48

2. Draw magnetic field lines around the conductors in Figure 9–49 and indicate direction with arrows.

FIGURE 9–49

3. Indicate the North and South poles of the current-carrying coil in Figure 9–50.

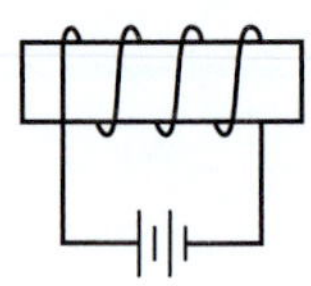

FIGURE 9–50

4. If a pole piece has a cross-sectional area of 0.005 m^2 and is emitting 1 μWb of flux:
    a. How many lines of flux are involved?
    b. What is the flux density, expressed in teslas?
5. How much induced emf results if a coil of 500 turns is cut at a rate of 10 μWb per second?
6. What is the magnetic field intensity (H) of a coil having 1,500 turns per meter when the current through that coil is 3 amperes?
7. How many lines of flux are represented by 4 Wb?

OUTLINE

OBJECTIVES

After studying this chapter, you should be able to:

1. List at least two key features of **digital multimeters (DMMs)**
2. Describe at least one advantage and one disadvantage of an **analog multimeter (VOM)**
3. Explain the meanings of the terms *autoranging* and *autopolarity*
4. Describe and calculate meter **loading effects** for specified measurement conditions
5. List at least two special purpose measuring devices
6. Define two basic methods of measuring voltage on a circuit having a ground reference
7. Describe the technique for making continuity checks on a 200-foot long cable
8. Define the purpose and function of meter protection circuits

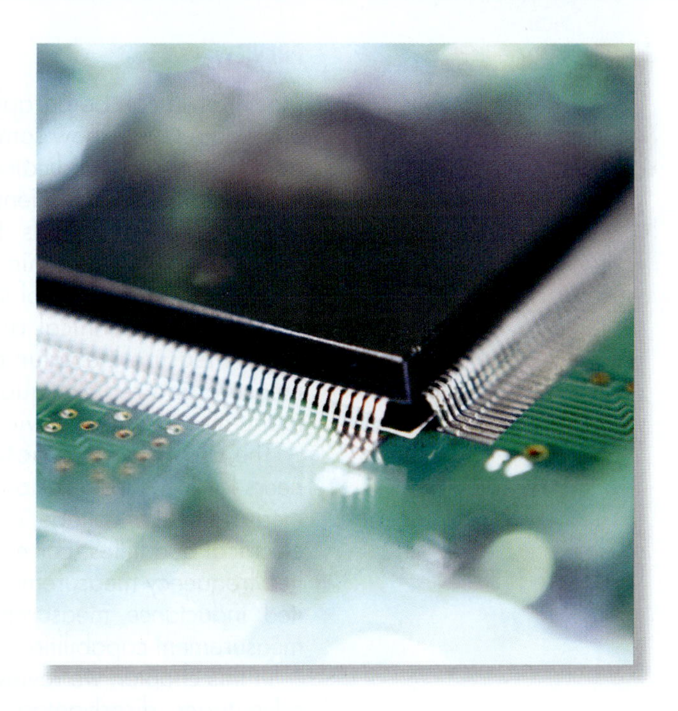

CHAPTER 10

Measuring Instruments

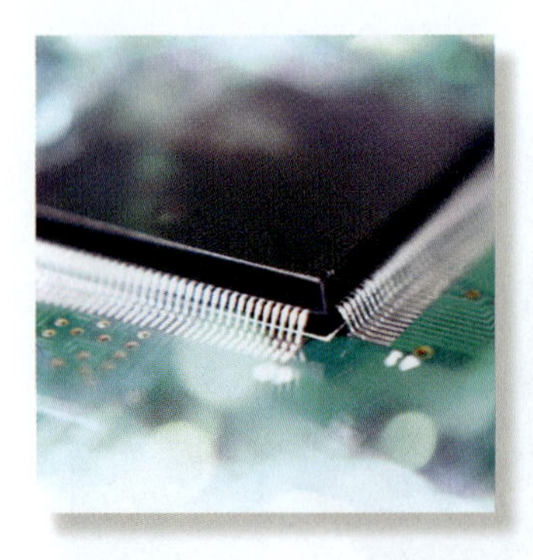

PREVIEW

Lord Kelvin can be paraphrased as saying, "If you can measure what you are speaking about, you know something about it. If you can't, your knowledge is very limited." Someone else has indicated that measuring instruments are like special eyes to technicians, engineers, and scientists. The dc measuring instruments you will study in this chapter are invaluable aids to people working in electronics. These instruments are indispensable for designing, troubleshooting, servicing, and monitoring equipment operation. Some samples of these instruments are in Figure 10–1.

There is a technical distinction between the terms *instrument* and *meter.* Instruments measure the value of present quantities. Meters, such as the watt-hour meter, measure and register quantities with respect to time. For convenience, we will refer to the measurement devices discussed using either term.

The **digital multimeter (DMM)** is the most-used device by technicians and engineers for measuring various electrical parameters. Many of these devices allow not only measurement of voltage, current, and resistance values, but a number of other important electrical quantities. For example, some allow measurement of capacitance values, limited frequency measurement, checking of semiconductors, and in some cases, even limited inductance measurements. You will greatly appreciate some of these special measurement capabilities as you proceed further in your training.

In this chapter, we review some of the basic operational features and list some of the advantages, disadvantages, and applications of both digital-type meters and analog-type meters.

Although the **analog multimeter (VOM** or **volt-ohm-milliammeter**) is used much less than the DMM today, there are still certain applications where it is more convenient to use than the digital-readout DMMs (e.g., tuning transmitters, or other scenarios where monitoring of smooth changes in parameters is helpful). Since analog multirange multimeters' operation depends on magnetism, series, parallel, and combination circuits that you have recently studied, we provide a very brief overview of how the analog meter works. You can readily see how this type of meter illustrates some of the practical applications of the theory you have previously studied.

At a more appropriate time in your studies, you will study the theory of operation of the DMM, which entails such things as oscillators, frequency counters, and other circuitry that you have not studied up to this time.

Information about *meter loading effects* (how the digital and analog-type movements affect the circuits they are testing) is also briefly discussed in this chapter. Finally, hints about how to effectively use these instruments for testing and troubleshooting applications are given.

KEY TERMS

Analog multimeter (VOM)
Damping (electrical)
Digital multimeter (DMM)
Loading effect
Multimeter

10–1 Digital Multimeters (DMMs)

The digital multimeter (DMM), Figure 10–1, is probably the piece of test equipment most used by technicians today. There are some applications where the analog multimeter (VOM) is still the best device to use. However, the DMM's accuracy, expanded measurement capabilities, ease of reading, and lighter loading of the circuit under test make it a prime choice in most cases.

Some DMMs provide polarity-indicated readouts and autoranging features. If the polarity being measured is negative, the readout indicates this. In contrast, a VOM would have its pointer hitting and/or pushing against the scale's left-side limit peg, if the polarity were opposite the way you connected the test leads. Certainly, meter damage could result. When used

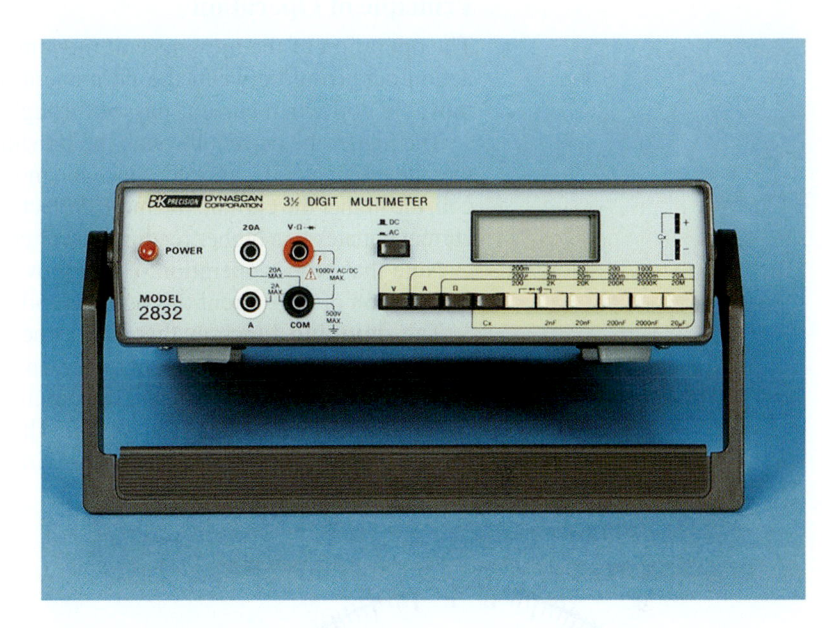

FIGURE 10–1 Examples of measuring instruments: (a) handheld digital multimeter *(Courtesy of John Fluke Mfg. Co.);* (b) a benchtop digital multimeter *(Courtesy of B & K Precision)*

in the autoranging mode, a DMM will automatically select the range with the best resolution, without the technician having to manually select the correct range.

Many DMMs also offer measurement capabilities beyond simply measuring voltage, current, or resistance. Some allow measurement of ac frequency, capacitor values, semiconductor diode condition, and so on. You will better understand the significance of these capabilities after you have studied ac signals and components used in ac.

Some popular DMMs also are capable of remembering important measurement data that you want to read at a later time. For example, there may be a "MIN MAX" mode on the meter. This means that the meter will remember the lowest and highest readings over a period of time, take the average, and provide that data in its readout. Also, some meters can store a reading in memory, then compare it to subsequent readings, and display the difference between the memorized reading and the subsequent reading. For safety's sake, some meters have a "hold" on the reading feature. This allows you to keep your eyes on the test probes when working in dangerous or difficult places for making measurements. You can then read the display when you have removed the test probes and are better able to safely fix your eyes on the meter readout.

Some of the advantages and disadvantages of the DMM are listed in the following table.

Advantages	**Disadvantages**
• Digital readout usually prevents a reading error	• Can be physically larger than equivalent VOM (not in all cases)
• Can have "autoranging" that prevents destroying a meter by having it on the wrong range	• Can be more costly to buy and operate
• Can have "autopolarity" that prevents problems from connecting meter to test circuit with wrong polarity	• Because of "step-type" digital feature rather than "smooth" analog action, are *not as good* for adjusting "tuned" circuits or "peaking" tunable responses
• Can have excellent accuracy (plus or minus 1%)	• Frequency limitations
• Portable and uses internal batteries	

10–2 Analog Multimeters (VOMs)

Principle of Operation

The operation of the moving-coil movement used in analog meters is based on the motor action occurring between the permanent magnet's field and the field established by the movable coil when current passes through it, as illustrated in Figures 10–2 and 10–3.

The higher the current through the moving coil, the stronger its magnetic field and the further the moving coil is moved. Since the movement indicator, or pointer, moves along with the coil, the higher the current, the more deflection of the pointer along the meter scale. Thus, more current equates to a higher reading; lower current equates to a lower reading on the meter.

Note that proper operation of this movement depends on dc current and proper polarity, or direction of the current through the coil.

The **damping** function that keeps the meter pointer from overshooting or oscillating above and below its final resting place on the scale is really a function explained by Lenz's law, which you studied earlier. The moving coil, when moved, induces a current in the aluminum frame upon which it is wound. The induced current creates a magnetic field opposing the motion that induced it, thus providing an electrical damping effect.

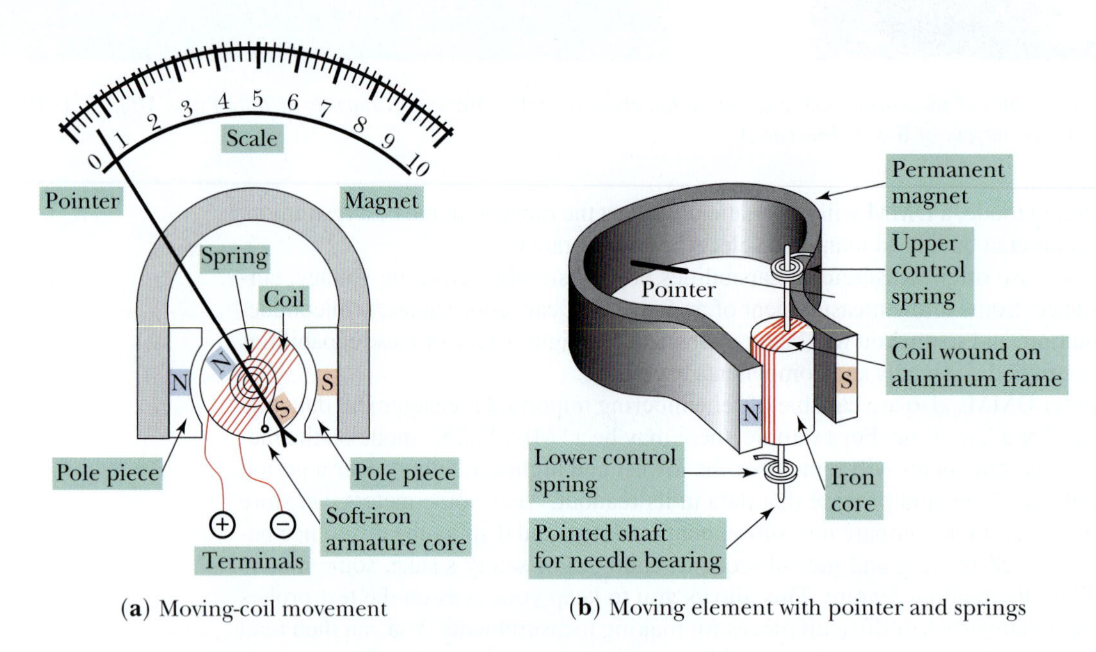

(**a**) Moving-coil movement (**b**) Moving element with pointer and springs

FIGURE 10–2 Key parts of a moving-coil analog meter movement

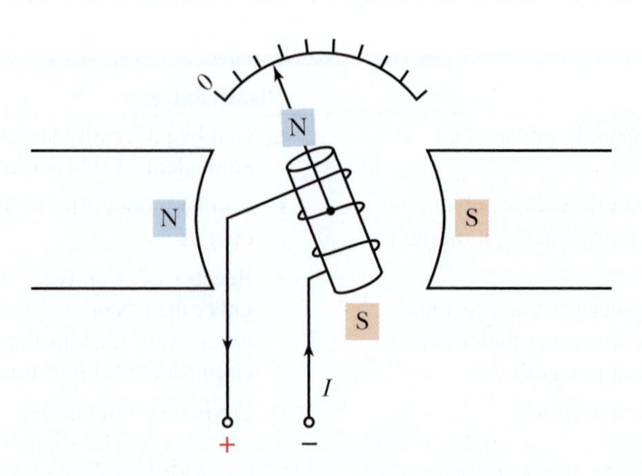

FIGURE 10–3 Polarity required for proper direction of motor action

FIGURE 10–4 The once-popular VOM (volt-ohm-milliammeter) *(Courtesy of Simpson Electric Co.)*

VOMs—Key Points Summary

The VOM or volt-ohm-milliammeter shown in Figure 10–4 was once a popular multimeter used for many years by technicians. Because the DMM has many advantages over the VOM, the VOM is used much less today. The advantages and disadvantages of the VOM are listed in the following table.

Advantages	Disadvantages
• Compact and portable • Does not require an external power source • Measures dc current, dc voltage resistance, and ac voltage • Typically has multiple ranges for each electrical quantity it measures • Is analog in nature, so is *good* for adjusting, tuning, or peaking tuned circuits	• Can cause significant meter loading effects on circuit being tested • Less accurate than a DMM • Analog readout scale can introduce "parallax" reading errors if pointer is viewed from an angle

10–3 Voltmeter Loading Effects

Because a voltmeter is connected in parallel with the component or the circuitry it is measuring, it would be ideal if the voltmeter circuitry had *infinite* resistance. This would mean none of the normal circuit operating conditions is altered with the voltmeter connected. But this is not the case! The voltmeter circuit has some finite value of resistance, and it alters circuit conditions! This situation is called **loading effect,** Figure 10–5.

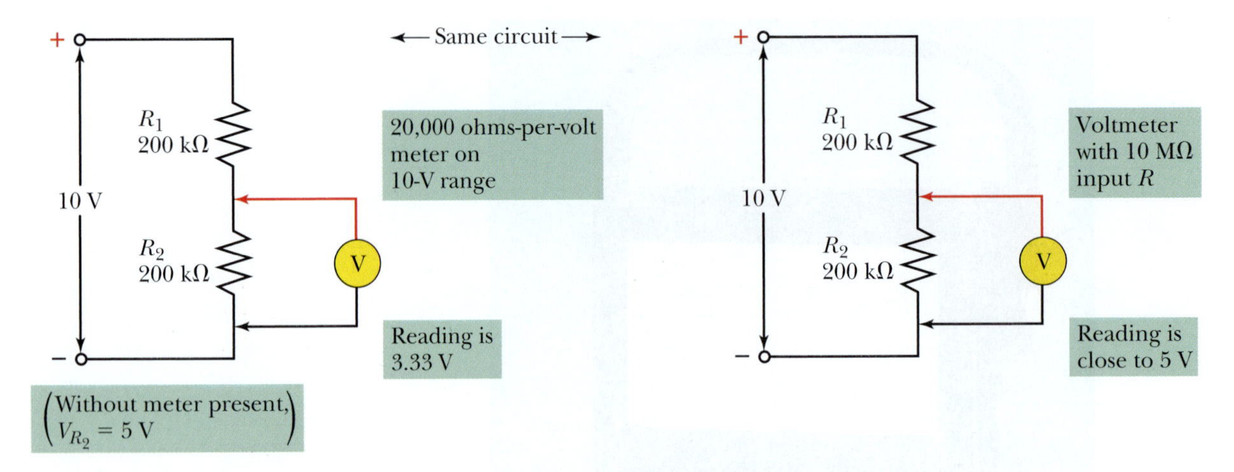

FIGURE 10–5 Voltmeter loading effect

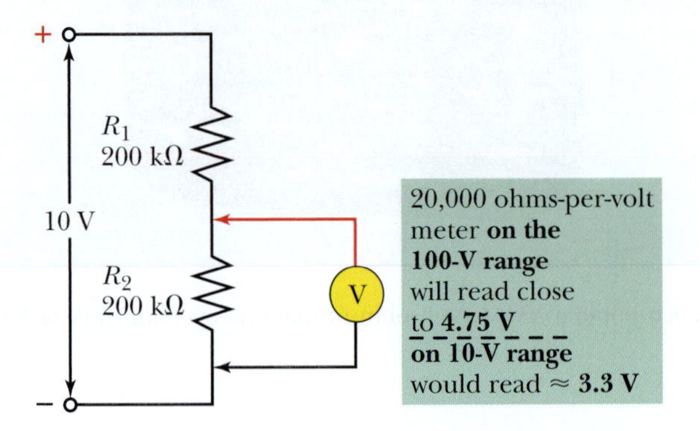

FIGURE 10–6 Sample conditions of loading effect

It is apparent that the higher the voltmeter circuit's resistance, the less loading effect it has on the tested circuit. That is, the higher the VOM voltage range selected, the less the loading effect.

A higher voltage range results in a higher meter resistance, which causes less voltmeter loading effect, Figure 10–6.

The greater the voltmeter circuit resistance ratio to the component or circuit resistance being tested, the more accurately the measurement reflects circuit conditions; that is, the less the circuit conditions change with the voltmeter attached. Stated another way, not only is the meter circuit resistance important, *but* the tested circuit's resistance characteristics are equally important! A higher circuit resistance before voltmeter connection results in greater disturbance of normal operating parameters when the meter is connected! In other words, low-resistance circuits are disturbed less by the meter connection than high-resistance circuits, Figure 10–7.

Practical Notes

One practical rule of thumb is to be sure the resistance of the voltmeter circuit (on the range being used) is at least 10 times the resistance of the circuit portion being tested by the voltmeter. For practical purposes, commercial voltmeters range from 20,000-Ω/V sensitivity ratings for common multimeters, or VOMs (volt-ohm-milliammeters), to 10 megohms, or higher input circuit resistances for the voltmeter circuitry common in digital measuring devices, Figure 10–8. Obviously, meters with multi-megohm input resistances cause negligible loading effect on lower-resistance circuits to be tested.

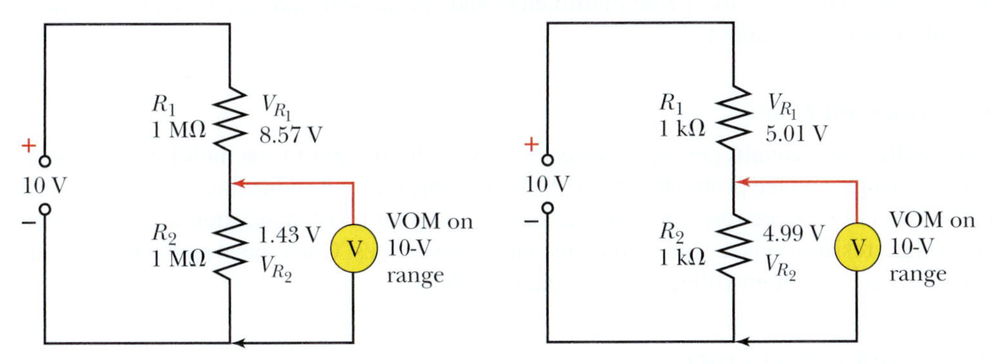

FIGURE 10–7 Variation in loading effect caused by differences in circuits being tested multiSIM

FIGURE 10–8 A DMM's higher meter circuitry resistance means less loading effect *(Courtesy of Simpson Electric Co.)*

Voltmeter Key Points Capsule

1. To measure voltage, the voltmeter is connected in parallel with the component or circuit portion where the difference of potential is to be measured. When measuring dc voltage, it is necessary to observe polarity. Also, care must be taken to assure that the voltage range of the metering circuit is sufficient to handle the voltage to be measured.
2. Because the voltmeter is connected in parallel, the higher the metering circuit resistance, the less it changes the circuit conditions, and the more accurately the voltmeter reading shows the actual circuit operating conditions.

10–4 Special Purpose Measuring Devices

There are special probes used with multimeters and special self-contained metering devices that you should know about.

High-Voltage Probe

Basic multimeters cannot generally measure very high voltages in the multikilovolt range; however, a high-voltage probe (Figure 10–9) will enable this to be done. In essence, the probe consists of a high-value resistance that acts as a multiplier resistance, external to the meter. The probe's resistor(s) is designed to withstand high voltages. One typical application of this probe is to measure high voltages on TV picture tubes.

Clamp-On Current Probe

Another accessory found for some multimeters is the clamp-on probe that measures ac parameters, Figure 10–10a. When ac current levels are in the ampere ranges, this device reacts to the magnetic fields surrounding the current-carrying conductors and translates this activity into appropriate readings. The key advantage to this device is that it measures current *without having to break the circuit* when using the metering circuit. The clamp-on probe simply surrounds the current-carrying conductor without breaking the circuit. NOTE: This device *does not measure dc* current!

Practical Notes

Measuring instruments are a vital tool to technicians when they are troubleshooting circuits that are operating abnormally. Electrical parameters indicate whether a component, circuit portion, or total circuit is operating normally or abnormally, or whether it is dead.

In a *qualitative* sense, the measurements indicate whether the values measured are normal, higher than normal, or lower than normal.

In a *quantitative* sense, the measurements indicate specific values for the parameters being measured. These values are then interpreted as normal or abnormal.

To troubleshoot any circuit or system, the technician *must know "what should be."* That is, he or she *must know the norms.* This is gained through experience, knowledge, or reference to appropriate technical documentation, such as schematics.

Although the following discussion is not exhaustive or comprehensive, the hints and techniques can be useful to you during your training and throughout your career as a technician. Study and apply these items, as appropriate.

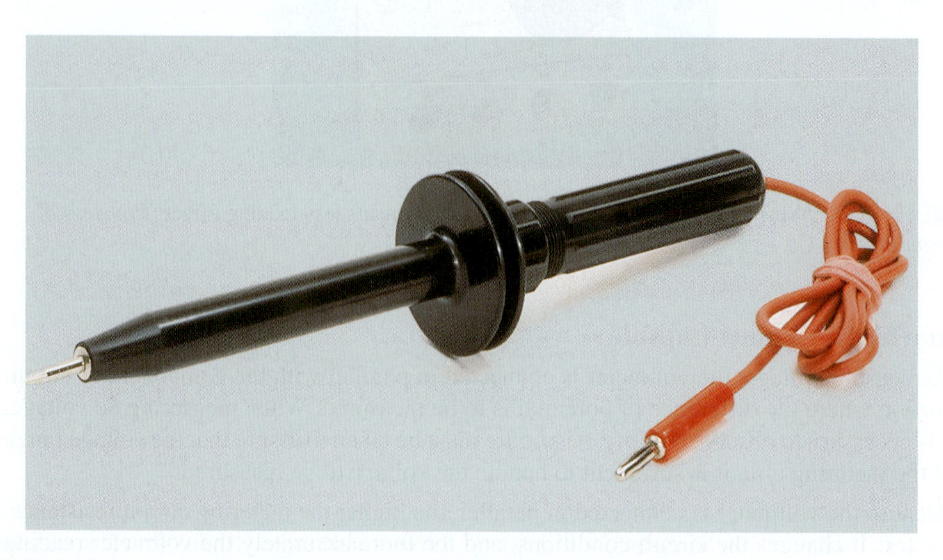

FIGURE 10–9 The high-voltage probe *(Courtesy of Amprobe Instruments)*

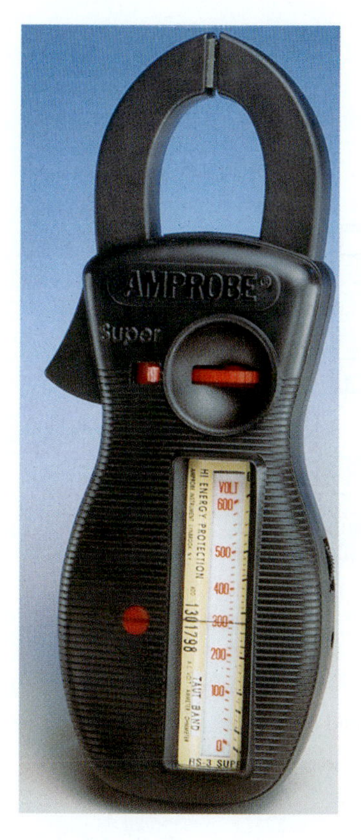

(a) (b)

FIGURE 10–10 Clamp-on probes *(Photos courtesy of Amprobe Instruments)*

Other Devices

There are a variety of special clamp-on measurement devices described in many electronic distributor catalogs. You will find the clamp-on meters we've been describing, as well as clamp-on volt-ohm-milliammeters (Figure 10–10b), clamp-on digital wattmeters, and "power factor" meters. (You will learn about "power factor" meters in later chapters.)

10–5 Troubleshooting Hints and Useful Measuring Techniques

General Information

Use test instruments properly:

1. **Voltmeters**—Using appropriate safety precautions to prevent shock, connect the meter in parallel with desired test points, making sure to observe proper range and polarity. For safety's sake, only use one hand at a time in making voltage measurements.
2. **Current meters**—*Turn power off circuit.* Insert meter in series, making sure to observe proper range and polarity. Turn power on circuit to be tested. Be sure the current metering circuit does not have too high a resistance, causing undue or unaccounted-for changes in circuit conditions.
3. **Ohmmeters**—*Turn power off circuit* where component or circuit portion to be measured resides. Disconnect at least one component lead. Connect test probes across component or circuit portion to be measured. Be sure ohmmeter batteries are good.

Techniques for Voltmeter Measurements

1. **Connecting one lead to common or ground**—Many electronic circuits have a common or ground reference point, such as the chassis or a common conductor "bus" where all

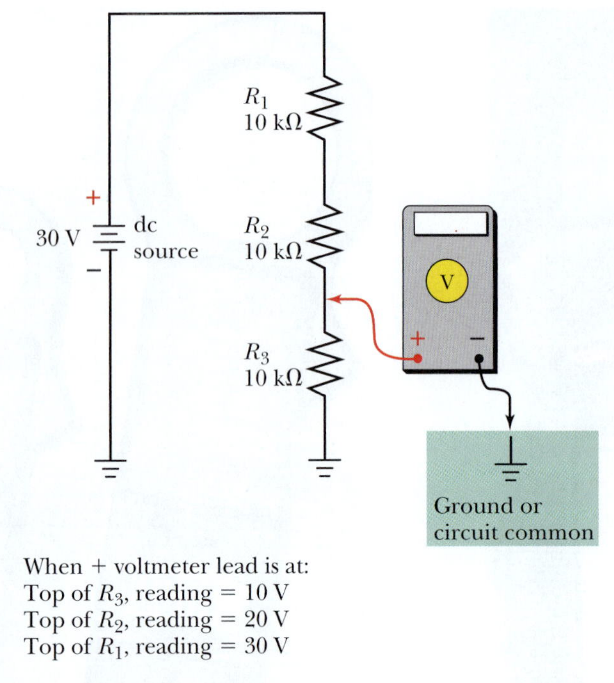

FIGURE 10–11 Making voltage measurements with respect to ground reference

voltages are referenced. By connecting one lead of the voltmeter (frequently the black or negative lead) to this ground reference, measurements are made throughout the circuit by moving the other lead to the various points where voltage measurements are desired, Figure 10–11. Frequently, the black lead on a meter has an alligator clip, making it easy to attach the lead to an electrically common point in the circuit.

2. **Using voltage checks to determine current**—Because it is often awkward and time-consuming to break a circuit when installing a current meter, technicians frequently use voltage measurements to determine current. For example, if the measured voltage drop across a 1-kΩ resistor is 15 V, it is simple (Ohm's law) to deduce that the current through the resistor must be 15 mA. That is,

$$I = \frac{V}{R} = \frac{15\ \text{V}}{1\ \text{k}\Omega} = 15\ \text{mA}$$

Obviously, Ohm's law enables you to determine current regardless of the R value where you are measuring the voltage. This technique prevents you from having to (1) turn off the circuit; (2) break the circuit; (3) insert a current meter; (4) make the measurement, then turn circuit off again; and (5) reconnect the circuit when the current meter is removed.

3. **Using voltage measurements to find opens**—Recall from a previous chapter that when an open is in a series circuit, the applied voltage appears across the open. This means the voltmeter is a very effective means to find the circuit's open portion. The series circuit applied voltage appears across the open component(s), and zero volts are measured across the remaining component(s), Figure 10–12.

4. **Using voltage measurements to find shorts**—Again recall that a zero-volt IR drop appears across a shorted component or circuit portion. Voltage measurements quickly show this.

Techniques for Ammeter Measurements

1. **Total current measurement as a circuit condition indicator**—Total current value gives a general indication of circuit condition. If the total current is higher than normal, a lower-than-normal total circuit resistance is indicated. If the total current is lower than normal,

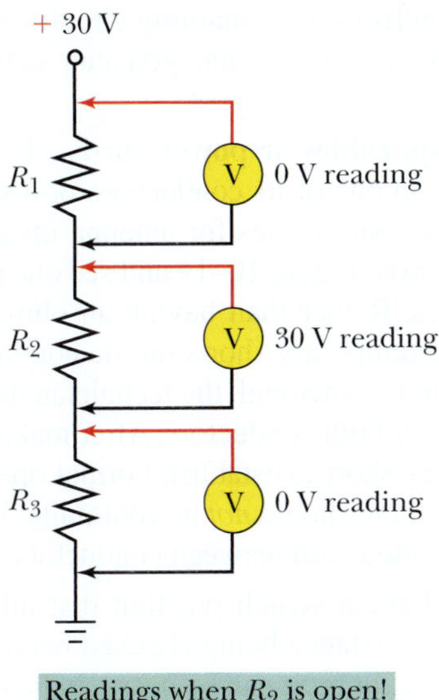

FIGURE 10–12 Voltage measurements can find "opens"

a higher-than-normal total circuit resistance is indicated. Additional isolation techniques are then used to find the problem.

2. **Special jacks for current measurement**—Some circuits or systems have built-in closed-circuit jacks where the ammeters are connected to measure current in the circuit, Figure 10–13.

   When the ammeter is not present, the jack is shorted, providing a closed-current path. But when the ammeter is present, the current must pass through the meter. (NOTE: This current-jack situation is uncommon.) More typically, voltage measurements determine current(s). In transmitters and other similar circuits where monitoring current is important, permanently mounted and installed ammeters are used.

3. **Clamp-on metering for ease of measurement**—Previously, we discussed the advantage of not breaking the circuit when using this type of current measurement. The limitations of this type of instrument are the limited available ranges and the limited physical situations where it is possible to connect such a meter.

(a)

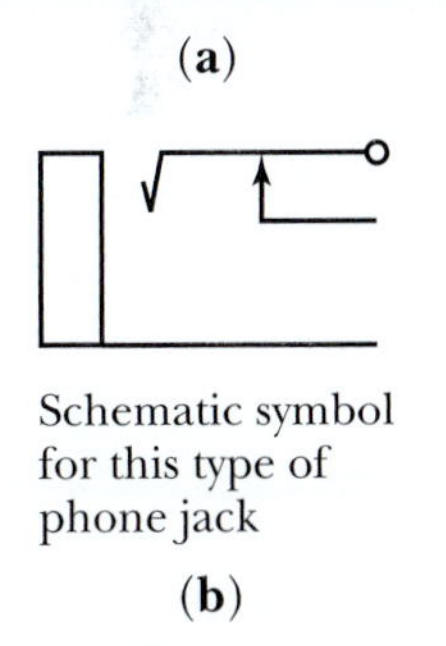

Schematic symbol for this type of phone jack

(b)

FIGURE 10–13 Shorting-type of jack

Techniques for Ohmmeter Measurements

1. **Continuity checks to find opens**—One frequently used check is the continuity check. The ohmmeter indicates ∞ ohms resistance or OL if there is no "continuity," or an open between the points where the probes are connected. Of course, an open is one extreme condition that can be measured.

   A continuity check is very useful for checking fuses, light bulbs, switches, circuit wires, and circuit board paths. For example, all of these items will show very low or almost zero ohms resistance across their terminals, if they are normal. If the measurement shows infinite ohms, the item is obviously open or blown.

   In fact, sometimes visual checks are used to confirm where the open is occurring. You can see the broken filament in an automobile light bulb and a broken copper path on the circuit board!

2. **Ohmmeter checks to find shorts**—Ohmmeter checks are valuable for finding shorted components or circuit portions. It is apparent when measuring a component or circuit resistance that when the resistance is zero ohms, there is a short. In the case of a closed switch, this is the normal situation. However, in cases where there should be a measurable resistance value, you have located a problem. By physically lifting one end of each

component or circuit portion from the remaining circuitry in order to electrically isolate it, and making appropriate measurements, you can determine which component is shorted.

3. **Checking continuity of long cables or power cords**—Frequently, it is handy to verify whether a cable has a break in one of its conductors, or a short between conductors. Examples are TV twin leads, coaxial cables for antenna installations, and cables in TV or computer network systems. Note Figure 10–14 and see one popular technique used to verify continuity in such cables. Rather than having an ohmmeter with 200-foot-long test leads, the technician creates temporary shorts on the conductors at one end of the cable. Then, using the ohmmeter at the other end, the technician determines if there is continuity throughout the cable length for both conductors. After making the measurement, the technician removes the temporary short to establish normal operation. Incidentally, when the temporary short is removed, there should *not* be continuity. If there is, this indicates an undesired low resistance, or shorted path between conductors.

   NOTE: Some multimeters have a switch position that allows for "audible" continuity checking. That is, when the resistance being checked between the test probes of the meter is very low, an audible tone, or buzz can be heard, indicating continuity. If you hear no sound, this indicates lack of continuity between the test probes.

4. **Measuring resistances with the ohmmeter**—Reiterating, *never* connect an ohmmeter to a circuit with power on it! Remove the power plug!

Watch Out for "Sneak Paths"!

One frequent error made by technicians is not sufficiently isolating the component or circuit portion to be tested from the remaining circuitry. Any unaccounted current path that remains electrically connected to the specific path you want to measure can change the measured R values from the "expected" to the "unexpected." Where possible, it is best to completely remove, or at least lift one end of the component being tested from the remaining circuit.

Another common "sneak path" occurs when the technician inadvertently allows his or her fingers to touch the test probes. This puts their body resistance in parallel with whatever is being measured. When it is impossible to isolate the component or circuit portion to be tested, but "in-circuit" tests are required, have a schematic diagram, or good knowledge about the circuitry so you can accurately interpret the measurements.

Naturally, when measuring resistances, use all of the normal operating procedures. That is, remove power from the circuit or component being tested, and use the proper meter range. If resistance measurements indicate improper values, proceed with additional isolating procedures, as appropriate.

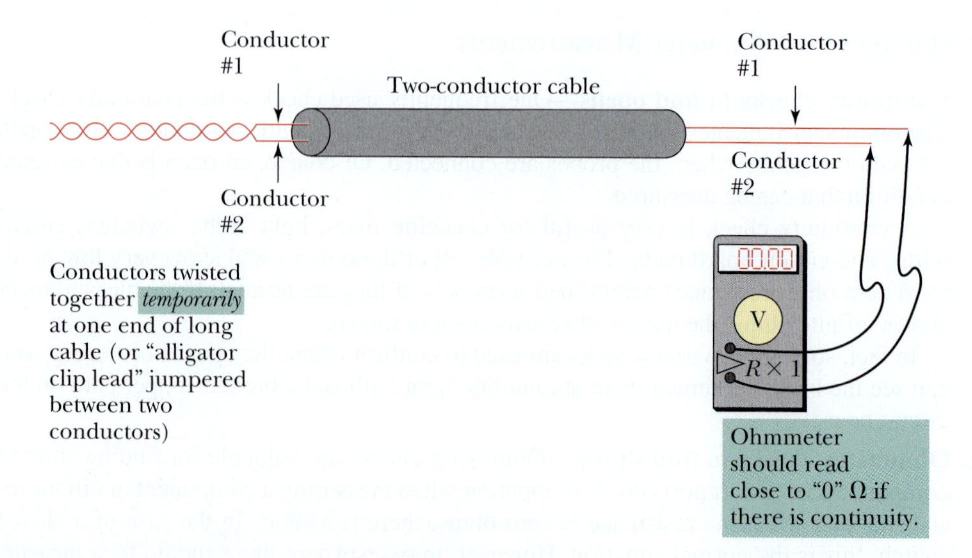

FIGURE 10–14 Checking continuity of long cables

10–6 Closing Comments about Measuring Instruments

In this chapter, we have shown dc measurements. We did mention, however, that most multimeters have additional built-in circuitry enabling the measurement of ac voltages.

Also you should know there are several other types of meter movement designs that react to ac quantities, which were not discussed in this chapter. These movements include the attraction-type moving-iron meter movement, the repulsion-type moving-iron movement, and the electrodynamometer movement (frequently used in wattmeters). It is beyond the scope of this chapter to discuss how each of these types operate, but we did want you to be aware of them.

As indicated throughout the chapter, meter movements are very fragile instruments. It is important that you give them the proper care to prevent physical abuse, and that you use proper operating procedures to prevent electrical damage.

Many instruments are manufactured with electrical protection circuitry built in. A variety of semiconductor circuits, from simple diode protection to more elaborate circuits, are examples of these circuits.

We have mentioned that technicians should be aware of factors affecting accuracy when making measurements. Meter loading effects were mentioned as one of these factors. Use simple series-parallel circuit analyses to determine the *voltmeter's* effect on the tested circuit. Use series circuit analyses to assess the *current meter's* effect on a given circuit.

Another point is the basic accuracy rating of the instrument itself. Typically, meters are rated from 1% to 3% accuracy at full-scale deflection for analog instruments and from 0.1% to 1% (or plus/minus one digit) on digital meter readouts. It is essential that technicians understand the limitations of the instruments used and have knowledge of the circuits or systems they are testing to properly interpret what is being measured. As a technician, you will find that *properly used and properly interpreted* circuit parameter measurements (enabled by the test instruments we have been discussing) are essential to the performance of your job. This is true whether the task is in designing, troubleshooting, servicing, or monitoring the operation of circuits and systems.

Summary

- Digital multimeters are definitely the most commonly used type of multimeter now.
- Digital multimeters have digital readouts that are much easier to read than analog-type multimeter scales, which require interpreting and interpolating.
- Digital multimeters have the advantage of causing far less "meter loading" effects on the circuits being tested.
- Digital multimeters are typically far superior in accuracy to analog types.
- An analog meter movement contains several key elements—permanent magnet, pole pieces, soft iron core, movable coil and support frame, spindle and bearings, springs, pointer and counterbalance, calibrated scale, and stops.
- The moving-coil (analog) movement depends on a force of repulsion between the permanent magnet's fixed magnetic field and the movable coil's magnetic field when current passes through the movable coil.
- Electrical damping prevents the instrument's pointer from oscillating above and below the actual reading; develops from the effects of induced current in the movable coil frame; and is analyzed by Lenz's law.
- Because voltmeters are connected in *parallel* with the portion of circuitry under test, the meter circuit's resistance alters the test circuit's operating conditions. This is called meter loading. The higher the meter circuit's input resistance, or Ω/V rating, the less loading effect it will cause.
- Factors to be observed when using dc instruments include using the proper range, connecting the meter to the circuit in a proper fashion, observing proper polarity, and, with the ohmmeter, making sure power is removed from the component or circuit being tested.
- Multimeters measure more than one type of electrical parameter, such as voltage, current, and resistance.
- A number of special probes and devices are available for electrical measurements. Some of these include clamp-on current probes, thermocouples, clamp-on multimeters, and high-voltage probes.
- Using measuring instruments to troubleshoot actually means using common sense, properly applying Ohm's law, and having some knowledge of how the tested circuit or system operates.

Review Questions

1. Name at least two principal advantages of the DMM over analog-type meters.
2. A typical moving-coil-type meter movement depends on the action of magnetic:
   a. attraction.
   b. repulsion.
   c. neutralization.
   d. all of the above.
3. Electrical damping in a meter movement refers to:
   a. moisture buildup in the meter if the meter is not sealed well.
   b. current passing through the meter's coil.
   c. induced current in the coil frame retarding oscillation of the pointer.
   d. induced current in the coil frame assisting the meter point movement.
4. List the procedural steps you would use to safely measure current in a circuit.
5. Why is it easier to use voltmeter checks, rather than current meter checks, when troubleshooting?
6. For what type situation is an analog-type instrument easier to use than a digital-type instrument?
7. The type of measuring instrument that causes less circuit loading effect for voltage measurements is:
   a. the VOM.
   b. the basic meter movement.
   c. the DMM.
8. When measuring current, it is desirable for the meter circuit to have:
   a. high resistance.
   b. low resistance.
   c. medium resistance.
   d. it doesn't matter.
9. The advantage of a clamp-on-type meter is that:
   a. it is small.
   b. you can measure with the power off.
   c. it is not necessary to break the circuit to use it.
   d. none of the above.

Analysis Questions

1. Explain why a current meter circuit should have a low resistance.
2. Draw a three-resistor parallel circuit with a dc source. Show where and how current meters are connected to measure I_T and the current through the middle branch. Indicate polarities, as appropriate.
3. Should a voltmeter circuit have a high or low resistance? Why?

OUTLINE

OBJECTIVES

After studying this chapter, you should be able to:

1. Draw a graphic illustrating an **ac** waveform
2. Define **cycle, alternation, period, peak-to-peak,** and **effective value (rms)**
3. Compute effective, peak, and peak-to-peak values of ac voltage and current
4. Explain average with reference to one-half cycle of sine-wave ac
5. Explain average with reference to nonsinusoidal waves
6. Define and calculate **frequency** and period
7. Draw a graph illustrating phase relationships of two **sine waves**
8. Describe the phase relationships of *V* and *I* in a purely resistive ac circuit
9. Label key parameters of nonsinusoidal waveforms
10. Use the computer to solve circuit problems

CHAPTER 11

Basic AC Quantities

PREVIEW

Previous chapters have discussed dc voltages, currents, and related quantities. As you have learned, direct current (dc) flows in one direction through the circuit. Your knowledge of dc and your practice in analyzing dc circuits will help you as you now investigate alternating current (ac)—current that periodically reverses direction.

In this chapter, you will become familiar with some of the basic terms used to define important ac voltage, current, and waveform quantities and characteristics. Knowledge of these terms and values, plus their relationships to each other, form the critical foundation on which you will build your knowledge and skills in working with ac, both as a trainee and as a technician.

Studying this chapter, you will learn how ac is generated; how it is graphically represented; how its various features are described; and how specific important values, or quantities are associated with these features.

In the next chapter, you will study a powerful test instrument, called the oscilloscope, which will enable you to visually observe the ac characteristics you will study in this chapter.

KEY TERMS

Ac
Alternation
Average value
Coordinate system
Cycle
Duty cycle
Effective value (rms)
Frequency *(f)*
Hertz (Hz)
Instantaneous value
Magnitude
Peak-to-peak value (*V*p-p)
Peak value (V_p or maximum value, V_{max})
Period *(T)*
Phase angle (θ)
Phasor
Quadrants
Rectangular wave
Sawtooth wave
Sine wave
Square wave
Vector

11–1 Background Information

The Difference between DC and AC

Whereas direct current is unidirectional and has a constant value, or magnitude, alternating current, as the name suggests, alternates in direction and is bidirectional. Also, ac differs from dc in that it is continuously changing in amplitude or value, as well as periodically changing in polarity.

In Figure 11–1, you see how both a simple dc circuit and a simple ac circuit may be represented diagrammatically. You are already familiar with the symbols in the dc circuit shown in Figure 11–1a. Notice the symbol used to represent the ac source in the ac circuit shown in Figure 11–1b. You will be seeing this symbol used throughout the remainder of the book. The wiggly line in the center of the symbol represents a sine-wave waveform, which you will study shortly.

Figure 11–1 also highlights the comparisons and contrasts between dc and ac voltage characteristics, using their graphical representations. As you proceed in the chapter, you will be focusing your attention on learning important specifics about the sine wave and other ac waveforms.

Alternating Current Used for Power Transmission

Alternating current **(ac)** has certain advantages over dc when transporting electrical power from one point to another point. A key advantage is its ability to be "stepped up" or "stepped down" in value. Again, you will study more about this later in the text. These advantages are why power is brought to your home in the form of ac rather than dc.

Some interesting contrasts between dc and ac are shown in Figure 11–2. This ac electrical power is produced by alternating-current generators. A simple discussion about generating ac voltage will be covered shortly. However, some groundwork will now be presented.

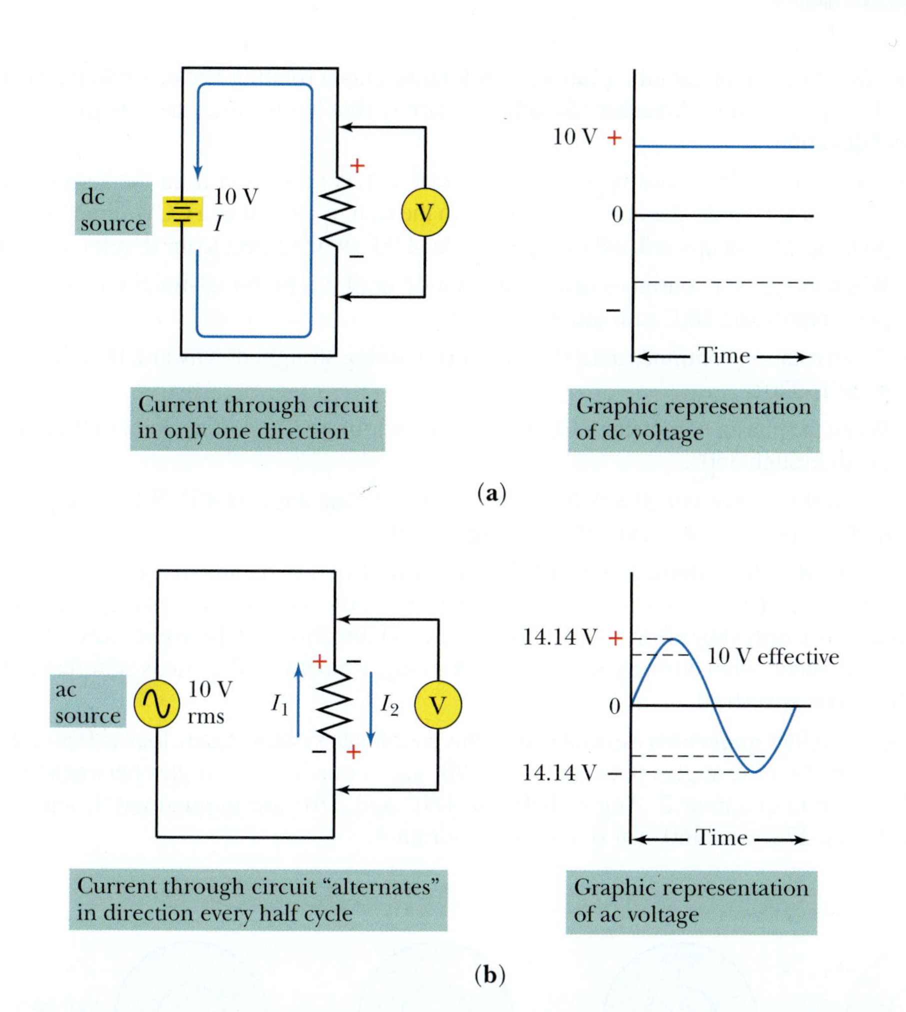

FIGURE 11–1 Graphic comparing dc and ac waveforms

CHARACTERISTIC DESCRIPTOR	IS CHARACTERISTIC OF: AC	DC
Current in one direction only	No	Yes
Current periodically alternates in direction—one-half the time in each direction	Yes	No
Voltage can be "stepped up" or "stepped down" by transformer action	Yes	No
Is an efficient means to transport power over long distances	Yes	No
Useful for all purposes	Yes	No

FIGURE 11–2 Similarities and differences between dc and ac power

Defining Angular Motion and the Coordinate System

As we deal with ac quantities, it is helpful to understand how angular relationships are commonly depicted. Refer to Figures 11–3a (angular motion) and 11–3b (the coordinate system) as you study the following discussions.

Mathematical and Graphic Concepts about AC

1. Angular motion is defined in terms of a 360° circle or a four-quadrant coordinate system.
2. The zero degree (starting or reference plane) is a horizontal line or arrow extending to the right, Figure 11–3a. NOTE: Other terms you might see for this line or arrow are a rotating

radius and a vector called a **phasor,** which represents a rotating vector's relative position with respect to time. A **vector** identifies a quantity that has (or illustrates) both magnitude *and* direction.

3. Angles increase in a positive, counterclockwise (CCW) direction from the zero reference point. In other words, the horizontal phasor to the right is rotated in a CCW direction until it is pointing straight upward, indicating an angle of 90° (with respect to the 0° reference point).
   - When the phasor continues in a CCW rotation until it is pointing straight to the left, that point represents 180° of rotation.
   - When rotation continues until the phasor is pointing straight downward, the angle at that point is 270°.
   - When the phasor completes a CCW rotation, returning to its starting point of 0°, it has rotated through 360°.
   - When the phasor is halfway between 0° and 90°, the angle is 45°. When the phasor is halfway between 90° and 180°, the angle is 135°.
4. The **coordinate system,** Figure 11–3b, consists of two perpendicular lines, each representing one of the two axes. The horizontal line is the x-axis. The vertical line is the y-axis. It is worth your effort to learn the names and directions of these two axes—they will be referenced often during your technical training. Remember the x-axis is horizontal and the y-axis is vertical.
5. There are four **quadrants** associated with this coordinate system. Again, look at Figure 11–3b and note that the angles between 0° and 90° are in quadrant 1. Angles between 90° and 180° are in quadrant 2. Angles between 180° and 270° are in quadrant 3; and angles between 270° and 360° (or 0°) are in quadrant 4.

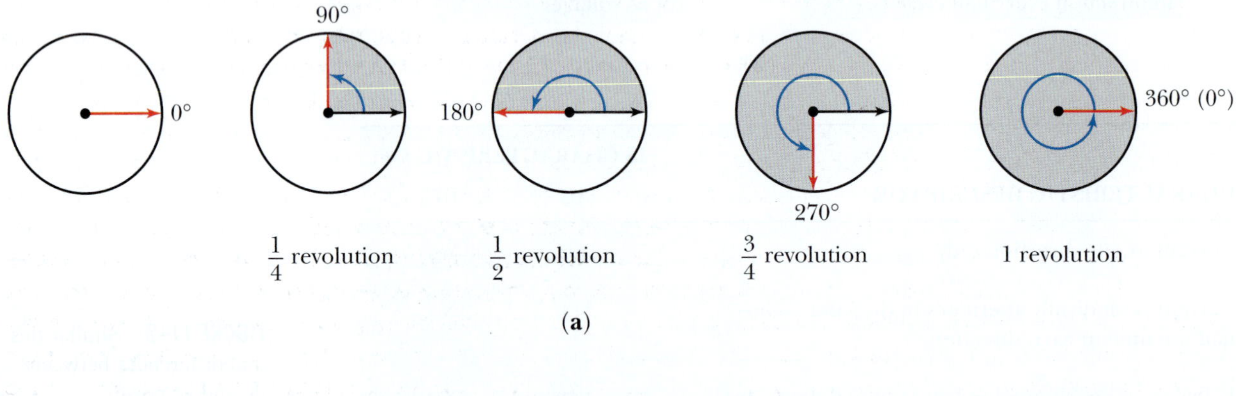

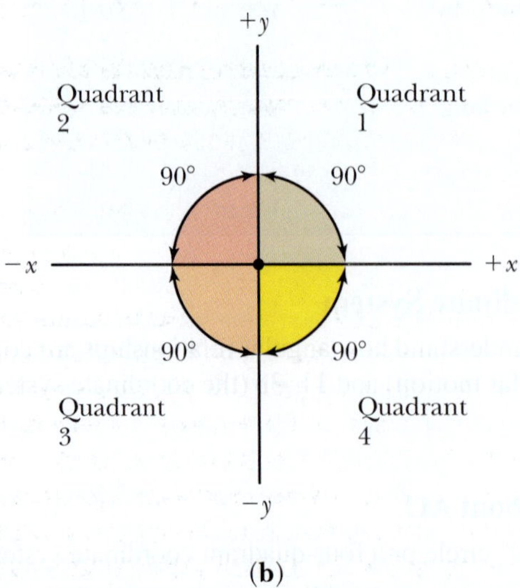

FIGURE 11–3 (a) Describing angular motion; (b) the coordinate system

6. When the phasor moves in a clockwise (CW) direction, the various angle positions are described as negative values. For example, the location halfway between 360° (0°) and 270° (–90°) in quadrant 4 (the same location as the +315° point) can be described as either –45° or +315°.

IN-PROCESS LEARNING CHECK 1

1. Define the difference between dc and ac. _____
2. Which direction is the reference, or 0° position, when describing angular motion? _____
3. Is the y-axis the horizontal or vertical axis? _____
4. Is the second quadrant between 0° and 90°, 90° and 180°, 180° and 270°, or 270° and 360°? _____
5. Define the term *vector.* _____
6. Define the term *phasor.* _____

Keeping in mind these thoughts about defining angles in degrees and a reference point on the coordinate system, let's move to a brief explanation of how an ac sine-wave voltage is generated.

11–2 Generating an AC Voltage

As a quick review, recall that generating a dc voltage is accomplished by means of rotating conductors (e.g., armature coil wires) cutting or linking lines of flux in a magnetic field. Observe that by means of a split-ring commutator and brushes, dc output voltage is delivered from the rotating armature to the load, and the dc is in the form of a pulsating dc, Figure 11–4a and b.

Refer to Figure 11–4c and note the elementary concepts of an ac generator. The simplified generator consists of a magnetic field through which a conductor is rotated. As you would expect when the maximum number of magnetic field flux lines are being cut (or linked) by the rotating conductor, maximum voltage is induced. Recall that if 10^8 flux lines are cut by a single conductor in one second, the induced voltage is one volt. Conversely, when minimum (or zero) flux lines are cut, minimum (or zero) voltage is induced.

Slip rings (sometimes called collector rings), are electrical contacting devices designed to make contact between stationary and rotating electrical junctures. As you can see in Figure 11–4c, the brushes are in a stationary position, and make contact between the external load resistance and the slip rings. The slip rings are connected to, and rotate along with, the rotating conductors. The outside surfaces of the slip rings, however, are in continuous sliding electrical contact with the brushes. The conductors, which are rotating through the magnetic field, are connected so that the same end of each conductor is always connected to the same slip ring.

Also, observe that the slip rings and brushes connect the external circuit load to the rotating loop so each end of the load is always connected to the same end of the rotating conductor. Thus, if the polarity of the induced voltage in the conductor changes, the voltage felt by the load through the brushes and the slip rings also changes.

Figure 11–4d shows the graphical representation of the changing magnitude and polarity of generator output felt by the load throughout one complete 360° rotation of the conductor loop(s). As we stated earlier, this graphic representation of the changing amplitude and polarity of the output voltage over time is called a waveform. In this case, the waveform is that of one full **cycle** of an ac sine-wave signal. A cycle is often defined as one complete sequence of a series of recurring events. There are many examples of cycles. The 24-hour day, the 12-hour clock face, one revolution of a wheel, and so on. In Figure 11–4d, you can see how at any given instant or point throughout the cycle, the signal's amplitude can be related to the angular degrees of rotation of the generator. Figure 11–5 again relates sine-wave amplitude and polarity to degrees throughout 360°. For a sine wave, you can see that maximum positive amplitude occurs at 90°. Maximum negative amplitude occurs at 270°. And, the sine-wave signal is at zero level at 0°, 180°, and 360° (assuming we start the waveform analysis at the zero point, with the signal progressing in a positive direction).

Practical Notes

As you will see in upcoming studies, we do not always have to analyze an ac sine-wave cycle as starting at zero amplitude level, and a cycle being completed at zero amplitude. We can define one complete cycle as starting at any point along the sine wave, and being completed at the next point along the waveform that is at the same amplitude and changing in the same direction. (That is, the point at which the sequence of recurring events starts again.)

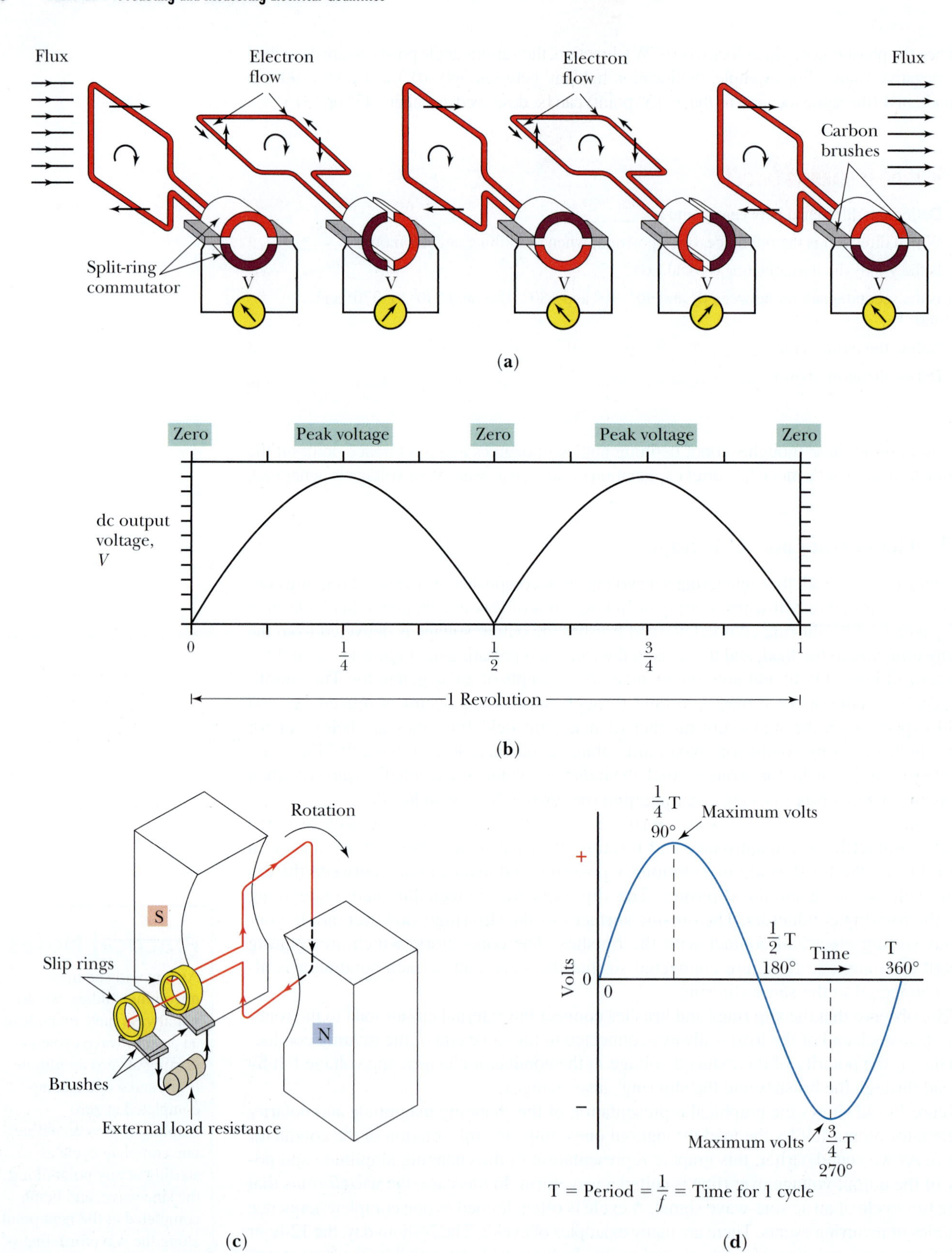

FIGURE 11–4 Basic concepts of an ac generator

The Sine Wave

This unique, single-frequency waveform reflects a quantity that constantly changes in amplitude and periodically reverses in direction.

Radians of Angular Measure

You have been introduced to the fact that sine-wave signals are often analyzed in terms of angular measure. The radian is another angular measure of importance, as many of the mathematics associated with ac analysis often use this measure.

As you might surmise from the word *radian* itself, there is a relationship between the term *radius* (of a circle) and the term *radian* (used in circular angular measure). One definition for a radian (sometimes abbreviated "rad") is: "In a circle, the angle included within an arc equal to the radius of the circle." In Figure 11–6, you can see that the angle included by an arc of this length is about 57.3°.

The thing that makes the radian of particular interest, mathematically, is that there are exactly 2π radians in 360°. If you divide 360° by 57.3°, you will see that it would take 6.28 (or 2π) radians to make up a complete 360° circle. Observe Figure 11–7 and note the relationships between the degrees and radians around the circle. It is important to remember that one complete cycle of ac (360°) equals 2π radians, that half a cycle (180°) equals 1π radian, and so on. See Figure 11–7b. The waveform in Figure 11–7b is called a **sine wave** because the amplitude (amount) of voltage at any given moment is directly related to the sine function, used in trigonometry. That is, the trigonometric sine (sin) value for any given angle (θ or theta), represented along the sine wave, relates directly to the voltage amplitude at that same

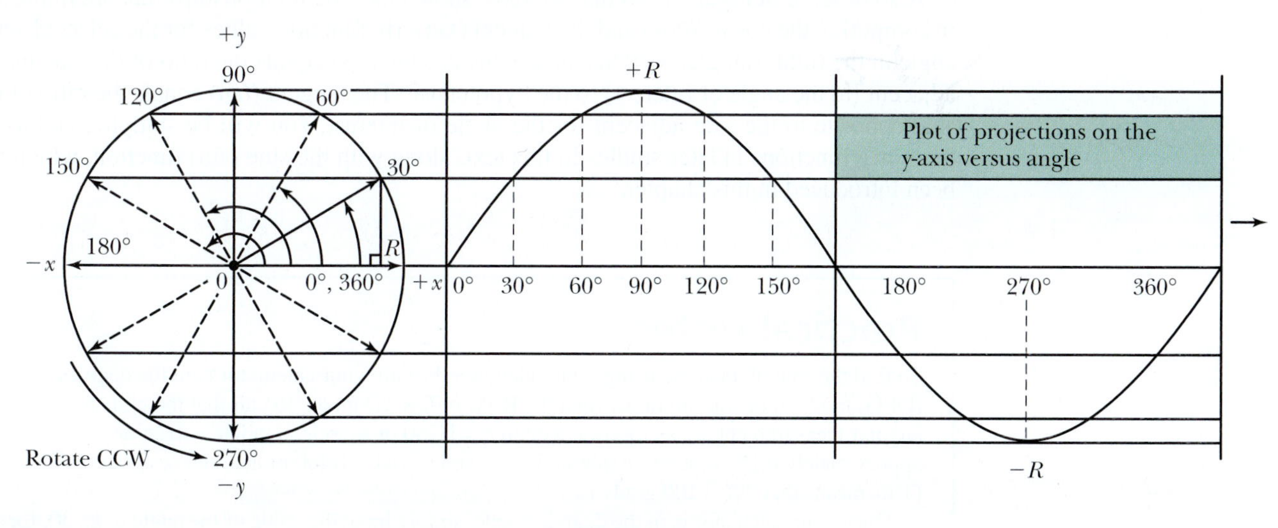

FIGURE 11–5 Projections of sine-wave amplitudes in relation to various angles

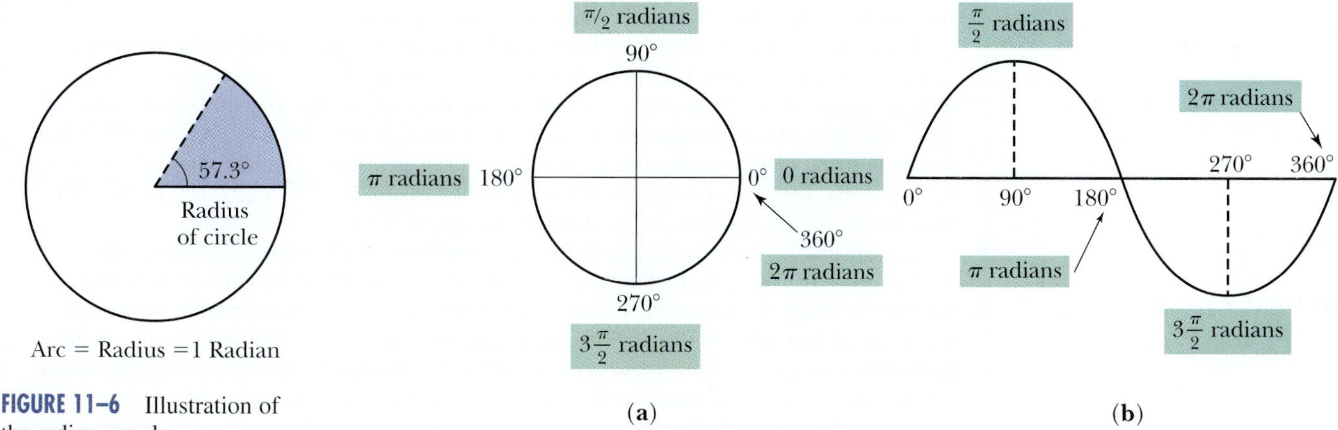

FIGURE 11–6 Illustration of the radian angular measurement

FIGURE 11–7 Degrees and radians of angular measure

For angle A:

1. The side opposite the right angle is called the *hypotenuse* (c).
2. The side opposite angle A is called the *opposite side* (a).
3. The side of angle A which is not the hypotenuse is called the *adjacent side* (b).

Sine (sin) of angle $= \dfrac{\text{opposite side}}{\text{hypotenuse}}$; $\sin A = \dfrac{a}{c}$

Cosine (cos) of angle $= \dfrac{\text{adjacent side}}{\text{hypotenuse}}$; $\cos A = \dfrac{b}{c}$

Tangent (tan) of angle $= \dfrac{\text{opposite side}}{\text{adjacent side}}$; $\tan A = \dfrac{a}{b} = \dfrac{\sin A}{\cos A}$

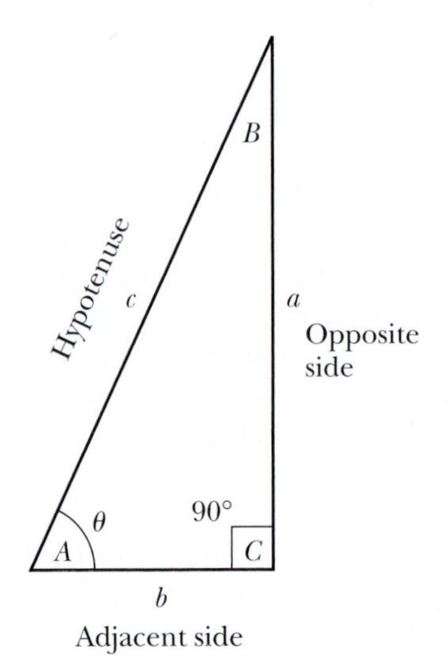

FIGURE 11–8 The right triangle

angle. NOTE: The sine of a given angle, using a right triangle as the basis of analysis, is equal to the ratio of the opposite side to the hypotenuse, Figure 11–8.

Also observe in Figure 11–8 that we have shown the side relationships that are important in computing the cosine (cos) and the tangent (tan) trig function values for the selected acute angle in the right triangle. As you can see, the cosine (cos) equals the ratio of the side that is adjacent (to the angle of interest) to the hypotenuse. The tangent (tan) equals the ratio of the side opposite to the side adjacent for the angle of interest. You will be studying and using these trig functions in later studies in this text, along with the sine (sin) function, which has been introduced in this chapter.

Practical Notes

To find the sine of an angle using your calculator, be sure your calculator is in the degrees (DEG) mode of operation, rather than the RAD or GRAD mode. Recall that the units for angular measurement are as follows: a degree = 1/360 of a circle; a radian = $1/2\pi$ or approximately 0.159 of a circle (about 57.3°); and a grad = 1/400 of a circle, or about 0.9°. (This means that 90° = 100 grads.)

Once your calculator is in the degrees mode, simply input the value of the angle (e.g., 30, for 30°), press the [SIN] function button, and read the value of the sine. In our example of 30°, the value of the sine is 0.5. The key sequence again is as follows: (the angle value), then the trig function key. The answer shows the value of the trig function.

To go the other way, that is, if you know that the ratio of the opposite side to the hypotenuse equals 0.5, and you want to find the angle whose sine value is 0.5: input 0.5, then press the [2nd] and [SIN^{-1}] keys. On some calculators, you press [INV], then [SIN]. By pressing the [2nd] and [SIN^{-1}] (or [INV], [SIN]) keys, you are telling the calculator you want to know the angle whose sin is 0.5. When you do this, you should see 30 in the readout of the calculator. This means the angle whose sine is 0.5 is 30°. The key sequence again is: (the trig function value), the [2nd] or [INV] key, then the trig function key (i.e., the [SIN^{-1}] or [SIN] key). The answer shows the value of the angle.

You can use the same techniques to find the value for cosines and tangents and/or the angles involved. That is, to find the trig function value when the angle is known, input the value of the angle, then press the correct trig function key to get your answer. In the case of the cosine, if you input 30 (for 30°), then press the [COS] button, the readout will indicate that the value of the cosine for an angle of 30° is 0.866. If you input 30, then press the [TAN] button, you find the tangent of an angle of 30° is 0.577. Going the other way, input the trig function value, then the [2nd] or [INV] key, then the appropriate trig function key (e.g., [SIN^{-1}] or [SIN] key, as appropriate). The answer is the angle involved.

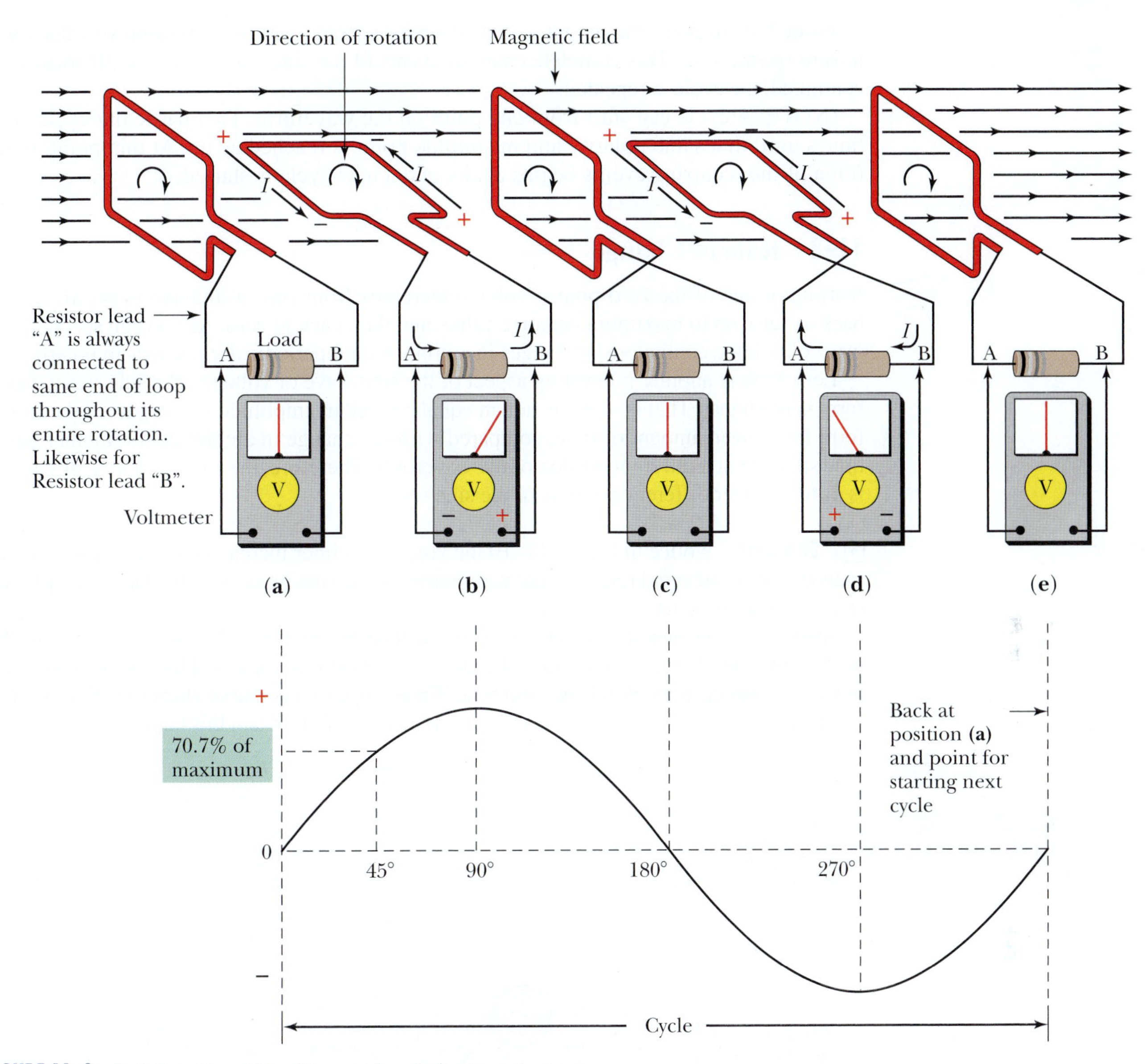

FIGURE 11–9 Relationship of flux lines cut in relation to output

Refer to Figure 11–9 as you read the following discussion. For simplicity, we will equate the angle of rotation from the starting point of the ac generator's mechanical rotation (where no flux lines are being cut by the rotating conductor) to the angle of the sine-wave output voltage waveform. For example, at the 0° point of rotation, no flux lines are being cut and the output voltage is zero (position a). If you find the sine value of 0° on a "trig" table, or use the "sin" function on a calculator, you will find it to be zero, and the output voltage at this time is zero volts.

At 45° into its rotation, the conductor cuts sufficient flux lines to cause output voltage to rise from 0 to 70.7% of its maximum value. The sine value for 45° is 0.707.

As the conductor rotates to the 90° position (position b), it cuts maximum lines of flux per unit time; hence, has maximum induced voltage. The sine function value for 90° is 1.00. For a sine wave, 100% of maximum value occurs at 90°.

As the conductor rotation moves from 90° toward 180°, fewer and fewer magnetic lines of force per unit time are being cut; hence, induced voltage decreases from maximum (positive) value back to zero (position c).

Continuing the rotation from 180° to 270°, the conductor cuts flux in the opposite direction. Therefore, the induced voltage is of the opposite polarity. Once again, voltage starts from zero value and increases in its opposite polarity **magnitude** until it reaches maximum negative value at 270° of rotation (position d).

From 270° to 360°, the output voltage decreases from its maximum negative value back to zero (position e). This complete chain of events of the sine wave is a cycle. If rotation is continued, the cycle is repeated.

(NOTE: A cycle can start from *any* point on the waveform. That particular cycle continues until that same polarity and magnitude point is reached again. At that point, repetition of the recurring events begins again, and a new cycle is started.)

11–3 Rate of Change

Starting at one of the zero points, voltage increases from zero to maximum positive, then back to zero, on to maximum negative value, and then back to zero. As you can see, the sine wave of voltage continuously changes in amplitude and periodically reverses in polarity.

Let's look at another interesting aspect of the sine wave of voltage. The sine-wave waveform is nonlinear. That is, there is not an equal change in amplitude in one part of the waveform for a given amount of time, compared with the change in amplitude for another equal time segment at a different portion of the waveform. Therefore, the *rate of change* of voltage is different over different portions of the sine wave.

◆ **EXAMPLE** Notice in Figure 11–10 the voltage is either increasing or decreasing in amplitude at its most rapid rate near the zero points of the sine wave (i.e., the largest amplitude *change* for a given amount of time).

Also, notice that near the maximum positive and negative areas of the sine wave, specifically, near the 90° and 270° points, the rate of change of voltage is minimum. This rate of change has much less *change* in amplitude per unit time. Knowledge of this rate of change (with respect to time) will be helpful to you in some later discussions, so try to retain this knowledge. ◆

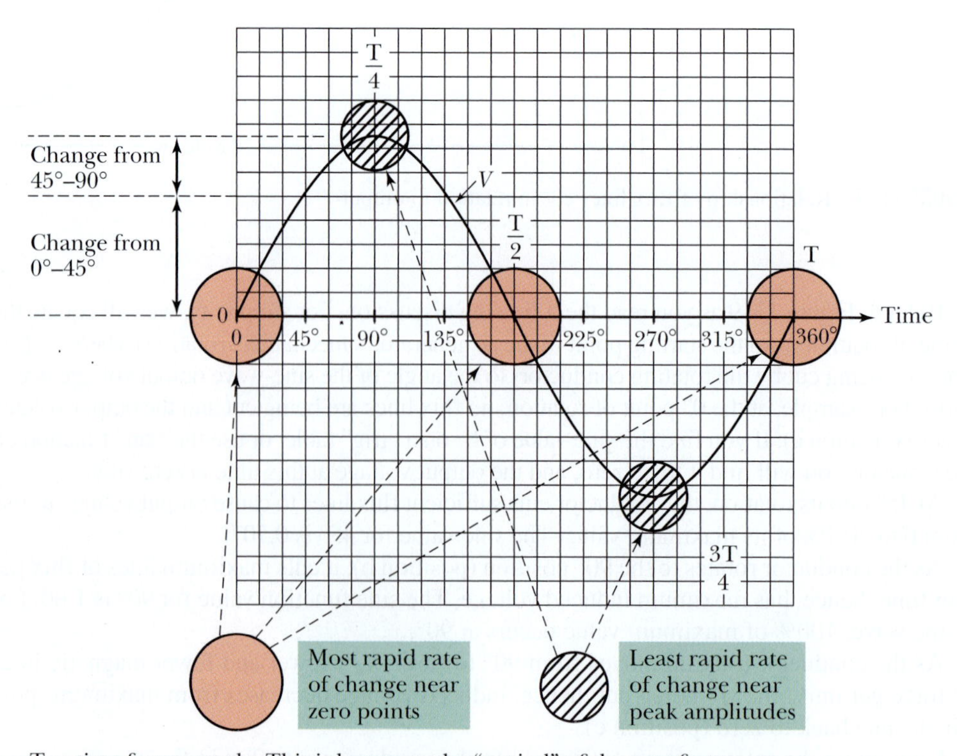

T = time for one cycle. This is known as the "period" of the waveform.

FIGURE 11–10 Different rates of change of voltage at various parts of sine wave

11–4 Introductory Information Summary Regarding AC Sine-Wave Signals

1. A sine-wave signal is an ac signal whose amplitude characteristics can be related to the trigonometric sine function.
2. An ac sine-wave signal continuously varies in amplitude and periodically changes polarity.
3. One complete cycle of an ac signal can be measured in terms of 360° and/or 2π radians (a radian = 57.3°).
4. When a signal's 0° starting point is considered to be at zero amplitude, it is common to think of that signal's maximum positive point occurring at 90°, the next zero point being at 180°, the maximum negative point being at 270°, and the cycle being completed at 360°, at zero amplitude, ready to start the next cycle.
5. One cycle of a sine-wave signal can be measured from any starting point along the waveform to the next point along the waveform that matches it, and starts another similar waveform sequence in terms of amplitude and direction variations.
6. The "rate of change" of the sine-wave waveform is greatest at the zero amplitude points, and least at the maximum amplitude points of the waveform.

PRACTICE PROBLEMS 1

1. Use a calculator or trig table and determine the sine of 35°.
2. Indicate on the following drawing at what point cycle 1 ends and cycle 2 begins, if the starting point for the first cycle is point A.

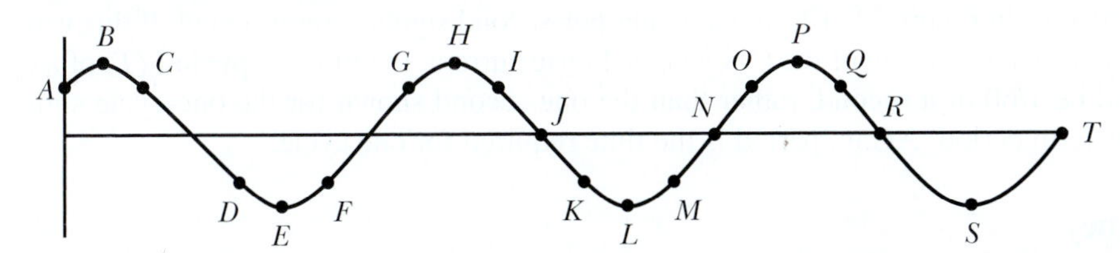

3. Which points on the drawing above represent areas of maximum rate of change of the sine wave(s)?

11–5 Some Basic Sine-Wave Waveform Descriptors

Refer to Figure 11–11. The horizontal line (x-axis) divides the sine wave into two parts—one above the zero reference line, representing positive values, and one below the zero reference line, representing negative values. Each half of the sine wave is an **alternation.** Naturally, the upper

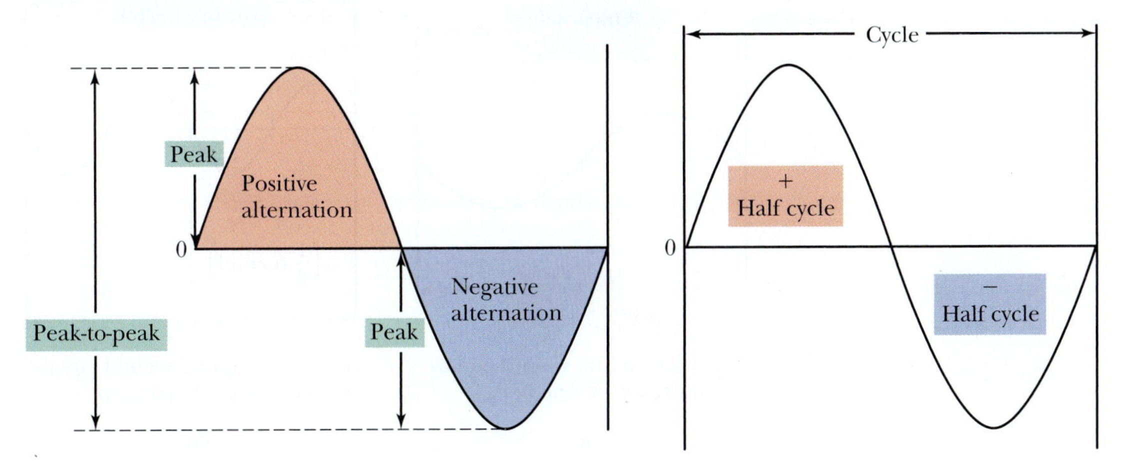

FIGURE 11–11 Several important descriptors of sine-wave ac

half is the "positive alternation," or positive half cycle. The lower half of the wave is the "negative alternation," or negative half cycle. With zero as the starting point, an alternation is the variation of an ac waveform from zero to a maximum value, then back to zero (in either polarity).

Examine Figure 11–11 again, and learn what is meant by positive peak, negative peak, and peak-to-peak values of the sine wave. Also, note in this illustration how a cycle is depicted.

Generalizing, a cycle, whether in electronics or any other physical phenomena, is one complete set of recurring events. Examples of this are one complete revolution of a wheel, and the 24-hour day (representing one complete revolution of the earth).

11–6 Period and Frequency

Periodic Waves

Periodic waves are those that repeat themselves regularly in time and form. Most ac signals and waves that you will study as a technician fall into this class. The period, or time of interest, for such waves is the time it takes for one complete cycle of that waveform. Let's see how the sine wave we have been discussing is defined in this regard.

Period

The ac sine-wave waveform has been described in terms of amplitude and polarity and has been referenced to various angles. It should be emphasized that the sine wave occurs over time—it does not instantaneously vary through all its amplitude values. The household ac, with which you are familiar, takes 1/60 of a second to complete a cycle. Look at Figure 11–12 and observe how the sine-wave illustrations have been labeled. Notice that time is along the horizontal axis. In Figure 11–12a, a one-cycle-per-second signal is represented. If this waveform represented one cycle of the 60-cycle ac house current, the time, or **period *(T)*** of a cycle would be 1/60 of a second, rather than the one second shown for the one-cycle signal. Learn the term period! Again, period is the time required for one cycle.

Frequency

The number of cycles occurring in one second is the **frequency** (f) of the ac. Refer again to Figure 11–12. The period of one cycle is inversely related to frequency. That is, the period equals the reciprocal of frequency, or $1/f$. The letter T represents period; that is, T is time of one cycle. Therefore:

FORMULA 11–1 $$T(\text{s}) = \frac{1}{f(\text{Hz})}$$

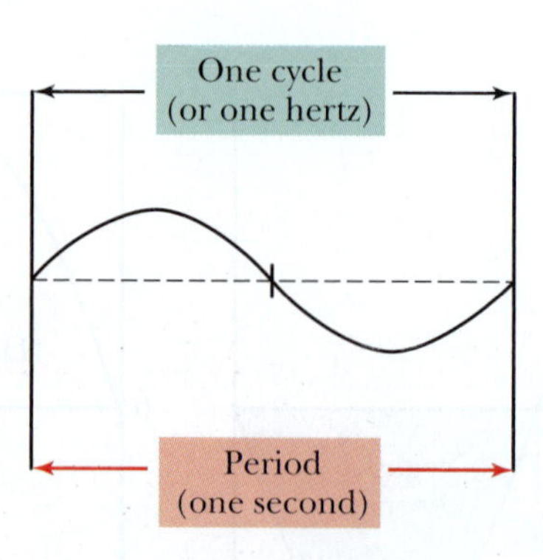

One cycle in one second equals a "frequency" of one hertz

(a)

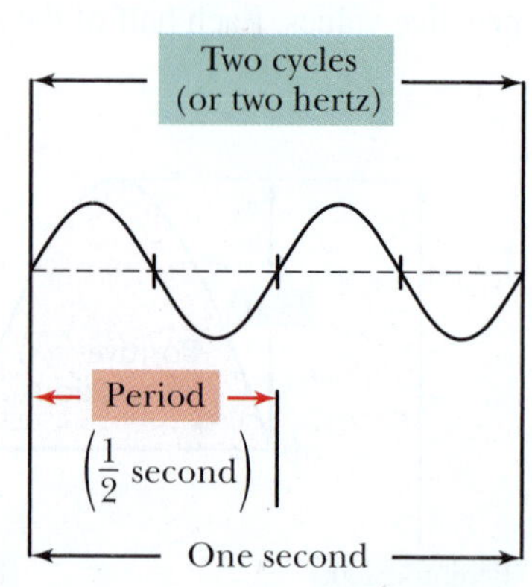

Two cycles in one second equals a "frequency" of two hertz

(b)

FIGURE 11–12 Comparing two frequencies

Restated, the period of one cycle (in seconds) equals 1 divided by the frequency in cycles per second. Rather than saying "cycles per second" when talking about frequency, we use a unit called the **Hertz (Hz),** named for the famous scientist Heinrich Hertz. For example, 1 Hz equals 1 cycle per second; 10 Hz equal 10 cycles per second; 1 kHz equals 1,000 cycles per second; and 1 MHz equals 1 million cycles per second. Notice the abbreviation for Hertz (Hz) *starts* with a capital H and is abbreviated Hz to prevent confusion with another electrical quantity you will study later—the henry, which is abbreviated with a capital H.

EXAMPLE

1. The period of a 100-Hz sine wave is $\frac{1}{100}$ of a second, $\left(T = \frac{1}{f}\right)$.

2. The period of a 1,000-Hz signal is $\frac{1}{1{,}000}$ of a second, $\left(T = \frac{1}{f}\right)$.

Relationship of Frequency and Period

Converse to calculating the period of one cycle by taking the reciprocal of the frequency (i.e., $T = 1/f$), the formula is transposed to find the frequency, if you know the time it takes for one cycle. The formula is:

FORMULA 11–2 $f\,(\text{Hz}) = \dfrac{1}{T\,(\text{s})}$

This indicates the period is inverse to the frequency, and vice versa. That is, the higher (or greater) the frequency, the less time required for one cycle (the shorter the period), or the longer (or greater) the period, the lower the frequency must be. In other words, if $T\uparrow$, then $f\downarrow$; and if $f\uparrow$, then $T\downarrow$. NOTE: Again the [1/x] function on your calculator is useful.

EXAMPLE

1. A signal with a period of 0.0001 seconds indicates a signal with a frequency of 10,000 Hz, $\left(f = \frac{1}{T}\right)$.

2. A signal with a period of 0.005 seconds indicates a signal with a frequency of 200 Hz, $\left(f = \frac{1}{T}\right)$.

Practical Notes

Use the [1/x] (or x^{-1}) reciprocal function of your calculator for this type of calculation.

IN-PROCESS LEARNING CHECK 2

1. Determine the frequency of an ac signal whose period is 0.0001 second. _____
2. What is the period for a frequency of 400 Hz? _____
3. What time does one alternation for a frequency of 10 kHz take? _____
4. Does the y-axis represent time or amplitude in waveforms? _____
5. For waveforms, at what *angular points* does the amplitude equal 70.7% of the positive peak value? At what angular points is the sine wave at a 70.7% level during the negative alternation? _____
6. In the coordinate system, in which quadrants are all the angles represented by the positive alternation of a sine wave? _____
7. What is the period of a frequency of 15 MHz? _____
8. What frequency has a period of 25 ms? _____
9. As frequency increases, does T increase, decrease, or remain the same? _____
10. The longer a given signal's period, the _____ the time for each alternation.

Practical Notes

Another meaningful measurement related to frequency is wavelength. (In this case, it is an inverse relationship; i.e., the higher the frequency the shorter the wavelength.) In your later studies, you will be likely to deal more with this measurement. You will be learning more details of the various frequency spectrums, such as audio frequencies, radio frequencies, optics, and x-rays.

Wavelength is the distance a given signal's energy travels in the time it takes for one cycle of that energy. Incidentally, the symbol used to represent wavelength is the Greek letter lambda (λ). Basically, two factors determine the wavelength of any given frequency. The speed at which the signal's energy moves through its propagation medium, and the length of time for one cycle (or the period) of the specific frequency. For example, radio waves and light waves travel through space at a speed of about 186,000 miles per second, which translates to about 300 million meters per second. Signals in the audio spectrum, however, travel at around 1,130 feet per second. As you might surmise, the higher the frequency, the less time there is in one period (one cycle); hence, the less distance the wave travels in that time, and the shorter its wavelength.

The formula that describes the above data is:

$$\lambda = \text{velocity/frequency, or } \lambda = v/f$$

For frequencies (radio frequencies, etc.) that travel at the speed of light, the formula is:

$$\lambda \text{ (wavelength)} = \frac{300 \times 10^6}{\text{Freq. (Hz)}} \text{ or } \frac{300}{f\text{(MHz)}}$$

For example, the wavelength of a 2-MHz signal would be 300/2 = 150 meters. You can also find a frequency, if you know its wavelength. That is, Frequency = v/λ. In our example, Frequency (in MHz) = $300 \times 10^6/150 = 2$ MHz.

11–7 Phase Relationships

Alternating current (ac) voltages and currents can be *in phase* or *out of phase* with each other by a difference in angle. This is called **phase angle** and is represented by the Greek letter theta (θ). When two sine waves are the same frequency and their waveforms pass through zero at different times, and when they do not reach maximum positive amplitude at the same time, they are out of phase with each other.

On the other hand, when two sine waves are the same frequency, and when their waveforms pass through zero and reach maximum positive amplitude at the same time, they are in phase with each other.

Look at Figure 11–13 and note the following:

1. Sine waves a and b are in phase with each other because they cross the zero points and reach their maximum positive levels at the same time (shown on the x-axis). Zero points are at times t_0, t_2, and t_4, and the positive peak at time t_1.
2. Sine waves b and c are out of phase with each other because they do not pass through the zero points and reach peak values at the same time. In fact, sine wave b reaches the peak positive point (at t_1) of its waveform 90° before sine wave c does (at t_2). It can be said that sine wave c "lags" sine wave b (and also sine wave a) by 90°. Conversely, it can be said that sine wave b "leads" sine wave c by 90°. In other words, when viewing waveforms of two *equal-frequency* ac signals, the signal reaching maximum positive first is leading the other signal, in phase. "First" refers to the peak that is closest to the y-axis; thus, showing an occurrence at an earlier time.

Obviously, ac signals can be out of phase by different amounts. This is illustrated by sine-wave waveforms and phasor diagrams, Figure 11–14. Figure 11–14a shows two signals that

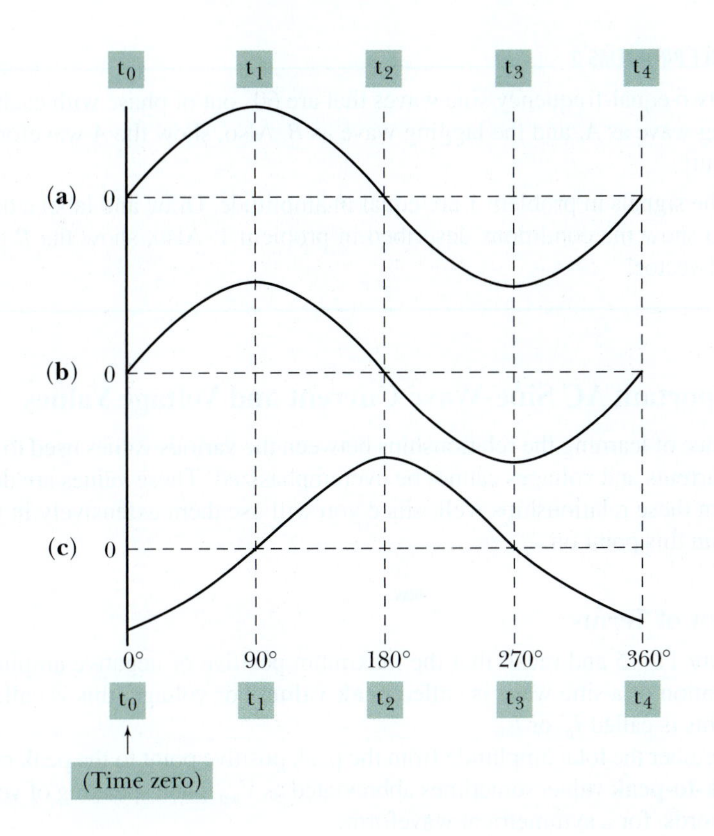

FIGURE 11–13 Phase relationships

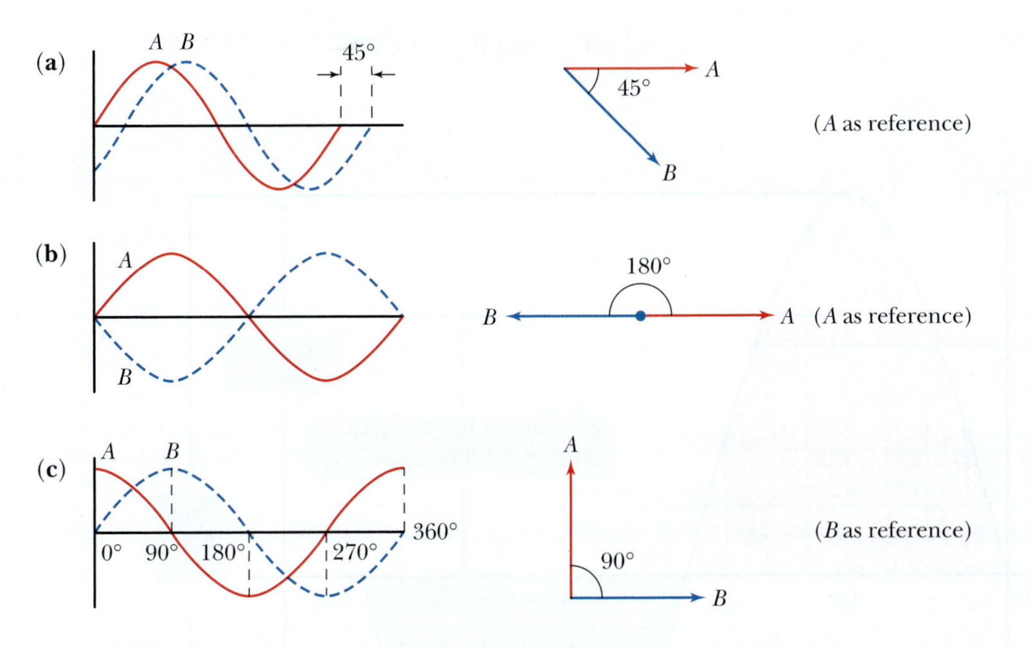

FIGURE 11–14 Diagrams depicting out-of-phase signals

are 45° out of phase. Which signal is leading the other one by 45°? The answer, of course, is *A* is leading *B* by 45°.

In Figure 11–14b, waveforms *A* and *B* are 180° out of phase. If these signals were of equal amplitude, they would cancel each other out.

In Figure 11–14c, you see two signals that are 90° out of phase.

Notice how the phasors in Figure 11–14 illustrate both magnitude and relative phase for several conditions. Observe that the *length* of the phasors can represent *amplitude* values of the two waveforms. Both phasors must represent the same relative point or quantity for each of the two waveforms being compared.

PRACTICE PROBLEMS 2

1. Illustrate two equal-frequency sine waves that are 60° out of phase with each other. Label the leading wave as *A*, and the lagging wave as *B*. Also, show the *A* waveform starting at the 0° point!
2. Assume the signals in problem 1 are equal in amplitude. Draw and label a diagram using phasors to show the conditions described in problem 1. Also, show the *B* phasor as the horizontal vector!

11–8 Important AC Sine-Wave Current and Voltage Values

The importance of learning the relationships between the various values used to measure and analyze ac currents and voltages cannot be overemphasized! These values are defined in this section. Learn these relationships well, since you will use them extensively in your training and work from this point on.

Brief Review of Terms

Look at Figure 11–15 and recall that the maximum positive or negative amplitude (height) of one alternation of a sine wave is called **peak value.** For voltage, this is called V_p, or V_M; for current, this is called I_p, or I_M.

Also, remember the total amplitude from the peak positive point to the peak negative point is called **peak-to-peak** value, sometimes abbreviated as $V_{p\text{-}p}$ when speaking of voltage values.

In other words, for a symmetrical waveform:

$$V_p = \frac{1}{2} \times V_{p\text{-}p},\ \text{and}\ V_{p\text{-}p} = 2 \times V_p$$

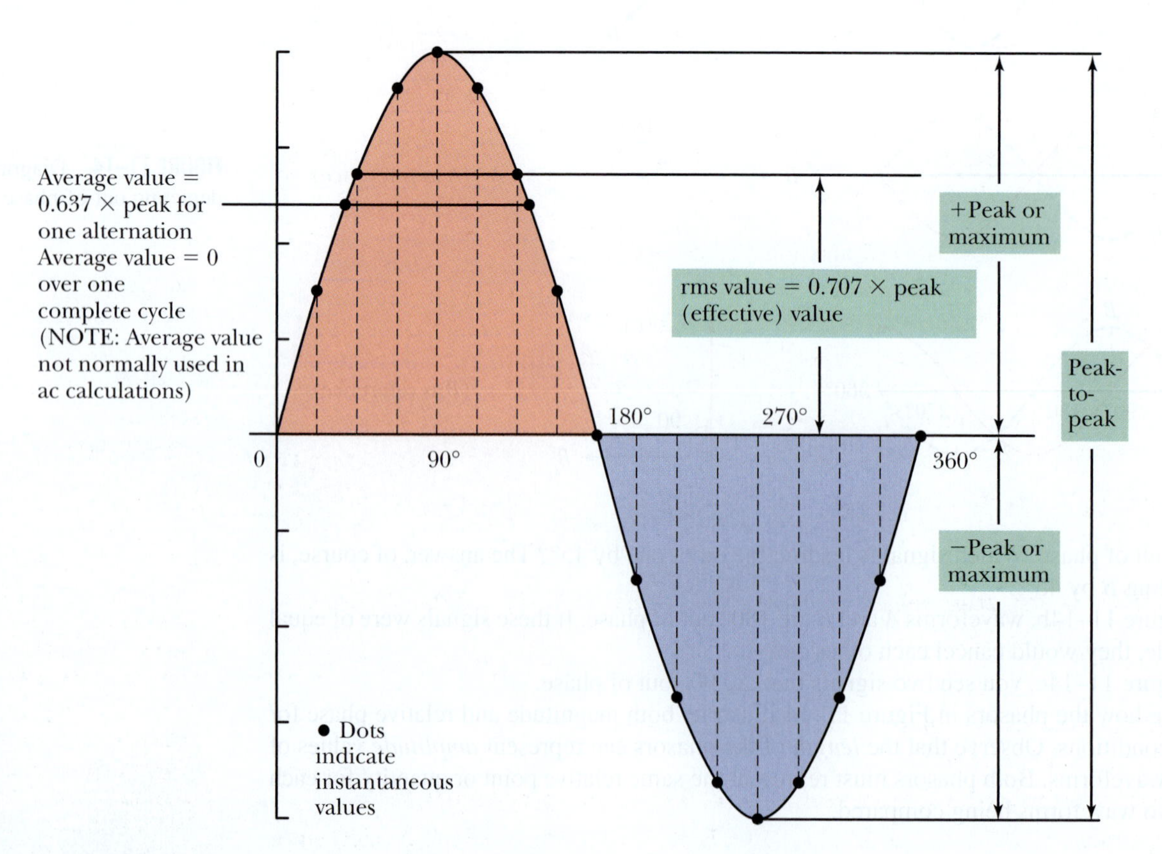

FIGURE 11–15 Important values related to sine-wave waveforms

Some New Terms and Values

Refer again to Figure 11–15 as you study the following discussions:

1. Since a given amount of current through a component in either direction causes the same amount of voltage drop, power dissipation, heating effect, and so on, the effects of an ac voltage or current are analyzed on a single alternation. For example, the positive alternation causes the same amount of current through a component (like a resistor) as the negative alternation, and vice versa.
2. The most frequently used value related to ac is called the **effective value (rms).** This is *that value of ac voltage or current producing the same heating effect as an equal value of dc voltage or current.* That is, it produces the same power:

$$I^2R \text{ or } \frac{V^2}{R}$$

For example, an effective value of ac current of two amperes through a given resistor produces the same heating effect as two amperes of dc current through that same resistor.

Effective value is also commonly termed the root-mean-square value. The root-mean-square term comes from the mathematical method used to find its value. It is derived from taking the square root of the mean (average) of all the squares of the sine values. Thus, effective value (rms) is 0.707 times maximum or peak value. Again, the reason effective value is greater than half of the maximum value relates to the nonlinearity of the rate of change of voltage or current throughout the alternation.

FORMULA 11–3 Effective value (rms) = 0.707 × Peak value

EXAMPLE Again, assuming 1-V peak value for the waveform in Figure 11–15, $V_{rms} = 0.707 \times 1 = 0.707$ V.

3. Although average value only has meaningful significance as it relates to the subject of rectification of ac (changing of ac to dc), we will simply introduce you to the concepts here. This awareness will then become useful in your later studies. If you compute the average amplitude or level under the waveform outline of *one-half cycle* or one alternation of an ac waveform, the result is the **average value** equals 0.637 (63.7%) of the maximum (or peak) value. Again, the reason the average value is more than 50% of the peak value is because the rate of change is slower near peak value than it is near the zero level. That means the voltage level is above 50% of maximum level for a longer time than it is below 50% of maximum level.

In other words, the sine wave is a nonlinear waveform. Therefore, the average is not half of the peak value. *Incidentally, when you "average" over the entire cycle, rather than for one alternation, the result is zero,* since the positive and negative alternations are symmetrical, equal in amplitude and time, and opposite in polarity. However, computing only one alternation or half cycle, the formula is:

FORMULA 11–4 Average value = 0.637 × Peak value

EXAMPLE If the peak value in Figure 11–15 were 1 V, V_{avg} (one-half cycle) = 0.637 × 1.0 = 0.637 V.

Summary of Relationships for Common AC Values (for Sinusoidal Waveforms)

1. Effective value (rms) = 0.707 times peak value, or 0.3535 times peak-to-peak value.
2. Average value = 0.637 times peak value. (NOTE: It also = 0.9 times effective value, since 0.637 is approximately 9/10 of 0.707.)

Practical Notes

Most voltmeters and current meters used by technicians are calibrated to measure effective, or rms, ac values. This includes VOMs, DMMs, and clip-on current and volt-meters.

In ac, power (heating effect) is computed in purely resistive components or circuits by using the effective voltage or current values. That is, $P = V_{rms} \times I_{rms}$, or $P = I^2R$, or $P = V^2/R$.

3. Peak value = maximum amplitude within one alternation. Since effective value (rms) is 0.707 of peak value, it follows that peak value equals 1/0.707, or is 1.414 times effective value. That is, $V_p = 1.414 \times$ rms value. (NOTE: It also equals 0.5 times peak-to-peak value.)
4. Peak-to-peak value = two times peak value and is the total sine-wave amplitude from positive peak to negative peak. It follows then, that effective value (rms) is 0.3535 of the peak-to-peak value (or half of 0.707). That is, $V_{p\text{-}p}$ equals 1/0.3535, or peak-to-peak value equals 2.828 times effective value.

With the knowledge of these various relationships, try the following practice problems.

PRACTICE PROBLEMS 3

1. The peak value of a sine-wave voltage is 14.14 V. What is the rms voltage value?
2. For the voltage described in question 1, what is the V_{avg} value over one alternation?
3. Approximately what percentage of effective value computed in question 1 is the average value computed in question 2?
4. Does frequency affect the rms value of a sine-wave voltage?
5. Are the results from calculating ac values from the negative alternation the same as ac values calculated from the positive alternation?
6. The peak-to-peak voltage of a sine-wave signal is 342 V. Calculate the following:
   a. ac rms (effective) voltage value __________
   b. peak voltage value __________
   c. instantaneous voltage value at 45° __________
   d. instantaneous voltage value at 90° __________
   e. ac average voltage value (over one-half cycle) __________

11–9 The Purely Resistive AC Circuit

Earlier, we discussed in-phase and out-of-phase ac quantities. In purely resistive ac circuits, the ac voltage across a resistance is in phase with the current through the resistance, Figure 11–16. It is logical that when maximum voltage is applied to the resistor, maximum current flows through the resistor. Also, when the instantaneous ac voltage value is at the zero point, zero current flows and so forth. Ohm's law is just as useful in ac circuits as it has been in dc circuits. (NOTE: **Instantaneous values** are expressed as *e* or *v* for instantaneous voltage and *i* for instantaneous current.)

EXAMPLE Applying Ohm's law to the circuit in Figure 11–16, if the rms ac voltage is 100 V, what is the effective value of ac current through the 10-Ω resistor? If you answered 10 A, you are correct! That is,

$$I = \frac{V}{R} = \frac{100 \text{ V}}{10\ \Omega} = 10 \text{ A}$$

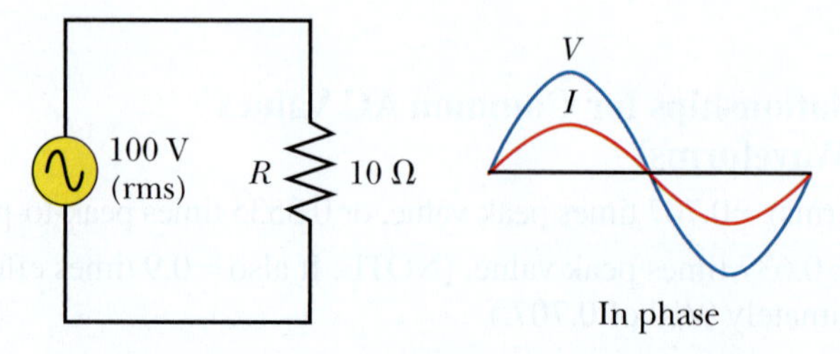

FIGURE 11–16 Basic ac resistive circuit

In ac circuits where voltage and current are in phase, the calculations to find *V, I,* and *P* are the same as they are in a dc circuit. Generally, effective (rms) values are used in these computations, unless otherwise indicated.

If you want to determine the instantaneous voltage at the 30° point of the sine wave, the value would equal the sine of 30° times the maximum value. In our sample case, the sine of 30° = 0.5. Therefore, if the maximum voltage (V_{max} or V_p) is 141.4 volts, the instantaneous value at 30° = 0.5 × 141.4 = 70.7 V.

FORMULA 11–5 $v = V_p \sin \theta$

where v = instantaneous voltage value
V_p = maximum voltage value
$\sin \theta$ = sine of the angle value

The analyses you have been using for dc circuits also applies to purely resistive circuits with more than one resistor. The only difference is that you typically will deal with ac effective values (rms). When you know only peak or peak-to-peak values, you have to use the factors you have been studying to find the rms value(s). In later chapters, you will study what happens when out-of-phase quantities are introduced into circuits via "reactive" components.

Try to apply your new knowledge of ac quantities and relationships to the following practice problems.

PRACTICE PROBLEMS 4

1. Refer to the series circuit in Figure 11–17 and determine the following parameters:
   a. Effective value of V_{R_1}
   b. Effective value of V_{R_2}
   c. I_p
   d. Instantaneous peak power
   e. P_{R_1}
2. Refer to the parallel circuit in Figure 11–18 and determine the following parameters:
   a. Effective value of I_1
   b. Effective value of I_2
   c. I_p
   d. $V_{p\text{-}p}$
3. Refer to the series-parallel circuit in Figure 11–19 and determine the following parameters:
   a. Effective value of V_{R_1}
   b. Effective value of V_{R_2}
   c. Effective value of V_{R_3}
   d. P_{R_1} (NOTE: Use appropriate rms values.)

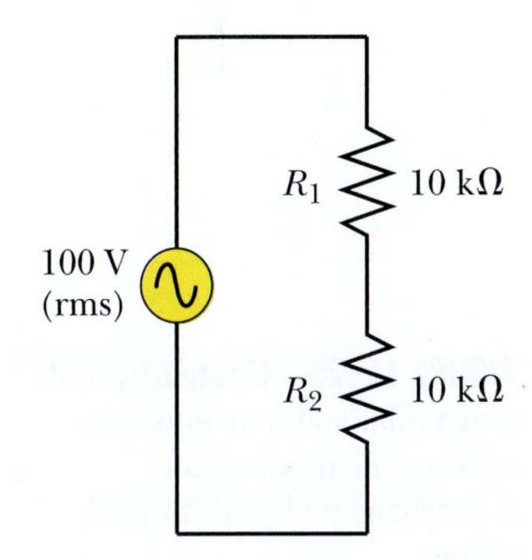

FIGURE 11–17 Purely resistive series ac circuit multiSIM

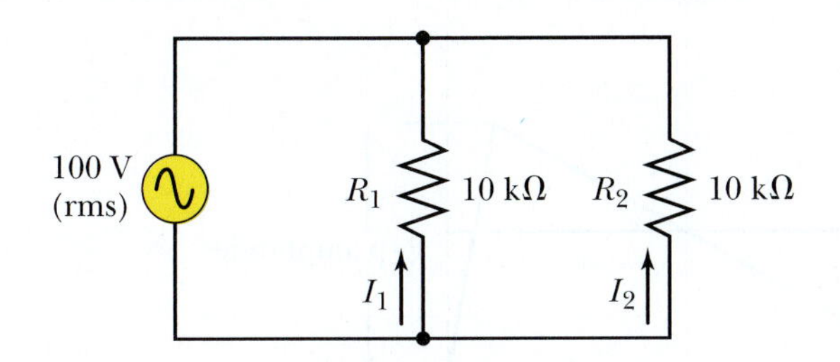

FIGURE 11–18 Purely resistive parallel ac circuit

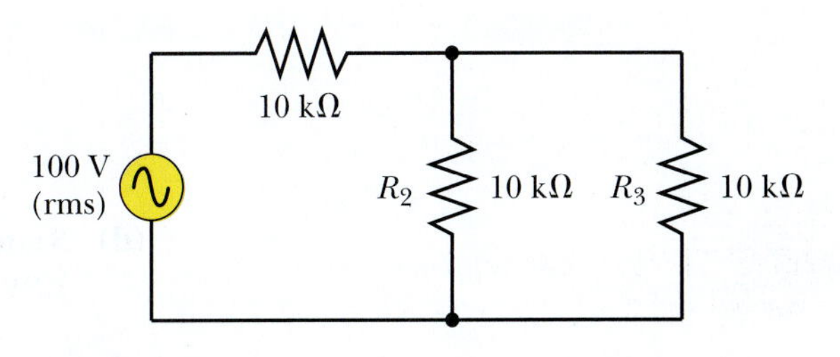

FIGURE 11–19 Purely resistive series-parallel ac circuit multiSIM

11–10 Other Periodic Waveforms

Important Waveshapes

Although the sine wave is a fundamental ac waveform, there are nonsinusoidal waveforms that are important for the technician to know and understand. Modern digital electronics systems, such as computers, data communications, radar, and pulse systems, and circuits requiring ramp waveforms, require that the technician become familiar with their features. It is not necessary in this chapter to deal in-depth with these various waveforms, but it is important to at least introduce you to them.

The most important waveforms include the **square wave,** the **rectangular wave,** and the **sawtooth** (ramp-shaped) **wave.** Refer to Figure 11–20 and note the comparative characteristic of each waveform with the sine wave in terms of period, wave shape, and peak-to-peak values.

How Nonsinusoidal Waveshapes Are Formed

It is interesting to note that a pure ac sine wave is comprised of one single frequency. Nonsinusoidal signal waveforms can be mathematically shown to be composed of a "fundamental" frequency sine wave *plus* a number of multiples of that frequency, called "harmonics." In other words, if the waveform is not sinusoidal, it indicates the presence of harmonics. For example, the square wave is composed of a fundamental sine-wave frequency and a large number of odd harmonic signals. The greater the number of odd harmonic signals that are added, or present, the closer to an ideal square wave the resultant complex waveform becomes (see Figure 11–21). Odd harmonics are those that represent

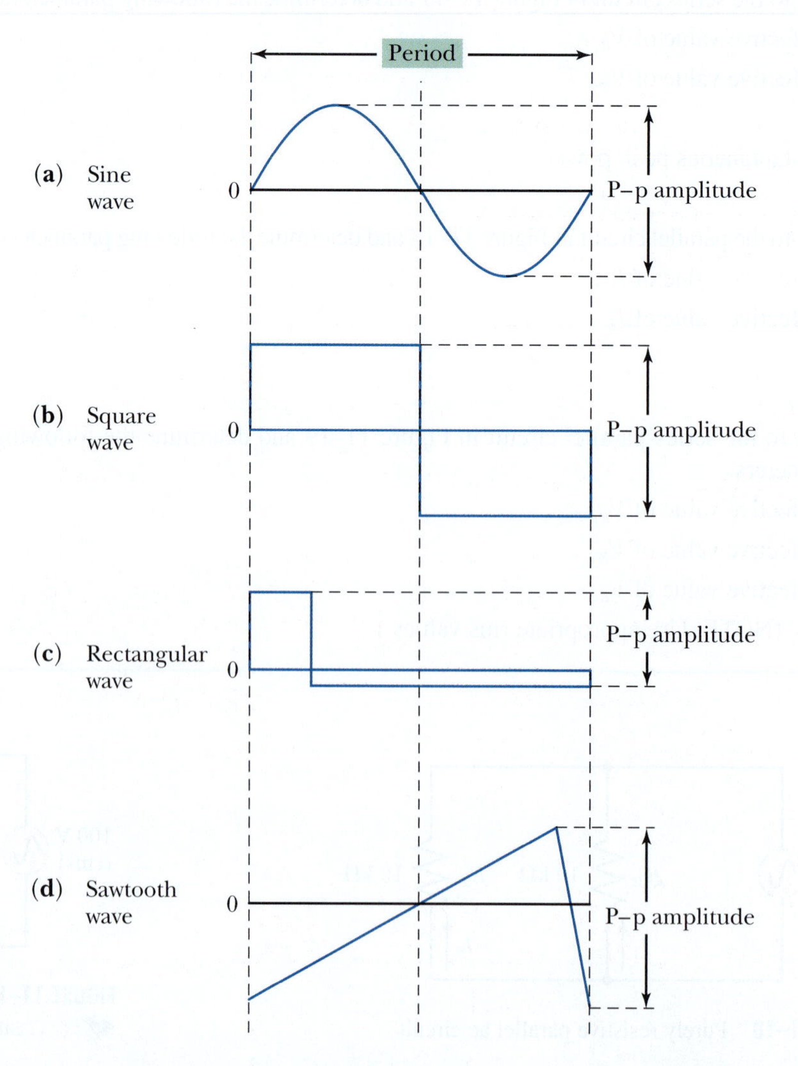

FIGURE 11–20 Comparison of sinusoidal and nonsinusoidal waveforms in terms of waveshape and peak-to-peak values

frequencies of 3 times, 5 times, 7 times, and so on, the fundamental frequency. For example, if the fundamental frequency is 100 Hz, then the third harmonic is 300 Hz, the fifth harmonic is 500 Hz, and so on.

Other nonsinusoidal waveforms may be created by the addition of even harmonic signals, or a mixture of even and odd harmonics, Figure 11–22. Even harmonics are even multiples of the fundamental frequency involved. That is, 200 Hz is the second harmonic of a fundamental frequency of 100 Hz; 400 Hz is the fourth harmonic frequency, and so on.

Important Parameters for Nonsinusoidal Waveforms

As you have already seen, the period of a nonsinusoidal waveform is determined in the same way the period is determined for a sine-wave signal; that is, the time it takes for one complete cycle of the signal, Figure 11–20. You have also seen that the peak-to-peak amplitude of nonsinusoidal signals is determined in a similar fashion to the sine-wave signal. That is, from the top-most point of the waveform to the bottom-most point on the waveform, Figure 11–20. Let's look at some more parameters of interest for these nonsinusoidal waveforms. NOTE: Some formulas used for sine-wave values are not useful for nonsinusoidal waves (i.e., $0.707 \times V_{PK}$, etc.).

Pulse-Width, Duty Cycle, and Average Values

Notice how the pulse width is identified in the series of cycles of rectangular waves shown in Figure 11–23. By comparing the pulse width with the time for one period, we can determine the

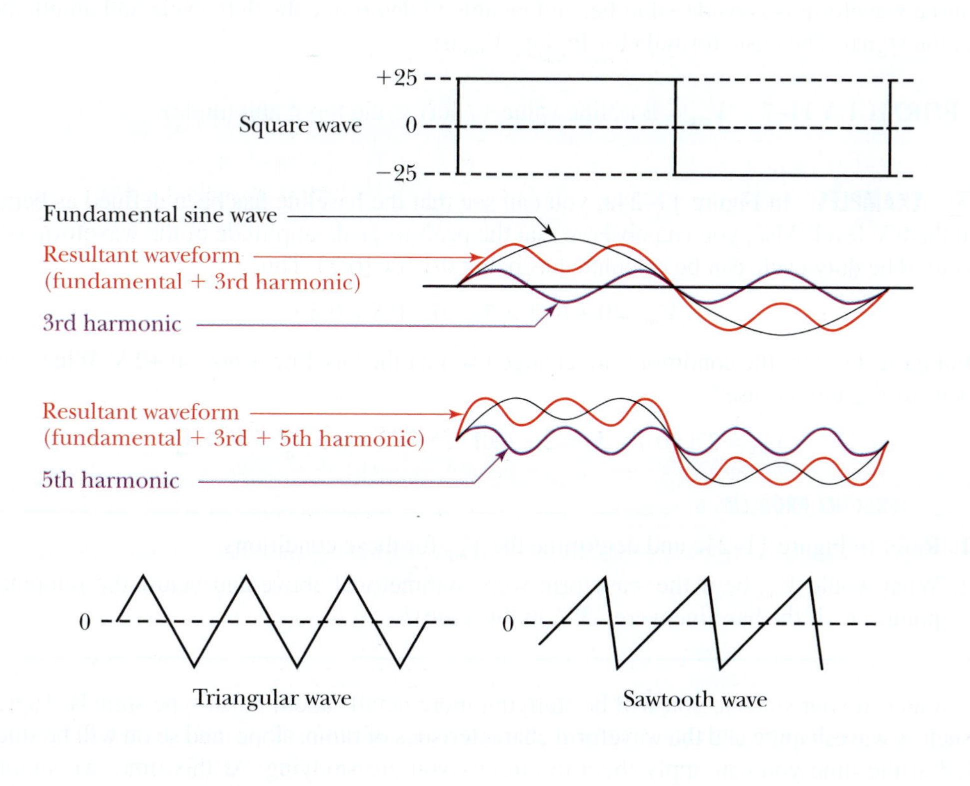

FIGURE 11–21 Graphic illustration of the forming of a square wave

FIGURE 11–22 Sample of other nonsinusoidal waveforms

FIGURE 11–23 Pulse width and period of a pulse-type waveform

duty cycle. That is, the *duty cycle equals the pulse width divided by the time for one period.* Since duty cycle is generally expressed in percentage, we then take this decimal number times 100% to find the percentage duty cycle:

FORMULA 11–6 $\% \text{ duty cycle} = \frac{t_w \text{ (Pulse width)}}{T \text{ (Time for 1 period)}} \times 100\%$

EXAMPLE If the time for one period of this repetitive pulse signal is 20 μs and the pulse width is 4 μs, then the percentage duty cycle is:

$$\% \text{ duty cycle} = \frac{4\ \mu\text{s}}{20\ \mu\text{s}} \times 100\% = 0.2 \times 100 = 20\% \text{ duty cycle}$$

PRACTICE PROBLEMS 5

What is the duty cycle of a repetitive pulse waveform if $T = 15$ μs and the $t_w = 1$ μs?

Finding Average Value for a Nonsinusoidal Waveform

In a number of situations in electronic circuits, these nonsinusoidal waveforms are not symmetrically balanced above and below the zero reference line. Many times a dc component is purposely added to "offset" the waveform, for operational purposes. To find the average value of voltage (V_{avg}) of such waveforms, it is necessary to understand where the "baseline" of the waveform is considered to be, and be able to determine the duty cycle and amplitude of the signal. The basic formula for finding V_{avg} is:

FORMULA 11–7 $V_{avg} = \text{Baseline value} + (\text{duty cycle} \times \text{p-p amplitude})$

EXAMPLES In Figure 11–24a, you can see that the baseline has been defined as being at the 0 V level. Also, you can observe that the peak-to-peak amplitude of the waveform is 5 volts. The duty cycle can be calculated as being 0.1 (or 10%). Thus,

$$V_{avg} = 0 + (0.1 \times 5) = 0 + 0.5 = 0.5 \text{ V}$$

In Figure 11–24b, the conditions are changed so that the baseline is now at +2 V. What happens to V_{avg} in this case?

$$V_{avg} = 2 + (0.1 \times 5) = 2 + (0.1 \times 5) = 2 + 0.5 = 2.5 \text{ V}$$

PRACTICE PROBLEMS 6

1. Refer to Figure 11–24c and determine the V_{avg} for these conditions.
2. What would V_{avg} be if the waveform were symmetrical above and below the reference point (i.e., if the baseline were –3 V, in this case)?

Later in your studies, you will be studying more details about pulse-type signals. Topics such as waveshaping and the waveform characteristics of ramp, slope, and so on will be studied at the time you can apply them to circuits you are studying. At this time, we simply wanted to introduce you to the fact that there are ac signals other than sine waves, and familiarize you with some of their basic characteristics.

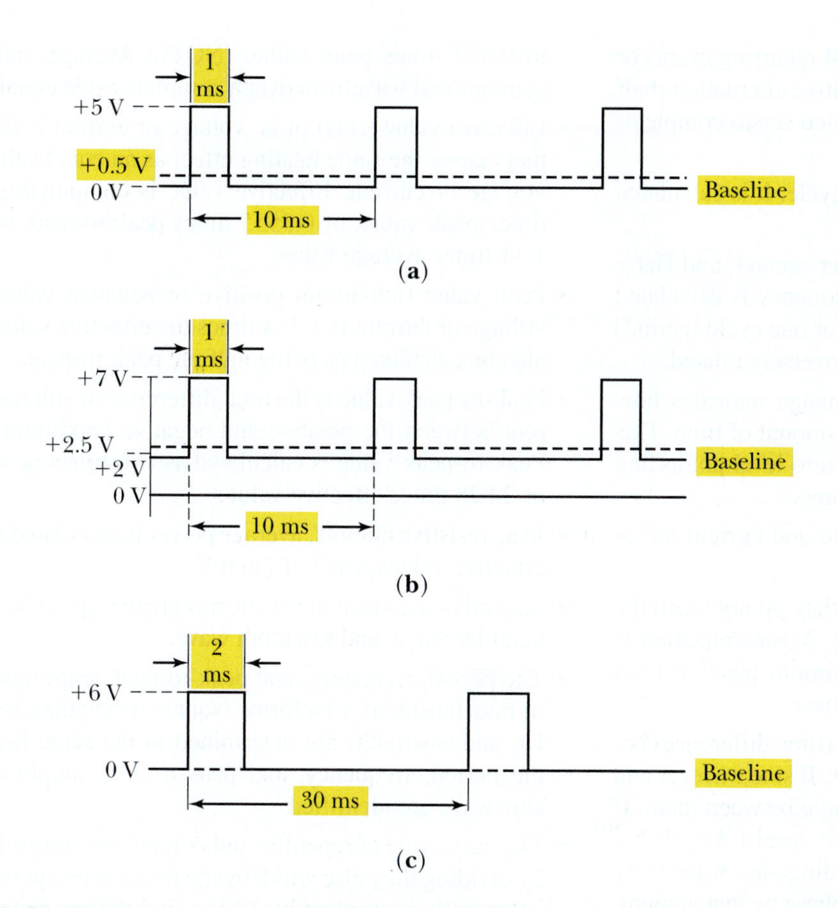

FIGURE 11–24 Finding V_{avg} for a rectangular waveform

Summary

- Alternating current (ac) varies continually in magnitude and periodically reverses in direction (polarity).
- Whereas ac is bidirectional, dc is unidirectional. Also, ac alternates in polarity, but dc has only one polarity. Furthermore, ac can be transformed or stepped up or down via transformer action, whereas dc cannot.
- The sine wave is related to angular motion in degrees, and any given point's value throughout the sine wave relates to the trigonometric sine function. Maximum values occur at 90° and 270° (–90°) points of the sine wave. Zero values occur at 0°, 180°, and 360° points of each cycle (assuming the wave is started at the zero level).
- A sine-wave voltage is generated by relative motion between conductor(s) and a magnetic field, assuming the conductor(s) links the flux lines via a rotating motion. The ac generator output is typically delivered to the load via slip rings and brushes.
- Quantities having magnitude and direction are frequently represented by lines or arrows called vectors. A vector's length and relative direction illustrate the quantity's value and angle.
- When representing angles relative to ac quantities, the zero reference point is horizontal and to the right. Angles increase positively from that reference plane in a counterclockwise (CCW) direction.
- When representing angles relative to the coordinate system, angles between 0° and 90° are in quadrant 1; angles between 90° and 180° are in quadrant 2; angles between 180° and 270° are in quadrant 3; and angles between 270° and 360° are in quadrant 4. These positive angles apply with CCW rotation of the phasor.

- A cycle represents one complete set of recurring events or values. In ac, the completion of a positive alternation (half cycle) and a negative alternation (or vice versa) completes a cycle.
- A period is the time it takes for one cycle. It is calculated by the formula $T = 1/f$.
- Frequency is the number of cycles per second, and Hertz (Hz) indicates cycles per second. Frequency is calculated using the formula $f = 1/T$, if the time for one cycle (period) is known. Frequency and period are inversely related.
- Referring to a sine wave, rate of change indicates how much amplitude changes for a given amount of time. This rate of change is maximum near the zero level points and minimum near the maximum level points.
- In purely resistive ac circuits, voltage and current are in phase with each other.
- When ac quantities are out of phase, they do not reach the same relative levels at the same time. A generalization is that the waveform that reaches maximum positive level first is leading the other quantity in phase.
- Phase angle is the difference in angle (time difference) between *two equal-frequency* sine waves. If sine waves A and B are in phase, there is zero phase angle between them. If sine wave A reaches maximum positive level 1/4 cycle before sine wave B, sine wave A is leading sine wave B by 90°. Thus, the two signals are out of phase by that amount. If sine wave A reaches maximum positive level 1/8 cycle before sine wave B, sine wave A is leading sine wave B in phase by 45°.
- Average value of ac voltage or current is the average level under one alternation's waveform. Average value is equal to 0.637 times peak value. (NOTE: Average value of a symmetrical waveform over a complete cycle equals zero.)
- Effective value (rms) of ac voltage or current is the value that causes the same heating effect as an equal value of dc voltage or current. Effective value is computed as 0.707 times peak value, or 0.3535 times peak-to-peak value, or 1.11 times average value.
- Peak value (maximum positive or negative value) of ac voltage or current is 1.414 times the effective value. It can also be calculated as being half the peak-to-peak value.
- Peak-to-peak value is the total difference in voltage or current between the positive and negative maximum values. Peak-to-peak value is calculated as two times peak value, or 2.828 times effective value.
- In ac resistive circuits, average power is calculated with the effective values (rms) of I and V.
- Several nonsinusoidal waveforms are the square wave, rectangular wave, and sawtooth wave.
- The period, frequency, and peak-to-peak amplitude values of nonsinusoidal waveforms (square, triangular, rectangular, and sawtooth) are determined in the same fashion as the period, frequency, and peak-to-peak amplitude of a sine wave are identified.
- The duty cycle of repetitive pulse-type waveforms is found by dividing the pulse width by the time for one period. This is typically multiplied by 100 to find the percentage duty cycle.
- The average voltage value (V_{avg}) of pulse-type waveforms is determined by adding the baseline voltage to the product of the duty cycle and the peak-to-peak amplitude of the waveform.

Formulas and Sample Calculator Sequences

Formula 11–1 *(To find the period of 1 cycle for a given frequency)*	$T(\text{s}) = \frac{1}{f(\text{Hz})}$ frequency in hertz, [1/x]
Formula 11–2 *(To find the frequency for a signal having a specified period time)*	$f(\text{Hz}) = \frac{1}{T(\text{s})}$ time for one cycle in seconds, [1/x]
Formula 11–3 *(To find effective value when peak value is known)*	Effective value (rms) = 0.707 × Peak value 0.707, [×], peak value of voltage, [=]
Formula 11–4 *(To find average value for one-half cycle when peak value is known)*	Average value = 0.637 × Peak value 0.637, [×], peak value of voltage, [=]
Formula 11–5 *(To find instantaneous voltage value when V peak is known)*	$v = V_p \sin \theta$ V peak value, [×], angle value, [SIN], [=]
Formula 11–6 *(To find percentage duty cycle of a pulse-type waveform)*	$\%\ \text{duty cycle} = \frac{t_w\ (\text{Pulse width})}{T\ (\text{Time for 1 period})} \times 100\%$ pulse-width time value, [÷], time for one cycle, [×], 100, [=]
Formula 11–7 *(To find the average voltage value for a pulse-type waveform)*	V_{avg} = Baseline value + (duty cycle × p-p amplitude) baseline value, [+], [(], duty cycle decimal value, [×], p-p value, [)], [=]

EXCEL AUTOMATED FORMULAS

Helpful problem-solving tools, such as the Excel automated formulas available on the CD, allow automatic calculations of formulas.

Using Excel

Basic AC Quantities Formulas

(Excel file reference: FOE11_01.xls)

DON'T FORGET! *It is not necessary to retype formulas once they are entered on the worksheet! Just input new parameters data for each new problem using that formula, as needed.*

- Use the Formula 11–2 spreadsheet sample and solve for frequency if the time for one period of the signal is 0.0025 seconds.

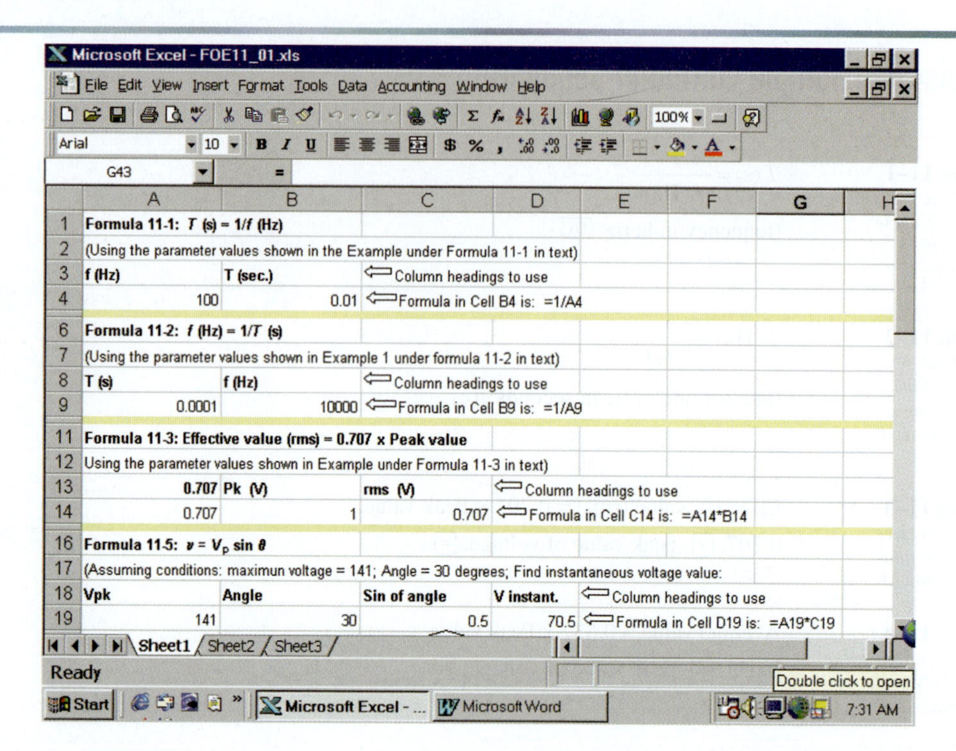

Using Excel

Basic AC Quantities Formulas

(Excel file reference: FOE11_01.xls)

DON'T FORGET! *It is not necessary to retype formulas once they are entered on the worksheet! Just input new parameters data for each new problem using that formula, as needed.*

- Use the Formula 11–5 spreadsheet sample and solve for the instantaneous voltage value assuming conditions of 200 *V* peak and an angle of 40°.

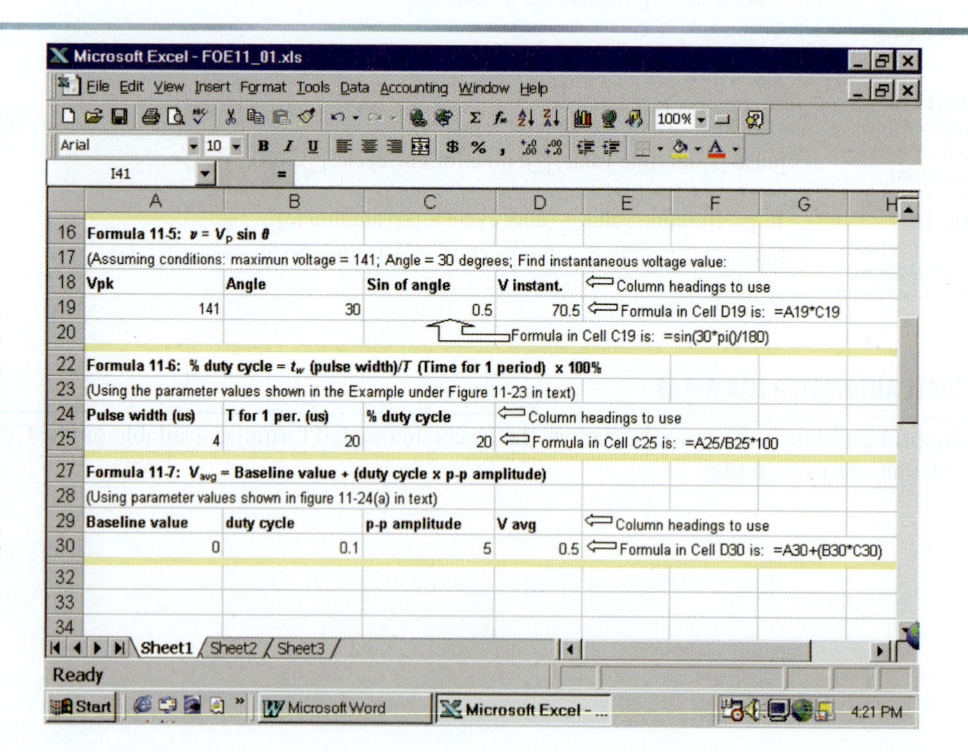

Review Questions

1. AC generator output is delivered via:
   - **a.** brushes alone.
   - **b.** commutator segments and brushes.
   - **c.** magnetic linkage.
   - **d.** slip rings and brushes.
2. The zero reference point often used to graphically represent angles for ac quantities on the coordinate system is:
   - **a.** vertical –90° upward from the horizontal plane.
   - **b.** horizontal, to the right.
   - **c.** horizontal, to the left.
   - **d.** vertical –90° downward from the horizontal plane.
3. Angles represented on the coordinate system increase positively from the zero reference point in a:
   - **a.** clockwise direction.
   - **b.** counterclockwise direction.
   - **c.** either CW or CCW direction.
4. When using the coordinate system, angles between 90° and 180° are represented:
   - **a.** in quadrant 1.
   - **b.** in quadrant 2.
   - **c.** in quadrant 3.
   - **d.** in quadrant 4.
5. In an ac sine wave, an alternation represents:
   - **a.** a full cycle.
   - **b.** a half cycle.
   - **c.** a quarter cycle.
   - **d.** two cycles.
6. A period is the time it takes for:
   - **a.** one-fourth cycle.
   - **b.** one-half cycle.
   - **c.** one cycle.
   - **d.** none of the above.
7. The formula for calculating a period is:
   - **a.** $T = \frac{1}{2}/f$.
   - **b.** $T = \frac{1}{4}/f$.
   - **c.** $T = 1/f$.
   - **d.** none of the above.
8. Frequency may be defined as:
   - **a.** the number of cycles per second, and is expressed in radians.
   - **b.** the number of cycles per second, and is expressed in Hertz.
   - **c.** the number of cycles per second, and is expressed in CPS.
   - **d.** none of the above.
9. As frequency is increased:
   - **a.** period increases.
   - **b.** period decreases.
   - **c.** period does not change.
   - **d.** none of the above.
10. The rate of change of an ac sinusoidal waveform is maximum at:
    - **a.** waveform maximum points.
    - **b.** waveform median points.
    - **c.** waveform zero points.
    - **d.** none of the above.
11. By convention, when two graphically represented sinewave waveforms are not in phase:
    - **a.** the waveform that reaches maximum positive level first is considered leading.
    - **b.** the waveform that reaches maximum negative level first is considered leading.
    - **c.** the waveform that reaches the zero point first is considered leading.
    - **d.** none of the above.
12. Phase angle between two equal-frequency sine waves represents the:
    - **a.** difference in amplitude.
    - **b.** difference in time.
    - **c.** difference in both amplitude and time.
    - **d.** none of the above.
13. The average value of ac voltage or current is its average level over:
    - **a.** one complete cycle.
    - **b.** two complete cycles.
    - **c.** one alternation.
    - **d.** two alternations.

14. The effective value of ac voltage or current may be calculated as:

a. 0.707 × peak value.

b. 0.3535 × peak-to-peak value.

c. both of the above.

d. none of the above.

15. Peak-to-peak value represents the total difference between:

a. zero and positive peak value.

b. zero and negative peak value.

c. positive peak to negative peak value.

d. both (a) and (b) above.

16. Explain the terms *fundamental* and *harmonic.*

17. Define the term *periodic wave.*

18. Wavelength of a signal is affected by:

a. the speed the signal travels and the frequency of the signal.

b. the speed the signal travels and the amplitude of the signal.

c. the time of day the signal is propagated and the signal strength.

d. none of the above.

19. A square wave is composed of:

a. a fundamental sine wave and a great number of even harmonics.

b. a fundamental sine wave and a great number of odd harmonics.

c. a distorted sine wave.

d. none of the above.

20. The period, frequency, and peak-to-peak amplitude values of square waves, triangular waves, and rectangular waveforms can be determined:

a. in a slightly different manner than for regular sine waves.

b. in a completely different manner than for regular sine waves.

c. in the same manner as for regular sine waves.

d. cannot be determined.

Problems

1. What is the period of a 2-kHz voltage?

2. What is the frequency of an ac having a period of 0.2 μs?

3. How much time does one alternation of 60-Hz ac require?

4. The peak value of a sine wave is 169.73 volts. What is the rms value?

5. What is the frequency of a signal having one-fourth the period of a signal whose T equals 0.01 seconds?

6. What is the peak-to-peak voltage across a 10-kΩ resistor that dissipates 50 mW?

7. If the rms value of an ac voltage doubles, the peak-to-peak value must _____.

8. The ac effective voltage across a given resistor doubles. What happens to the power dissipated by the resistor?

9. The period of an ac signal triples. What must have happened to the frequency of the signal?

10. The rms value of a given ac voltage is 75 volts. What is the peak value? What is the peak-to-peak value?

11. What time is needed for one alternation of a 500-Hz ac signal?

12. One alternation of a signal takes 500 μs. What is its frequency? What is the time of two periods? (Use a calculator to find answers.)

13. Draw the diagram of a three-resistor series-parallel circuit having a 10-kΩ resistor (labeled R_1) in series with the source, and two 100-kΩ resistors (labeled R_2 and R_3) in parallel with each other and in series with R_1. Assume a 120-V ac source. Calculate the following parameters and appropriately label them on the diagram.

a. V_{R_1}

b. I_{R_1}

c. P_{R_1}

d. I_{R_2}

e. V_{R_3}

f. θ between V applied and I total

14. What is the wavelength of a radio signal with a frequency of 1 MHz?

15. If the frequency of a signal is lowered to one-third its original frequency, what happens to its wavelength? Be specific.

Analysis Questions

1. What are two differences between ac and dc?
2. Draw an illustration showing one cycle of a 10-V (rms), 1-MHz ac signal. Identify and appropriately label the positive alternation, the period, the peak value, the peak-to-peak value, the effective value, and points on the sine wave representing angles of 30°, 90°, and 215°.
3. The rms value of a given ac voltage is 100 V. What is the peak value? What is the peak-to-peak value?

Refer to Figure 11–25 and answer questions 4 to 8.

4. Which waveform illustrates the signal with the longest period?
5. What is the peak-to-peak voltage value of waveform B?
6. What is the effective value of voltage waveform A?
7. If the period of waveform A equals 0.002 ms, what is the frequency represented by waveform B?
8. If the peak-to-peak value of waveform A changes to 60 V, what is the new effective voltage value?
9. **a.** Draw a graphic illustration of one and one-half cycles of a rectangular wave. Assume the amplitude is 10 V, the baseline is +3 V, the period is 25 ms, and the duty cycle is 25%. Label all appropriate parameters on the drawing.
   b. Determine the V_{avg} for the signal described in question 9a.
10. Draw one cycle of a sine wave and a cosine wave on the same "baseline." Assume each maximum value is 10 units. Appropriately label the diagram to show the following information:
    a. Angles of 0°, 45°, 90°, 135°, 180°, 225°, 270°, 315°, and 360°, marked on the x-axis.
    b. Amplitude values of the sine wave at 45° and 90°, respectively. (Use a calculator to determine values.)
    c. Amplitude values of the cosine wave at 0°, 45°, and 90°, respectively. (Use a calculator to determine values.)
    d. Indicate which wave is leading in phase, and by how much.

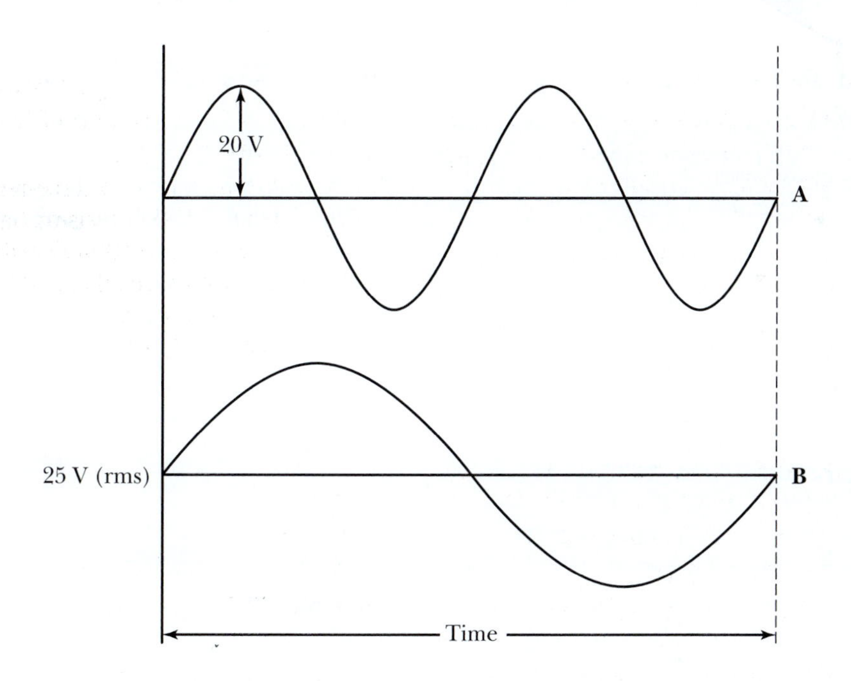

FIGURE 11–25

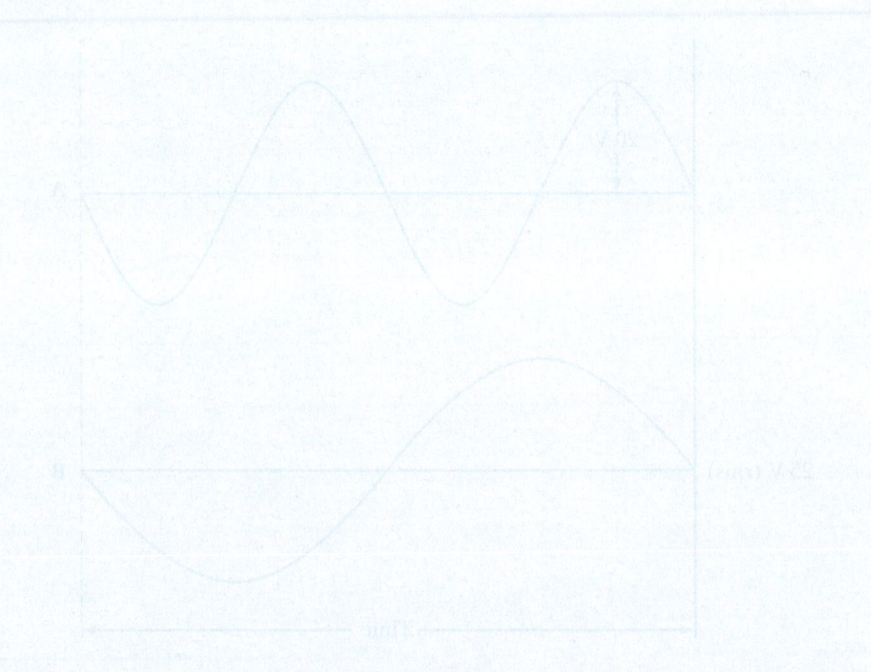

OUTLINE

OBJECTIVES

After studying this chapter, you should be able to:

1. List the key sections of the **oscilloscope**
2. List precautions when using scopes
3. List procedures when measuring voltage with a scope
4. List procedures to display and interpret waveforms
5. List procedures relating to phase determination
6. List procedures when checking frequency with a scope
7. Use the computer to solve circuit problems
8. Use the SIMPLER troubleshooting sequence to solve the Troubleshooting Challenge problem

CHAPTER 12

The Oscilloscope

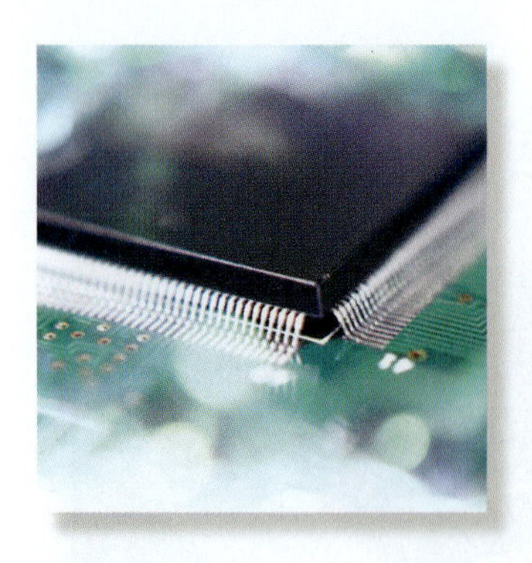

PREVIEW

The **oscilloscope,** often called a scope, is without doubt one of the most versatile tools available to technicians and engineers. It can display waveforms; help in determining voltages, currents, and phase relationships; and assist in computing and comparing frequencies.

In your studies, you have been introduced to other measuring instruments and to some important ac quantities. This chapter will increase your knowledge about measuring techniques. This chapter will also reinforce your knowledge about the important ac quantities you will measure and analyze throughout your remaining studies and in your career as a technician.

The purpose of this chapter is not to discuss the details of circuitry involved in making a scope function, nor to instruct you on subjects that are best learned with actual experience. However, this chapter will, perhaps, inspire you to become familiar with this valuable tool and will be a good background, as a first step in helping you to eventually become proficient in using this device.

KEY TERMS

Cathode-ray tube (CRT)
Oscilloscope
Position control(s)
Sweep frequency control(s)
Synchronization control(s)
Vertical amplifier
Vertical volts/div control(s)

12–1 Background Information

Scopes come in a variety of brands and complexities. Two general categories of scopes include those used for general-purpose benchwork applications, and the more complex (and expensive) laboratory-quality scopes, Figure 12–1a and b. A portable scope is shown in Figure 12–1c.

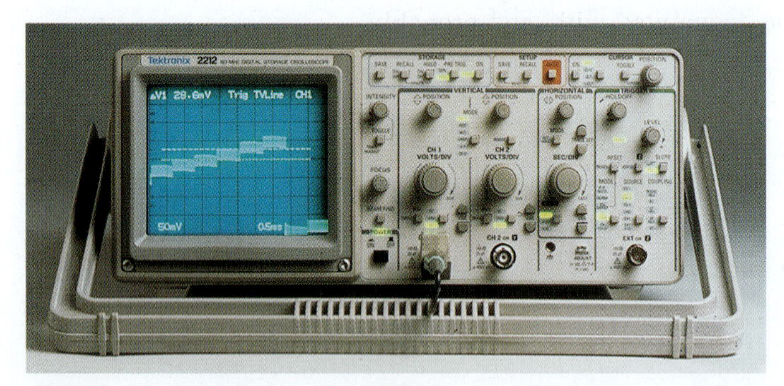

(a)

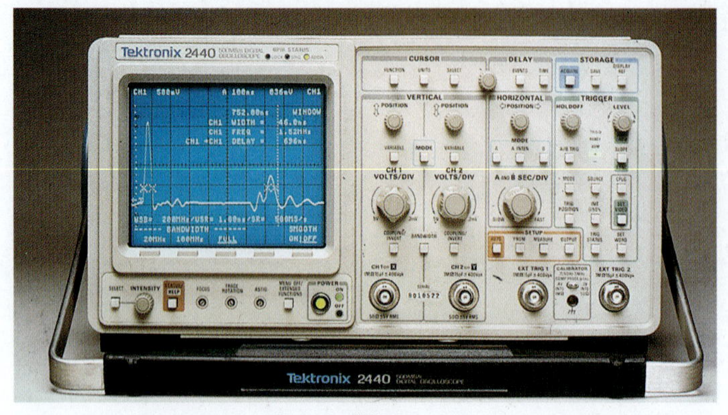

(b)

(c)

FIGURE 12–1 (a) General-purpose oscilloscope; (b) lab-quality oscilloscope; (c) portable oscilloscope *(Photos courtesy of Tektronix, Inc.)*

In Figure 12–2a, a "digital storage" oscilloscope (DSO) is shown. In today's world, you will find these quite commonly used. These scopes have the ability to store signal waveforms in memory for either immediate or later retrieval and display. This type of scope makes possible such applications as waveform comparisons at different times or locations, complex, long-duration and/or short-duration signal change analysis, and other unique types of signal comparisons and analyses. These are analyses that would be difficult or impossible to perform using standard analog real-time (ART) scopes. The "real-time" identifier simply indicates that what you see on the scope is happening now. To store signals, this scope "digitizes" the analog signals by sampling signal levels many times over the duration of the signal waveform. The digital data is then placed in memory for retrieval at the desired time.

Another generation in scopes, advertised by at least one well-known oscilloscope manufacturer, are those with "automatic measurements" capability, graphical interfaces, saving of image files to storage media disks, connectivity to computers, and other amazing capabilities. These scopes are advertised as having ability to display, store, and analyze complex signals in real-time, using *three dimensions* of signal information: *amplitude, time,* and *distribution of amplitude over time.* These abilities make this type of scope go beyond analog real-time (ART) and digital storage oscilloscopes' (DSO) capabilities. See Figure 12–2b for an example of this type of scope.

Many modern digital oscilloscopes make difficult waveform measurements much easier than was true of the old analog scopes. Front panel buttons and on-screen menus make your selection of automated measurements very intuitive. Measurements of amplitude, period, rise, or fall times are all easily done. Some of these modern scopes also can perform some mathematics for you in terms of determining mean and RMS calculations, as well as duty cycle computations, and so on. Automated measurements with these scopes appear as on-screen alphanumeric readouts, which are typically more accurate than you could perform by trying to interpret values from the scope graticule. Also, many modern scopes can produce outputs that can be coupled to computers and printers, which adds further to their flexibility and usefulness.

For our purposes in this chapter, we will simply introduce you to some basic functions that you will initially encounter in your training environment. As you get out into the industry, you will have the opportunity to become acquainted with, and probably use, some of the more sophisticated varieties of scopes we have mentioned.

Two general features that are common to virtually all scopes include the following:

1. All scopes have a cathode-ray tube (CRT) where the visual displays are viewed.
2. All scopes have controls and related circuits that adjust the display to help you analyze voltage, time, waveform, and frequency parameters of signal(s) under test. In some cases, these controls are manipulated manually; in other cases, they may be automated, depending upon the type of oscilloscope you are using.

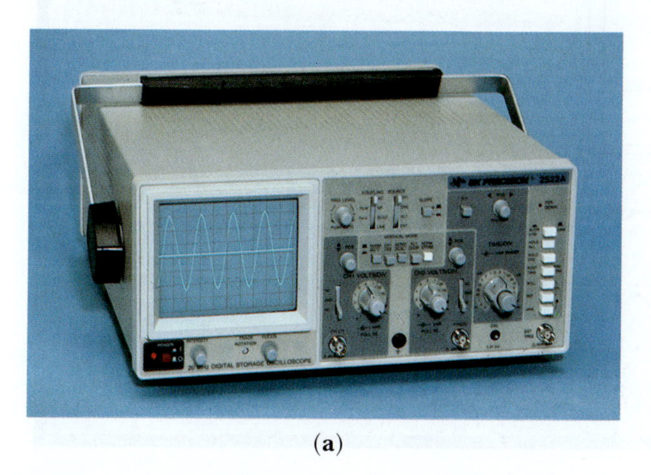

(a)

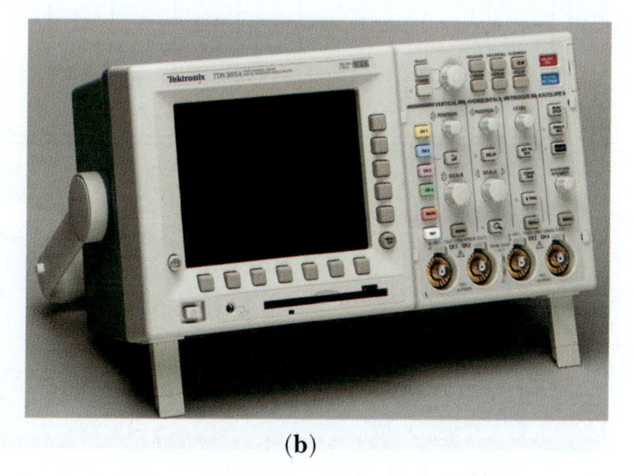

(b)

FIGURE 12–2 (a) A digital storage oscilloscope (DSO) *(Courtesy of B & K Precision);* (b) a digital phosphor "high tech" oscilloscope *(Courtesy of National Instruments)*

12–2 Key Sections of the Scope

A simplified sample scope front panel layout is shown in Figure 12–3. Refer to this figure as you read the following list and descriptions of the various controls and related circuits involved.

The key sections of a scope that relate to the controls shown in Figure 12–3 are the following:

The **cathode-ray tube (CRT),** which consists of a screen where signals are viewed, and the elements within the CRT, which generate and control a stream of electrons that strike the back (inside) of the CRT screen, producing the illumination you view on the outside of the CRT screen. For informational purposes only, note the simplified diagram in Figure 12–4, showing some of these CRT elements. NOTE: It is not necessary for you to learn details regarding these elements at this time in your training.

Intensity and focus controls, which allow the users to adjust the brightness, size, clarity, and focus of the spot or trace on the CRT screen caused by the electron beam (see Figure 12–3).

Position controls (vertical and horizontal), which allow adjusting voltages that control the position of the trace on the CRT screen. In Figure 12–4, you can see the CRT "deflection plates" that control the position and movement of the electron beam coming from the opposite end of the tube and striking the back of the screen. The position where the electron beam strikes the back of the CRT screen is controlled by the electrostatic field set up between the plates by the difference of potential between them. The reactions of the electron beam to these fields are illustrated in Figure 12–5. Notice the electron beam is attracted to the plate(s) having a positive charge and repelled from any deflection plates having a negative potential, or charge. Therefore, the position where the electron beam strikes the screen can be controlled by the voltages present at the deflection

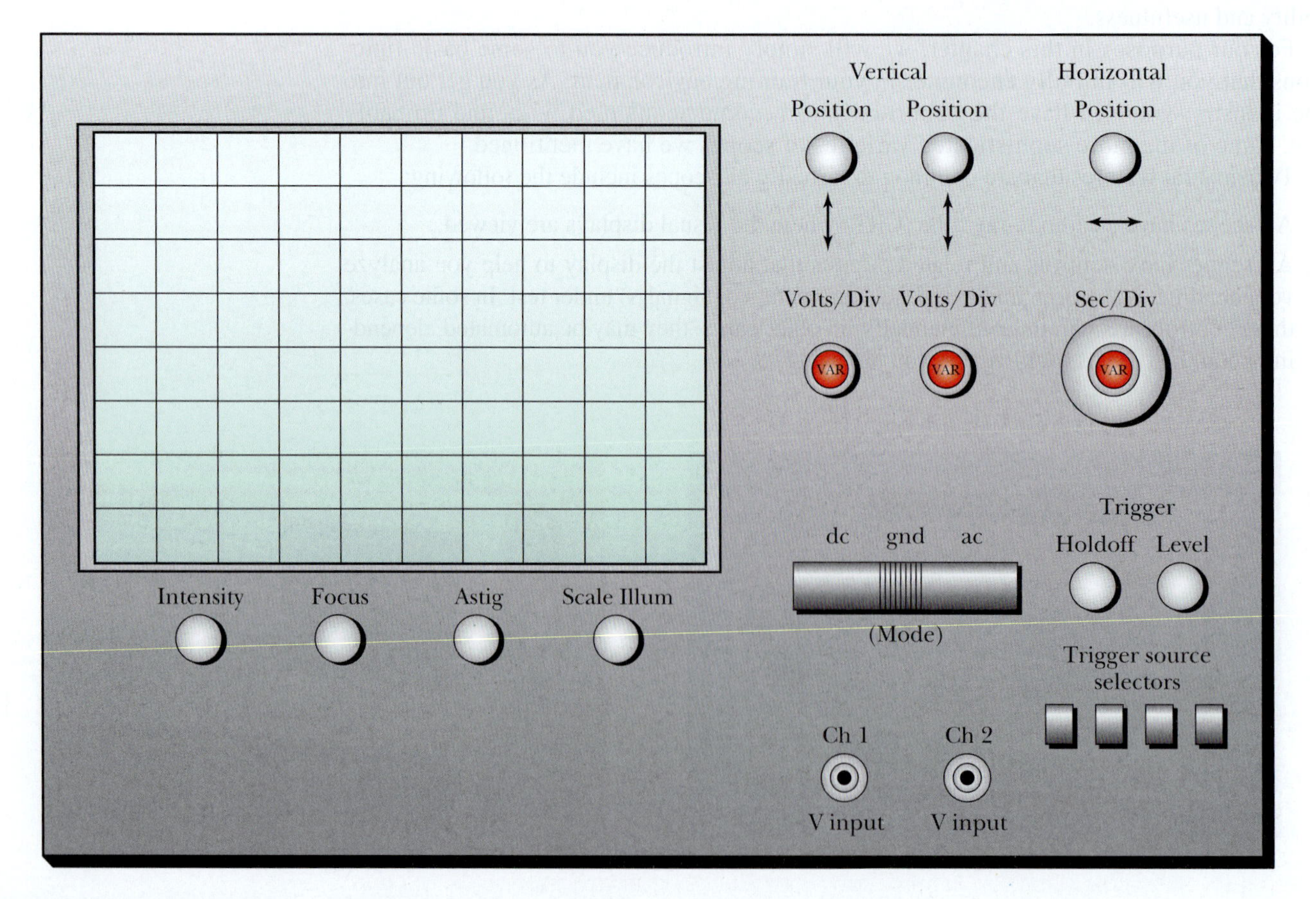

FIGURE 12–3 Typical controls on a dual-trace scope (simplified)

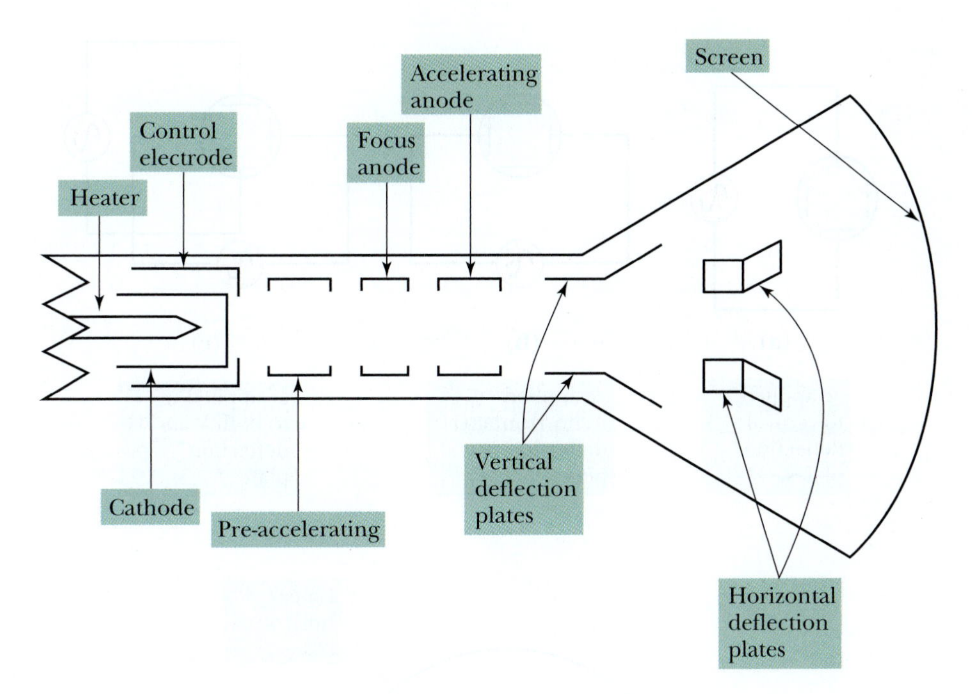

FIGURE 12–4 Elements in a typical cathode-ray tube (CRT)

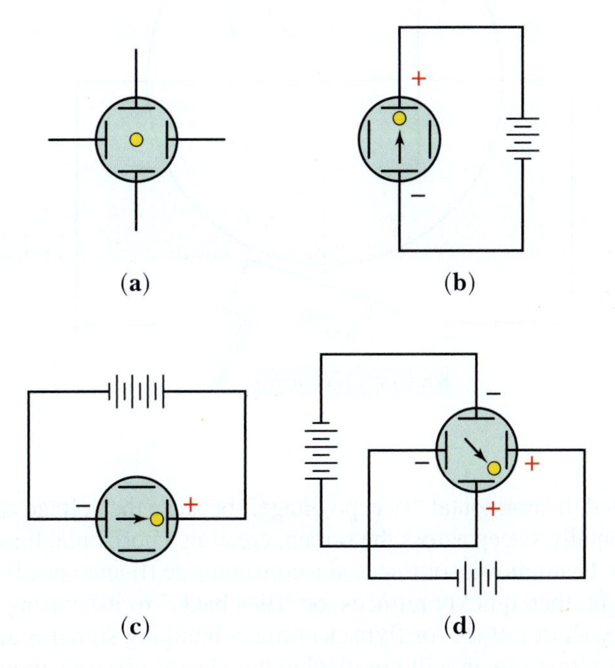

FIGURE 12–5 Electron-beam movement caused by dc voltages on deflection plates

plates. Knowing that deflection of the electron beam is toward positive deflection plates and away from negative deflection plates, you can understand the position controls; simply make the appropriate deflection plates either more positive or more negative, depending on which way you want the electron beam to move on the CRT face.

Also, observe what happens when an ac signal is applied to the various deflection plates, as shown in Figure 12–6a, b, and c.

The **horizontal sweep frequency controls** are used to control the linear trace speed and repetition rate of the horizontal trace of the electron beam across the CRT screen. A sawtooth (or ramp-type) waveform applied to the horizontal deflection plates creates a horizontal trace on the screen, Figure 12–7. The voltage applied to the horizontal plates is

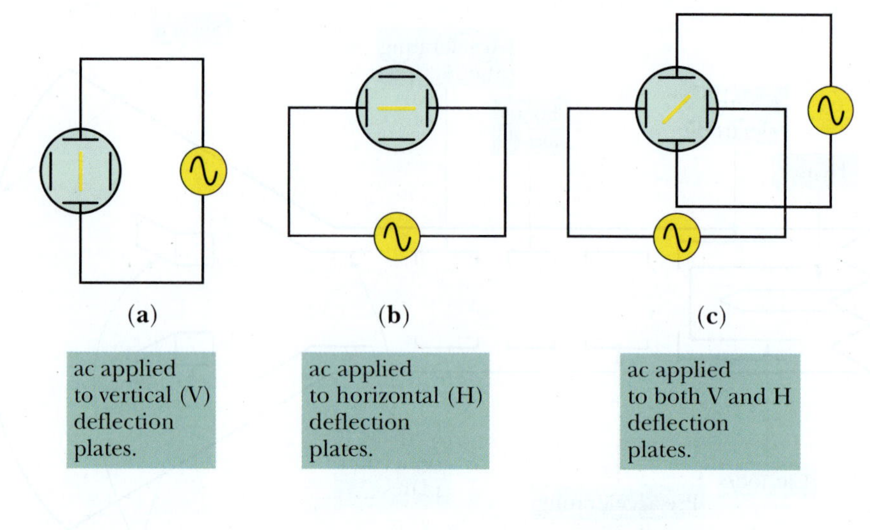

(a)
ac applied to vertical (V) deflection plates.
(b)
ac applied to horizontal (H) deflection plates.
(c)
ac applied to both V and H deflection plates.

FIGURE 12–6 Electron-beam movement caused by ac voltages on deflection plates

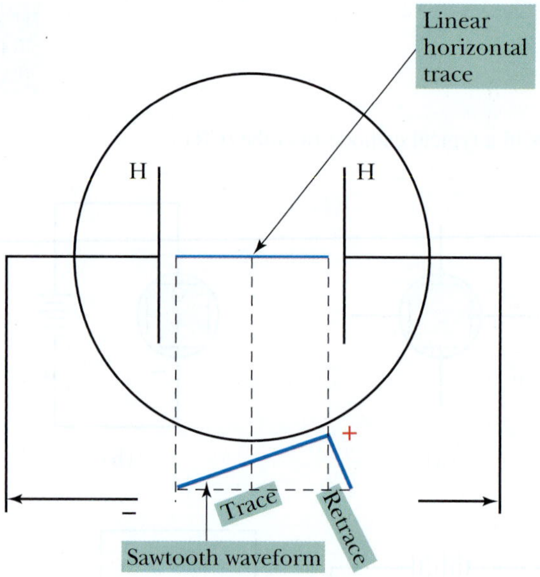

FIGURE 12–7 Linear trace caused by sawtooth voltage on horizontal (H) plates

sometimes called the horizontal "sweep voltage," because the voltage causes the electron beam to horizontally sweep across the screen, creating a horizontal trace, or line.

The electron beam moves, or traces at a constant rate (linear speed) across the screen from left to right, then quickly retraces, or "flies back," to its starting point on the left. During the very short retrace (or flyback) time, a blanking signal is applied to the system so that the scope screen will not display the electron beam's movement across the screen (from right to left) during the retrace time.

Because the electron beam is moving at a constant speed across the screen during the left-to-right trace time, a horizontal "time base" has been established. In other words, we know how much time it takes for the beam to move from one horizontal point to another on the screen. Because of this, we can measure times and calculate frequencies of signal waveforms that are displayed, as you will see later.

The number of times the beam traces across the screen in a second is determined by the frequency of the sweep (sawtooth) voltage. You see a horizontal line on the screen when this happens at a rapid rate. Two reasons for this are that (1) the CRT screen material has a persistance that causes continual emission of light from the screen for a short time after the electron beam has stopped striking that area; and (2) the retinas in our eyes also have a persistence feature. This is why you don't see flicker between frames in movies or on TV.

The **sweep frequency controls**—horizontal time variable, "VAR," and horizontal sec/div (seconds per division)—adjust the frequency of the horizontal sweep circuitry in the oscilloscope. This controls the number of times per second the beam is traced horizontally across the CRT screen. The horizontal frequency control(s) allows—us to view signals of different frequencies.

Having accumulated this information, let's now see how the electron beam is further controlled, amplified, or attenuated to provide meaningful displays on the CRT screen.

Vertical section is where signals to be analyzed are input into the scope, amplified, or attenuated, as needed for proper viewing. Key elements in this section are the vertical input jack(s) (sometimes called Y input jacks), and the **vertical attenuator** and **vertical amplifier,** with related controls. When the scope is a single-trace scope, there is only one vertical input jack. When a dual-trace scope is involved, there are two vertical input jacks. (See Chapter 1 and Chapter 2 V input jacks on drawing for Figure 12–3.)

The vertical attenuator and amplifier circuitry, and associated controls, enable decreasing or increasing the amplitude of the signals that are applied to the scope via the vertical input jack(s). View Figure 12–3 again, and note the calibrated **vertical volts/div** and the **variable control(s),** which are used to adjust the scopes vertical sensitivity, giving signal level control. Using these controls allows you to adjust for larger or smaller vertical deflection of the CRT trace with a given vertical input signal amplitude.

Horizontal section allows control of voltages and signals applied to the horizontal deflection plates. We have briefly discussed some of the important horizontal deflection elements and controls. One of these controls adjusts the frequency of the internally generated horizontal sweep signal.

Other possible elements related to the horizontal trace may include a horizontal **gain control** that changes the length of the horizontal trace line, and a jack (often marked "Ext X" input) that enables inputting a signal from an external source to the horizontal deflection system, in lieu of using the internally generated sweep signal.

Synchronization controls are used to synchronize the signal you want to observe with the horizontal trace; thus, the waveform appears stationary (assuming the signal has a periodic waveform). The effect is much like a strobe light that provides "stop action" when timing an automobile, for example. Also, you have probably observed that spinning fan blades can appear to be standing still if the light shining on them is blinking at the appropriate rate.

It is not necessary to go into detail regarding the circuitry, but it is sufficient to say that by starting or "triggering" the trace (left-to-right horizontal sweep) of the electron beam in proper time relationship to the signal to be observed on the scope (signal fed to the vertical deflection plates), a stable waveform is displayed.

Controls and jacks associated with this synchronization process include the following:

1. Trigger source selector switches
2. "Trigger holdoff and level" controls, which help set the appropriate level for the triggering signal to work best. (Again, look at Figure 12–3 to observe these controls.)

Practical Notes

Caution! One thing you should learn when operating a scope is that it is not good to leave a bright spot at one position on the CRT. The CRT screen material can be burned or damaged if this occurs for any great length of time. Never have the trace bright enough to cause a "halo" effect on the screen.

■ **IN-PROCESS LEARNING CHECK 1**

1. The part of the oscilloscope producing the visual display is the _______ _______ _______.
2. The scope control that influences the brightness of the display is the _______ control.
3. A waveform is moved up or down on the screen by using the _______ _______ control.
4. A waveform is moved left or right on the screen by using the _______ _______ control.
5. For a signal fed to the scope's vertical input, the controls that adjust the number of cycles seen on the screen are the horizontal _______ and the horizontal _______ control.
6. The control(s) that help keep the waveform from moving or jiggling on the display are associated with the _______ circuitry.

12–3 Combining Horizontal and Vertical Signals to View a Waveform

Having an electron beam cause a horizontal line across the CRT screen by applying voltage(s) to the horizontal deflection plates alone is not useful for displaying and analyzing waveforms. Causing an electron beam to vertically move by applying voltage(s) to the vertical deflection plates alone also is not useful for displaying and analyzing waveforms. However, useful waveforms appear when appropriate voltages are simultaneously applied to both the horizontal and vertical deflection plates.

Look at Figure 12–8. Notice how a sinusoidal waveform is displayed when a sine-wave voltage is applied to the vertical deflection plates, and when a sawtooth voltage of the same frequency (from internal sweep circuitry) is simultaneously applied to the horizontal deflection plates of the CRT. In this case, the frequency is 60 Hertz.

Again refer to Figure 12–8 and study the following discussion. Look at the conditions at times t_0, t_1, t_2, t_3, and t_4.

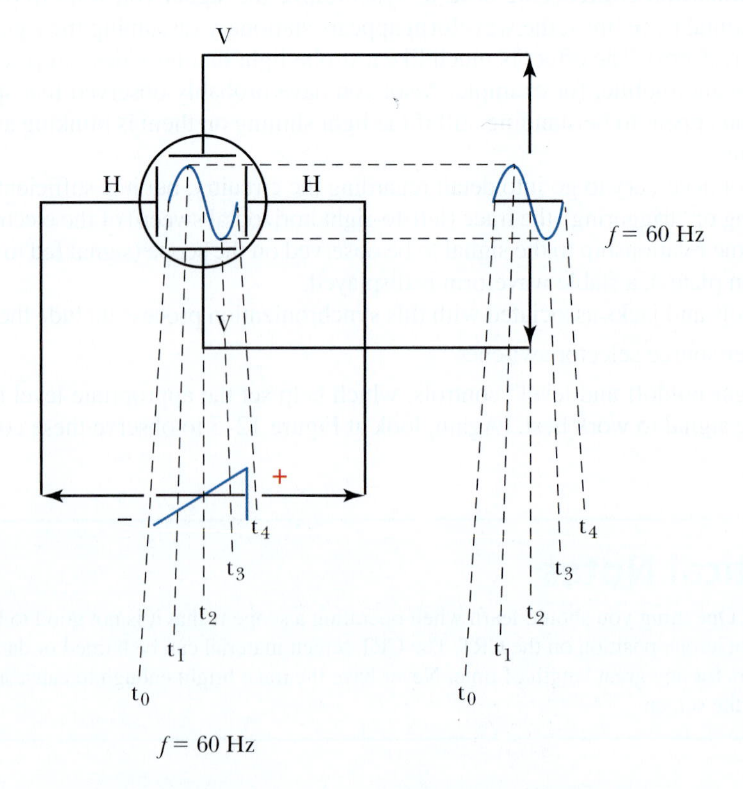

FIGURE 12–8 Waveform with sawtooth on horizontal (H) plates and sine wave on vertical (V) plates

1. The sawtooth voltage is applied to the horizontal deflection plates so that at the beginning of the sawtooth, the electron beam is at the left side of the screen (viewed by the person looking at the front of the scope). Time is t_0.
2. As the sawtooth voltage increases in a positive direction, the right-hand deflection plate becomes increasingly positive. Thus, the beam moves from the screen's left side toward its right side. Time is t_1.
3. The 60-Hz sine-wave voltage is applied to the vertical deflection plates. At time zero (t_0), the beam is horizontally on the screen's left side. Since no vertical voltage is at t_0 of the sine wave, a spot appears on the screen's left side.
4. As the horizontal sawtooth signal causes the beam to begin its horizontal trace to the right, the voltage on the vertical deflection plates causes the beam to simultaneously move to its maximum upward position. Time is t_1.
5. By the time the vertical signal has decreased from maximum positive value back to zero (180° into the sine wave), the horizontal signal has caused the beam to arrive at the center of the screen. In other words, the positive alternation of the sine wave has been traced and displayed on the screen at t_2.
6. The same rationale is used when tracing the negative alternation's waveform. That is, as the beam continues its horizontal trek across the screen, the signal applied to the vertical deflection plates is causing the beam to be sequentially moved to the maximum downward position, then back to the zero voltage position, vertically (center of the screen vertically). During this sequence, the negative alternation of the sine wave has been displayed.
7. Throughout the cycle, one cycle of the sawtooth waveform applied to the horizontal plates has caused one horizontal trace. At the same time, the electron beam has been moved up and down, corresponding to the voltage throughout one sine-wave signal cycle applied to the vertical plates. The result is a sine-wave waveform viewed on the CRT.
8. Because these events have occurred 60 times in 1 second, our eyes perceive a single sine-wave waveform that traces over itself 60 times in 1 second.

EXAMPLE A horizontal sweep frequency of 60 Hz and a vertical signal of 120 Hz result in two cycles of vertical deflections occurring at the same period that one sweep occurred. The display shows two full cycles of a sine wave, Figure 12–9.

PRACTICE PROBLEMS 1

Assume the vertical frequency is four times the horizontal frequency. What does the waveform look like?

By looking at the electron beam position at a number of instantaneous values of the applied horizontal and vertical signals, you can determine what type of waveform will appear on the scope screen, Figure 12–10. These concepts help interpret what we see on the CRT screen. Later, you will see how these concepts help us determine frequency and phase of ac signals.

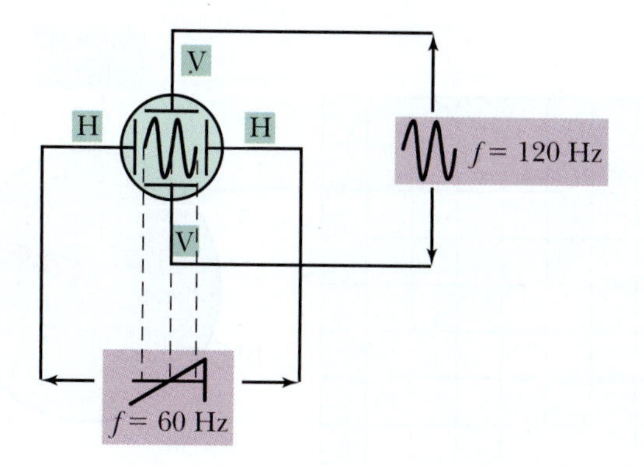

FIGURE 12–9 Result of two cycles of sine-wave signal per one horizontal sweep

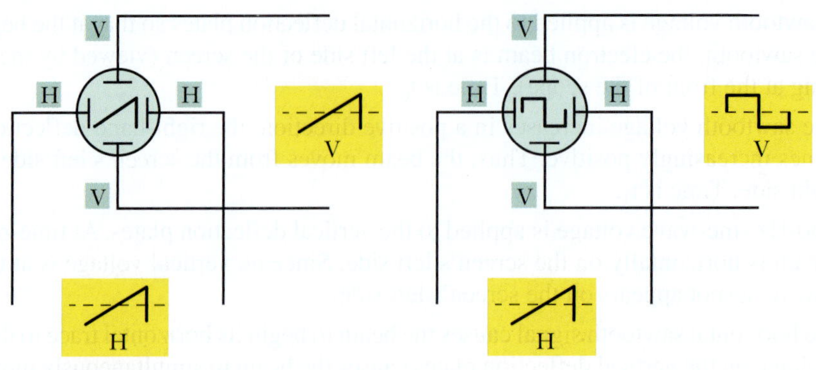

FIGURE 12–10 Illustration of how different waveforms are traced

12–4 Measuring Voltage and Determining Current with the Scope

General Information

Both dc and ac voltage measurements can be made with oscilloscopes. It is useful to know how to make these measurements and to interpret waveforms in terms of voltage amplitudes. One common method used to measure voltages with the scope is to interpret the scope display in reference to the "volts/division" control setting when the "V VAR" (variable) control is set at the "Cal" position. This technique is used frequently.

As indicated earlier, learning to use the scope in a practical way *only comes through "hands-on" experience.* The descriptions in this chapter provide only conceptual knowledge. Therefore, you will have to expand on these concepts by actual laboratory experiences—be sure you do!

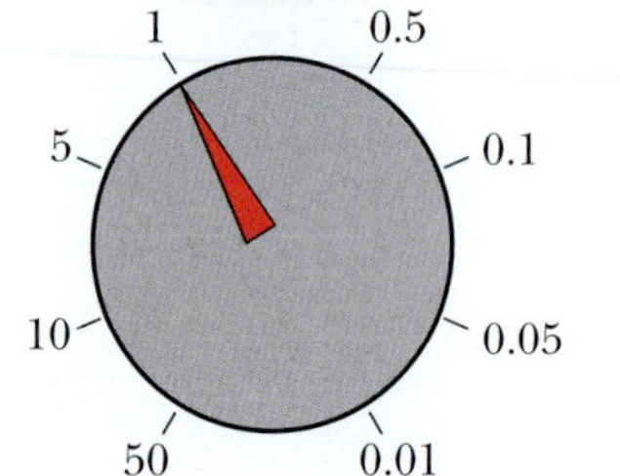

FIGURE 12–11 Using the calibrated volts/div switch setting to determine voltage

Volts-per-Division Control Setting Approach

By making sure that the "V-Variable" control is set in the "Cal" position, you can directly interpret voltage readings of scope patterns by reference to the vertical volts/div control setting. In Figure 12–11, you can see that our example control can be set at eight different positions. The beauty in this is that you can select the position that makes the signal display the most readable amplitude for interpreting the signal being measured. The interpreted signal voltage value equals the volts/div setting times the number of divisions of deflection.

EXAMPLE Observe in Figure 12–12 that the volts/div control is set at the 0.1 position. With the "V-Variable" control in the "Cal" position, each tenth-volt amplitude of the signal will cause one division displacement of the trace on the scope display. In our sample then, there are six divisions of deflection caused by the peak-to-peak amplitude of the signal. The interpretation is that the signal being measured must be 0.6 volts peak-to-peak. The rms value is 0.3535×0.6, or 0.21 V. If the tested voltage were a dc voltage of 0.6 V, the trace would

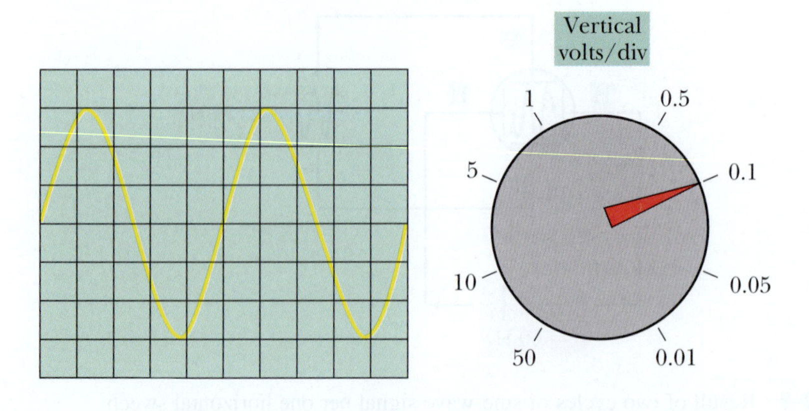

FIGURE 12–12 Example of volts/div voltage measurement approach

have jumped upward or downward from center by six divisions, depending on the polarity applied to the scope input jacks. This would cause the trace to be "off-screen." You would have to move the 0-V reference trace off center by two or more vertical divisions in order to see the 0.6-V dc deflection.

An alternative approach would be to set the **volts/div control** to the next higher setting (0.5 V/div). Then, the display would only jump one and one-fifth squares, which would be viewable. Of course, you would interpret the value based on the new volts/div setting. ◆

Practical Notes

Once you have calculated the rms value for a given measurement, you don't need to recalculate to find the rms value for other measurements if you DO NOT change the vertical control settings. In the preceding example, the scope was set up so that 0.6 V peak-to-peak, or 0.21 V rms, caused six divisions of trace deflection. If the settings are not changed, you know that if you had three divisions of deflection, the rms must be one-half of 0.21 V rms, or 0.105 V rms. In other words, you would know that since 0.21 V rms caused six divisions of deflection, then each division must equal 0.21/6 V rms, 0.035 volts rms per division.

PRACTICE PROBLEMS 2

1. If the vertical volts/div control in Figure 12–12 were set on the "5" position and the scope display were the same as shown, what would be the peak-to-peak and rms values of the signal being checked?
2. What is the rms volts-per-division sensitivity setup used in question 1?
3. If you were going to measure a signal that you knew was about twice as great in amplitude as the one measured in question 1, which setting of the vertical volts/div control would you use?

Attenuator Test Probes

Another technique often used with many scopes is a test probe that attenuates the signal by a given factor, for example 10 times attenuation, Figure 12–13. Thus, on the 10× switch position, you multiply by 10 times what you are reading on the scope to find the actual value of voltage applied to the probe. This probe often has another purpose—it prevents the circuit being tested from being disturbed by the scope's input circuitry, or test cable. Also, the cable is a shielded cable, which minimizes stray signal pickup.

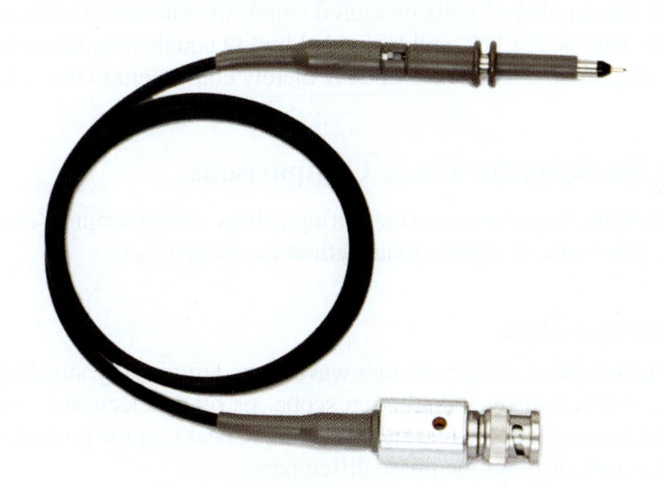

FIGURE 12–13 An example of a 10× test probe. *(Courtesy of Probe Master, Inc.)*

FIGURE 12–14

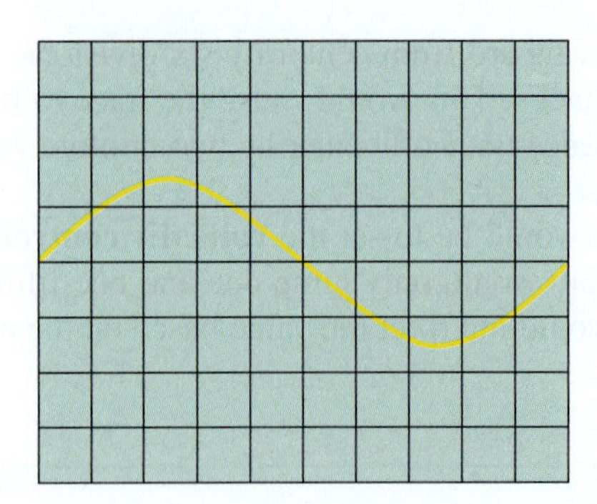

Practical Notes

There are numerous types of test probes and associated cables available for use with oscilloscopes. Some are switchable and have one switch position for measuring low frequencies and another switch position for measuring high frequencies. Many have switchable attenuation, for example, ×1 (times 1) and ×10 (times 10). Sometimes you will see these written as 1:1 and 10:1 modes. As indicated earlier, the ×10 mode indicates that the signal has been attenuated 10 times; therefore, you would multiply your scope reading by 10 to find the actual signal level input to the test probe. There are also "current probes" available for use with scopes; however, the voltage probes are the more common types found in practice. As you proceed in your studies, you will be learning other parameters related to scope input circuit characteristics and matching of test probes/cables to the circuit conditions.

PRACTICE PROBLEMS 3

1. Refer to the waveform shown in Figure 12–14 and determine the peak-to-peak and rms values of the signal. Assume that the volts/div setting of the vertical input signal controls is set at 0.5 V/div and that you are using a 1:1-type test probe.
2. Assume that the same waveform pattern is showing on the scope as is shown in Figure 12–14; however, this time the volts/div setting is 100 mV/div and you are using a 10:1 test probe. What are the signal's peak-to-peak and rms voltage values?

Measuring Voltage to Find Current

Current can be calculated by the Ohm's law technique you learned in previous chapters. If you use the scope to determine the rms voltage across a 1-kΩ resistor, for example, you automatically know the number of volts measured equals the number of milliamperes through the 1-kΩ resistor. That is, $I = V/R$, and V divided by kΩ equals mA. Obviously, Ohm's law applies to any value resistor. The 1-kΩ value is merely convenient to use, where possible.

12–5 Using the Scope for Phase Comparisons

In addition to exhibiting waveforms and measuring voltage (or determining current), the scope can compare the phase of two signals. One method for doing this is:

The Overlaying Technique

The overlaying technique is where the two waveforms are visually superimposed on each other. This is accomplished by a dual-trace scope, or by an electronic switching device. Therefore, the displacement of the two signals' positive peaks, or the points where they cross the zero reference axis, displays the phase difference.

EXAMPLE Adjust the horizontal sweep frequency so that exactly one cycle of the signal applied to the A (or #1) input covers the screen. (NOTE: Ten divisions are typical.)

The second signal is applied to the scope B (or #2) input. Since the screen is divided into 10 divisions, and 10 divisions represent 360°, then each division represents 36°. By noting the amount of offset between the two waveform traces and knowing that each division difference equals 36°, you can easily interpret the phase difference between the two signals.

PRACTICE PROBLEMS 4

Suppose the horizontal frequency in our example were adjusted so there were exactly 2 cycles across the 10 divisions of the screen. How many degrees would each major division on the screen represent?

Summarizing the Dual-Trace Scope Approach:

1. Two horizontal traces are obtained on the scope. They are then superimposed at the center of the screen by using the positioning controls, as appropriate.
2. Signal #1 is applied to one channel's vertical input jack. For convience, the scope sweep frequency is adjusted for one or two cycles on the screen. Vertical level controls are adjusted to obtain a convenient size waveform for viewing purposes.
3. Signal #2 is connected to the other channel's vertical input jack. Again, appropriate adjustments are made to appropriately display signal #2. The signals should be adjusted to the same vertical size and should be the same frequency.
4. Look at Figure 12–15 to see one method that can be used to interpret the display in terms of phase difference between the signals.

12–6 Determining Frequency with the Scope

Today's technology has given us devices called digital frequency counters that are excellent for making frequency measurements. However, there are still many times when you have a scope connected to the circuit under test, and you do not want or need to pull out another piece of test equipment when you already have a pattern on the scope that you can interpret for frequency measurement.

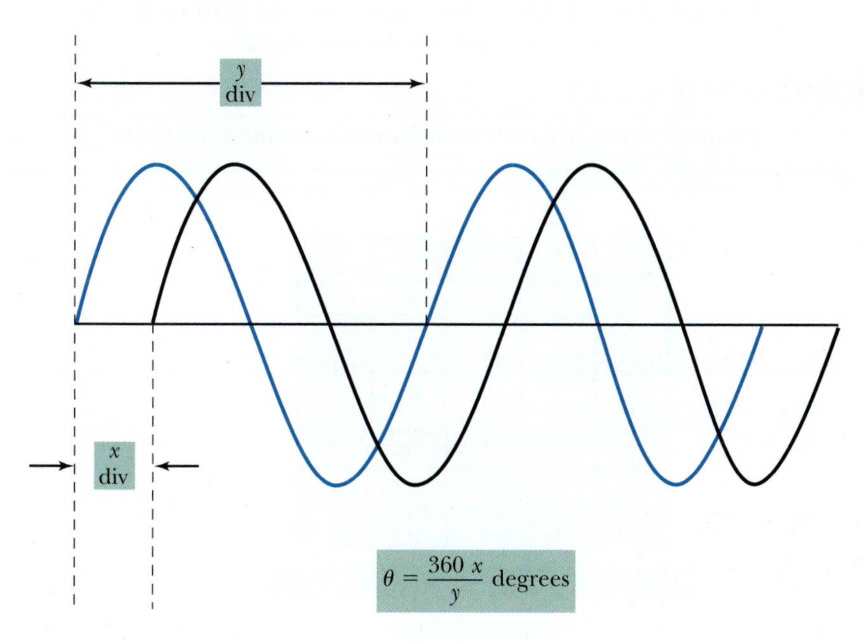

FIGURE 12–15 Overlay technique for phase measurement

Let's take a quick look at how this is done. You have already learned that the sweep frequency controls allow us to have a "linear horizontal time base sweep" whose time per division we can easily determine. You have learned in an earlier chapter that a signal's period (time for one cycle) and its frequency are inversely proportional. That is $f = 1/T$ and $T = 1/f$. Add your calculator's 1/x (reciprocal) function and you have all the tools you need to find the frequency quickly from a scope pattern.

EXAMPLE Observe Figure 12–16. The sweep frequency control is set on 5 ms/div. This means the horizontal trace takes 5 milliseconds for each division of horizontal movement across the scope screen. With this knowledge, you can see that the signal being looked at is taking about 50 milliseconds for one cycle, or the period equals 50 ms. Using your calculator [1/x] function and inputting 50, [EE] (or [EXP]), [±], 3, then using the [1/x] function key, the answer is shown as 20. This means the frequency is 20 Hz. Of course, you probably could have easily interpreted in your head that 50 milliseconds = 0.05 seconds and simply used the reciprocal function on your calculator for 0.05 to get to the answer of 20.

PRACTICE PROBLEMS 5

1. Assume the s/div control is set on the 0.1 s/div setting. Also assume that the scope pattern shows exactly 4 cycles of the signal to be measured across its 10 horizontal divisions.
   a. What is the period of the signal being measured?
   b. What is the frequency of the signal being measured?
2. Assume the s/div control is set on the 50 μs/div setting. This time, the pattern shows exactly 10 cycles of the signal under test.
   a. What is the signal's period?
   b. What is the signal's frequency?

Practical Notes

Translating the $f = 1/T$ formula into practical terms relating to the scope settings and waveform being viewed:

f = 1 cycle display ÷ (number of horizontal squares for that 1 cycle × the "time per div" horizontal sweep setting)

Thus, for Figure 12–16:

$1/(5 \text{ ms} \times 10 \text{ divisions}) = 1/50 \text{ ms}$; therefore, $f = 1/0.05 \text{ s} = 20 \text{ Hz}$

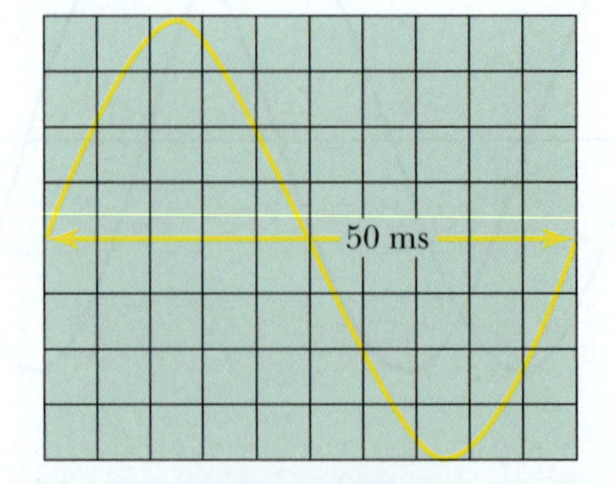

FIGURE 12–16 Example of frequency determination

Final Comments about the Oscilloscope

Certainly, reading this chapter has not made you a professional oscilloscope user! We only hope that by briefly discussing the remarkable versatility of the oscilloscope, you will desire to become knowledgeable and practiced in applying this instrument throughout your studies and career. The ability of the scope to display waveforms so that increments of time are measured and interpreted, and its ability to measure relative amplitudes makes this instrument one of the technician's most valuable assets. Learn to use it!

Figure 12–17 shows a very simplified block diagram of the elemental parts of a scope. As you can see, the vertical input signal is processed through the vertical amplifier/attenuator system before being applied to the vertical deflection plates of the CRT. The processing can either be in the form of amplification or attenuation of the signal to be displayed by the scope. Generally, the signal applied to the vertical input system is the signal you are trying to analyze or check.

The horizontal amplifier system has two possible inputs. The one most often used is the input coming from the horizontal timebase generator (called the sweep system in our diagram). The output of this horizontal amplifier system is then fed to the horizontal deflection plates of the CRT. This produces the horizontal trace on the scope. Sometimes, it is desired to use a horizontal signal source other than the internal sweep generator circuitry. In that case, the horizontal signal is input to the horizontal amplifier system via the H input jack (sometimes labeled the "x" input jack).

Of course, the dc power supply shown in our diagram supplies the appropriate dc voltages to operate the CRT, scope amplifier circuits, timebase generator circuits, and so on throughout the oscilloscope circuitry.

A Good Idea

When creating brackets for troubleshooting, place one of the brackets at the point where there is an abnormal "output indicator"! Place the other bracket at the first point in an earlier part of the flow path where things are known to be normal!

Practical Notes

Some precautions: When using a scope, the technician should use common sense!

1. Do not leave a "bright spot" on the screen for any length of time.
2. Do not apply signals that exceed the scope's voltage rating.
3. Do not try to make accurate measurements on signals whose frequency is outside the scope's frequency specifications.
4. Be aware that the scope's input circuitry (including the test probe) can cause loading effects on the circuitry under test—use the correct probe for the job!
5. Check the attenuator probe setting (1:1 or 10:1) to interpret the scope display properly.

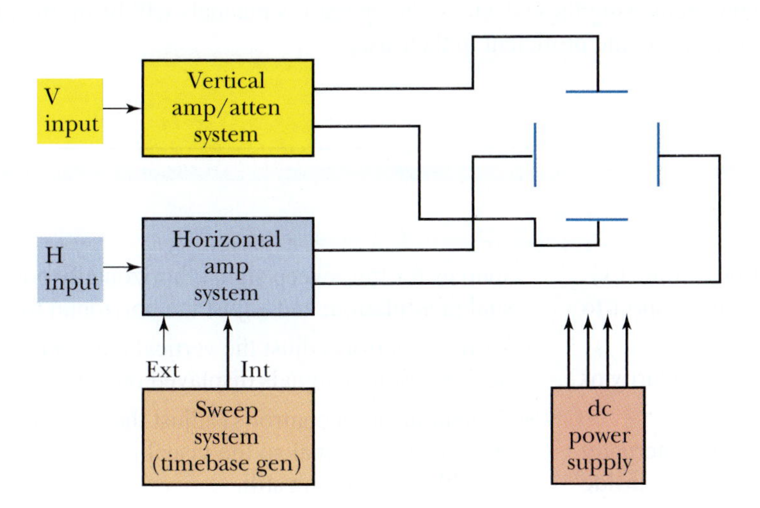

FIGURE 12–17 Simple oscilloscope functional block diagram

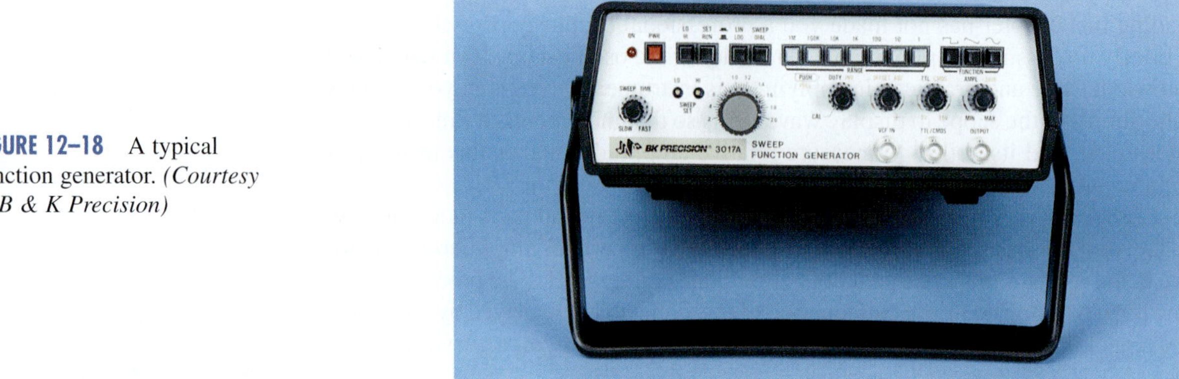

FIGURE 12–18 A typical function generator. *(Courtesy of B & K Precision)*

12–7 A Test Device Often Used with Circuits and Scopes

Function Generators (or Signal Generators)

It is frequently necessary when checking a circuit's operation to supply a signal to the input of that circuit that has the proper signal shape (waveform), frequency, and amplitude. The modern "function generator" has the capability of being such a signal source. See Figure 12–18.

Note that there are controls (switches or knobs) that allow you to control the wave shape, frequency, and amplitude value of the signal available from the generator's output terminals, or connector. For example, you can choose to have the output be a sine wave, a square wave, a triangular wave, or a ramp-shaped wave. Using the frequency adjusting controls, you can set the output frequency over a wide spectrum of frequencies. The output controls often allow controlling outputs from 0 volts to 7 volts (rms), or greater.

You will have the opportunity to get used to these controls in a "hands-on" fashion while performing laboratory projects during training, and in the workplace after your training. In "lab" you will often connect the signal source to the input of the circuit being tested; then, use the oscilloscope to monitor signal conditions at various points in the circuit to aid in verifying circuit operation, troubleshooting, and/or calibrating circuit settings.

Sometimes, these function generators have a built-in "frequency counter" and a digital readout so you can more accurately set the frequency output. Many generators depend on the user interpreting the frequency dial on the knob, along with the frequency multiplier switches to set output to a close approximation of the desired frequency. Also, these built-in frequency counters can be used to measure frequencies outside the signal generator itself. Other times, frequency counters are separate pieces of equipment. The price range of the equipment often determines the frequency range over which it measures, and the accuracy of the measurements. As you begin using these devices, the operator's manuals will be of great use to you, and will help you become proficient in their use.

Summary

- The oscilloscope (scope) is an instrument providing a visual presentation of electrical parameters with respect to time and amplitude.
- Intensity and focus controls adjust brightness and clarity of the visual presentation.
- Horizontal controls adjust the frequency of the internally generated horizontal sweep signal; select outside or inside sources for this sweep signal; horizontally position the visual presentation; and adjust the horizontal trace length.
- Vertical controls adjust the vertical amplitude and the vertical position of signals displayed on the screen.
- Synchronization control(s) adjust the start time of the horizontal sweep signal so the visual presentation of a signal is stable, or standing still.

- Scope presentations display the peak-to-peak values of the waveform. Other significant values are interpreted from the waveform by appropriate interpretation factors. For example, rms value = 0.707 × peak, or 0.3535 × peak-to-peak value.
- To measure voltage with an oscilloscope, the appropriate vertical controls are adjusted so the user knows the vertical sensitivity of the scope. That is, what voltage is represented by each division of vertical deflection on the calibrated screen.
- To determine the phase difference between two signals with an oscilloscope, a dual-trace scope is used so each signal is applied to a separate vertical input. Then, by appropriate adjustment, the two signals are superimposed on the visual presentation. Thus, the technician can interpret the phase differential in terms of the two signals.

Review Questions

1. What is the name and abbreviation of the oscilloscope component that displays signals on its screen?
2. What are four practical uses of the oscilloscope?
3. Why should the intensity control be set to provide the minimum brightness that allows good readability of the presentation?
4. What control(s) set the sweep frequency, therefore determining the number of cycles of a given signal viewed?
5. To what input terminal(s) is the signal to be viewed normally applied?
6. To what terminal(s) is an external sweep voltage applied?
7. To what terminal(s) is an external synchronizing signal applied?
8. Along what axis is deflection caused by a signal applied to the vertical deflection system?
9. Why should the setting of the scope vertical gain control and vertical input switch remain unchanged after initial calibration, if using the scope for voltage measurement?
10. What is an attenuator probe?

Problems

1. The time/div control and horizontal gain control settings are set so the horizontal sweep line represents 0.1 ms per division. What is the frequency of a signal if one displayed cycle uses four horizontal divisions on the horizontal sweep line?
2. If the controls are set so that the horizontal sweep line represents 2.5 ms per division, what is the frequency of a signal if a single displayed cycle uses 8 horizontal divisions on the horizontal sweep line?
3. For a frequency that is double that calculated in question 2, how many horizontal divisions would a single cycle take if the sweep controls were set so that each horizontal division represented 5 ms?

Refer to Figure 12–19 and answer questions 4 to 8:

4. What is the peak-to-peak value of voltage indicated by the waveform display?
5. What is the value of T for the signal displayed?
6. If you want the display to show only one cycle of the signal, which controls do you adjust on the scope?
7. What is the time-per-division setting to accomplish a one-cycle display?
8. The vertical sensitivity is set for 5 V per division (calibrated). What are the rms, peak, and peak-to-peak values of voltage of the signal?

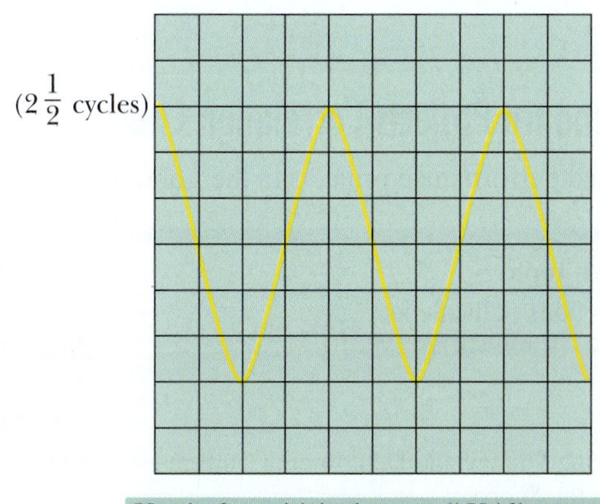
($2\frac{1}{2}$ cycles)
Vertical sensitivity is set at 5 V/div
Horizontal time/div is set at 0.001 s/div

FIGURE 12–19

9. If a scope is set with the vertical sensitivity at 2 V per division, and in measuring the voltage across a 4.7-kΩ resistor with the scope the peak-to-peak deflection is 2.5 divisions, what is the value of rms current through the resistor?
10. How many cycles of a 400-Hz signal would appear across a scope having 10 horizontal divisions if the horizontal sweep controls were set at 0.5 ms/div?

Analysis Questions

Since the best method to familiarize yourself with an oscilloscope is to see one, the following questions pertain to observing the oscilloscope you may have available in your training environment. Since the best way to learn how to use an oscilloscope is through hands-on experience, we further recommend that you also perform "hands-on" laboratory projects(s) related to using the scope. (See the end-of-chapter "Performance Projects Correlation Chart" for suggestions along this line.)

1. List the precise names of the scope controls observed on your training program's oscilloscope that:
   - **a.** control the trace focus.
   - **b.** control the trace intensity.
   - **c.** control the vertical position of the trace(s).
   - **d.** control the horizontal position of the trace(s).
   - **e.** allow for scope adjustment and measurement of vertical input signal amplitude(s).
   - **f.** provide for measuring vertical input signal frequency(ies).
2. Observe the switches and controls on your training program's oscilloscope, and list the following data:
   - **a.** Maximum possible vertical input signal sensitivity setting (the smallest volts/div setting)
   - **b.** Minimum possible vertical input signal sensitivity
3. Refer to Figure 12–20 and for each pattern shown (except for the model waveform pattern) list the oscilloscope controls used to cause the waveform shown to match the model waveform. (NOTE: a = model waveform.)

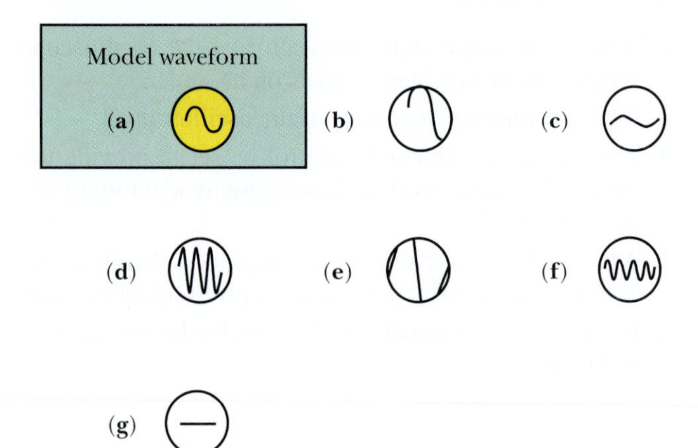

FIGURE 12–20

Performance Projects Correlation Chart

Suggested performance projects in the Laboratory Manual that correlate with topics in this chapter are:

Chapter Topic	Performance Project	Project Number
Key Sections of the Scope	Basic Operation: Familiarization	33
	Basic Operation: Controls Manipulation	34
	Basic Operation: Vertical Controls and DCV.	35
	Basic Operation: Observing Various Waveforms	36
Measuring Voltage and Determining Current with the Scope	Voltage Measurements	37
Using the Scope for Phase Comparisons	Phase Comparisons	38
Determining Frequency with the Scope	Determining Frequency	39

NOTE: It is suggested that after completing the above projects, the student should be required to answer the questions in the "Summary" at the end of this section of projects in the Laboratory Manual.

Troubleshooting Challenge

CHALLENGE CIRCUIT 7

(Follow the SIMPLER sequence by referring to inside front cover.)

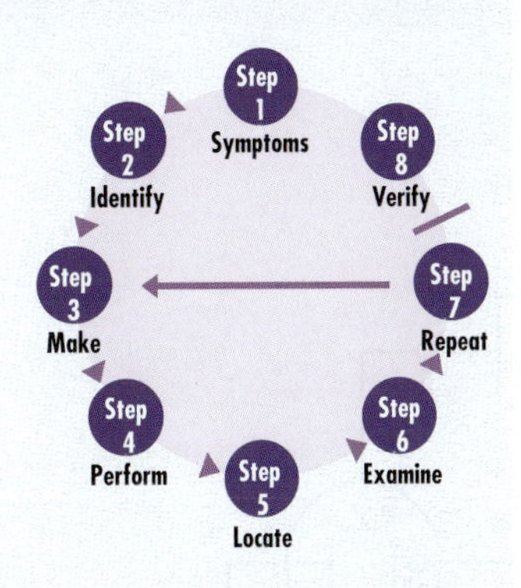

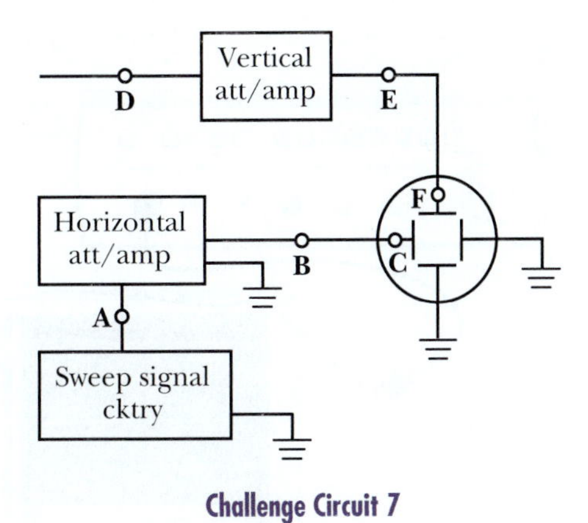

Challenge Circuit 7

STARTING POINT INFORMATION

1. Simplified block diagram
2. Vertical input signal is 300-Hz ac sine wave
3. Output of sweep signal circuitry is a 150-Hz sawtooth waveform
4. Horizontal sweep = 150 Hz
5. Trace seen on CRT is a horizontal line from left to right

TEST	Results in Appendix C
Signal/Voltage Checks	
Point A to Ground	(70)
Point B to Ground	(2)
Point C to Ground	(113)
Point D to Ground	(57)
Point E to Ground	(23)
Point F to Ground	(106)

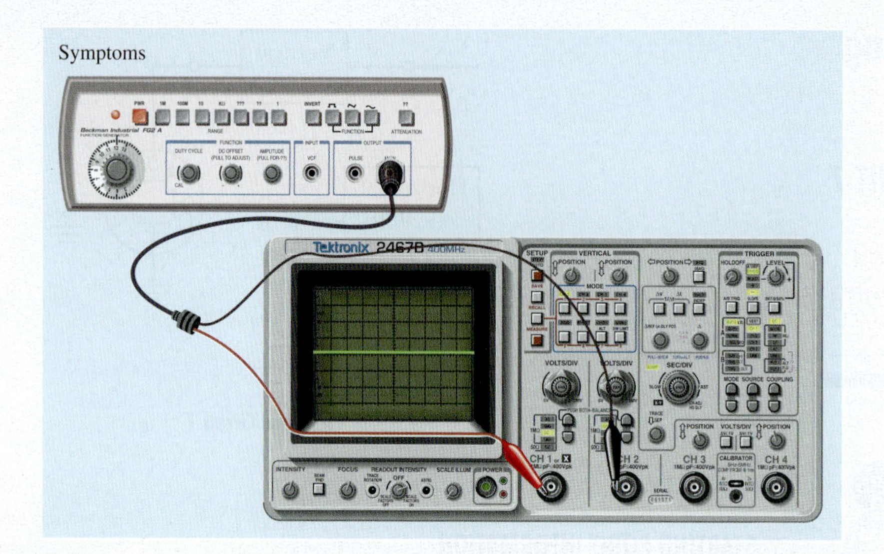

CHALLENGE CIRCUIT 7

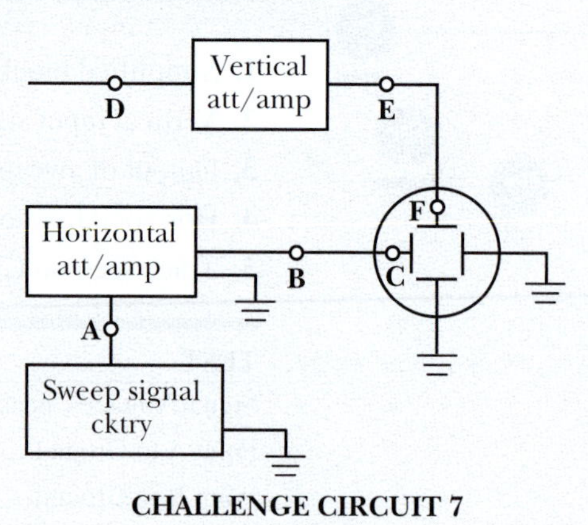

CHALLENGE CIRCUIT 7

STEP [1]

SYMPTOMS Although there should be a waveform on the CRT screen, there is only a straight horizontal line. This implies there is a horizontal signal being fed to the CRT horizontal plates, but no vertical signal is reaching the CRT vertical plates.

STEP [2]

IDENTIFY initial suspect area: The initial suspect area includes all elements involved in the vertical signal path to the CRT vertical plates. That is, the signal input to the vertical amplifier, the vertical amplifier, and the path from the vertical amplifier output to the CRT vertical plates.

STEP [3]

MAKE test decision: We'll use the "inputs-outputs" technique and first check the input of the vertical amplifier to see if an input signal is present.

STEP [4]

PERFORM **1st Test:** Check the presence of an input signal for the vertical amplifier at point D. The input signal is present.

STEP [5]

LOCATE Locate new suspect area: The new suspect area is the remainder of the vertical signal path to the CRT vertical plates.

STEP [6]

EXAMINE available data.

STEP [7]

REPEAT analysis and testing:

2nd Test: Check the signal "out of" the vertical amplifier at point E to ground. No signal is present, indicating a problem in the vertical amplifier circuitry. We have performed block-level troubleshooting. At this point, the vertical amplifier requires component-level troubleshooting.

STEP [8]

VERIFY **3rd Test:** After component-level troubleshooting has been completed and the needed repairs performed, test the system with the same signals. The result is that two sine-wave cycles are seen on the CRT screen.

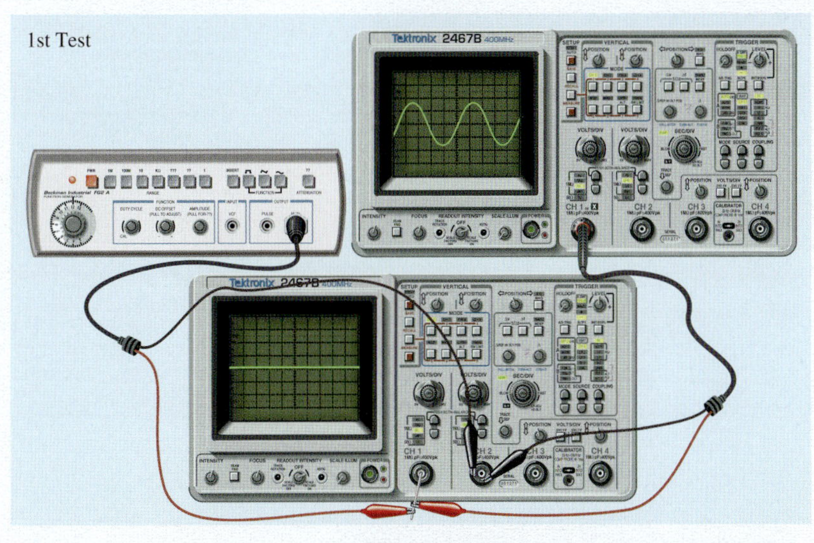
1st Test
Tektronix 2467B 400MHz

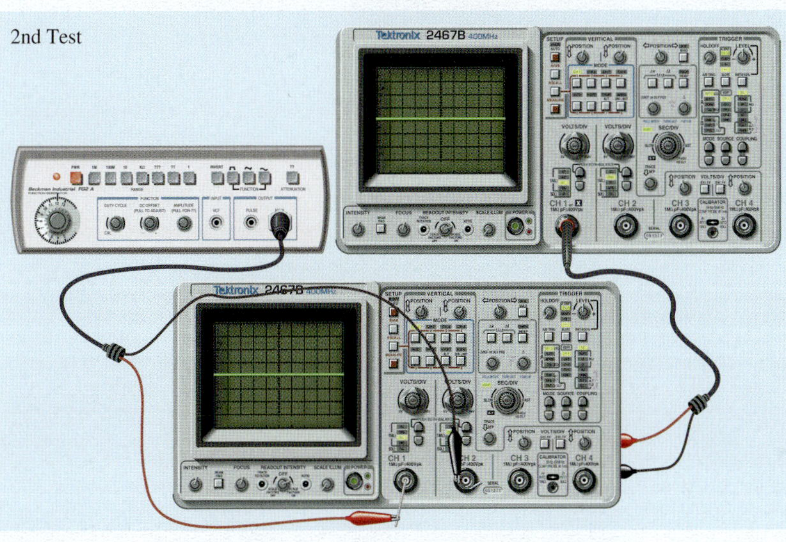
2nd Test
Tektronix 2467B 400MHz

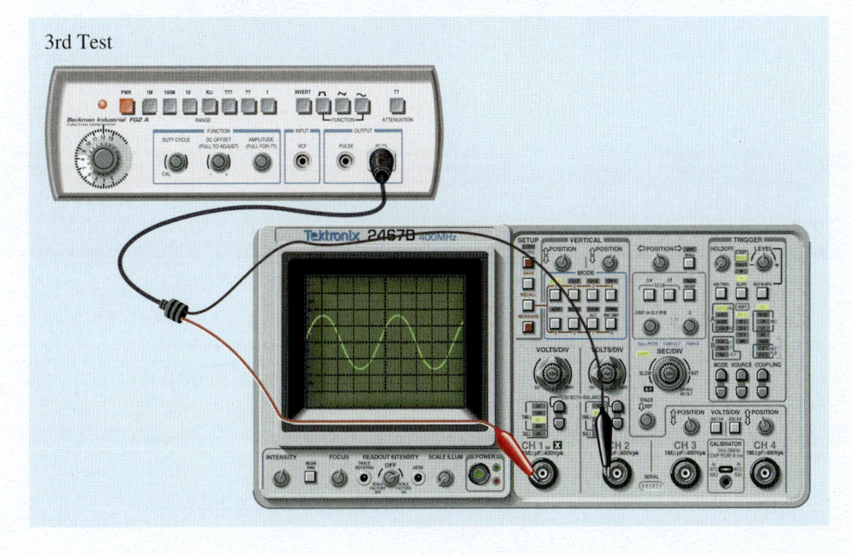
3rd Test
Beckman Industrial FG2 A
Tektronix 2467B 400MHz

PART IV

Reactive Components

OUTLINE

OBJECTIVES

After studying this chapter, you should be able to:

1. Define **inductance** and **self-inductance**
2. Explain **Faraday's** and **Lenz's laws**
3. Calculate induced **cemf** values for specified circuit conditions
4. Calculate inductance values from specified parameters
5. Calculate inductance in series and parallel
6. Determine energy stored in a magnetic field
7. Draw and explain time-constant graphs
8. Calculate time constants for specified circuit conditions
9. Use ϵ to aid in finding circuit quantities in dc *RL* circuits
10. List common problems of inductors
11. Use the computer to solve circuit problems

CHAPTER 13

Inductance

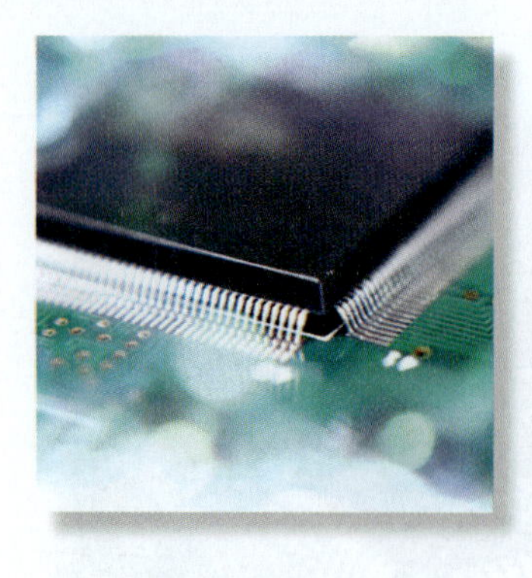

PREVIEW

The electrical property of resistance and its component, a resistor, are important elements in your understanding of electronics. Likewise, the property of inductance and its related component, an inductor, are equally important. A significant contrast between resistors and inductors is that resistors dissipate electrical energy and inductors have the ability to temporarily store electrical energy. Even as you applied your newly gained insights about magnetism and electromagnetism as you studied dc meters and generator action, you will again apply your learning about electromagnetism, as you consider the characteristics of inductors.

Inductors have many uses in electronics. For example, they are used in "tuned" circuits, which enable the selection of a desired radio or television station. They are used as filter elements to smooth waveforms or to suppress unwanted frequencies. Inductance is also used in transformers, which are devices that increase or decrease ac voltages, currents, and impedances. The inductor (or coil) in automobiles has been used for years to help develop high voltages, which create the spark across the spark plug gaps. These applications and many others use this unique electrical property of inductance.

In this chapter, we will discuss the property of inductance as it relates to dc circuit action. In the following chapters, you will study the action of this electrical property under ac conditions. An understanding of both will be very useful to you!

KEY TERMS

counter-emf or cemf
ϵ
Faraday's law
Henry (H)
Inductance (self-inductance)
Joule
Lenz's law
L, L_T
L/R time constant
τ

13–1 Background Information

From previous *chapters,* you know that electrical resistance *(R)* is that property in a circuit that *opposes current flow.* It is also important for you to know that **inductance (*L*)** is the property in an electrical circuit that *opposes a change in current.*

Inductors come in a variety of physical and electrical sizes, Figure 13–1. The forms you see here are components that have been designed to oppose changes in current. It should be pointed out that even a straight piece of wire exhibits a small amount of inductance. Sometimes this "stray" (many times undesired) inductance presents conditions that must be contended with in circuit designs and layouts. This is particularly true in high-frequency circuits, which you will study later in your training. Again, remember that inductance does not oppose current flow, as such, but opposes a *change* in current flow.

Another important characteristic of inductance and inductors is that they store electrical energy in the form of the magnetic field surrounding them. As current increases through the inductor, a resulting magnetic field expands, surrounding the coil, Figure 13–2a. When current tries to decrease, the expanded field collapses and tries to prevent the current from decreasing, Figure 13–2b. (Recall Lenz's law!) As the field expands, it absorbs and stores energy. When the field collapses, it effectively returns that stored energy to the circuit.

Another feature of inductance and inductors is that an emf (electromotive force or voltage) is induced when a current *change* occurs through them. This induced emf is due to the expanding or collapsing magnetic field as it cuts the conductors of the coil. This self-induced emf is sometimes called **counter-emf,** or **cemf,** because its polarity opposes the change that

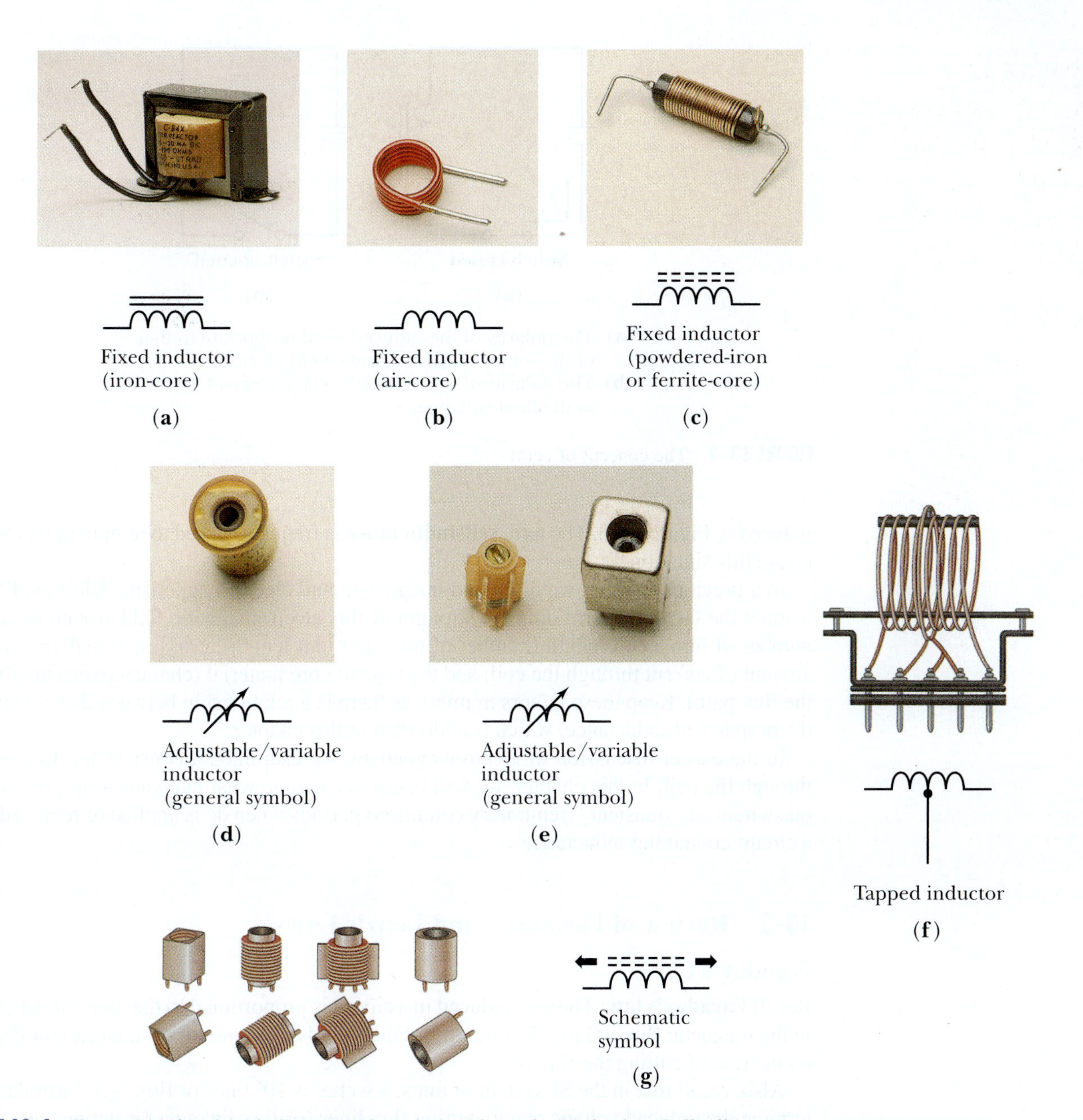

FIGURE 13–1 Examples of various inductors and symbols *(Photos by Michael A. Gallitelli)*

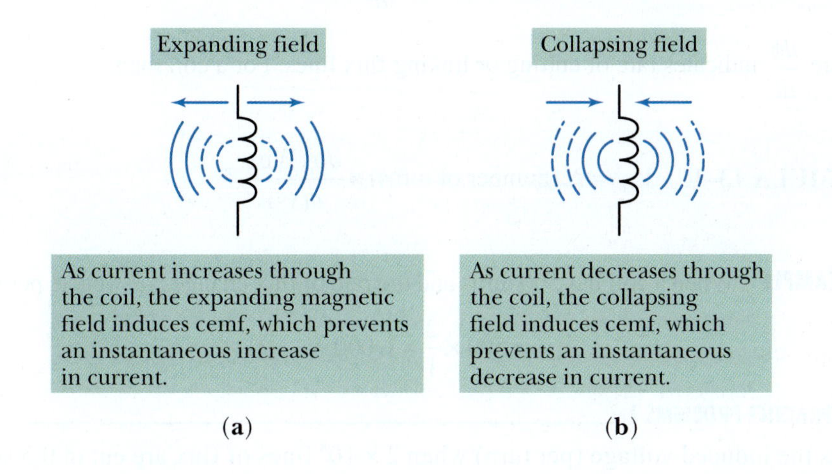

Expanding field
As current increases through
the coil, the expanding magnetic
field induces cemf, which prevents
an instantaneous increase
in current.
(a)
Collapsing field
As current decreases through
the coil, the collapsing
field induces cemf, which
prevents an instantaneous
decrease in current.
(b)

FIGURE 13–2 Expanding and collapsing fields

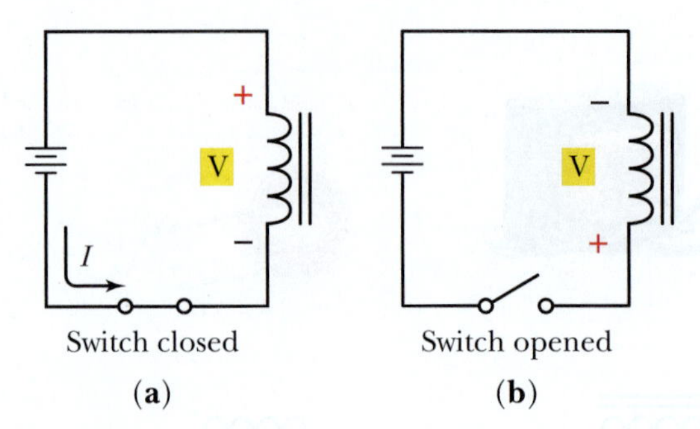

(**a**) The polarity of the induced cemf is opposite to that of the source as the magnetic field expands.
(**b**) The polarity of the induced voltage reverses as the field collapses.

FIGURE 13–3 The concept of cemf

induced it, Figure 13–3. The term **self-inductance** is frequently used to refer to inductors that create this situation.

In a previous chapter, we discussed magnetism and electromagnetism. Additionally, you learned the factors determining the strength of this electromagnetic field around the coil—number of turns; coil length (number of turns per unit length); cross-sectional area of coil; amount of current through the coil; and the type of core material (characteristics involved in the flux path). Keep these factors in mind, as there is a relationship between these items and the property of inductance, which is addressed in this chapter.

In the earlier discussion of electromagnetism, we examined a constant level dc current through the coil. In this chapter, we will begin to examine what happens during the switch-on/switch-off "transient" (temporary condition) periods when dc is applied or removed from a circuit containing inductance.

13–2 Review of Faraday's and Lenz's Laws

Faraday's Law

Recall **Faraday's law:** The emf induced in a circuit is proportional to the time-rate of change in the magnetic flux linking the circuit. In other words, the amount of induced emf depends on the rate of cutting the flux (ϕ).

Also, recall that in the SI system of units, a weber is 10^8 lines of flux. The formula to determine the induced voltage due to cutting flux lines (or flux linkages) is defined as:

$$V_{ind} = \frac{d\phi}{dt}$$

where $\frac{d\phi}{dt}$ indicates rate of cutting or linking flux lines. For a coil then:

FORMULA 13–1 $V_{ind} = N\,(\text{number of turns}) \times \frac{d\phi\,(\text{Wb})}{dt\,(\text{s})}$

EXAMPLE When a coil has 500 turns and the rate of flux change is 2 webers per second,

$$V_{ind} = 500 \times \frac{2}{1} = 1{,}000 \text{ V}$$

PRACTICE PROBLEMS 1

What is the induced voltage (per turn) when 2×10^8 lines of flux are cut in 0.5 second?

Lenz's Law

Recall **Lenz's law:** The direction of an induced voltage (or current) opposes the change that caused it. That is, when the amount of magnetic flux linking an electric circuit changes, an emf is induced that produces a current opposing this flux change.

To summarize, Faraday's law describes how much voltage is induced by the parameters affecting it, and Lenz's law describes the polarity and nature of this induced voltage and the current produced. Let's now begin to correlate these laws with some new information as we examine inductance and its dc characteristics.

13–3 Self-Inductance

In addition to inductance being defined as the property of a circuit that opposes a change in current, another definition for self-inductance (or inductance) is *the property of an electric circuit where a voltage is induced when the current flowing in the circuit changes.* You have learned the symbol for inductance, ***L.*** The unit of inductance is the **henry.** Named in honor of Joseph Henry, this unit of inductance is abbreviated **H.**

As you might suspect, the amount or magnitude of induced emf produced is related to the amount of inductance *(L)* in henrys (H) and the rate of current change. A formula expressing this is:

FORMULA 13–2 $V_L = L\frac{di \text{ (change in current, in amperes)}}{dt \text{ (change in time, in seconds)}}$

NOTE: The terms *di* and *dt* are the same as saying Δi and Δt or delta *i* and delta *t.* Δ or *d* meaning "the change in . . ."

where L = inductance in henrys

$$\frac{\Delta i}{\Delta t} \text{ or } \frac{di}{dt} = \text{rate of change of current} \left(\frac{\text{A}}{\text{s}}\right)$$

It should be noted that the polarity of the induced voltage is opposite to the source voltage that caused the current through the coil.

From the above, a definition for a henry of inductance states that *a circuit has an inductance of one henry when a rate of change of 1 ampere per second causes an induced voltage of 1 volt.* The induced voltage is in the opposite direction to the source voltage; therefore, it is called *counter-emf* or *back-emf.* Again, see Figure 13–3.

◆ **EXAMPLE** When the rate of current change is 5 amperes per second through an inductance of 10 henrys, what is the magnitude of induced emf?

$$V_L = L\frac{di \text{ (change in current, in amperes)}}{dt \text{ (change in time, in seconds)}}$$

$$V_L = 10 \times 5 = 50 \text{ V (Or, induced cemf} = 50 \text{ V)}$$ ◆

PRACTICE PROBLEMS 2

What is the counter-emf produced by an inductance of 5 henrys if the rate of current change is 3 amperes per second?

From these definitions and formulas, a useful formula to calculate inductance value is:

FORMULA 13–3 $L = \frac{V_L}{\frac{di}{dt}} \text{ or } \frac{V_L}{1} \times \frac{dt}{di}$

where L = inductance in henrys
V_L = induced voltage across inductor
$\frac{di}{dt}$ = rate of current change $\left(\frac{A}{s}\right)$

EXAMPLE What is the circuit inductance if a current change of 12 amperes in a period of 6 seconds causes an induced voltage of 5 volts?

$$L = \frac{V_L}{\frac{di}{dt}} = \frac{V_L}{1} \times \frac{dt}{di} = \frac{5}{1} \times \frac{6}{12} = 5 \times 0.5 = 2.5 \text{ H}$$

13–4 Factors That Determine Inductance Value of Inductors

As implied earlier, some of the same factors that affect the strength of a coil's electromagnetic field are also related to the amount of coil inductance.

Look at Figure 13–4. The physical properties of an inductor that affect its inductance include the following:

1. *Number of turns.* The greater the number of turns, the greater the inductance value. In fact, inductance is proportional to the square of the turns. That is, when you double the number of turns in a coil of a given length and diameter, the inductance quadruples, assuming the core material is unchanged.
2. *Cross-sectional area (A) of the coil.* This feature relates to the square of the coil diameter. The greater the cross-sectional area, the greater the inductance value. When the cross-sectional area doubles, the inductance *(L)* doubles.
3. *Length of the coil.* The longer the coil for a given diameter and number of turns, the *less* the inductance value. This is due to a less concentrated flux, thus, less flux cut per unit time and the less the induced voltage produced.
4. *Relative permeability of the core.* For air-core inductors, the relative permeability (μ_r) is virtually one. The higher the permeability of the inductor's core material, the greater the inductance value. Recall that permeability relates to the magnetic path's ability to concentrate magnetic flux lines.

As a technician you probably will not design inductors. However, Formula 13–4 illustrates the relationship of inductance to the various factors we have been discussing. In SI units, the formula is:

FORMULA 13–4 $L = 12.57 \times 10^{-7} \times \frac{\mu_r N^2 A}{l}$

where L = inductance in henrys
μ_r = relative permeability (for air or vacuum = 1)
N = number of turns
A = cross-sectional area in square meters
l = length of coil in meters
12.57×10^{-7} = a constant representing the absolute permeability of air (in SI units)

EXAMPLE An air-core coil that is 0.01 meters in length with a cross-sectional area of 0.001 square meters has 2,000 turns. What is its inductance in henrys?

$$L = 12.57 \times 10^{-7} \times \left(\frac{1 \times (2{,}000)^2 \times 0.001}{0.01}\right)$$

$$L = 12.57 \times 10^{-7} \times \left(\frac{(40 \times 10^5) \times 1 \times 10^{-3}}{10 \times 10^{-3}}\right)$$

$$L = 12.57 \times 10^{-7} \times 4 \times 10^5 = \text{about 0.5 H, or 500 mH}$$

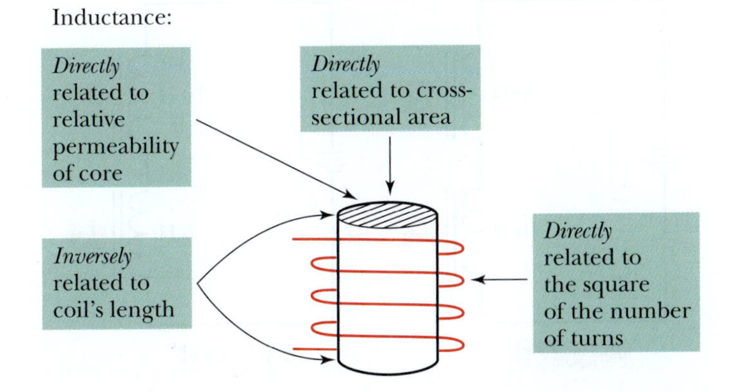
Inductance:
Directly related to relative permeability of core
Directly related to cross-sectional area
Inversely related to coil's length
Directly related to the square of the number of turns

FIGURE 13–4 Factors that affect the amount of inductance

Since the μ_r of air is one, the formula is modified for *air-core* inductors and stated as:

FORMULA 13–5 $L = 12.57 \times 10^{-7} \times \frac{N^2 A}{l}$

■ IN-PROCESS LEARNING CHECK 1

Fill in the blanks as appropriate.

1. All other factors remaining the same, when an inductor's number of turns increases four times, the inductance value _____ by a factor of _____.
2. All other factors remaining the same, when an inductor's diameter triples, the inductance value _____ by a factor of _____.
3. All other factors remaining the same, when an inductor's length increases, the inductance value _____.
4. When the core of an air-core inductor is replaced by material having a permeability of 10, the inductance value _____ by a factor of _____.

13–5 Inductors in Series and Parallel

For our purposes, we will assume the inductors in this section do not have any *coupling* between them. That is, there are no flux lines from one inductor linking the turns of another inductor. In technical terms, no *mutual inductance* exists between coils.

Inductance in Series

Finding the total inductance of inductors in *series,* Figure 13–5, is simple. Since they are in series, the current change through each inductor is the same. Thus, when they are equal-value inductors, the induced voltage in each is equal, and the total inductance is two times the value of either inductor. This implies that when the inductors are not equal in value, the total inductance **(L_T)** is the sum of the inductance values in series. That is, $L_T = L_1 + L_2$.

Stated another way: *Inductors in series add inductances in the same way that resistors in series add resistances!*

FORMULA 13–6 $L_T = L_1 + L_2 \ldots + L_n$

REMINDER: This formula holds true as long as there is no coupling of flux lines from one inductor to another; that is, there is no coupling between inductors.

EXAMPLE When the inductors in Figure 13–5a are tripled in value, the total inductance of the circuit on the left is 60 H, rather than 20 H, and the total inductance of Figure 13–5b is 45 H, rather than 15 H.

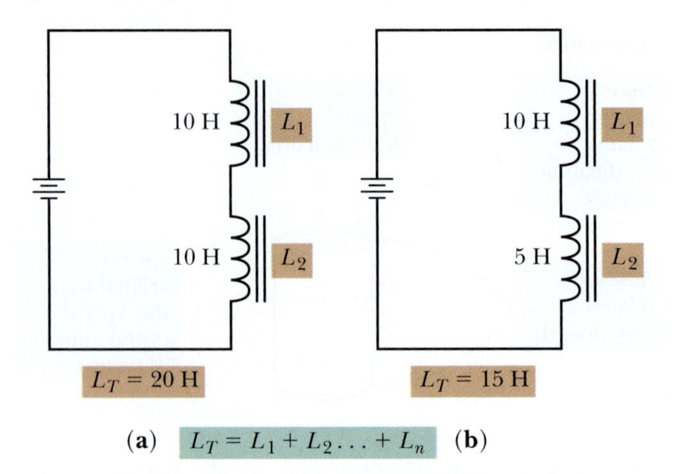

FIGURE 13–5 Total inductance of series inductors

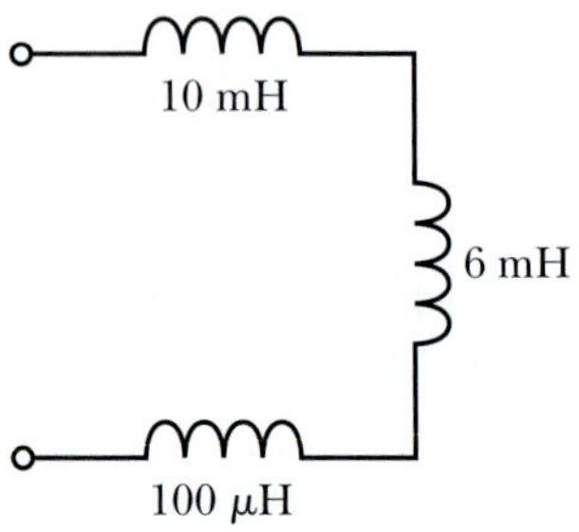

10 mH

6 mH

100 μH

FIGURE 13–6

PRACTICE PROBLEM 3

Determine the total inductance of the circuit shown in Figure 13–6.

Inductance in Parallel

As you might expect, the total inductance of inductors in parallel is less than the least value inductor in parallel. This is analogous to resistors in parallel. The general formulas used to compute total inductance for inductors in parallel, Figure 13–7, are the reciprocal formulas: (NOTE: Don't forget to use the reciprocal function of your calculator to advantage when dealing with these formulas.)

FORMULA 13–7 $$\frac{1}{L_T} = \frac{1}{L_1} + \frac{1}{L_2} + \frac{1}{L_3} \cdots + \frac{1}{L_n}$$

FORMULA 13–8 $$L_T = \frac{1}{\frac{1}{L_1} + \frac{1}{L_2} + \frac{1}{L_3} \cdots + \frac{1}{L_n}}$$

Practical Notes

Be aware that Formula 13–8 ("the reciprocal of the summed reciprocals") is simply displaying what you would need to do to find L_T, after you use Formula 13–7 and find $1/L_T$. That is, to find L_T, you take the reciprocal of $1/L_T$, or 1 over $1/L_T$. This then gives you the total inductance of the parallel inductors used in the calculations.

(NOTE: If all units are in henrys, the answer is in henrys.) Also, the product-over-the-sum approach, used in Figure 13–8, works for two inductors in parallel:

FORMULA 13–9 $$L_T = \frac{L_1 \times L_2}{L_1 + L_2}$$

◆ **EXAMPLE** You have probably already surmised that the same time-saving techniques can be used for inductance calculations as you used in resistance calculations. For example, if the two inductors in parallel are of equal value, the total inductance equals half of one inductor's value. This is similar to two 10-Ω resistors in parallel having an equivalent total resistance of 5 Ω. ◆

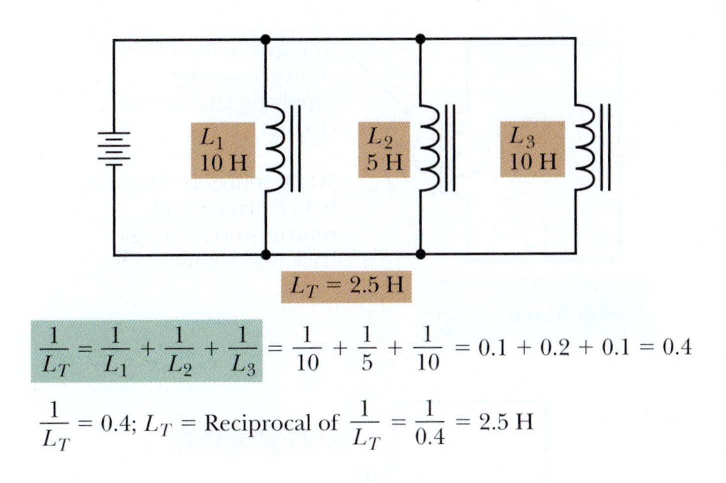

FIGURE 13–7 Finding L_T using the reciprocal formula

$$\frac{1}{L_T} = \frac{1}{L_1} + \frac{1}{L_2} + \frac{1}{L_3} = \frac{1}{10} + \frac{1}{5} + \frac{1}{10} = 0.1 + 0.2 + 0.1 = 0.4$$

$$\frac{1}{L_T} = 0.4;\ L_T = \text{Reciprocal of } \frac{1}{L_T} = \frac{1}{0.4} = 2.5 \text{ H}$$

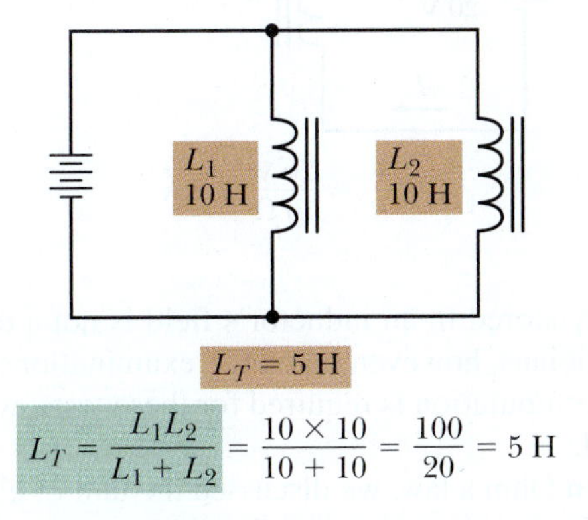

FIGURE 13–8 Finding L_T using the product-over-the-sum formula

$$L_T = \frac{L_1 L_2}{L_1 + L_2} = \frac{10 \times 10}{10 + 10} = \frac{100}{20} = 5 \text{ H}$$

PRACTICE PROBLEMS 4

1. What is the total inductance of three 15-H inductors that are connected in parallel and have no mutual inductance?
2. Use your calculator and the reciprocal method and find the total inductance of four inductors in parallel. Their individual values are: 10 mH, 12 mH, 15 mH, and 20 mH. Assume no coupling between them.

13–6 Energy Stored in the Inductor's Magnetic Field

The magnetic field surrounding an inductor when current passes through the inductor is a form of stored energy, Figure 13–9. The current producing the electromagnetic field is provided by the circuit's source. Hence, this stored energy is supplied by the source to the magnetic field.

When the source is removed from the circuit, the field collapses and returns the stored energy to the circuit in the form of induced emf. This induced emf tries to keep the current through the coil from decaying.

If the inductor were a perfect inductor with no resistance, no I^2R loss would occur; therefore, no power would be dissipated in this energy storing-and-returning process. NOTE: In reality, inductors have some resistance (the resistance of the wire), and they dissipate power; however, for the present discussion, we won't analyze the details of this fact. Note also that in dc circuits, the inductor's resistance (plus any other resistance in series with the inductor) and the amount of voltage applied determine the amount of current through the coil (Ohm's law).

Notice in Figure 13–10, if a coil has 10 ohms of resistance, and a dc voltage of 20 V is applied, the dc current through the coil will be 20 V/10 Ω, or 2 amperes, after the field has finished building up.

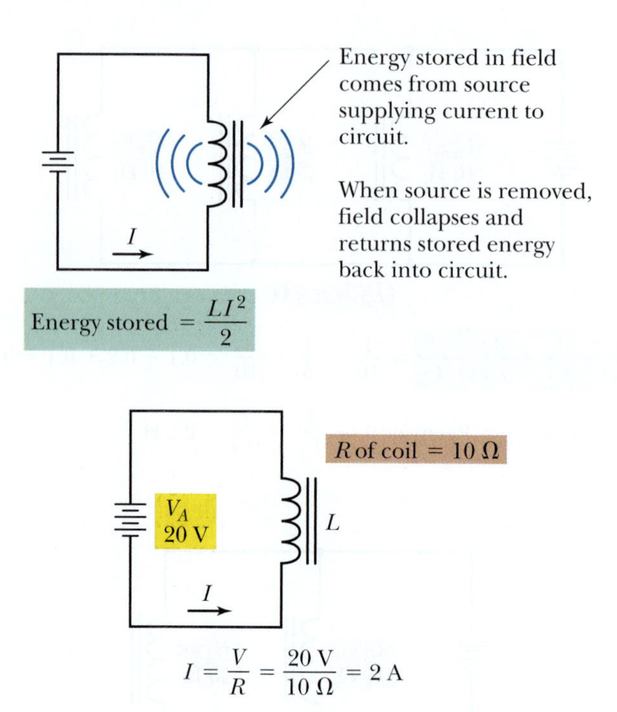
Energy stored in field comes from source supplying current to circuit.
When source is removed, field collapses and returns stored energy back into circuit.
I
Energy stored = $\frac{LI^2}{2}$
R of coil = 10 Ω
V_A
20 V
L
I
$I = \frac{V}{R} = \frac{20\text{ V}}{10\ \Omega} = 2\text{ A}$

FIGURE 13–9 Concept of energy being stored in inductor's magnetic field

FIGURE 13–10 Direct current (dc) through inductor found by using Ohm's law

Computing the energy stored in an inductor's field is not a daily calculation for most technicians. Many technicians, however, take FCC examinations to obtain commercial radio licenses. Since this computation is required for these tests, we will show you how this calculation is performed.

In an earlier chapter on Ohm's law, we discussed the unit of electrical energy, the **joule.** A joule is sometimes defined as the amount of electrical energy required to move 1 coulomb of electrical charge between two points having a potential difference of 1 volt. Also remember that the watt of power relates to the joule since the performance of electrical work at a rate of 1 joule per second equals 1 watt.

Relating all of these factors, it can be shown that the stored energy correlates to the average power supplied to the inductor circuit and the time involved. The amount of energy stored, in turn, relates to the amount of inductance and current involved.

Accumulating these facts, then:

FORMULA 13–10 E (energy in joules) $= 0.5\ LI^2$

where L = inductance in henrys
I = current in amperes
Energy = joules (or watt-seconds)

EXAMPLE What is the energy stored by an inductor of 20 henrys that has a current of 5 amperes passing through it?

Substituting into the formula:

$$E \text{ (energy in joules)} = 0.5\ LI^2 \text{ or } \frac{LI^2}{2}$$

$$E \text{ (energy in joules)} = \frac{20 \times (5)^2}{2} = \frac{20 \times 25}{2} = \frac{500}{2} = 250\text{ J}$$

PRACTICE PROBLEMS 5

A dc voltage of 100 V is applied to a 1-H inductor that has 20 Ω of dc resistance. (a) What is the dc current through the coil? (b) How many joules of energy are stored in the inductor's magnetic field?

13–7 The *L/R* Time Constant

We earlier established that an inductor opposes a change in current flow due to the expanding or contracting magnetic field as it cuts the turns of the coil. The induced voltage opposes the change that originally induced it (Lenz's law). This implies that even when dc voltage is first applied to a coil, the current cannot instantly rise to its maximum value (the *V/R* value), which is true!

Let's examine the transient conditions, or the temporary state caused by a sudden change in circuit conditions in a simple inductive circuit when dc is first applied. Refer to Figures 13–11 and 13–12 as you study the following discussion.

1. In Figure 13–11, you see the conditions that exist in a purely resistive circuit. When the switch is closed, the current immediately reaches its dc (Ohm's law) value determined by the voltage applied and the resistance involved.
2. In Figure 13–12a, the inductor in the resistor-inductor *(RL)* circuit prevents current from rising to its maximum value during the first moments after the switch is closed. This is because the current increases from zero to some value, causing an expansion of the magnetic field in the inductor. The expanding field cuts the turns of the coil, inducing a back-emf (cemf) that opposes the current change. Therefore, the current does not abruptly rise from zero to the final (stable) dc value.
3. In Figure 13–12b, at time t_1, the current has arrived at only 63.2% of its final value. The time from t_0 to t_1 is equal to the value of *L/R*, where *L* is in henrys and *R* is the circuit resistance value in ohms. This value ***(L/R)*** is known as one **time constant.** The symbol for the time constant is the Greek letter tau (τ). In a circuit containing *L* and *R*, the formula for time constant is:

FORMULA 13–11 $\tau = \frac{L}{R}$

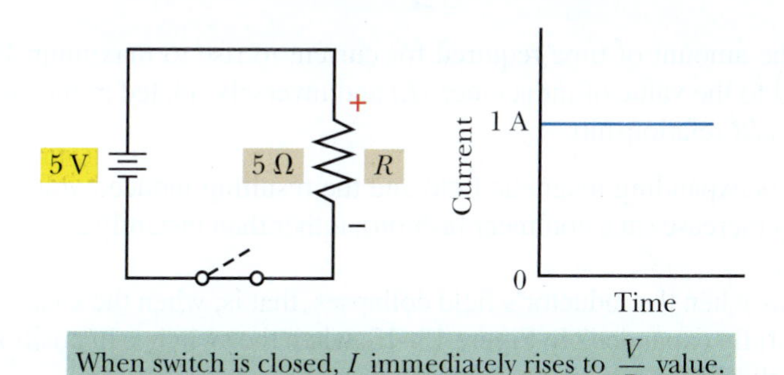

FIGURE 13–11 Direct current (dc) change in a purely resistive circuit

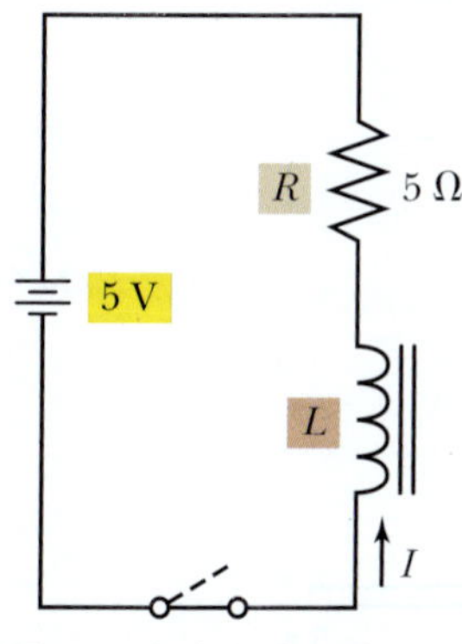

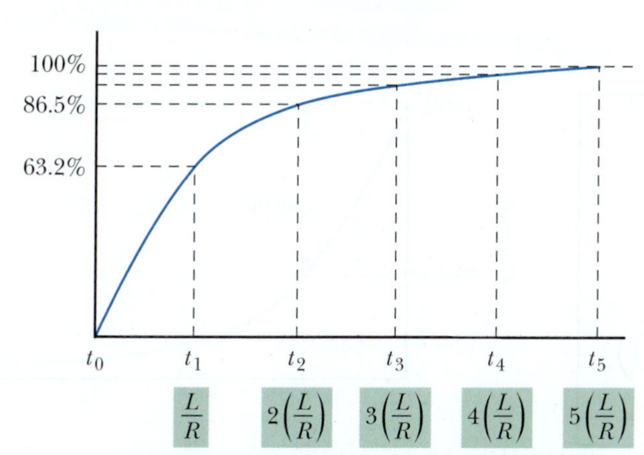

FIGURE 13–12 Current change in a circuit containing inductance multiSIM

Let's think about Formula 13–11 for a moment. One time constant (τ) equals the value of the inductance *(L)* in henrys divided by the resistance *(R)* in ohms. The amount of time for one time constant is *directly* related to the value of inductance. The time for one time constant is inversely related to the resistance value. Why might this be true?

You have already learned that the greater the inductance value, the greater will be the amount of counter-emf generated for a given amount of current change ($V_L = L \times d\phi/dt$). This indicates that the larger the *L* value, the greater the opposition to current change. This means it will take more time for the current (and field) to build up to their maximum, or "steady-state" level, which occurs after five time constants. In other words: As *L* increases, the time for one time constant (τ) increases. Conversely, as *L* decreases, τ decreases.

In accordance with Ohm's law, the smaller the resistance value, the higher the final current will be ($I = V/R$); thus, the greater the required change in current to arrive at the higher current value, and the longer the time involved for one time constant. In other words: As *R* decreases, the time for τ increases. Conversely, as *R* increases, τ decreases.

With these thoughts in mind, continue examining the graphing of the rise of current through the *RL* circuit, with dc applied. Incidentally, this type of graph is often referred to as the "Universal Time-Constant Chart" or graph.

4. In Figure 13–12b, at time t_2 (where the time from t_1 to t_2 represents a second time constant), the current has risen 63.2% of the remaining distance to the final dc current level. That is, current rises 63.2% of the remaining 36.8% to be achieved, or 63.2% of 36.8% equals an additional 23.3%. Thus, in two time constants the current achieves 63.2% + 23.3%, or 86.5% of its final maximum value.
5. As time continues, the current rises another 63.2% of the remaining value for each time constant's worth of time. After *five time constants,* the current has virtually arrived at its final dc value. In this case,

$$\frac{5\text{ V}}{5\ \Omega} = 1\text{ A}$$

Therefore, the amount of time required for current to rise to maximum *V*/*R* value is directly related to the value of inductance *(L)* and inversely related to the circuit *R*, as indicated by the *L*/*R* relationship.

The inductor's expanding magnetic field and the resulting induced voltage cause current to exponentially increase (in a nonlinear fashion), rather than instantly as it would in a purely resistive circuit.

What happens when the inductor's field collapses, that is, when the source is removed after the field has fully expanded? In Figure 13–13, when the switch is in position A, the source is connected to the *RL* circuit.

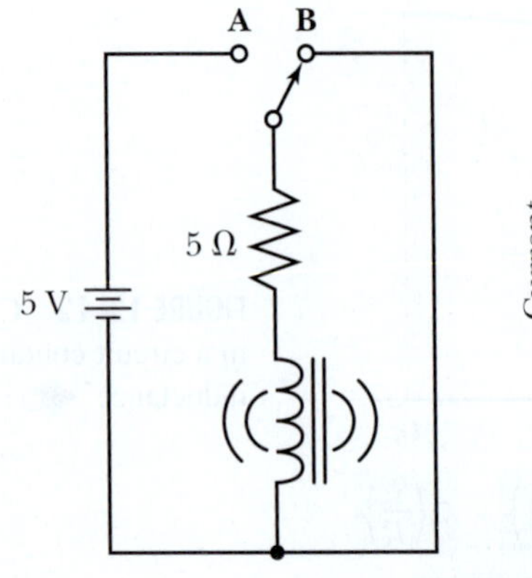

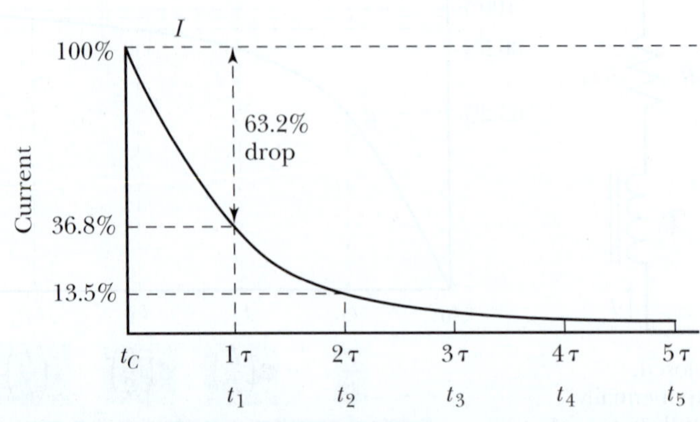

FIGURE 13–13 Circuit with same *L*/*R* time constant for current decay as for current buildup

When the switch is in position B, the source is disconnected from the *RL* circuit. However, a current path is provided for the *RL* circuit, even when the source is disconnected.

Assume the switch was in position A for at least five time constants and the current has reached its maximum level. Therefore, the magnetic field is fully expanded. When the switch is moved to position B, the expanded field collapses, inducing an emf into the coil, which attempts to keep the current from changing. The inductor's collapsing field and induced emf have now, in effect, become the source. Obviously, as the field fully collapses, current decays and eventually becomes zero. The following comments describe what happens during this declining current time. Figure 13–14, the Universal Time-Constant Chart, shows a graphic display of both the current buildup and decay. Refer to it while reading the following discussion.

1. The current decays 63.2% from maximum current value toward zero in the first time constant. In other words, it falls to 36.8% of maximum value.
2. During the second τ, it falls 63.2% of the remaining distance, or another 23.3% for a total of 86.5% decay. In other words, the current is now at 13.5% of its maximum value.
3. This continues, until at the end of the five *L/R* time constants, the current is virtually back at the zero level.
4. Given the same *L/R* conditions, the current decay is the mirror image (or the inverse) of the current-rise condition, described earlier.

Let's summarize. In a dc circuit composed of an inductor and a series resistance (*RL* circuit), it takes five *L/R* time constants for current to change *from one static level to a "changed" new level,* whether it be an increase or a decrease in current levels. (NOTE: This is true whether the change is from 0 to 1 ampere, from 1 to 5 amperes, or from 4 amperes to 2 amperes, and so forth.) This is an important concept, so remember it!

Observe Figure 13–15 and notice that in a *dc RL* circuit, the sum of V_L and V_R must equal *V* applied. This means that as current increases through the *RL* circuit, V_R is increasing and V_L is decreasing proportionately until V_R equals approximately *V* applied and V_L equals approximately 0 V when the current has finally reached a steady-state value (assuming the *R* of the inductor is close to zero ohms). Conversely, when current decays, V_R also decays at the same rate to the steady-state minimum value the circuit is moving toward. The τ curves are one convenient method to determine what values V_R and V_L have after specified times.

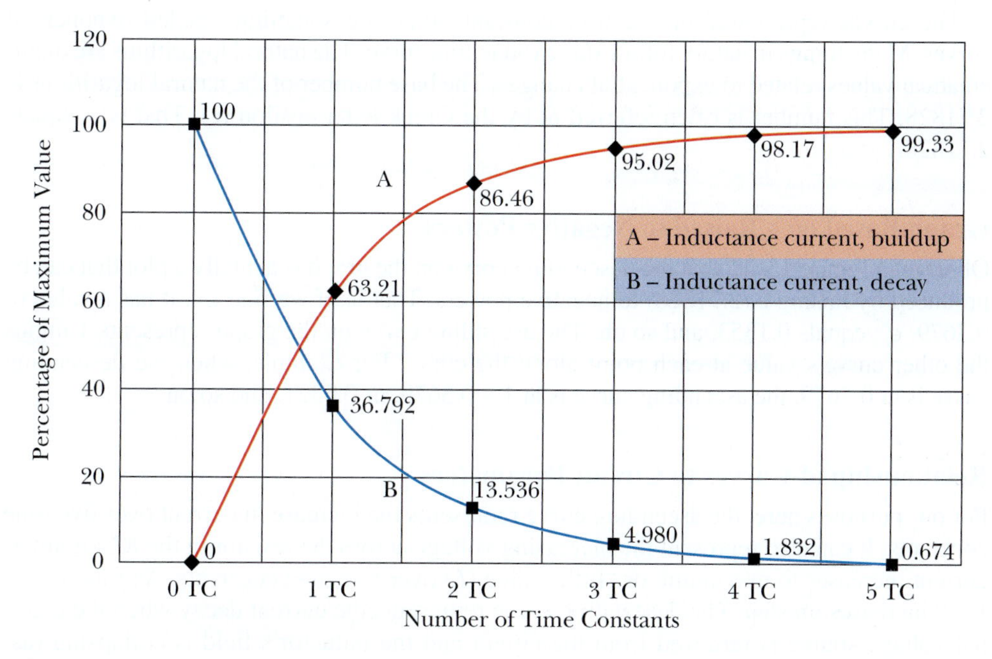

FIGURE 13–14 Universal Time-Constant Chart

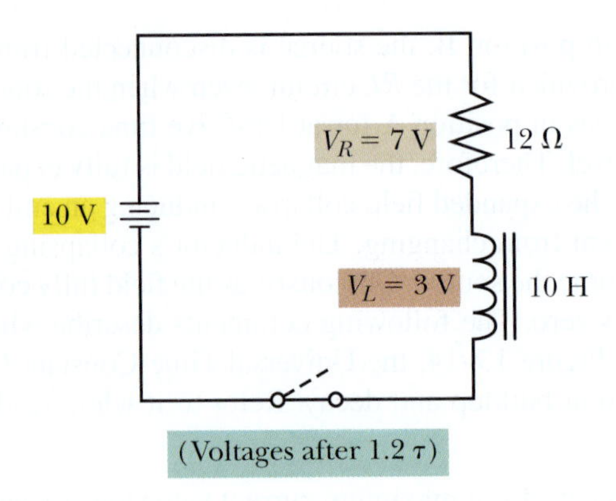

FIGURE 13–15 Simple dc *RL* circuit multiSIM

EXAMPLE Refer to Figure 13–15. What is the value of V_R 1 second after the switch closed? Answer: First, compute how many time constants are represented by 1 second:

$$\tau = \frac{L}{R} = \frac{10\text{ H}}{12\ \Omega} = 0.833 \text{ seconds}$$

Therefore, 1 second = 1/0.833 time constants, or 1 second = 1.2 τ.

Next, refer to the τ curves in Figure 13–14 and note that at 1.2 τ, the current rises to 70% of maximum value. Therefore, $V_R = 0.7 \times V$ applied. Thus, V_R at that instant equals 0.7 × 10 V = 7 V and V_L is 3 V.

PRACTICE PROBLEMS 6

Using the same circuit as in Figure 13–15, determine the V_R and V_L values 2 seconds after the switch is closed.

13–8 Using Epsilon (ϵ) to Help Calculate Circuit Parameters

As you have seen, you can come fairly close in approximating circuit conditions and values from the exponential Universal Time-Constant Chart. Another more precise way to calculate circuit parameters at any given instant is to use natural logarithms and epsilon, rather than graphical means.

The curves represented on the time-constant graph are sometimes called exponential curves. Many things in nature follow this exponential curve. The natural logarithms are mathematical values related to exponential changes. The base number of the natural logarithms is 2.71828. This number is often referred to by the Greek letter epsilon (ϵ). That is, ϵ equals 2.71828.

Graph Based on ϵ Raised to Negative Powers

Observe in Figure 13–16 that the descending curve on the graph is actually a plot that can be produced by raising ϵ (2.71828) to negative powers. That is, if we find ϵ^{-1}, it has a value of 0.3679, ϵ^{-2} equals 0.1353, and so on. The ascending curve on the graph represents 1 minus the other curve's value at each point along the curve. For example, when the descending curve is at 0.3679, the ascending curve is at 1 – 0.3679, or 0.6321, and so on.

Relationship of Curves to Circuit Parameters

For our purposes here, the ascending curve represents the increase in current over five time constants. It can also represent the increasing voltage across the resistor in the *RL* circuit as current increases to maximum, or "full," value (I_f) over five time constants. As I increases, $I \times R$ increases, in step. The descending curve represents the current decay when the external voltage source is removed from the circuit and the inductor's field is collapsing (assuming there is a current path when the external source is removed, Figure 13–13). Also,

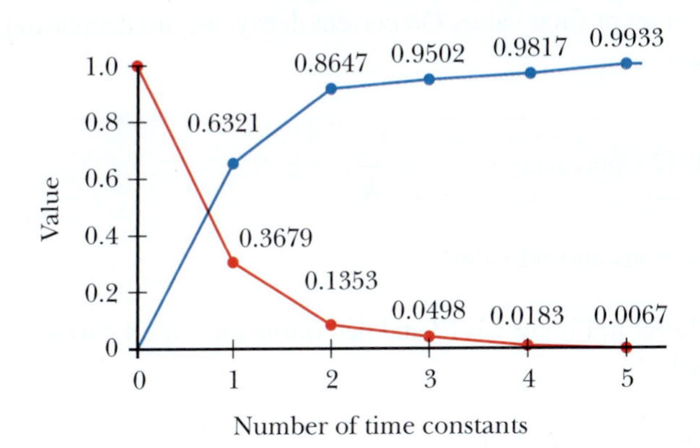

FIGURE 13–16 Graph based on ε raised to negative powers

the descending curve can represent the decay in voltage drop across the resistor, since its $I \times R$ drop is going to be in step with the current conditions.

As you know, in the dc circuit, the $V_R + V_L$ must add up to equal V_{applied}. At any given point along these curves, the decimal values of the descending and ascending curves must add up to 1, representing 100% of the maximum, or final, state value.

Relationship of Negative Powers for Epsilon and *L/R* Time Constant

As stated, the negative power to which epsilon (2.71828) is raised, determines where we are on the descending exponential curve. How does this curve relate to time constants?

You have already learned that after five *L/R* time constants, the circuit parameters will have changed from wherever they are to the new "stable state" conditions. The relationship of importance to us is: *how much time is being allowed for the change versus the time for one time constant?*

If the time being allowed for the change is five time constants or greater, (1) current will be at its final value, if it is building up, or at zero value, if decaying; (2) V_R will be equal to V_{applied} if current is building up, or at zero value, if current is decaying; (3) V_L will be at zero value in either case.

If the time being allowed for the change is only one time constant, then current will be at 63.2% of its new maximum, if current is building up, or at 36.8% of maximum, if current is decaying.

The bottom line is, if we know the amount of time being allowed and know, or can calculate, the *L/R* time constant, we know how many time constants are being allowed for the change. *This ratio of time allowed to time for one time constant is the number we want to use as the negative power of epsilon (ϵ) in calculating circuit parameters!*

The method of finding the number of time constants allowed is:

$$\text{Number of time constants} = \frac{\text{Time allowed for change } (t)}{\text{Time for one } L/R \text{ time constant}}$$

$$\text{Number of } \tau = \frac{\text{t}}{\frac{L}{R}} = \frac{t}{1} \times \frac{R}{L} = \frac{Rt}{L}$$

So the negative power we will want to raise epsilon to in our calculations related to *L/R* time constants is *Rt/L*.

Of course, your scientific calculator has either the natural log function key [e^x], or a key labeled [$\ln_x$], which will greatly simplify calculations involving raising epsilon to the various powers of interest. In our examples, we will use the [e^x] key.

Armed with these facts, let's introduce the formulas used to solve for the instantaneous value of current. Of course, if you know the current, you can calculate the $I \times R$ resistor voltage drop. If you know the applied voltage, you can then subtract V_R from V_S to find V_L at that point in time.

The circuit conditions assumed here are as follows: On current buildup, we are increasing from zero to maximum or final value. On current decay, we are decreasing from the present level to zero current.

FORMULA 13–12 (Increasing I) $i = \frac{V_S}{R}(1 - \epsilon^{-Rt/L})$

where i = instantaneous current value

EXAMPLES Refer to Figure 13–17 and determine the circuit current 1 millisecond after the switch is closed.

$$i = \frac{V_S}{R}(1 - \epsilon^{-Rt/L})$$

$$i = \frac{10 \text{ V}}{10 \text{ k}\Omega}(1 - \epsilon^{-10 \text{ k}\Omega \times 1 \text{ ms}/10 \text{ H}})$$

$$i = 1 \text{ mA }(1 - \epsilon^{-1}) = 1 \text{ mA }(1 - 0.3678) = 1 \text{ mA} \times 0.6322$$

$$i = 0.0006322 \text{ A or } 0.6322 \text{ mA}$$

This checks out against our exponential graph. The time allowed for change was one time constant and the current increased from zero to about 63% of its maximum possible value of 1 mA.

$$V_R = 0.63 \text{ mA} \times 10 \text{ k}\Omega, \text{ or } 6.3 \text{ V}; V_L = 10 \text{ V} - 6.3 \text{ V} = 3.7 \text{ V}$$

On many calculators, the keystrokes for finding $\epsilon^{-Rt/L}$ for our example are: 10, [EE], 3, [×], 1, [EE], [+/–], 3, [÷], 10, [=], [+/–], [2nd], [e^x]. This yields the 0.3678, which easily plugs into the formula in place of the ϵ^{-1} from our example.

The complete sequence for many calculators for this problem would be: 10, [EE], 3, [×], 1, [EE], [+/–], 3, [÷], 10, [=], [+/–], [2nd], [e^x], [+/–], [+], 1, [=], [×], 10, [÷], 10, [EE], 3, [=].

Observe that this sequence indicates that first you find the product of $R \times t$, then divide it by the L value, yielding the value of Rt/L (the number of time constants being allowed for the change). Then use that result as the negative power to raise epsilon to in order to find out where we are on the exponential curve. Then, make that result negative with the [+/–] button in order to subtract it from a + 1. The resultant decimal value is then multiplied times the value of maximum current found by V/R. The result, for our example, is the circuit current present after the one time constant allowed for the current change from zero.

PRACTICE PROBLEM 7

Suppose for the example circuit in Figure 13–17, we want to compute the current 2.5 milliseconds after the switch was closed. What would be the value of current? (Use the calculator sequence, then see if the result makes sense by checking against the τ chart.)

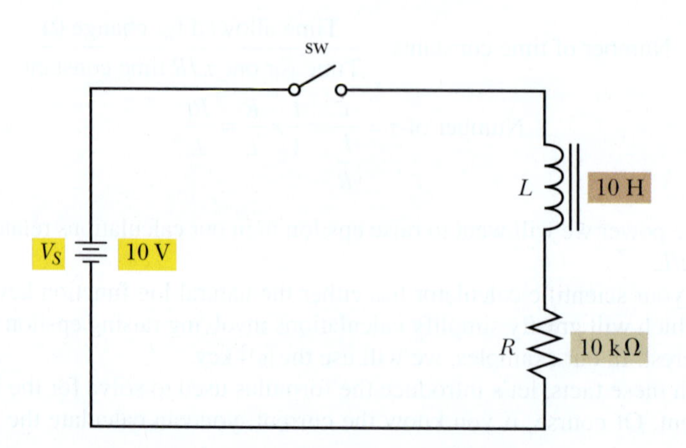

FIGURE 13–17

To find the value of decaying (decreasing) current, it is not necessary to subtract from 1, since we are not trying to find the value of the ascending curve, but rather we want the descending curve value. The formula for instantaneous value of current during decreasing current is:

FORMULA 13–13 (Decreasing I) $i = \frac{V_S}{R}(\epsilon^{-Rt/L})$

EXAMPLE Refer to the circuit shown in Figure 13–18. Assume that the current is at maximum, final-state value; then the switch is moved from point A to point B. What is the value of circuit current after 68 microseconds?

$$i = \frac{V_S}{R}\epsilon^{-Rt/L}$$

$$i = \frac{6 \text{ V}}{2.2 \text{ k}\Omega} \times \epsilon^{-(2.2 \text{ k}\Omega \times 68 \text{ }\mu\text{s})/50 \text{ mH}} = 2.72 \text{ mA} \times \epsilon^{-2.992} = 2.72 \text{ mA} \times 0.05$$

$$= 0.136 \text{ mA}$$

This checks out against our τ chart. After about three time constants, the current should have decayed to approximately 0.0498 times maximum current. Our calculations indicated that after 2.992 time constants, the current is at 0.05 times its original value. $V_R = 0.136 \text{ mA} \times 2.2 \text{ k}\Omega = 0.299$ volts at this point in time.

Practical Notes

Adaptations of the current formulas can be made to find resistor and inductor voltage at any instant during current rise or at any instant during current decay. The formulas that may be used are as follows:

Voltage across resistor during current buildup:

$$V_R = V_S \times (1 - \epsilon^{-Rt/L})$$

Voltage across resistor during current decay

$$V_R = V_S \times (\epsilon^{-Rt/L})$$

Voltage across inductor during current buildup or decay:

$$V_L = V_S \times (\epsilon^{-Rt/L})$$

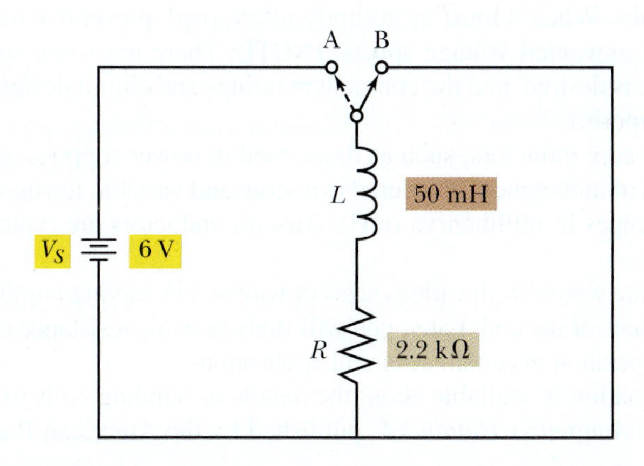

FIGURE 13–18

A Rapidly Collapsing Field Induces Very High Voltage(s)

In a circuit, such as in Figure 13–15, when the source voltage is disconnected, there is no low-resistance current path, as was shown earlier when explaining the current decay events. For purposes of explanation, let's assume the resistance across the open-switch contacts is extremely high. What happens to the total *R* in the current path when the switch opens? The *R* drastically increases (approaching infinity). What impact does this have on the *L*/*R* time constant? Since *R* drastically increases and *L* remains the same, the time constant must greatly decrease!

The result of all this activity is that the expanded field collapses almost instantly. This means many flux lines cut the turns of the inductor in a very short time; thus, a very high voltage is induced in the coil. (NOTE: This phenomenon produces high voltage and high-voltage "arcs" jumping the gaps of spark plugs in your automobile.) In Figure 13–15, the induced voltage could be high enough to cause an arc across the open-switch contacts. The stored magnetic energy represented by the collapsing field is thus dissipated by the energy used in producing the arc.

ATTENTION: Please be informed that the problem of rapidly collapsing fields producing very high voltages is a common danger and can be a creator of many problems in industrial control and other circuits used in industry. These circuits are often controlling very high currents through very large inductances (including motors, etc.). For that reason it is a topic that you should consider very significant! In many cases, there may be some "protective" circuits used to manage these dangers; however, we want you to understand how important it is for you to learn about these circuits at the appropriate time in your studies. The bottom line is that you should not only become aware of the dangers of rapidly collapsing fields, but should endeavor to fully learn the circuits and safety precautions needed when working with such circuits.

13–9 Summary Comments about Inductors

You have studied that inductors have properties of inductance (which oppose a *change* in current) and resistance in the wire from which the inductor is wound (which opposes current flow). The amount of inductance, *L* (measured in henrys), relates to the physical properties of the coil, including the length, cross-sectional area, number of turns, and type of core. Also, there is a relationship between the amounts of inductance and induced cemf that develops for any given rate of current change through the coil.

Also, you know that when a dc voltage is first applied to a circuit with an inductor present, current does not instantly rise to the value dictated by the dc resistance of the circuit divided by the applied voltage. This delay is due to the inductor's opposing any current change. It takes five *L*/*R* time constants for the complete change from one current level (even if it is zero amperes) to a new current level. Under opposite conditions (i.e., if the dc source is removed from the inductor circuit) current will not abruptly decrease to zero, since the inductor opposes the current change and the collapsing field induces an emf in the coil to attempt to prevent the current from changing. When the time constant is very small during the decay sequence, the field quickly collapses, inducing a very high voltage in the coil. Wise circuit designers, therefore, make sure that current is not abruptly interrupted through the inductors in general dc circuits. When it must be abruptly interrupted, preventive measures are generally used to avoid unwanted voltage spikes. (NOTE: There are some special applications where high voltage is desired, and the component ratings and circuit design are appropriately chosen for this purpose.)

Laminated iron-core inductors, such as those used in power supplies and audio circuitry, are often in henrys of inductance. Powdered iron-core and variable ferrite-core inductors can have inductance ranges in millihenrys (mH). Air-core inductors are typically in the microhenrys (μH) range.

Also, inductors are wound with various sizes of wire and in varying numbers of turns, which affect the dc resistance of the coil. Later, you will study how the resistance to inductance ratios affects inductors' operation in certain ac circuit applications.

Excellent information is available about the details of winding coils to desired specifications in *The Radio Amateur's Handbook,* published by the American Radio Relay League

A Good Idea

Watch out for "inductive kick" when breaking a dc circuit containing a large inductance! Use appropriate safety precautions to avoid being shocked!

(ARRL). Also, there are numerous inexpensive, special slide rules that allow you to establish required parameters and to read the number of turns, coil length, coil diameter, and so on, needed for the desired results.

13–10 Troubleshooting Hints

Typical Troubles

Troubles in inductors are usually associated with either opens or shorts. For this reason, the ohmmeter is a useful instrument to check the condition of an inductor—*provided* you know the normal resistance for the coil being checked.

Usual values range from 1 Ω to several hundred ohms for iron-core "chokes" and inductors with many turns and inductance values in the henrys range. Inductors in the millihenrys and microhenrys ranges typically have dc resistances from less than 1 Ω to something less than 100 Ω.

CAUTION! Care should be taken when removing ohmmeter test leads from large inductances. The "inductive kick" caused by the quick collapse of the inductor's field when the ohmmeter is removed can give you a nasty shock! This field is created by current produced through the inductor by the ohmmeter's dc source during testing. Be sure your hands or fingers are not touching or near the metallic tips of the test probes when removing the test leads from the inductor connections.

Open Coil

If an inductor is open, it is obvious with an ohmmeter check that it measures infinite ohms. **CAUTION!** Be sure there are no sneak paths when you measure. If it is easy to do, it is good to isolate the component being measured from the rest of the circuitry when checking it. Probably the most common trouble is this open condition. There might be a broken turn in the coil, a bad connection to the coil, and so forth.

Shorted Turns

Because each coil turn must be electrically insulated from all other coil turns for the inductor to operate properly, when the varnish or insulation on adjacent turns breaks down, these turns can short together. This is more difficult to detect with an ohmmeter test. The key is to know what R value the given coil should read. If there are shorted turns, the R value will read lower than normal. (NOTE: If only one or two turns are shorted, it may be impossible to detect by simple ohmmeter measurements!)

Obviously, the more turns that are shorted, the greater is the difference between the normal value and that which is measured. It is also possible in iron-core inductors to have turns shorting to the core, or being shorted by the core.

Another possibility is the whole inductor is being shorted from one end to the other. The reading will then be 0 Ω. This condition, however, is less apt to happen than simply having shorted turns. Many times, an inductor with shorted turns will give the technician a clue by the burned insulation smell or visual effects shown by an extremely overheated component.

Practical Notes

When selecting inductors for replacement or new designs, be sure to notice the important parameter ratings. Four important ratings you should note are:

1. the inductance value;
2. the current rating;
3. the dc resistance of the inductor; and
4. any voltage and insulation ratings or limitations related to the inductor's use.

Summary

- Inductance or self-inductance is the property of a circuit that opposes any *change* in current flow. Any change in current creates moving magnetic flux lines that, when linked (or cutting conductor[s]), induces an emf opposing the current change that initially produced it (Lenz's law).
- An inductor is a component that has the property of inductance. Any current-carrying conductor possesses some amount of inductance. However, most inductors are created by forming a conductor into loops of wire wound around some sort of structure. The amount of inductance created depends on the number of turns, the cross-sectional area, the length of the coil, and the permeability (magnetic characteristics) of the core material.
- Inductors come in a variety of physical and electrical sizes. Iron-core inductors typically have inductances in the henrys ranges. Powdered iron-core and variable ferrite-core inductors have values in the millihenrys range. Air-core inductors are typically in the microhenrys range.
- The amount of back-emf induced in an inductor relates to the amount of inductance involved and the rate of current change. That is,

$$V_L = L\frac{di}{dt}$$

where L = inductance in henrys

$\frac{di}{dt}$ = the rate of current change in amperes per second

- Inductance value is calculated when the inductor's physical characteristics and the core's magnetic properties (permeability) are known. That is,

$$L = (12.57 \times 10^{-7}) \times \frac{(\mu_r N^2 A)}{\text{length}}$$

where 12.57×10^{-7} is the absolute permeability of air, μ_r is the relative permeability of the core, N is the number of turns, and A is the core's cross-sectional area in square meters and length is in meters.

- Total inductance of inductors in series (with no mutual coupling) is calculated by adding the individual inductance values. That is, $L_T = L_1 + L_2 \ldots + L_n$. This method is similar to finding total resistance of series resistors.
- Total inductance of inductors in parallel (with no mutual coupling) is calculated in a similar fashion to that used when calculating equivalent resistance of parallel resistors. That is, either the reciprocal formula or the product-over-the-sum formula may be used. These methods are:

$$\frac{1}{L_T} = \frac{1}{L_1} + \frac{1}{L_2} \cdots + \frac{1}{L_n}$$

$$\text{and } L_T = \frac{L_1 \times L_2}{L_1 + L_2}$$

- Energy is stored in the inductor magnetic field when there is current flowing through the inductor. The amount of energy stored is calculated in joules. That is,

$$E\ (\textit{energy in joules}) = \frac{LI^2}{2}\ \left(\text{or } \frac{1}{2}LI^2\right)$$

where E is energy in joules, L is inductance in henrys, and I is current in amperes.

- Current through an inductor does not instantly rise to its maximum value. Neither does it instantly fall from the value it is at to the minimum value where it decays. Current takes five time constants to change from the level it is at to a new stable level. A time constant (τ) is equal to the inductance in henrys, divided by the R in ohms, or $\tau = L/R$. (NOTE: The L/R time for the increasing current condition may be different than the L/R time for the decaying current condition depending on the circuitry involved.)
- The Universal Time-Constant Chart is a graph of the exponential growth and decay of many things in nature. Looking at the curve of increasing values, after one time constant (τ), the quantity reaches 63.2% of maximum value. After two time constants, the quantity has reached 86.5% of maximum value and so on. After five time constants, the quantity has virtually reached 100% of the maximum value to be achieved. Looking at the decay curve, after one time constant, the value has fallen from 100% of maximum to 36.8% of maximum (or a decay of 63.2%). After two time constants, the quantity has fallen to 13.5% of the original maximum value, and so forth. After five time constants, the quantity has fallen to the minimum value involved.
- Natural logarithms and epsilon can be used to mathematically determine instantaneous values of dc transient circuit quantities, such as instantaneous current (i) in dc RL circuits.

$$\text{For increasing current: } i = \frac{V_S}{R} \times (1 - \epsilon^{-Rt/L})$$

$$\text{For decaying current: } i = \frac{V_S}{R} \times \epsilon^{-Rt/L}$$

- When an inductor's magnetic field rapidly collapses, a high voltage is induced. Therefore, in dc circuits, precautions should be taken to prevent abrupt interruptions of current through large inductances. Also, circuit design should minimize the effects of such changes.
- Typical problems with inductors are opens or shorts. The ohmmeter is useful to check for these problems, provided the technician has knowledge of normal resistance values for the component being tested. Shorted turns are more difficult to find with ohmmeter checks.

Formulas and Sample Calculator Sequences

Formula 13–1
(To find induced voltage)

$$V_{ind} = N\text{ (number of turns)} \times \frac{d\phi\text{ (Wb)}}{dt\text{ (s)}}$$

number of turns, [×], [(], number of webers cut, [÷], amount of time, [)], [=]

Formula 13–2
(To find induced voltage across an inductor)

$$V_L = L\frac{di}{dt}$$

inductance value, [×], [(], change in current, [÷], amount of time, [)], [=]

Formula 13–3
(To find inductance value from V_L and rate of change of current)

$$L = \frac{V_L}{\frac{di}{dt}}$$

V_L value, [÷], [(], change in current, [÷], amount of time, [)], [=]

Formula 13–4
(To find inductance value from inductor physical factors)

$$L = 12.57 \times 10^{-7} \times \frac{\mu_r N^2 A}{l}$$

12.57, [EE] or [EXP], [+/–], 7, [×], [(], μ_r value, [×], number of turns, [x^2], [×], area of coil in square meters, [)], [÷], length of coil in meters, [=]

Formula 13–5
(To find inductance value from inductor physical factors)

$$L = 12.57 \times 10^{-7} \times \frac{N^2 A}{l}$$

12.57, [EE] or [EXP], [+/–], 7, [×], [(], number of turns, [x^2], [×], area of coil in square meters, [)], [÷], length of coil in meters, [=]

Formula 13–6
(To find total inductance of series inductors)

$$L_T = L_1 + L_2 \ldots + L_n$$

L_1 value, [+], L_2 value, . . . , [=]

Formula 13–7
(To find the reciprocal of total inductance of inductors in parallel)

$$\frac{1}{L_T} = \frac{1}{L_1} + \frac{1}{L_2} + \frac{1}{L_3} \ldots + \frac{1}{L_n}$$

L_1 value, [1/x], [+], L_2 value, [1/x], [+], L_3 value, [1/x], . . . , [=]

Formula 13–8
(To find the total inductance of inductors in parallel)

$$L_T = \frac{1}{\frac{1}{L_1} + \frac{1}{L_2} + \frac{1}{L_3} \ldots + \frac{1}{L_n}}$$

L_1 value, [1/x], [+], L_2 value, [1/x], [+], L_3 value, [1/x], [=], [1/x]

Formula 13–9
(To find the total inductance of two inductors in parallel)

$$L_T = \frac{L_1 \times L_2}{L_1 + L_2}$$

L_1 value, [×], L_2 value, [÷], [(], L_1 value, [+], L_2 value, [)], [=]

Formula 13–10
(To find the energy stored by an inductor)

$$E\text{ (energy in joules)} = 0.5\, LI^2$$

0.5, [×], [(], L value, [×], I value, [x^2], [)], [=]

Formula 13–11
(To find the time of one time constant for given L and R values)

$$\tau = \frac{L}{R}$$

inductance value (henrys), [÷], resistance value (ohms), [=]

Formula 13–12
(To find the instantaneous value of current on current buildup through a given L-R circuit)

$$i = \frac{V_S}{R} \times (1 - \epsilon^{Rt/L})$$

(NOTE: This is the formula for increasing current.)

R value, [×], time value, [÷], inductance value, [=], [+/–], [2nd], [e^x], [+/–], [+], 1, [=], [×], V_s value, [÷], R value, [=]

Formula 13–13
(To find the instantaneous value of current on current decay through a given L-R circuit)

$$i = \frac{V_S}{R} \times \epsilon^{-Rt/L}$$

(NOTE: This is the formula for decreasing current.)

R value, [×], time value, [÷], inductance value, [=], [+/–], [2nd], [e^x], [×], V_s value, [÷], R value, [=]

EXCEL AUTOMATED FORMULAS

Helpful problem-solving tools, such as the Excel automated formulas available on the CD, allow automatic calculations of formulas.

Using Excel

Inductance Formulas

(Excel file reference: FOE13_01.xls)

DON'T FORGET! *It is not necessary to retype formulas once they are entered on the worksheet! Just input new parameters data for each new problem using that formula, as needed.*

- Use the Formula 13–2 spreadsheet sample and solve for the value of voltage across the inductor if the inductor value is 5 H, and the rate of current change is 3 amperes per second. Check your answer against the Appendix answer for Practice Problems 2.
- Use the Formula 13–13 spreadsheet sample and solve for instantaneous current value for the circuit of Figure 13–17. Assume the switch closed for a time of 2.5 milliseconds. Check your answer against the Appendix answer for Practice Problems 7.

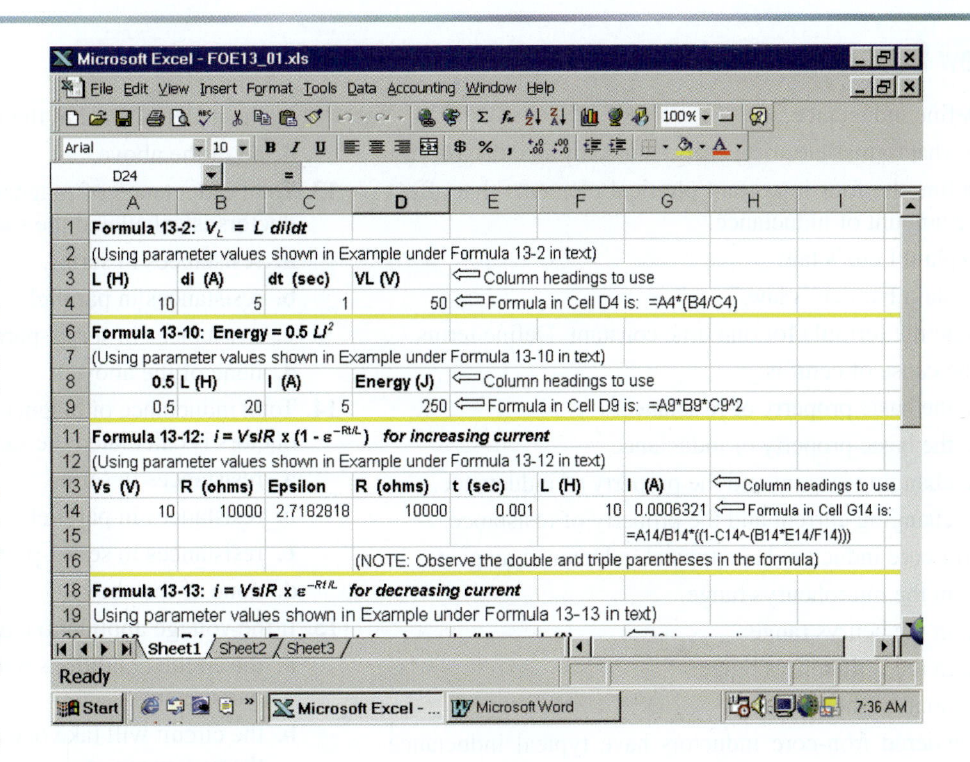

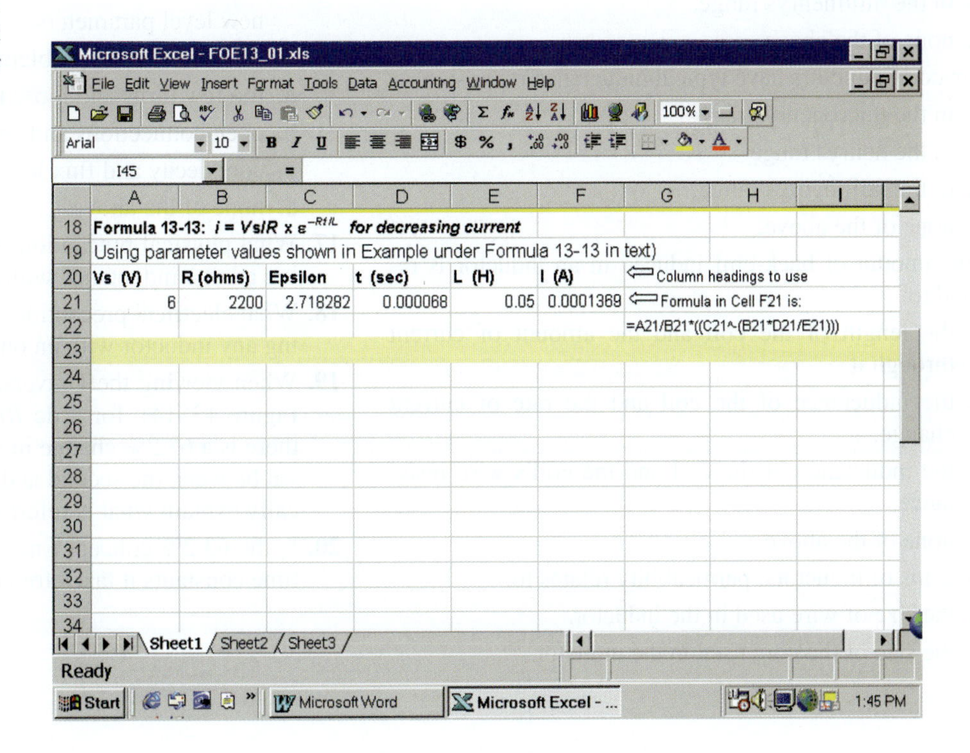

Review Questions

1. Define inductance.
2. In what form is electrical energy stored by an inductor?
3. Define the four important physical elements that affect the amount of inductance.
4. Explain Lenz's law.
5. Explain Faraday's law.
6. State the formula for one time constant. Define terms.
7. The cause of cemf is:
   a. the basic property of resistance.
   b. the basic property of inductance.
   c. changing current and the property of inductance.
   d. changing current and the property of resistance.
8. Iron-core inductors have typical inductance ranges:
   a. in the microhenrys range.
   b. in the henrys range.
   c. in the millihenrys range.
   d. none of the above.
9. Powdered iron-core inductors have typical inductance ranges:
   a. in the microhenrys range.
   b. in the henrys range.
   c. in the millihenrys range.
   d. none of the above.
10. Air-core inductors have typical inductance ranges:
    a. in the microhenrys range.
    b. in the henrys range.
    c. in the millihenrys range.
    d. none of the above.
11. The amount of back-emf induced in an inductor is related to:
    a. the length of the coil and the amount of current through it.
    b. the inductance of the coil and the rate of current change.
    c. the inductance of the coil and the coil's wire resistance.
    d. none of the above.
12. In terms of inductors, permeability relates to:
    a. the size of wire used in the inductor.
    b. the size of the core used in the inductor.
    c. the material used in the inductor core.
    d. all of the above.
13. Total inductance of magnetically noncoupled series inductors is calculated the same as:
    a. resistances in series.
    b. resistances in parallel.
    c. resistances in series-parallel.
    d. none of the above.
14. Total inductance of magnetically noncoupled parallel inductors is calculated the same as:
    a. resistances in series.
    b. resistances in parallel.
    c. resistances in series-parallel.
    d. none of the above.
15. If the voltage applied to a dc *RL* circuit changes level:
    a. the circuit conditions will instantly be at the new level parameters.
    b. the circuit will take one *L*/*R* time constant to arrive at the new parameters.
    c. it will take more than one *L*/*R* time constant to arrive at the new parameters.
    d. the circuit will take 10 time constants to arrive at the new level parameters.
16. The most common problems found with inductors are:
    a. too few turns and loose turns.
    b. open connections and shorted turns.
    c. core decay and flux leakage.
    d. none of the above.
17. What physical precaution should be taken when checking a large inductance with an ohmmeter?
18. What electrical precaution should be taken when checking any inductor with an ohmmeter?
19. When viewing the Universal Time-Constant Chart (see Figure 13–14), for a dc *RL* circuit, it can be said that there is a 63.2% change in current rise in the time interval between the second and third time constant. Specifically explain what is referred to by this 63.2% term.
20. Is the 63.2% concept true between each of the five *L*/*R* time constants it takes for complete change?

Problems

1. When the number of turns on an inductor that has an inductance of 100 mH doubles, while keeping its length, cross-sectional area, and core material constant, what is its new inductance value? (Express your answer in three ways: μH, mH, and H.)
2. Current through an inductor is changing at a rate of 100 mA/s and the induced voltage is 30 mV. What is the value of inductance?
3. What is the new inductance for a 100-μH air-core coil whose core is replaced with one having a relative permeability of 500?
4. How much energy is stored in the magnetic field of a 15-H inductor that is carrying current of 2 A?
5. A 100-mH, a 0.2-H, and a 1,000-μH inductor are connected in series. Assuming no mutual coupling exists between them, what is the circuit's total inductance? (Express your answer in mH.)
6. A 10-H and a 15-H inductor are connected in parallel. Assuming no mutual coupling exists between them, what is the resultant total inductance?
7. A series *RL* circuit consists of a 250-mH inductor and a resistor having a resistance of 100 Ω. What is the value of one time constant? How long will it take for current to change completely from one level to another?
8. Refer to the Universal Time-Constant Chart, Figure 13–14, and determine what percentage of applied voltage appears across the resistor in a series *RL* circuit after four time constants. What percentage of the applied voltage is across the inductor at this same time?
9. A 500-turn inductor has a total inductance of 0.4 H. What is the inductance when connections are made at one end of the coil and at a tap point connection at the 250th turn?
10. How long will it take the voltage across the resistance in a series *RL* circuit consisting of 5 H of inductance and 10 Ω of resistance to achieve 86.5% of *V* applied? (Assume the inductor has negligible resistance.)

Perform the calculations for questions 11 and 12 using your calculator and the e^x function, as appropriate.

11. Refer to Figure 13–19. What is the voltage across the inductor after the switch has been closed for 0.6 μs?
12. Refer to Figure 13–20. How much time will it take for the voltage across the resistor to reach 38 V? (Round off, as appropriate.)

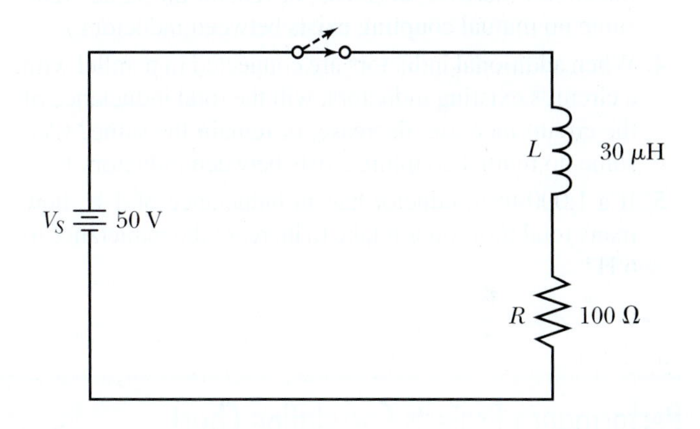

FIGURE 13–19

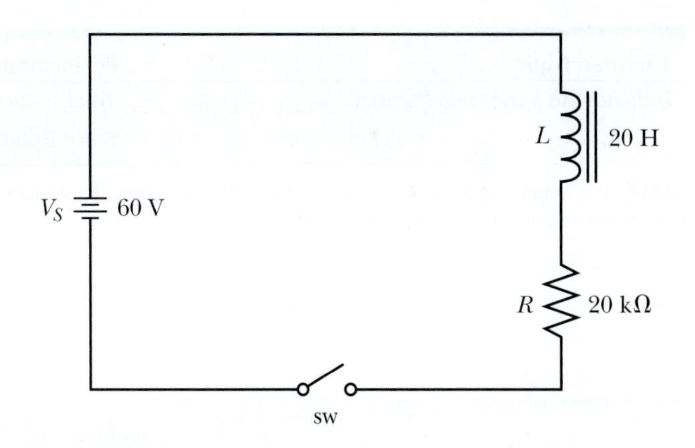

FIGURE 13–20

Analysis Questions

1. If the iron core is removed from an inductor so that it becomes an air-core inductor, will the coil's inductance increase, decrease, or remain the same?
2. When the current through an inductor has reached its steady-state condition and the current through the coil remains the same, is there an induced back-emf?
3. When additional inductors are connected in series with a circuit's existing inductors, will the total inductance of the circuit increase, decrease, or remain the same? (Assume no mutual coupling exists between inductors.)
4. When additional inductors are connected in parallel with a circuit's existing inductors, will the total inductance of the circuit increase, decrease, or remain the same? (Assume no mutual coupling exists between inductors.)
5. If a 1,000-turn inductor has an inductance of 4 H, how many total turns does it take to increase the inductance to 6 H?
6. Draw a graph of the V_R and V_L voltages with respect to V applied for an inductor circuit consisting of a 10-H inductor, a series 10-Ω resistance, and a source voltage of 20 V. Identify the specific voltages across L and R at the 1τ, 2τ, 3τ, and 4τ points on your graph.
7. Explain why a 1,000-turn coil, consisting of wire that has a varnish insulation with a voltage breakdown rating of only 100 V, can safely handle an applied voltage across the coil at 2,000 V.
8. What is the difference between the current-limiting effects of an inductor's resistance and its property of inductance?
9. Explain why the joules of energy stored by an inductor is directly related to the amount of inductance the inductor has.
10. Explain why there is a direct relationship between the current through a coil and the amount of energy stored by that coil.

Performance Projects Correlation Chart

Suggested performance projects in the Laboratory Manual that correlate with topics in this chapter are:

Chapter Topic	Performance Project	Project Number
Inductors in Series and Parallel	Total Inductance in Series and Parallel	40
	Story Behind the Numbers: Inductance	

NOTE: It is suggested that after completing the above projects, the student should be required to answer the questions in the "Summary" at the end of this section of projects in the Laboratory Manual.

OUTLINE

OBJECTIVES

After studying this chapter, you should be able to:

1. Illustrate *V-I* relationships for a purely resistive ac circuit
2. Illustrate *V-I* relationships for a purely inductive ac circuit
3. Explain the concept of **inductive reactance**
4. Write and explain the formula for inductive reactance
5. Use Ohm's law to solve for X_L
6. Use the X_L formula to solve for inductive reactance at different frequencies and with various inductance values
7. Use the X_L formula to solve for unknown *L* or *f* values
8. Determine X_L, I_L, and V_L values for series- and parallel-connected inductances
9. Use the computer to solve circuit problems
10. Use the SIMPLER troubleshooting sequence to solve the Troubleshooting Challenge problem

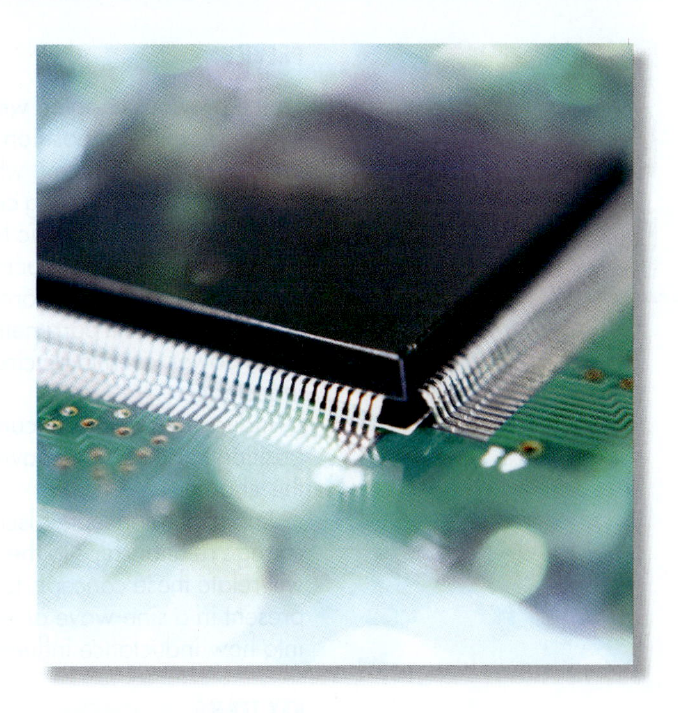

CHAPTER 14

Inductive Reactance in AC

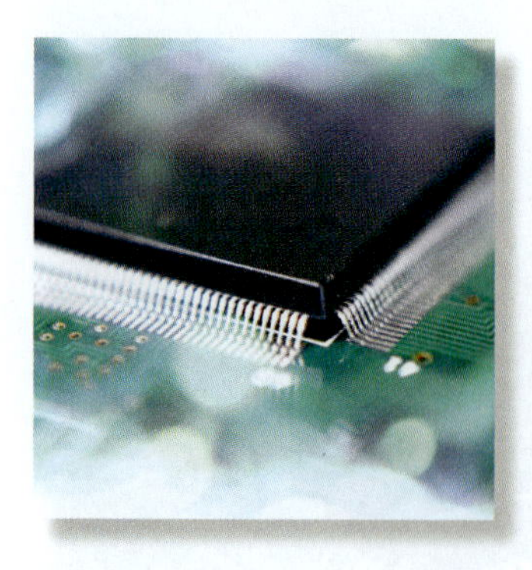

PREVIEW

In the previous chapter, we learned that an inductor opposes a current change but provides very little opposition to a steady (dc) current. The dc winding resistance is the only opposition to dc current; whereas the opposition to the current change is because an inductor reacts to changing current by producing a back-emf. You saw that energy is stored in the inductor's magnetic field during current buildup and returned to the circuit during current decay. If the inductor were "perfect" (without dc wire resistance or other losses), it would return all the stored energy to the circuit, dissipate no power, and have no effect on dc circuit parameters. Practical inductors have some resistance and dissipate some power, even in dc circuits. However, these losses are not caused by the property of inductance.

Inductors react to ac current, which is constantly varying. Also, they produce an opposition to ac current known as inductive reactance, the reactance that will be studied in this chapter.

Important concepts discussed earlier were the inductance value and rate of current change relationships to the induced voltage appearing across the coil. In this chapter, we will relate these concepts to the ac circuit and explore what happens when inductance is present in a sine-wave ac environment. We will show how these characteristics translate into how inductance influences parameters in ac circuits.

KEY TERMS

Inductive reactance
Leading in phase
Phase angle
Q (figure of merit)

14–1 *V* and *I* Relationships in a Purely Resistive AC Circuit

For a quick review, observe Figure 14–1. Note in a purely resistive circuit, the voltage across a resistor is in phase with the current through it. Both the voltage and current pass through the zero points and reach the maximum points of the same polarity at the same time.

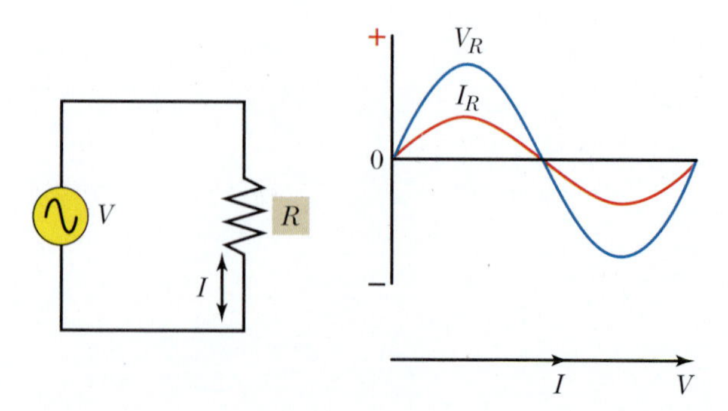

FIGURE 14–1 *V* and *I* relationships in a purely resistive ac circuit.

14–2 *V* and *I* Relationships in a Purely Inductive AC Circuit

For purposes of explanation, let's assume the inductor is an ideal inductor, having only inductance and no resistance. What phase relationship will the current through the coil and the voltage across the coil have with respect to each other? Examine Figure 14–2 as you study the following explanations and comments.

1. In the circuit shown, the inductor is connected across the source output terminals. Therefore, recalling the rule about voltage in parallel circuits, you know the impressed voltage across the coil and the source applied voltage are the same.
2. In the circuit shown, there is only one current path. Therefore, recalling the rule about current in series circuits, you know the current through all circuit parts is the same. That is, the circuit current *(I)* and the current through the inductor (I_L) are the same.
3. Recall the induced voltage (cemf) across the inductor is maximum when the rate of current change is maximum and minimum when the rate of current change is minimum, Figure 14–3.
4. Also, remember for a sine wave the maximum rate of change occurs as the waveform crosses the zero points. The minimum rate of change occurs at the maximum positive and negative peaks of the sine wave, Figure 14–3.

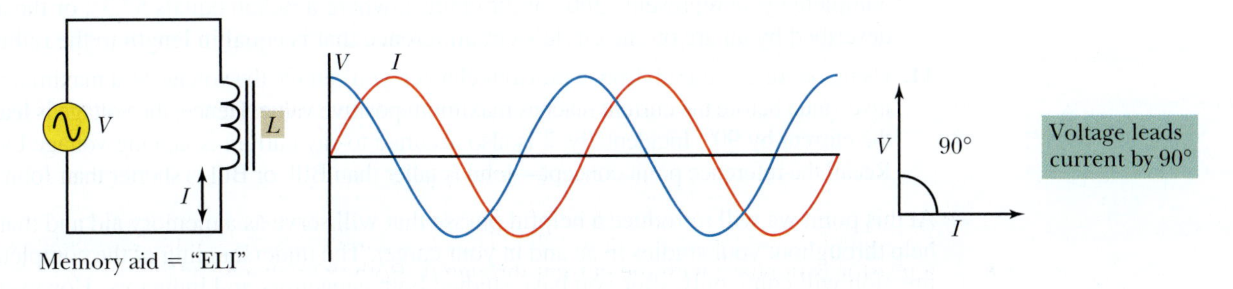

FIGURE 14–2 *V* and *I* relationships in a purely inductive ac circuit

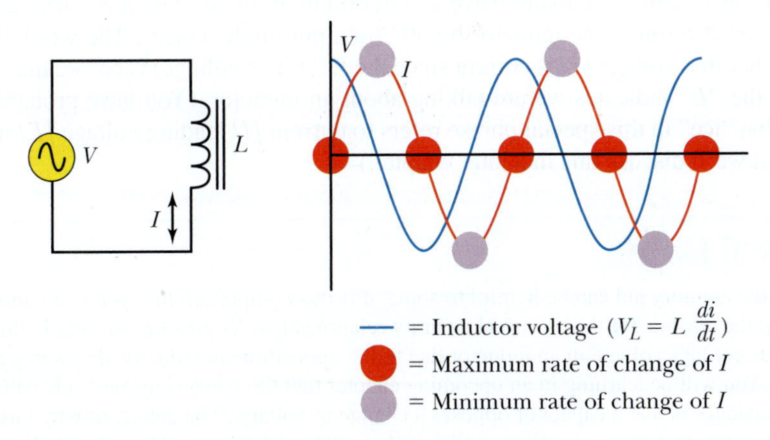

FIGURE 14–3 Inductor voltage related to rate of change of current

5. The implication is that maximum induced V occurs across the inductor when the current sine wave is either increasing from (or through) zero in the positive direction or decreasing from (or through) zero in the negative direction. See Figure 14–3 again.
6. The *amplitude* of the induced voltage is determined by the rate of current change and the value of L.
7. Two factors determine the *polarity* of the induced voltage:
   a. the direction of the current flow through the coil, and
   b. whether the current is trying to increase or decrease.
8. Because the induced voltage opposes any current change, when the current tries to increase, the cemf will oppose it. When the current tries to decrease, the cemf will again oppose it (Lenz's law).
9. Figures 14–2 and 14–3 illustrate the results of these actions and reactions. In essence, because the induced back-emf is maximum when the rate of current change is maximum and minimum when the rate of current change is minimum, the current lags behind the impressed voltage. The time lag, or delay of current behind voltage, has obviously been caused by the inductance's opposition to changes in current flow. The current and voltage are out of phase. The 90° **phase angle** is due to the fact that at the moment when the current increases or decreases, the inductor instantly develops the cemf, which opposes the change in current.
10. Since the V and I are one-fourth cycle apart, they are 90° out of phase with each other. A complete cycle represents 360° or 2π radians, where a radian equals 57.3°, or the angle described by an arc on the circle's circumference that is equal in length to the radius.
11. Using our rule of thumb from a previous chapter, you can see the voltage is at maximum positive value before the current reaches maximum positive value. Hence, the voltage is **leading** the current by 90°. Incidentally, it is also accurate to say current is lagging voltage by 90°. Recall the reference point concept—John is taller than Bill, or Bill is shorter than John.

At this point we will introduce a helpful phrase that will serve as a memory aid and that will help throughout your studies in ac and in your career. The understanding of the complete expression will come only after you have studied both capacitors and inductors. However, the phrase can also be helpful to you now.

Here's the phrase: *Eli the ice man.* It's a short saying, so you should not have trouble memorizing it, particularly after you understand the implications!

You have just learned in a sine-wave ac circuit environment, voltage across an inductor leads the current through the inductor by 90° (in a pure inductance). The word "Eli" helps you remember that voltage leads current since the E (emf or voltage) is before the I (current). Obviously the *"L"* indicates we are talking about an inductor. (You have probably already surmised that "ice" in this special phrase refers to current *[I]* leading voltage *[E]* in a capacitor *[C]*, but we'll discuss this in a later chapter.)

Practical Notes

Although the memory aid can be helpful to some, it is more important that you learn and understand the reasons for the electrical quantity relationships. As you are learning in this chapter, current lags voltage in an inductor due to the opposition an inductor shows to a change in current. You will be learning in an upcoming chapter that the reason current leads voltage across a capacitor is that a capacitor opposes a change in voltage. The details of why this is true will become apparent as you continue studying these subjects. We simply want to indicate that the memory aid should not replace your understanding of these matters.

14–3 Concept of Inductive Reactance

Because an inductor reacts to the constantly changing current in an ac circuit by producing an emf that opposes current changes, inductance displays opposition to ac current, Figure 14–4. Recall the inductor did not show opposition to dc current, except for the small resistance of the wire. This lack of opposition is because current must be changing to cause a changing flux that induces the back-emf we've been discussing. In dc the current is not changing, except for the temporary switch-on and switch-off conditions discussed earlier.

This special opposition an inductor displays to ac current is **inductive reactance.** The symbol "X" denotes reactance. Because it is inductive reactance, the correct abbreviation or symbol is X_L. Because X_L is an opposition to current, the *unit of measure for inductive reactance* (X_L) *is the ohm!*

14–4 Relationship of X_L to Inductance Value

We have already stated that the amount of opposition inductance shows to ac current is directly related to the amount of back-emf induced. For a given rate of current change, the amount of back-emf induced is directly related to the value of coil inductance (i.e., the greater the L, the greater the back-emf produced). Therefore, the higher the value of L, the higher will be the inductive reactance (X_L) value. In other words, if the L value doubles, the number of ohms of X_L doubles; if the L in the circuit decreases to half its original value, X_L decreases to half, and so on, Figure 14–5.

EXAMPLE If a purely inductive circuit has a 5-H inductor with a second 5-H inductor connected in series, what happens to the circuit current? (Assume all other circuit parameters remain the same.)

Answer: Current decreases to half the original value, since the total opposition to current (X_{L_T}) has doubled when inductance was doubled.

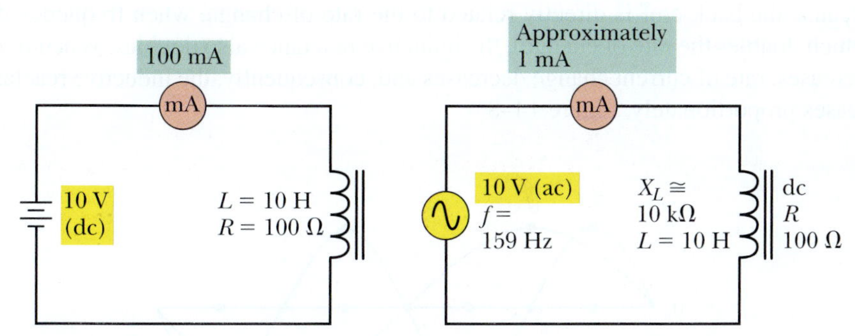

FIGURE 14–4 Concept of inductive reactance to ac

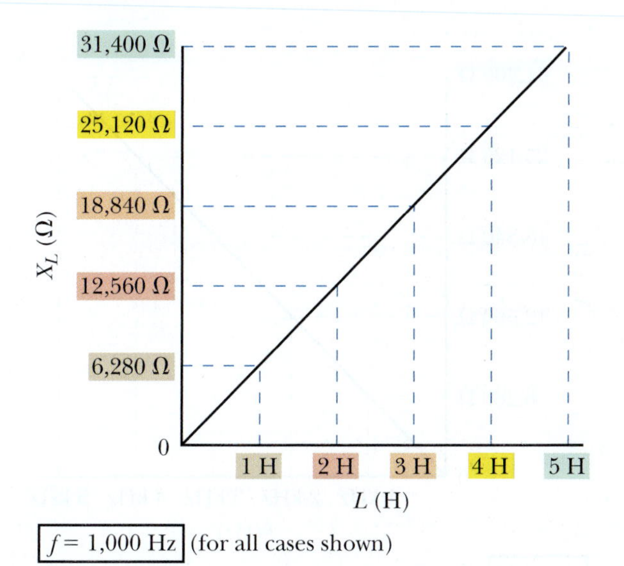

FIGURE 14–5 Relationship of X_L to L

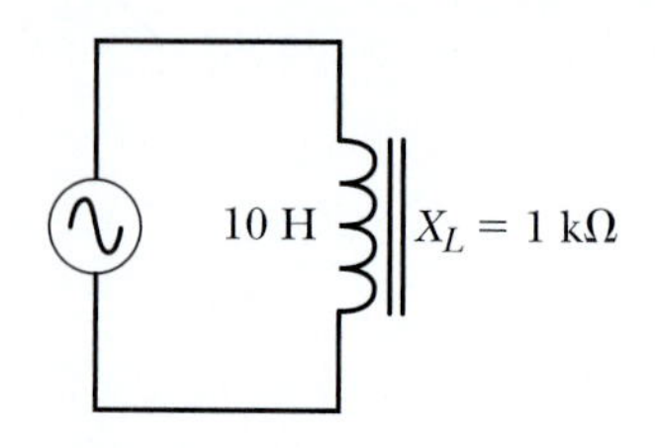

FIGURE 14–6

PRACTICE PROBLEMS 1

1. For the circuit in Figure 14–6, what is the value of total X_L when two more equal-value inductors are added in series with the existing inductor, thus making $L_T = 30$ H? (Assume all other circuit conditions remain the same.)
2. If the three 10-H inductors in question 1 are parallel-connected rather than series-connected, what is the total X_L?

14–5 Relationship of X_L to Frequency of AC

Rate of Change Related to Frequency

It has been previously established that the amount of back-emf produced in an inductor relates to both the value of inductance and to the rate of current change. As you know, the faster the rate of change, the greater the induced emf produced. And the greater the back-emf, the higher the value of opposition or X_L.

Let's take a look at how changing the frequency of an ac signal affects this rate of change. Refer to Figure 14–7 as you study the following discussion.

1. Only one cycle is shown for the lower of the two frequency signals displayed. Two cycles of the higher frequency signal are shown.
2. The two signals have the same maximum (peak) amplitude. You can see that the higher-frequency signal must have a higher rate of change of current to reach its maximum amplitude level in half the time. Carrying this thought throughout the whole cycle, it is apparent that the higher-frequency signal has twice the rate of change compared to the lower-frequency signal.
3. Because the back-emf is directly related to the rate of change, when frequency doubles (which doubles the rate of change), the inductive reactance also doubles. When frequency decreases, rate of current change decreases and, consequently, the inductive reactance decreases proportionately, Figure 14–8.

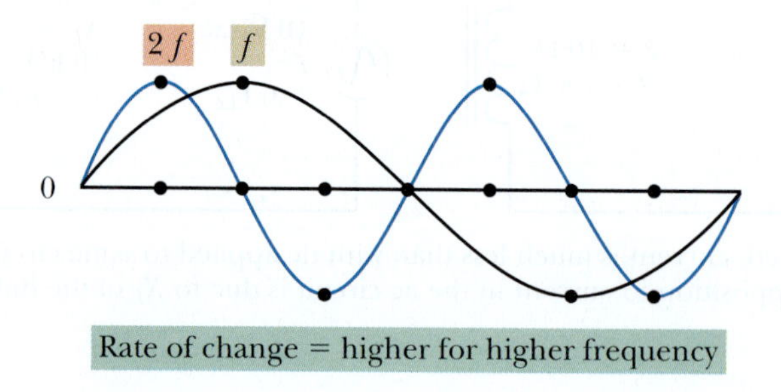

FIGURE 14–7 Comparison of rate of change for two frequencies

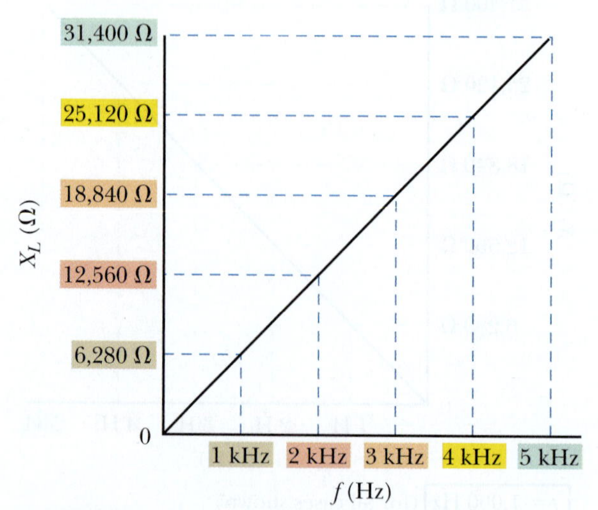

FIGURE 14–8 Relationship of X_L to f

Rate of Change Related to Angular Velocity

In an earlier chapter, a rotating vector to describe a sine wave was introduced. At this point, we want to indicate that a relationship exists between frequency and the angular velocity of the quantity represented by a rotating vector.

Angular velocity is the speed at which a vector rotates about its axis. The symbol representing angular velocity is the lowercase Greek letter omega, ω. Thus, the relationship of angular velocity to frequency is ω equals $2\pi f$ (radians per second). (Recall that 2π radians equal 360°, or one complete cycle.) We introduce angular velocity and its mathematical representation at this point to acquaint you with the expression "$2\pi f$," which is used in formulas dealing with ac quantities. You will see this expression frequently as you continue in your studies and in your work as a technician.

■ IN-PROCESS LEARNING CHECK 1 Fill in the blanks as appropriate.

1. When current increases through an inductor, the cemf (helps, hinders) _____ the current increase.
2. When current decreases through an inductor, the cemf (helps, hinders) _____ the current decrease.
3. In a pure inductor, the _____ leads the _____ by 90°.
4. What memory aid helps in recalling the relationship described in question 3? _____
5. The opposition that an inductor shows to ac is termed _____ _____.
6. The opposition that an inductor shows to ac (increases, decreases, remains the same) _____ as inductance increases.
7. The opposition that an inductor shows to ac (increases, decreases, remains the same) _____ as frequency decreases.
8. X_L is (directly, inversely) _____ related to inductance value.
9. X_L is (directly, inversely) _____ related to frequency.

14–6 Methods to Calculate X_L

The X_L Formula

Two common methods are used to calculate the value of X_L. One method is to use Ohm's law. The other method is the inductive reactance formula,

FORMULA 14–1 $X_L = 2\pi fL$

where X_L = inductive reactance in ohms
2π = 6.28 (approximately)
f = frequency in hertz
L = inductance in henrys

You can see from the formula and previous discussions that inductive reactance is directly related to both frequency and inductance. That is, when *either* frequency *or* inductance increases, inductive reactance increases. When *either* frequency *or* inductance decreases, X_L decreases. That is, when one factor doubles while the other one halves, the reactance stays the same, and so on.

The X_L formula can also be transposed to solve for L or f, if the other two factors are known.

FORMULA 14–2 $L = \dfrac{X_L}{2\pi f}$

$$\text{That is, } L = \frac{X_L}{6.28f} \text{ henrys or } \frac{X_L}{2\pi f}$$

and

FORMULA 14–3 $f = \dfrac{X_L}{2\pi L}$

$$\text{That is, } f = \frac{X_L}{6.28\ L} \text{ or } \frac{X_L}{2\pi L}$$

◆ **EXAMPLE** What is the inductive reactance of an iron-core choke having 5 H of inductance that is in a 60-Hz ac circuit?

$$X_L = 2\pi fL = 6.28 \times 60 \times 5 = 1{,}884\ \Omega$$ ◆

◆ **EXAMPLE** What is the inductance of an inductor that offers 2 kΩ of inductive reactance at a frequency of 1.5 MHz?

$$L = \frac{X_L}{2\pi f} = \frac{2 \times 10^3}{6.28 \times (1.5 \times 10^6)} = 212\ \mu\text{H}$$ ◆

◆ **EXAMPLE** What frequency of operation would cause a 30-μH inductor to produce 4 kΩ of inductive reactance?

$$f = \frac{X_L}{2\pi L} = \frac{4 \times 10^3}{6.28 \times (30 \times 10^{-6})} = 21.23\ \text{MHz}$$ ◆

Practical Notes

When using Ohm's law, as indicated, it is common to use the (rms) or effective values of voltage and current in your calculations, unless otherwise indicated!

PRACTICE PROBLEMS 2

1. What is the X_L of a 10-mH inductor that is operating at a frequency of 100 kHz?
2. What is the inductance in a purely inductive circuit, if the frequency of the applied voltage is 400 Hz and the inductive reactance is 200 Ω?
3. What frequency causes a 100-mH inductor to exhibit 150 Ω of inductive reactance?

Using the Ohm's Law Approach

The ac current through an inductor is also calculated by using the Ohm's law expression: $I_L = V_L/X_L$. We can transpose this formula to find any one of the three factors, provided we know the other two factors. (We have previously done this using Ohm's law expressions.) This means that:

$$X_L = \frac{V_L}{I_L}$$

$$V_L = I_L X_L$$

$$I_L = \frac{V_L}{X_L}$$

PRACTICE PROBLEMS 3

Use Ohm's law and answer the following.

1. What is the inductive reactance of an inductor whose voltage measures 25 V and through which an ac current of 5 mA is passing?
2. What is the voltage across a 2-H inductor passing 3 mA of current? Assume the frequency of operation is 400 Hz. (Hint: Calculate X_L using the reactance formula first.)
3. What is the current through an inductor having 250 Ω of inductive reactance if the voltage across the inductor measures 10 V?

14–7 Inductive Reactances in Series and Parallel

You are aware that inductances in series are additive just like series resistances are additive. Likewise, you have already learned that inductive reactance, or X_L, is directly related to the amount of inductance. Therefore, the ohmic values of inductive reactances of series inductors add like the ohmic values of series resistances add. In other words, $X_{L_T} = X_{L_1} + X_{L_2} \ldots + X_{L_n}$, Figure 14–9.

Inductance values in parallel are also treated in the same manner that parallel resistance values are treated. That is, the reciprocal or product-over-the-sum approaches are used to find the total or equivalent value(s). Because inductive reactance is directly related to inductance value, total X_L is again correlated to the total inductance of parallel-connected inductors like it was to the total inductance of series-connected inductors, Figure 14–10.

Useful formulas to find the total inductive reactance of parallel-connected inductors are:

FORMULA 14–4 $$\frac{1}{X_{L_T}} = \frac{1}{X_{L_1}} + \frac{1}{X_{L_2}} \ldots + \frac{1}{X_{L_n}}$$

or

FORMULA 14–5 $$X_{L_T} = \frac{1}{\frac{1}{X_{L_1}} + \frac{1}{X_{L_2}}}, \text{ etc.}$$

Practical Notes

Once again, the reciprocal ([1/x]) function on your calculator is found to be very useful. You can solve for total inductive reactance of parallel-connected inductive reactances using the same techniques you have previously used to solve for total inductance of parallel-connected inductances. (NOTE: Assume no coupling between inductances.)

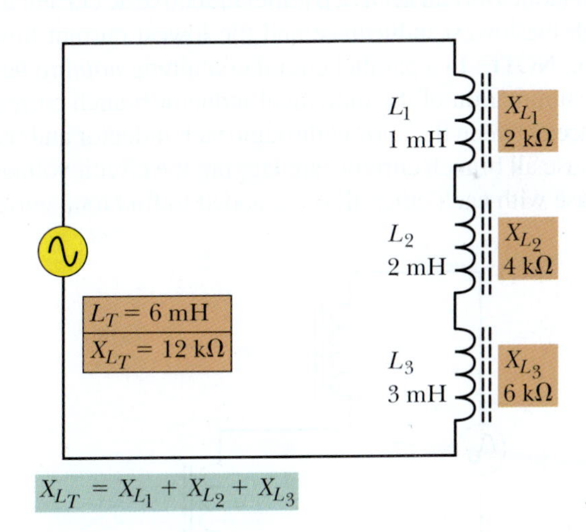

FIGURE 14–9 Inductive reactances in series
multiSIM

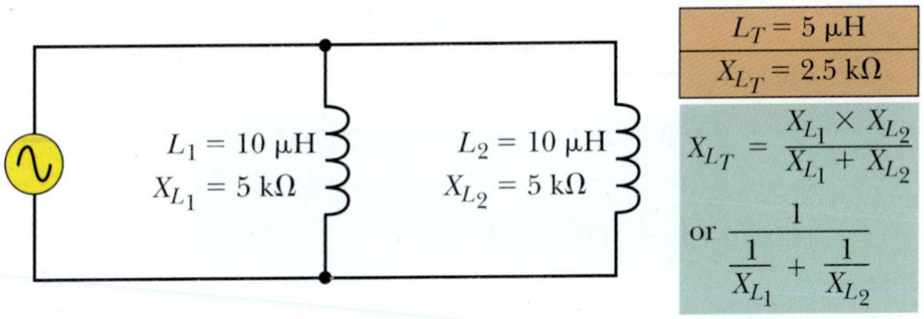

$L_1 = 10\ \mu H$, $X_{L_1} = 5\ k\Omega$

$L_2 = 10\ \mu H$, $X_{L_2} = 5\ k\Omega$

$L_T = 5\ \mu H$

$X_{L_T} = 2.5\ k\Omega$

$$X_{L_T} = \frac{X_{L_1} \times X_{L_2}}{X_{L_1} + X_{L_2}} \text{ or } \frac{1}{\frac{1}{X_{L_1}} + \frac{1}{X_{L_2}}}$$

FIGURE 14–10 Inductive reactances in parallel
multiSIM

For two inductors in parallel, you can use:

FORMULA 14–6 $X_{LT} = \dfrac{X_{L_1} \times X_{L_2}}{X_{L_1} + X_{L_2}}$

Where appropriate, you can use all the "shortcuts" you have previously learned about parallel circuits—two 25-ohm reactances in parallel yield a total equivalent reactance of 12.5 ohms, or half of either reactor's value, and so forth.

14–8 Summarizing Comments about Inductors Related to Inductive Reactance

Collecting what you now know about inductance and reactance, the following comments can be made:

1. The inductive reactance of an inductor limits or opposes the ac current flow through the inductor. That is, given an inductor in a circuit with a specified value of dc applied voltage, if that same inductor is put into a circuit with the same ac voltage (rms value), the current will be less in the ac circuit than in the dc circuit. This situation is due to the opposition of the inductive reactance to ac current, Figure 14–4.
2. Connecting inductances in series causes total inductance to increase, total inductive reactance to increase, and current to decrease (assuming a constant V applied), Figure 14–11.
3. Series-connected inductors may act as a series, inductive ac voltage divider with the greatest voltage across the largest inductance and the smallest voltage across the smallest inductance, and so on. NOTE: In a series circuit containing *nothing but inductors,* the total voltage equals the sum of the individual voltages across the inductors. The current through the inductors is 90° out of phase with the voltage across each inductor. Each of the inductor's voltages is in phase with the other inductors' voltages, thus are added to obtain the total voltage, Figure 14–12.
4. Connecting inductances in parallel causes total inductance to decrease, total inductive reactance to decrease, and current to increase (assuming a constant V applied), Figure 14–13.
5. Parallel-connected inductors can act as a parallel inductive ac current divider with the highest current through the lowest inductance and the lowest current through the highest inductance and so on. NOTE: In a parallel circuit containing *nothing but inductors,* the total current equals the summation of the individual inductor branch currents. Again, there is a 90° phase difference between the current through each inductor and the voltage across that inductor. But because all branch currents are lagging the circuit voltage by 90°, the branch currents are in phase with each other, thus are added to find total current, Figure 14–14.

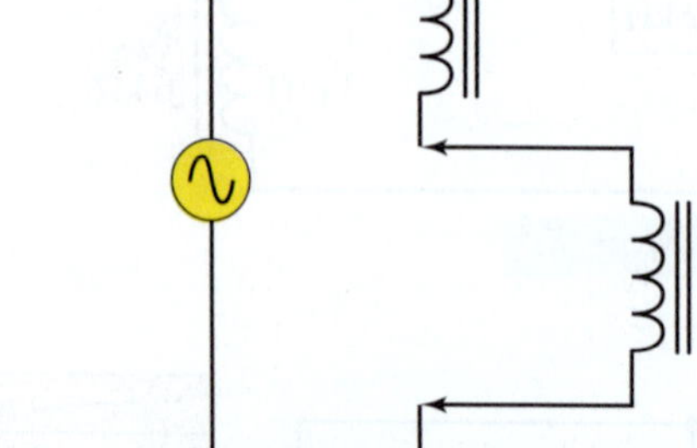

Connecting inductances in series causes:
L_T ↑
X_{L_T} ↑
I_T ↓

FIGURE 14–11 Effects of connecting inductances in series

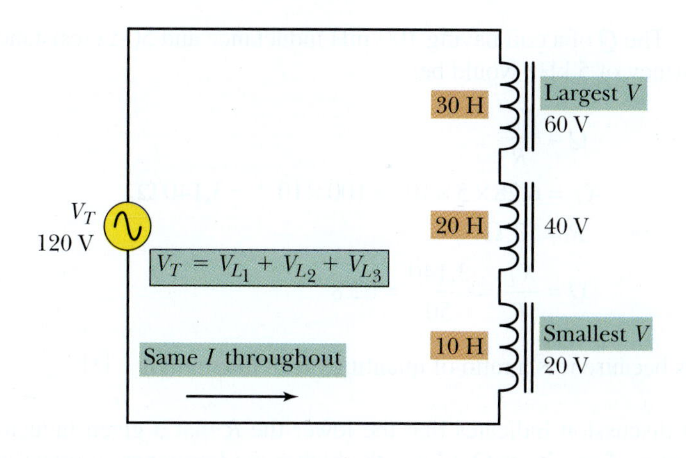

FIGURE 14–12 Voltage division across series-connected inductors

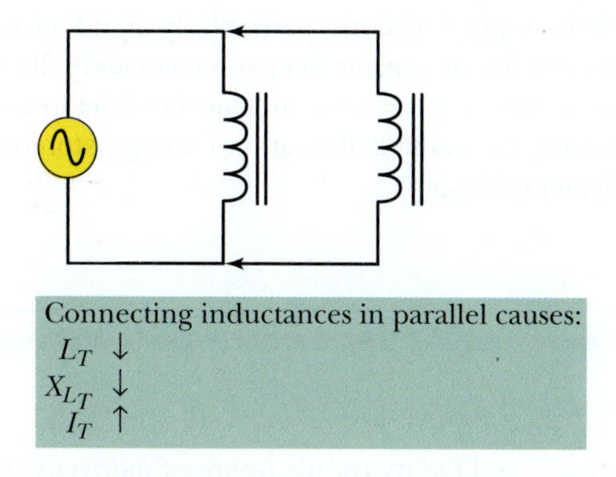

FIGURE 14–13 Effects of connecting inductors in parallel

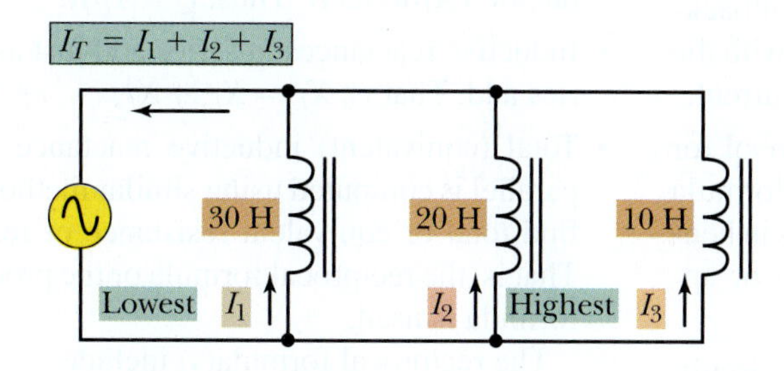

FIGURE 14–14 Current division through parallel-connected inductors

14–9 Inductor Q Factor

We have already implied that a perfect inductor would have zero resistance, and that the only opposition this inductor would show to ac current flow would be the inductive reactance. It has also been stated that every coil has some resistance in the wire making up the coil. In many cases, this resistance is so small that it has negligible effect on the circuit operation. In other cases, this is not true. In fact, under certain conditions it may have considerable effect. Therefore, inductors sometimes have a Q rating, which is a ratio describing the energy the inductor stores in its field compared to the energy it dissipates due to I^2R losses.

In essence, the ratio of the X_L that the inductor shows (at a given frequency) to the R that it has, depicts this ratio. That is:

FORMULA 14–7 $Q = \frac{X_L}{R}$

EXAMPLE The Q of a coil having 100-mH inductance and 50-Ω resistance that is operating at a frequency of 5 kHz would be:

$$Q = \frac{X_L}{R}$$

$$X_L = 6.28 \times 5 \times 10^3 \times 100 \times 10^{-3} = 3{,}140\ \Omega$$

$$R = 50\ \Omega$$

$$Q = \frac{X_L}{R} = \frac{3{,}140}{50} = 62.8$$

(Q has no units because it is a ratio of quantities with like units.)

The preceding discussion indicates that the lower the R that a given inductor (L) has, the higher is its **figure of merit,** or **Q.** (Also, the higher the frequency of operation, the greater the X_L and the higher the Q for a given inductor.)

It is generally desirable to use high Q coils, particularly at radio frequencies. A coil is considered to be high Q if its Q value is greater than approximately 20 to 25. Of course, this is an arbitrary value. It does, however, give some indication of the inductance-to-resistance ratio of the inductor. Therefore, the value is the ratio of energy stored to energy dissipated or the X_L/R ratio we have been examining.

Summary

- Inductive reactance is the opposition that inductance displays to ac (or pulsating) current.
- Inductive reactance is associated with the amount of back-emf generated by the inductor, which is associated with the value of inductance (L) and the rate of change of current.
- Inductive reactance is measured in ohms. The symbol for inductive reactance is X_L. X_L is calculated by the formula $X_L = 2\pi fL$, where X_L is in ohms, f is in hertz, and L is in henrys. The 2π is the constant equal to two times pi or approximately 6.28.
- Voltage leads current by 90° in a purely inductive circuit. (Voltage and current are in phase for a purely resistive circuit.)
- The first part of the memory-aid phrase helps you remember the phase relationship of voltage and current relative to inductors. The phrase is *Eli the ice man.* The word *Eli* indicates that E comes before I in an inductor (L).
- Inductive reactance is *directly proportional* to both inductance (L) *and* to frequency (f).
- Quality (or the figure of merit) of an inductor is sometimes described as its Q. This factor shows the ratio of the inductor's X_L to its R. Thus, $Q = X_L/R$.
- Inductive reactances in series add just as resistances in series add. That is, $X_{L_T} = X_{L_1} + X_{L_2} \ldots + X_{L_n}$.
- Total (equivalent) inductive reactance of inductances in parallel is computed using similar methods to those used to find total or equivalent resistance of resistors in parallel. That is, the reciprocal formula or the product-over-the-sum formula is used.

  The reciprocal formula(s) include:

  $$\frac{1}{X_{LT}} = \frac{1}{X_{L1}} + \frac{1}{X_{L2}} + \ldots \frac{1}{X_{Ln}} \text{ or } X_{LT} = \frac{1}{\frac{1}{X_{L1}} + \frac{1}{X_{L2}}}$$

  The product-over-the-sum formula is:

  $$X_{LT} = \frac{X_{L1} \times X_{L2}}{X_{L1} + X_{L2}}$$

Formulas and Sample Calculator Sequences

Formula 14–1

(To find inductive reactance when frequency and inductance values are known)

$X_L = 2\pi fL$

6.28, [×], frequency value, [×], inductance value, [=]

Formula 14–2

(To find inductance value when inductive reactance and frequency values are known)

$L = \frac{X_L}{2\pi f}$

X_L value, [÷], [(], 6.28, [×], frequency value, [)], [=]

Formula 14–3

(To find frequency when inductance and inductive reactance values are known)

$f = \frac{X_L}{2\pi L}$

X_L value, [÷], [(], 6.28, [×], inductance value, [)], [=]

Formula 14–4

(To find the reciprocal of inductive reactance of parallel inductors)

$\frac{1}{X_{L_T}} = \frac{1}{X_{L_1}} + \frac{1}{X_{L_2}} \cdots + \frac{1}{X_{L_n}}$

X_{L_1} value, [1/x], [+], X_{L_2} value, [1/x], . . . , [=]

Formula 14–5

(To find total inductive reactance of parallel inductances)

$X_{L_T} = \frac{1}{\frac{1}{X_{L_1}} + \frac{1}{X_{L_2}}, \text{etc.}}$

X_{L_1} value, [1/x], [+], X_{L_2} value, [1/x], [+], . . . , [=], [1/x]

Formula 14–6

(To find total inductive reactance of two parallel inductances)

$X_{L_T} = \frac{X_{L_1} \times X_{L_2}}{X_{L_1} + X_{L_2}}$

X_{L_1} value, [×], X_{L_2} value, [÷], [(], X_{L_1} value, [+], X_{L_2} value, [)], [=]

Formula 14–7

(To find circuit Q factor when X_L and R are known)

$Q = \frac{X_L}{R}$

X_L value, [÷], R value, [=]

EXCEL AUTOMATED FORMULAS

Helpful problem-solving tools, such as the Excel automated formulas available on the CD, allow automatic calculations of formulas.

Using Excel

Inductive Reactance in AC Formulas

(Excel file reference: FOE14_01.xls)

DON'T FORGET! *It is not necessary to retype formulas once they are entered on the worksheet! Just input new parameters data for each new problem using that formula, as needed.*

- Use the Formula 14–1 spreadsheet sample and solve for the value of inductive reactance for the parameters given in Practice Problems 2, question 1. Check your answer against the answer given in the Appendix for this question.
- Use the Formula 14–2 spreadsheet sample and solve for the value of inductance involved for the parameters given in Practice Problems 2, question 2. Check your answer against the answer given in the Appendix for this question.

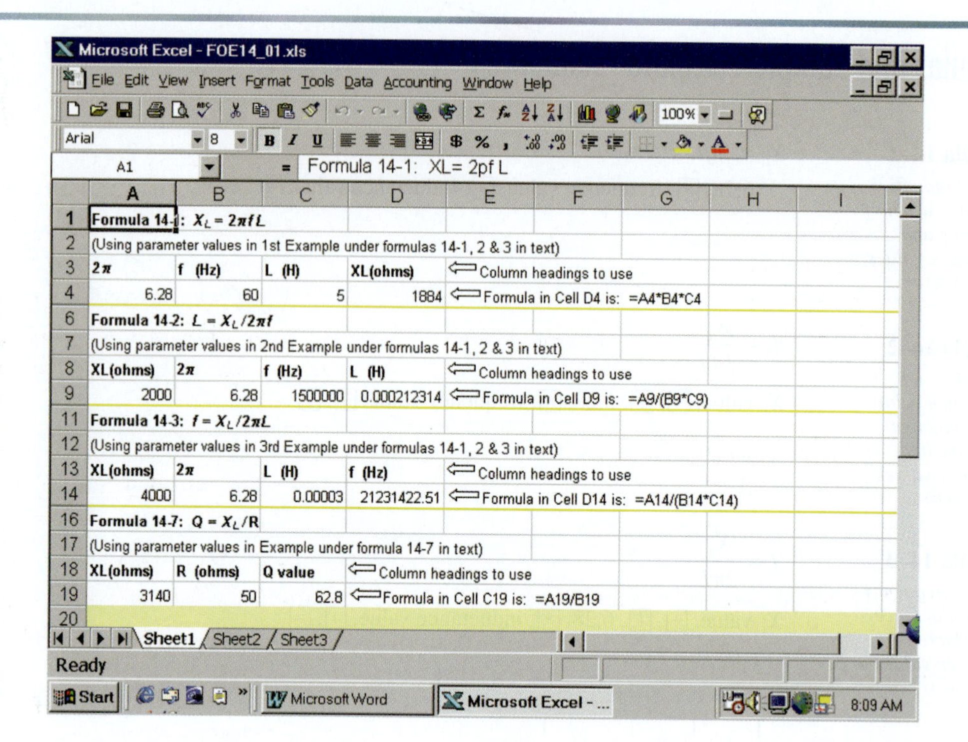

Microsoft Excel - FOE14_01.xls

A1 = Formula 14-1: XL= 2pf L

	A	B	C	D	E
1	Formula 14-1: $X_L = 2\pi f L$				
2	(Using parameter values in 1st Example under formulas 14-1, 2 & 3 in text)				
3	2π	f (Hz)	L (H)	XL(ohms)	⇐Column headings to use
4	6.28	60	5	1884	⇐Formula in Cell D4 is: =A4*B4*C4
6	Formula 14-2: $L = X_L/2\pi f$				
7	(Using parameter values in 2nd Example under formulas 14-1, 2 & 3 in text)				
8	XL(ohms)	2π	f (Hz)	L (H)	⇐Column headings to use
9	2000	6.28	1500000	0.000212314	⇐Formula in Cell D9 is: =A9/(B9*C9)
11	Formula 14-3: $f = X_L/2\pi L$				
12	(Using parameter values in 3rd Example under formulas 14-1, 2 & 3 in text)				
13	XL(ohms)	2π	L (H)	f (Hz)	⇐Column headings to use
14	4000	6.28	0.00003	21231422.51	⇐Formula in Cell D14 is: =A14/(B14*C14)
16	Formula 14-7: $Q = X_L/R$				
17	(Using parameter values in Example under formula 14-7 in text)				
18	XL(ohms)	R (ohms)	Q value	⇐Column headings to use	
19	3140	50	62.8	⇐Formula in Cell C19 is: =A19/B19	
20					

Review Questions

1. Inductive reactance is the opposition that an inductor shows:

a. to direct current.
b. to dc and ac.
c. to alternating current.
d. all of the above.

2. The higher the back-emf produced by an inductor:

a. the higher its X_L.
b. the lower its X_L.
c. the higher its resistance.
d. back-emf has no relationship to either resistance or reactance.

3. X_L is:

a. directly related to only frequency.
b. directly related to only inductance.
c. directly related to both frequency and inductance.
d. none of the above.

4. Inductive reactance is indicated in:

a. siemens.
b. ohms.
c. ampere turns.
d. impedance.

5. In a purely inductive circuit:

a. voltage lags current by 90°.
b. current lags voltage by 90°.
c. voltage and current are in phase.
d. none of the above.

6. Total inductive reactance for series inductors:

a. is calculated similarly to determining R_{equiv} of parallel resistors.
b. is calculated similarly to determining total inductance of series inductors.
c. is calculated using the reciprocal formula.
d. none of the above.

7. The Q of an inductor is equal to:

a. the ratio of its turns to its length.
b. the ratio of its length to its turns.
c. the ratio of its resistance to its inductive reactance.
d. the ratio of its inductive reactance to its resistance.

8. The characteristic of an inductor that causes voltage to lead current is:

a. its resistance and its related impedance.
b. its inductance and its related impedance.
c. its inductance and its related reactance.
d. all of the above.

9. In a series circuit composed of only inductors:

a. V_T equals the difference between the maximum and minimum V drops.
b. V_T equals the sum of all the individual voltage drops.
c. V_T equals the sum of the maximum and minimum V drops.
d. none of the above.

10. In a parallel circuit composed of only inductors:

a. total current is the same throughout each inductor.
b. total current is the sum of all the branch currents.
c. both a and b.
d. none of the above.

Problems

1. What is the inductance of an inductor that exhibits 376.8-Ω reactance at a frequency of 100 Hz? What is inductive reactance at a frequency of 450 Hz?
2. What is the X_L of a 5-H inductor at 120 Hz? What is the value of X_L at a frequency of 600 Hz?
3. What is the total inductance of the circuit in Figure 14–15? When the voltage applied is 100 V and the current through L_1 is 100 mA, what is the frequency of the applied voltage? (Use one decimal place in your calculations.)
4. Find L_T for the circuit in Figure 14–16. At approximately what frequency does X_{L_T} equal 5,000 ohms?
5. Find the Q of the coil in the circuit shown in Figure 14–17. When the frequency triples and the R of the inductor doubles, does Q increase, decrease, or remain the same? If the applied voltage were dc instead of ac, would Q change? Explain.
6. Draw a waveform showing two cycles that illustrates the circuit V and I for the circuit in Figure 14–18. Appropriately label the peak values of V and I, assuming the source voltage is shown in rms value.
7. Find V_T, V_{L_1}, and I_T for the circuit in Figure 14–19.
8. What is the branch current through L_2 for the circuit in Figure 14–20?
9. What is the value of the source voltage in the circuit of Figure 14–20?
10. Determine the total inductance of the circuit in Figure 14–21.

Refer again to Figure 14–21 and answer questions 11 and 12 with "I" for increase, "D" for decrease, or "RTS" for remain the same.

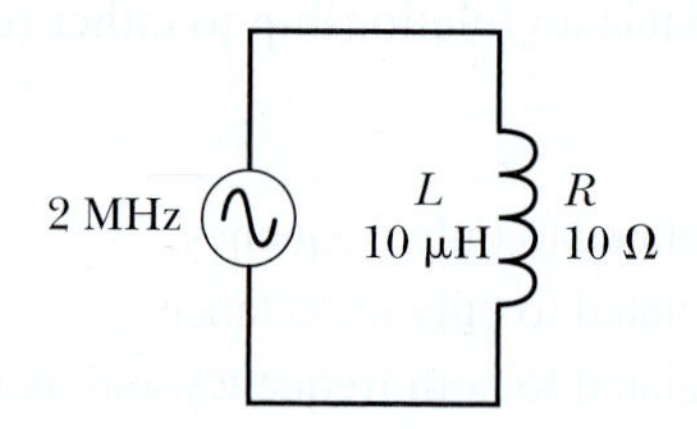

FIGURE 14–17

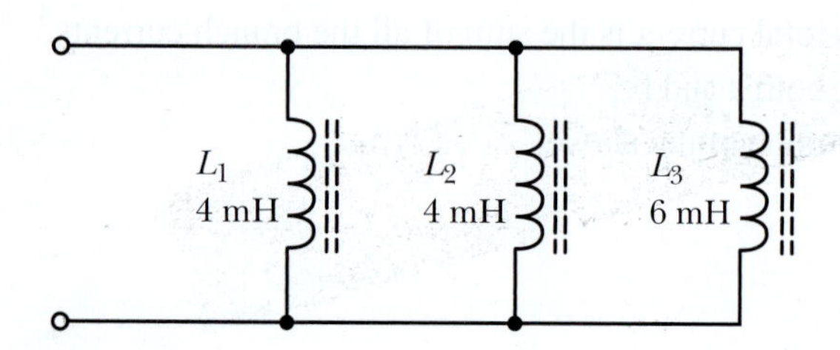

FIGURE 14–15

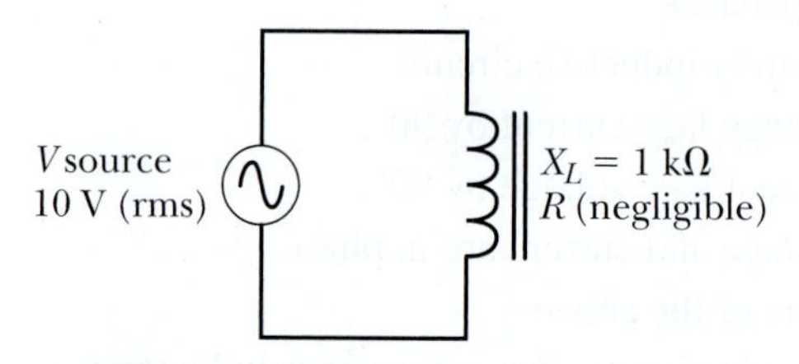

FIGURE 14–18

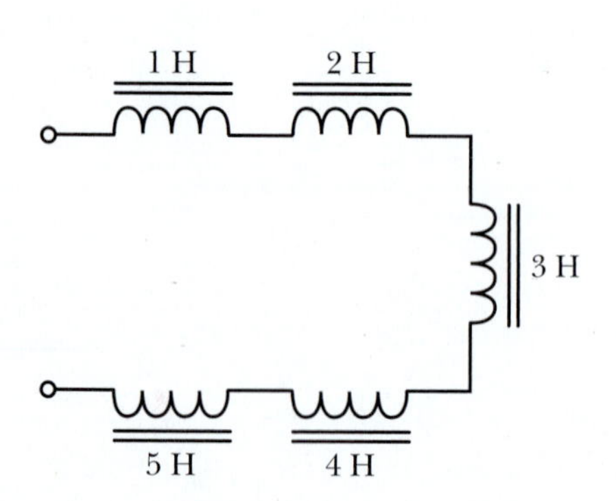

FIGURE 14–16

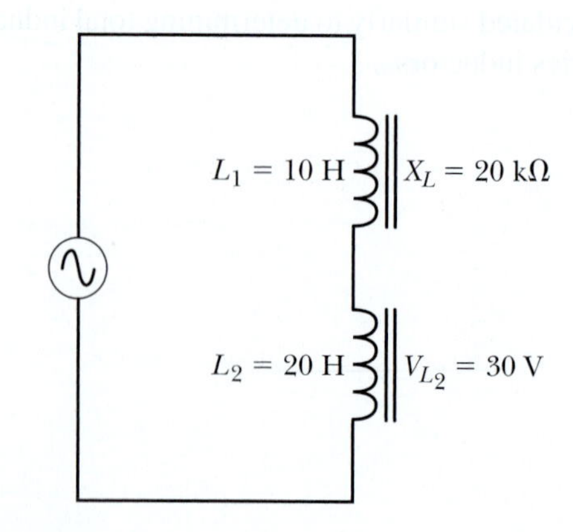

FIGURE 14–19

11. If frequency doubles and V applied stays the same,
L_T will _____.
X_{L_T} will _____.
I_T will _____.
V_{L_4} will _____.
I_{L_1} will _____.

12. If L_2 changes to a 1-H inductor,
L_T will _____.
X_{L_1} will _____.
I_T will _____.
V_{L_1} will _____.
f will _____.

13. What value of inductor will produce 2,000 ohms of X_L at a frequency of 1 kHz?

14. Refer to the circuit in Figure 14–22. Assuming that L_1 = 10 H, L_2 = 15 H, and L_3 = 20 H, what is the total inductance of the circuit?

15. For the circuit of Figure 14–22, what is the total inductive reactance value?

16. For the circuit of Figure 14–22, what is:
a. the circuit total current?
b. the current through each inductor?

Referring to the circuit of Figure 14–22, answer questions 17 through 20 with "I" for increase, "D" for decrease, or "RTS" for remain the same, as appropriate for each question.

17. If L_2 were removed from the circuit, would the circuit total current increase, decrease, or remain the same?

18. If the source frequency were increased, would the current through L_3 increase, decrease, or remain the same?

19. What would have to happen to the circuit applied voltage to keep the circuit current constant if the frequency of the source were decreased to one-third its original value? Be specific.

20. If the three inductors were rewired to be in series, rather than in parallel, what would the new circuit total current be? (Assume f and V source remained the same.)

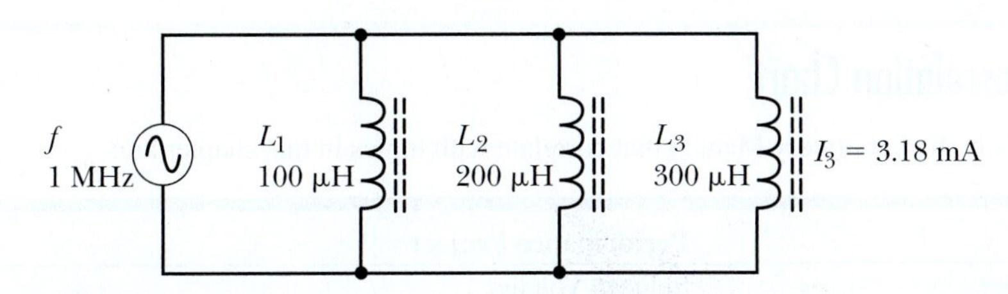

FIGURE 14–20

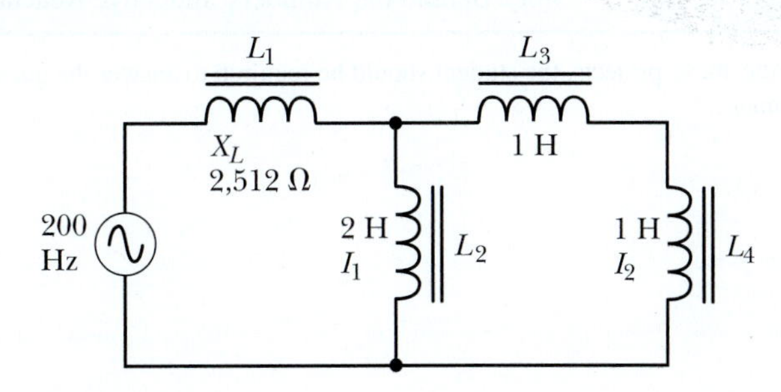

FIGURE 14–21

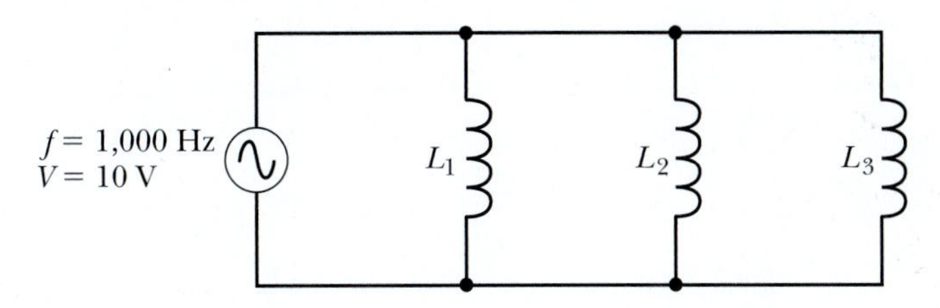

FIGURE 14–22

Analysis Questions

1. If a circuit's parameters are changed so that frequency triples and the amount of inductance doubles, how many times greater will the new X_L be compared with the original value?
2. Assume circuit conditions change for a given circuit so that frequency is halved and inductance is tripled. What is the relationship of the new X_L to the original value?
3. In a given circuit containing inductance, when frequency decreases, will circuit current increase, decrease, or remain the same?
4. When the frequency of applied voltage for a given inductive circuit doubles and the circuit voltage increases by three times, will the circuit current increase, decrease, or remain the same?
5. When the frequency of applied voltage to a given inductor decreases, will the Q of the coil change? If so, in what way?
6. Does the R of a given inductor change when frequency is changed?
7. Explain why you think, in most cases, it is best to have an inductor with a high Q factor.
8. For a theoretically "perfect" inductor:
   a. current and voltage are in phase.
   b. current and voltage are 45° out of phase.
   c. current and voltage are 90° out of phase.
   d. current and voltage are 180° out of phase.
9. When current through an inductor is increasing, the induced voltage in the inductor:
   a. aids the increase in current.
   b. opposes the increase in current.
   c. has no effect on the increase in current.
   d. none of the above.

Performance Projects Correlation Chart

Suggested performance projects in the Laboratory Manual that correlate with topics in this chapter are:

Chapter Topic	Performance Project	Project Number
Concept of Inductive Reactance	Induced Voltage	41
Relationship of X_L to Inductance Value	Relationship of X_L to L and Frequency	42
Relationship of X_L to Frequency of AC	Story Behind the Numbers: Inductive Reactance	

NOTE: It is suggested that after completing the above projects, the student should be required to answer the questions in the "Summary" at the end of this section of projects in the Laboratory Manual.

Troubleshooting Challenge

CHALLENGE CIRCUIT 8

(Follow the SIMPLER sequence by referring to inside front cover.)

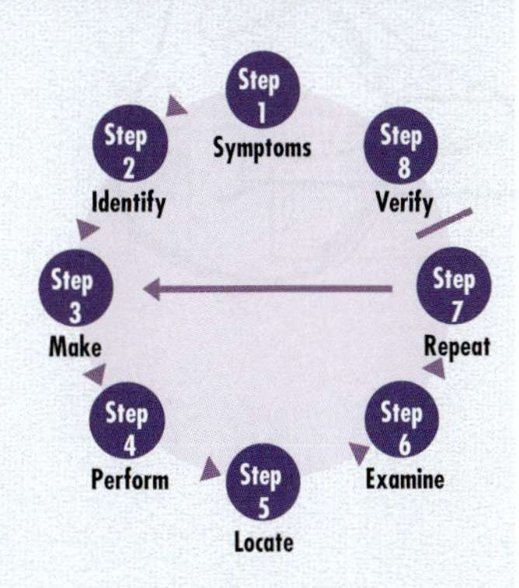

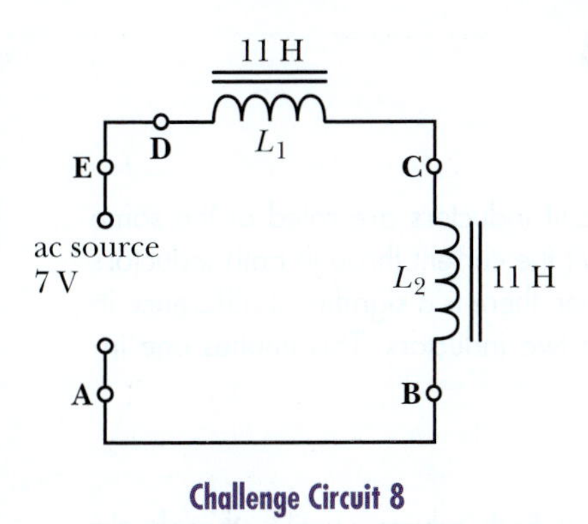

Challenge Circuit 8

STARTING POINT INFORMATION

1. Circuit diagram
2. Each inductor is rated as 11-H, 400-Ω dc resistance
3. The ac voltage measured across L_1 is significantly lower than the voltage measured across L_2

TEST	Results in Appendix C
V_{A-B}	(66)
V_{B-C}	(31)
V_{C-D}	(4)
V_{D-E}	(82)
V_{E-A}	(43)
R_{B-C}	(17)
R_{C-D}	(114)

CHALLENGE CIRCUIT 8

STEP 1

SYMPTOMS The series circuit inductors are rated at the same value, and in a series circuit the current through both inductors should be the same. However, there is a significant difference in the $I \times X_L$ drops across the two inductors. This implies one inductor has a problem.

STEP 2

IDENTIFY initial suspect area: Both inductors until further checks are made.

STEP 3

MAKE test decision: It is less likely that an inductor would increase rather than decrease in inductance (due to shorted turns or change in magnetic path). So let's start by examining the inductor with the lowest voltage drop (L_1). Our assumption is L_2 has remained at the rated value of L, while L_1 may have decreased. This would fit the symptom information regarding L_1's voltage drop being low.

STEP 4

PERFORM **1st Test:** Check the dc resistance of L_1 versus its rated value to see if there might be a hint of shorted turns. (NOTE: Source is removed from the circuit.) R is 297 Ω and the rated value is 400 Ω, indicating the possibility of some shorted turns.

STEP 5

LOCATE new suspect area: L_1 is now a strong suspect area. But we still need to check L_2's parameters to compare and verify our suspicions about L_1.

STEP 6

EXAMINE Examine the available data.

STEP 7

REPEAT analysis and testing: A resistance check of L_2 might be necessary.

2nd Test: Check R_{B-C}. R_{B-C} is 401 Ω, which agrees with the rating. Replace L_1 with the appropriately rated inductor. Then check the voltage across L_1.

3rd Test: V_{C-D} is now close to 3.5 V, which is half of the source voltage. This is normal for this circuit with equal inductor values.

4th Test: Check the voltage across L_2 to verify that it is approximately equal to the voltage across L_1.

STEP 8

VERIFY When checked, the result is also about 3.5 V. The trouble is L_1 has shorted turns. Replacing L_1 restores the circuit to normal.

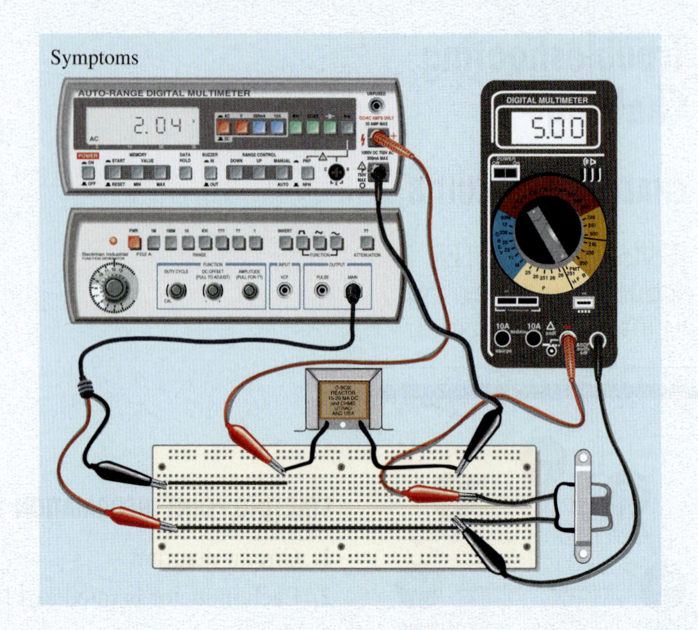

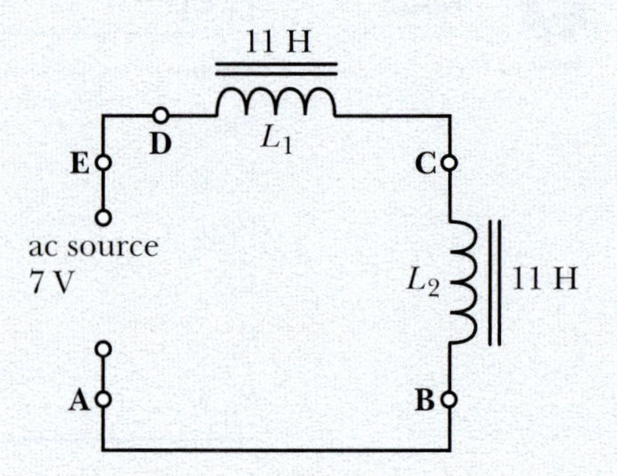

CHALLENGE CIRCUIT 8

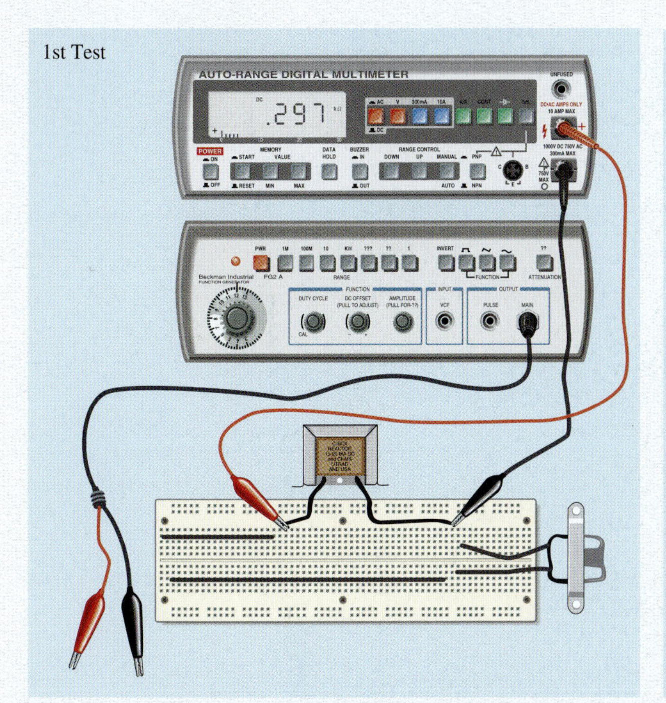
1st Test
AUTO-RANGE DIGITAL MULTIMETER
.297
POWER

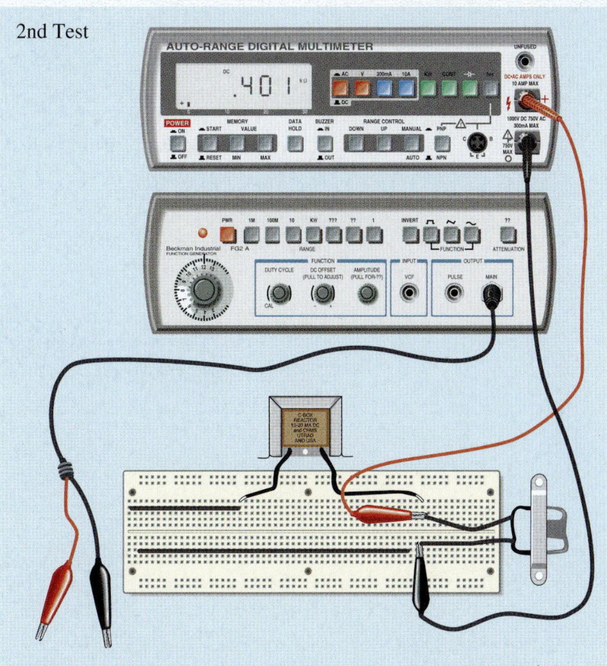
2nd Test
AUTO-RANGE DIGITAL MULTIMETER
.401
POWER

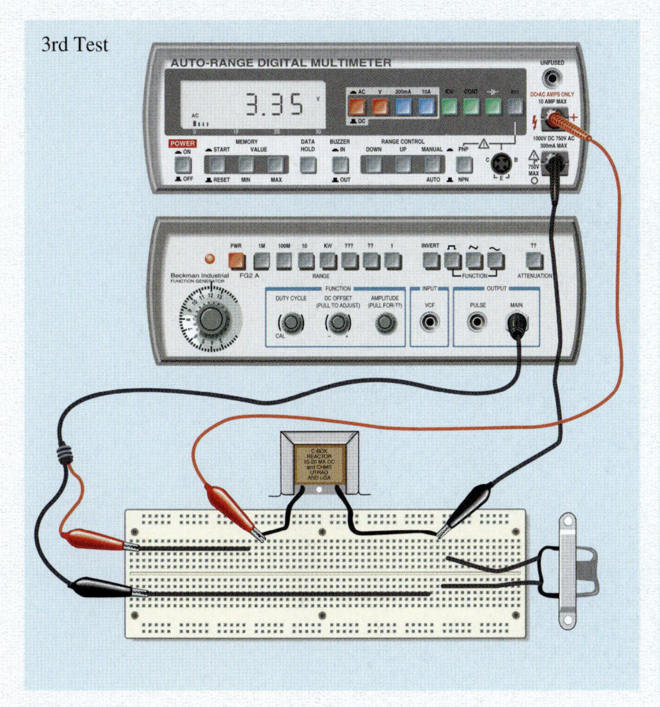
3rd Test
AUTO-RANGE DIGITAL MULTIMETER
3.35
POWER

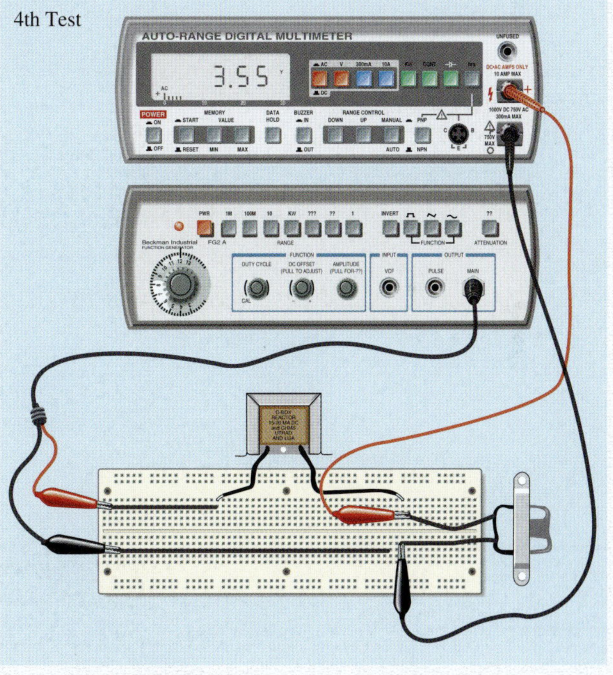
4th Test
AUTO-RANGE DIGITAL MULTIMETER
3.55
POWER

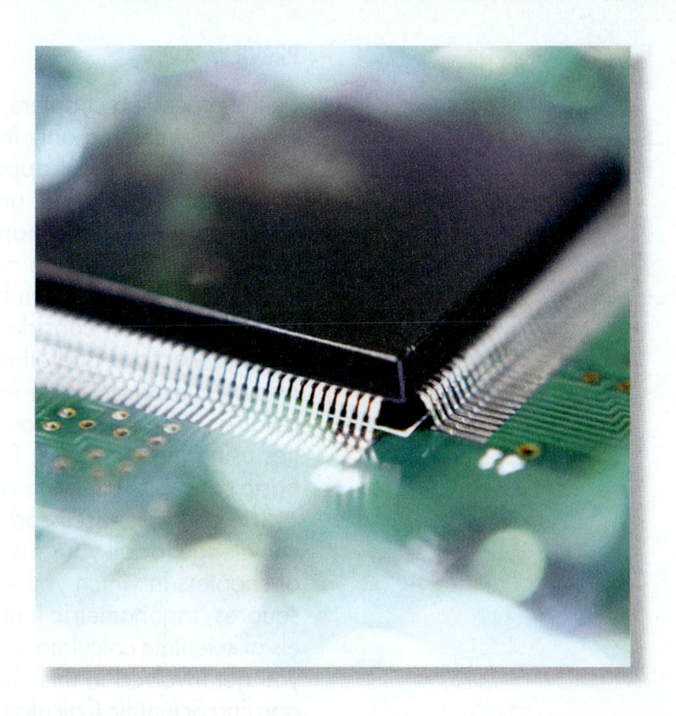

CHAPTER 15

RL Circuits in AC

OUTLINE

OBJECTIVES

After studying this chapter, you should be able to:

1. Use **vectors** to determine magnitude and direction
2. Determine circuit impedance using the **Pythagorean theorem**
3. Determine V_T and I_T using the Pythagorean theorem
4. Determine ac circuit parameters using trigonometry
5. Calculate ac electrical parameters for series *RL* circuits
6. Calculate ac electrical parameters for parallel *RL* circuits
7. List at least three practical applications of inductive circuits
8. Use the computer to solve circuit problems
9. Use the SIMPLER troubleshooting sequence to solve the Troubleshooting Challenge problem

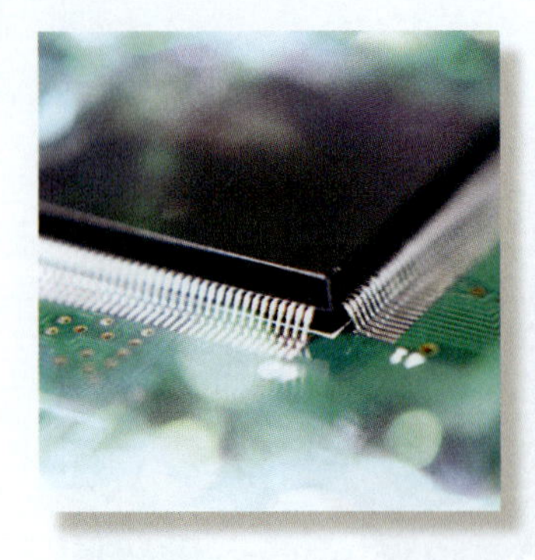

PREVIEW

In the preceding chapters, you have studied the circuit characteristics of both dc and ac circuits containing only inductance. You have learned basic concepts related to inductors, such as inherent opposition to current changes, the self-induced back-emf, phase relationships of *V* and *I*, and other characteristics of inductance.

The purpose of this chapter is to introduce you to basic facts and circuit analysis techniques used to examine ac circuits containing both resistance and inductance (*RL* circuits). Impedance, a combination of resistance and reactance, will be discussed and analyzed. Further considerations about **phase angle** will be given. Primary methods of using **vectors** and other simple math techniques to analyze *RL* circuit parameters will also be introduced. Later in the text, you will study vector analysis techniques that involve polar and rectangular coordinate systems, and methods of expressing electrical quantities using these systems. For this chapter, however, we simply want to introduce you to basic concepts related to analyzing *RL* circuits in ac.

The concepts discussed in this chapter are important because they provide valuable knowledge for all your future ac circuit analyses endeavors. This chapter begins a series of chapters in which you will be performing calculator operations such as square roots, squares, trigonometric functions, and so on. We want to remind you that different models of scientific calculators may use slightly different keystrokes for certain operations. As you get involved in this chapter, study and/or refer back to the "Special Notations Regarding Scientific Calculators" chart found on the CD provided with this text, as required.

KEY TERMS

Cosine function
Impedance
Phase angle (θ)
Pythagorean theorem
Scalar
Sine function
Tangent function
Vector

15–1 Review of Simple *R* and *L* Circuits

Simple *R* Circuit

The ac current through a resistor and the ac voltage across a resistor are in phase, as you have already learned. You know this may be shown by a waveform display or phasor diagram, Figure 15–1.

In the purely resistive *series* ac circuit, Figure 15–2a, you simply add the individual series resistances to find total circuit opposition to current. Or you add voltage drops across resistors to find total circuit voltage, since all voltage drops are in phase with each other.

In purely resistive *parallel* circuits, Figure 15–2b, total resistance is found by using the product-over-the-sum or reciprocal methods. The total circuit current is found by adding the branch currents, because all branch currents are in phase with each other.

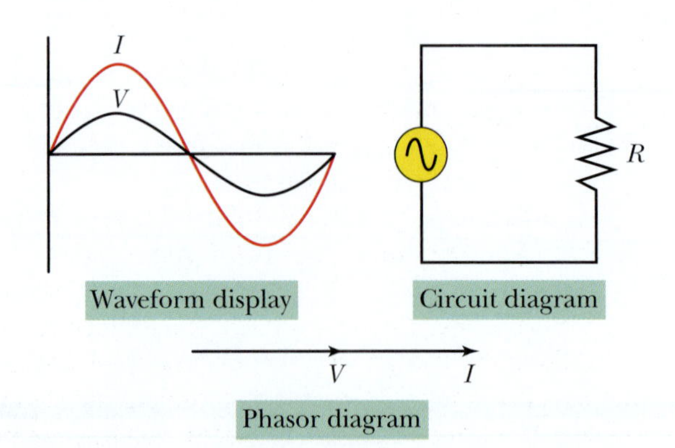

FIGURE 15–1 *V* and *I* in a simple *R* circuit

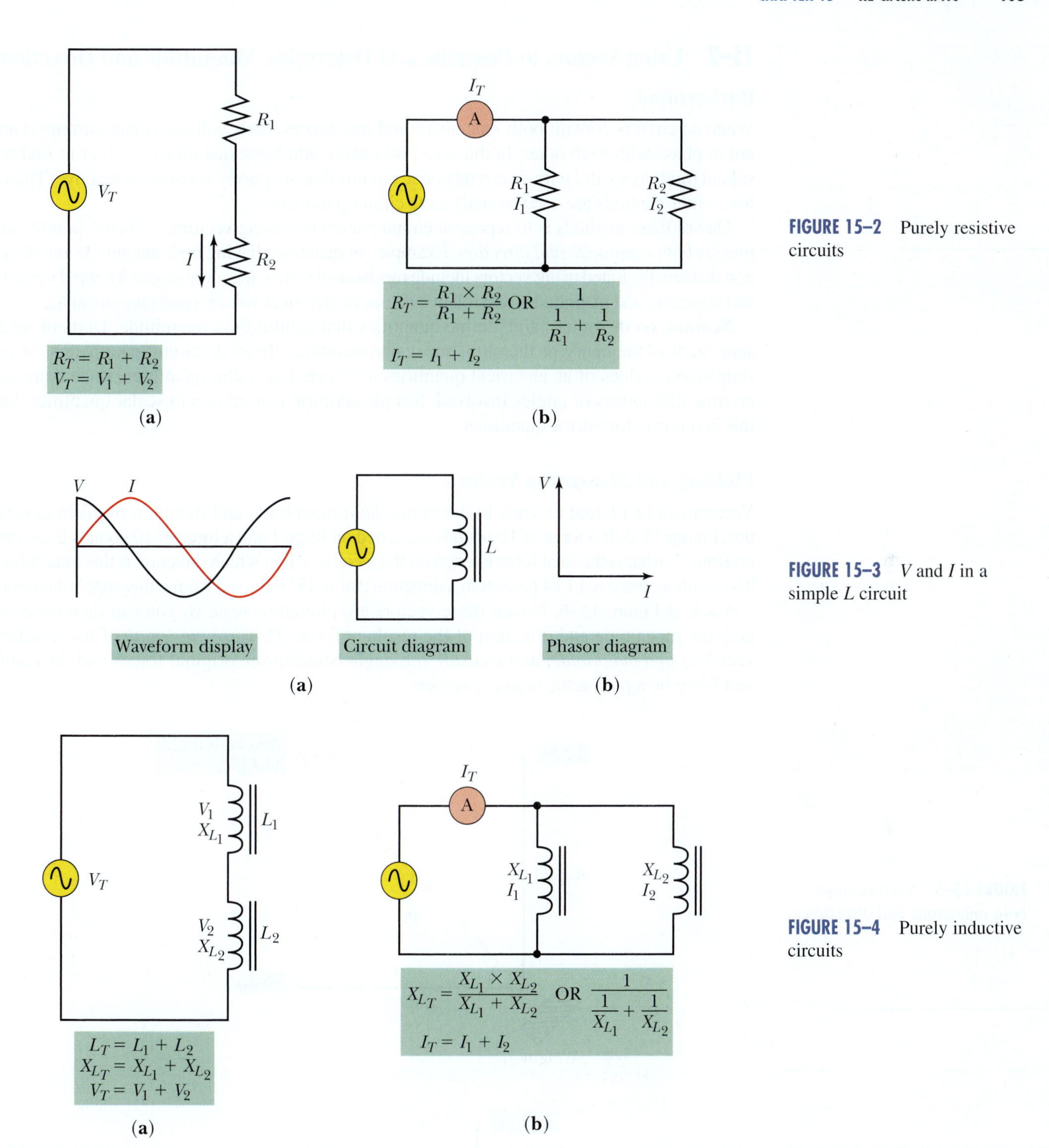

FIGURE 15–2 Purely resistive circuits

FIGURE 15–3 V and I in a simple L circuit

FIGURE 15–4 Purely inductive circuits

Simple *L* Circuit

In a purely inductive ac circuit, you know that current through the inductor and voltage across the inductor are 90° out of phase with each other. Again, this is illustrated by a waveform display, Figure 15–3a, or a phasor diagram, Figure 15-3b.

You know that in purely inductive series circuits, total inductance is the sum of the individual inductances; total inductive reactance (X_{L_T}) is the sum of all X_Ls; and total voltage equals the sum of all V_Ls, since each V_L is in phase with the other V_Ls although they are not in phase with the current, Figure 15–4a. In purely inductive parallel circuits, total circuit opposition to current (X_{L_T}) is found by using reciprocal or product-over-the-sum approaches. Also, total circuit current equals the sum of the branch currents, since branch currents are in phase with each other although not in phase with the circuit voltage, Figure 15–4b.

15–2 Using Vectors to Describe and Determine Magnitude and Direction

Background

When ac circuits contain both reactances and resistances, their voltage(s) and current(s) are not in phase with each other. In this case, we cannot add these quantities together to find resultant totals as we did with dc circuits or ac circuits that are purely resistive or reactive. Therefore, other methods are used to analyze ac circuit parameters.

One of these methods is to represent circuit parameters using **vectors.** A *vector quantity expresses both magnitude and direction.* Examples of quantities having both magnitude and direction that are illustrated using vectors include mechanical forces, the force of gravity, wind velocity and direction, and magnitudes and relative angles of electrical voltages and currents in ac.

Scalars, on the other hand, define quantities that exhibit only magnitude. Units of measure, such as the henry or the ohm, are scalar quantities. Even dc circuit parameters, or instantaneous values of ac electrical quantities are treated as scalar quantities, since there are no time differences or angles involved. Simple addition is used to add scalar quantities, but this is not true for vector quantities.

Plotting and Measuring Vectors

Vectors can be plotted to scale to determine *both* magnitude and direction of given quantities, Figure 15–5. If a force of 10 pounds is exerted on Rope 1 and a force of 10 pounds is exerted on Rope 2, what is the total force exerted on the weight, and in which direction is this total force? The resultant force is 14.14 pounds in a direction that is 45° removed from either rope's direction.

Look at Figure 15–6. Notice these vectors are plotted to scale so you can then *measure* both the magnitude and direction of the resultant force. *Measure the length* of the resultant vector *to find magnitude,* and *measure the angle* between the original forces and the resultant force by a protractor *to find direction.*

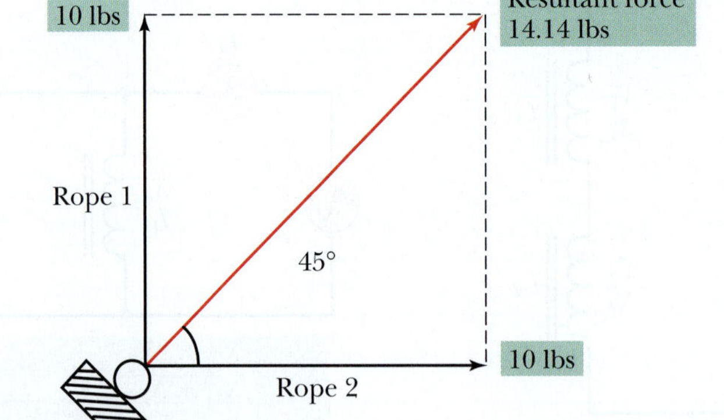

FIGURE 15–5 Vectors show both magnitude and direction.

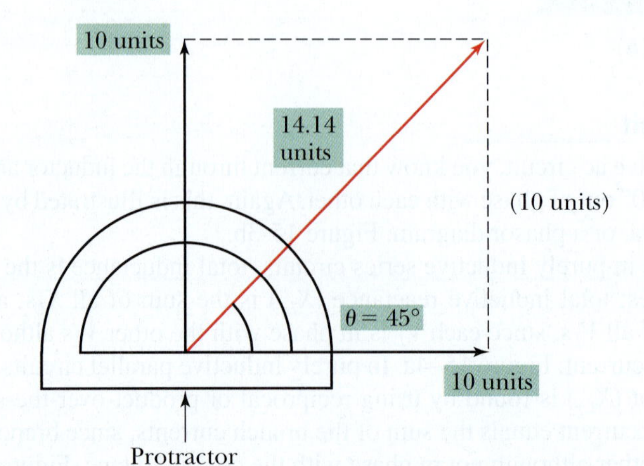

FIGURE 15–6 Plotting vectors to scale

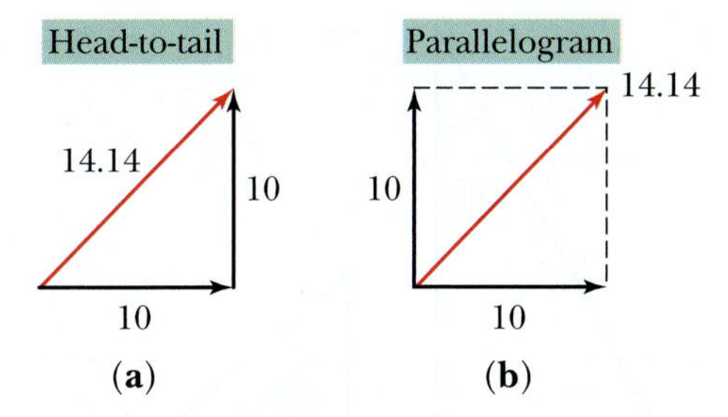

FIGURE 15–7 Methods of formatting vectors

Observe Figure 15–7a, which illustrates how vectors are laid end-to-end or head-to-tail to determine resultant values. Also, notice in Figure 15–7b that a parallelogram is drawn that projects the same magnitudes as the end-to-end approach. Although either method can be used to find resultant values, we will use the parallelogram approach most frequently.

Later in the chapter you will see how this plotting and measuring technique is used to analyze *RL* circuit parameters. Although it is not always convenient to have graph paper and protractor handy, or to draw circuit parameters to scale when analyzing a circuit, the plotting approach can be used. Also, it is a useful tool to help introduce you to vectors.

Even when not plotted to scale for the direct measurement method, vectors are useful to illustrate ac circuit parameters and show relative magnitudes and angles. Sketching this diagram helps you understand the situation and verify your computations.

Practical Notes

Even when not precisely measuring vectors to determine resultant values, get in the habit of approximating the *relative* lengths and directions of vectors when you draw them to solve ac problems. This allows you to visually determine if your computations are making sense! A good practice is to draw simple vector diagrams whenever you are analyzing ac circuits—they help visualize the situation.

PRACTICE PROBLEMS 1

Assume the forces on the ropes in Figure 15–5 are 3 pounds for Rope 1 and 4 pounds for Rope 2. What are the magnitude and direction of the resultant force on the weight? (NOTE: Use the plotting-to-scale approach and remember to keep things relative to the scale you use.)

IN-PROCESS LEARNING CHECK 1

1. In a purely resistive ac circuit, the circuit voltage and current are _____ phase.
2. In a purely inductive ac circuit, the current through the inductance and the voltage across the inductance are _____ degrees out of phase. In this case, the _____ leads the _____.
3. A quantity expressing both magnitude and direction is a _____ quantity.
4. The length of the vector expresses the _____ of a quantity.

15–3 Introduction to Basic AC Circuit Analysis Techniques

Most ac circuit analyses solve for quantities that are illustrated with vectors in right-triangle form. Recall from our previous discussions that a right triangle has one angle forming 90° and two angles that total another 90°, Figure 15–8. Many electrical parameters have this 90° phase difference between quantities. For example, in perfect inductors and capacitors, the voltages and currents are 90° out of phase. Also, ac circuit resistive and reactive voltage drops or currents are often represented as 90° out of phase with each other.

Practical Notes

NOTE: The calculations are easy if you have a commonly used scientific calculator with square, square root, and "trig" function keys. You probably own or presently have access to such a calculator. If not, you might want to gain access to one and learn to use these functions as you proceed through the remainder of the chapter. If this device is not available, paper and pencil will work just fine! NOTE: *Be aware* that some calculators only accept radian measure! Therefore, to convert degrees to number of radians, divide the number of degrees by 57.3; and to convert radians to number of degrees, multiply the number of radians by 57.3. Recall that 2π radians equal 360°; that is, 6.28 times 57.3° equals 360°.

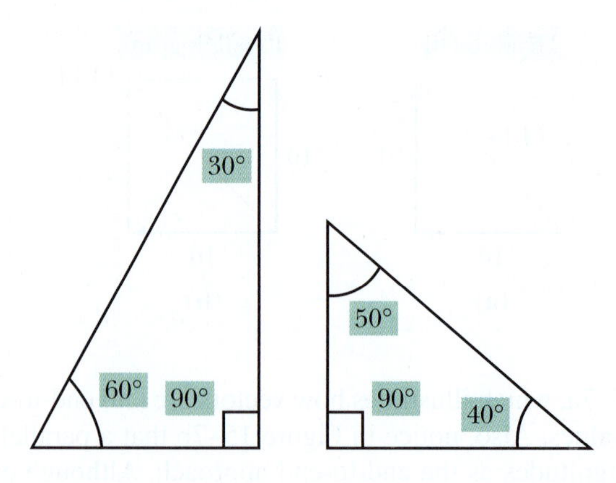

FIGURE 15–8 Examples of right triangles

We will discuss two basic techniques used to analyze these right-triangle problems. The first method is the **Pythagorean theorem,** which is useful to determine the *magnitude* of quantities. The other method briefly addressed in this chapter is trigonometric functions that determine the circuit phase angles. Also, trigonometry can be used to find the magnitude of various electrical quantities, if the phase angle is known.

Pythagorean Theorem

The Pythagorean theorem states that the square of the hypotenuse of a right triangle equals the sum of the squares of the other two sides.

FORMULA 15–1 $c^2 = a^2 + b^2$

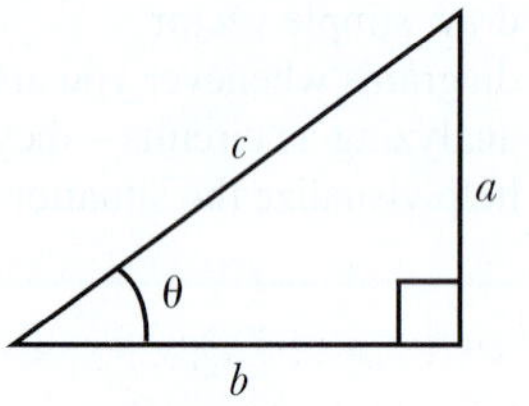

$c^2 = a^2 + b^2$
$\sqrt{c^2} = \sqrt{a^2 + b^2}$
$c = \sqrt{a^2 + b^2}$

FIGURE 15–9 Pythagorean theorem

Refer to Figure 15–9 and review the equations related to the right triangle and Pythagorean theorem.

Now, let's transfer this information into finding the resultant value, or magnitude, of two voltages that are displaced from each other by 90° and vectorially added.

EXAMPLE Refer to Figure 15–10 and observe the following steps: In the diagram shown,

1. V_S represents the hypotenuse of the right triangle.
2. V_1 represents one of the other two sides of the right triangle.
3. V_2 represents the remaining side of the right triangle.
4. Transferring this information into the Pythagorean formula:

$$V_S = \sqrt{V_1^2 + V_2^2}$$
$$V_S = \sqrt{25^2 + 25^2}$$
$$V_S = \sqrt{1{,}250} = 35.35 \text{ V}$$

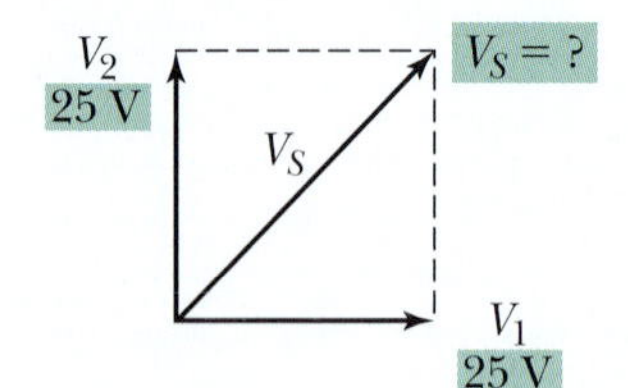

FIGURE 15–10

PRACTICE PROBLEMS 2

Refer to Figure 15–11. Use the Pythagorean theorem and solve for the magnitude of side V_S.

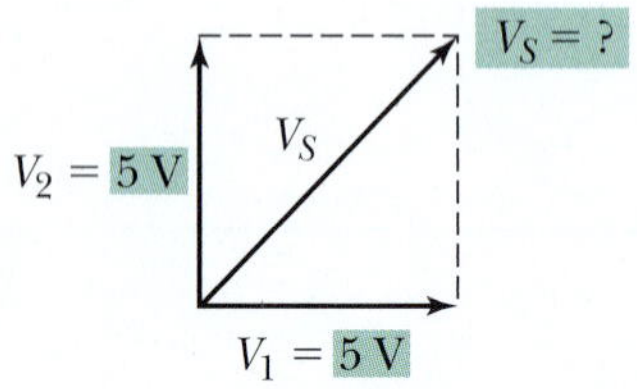

FIGURE 15–11

Trigonometric Functions

Trigonometric functions result from special relationships existing between the sides and angles of right triangles. "Trig" is a common shortened term for trigonometry or trigonometric. We will only discuss three basic trig functions used for your study of ac circuits—**sine, cosine,** and **tangent functions.**

Notice the side opposite (opp), the side adjacent (adj), and the hypotenuse (hyp) in Figure 15–12. As you can see, the relationship of each side to the "angle of interest," θ, determines whether it is called the opposite or the adjacent side. The hypotenuse is *always* the longest side of the triangle.

Refer to Figure 15–12 to review the basic trig functions.

If we assume a rotating vector with a unit length of one starts at 0° and rotates counterclockwise to the 90° position, the values of sine, cosine, and tangent vary within limits. That is:

$\sin\theta$ increases from 0 to 1.0 (with angles from 0° to 90°)
$\cos\theta$ decreases from 1.0 to 0 (with angles from 0° to 90°)
$\tan\theta$ increases from 0 to ∞ (with angles from 0° to 90°)

See if you can relate this to the sine and cosine waves in Figure 15–13. Assume the maximum (peak) value is one volt for each waveform. Relate the zero points and the maximum points to the angles. Do they compare?

Figure 15–14 carries this idea one step further. Notice how these functions vary in the four quadrants of the coordinate system we discussed in an earlier chapter. Again, try relating the sine and cosine waves in Figure 15–13 to the statements in Figure 15–14.

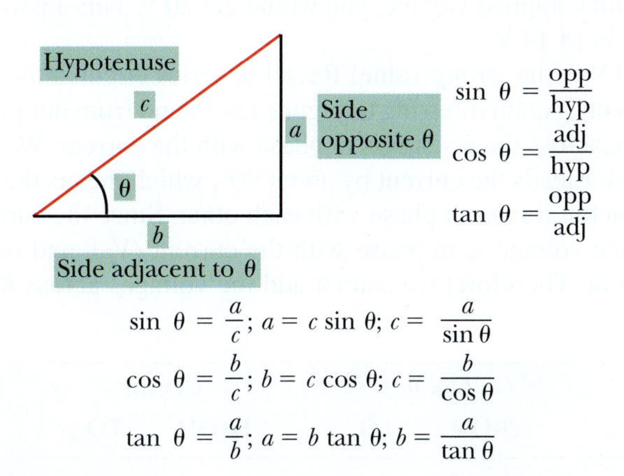

FIGURE 15–12 Labeling sides of a right triangle

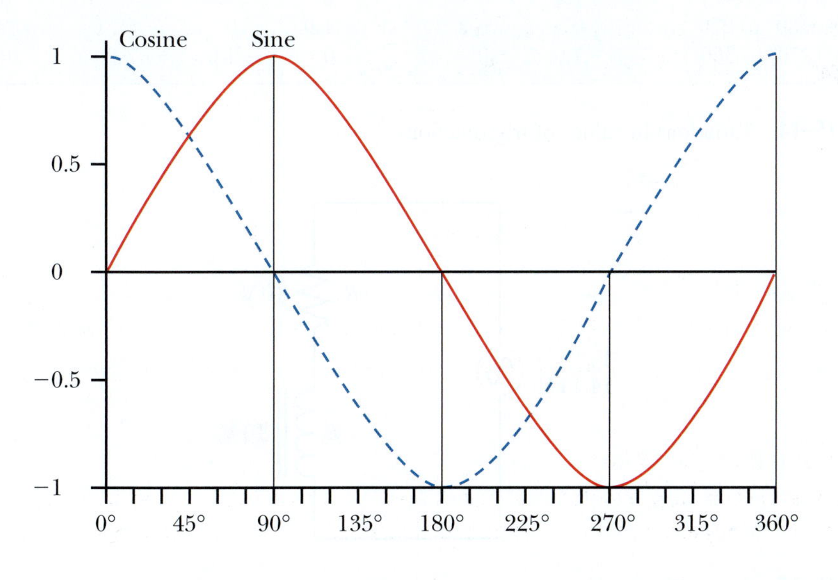

FIGURE 15–13 Sine and cosine waves

EXAMPLE What is the described angle if the side opposite is 2 units and the hypotenuse is 2.828 units?

Referring to the trig formulas for sine, cosine, and tangent, the trig formula containing the side opposite and the hypotenuse is the formula for the sine of the angle. Therefore,

$$\sin\theta = \frac{2}{2.828} = 0.707$$

The $\sin^{-1}$ of 0.0707 = 45°.

Now try the following practice problems.

PRACTICE PROBLEMS 3

1. What is the described angle if the side adjacent has a magnitude of 5 units and the hypotenuse has a magnitude of 10 units?
2. What is the tangent of 45°?
3. What is the sine of 30°?

15–4 Fundamental Analysis of Series *RL* Circuits

Now we will take the knowledge you have acquired and apply it to some practical situations and problems, so you can see the benefit of this knowledge!

Refer to Figure 15–15. Notice that if you added the voltages across the series resistor and the inductor to find total applied voltage, you would get 20 V. This answer is wrong! The actual applied voltage is 14.14 V.

Let's see why 20 V is the wrong value! Recall in series circuits, the current is the same through all components. Again referring to Figure 15–15 and from our previous discussions, we know the voltage across the resistor is in phase with the current. We also know the voltage across the inductor leads the current by about 90°, which means the voltages across the resistor and the inductor are *not* in phase with each other. Since the current through both is the same current, one voltage is in phase with the current (V_R), and one voltage is not in phase with the current. Therefore, we cannot add the voltages across R and L to find total

QUADRANT	SIN θ FROM	SIN θ TO	COS θ FROM	COS θ TO	TAN θ FROM	TAN θ TO
I (0° to 90°)	0	1.0	1.0	0	0	∞
II (90° to 180°)	1.0	0	0	−1.0	−∞	0
III (180° to 270°)	0	−1.0	−1.0	0	0	∞
IV (270° to 360°)	−1.0	0	0	1.0	−∞	0

FIGURE 15–14 Variations in values of trig functions

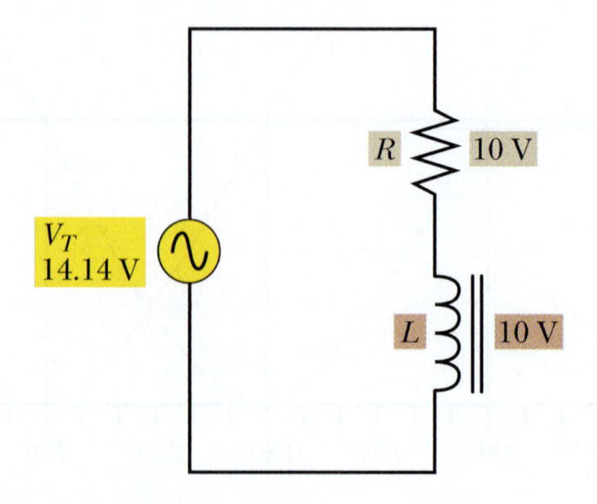

FIGURE 15–15 Series *RL* circuit voltages

voltage, as we would do in a simple dc circuit. Rather, we must use a special method of addition called vector addition to find the correct answer. The voltages must be added "vectorially," not "algebraically."

Let's try the two methods previously discussed to find total voltage in series *RL* circuits.

Using the Pythagorean Theorem to Analyze Voltage in Series *RL* Circuits

Figure 15–16 illustrates how V_R, V_L, and V_T are typically represented in a right triangle by phasors or vectors.

EXAMPLE Let's transfer the Pythagorean theorem formula in Figure 15–9, $c = \sqrt{a^2 + b^2}$, to this situation. As you can see, the longest side is V_T. This is equivalent to the *c* in the original Pythagorean formula. The other two quantities (V_R and V_L) are the other quantities in the formula. Thus, the formula to solve for V_T becomes:

FORMULA 15–2 $V_T = \sqrt{V_R^2 + V_L^2}$

Therefore, $V_T = \sqrt{10^2 + 10^2} = \sqrt{200} = 14.14$ volts.

EXAMPLE Assume the voltage values in Figure 15–15 change so that V_R is 30 volts and V_L is 40 volts. What is the new *V* applied value?
Answer: The V_T is 50 volts, Figure 15–17.

PRACTICE PROBLEMS 4

In a simple series *RL* circuit, V_R is 120 V and V_L is 90 V. What is the *V* applied value? SUGGESTION: Use your calculator.

You can see that, using the Pythagorean theorem, it is easy to find the magnitude of total voltage in a series *RL* circuit—simply substitute the appropriate values in the formula. We are not able to find the resultant angular direction by this method. However, the trig functions help us.

Practical Notes

For *series circuits,* since current is common to all parts, *current is commonly the reference phasor* and is plotted on the reference phasor 0° axis.

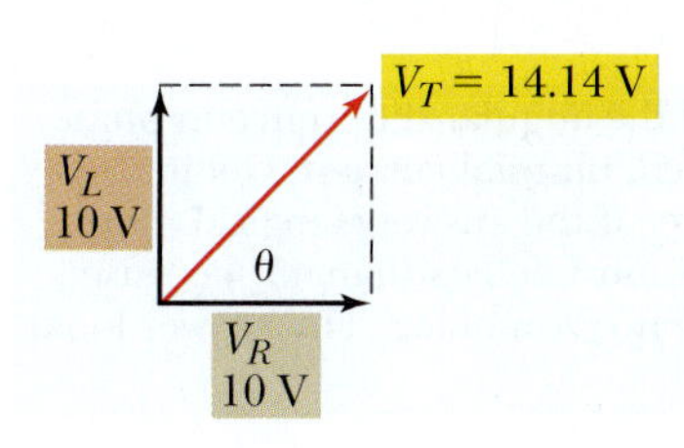

FIGURE 15–16 Voltage vector diagram

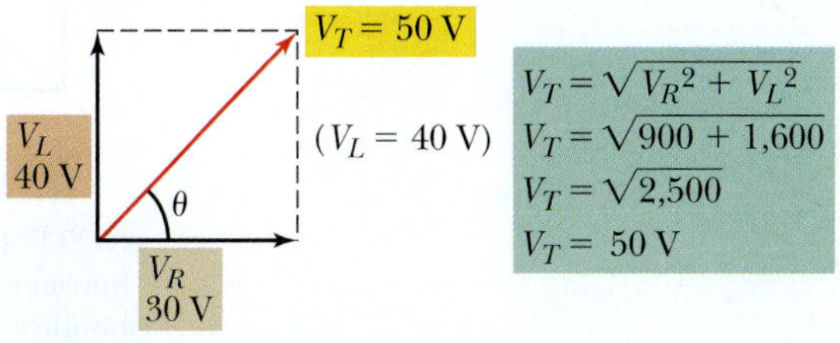

$V_T = 50$ V

V_L 40 V

($V_L = 40$ V)

θ

V_R 30 V

$V_T = \sqrt{V_R^2 + V_L^2}$
$V_T = \sqrt{900 + 1{,}600}$
$V_T = \sqrt{2{,}500}$
$V_T = 50$ V

(NOTE: V_T is *not* equal to the sum of V_R and V_L. It is greater than either one alone, but not equal to their sum.)

FIGURE 15–17

Using Trig to Solve for Angle Information in Series *RL* Circuits

Look at Figure 15–18 as you study the following statements.

1. To find the desired angle information using trig functions, identify the sides of the right triangle in terms of the circuit parameters of interest. Note in Figure 15–18 that V_L is in the upward (+90°) direction, and V_R is at the 0° reference direction. In *series* circuits, the current is the same throughout the circuit and is in phase with the resistive voltage. Therefore, *I* is also on the 0° axis.

   Because these directions are important, let's restate them! They are general rules of thumb that you will use from now on! V_L is vertical (up), V_R is horizontal to the right, and *I* is in phase with V_R.

2. Referring to the values in Figure 15–18, let's find the *circuit phase angle* that is defined as *the phase angle between circuit applied voltage and circuit total current.*

   a. Consider the angle between V_T and I_T the angle of interest. Thus, the side opposite to this angle is the V_L value and the hypotenuse is V_T.

   b. Use the information relating these two sides in the sine function formula to find the angle. (Recall that sin θ= opp/hyp.) Therefore, in our example, sin θ= 30/50 = 0.60.

   c. Using a calculator, computer, or trig table, to find the angle having a sine function value of 0.60, we find that it is 36.9°.

Practical Notes In ac circuits containing both resistance and reactance, the phase difference in degrees between applied voltage and total current must fall between 0° (indicating a purely resistive circuit) and 90° (indicating a purely reactive circuit). Since both resistance and reactance are present, the circuit cannot act purely resistive or purely reactive.

Practical Notes Rather than having to say in words "the angle having a tangent value equal to . . . ," it is common to use the mathematical term "arc tangent," or "arctan." In other words, if I find the arctan of 0.75, it is an angle of 36.9°. You will also see arctan written as "$\tan^{-1}$." To find the arctan on a calculator, input the ratio of the opposite side to the adjacent side. That is: (opposite side value, [÷], adjacent side value, [=]), then press the [2nd] and [TAN-1] function keys (in that order). You will then see the $\tan^{-1}$ (arctan, or angle value) displayed in the readout.

Naturally, you can find the arc sine or arc cosine using the same technique. For the sine, the values to input are the opposite side divided by the hypotenuse. For the cosine, the values to input are the adjacent side divided by the hypotenuse. Once you have the sine or cosine value, use the [2nd] and [SIN-1] function keys or the [2nd] and [COS-1] function keys as appropriate to find the arc sine or arc cosine, respectively. (NOTE: This gives you the angle value.)

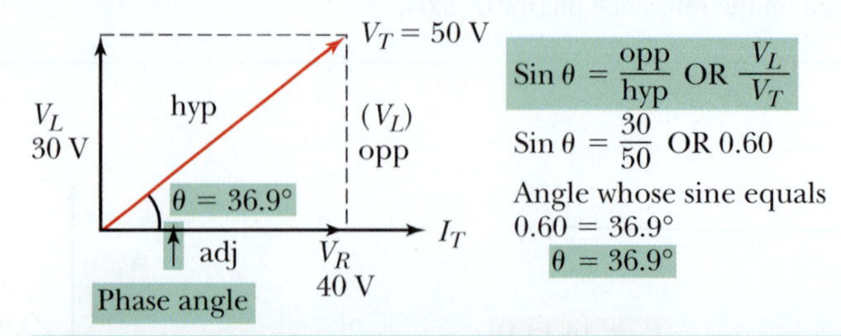

NOTE: Phase angle is considered the angular difference in phase between circuit total voltage and circuit total current. Get in the habit of looking at the sketch to see if the answer is logical. It is obvious that the angle in this case must be less than 45° because the vertical leg is shorter than the horizontal leg. The answer looks and seems logical!

FIGURE 15–18 Using trig to find angle information

d. In this example, since we know all sides, we can also use the other two trig functions. That is,

$$\cos\theta = \frac{\text{adj}}{\text{hyp}} = \frac{V_R}{V_T} = \frac{40}{50} = 0.8$$

The angle with a cosine value of 0.8 is 36.9°.

$$\tan\theta = \frac{\text{opp}}{\text{adj}} = \frac{V_L}{V_R} = \frac{30}{40} = 0.75$$

The angle with a tangent value of 0.75 is 36.9°.

e. Use the sine, cosine, or tangent functions to solve for the unknown angle depending on the information available. Obviously, when you know the values for the opposite side and the hypotenuse, use the sine function. When you know the values for side adjacent and the hypotenuse, use the cosine function. When you know only the values for side opposite the angle of interest and the side adjacent to the angle of interest, use the tangent function.

Stated another way, select and use the formula containing the two known values and the value you are solving for.

PRACTICE PROBLEMS 5

1. Draw the vector diagram for the voltages in Figure 15–19.
2. Use trig and solve for the phase angle of the circuit in Figure 15–19.
3. Which trig function did you use in this case?
4. If the voltages had been reversed (i.e., $V_R = 20$ V and $V_L = 30$ V), would the phase angle have been greater, smaller, or the same?

Using Trig to Solve for Voltages in Series *RL* Circuits

You have seen how trig helps find angle information. You have also learned that the Pythagorean theorem is useful to find magnitude or voltage value in series *RL* circuits. Can trig also be used to find magnitude? Yes, it can! Let's see how!

EXAMPLE If the angle information is known, determine the sine, cosine, or tangent values for that angle. Next, use the appropriate, available sides information, and find the value of the unknown side. Look at Figure 15–20 and let's see how to do this.

1. Start by making sure the formula for the trig function used includes the unknown side of interest, plus a known side. This allows us to solve for the unknown by using equation techniques.

$$\text{Sin }\theta = \frac{V_L}{V_T}$$

$$\text{Sin }\theta = \frac{V_L}{100}$$

$$\text{Sin } 60° = \frac{V_L}{100}$$

$$0.866 = \frac{V_L}{100}$$

$$86.6\text{ V} = V_L$$

$$\text{Cos }\theta = \frac{V_R}{V_T}$$

$$0.5 = \frac{V_R}{100}$$

$$50\text{ V} = V_R$$

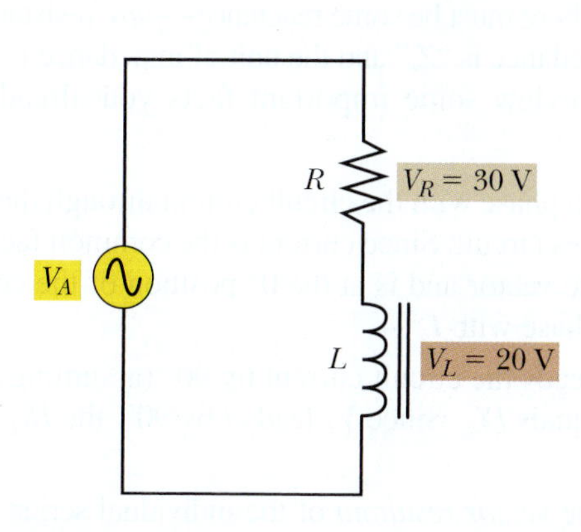

FIGURE 15–19 multiSIM

FIGURE 15–20

2. To solve for V_L, use the sine function since sin θ uses the opposite side divided by the hypotenuse. V_L is the unknown side of interest and is the side opposite the known angle. Since we know the hypotenuse value, we have our starting point.
3. Sin θ = opp/hyp, thus sin 60° = V_L/100. Now we determine the sine of 60° and substitute in the equation. Use the trig table, calculator, or computer and find the sine of 60° is 0.866. (NOTE: Round off to 0.87) Put this in the formula, 0.87 = V_L/100. Transpose and solve: $V_L = 100 \times 0.87 = 87$ V.
4. To solve for V_R, use either the cosine or tangent functions with the information now known. Cosine uses the adjacent divided by the hypotenuse, and tangent uses the opposite divided by the adjacent. In either case, a known and the desired unknown (the adjacent) values appear in the formulas.
5. Assuming we do not know the V_L value, we use the cosine function. Cos θ = adj/hyp, therefore cos 60° = V_R/100. Next, determine the cosine of 60°, which equals 0.5. Substitute in the formula, 0.5 = V_R/100. Transpose and solve: $V_R = 100 \times 0.5 = 50$ V.

Now, let's see if these results make sense with respect to our vector diagram in Figure 15–20. Since the angle is greater than 45°, we know V_L must be greater than V_R and it is!

Now we can use the tangent function to check our results, since it uses the opposite divided by the adjacent sides. Insert the values found for V_L and V_R into the tan θ = opp/adj formula to find that 86.6/50 = 1.732. The angle whose tangent is 1.732 is 60°; that is, the arc tan ($\tan^{-1}$) of 1.732 = 60°. It checks out! Now, you try some problems. ◆

PRACTICE PROBLEMS 6

1. A series *RL* circuit has one resistor and one inductor. What is the V_R value if the applied voltage is 50 V and the circuit phase angle is 45°?
2. A series *RL* circuit composed of a single resistor and inductor has an applied voltage value of 65 V and the circuit phase angle is 60°.
   a. What is the value of V_R?
   b. What is the value of V_L?
3. A series *RL* circuit composed of a single resistor and inductor has values where $V_L = 36$ V and $V_R = 24$ V.
   a. What is the value of the angle between V_R and V_T?
   b. What is the value of V_T?

Analyzing Impedance in Series *RL* Circuits

You have learned how to analyze various voltages in series *RL* circuits by measuring vectors, using the Pythagorean theorem, and applying basic trig functions. Now, let's look at another important ac circuit consideration: impedance.

Impedance is the total opposition a circuit offers to ac current flow at a given frequency. This means impedance is the combined opposition of all *resistances and reactances* in the circuit. Since impedance is sensitive to frequency, there must be some reactance—*pure* resistance is not sensitive to frequency. The symbol for impedance is "Z," and the unit of impedance is the ohm.

Before discussing impedance, let's review some important facts you already know, Figure 15–21.

1. The voltage drop across the resistor is in phase with the circuit current through the resistor, and V_R equals *IR*. Note that this is a series circuit. Since current is the common factor in series circuits, the current is the reference vector and is at the 0° position of the coordinate system. Also, V_R is at 0° since it is in phase with *I*.
2. The voltage drop across the inductor leads the circuit current by 90° (assuming a perfect inductor with no resistance), and V_L equals IX_L. Since V_L leads *I* by 90°, the IX_L vector is leading the *I* vector by 90°.
3. As you know, total circuit voltage is the *vector resultant* of the individual series voltages. Since the circuit is not purely resistive, nor purely inductive, the resultant circuit phase an-

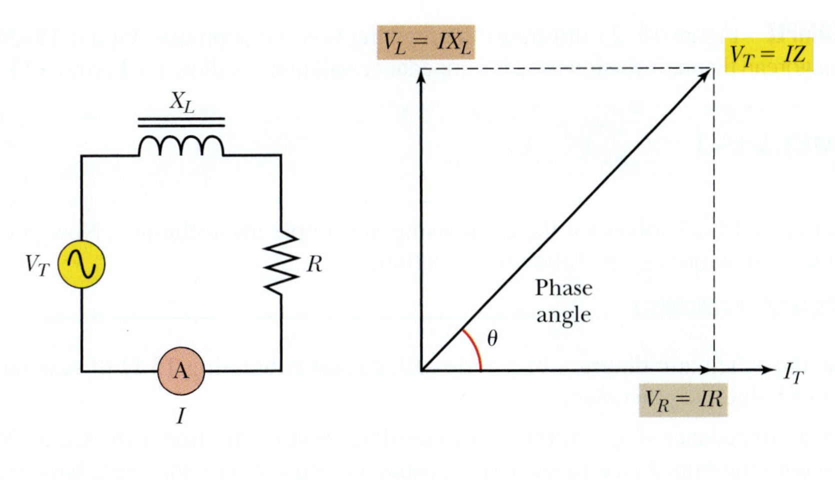

FIGURE 15–21 Series *RL* circuit quantities

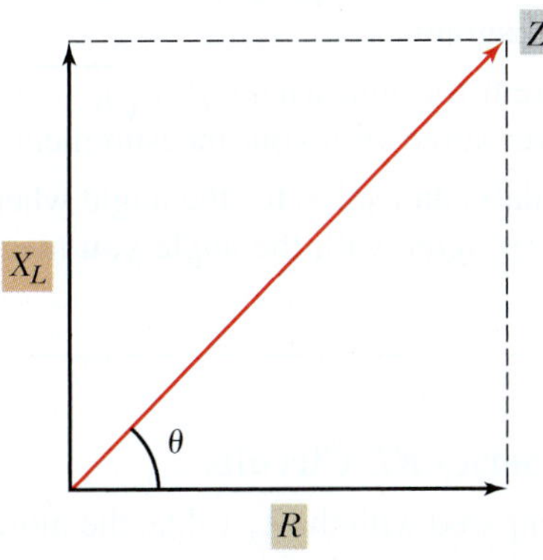

FIGURE 15–22 An impedance diagram

gle between circuit *V* and circuit *I* is between 0° and 90°. Also, as you can see from the diagram, total circuit voltage equals *IZ*, where *Z* is the total opposition to ac current from the combination of resistance and reactance. Since *I* is at 0° and *IZ* is at an angle between 0° and 90°, when we plot *Z*, it also will be between 0° and 90°.

Let's pursue these thoughts. Refer to Figure 15–22 as you study the following discussion about how to plot a diagram of *R*, X_L, and *Z* (impedance diagram) for series *RL* circuits.

1. The V_R equals *IR*, and *I* is plotted at the 0° reference vector position. Also, since the resistor voltage is in phase with the current through the resistor, *R* is plotted at the 0° reference position when plotting an impedance diagram of *R*, X_L, and *Z*.

2. Since V_L equals IX_L and V_L is plotted at 90° from the reference vector position, it is logical that X_L is plotted at the 90° position when plotting *R*, X_L, and *Z* on an impedance diagram. Thus, the result of *I* times X_L is plotted at 90°.

3. The V_T equals *IZ*, and V_T is plotted at an angle (depending on the vector resultant of V_R and V_L) between 0° and 90°. Then it is logical to plot *Z* at the same angle in an impedance diagram where V_T is plotted in the voltage diagram of the same series *RL* circuit. In fact, the angle between *Z* and *R* is the same as the circuit phase angle between V_T and I_T.

Knowing all this, how can the techniques to analyze voltages in series *RL* circuits be applied to impedance? As you can see by comparing the diagrams of Figures 15–21 and 15–22, the layout of the impedance diagram looks just like the diagram used in the voltage vector diagram analysis. This implies the techniques used to find impedance *(Z)* are those techniques used to find V_T. Recall that these methods include:

1. Draw to scale and measure to find magnitude and direction.
2. Apply the Pythagorean theorem to find magnitudes.
3. Apply trig functions to find angles or magnitudes.

EXAMPLE Figure 15–23 illustrates the drawing-to-scale approach. Figure 15–24 shows the Pythagorean theorem version used for impedance solutions as shown in Formula 15–3.

FORMULA 15–3 $Z = \sqrt{R^2 + X_L^2}$

Finally, Figure 15–25 solves for the angle using the simple trig technique. Now you try applying these techniques in the following situation.

PRACTICE PROBLEMS 7

1. Draw the schematic diagram of a series *RL* circuit containing 15 Ω of resistance and 20 Ω of inductive reactance.
2. Draw an impedance diagram for the circuit described in question 1 (to scale). Measure and record the impedance value. Use a protractor and measure the angle between *R* and *Z*. Label this angle on the diagram.
3. Use the Pythagorean theorem formula where $Z = \sqrt{R^2 + X_L^2}$ and calculate and record the *Z* value. Does the answer agree with your measurement results?
4. Use appropriate trig formula(s) and solve for the angle where *Z* should be plotted. Show your work. Does this answer agree with the angle you measured using the protractor in question 2?

Summary of Analyses of Series *RL* Circuits

1. The larger the *R* value compared with the X_L value, the more resistive the circuit will act. Conversely, the larger the X_L compared with the *R*, the more inductive the circuit will act.
2. The more resistive the circuit, the closer the circuit phase angle is to 0°. The more inductive the circuit, the closer the circuit phase angle is to 90° because circuit voltage is lead-

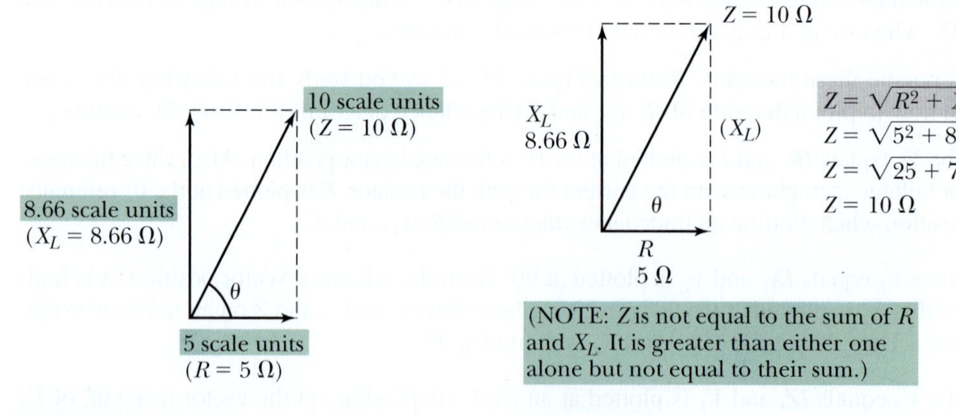

FIGURE 15–23

FIGURE 15–24

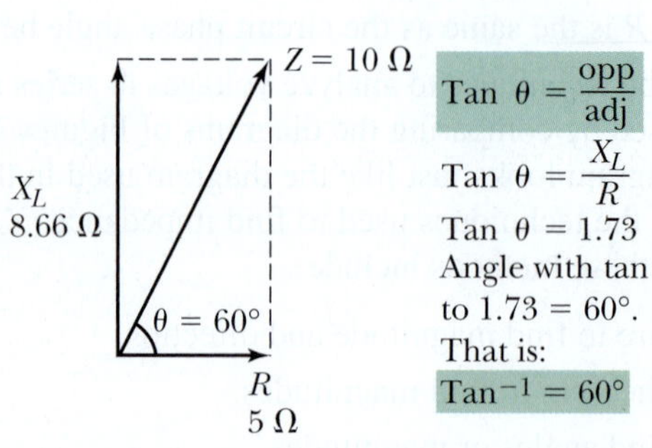

$Z = 10\ \Omega$

X_L 8.66 Ω

$\theta = 60°$

R 5 Ω

$\text{Tan } \theta = \frac{\text{opp}}{\text{adj}}$

$\text{Tan } \theta = \frac{X_L}{R}$

$\text{Tan } \theta = 1.73$

Angle with tan equal to 1.73 = 60°.

That is:

$\text{Tan}^{-1} = 60°$

FIGURE 15–25

ing circuit current, which is the 0° reference vector in series circuits. Recall in a purely resistive series circuit, phase angle equals 0°. In a purely inductive series circuit, phase angle equals 90°. Refer to Figure 15–26 for some examples of R, X_L, Z, V_R, V_L, V_T, and phase angle (θ) relationships for series *RL* circuits.

3. Since current (I) is the same through all parts of a series circuit, the ratio of R and X_L is the same as the relationship (or ratio) of V_R and V_L. That is, $V_R = IR$ and $V_L = IX_L$. For example, when the R value is 2 times the X_L value, the V_R value is twice the V_L value.
4. Total circuit impedance is the result of both factors, resistance and reactance. Impedance is greater than each factor but *not* equal to their sum. Right-triangle methods are used to find its value. Refer again to Figure 15–24.
5. Total circuit voltage is the vector resultant of the resistive and reactive voltages in series. This voltage is greater than each factor but *not* equal to their sum. Right-triangle methods are used. Refer again to Figure 15–17.
6. Phase angle for series *RL* circuits is found using sine, cosine, or tangent functions. For example:

$$\sin\theta = \frac{V_L}{V_T}$$

$$\cos\theta = \frac{V_R}{V_T}$$

$$\tan\theta = \frac{V_L}{V_R}$$

NOTE: The tangent function is useful for many ac circuit problems because it does not require knowing the hypotenuse value.

For the impedance diagram:

$$\sin\theta = \frac{X_L}{Z}$$

$$\cos\theta = \frac{R}{Z}$$

$$\tan\theta = \frac{X_L}{R}$$

7. A greater R or X_L value results in greater circuit impedance and lower current value for a given applied voltage.
8. The circuit current value is found by using Ohm's law, where

$$I_T = \frac{V_T}{Z_T}$$

R Ω	V_R (IR)	X_L Ω	V_L (IX_L)	ZΩ ($\sqrt{R^2 + X_L^2}$)	V_T (IZ) VOLTS	θ (PHASE ANGLE) Arc Tan of $\frac{X_L}{R}$ $\left(\text{or } \frac{V_L}{V_R}\right)$
10	10 V	5	5 V	11.18	11.18	26.56°
10	10 V	10	10 V	14.14	14.14	45°
5	5 V	10	10 V	11.18	11.18	63.43°
100	100 V	10	10 V	100.5	100.5	5.71°
10	10 V	100	100 V	100.5	100.5	84.29°

Assume an I_T value of one ampere in each case. (This means Vapplied is changing.)

FIGURE 15–26 Examples of parameter relationships in series *RL* circuits

9. *Impedance diagrams can only be drawn and used to analyze series ac circuits!* Also, they can be used to analyze series portions of more complex circuits but *should not* be drawn to analyze parallel circuits.
10. An impedance diagram is not a phasor diagram, since the quantities shown are not sinusoidally varying quantities. However, because Z is the vector resulting from R and X, it can be graphed.

15–5 Fundamental Analysis of Parallel *RL* Circuits

Basically, the same strategies we have been using in the series circuit V, I, and θ solutions can be applied to parallel circuits. That is, we can use vectors, the Pythagorean theorem, and trig functions. The differences are in *which* parameters are used on the diagrams and in the formulas.

Because *voltage* is the common factor in parallel circuits, it becomes the reference vector at the 0° position on our diagrams. Since I_T equals the vector sum of the branch currents in parallel circuits, our diagrams illustrate the branch currents in appropriate right-triangle form. (Recall that in series circuits, current was the reference and voltages are plotted in right-triangle format.)

Another contrast between parallel *RL* circuit analysis and series *RL* circuit analysis is that we *do not* use impedance diagrams to analyze parallel *RL* circuits.

With these facts in mind, let's apply the principles you have learned to some parallel *RL* circuit analysis. Refer to Figure 15–27 as you study the following comments.

1. V_T is at the reference vector position because it is common to both the resistive and the reactive branches.
2. Since I_R (resistive branch current) is in phase with V_R (or V_T), it is also at the 0° position.
3. We know the current through a coil lags the voltage across the coil by 90°. Thus, I_L is lagging V_T (or V_L) by 90°. (Remember in the four-quadrant system that –90° is shown as straight down.)
4. The total current is the vector resultant of the resistive and the reactive branch currents. Thus, it is between 0° and –90°.
5. The higher the resistive branch current relative to the reactive branch current, the more resistive the circuit will act. This means that the phase angle is smaller between V_T and I_T. Also, the greater the inductive branch current relative to the resistive branch current, the more inductive the circuit will act. This means the phase angle is greater between V_T and I_T.

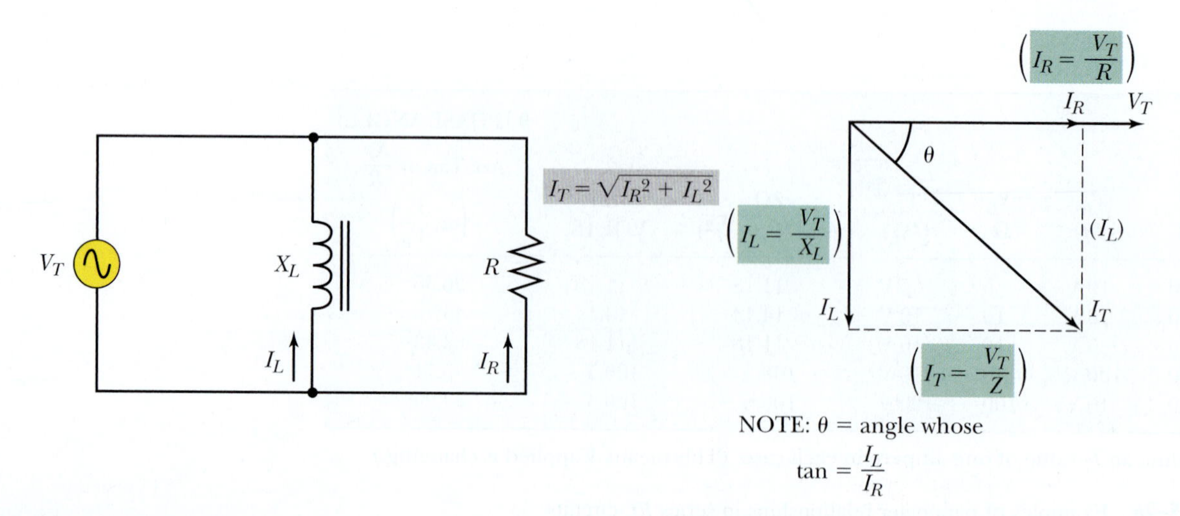

FIGURE 15–27 Parallel *RL* circuit analysis

6. In a parallel *RL* circuit, total impedance is the resultant opposition offered to current by the combined circuit resistance and reactance. That is,

$$Z = \frac{V_T}{I_T}$$

Because of the phase relationships in a circuit with both resistance and reactance, the impedance is less than the impedance of any one branch. However, the shortcut methods used in purely resistive or reactive circuits will not work. For example, if there are 10 Ω of resistance in parallel with 10 Ω of reactance, the impedance will *not* be 5 Ω but 7.07 Ω. This is because the total current is not the simple arithmetic sum of the branch currents, as it is in purely resistive or reactive circuits. Rather, total current is the vector sum.

7. Branch currents are found using Ohm's law, where

$$I_R = \frac{V_R}{R} \text{ and } I_L = \frac{V_L}{X_L}$$

8. Total current is found using the Pythagorean formula, where:

FORMULA 15–4 $I_T = \sqrt{I_R^2 + I_L^2}$

EXAMPLE Parameters for the circuit in Figure 15–28 are $V_T = 150$ V, $R = 50$ Ω, and $X_L = 30$ Ω. Find I_T, Z, and the circuit phase angle (θ).

1. Label the known values on the diagram, Figure 15–29.
2. Determine the branch currents:

$$I_R = \frac{V_T}{R} = \frac{150 \text{ V}}{50\ \Omega} = 3 \text{ A}$$

$$I_L = \frac{V_T}{X_L} = \frac{150 \text{ V}}{30\ \Omega} = 5 \text{ A}$$

3. Draw a current vector diagram and label knowns, Figure 15–30.

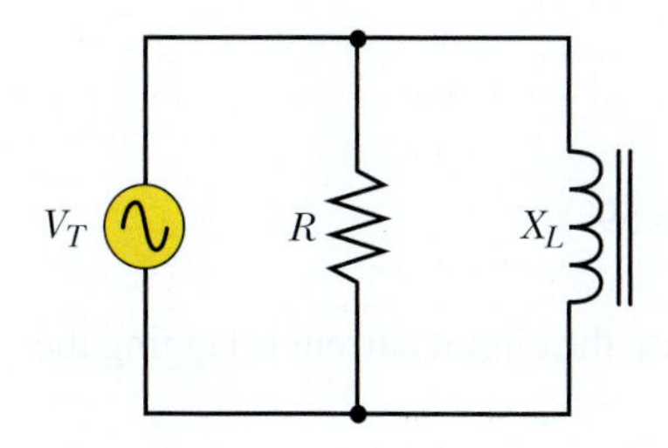

FIGURE 15–28

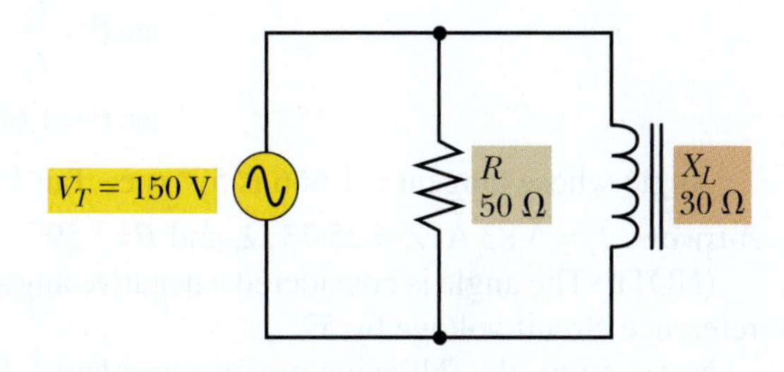

FIGURE 15–29 multiSIM

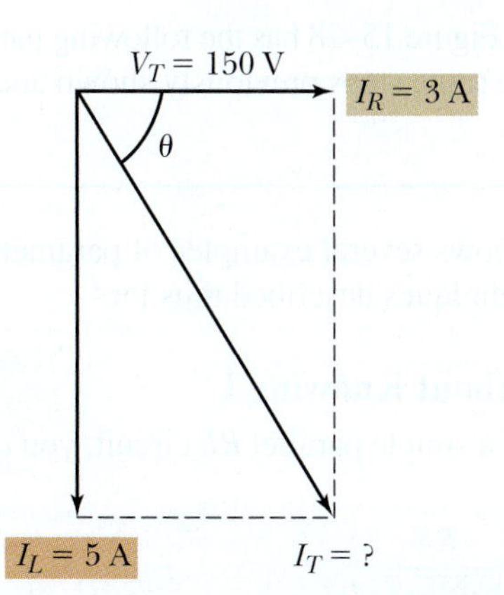

FIGURE 15–30

R Ω	I_R $\left(\frac{V_T}{R}\right)$	X_L Ω	I_L $\left(\frac{V_T}{X_L}\right)$	I_T $(\sqrt{I_R^2 + I_L^2})$	Z $\left(\frac{V_T}{I_T}\right)$	θ (PHASE ANGLE) Arc Tan of $\frac{I_L}{I_R}$
10	5 A	5	10 A	11.18 A	4.47 Ω	−63.43°
10	5 A	10	5 A	7.07 A	7.07 Ω	−45°
5	10 A	10	5 A	11.18 A	4.47 Ω	−26.56°
100	0.5 A	10	5 A	5.02 A	9.96 Ω	−84.29°
10	5 A	100	0.5 A	5.02 A	9.96 Ω	−5.7°
						NOTE: The larger the (−) phase angle, the more inductive the circuit is acting.

Assume a voltage applied (V_T) of 50 V for each case. (This means I_T will be changing.)

FIGURE 15–31 Examples of parameter relationships in parallel *RL* circuits

4. Use the Pythagorean theorem and solve for I_T:

$$I_T = \sqrt{I_R^2 + I_L^2}$$
$$I_T = \sqrt{9 + 25}$$
$$I_T = \sqrt{34}$$
$$I_T = 5.83\text{ A}$$

5. Use Ohm's law and solve for *Z:*

$$Z = \frac{V_T}{I_T}$$
$$Z = \frac{150\text{ V}}{5.83\text{ A}}$$
$$Z = 25.73\ \Omega$$

6. Use the trig formula and solve for the magnitude of the phase angle:

$$\tan\theta = \frac{I_L}{I_R}$$
$$\tan\theta \approx 1.666$$

Angle whose tangent = 1.666 is 59° (i.e., Tan^{-1} 1.666 = 59°)

Answers: I_T = 5.83 A, Z = 25.73 Ω, and θ = −59°

(NOTE: The angle is considered a negative angle because the circuit current is lagging the reference circuit voltage by 59°.)

Now you try the following practice problem.

PRACTICE PROBLEMS 8

A circuit similar to the one in Figure 15–28 has the following parameters: V_T = 300 V, R = 60 Ω, and X_L = 100 Ω. Perform the same steps previously shown and find (1) I_T, (2) *Z,* and (3) the phase angle *(θ).*

The chart in Figure 15–31 shows several examples of parameter relationships in parallel *RL* circuits using calculation techniques described thus far.

A Formula to Find Z without Knowing I

To find the impedance *(Z)* of a simple parallel *RL* circuit, you can use the following formula:

FORMULA 15–5 $Z = \dfrac{RX_L}{\sqrt{R^2 + X_L^2}}$

EXAMPLE Using the same values of R and X_L just used in our preceding example:

$$Z = \frac{RX_L}{\sqrt{R^2 + X_L^2}} = \frac{50 \times 30}{\sqrt{50^2 + 30^2}} = \frac{1,500}{\sqrt{3,400}} = 25.72\ \Omega$$

As you can see, it checks out with our previous solution in which we used Ohm's law to solve for Z. Now you try one using this Z formula technique.

PRACTICE PROBLEMS 9

Assume a simple parallel *RL* circuit in which the resistance equals 47 Ω and the inductive reactance equals 100 Ω. What is the impedance value of this circuit?

Another Method for Finding Parallel *RL* Circuit Parameters

Recall that a conductance-related method for calculating the circuit parameters in purely resistive *parallel circuits* was introduced in an earlier chapter in the text. Another method, somewhat similar to the conductance approach, can also be used for determining circuit parameters in *parallel RL ac circuits.*

Conductance

As you will remember, *conductance* is the ease with which current can flow through a resistive circuit, and is the reciprocal of a circuit's opposition to current. The symbol for conductance is G. The unit of measure for conductance is siemens. The formula to find conductance value is:

FORMULA 15–6 $G = \frac{1}{R}$

where: **G** is conductance in siemens
R is resistance in ohms

Inductive Susceptance

A similar concept to that of resistive component conductance (G) is the concept of a reactive component's *susceptance.* For inductors in ac circuits this is called *inductive susceptance.* The general symbol for *susceptance* is B. The specific symbol for inductive susceptance is B_L. The unit of measure for susceptance is the siemens. The formula to find inductive susceptance is:

FORMULA 15–7 $B_L = \frac{1}{X_L}$

where: X_L is inductive reactance in ohms
B_L is inductive susceptance in siemens

As you can see, inductive susceptance is the reciprocal of inductive reactance, even as conductance is the reciprocal of resistance.

Admittance

When an ac circuit is composed of both resistive and reactive components, it is called *admittance.* In parallel circuits, this is the vector resultant of both conductance and susceptance. The symbol for admittance is Y. The unit used for admittance is also the siemens.

The basic formula for admittance is:

FORMULA 15–8 $Y = \frac{1}{Z}$

From this formula you see that admittance is the reciprocal of impedance. It follows then that impedance is the reciprocal of admittance ($Z = 1/Y$). Other Ohm's law formulas derived from

FIGURE 15–32

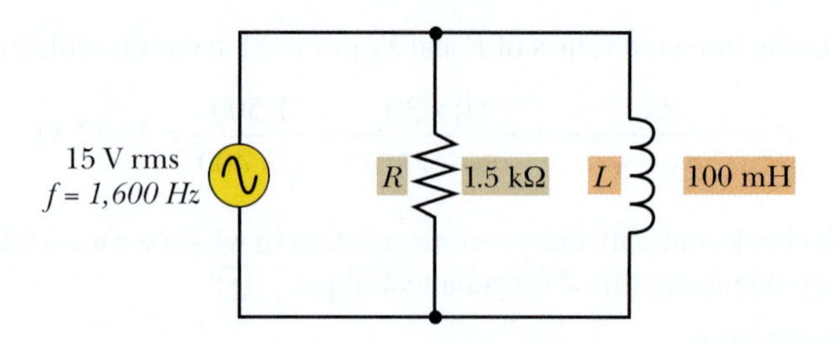

these admittance formulas include: To find total circuit current: ($I = VY$) and to find circuit voltage: ($V = I/Y$).

As you know, in *series* ac circuits containing both resistance and reactance, circuit impedance is a vector resultant of R and X. Likewise, admittance for *parallel* ac circuits containing both resistive and reactive branches is computed as the vector resultant of resistive branch conductance (G) and reactive branch susceptance. In a parallel RL circuit, then:

FORMULA 15–9 $Y = \sqrt{G^2 + B_L{}^2}$

EXAMPLE Calculate the conductance, susceptance, admittance, impedance, and total current for the circuit of Figure 15–32.

Conductance: of the R branch, (G) = $1/R$ = 1/1500 = 0.000666 S or 0.666 mS
X_L: of the inductive branch = $2\pi fL$ = 6.28 × 1600 × 0.1 = 1005 ohms, therefore:
Inductive susceptance: $B_L = 1/X_L$ = 1/1005 ohms = 0.000995 S or 0.995 mS

Admittance: $(Y_{tot}) = \sqrt{G^2 + B_L{}^2} = \sqrt{(0.666 \text{ mS})^2 + (0.995 \text{ mS})^2} = 1.197 \text{ mS}$

Using the Ohm's law formulas, from the admittance data:

$$Z = 1/Y = 1/1.197 \text{ mS} = \text{approximately } 835 \text{ ohms}$$
$$I_T = VY \text{ or } IT = 15 \text{ V} \times 1.197 \text{ mS} = 17.95 \text{ mA}$$
$$V = I/Y = 17.95 \text{ mA}/1.197 \text{ mS} = 14.99 \text{ V}$$
$$Y = I/V = 17.95 \text{ mA}/15 \text{ V} = 1.197 \text{ mS}$$

Comparing these results with the ones that we can get by using the approaches discussed earlier in the chapter:

$$I_R = V/R = 15 \text{ V}/1500 \text{ ohms} = 10 \text{ mA}$$
$$I_L = V/X_L = 15 \text{ V}/1005 \text{ ohms} = 14.93 \text{ mA}$$
$$I_T = \sqrt{I_R{}^2 + I_L{}^2} = \sqrt{(10 \text{ mA})^2 + (14.93 \text{ mA})^2} = 17.96 \text{ mA}$$
$$Z = V/I_T = 15 \text{ V}/17.96 \text{ mA} = 835 \text{ ohms}$$
$$V = I \times Z = 17.96 \text{ mA} \times 0.835 \text{ k}\Omega = 14.99 \text{ V}$$

As you can see, the answers from the two approaches correlate well!

Another Method to Find Phase Angle (θ) for Parallel RL Circuits

Recall that we have been solving for the magnitude of phase angles in parallel RL circuits by using the arc tangent function ($\tan^{-1}$). We use the inductive branch current to represent the opposite side and the resistive branch current to represent the adjacent side for the tangent function formula. So, to find the magnitude of the phase angle value, we have been using: $\theta(\text{magnitude}) = \tan^{-1}(I_{XL}/I_R)$. Since the current through each branch of a parallel circuit is inversely related to the opposition of that branch to current, the result of R over X_L, that is R/X_L turns out to be the same result as the I_{XL} over I_R. This means that the arc tangent ($\tan^{-1}$) of R/X_L will also give us the magnitude of the circuit phase angle.

FORMULA 15–10 $\theta(\text{magnitude}) = \tan^{-1}\dfrac{R}{X_L}$

Equivalent series *RL* circuit *Z* diagram:

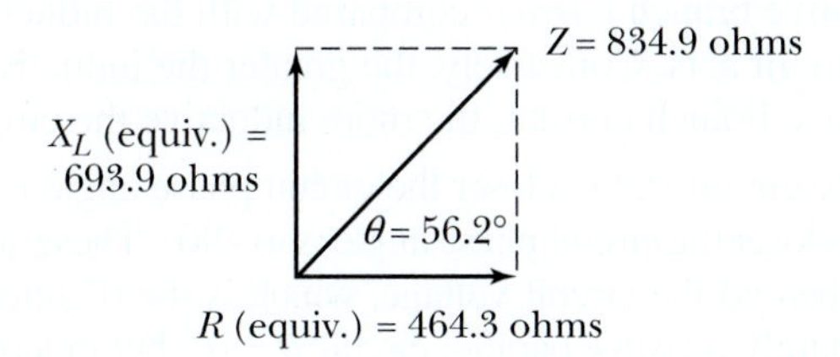

FIGURE 15–33

In our previous example: I_{XL}/I_R = 14.93 mA/10 mA = 1.49 and the $\tan^{-1}$ of I_{XL}/I_R = 56.18°, *which also represents the angle between circuit voltage and total current. In this case, the voltage leads circuit current 56.18°.*

R/X_L = 1500 Ω/1005 Ω = 1.49 and the $\tan^{-1}$ of R/X_L = 56.18°

Conversion from a Parallel *RL* Circuit to an Equivalent Series *RL* Circuit

Objective of conversion:

To find a series *RL* circuit whose parameters will present the same impedance and magnitude of phase angle to the circuit source as the parallel *RL* circuit.

Step 1: Find the *Z* and *θ* of the parallel circuit.

Step 2: Construct the equivalent series circuit *Z* diagram, where:

Equivalent series circuit *R* = parallel circuit *Z* × cos *θ* and the equivalent series circuit X_L = parallel circuit *Z* × sin *θ*.

EXAMPLE Using the parameters from our previous parallel *RL* circuit admittance calculations (where the parallel *RL* circuit *Z* = 835 ohms and the parallel *RL* circuit magnitude of phase angle (*θ*) = 56.18°): (see Figure 15–33)

Equivalent series *RL* circuit *R* = parallel circuit *Z* × cos 56.18°
R = 835 ohms × 0.556
R = 464.3 ohms

Equivalent series *RL* circuit X_L = parallel circuit *Z* × sin 56.18°
X_L = 835 ohms × 0.831
X_L = 693.9 ohms

Verification that conversion works to present a series circuit having the same *Z* and magnitude of phase angle (*θ*) as the parallel *RL* circuit.

Using series *RC* circuit analysis:

$$Z = \sqrt{R^2 + X_L^{\,2}} = \sqrt{464.3^2 + 693.9^2} = 834.9 \text{ohms}$$

θ (equiv. series *RL* circuit) = $\tan^{-1}$ of X_L/R = $\tan^{-1}$ 693.9 Ω/464.3 Ω = 56.2°

PRACTICE PROBLEMS 10

Referring to Figure 15–28 (where *R* = 60 Ω, X_L = 100 Ω, V_T = 300V):

1. Determine the value of conductance (*G*) in this circuit.
2. Determine the value of inductive susceptance (B_L) in this circuit.
3. Determine the value of circuit admittance (*Y*).
4. Determine the value of *Z* for this circuit using the *Z* = 1/*Y* formula.
5. Determine the value of I_T for this circuit, using the I_T = *V* × *Y* formula.
6. Confirm the value of V_T using the *V* = *I*/*Y* formula.
7. Confirm the value of *Y* using the *Y* = *I*/*V* formula.
8. Use the R/X_L approach to solve for the magnitude of the circuit phase angle.

Summary of Analysis of Parallel *RL* Circuits

1. The greater the resistive branch *current* compared with the inductive branch *current,* the more resistive the circuit acts. Conversely, the greater the inductive branch *current* compared with the resistive branch current, the more inductive the circuit acts.
2. The more resistive the circuit is, the closer the circuit phase angle is to 0°. The more inductive the circuit is, the closer the circuit phase angle is to –90°. These situations result from circuit current lagging behind the circuit voltage, which is the 0° reference vector in parallel circuits. Recall in a purely *resistive* parallel circuit $\theta = 0°$, but in a purely inductive *parallel* circuit $\theta = -90°$. Refer to Figure 15–31 to see some examples of the relationships of R, X_L, Z, I_R, I_L, I_T, and phase angle (θ) with selected circuit parameters purposely varied.
3. Branch current is *inverse* to the resistance or reactance of a given branch. For example, when R is two times X_L, current through the *resistive* branch is half the current through the *reactive* branch.
4. Total circuit impedance of a given parallel *RL* circuit is less than the opposition of any one branch. However, impedance cannot be found with product-over-the-sum or reciprocal formulas used in dc circuits. Neither can impedance be found using an impedance diagram approach. Ohm's law, admittance, or other methods can be used.
5. Total circuit current is the vector resultant of the resistive and reactive branch currents. Current is greater than any one branch current but *not* equal to their sum. Right-triangle methods are used. Refer again to Figure 15–27.
6. Phase angle for parallel *RL* circuits is found using sine, cosine, or tangent trig functions. For example:

$$\sin\theta = \frac{I_L}{I_T}$$

$$\cos\theta = \frac{I_R}{I_T}$$

$$\tan\theta = \frac{I_L}{I_R}$$

   NOTE: The tangent function is useful for many ac circuit problems because it does not require knowing the hypotenuse value.

7. A greater R or X_L value results in greater circuit impedance and lower circuit current value for a given applied voltage.
8. The circuit impedance value is found by using Ohm's law, or admittance, where

$$Z_T = \frac{V_T}{I_T} \text{ or } Z = \frac{1}{Y}$$

9. *Impedance diagrams can only be drawn and used to analyze series ac circuits!* Also, they can be used to analyze series portions of more complex circuits but *SHOULD NOT* be drawn to analyze parallel circuits. Obviously, since the impedance of a parallel *RL* circuit is smaller than any one branch, the hypotenuse cannot be shorter than one of the sides when using the right-triangle techniques. Therefore, an impedance diagram cannot be drawn for parallel *RL* circuits.

15–6 Examples of Practical Applications for Inductors and *RL* Circuits

Inductors and inductances have a variety of applications. Some of these applications use the property of inductance or mutual inductance. Other applications involve using inductance with resistance or capacitance, or both. Inductance can be used in circuits operating at powerline (low) frequencies; circuits operating in the audio-frequency range; and circuits operating at higher frequencies, known as radio frequencies.

The following is a brief list of applications.

1. Power-frequency applications, Figure 15–34.

  a. Power transformers

  b. Filter circuits (both powerline and power-supply filters)

  c. Electromagnets, solenoids, relays

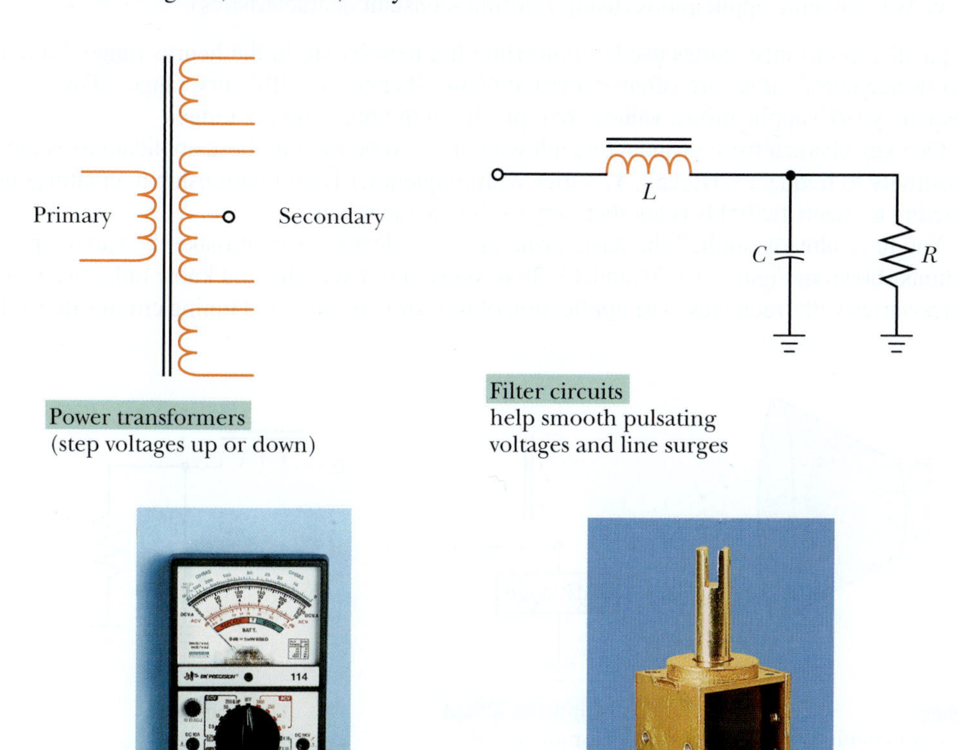

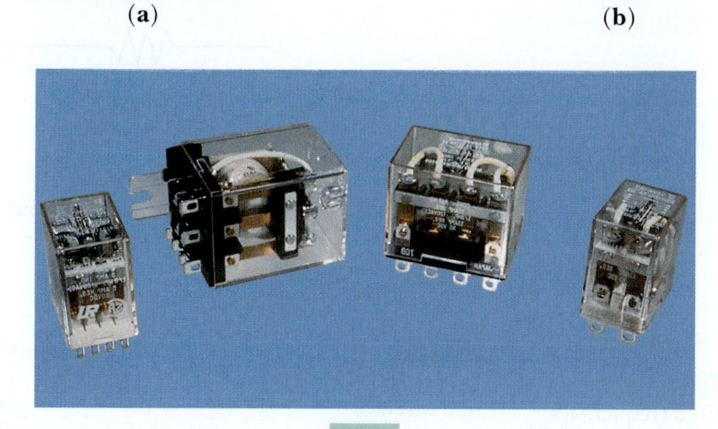

FIGURE 15–34 Examples of applications at low ac (or pulsating dc) frequencies *(Photo **a** courtesy of B & K Precision; photos **b** and **c** courtesy of Guardian Electric Mfg. Co.)*

2. Audio-frequency applications, Figure 15–35.
   a. Audio transformers (used for coupling and Z-matching)
   b. Audio chokes (as loads for amplifiers)
   c. Filter circuits
3. Radio-frequency applications, Figure 15–36.
   a. Tuning circuit applications
   b. Radio frequency chokes/filters
   c. Waveshaping applications (using L/R time-constant characteristics)

Typically, inductance values used at powerline frequencies are in the henrys range. For audio frequencies, values are often in the tenths of a henry, or millihenry ranges. For radio-frequency *(RF)* applications, values are typically in the microhenrys range.

One key characteristic of inductors allowing them to be used in these applications is their sensitivity to frequency. (Recall, X_L varies with frequency.) Their relationship to creating and reacting to magnetic fields is another very useful characteristic.

You have already studied the basic principles of inductors, which make the varied applications shown in Figures 15–35 and 15–36 possible. For example, you know inductive reactance varies with frequency. The application of inductors in filter and tuning circuits depends

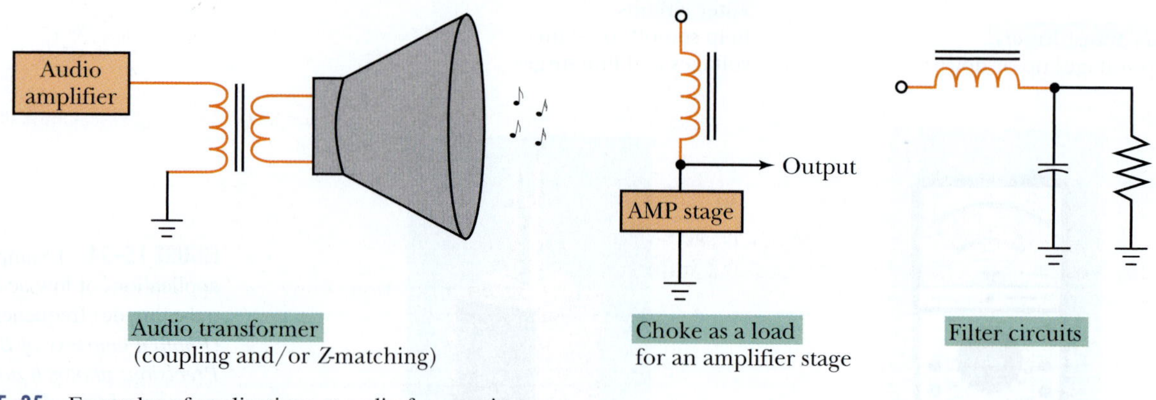

FIGURE 15–35 Examples of applications at audio frequencies

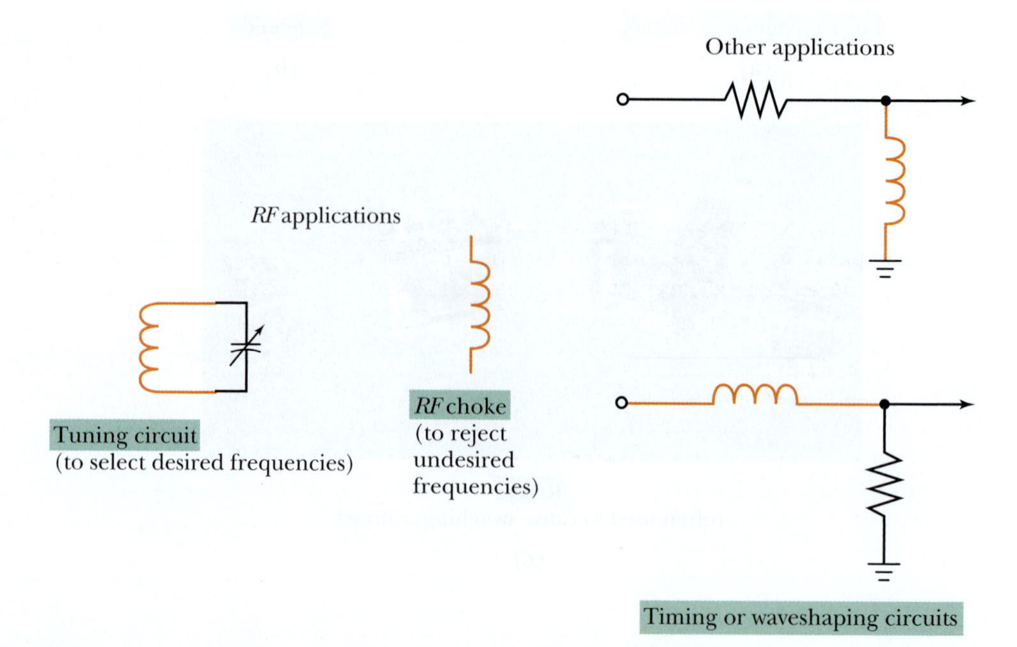

FIGURE 15–36 Examples of radio-frequency *(RF)* and other applications

on this principle. Also, you studied the principle for waveshaping when you studied L/R time constants.

As you can see, inductors, the property of inductance, and inductive reactance are valuable tools in the field of electronics. Your understanding of inductors will be of great help in your career.

Summary

- Phase angle is the difference in phase between the circuit applied voltage (V_T) and the circuit current (I_T).
- Circuit current and voltage are in phase in purely resistive ac circuits; 90° out of phase in purely inductive circuits; and between 0° and 90° out of phase in circuits containing both resistance and inductance.
- For *series* ac circuits containing both resistance and inductive reactance, total circuit voltage and impedance cannot be found by adding voltage values or oppositions to current, as is done in dc circuits. Rather, values must be determined by vectors or right-triangle techniques.
- For voltage-current vector diagrams relating to *series RL* circuits, *current* is the *reference vector* positioned at the 0° location since it is common throughout the circuit. Resistive voltage drop(s) are also at the 0° position, since they are in phase with current. Inductive voltages are shown at 90° (since V_L leads I by 90°). Total circuit voltage is plotted as the vector resultant of the resistive and inductive voltages.
- For impedance diagrams (used *only* for *series* circuits), the R values are plotted at the 0° reference position. The X_L value is plotted vertical (up) at 90° (assuming a perfect inductor). Impedance is plotted at a position that is the vector resultant of the R and X_L vectors.
- The Pythagorean theorem applied to *series RL* circuits is used to calculate the total voltage value where $V_T = \sqrt{V_R^2 + V_L^2}$. It is also useful to determine the impedance (Z) value where $Z = \sqrt{R^2 + X_L^2}$.
- The trig functions of sine, cosine, and tangent are used to determine phase angle, if two sides are known. Also, these functions can be used to determine voltage and impedance values of *series RL* circuits, if the phase angle and one side are known.
- For *parallel* ac circuits containing both resistance and inductive reactance, total circuit current cannot be found by adding branch currents, as is done in dc circuits. Right-triangle methods are again used.
- For voltage-current vector diagrams relating to *parallel RL* circuits, *voltage* is the *reference vector* at the 0° location since it is common throughout the circuit. Resistive branch current(s) are also at the 0° position, since they are in phase with the voltage. Inductive branch currents are at –90° since I_L lags V by 90°. Total circuit current is plotted as the vector resultant of the resistive and inductive branch currents.
- The Pythagorean theorem applied to *parallel RL* circuits is used to calculate the total current value where $I_T = \sqrt{I_R^2 + I_L^2}$.
- A formula that may be used to find Z in a simple parallel *RL* circuit is:

$$Z = \frac{RX_L}{\sqrt{R^2 + X_L^2}}$$

- The trig functions (sine, cosine, and tangent) are also used to determine phase angle in parallel circuits, as well as series circuits, if two sides are known. They can be used to determine current values in *parallel RL* circuits, if the phase angle and one side are known.
- Plotting vectors to scale and then measuring lengths and angles is one method to determine parameters of interest in *RL* circuits. For series *RL* circuits, the resistive and inductive voltage values are plotted to find V_T and θ. For parallel *RL* circuits, the resistive and inductive branch current values are plotted to find I_T and θ.
- Impedance is the total opposition offered to ac current by a combination of resistance and reactance. The symbol for impedance is "Z," and the unit of impedance is the ohm.
- Inductors, either alone or in conjunction with other components (R or C), are used in many applications. General uses include transformers, tuning circuits, filter circuits, and waveshaping circuits.

Formulas and Sample Calculator Sequences

FORMULA 15–1
(To find the value of the hypotenuse, squared)

$c^2 = a^2 + b^2$

a value, [x^2], [+], b value, [x^2], [=]

FORMULA 15–2
(To find the value of total voltage in a series RL circuit)

$V_T = \sqrt{V_R^2 + V_L^2}$

V_R value, [x^2], [+], V_L value, [x^2], [=], [$\sqrt{x}$]

FORMULA 15–3
(To find the value of circuit impedance in a series RL circuit)

$Z = \sqrt{R^2 + X_L^2}$ (for series circuits)

R value, [x^2], [+] X_L value, [x^2], [=], [$\sqrt{x}$]

FORMULA 15–4
(To find the value of total circuit current in a parallel RL circuit)

$I_T = \sqrt{I_R^2 + I_L^2}$ (for parallel circuits)

I_R value, [x^2], [+], I_L value, [x^2], [=], [$\sqrt{x}$]

FORMULA 15–5
(To find the value of circuit impedance in a simple parallel RL circuit)

$Z = \dfrac{RX_L}{\sqrt{R^2 + X_L^2}}$ (for parallel circuits)

R value, [×], X_L value, [÷], [(], R value, [x^2], [+], X_L value, [x^2], [)], [$\sqrt{x}$], [=]

Formula 15–6
(To find conductance of a resistive branch)

$G = \dfrac{1}{R}$ (Parallel RL circuits)

R value, [1/x]

Formula 15–7
(To find inductive susceptance of an inductive branch)

$B_L = \dfrac{1}{X_L}$ (Parallel RL circuits)

X_L value, [$\sqrt{x}$]

Formula 15–8
(Ohm's law style formula to find admittance value)

$Y = \dfrac{1}{Z}$

Formula 15–9
(To find admittance of a parallel RL circuit)

$Y = \sqrt{G^2 + B_L^2}$ (Parallel RL circuits)

G value, [x^2], [+], B_C value, [x^2], [=], [$\sqrt{x}$]

Formula 15–10
(To find magnitude of the phase angle using resistance and reactance values)

θ (magnitude) $= \tan^{-1} \dfrac{R}{X_L}$ (Parallel RL circuits)

R value, [÷], X_L value, [=], [2nd], [TAN-1]

EXCEL AUTOMATED FORMULAS

Helpful problem-solving tools, such as the Excel automated formulas available on the CD, allow automatic calculations of formulas.

Using Excel

RL *Circuits in AC* Formulas

(Excel file reference: FOE15_01.xls)

DON'T FORGET! It is not necessary to retype formulas once they are entered on the worksheet! Just input new parameters data for each new problem using that formula, as needed.

- Use the Formula 15–2 spreadsheet sample and solve for the value of total circuit voltage in a series *RL* circuit with the parameters given in Practice Problems 4. Check your answer against the answer given in the Appendix for this problem.
- Use the Formula 15–5 spreadsheet sample and solve for the value of impedance of a simple parallel *RL* circuit having the parameters described in Practice Problems 9. Check your answer against the answer given in the Appendix for this problem.

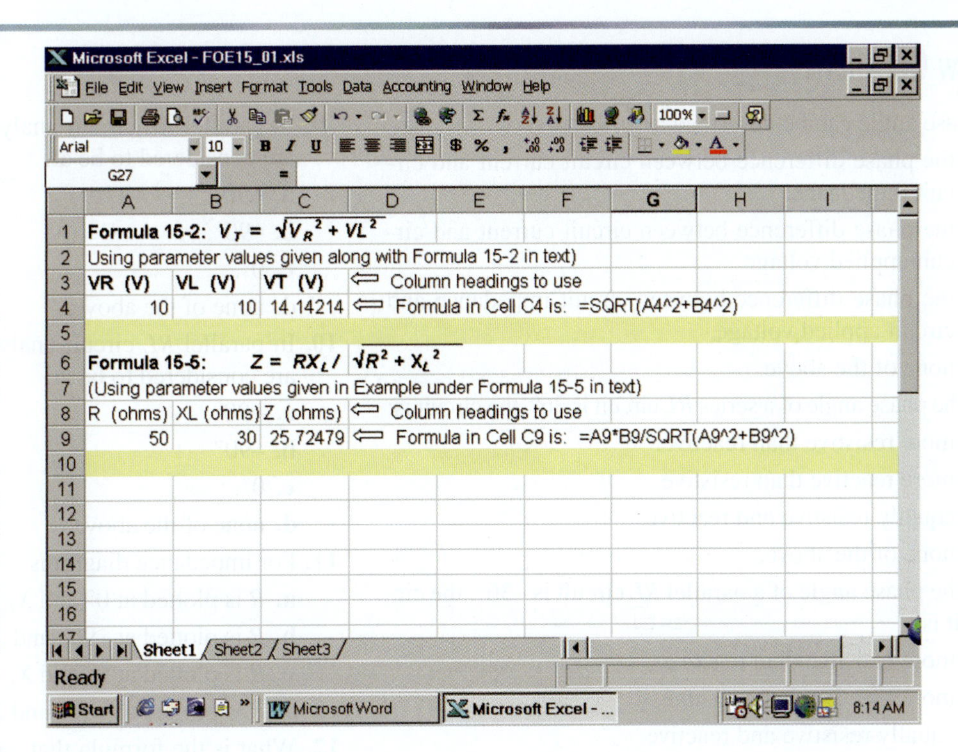

Review Questions

1. Phase angle can be described as:
   - **a.** the phase difference between circuit current and circuit impedance.
   - **b.** the phase difference between circuit current and circuit applied voltage.
   - **c.** the phase difference between circuit impedance and circuit applied voltage.
   - **d.** none of the above.
2. If the phase angle of a series *RL* circuit is 36°, the circuit is:
   - **a.** more resistive than reactive.
   - **b.** more reactive than resistive.
   - **c.** equally resistive and reactive.
   - **d.** none of the above.
3. If the phase angle of a parallel *RL* circuit is –36°, the circuit is:
   - **a.** more resistive than reactive.
   - **b.** more reactive than resistive.
   - **c.** equally resistive and reactive.
   - **d.** none of the above.
4. Impedance vector diagrams should not be used for analyzing:
   - **a.** series *RL* circuits.
   - **b.** parallel *RL* circuits.
   - **c.** series-parallel *RL* circuits.
   - **d.** can be used for all the above circuits.
5. For analyzing series *RL* circuits, the reference vector is:
   - **a.** current.
   - **b.** voltage.
   - **c.** impedance.
   - **d.** none of the above.
6. For analyzing parallel *RL* circuits, the reference vector is:
   - **a.** current.
   - **b.** voltage.
   - **c.** impedance.
   - **d.** none of the above.
7. In series *RL* circuit analysis, resistive voltage drops are considered to be at:
   - **a.** +90°.
   - **b.** –90°.
   - **c.** 0°.
   - **d.** none of the above.
8. In series *RL* circuit analysis, reactive voltage drops are considered to be at:
   - **a.** +90°.
   - **b.** –90°.
   - **c.** 0°.
   - **d.** none of the above.
9. In parallel *RL* circuit analysis, resistive branch currents are considered to be at:
   - **a.** +90°.
   - **b.** –90°.
   - **c.** 0°.
   - **d.** none of the above.
10. In parallel *RL* circuit analysis, reactive branch currents are considered to be at:
    - **a.** +90°.
    - **b.** –90°.
    - **c.** 0°.
    - **d.** none of the above.
11. For impedance diagrams:
    - **a.** R is plotted at 0° and X_L is plotted at –90°.
    - **b.** R is plotted at –90° and X_L is plotted at 0°.
    - **c.** R is plotted at 0° and X_L is plotted at +90°.
    - **d.** R is plotted at +90° and X_L is plotted at 0°.
12. What is the formula that can be used to find V_T in a series *RL* circuit?
13. What is the formula that can be used to find Z in a series *RL* circuit?
14. What is the formula that can be used to find I_T in a parallel *RL* circuit?
15. What is the formula that can be used to find Z for a simple parallel *RL* circuit?
16. Impedance is the:
    - **a.** algebraic sum of resistance and reactance in an ac circuit.
    - **b.** arithmetic sum of resistance and reactance in an ac circuit.
    - **c.** vector sum of resistance and reactance in a series *RL* ac circuit.
    - **d.** vector sum of resistance and reactance in a parallel *RL* ac circuit.
17. For a series *RL* circuit:
    - **a.** the greater the resistance compared with the inductive reactance, the greater the circuit phase angle will be.
    - **b.** the greater the resistance compared with the inductive reactance, the smaller the circuit phase angle will be.
    - **c.** the greater the inductive reactance compared with the resistance, the smaller the circuit phase angle will be.
    - **d.** none of the above.
18. For a parallel *RL* circuit:
    - **a.** the greater the resistance compared with the inductive reactance, the greater the circuit phase angle will be.
    - **b.** the greater the resistance compared with the inductive reactance, the smaller the circuit phase angle will be.
    - **c.** the greater the inductive reactance compared with the resistance, the greater the circuit phase angle will be.
    - **d.** none of the above.

19. For an impedance diagram:

a. $\sin\theta = X_L/Z$.

b. $\sin\theta = R/Z$.

c. $\sin\theta = X_L/R$.

d. none of the above.

20. For a parallel *RL* circuit:

a. $\cos\theta = I_L/I_T$.

b. $\cos\theta = I_L/I_R$.

c. $\cos\theta = I_R/I_T$.

d. none of the above.

Problems

1. A series *RL* circuit contains a resistance of 20 Ω and an inductive reactance of 30 Ω. Find the following parameters using the specified techniques.

a. Use the plot-and-measure technique to find Z. (Round off to the nearest whole number.) Measure the θ with a protractor. (Round off to the nearest degree.)

b. Use the Pythagorean theorem to find Z.

c. Use trig to find the angle between *R* and *Z*.

2. A parallel *RL* circuit has one branch consisting of a 200-Ω resistance, and another branch consisting of an inductor with X_L equal to 150 Ω. The applied voltage is 300 V. Find the values for the following parameters. Draw and label the schematic and vector diagrams.

a. I_R

b. I_L

c. I_T

d. θ

e. Z

3. Draw the circuit diagram, voltage-current vector diagram, and impedance diagram for a series *RL* circuit having one *R* and one *L*. Voltage drop across the 10-kΩ resistor is 5 V, and circuit phase angle is 45°. Label all components and vectors, and show all calculations.

4. Draw the circuit diagram and voltage-current vector diagram of a parallel *RL* circuit having one *R* and one *L*. Assume a total current of 12 A and a phase angle of –75°. Label all components and vectors in the diagrams, and show all calculations. (Round answers to the nearest whole number and the nearest degree values.)

5. Refer to Figure 15–37 and solve for the following:

a. Z

b. V_R

c. θ

6. Refer to Figure 15–38 and solve for the following:

a. X_L

b. V_L

c. V_R

d. θ

7. Refer to Figure 15–39 and solve for the following:

a. I

b. Z

c. V_L

d. X_L

e. L

8. Draw the circuit diagram and all vector diagrams of the series *RL* circuit described below to find and illustrate all voltage, current, impedance, and phase-angle parameters. Show all calculations.

Given: Inductance = 10 H

Voltage leads current by 60°

Frequency = 1 kHz

Total voltage (V_T) = 72.5 V

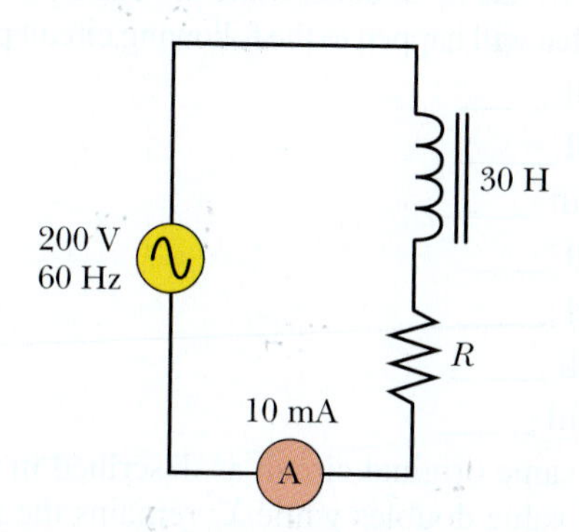

FIGURE 15–38

FIGURE 15–37

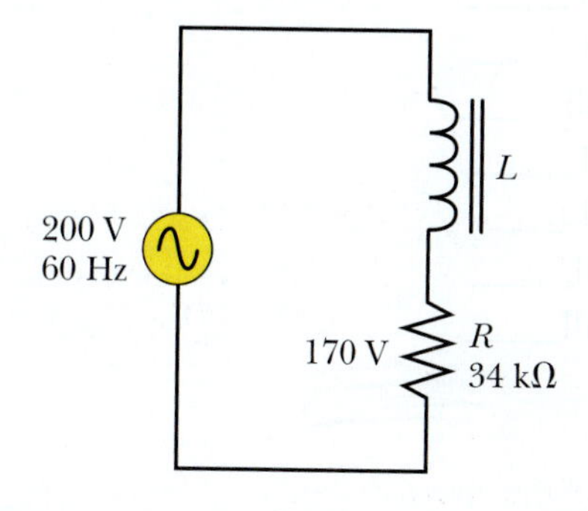

FIGURE 15–39

9. Draw the circuit diagram and all vector diagrams for the parallel RL circuit described below to find and illustrate all voltage, current, and phase-angle parameters. Show all calculations.

   (NOTE: Rounding off is acceptable for this problem.)

   Given: Inductance = 1.99 H

   Frequency = 400 Hz

   Resistance = 7 kΩ

   $V_T = 21$ V

10. Draw the circuit diagram for a series RL circuit containing two 4-mH inductances and two 5-kΩ resistors. The circuit has a frequency of 200 kHz. Find I_T if the applied voltage is 140 V. (Round off to the nearest kilohm, volt, and milliampere, as appropriate.)

Analysis Questions

Answer the following with "I" for increase, "D" for decrease, and "RTS" for remain the same.

1. If the frequency applied to a parallel RL circuit is doubled, and the circuit has one branch consisting of a 200-Ω resistance, and another branch consisting of an inductor with an X_L of 150-Ω, what will happen to the following circuit parameters?
   a. R will _____
   b. L will _____
   c. X_L will _____
   d. Z will _____
   e. θ will _____
   f. I_T will _____
   g. V_T will _____
2. If the inductance in a series RL circuit consisting of a 10-kΩ R and a 10-kΩ X_L doubles while the circuit R remains the same, what will happen to the following circuit parameters?
   a. R will _____
   b. L will _____
   c. X_L will _____
   d. Z will _____
   e. θ will _____
   f. I_T will _____
   g. V_T will _____
3. For the same original circuit as described in question 2, if the R value doubles while X_L remains the same, what will happen to the circuit parameters?
   a. R will _____
   b. L will _____
   c. X_L will _____
   d. Z will _____
   e. θ will _____
   f. I_T will _____
   g. V_T will _____
4. For the parallel RL circuit you drew in Practice Problems, question 4, if inductance doubles while R remains the same, what will happen to the circuit parameters?
   a. R will _____
   b. L will _____
   c. X_L will _____
   d. Z will _____
   e. θ will _____
   f. I_T will _____
   g. V_T will _____
5. For the parallel RL circuit you drew in Practice Problems, question 4, if the R value doubles while X_L remains the same, what will happen to the circuit parameters?
   a. R will _____
   b. L will _____
   c. X_L will _____
   d. Z will _____
   e. θ will _____
   f. I_T will _____
   g. V_T will _____
6. For the parallel RL circuit you drew in Practice Problems, question 4, if both R and X_L values are halved, what will happen to the circuit parameters?
   a. R will _____
   b. L will _____
   c. X_L will _____
   d. Z will _____
   e. θ will _____
   f. I_T will _____
   g. V_T will _____
7. What calculator keystrokes would you use to find the sine of 57°?
8. What calculator keystrokes would you use to find the $\tan^{-1}$ related to a ratio of 0.80?

9. What calculator keystrokes would you use to find the cosine of 34°?
10. What calculator keystrokes would you use to find the arcsine for a ratio of 0.75?
11. What, if any, are the advantages of using the tangent function for ac circuit analysis?
12. How many radians are represented by 75°?

MultiSIM Exercise for *RL* Circuits in AC

1. Use the MultiSIM program to set up a *series RL* circuit with the following parameters: Source = *f* of 1,000 Hz, 50-V rms; $R = 10\ \text{k}\Omega$; and $L = 1.59$ H. Don't forget to connect ground reference at the bottom of the source.
2. Measure and record the values of V_L and V_{R_1}. Double the frequency of the source; then measure and record the V_L and V_{R_1} parameters again.
3. Across which component did the voltage drop increase? Across which component did the voltage drop decrease? Since the source frequency was doubled, did the voltage drop across the component with the increase have twice its previous *V* drop? Explain this finding.
4. Did the phase angle change from the original circuit 45°? If so, is the phase angle greater or less with the 2000-signal?
5. With a dual trace scope simulator, determine the circuit phase angle when the circuit is operating at a frequency of 2000 Hertz. (NOTE: Since circuit has a ground you do not need to use the ground terminal on the scope simulator.)

Performance Projects Correlation Chart

Suggested performance projects in the Laboratory Manual that correlate with topics in this chapter are:

Chapter Topic	Performance Project	Project Number
Fundamental Analysis of Series *RL* Circuits	*V, I, R, Z*, and θ Relationships in a Series *RL* Circuit	43
Fundamental Analysis of Parallel *RL* Circuits	*V, I, R, Z*, and θ Relationships in a Parallel *RL* Circuit	44
	Story Behind the Numbers: *RL* Circuits in AC: Series Circuit	
	Story Behind the Numbers: *RL* Circuits in AC: Parallel Circuit	

NOTE: It is suggested that after completing the above projects, the student should be required to answer the questions in the "Summary" at the end of this section of projects in the Laboratory Manual.

Troubleshooting Challenge

CHALLENGE CIRCUIT 9

(Follow the SIMPLER sequence by referring to inside front cover.)

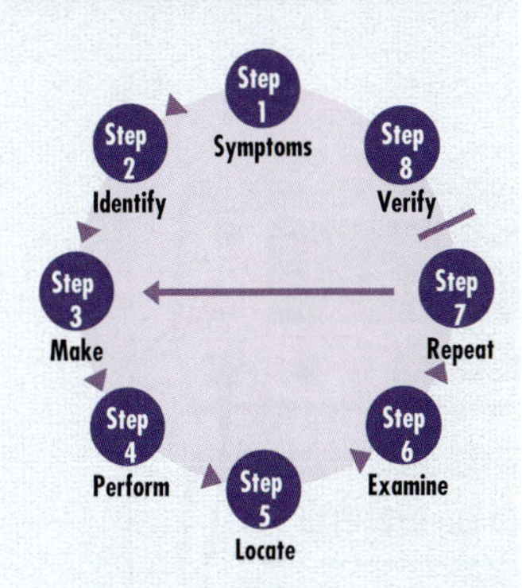

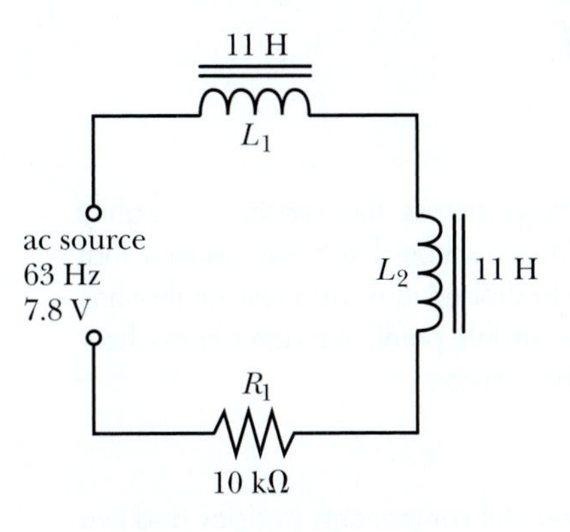

Challenge Circuit 9 multiSIM

STARTING POINT INFORMATION

1. Circuit diagram
2. V_{R_1} is higher than it should be
3. V source checked as being 7.8 V

TEST	Results in Appendix C
V_{R_1}	(34)
V_{L_1}	(89)
V_{L_2}	(48)
R_1	(109)

CHALLENGE CIRCUIT 9

STEP 1

SYMPTOMS The voltage drop across the resistor is slightly higher than it should be. This could be due to an inductor that has decreased in value due to shorted turns or a resistor that has increased or opened since, at this point, we don't know how high the voltage is across the resistor.

STEP 2

IDENTIFY initial suspect area: All components (resistor and two inductors) are in the initial suspect area.

STEP 3

MAKE test decision: Since inductors typically open more frequently than they short and since we don't know how high the voltage is across the resistor, it makes sense to check the voltage across R_1.

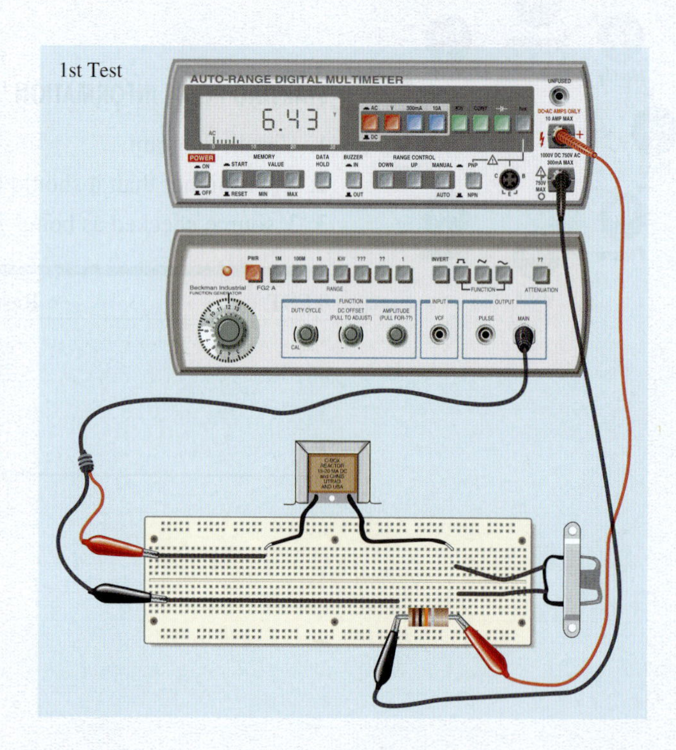

STEP 4

PERFORM **1st Test:** V_{R_1} is 6.4 V. If everything were normal, the resistor would drop 5.9 V, and each inductor would drop about 2.6 V, since their inductive reactances at this frequency would be close to 4.35 kΩ each. The total X_L is about 8.7 kΩ and the rated R is 10 kΩ. Z would then equal about 13.3 kΩ, and circuit current would be approximately 0.59 mA. Thus, the vector sums of the resistive and inductive voltage drops would equal approximately 7.8 V.

STEP 5

LOCATE new suspect area: The new suspect area stays the same as the initial suspect area since we do not have enough information to eliminate the inductors or the resistor.

STEP 6

EXAMINE available data.

STEP 7

REPEAT analysis and testing: It would be good to look at the inductor voltage drops so we can draw some conclusions.
2nd Test: Measure V_{L_1}. V_{L_1} is close to 2.3 volts, which is slightly lower than normal.
3rd Test: Measure V_{L_2}. V_{L_2} is close to 2.36 volts, which is slightly lower than normal.
4th Test: Measure R_1. R_1 is measured with power removed and equals 12 kΩ, which is higher than the 10 kΩ it should be.

STEP 8

VERIFY **5th Test:** Since R_1 resistance is high, replace R_1 with a new 10-kΩ resistor. After replacing the resistor, all the circuit parameters become normal.

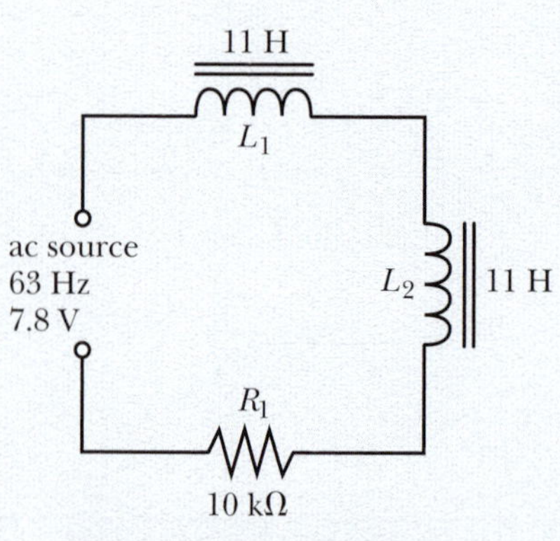

CHALLENGE CIRCUIT 9

multiSIM

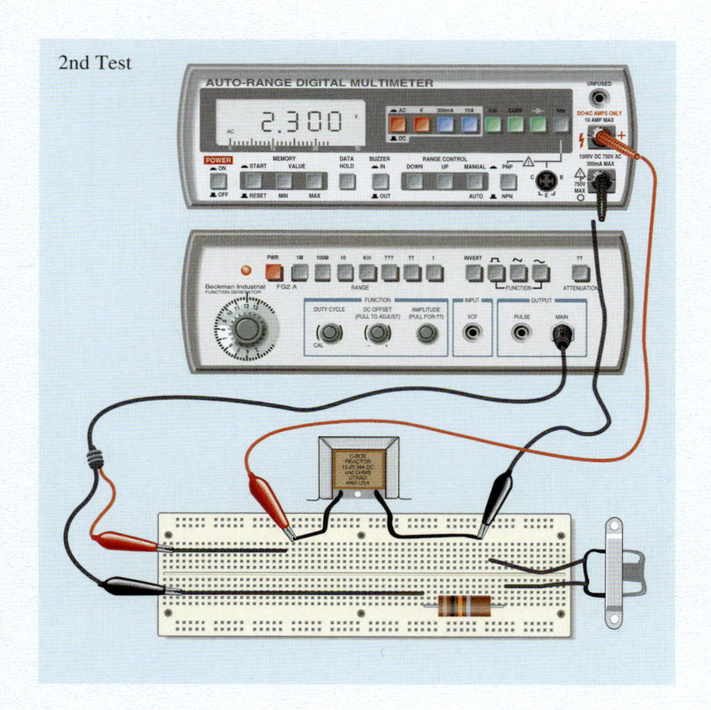
2nd Test
AUTO-RANGE DIGITAL MULTIMETER
2.300
POWER

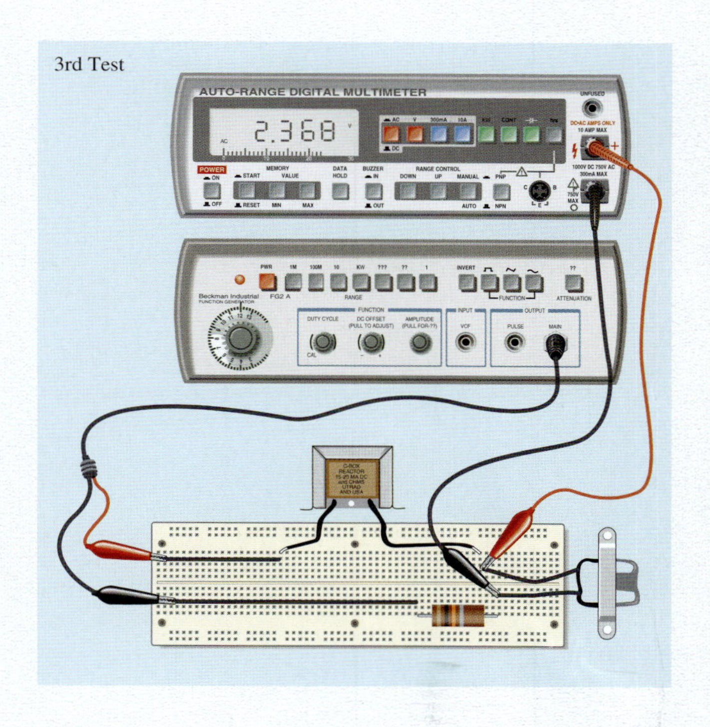
3rd Test
AUTO-RANGE DIGITAL MULTIMETER
2.368
POWER

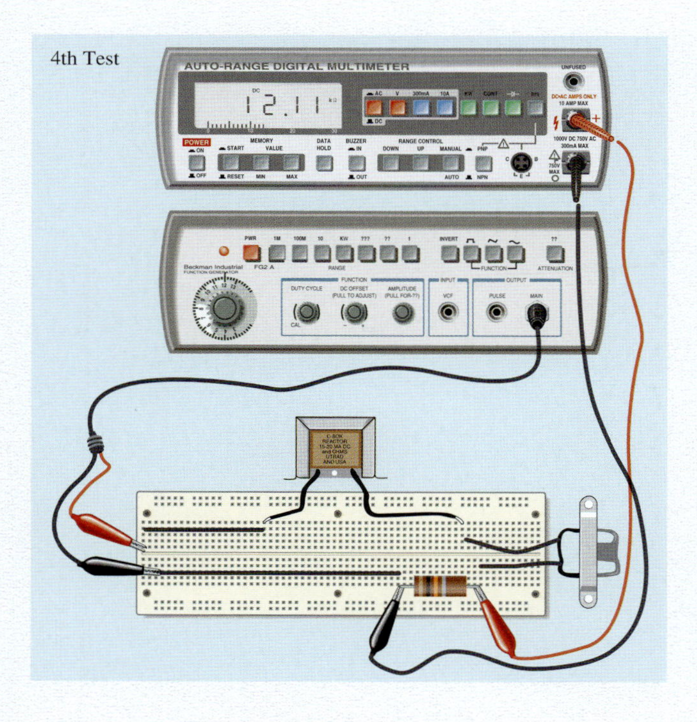
4th Test
AUTO-RANGE DIGITAL MULTIMETER
12.11
POWER

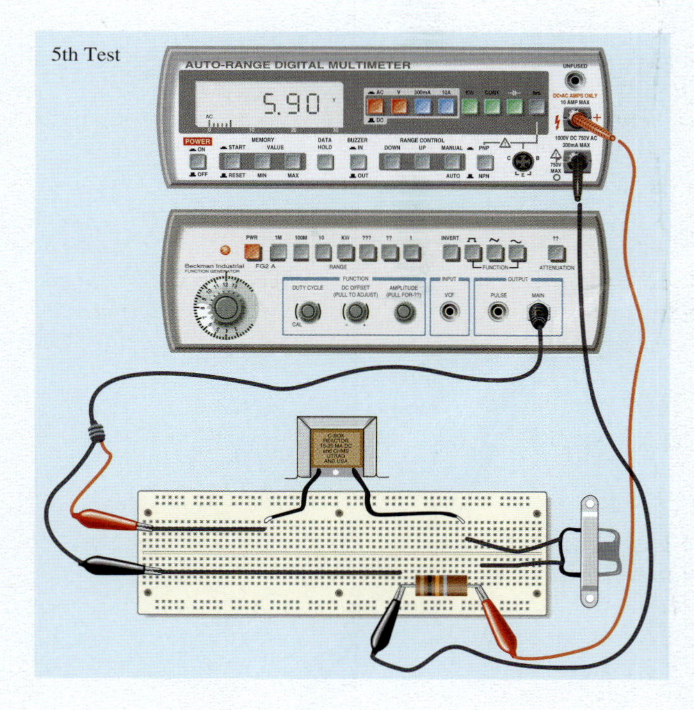
5th Test
AUTO-RANGE DIGITAL MULTIMETER
5.90
POWER

OUTLINE

OBJECTIVES

After studying this chapter, you should be able to:

1. Define **mutual inductance**
2. Calculate mutual inductance values
3. Calculate **coefficient of coupling** values
4. Calculate **turns, voltage, current,** and **impedance ratios**
5. List, draw, or explain physical, magnetic, electrical, and schematic characteristics of various **transformers**
6. List common transformer color codes
7. Define at least two types of **core losses**
8. List common problems found in transformers
9. List troubleshooting procedures
10. Use the computer to solve circuit problems

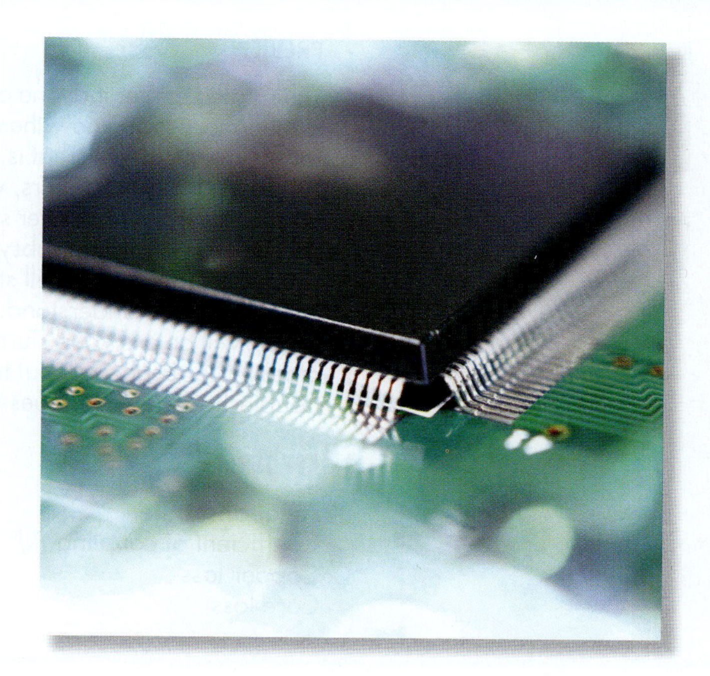

CHAPTER 16

Basic Transformer Characteristics

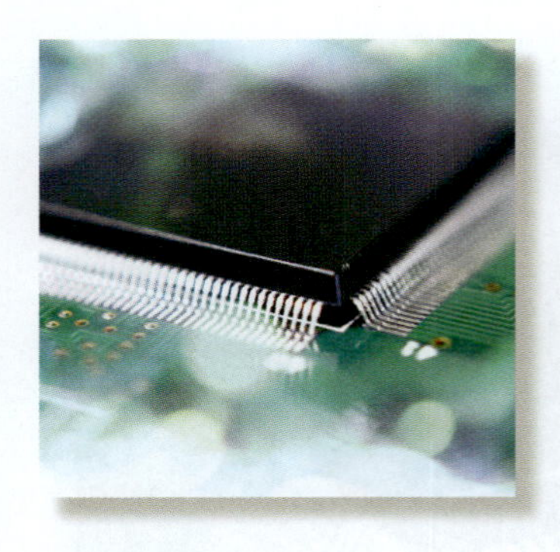

PREVIEW

We stated earlier that one application of electromagnetic induction is the transformer. Furthermore, you know these devices have application in different frequency ranges and for various purposes. That is, there are power transformers, audio transformers, and radio-frequency (rf) transformers, which all find various applications in many types of electronic circuits and systems. Power supplies, audio amplifiers, radio and TV receivers, transmitters, and other systems and subsystems use the unique characteristics of these devices.

In this chapter, you will study the basic features and types of transformers that a technician needs to understand. Practical information, such as schematic symbols and color codes, will be provided. Turns ratios, voltage ratios, current ratios, impedance ratios, and other major aspects about transformers will be studied. Finally, typical malfunctions and troubleshooting techniques will be examined.

KEY TERMS

Autotransformer
Coefficient of coupling *(k)*
Copper loss
Core loss
Current ratio
Impedance ratio
Isolation transformer
Mutual inductance (*M* or L_M)
Transformers
Turns ratio *(TR)*
Voltage ratio

16–1 Background Information

You know when a changing current passes through a conductor wire or a coil, an expanding or collapsing magnetic field is produced. This expanding or collapsing field, in turn, induces a counter-emf (cemf). This process of producing voltage through a changing magnetic field is known as electromagnetic induction. When the induced voltage is created across the current-carrying conductor or coil itself, it is called self-induction. As you know, the symbol for this self-inductance is *L*.

Another important kind of induction is **mutual induction.** The unit of mutual inductance is the henry and the symbol is ***M***. (NOTE: Sometimes you will also see the abbreviation L_M.)

One definition for mutual inductance is the property of inducing voltage in one circuit by varying the current in another circuit. Refer to Figure 16–1 and note that Coil A is connected to an ac source. Coil B is not connected to any source. However, it is located close to coil A so that most of the flux produced in Coil A by current from the source cuts or "links" Coil B. Since a *conductor, magnetic field,* and *relative motion* will induce voltage, a voltage is induced in Coil B because of the changing current in Coil A. By definition, there must be mutual inductance. These two circuits (Coil A and Coil B) are positioned so that energy is transferred by *magnetic linkage.* This coupling between circuits is also sometimes called *inductive coupling.* Inductive coupling is due to flux linkages.

16–2 Coefficient of Coupling

In Figure 16–2 you can see smaller and greater amounts of flux linkage between coils. A term expressing this relationship by the fractional amount of total flux linking the two coils is

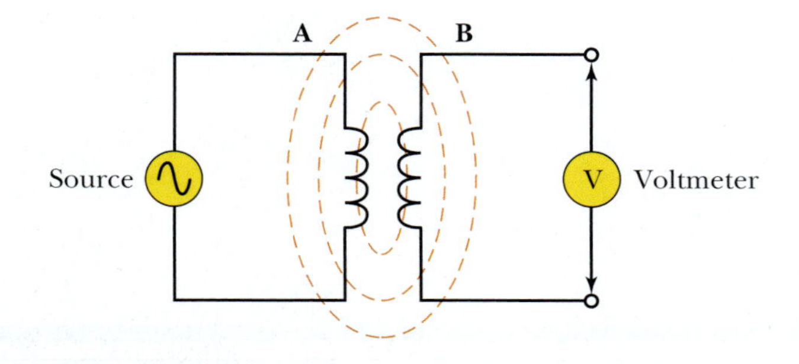

FIGURE 16–1 Mutual induction by flux linkage

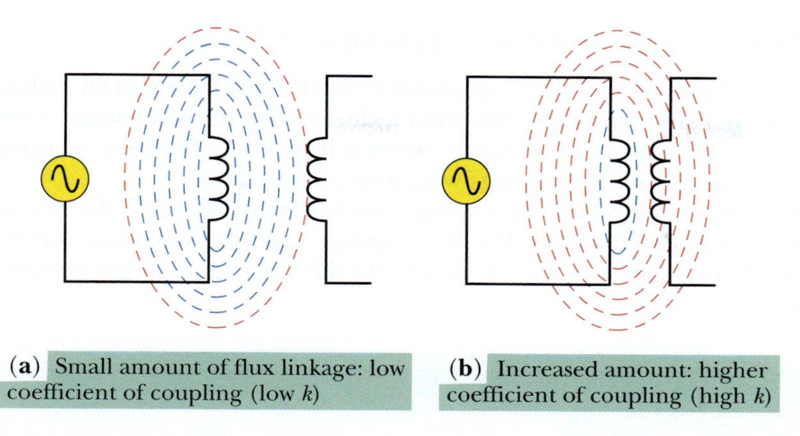

(a) Small amount of flux linkage: low coefficient of coupling (low k)

(b) Increased amount: higher coefficient of coupling (high k)

FIGURE 16–2 Coefficient of coupling related to degree of flux linkage

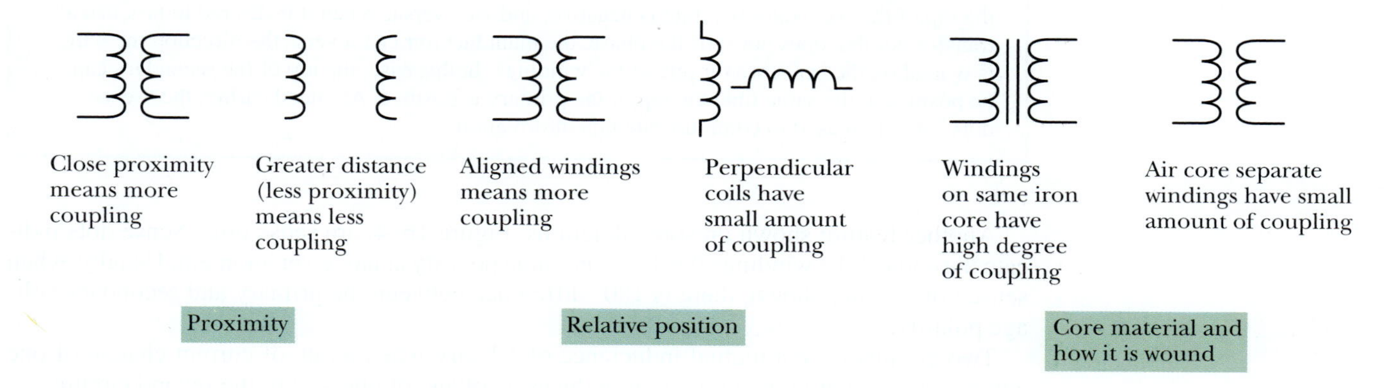
Close proximity means more coupling
Greater distance (less proximity) means less coupling
Aligned windings means more coupling
Perpendicular coils have small amount of coupling
Windings on same iron core have high degree of coupling
Air core separate windings have small amount of coupling
Proximity
Relative position
Core material and how it is wound

FIGURE 16–3 Factors influencing coefficient of coupling (k)

coefficient of coupling, represented by the letter ***k*.** Coefficient of coupling and mutual inductance are *not* the same. However, they are related. Figure 16–2b shows the coils with the higher coefficient of coupling. The coefficient of coupling is therefore computed as the fractional amount of the total flux linking the two circuits. When *all* the flux from one circuit links the other, coefficient of coupling (k) equals one. When two-thirds of the flux links the other circuit, k equals 0.66, and so forth.

See Figure 16–3 for examples of factors that influence the coefficient of coupling. These factors include the proximity of coils (the closer they are the higher the k); the relative positions of coils with respect to each other (parallel coils at a given distance have greater k than perpendicular coils); and other factors, such as when the coils are wound on the same core. Coils wound on the same iron core have a coefficient of coupling almost equal to unity (1). Virtually all the flux produced by one coil links the other coil, with little leakage flux occurring. Air-core coils (often used in rf circuits) have a k that indicates a percentage of coupling from less than 5% $(k = 0.05)$ to about 35% $(k = 0.35)$. Obviously, air-core coils have more leakage flux.

■ IN-PROCESS LEARNING CHECK 1

Fill in the blanks as appropriate.

1. Producing voltage via a changing magnetic field is called electromagnetic _____.
2. Inducing voltage in one circuit by varying current in another circuit is called _____ _____.
3. The fractional amount of the total flux that links two circuits is called the _____ of coupling, which is represented by the letter _____. When 100% of the flux links the two circuits, the _____ of coupling has a value of _____.
4. The closer the coils are, the _____ the coupling factor produced. Compared with parallel coils, perpendicular coils have a _____ degree of coupling.

16–3 Mutual Inductance and Transformer Action

As stated, when two circuits are coupled magnetically, they have mutual inductance. Such is the case for transformers. **Transformers** are devices that transfer energy from one circuit to another by electromagnetic induction (mutual induction). Typically, transformers have their *primary* and *secondary* windings wound on a single core.

Refer to Figure 16–4. Notice the winding connected to the source is the primary winding. The winding connected to the load is the secondary winding. Some transformers can have more than one secondary winding. Also, observe the schematic symbol(s) used for various transformers.

Practical Notes

Most transformers shift the phase of voltage by 180° from input to output due to the electromagnetic induction process. That is, when the top of the primary winding is positive, the top of the secondary winding is negative, and vice versa. When it is desired to produce a transformer that does not shift the phase, the manufacturer can reverse the direction the wire is wound on the iron core for one of the windings. In this case, the top of the secondary can be positive at the same time the top of the primary is positive. As stated earlier, the "sense dots" are often used to communicate this information.

Another feature shown on some diagrams, Figure 16–4, are sense dots. Sense dots indicate the ends of the windings that have the same polarity at any given moment. Usually, when sense dots are *not* shown, there is 180° difference between the primary and secondary voltage polarity.

Two circuits have a mutual inductance of 1 henry when a rate of current change of one ampere per second in the first circuit induces a voltage of one volt in the second circuit.

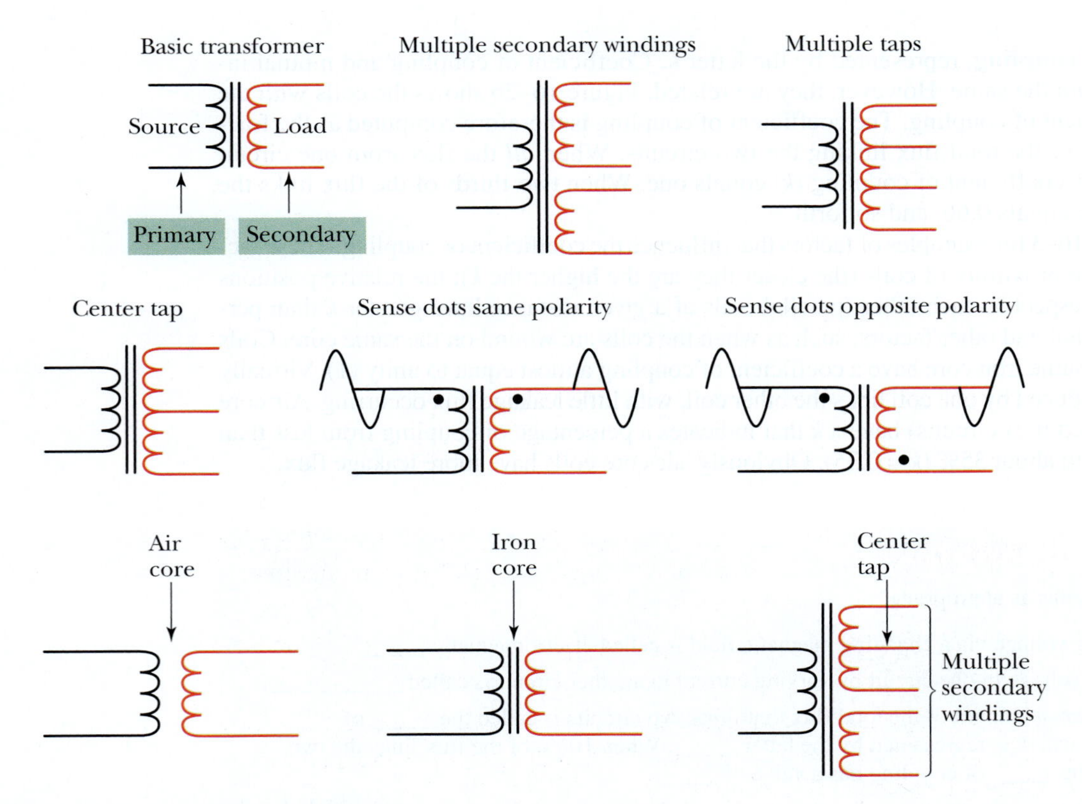

FIGURE 16–4 Primary and secondary transformer windings

Mutual inductance between windings is directly related to the inductance of each winding and the coefficient of coupling. The formula is:

FORMULA 16–1 $M = k\sqrt{L_1 \times L_2}$

where M = mutual inductance in henrys
L_1 = self-inductance of the primary and L_2 = self-inductance of the secondary (for transformers) where Ls are in henrys

EXAMPLE A transformer has a k of 1 (100% coupling of flux lines), and the primary and secondary each have inductances of 3 H. The mutual inductance (M or L_M) is computed as $M = 1\sqrt{3 \times 3} = 1\sqrt{9} = 1\sqrt{9} = 3$ H.

PRACTICE PROBLEMS 1

What is the mutual inductance of a transformer where the coefficient of coupling is 0.95, the primary inductance is 10 H, and the secondary inductance is 15 H?

16–4 Mutual Inductance between Coils Other Than Transformers

We have been discussing mutual inductance in transformers. How does mutual inductance (coupled inductance) affect the total inductance of coils other than transformers that are located where their fields can interact?

Series Coils

When two coils are connected in *series* so that their fields *aid* each other, total inductance is calculated from the formula:

FORMULA 16–2 $L_T = L_1 + L_2 + 2M$

where L_T = total inductance (H)
L_1 = inductance of one coil (H)
L_2 = inductance of other coil (H)
M = mutual inductance (H)

When two coils are connected in *series* so their fields oppose each other, the formula becomes:

FORMULA 16–3 $L_T = L_1 + L_2 - 2M$

Therefore, the general formula for series inductors is:

FORMULA 16–4 $L_T = L_1 + L_2 \pm 2M$

EXAMPLE Two inductors are series connected so their fields aid one another. Inductor 1 has an inductance of 8 H. Inductor 2 also has an inductance of 8 H. Their mutual inductance is 4 H. What is the total inductance of the circuit?

$$L_T = L_1 + L_2 + 2M = 8 + 8 + 8 = 24 \text{ H}$$

PRACTICE PROBLEMS 2

What is the total inductance of two series-connected 10-H inductors whose mutual inductance is 5 H and whose magnetic fields are opposing?

Parallel Coils

When two inductors are in *parallel* with their fields *aiding* each other, the formula is:

FORMULA 16–5
$$L_T = \frac{1}{\frac{1}{L_1 + M} + \frac{1}{L_2 + M}}$$

When two inductors are in *parallel* with their fields *opposing* each other, the formula becomes:

FORMULA 16–6
$$L_T = \frac{1}{\frac{1}{L_1 - M} + \frac{1}{L_2 - M}}$$

PRACTICE PROBLEMS 3

1. What is the total inductance of two parallel coils whose fields are aiding, if $L_1 = 8$ H, $L_2 = 8$ H, and $M = 4$ H?
2. What is the total inductance of two parallel coils whose fields are opposing, if $L_1 = 8$ H, $L_2 = 8$ H, and $M = 4$ H?

Relationship of *k* and *M*

The coefficient of coupling and mutual inductance are related by the following formulas for *k* and *M*.

FORMULA 16–7
$$k = \frac{M}{\sqrt{L_1 \times L_2}}$$

Recall Formula 16-1 is $M = k\sqrt{L_1 \times L_2}$. As you can see from these two formula, mutual inductance and coefficient of coupling are directly related.

16–5 Important Transformer Ratios

We will begin this section by making the following assumptions.

1. With *no load* connected to the secondary winding, the primary winding acts like a simple inductor.
2. Current flowing in the primary depends on applied voltage and inductive reactance of the primary.

$$I_p = \frac{V}{X_P}$$

3. The primary current produces a magnetic field that cuts both the primary and secondary turns.
4. A cemf is produced in the primary.
5. A voltage is produced in *each turn* of the secondary, which tends to equal the voltage induced in *each turn* of the primary because of mutual inductance. This means when there are equal turns on the secondary and the primary, their voltages are virtually equal.

Using these assumptions, let's look at the ratio between electrical output and input parameters under specified conditions. When 100% of the primary flux links the secondary winding (when $k = 1$), the ratio of the induced voltages (i.e., the ratio of primary voltage to the secondary voltage) is the same as the ratio of turns on the primary compared with the turns on the secondary. Thus,

$$\frac{V_P}{V_S} = \frac{N_P}{N_S}$$

where V_P is primary induced voltage; V_S is secondary induced voltage; N_P is the number of turns on the primary winding; and N_S is the number of turns on the secondary winding.

Observe Figure 16–5 and notice how the terms *step up* and *step down* relate to transformers. When the secondary (output) voltage is higher than the primary (input) voltage, the transformer is a step-up transformer. Conversely, when the secondary voltage is lower than the primary voltage, the transformer is a step-down transformer.

Turns Ratio

You will often find the **turns ratio (TR)** of a transformer expressed in terms of the ratio of primary turns to secondary turns (N_P/N_S). For example, when the primary has 250 turns and the secondary has 1,000 turns, the turns ratio is:

$$\frac{N_P}{N_S} = \frac{250}{1{,}000} = 1{:}4$$

NOTE: A number of texts also show the opposite approach (i.e., secondary-to-primary ratio). *Either approach works as long as formulas are consistently set up using the same approach.* In any case, the technician should be careful in making transformer calculations! Be *consistent* in the way you set up the various ratios. We will use *both* ratios to familiarize you with both approaches.

Notice that in Figure 16–6a the turns ratio is shown for both approaches. Also, in Figure 16–6b, the **voltage ratio** is shown for both approaches.

Voltage Ratio

As you learned with the ideal transformer, the voltage ratio has the same ratio as the turns ratio; again see Figure 16–6b. That is:

FORMULA 16–8 $\dfrac{V_P}{V_S} = \dfrac{N_P}{N_S}$

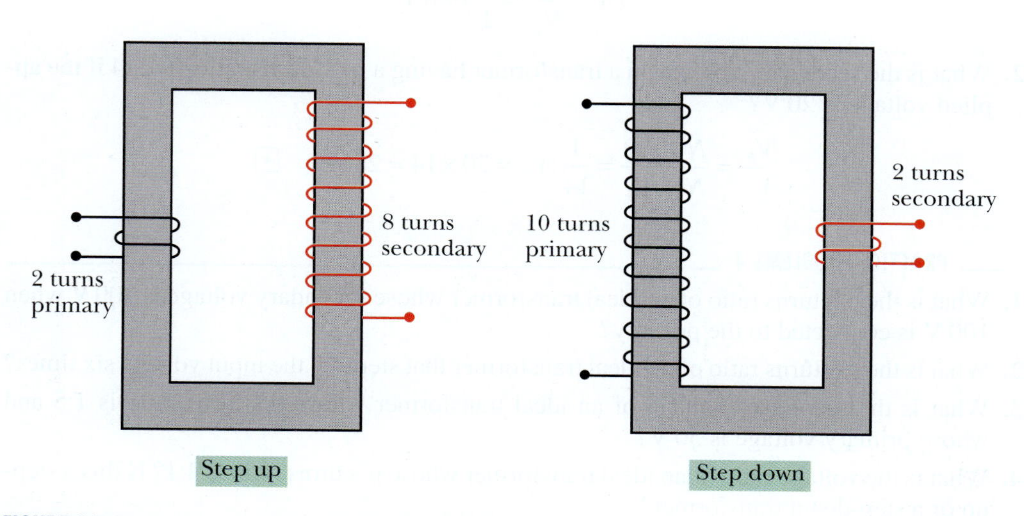

FIGURE 16–5 Step-up and step-down transformers

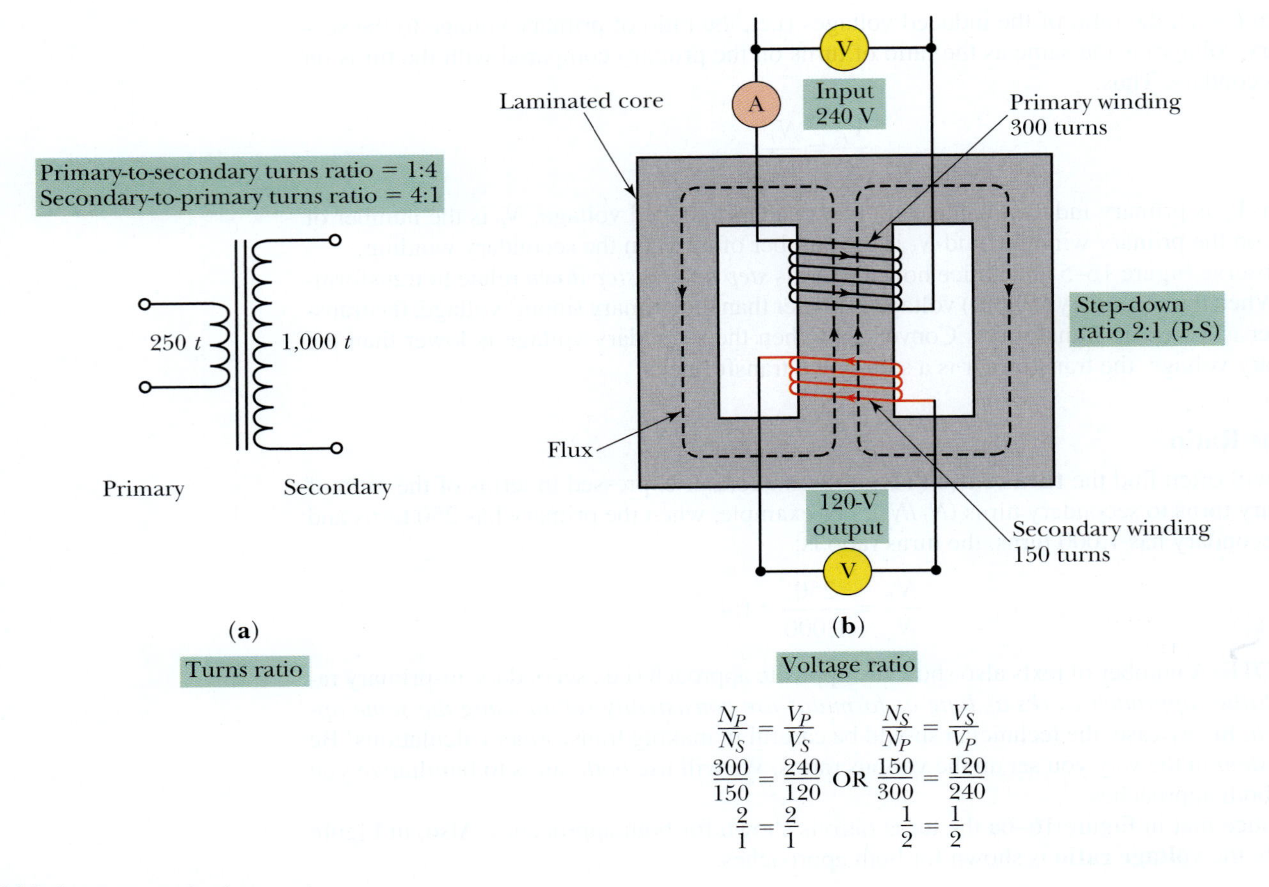

$$\frac{N_P}{N_S} = \frac{V_P}{V_S} \qquad \frac{N_S}{N_P} = \frac{V_S}{V_P}$$

$$\frac{300}{150} = \frac{240}{120} \quad \text{OR} \quad \frac{150}{300} = \frac{120}{240}$$

$$\frac{2}{1} = \frac{2}{1} \qquad \frac{1}{2} = \frac{1}{2}$$

FIGURE 16–6 Turns and voltage ratios

EXAMPLE

1. What is the primary-to-secondary turns ratio of a transformer having eight times as many turns on the primary as on the secondary?

$$TR = \frac{N_P}{N_S} = \frac{8}{1} \text{ or } 8{:}1$$

2. What is the secondary voltage on a transformer having a (p-s) turns ratio of 1:14 if the applied voltage is 20 V?

$$\frac{V_P}{V_S} = \frac{N_P}{N_S}; \frac{20}{V_S} = \frac{1}{14}; V_S = 20 \times 14 = 280 \text{ V}$$

PRACTICE PROBLEMS 4

1. What is the p-s turns ratio of an ideal transformer whose secondary voltage is 300 V when 100 V is connected to the primary?
2. What is the p-s turns ratio of an ideal transformer that steps up the input voltage six times?
3. What is the secondary voltage of an ideal transformer whose p-s turns ratio is 1:5 and whose primary voltage is 50 V?
4. What is the voltage ratio of an ideal transformer whose p-s turns ratio is 4:1? Is this a step-up or a step-down transformer?

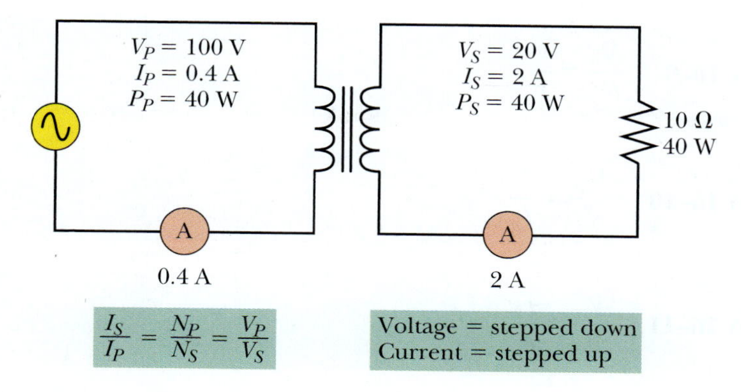

FIGURE 16–7 Current ratio multiSIM

Current Ratio

Up to now, our discussions have been about transformers with no load connected to the secondary(ies). Notice in Figure 16–7 that when a load is connected to the secondary, current flows through the load and the secondary. This current produces a magnetic field related to the secondary current. The polarity produced by the secondary current cancels some of the primary field. Hence, the cemf is reduced in the primary, which reduces the impedance of the primary to current flow. This reduced impedance causes primary current to increase. The interaction between the secondary and the primary is termed *reflected impedance.* In other words, the impedance reflected across the input or primary results from current in the output or secondary.

In Figure 16–7, the power available to the load from the transformer secondary must come from the source that supplies the primary, since a transformer does not *create* power, but *transfers* power from the primary circuit to the secondary circuit. Transformers are not 100% efficient. Therefore, the power delivered to the primary by its source is slightly higher than the power delivered by the secondary to its load. Since transformers are highly efficient (generally ranging from 90% to 98% efficient), Figure 16–7 has assumed 100% efficiency to illustrate this principle. Later in this chapter, we will look at some losses that prevent actual 100% efficiency.

Also notice in Figure 16–7 that the power delivered to the load is 40 W, where $V_S = 20$ V, $R_S = 10\ \Omega$, and $I_S = 2$ A. The power delivered to the primary is 40 W, where $V_P = 100$ V and $I_P = 0.4$ A. This is a step-down transformer where the voltage has been stepped down five times. That is,

$$\frac{N_P}{N_S} = 5{:}1 \text{ and } \frac{V_P}{V_S} = 5{:}1$$

What about the **current ratio?** Observe that the current in the secondary is five times higher than the current in the primary. Using our primary-to-secondary principle, the current ratio is 1:5, or the inverse of the voltage ratio.

This makes sense, if we assume that the power in the secondary is the same as the power in the primary. In other words, $V_S I_S = V_P I_P$. If voltage were decreased by five times, the current would have to be increased by five times so the V times I (or P) would be the same in both primary and secondary.

(NOTE: Power supplied to the primary must equal power delivered to the secondary plus the power lost.)

Now let's relate current ratio to the turns and voltage ratios. Thus,

$$\frac{I_S}{I_P} = \frac{N_P}{N_S} = \frac{V_P}{V_S}$$

Note the inversion in the current ratio. Since quantities equaling the same thing are equal to each other, we can list three useful equations from the preceding information.

FORMULA 16–9 $\frac{N_P}{N_S} = \frac{V_P}{V_S}$

FORMULA 16–10 $\frac{N_P}{N_S} = \frac{I_S}{I_P}$

FORMULA 16–11 $\frac{V_P}{V_S} = \frac{I_S}{I_P}$

If you know any three of the quantities in any of these equations, you can solve for the unknown quantity, assuming 100% efficiency.

◆ EXAMPLES

1. A transformer has 250 turns on the primary, 1,000 turns on the secondary, and the primary voltage is 120 V. What is the secondary voltage value?

$$\frac{N_P}{N_S} = \frac{V_P}{V_S}$$

$$\frac{250}{1{,}000} = \frac{120}{V_S}$$

$$250\,V_S = 120{,}000$$

$$V_S = \frac{120{,}000}{250} = 480 \text{ V}$$

2. A transformer has a primary to secondary turns ratio of 5:1. What is the primary current when the secondary load current is 0.5 A?

$$\frac{N_P}{N_S} = \frac{I_S}{I_P}$$

$$\frac{5}{1} = \frac{0.5}{I_P}$$

$$5I_P = 1 \times 0.5$$

$$I_P = \frac{0.5}{5} = 0.1 \text{ A}$$

3. The primary-to-secondary current ratio of a transformer is 4:1, and the secondary voltage is 60 V. What is the voltage on the primary?

$$\frac{V_P}{V_S} = \frac{I_S}{I_P}$$

$$\frac{V_P}{60} = \frac{1}{4}$$

$$4V_P = 60 \times 1$$

$$V_P = \frac{60}{4} = 15 \text{ V}$$ ◆

Impedance Ratio

The **impedance ratio** is related to the turns ratio, Figure 16–8. Also, since inductance and thus the inductive reactance of a coil, relates to the square of the turns, the impedance also has a similar relationship. A load of a given impedance connected to the transformer sec-

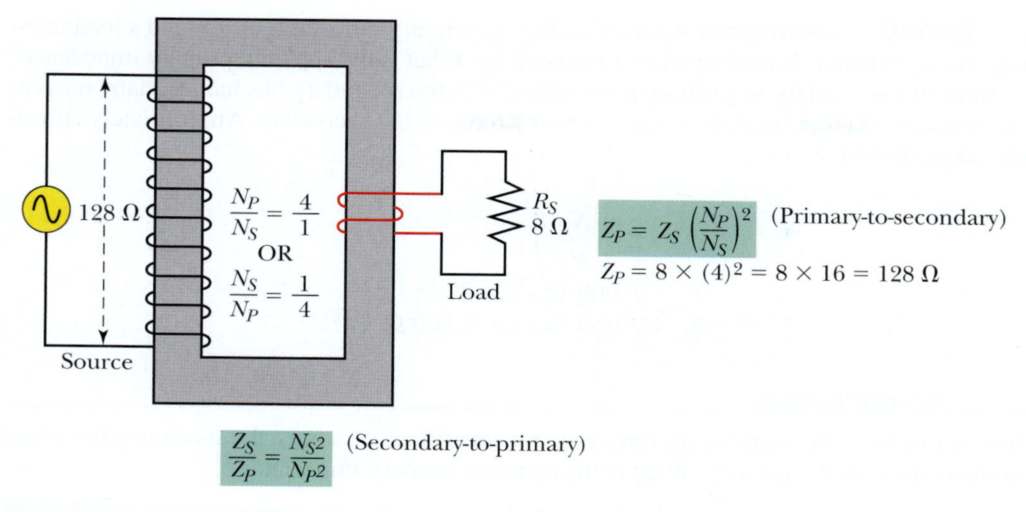

FIGURE 16–8 Impedance ratio

ondary is transformed to a different value by what the source sees "looking into" the primary (unless the turns ratio is 1:1). In fact, for an ideal (zero loss) transformer, the impedance transformation (secondary-to-primary) is proportional to the square of the secondary-to-primary turns ratio. That is,

FORMULA 16–12 $\left(\frac{N_S}{N_P}\right)^2 = \frac{Z_S}{Z_P}$

where Z_S = the impedance of the load connected to the secondary
Z_P = the impedance looking into the primary from source
$\frac{N_S}{N_P}$ = the secondary-to-primary turns ratio

Another way of thinking about impedance ratios and their relationship to turns ratios is to consider that impedance can be regarded as the ratio of V to I. That is $Z = V/I$. For example, if we consider a step-up transformer that steps up voltage by two times, we understand the current is "stepped down" by two times (or is one-half). Then our ratio of V/I has become $2V/0.5I$, which translates to a ratio of 4. Thus, the impedance of the winding with the higher turns is four times that of the winding with the lesser turns. Of course, the converse is true if the transformer is a step-down transformer. The winding with the lesser turns would have half the voltage and twice the current; thus, the $0.5V/2I$ ratio would indicate that the winding with the lower number of turns has an impedance that is one-fourth that of the winding with the greater number of turns. Therefore, we see again, that the impedance ratio is related to the square of the turns ratio.

EXAMPLE When the N_S/N_P (secondary-to-primary) turns ratio of a transformer is 1:4 and the impedance connected to the secondary is 8 Ω, the primary impedance is related to the square of the turns ratio. Hence, Z_P is 16 times Z_S. That is $Z_P = 8 \times 16 = 128\ \Omega$, Figure 16–8. A practical formula expressing this in terms of *primary-to-secondary* turns ratio is:

FORMULA 16–13 $Z_P = Z_S\left(\frac{N_P}{N_S}\right)^2$

◆ **EXAMPLE** A transformer has a secondary-to-primary turns ratio of 0.5, and a load resistance of 2,000 ohms is connected to the secondary. What is the apparent primary impedance?

Since the secondary-to-primary turns ratio is 0.5, the secondary has half as many turns as the primary, or the primary has twice as many turns as the secondary. Appropriately substitute in the formula:

$$Z_P = Z_S\left(\frac{N_P}{N_S}\right)^2$$
$$Z_P = 2{,}000\ \Omega \times (2)^2$$
$$Z_P = 2{,}000\ \Omega \times 4 = 8{,}000\ \Omega$$ ◆

PRACTICE PROBLEMS 5

In the prior example, suppose the turns ratio reverses. In other words, the secondary has twice as many turns as the primary. What is the apparent primary impedance?

These discussions indicate the primary impedance of a very efficient transformer (as it looks to the power source feeding the primary) is determined by the load connected to the secondary and the turns ratio. Because this is true, transformers are often used to match impedances. That is, a transformer is used so that a secondary load of given value is transformed to the optimum impedance value into which the primary source should be looking. (The source may be an amplifier stage, another ac signal, or power source.) Therefore, a useful formula is:

FORMULA 16–14 $\frac{N_P}{N_S} = \sqrt{\frac{Z_P}{Z_S}}$

where $\frac{N_P}{N_S}$ = the required *primary-to-secondary* turns ratio
Z_P = the primary impedance desired (required)
Z_S = the impedance of the load connected to the secondary

◆ **EXAMPLE** An amplifier requires a load of 1,000 ohms for best performance. The amplifier output is to be connected to a loudspeaker having an impedance of 10 ohms. What must the turns ratio *(primary-to-secondary)* be for a transformer used for impedance matching?

$$\frac{N_P}{N_S} = \sqrt{\frac{Z_P}{Z_S}}$$
$$\frac{N_P}{N_S} = \sqrt{\frac{1{,}000}{10}} = \sqrt{100} = 10$$

This indicates the primary must have 10 times as many turns as the secondary. ◆

PRACTICE PROBLEMS 6

1. What is the primary-to-secondary impedance ratio of a transformer that has 1,000 turns on its primary and 200 turns on its secondary?
2. What impedance would the primary source see looking into a transformer where the primary-to-secondary turns ratio is 16 and the load resistance connected to the secondary is 4 Ω?

16–6 Transformer Losses

Because of energy losses, practical transformers do not meet all the criteria for the ideal, 100% efficient devices we have been discussing. These transformer losses consist primarily of two categories. One category is **copper loss,** caused by the I^2R copper wire losses that make up the primary and secondary windings. Current passing through the resistance of the wire produces energy loss in the form of heat.

The other category of transformer losses is termed **core loss.** Core losses result from two factors. One factor is called *eddy current* losses. Because of the changing magnetic fields and the iron-core conductivity, currents are produced in the iron core that do not aid transformer output. These eddy currents produce an I^2R core loss related to the core material resistance and the current amount. To reduce the current involved, the iron core is usually constructed of thin sheets of metal called *laminations.* These laminations are electrically insulated from each other by varnish or shellac. Reducing the cross-sectional area of the conductor by laminating significantly increases the core resistance, which decreases the current, and greatly reduces the I^2R loss.

The other core loss is called *hysteresis loss.* Recall magnetic domains in the magnetic core material require energy to rearrange. (Recall in an earlier chapter, the lagging effect of the core magnetization behind the magnetizing force was illustrated by the hysteresis loop.) Since ac current constantly changes in magnitude and direction, the tiny molecular magnets within the core are constantly being rearranged. This process requires energy that causes some loss.

Hysteresis loss increases as the frequency of operation is increased. For this reason, iron-core transformers, such as those used at power frequencies, are not used in radio-frequency applications.

Copper losses, core losses, leakage flux, and other factors prevent a transformer from being 100% efficient. Efficiency percentage is calculated as output power divided by input power times 100.

FORMULA 16–15 $\text{Efficiency } \% = \frac{P_{out}}{P_{in}} \times 100$

To minimize these various losses, the physical features of coils and transformers used for applications in power, audio, and rf frequency ranges differ slightly. We will investigate some of these characteristics in the next section.

16–7 Characteristics of Selected Transformers

Power Transformers

Power transformers, Figure 16–9, typically have the following features:

1. They are heavy (because of their iron core) and can be very large, like the transformers on telephone poles. However, power transformers in electronic devices are much smaller and vary considerably in size.
2. They have a laminated iron core to reduce eddy-current losses.

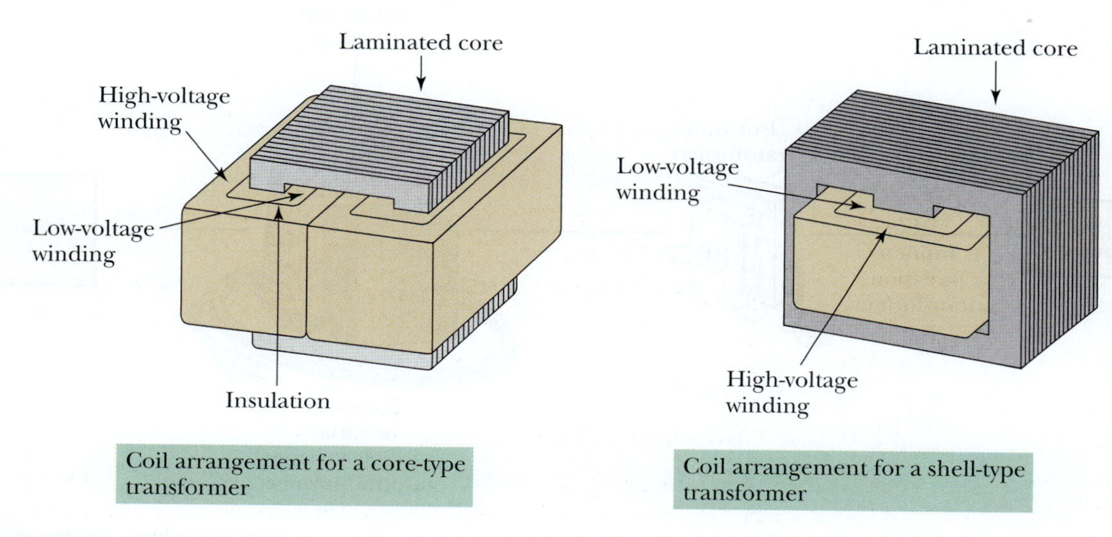

FIGURE 16–9 Power transformers

3. The flux path is as short as possible to reduce leakage flux and minimize magnetizing energy needed.
4. They use one of two core shapes—core type (primary and secondary windings on separate legs of the core) and shell type (both primary and secondary windings on the center leg), again see Figure 16–9.
5. They have one or more primary and secondary windings.
6. They may have *one* tapped winding, such as the **autotransformer,** Figure 16–10. The autotransformer is a special single-winding transformer.

When the input source is connected from one end to the tap while the output is across the whole coil, the voltage is stepped up. If the input voltage is applied across the whole coil and the output taken from the tap to one end, the voltage is stepped down.

Variacs or *Powerstats* are trade names for autotransformers having a movable wiper-arm tap that allows varying voltage from 0 V to the maximum level. They are often used as a variable ac source in experimental setups.

7. They isolate the circuitry connected to its secondary from the primary ac source (e.g., the ac powerline). A 1:1 turns ratio transformer is specifically designed for isolation applications, Figure 16–11. (NOTE: All standard transformers isolate secondary from primary source regardless of turns ratios, **except for autotransformers**.)

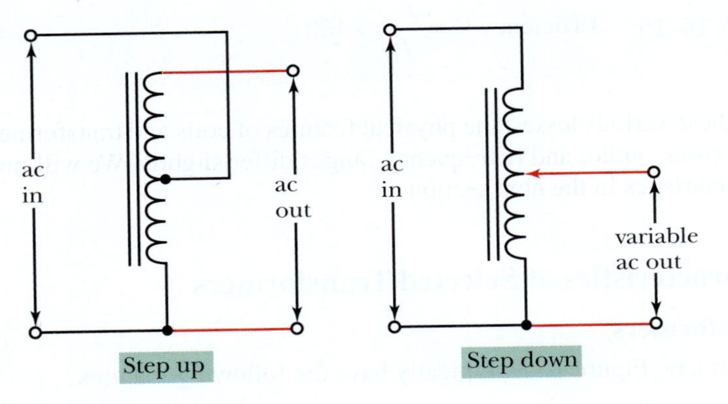

FIGURE 16–10 Autotransformer

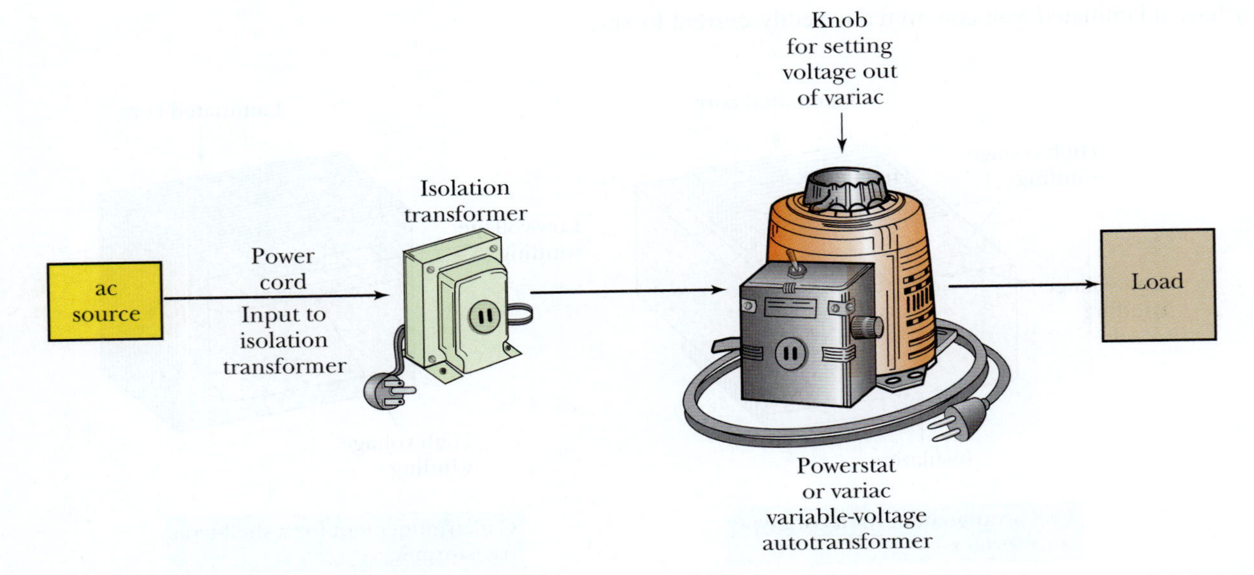

FIGURE 16–11 Using an isolation transformer for safety when using a powerstat or variac-type variable voltage autotransformer

When using this special 1:1 turns-ratio isolation transformer, the load on the secondary has no *direct* connection to the primary source. This provides a safety measure for people working with the circuitry connected to the isolation transformer secondary.

8. They cannot operate at high frequency because of excessive losses that would be created.
9. Just two examples of the many color codes in use are shown in Figures 16–12a and b.

Safety Hints

Caution: Bad shock can occur if no **isolation transformer** is used on the autotransformer input side, because there is no isolation from the raw line voltage and ground connections in the autotransformer circuit. For this reason, do *not* use autotransformers with a movable wiper-arm tap unless you plug the autotransformer into an isolation transformer output, which isolates it from the raw ac power source. ■

Practical Notes

Some power transformers are used when it is necessary to be able to select either a 120- or 240-V primary input source for the transformer. Therefore, there are transformers with two primary windings. These primary windings may be connected in series, when 240 V is the source (i.e., the end of one winding is connected to the beginning of the other winding, being careful to pay attention to phasing of the two windings). Alternatively, by connecting the two primaries in parallel (again being careful of phasing), the transformer primary can be fed from a 120-V source. By using these techniques, the secondary voltage output will be the same with 120- or 240-V inputs, as appropriate.

Audio Transformers

Audio transformers are similar to power transformers in their construction. They typically are:

1. smaller and lighter than most power transformers; and
2. unable to operate at frequencies above the audio range without excessive losses.

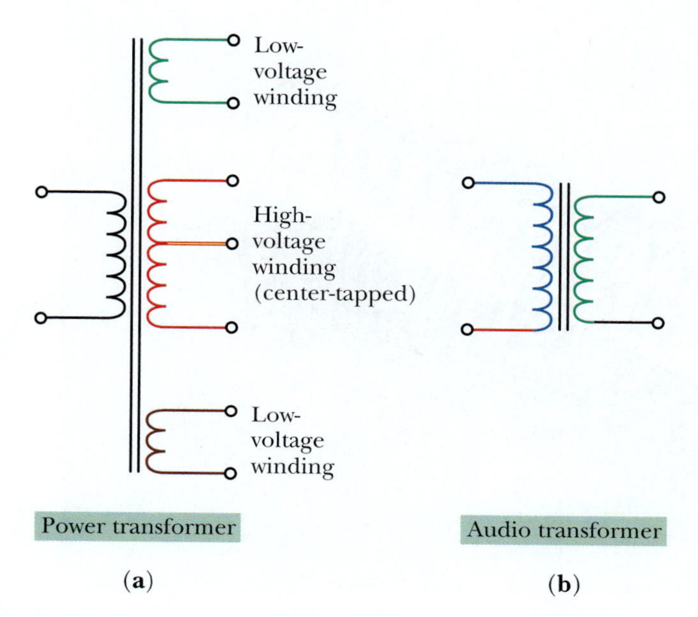

FIGURE 16–12 Typical power and audio transformer color-coding systems

RF Transformers and Coils

Because the flux is changing *very rapidly, RF* transformers or coils, Figures 16–13 and 16-14, have different features compared with power or audio transformers. Some features include the following:

1. Cores are air core, or powdered iron core, depending on frequency and application. Air-core coils have virtually zero hysteresis or eddy-current losses. Powdered iron cores are made by grinding metal materials and nonconductive materials into granules, which are pressed into a cylindrical-shaped "slug." Each metal particle is therefore insulated, greatly reducing losses such as eddy currents. See Figure 16–13a for examples of this coil.

   Also, nonconductive ferrite materials can be used in *RF* coils or transformers. As you recall, ferrite materials are special magnetic materials that provide easy paths for magnetic flux

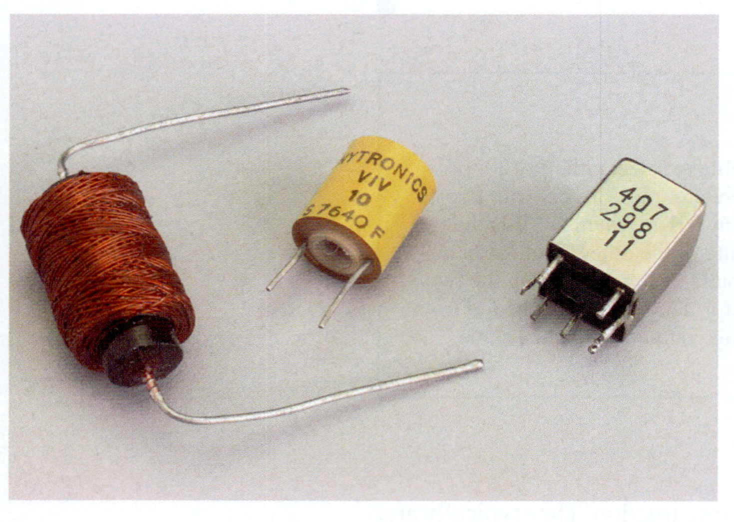

(a)

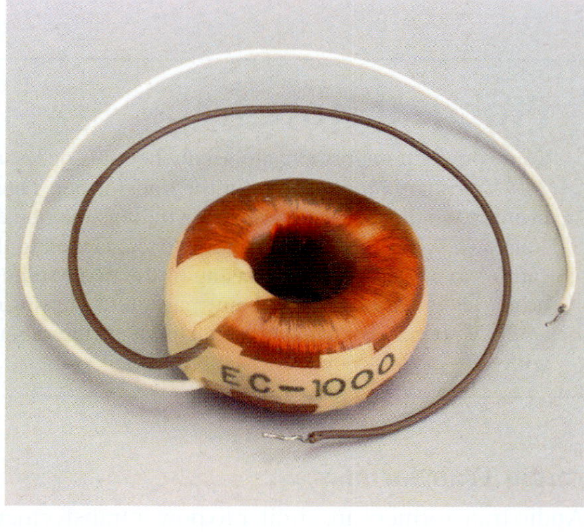

(b)

FIGURE 16–13 Examples of various types of coils and cores: (a) powdered iron-core-type coils; (b) toroidal core

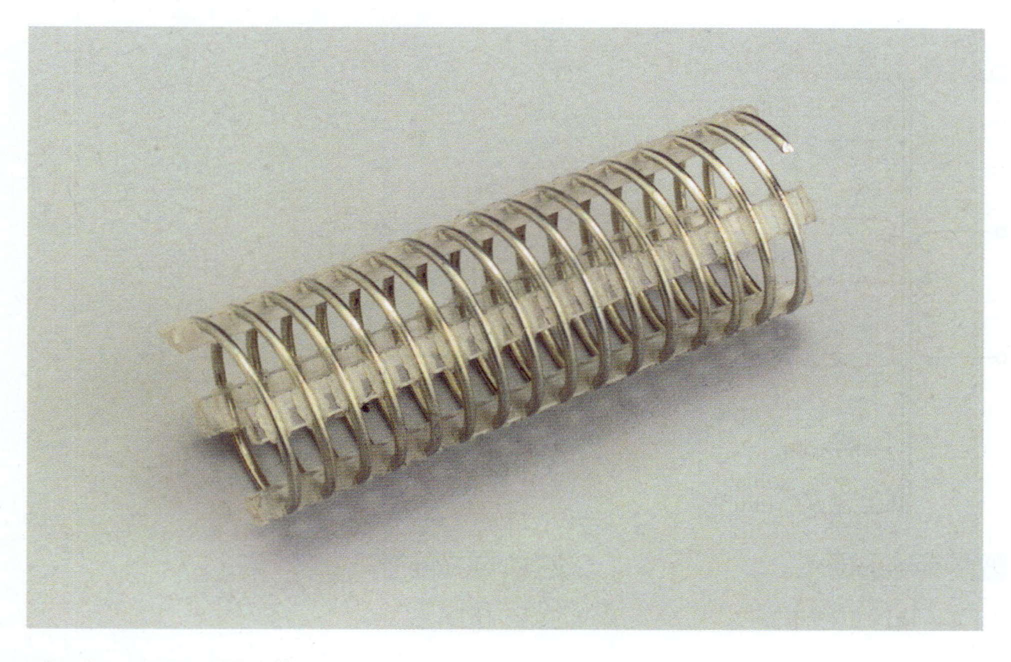

FIGURE 16–14 Example of an air-core *RF* coil

but are nonconductive. Often, toroid-type cores are used in radio-frequency circuits and lower-frequency applications, Figure 16–13b.

2. *RF* coils and transformers typically have very low inductance values. Generally, these values are in the microhenry or millihenry ranges.
3. Figure 16–14 shows an example of an air-core coil found in transmitter applications.

Color Coding

Today, many power, audio, and *RF* transformers are designed and manufactured to be used and mounted on printed circuit (PC) boards. Consequently, their connector terminals are designed for that technique. However, some transformers that have wire leads by which connection is made to the various windings are still available and in use. In that case, the leads may be color coded to identify the various windings and their connection points. Although there may be variations from these, frequently the colors used include the following:

- *Power transformers:* Typically primary leads are black, high-voltage secondary is red, and if center-tapped, the center-tap is red with yellow stripe; low-voltage windings may be yellow for one winding, green for another, brown for another and slate for another, depending on the number of separate low-voltage windings the transformer has, and the type of circuitry involved.
- *Audio transformers:* Typically, one primary winding lead is blue and connects to the output electrode of the device feeding the transformer primary, the other end of the primary winding is red, and connects to the circuit source; the secondary has one lead that is green and feeds the input to the stage connected to the transformer secondary, the other secondary lead (return) is typically black.
- *Radio frequency (RF)* interstage transformers, such as intermediate-frequency (i.f.) transformers used in radio receivers, use colors similar to the audio transformers. That is, primary leads are blue and red, and secondary leads are green and black.

Important Ratings to Consider When Replacing Transformers

A listing of practical transformer ratings, which should be considered when replacing transformers, is shown below. Each of these parameters can be very important considerations when replacing a transformer.

Practical Transformer Ratings	
POWER TRANSFORMERS	AUDIO TRANSFORMERS
Input volts	Primary impedance
Output *V, I,* and kVA ratings	Secondary impedance
Insulation ratings	Maximum *V, I,* and *P*
Size	Size
Mounting dimensions	Mounting dimensions
Weight	

16–8 Troubleshooting Hints

Transformer problems are virtually the same problems as those described in the chapter on inductance. Two basic problems are opens or shorts. The primary test to detect these defects is the ohmmeter.

Safety Hints

Once again, we want to remind you that when checking large inductances (including transformer windings) with an ohmmeter, take precautions not to get shocked by the "inductive kick" of the winding's collapsing field when removing the ohmmeter leads from the winding under test. ■

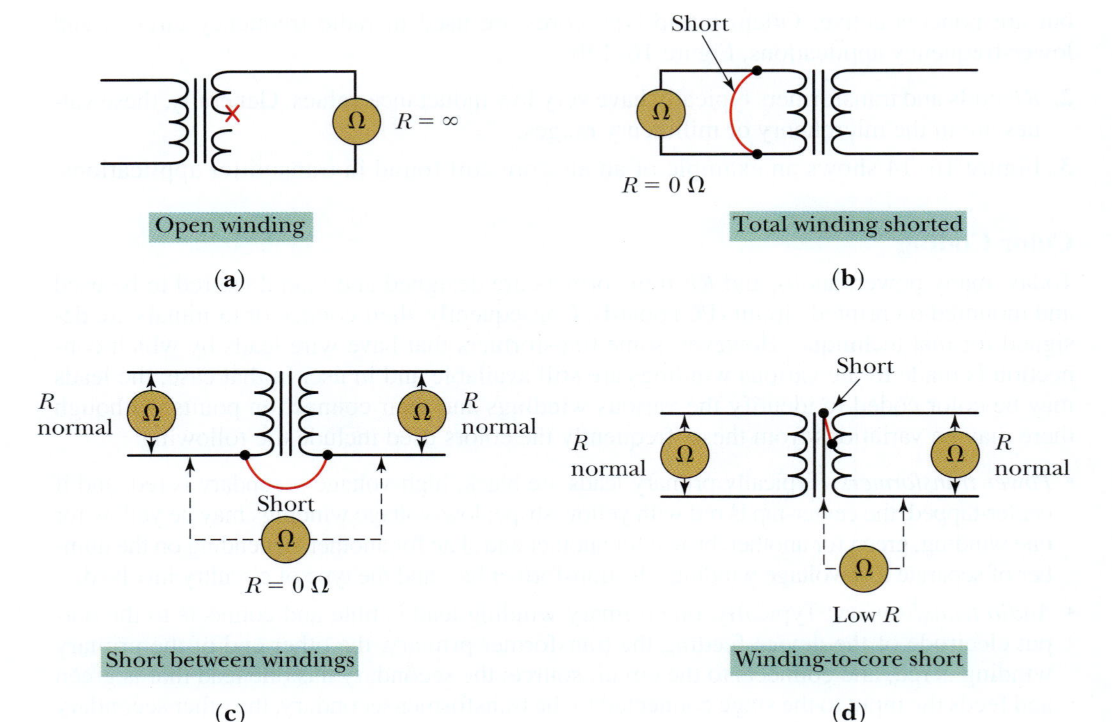

FIGURE 16–15 Typical transformer troubles

Open Winding(s)

With an open winding, the obvious result is zero secondary output voltage, even when source voltage is applied to the primary. When the primary is *not* open and only one secondary of a multiple-secondary transformer is open, only the open winding shows zero output. To check these conditions, remove the power and use the ohmmeter.

Refer to Figure 16–15a to see an open winding. In this case, the ohmmeter registers an infinite resistance reading, indicating the winding under test has an open. Both the primary and secondary windings should be checked! Normal readings are the specified dc resistance values for each winding.

Shorted Winding(s)

Shorted turns or a shorted winding can cause a fuse in the primary circuit to blow (open) from excessive current. Another possible clue to this condition is if the transformer operates at a hotter than normal temperature. Also, another hint of shorted turns is lower-than-normal output voltage. Actually, partially shorted windings or windings with very low *normal* resistances are hard to check when only using an ohmmeter. Therefore, watch for these clues when troubleshooting transformer problems.

However, the ohmmeter is useful to detect certain types of transformer short conditions. Figure 16–15b shows a total-winding short condition. In this case, the ohmmeter reads virtually zero ohms when it should be reading the normal winding resistance value, if there were no short.

Figure 16–15c illustrates the condition when the primary and secondary windings are shorted together. In this case, each winding resistance might be normal; however, there would be a low resistance reading between windings. When conditions are normal, "infinite ohms" is measured between primary and secondary windings.

Another possible short condition is when one or both windings are shorted to the metal core, Figure 16–15d. Again, if things are normal, an infinite ohms reading will be present between either winding and the core material.

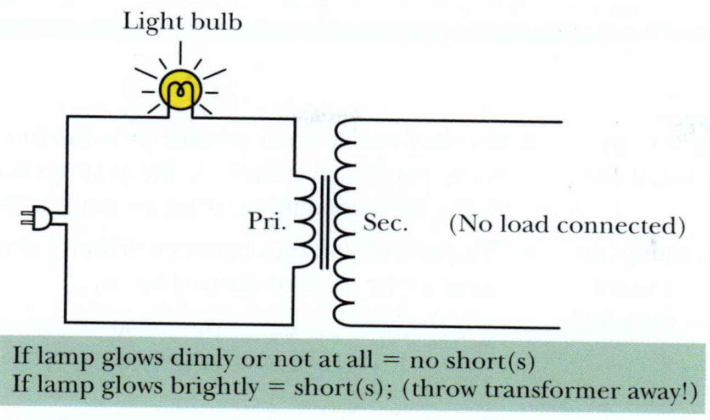
Light bulb
Pri.
Sec.
(No load connected)
If lamp glows dimly or not at all = no short(s)
If lamp glows brightly = short(s); (throw transformer away!)
NOTE: Use small bulb rating for small transformers;
use larger bulb rating for large transformers.

FIGURE 16–16 One way to find shorted turns

*The requirements for bulb size and its electrical ratings are as follows:

1. The voltage rating should equal the same voltage rating as the transformer primary.
2. For small transformers, a small-wattage lamp should be used.
3. For large transformers, a large-wattage lamp should be used.

Be aware that choosing the correct bulb size and interpreting the results come only with experience. This test is simply mentioned to show you some innovative ways technicians find out *what they need to know,* even in difficult situations.

Practical Notes

*The following discussion describes one practical approach used to determine whether a short in a *power transformer* exists. This approach is a simple test involving the transformer in question and a light bulb. In Figure 16–16, you can see a light bulb is placed in series with the transformer primary winding. When a short is present in *either* the primary or secondary, the lamp glows brightly because of the higher-than-normal primary current. When no short exists, the lamp does *not* glow or may glow dimly.

Summary

- When the magnetic flux of two coils link and interact, mutual inductance and inductive coupling occur between the two circuits.
- When mutual inductance is present, current changing in one circuit causes an induced voltage in the other circuit. The symbol for mutual inductance is M or L_M, and the unit of measure is the henry.
- One henry of mutual inductance is present when a change of 1 ampere per second in one circuit induces 1 volt in the mutually coupled circuit.
- Expressing the fraction of total flux linking one circuit to another circuit is called coefficient of coupling. The abbreviation for coefficient of coupling is k. NOTE: Multiplying k by 100 yields the percentage of total flux lines linking the circuits.
- The relationship between mutual inductance and coefficient of coupling is shown by the formulas:

$$M = k\sqrt{L_1 \times L_2} \text{ and } k = \frac{M}{\sqrt{L_1 \times L_2}}$$

where M = mutual inductance in henrys
k = coefficient of coupling
L_1 = self-inductance of coil or transformer winding 1
L_2 = self-inductance of coil or transformer winding 2

- Mutual inductance can exist between coils that are not transformer windings but that are close enough for flux linkage to occur. This can affect total inductance for the circuit.
- Transformers are composed of separate coils or windings, having inductive coupling. The primary winding is connected to an ac source. The load(s) are connected to the secondary winding(s).
- One feature of a two-winding transformer is that the primary and secondary are magnetically coupled, yet electrically isolated from each other. This feature can keep circuits and loads from being directly connected to the power lines, which increases safety.
- One key transformer parameter is the turns ratio. Turns ratio is expressed either by the primary-to-secondary ratio N_P/N_S, or secondary-to-primary ratio N_S/N_P.
- The ratio of voltages between primary and secondary is the same as the ratio of turns. That is,

$$\frac{N_P}{N_S} = \frac{V_P}{V_S} \text{ or } \frac{N_S}{N_P} = \frac{V_S}{V_P}$$

- The current ratio of a transformer is the inverse of the voltage ratio. When voltage is stepped up from primary to secondary, current is stepped down by a proportionate factor:

$$\frac{V_P}{V_S} = \frac{I_S}{I_P} \text{ or } \frac{V_S}{V_P} = \frac{I_P}{I_S}$$

- The impedance ratio of a transformer is related to the square of the turns ratio. That is,

$$\frac{Z_P}{Z_S} = \frac{N_P^2}{N_S^2}$$

- When a transformer has 100% efficiency, the power delivered to the load by the secondary equals the power supplied to the primary by the ac source. The efficiency of iron-core transformers is usually in the 90% to 98% range.
- Two categories of transformer losses are copper losses and core losses. Copper losses relate to the I^2R losses of the copper wire in the primary and secondary windings. Core losses result from eddy currents being induced in the core by the changing magnetic flux, and the energy wasted due to hysteresis losses. Eddy currents in the core material cause I^2R losses.
- Transformers have various applications at different frequencies. Three transformer categories include power transformers, audio transformers, and RF transformers.
- Power and audio transformers generally have laminated iron cores, for efficiency's sake. RF transformers usually use air cores, special powdered iron, or ferrite in their core material.
- Transformers can develop abnormal operations because of opens or shorts. Opens are evident by lack of output. Shorts between turns, from winding-to-winding or from winding-to-core, can be seen by overheating, blown fuses, and so forth.

Formulas and Sample Calculator Sequences

Formula 16–1 *(To find the value of mutual inductance)*	$M = k\sqrt{L_1 \times L_2}$ k value, [×], [(], L_1 value, [×], L_2 value, [)], [$\sqrt{x}$], [=]
Formula 16–2 *(To find the value of total inductance for two-series inductors having aiding fields)*	$L_T = L_1 + L_2 + 2M$ L_1 value, [+], L_2 value, [+], 2, [×], M value, [=]
Formula 16–3 *(To find the value of total inductance for two-series inductors having opposing fields)*	$L_T = L_1 + L_2 - 2M$ L_1 value, [+], L_2 value, [−], 2, [×], M value, [=]
Formula 16–4 *(The general formula for total inductance of two inductors having interacting fields)*	$L_T = L_1 + L_2 - 2M$ L_1 value, [+], L_2 value, [+] or [−] as appropriate, 2, [×], M value, [=]
Formula 16–5 *(To find the total inductance of two parallel inductors having aiding fields)*	$L_T = \dfrac{1}{\dfrac{1}{L_1 + M} + \dfrac{1}{L_2 + M}}$ L_1 value, [+], M value, [=], [1/x], [+], [(], L_2 value, [+], M value, [)], [1/x], [=], [1/x]
Formula 16–6 *(To find the total inductance of two parallel inductors having opposing fields)*	$L_T = \dfrac{1}{\dfrac{1}{L_1 - M} + \dfrac{1}{L_2 - M}}$ L_1 value, [−], M value, [=], [1/x], [+], [(], L_2 value, [−], M value, [)], [1/x], [=], [1/x]
Formula 16–7 *(To find the coefficient of coupling between two inductors)*	$k = \dfrac{M}{\sqrt{L_1 \times L_2}}$ M value, [÷], [(], L_1 value, [×], L_2 value, [)], [$\sqrt{x}$], [=]
Formula 16–8 *(To show the relationship of the voltage ratio of a transformer to its turns ratio)*	$\dfrac{V_P}{V_S} = \dfrac{N_P}{N_S}$ V_P value, [÷], V_S value, [=], (should equal) N_P value, [÷], N_S value, [=]

Formula 16–9
(To show the relationship of the turns ratio of a transformer to its voltage ratio)

$$\frac{N_P}{N_S} = \frac{V_P}{V_S}$$

N_P value, [÷], N_S value, [=] (should equal) V_P value, [÷], V_S value, [=]

Formula 16–10
(To show the relationship of the turns ratio of a transformer to its current ratio)

$$\frac{N_P}{N_S} = \frac{I_S}{I_P}$$

N_P value, [÷], N_S value, [=] (should equal) I_S value, [÷], I_P value, [=]

Formula 16–11
(To show the relationship of the voltage ratio of a transformer to its current ratio)

$$\frac{V_P}{V_S} = \frac{I_S}{I_P}$$

V_P value, [÷], V_S value, [=] (should equal) I_S value, [÷], I_P value, [=]

Formula 16–12
(To show the relationship of the turns ratio of a transformer to its impedance ratio)

$$\left(\frac{N_S}{N_P}\right)^2 = \frac{Z_S}{Z_P}$$

N_S value, [÷], N_P value, [=], [x^2], [=] (should equal) Z_S value, [÷], Z_P value, [=]

Formula 16–13
(To find apparent primary impedance of a transformer with a known secondary load impedance, when the transformer turns ratio is known)

$$Z_P = Z_S\left(\frac{N_P}{N_S}\right)^2$$

Z_S value, [×], [(], N_P value, [÷], N_S value, [)], [x^2], [=]

Formula 16–14
(To find the needed transformer turns ratio to match given impedance conditions)

$$\frac{N_P}{N_S} = \sqrt{\frac{Z_P}{Z_S}}$$

N_P value, [÷], N_S value, [=] (should equal), [(], Z_P value, [÷], Z_S value, [)], [$\sqrt{x}$]

Formula 16–15
(To find transformer efficiency)

$$\text{Efficiency } \% = \frac{P_{out}}{P_{in}} \times 100$$

P_{out} value, [÷], P_{in} value, [×], 100, [=]

EXCEL AUTOMATED FORMULAS

Helpful problem-solving tools, such as the Excel automated formulas available on the CD, allow automatic calculations of formulas.

Using Excel

Basic Transformer Characteristics Formulas

(Excel file reference: FOE16_01.xls)

DON'T FORGET! *It is not necessary to retype formulas once they are entered on the worksheet! Just input new parameters data for each new problem using that formula, as needed.*

- Use the Formula 16–13 spreadsheet sample and solve for the value of primary impedance, assuming that the parameters are those called for in Practice Problems 5. Check your answer out against the answer given in the Appendix for this problem.
- Again, use the Formula 16–13 spreadsheet sample and the same turns ratio as used in the problem above. What would the primary impedance value be if the secondary load were 16 ohms?

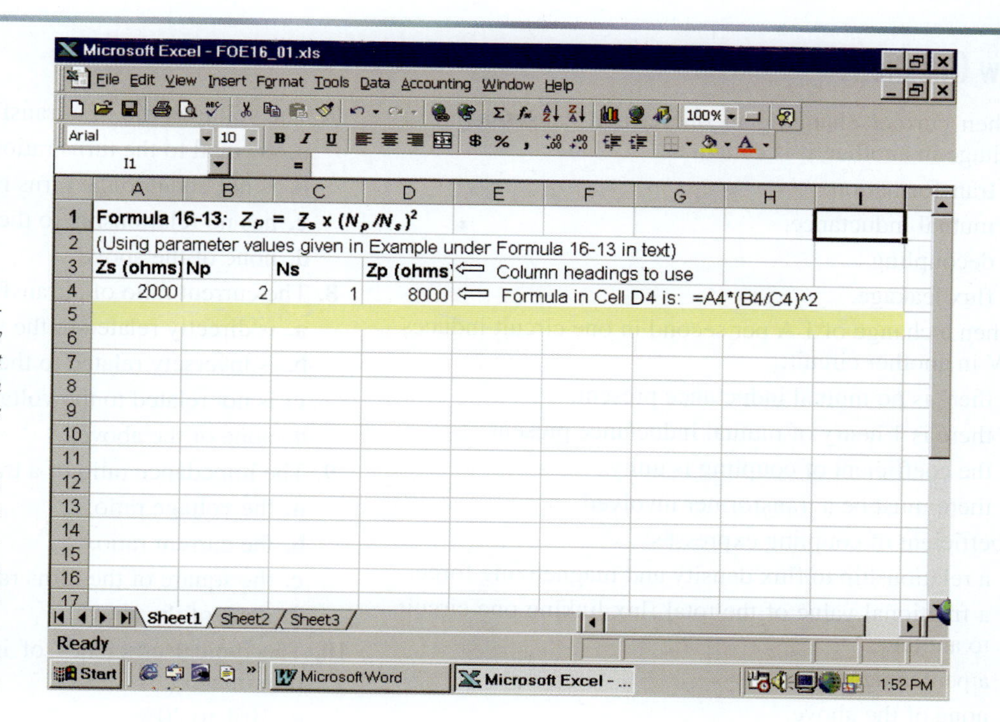

Review Questions

1. When current changing in one circuit causes induced voltage in another, it is called:
   - **a.** transformation.
   - **b.** mutual inductance.
   - **c.** decoupling.
   - **d.** flux leakage.
2. When a change of 1 A per second in one circuit induces 1 V in another circuit:
   - **a.** there is no mutual inductance present.
   - **b.** there is 1 henry of mutual inductance present.
   - **c.** the coefficient of coupling is unity.
   - **d.** there must be a transformer involved.
3. Coefficient of coupling expresses:
   - **a.** a relationship to flux density and magnetizing force.
   - **b.** a fractional value of the total flux linking one circuit to another.
   - **c.** a percentage.
   - **d.** none of the above.
4. In a transformer:
   - **a.** the secondary is connected to the source.
   - **b.** the secondary is always connected to a load.
   - **c.** the primary is connected to the source.
   - **d.** none of the above.
5. In a transformer:
   - **a.** the primary and secondary are electrically connected to each other.
   - **b.** the primary and secondary are only magnetically coupled.
   - **c.** the primary and secondary are both electrically and magnetically coupled.
   - **d.** none of the above.
6. Turn ratios for transformers may be expressed:
   - **a.** only as a ratio of primary turns-to-secondary turns.
   - **b.** only as a ratio of secondary turns-to-primary turns.
   - **c.** either as primary-to-secondary or secondary-to-primary ratios.
   - **d.** none of the above.
7. The voltage ratio of a transformer:
   - **a.** is equal to the turns ratio.
   - **b.** is not equal to the turns ratio.
   - **c.** has no relationship to the turns ratio.
   - **d.** none of the above.
8. The current ratio of a transformer:
   - **a.** is directly related to the voltage ratio.
   - **b.** is inversely related to the voltage ratio.
   - **c.** is not related to the voltage ratio.
   - **d.** none of the above.
9. The impedance ratio of a transformer is equal to:
   - **a.** the voltage ratio.
   - **b.** the current ratio.
   - **c.** the square of the turns ratio.
   - **d.** none of the above.
10. The typical efficiency of iron-core transformers is approximately:
    - **a.** 10% to 20%.
    - **b.** 80% to 88%.
    - **c.** 90% to 98%.
    - **d.** none of the above.
11. Copper losses in a transformer come from:
    - **a.** resistance of the iron in the core material.
    - **b.** eddy currents.
    - **c.** resistance of the wire in the windings.
    - **d.** none of the above.
12. Core losses in a transformer may come from:
    - **a.** hysteresis and eddy currents.
    - **b.** winding losses and metal core losses.
    - **c.** leakage flux alone.
    - **d.** all of the above.
13. Iron cores are typically laminated, rather than solid:
    - **a.** to decrease their resistance to current.
    - **b.** to increase their resistance to current.
    - **c.** to make the cores lighter in weight.
    - **d.** none of the above.

14. Iron cores are usually used for:
  a. audio and *RF* transformers.
  b. power and *RF* transformers.
  c. power and audio transformers.
  d. power and large *RF* transformers.

15. The most common problems in transformers are:
  a. shorted turns and lamination buzz.
  b. shorted turns and overheating.
  c. opens and shorts.
  d. opens between windings and core.

16. A blown fuse in a transformer circuit may indicate:
  a. an open transformer winding.
  b. a short between windings.
  c. an open in a single winding.
  d. none of the above.

17. A lack of output from a transformer may indicate:
  a. a low resistance.
  b. an open.
  c. excessive leakage flux.
  d. none of the above.

18. The ohmmeter can be used to find transformer troubles, such as:
  a. open windings.
  b. shorts between windings.
  c. shorts between core and windings.
  d. all of the above.

Problems

1. What is the primary current of a transformer whose source is 120 V and secondary is supplying 240 V at 100 mA?

2. What is the secondary voltage of a transformer having a secondary-to-primary turns ratio of 5.5:1, if the source voltage is 100 V?

3. What is the primary-to-secondary turns ratio of a transformer whose primary voltage is 120 V and secondary voltage is 6 V?

4. If a transformer has a primary-to-secondary turns ratio of 5:1 and the load resistance connected to the secondary is 8 Ω, what impedance is seen looking into the primary?

5. What is the mutual inductance of two coils having a coefficient of coupling of 0.6, if each coil has a self-inductance of 2 H?

6. What is the total inductance of two coils in series having a 0.3 k, if each of the coils has a self-inductance of 10 mH? (Assume their fields are aiding.)

7. What is the total inductance of two coils in parallel having a 0.5 k, if each coil has a self-inductance of 10 H. (Assume their fields are opposing.)

8. What is the secondary voltage of an ideal transformer whose secondary-to-primary turns ratio is 15 and primary voltage is 12 V?

9. How many volts per turn are in a transformer having a 500-turn primary and a 100-turn secondary, if the primary voltage is 50 V?

10. A 6:1 step-down transformer has a secondary current of 0.1 A. Assuming an ideal transformer, what is the primary current?

11. What is the number of turns in the primary winding of the transformer shown in Figure 16–17?

12. Assuming an ideal transformer, what is the power delivered to R_L in Figure 16–18?

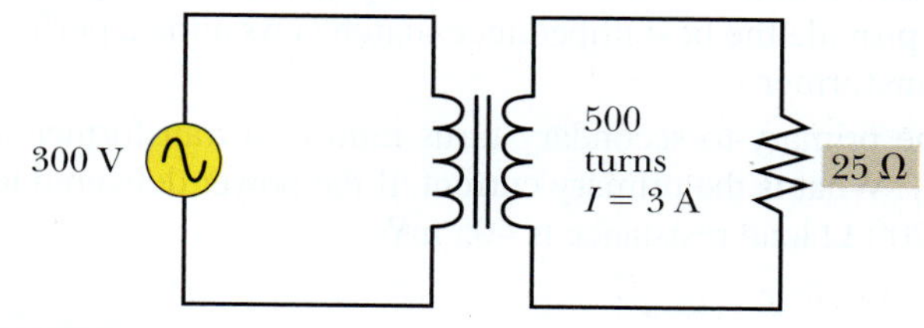

FIGURE 16–17

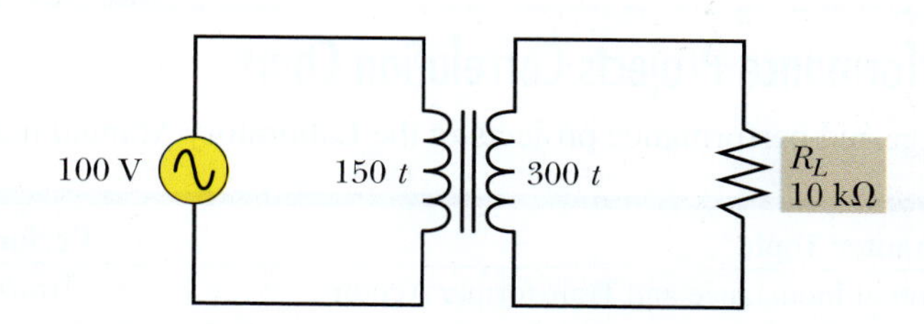

FIGURE 16–18

13. What is the primary-to-secondary turns ratio in Figure 16–19?
14. What is the primary current in the transformer circuit in Figure 16–20? (Assume an ideal transformer.)
15. What is the voltage applied to the primary of the circuit in Figure 16–21?

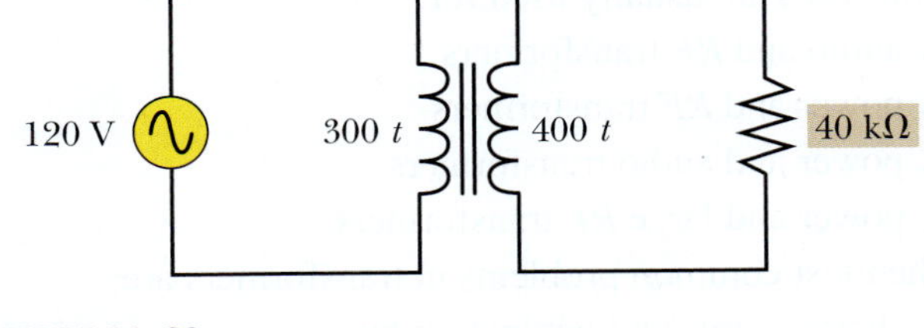

FIGURE 16–20

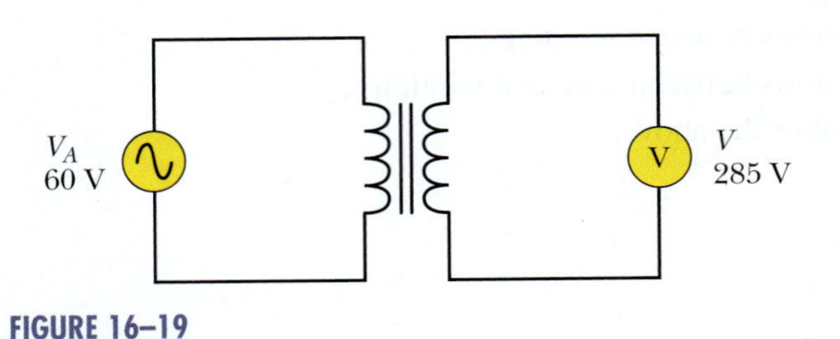

FIGURE 16–19

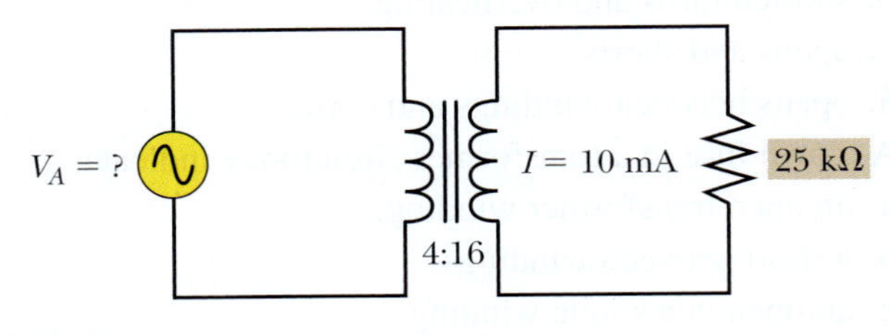

FIGURE 16–21

Analysis Questions

1. Research the term *flyback transformer.* Indicate where it is used, and whether its output is a high or low voltage. Also, list the frequency at which a TV flyback transformer is operating.
2. Why are the cores of transformers laminated iron rather than solid iron?
3. An amplifier having an output impedance of 5,000 Ω is feeding its output signal to a matching transformer. If the speaker connected to the secondary is a 16-Ω speaker and the number of turns on the secondary is 200 turns, what number of turns should the primary winding have to provide the best impedance match? (Assume a perfect transformer.)
4. The primary-to-secondary turns ratio of a transformer is 8:1. What is the primary current, if the power delivered to a 200-Ω load resistance is 405 mW?
5. If the impedance ratio (primary-to-secondary) of an ideal transformer is 36:1 and the number of turns on the primary is 1,200, how many turns are on the secondary winding?
6. A TV receiver contains a transformer in its power supply. If you unplug the receiver from its 120-V, 60-Hz wall plug (ac source), turn the on/off switch to the on position, and connect an ohmmeter to each lead of the ac cord's plug, the ohmmeter reads 10 Ω. Using Ohm's law, you think the current supplied to the transformer is 12 A, since $I = V/R$, or 120 V/10 Ω. Explain why the circuit is fused with only a 5-A fuse, yet, when the TV receiver is turned on, the fuse does not blow!

Performance Projects Correlation Chart

Suggested performance projects in the Laboratory Manual that correlate with topics in this chapter are:

Chapter Topic	Performance Project	Project Number
Mutual Inductance and Transformer Action	Turns, Voltage, and Current Ratios	45
Important Transformer Ratios	Turns Ratio(s) versus Impedance Ratios	46
	Story Behind the Numbers: Basic Transformer Characteristics	

NOTE: It is suggested that after completing the above projects, the student should be required to answer the questions in the "Summary" at the end of this section of projects in the Laboratory Manual.

OUTLINE

OBJECTIVES

After studying this chapter, you should be able to:

1. Define **capacitor, capacitance, dielectric, dielectric constant, electric field,** farad, ***RC* time constant,** and leakage resistance
2. Describe **capacitor charging** action and **discharging** action
3. Calculate charge, voltage, capacitance, and stored energy, using the appropriate formulas
4. Determine total capacitance in circuits with more than one capacitor (series and parallel)
5. Calculate circuit voltages using appropriate RC time constant formulas
6. List and describe the physical and electrical features of at least four types of capacitors
7. List typical capacitor problems and describe troubleshooting techniques
8. Use the computer to solve circuit problems
9. Use the SIMPLER troubleshooting sequence to solve the Troubleshooting Challenge problem

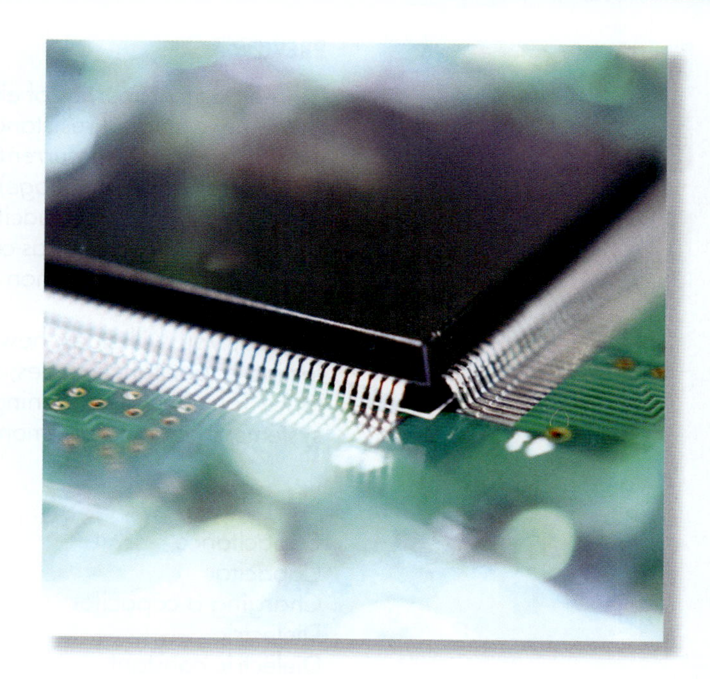

CHAPTER 17

Capacitance

PREVIEW

Three basic properties of electrical circuits are resistance, inductance, and capacitance. You have studied resistance (opposition to current flow) and inductance (property opposing a *change in current*). In this chapter, you will study capacitance (property opposing a *change in voltage*).

You will learn that capacitors have the unique ability to store electrical energy because of an electric field. Even as current produces a magnetic field, voltage establishes an electric field. This phenomenon and other important dc characteristics of capacitors will be analyzed.

Also, you will study how capacitors charge and discharge; the physical factors affecting capacitance values; some examples of capacitors; how to calculate total capacitance for circuits containing more than one capacitor; examples of capacitor coding systems; and some common capacitor problems and troubleshooting techniques.

KEY TERMS

Capacitance
Capacitor
Charging a capacitor
Dielectric
Dielectric constant
Dielectric strength
Discharging a capacitor
Electrostatic (electric) field
RC time constant

17–1 Definition and Description of a Capacitor

In *physical* terms, a **capacitor** is an electrical component, generally consisting of two conducting surfaces (often called capacitor plates) that are separated by a nonconductor (called the dielectric), Figure 17–1.

The nonconducting material, or **dielectric,** can be any of a number of materials. Some examples of dielectric materials are vacuum, air, waxed paper, plastic, glass, ceramic material, aluminum oxide, and tantalum oxide.

The schematic symbol for the capacitor plates is one straight line and one curved line, each perpendicular to the lines representing the wires or circuit conductors connected to the plates; again see Figure 17–1.

In *electrical* terms, a capacitor is an electrical component that stores electrical charge when voltage is applied.

17–2 The Electrostatic Field

You know that unlike charges attract and like charges repel. Refer to Figure 17–2 as you study the following discussion.

Lines are used to illustrate **electrostatic fields** between charged bodies, much like lines illustrate magnetic flux between magnetic poles. Direction of the electrostatic field lines is usually shown from the positive charged body to the negative charged body. (NOTE: An electron placed in this field travels in the opposite direction from the electrostatic field. In other words, the electron is repelled from the negative body and attracted to the positive body.)

The force between charged bodies is *directly* related to the product of the charges on the bodies and *inversely* related to the square of the distance between them.

$$F = \frac{kQ_1 \times Q_2}{d^2} \quad \text{(coulomb's law)}$$

This means the closer together a capacitor's plates are for a given charge, the stronger the electric field produced. Also, the greater the charge (in coulombs) stored on the plates, the stronger the field produced.

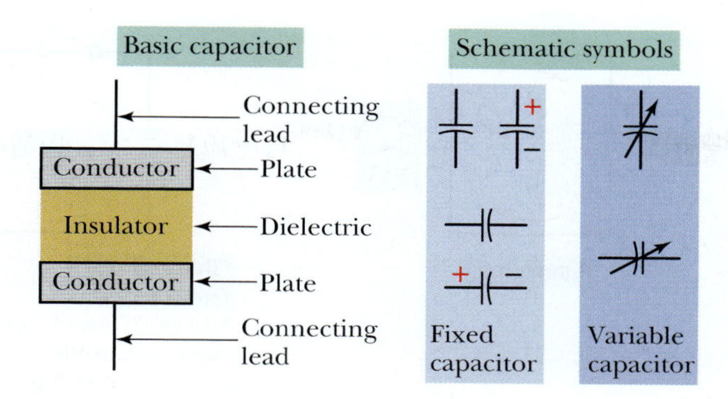

FIGURE 17–1 Basic capacitor and symbols

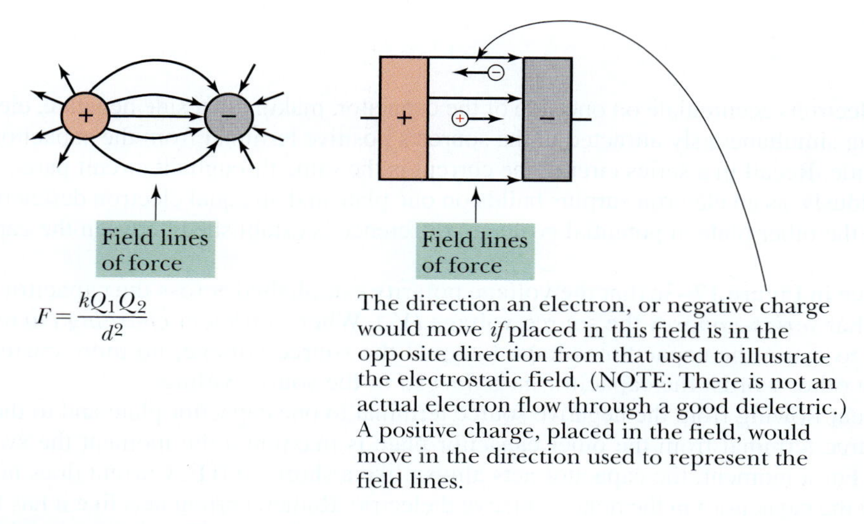

Field lines of force
$F = \frac{kQ_1Q_2}{d^2}$
Field lines of force
The direction an electron, or negative charge would move *if* placed in this field is in the opposite direction from that used to illustrate the electrostatic field. (NOTE: There is not an actual electron flow through a good dielectric.) A positive charge, placed in the field, would move in the direction used to represent the field lines.

FIGURE 17–2 The electrostatic field

The electrostatic field represents the storage of electric energy. This energy originates with a source and can be returned to a circuit when the source is removed. (This is analogous to energy stored in the buildup of an inductor's magnetic field, which may be returned to the circuit when the magnetic field collapses.)

17–3 Charging and Discharging Action

Figure 17–3a illustrates a capacitor with no *charge,* that is, no electrostatic field exists between the plates. The plates and dielectric between them are electrically *neutral.*

Charging Action

In Figure 17–3b, the circuit switch is closed and electrons move for a short time, **charging the capacitor.** Let's look at this action.

As soon as the switch is closed, neither capacitor plate has an electron excess or deficiency. Also, there is no voltage across the plates or electric field across the dielectric to oppose electron movement to or from the plates.

Electrons leave the negative terminal of the battery and travel to the capacitor plate, which is electrically connected to it. The electrons do not travel through the nonconductive dielectric material between the capacitor plates. Rather, the electrons accumulate on that side of the capacitor, establishing an electron excess and a negative charge at that location. Incidentally, the electron orbital paths in the atoms of the dielectric material (unless it is vacuum) are distorted, since they are repelled by the negative charged plate of the capacitor, Figure 17–4. (NOTE: The distorted orbits represent storage of some electrical energy.)

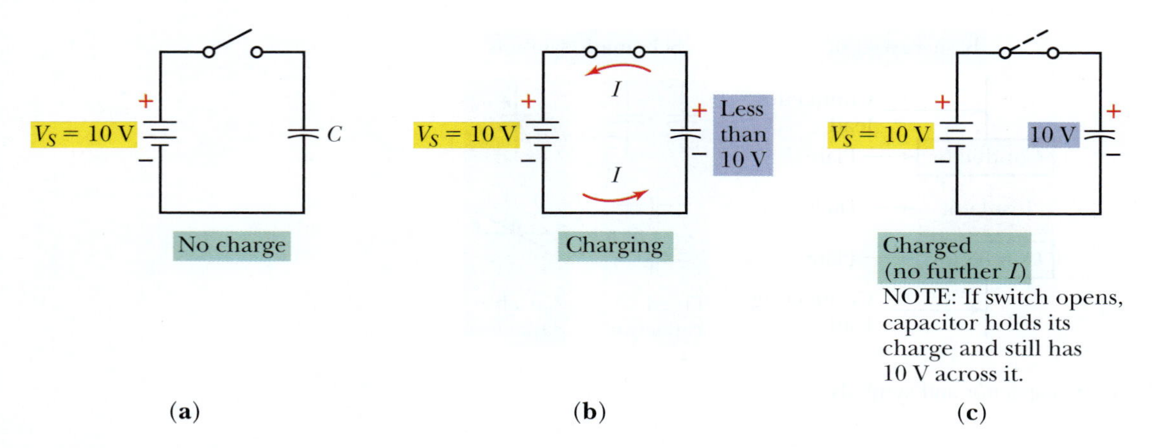

FIGURE 17–3 Capacitor charging action

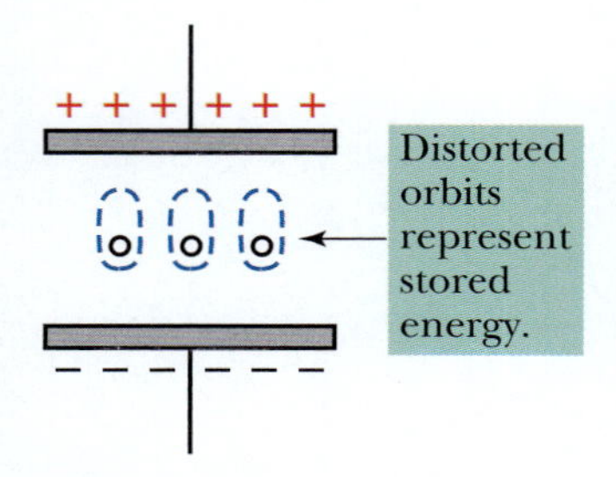

FIGURE 17–4 Distorted orbits represent stored electrical energy.

As electrons accumulate on one side of the capacitor, making that side negative, electrons are being simultaneously attracted to the source's positive terminal from the capacitor's opposite side. Recall in a series circuit, the current is the same through all circuit parts.

Obviously, as an electron surplus builds on one plate and an equal electron deficiency occurs on the other plate, a potential (voltage) difference is established between the capacitor plates.

Notice in Figure 17–3c that the voltage polarity established across the capacitor plates is one that *series-opposes* the source voltage (V_S). When sufficient *charging current* has flowed to cause the capacitor voltage to equal the source voltage, no more current can flow. In other words, the capacitor has *charged* to the source voltage.

Current flowing from the negative source terminal to one capacitor plate and to the positive source terminal from the other capacitor plate is maximum the moment the switch is closed. For a moment, the capacitor acts almost like a short. NOTE: Current does not flow *through* the capacitor via the nonconductive dielectric. Rather, current acts like it has flowed because of the electron movement on either side of the capacitor. Charging current is maximum the first instant the switch is closed and decreases to zero as the capacitor becomes *charged* to the source voltage value.

When the source voltage increases to a higher voltage, current again flows until the capacitor becomes charged to this new voltage level. Once the capacitor has charged to the source voltage, it acts like an open and blocks dc, again see Figure 17–3c.

⚠ Safety Hints

Caution: ***Always discharge circuit capacitors* after power has been removed and before working on circuits containing them!** Technicians have been killed by accidentally becoming the discharge path! By providing an appropriate discharge path to bleed the charge from circuit capacitors before working on the circuit, this danger is avoided.

Many power supplies have a "bleeder resistor," which is intended to bleed off the charge from power supply capacitors when the power is turned off. IT IS NOT SAFE to depend on the bleeder working, as they occasionally become open. A number of technicians have been killed, assuming the bleeder resistor would protect them!

Some dangerous equipment have a "shorting stick" mounted inside the cabinet of the equipment. This is a long, nonconductive, wooden or plastic stick with a large braided conductor attached at one end. The other end of the braided conductor is permanently connected to the system ground. The technician holds the end of the stick at the opposite end from which the conductor is attached to the stick, then positions the conductor end of the stick to touch the capacitor terminals to be discharged. This of course discharges the capacitor through the braid to ground. Caution must be used in doing this due to the large arc that can result. All operator safety precautions recommended by the equipment manufacturer and common sense MUST be followed when using this procedure.

The point to be made here is NEVER ASSUME LARGE CAPACITORS IN SYSTEMS ARE DISCHARGED JUST BECAUSE THE POWER HAS BEEN TURNED OFF! ■

In summary, maximum charging current flows when voltage is first applied to the capacitor. As the capacitor charges, voltage across the capacitor increases to the source voltage value and charging current decreases. When the capacitor is charged to V_S, no further current flows and the capacitor is holding a charge, which causes a voltage equal to V_S across its plates. The electrical energy has been stored in the electrostatic (dielectric) field of the capacitor.

When we open the switch in the charged capacitor circuit, Figure 17–5, and we assume the dielectric has infinite resistance to the current (i.e., no leakage current flows between the plates through the dielectric), the capacitor remains charged *as long* as no external *discharge* path for current is provided.

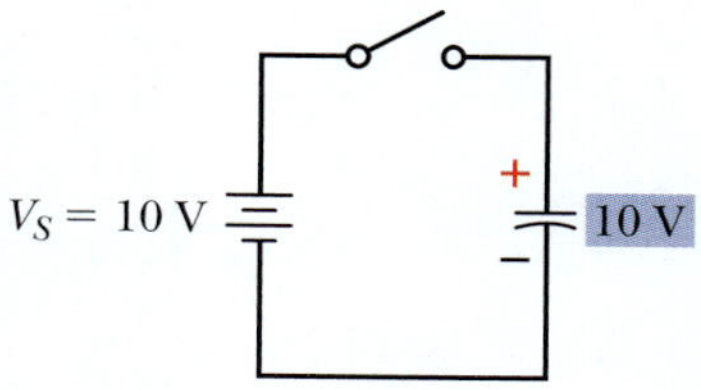

FIGURE 17–5 Capacitors retain charge until it bleeds off either through leakage R of dielectric or through some external current path.

■ IN-PROCESS LEARNING CHECK 1

Fill in the blanks as appropriate.

1. A capacitor is an electrical component consisting of two conducting surfaces called capacitor plates that are separated by a nonconductor called the dielectric.
2. A capacitor is a device that stores electrical charge when voltage is applied.
3. Electrons do not travel through the capacitor dielectric.
4. During charging action, one capacitor plate collects _____, making that plate _____. At the same time, the other plate is losing _____ to become _____.
5. Once the capacitor has charged to the voltage applied, it acts like an _____ circuit to dc. When the level of dc applied increases, the capacitor _____ to reach the new level. When the level of dc applied decreases, the capacitor _____ to reach the new level.
6. When voltage is first applied to a capacitor, _____ charge current occurs.
7. As the capacitor becomes charged, current _____ through the circuit in series with the capacitor.

Discharging Action

Figure 17–6 illustrates **discharging a capacitor.** An external path for electron movement has been supplied across the capacitor plates. The excess electrons on the bottom plate quickly move through this path to the top plate, which has an electron deficiency. The charge difference between the plates is neutralized, and the potential difference between the plates falls to zero. When full neutralization occurs and the potential difference between plates is zero, we have completely discharged the capacitor.

17–4 The Unit of Capacitance

You have seen that **capacitance** relates to the *capacity* of a *capacitor* to store electrical charge. As you might expect, the more charge a capacitor stores for a given voltage, the larger its capacitance value must be. The unit of capacitance is the farad, named in honor of Michael Faraday. The farad is that amount of capacitance where a charge of 1 coulomb develops a potential difference of 1 volt across the capacitor plates, or terminals. The farad is abbreviated as capital F.

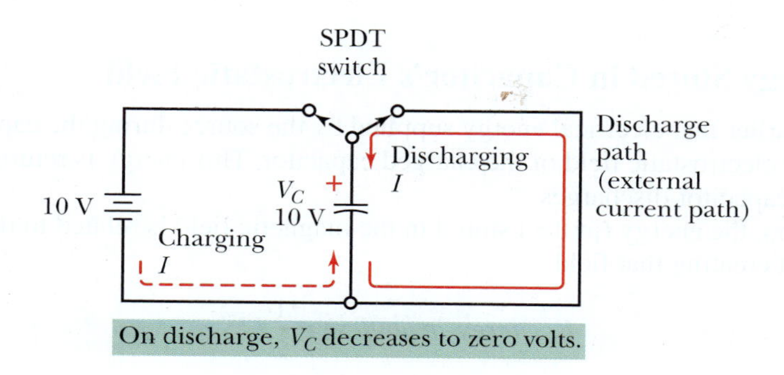

FIGURE 17–6 Discharging a capacitor

From this definition, relationships between the electrical parameters of capacitance *(C)*, voltage *(V)*, and charge *(Q)* are shown as:

FORMULA 17–1 $C = \dfrac{Q}{V}$

FORMULA 17–2 $Q = CV$

FORMULA 17–3 $V = \dfrac{Q}{C}$

where V = volts
Q = coulombs
C = capacitance in farads

As you can see from Formula 17–2, the amount of charge a capacitor stores is *directly related to* the amount of *capacitance and* the *voltage* across the capacitor.

EXAMPLE When 10 coulombs of charge are stored on a capacitor charged to 2 V, the capacitance (in farads) is calculated as:

$$C\text{ (farads)} = \frac{Q\text{ (coulombs)}}{V\text{ (volts)}}$$

$$C(F) = \frac{10\text{ C}}{2\text{ V}}$$

$$C = 5\text{ F}$$

NOTE: The unit of capacitance *(C)* is the farad (F). The unit of charge *(Q)* is the coulomb, also abbreviated as "C."

It should be noted that the farad is a large amount of capacitance. Practical capacitors are usually in the range of millionths of a farad (μF) or millionths of a millionth of a farad, formerly called micro-micro farad (μμF). This value is now called picofarad, pF. Using powers of 10, these values are expressed as:

$$\text{One microfarad} = 1\ \mu\text{F} = 1 \times 10^{-6}\text{ F}$$

$$\text{One picofarad} = 1\text{ pF} = 1 \times 10^{-12}\text{ F}$$

PRACTICE PROBLEMS 1

1. What is the capacitance of a capacitor that develops 25 V across its plates when storing a 100-μC charge? (Express answer in microfarads.)
2. What is the charge on a 10-μF capacitor when the capacitor is charged to 250 V? (Express answer in μC.)
3. What voltage develops across the plates of a 2-μF capacitor with a charge of 50 μC?

17–5 Energy Stored in Capacitor's Electrostatic Field

It was stated earlier that electrical energy supplied by the source during the capacitor charge is stored in the electrostatic field of the charged capacitor. This energy is returned to the circuit when the capacitor discharges.

For inductors, the energy (joules) stored in the magnetic field is related to the inductance and the current creating that field

$$E(\textit{energy in joules}) = \frac{1}{2}LI^2$$

However, for capacitors, the joules of stored energy are related to the capacitance and the voltage, or potential difference, between capacitor plates created by the electrostatic field. The formula for energy (in joules) is:

FORMULA 17–4 $E \text{ (energy in joules)} = \frac{1}{2}CV^2 \text{ or } 0.5\ CV^2$

EXAMPLE How many joules are stored in a 5-μF capacitor that is charged to 250 V?

$$\text{Energy } (J) = \frac{1}{2}CV^2$$

$$\text{Energy} = \frac{5 \times 10^{-6} \times (250)^2}{2} = \frac{0.3125}{2} = 0.15625 \text{ J}$$

PRACTICE PROBLEMS 2

How much electrical energy is stored in a 10-μF capacitor that is charged to 100 V?

17–6 Factors Affecting Capacitance Value

Capacitance value is related to how much charge is stored per unit voltage. The physical factors influencing the capacitance of parallel-plate capacitors are size of plate area, thickness of dielectric material (spacing between plates), and type of dielectric material, Figure 17–7. Let's examine these factors.

- **Plate area.** Obviously, the larger the area of the plates facing each other, the greater the number of electrons (and charge) that can be stored. Therefore in Figure 17–7, C is directly related to plate area. For example, when the plate area doubles, there is room for twice as many electric lines of force with the same potential difference between plates. Thus, the amount of charge the capacitor holds per given voltage has doubled and the capacitance has doubled. The area of the plates directly facing each other equals the area of one plate. Using SI units, this area is expressed in square meters.
- **Distance between plates (dielectric thickness).** From previous discussions, it is logical to assume the further apart the plates, the less the electric field strength established for a given charge and potential difference between plates, and the less the electrical energy or charge stored in this electric field. In other words, the *capacitance value is inverse to the spacing between plates.* Conversely, the closer the plates (the thinner the dielectric material), the stronger the electric field intensity per given voltage and the higher the capacitance value. In SI units, this distance (thickness of dielectric) is expressed in meters.
- **Type of dielectric material.** Different dielectric materials exhibit different abilities to concentrate electric flux lines. This property is sometimes called *permittivity,* which is analogous to conductivity of conductors. Vacuum (or air) has a permittivity of one. The relative permittivity of a given dielectric material is the ratio of the absolute permittivity (ϵ_o) of the

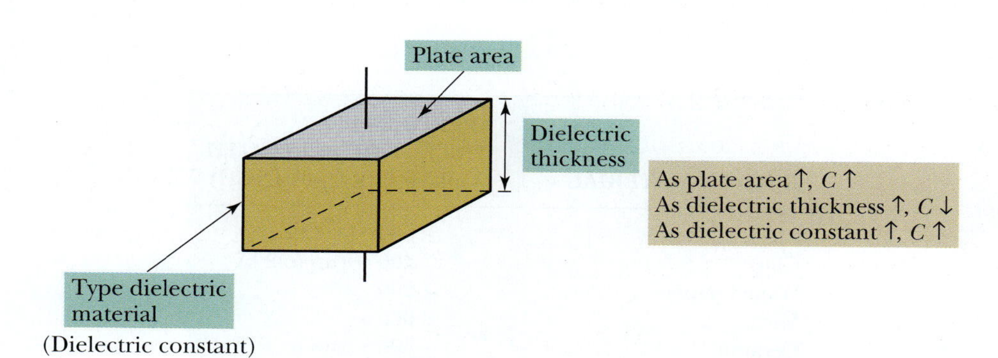

FIGURE 17–7 Physical factors affecting capacitance value

given dielectric to the absolute permittivity of free space (vacuum) (ϵ_v). This relative permittivity is called the **dielectric constant** *(k)* of the material.

Dielectric Constants

Refer to the chart in Figure 17–8 for examples of the average dielectric constant of various materials. The ratio reflecting the dielectric constant is shown as:

FORMULA 17–5 $k = \frac{\epsilon_O \text{ (absolute permittivity of dielectric materials)}}{\epsilon_v \text{ (absolute permittivity of vaccum)}}$

(NOTE: Sometimes the symbol K_ϵ is used in lieu of k to represent dielectric constant.)

The higher the dielectric constant is, the greater the density of electric flux lines established for any given plate area and spacing between plates. Therefore, capacitance is directly related to the dielectric constant. For example, when the average dielectric constant (k or K_ϵ) of a given material is five, the capacitance is five times as great with this dielectric material compared with a dielectric of air or vacuum.

Dielectric Strength

The electric field intensity is increased by increasing the voltage across a dielectric. As we continue increasing the potential difference across the capacitor plates, eventually a point is reached where the electron orbits in the dielectric material are stressed (distorted in their orbital paths), causing electrons to be torn from their orbits. At this time, the dielectric material breaks down and typically punctures or arcs, becoming a conductor. The material is no longer functioning as a nonconducting dielectric material (unless it is air or vacuum, which are self-healing). The breakdown value relates to the **dielectric strength** of the material. The practical measure of dielectric strength is called the *breakdown voltage.* The breakdown voltage depends on the kind of material and its thickness. Figure 17–9 lists some materials and their approximate average dielectric strength ratings in terms of volts per mil (thousandths of an inch) breakdown voltage ratings. Be aware these numbers will vary, depending on manufacture, frequency of operation, and so on.

FIGURE 17–8 Sample dielectric constants

KIND OF MATERIAL	APPROXIMATE *K*
Air	1.0
Glass	8.0
Waxed paper	3.5
Mica	6.0
Ceramic	100.0 +

NOTE: Values vary depending on quality, grade, environmental conditions, and frequency.

FIGURE 17–9 Examples of dielectric strengths

KIND OF MATERIAL	APPROXIMATE DIELECTRIC STRENGTH (VOLTS PER 0.001 INCH)
Air	80
Glass	200 + (up to 2 kV)
Waxed paper	1,200
Mica	2,000
Ceramic	500 + (up to ≈ 1 kV)

17–7 Capacitance Formulas

As you have learned, capacitance depends on the surface dimensions of plates facing each other, and the spacing and the dielectric constant of the material between plates. Using the appropriate factors to compute capacitance in farads using the SI units, the formula for capacitance of a parallel-plate capacitor with air (or vacuum) as the dielectric is:

FORMULA 17–6 $C = \frac{8.85A}{10^{12}s}$

where C = capacitance in farads
A = area of the plates facing each other, in square meters
s = spacing between plates, in meters the 8.85 constant relates to the absolute permittivity of air or vacuum

NOTE: If the plates have different dimensions, *use the smaller plate area for "A"* in the formula.

To calculate capacitance when dielectric materials other than air or vacuum are used, use this formula and insert the dielectric constant *(k)* in the formula. Thus:

FORMULA 17–7 $C = \frac{8.85kA}{10^{12}s}$

Again, C = capacitance in farads
A = area of the plates facing each other in square meters
k = the dielectric constant
s = the spacing between the parallel plates in meters

For your information, a practical formula used to determine the capacitance in *microfarads* (μF) of multiplate capacitors is in Figure 17–10. NOTE: This formula uses inches and square inches, rather than meters and square meters.

FORMULA 17–8 $C = 2.25\frac{kA}{10^{7}s}(n-1)$

where C = capacitance in μF
k = dielectric constant in material between plates
A = area of one plate in square inches
s = separation between plate surfaces in inches
n = number of plates

EXAMPLES

1. A two-plate capacitor has a dielectric with a dielectric constant of 3. The plates are each 0.01 square meters in area and spaced 0.001 meter apart. What is the capacitance?

$$C = \frac{8.85kA}{10^{12}s} = \frac{8.85 \times 3 \times 0.01}{10^{12} \times 0.001}$$

$$C = \frac{0.2655}{1 \times 10^{9}} = 0.0000000002655 \text{ F}$$

$$C = 265.5 \text{ pF}$$

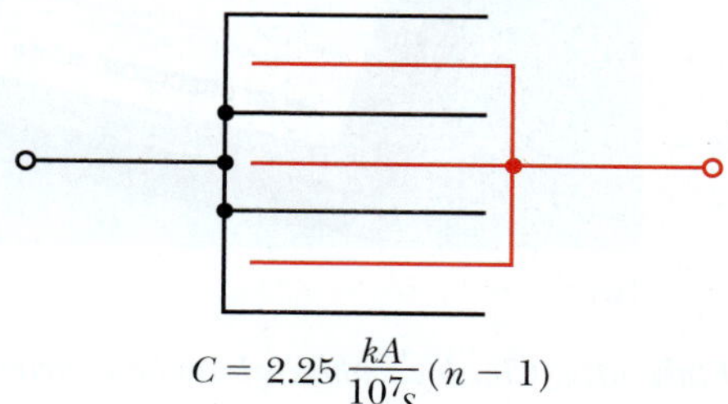

where C = capacitance in μF
k = dielectric constant of dielectric
A = area of one plate in square inches
s = separation between plates in inches
n = number of plates

FIGURE 17–10 Multiplate capacitor

Practical Notes

As a technician, you will rarely, if ever, have to actually measure capacitor plate dimensions, plate-spacing, and so on, and use the formulas we have shown you to calculate capacitance value. These formulas were introduced as a reinforcement tool to help you understand the important physical factors that influence the value of capacitance.

Technicians do, however, have to check and measure capacitances quite routinely. There are several pieces of test equipment that can be used for doing this (see Figure 17–11):

1. DMMs that have capacity-checking range(s)
2. "Dedicated" digital capacitance meters or checkers
3. LCR meters that can check inductance, resistance, and/or capacitance
4. Other (such as bridges and dip-meters)

2. What is the capacitance (in microfarads) of a capacitor if the plate area of one plate is 2 square inches, and the dielectric has a k of 100 and is 0.005 inches thick? (Assume a two-plate capacitor.)

$$C = 2.25\frac{kA}{10^7 s}(n-1)$$

$$C = \frac{2.25 \times 100 \times 2}{10^7 \times 0.005} \times (2-1) = \frac{450}{50{,}000} \times 1 = 0.009\ \mu\text{F}$$ ◆

PRACTICE PROBLEMS 3

1. A capacitor has four plates of equal size. Each plate has an area of 1 square inch. This capacitor uses a dielectric having a k of 10. The dielectric thickness between each pair of plates is 0.1 inch. What is the capacitance value in microfarads?
2. Each plate of a two-plate capacitor has an area of 0.2 square meters. The plates are uniformly spaced 0.005 meters apart and the dielectric is air. What is the capacitance in picofarads?

(a)

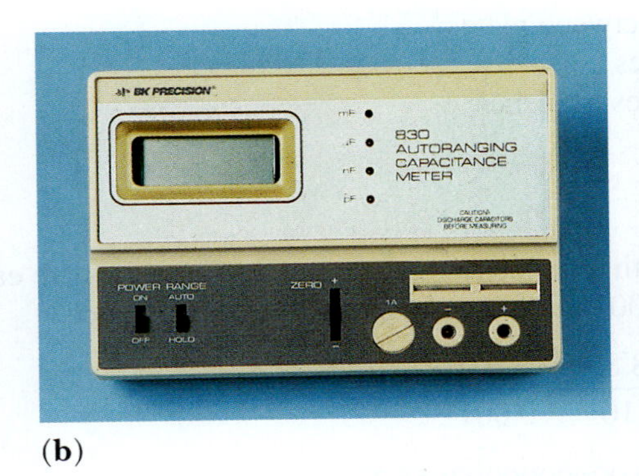

(b)

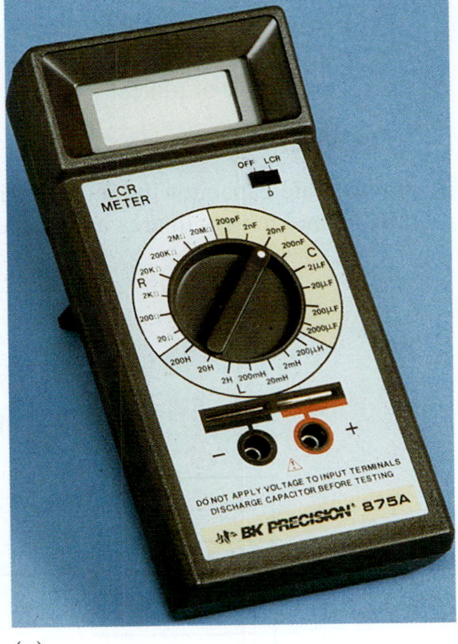

(c)

FIGURE 17–11 Test instruments used to measure capacitance *(Photo **a** courtesy of John Fluke Mfg. Co., Inc.; photos **b** and **c** courtesy of B & K Precision)*

17-8 Total Capacitance in Series and Parallel

As a technician, you may have to determine circuit parameters with capacitors connected in series or parallel. The following information will help you learn to perform such required circuit analysis.

Capacitance of Series Capacitors

Refer to Figure 17–12 as you study the following comments.

1. Connecting two capacitors in series increases the dielectric thickness, while the plate areas connected to the source remain the same, Figure 17–12. From our previous discussions, you can surmise that connecting capacitors in series *decreases* the resultant total capacitance.
2. Finding total capacitance of *capacitors in series* is analogous to finding total resistance of *resistors in parallel.*
3. The formulas to determine total capacitance (C_T) of series-connected capacitors are as follows:

   a. For two capacitors in series:

FORMULA 17–9 $$C_T = \frac{C_1 \times C_2}{C_1 + C_2}$$

   b. For equal-value capacitors in series:

FORMULA 17–10 $$C_T = \frac{C}{N}$$

   where C is the value of one of the equal-value capacitors and N is the number of capacitors in series.

   c. The general formula for any number and value of capacitors in series:

FORMULA 17–11 $$C_T = \frac{1}{\frac{1}{C_1} + \frac{1}{C_2} + \frac{1}{C_3} \cdots + \frac{1}{C_n}}$$

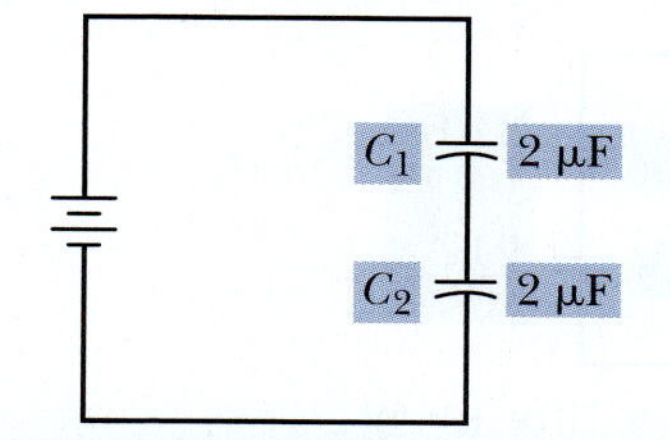

For two series capacitors:

$$C_T = \frac{C_1 \times C_2}{C_1 + C_2}$$

For equal value C's:

$$C_T = \frac{C_1}{N}$$

Generalized formula:

$$C_T = \frac{1}{\frac{1}{C_1} + \frac{1}{C_2} \cdots + \frac{1}{C_n}}$$

Equivalent: Dielectric thicknesses add and effective plate area remains the same; therefore, $C\downarrow$.

FIGURE 17–12 Total capacitance of series capacitors

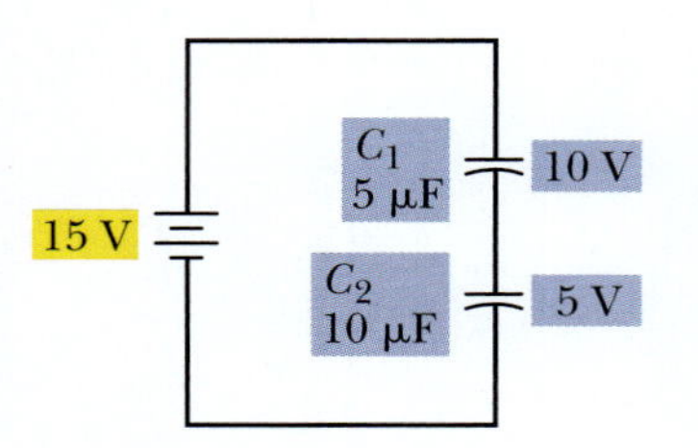

FIGURE 17–13 Voltage distribution across series capacitors

Voltage Distribution Related to Series Capacitors

Because the current throughout a series circuit is the same at any point, capacitors connected in series have the same number of coulombs (C) of charge *(Q)*. The potential difference between plates of equal-value capacitors for a given charge is equal. But for different-value capacitors, potential difference between plates is not equal. Recall the $V = Q/C$ formula, where V is directly related to Q and inversely related to C. This indicates in a series circuit (since the Q is the same), the voltage across any given capacitor is *inversely* proportional to its C value. For example, if we have a 5-μF capacitor and a 10-μF capacitor in series, the voltage across the 5-μF capacitor is two times the voltage across the 10-μF capacitor, Figure 17–13. To obtain capacitances in the ranges described (i.e., multi-microfarad), the electrolytic capacitor is typically used. You will study this capacitor later in the chapter. However, you should take note of the preceding comments and following examples about voltage distribution.

EXAMPLES

1. What is the total capacitance of the circuit in Figure 17–14?

$$C_T = \frac{C_1 \times C_2}{C_1 + C_2} = \frac{3\ \mu\text{F} \times 6\ \mu\text{F}}{3\ \mu\text{F} + 6\ \mu\text{F}} = \frac{18\ \mu\text{F}}{9} = 2\ \mu\text{F}$$

2. What is the total capacitance of three 10-μF capacitors connected in series?

$$C_T = \frac{C}{N} = \frac{10\ \mu\text{F}}{3} = 3.33\ \mu\text{F}$$

3. When the voltage applied to the circuit in Figure 17–14 is 90 V, what are the V_{C_1} and V_{C_2} values?

Because the voltage distribution is *inverse* to the capacitance values, V_{C_1} equals twice the value of V_{C_2}, or two-thirds of the applied voltage. Thus,

$$V_{C_1} = \frac{2}{3} \times 90 = 60\ \text{V, and } V_{C_2} = \frac{1}{3} \times 90 = 30\ \text{V}$$

> **Practical Notes**
>
> Because of the large tolerances typical of electrolytics, the voltage distribution may not "track" the theoretical values computed by using the "marked values" on the capacitors. If the capacitors' values are precisely as marked, the resulting voltage distribution will be as indicated in the preceding discussion.

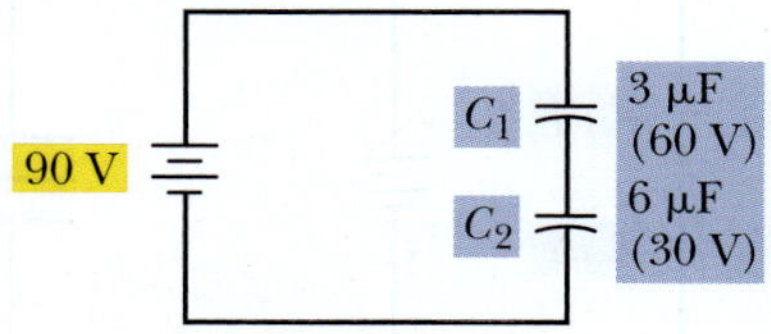

90 V
C_1 3 μF (60 V)
C_2 6 μF (30 V)

$$C_T = \frac{(3 \times 10^{-6}) \times (6 \times 10^{-6})}{(3 \times 10^{-6}) + (6 \times 10^{-6})}$$

$$C_T = \frac{18}{9} = 2\ \mu\text{F}$$

$$V_{C_1} = \frac{2}{3} \times V_T$$

$$V_{C_1} = 60\ \text{V}$$

$$V_{C_2} = \frac{1}{3} \times V_T$$

$$V_{C_2} = 30\ \text{V}$$

$$Q_T = V_T \times C_T$$

$$Q_T = 90\ \text{V} \times 2\ \mu\text{F} = 180\ \mu C$$

FIGURE 17–14

An alternate solution is to find Q by using the total capacitance value in the formula $Q = C_T \times V_T$. Then solve for each capacitor voltage value using $V = Q/C$. Therefore, since the total capacitance is 2 μF:

$$Q = 2 \times 10^{-6} \times 90 = 180\ \mu C$$

$$V_{C_1} = \frac{180\ \mu C}{3\ \mu F} = 60\ V$$

$$V_{C_2} = \frac{180\ \mu C}{6\ \mu F} = 30\ V$$

PRACTICE PROBLEMS 4

1. What is the total capacitance of the circuit in Figure 17–15?
2. What is the total capacitance of four 20-μF capacitors connected in series?
3. What are the voltages across each capacitor in Figure 17–15?

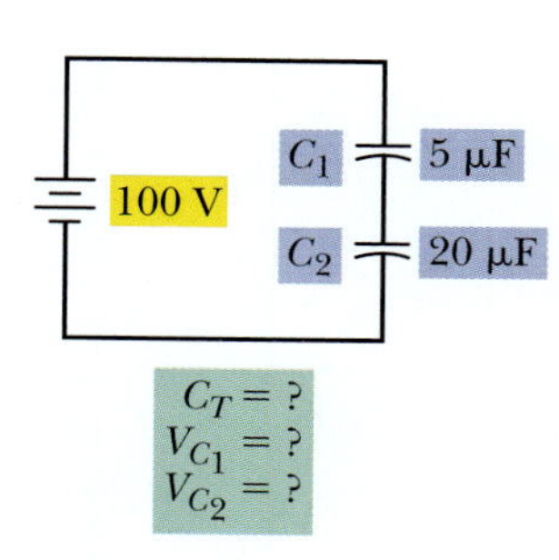

FIGURE 17–15

17–9 Finding Voltage When Three or More Capacitors Are in Series

A handy formula used to find voltage across a given capacitor when three or more capacitors are in series is (see Figure 17–16):

FORMULA 17–12 $V_X = V_S \dfrac{C_T}{C_X}$

where V_X = voltage across capacitor "x"
V_S = the dc source voltage
C_T = total series capacitance
C_X = value of capacitor "x"

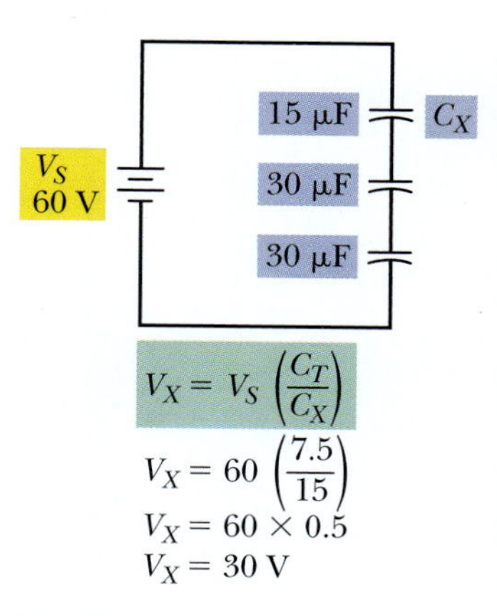

FIGURE 17–16 Series capacitors voltage formula

Capacitance of Parallel Capacitors

Refer to Figure 17–17 as you study the following information. Since the top and bottom plates of all the parallel capacitors are electrically connected to each other, the combined plate area equals the sum of the individual capacitor plate areas. However, the dielectric thicknesses do not change and are not additive.

Increasing the effective plate area while keeping the dielectric thickness(es) the same results in the total capacitance of the parallel-connected capacitors equaling the sum of the individual capacitances. In other words, *capacitances in parallel add like resistances add in series.* The formula for total capacitance (C_T) of parallel-connected capacitors is:

FORMULA 17–13 $C_T = C_1 + C_2 \ldots C_n$

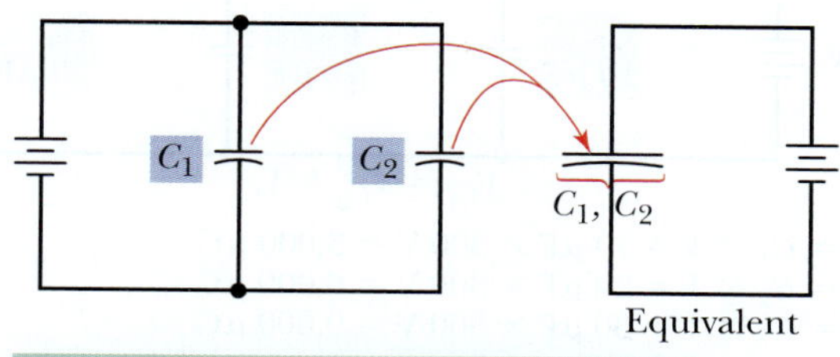

FIGURE 17–17 Capacitance of parallel capacitors

Charge Distribution Related to Parallel Capacitors

Because the voltage across components connected in parallel is the same, the charge (in coulombs) contained by each capacitor in a parallel-capacitor circuit is *directly* related to each capacitance value. That is, the larger the capacitance value, the greater the charge for a given voltage, $Q = CV$. For example, a 10-µF, 20-µF, and 30-µF capacitor are in parallel, and the voltage applied to this circuit is 300 V. Thus, the charge *(Q)* on the 10-µF capacitor is half that on the 20-µF capacitor and one-third that on the 30-µF capacitor ($Q = CV$). The Q on the 10-µF capacitor equals 10 µF × 300 V =3,000 µC. The charge on the 20-µF capacitor equals 20 µF × 300 V = 6,000 µC. The charge on the 30-µF capacitor equals 30 µF × 300 V = 9,000 µC.

NOTE: Because total charge equals the sum of the charges, it can also be computed from the $Q_T = C_T \times V_T$ formula. Therefore, total charge = 60 µF × 300 V, so Q_T = 18,000 µC, Figure 17–18.

EXAMPLES

1. What is the total capacitance of a parallel circuit containing one 15-µF capacitor, two 10-µF capacitors, and one 3-µF capacitor?

$$C_T = C_1 + C_2 + C_3 + C_4 = 15 + 10 + 10 + 3\ \mu\text{F} = 38\ \mu\text{F}$$

2. What is the charge on the 3-µF capacitor, if the voltage applied to the circuit in question 1 is 50 V?

$$Q = CV = 3 \times 10^{-6} \times 50 = 150\ \mu\text{C}$$

PRACTICE PROBLEMS 5

1. What is the total capacitance of five 12-µF capacitors connected in parallel?
2. If V_T equals 60 V, what is the charge on each capacitor in problem 1? What is the total charge on all capacitors?

17–10 The *RC* Time Constant

At the beginning of the chapter, we said a capacitor opposes a *change* in voltage. You have seen how charge current is required to establish a potential difference between the capacitor plates. Obviously, the rate of charge for a given capacitor is directly related to the charge current. That is, the higher the current, the quicker the capacitor is charged to V_S.

You also know current allowed to flow is *inversely* related to the resistance in the current path. Therefore, the larger the resistance value in the charge (or discharge) path of the capacitor, the longer the time required for voltage to reach V_S across the capacitor during charge, or to reach 0 V during discharge.

The larger the capacitance *(C)* value, the more charge is required to develop a given potential difference between its plates, $V = Q/C$. Because a capacitor prevents an instantaneous

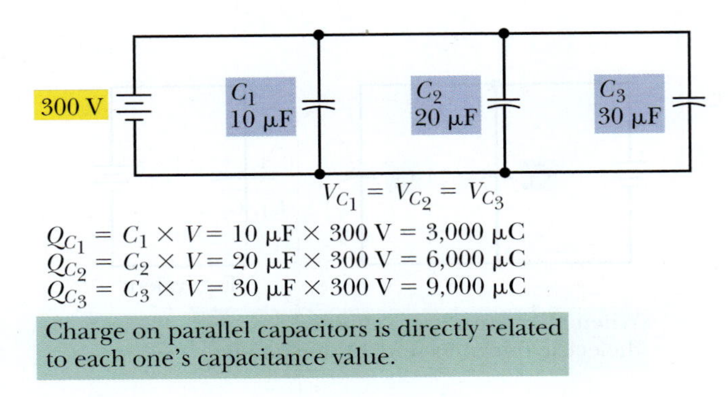

FIGURE 17–18 Charge distribution on parallel capacitors

voltage change across its plates (if there is any resistance in the charge path or discharge path), a capacitor is said to oppose a *change* in voltage across itself.

Recall that when inductors were discussed in a previous chapter, there was an L/R time constant relating the time required to change current through the coil. You learned that five L/R time constants are needed for a complete change from one circuit current value to the new level.

In like manner, an ***RC* time constant** relating the time required to charge (or discharge) a capacitor exists. *To completely change from one voltage level to a "new level" of source voltage,* the capacitor needs five *RC* time constants. *One* RC *time constant* (τ) = R *(in ohms)* × C *(in farads)*. That is:

FORMULA 17–14 $\tau = R \times C$

Refer to Figure 17–19 as you study the following information about a typical series *RC* circuit.

The sum of the dc voltages across the capacitor and the resistor equals the source voltage (V_S) at all times.

As the capacitor is charged, the voltage across the capacitor is series-opposing the source voltage. Therefore, the voltage across the resistor at any given moment equals the source voltage minus the capacitor voltage. That is, $V_R = V_S - V_C$. Thus, as the capacitor charges to the V_S value, the resistor voltage decreases toward 0 V. When the capacitor is fully charged, V_C equals V_S and zero current flows, and no *IR* drop occurs across the resistor.

When the switch is initially closed, the capacitor has zero charge and zero voltage. Therefore, the source voltage is dropped by the resistor *(R)*. Because this is a series circuit, and the current is the same throughout a series circuit, Ohm's law reveals what the charging current is at that moment.

$$I = \frac{V_R}{R}$$

In other words, at that moment, the capacitor behaves like a short; therefore, only the resistor limits current.

As the charging current flows, a series-opposing potential difference across the capacitor develops, and the charging current diminishes at a nonlinear rate (toward zero). The *IR* drop across the resistor drops accordingly. For example, after a certain interval of time, the capacitor is charged to half of V_S. At that time, half the source voltage is across the resistor, and one-half the source voltage is across the capacitor.

The percentage of voltage across the capacitor at any given moment relates to the *RC* time constant, where one time constant *(in seconds)* = R (Ω) × C (farads). Often the Greek lowercase letter tau (τ) is used to represent one time constant. Note also that τ (sec) = R (megohms) × C (microfarads). Also, observe that τ *(μsec)* = R (Ω) × C (μF).

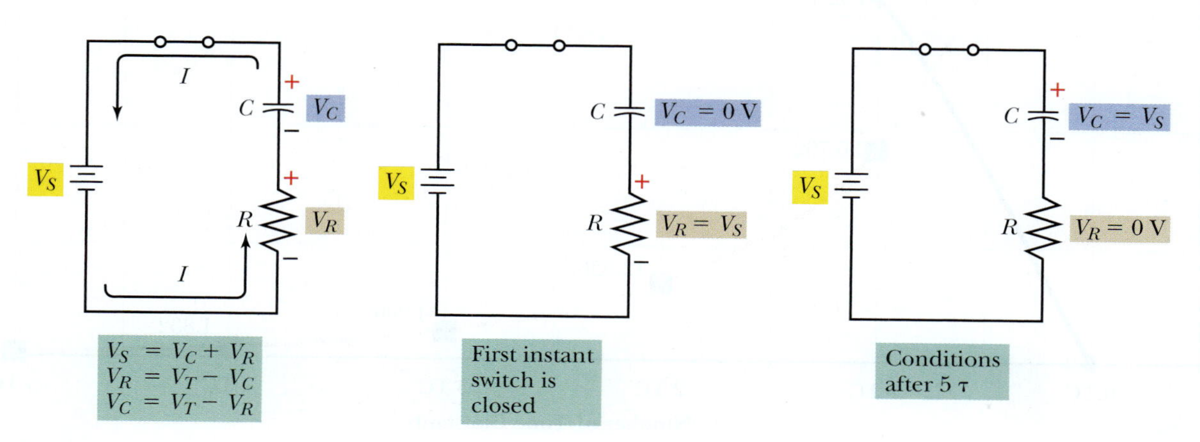

FIGURE 17–19 Voltages in an *RC* circuit

Refer to Figure 17–20 (curve A) and notice that after one time constant *(τ), the capacitor voltage (V_C)* equals 63.2% of V_S. After 2 τ, V_C equals 86.5% of V_S; after 3 τ, V_C equals 95%; after 4 τ, V_C equals 98.5%; and after 5 τ, V_C equals more than 99% of V_S; therefore, it is considered fully charged. Are these percentages and corresponding numbers familiar? Of course! They are the same percentages and numbers used when the *L/R* time constant was discussed. In other words, the same exponential curves (or time-constant chart curves) are applicable for analyzing voltages in a capacitor circuit as were used to analyze current in an inductor circuit.

Note in Figure 17–20 that curve B illustrates what happens to the resistor voltage as the capacitor is charged. That is, at the first instant V_R equals V_S. As the capacitor charges, V_R decreases toward zero. In effect, the resistor voltage and the circuit current are inverse to the capacitor voltage. As the capacitor voltage increases, the circuit current decreases, and thus the voltage across the resistor also decreases.

Using Exponential Formulas to Find Voltage(s)

Values along the exponential curves on the time-constant chart in Figure 17–20 are more precisely found by using some simple formulas. The values along the descending exponential curve relate to the base of *natural logarithms,* called epsilon (ϵ), which has a value of 2.718. (Recall this number of natural growth was discussed in an earlier chapter.)

When a capacitor is charging in a series *RC* circuit, the voltage across the resistor during the first instant is maximum, then decreases exponentially as the capacitor charges to the new voltage level. Thus, the resistor voltage follows the descending exponential curve shown in the time-constant chart. The level of v_R for any given point of time relates to the length of time the capacitor has charged relative to the circuit's *RC* time constant. That is, the time allowed *(t)* divided by the time of 1 *RC* time constant (τ), or *t*/τ. Obviously, when the *t* (time allowed) equals

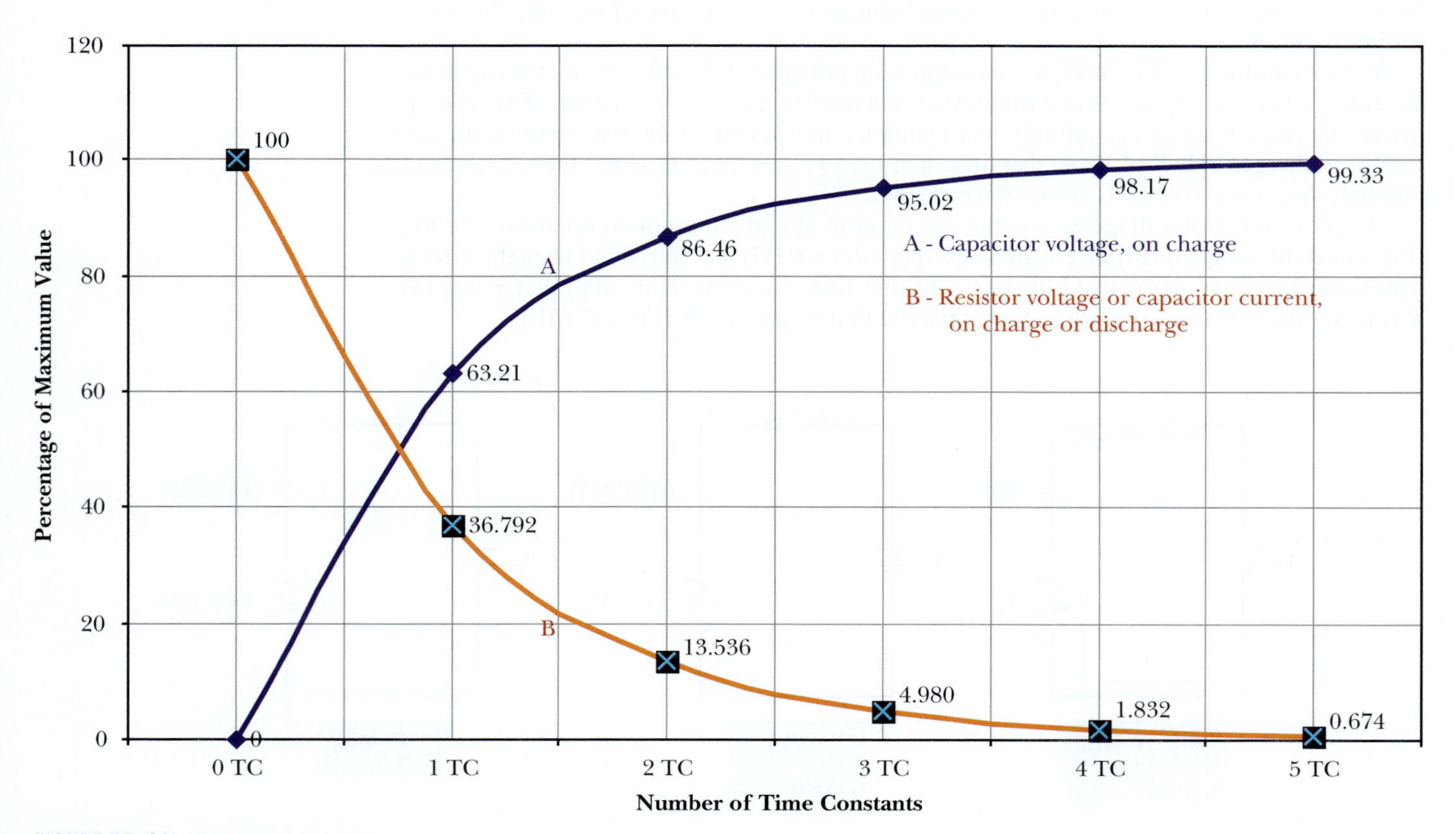

FIGURE 17–20 Time-constant chart

5 τ or more, then the capacitor has fully charged, no current is flowing, and v_R equals zero. When t is less than 5 τ, then v_R equals a value that is found by the formula:

FORMULA 17–15 $v_R = V_S \times \epsilon^{-t/\tau}$

NOTE: The small letter v denotes an instantaneous voltage value for a given moment in time.

where ϵ = epsilon (or 2.71828)
t = time allowed in seconds
$\tau = R \times C$ (or 1 time constant)

EXAMPLE For the circuit in Figure 17–21 find v_R after 10,000 μs.

$$v_R = V_S\, \epsilon^{-t/\tau}$$
$$v_R = 50 \times 2.718^{-10{,}000/(100{,}000 \times 0.1)}$$
$$v_R = 50 \times 2.718^{-1}$$
$$v_R = 50 \times 0.3679$$
$$v_R = 18.395 \text{ V}$$

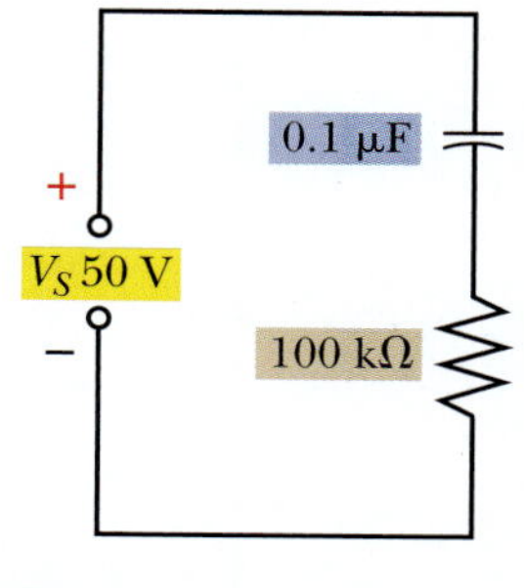

FIGURE 17–21 multiSIM

EXAMPLE For the circuit in Figure 17–21, what is the V_R after 1.8 time constants?

$$v_R = V_S \times \epsilon^{-t/\tau}$$
$$v_R = 50 \times 2.718^{-1.8}$$
$$v_R = 50 \times 0.1653$$
$$v_R = 8.265 \text{ V}$$

The calculator keystrokes that can be used to perform the calculations in this example are:

1.8, [+/–], [2nd], [e^x], [×], 50, [=]

Practical Notes

Using a calculator to find the ε value raised to the $-t/\tau$ power is easy. You simply enter the result of dividing t by τ and then press the [+/–] key to make it negative. Following that, you press the [2nd] key, then the [e^x] key to get the natural antilogarithm of the displayed number.

For the first Figure 17–21 example, the keystrokes to find v_R would be:

1, [+/–], [2nd], [e^x], [×], 50, [=]

The 1.8, [+/–], [2nd], [e^x], sequence gives the result of raising 2.718 to the negative 1.8th power, which is 0.1653. This result times the 50 V (V_A) provides the value of voltage across the resistor (8.265 V).

The capacitor voltage is the difference between the resistor voltage and the new voltage applied or source voltage for any given instant. A formula for v_C is:

FORMULA 17–16 $v_C = V_S\,(1 - \epsilon^{-t/\tau})$

For the first example then:

$$v_C = 50 \times (1 - 0.3679)$$
$$v_C = 50 \times 0.6321$$
$$v_C = 31.605 \text{ V}$$

To check, add the V_R and the v_C (18.395 + 31.605); the result is 50 V (our applied voltage). It checks out!

For the second example (see Figure 17–21):

$$v_C = 50 \times (1 - 0.1653)$$
$$v_C = 50 \times 0.8347$$
$$v_C = 41.735 \text{ V}$$

When we add v_R (8.265 V) + v_C (41.735 V), we get 50 V. It checks! ◆

PRACTICE PROBLEMS 6

Use either the time-constant chart or the formulas and a scientific calculator to solve the following.

1. What is the voltage across the capacitor of the circuit in Figure 17–22, 20 μs after the switch is closed?
2. What is the voltage across the resistor in Figure 17–22, 15 μs after the switch is closed? What is the charging current value at that moment?
3. What is the value of v_C 30 μs after the switch is closed?
4. What is the value of v_R 25 μs after the switch is closed?

17–11 Types of Capacitors

General Classification

There are two basic classifications of capacitors—*fixed* and *variable*. These capacitors come in various configurations, sizes, and characteristics. As the two names imply, fixed capacitors have one fixed value of capacitance and variable capacitors have a range of capacitance values that are available by adjusting a movable section of the capacitor.

Figure 17–23a shows a variable (tuning) capacitor. By changing the position of the *rotor* plates with respect to the fixed *stator* plates, the plate areas facing each other are changed.

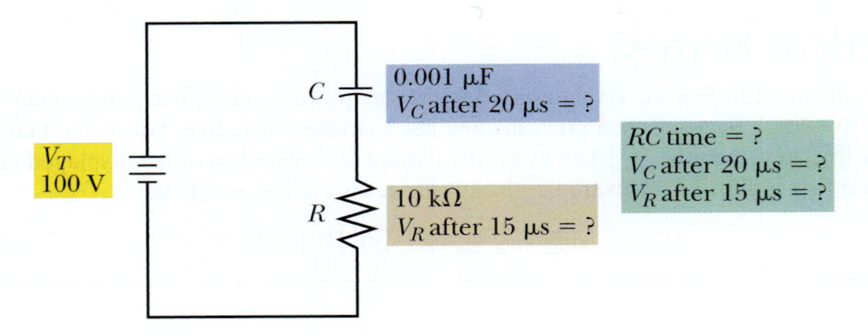

FIGURE 17–22 multiSIM

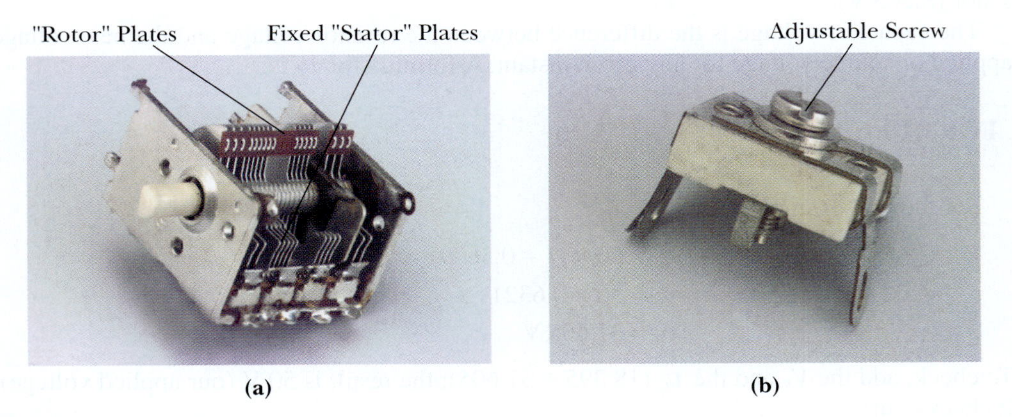

FIGURE 17–23 Variable capacitors

To move the rotor plates, the shaft upon which the rotor plates are mounted is changed in position. As you can see in the picture, the shaft protrudes through the front of the capacitor frame so that a knob could be attached to the shaft. By turning the knob, we can move the rotor plates anywhere from being fully meshed with the stator plates (maximum capacitance position) to being fully unmeshed (minimum capacitance position).

Figure 17–23b shows a small variable capacitor (sometimes called a trimmer capacitor). This small compression-type capacitor has only one movable plate. By turning the adjustable screw in, the spring-like compression plate can be squeezed tightly against the bottom fixed plate (maximum capacitance position, as shown in the picture). If the screw is turned out, or loosened, the compression-type plate follows the screw outward. This widens the gap between the compression-type plate and the fixed bottom plate, decreasing the capacitance value. The wider the gap, the less the capacitance.

Capacitors and Their Characteristics

Capacitors are generally referred to in terms of their dielectric material. The following discussion describes several common capacitors and their features.

1. *Paper and plastic capacitors* use a variety of dielectric materials. Some of these include mylar, polystyrene and polyethylene for the plastic types, and waxed or oiled paper for the older, less expensive paper type. Typically, the plates are long strips of tinfoil separated by the dielectric material. The foil and dielectric material are commonly rolled into a cylindrical component, Figure 17–24. It is interesting to note that because of their construction, each plate has two active surfaces. This means to calculate the plate areas, use twice the area of one plate, rather than only the area of one plate. Typical capacitance values for this type capacitor range from about 0.001 μF to 1 μF or more. Paper and plastic capacitors are typically used in fairly low-frequency applications such as in audio amplifiers.
2. *Mica capacitors* use mica for the dielectric material. Because the mica dielectric has a high breakdown voltage, these low-capacitance, high-voltage capacitors are frequently found in high-voltage circuits. Often, their construction is alternate layers of foil with mica that is molded into a plastic case, Figure 17–25. They are compact, moisture-proof, and durable. Voltage ratings are in thousands of volts. Typical capacitance values range from about 5 to 50,000 picofarads, depending on voltage ratings.
3. *Ceramic capacitors* are typified by their small size and high-dielectric strength. Ceramic capacitors generally come in the shape of a flat disk (disk-ceramics), Figure 17–26a, or in cylindrical shapes. These capacitors are also compact, moisture-proof, and durable. Typical available ranges having 1,000-V ratings are from approximately 5 pF to about 5,000 pF. At lower voltage ratings, higher capacitance values are available. Because of good dielectric characteristics, both mica and ceramic capacitors can be used in applications from the audio-frequency range up to several hundred megahertz. Figure 17–26b shows a sample of typical coding used with disk ceramic capacitors.

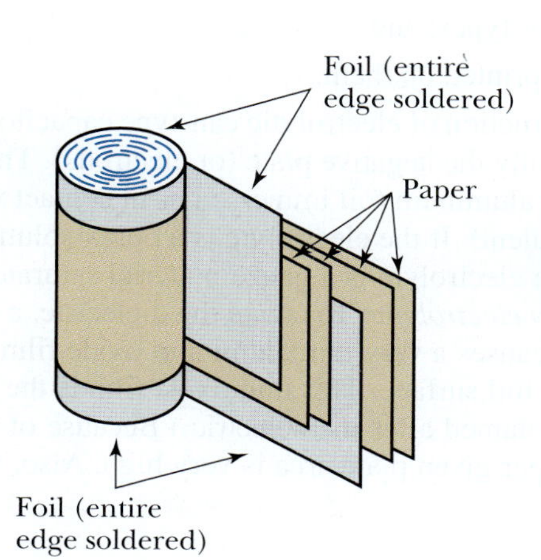

Typical available capacitance values: Approximately 0.001 to 1.0 μF

Typical working voltage dc (WVDC): 100 to 1,500 V

FIGURE 17–24 Paper-type capacitor

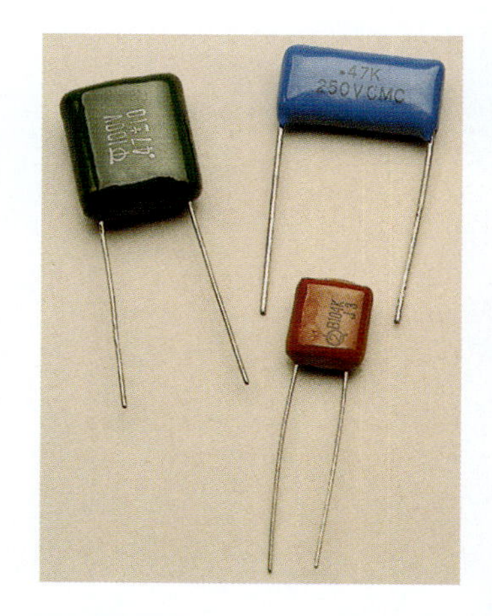

FIGURE 17–25 Mica capacitors *(Photo by Michael A. Gallitelli)*

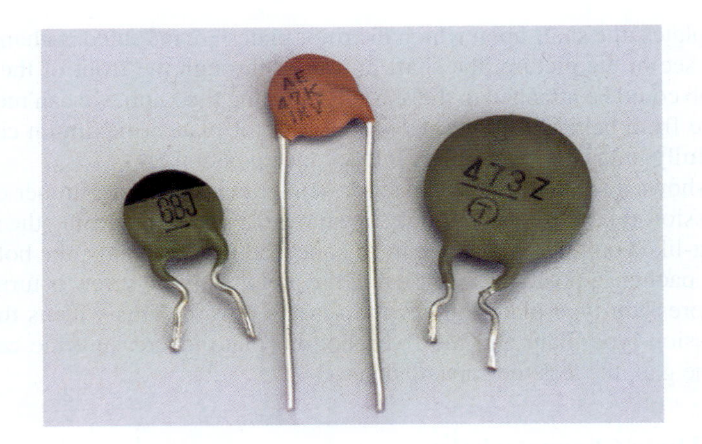

(a)

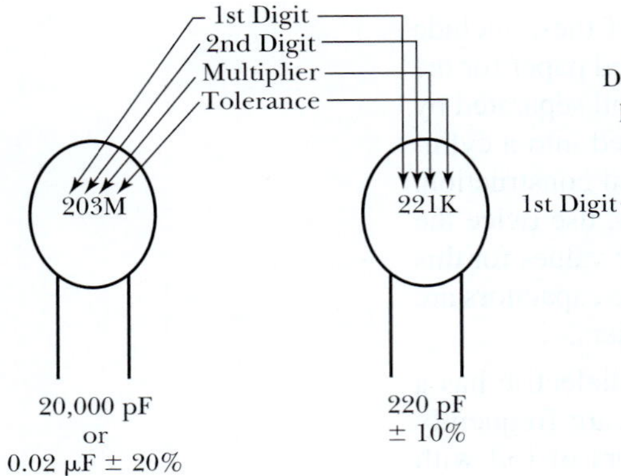

Third-Digit Multiplier

Number	Multiply by
0	1
1	10
2	100
3	1,000
4	10,000

Letter Tolerance Code

Under 10 pF Values:

Letter	Tolerance
B	± 0.1 pF
C	± 0.25 pF
D	± 0.5 pF
F	± 1.0 pF

Over 10 pF Values:

Letter	Tolerance
E	± 25%
F	± 1%
G	± 2%
H	± 2.5%
J	± 5%
K	± 10%
M	± 20%
P	− 0%, + 100%
S	− 20%, + 50%
W	− 0%, + 200%
X	− 20%, + 40%
Z	− 20%, + 80%

(b)

FIGURE 17–26 (a) Ceramic capacitors; (b) sample ceramic disk capacitors coded markings

4. *Electrolytic capacitors* have several prominent characteristics, including:
   - **a.** high capacitance-to-size ratio;
   - **b.** polarity sensitivity and terminals marked + and –;
   - **c.** allowing more leakage current than other types; and
   - **d.** having their *C* value and voltage rating printed on them.

   Refer to Figure 17–27 for the typical construction of electrolytic can-type capacitors. The external aluminum can or housing is typically the negative plate (or electrode). The positive electrode external contact is generally aluminum foil immersed or in contact with an electrolyte of ammonium borate (or equivalent). If the electrolyte is a borax solution, the capacitor is called a *wet electrolytic*. If the electrolyte is a gauze material saturated with borax solution, the capacitor is called a *dry electrolytic*. To create the dielectric, a dc current is passed through the capacitor. This causes a very thin aluminum oxide film (about 10 microcentimeters thick) to form on the foil surface. This thin oxide film is the dielectric. (NOTE: This type of capacitor is *not* named after its dielectric.) Because of the extremely thin dielectric, capacitance value per given plate area is very high. Also, voltage

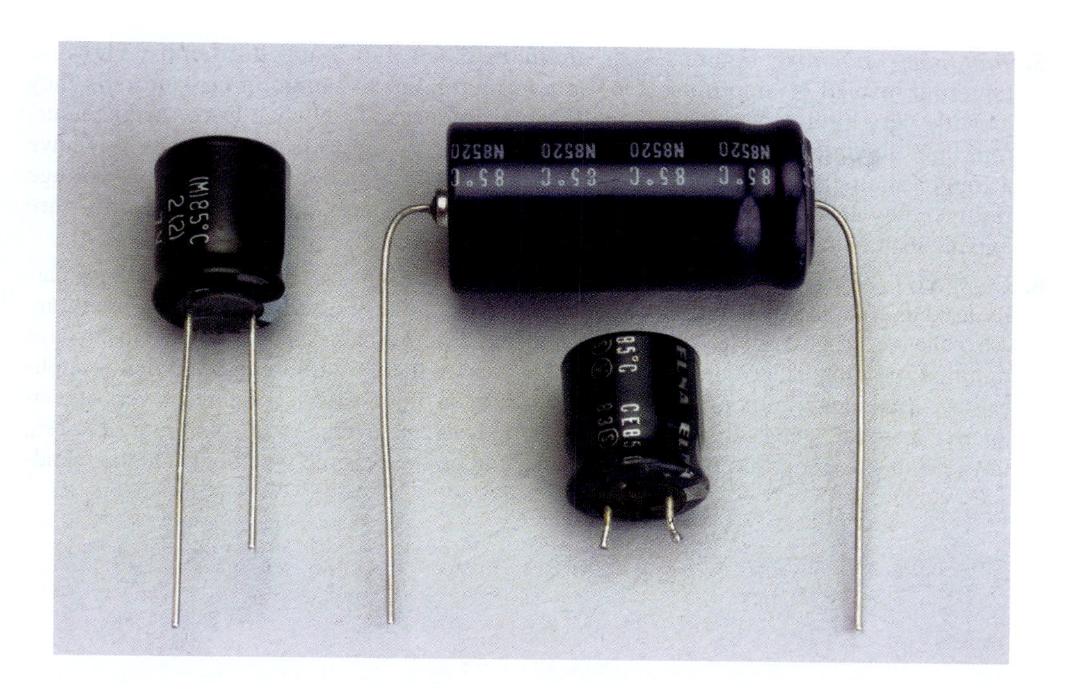

FIGURE 17–27 Aluminum-type electrolytic capacitors

Safety Hints

A **very important** fact about electrolytics is that you must ***observe polarity*** when connecting them into a circuit. If current is passed in the wrong direction through the capacitor, the chemical action that created the dielectric layer will be reversed and will destroy the capacitor. A shorted capacitor is actually created. Gas may build up in the component, causing an explosion! Depending on the voltage ratings, common values of electrolytic capacitors range from 2 μF to several hundred μF. Usual voltage ratings for this capacitor range from 10 V to about 500 V. ■

Practical Notes

Some specially made "nonpolarized" electrolytic capacitors are used in ac applications. One application for this type capacitor is as a motor-starting capacitor. This application requires a large capacitance value that must operate in an ac environment. These capacitors are actually constructed by putting two electrolytic capacitors "back-to-back" in one package. To function properly, they are connected with series-opposing polarities.

breakdown levels are relatively low. Furthermore, since the dielectric is very thin, some leakage current occurs. Typically, this leakage is a fraction of a milliampere for each microfarad of capacitance.

The main advantage of the electrolytic capacitor is the large capacitance-per-size factor. Two obvious disadvantages are the polarity, which must be observed, and the higher leakage current feature. Also, in many capacitors of this type, the electrolyte can dry out with age and depreciate the capacitor quality or render it useless. Because of the losses of the dielectric at higher frequencies, electrolytic capacitor applications are generally limited to power-supply circuits and audio-frequency applications.

5. *Tantalum capacitors,* Figure 17–28, are in the electrolytic capacitor family. They use tantalum instead of aluminum. One important quality of tantalum capacitors is they provide very high capacities in small-sized capacitors. They have lower leakage current than the older electrolytics. Also, they do not dry out as fast, therefore they have a longer shelf-life. Because they are generally only manufactured with low-voltage ratings, they are used in low-voltage semiconductor circuitry. Tantalum capacitors are expensive.
6. *Chip (SMT) capacitors,* Figure 17–29a and b. Very small in size, these devices are primarily used as compact components on PC board–type circuitry where space is at a premium. Also, the construction style, with virtually no lead length, enables these components to have minimal stray inductance or capacitance. This makes them useful in high-frequency applications. Materials commonly used in the construction of these devices include layers of conductive material, using ceramic as the dielectric between layers. As seen in Figure 17–29a, the general exterior portions of the chip capacitor are the ceramic-type body and the metal-

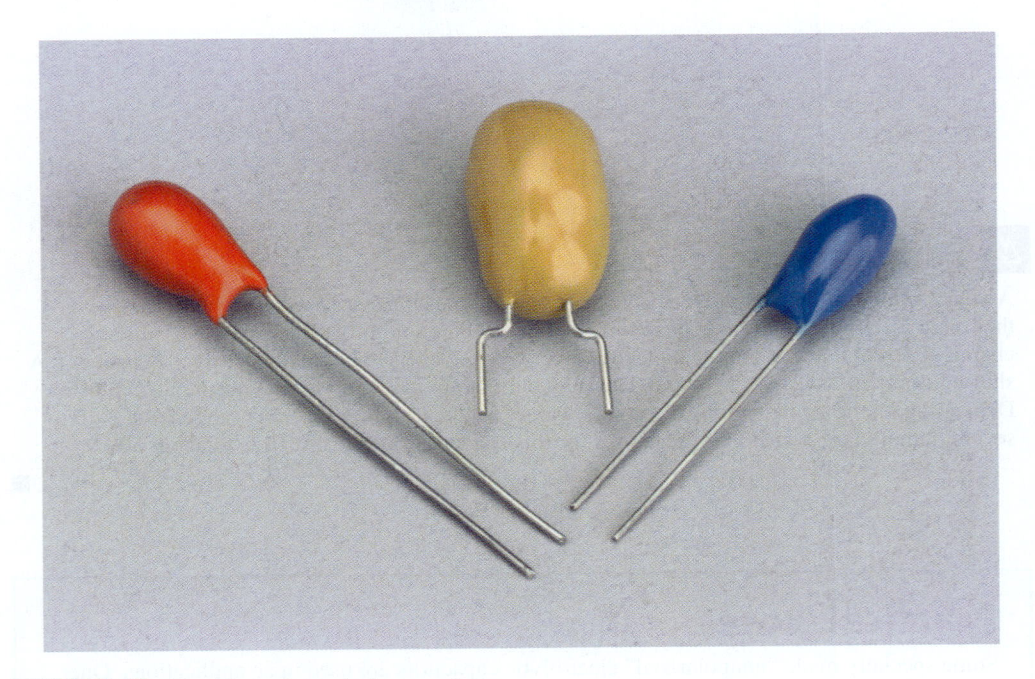

FIGURE 17–28 Tantalum electrolytic capacitors

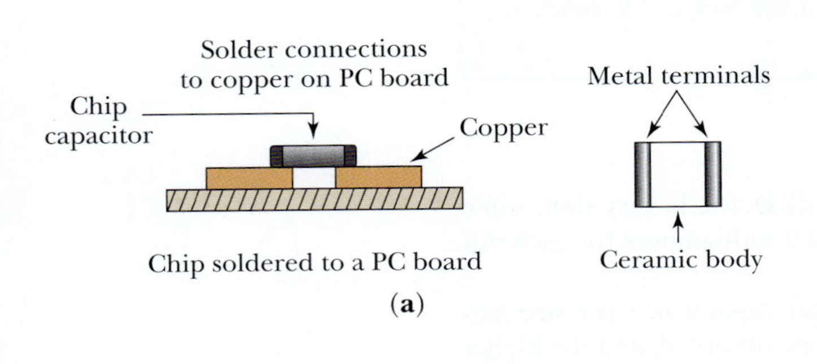

(b)

FIGURE 17–29 "Chip" capacitors

lic end contacts. These metallic end pieces are the electrical contact points to the capacitor. Their miniature size makes marking their value difficult. Therefore, they are typically marked with a short two-character code (which you may need a magnifying glass to read). This two-letter code consists of a letter, which indicates the significant digits in its value, and a number, which indicates the multiplier (or how many zeroes should follow the significant digits). This coding indicates the capacitor value in pF.

Observe Figure 17–30 to see the meaning of the letters and numbers in this two-character coding system.

Ratings

Several important factors should be observed when selecting or replacing capacitors in various applications. Some of these include:

- physical size and mounting characteristics;
- capacitance value (in microfarads or picofarads);
- capacitance tolerance (in percentage of rated value);
- voltage ratings (e.g., working volts dc or WVDC);

Surface-Mount Technology (SMT) "Chip" Capacitor Coding

SMT Capacitor Significant Figures Letter Code

Character	Significant Figures	Character	Significant Figures
A	1.0	R	4.3
B	1.1	S	4.7
C	1.2	T	5.1
D	1.3	U	5.6
E	1.5	V	6.2
F	1.6	W	6.8
G	1.8	X	7.5
H	2.0	Y	8.2
J	2.2	Z	9.1
K	2.4	a	2.5
L	2.7	b	3.5
M	3.0	d	4.0
N	3.3	e	4.5
P	3.6	f	5.0
Q	3.9	m	6.0
		n	7.0
		t	8.0
		y	9.0

SMT Capacitor Multiplier Code

Number	Decimal Multiplier
0	1
1	10
2	100
3	1,000
4	10,000
5	100,000
6	1,000,000
7	10,000,000
8	100,000,000
9	0.1

(a)

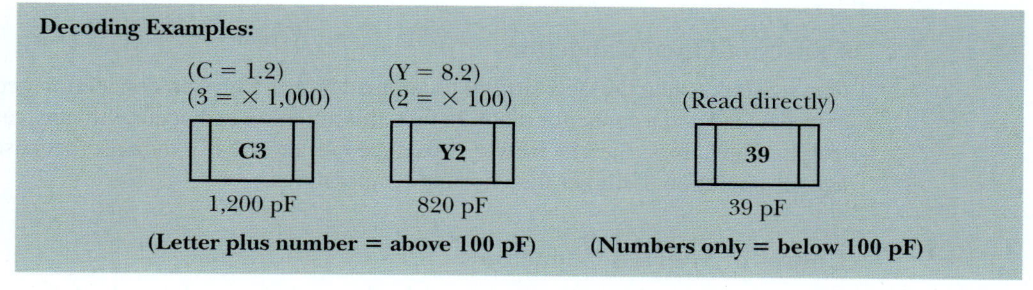

(b)

FIGURE 17–30

TYPICAL RANGE OF RATINGS FOR VARIOUS CAPACITORS
(Shown in standard electronic parts catalogs)

CAPACITOR TYPE	CAPACITANCE RATINGS	VOLTAGE RATINGS	TOLERANCE RATINGS	OPER. TEMPS	TEMP. COEFF.
Paper	0.001–1.0 μF	100–1,500 V	±10% (common)	−40 to +85°C	N/A
Mica	5–50,000 pF	600 to several kV	±1–5%	−55 to +125°C	N/A
Ceramic	1–10,000 pF	1 kV–6 kV	±10–20%	−55 to +85°C	N or P 0–750
Electrolytic	10–1,000 μF	5–500 V	−10–+50%	−40 to +85°C	N/A

NOTE: Temperature coefficient ratings are generally given for ceramic-type capacitors that are designed to increase or decrease in capacitance with temperature change. The rating is normally given in "parts-per-million per degree Celsius." For example, if a capacitor decreases 500 ppm per degree Celsius, it would be labeled "N500."

FIGURE 17–31 Summary of typical ratings

- safe temperature operating range;
- temperature coefficient (for temperature-compensating ceramic capacitors used in tuned RF circuits);
- power factor of the capacitor (expressing the loss characteristics of dielectric used in capacitor); and
- inductance characteristics (when used at high frequencies).

Taking into consideration these factors, Figure 17–31 shows examples of typical ratings for various types of capacitors.

17–12 Typical Color Codes

Capacitors are color coded with various systems. It is not critical to learn these systems at this time. However, you should be aware that they exist.

Appendix A displays some common methods of color coding for mica, ceramic disk, ceramic tubular, and paper capacitors. As in other color-coding systems, the position and color of the dots or bands conveys the desired information. (NOTE: Electrolytic capacitors are generally large enough so that the information is printed on their body, rather than using color codes. Also, ceramic capacitors often have the critical information printed on them.)

17–13 Typical Problems and Troubleshooting Techniques

Capacitors develop problems more frequently than resistors or inductors. As with the other components, troubles may be related to a shorted capacitor, an open capacitor, or degenerated performance caused by excessive leakage, age, and other causes.

Causes of Open Capacitors

One reason a capacitor acts similar to an open is that a broken connection occurs between an external lead and a capacitor plate. In fact, this situation can occur with any capacitor! In electrolytic capacitors, the electrolyte resistance can drastically increase because of drying out, age, or operation at higher than normal temperatures.

Causes of Shorted Capacitors

Shorts can occur because of external leads touching or plates shorted internally. Internal shorts are most common with paper capacitors and electrolytic capacitors. Common causes for the dielectric material breaking down are age and usage. Again, operating under high temperature conditions accelerates this aging process. In paper electrolytic capacitors, the dielectric material sometimes deteriorates with age, which decreases leakage resistance. The resulting increase in leakage current is equivalent to a *partial short* condition.

Troubleshooting Techniques

Two general methods for testing capacitors are *in-circuit* testing and *out-of-circuit* testing. There are capacitance testers that measure capacitance values, leakage characteristics, and so on; however, the ohmmeter is often useful for making out-of-circuit general-condition tests. Because of the capacitor charging action, there typically is a current surge when a voltage is first applied to the capacitor, and the current decreases to zero (or to leakage-current level) when the capacitor is fully charged. Often, this action is monitored with the ohmmeter mode on a VOM or DMM.

Using the Ohmmeter for General Tests

General Precautions

When using the ohmmeter to check electrolytic capacitors, you should *observe polarity.* First, determine the voltage polarity at the test leads of the ohmmeter when in the ohms function. Frequently, the red lead is positive and the black lead is negative *but not* for all brands and models of instruments—check with the instrument's literature or another voltmeter. Next, observe the polarity markings on the capacitor, as appropriate. Also, to protect the meter, *make sure the capacitor to be checked is discharged* before testing.

A Good Idea

Before working on a circuit, make sure all appropriate power supply capacitors are discharged, even though the power has been removed! Sometimes the "bleeder resistor" fails, and someone can get injured or killed!

Action with Good Capacitors

An out-of-circuit test can be performed with a DMM or a VOM where one capacitor lead is disconnected from the circuit, or the capacitor is physically removed from the circuit.

When checking a capacitor with a VOM, be sure a high range (e.g., $R \times 1$ meg range selector position) is used. The auto-ranging DMMs make this selection unnecessary. Just be sure that you have selected the Ohms mode for either type meter.

As you can see in Figure 17–32, after you have made sure the capacitor to be checked HAS BEEN DISCHARGED, prior to testing, you simply connect the meter leads to the capacitor leads, observing polarity, as appropriate. For electrolytic capacitors, be sure you connect the ohmmeter's test voltage so that the + polarity meter test voltage is connected to the + polarity of the capacitor. In some cases the meter's black and red leads might fool you. That is, the positive side of the ohmmeter's source may be at the black lead, rather than the red. You should check this out with another voltmeter before connecting the ohmmeter to the capacitor to be checked.

When you connect the meter leads to the capacitor, you will observe that at first, the resistance reading is low; then, it climbs higher and higher in value as the capacitor charges (series-opposing) to the ohmmeter's source voltage. Eventually, the reading should come to rest at a high resistance value, if the capacitor is good. For small-value capacitors, the reading can be close to infinite ohms. For electrolytic capacitors, the leakage resistance by this method varies from about 0.25 to 1 megohm, depending on the capacitor value and condition. When the R reading is less than this, the capacitor probably needs to be replaced.

For small-value capacitors, the charging action deflection (on a VOM), or reading changes on a DMM, are very quick, and little pointer or reading change can be seen. For larger value capacitors, the changes are often quite observable.

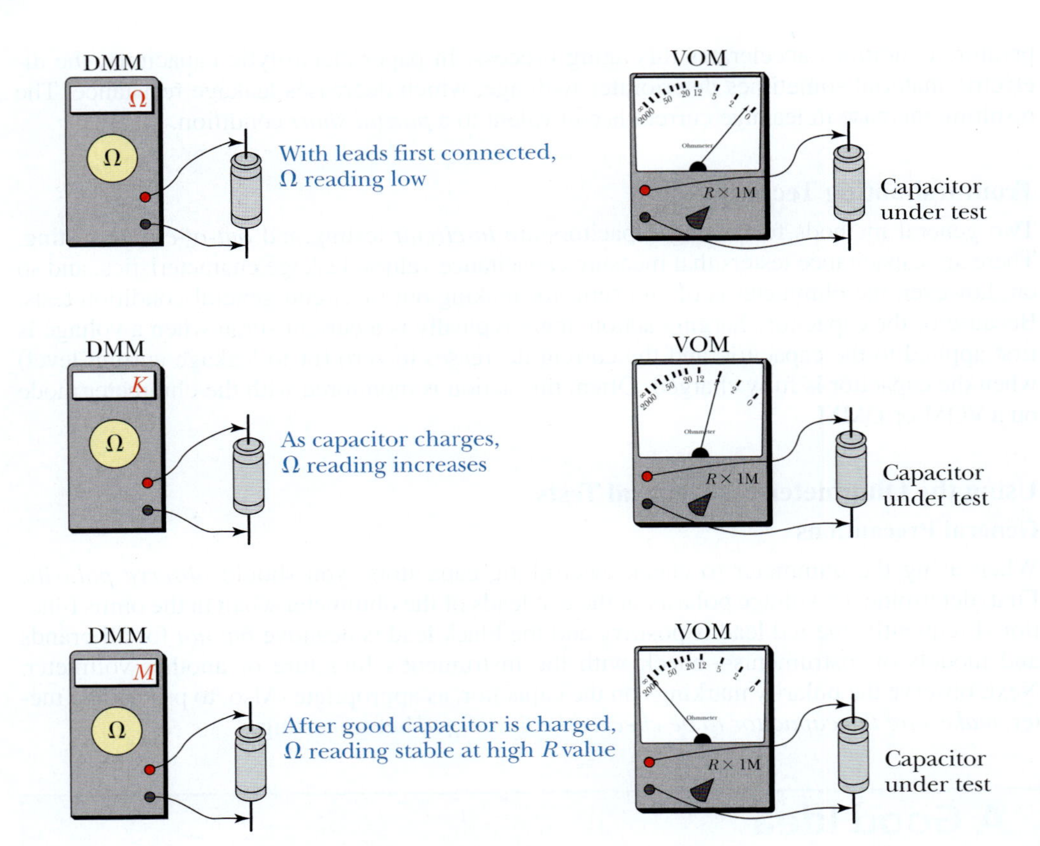

FIGURE 17–32 Checking capacitors with an ohmmeter

Open Capacitors

When an ohmmeter is connected across an open capacitor, the meter reading does not change (or an infinite Ω reading is shown), and no charging action is observed. Of course, the technician should be aware of the capacitor type and value being checked. For example, small-value capacitors (such as mica capacitors) may not provide an observable reading change, while large-value capacitors should provide some discernable charging action.

Shorted Capacitors

When an ohmmeter is connected across a shorted capacitor, the meter reading is 0 Ω and remains there. Sometimes, a variable capacitor short is remedied by locating the plate(s) that touches. In this case, simply bend the plates apart. Fixed capacitors indicating a shorted condition must be replaced.

Excessive Leakage Indication

When an ohmmeter is connected across a capacitor having excessively high leakage current (excessively low leakage resistance), the obvious result is a lower-than-normal resistance reading. The capacitor shows a charging action; however, the leakage resistance is lower than it should be.

Replacing Electrolytic Capacitors

Electrolytic capacitors definitely deteriorate with age. Their electrolyte may dry out, causing an increase in resistance and degeneration of operation. Also, the dielectric resistance greatly decreases from constant use causing excessive leakage current and reduced effectiveness in operation. When selecting an electrolytic capacitor for replacement, pay attention to the age of the capacitor—shelf-life is an important consideration.

Closing Comments

Capacitors are tremendously useful components found in most electronic circuits. Their characteristics of opposing a change in voltage, of blocking dc current once charged, and of storing electrical energy make them useful in numerous applications.

You should be aware that resistance, inductance, and capacitance are present in each of the three basic components you have studied thus far. Also, there is stray, or "distributed," capacitance and inductance present in most circuits. For example, wires and conductors that carry current exhibit some inductance, even though it may be a small amount. The leads and foil of capacitors have some inductance, as well. There is capacitance between the turns of an inductor. Distributed capacitance is present between wires and chassis, or ground. Both inductance and capacitance are present along the length of antenna elements. Coaxial cables have capacitance between conductors and inductance along their length. A resistor has a small degree of both inductance and capacitance present because of its structure.

These stray, or distributed, capacitances and/or inductances normally do not affect circuits operating at low frequencies. However, in your later studies of high-frequency components and circuits, you will see that they can play a part in proper or improper operation of circuits and systems. You will learn that the effects of some of these stray electrical quantities can be minimized by using shielding and/or using proper location of components and wise routing of wires. In summary, then, the frequency where components are expected to operate determines how critical these stray or distributed *R, L,* and *C* values are to circuit operation. (Look ahead to the illustration in Figure 21–21 to see how these stray electrical quantities are illustrated.)

Summary

- Capacitance is the circuit property that opposes a voltage *change* caused by the storing of electrical energy (charge) in an electrostatic field. The unit of capacitance is the farad (F). One farad of capacitance stores one coulomb of charge when 1 V is applied or is present across its plates. Practical subunits of the farad are the microfarad (μF) and the picofarad (μμF or pF). The symbol for capacitance is *C.*
- A capacitor is a device that possesses the property of capacitance. Capacitance is present when there are two conductors separated by an insulator (nonconductor or dielectric). Capacitance can be lumped as in the capacitor component, or distributed as the capacitance between wires and a chassis.
- Capacitors come in various sizes and shapes and in various capacitance values. Capacitors are often named after the type of dielectric used in their construction (e.g., paper, mica, and ceramic). Electrolytic capacitors are an exception and are *not* named after their aluminum oxide dielectric. This capacitor is also the only basic category that is polarity sensitive.
- Factors affecting the capacitance amount include conductive plate areas, spacing between the plates (dielectric thickness), and the type of dielectric material. The formula is

$$C = \frac{8.85kA}{10^{12}s}$$

The greater the plate areas, the greater the capacitance value. The greater the spacing between plates (the thicker the dielectric), the less the capacitance value. Also, the higher the dielectric constant *(k)*, the higher the capacitance value.

- The capacitance, charge, and voltage relationships for a capacitor are expressed as:

$$Q = CV,\ C = \frac{Q}{V},\ \text{and}\ V = \frac{Q}{C}$$

where Q is charge in coulombs; C is capacitance in farads; and V is voltage across the capacitor in volts.

- Dielectric constant (k or K_e) is the comparative charge that different materials store relative to that charge stored in vacuum or air. That is, vacuum and air have a reference k of one.
- When capacitors charge, they are accumulating electrons on their negative plate(s) and losing electrons from their positive plate(s). As this happens, the electron orbits in the dielectric material are stressed, or distorted. This stored energy is returned to the circuit when the capacitor discharges. When the capacitor discharges, the excess electrons on the negative plate(s) travels to the positive plate(s), eventually neutralizing the charge and the potential difference between plates.
- Leakage resistance is related to the resistance of the dielectric material to current flow. If the dielectric were a perfect nonconductor, the leakage resistance would be infinite. Most capacitors have very high leakage resistance, unless their voltage ratings are exceeded. Electrolytic capacitors have a lower leakage resistance than the other types because of the thinness and type of dielectric.
- When capacitors are connected in series, the total capacitance is less than the least capacitance in series. That is, capacitances in series add like resistances in parallel.
- When capacitors are in parallel, total capacitance equals the sum of all the individual capacitances. That

is, capacitances in parallel add like resistances in series, where $C_T = C_1 + C_2 + \ldots C_n$.

- Capacitors take time to charge and discharge. The time depends on the amount of capacitance storing the charge and the amount of resistance limiting the charging or discharging current. One *RC* time constant equals *R* (in ohms) times *C* (in farads). Five *RC* time constants are required for the capacitor to completely charge or discharge to a new voltage level.
- Basic types of capacitors are paper, mica, ceramic, and electrolytic types that come in a variety of combinations. Some of these variations are epoxy-dipped tantalum capacitors, epoxy-coated polyester film capacitors, dipped-mica capacitors, polystyrene film capacitors, dipped-tubular capacitors, "chip" capacitors, and so on.
- Capacitors have various color codes or coding systems. Electronic Industries Association (EIA) and MIL (military) color codes are common. Tubular capacitors often use a banded color-code system (typically six bands). Older types of mica capacitors are frequently color coded with either a five-dot, or six-dot system. Ceramic capacitors may sometimes be color coded with MIL color codes, using three-, four-, or five-color dots or bands. Chip capacitors use a two-character coding system.
- Typical problems of capacitors include opens, shorts, or degeneration in operation due to low leakage resistance, or high electrolyte resistance.
- An ohmmeter check of a good capacitor will indicate a charging action occurring when the ohmmeter is first connected to the capacitor. The capacitor is charging to the ohmmeter's internal voltage source value. This is displayed by the ohmmeter reading starting at a low resistance reading, then moving to a very high resistance reading once the capacitor has charged.
- Typical *R* values after charging are 100 or more megohms. Electrolytic capacitors have a lower leakage resistance in the range of 1/2 to 1 megohm.

Formulas and Sample Calculator Sequences

FORMULA 17–1
(To find capacitance value when capacitor charge and voltage are known)

$$C = \frac{Q}{V}$$

charge value in coulombs, [÷], voltage value, [=]

FORMULA 17–2
(To find charge value when capacitance and the voltage across the capacitor are known)

$$Q = CV$$

capacitance value in farads, [×], voltage value, [=]

FORMULA 17–3
(To find voltage value when capacitance and charge values are known)

$$V = \frac{Q}{C}$$

charge value in coulombs, [÷], capacitance value in farads, [=]

FORMULA 17–4
(To find the value of energy stored in a given capacitance with given voltage across it)

$$E \text{ (energy in joules)} = \frac{1}{2}CV^2 \text{ or } 0.5\ CV^2$$

capacitance value in farads, [×], voltage value, [x^2], [÷], 2, [=]

FORMULA 17–5
(To find the value of dielectric constant)

$$k = \frac{\epsilon_O \textit{ (absolute permittivity of dielectric material)}}{\epsilon_V \textit{ (absolute permittivity of vaccum)}}$$

ϵ_o value, [÷], ϵ_V value, [=]

FORMULA 17–6
(To find the value of capacitance for a capacitor having given physical dimensions)

$$C = \frac{8.85A}{10^{12}s} \text{ (where } A \text{ is in square meters and } s \text{ is in meters)}$$

8.85, [×], area of plates, [÷], [(], 1, [EE], 12 , [×], spacing, [)], [=]

FORMULA 17–7
(To find the value of capacitance for a capacitor having given physical dimensions)

$C = \frac{8.85kA}{10^{12}s}$ (where A is in square meters and s is in meters)

8.85, [×], k value, [×], area of plates, [÷], [(], 1, [EE], 12, [×], spacing, [)], [=]

FORMULA 17–8
(To find the value of capacitance for a capacitor having given physical characteristics)

$C = 2.25\frac{kA}{10^{7}s}(n-1)$ (where A is in square inches and s is in inches)

2.25, [×], k value, [×], A value, [÷], [(], 1, [EE], 7, [×], s value, [)], [×], [(], n value, [−], 1, [)], [=]

FORMULA 17–9
(To find the total capacitance for two capacitors in series)

$C_T = \frac{C_1 \times C_2}{C_1 + C_2}$ (series capacitors)

C_1 value, [×], C_2 value, [÷], [(], C_1 value, [+], C_2 value, [)], [=]

FORMULA 17–10
(To find the total capacitance of multiple equal-value capacitors in series)

$C_T = \frac{C}{N}$ (series capacitors)

capacitor value of one of the equal-value capacitors, [÷], number of capacitors, [=]

FORMULA 17–11
(To find the total capacitance of multiple capacitors in series)

$C_T = \frac{1}{\frac{1}{C_1} + \frac{1}{C_2} + \frac{1}{C_3} \cdots + \frac{1}{C_n}}$ (series capacitors)

C_1 value, [1/x], [+], C_2 value, [1/x], . . . , [=], [1/x]

FORMULA 17–12
(To find the voltage across a given capacitor when several capacitors are in series)

$V_X = V_S\frac{C_T}{C_X}$

V_S value, [×], [(], C_T value, [÷], C_X value, [)], [=]

FORMULA 17–13
(To find the total capacitance of parallel capacitors)

$C_T = C_1 + C_2 \ldots + C_n$ (parallel capacitors)

C_1 value, [+], C_2 value, . . . , [=]

FORMULA 17–14
(To find the time constant of a simple RC circuit)

$\tau = R$ (ohms) $\times$ C (farads)

R value, [×], C value, [=]

FORMULA 17–15
(To find the value of instantaneous voltage across the resistor in an RC circuit when the capacitor is charging)

$\upsilon_R = V_S \times \epsilon^{-t/\tau}$

time allowed value, [÷], [(], R value, [×], C value, [)], [=], [+/−], [2nd], [e^x], [×], V_S value, [=]

FORMULA 17–16
(To find the value of instantaneous voltage across the capacitor in an RC circuit when the capacitor is charging)

$V_C = V_S(1 - \epsilon^{-t/\tau})$

time allowed value, [÷], [(], R value, [×], C value, [)], [=], [+/−], [2nd], [e^x], [+/−], [+], 1, [=], [×], V_S value, [=]

EXCEL AUTOMATED FORMULAS

Helpful problem-solving tools, such as the Excel automated formulas available on the CD, allow automatic calculations of formulas.

Using Excel

Capacitance Formulas

(Excel file reference: FOE17_01.xls)

***DON'T FORGET!** It is not necessary to retype formulas once they are entered on the worksheet! Just input new parameters data for each new problem using that formula, as needed.*

- Use the Formula 17–15 spreadsheet sample and the parameters given in Practice Problems 6, question 2. Solve for the value voltage across the resistor of Figure 17–22, 15 microseconds after the start of the charge time, as stated in the question. Check your answer against the answer for this question in the Appendix.
- Use the Formula 17–16 spreadsheet sample and solve for the value of capacitor voltage, using the parameters from Practice Problems 6, question 3. Again, check your answer against the answer given in the Appendix for this question.

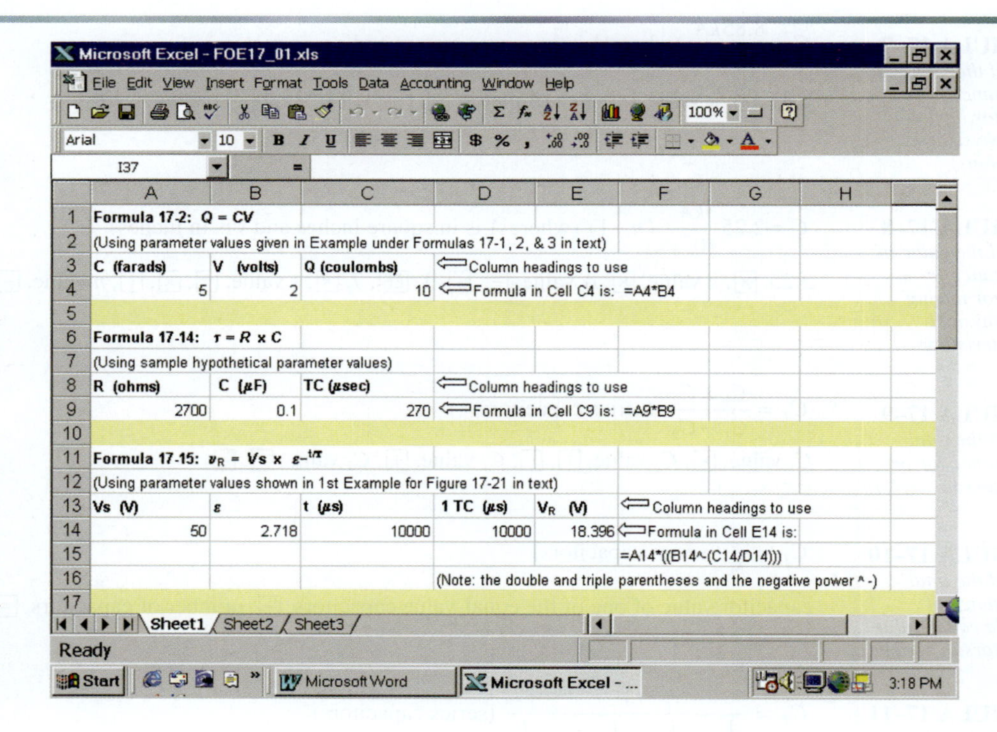

	A	B	C	D	E	F
1	**Formula 17-2: Q = CV**					
2	(Using parameter values given in Example under Formulas 17-1, 2, & 3 in text)					
3	**C (farads)**	**V (volts)**	**Q (coulombs)**	⇐Column headings to use		
4	5	2	10	⇐Formula in Cell C4 is: =A4*B4		
5						
6	**Formula 17-14: $\tau = R \times C$**					
7	(Using sample hypothetical parameter values)					
8	**R (ohms)**	**C (µF)**	**TC (µsec)**	⇐Column headings to use		
9	2700	0.1	270	⇐Formula in Cell C9 is: =A9*B9		
10						
11	**Formula 17-15: $v_R = Vs \times \varepsilon^{-t/T}$**					
12	(Using parameter values shown in 1st Example for Figure 17-21 in text)					
13	**Vs (V)**	**ε**	**t (µs)**	**1 TC (µs)**	**V_R (V)**	⇐Column headings to use
14	50	2.718	10000	10000	18.396	⇐Formula in Cell E14 is:
15						=A14*((B14^-(C14/D14)))
16				(Note: the double and triple parentheses and the negative power ^ -)		
17						

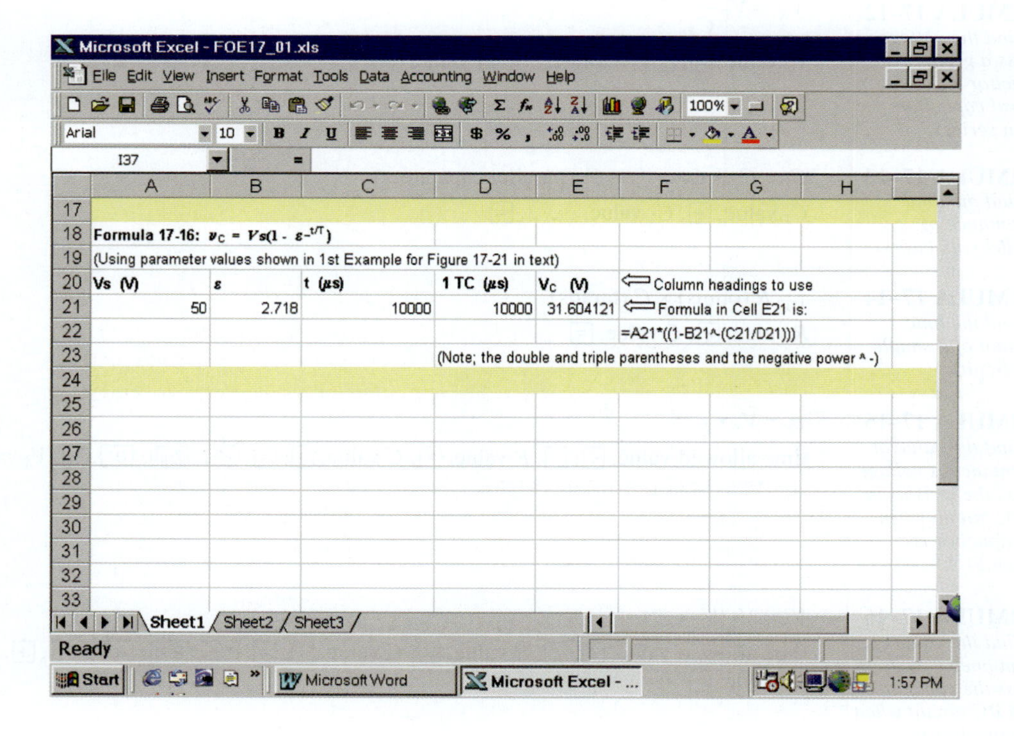

	A	B	C	D	E	F
17						
18	**Formula 17-16: $v_C = Vs(1 - \varepsilon^{-t/T})$**					
19	(Using parameter values shown in 1st Example for Figure 17-21 in text)					
20	**Vs (V)**	**ε**	**t (µs)**	**1 TC (µs)**	**V_C (V)**	⇐Column headings to use
21	50	2.718	10000	10000	31.604121	⇐Formula in Cell E21 is:
22						=A21*((1-B21^-(C21/D21)))
23				(Note; the double and triple parentheses and the negative power ^ -)		
24						

Review Questions

1. Capacitors store energy in the form of:
   a. an electromagnetic field.
   b. an electrostatic field.
   c. an electromagnetic charge.
   d. none of the above.
2. The property of capacitance requires:
   a. two or more insulators connected by conductors.
   b. two or more conductors with electrical connections.
   c. two or more conductors separated by insulator(s).
   d. all of the above.
3. Various types of capacitors are often named:
   a. after the type of dielectric used in them.
   b. after the type of conductor used in them.
   c. only after their size and shape.
   d. none of the above.
4. Capacitance value is:
   a. directly related to dielectric thickness.
   b. inversely related to dielectric thickness and plate area.
   c. inversely related to the dielectric constant.
   d. none of the above.
5. Dielectric constant relates to:
   a. comparative charge stored for a given material versus air.
   b. the fact that the dielectric does not change characteristics.
   c. the insulating properties of the nonconducting material.
   d. none of the above.
6. When a capacitor charges:
   a. it is returning energy to the circuit from which it got the energy.
   b. it is losing electrons to the source.
   c. it is building up an electrostatic field.
   d. none of the above.
7. Because dielectric materials are nonconductive:
   a. a capacitor's leakage resistance is infinite.
   b. a capacitor's leakage resistance is usually low.
   c. a capacitor cannot pass current through the circuit.
   d. none of the above.
8. Capacitances in series add like:
   a. resistances in series.
   b. inductances in series.
   c. resistances in parallel.
   d. none of the above.
9. The total capacitance of series capacitors is:
   a. greater than any one of the capacitors in series.
   b. less than any one of the capacitors in series.
   c. equal to the capacitance of one, divided by the number of capacitors.
   d. none of the above.
10. Capacitances in parallel:
    a. add like resistances in series.
    b. add like inductances in parallel.
    c. add like resistances in parallel.
    d. none of the above.
11. Charge on a capacitor is directly related to:
    a. voltage and inversely related to capacitance.
    b. capacitance and inversely related to voltage.
    c. both capacitance and voltage.
    d. none of the above.
12. The type of capacitor most likely to deteriorate with age is the:
    a. mica capacitor.
    b. ceramic capacitor.
    c. paper capacitor.
    d. electrolytic capacitor.
13. List one very impotant precaution that must be observed when connecting electrolytic capacitors into a circuit.
14. List the main advantage of an electrolytic capacitor.
15. What type of capacitor is noted for its high-voltage breakdown characteristic?
16. What are two basic classifications for capacitors?
17. List two advantages of "chip" capacitors.
18. What rule of thumb is used to indicate the amount of time it will take a capacitor to charge to the source voltage, if there is a resistor in series with the capacitor?
19. In a series dc circuit containing both a resistor and a capacitor, as the capacitor charges:
    a. the voltage across the resistor increases.
    b. the voltage across the resistor decreases.
    c. the voltage across the resistor remains the same.
    d. the voltage across the capacitor decreases.
20. Name at least two test devices that may be used to check capacitance value.

Problems

1. The capacitor plate area doubles while the dielectric thickness is halved. What is the relationship of the new capacitance value to the original capacitance value? (Assume the same dielectric material.)
2. What charge is a 100-pF capacitor storing, if it is charged to 100 V?
3. What is the capacitor value that stores 200 pC when charged to 50 V?
4. What voltage is present across a 10-μF capacitor having a charge of 2,000 μC?
5. What is the total capacitance of a circuit containing a 100-pF, a 400-pF, and a 1,000-pF capacitor connected in series? If the capacitors were connected in parallel, what would be the total capacitance?
6. A 10-μF, a 20-μF, and a 50-μF capacitor are connected in series across a 160-V source. What voltages appear across each capacitor after it is fully charged? What value of charge is on each capacitor?
7. A 10-μF, a 20-μF, and a 50-μF capacitor are connected in parallel across a 100-V source. What is the charge on each capacitor? What is the total capacitance and total charge for the circuit?
8. Using

$$V_X = V_S \frac{C_T}{C_X}$$

determine the voltage across the 20-μF capacitor in a series circuit containing a 20-μF, 40-μF, and 60-μF capacitor across a dc source of 100 V. (Show all work.)

9. Use the appropriate exponential formula(s) and find the v_R value in a series *RC* circuit after 1.2 τ if V_S is 200 V. What is the v_C value? (Use a calculator with the e^x or l_{nx} key, if possible.)

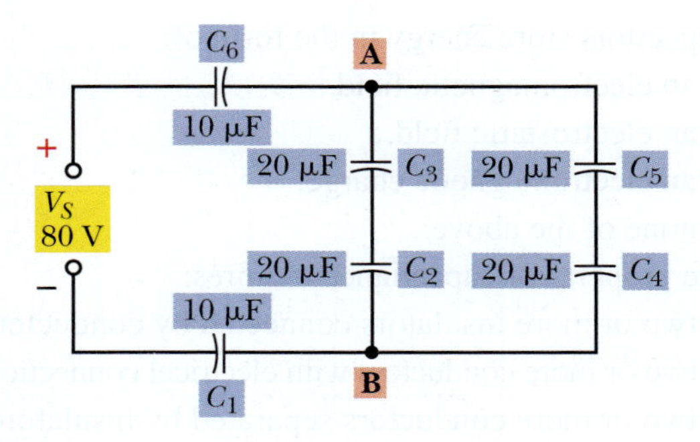

FIGURE 17–33

10. What is the total capacitance of the circuit in Figure 17–33?
11. What is the voltage across C_5 in question 10?
12. What is the charge on C_6 in question 10?
13. Assume each capacitor in question 10 has a breakdown voltage rating of 300 V. What voltage should not be exceeded between points A and B?
14. For the voltage described between points A and B in question 13, what is the V_S value that causes that condition to exist?
15. What is the total energy stored in the circuit of question 10 with 80-V V_S?
16. Assume a 15-μF capacitor is charged from a 150-V source and another 15-μF capacitor is charged from a separate 300-V source. If the capacitors are disconnected from their sources after being charged and connected in parallel with positive plate to positive plate and negative plate to negative plate, what is the resultant *voltage* and *charge* on each capacitor?
17. For the conditions described in question 16, what is the total stored energy after the capacitors are connected?

Analysis Questions

1. Research and describe the following:
   a. The procedure for using a DMM to measure capacitance values.
   b. For the particular meter you research, define the ranges of capacitance the meter can measure.
2. Research and describe the following:
   a. The procedure for using a dedicated capacitance meter to measure capacitance values.
   b. For the particular meter you research, define the ranges of capacitance the meter can measure.
3. Research and describe the following:
   a. The procedure for using an LCR meter to measure capacitance values.
   b. For the particular meter you research, define the ranges of capacitance the meter can measure.
4. Research and describe the following:
   a. Any other types of measurements, if any, that can be made relative to capacitors' characteristics or conditions.
   b. By what test equipment and/or means these other measurements might be made.

A Good Idea

After making your first test in troubleshooting, move one of the brackets you used to define the "area of uncertainty" so that the bracketed area becomes smaller! (If at all possible!) Then, make new tests only in the new, smaller bracketed area!

Performance Projects Correlation Chart

Suggested performance projects that correlate with topics in this chapter are:

Chapter Topic	Performance Project	Project Number
Charging and Discharging Action	Charge and Discharge Action and *RC* Time	47
The *RC* Time Constant		
Total Capacitance in Series and Parallel	Total Capacitance in Series and Parallel	48
	Story Behind the Numbers: Capacitance (in DC)	

NOTE: It is suggested that after completing the above projects, the student should be required to answer the questions in the "Summary" at the end of this section of projects in the Laboratory Manual.

Troubleshooting Challenge

CHALLENGE CIRCUIT 10

(Follow the SIMPLER sequence by referring to inside front cover.)

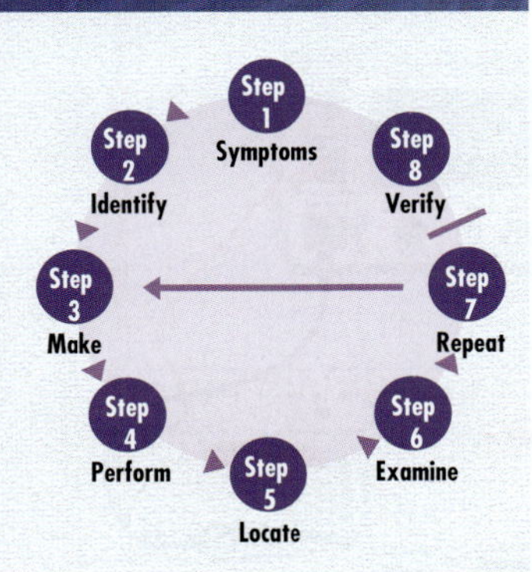

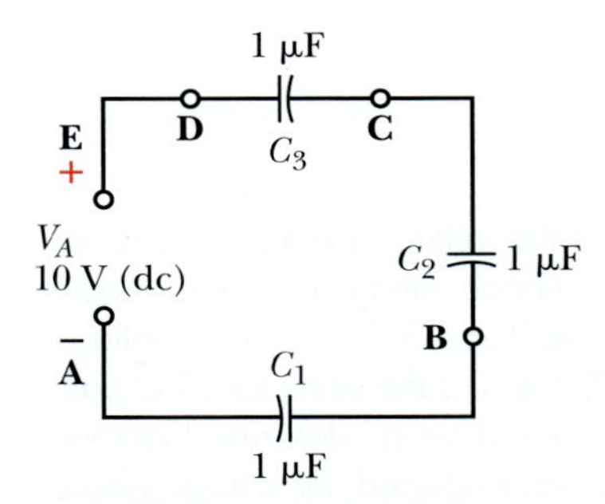

CHALLENGE CIRCUIT 10

STARTING POINT INFORMATION

1. Circuit diagram
2. V_{C_2} is lower than it should be

TEST	Results in Appendix C
V_{C_1}	(116)
V_{C_2}	(15)
V_{C_3}	(94)

CHALLENGE CIRCUIT 10

STEP 1

SYMPTOMS All we know is that the voltage across C_2 is lower than it should be. If things were normal, voltage across C_1 would be 3.3 V; the voltage across C_2 would be 3.3 V; and the voltage across C_3 would also be 3.3 V. Since Q is the same for all capacitors, the voltage distributions should be as described above. Also, we could predict that if C_2 were opened, its voltage would be 10 V (which does not agree with our symptom information); if C_2 were shorted or one of the other capacitors were opened, V_{C_2} would be 0 V; and if V_{C_2} were leaky, voltage would also be close to 0 V.

STEP 2

IDENTIFY initial suspect area: C_2 is the initial suspect area. However, we can't disregard the other capacitors since we have limited information.

STEP 3

MAKE test decision: Let's first check C_2's voltage to see where we are.

STEP 4

PERFORM **1st Test:** Check V_{C_2}. V_{C_2} is close to 0 V, which is lower than normal.

STEP 5

LOCATE new suspect area: If either one of the other capacitors were open, the voltage for C_2 would still be 0 V. If either one of the other two capacitors were leaky or shorted, C_2's voltage would be higher than normal.

STEP 6

EXAMINE available data.

STEP 7

REPEAT analysis and testing: A check of V_{C_1} or V_{C_3} might be necessary.

2nd Test: Check V_{C_3}. V_{C_3} is about 5 V, which is higher than normal, but C_3 is certainly not open or shorted. It looks like C_2 might be leaky. Let's take a look at C_2.

3rd Test: Lift one of C_2's leads and check the capacitor's dc resistance. The test reveals a lower-than-normal R.

STEP 8

VERIFY **4th Test:** Replace C_2 and check the circuit operation. The circuit checks out normal. C_2 was leaky.

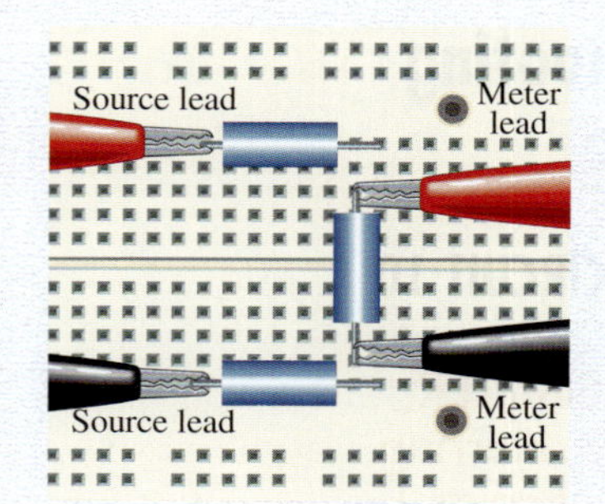

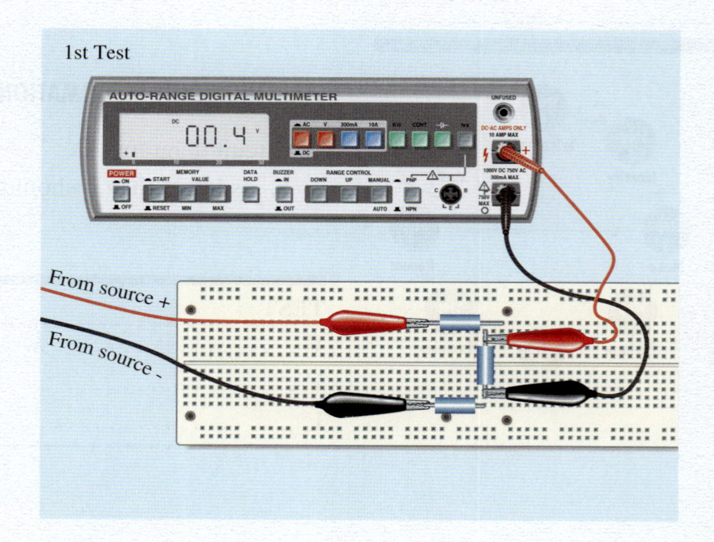

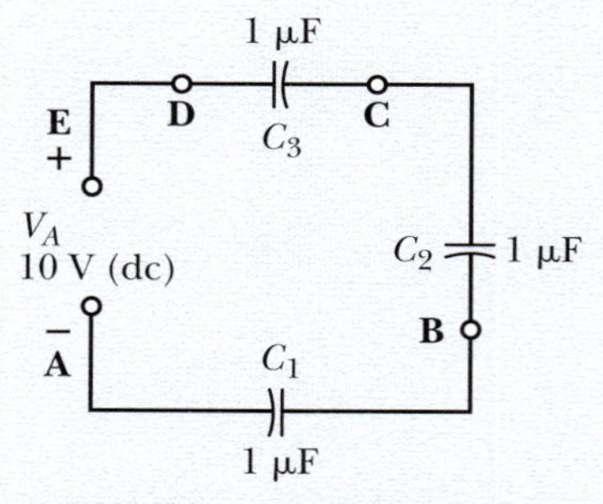

CHALLENGE CIRCUIT 10

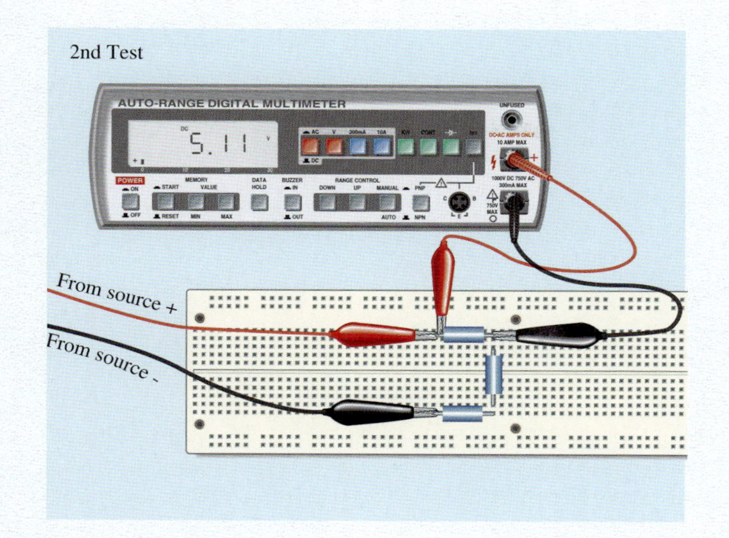
2nd Test
AUTO-RANGE DIGITAL MULTIMETER
5.11
From source +
From source -

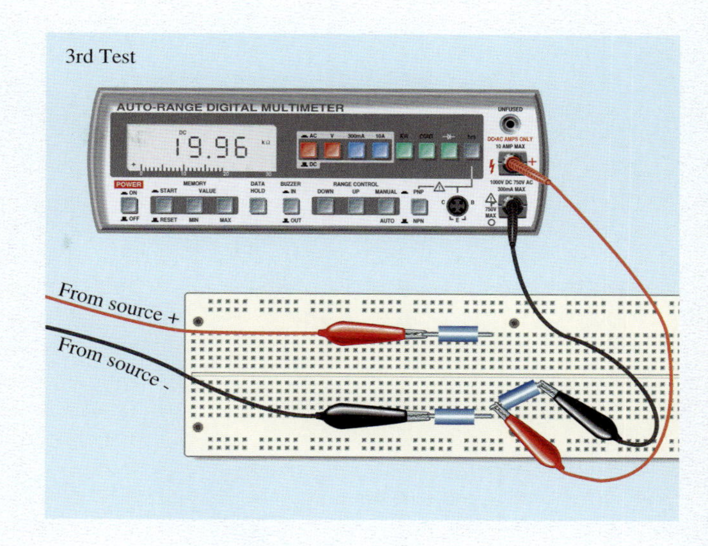
3rd Test
AUTO-RANGE DIGITAL MULTIMETER
19.96
From source +
From source -

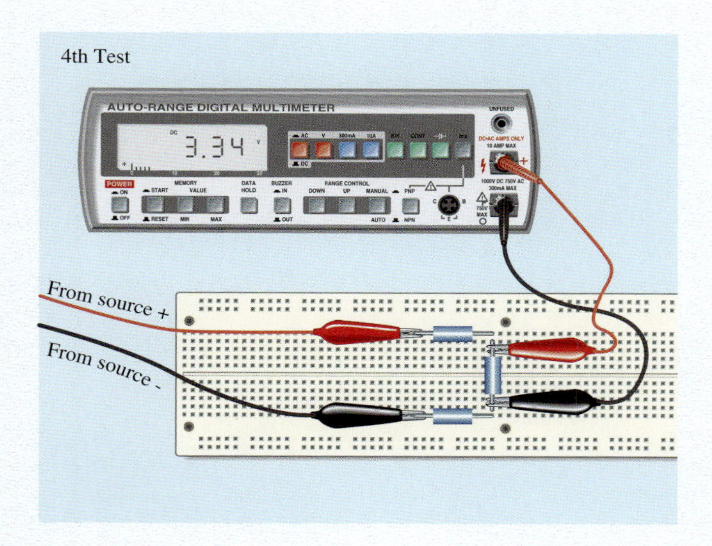
4th Test
AUTO-RANGE DIGITAL MULTIMETER
3.34
From source +
From source -

OUTLINE

OBJECTIVES

After studying this chapter, you should be able to:

1. Illustrate V-I relationships for purely resistive and purely capacitive circuits
2. Explain **capacitive reactance**
3. Use Ohm's law to solve for X_C value(s)
4. Use the capacitive reactance formula to solve for X_C value(s)
5. Use the X_C formula to solve for unknown C and f values
6. Use Ohm's law and **reactance** formulas to determine circuit reactances, voltages, and currents for series-connected and parallel-connected capacitors
7. List two practical applications for X_C
8. Use the computer to solve circuit problems
9. Use the SIMPLER troubleshooting sequence to solve the Troubleshooting Challenge problem

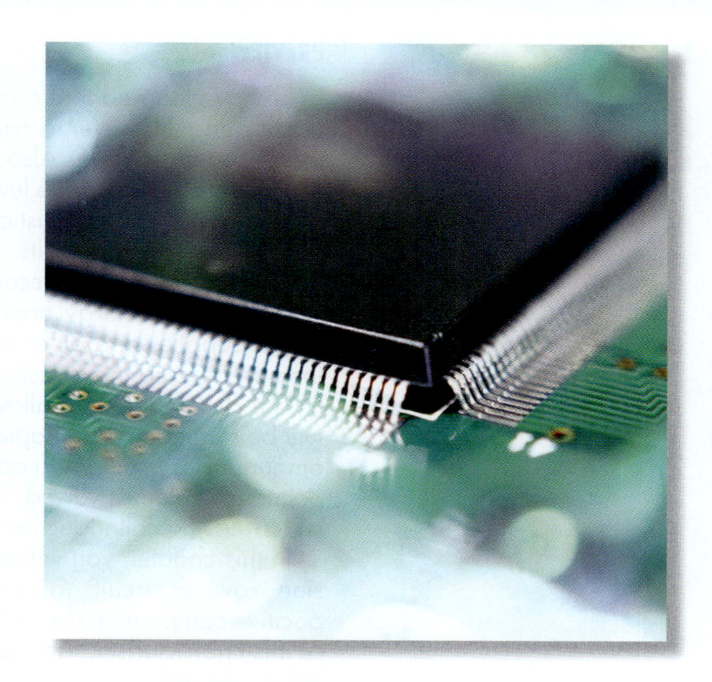

CHAPTER 18

Capacitive Reactance in AC

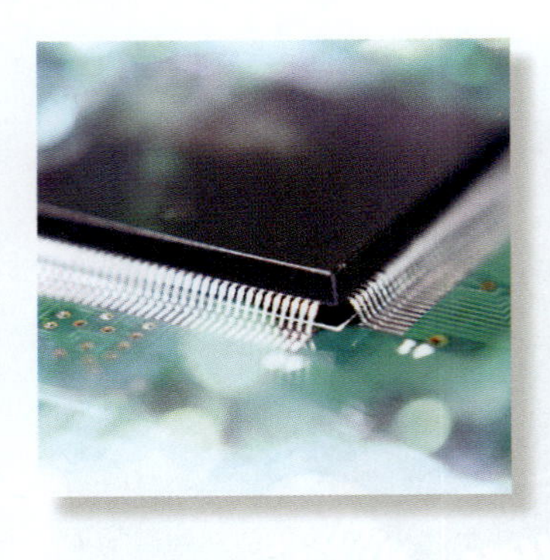

PREVIEW

You have learned that a capacitor allows current flow, provided it is charging or discharging. However, a capacitor does not allow dc current to continue once charged to the source voltage. Also, you learned a capacitor does not instantly charge or discharge to a new voltage level; it requires current flow and time.

The unique characteristics of capacitors allow them to be used for a number of purposes in electronic circuits. Their ability to block dc current (once charged) makes them useful in coupling and decoupling applications for amplifiers, and so on. Their ability to store electrical energy provides great value when they are used for filtering circuits in power supplies and other circuits. Their charging and discharging characteristics allow them to be used in waveshaping applications and for timing circuits of all kinds.

The characteristic of allowing ac current due to the charging and discharging current will be studied in this chapter. You will learn there is an inverse relationship between the amount of *opposition* a capacitor gives to ac current and the circuit frequency and capacitance values involved. Capacitor opposition to ac current is called **capacitive reactance.** (Recall inductive reactance describes inductor opposition to ac current.)

In this chapter, you will examine capacitance as it relates to the capacitor action in a sine-wave ac circuit. You will study the phase relationship of voltage and current for capacitive components; learn to compute capacitive reactance; discover some applications of this characteristic; and learn other practical information about this important electrical parameter.

KEY TERMS

Capacitive reactance *(X_C)*
Rate of change of voltage
Reactance

18–1 *V* and *I* Relationships in a Purely Resistive AC Circuit

As a quick review, look at Figure 18–1. Note that *V* and *I* are in phase in a purely resistive ac circuit.

18–2 *V* and *I* Relationships in a Purely Capacitive AC Circuit

Observe Figure 18–2 as you study the following statements about the relationship of *V* and *I* in a purely capacitive ac circuit.

1. To develop a voltage (potential difference or p.d.) across capacitor plates, charge must move. That is, charging current must flow first.
2. When a sine-wave voltage is applied across a capacitor that is effectively in parallel with the source terminals, voltage across the capacitor must be the same as the applied voltage.
3. For the voltage across the capacitor to change, there must be alternate charging and discharging action. This implies charging and discharging current flow. The value of instantaneous capacitor current (i_C) directly relates to the amount of capacitance and the rate of change of voltage. (This principle fits well with the concept learned earlier where $Q = CV$.)
4. The formula that expresses the direct relationship of i_C to C and rate of change of voltage is:

FORMULA 18–1 $i_C = C\frac{dv}{dt}$

where dv/dt indicates change in voltage divided by amount of time over which the change occurs, or the rate of change of voltage.

For example, if the voltage across a 100-pF capacitor changes by 100 V in 100 μs, then:

$$i_C = C\frac{dv}{dt}$$

$$i_C = 100 \times 10^{-12} \times \frac{100}{100 \times 10^{-6}}$$

$$i_C = \frac{10{,}000 \times 10^{-12}}{100 \times 10^{-6}}$$

$$i_C = 100 \times 10^{-6}, \text{ or } 100\ \mu\text{A}$$

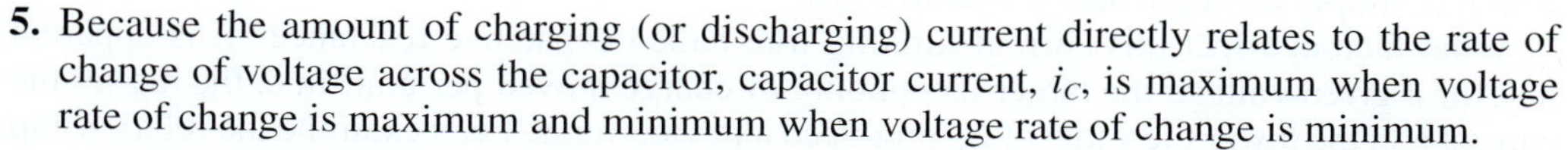

5. Because the amount of charging (or discharging) current directly relates to the rate of change of voltage across the capacitor, capacitor current, i_C, is maximum when voltage rate of change is maximum and minimum when voltage rate of change is minimum.

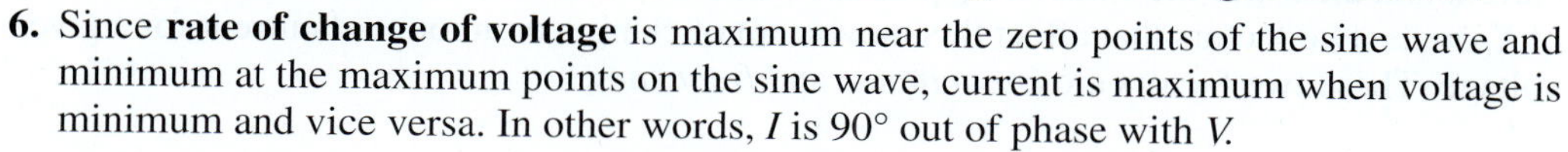

6. Since **rate of change of voltage** is maximum near the zero points of the sine wave and minimum at the maximum points on the sine wave, current is maximum when voltage is minimum and vice versa. In other words, I is 90° out of phase with V.

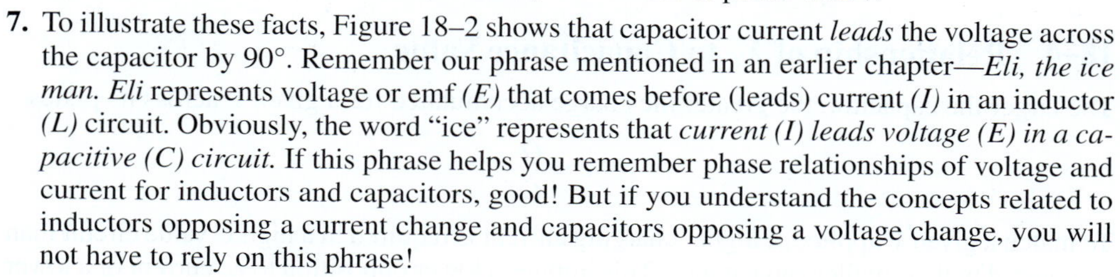

7. To illustrate these facts, Figure 18–2 shows that capacitor current *leads* the voltage across the capacitor by 90°. Remember our phrase mentioned in an earlier chapter—*Eli, the ice man. Eli* represents voltage or emf *(E)* that comes before (leads) current *(I)* in an inductor *(L)* circuit. Obviously, the word "ice" represents that *current (I) leads voltage (E) in a capacitive (C) circuit.* If this phrase helps you remember phase relationships of voltage and current for inductors and capacitors, good! But if you understand the concepts related to inductors opposing a current change and capacitors opposing a voltage change, you will not have to rely on this phrase!

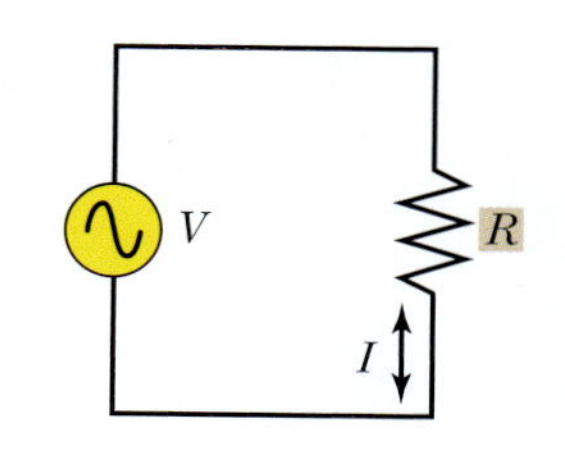

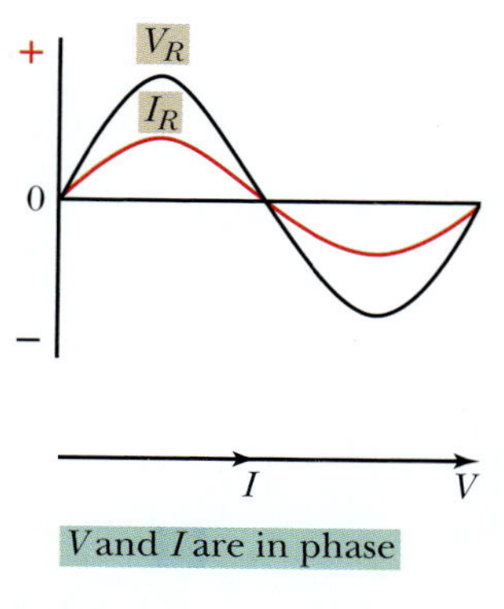

FIGURE 18–1 V and I phase relationships in a purely resistive circuit

■ IN-PROCESS LEARNING CHECK 1

Fill in the blanks as appropriate.

1. For a capacitor to develop a potential difference between its plates, there must be a _____ current.
2. For the potential difference between capacitor plates to decrease (once it is charged), there must be a _____ current.
3. When ac is applied to capacitor plates, the capacitor will alternately _____ and _____.
4. The value of instantaneous capacitor current directly relates to the value of _____ and rate of change of _____.
5. The amount of charging or discharging current is maximum when the rate of change of voltage is _____. For sine-wave voltage, this occurs when the sine wave is near _____ points.
6. Capacitor current _____ the voltage across the capacitor by _____ degrees. The expression that helps you remember this relationship is "Eli, the _____ man."

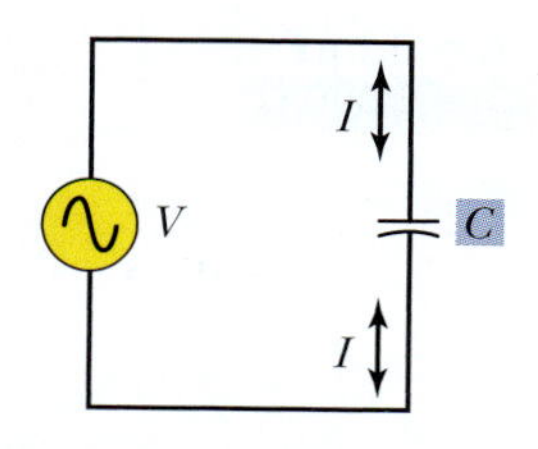

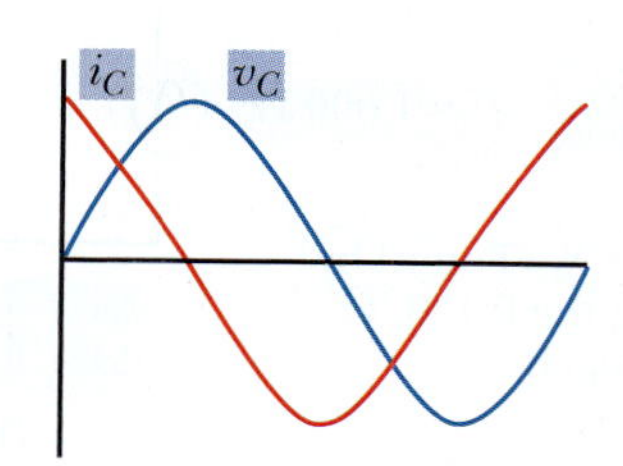

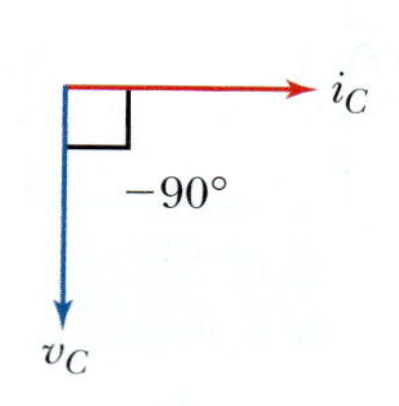

FIGURE 18–2 V and I phase relationships in a purely capacitive circuit

18–3 Concept of Capacitive Reactance

As you have just seen, a sine-wave current develops in a capacitor circuit by the charge and discharge action of the capacitor when a sine-wave voltage is applied. This sine-wave capacitor current leads the sine-wave capacitor voltage by 90°.

In a purely capacitive circuit, the factor that limits the ac current is called capacitive reactance. The symbol used to represent this quantity is X_C. It is measured in ohms. As you would anticipate, the higher the capacitive reactance, the lower the amount of current produced for a given voltage. That is,

$$I_C = \frac{V_C}{X_C}$$

which is simply an application of Ohm's law.

What factors affect this current-limiting trait called capacitive reactance? It is apparent that for a given voltage the larger the amount of charge moved per unit time, the higher the current and the lower the value of X_C produced and vice versa. Let's examine the relationship of X_C to the two factors that affect this movement of charge, or amount of current for a given voltage. These two factors are *capacitance* and *frequency.*

18–4 Relationship of X_C to Capacitance Value

The larger the capacitor, the greater the charge accumulated for a given V across its plates

$$V = \frac{Q}{C}$$

To match a given V applied, a higher charging current is required in a high C value circuit than is required with a smaller capacitance. This implies a lower opposition to ac current or a lower capacitive reactance *(X_C)*. This means X_C *is inversely proportional to* C. In other words, the larger the C, the lower the X_C present; and the smaller the C, the higher the X_C present.

You will soon be introduced to a formula for calculating X_C; however, for now, notice in Figure 18–3a that for a given frequency of operation (100 Hz, in this case), the amount of

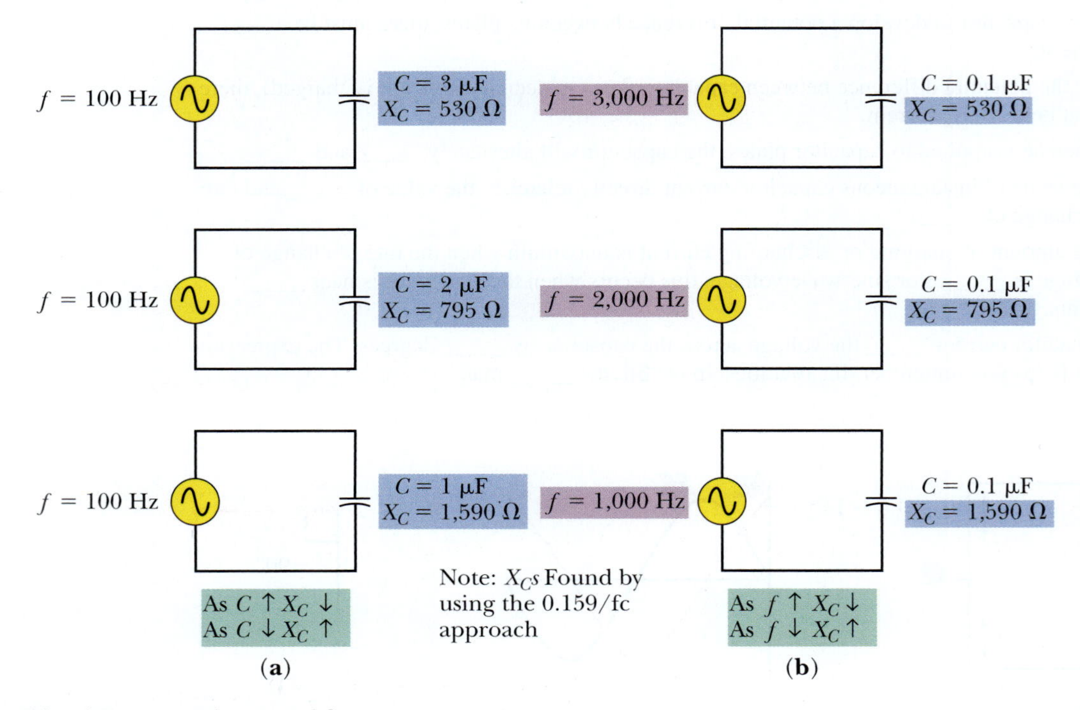

FIGURE 18–3 Relationship of X_C to capacitance and frequency

capacitive reactance in ohms is *inversely related* to the capacitance value, which illustrates our previous statement. We will ask you to confirm the values shown in the figure after you study the X_C formula. For now, assume the values are valid. The point here is that the greater the C value, the less the X_C value, and vice versa.

18–5 Relationship of X_C to Frequency of AC

If the capacitor voltage is tracking the source V, as in the circuit of Figure 18–3b, then the higher the frequency and the less available time there is for the required amount of charge (Q) to move to and away from the respective capacitor plates during charge and discharge action. As you know, it takes a given amount of Q to achieve a specific voltage across a given capacitor ($V = Q/C$).

The higher the frequency of applied voltage, the faster is the rate of change of voltage for any given voltage level. As you learned earlier, the i_C directly relates to the voltage rate of change (dV/dt). Since higher frequencies produce a greater charge and discharge current for any given voltage level, as frequency increases opposition to current decreases. More current for a given V applied indicates there is *less* opposition, meaning capacitive reactance (X_C) decreases as frequency increases. That is, *X_C is inversely proportional to f.* In other words, the higher the frequency, the lower the X_C present; and the lower the frequency, the greater the X_C present. See again Figure 18–3b.

18–6 Methods to Calculate X_C

Ohm's Law Method

Ohm's law is one practical method to calculate X_C, where

$$X_C = \frac{V_C}{I_C}$$

For example, if the effective ac voltage applied to a capacitive circuit is 100 volts, and the I_C is 1 ampere (rms), capacitive reactance (X_C) must equal 100 ohms. NOTE: Unless told otherwise, assume effective (rms) values of ac voltage and current in your measurements and computations relating to V_C, I_C, and X_C. As an electronics student and technician, you will frequently make such measurements and calculations. Remember, the Ohm's law formula can also be transposed to yield

$$V_C = I_C \times X_C, \text{ and } I_C = \frac{V_C}{X_C}$$

The Basic X_C Formula

From the previous discussions, it is easy to understand the concepts illustrated by the formula to calculate X_C when only C and f are known. The formula states that:

FORMULA 18–2 $X_C \text{ (ohms)} = \dfrac{1}{2\pi fC}$

where f is frequency in hertz
C is capacitance in farads

Notice that X_C is *inversely* proportional to both f and C. That is, if either f or C (or both) are increased, the resulting larger product in the denominator divided into one results in a smaller

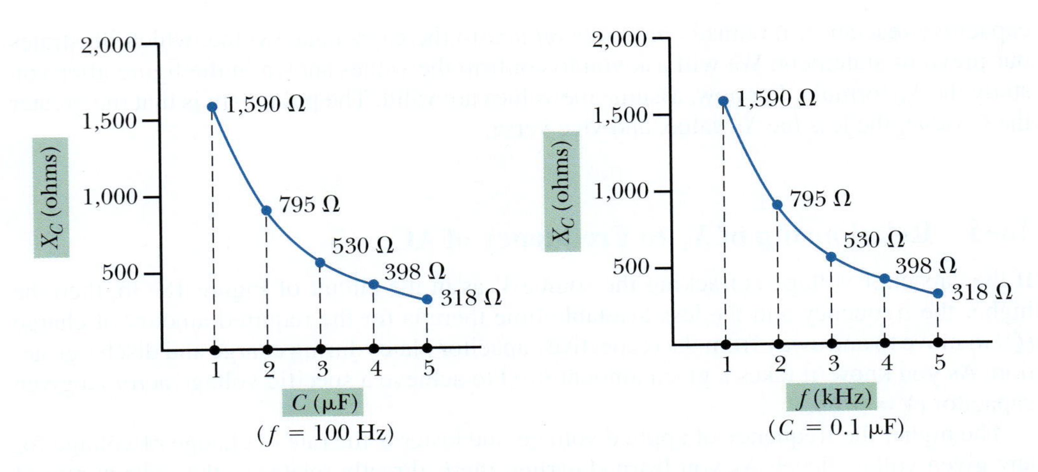

FIGURE 18–4 Examples of the inverse relationship of X_C to C and to f

value of X_C. In fact, if either f or C is doubled, X_C is halved. If either f or C is halved, X_C doubles, and so forth, Figure 18–4.

Stated another way, larger C means more charging and discharging current flows to charge to and discharge from a given voltage. Thus, a larger C results in a lower X_C. Since a higher f means the voltage rate of change is faster, more charging and discharging current must flow in the time allowed by the higher frequency; hence, a higher f results in a lower X_C.

Recall that the "$2\pi f$" in the formula relates to angular velocity, which relates to sine-wave frequency. The X_C formula is the second formula where you have found this angular velocity expression—the other formula was for inductive reactance where $X_L = 2\pi fL$.

Practical Notes

Some people find it convenient to alter the formula

$$X_C = \frac{1}{2\pi fC} \text{ to read } \frac{0.159}{fC}$$

This can be done because the value of 2π does not change. That is, it is a "constant." One divided by $2\pi = 0.159$; therefore, the formula can be simplified to read $0.159/(fC)$. When X_C calculations must be performed without a calculator, this modified formula may be easier to use. When a calculator is available, either formula format is easily used.

EXAMPLE

1. What is the capacitive reactance of a 2-μF capacitor operating at a frequency of 120 Hz?

$$X_C = \frac{1}{2\pi fC} = \frac{1}{6.28 \times 120 \times (2 \times 10^{-6})} = \frac{1}{1{,}507.2 \times 10^{-6}} = 663.5\Omega$$

2. The frequency of applied voltage doubles while the C remains the same in a capacitive circuit. What happens to the circuit X_C?

$$X_C = \frac{1}{2\pi fC}, \text{ where } X_C \text{ for stated conditions} = \frac{1}{2\pi(2f)\,C}$$

Therefore, the denominator's product is twice as large as that divided into the unchanged numerator. That is, the X_C is half of the original frequency X_C.

3. In a capacitive circuit, if X_C quadruples while the frequency of the applied voltage remains the same, what must have happened to the C value in the circuit?

Since X_C is inversely proportional to f and C and f remains unchanged, C must have decreased to one-fourth its original value. That is,

$$4 \times \text{original } X_C = \frac{1}{2\pi f(0.25C)}$$

PRACTICE PROBLEMS 1

1. Use the X_C formula and verify values of X_C shown in Figure 18–3. Do the formula results validate the values of X_C shown? (NOTE: X_Cs in Figure 18–3 found using the $0.159/fC$ approach.)
2. What is the capacitive reactance of a 5-μF capacitor operating at a frequency of 200 Hz?
3. What happens to the X_C in a circuit when f doubles and the capacitance value is halved at the same time?

Rearranging the X_C Formula to Find f and C

By rearranging the X_C equation, we can solve for frequency if both C and X_C are known, or solve for C if both f and X_C are known. When the equation is transposed, the new equations are as follows:

FORMULA 18–3 $$f = \frac{1}{2\pi C X_C}$$

and

FORMULA 18–4 $$C = \frac{1}{2\pi f X_C}$$

Try these formulas on the following practice problems.

PRACTICE PROBLEMS 2

1. What is the frequency of operation of a capacitive circuit where the total X_C is 1,000 Ω and the total capacitance is 500 pF?
2. What is the total circuit capacitance of a circuit exhibiting an X_C of 3,180 Ω at a frequency of 2 kHz?

18–7 Capacitive Reactances in Series and Parallel

In the last chapter, you learned that capacitances in series add like resistances in parallel, and capacitances in parallel add like resistances in series. It is interesting that when we consider capacitive *reactances* in series and parallel, the converse is true.

Capacitive Reactances in Series

Since capacitive reactance (X_C) is an opposition to current flow, measured in ohms, capacitive reactances (current oppositions) in series add like resistances in series. That is, $X_{C_T} = X_{C_1} + X_{C_2} + X_{C_3} \ldots + X_{C_n}$, Figure 18–5.

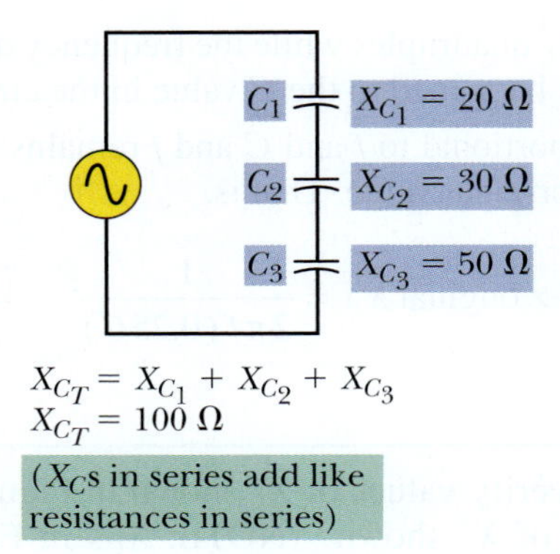

FIGURE 18–5 X_Cs in series

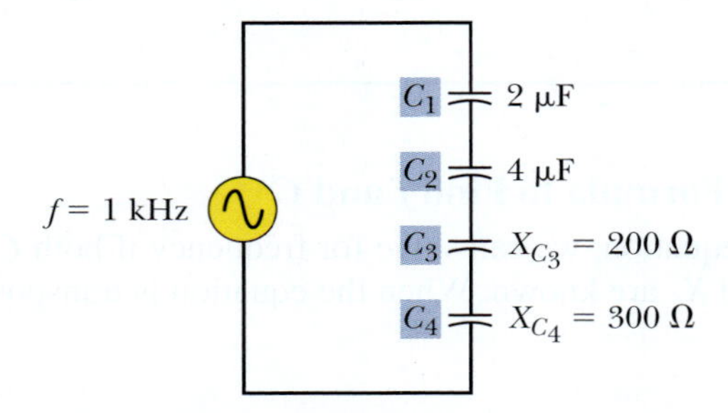

FIGURE 18–6

EXAMPLE What is the total X_C of a circuit containing three series-connected 3-μF capacitors operating at a frequency of 100 Hz?

Approach 1: Find total C, then use the X_C formula:

C_T for three equal-value capacitors in series equals the C value of one capacitor divided by the number of equal-value capacitors ($C_T = C_1/N$).

$$C_T = \frac{3\ \mu\text{F}}{3} = 1\ \mu\text{F}$$

$$X_{C_T} = \frac{1}{2\pi f C_T} = \frac{1}{6.28 \times 100 \times 1 \times 10^{-6}} = 1{,}592\ \Omega$$

Approach 2: Find X_C of a 3-μF capacitor, then multiply by 3:

$$X_{C_1} = \frac{1}{2\pi f C_1} = \frac{1}{6.28 \times 100 \times 3 \times 10^{-6}} = 530.7\ \Omega$$

$$X_{C_T} = 3 \times 530.7\ \Omega = 1{,}592\ \Omega$$

PRACTICE PROBLEMS 3

What is the total X_C of the circuit in Figure 18–6?

Capacitive Reactances in Parallel

Because capacitive reactance is an opposition to ac current flow, when we supply more current paths by connecting X_Cs in parallel, the resulting total opposition decreases just as it does when we parallel resistances.

Useful formulas to solve for total capacitive reactance of capacitors in parallel are as follows:

- General formula for X_Cs in parallel:

FORMULA 18–5 $$X_{C_T} = \frac{1}{\frac{1}{X_{C_1}} + \frac{1}{X_{C_2}} + \frac{1}{X_{C_3}} \cdots + \frac{1}{X_{C_n}}}$$

- Formula for two X_Cs in parallel:

FORMULA 18–6 $$X_{C_T} = \frac{X_{C_1} \times X_{C_2}}{X_{C_1} + X_{C_2}}$$

- Formula for equal-value X_Cs in parallel:

FORMULA 18–7 $$X_{C_T} = \frac{X_{C_1}}{N}$$

where N = number of equal-value X_Cs in parallel

EXAMPLE What is the total X_C of a circuit containing three parallel-connected capacitor branches with values of 2 μF, 3 μF, and 5 μF, respectively? Assume the circuit is operating at a frequency of 400 Hz.

Two possible approaches are:

1. Find total C, then use X_C formula to find X_{C_T}.
2. Find X_C of each capacitor using X_C formula, then use the general formula to solve for X_{C_T}.

Of these two approaches, approach 1 is easiest for these circumstances. Therefore,

$$C_T = C_1 + C_2 + C_3$$
$$C_T = 2\ \mu F + 3\ \mu F = 5\ \mu F = 10\ \mu F$$
$$X_{C_T} = \frac{1}{2\pi f C_T} = \frac{1}{6.28 \times 400 \times 10 \times 10^{-6}} = \frac{1}{0.02512} = 39.8\ \Omega$$

PRACTICE PROBLEMS 4

1. What is the total capacitive reactance of a circuit containing three parallel-connected capacitors of 1 μF, 2 μF, and 3 μF, respectively? (Assume an operating frequency of 1 kHz.)
2. What is the total capacitive reactance of two parallel-connected capacitors where one capacitance is 5 μF and the other capacitance is 20 μF? (Circuit is operating at a frequency of 500 Hz.)
3. What is the total capacitive reactance of four 1,600-Ω capacitive reactances connected in parallel?

18–8 Voltages, Currents, and Capacitive Reactances

Our computations have concentrated on finding the capacitive reactances in capacitive circuits. Let's discuss voltage distribution throughout series capacitive circuits and current distribution throughout parallel capacitive circuits.

Capacitive reactances represent opposition to ac current (in ohms). When dealing with V, I, and X_C, we treat ohms of opposition like we treat resistances in resistive circuits. That is, voltage drops are calculated by using $I \times X_C$; and currents are calculated by using V/X_C, and so on.

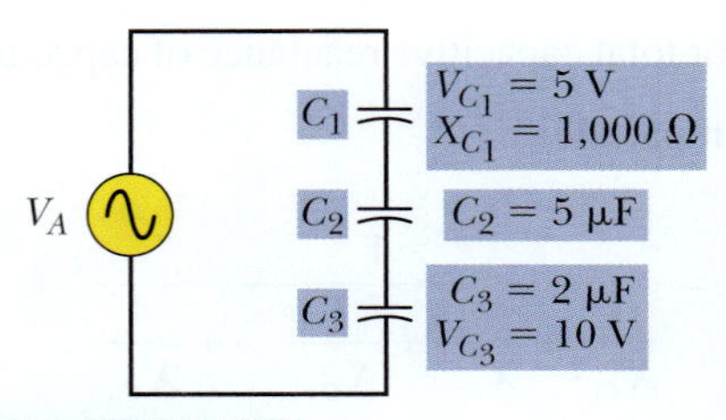

Find f; C_1; I; X_{C_2}; X_{C_3}; X_{C_T}; V_{C_2}; and V_A.

(Since we know two things about C_1, let's start there.)

$$I = \frac{V_{C_1}}{X_{C_1}} = \frac{5}{1{,}000} = 0.005\ A,\text{ or } I = 5\text{ mA}.$$

Since V_{C_3} is double V_{C_1}, its X_C must be double that of C_1 because they have the same I and $V = I \times X_C$. Thus, $X_{C_3} = 2{,}000\ \Omega$.

Since V_{C_1} is half V_{C_3}, C_1 must have twice the capacitance and half the X_C of C_3. Thus, $C_1 = 4\ \mu$F.

FIGURE 18–7 Series capacitive circuit analysis

Since C_2 is 2.5 times the C value of C_3 and X_C is inversely related to C, X_{C_2} must equal $\frac{X_{C_3}}{2.5}$ or $\frac{2{,}000}{2.5}$. Thus, $X_{C_2} = 800\ \Omega$.

$$V_{C_2} = I \times X_{C_2} = 5\text{ mA} \times 800\ \Omega$$
$$V_{C_2} = 4\text{ V}$$
$$X_{C_T} = X_{C_1} + X_{C_2} + X_{C_3} = 1{,}000 + 800 + 2{,}000$$
$$X_{C_T} = 3{,}800\ \Omega$$
$$C_T = \frac{1}{\left(\frac{1}{4} + \frac{1}{5} + \frac{1}{2}\right)}$$
$$C_T = 1.05\ \mu\text{F}$$
$$f = \frac{1}{2\pi C_T X_{C_T}} = \frac{1}{6.28 \times 1.05 \times 10^{-6} \times 3{,}800}$$
$$f = \frac{1}{0.02505} = 39.9\text{ Hz}$$
$$f = 39.9\text{ Hz or about 40 Hz}$$
$$V_A = I \times X_{C_T} = 5\text{ mA} \times 3.8\text{ k}\Omega$$
$$V_A = 19\text{ V}$$
$$(\text{Also, } V_1 + V_2 + V_3 = 19\text{ V})$$

Refer to Figure 18–7 to see series circuit parameters and computations illustrated. Then refer to Figure 18–8 and observe how parallel circuit parameters and computations are made. After studying these figures, try the following practice problems.

PRACTICE PROBLEMS 5

1. For the circuit in Figure 18–9, find X_{C_1}, X_{C_2}, X_{C_3}, X_{C_T}, C_2, C_3, V_2, and I_T.
2. For the circuit in Figure 18–10, find C_1, X_{C_1}, X_{C_2}, X_{C_3}, I_2, and I_T.

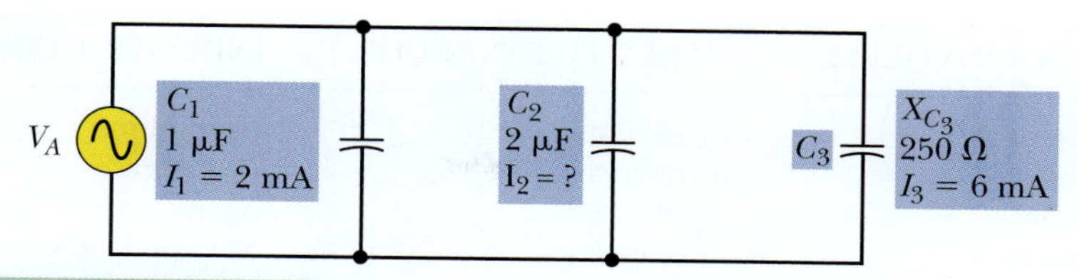

Find V_A; f; X_{C_2}; X_{C_1}; I_2; I_T.

Starting with information regarding C_3,

$$V_A = V_3 = I_3 \times X_{C_3}$$
$$V = 6 \text{ mA} \times 0.25 \text{ k}\Omega$$
$$V_A = 1.5\text{V}$$

Since I_3 is triple I_1, X_{C_3} must be one-third X_{C_1}. Therefore, C_3 must be triple C_1.

$$C_3 = 3\ \mu\text{F}$$
$$X_{C_1} = 3 \times X_{C_3} = 3 \times 250\ \Omega$$
$$X_{C_1} = 750\ \Omega$$

$$X_{C_2} = \frac{1}{2} X_{C_1} = \frac{1}{2} \text{ of } 750\ \Omega$$
$$X_{C_2} = 375\ \Omega$$
$$I_2 = \frac{V_2}{X_{C_2}} = \frac{1.5 \text{ V}}{375\ \Omega}$$
$$I_2 = 4 \text{ mA}$$
$$I_T = I_1 + I_2 + I_3 = 2 \text{ mA} + 4 \text{ mA} + 6 \text{ mA}$$
$$I_T = 12 \text{ mA}$$

f is found by using any one of the branches information or by using the total C and X_C parameters. That is,

$$f = \frac{1}{2\pi C X_C} = \frac{1}{6.28 \times 1 \times 10^{-6} \times 750}$$
$$f = \frac{1}{0.00471}$$
$$f = 212 \text{ Hz}$$

FIGURE 18–8 Parallel capacitive circuit analysis

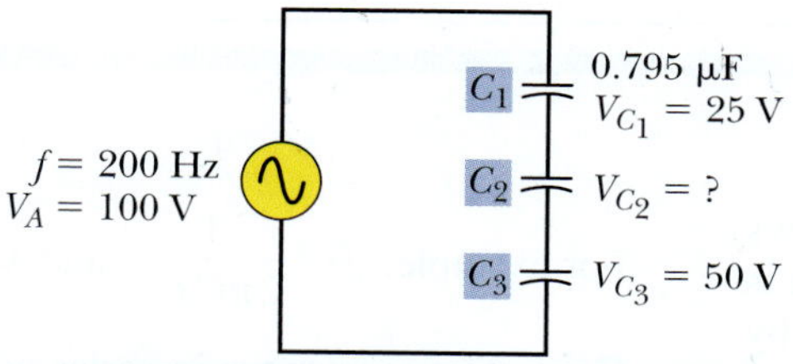

$f = 200$ Hz
$V_A = 100$ V
C_1 0.795 μF $V_{C_1} = 25$ V
C_2 $V_{C_2} = ?$
C_3 $V_{C_3} = 50$ V

Find X_{C_1}; X_{C_2}; X_{C_3}; X_{C_T}; C_2; C_3; V_2; and I_T.

FIGURE 18–9 

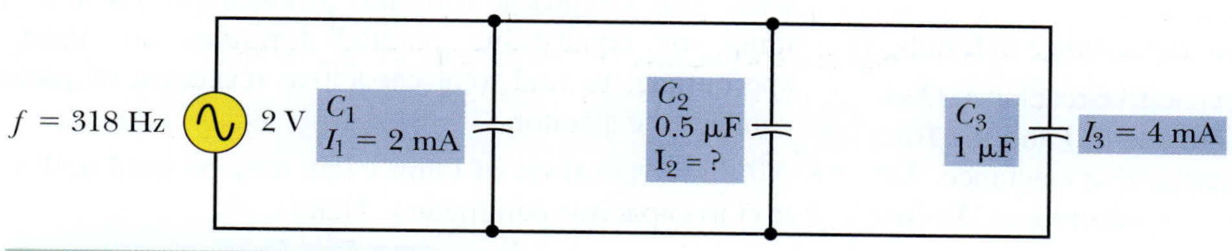

Find C_1; X_{C_1}; X_{C_2}; X_{C_3}; I_2; and I_T.

FIGURE 18–10 

PARAMETER INVOLVED	CAPACITIVE CIRCUIT	INDUCTIVE CIRCUIT
Opposition to I	X_C in ohms	X_L in ohms
As f increases	X_C decreases	X_L increases
As f decreases	X_C increases	X_L decreases
As C increases	X_C decreases	Not applicable
As C decreases	X_C increases	Not applicable
As L increases	Not applicable	X_L increases
As L decreases	Not applicable	X_L decreases
Voltage across component	$I \times X_C$	$I \times X_L$
Instantaneous values	$i_C = C \times dv/dt$	$v_L = L \times di/dt$
Phase of V and I	I leads V by 90 degrees	V leads I by 90 degrees
To calculate X	$X_C = 1/2\pi fC$	$X_L = 2\pi fL$
Series-connected	$C_T <$ least C	$L_T =$ sum of L's
Parallel-connected	$C_T =$ sum of C's	$L_T <$ least L
Series-connected	$X_{C_T} =$ sum of X_C's	$X_{L_T} =$ sum of X_L's
Parallel-connected	$X_{C_T} <$ least X_C	$X_{L_T} <$ least X_L

FIGURE 18–11 Comparisons of C and L circuit characteristics

18–9 Final Comments about Capacitors and Capacitive Reactance

Having studied capacitive reactance, you can more clearly see how capacitors are used. As you proceed in your studies and career, you should *keep in mind* the inverse relationship of capacitance and capacitive reactance. Also, remember capacitances in series add like resistances in parallel. Keeping these ideas in mind will help prevent errors when analyzing circuits containing both inductances and capacitances.

A list of similarities and differences between capacitive and inductive circuits is shown in Figure 18–11. Review what you have learned about the operating features of these two electronic components. It is important that you thoroughly understand each component to analyze circuits that combine these components in various configurations.

Summary

- Capacitive reactance is the opposition a capacitor displays to ac or pulsating dc current. The symbol is X_C and it is measured in ohms. Capacitive reactance is calculated by the formula

$$X_C = \frac{1}{2\pi fC}$$

where f is frequency in hertz and C is capacitance in farads.

- Factors influencing the amount of capacitive reactance *(X_C)* in a circuit are the amount of capacitance *(C)* and the frequency *(f)*. Voltage does *not* affect capacitive reactance. As C *or* f increases, X_C decreases. As C *or* f decreases, X_C increases proportionately.
- In a purely capacitive circuit, current leads voltage by 90°. (Charging current must flow before a potential difference develops across the capacitor plates.)
- Capacitive reactance is calculated by using Ohm's law or the capacitive reactance formula, where

$$X_C = \frac{V_C}{I_C}, \text{ or } X_C = \frac{1}{2\pi fC}$$

- The capacitive reactance formula can be rearranged to find either f or C, if all other factors are known.

For example, $f = \dfrac{1}{2\pi C X_C}$, and $C = \dfrac{1}{2\pi f X_C}$

- Capacitive reactances in series add like resistances in series. That is, $X_{C_T} = X_{C_1} + X_{C_2} + X_{C_3} \ldots + X_{C_n}$.
- Capacitive reactances in parallel add like resistances in parallel. The reciprocal formula, product-over-the-sum formula, or equal-value parallel formulas are used, as appropriate, to find total capacitive reactance of parallel-connected capacitors.
- All three variations of Ohm's law may be used with relation to capacitor parameters. That is,

$$I_C = \frac{V_C}{X_C};\ V_C = IX_C;\ \text{and } X_C = \frac{V_C}{I_C}$$

- Capacitance and capacitive reactance exhibit converse characteristics. In a *series* circuit, the *smallest* C has the largest voltage drop and the largest C the least voltage drop. In this same series circuit, the largest X_C drops the most voltage (because the smallest C has the highest X_C), and so forth. In *parallel* circuits, the largest C holds the largest amount of charge *(Q)* and the smallest C has the least Q.

Formulas and Sample Calculator Sequences

FORMULA 18–1
(To find the instantaneous value of current when the value of capacitance and the rate of change of voltage across the capacitor are known)

$$i_C = C\frac{dv}{dt}$$

Cap value, [×], voltage change value, [÷], amount of time, [=]

FORMULA 18–2
(To find the value of capacitive reactance when the values of frequency and capacitance are known)

$$X_C \text{ (ohms)} = \frac{1}{2\pi fC}$$

6.28, [×], frequency value, [×], capacitance value, [=], [1/x]

FORMULA 18–3
(To find the frequency involved that produces a given capacitive reactance with a given capacitance value)

$$f = \frac{1}{2\pi CX_C}$$

6.28, [×], capacitance value, [×], X_C value, [=], [1/x]

FORMULA 18–4
(To find the capacitance value that produces a given capacitive reactance at a specified frequency)

$$C = \frac{1}{2\pi fX_C}$$

6.28, [×], frequency value, [×], X_C value, [=], [1/x]

FORMULA 18–5
(To find the total capacitive reactance of parallel capacitors whose capacitive reactances are known)

$$X_{C_T} = \frac{1}{\frac{1}{X_{C_1}} + \frac{1}{X_{C_2}} + \frac{1}{X_{C_3}} \cdots \frac{1}{X_{C_n}}}$$

X_{C_1} value, [1/x], [+], X_{C_2} value, [1/x], [+], X_{C_3} value, [1/x], [=], [1/x]

FORMULA 18–6
(To find the total capacitive reactance of two parallel capacitors whose capacitive reactances are known)

$$X_{C_T} = \frac{X_{C_1} \times X_{C_2}}{X_{C_1} + X_{C_2}}$$

X_{C_1} value, [×], X_{C_2} value, [÷], [(], X_{C_1} value, [+], X_{C_2} value, [)], [=]

FORMULA 18–7
(To find the total capacitive reactance of equal value capacitive reactances in parallel)

$$X_{C_T} = \frac{X_{C_1}}{N}$$

X_{C_1} value, [÷], number of equal-value X_Cs in parallel, [=]

EXCEL AUTOMATED FORMULAS

Helpful problem-solving tools, such as the Excel automated formulas available on the CD, allow automatic calculations of formulas.

Using Excel

Capacitive Reactance in AC Formulas

(Excel file reference: FOE18_01.xls)

DON'T FORGET! *It is not necessary to retype formulas once they are entered on the worksheet! Just input new parameters data for each new problem using that formula, as needed.*

- Use the Formula 18–2 spreadsheet sample and find the answer to Practice Problems 1, question 2. Check your answer against that given in the Appendix for this problem.
- Use the Formula 18–4 spreadsheet sample and solve for the value of capacitance, as stipulated in Practice Problems 2, question 2. Again, check your answer against that given in the Appendix for this question.

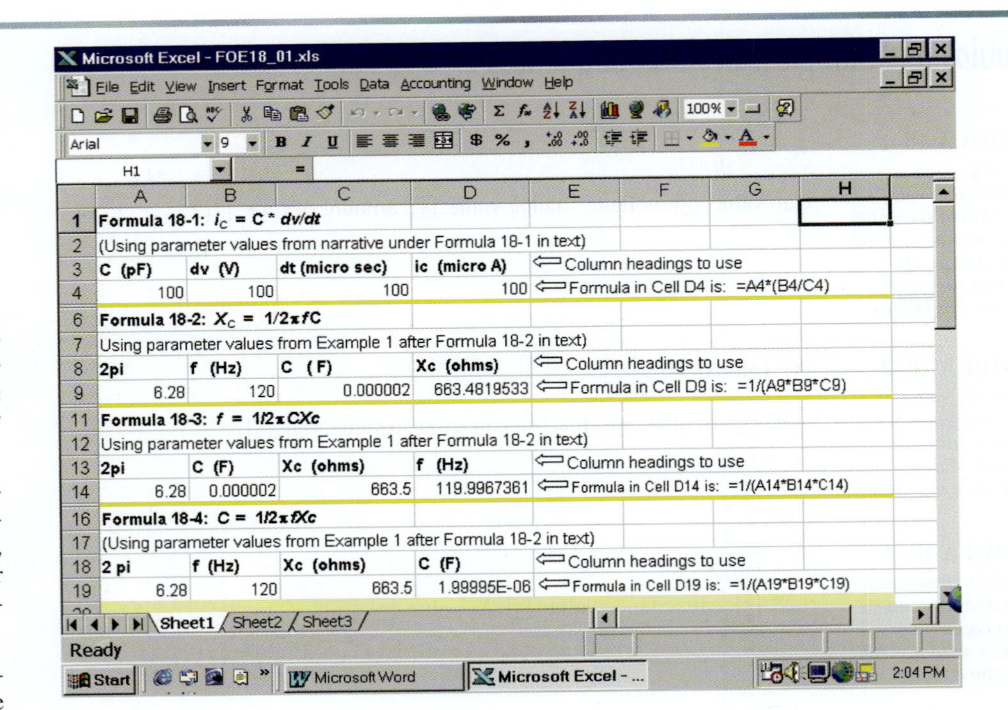

Review Questions

1. Capacitive reactance is:
   a. the opposition a capacitor shows to dc only.
   b. the opposition a capacitor shows to ac only.
   c. the opposition a capacitor shows to ac or pulsating dc.
   d. none of the above.
2. The amount of capacitive reactance shown by a capacitor can be influenced by:
   a. circuit voltage only.
   b. circuit voltage and current.
   c. frequency and resistance.
   d. frequency and capacitance.
3. Two commonly used methods used to compute capacitive reactance values are:
   a. Ohm's law and Kirchhoff's law.
   b. Ohm's law and the capacitive reactance formula.
   c. Ohm's law and epsilon natural log approach.
   d. none of the above.
4. The symbol used to represent capacitive reactance is:
   a. Z.
   b. R.
   c. X_C.
   d. none of the above.
5. Total capacitive reactance of parallel capacitors can be calculated like resistances in:
   a. series.
   b. parallel.
   c. series-parallel.
   d. none of the above.
6. Total capacitive reactance of series capacitors can be calculated like resistances in:
   a. series.
   b. parallel.
   c. series-parallel.
   d. none of the above.
7. In a series capacitive circuit, the smallest capacitance has:
   a. the lowest voltage drop.
   b. the highest voltage drop.
   c. the same voltage drop as all other capacitors.
   d. none of the above.
8. In a parallel capacitive circuit, the largest capacitance has:
   a. the lowest charge value.
   b. the highest charge value.
   c. the same Q as the other capacitors in the circuit.
   d. none of the above.
9. If you know the frequency and capacitive reactance:
   a. the value of C can be found by the formula $1/2\pi fC$.
   b. the value of C can be found by the formula $0.159/fC$.
   c. the value of C can be found by the formula $1/2\pi fX_C$.
   d. the value of C cannot be found with just this amount of information.
10. If four capacitors of equal value are in series:
    a. C_T = four times one C's value, and X_{C_T} = one-fourth each capacitor's reactance.
    b. X_{C_T} = four times one C's reactance and C_T = one-fourth of one of the Cs.
    c. total capacitance and total reactance are four times each individual C's values.
    d. none of the above.
11. If four capacitors of equal value are in parallel:
    a. C_T is four times one C's value, and X_{C_T} is one-fourth each capacitor's reactance.
    b. X_{C_T} = four times one C's reactance, and C_T = one-fourth each capacitor's C.
    c. total capacitance and total reactance are four times each individual C's values.
    d. none of the above.
12. The voltage rating of the smallest capacitor in a series C circuit needs:
    a. to be higher than the other capacitors' voltage ratings, if the others are operating at their voltage rating.
    b. to be lower than the other capacitor's voltage ratings, if the others are operating at their voltage rating.
    c. to be the same as the other capacitors' voltage ratings, if the others are operating at their voltage rating.
    d. no special consideration.
13. In a purely capacitive circuit, the current:
    a. leads the voltage by 45°.
    b. lags the voltage by 45°.
    c. leads the voltage by 90°.
    d. lags the voltage by 90°.
14. Capacitive reactance and inductive reactance:
    a. vary in the same way as frequency of operation is changed.
    b. vary in opposite ways as frequency of operation is changed.
    c. both remain the same as frequency of operation is changed.
    d. none of the above.
15. Capacitive reactance is related to the rate of change of:
    a. current.
    b. voltage.
    c. frequency.
    d. none of the above.

16. When computing X_C, the constant 0.159, when used, is equivalent to:
   a. 2π.
   b. $1/2\pi$.
   c. $2\pi fC$.
   d. none of the above.
17. The process of developing a voltage across capacitor plates may be called:
   a. rate of change of voltage.
   b. charging.
   c. discharging.
   d. none of the above.
18. When the rate of change of voltage is maximum, capacitor charging or discharging current is:
   a. maximum.
   b. minimum.
   c. not changing.
   d. none of the above.
19. When the sine-wave voltage applied to capacitor circuit is near maximum level, i_C is
   a. maximum.
   b. minimum.
   c. not changing.
20. What formula would you use to find f if you know the C and X_C?

Problems

1. What is the capacitive reactance of a 0.00025-µF capacitor operating at a frequency of 5 kHz?
2. What capacitance value exhibits 250 Ω of reactance at a frequency of 1,500 Hz?
3. The frequency of applied voltage to a given circuit triples while the capacitance remains the same. What happens to the value of capacitive reactance?
4. Series-circuit capacitors are all exchanged for Cs, each having twice the capacitance of the capacitor it replaces. What happens to total circuit X_C and total C?
5. For the circuit in Figure 18–12 find C_T, X_{C_T}, V_A, C_1, C_2, and X_{C_2}.
6. For the circuit in Figure 18–12, if C_1 and C_2 were equal in value, what would be the new voltages on C_1 and C_2?
7. Assume the frequency doubles in the circuit in Figure 18–12. Answer the following statements with "I" for increase, "D" for decrease, and "RTS" for remain the same.
   a. C_1 will _____
   b. C_2 will _____
   c. C_T will _____
   d. X_{C_1} will _____
   e. X_{C_2} will _____
   f. X_{C_T} will _____
   g. Current will _____
   h. V_{C_1} will _____
   i. V_{C_2} will _____
   j. V_A will _____
8. For the circuit in Figure 18–13 find C_T, I_1, I_2, X_{C_T}, X_{C_1}, X_{C_2}, and I_T.
9. For the circuit in Figure 18–13, if C_2 is made to equal C_1, what is the I_T?

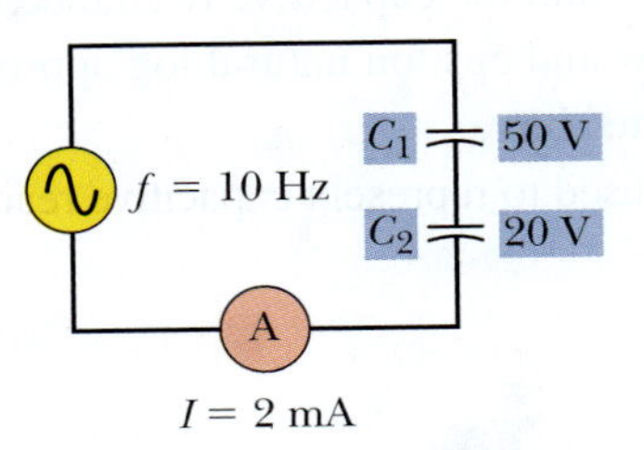

FIGURE 18–12

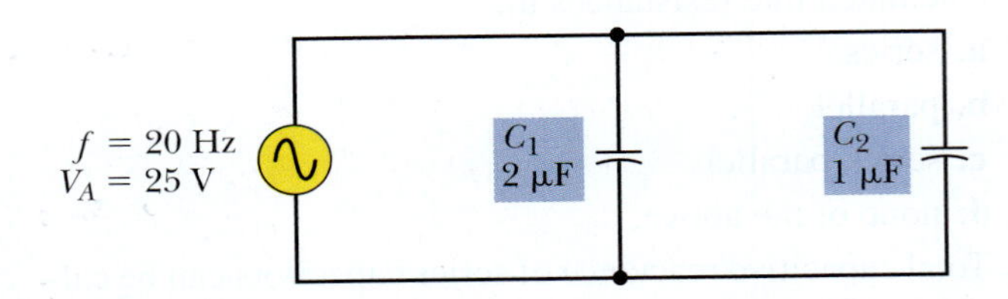

FIGURE 18–13

10. Assume the frequency doubles in the circuit in Figure 18–13. Answer the following statements with "I" for increase, "D" for decrease, and "RTS" for remain the same.
   a. C_1 will _____
   b. C_2 will _____
   c. C_T will _____
   d. X_{C_1} will _____
   e. X_{C_2} will _____
   f. X_{C_T} will _____
   g. Current will _____
   h. V_{C_1} will _____
   i. V_{C_2} will _____
   j. V_A will _____

Refer to Figure 18–14 and answer questions 11 to 15:

11. What is the X_C of C_1?

12. What is the X_{C_T}?

13. What is the value of C_T?

14. If f were changed to 5 kHz, what would be the new value of circuit X_{C_T}?

15. If C_2 were replaced by a larger capacitance value capacitor, would the voltage across C_4 increase, decrease, or remain the same?

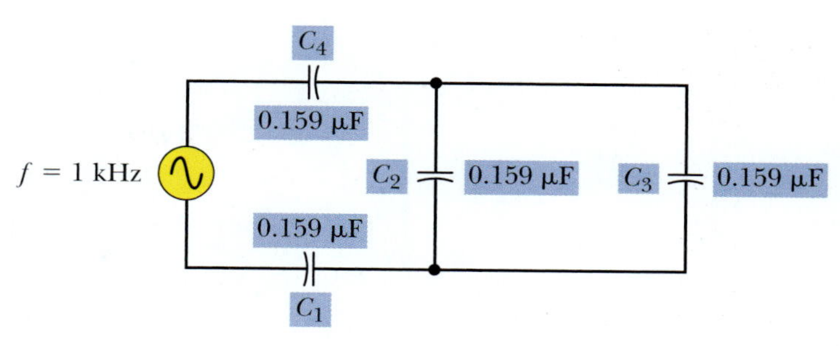

FIGURE 18–14

Analysis Questions

1. In your own words, explain when and why a capacitor might be said to be an "open circuit" with respect to dc, or that a capacitor "blocks" dc.

2. In your own words, explain how a capacitor can "pass" ac.

3. Draw a capacitive voltage divider circuit where the following conditions are found. (Label the diagram with all component and parameter values.)

a. Three capacitors are used.

b. The voltage across the middle capacitor will be double that found across the capacitors on either side of it.

c. The frequency of operation is 2 kHz.

d. The capacitor with maximum X_C has a reactance value of 79.5 Ω.

e. $V_A = 159$ V.

4. Draw a circuit configuration where the total capacitance is 0.66 μF, using six 1-μF capacitors.

Performance Projects Correlation Chart

Suggested performance projects that correlate with topics in this chapter are:

Chapter Topic	Performance Project	Project Number
V & I Relationships in a Purely Capacitive AC Circuit	Capacitance Opposing a Change in Voltage	49
Relationship of X_C to Capacitance Value and Relationship of X_C to Frequency of AC	X_C Related to Capacitance and Frequency	50
Methods to Calculate X_C	The X_C Formula	51
	Story Behind the Numbers: Capacitive Reactance in AC	

NOTE: It is suggested that after completing the above projects, the student should be required to answer the questions in the "Summary" at the end of this section of projects in the Laboratory Manual.

Troubleshooting Challenge

CHALLENGE CIRCUIT 11

(Follow the SIMPLER sequence by referring to inside front cover.)

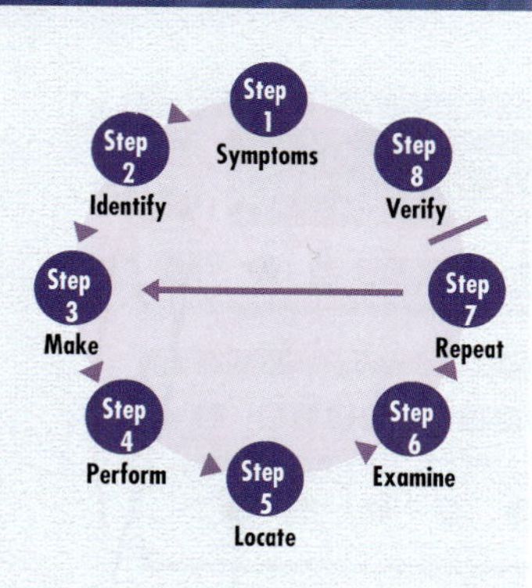

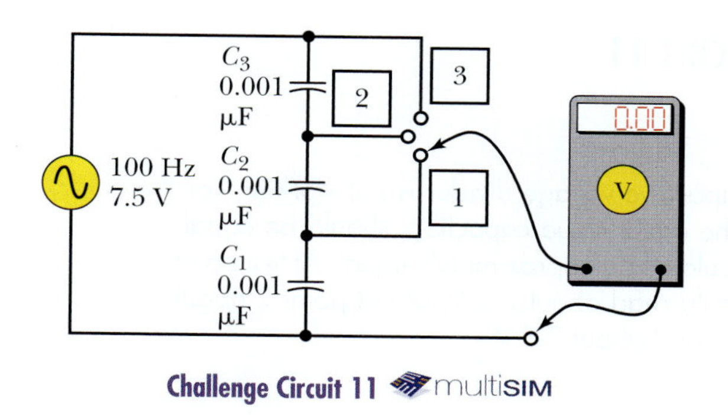

Challenge Circuit 11 multiSIM

STARTING POINT INFORMATION

1. Circuit diagram
2. When testing at test point "1," voltmeter reading is 0 volts.
3. When testing at test point "2," voltmeter reading is 7.5 volts.
4. When testing at test point "3," voltmeter reading is 7.5 volts.

TEST	Results in Appendix C
V_{C_1}	(28)
V_{C_2}	(67)
V_{C_3}	(102)
V_T	(79)
"Bridging" with known good 0.001-μF capacitor:	
Bridge C_1	(71)
Bridge C_2	(24)
Bridge C_3	(111)

CHALLENGE CIRCUIT 11

STEP 1

SYMPTOMS The capacitive voltage divider is not working correctly. The X_Cs of the equal-value capacitors should be equal. The $I \times X_C$ drops should all be approximately equal. At test point 1, the voltmeter should read about 2.5 V; at test point 2 about 5.0 V; and at test point 3 about 7.5 V.

STEP 2

IDENTIFY initial suspect area: If C_1 were shorted, the V at test point 1 would be 0 V. The V at test point 2 would be 3.75 V rather than 2.5 V. If C_3 were shorted, the V at switch position 1 would be 3.75 V, not 0 V. If C_1 were open, the V at switch position 1 would be 7.5 V. If C_3 were open, the V at switch position 2 would be 0 V. The normal voltage at switch position 3 is 7.5 V (the source voltage), which is what is measured. Without making our first test, we have narrowed the initial suspect area to C_2. If it were open, the circuit would operate similar to the symptom descriptions.

STEP 3

MAKE test decision: Bridge the suspected open capacitor with a known good one of the same ratings and note any changes in circuit operation. We'll try the quick bridging technique. CAUTION: Make bridging connections with the power off. Then turn power on and make the necessary observations.

STEP 4

PERFORM **1st Test:** Bridge C_2 while monitoring voltage at switch position 2. When this is done, V measures about 4.8 V, close to the nominal 5-V normal operation.

STEP 5

LOCATE new suspect area: This is not necessary since we've found our problem.

STEP 6

EXAMINE available data.

STEP 7

REPEAT analysis and testing.

STEP 8

VERIFY that the circuit operates normally by checking the voltage at each switch position **(2nd, 3rd,** and **4th Tests).** C_2 was open.

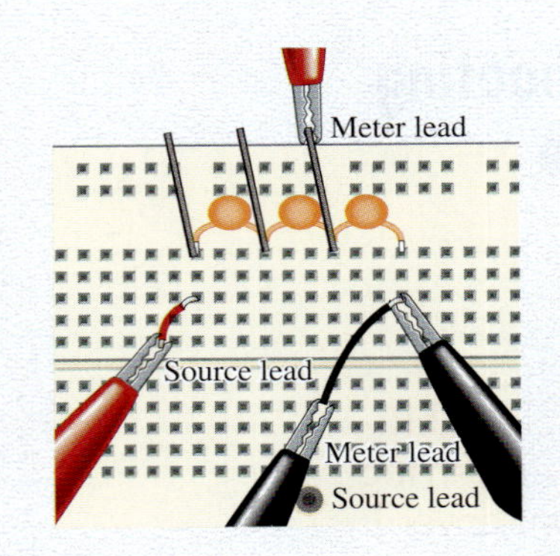

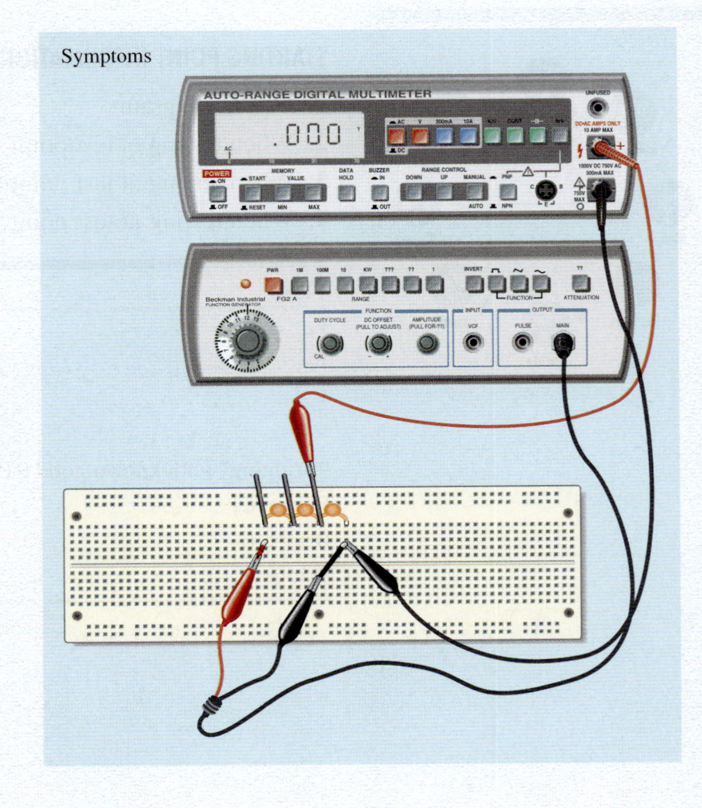

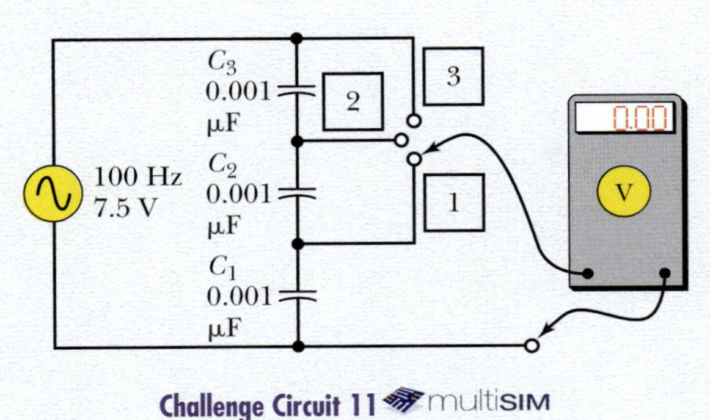

Challenge Circuit 11 multiSIM

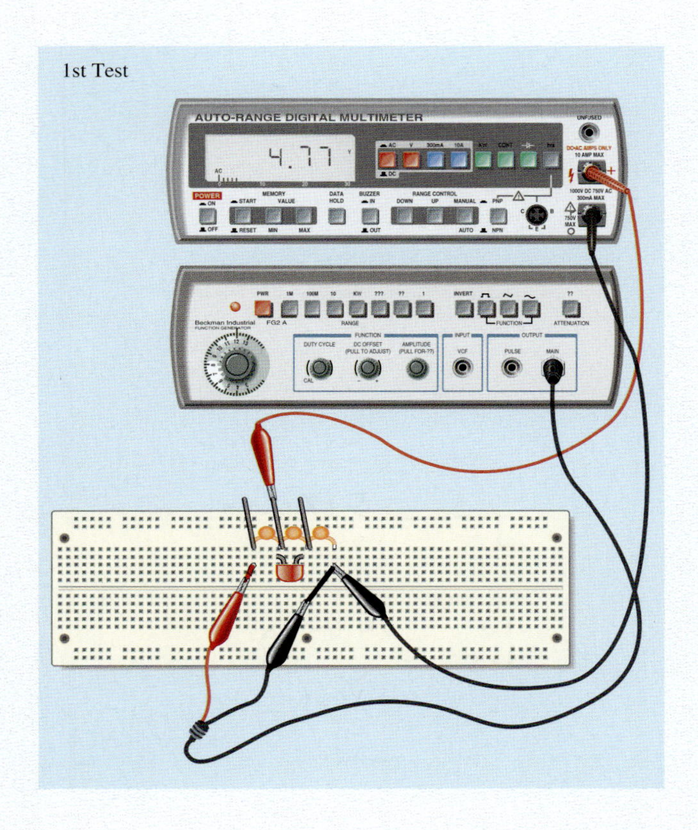
1st Test
AUTO-RANGE DIGITAL MULTIMETER
4.77
Beckman Industrial
FG2 A

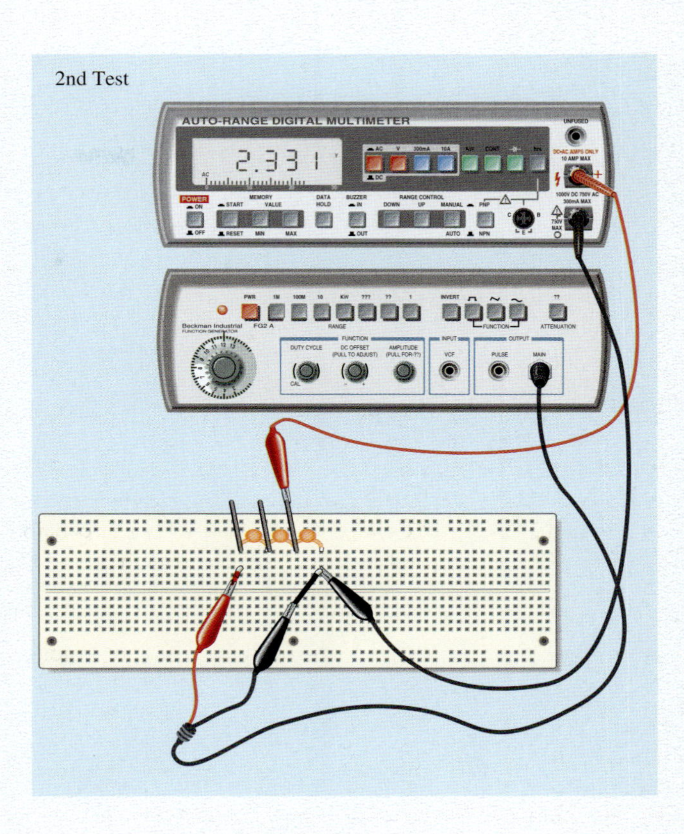
2nd Test
AUTO-RANGE DIGITAL MULTIMETER
2.331
Beckman Industrial
FG2 A

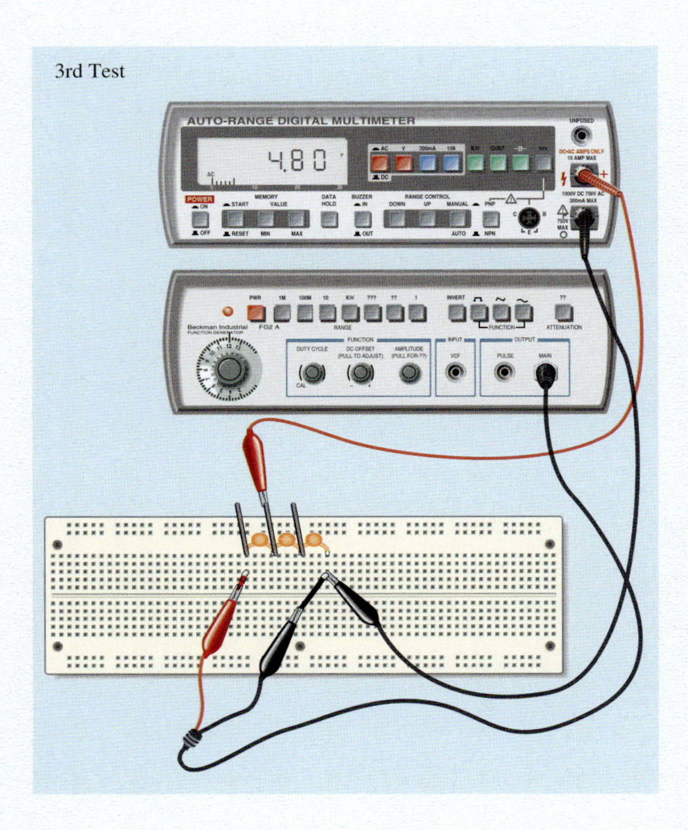
3rd Test
AUTO-RANGE DIGITAL MULTIMETER
4.80
Beckman Industrial
FG2 A

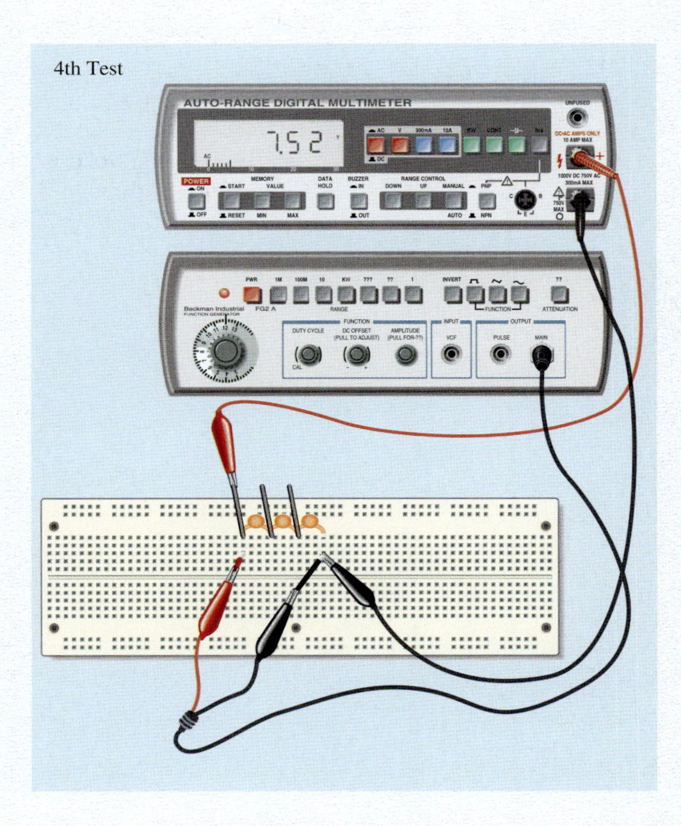
4th Test
AUTO-RANGE DIGITAL MULTIMETER
7.52
Beckman Industrial
FG2 A

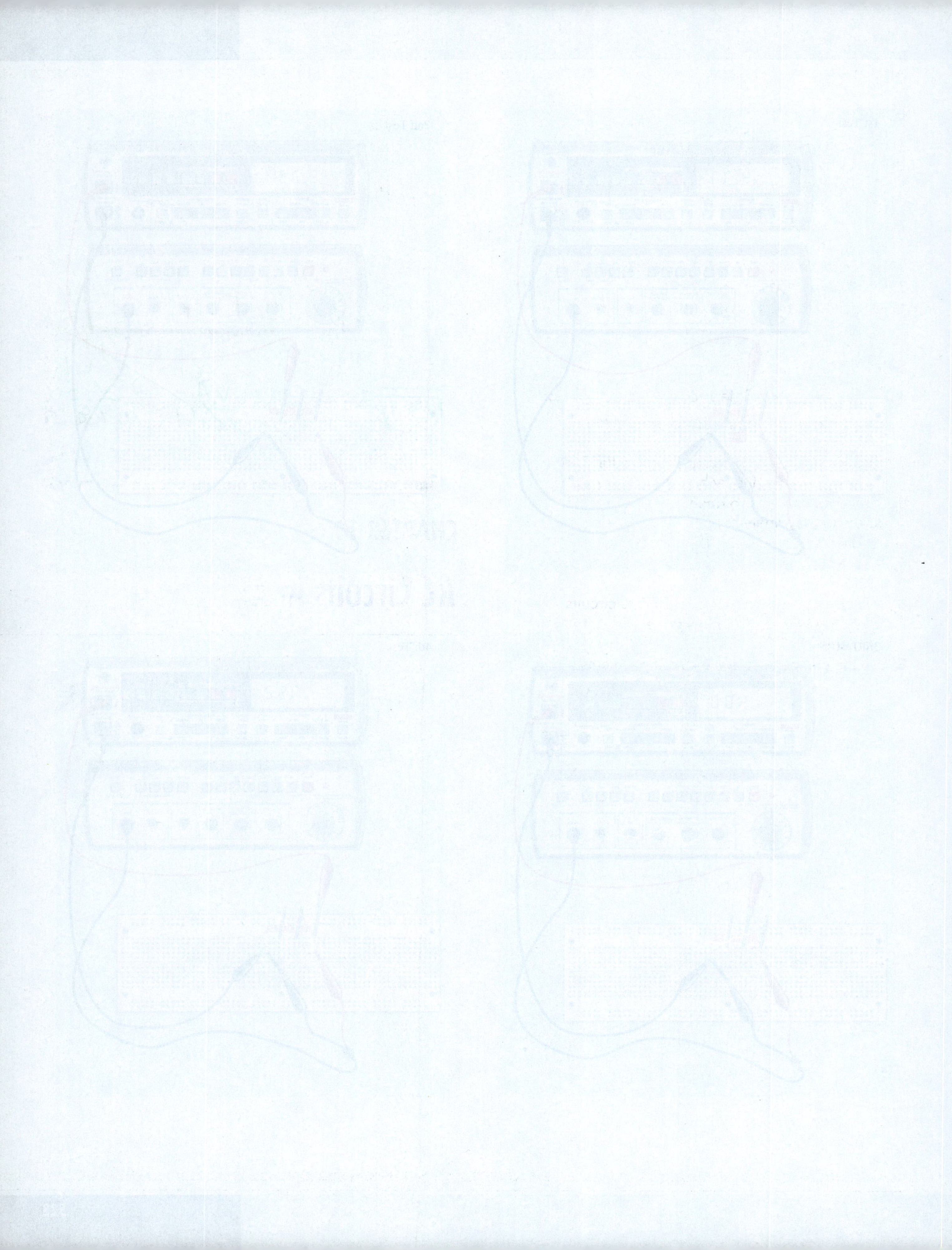

OUTLINE

OBJECTIVES

After studying this chapter, you should be able to:

1. Draw or describe operation of simple *R* and *C* circuits
2. Analyze appropriate series and parallel *RC* circuit parameters using the Pythagorean theorem
3. Use vector analysis to analyze series and parallel *RC* circuit parameters
4. List differences between *RC* and *RL* circuits
5. Predict output waveform(s) from a waveshaping network
6. List two practical applications for *RC* circuits
7. Apply troubleshooting hints to help identify problems
8. Use the computer to solve circuit problems

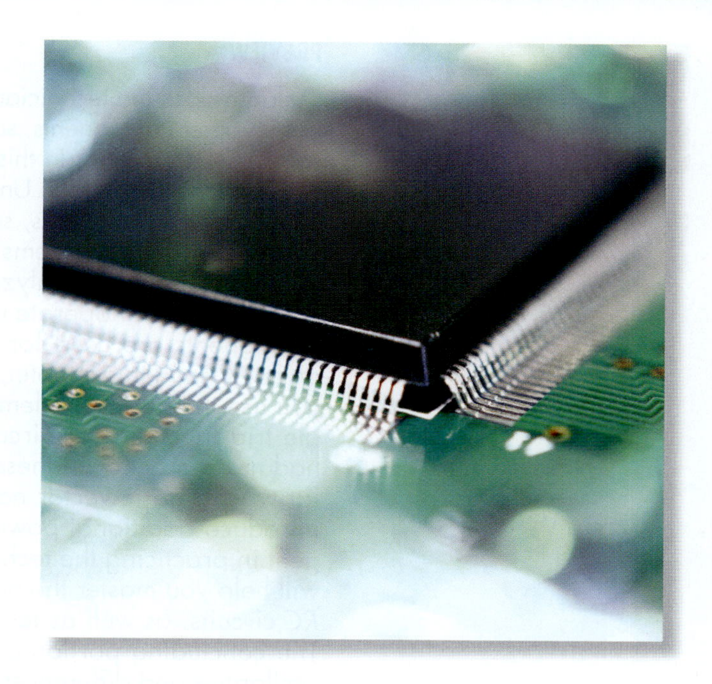

CHAPTER 19

RC Circuits in AC

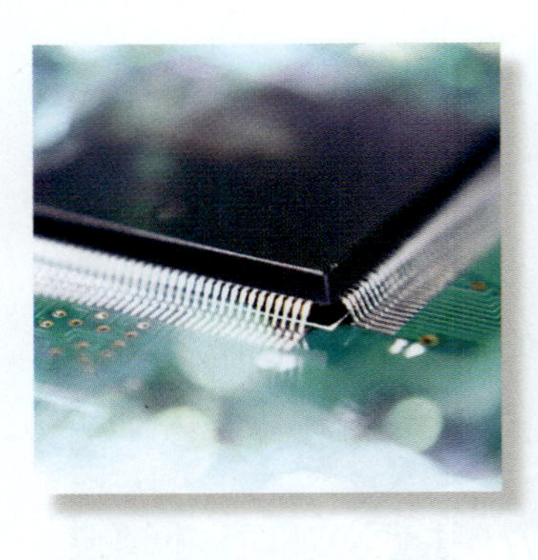

PREVIEW

Seldom does the technician find capacitors alone in a circuit. They are generally used with other components, such as resistors, inductors, transistors, and so on, to perform some useful function. In this chapter, we will investigate simple circuits containing both resistors and capacitors. Understanding *RC* circuits is foundational to dealing with practical circuits and systems, such as amplifiers, receivers, transmitters, computers, power supplies, and other systems and subsystems.

You will learn to analyze both series and parallel combinations of resistance and capacitance that operate under sine-wave ac conditions. An overview of how these components are used for waveshaping in nonsinusoidal applications will also be given. Also in this chapter, as in the earlier chapter "*RL* Circuits in AC," we will have you use the basic, or elemental, approaches of the Pythagorean theorem, and simple trig functions for circuit analysis. After completing this chapter, you will have had more practice in these techniques. Since after this chapter you will have familiarity with basic vector notations for both *RL* and *RC* circuits, the next chapter will introduce additional powerful vector analysis and notation techniques. So be diligent in practicing the techniques and calculator usage afforded in this chapter. This will help you master the next chapter, which combines all your knowledge of *RL* and *RC* circuits, as well as teaching you some additional powerful analysis techniques. The concluding portion of this chapter will provide a summary and review of the similarities and differences between *RC* and *RL* circuits and supply some troubleshooting hints for *RC* circuits.

KEY TERMS

Current leading voltage
Long time-constant circuit
Short time-constant circuit

19–1 Review of Simple *R* and *C* Circuits

Although you have studied purely resistive and capacitive circuits, review Figure 19–1 as a quick refresher.

19–2 Series *RC* Circuit Analysis

As Figure 19–1 has just revealed, current and voltage are in phase for purely resistive ac circuits. Circuit **current leads voltage** by 90° in purely capacitive circuits. Therefore, when resistance and capacitance are combined, the overall difference in angle between circuit voltage and current is an angular difference between 0° and 90°, Figure 19–2. The actual angle depends on the relative values of the components' electrical parameters and the frequency of operation.

Positional Information for Phasors

In series circuits, the current is used as the zero-degree reference vector for *V-I* vector diagrams because *current is common throughout a series circuit.* The *V-I* vector diagram, therefore, shows the resultant circuit voltage lagging behind the current by some angle between 0° and –90°, again see Figure 19–2. Remember, this is in the fourth quadrant of the coordinate system.

In *series RC* circuits (and *only* for *series* circuits), we can also develop an impedance diagram as well as a *V-I* vector diagram, Figure 19–3. Since the resistive voltage drop ($I \times R$)

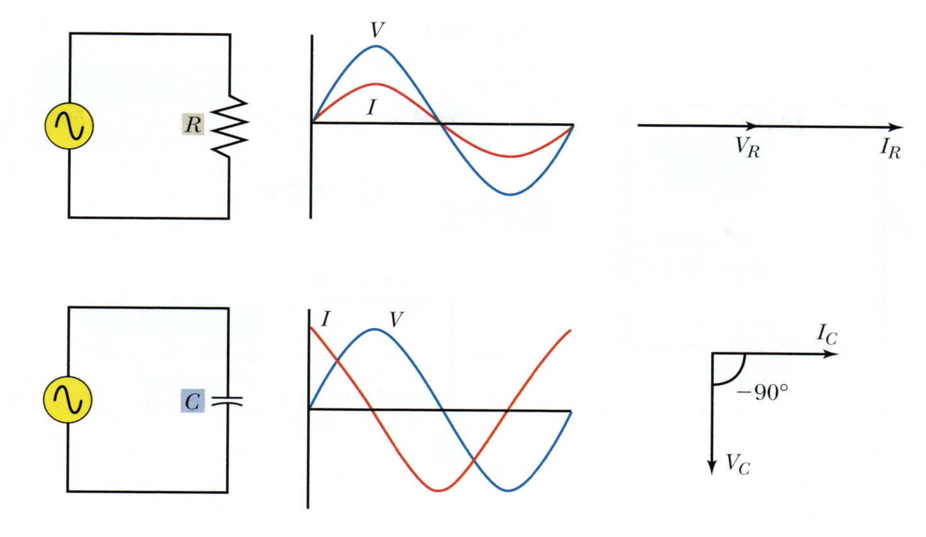

FIGURE 19–1 Voltage-current relationships in purely R and purely C circuits

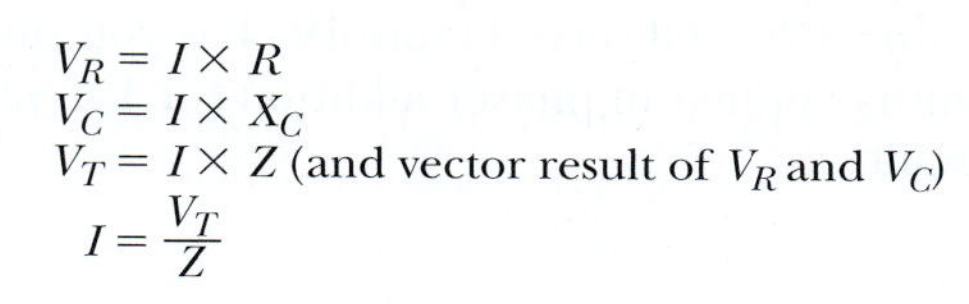

$V_R = I \times R$
$V_C = I \times X_C$
$V_T = I \times Z$ (and vector result of V_R and V_C)
$I = \frac{V_T}{Z}$

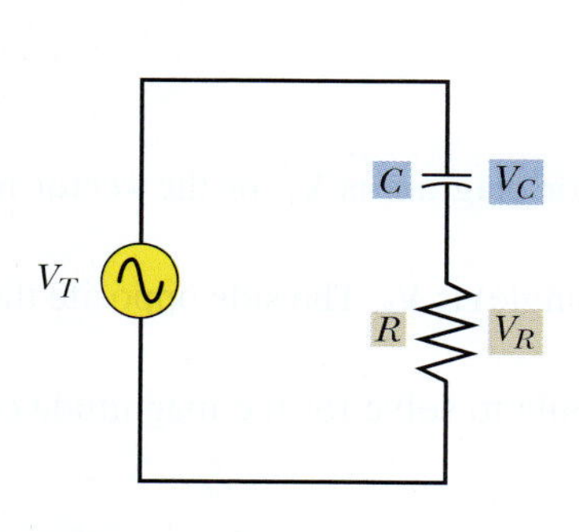

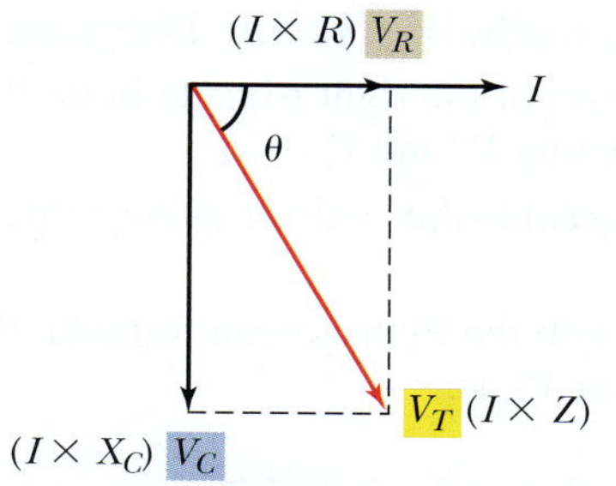

Angular difference between V_T and I (θ) is greater than 0° but less than 90°.

FIGURE 19–2 Positional information for phasors in series RC circuits

vector must indicate the resistor voltage is in phase with the circuit current, the impedance diagram line representing R is shown at the 0° position. Because the capacitor voltage ($I \times X_C$) lags the capacitor (and circuit) current by 90°, the X_C on the impedance diagram is shown downward at the –90° position.

Summarizing for *series RC* circuits, I and V_R are shown at 0°, and V_C is shown at –90° in V-I vector diagrams. The R is shown at 0° and X_C at – 90° for impedance diagrams. ***The angle between V_T and I (in V-I vector diagrams) and between Z and R (in impedance diagrams) is the value of the circuit phase angle.*** With this information, let's proceed with some analysis techniques.

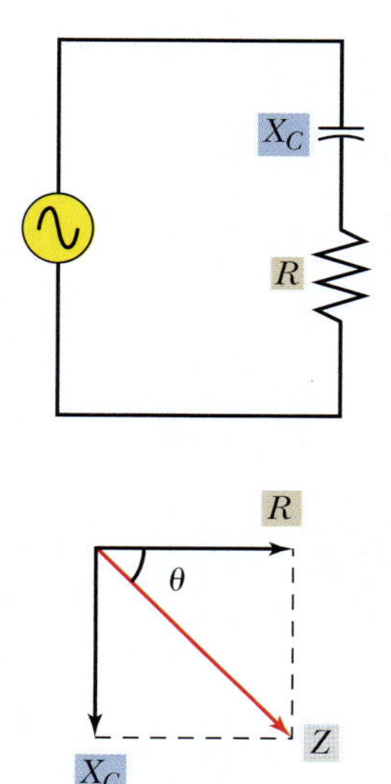

FIGURE 19–3 Impedance diagram for a series RC circuit

Solving Magnitude Values by the Pythagorean Theorem

As in the chapter on RL circuits, we can use the Pythagorean theorem to solve for magnitudes of voltages, currents, and circuit impedance in series RC circuits.

Recall that the Pythagorean theorem deals with right triangles and the relationship of triangle sides. The hypotenuse (c) relates to the other sides by the formula $c^2 = a^2 + b^2$,

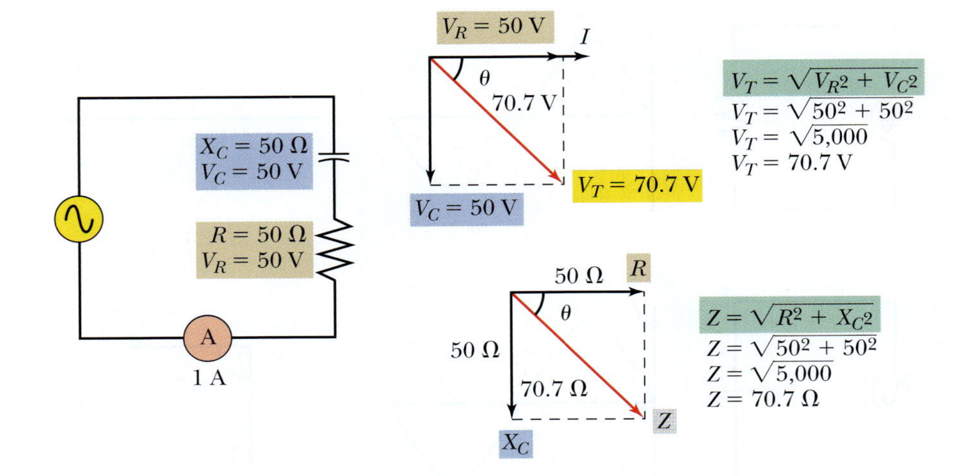

FIGURE 19–4 Using the Pythagorean theorem approach

multiSIM

or $c = \sqrt{a^2 + b^2}$. Observe Figure 19–4 as you study the following statements to see how this theorem is applied in phasor addition for *V-I* vector and impedance diagrams for a series *RC* circuit.

Solving for V_T in the *V-I* Vector Diagram

1. The hypotenuse of the right triangle in the *V-I* vector diagram is V_T, or the vector resultant of combining V_R and V_C.
2. The side adjacent to the angle of interest (the phase angle) is V_R. The side opposite the angle is V_C.
3. Substituting into the Pythagorean formula, the formula to solve for the magnitude of the hypotenuse, or V_T, is:

FORMULA 19–1 $V_T = \sqrt{V_R^{\ 2} + V_C^{\ 2}}$

EXAMPLE Thus, in Figure 19–4, $V_T = \sqrt{50^2 + 50^2} = \sqrt{2{,}500 + 2{,}500} = \sqrt{5{,}000} = 70.7$ V.

Solving for Z in the Impedance Diagram

1. The hypotenuse of the right triangle in the impedance diagram is Z (impedance).
2. The side adjacent to the angle representing the phase angle is R. The side opposite the angle represents X_C.
3. Substituting into the Pythagorean formula, the formula to solve for the magnitude of the hypotenuse, or Z, is:

FORMULA 19–2 $Z = \sqrt{R^2 + X_C^{\ 2}}$

EXAMPLE Thus, in Figure 19–4, $Z = \sqrt{50^2 + 50^2} = \sqrt{2{,}500 + 2{,}500} = \sqrt{5{,}000} = 70.7\ \Omega$.

Try applying the Pythagorean formula to the following problems.

PRACTICE PROBLEMS 1

1. For the circuit in Figure 19–5, use the Pythagorean approach and solve for V_T. Draw the *V-I* vector diagram and label all elements. (NOTE: You may want to try the drawing-to-scale approach learned in an earlier chapter to check your results.)
2. For the circuit in Figure 19–5, use the Pythagorean approach and solve for *Z*. Draw and label an appropriate impedance diagram.

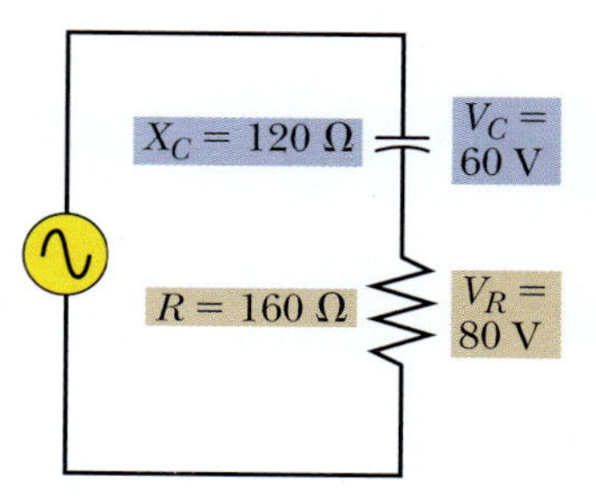

FIGURE 19–5

Solving for Phase Angle Using Trigonometry

Recall from the *RL* circuits chapter that we can use simple trig functions to solve for the angles of a right triangle. Remember that

$$\sin\theta = \frac{\text{opp}}{\text{hyp}},\ \cos\theta = \frac{\text{adj}}{\text{hyp}},\ \text{and}\ \tan\theta = \frac{\text{opp}}{\text{adj}}$$

Because we have already identified which sides are the side adjacent, side opposite, and hypotenuse in the series *RC* circuit *V-I* vector and impedance diagrams, substitute into the trig formulas to solve for angle information.

Note that the trig formula that does not require knowing the magnitude of the hypotenuse to solve for the angle information is the tangent function. Therefore, this function is often used.

Practical Notes

Reminder: It is a good habit to sketch diagrams for any problem you are solving even if you don't draw them precisely. By sketching, you can approximate to see if the answers you get are logical.

Practical Notes

With the accessibility of scientific calculators, finding phase angles (and other circuit parameters) in ac circuits has become increasingly simple. In this chapter, we are having you use, practice, and continue to learn the rudimentary concepts related to trig functions for phase-angle solutions.

In the next chapter, you will have the opportunity to learn rectangular and polar vector notation techniques and simple calculator keystrokes that allow conversion between these notation methods. As you will see, these notation methods and the calculator can take much of the "grief" out of ac circuit calculations.

The foundations you have built in previous chapters, and are building in this chapter, will help you to better understand and use these techniques as you progress in your studies. We encourage you to try to understand concepts, rather than simply memorizing formulas!

Solving the Angle from the *V-I* Vector Diagram

EXAMPLE Refer again to Figure 19–4. Since we know all three sides, we can use any of the three basic trig formulas. However, let's use the tangent function.

1. To find the phase angle, or the angle between V_T and *I*, the V_C and V_R values can be used.
2. Substitute these values into the tangent formula.

$$\tan\theta = \frac{\text{opp}}{\text{adj}} = \frac{V_C}{V_R} = \frac{50}{50} = 1$$

3. Using a calculator, we find that the arctan θ, or the angle whose tangent equals 1 is 45°.
4. Therefore, the angle value is 45°. This makes sense since both legs of the triangle are equal in length. If you draw this diagram to scale and use a protractor to measure the angle, the result should verify a 45° angle.
5. Since V_T is in the fourth quadrant, the phase angle is considered negative; thus, $\theta = -45°$.

Solving the Angle from the Impedance Diagram

EXAMPLE Refer to Figure 19–6 and use the tangent function to find the phase angle.

1. The angle between Z and R is the same value as the phase angle (the angle between the circuit voltage and circuit current).
2. Using the tangent function to find this angle requires the X_C and R values.
   a. Add the resistances to get the R value.

$$R = 270\ \Omega + 270\ \Omega = 540\ \Omega$$

   b. Add the X_Cs to get the X_C value.

$$X_C = 300\ \Omega + 100\ \Omega = 400\ \Omega$$

3. Substitute appropriate values into the tangent formula.

$$\tan\theta = \frac{\text{opp}}{\text{adj}} = \frac{X_C}{R} = \frac{400}{540} = 0.74$$

4. Find the $\tan^{-1}$ of 0.74 to get the angle.

$$\tan^{-1} 0.74 = 36.5°;\ \text{thus, } \theta = 36.5°$$

NOTE: When impedance for series RC circuits is plotted on an impedance vector diagram, the location of the Z vector is in the fourth quadrant, somewhere between 0° and –90° removed from the reference vector (R). This means the phase angle is considered a negative angle. Thus, $\theta = -36.5°$, in this case.

Apply these principles to solve the following problems.

PRACTICE PROBLEMS 2

1. Refer to Figure 19–7. Draw a V-I vector diagram for the circuit shown, and find the phase angle using a basic trig function. (Round off to the nearest volt.)
2. Draw an impedance diagram for the circuit in Figure 19–7. Then solve for the value of θ, using an appropriate trig function.

FIGURE 19–6

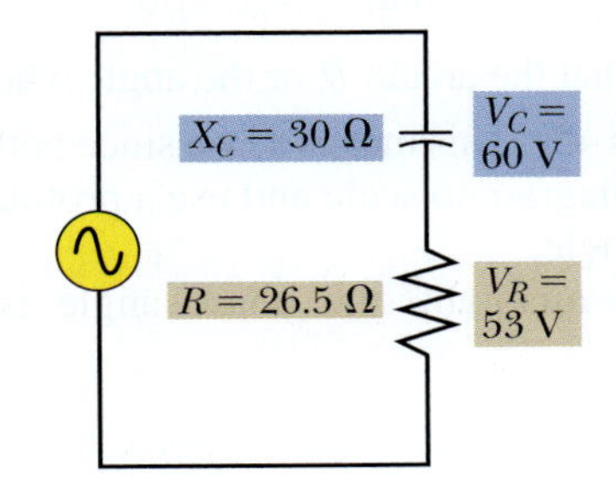

FIGURE 19–7 multiSIM

19–3 Phase-Shift Applications of Series *RC* Networks

Background Information

Phase-shift networks are frequently needed and used in electronic circuits and electronically controlled systems. The series *RC* network is one of the simpler networks for achieving a desired phase shift (or phase difference). The network is used to create and/or control the phase difference between the voltage into the network, and the voltage being taken as output voltage from the network. These phase-shift networks are sometimes called *RC* "lead" or "lag" circuits.

The principle on which these *RC* phase-shift networks operate is based on the fact that a capacitor opposes a change in voltage. That is, it takes time for the capacitor to change from one voltage level to another (because of charging and discharging time required). You have already learned that in a series *RC* circuit, the circuit current may lead the circuit applied voltage (V_{in}) anywhere from 0° to 90°, depending on the relative values of R and X_C. Also, you have learned that the voltage across the resistor is in phase with the circuit current.

We can control the relative R and X_C values by choosing or varying the component values. For example, we can use a variable resistor in the network to control the amount of phase shift. We can also select which component we will take the output voltage across. By selecting the appropriate component, we can choose whether the output voltage leads or lags the input voltage. A "lead" network is defined as one in which the *output voltage leads input voltage.* A "lag" network is one in which the *output voltage lags the input voltage.*

EXAMPLE (AN *RC* LEAD NETWORK) Refer to Figure 19–8 as you study the following comments.

1. For the lead network, the output is taken across R.
2. For the lead network, then, V_R and V_{out} are the same voltage.
3. For our vector diagram, we will use V_{out} as the reference vector.
4. Since the capacitor causes the circuit current to "lead" the circuit voltage and the output voltage is taken across the resistor, the output voltage is leading the input voltage.
5. The amount of lead is controllable by the values of C and R, which control the relative values of X_C and R.
6. The circuit phase angle $(\theta) = \arctan \dfrac{X_C}{R}$

$$X_C = \frac{1}{2\pi fC} = \frac{1}{6.28 \times 400 \times 0.5 \times 10^{-6}} = 796\ \Omega$$

$$\frac{X_C}{R} = \frac{796}{1{,}000} = 0.796$$

$$\arctan 0.796 = 38.5°$$

7. For this network, the output voltage leads the input voltage by 38.5°.

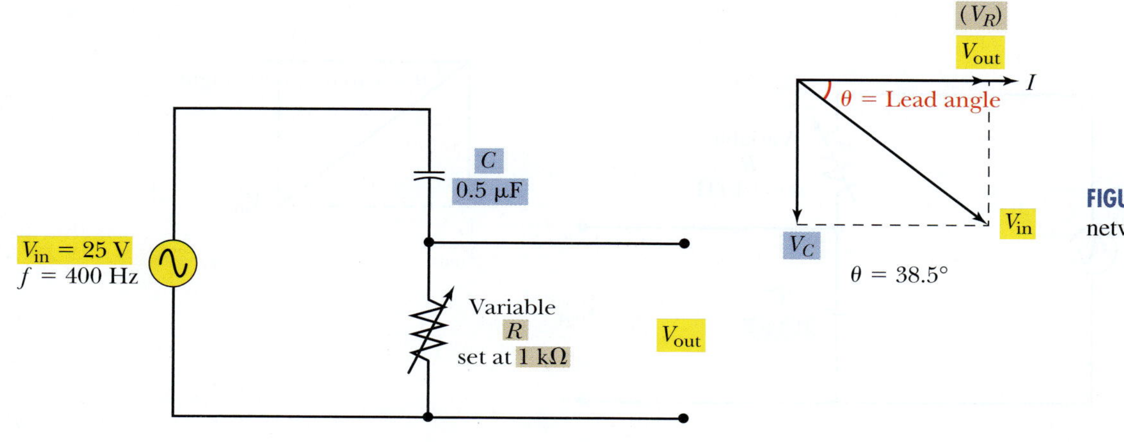

FIGURE 19–8 An *RC* "lead" network

8. The value of V_{out} is determined by the voltage divider action of the *RC* network. In essence, the output voltage (in this case V_R) relates to the ratio of the *R* to the circuit impedance. That is, $V_{out} = V_{in} \times R/Z$. (Of course *Z* in this circuit equals $\sqrt{R^2 + X_C^{\ 2}}$.)

$$V_{out} = V_{in}\ (R/Z)$$
$$\text{where } Z = \sqrt{R^2 + X_C^{\ 2}} = \sqrt{1{,}000^2 + 796^2} = 1{,}278\ \Omega$$
$$V_{out} = 25\text{ V }(1{,}000/1{,}278) = 25\text{ V} \times 0.78 = 19.5\text{ V}$$

PRACTICE PROBLEMS 3

What would be the angle of "lead" should the variable resistor in Figure 19–8 be set at 500 Ω?

EXAMPLE (AN *RC* LAG NETWORK) Refer to Figure 19–9 as you study the following information.

1. For the "lag" network, the output is taken across *C*.
2. For the lag network, then, V_C and V_{out} are the same voltage.
3. For our vector diagram, we will use V_R as the reference vector.
4. Since the capacitor causes the circuit current to lead the circuit voltage and the output voltage is taken across the capacitor, the output voltage is lagging the input voltage.
5. The circuit phase angle and the amount of lag are NOT the same thing, in this case. The circuit phase angle is the phase difference between V_R and V_{in}. Since the output is taken across the capacitor, the *lag angle is the difference in degrees between* V_C *and* V_{in}.
6. The circuit phase angle and the amount of lag are controllable by the values of *C* and *R*, which, in turn, control the relative values of X_C and *R*.
7. The circuit phase angle $(\theta) = \arctan \dfrac{X_C}{R}$

$$X_C = \frac{1}{2\pi fC} = \frac{1}{6.28 \times 400 \times 0.5 \times 10^{-6}} = 796\ \Omega$$
$$\frac{X_C}{R} = \frac{796}{1{,}000} = 0.796$$
$$\arctan 0.796 = 38.5^\circ$$

8. For this network, V_{in} lags V_R by this 38.5° (the circuit phase angle); however, the amount of lag between V_{out} compared with V_{in} equals 90° minus 38.5° or 51.5° of lag.

Lag angle = 90° − circuit phase angle

9. To vary the amount of lag in this circuit, we can vary the *R* value. If the *R* value is decreased, the circuit becomes more capacitive and the phase angle will increase; however, the amount of lag will decrease. Remember that the lag angle is the angular difference

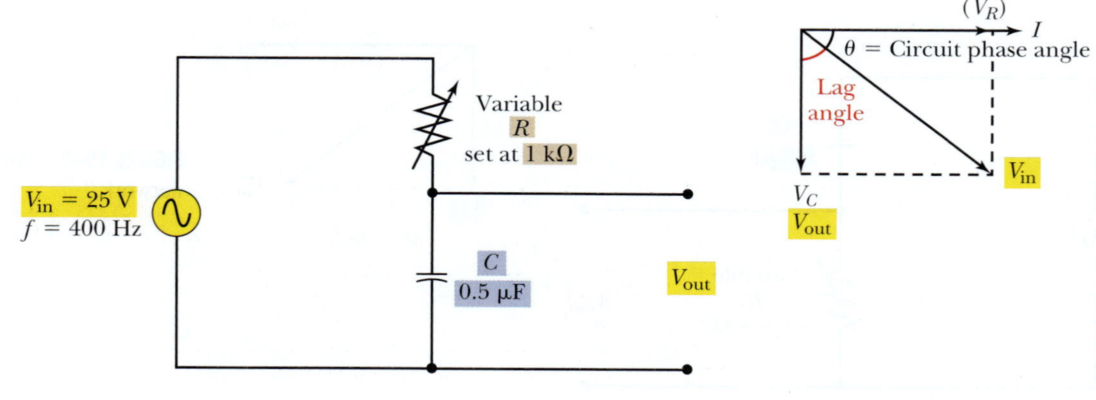

FIGURE 19–9 An *RC* "lag" network

between V_{in} and V_C (the output voltage). If the R value is increased, the circuit phase angle will decrease; however, the lag angle will increase.

10. The value of V_{out} is determined by the voltage-divider action of the RC network. In essence, the output voltage (in this case V_C) relates to the ratio of the X_C to the circuit impedance. That is, $V_{out} = V_{in} \times X_C/Z$. (Of course Z, in this circuit, equals $\sqrt{R^2 + X_C^2}$.

$$V_{out} = V_{in}\left(\frac{X_C}{Z}\right)$$

$$\text{where } Z = \sqrt{R^2 + X_C^2} = \sqrt{1{,}000^2 + 796^2} = 1{,}278\ \Omega$$

$$V_{out} = 25\text{ V}\left(\frac{796}{1{,}278}\right) = 25\text{ V} \times 0.62 = 15.5\text{ V}$$ ◆

PRACTICE PROBLEMS 4

What would be the angle of lag should the R be set at 750 Ω in the circuit of Figure 19–9?

Summary of Series RC Circuits

Figure 19–10 shows examples of series RC circuit parameters for various circuit conditions. These examples help reinforce the following generalizations about series RC circuits.

1. Ohm's law formulas can be used at any time to solve for individual component or total circuit voltages, currents, resistances, reactances, and impedances, as long as the appropriate parameter values are substituted into the formulas. That is,

$$V_C = I_C \times X_C,\ Z = \frac{V_T}{I_T},\ \text{and so forth}$$

CAPACITANCE CHANGED BY A FACTOR OF 2 (FROM 2 μF TO 1 μF)

f (Hz)	C (μF)	X_C (Ω)	R (Ω)	Z (Ω)	I (amps)	V_C (IX_C)	V_R (IR)	V_T (IZ)	θ deg
100	2	795	7,950	7,989.65	0.025	19.90	199.01	200	−5.71
100	1	1,590	7,950	8,107.44	0.0246	39.22	196.12	200	−11.31

(C to half; X_C doubled; Z increased; I decreased; V_C increased; V_R decreased; and θ increased)

(a)

RESISTANCE CHANGED BY A FACTOR OF 2 (FROM 7,950 Ω TO 3,975 Ω)

f (Hz)	C (μF)	X_C (Ω)	R (Ω)	Z (Ω)	I (amps)	V_C (IX_C)	V_R (IR)	V_T (IZ)	θ deg
100	1	1,590	7,950	8,107.44	0.0246	39.22	196.12	200	−11.31
100	1	1,590	3,975	4,281.21	0.0467	74.28	185.70	200	−21.80

(R to half; Z decreased; I increased; V_C increased; V_R decreased; and θ increased)

(b)

FREQUENCY CHANGED BY A FACTOR OF 10 (FROM 100 Hz TO 10 Hz)

f (Hz)	C (μF)	X_C (Ω)	R (Ω)	Z (Ω)	I (amps)	V_C (IX_C)	V_R (IR)	V_T (IZ)	θ deg
100	1	1,590	3,975	4,281.21	0.0467	74.28	185.7	200	−21.80
10	1	15,900	3,975	16,389.34	0.0122	194.03	48.51	200	−75.96

(f to 1/10; X_C increased ten times; Z increased; I decreased; V_R decreased; and θ increased greatly)

(c)

FIGURE 19–10 Sample series RC circuit parameters charts

2. The circuit voltage lags the circuit current by some angle between 0° and –90°. Conversely, the circuit current leads the circuit voltage by that same angular difference.
3. Since current is common throughout series circuits, current is the reference vector in phasor diagrams of *V* and *I*.
4. The larger the *R* value is with respect to the X_C value, the more resistive the circuit is. That is, the phase angle is smaller or closer to 0°. Conversely, the larger the X_C value is with respect to the circuit *R* value, the greater is the phase angle and the closer it is to –90°.
5. Impedance of a series *RC* circuit is greater than either the *R* value alone or the X_C value alone but *not* as great as their arithmetic sum. Impedance equals the vector sum of *R* and X_C.
6. When either *R* or X_C increases in value, circuit impedance increases and circuit current decreases.
7. Total voltage of a series *RC* circuit is greater than either the resistive voltage drops or the reactive voltages alone but *not* as great as their arithmetic sum. The voltages must be added vectorially to find the resulting V_T value.
8. For series *RC* circuits, both *V-I* vector and impedance diagrams are used. The quantities used in the *V-I* vector diagram are the circuit current as a reference vector at 0°; V_R (also at the 0° reference position); and V_C at –90°. V_T (or V_A) is plotted as the vector resulting from the resistive and reactive voltage vectors. For the impedance diagram, *R* is the reference vector and X_C is plotted downward at the –90° position. Circuit impedance *(Z)* is the vector sum resulting from *R* and X_C.
9. In series *RC* circuits, the V_T for *V-I* vector diagrams and the *Z* for *Z* diagrams are in the fourth quadrant, so the phase angle is considered negative.

■ IN-PROCESS LEARNING CHECK 1

Fill in the blanks as appropriate.

1. In series *RC* circuits, the _____ is the reference vector, shown at 0°.
2. In the *V-I* vector diagram of a series *RC* circuit, the circuit voltage is shown _____ the circuit current.
3. In series *RC* circuits, an impedance diagram (can, cannot) _____ be drawn.
4. Using the Pythagorean approach, the formula to find V_T in a series *RC* circuit is _____.
5. Using the Pythagorean approach, the formula to find *Z* in a series *RC* circuit is _____.
6. The trig function used to solve for phase angle when you do not know the hypotenuse value is the _____ function.

19–4 Parallel *RC* Circuit Analysis

In parallel *RC* circuits, the voltage is common to all parallel branches. Therefore, *voltage* is the reference vector for the *V-I* vector diagrams, Figure 19–11a. Again, you *will not* draw impedance diagrams for *parallel* circuits.

In drawing a *V-I* vector diagram, branch currents are plotted relative to the voltage reference vector. I_R is plotted at the zero-degree position and is the side adjacent to the angle of interest. I_C is plotted at the +90° position because it leads the voltage by 90° and is the side opposite the angle for computational purposes. Again, refer to Figure 19–11a.

Solving for I_T Using the Pythagorean Approach

For parallel *RC* circuits, the Pythagorean formula to solve for I_T becomes:

FORMULA 19–3 $I_T = \sqrt{I_R^2 + I_C^2}$

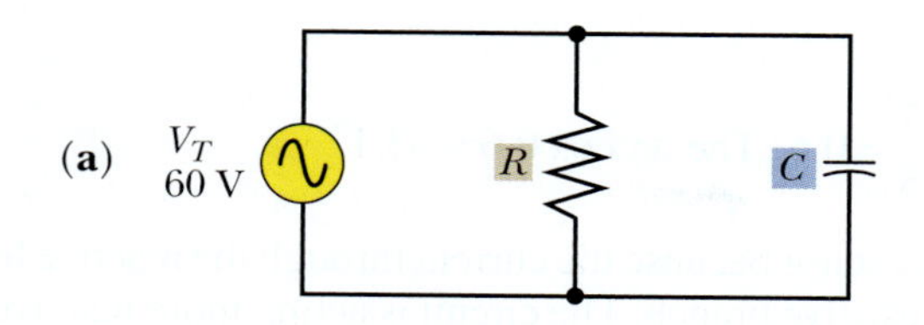

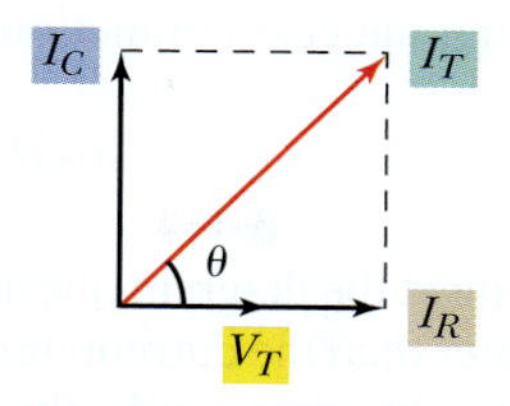

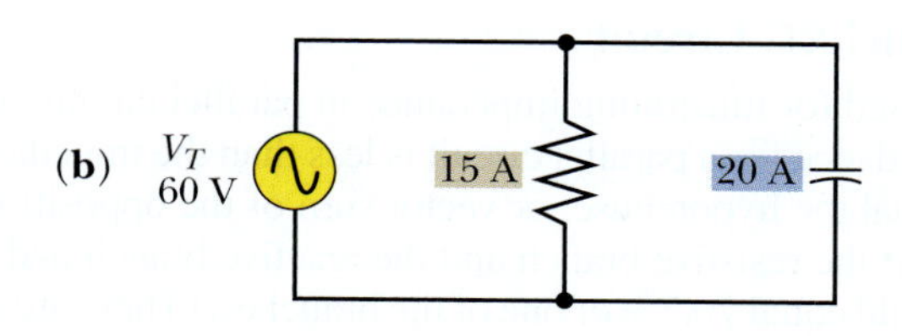

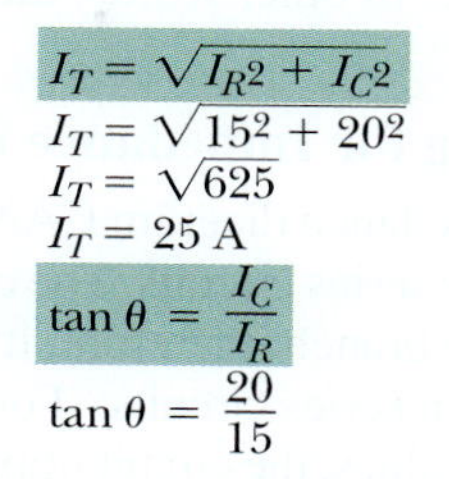

tan θ = 1.33
Angle whose tan = 1.33 is 53.1°

(c) V_T 60 V; 15 A, 4 Ω; 20 A, 3 Ω; $Z = 2.4\ \Omega$; A 25 A

$$Z = \frac{V_T}{I_T} = \frac{60}{25} = 2.4\ \Omega$$

$$R = \frac{V_R}{I_R} = \frac{60}{15} = 4\ \Omega$$

$$X_C = \frac{V_C}{I_C} = \frac{60}{20} = 3\ \Omega$$

(NOTE: Z is less than either of the branch ohms.)

FIGURE 19–11 Parallel *RC* circuit analysis

◆ **EXAMPLE** Use the quantities in Figure 19–11b, and substitute into the preceding formula. Thus,

$$I_T = \sqrt{I_R^2 + I_C^2} = \sqrt{15^2 + 20^2} = \sqrt{625} = 25\text{ A}$$ ◆

Solving for Phase Angle Using Trigonometry

The sine, cosine, and tangent functions, as related to a parallel *RC* circuit, are stated as:

$$\sin\theta = \frac{\text{opp}}{\text{hyp}} \text{ or } \sin\theta = \frac{I_C}{I_T}$$

$$\cos\theta = \frac{\text{adj}}{\text{hyp}} \text{ or } \cos\theta = \frac{I_R}{I_T}$$

$$\tan\theta = \frac{\text{opp}}{\text{adj}} \text{ or } \tan\theta = \frac{I_C}{I_R}$$

It should be noted that due to the fact that the tangent function does not require solving for the hypotenuse (I_T, in this case), it is often used to solve for phase angles.

◆ **EXAMPLES** Referring to the information in Figure 19–11b, trig functions can be used to solve for the phase angle (θ) as follows:

1. Using the tangent function:

$$\tan\theta = \frac{I_C}{I_R} = \frac{20}{15} = 1.33. \text{ The arctan } 1.33 = 53.1^\circ$$

2. Using the sine function:

$$\sin\theta = \frac{I_C}{I_T} = \frac{20}{25} = 0.8. \text{ The arcsin } 0.8 = 53.1^\circ$$

3. Using the cosine function:

$$\cos\theta = \frac{I_R}{I_T} = \frac{15}{25} = 0.6. \text{ The arccos } 0.6 = 53.1^\circ$$

Looking at the diagram, the angle makes sense because the current through the reactive branch is greater than the current through the resistive branch. The circuit is acting more reactive than resistive. In other words, the angle is greater than 45°. ◆

Solving for Impedance in a Parallel *RC* Circuit

An impedance diagram CANNOT be used for illustrating impedance in parallel circuits as it is used for series circuits. The circuit impedance for a parallel circuit is less than the impedance of any one branch; therefore, it cannot equal the hypotenuse, or vector sum of the oppositions, as it does in series circuits. (For example, if the resistive branch and the reactive branch had equal ohmic values, the circuit impedance would equal 70.7% of one of the branches.) There are a couple of simple ways, however, to solve for the value of impedance in a basic parallel *RC* circuit.

1. Use Ohm's law, where $Z = V_T/I_T$.
2. Use the following formula:

FORMULA 19–4 $$Z = \frac{RX_C}{\sqrt{R^2 + X_C^2}}$$

Let's try these out!

◆ **EXAMPLES** Refer again to Figure 19–11b.

1. Using Ohm's law:

$$Z = \frac{V_T}{I_T}$$

$$Z = \frac{60\text{ V}}{25\text{ A}}$$

$$Z = 2.4\ \Omega$$

2. Using the special formula:

$$Z = \frac{RX_C}{\sqrt{R^2 + X_C^2}}$$

To use this formula, we need to know the R and X_C values. Since the voltage is known and each branch current is known, we can use Ohm's law to find these.

$$R = \frac{60\text{ V}}{15\text{ A}} = 4\ \Omega;\ X_C = \frac{60\text{ V}}{20\text{ A}} = 3\ \Omega$$

$$Z = \frac{4 \times 3}{\sqrt{4^2 + 3^2}} = \frac{12}{5} = 2.4\Omega$$ ◆

PRACTICE PROBLEMS 5

Refer to Figure 19–12 and solve the following problems.

1. Use the capacitive reactance formula, Ohm's law, and the Pythagorean approach, as appropriate, to solve for I_T.
2. Draw the *V-I* vector diagram for the circuit and determine the phase angle, using appropriate trig formula(s). Label all the elements in the diagram.
3. Determine the circuit impedance.

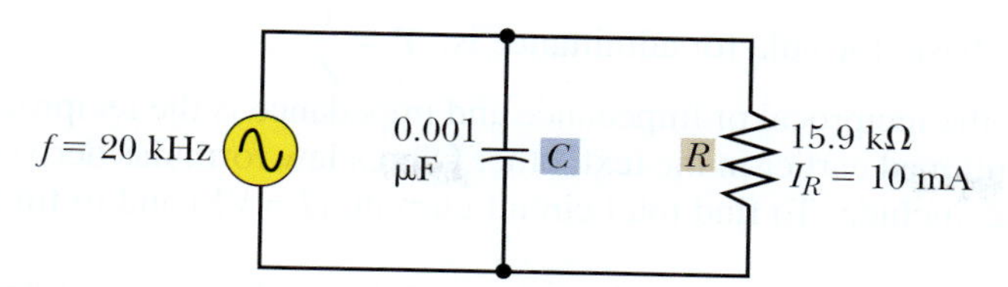

FIGURE 19–12 multiSIM

IN-PROCESS LEARNING CHECK 2

1. In parallel *RC* circuits, _____ is the reference vector in *V-I* vector diagram.
2. In parallel *RC* circuits, current through the resistor(s) is plotted (in phase, out of phase) _____ with the circuit current vector.
3. In parallel *RC* circuits, current through capacitor branch(es) is plotted at _____ with respect to circuit applied voltage.
4. The Pythagorean formula to find circuit current in parallel *RC* circuits is _____.
5. Is the Pythagorean formula used to find circuit voltage in a parallel *RC* circuit? _____
6. Can the tangent function be used to find phase angle in a parallel *RC* circuit? _____
7. Can an impedance diagram be used to find circuit impedance in a parallel *RC* circuit? _____

Finding Conductance, Susceptance, and Admittance in Parallel RC Circuits

Conductance

As discussed earlier in the text, *conductance* is the ease with which current can flow through a resistive circuit, and is the reciprocal of a circuit's opposition to current. The symbol for conductance is *G*. The unit of measure for conductance is siemens. The formula to find conductance value is:

FORMULA 19–5 $G = \frac{1}{R}$

Where: G is conductance in siemens
R is resistance in ohms

Capacitive Susceptance

Recall that for resistive components, there is conductance (G), which is the reciprocal of resistance, and for reactive components, there is *susceptance,* which is the reciprocal of reactance. For capacitors in ac circuits this is called *capacitive susceptance.* The general symbol for *susceptance* is *B*. The specific symbol for capacitive susceptance is B_C. The unit of measure for susceptance is the siemens. The formula to find capacitive susceptance is:

FORMULA 19–6 $B_C = \frac{1}{X_C} = 2\pi fC$

Where: X_C is capacitive reactance in ohms
B_C is capacitive susceptance in siemens.

Again, you see that capacitive susceptance is the reciprocal of capacitive reactance, even as conductance is the reciprocal of resistance.

Admittance

Admittance in parallel circuits containing both conductance and susceptance is calculated as the vector resulting from these two. The symbol for admittance is *Y*. The unit used for admittance is also the siemens.

Recall that the basic formula for admittance is: $Y = \frac{1}{Z}$

Admittance is the reciprocal of impedance and impedance is the reciprocal of admittance ($Z = 1/Y$). As mentioned earlier in the text, other Ohm's law formulas derived from these admittance formulas include: To find total circuit current: ($I = VY$) and to find circuit voltage: ($V = I/Y$).

Because admittance for *parallel* ac circuits containing both resistive and reactive branches is computed as the vector resultant of resistive branch conductance (G) and reactive branch susceptance (B). In a parallel RC circuit, then:

FORMULA 19–7 $Y = \sqrt{G^2 + B_C^2}$

EXAMPLE Calculate the conductance, susceptance, admittance, impedance, and total current for the circuit of Figure 19–13:

Conductance: of the R branch, $(G) = 1/R = 1/1500 = 0.000666$ S or 0.666 **mS**

X_C: of the capacitive branch $= 1/2\pi fC = 1/(6.28 \times 1600 \times 0.1 \times 10^{-6}) = 995$ **ohms,** therefore:

Capacitive Susceptance: $B_C = 1/X_c = 2\pi fC = 0.001$ or 1 mS

Admittance: $(Y_{tot}) = \sqrt{G^2 + B_C^2} = \sqrt{(0.66\text{ mS})^2 + (1\text{ mS})^2} = 1.2$ **mS**

Using the Ohm's law formulas, from the admittance data:

$Z = 1/Y = 1/1.2$ mS = approximately 833 **ohms**

$I_T = VY$ or $I_T = 15\text{ V} \times 1.2\text{ mS} = 18$ **mA**

$V = I/Y = 18\text{ mA}/1.2\text{ mS} = 15$ **V**

$Y = I/V = 18\text{ mA}/15\text{ V} = 1.2$ **mS**

Comparing these results with the ones that we can get by using the approaches discussed earlier in the chapter:

$I_R = V/R = 15\text{ V}/1500\text{ ohms} = 10$ **mA**

$I_C = V/X_C = 15\text{ V}/995\text{ ohms} = 15$ **mA**

$I_T = \sqrt{I_R^2 + I_C^2} = \sqrt{(10\text{ mA})^2 + (15\text{ mA})^2} = 18$ **mA**

$Z = V/I_T = 15\text{ V}/18\text{ mA} = 833$ **ohms**

$V = I \times Z = 18\text{ mA} \times 0.833\text{ k}\Omega = 14.99$ **V**

As you can see, the answers from the two approaches correlate well!

Another Method to Find Phase Angle (θ) for Parallel RC Circuits

Recall that we have been solving for the magnitude of phase angles in parallel RC circuits by using the arc tangent function ($\tan^{-1}$). We use the capacitive branch current to represent the opposite side and the resistive branch current to represent the adjacent side for the tangent

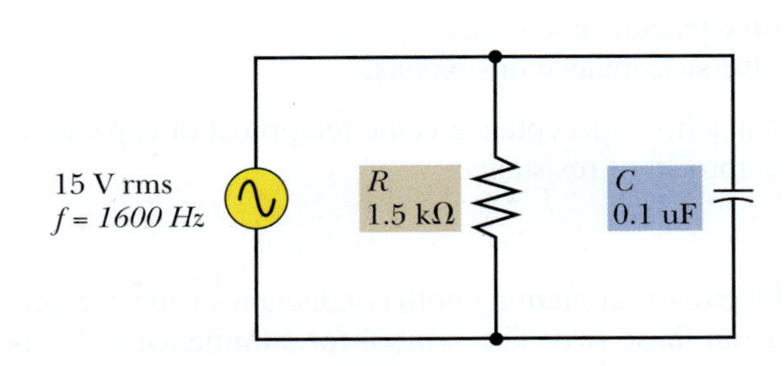

FIGURE 19–13

function formula. So, to find the magnitude of the phase angle, we have been using: θ(magnitude) = $\tan^{-1}(I_{XC}/I_R)$. Since the current through each branch of a parallel circuit is inversely related to the opposition of that branch to current, the result of R over X_C, that is R/X_C, turns out to be the same result as the I_{XC} over I_R. This means that the arc tangent ($\tan^{-1}$) of R/X_C will also give us the magnitude of the circuit phase angle.

FORMULA 19–8 $\theta(\text{magnitude}) = \tan^{-1}\dfrac{R}{X_C}$

From our previous example: I_{XC}/I_R = 15.075 mA/10 mA = 1.5075 and the $\tan^{-1}$ of I_{XC}/I_R = 56.44° (which also represents the angle between circuit voltage and total current. As indicated, in this case circuit, voltage lags circuit current by 56.44°)

$R/X_C = 1500\Omega/995\ \Omega = 1.5075$

and the $\tan^{-1}$ of $R/X_C = 56.44°$

Conversion from a Parallel *RC* Circuit to an Equivalent Series *RC* Circuit

Objective of conversion:

To find a series *RC* circuit whose parameters will present the same impedance and magnitude of phase angle to the circuit source as does the parallel *RC* circuit.

Step 1. Find the Z and θ of the parallel circuit.

Step 2. Construct the equivalent series circuit Z diagram, where:
Equivalent series circuit R = parallel circuit Z x cos θ
and the equivalent series circuit X_C = parallel circuit $Z \times$ sin θ.

Example conversion: (See Figure 9–14.)

Using the parameters from our previous parallel *RC* circuit admittance calculations (where the parallel *RC* circuit Z = 829 ohms and the parallel *RC* circuit magnitude of θ= 56.44°):

Equivalent series *RC* circuit R = parallel circuit $Z \times$ cos 56.44°

$R = 829 \text{ ohms} \times 0.553$

$R = 458 \text{ ohms}$

Equivalent series *RC* circuit X_C = parallel circuit $Z \times$ sin 56.44°

$X_C = 829 \text{ ohms} \times 0.833$

$X_C = 691 \text{ ohms}$

Equivalent series *RC* circuit Z diagram:

R (equiv.) = 458 ohms

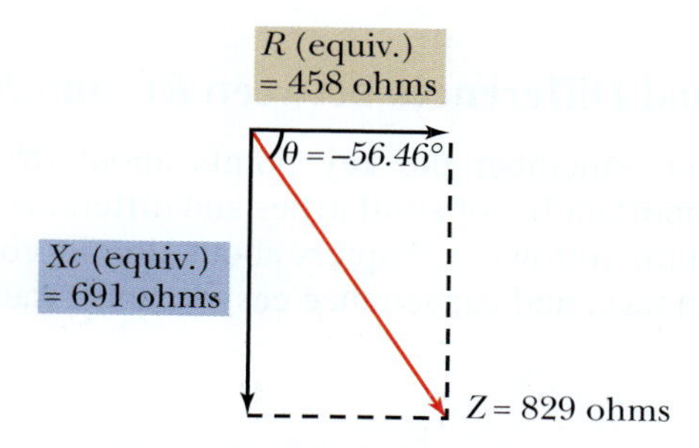

Equivalent series *RC* circuit *Z* diagram

FIGURE 19–14

Verification that conversion works to present a series circuit having the same Z and θ as the parallel RC circuit.

Using series RC circuit analysis:

$$Z = \sqrt{R^2 + X_C^{\,2}} = \sqrt{458^2 + 691^2} = 829 \text{ ohms}$$

$$\theta(\text{magnitude}) = \text{Tan}^{-1} \text{ of } 691/458 = 56.46°$$

PRACTICE PROBLEMS 6

Assume a parallel RC circuit where: $R = 4\Omega$, $X_C = 3\Omega$, and $V_T = 60$ V:

1. Determine the value of conductance (G) in this circuit?
2. Determine the value of capacitive susceptance (B) in this circuit?
3. Determine the value of circuit admittance (Y).
4. Determine the value of Z for this circuit using the $Z = 1/Y$ formula.
5. Determine the value of I_T for this circuit, using the $I_T = V \times Y$ formula.
6. Confirm the value of V_T using the $V = I/Y$ formula
7. Confirm the value of Y using the $Y = I/V$ formula
8. Use the R/X_C approach to solve for the circuit phase angle magnitude.

Summary of Parallel *RC* Circuits

Figure 19–15 provides several examples of parallel *RC* circuit parameters that illustrate the following facts. Study and understand these general principles before proceeding in the chapter.

1. Ohm's law formulas are used at any time to solve for individual component or overall circuit parameters, providing you substitute proper quantities into the formulas.
2. The applied voltage (V_T) is the reference vector in phasor diagrams of parallel *RC* circuits, since V is common to all branches.
3. The total circuit current leads the circuit voltage by some angle between 0° and + 90°.
4. The *smaller* the R value with respect to the X_C value, the more resistive the circuit is. That is, the closer to 0° the phase angle is. Conversely, the larger the R value is with respect to the X_C value, the more capacitive the circuit is.
5. Impedance in a parallel *RC* circuit is less than the ohmic opposition of any one branch and must be found through Ohm's law techniques or a special formula. Impedance for a parallel *RC* circuit *cannot* be found by drawing an impedance diagram.
6. When either R or X_C increases, impedance increases and circuit current decreases.
7. Total current in a parallel *RC* circuit is greater than any one branch current. However, total current is *not* as great as the sum of the branch currents. The branch currents must be added vectorially to find the resulting I_T value.
8. For a *V-I* vector diagram of a parallel *RC* circuit, voltage applied is the reference vector, I_R is plotted at 0°, and I_C is plotted at + 90°. I_T is plotted as the vector resulting from the resistive and reactive branch currents.

19–5 Similarities and Differences between *RC* and *RL* Circuits

So that you can see and remember the key points about *RL* and *RC* circuits, Figure 19–16 provides an important list of similarities and differences. If you understand the facts listed, you will find the following chapters about circuits containing all three quantities of resistance, inductance, and capacitance easy to understand.

CAPACITANCE CHANGED BY A FACTOR OF 2 (FROM 2 μF TO 1 μF)									
f (Hz)	V_T (V)	C (μF)	X_C (Ω)	R (Ω)	I_C (amps)	I_R (amps)	I_T (amps)	Z (ohms)	θ deg
100	100	2	795	5,000	0.1258	0.0200	0.1274	758.14	80.97
100	100	1	1,590	5,000	0.0629	0.0200	0.0660	1,515.2	72.36

(C to half; X_C doubled; I_C halved; I_R = same; I_T decreased; and θ decreased)

(a)

RESISTANCE CHANGED BY A FACTOR OF 2 (FROM 5,000 Ω TO 2,500 Ω)									
f (Hz)	V_T (V)	C (μF)	X_C (Ω)	R (Ω)	I_C (amps)	I_R (amps)	I_T (amps)	Z (ohms)	θ deg
100	100	1	1,590	5,000	0.0629	0.0200	0.0660	1,515.23	73.36
100	100	1	1,590	2,500	0.0629	0.0400	0.0745	1,341.64	57.54

(R to half; X_C = same; I_C = same; I_R doubled; I_T increased; and θ decreased)

(b)

FREQUENCY CHANGED BY A FACTOR OF 10 (FROM 100 Hz TO 10 Hz)									
f (Hz)	V_T (V)	C (μF)	X_C (Ω)	R (Ω)	I_C (amps)	I_R (amps)	I_T (amps)	Z (ohms)	θ deg
100	100	1	1,590	2,500	0.0629	0.0400	0.0745	1,341.64	57.54
10	100	1	15,900	2,500	0.0063	0.0400	0.0405	2,469.66	8.94

(f to $1/10$; X_C increased 10x; I_C decreased to $1/10$; I_R = same; I_T decreased; and θ decreased greatly)

(c)

RESISTANCE CHANGED TO MAKE EQUAL WITH X_C									
f (Hz)	V_T (V)	C (μF)	X_C (Ω)	R (Ω)	I_C (amps)	I_R (amps)	I_T (amps)	Z (ohms)	θ deg
10	100	1	15,900	2,500	0.0063	0.0400	0.0405	2,469.66	8.94
10	100	1	15,900	15,900	0.0063	0.0063	0.0089	11,243.00	45

(R made = X_C; X_C = same; I_C = same; I_R decreased greatly; I_T decreased greatly; and $\theta = 45°$ [since I_C and I_R are equal])

(d)

V_T IS DOUBLED									
f (Hz)	V_T (V)	C (μF)	X_C (Ω)	R (Ω)	I_C (amps)	I_R (amps)	I_T (amps)	Z (ohms)	θ deg
10	100	1	15,900	15,900	0.0063	0.0063	0.0089	11,243.00	45
10	200	1	15,900	15,900	0.0126	0.0126	0.0178	11,243.00	45

(V_T doubled; all currents doubled; X_C, R, Z, and θ did not change)

(e)

FIGURE 19–15 Sample parallel *RC* circuit parameters charts

ITEM DESCRIPTION	SERIES *RC* CIRCUIT	PARALLEL *RC* CIRCUIT	SERIES *RL* CIRCUIT	PARALLEL *RL* CIRCUIT
Formulas	$V_T = \sqrt{V_R^2 + V_C^2}$ $I_T = I_R = I_C = \frac{V_T}{Z}$ $Z = \sqrt{R^2 + X_C^2}$ $\tan\theta = \frac{X_C}{R}$ or $\frac{V_C}{V_R}$ $X_C = \frac{0.159}{fC}$ or $\frac{V_C}{I_C}$	$V_T = I_T \times Z$ $I_T = \sqrt{I_R^2 + I_C^2}$ $Z = \frac{V_T}{I_T}$ $\tan\theta = \frac{I_C}{I_R}$ $X_C = \frac{0.159}{fC}$ or $\frac{V_C}{I_C}$	$V_T = \sqrt{V_R^2 + V_L^2}$ $I_T = I_R = I_L = \frac{V_T}{Z}$ $Z = \sqrt{R^2 + X_L^2}$ $\tan\theta = \frac{X_L}{R}$ or $\frac{V_L}{V_R}$ $X_L = 2\pi fL$ or $\frac{V_L}{I_L}$	$V_T = I_T \times Z$ $I_T = \sqrt{I_R^2 + I_L^2}$ $Z = \frac{V_T}{I_T}$ $\tan\theta = \frac{I_L}{I_R}$ $X_L = 2\pi fL$ or $\frac{V_L}{I_L}$
As $f\uparrow$	$X_C\downarrow Z\downarrow I\uparrow \theta\downarrow$	$X_C\downarrow Z\downarrow I\uparrow \theta\uparrow$	$X_L\uparrow Z\uparrow I\downarrow \theta\uparrow$	$X_L\uparrow Z\uparrow I\downarrow \theta\downarrow$
As $C\uparrow$	$X_C\downarrow Z\downarrow I\uparrow \theta\downarrow$	$X_C\downarrow Z\downarrow I\uparrow \theta\uparrow$	N/A	N/A
As $L\uparrow$	N/A	N/A	$X_L\uparrow Z\uparrow I\downarrow \theta\uparrow$	$X_L\uparrow Z\uparrow I\downarrow \theta\downarrow$
As $R\uparrow$	$Z\uparrow I\downarrow \theta\downarrow$	$Z\uparrow I\downarrow \theta\uparrow$	$Z\uparrow I\downarrow \theta\downarrow$	$Z\uparrow I\downarrow \theta\uparrow$

FIGURE 19–16 Similarities and differences between *RL* and *RC* circuits

19–6 Linear Waveshaping and Nonsinusoidal Waveforms

Many electronic circuits depend on the ability to change nonsinusoidal circuit waveforms (such as square or rectangular waveforms) into different shapes for different purposes. For example, in radar systems it is not uncommon to need sharp or pointed waveforms to act as *trigger pulses* for circuits where timing action is performed.

In Figure 19–17, a square-wave input fed to a series *RC* network is shaped to be somewhat sharp or peaked across the resistor. This output is called the *differentiated output.* By definition, a differentiating circuit is one whose output relates to the rate of change of voltage or current of the input.

On the other hand, across the capacitor an output called an *integrated output* is taken. In this case, the waveform is rounded off or flattened out. This seems logical since a capacitor opposes a voltage change. Therefore, the voltage can't make sudden or "sharp" changes across the capacitor. By definition, an integrating circuit output relates to the integral of the input signal.

The amount of change of the input signal waveshape at the output relates to the amount of time allowed for the capacitor to charge or discharge during one alternation of the input signal. If the capacitor has more than ample time (over five time constants) to make its voltage change, the circuit is a **short time-constant circuit.** By definition, a short time-constant circuit is one where the time allowed (t) is 10 times or more than one *RC* time constant; that is, $t/RC = 10$ or more. *RC* time is short compared with t.

Conversely, if the *RC* time constant is such that the time allowed by the input signal alternation is short compared with the time needed for complete change across the *RC* network, it is a **long time-constant circuit.** That is, $t/RC = 0.1$ or less. *RC* time constant is long compared with the time allowed to charge or discharge.

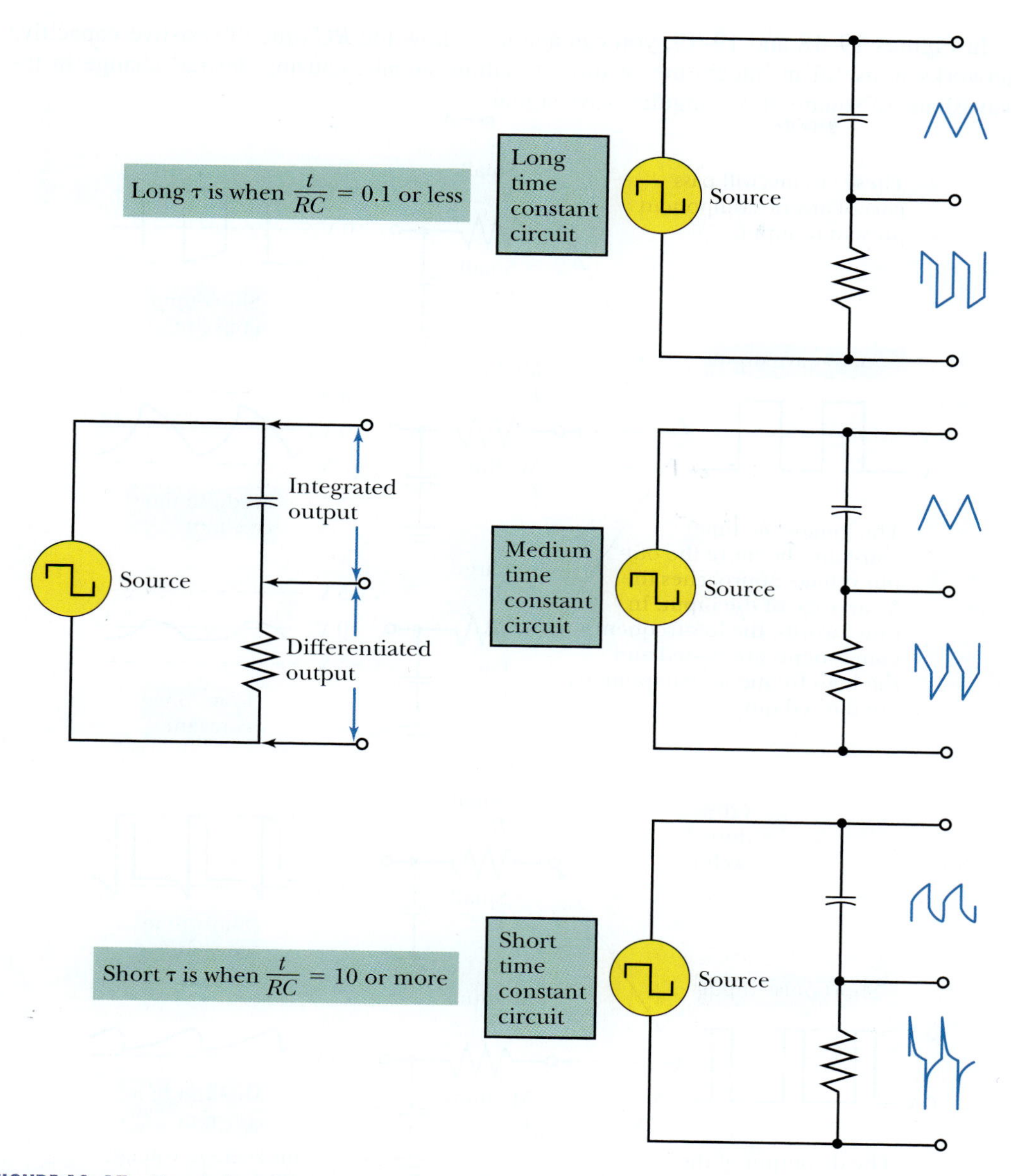

FIGURE 19–17 Examples of *RC* circuit waveshaping

Refer again to Figure 19–17. Notice both the differentiated and integrated waveforms present across the resistor and capacitor for long, medium, and short time-constant conditions. You will see practical applications of such circuits as you study a variety of waveshaping circuit applications.

In Figures 19–18 and 19–19, you can again see how the *RC* time of resistive-capacitive networks is useful in integrating or differentiating signals, causing desired change in the waveshape of square or rectangular wave signals.

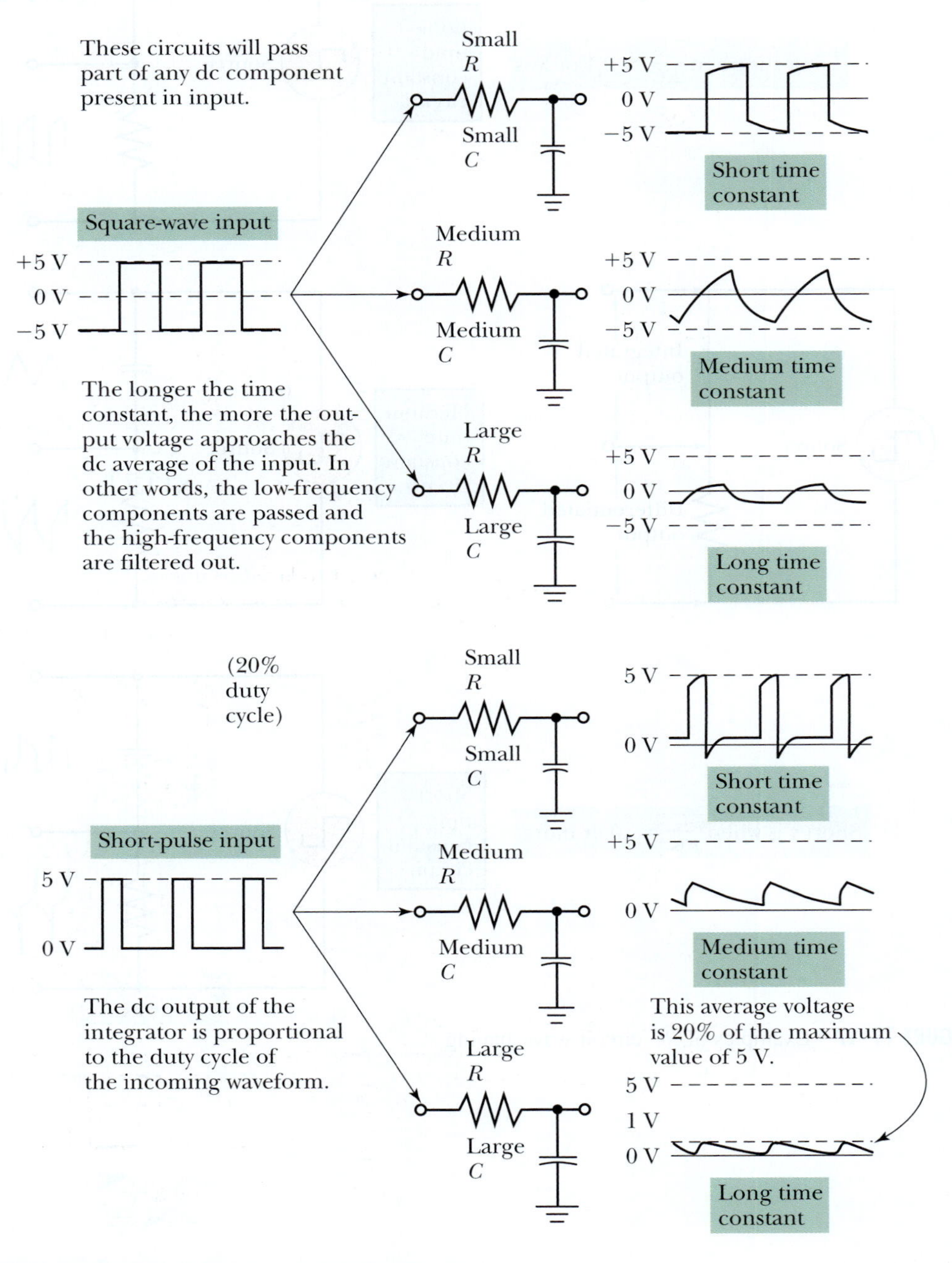
These circuits will pass part of any dc component present in input.
Small R
Small C
+5 V
0 V
−5 V
Short time constant
Square-wave input
Medium R
Medium C
Medium time constant
The longer the time constant, the more the output voltage approaches the dc average of the input. In other words, the low-frequency components are passed and the high-frequency components are filtered out.
Large R
Large C
Long time constant
(20% duty cycle)
5 V
0 V
Short-pulse input
The dc output of the integrator is proportional to the duty cycle of the incoming waveform.
This average voltage is 20% of the maximum value of 5 V.
1 V

FIGURE 19–18 Waveshaping—*RC* integrator circuits

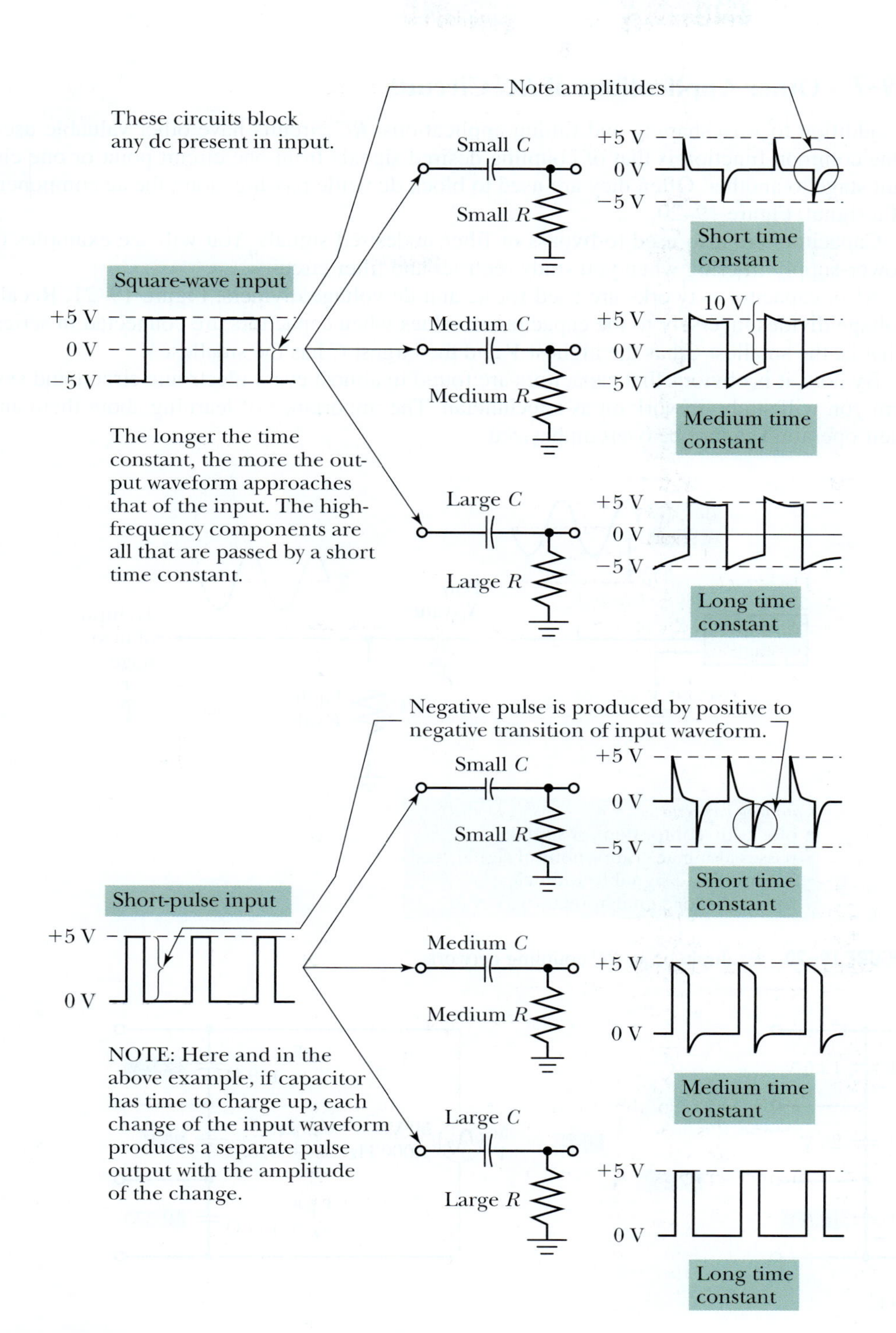

FIGURE 19–19 Waveshaping—*RC* differentiator circuits

19–7 Other Applications of *RC* Circuits

In addition to waveshaping and timing applications, *RC* circuits have other valuable uses. One common function is that of coupling desired signals from one circuit point or one circuit stage to another. Often they are used to block dc while passing along the ac component of a signal, Figure 19–20.

Capacitors are also used to bypass or filter undesired signals. You will see examples of power-supply filtering when you study rectifier and filter circuits.

Also, capacitor networks are used for ac and dc voltage dividers, Figure 19–21. Recall, voltage divides inversely to the capacitance values when capacitors are connected in series. That is, the smallest C has the highest V and the largest C has the smallest V.

By now, it is obvious that capacitors are found in almost every electronic circuit and system you will study or work on as a technician. The importance of learning about them and their operation cannot be overemphasized.

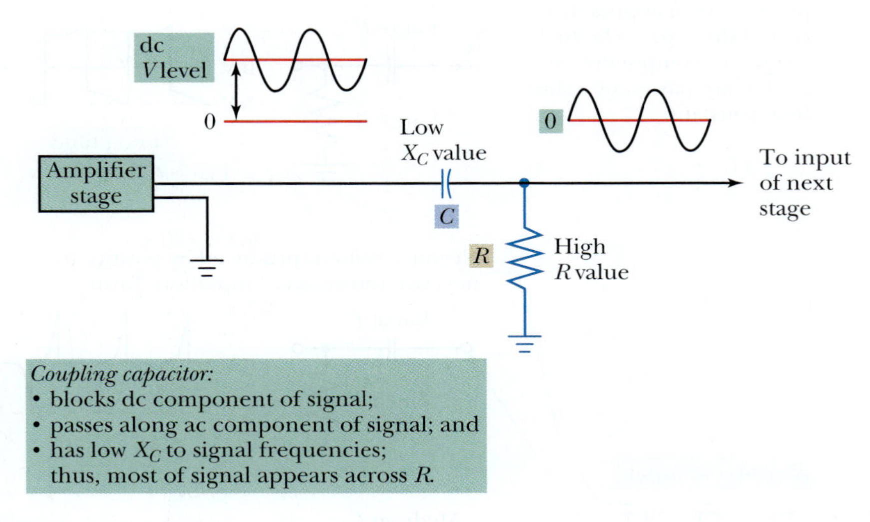

FIGURE 19–20 Example of an *RC* coupling network

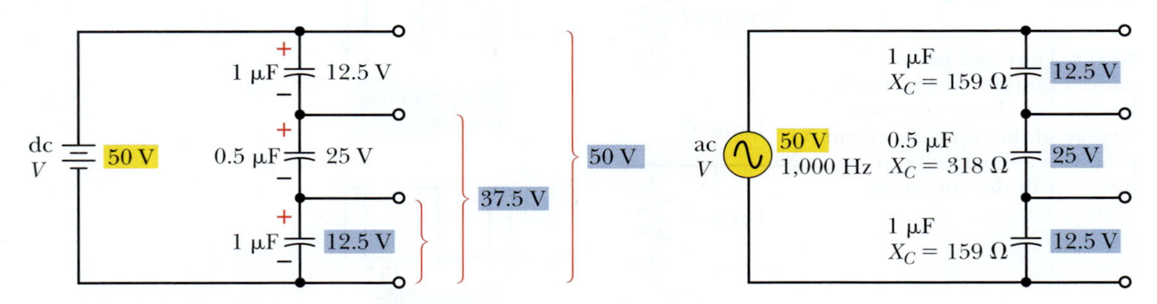

FIGURE 19–21 Capacitor voltage dividers

19–8 Troubleshooting Hints and Considerations for *RC* Circuits

Based on information supplied earlier regarding typical troubles with capacitors, one can surmise some of the problems that can occur for the various applications where capacitors are used. Some examples follow.

When used as a filtering device—for example, in power supplies or in decoupling circuits—either an excessive leakage condition or excessive internal resistance condition will prevent proper filtering action. As you will learn, in such systems as amplifiers, radio or TV receivers a range from excessive hum to high-pitched squeals can be heard when such problems exist.

The only cure for this situation is to isolate the bad capacitor and *replace it!* Sometimes, bridging a degenerated capacitor with a good capacitor displays sufficient difference to help isolate the bad capacitor.

Practical Notes

1. If a capacitor is used in a *timing circuit,* one symptom of a possible bad C (or R) is an improper time period.
2. If the capacitor is used in a waveshaping function, obvious clues are noted if an oscilloscope is used to analyze the normality or abnormality of the waveform(s).
3. If a capacitor is used in a *coupling circuit* where it is supposed to block dc and pass ac, the presence of measurable dc across the network resistor in series with the coupling capacitor shows the capacitor is probably leaking dc.

The primary tools used to help find defective capacitors in these various functions are the oscilloscope, digital multimeters, and capacitor testers. Another useful device is a component *substitution box,* which enables selection of an appropriate-value component. This component is temporarily substituted in the suspected area of trouble to see if the problem disappears with the substitute part.

As you study power supplies, amplifiers, oscillators, pulse and digital circuits, you will gain more familiarity with common symptoms and troubleshooting techniques associated with *RC* circuits. Perhaps these few comments will whet your appetite to learn more.

Safety Hints

1. To avoid shock, don't make connections with the circuit on.
2. In a defective circuit, capacitors may not always be discharged. BE CAREFUL when inserting test equipment. Check that capacitors are discharged by making appropriate voltage measurements, and/or using safe discharge techniques. ■

Summary

- Circuit current and circuit voltage are in phase in purely resistive ac circuits. They are 90° out of phase in purely capacitive circuits. They are somewhere between 0° and 90° out of phase with each other in circuits containing both resistance and capacitance.
- To calculate total circuit voltage in a series *RC* circuit, vectorially add the resistive and capacitive voltages, as opposed to adding them with simple arithmetic.
- Since current is common throughout any series circuit, the reference vector for phasor diagrams of series *RC* circuits is the circuit current. The various circuit voltages (V_C and V_R) are then plotted with respect to the reference vector. V_C is shown lagging the current by 90°, and V_R is in phase with the current.
- The reference vector for phasor diagrams of parallel *RC* circuits is the circuit voltage. The various parallel branch currents are plotted with respect to the reference vector. I_C is shown leading the voltage by 90°, and I_R is in phase with the circuit voltage.
- Impedance is the opposition shown to ac current by the combination of resistance and reactance.
- Impedance diagrams can only be used for *series* circuits. They should not be used for analyzing parallel *RC* (or *RL*) circuits.
- The Pythagorean theorem approach is used to solve for the magnitudes of various phasors. In series *RC* circuits, use the formula to calculate Z and V_T. For parallel *RC* circuits, use the formula to calculate I_T.
- Trig functions are used to solve for both phase angles and magnitudes of various sides. One commonly used trig function is the tangent function, used for both series and parallel resistive and reactive circuits.
- In series *RC* circuits, the larger the C is, the smaller the X_C present, and the less the capacitance controls the circuit parameters. In series circuits, the smaller the X_C is with respect to the R, the less the $I \times X_C$ drop is compared with the $I \times R$ drop. Hence, the less capacitive the circuit acts, and the smaller the phase angle will be.
- In parallel *RC* circuits, the larger the C is, the smaller the X_C present, and the more the capacitance controls the circuit parameters. In parallel circuits, the smaller the X_C is with respect to the R, the higher is the current through the capacitive branch. Hence, the more capacitive the circuit acts, and the larger the phase angle will be.
- Any frequency change changes the X_C. Therefore, frequency change affects circuit current, component voltages, impedance, and phase angle for series *RC* circuits. Frequency changes also affect branch currents, impedance, and phase angle in parallel *RC* circuits.
- Capacitors and *RC* circuits have various applications, including filtering, coupling, voltage-dividing, waveshaping, and timing. These uses take advantage of the capacitor's sensitivity to frequency, ability to block dc while passing ac and to store charge.
- Because electrolytic capacitors age faster than other components, such as resistors or inductors, periodically replace them to assure optimum operation of the equipment where they are used.
- The relative ease that current flows through a resistance is called conductance, where $G = 1/R$. A comparable term for reactance is susceptance, where $B = 1/\pm X$. When resistance and reactance are combined in a circuit, the term for ease of current flow is admittance (Y). Admittance is the reciprocal of impedance, where $Y = 1/Z$. The unit of measure for all these terms is siemens (S).
- Some important relationships between admittance, susceptance, and conductance are $Y = G \pm jB$, $Y = \sqrt{G^2 + B^2}$, and $\theta =$ the angle whose tangent $= B/G$.

Formulas and Sample Calculator Sequences

FORMULA 19–1
(To find the total voltage in a series RC *circuit)*

$V_T = \sqrt{V_R{}^2 + V_C{}^2}$ (Series *RC* circuits)

V_R value, [x^2], [+], V_C value, [x^2], [=], [$\sqrt{x}$]

FORMULA 19–2
(To find the impedance in a series RC *circuit)*

$Z = \sqrt{R^2 + X_C{}^2}$ (Series *RC* circuits)

R value, [x^2], [+], X_C value, [x^2], [=], [$\sqrt{x}$]

FORMULA 19–3
(To find the total current in a parallel RC *circuit)*

$I_T = \sqrt{I_R{}^2 + I_C{}^2}$ (Parallel *RC* circuits)

I_R value, [x^2], [+], I_C value, [x^2], [=], [$\sqrt{x}$]

FORMULA 19–4 *(To find the impedance in a simple parallel RC circuit)*

$$Z = \frac{RX_C}{\sqrt{R^2 + X_C{}^2}} \text{ (Parallel } RC \text{ circuits)}$$

R value, [×], X_C value, [÷], (R value, [x^2], [+], X_C value [x^2],), [$\sqrt{x}$], [=]

FORMULA 19–5 *(To find conductance of a resistive branch)*

$$G = \frac{1}{R} \text{ (Parallel } RC \text{ circuits)}$$

R value, [$1/x$]

FORMULA 19–6 *(To find capacitive susceptance of a capacitive branch)*

$$B_C = \frac{1}{X_C} = 2\pi fC \text{ (Parallel } RC \text{ circuits)}$$

X_C value, [$1/x$]

FORMULA 19–7 *(To find admittance of a parallel RC circuit)*

$$Y = \sqrt{G^2 + B_C{}^2} \text{ (Parallel } RC \text{ circuits)}$$

G value, [x^2], [+], B_C value, [x^2], [=], [$\sqrt{x}$]

FORMULA 19–8 *(To find phase angle value using resistance and reactance values)*

$$\theta\text{(magnitude)} = \tan^{-1}\frac{R}{X_C} \text{ (Parallel } RC \text{ circuits)}$$

R value, [÷], X_C value, [=], [2nd], [TAN-1]

NOTE: If the circuit capacitance value is in μF, use: Cap. value number, [EE], [+/–], 6 to express the "Cap. value" in microfarads for the calculator sequence.

TRIG FUNCTIONS:

sin θ = opp/hyp = V_C/V_T, or X_C/Z (series RC circuits)

sin θ = opp/hyp = I_C/I_T (parallel RC circuits)

opp side value, [÷], hyp value, [=]

cos θ = adj/hyp = V_R/V_T, or R/Z (series RC circuits)

cos θ = adj/hyp = I_R/I_T (parallel RC circuits)

adj side value, [÷], hyp value, [=]

tan θ = opp/adj = V_C/V_R, or X_C/R (series RC circuits)

tan θ = opp/adj = I_C/I_R (parallel RC circuits)

opp side value, [÷], adj side value, [=]

arctan θ = the angle represented by the tangent value

tan value, [2nd], [TAN-1]

EXCEL AUTOMATED FORMULAS

Helpful problem-solving tools, such as the Excel automated formulas available on the CD, allow automatic calculations of formulas.

Using Excel

RC Circuits in AC Formulas

(Excel file reference: FOE19_01.xls)

DON'T FORGET! *It is not necessary to retype formulas once they are entered on the worksheet! Just input new parameters data for each new problem using that formula, as needed.*

- Use the Formula 19–1 spreadsheet sample and find total voltage for the circuit parameters given in Figure 19–5.
- Use the Formula 19–4 spreadsheet sample and solve for the value of circuit impedance for the parameters given in Figure 19–12.

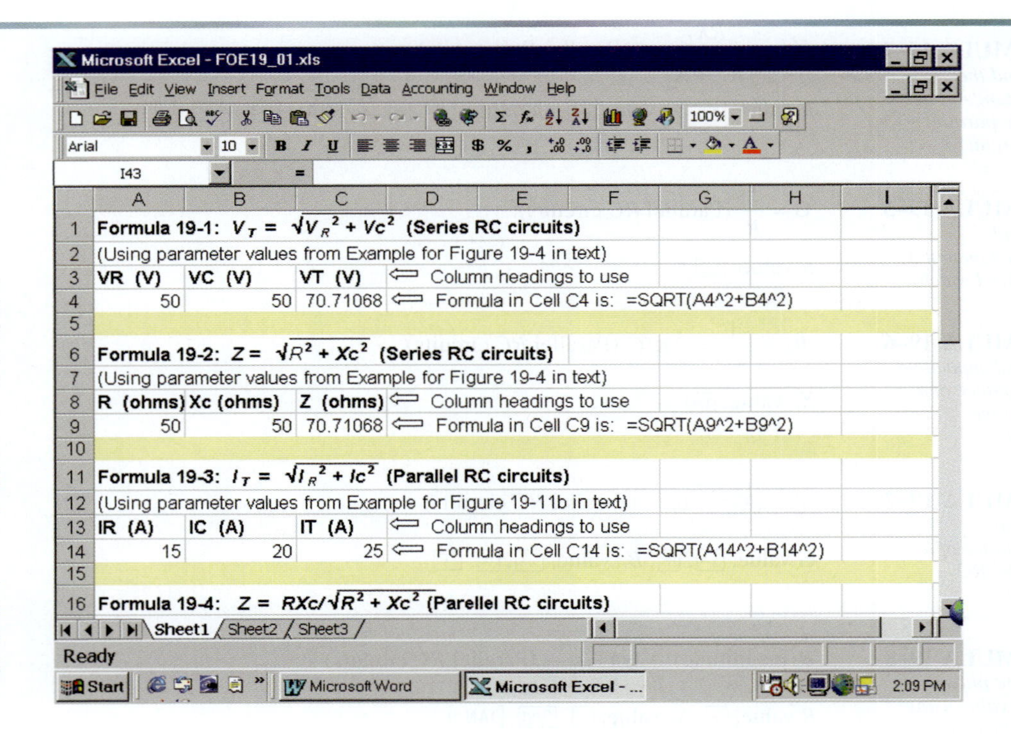

Using Excel

RC Circuits in AC Formulas

(Excel file reference: FOE19_01.xls)

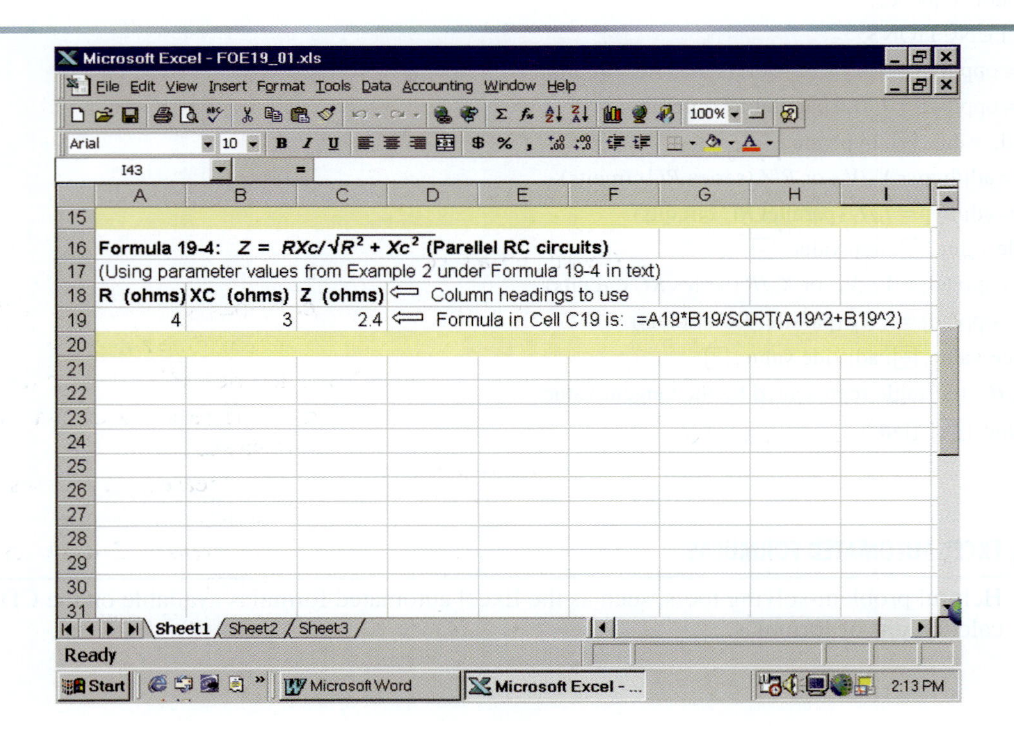

Review Questions

1. In *RC* circuits:

a. circuit phase angle is 90°.
b. circuit phase angle is 0°.
c. circuit phase angle is 45°.
d. circuit phase angle is between 0° and 90°.

2. Total voltage in a series *RC* circuit is found by:

a. algebraic addition.
b. vector addition.
c. mathematical addition.
d. none of the above.

3. The reference vector, when analyzing parallel *RC* circuits is:

a. circuit current.
b. circuit voltage.
c. circuit impedance.
d. none of the above.

4. The quantity(ies) that is(are) plotted with respect to the reference vector in parallel *RC* circuits:

a. are the circuit currents.
b. is the circuit voltage.
c. are the circuit reactances.
d. none of the above.

5. In a series *RC* circuit, the larger the *C* value for a given *R* value:

a. the larger the circuit phase angle.
b. the smaller the circuit phase angle.
c. the value of *C* does not affect phase angle.
d. the more "capacitive" the circuit acts.

6. In parallel *RC* circuits, the larger the *C* value for a given *R* value:

a. the larger the circuit phase angle.
b. the smaller the circuit phase angle.
c. the value of *C* does not affect phase angle.
d. the less "capacitive" the circuit acts.

7. If frequency of operation is increased for a series *RC* circuit:

a. the circuit will act more capacitively.
b. the circuit will act less capacitively.
c. the circuit conditions will remain the same.
d. none of the above.

8. If the frequency of operation is decreased for a parallel *RC* circuit:

a. the circuit will act more capacitively.
b. the circuit will act less capacitively.
c. the circuit conditions will remain the same.
d. none of the above.

9. *RC* circuits can act as filters because:

a. capacitors oppose a change in current.
b. capacitors oppose a change in voltage.
c. resistors react to changes in frequency.
d. none of the above.

10. *RC* circuits can act as signal coupling networks because:

a. capacitors block ac and pass dc.
b. capacitors block both ac and dc.
c. capacitors block dc and pass ac.
d. none of the above.

11. A series *RC* circuit's impedance increases as:

a. frequency increases.
b. frequency decreases.
c. phase angle increases.
d. phase angle decreases.

12. A parallel *RC* circuit's phase angle increases as:

a. frequency increases.
b. frequency decreases.
c. impedance increases.
d. none of the above.

13. Capacitive by-pass filters:

a. by-pass or filter desired signal frequencies.
b. by-pass or filter undesired signal frequencies.
c. by-pass or filter undesired electronic components.
d. none of the above.

14. In a series *RC* circuit, as frequency is increased:

a. X_C decreases, Z decreases, I decreases, and θ decreases.
b. X_C increases, Z increases, I decreases, and θ increases.
c. X_C decreases, Z decreases, I increases, and θ decreases.
d. none of the above.

15. For a *V-I* vector diagram for a parallel *RC* circuit:

a. V_A is the reference, I_R is at 0°, I_C is at –90°, and I_T is between 0° and +90°.

b. V_A is the reference, I_R is at 0°, I_C is at +90°, and I_T is between 0° and +90°.

c. I_T is the reference, V_R is at 0°, V_C is at –90°, and V_A is between 0° and +90°.

d. I_T is the reference, V_R is at 0°, V_C is at +90°, and V_A is between 0° and +90°.

16. For a *V-I* vector diagram for a series *RC* circuit:

a. V_A is the reference, I_R is at 0°, I_C is at –90°, and I_T is between 0° and –90°.

b. V_A is the reference, I_R is at 0°, I_C is at +90°, and I_T is between 0° and –90°.

c. I_T is the reference, V_R is at 0°, V_C is at –90°, and V_A is between 0° and –90°.

d. I_T is the reference, V_R is at 0°, V_C is at +90°, and V_A is between 0° and –90°.

17. For a series *RC* lead (phase shift) network, the output voltage is taken across:

a. the capacitor.

b. the resistor.

c. both the resistor and the capacitor.

d. neither the resistor nor the capacitor.

18. For a series *RC* lag (phase shift) network, the output voltage is taken across:

a. the capacitor.

b. the resistor.

c. both the resistor and the capacitor.

d. neither the resistor nor the capacitor.

19. What two formulas might be used to find total current in a parallel *RC* circuit?

20. Given a parallel *RC* circuit's *R* and X_C values, what formula would you use to find the circuit's impedance?

Problems

Answer questions 1–5 in reference to Figure 19–22.

1. What is the circuit impedance?

2. What are the V_C, V_R, and V_T values?

3. What is the *C* value?

4. Is circuit voltage leading or lagging circuit current? By how much?

Answer the following with "I" for increase, "D" for decrease, and "RTS" for remain the same.

5. If the frequency of the circuit in Figure 19–22 doubles and component values remain the same:

a. *R* will _____

b. X_C will _____

c. *Z* will _____

d. I_T will _____

e. θ will _____

f. V_C will _____

g. V_R will _____

h. V_T will _____

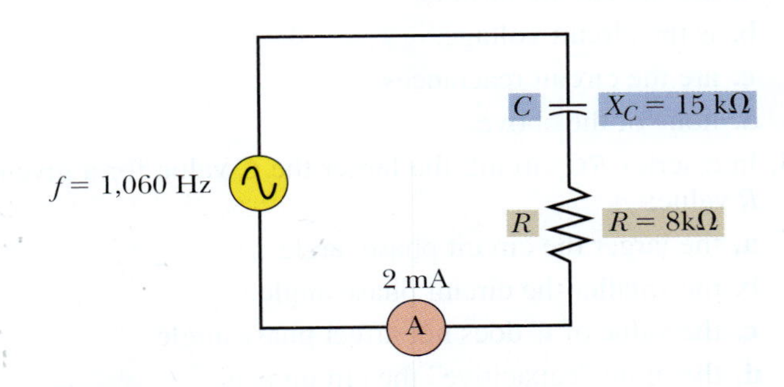

FIGURE 19–22

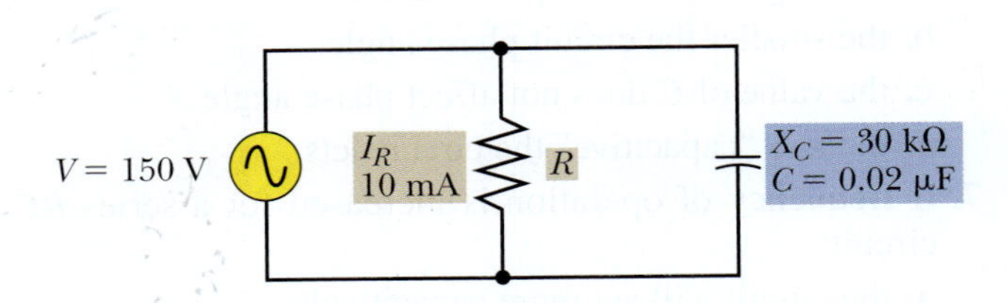

FIGURE 19–23

Answer questions 6 to 10 in reference to Figure 19–23.

6. What are the I_C and I_T values?

7. What is the *R* value?

8. What is the *Z* value?

9. What is the frequency of operation?

10. Answer the following with "I" for increase, "D" for decrease, and "RTS" for remain the same. If the frequency of the circuit in Figure 19–23 doubles and component values remain the same:
   a. R will _____
   b. X_C will _____
   c. Z will _____
   d. I_T will _____
   e. θ will _____
   f. I_C will _____
   g. I_R will _____
   h. V_T will _____

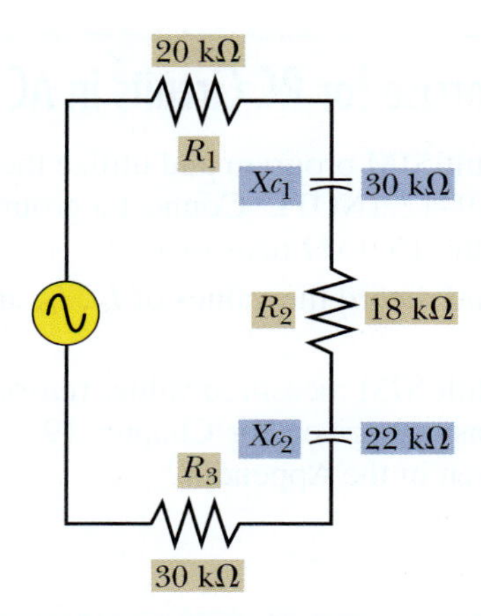

FIGURE 19–24

Answer questions 11 to 15 in reference to Figure 19–24.

11. Find R_T
12. Find X_{C_T}
13. Find Z
14. Find θ
15. If C_1 equals 0.1 µF, what is the frequency of the source voltage?

Answer questions 16 to 20 in reference to Figure 19–25.

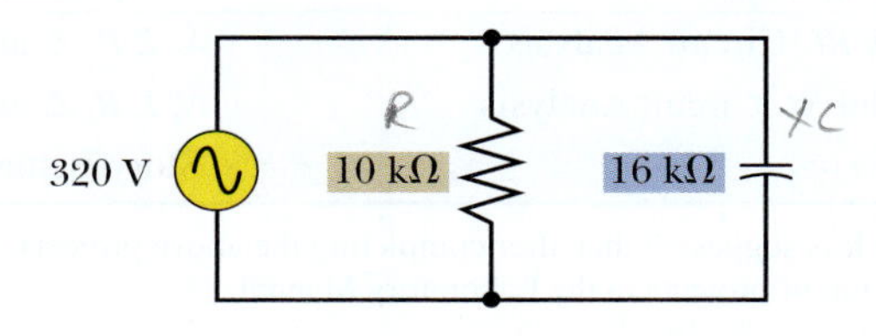

FIGURE 19–25

16. Find I_R
17. Find I_C
18. Find I_T
19. Find θ
20. Find Z_T
21. A two-component series *RC* circuit is operated at a frequency of 3,975 Hz. The circuit is composed of a 0.04-µF capacitor and an undisclosed-value resistance. The circuit impedance is 1,250 Ω. The applied voltage is 25 V. Determine the following circuit parameters: R, X_C, I, V_R, V_C, and θ.
22. A two-branch parallel *RC* circuit has a total current of 5 mA. The circuit's capacitive reactance is 2 kΩ. The applied voltage is 5 V. Determine the following circuit parameters: R, I_C, I_R, Z, and θ.

Analysis Questions

1. In your own words, describe two basic circuit operational differences between a series *RC* circuit and a series *RL* circuit.
2. In your own words, describe two basic circuit operational differences between a parallel *RC* circuit and a parallel *RL* circuit.
3. Are there logical comparisons between a series *RC* circuit and a parallel *RL* circuit? Explain.
4. Draw the circuit diagram of a phase lead *RC* circuit that will produce a 45° lead angle. The components you can use include a variable resistor whose range of R is from 0 to 10,000 Ω, and a 0.25-µF capacitor. Assume the frequency of operation is 120 Hz.

MultiSIM Exercise for RC Circuits in AC

1. Use the MultiSIM program and utilize the circuit shown in Figure 19–12. (NOTE: Connect a ground reference to bottom of the 15.9 kΩ resistor.)
2. Measure and record the values of I_C, I_R, and total circuit current.
3. Do your MultiSIM measured values reasonably compare with the answers given for Chapter 19—Practice Problems 5, given in the Appendix?
4. Using MultiSIM, determine by measurement, and record what the values of I_C, I_R, and total circuit current would be if the source voltage remained the same, but the frequency were changed to 10 kHz.
5. Would the circuit phase angle change? If so, would it increase or decrease?

Performance Projects Correlation Chart

Suggested performance projects that correlate with topics in this chapter are:

Chapter Topic	Performance Project	Project Number
Series *RC* Circuit Analysis	*V, I, R, Z,* and θ Relationships in a Series *RC* Circuit	52
Parallel *RC* Circuit Analysis	*V, I, R, Z,* and θ Relationships in a Parallel *RC* Circuit	53
	Story Behind the Numbers: *RC* Circuits in AC	

NOTE: It is suggested that after completing the above projects, the student should be required to answer the questions in the "Summary" at the end of this section of projects in the Laboratory Manual.

OUTLINE

OBJECTIVES

After studying this chapter, you should be able to:

1. Solve *RLC* circuit problems using the Pythagorean approach and trig functions
2. Define and illustrate ac circuit parameters using both **rectangular** and **polar form notation**
3. Define **real numbers** and **imaginary numbers**
4. Define real power, **apparent power, power factor,** and **voltampere-reactive**
5. Calculate values of real power, apparent power, and power factor, and draw the power triangle
6. Analyze *RLC* circuits and state results in rectangular and polar forms
7. Use the computer to solve circuit problems
8. Use the SIMPLER troubleshooting sequence to solve the Troubleshooting Challenge problem

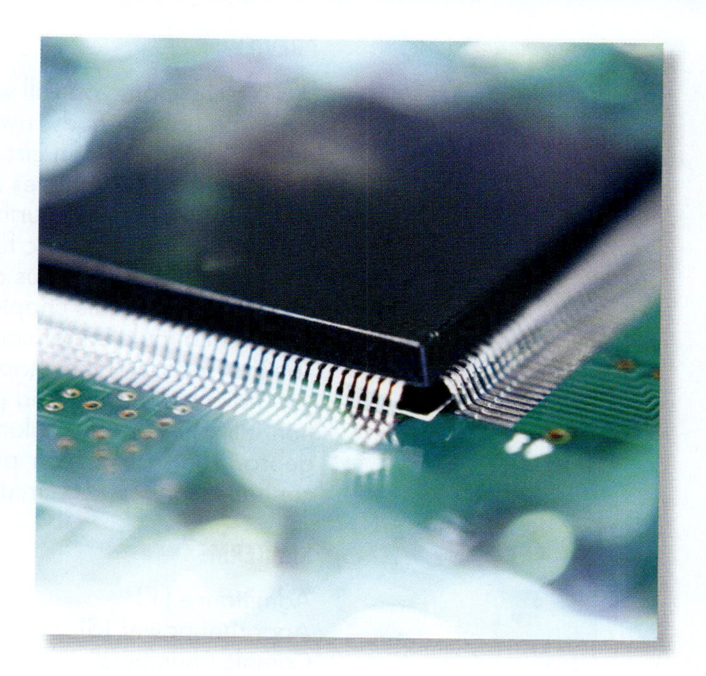

CHAPTER 20

RLC Circuit Analysis

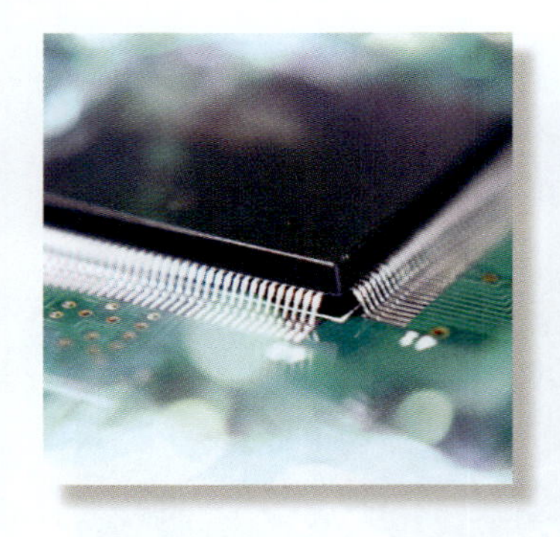

PREVIEW

As a technician, you will work with many circuits that contain combinations of *R, L,* and *C*. You have already investigated the property of inductance in ac circuits, in pure inductive circuits, and in circuits with inductance and resistance. Likewise, you have seen how capacitance behaves in purely capacitive circuits and in circuits with both capacitance and resistance. Furthermore, you have learned to use the Pythagorean theorem and simple trigonometric functions to analyze these circuits. In this chapter, you will analyze circuits with various combinations of *R, L,* and *C*.

The outline of this chapter accommodates either programs that teach vector algebra and its manipulation of complex numbers or only the approaches learned to this point. The first section of the chapter will present *RLC* circuits and the basic analysis techniques you have been using and provide a discussion of power in ac circuits. The second section introduces rectangular and polar form vector analysis, sometimes called vector algebra. The last portion provides an opportunity to apply vector algebra to various configurations of *RLC* circuits.

KEY TERMS

Admittance (*Y*)
Apparent power (*S*)
Complex numbers
Imaginary numbers
Polar form notation
Power factor (*p.f.*)
Reactive power (*Q*)
Real numbers
Rectangular form notation
Susceptance (*B*)
True (or real) power (*P*)
Voltampere-reactive (VAR)

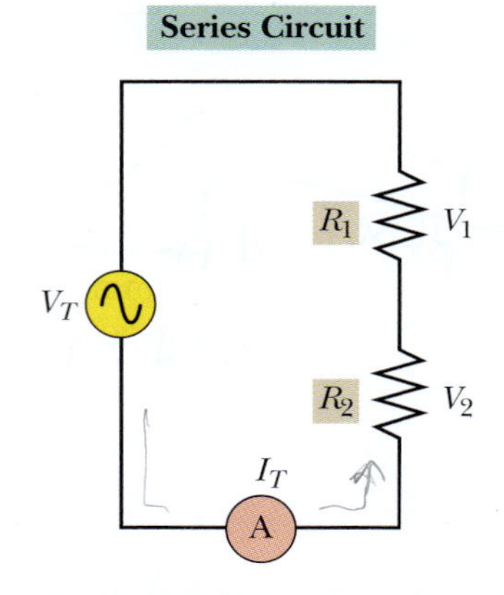

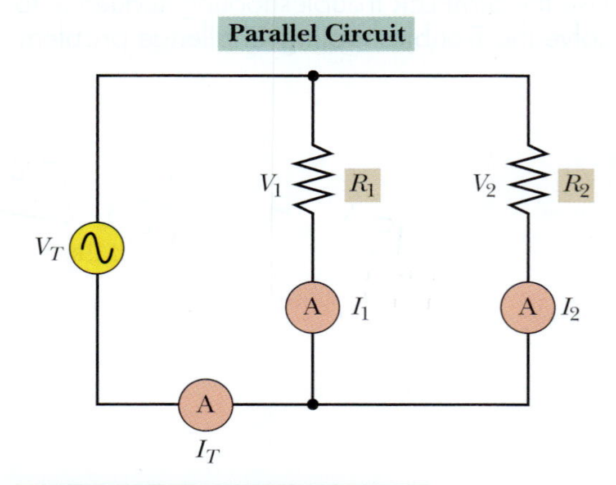

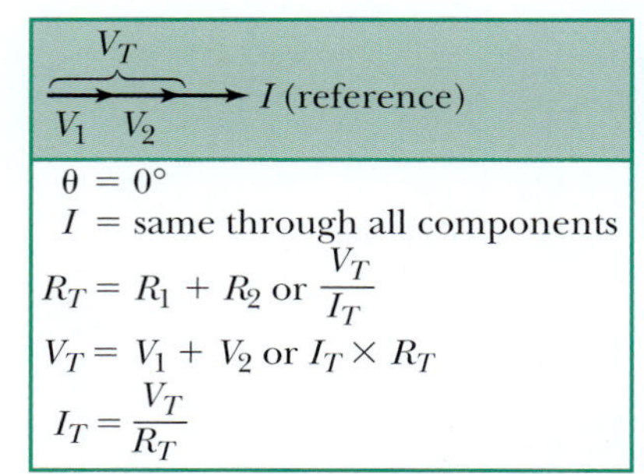

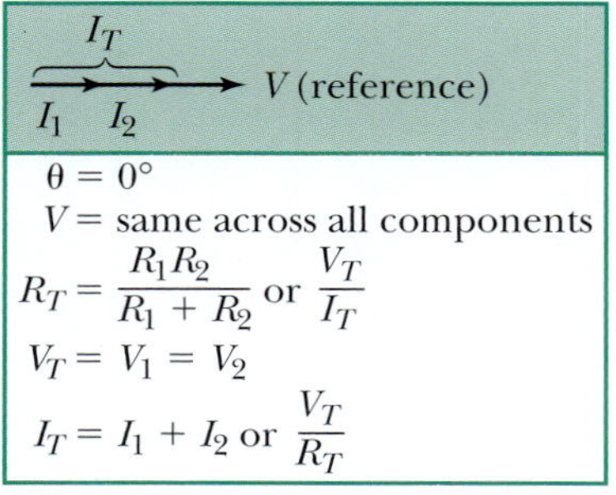

FIGURE 20–1 Purely resistive circuits

20–1 Basic *RLC* Circuit Analysis

Review of Resistive Circuits

In studying circuits containing only resistances, you learned resistance in ac is treated the same as resistance in dc. Key facts about series and parallel resistive circuits are in Figure 20–1. Take a few minutes to review these facts.

Review of Inductive Circuits

As you know, inductance opposes a change in current. This feature causes the voltage to lead current by 90° in purely inductive circuits and by some angle between 0° and 90° in circuits with both L and R. Examine Figures 20–2 and 20–3 as you review inductive

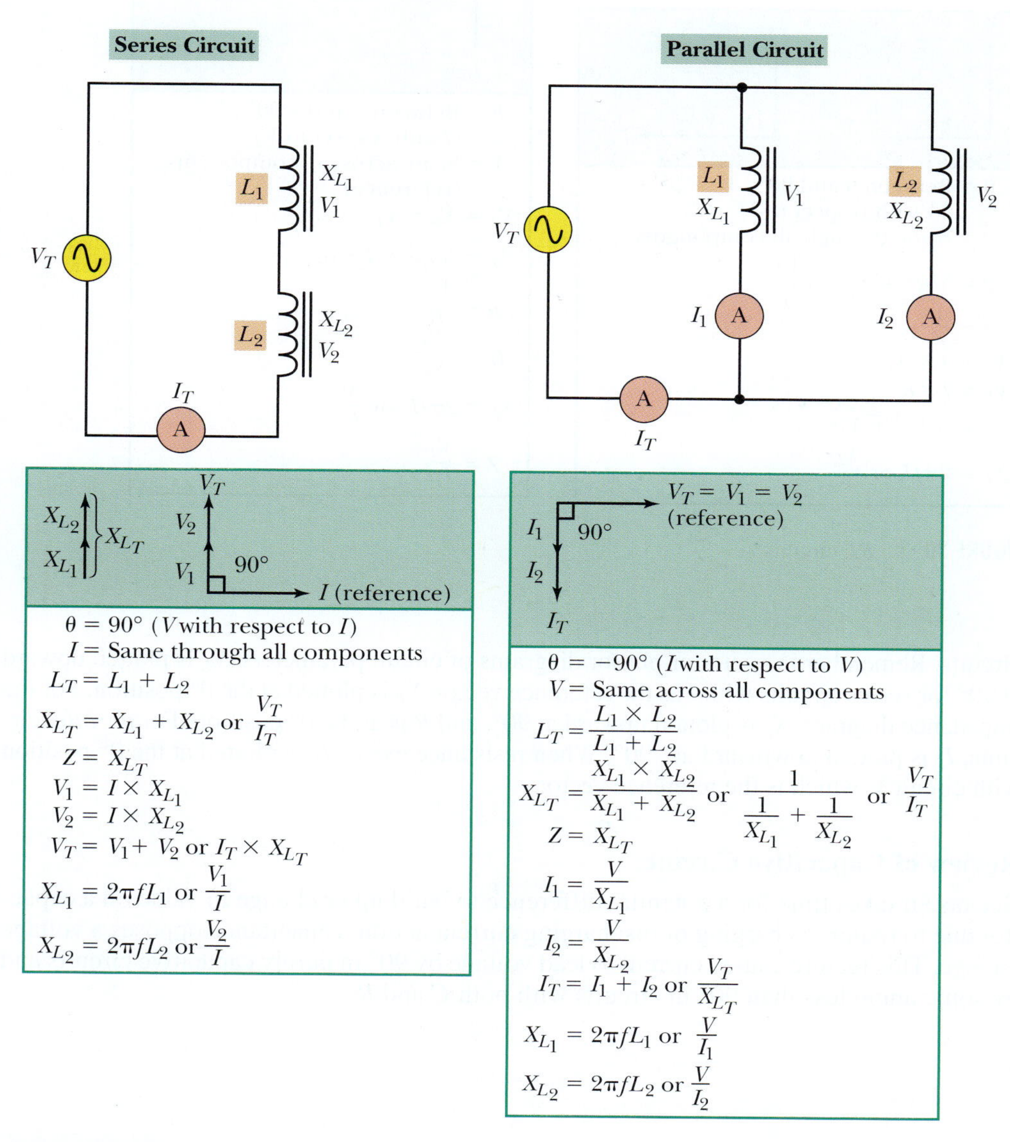

FIGURE 20–2 Purely inductive circuits

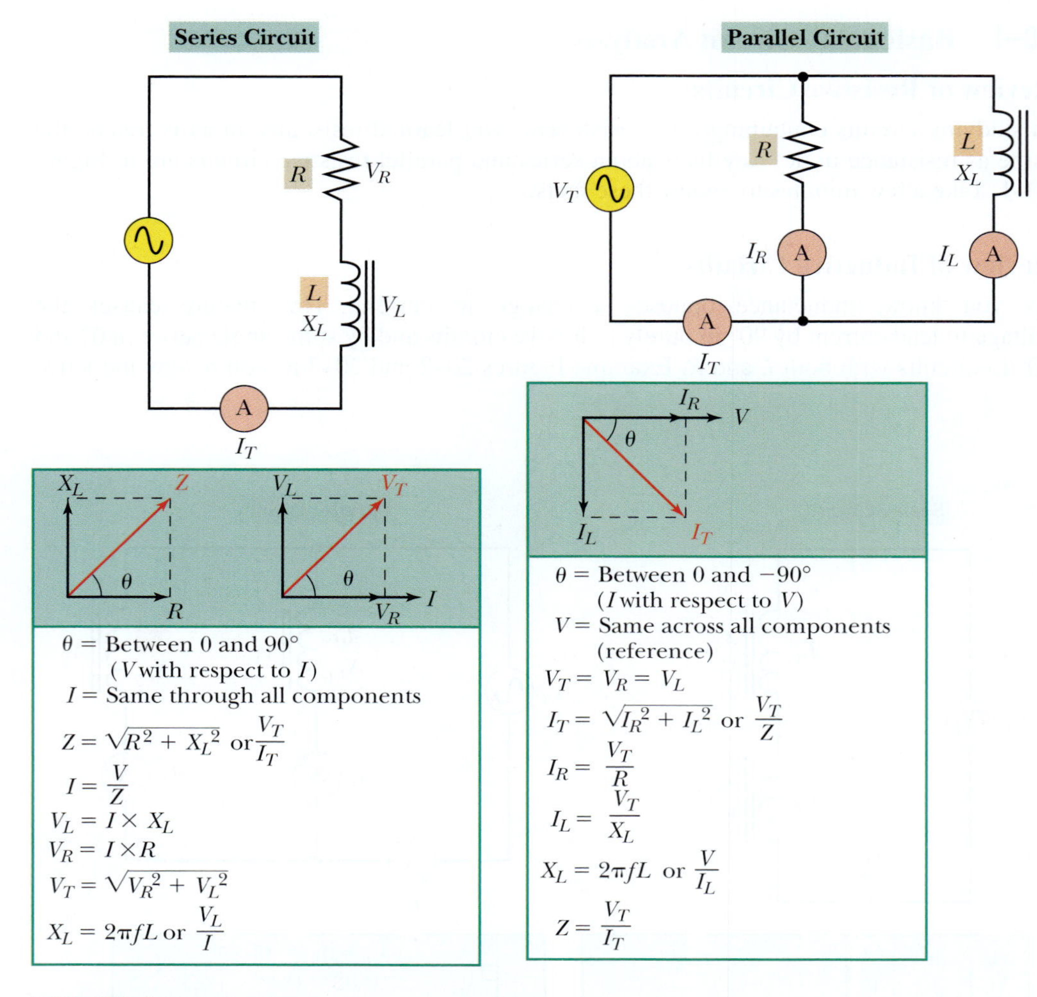

FIGURE 20–3 *RL* circuits

circuits. Remember in vector or phasor diagrams of circuit parameters, V_L is plotted upward at 90° for *series* circuits with I as the reference vector. V_R is plotted at the 0° position. For the impedance diagram, X_L is plotted upward at 90°, and R is at the 0° position. For *parallel* circuits, I_L is plotted downward at –90°. When resistance exists, I_R is plotted at the 0° position with circuit V, which is the reference vector.

Review of Capacitive Circuits

Because it takes time for a potential difference to build up or change in value on a capacitor due to required charging or discharging current action, capacitance opposes a voltage change. This feature causes current to lead voltage by 90° in purely capacitive circuits and by some angle less than 90° in circuits with both C and R.

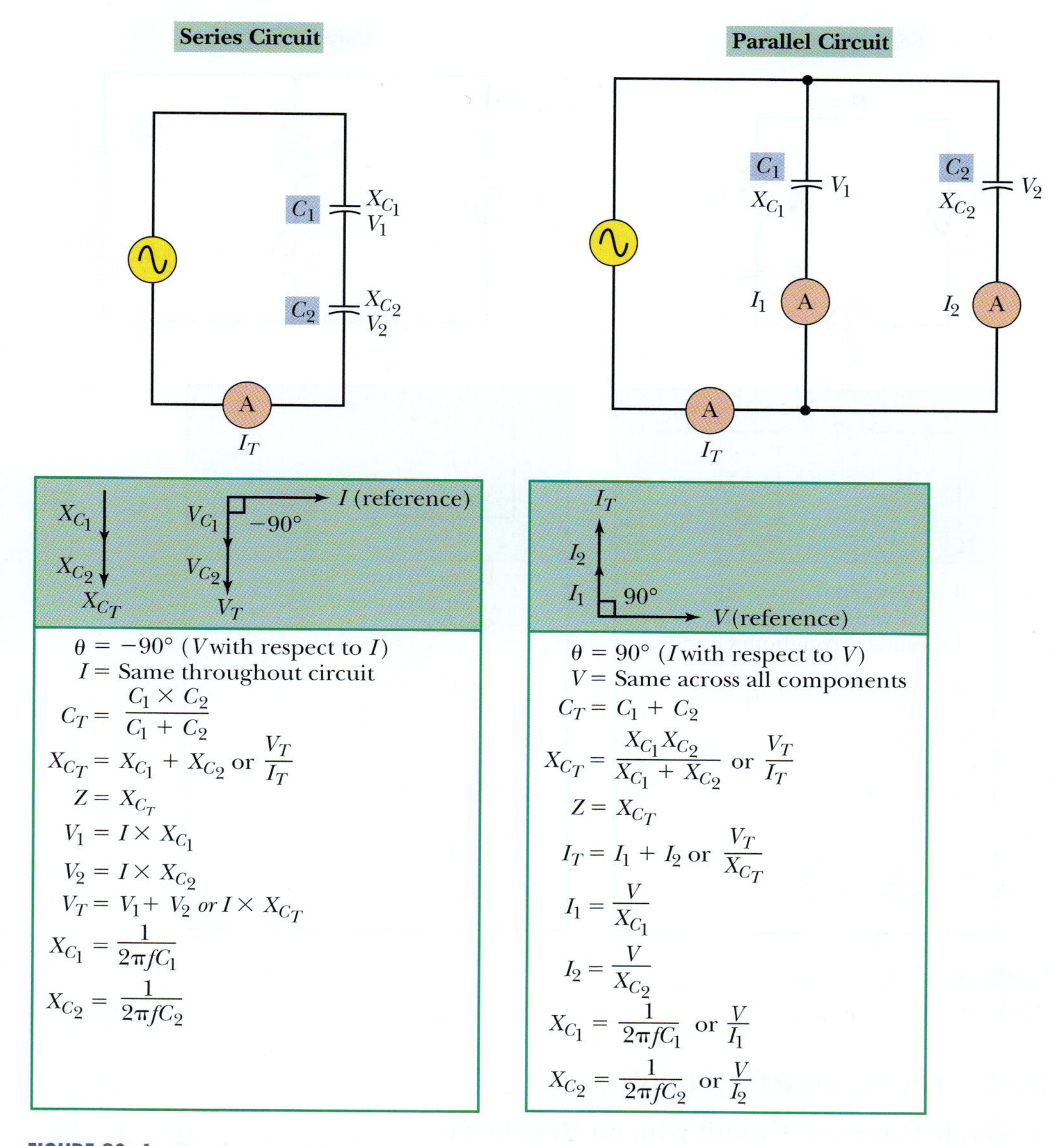

FIGURE 20–4 Purely capacitive circuits

Refer to Figures 20–4 and 20–5. Notice for series circuit V-I vector diagrams, V_C is plotted downward at –90°. For Z diagrams, X_C is plotted downward at –90°. When resistance exists, V_R is plotted at the 0° position with the circuit I, which is the reference vector. X_C is plotted downward and R is plotted at 0° for series circuit impedance diagrams.

For parallel circuit V-I diagrams, I_C is plotted upward at 90°, and I_R is plotted at the 0° position along with circuit V, which is the reference vector. Remember all of these facts as we move to the analysis of circuits containing all three components—R, L, and C.

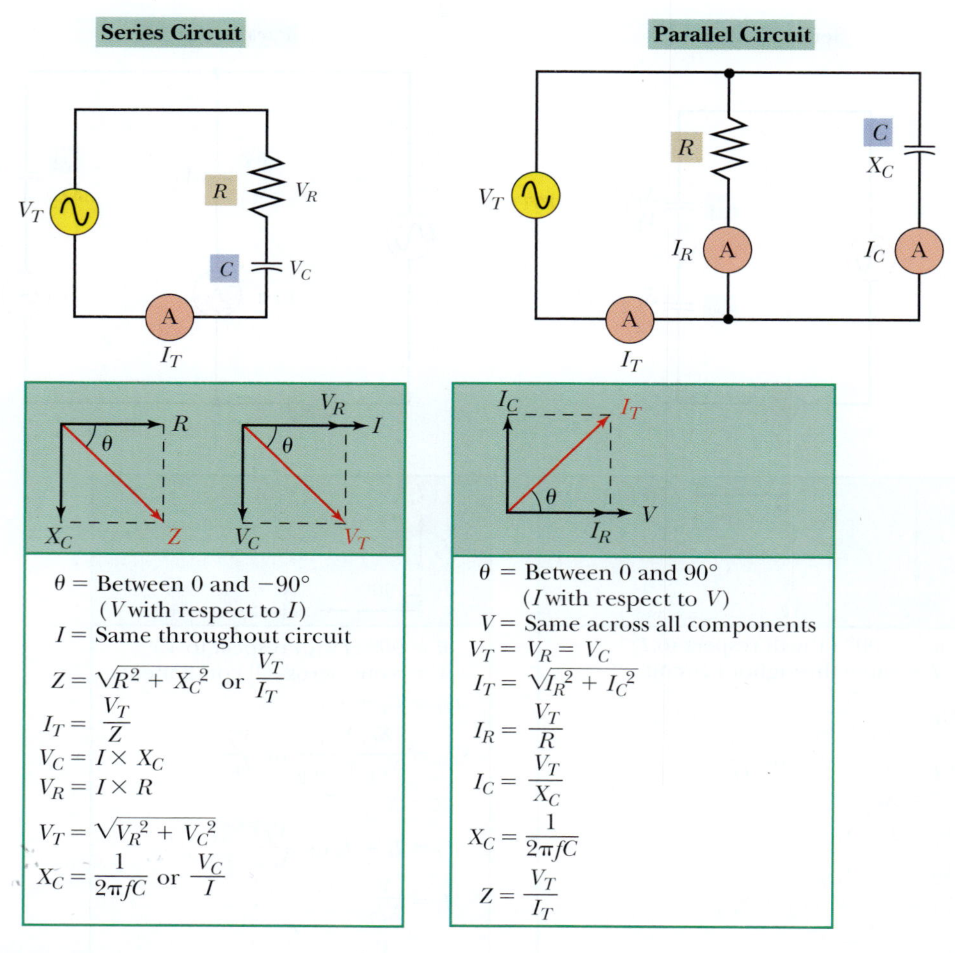

FIGURE 20–5 *RC* circuits

20–2 The Series *RLC* Circuit

A Simple Series *LC* Circuit without Resistance

Let's look at what happens when we combine inductive reactance (X_L) and capacitive reactance (X_C) in a simple series circuit. Refer to Figure 20–6 as you study the following statements.

1. Since X_L is plotted upward and X_C is plotted downward, they *cancel* each other.
2. The result of combining them is that any remaining, or *net* reactance is the difference between the two reactances. That is, if X_L is greater than X_C, the net reactance is some value of X_L. If X_C is greater than X_L, the net circuit reactance is some value of X_C.
3. Apply the same rationale to voltages. Since the voltages are plotted 180° out of phase with each other and calculated by multiplying the common current (I) times each reactance, the resulting voltage of the two opposite-reactive voltages is the difference between them. In other words, V_{X_T} = largest V_X – smaller V_X.
4. Since the resulting reactance ohms are less than either reactance (because of the cancelling effect), the current is higher than it would be with either reactance alone. This means the $I \times X_L$ or $I \times X_C$ voltages are higher than the V applied! Obviously, since the opposite reactive voltages also cancel each other, the net voltage equals the applied voltage V_T.

Series *RLC* Circuit Analysis

Refer to Figure 20–7 and note the *RLC* circuit analysis as you study the following procedures.

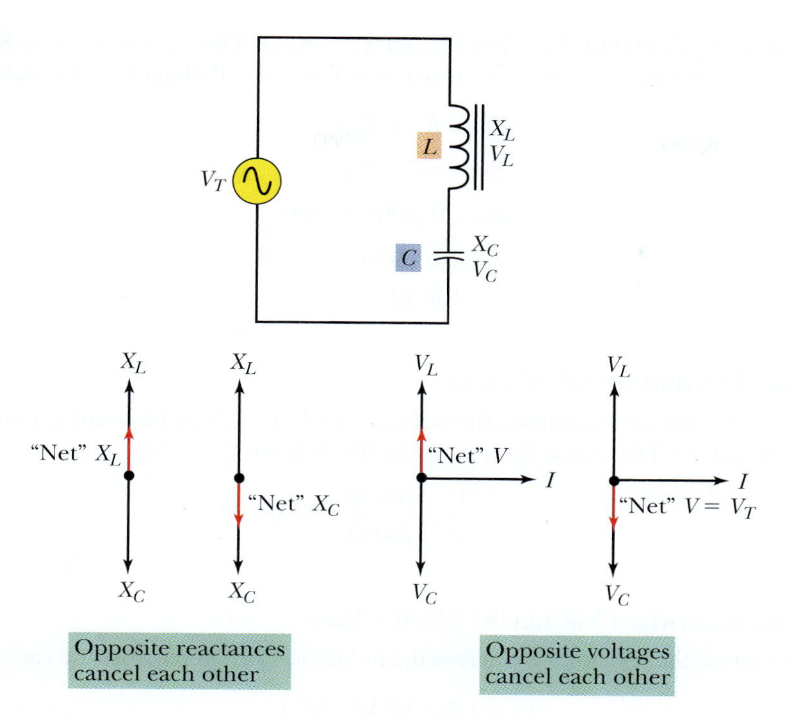

FIGURE 20–6 Series *LC* circuit without resistance

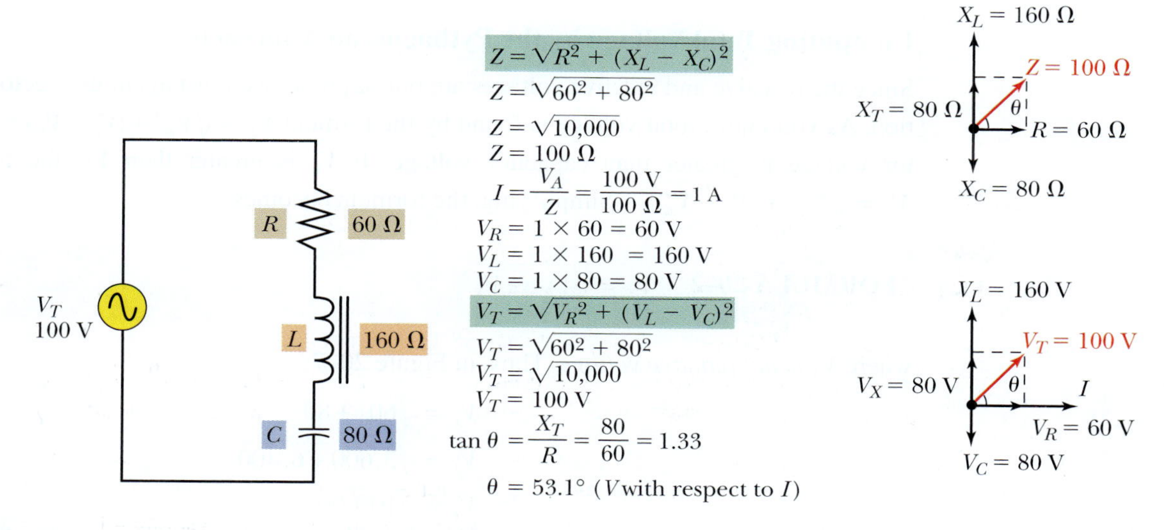

FIGURE 20–7 Series *RLC* circuit

Computing Impedance by the Pythagorean Approach

Using the concepts previously presented, the impedance must equal the vector sum of *R* and the remaining reactance. The formula is $Z = \sqrt{R^2 + (X_L - X_C)^2}$ when *XL* is greater than *XC*. If *XC* is greater than *XL*, then $Z = \sqrt{R^2 + (X_C - X_L)^2}$. In essence,

FORMULA 20–1 $Z = \sqrt{R^2 + X^2}$

where net reactance (*X*) depends on the relative reactance values, that is

$$X = X_L - X_C$$
$$X = X_C - X_L$$

In Figure 20–7, $R = 60\ \Omega$, $X_L = 160\ \Omega$, and $X_C = 80\ \Omega$. The net reactance is 80 Ω of inductive reactance in series with 60 Ω of resistance. Using the Pythagorean formula to find Z:

$$Z = \sqrt{R^2 + X^2}$$
$$Z = \sqrt{60^2 + 80^2}$$
$$Z = \sqrt{3{,}600 + 6{,}400}$$
$$Z = \sqrt{10{,}000}$$
$$Z = 100\ \Omega$$

Computing Current by Ohm's Law

In series circuits, current is common throughout the circuit. Circuit current is found through Ohm's law where $I = V/Z$. Since V_A is given as 100 V, then

$$I = \frac{V}{Z} = \frac{100\ \text{V}}{100\ \Omega} = 1\ \text{A}$$

Computing Individual Voltage by Ohm's Law

The voltage across the resistor = IR, across the inductor = IX_L, and across the capacitor = IX_C.

$$V_R = 1\ \text{A} \times 60\ \Omega = 60\ \text{V}$$
$$V_L = 1\ \text{A} \times 160\ \Omega = 160\ \text{V}$$
$$V_C = 1\ \text{A} \times 80\ \Omega = 80\ \text{V}$$

Computing Total Voltage by the Pythagorean Approach

Since the resistive and reactive voltages are out of phase, we must again use vector summation. As you know, total voltage is found by the formula $V_T = \sqrt{V_R^2 + (V_L - V_C)^2}$, if inductor voltage is greater than capacitor voltage. If V_C is greater than V_L, the formula is $V_T = \sqrt{V_R^2 + (V_C - V_L)^2}$. Simplifying, the formula becomes

FORMULA 20–2 $V_T = \sqrt{V_R^2 + V_X^2}$

where V_X *is net reactive voltage.* Thus, in Figure 20–7:

$$V_T = \sqrt{60^2 + 80^2}$$
$$V_T = \sqrt{3{,}600 + 6{,}400}$$
$$V_T = \sqrt{10{,}000}$$
$$V_T = 100\ \text{V}$$

Computing Phase Angle from Impedance and Voltage Data

Using impedance data, the formula is:

FORMULA 20–3 $\text{tangent of } \theta = \dfrac{X\ \text{(net reactance)}}{R}$

$$\tan\theta = \frac{80}{60}$$
$$\tan\theta = 1.33$$

The angle whose tangent = 1.33 is 53.1°. Since the net reactance is inductive reactance (plotted upward), the angle is +53.1°. NOTE: If the net reactance were capacitive reactance (plotted downward) of the same value, the phase angle would be –53.1° rather than +53.1°.

Using voltage data, the formula is:

FORMULA 20–4 tangent of $\theta = \dfrac{V_X}{V_R}$

$$\tan\theta = \frac{80}{60}$$
$$\tan\theta = 1.33$$

The angle whose tangent = 1.33 is 53.1°.

Since the net reactive voltage is inductive voltage, which leads current (the reference vector), the angle is +53.1°. NOTE: If the net reactive voltage were V_C (plotted downward) of the same value, the phase angle would be –53.1° rather than +53.1°.

EXAMPLE Refer to the series *RLC* circuit in Figure 20–8 and calculate Z, V_T, V_{L_1}, V_{C_1}, V_{R_1}, and θ.

In our example circuit, $R_1 = 20\ \Omega$ and $R_2 = 10\ \Omega$, so circuit $R = 30\ \Omega$.

$X_{L_1} = 50\ \Omega$ and $X_{L_2} = 40\ \Omega$, so circuit $X_L = 90\ \Omega$.

$X_{C_1} = 80\ \Omega$ and $X_{C_2} = 40\ \Omega$, so circuit $X_C = 120\ \Omega$.

Net reactance = $120\ \Omega\ (X_C) - 90\ \Omega\ (X_L) = 30\ \Omega$ of capacitive reactance.

$Z = \sqrt{R^2 + X^2} = \sqrt{30^2 + 30^2} = \sqrt{900 + 900} = \sqrt{1{,}800} = 42.42\ \Omega$

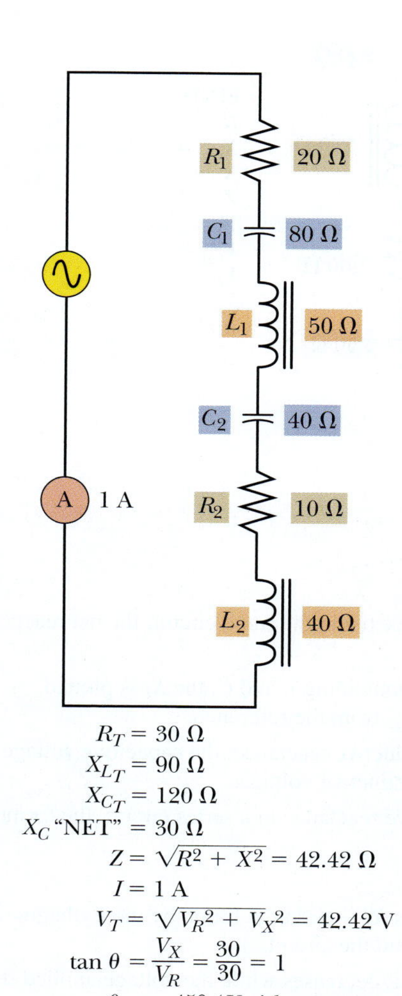

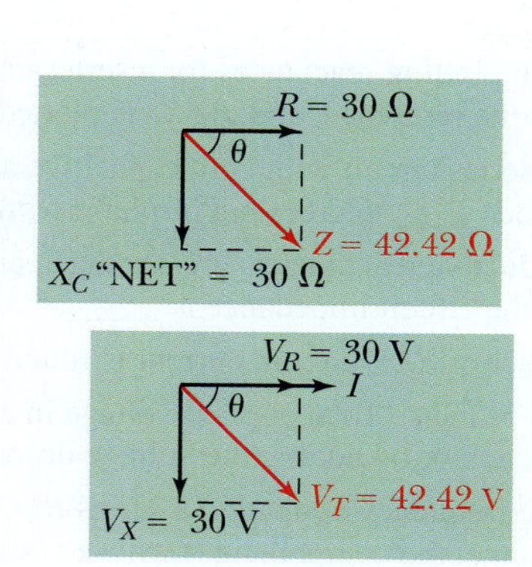

FIGURE 20–8 multiSIM

Practical Notes

With more than one resistor or one reactance of each kind in the circuit, combine resistances to one total R value; inductive reactances to one total X_L value; and capacitive reactances to one total X_C value. Then solve the problem with the standard procedures to solve series RLC circuits.

I is given as 1 ampere.

$V_R = IR = 1 \times 30 = 30 \text{ V}$

$V_L = IX_L = 1 \times 90 = 90 \text{ V}$

$V_C = IX_C = 1 \times 120 = 120 \text{ V}$

$V_T = \sqrt{V_R^{\,2} + V_X^{\,2}} = \sqrt{30^2 + 30^2} = \sqrt{900 + 900} = \sqrt{1{,}800} = 42.42 \text{ V}$

$V_{L_1} = IX_{L_1} = 1 \text{ A} \times 50\ \Omega = 50 \text{ V}$

$V_{C_1} = IX_{C_1} = 1 \text{ A} \times 80\ \Omega = 80 \text{ V}$

$V_{R_1} = IR_1 = 1 \text{ A} \times 20\ \Omega = 20 \text{ V}$

$$\tan\theta = \frac{X}{R},\ \text{or}\ \frac{V_X}{V_R}$$

$$\tan\theta = \frac{30}{30} = 1$$

The angle whose tangent = 1 is 45°. ◆

Now that you have seen how this is done, try the following practice problem.

PRACTICE PROBLEMS 1

Refer to Figure 20–9 and calculate R, X_{C_T}, X_L, Z, V_T, V_{C_2}, V_L, V_R, and θ.

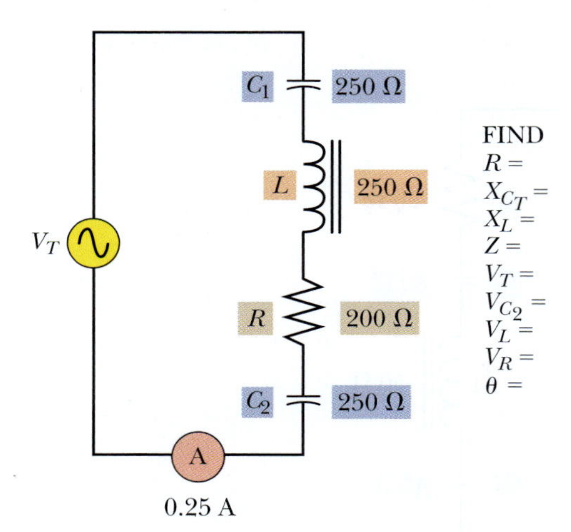

FIGURE 20–9

■ IN-PROCESS LEARNING CHECK 1

Fill in the blanks as appropriate.

1. When both capacitive and inductive reactance exist in the same circuit, the net reactance is _____.
2. When plotting reactances for a series circuit containing L and C, the X_L is plotted _____ from the reference, and the X_C is plotted _____ from the reference.
3. In a series circuit with both capacitive and inductive reactance, the capacitive voltage is plotted _____ degrees out of phase with the inductive voltage.
4. If inductive reactance is greater than capacitive reactance in a series circuit, the formula to find the circuit impedance is _____.
5. In a series RLC circuit, current is found by _____.
6. True or False. Total applied voltage in a series RLC circuit is found by the Pythagorean approach or by adding the voltage drops around the circuit.
7. If the frequency applied to a series RLC circuit decreases while the voltage applied remains the same, the component(s) whose voltage will increase is/are the _____.

20–3 The Parallel *RLC* Circuit

A Simple *LC* Parallel Circuit

Refer to Figure 20–10 as you study the following discussion on combining inductive and capacitive reactances in parallel.

1. Since I_L is plotted downward and I_C is plotted upward, obviously the reactive branch currents *cancel* each other.
2. Combining the reactive branch currents produces the resulting reactive current that is the difference between the opposite-reactance branch currents. That is, the resulting reactive current (I_X) is the larger reactive branch current minus the smaller reactive branch current.
3. In a parallel circuit, the voltage is the same across all branches; that is, V_T.

Parallel *RLC* Circuit Analysis

As was just stated, the *opposite-direction phasors* cancelling effect is also applicable to parallel circuits. But this cancellation effect is different between series and parallel circuits because of the circuit parameters involved. In series circuits, the current is common and the reactive voltage opposite-direction phasors cancel. However, in parallel circuits, the voltage is common and the reactive branch currents are opposite-direction phasors. Look at Figure 20–11 as you study the following discussion.

1. I_T is the vector sum of the circuit resistive and reactive branch currents.
2. The resulting reactive branch current value used to calculate total current is found by subtracting the smaller reactive branch current value from the larger reactive branch current value. This value (I_X) is then used to calculate I_T.

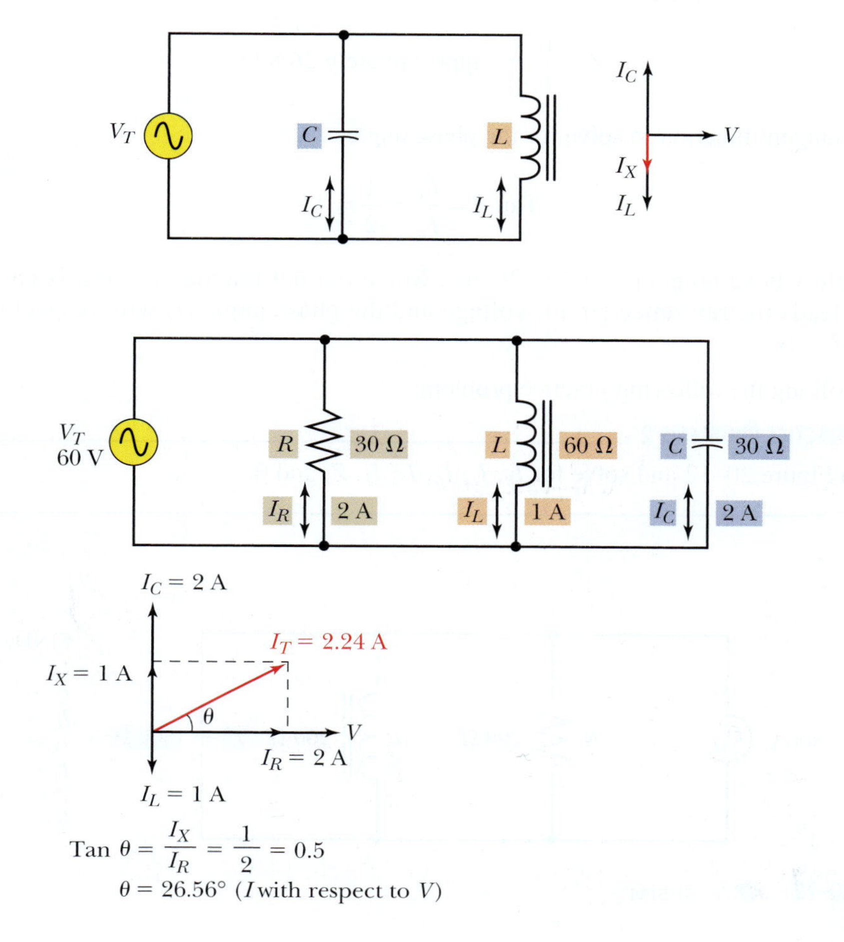

FIGURE 20–10 Parallel *LC* circuit without resistance

FIGURE 20–11 Parallel *RLC* circuit

3. The general formula for I_T is $I_T = \sqrt{I_R^2 + (I_L - I_C)^2}$, where I_L is greater than I_C. If I_C is greater than I_L, the formula becomes $I_T = \sqrt{I_R^2 + (I_C - I_L)^2}$. Using the *net reactive current,* the formula becomes

FORMULA 21–5 $I_T = \sqrt{I_R^2 + I_X^2}$

Let's apply this knowledge and Ohm's law to solve a problem.

EXAMPLE Refer again to the circuit in Figure 20–11 and find I_R, I_L, I_C, I_X, I_T, Z, and θ. Use Ohm's law to solve for each branch current:

$$I_R = \frac{V_T}{R} = \frac{60}{30} = 2 \text{ A}$$

$$I_L = \frac{V_T}{X_L} = \frac{60}{60} = 1 \text{ A}$$

$$I_C = \frac{V_T}{X_C} = \frac{60}{30} = 2 \text{ A}$$

Subtract the smaller reactive current from the larger to solve for I_X:

$$I_X = I_C - I_L = 2 - 1 = 1 \text{ A}$$

Use the Pythagorean approach to solve for I_T:

$$I_T = \sqrt{I_R^2 + I_X^2} = \sqrt{2^2 + 1^2} = \sqrt{4+1} = \sqrt{5} = 2.24 \text{ A}$$

Use Ohm's law to solve for Z:

$$Z = \frac{V_T}{I_T} = \text{ approximately } 26.8\ \Omega$$

Use the tangent function to solve for the phase angle:

$$\text{Tan } \theta = \frac{I_X}{I_R} = \frac{1}{2} = 0.5$$

The angle whose tangent = 0.5 is 26.56°. Since the net reactive current is capacitive, current leads the reference circuit voltage and the phase angle (I_T with respect to V_T) is + 26.56°.

Try solving the following practice problem.

PRACTICE PROBLEMS 2

Refer to Figure 20–12 and solve for I_R, I_L, I_C, I_X, I_T, Z, and θ.

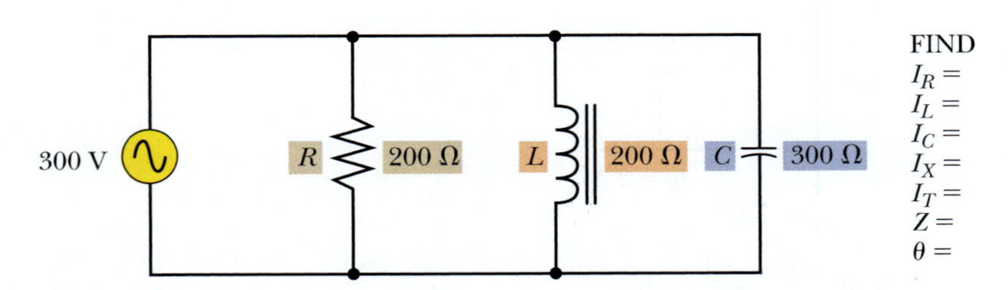

FIGURE 20–12 multiSIM

IN-PROCESS LEARNING CHECK 2

1. In parallel *RLC* circuits, capacitive and inductive branch currents are _____ out of phase with each other.
2. The formula to find total current in a parallel *RLC* circuit when I_C is greater than I_L is _____.
3. True or False. The quickest way to find the circuit impedance of a parallel *RLC* circuit is to use an impedance diagram. _____
4. If the frequency applied to a parallel *RLC* circuit increases while the applied voltage remains the same, the branch current(s) that will increase is/are the _____ branch(es).

20–4 Power in AC Circuits

Background

In dc circuits, you know how to find the power expended by a circuit through the resistive power dissipation. That is,

$$I^2R, \frac{V^2}{R}, \text{ and } V \times I$$

In ac circuits with both resistance and reactance, the apparent power supplied by the source is not the real or true power dissipated in the form of heat. Obviously, this is because the reactances take power from the circuit during one portion of the cycle and return it to the circuit during another portion of the cycle. As you know, circuits with reactances cause the voltage and current to be out of phase with each other. Therefore, the product of the out-of-phase circuit *V* and *I* does not indicate the actual circuit power dissipation.

Power in a Purely Resistive Circuit

Look at Figure 20–13. The power is dissipated during both half cycles. During the positive alternation, the power graph is represented by the products of a series of instantaneous values

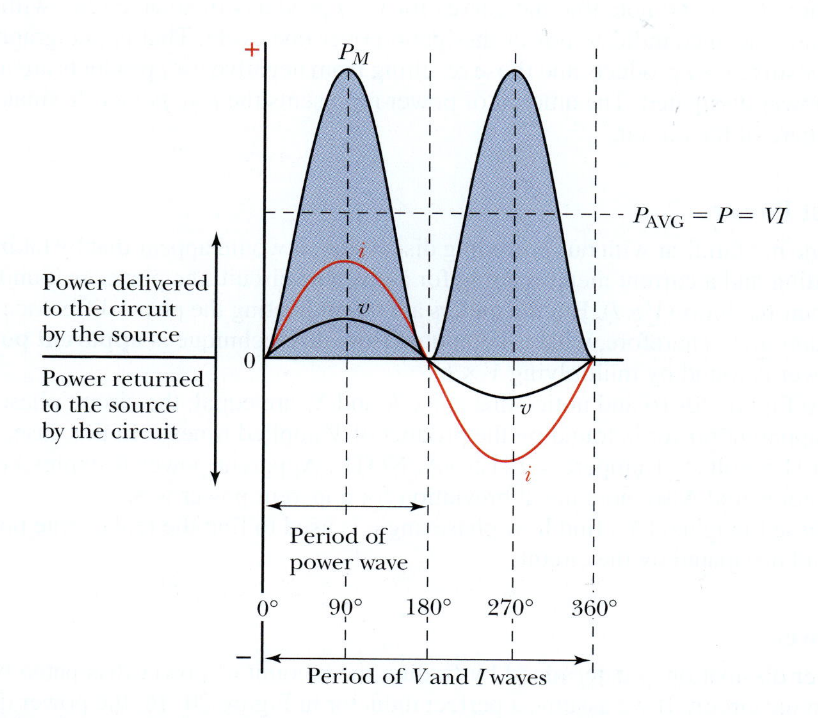

FIGURE 20–13 Instantaneous power graphs in a purely resistive circuit

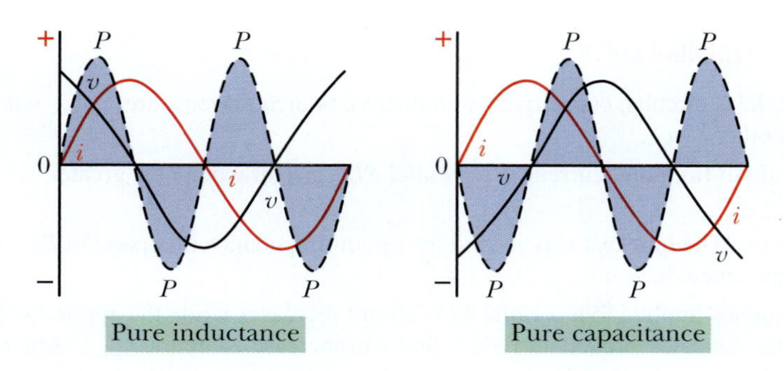

FIGURE 20–14 Instantaneous power graphs in purely reactive circuits

of *i* and *v* along the waveform. During the negative alternation, the power graph represents the products of the series of instantaneous values of negative *i* and negative *v* along the waveform. The product of a negative times a negative, of course, comes out positive. In other words, the resistor dissipates power during both alternations. Over the whole cycle, an average power dissipation equal to the product of current and voltage results. It can be mathematically shown that the average power dissipated equals the effective (rms) voltage value times the effective (rms) current value.

Power in Purely Reactive Circuits

Phase differences between voltage and current for purely inductive or capacitive circuits cause the power graphs, Figure 20–14. Because reactances take power from the circuit for one-quarter cycle and return it the next quarter cycle, you see a power graph of twice the source frequency. Notice, too, the average power over the complete cycle is zero because the reactive instantaneous *v* times *i* products are positive half the time and negative half the time. This fact implies that purely inductive or capacitive reactances do not dissipate power. In reality, inductors (because of their wire resistance) have some resistive component, which dissipates some power in the coil. Good capacitors have very low effective resistance and typically dissipate almost zero power.

Combining Resistive and Reactive Power Features

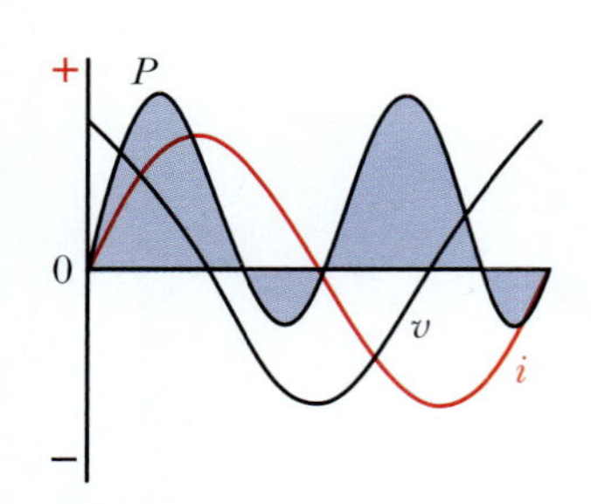

FIGURE 20–15 Instantaneous power graphs—*R* and *X* combined

View Figure 20–15 and note that the curves for $v \times i$ products in an ac circuit with both resistance and reactance indicate power dissipation over one cycle. That is, the graphs resulting from positive $v \times i$ products and those resulting from negative $v \times i$ products are not equal. There is power dissipated. The amount of power represents the *true power* dissipated by the resistive share of the circuit.

Apparent Power

To someone not familiar with our preceding discussion, it would appear that by taking a voltmeter reading and a current meter reading for a given ac circuit, the power is found by multiplying their readings ($V \times I$). But the meters are *not* indicating the phase difference between the two quantities. Therefore, what is computed from this technique is **apparent power.** Apparent power is found by multiplying $V \times I$.

Refer to Figure 20–16 and notice that since R and X_L are equal, the circuit phase angle is 45°. The apparent power is found by the product of V applied times I. In this case, apparent power = 141.5 volts × 1 ampere = 141.5 VA. NOTE: Apparent power is expressed as *volt-amperes,* not watts! Also, note the abbreviation for apparent power is ***S***.

With these thoughts, let's find how phase angle is used to find the real or true power supplied to and dissipated by the circuit.

True Power

True power dissipation is determined by finding the amount of power dissipated by the resistance in the circuit. If we assume a perfect inductor in Figure 20–16, the power dissipated by the resistor equals the true power dissipation. Because the I_R and V_R are in phase with each other, the formula $V_R \times I_R$ is used. Therefore, true power = 100 V × 1 A = 100 W.

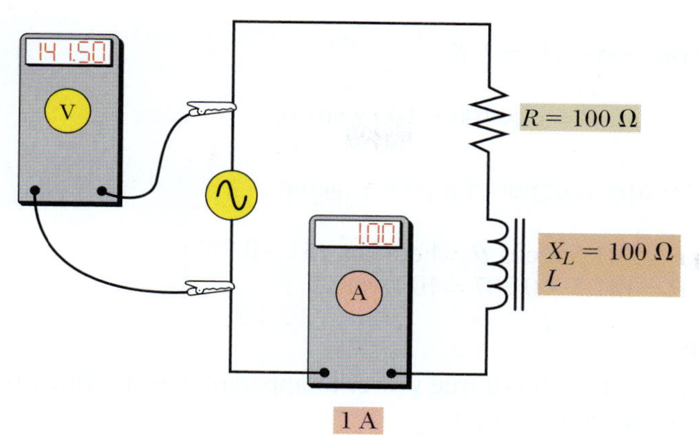

$p.f. = \cos\theta = \cos 45° = 0.707$

$p.f. = \frac{R}{Z}$ or $\frac{V_R}{V_T}$

$S = VI = 141.5$ VA (apparent power)
$P = VI\cos\theta = I^2R = 100$ W (true power)
$Q = VI\sin\theta = 100$ VAR (reactive power)

FIGURE 20–16 Power and power factor in a series ac circuit

The I^2R Method

The preference is to say that true power equals I^2R, where R represents all circuit resistive quantities. This also includes resistor components and resistance values in reactive components. These R values are easily measured. It is not as easy (for example) to separate reactive and resistive voltage components in an inductor by measuring V_L.

The Power Factor Method

The power factor method considers the phase angle between current and voltage, created by the reactance. The value of the circuit **power factor *(p.f.)*** reflects how resistive or reactive the circuit is acting. A power factor of one indicates a purely resistive circuit. A power factor of zero represents a purely reactive circuit. Obviously, a power factor between zero and one indicates the circuit has both resistance and reactance, and true power is less than apparent power. Power factor is equal to the cosine of the phase angle.

Practical Notes

A power factor of one indicates a circuit with a purely resistive load that uses all the energy delivered to it by the source. Power factors of less than one indicate some reactance is present in the circuit; and hence, the source must deliver more energy (or current) than is actually being put to work as "true power." In practice then, the power company likes industries and people (who are their "loads") to keep the power factor as close to one as possible. This means the power company's electrical generators have to produce less electrical energy, and their wires have to carry less current than when the power factors indicate reactive loads. Having smaller generator requirements and being able to transport the energy via smaller conductors can save the power company a lot of money.

Since many loads are inductive (such as motors), many industries use reactive compensating components to offset the reactance of their loads. For example, capacitors can be used to produce opposite reactances to help offset the inductance of inductive loads, such as large motors used in industry.

FORMULA 20–6 Power factor $(p.f.) = \cos\theta$

Then for our sample circuit in Figure 20–16:

FORMULA 20–7 Apparent power $(S) = V \times I$

$$S = 141.5 \times 1 = 141.5 \text{ volt-amperes}$$

FORMULA 20–8 True power $(P) = I^2R$

$$P = 1^2 \times 100 = 100 \text{ watts or}$$

FORMULA 20–9 P = Apparent power × power factor

$$P = 141.5 \times \cos\theta \text{ (where } \cos 45° = 0.707\text{)}$$
$$P = 141.5 \times 0.707 = 100 \text{ watts}$$

Power Factor Formulas

Power factor really expresses the ratio of true power to apparent power (which is the same ratio expressed by the cos θ, Formula 20–6).

FORMULA 20–10 $p.f. = \dfrac{\text{True power } (P)}{\text{Apparent power } (S)}$

$$\frac{VI \times \cos\theta}{VI} \text{ or } \frac{R}{Z} \text{ (Series)} = \frac{I_R}{I_T} \text{ (Parallel)}$$

EXAMPLE Recall in *series circuits*

$$\cos\theta = \frac{R}{Z}\left(\text{or } \frac{V_R}{V_T}\right)$$

For Figure 20–16, our series circuit example:

$$Z = \frac{V_T}{I_T} = \frac{141.5 \text{ V}}{1 \text{ A}} = 141.5\ \Omega$$

$$\cos\theta = \frac{R}{Z} = \frac{100}{141.5} = 0.707. \text{ Therefore, } p.f. = 0.707$$

(Incidentally, the angle whose cosine = 0.707 is 45°, as you know.)

In *parallel circuits* the *power factor* = cos $\theta = I_R/I_T$. Refer to Figure 20–17 and notice the calculations for power factor using the resistive and total current values. In this case, the voltage applied is 100 V. Thus:

$$I_R = \frac{100}{100} = 1 \text{ A}$$
$$I_L = \frac{100}{100} = 1 \text{ A}$$
$$I_T = \sqrt{I_R^2 + I_L^2} = \sqrt{1^2 + 1^2} = \sqrt{2} = 1.414$$
$$p.f. = \frac{I_R}{I_T} = \frac{1}{1.414} = 0.707$$

Again, in this case $\theta = 45°$.

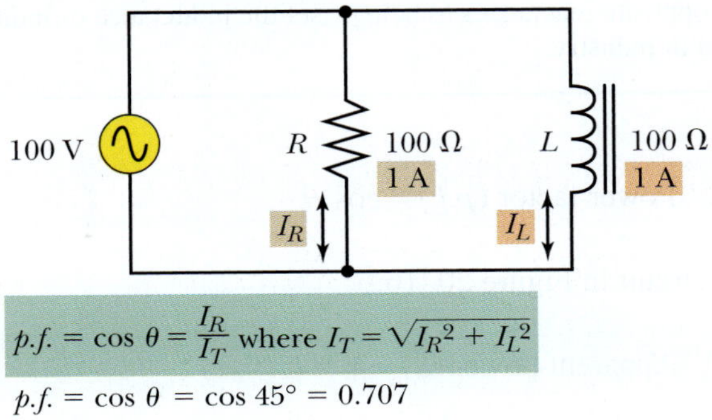

FIGURE 20–17 Power factor in a parallel ac circuit

The Power Triangle

The relationship between true and apparent power is sometimes represented visually by a *power triangle* diagram.

Figure 20–18a is a power triangle for a series *RL* circuit where $R = X_L$. Apparent power (S) in volt-amperes is plotted on a line at an angle equal to θ from the x-axis. Its length, representing magnitude or value, is $V \times I$. Notice the apparent power is the result of two components:

1. The **true power *(P)*,** sometimes called *resistive power,* is plotted on the x-axis (horizontal) and has a magnitude of $VI \times \cos\theta$ watts.
2. The **reactive power *(Q)*,** often called **voltampere-reactive (VAR),** is plotted on the y-axis (vertical) and has a magnitude of $VI \times \sin\theta$.

Figure 20–18b is a power triangle for a circuit with resistance and capacitive reactance. From your knowledge of the Pythagorean theorem and right-triangle analysis, you can see apparent power is computed from the formula:

FORMULA 20–11 $S = \sqrt{P^2 + Q^2}$

where S = apparent power (measured in volt-amperes, VA)
P = true power (measured in watts, W)
Q = net reactive power (measured in volt-amperes reactive, VAR)

(NOTE: Power triangles are often used to analyze ac power distribution circuits.)

Practical Notes

This formula is applicable to both series and parallel circuits. To summarize:

1. Knowing only the value of voltage or current does not indicate the amount of power actually being dissipated by the circuit.
2. A power triangle (different from *V-I* vector diagram) is used to diagram the various types of power in the ac circuit. Because the power waveform is a different frequency sinusoidal representation compared with the *V* and *I* quantities, we cannot plot the power values on the same vector diagram with *V* and *I*. Thus, the right-triangle (Pythagorean) approach displays the various types of power in the ac circuit.
3. The vertical element in the power triangle is the reactive power. The horizontal element in the triangle is the true or resistive power. Therefore, the hypotenuse of the triangle is the apparent power, which is the result of reactive and resistive power.

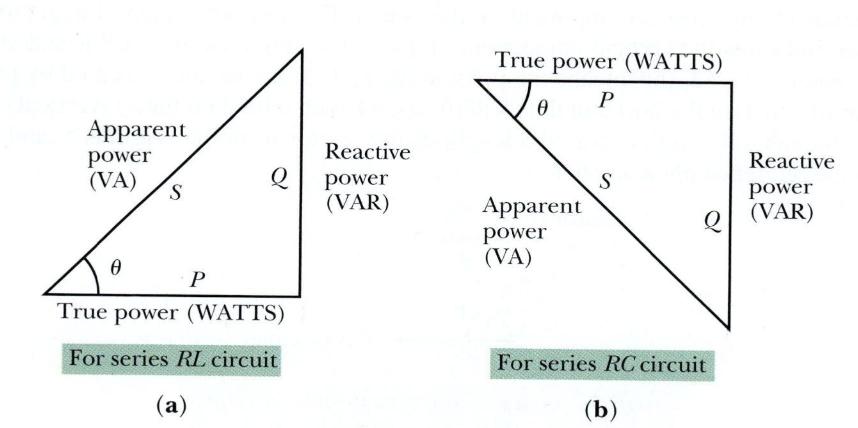

FIGURE 20–18 The power triangle

■ **IN-PROCESS LEARNING CHECK 3**

1. Power dissipated by a perfect capacitor over one whole cycle of input is _____.
2. The formula for apparent power is _____, and the unit expressing apparent power is _____.
3. True power in *RLC* circuits is that power dissipated by the _____.
4. The formula for power factor is _____.
5. If the power factor is one, the circuit is purely _____.
6. The lower the power factor, the more _____ the circuit is acting.
7. In the power triangle, true power is shown on the _____ axis, reactive power (VAR) is shown on the _____ axis, and apparent power is the _____ of the right triangle.

20–5 Rectangular and Polar Vector Analysis

Overview of Rectangular and Polar Forms

In studying dc circuits, you learned that adding, subtracting, multiplying, and dividing electrical quantities was all that was needed to analyze circuits. When you studied ac circuits, you learned that out-of-phase quantities need a different treatment. You learned several approaches—graphic analysis (plotting and measuring vectors or phasors and angles); the Pythagorean theorem (to find resulting magnitudes or values); and simple trigonometry (to find related angles). Also, you learned trigonometry can be used to find magnitude information.

In this section, you will learn about algebraic vector analysis that involves both rectangular and polar forms to analyze vector quantities. You will learn how the *j* operator is used in complex numbers to distinguish the vertical vector from other components. Actually, these techniques are only a small extension of what you already know and use!

Phasors

You know that phasors represent sinusoidal ac quantities when they are the *same* frequency. Of course, they must be the same frequency so that when using phasor drawings the quantities compared maintain the same phase differential throughout the ac cycle. The normal direction of rotation of the constant-magnitude rotating radius vector (phasor) is counterclockwise from the 0° position (reference point). The relative position of the phasor shows its relationship to the reference (0°) position and to other phasors in the diagram. The phasor length indicates the magnitude, or relative value of the quantity represented, Figure 20–19. In ac, this is the effective (rms) value of *V*, *I*, and so forth.

Rectangular Notation

In **rectangular form notation,** the magnitude and direction of a vector relates to both the *x*-axis (horizontal) and the *y*-axis (vertical) of the rectangular coordinate system, Figure 20–20. One way to find the location and magnitude of a given vector on this coordinate system is by **complex numbers.** A complex number locates a positional point with respect to two axes.

A complex number has two parts—**real** and **imaginary.** The first part, real term or number, indicates the horizontal component on the *x*-axis. The second element, imaginary term or number, indicates the vertical component on the *y*-axis. For example, if it is stated that a specific vector = $4 + j3$, this means the point at the end of that vector is located by projecting upward from 4 on the horizontal axis until a point even with 3 on the $+j$ (vertical) axis is reached. (Incidentally, in this case, the length of this vector from the origin is 5, and its angle from the reference plane is 36.9°.)

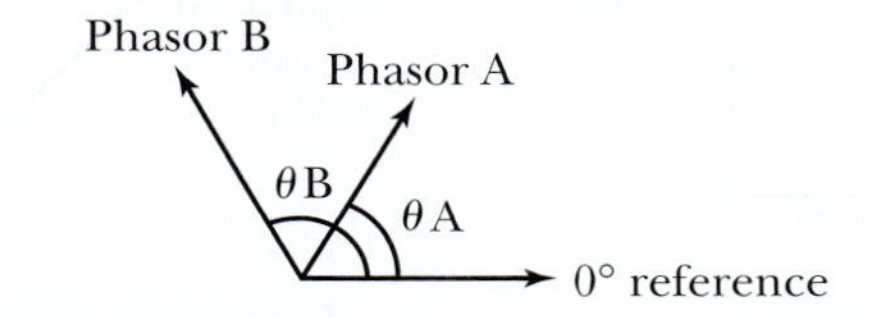

Length of phasor denotes magnitude or value.
Angle of phasor denotes relative direction.

FIGURE 20–19 Phasor illustration

Notice in Figure 20–20 the horizontal axis is also called the *x*-axis, the real axis, or the resistance axis. The vertical axis is also called the imaginary axis, or the reactance axis. The *j* operator designates values on that axis. The +*j* indicates an upward direction, while –*j* signifies a downward direction. When a number in a formula is preceded by the letter *j*, the number is a vertical (or *y*-axis) component.

Since the *j* operator is so important in the rectangular form of vector algebra, let's define the *j* operator more specifically:

$$j = \sqrt{-1}$$
$$j^2 = (\sqrt{-1})(\sqrt{-1}) = -1$$
$$j^3 = (j^2)(j) = (-1)(j) = -j$$
$$j^4 = (j^2)(j^2) = (-1)(-1) = +1$$

Look at Figure 20–21 to relate these equations to the rectangular coordinate format. Notice the various powers of the *j* operator rotate a phasor as follows:

- +*j* is a 90° operator, rotating a phasor 90° in the counterclockwise (CCW) direction, or +90°.
- –*j* rotates a phasor 90° in the clockwise (CW) direction, or –90°.
- j^2 rotates the phasor 180°.
- j^3 rotates the phasor to the +270° position, or –90°.
- j^4 is +1, rotating the vector back to the starting point (0°).

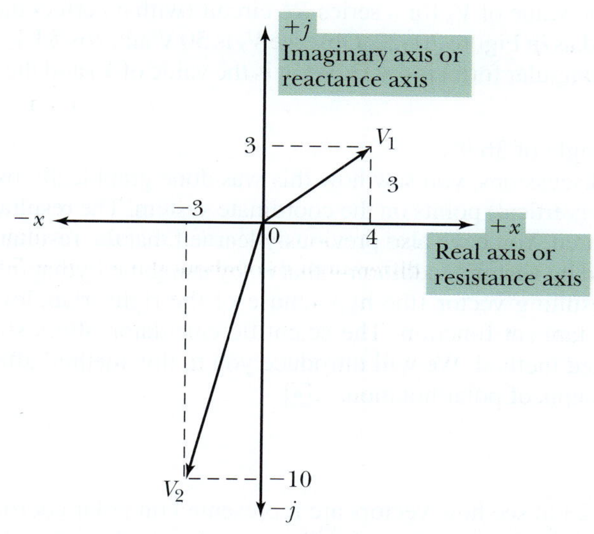

FIGURE 20–20 Illustration of rectangular notation form

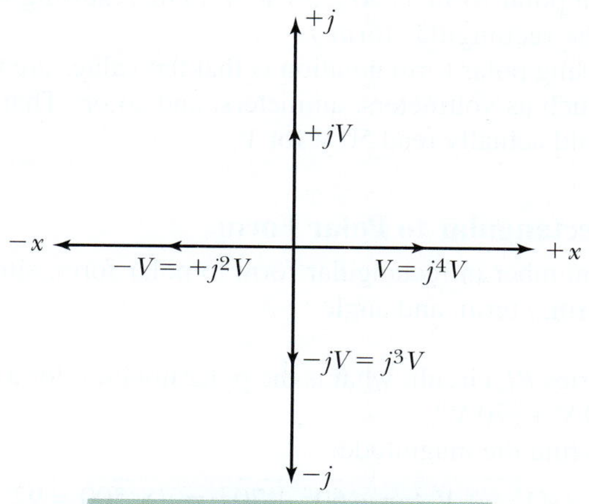

FIGURE 20–21 The *j* operator

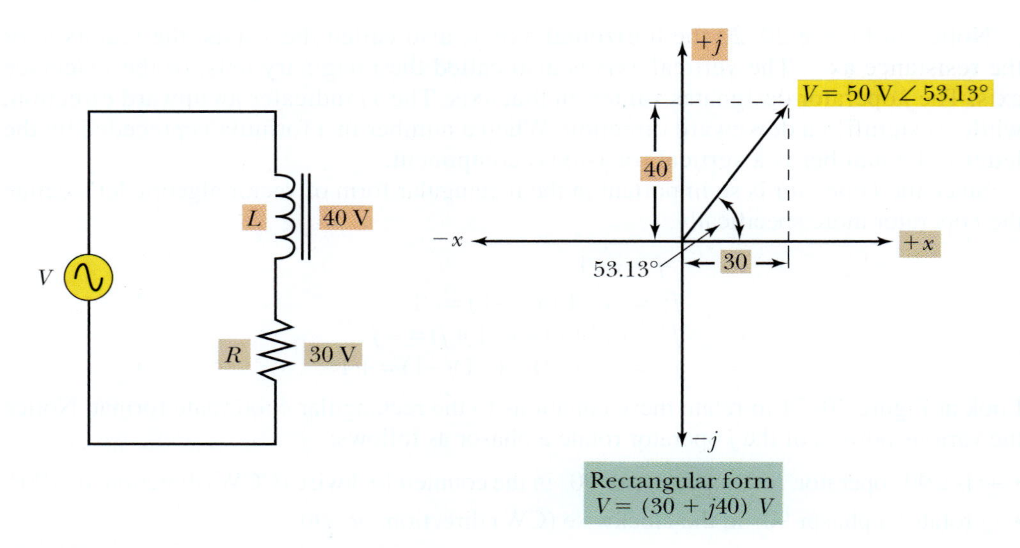

FIGURE 20–22 Example of rectangular notation

EXAMPLE If the value of V_T for a series *RL* circuit (with a perfect inductor) is $30 + j40$, it might be illustrated as in Figure 20–22. Observe V_T is 50 V and θ is 53.1°. Now suppose that V_T, expressed in rectangular form, is $4 + j3$. What is the value of V_T and the angle? This should be familiar to you.

Answer: 5 V at an angle of 36.9°.

In our previous discussions, you saw how this was done graphically by projecting from 4 (horizontal) and $+j3$ (vertical) points on the coordinate system. The resultant vector and angle could then be measured. You have also previously learned that the resultant of these two projections can be found by vectorial addition—that is, by using the Pythagorean theorem to find the length of the resulting vector (the hypotenuse of the right triangle). The angle can be found by using the tangent function. The scientific calculator offers still another method, which is the preferred method. We will introduce you to this method after you are familiarized with basic concepts of polar notation.

Polar Notation

Refer to Figure 20–23 to see how vectors are represented on polar coordinates.

Whereas the rectangular form expresses vectors by the horizontal and vertical axes, the **polar form notation** expresses the length and angle of the resulting vector to define a vector. Previously, we stated one value of V_T is $30 + j40$ in rectangular notation. This same V_T is expressed in polar form as $50 \angle 53.1°$ V. (The resulting vector of the other two vectors described in the rectangular form.)

The advantage for using polar form notation is that the values are the same as those measured by instruments such as voltmeters, ammeters, and so on. That is, in the previous example, a voltmeter would actually read 50 V for V_T.

Converting from Rectangular to Polar Form

To convert a complex number in rectangular form to polar form, simply find the vector resulting from the real term, *j* term, and angle.

EXAMPLE In a series *RL* circuit, what is the polar notation for a value expressed in rectangular notation as $60\text{ V} + j70\text{ V}$?

Use phasor addition to find the magnitude:

$$V_T = \sqrt{V_R^2 + V_L^2} = \sqrt{60^2 + 70^2} = \sqrt{8{,}500} = 92.2\text{ V}$$

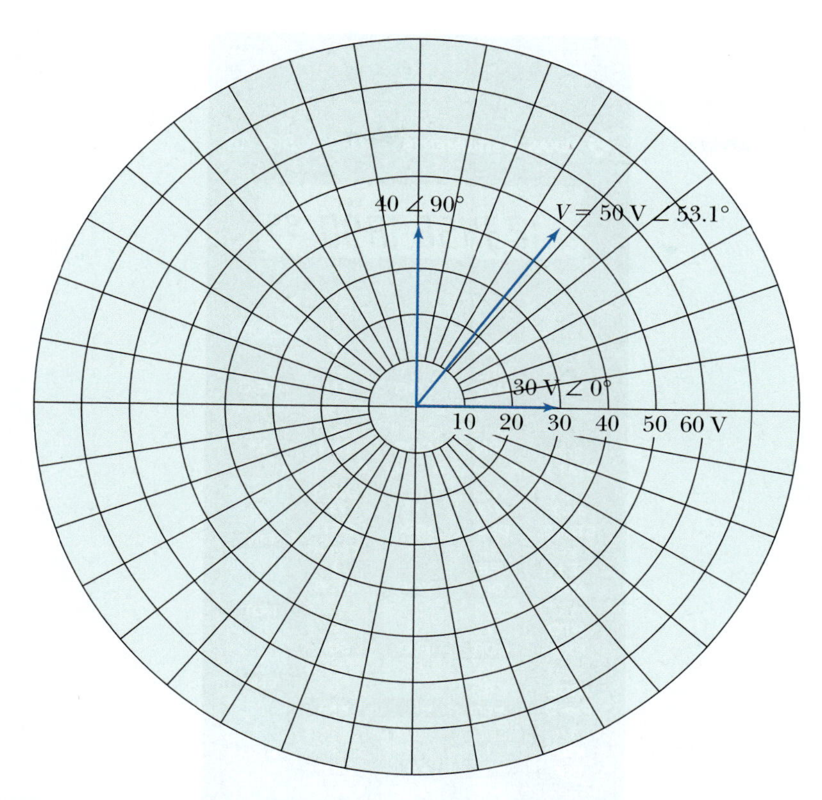

FIGURE 20–23 Illustration of vectors represented on polar coordinate system

Use the tangent function to solve for the angle:

$$\tan\theta = \frac{V_L}{V_R} = \frac{70}{60} = 1.166$$

The angle whose tangent = 1.166 is 49.3°

Polar form = 92.2 ∠49.3° V

You get the idea! To convert, solve for the magnitude by vectorially adding the real and the *j* terms; then solve for the angle, using the arctangent, as appropriate. The resulting magnitude and angle are then combined to express the polar notation for the vector. ◆

Using a Scientific Calculator to Convert from Rectangular to Polar Form

In Figure 20–24, you see one example of a scientific calculator. Let's see the steps you can use to perform the conversion of values expressed in rectangular form to find the equivalent in polar form.

Keys of particular interest will be the [x↔y] key, and the [R▸P] key. Let's use the values given in our previous example and perform the conversion using the calculator.

◆ **EXAMPLE**

1. Enter the real term value, in this case 60.
2. Press the [x↔y] key. (This enters the real term portion.)
3. Enter the imaginary term value, in this case 70.
4. Press the [3rd] key, then the [R▸P] key (rectangular to polar key). By pressing the [3rd] and [R▸P] key in sequence, you have invoked the rectangular-to-polar conversion mode of the calculator.

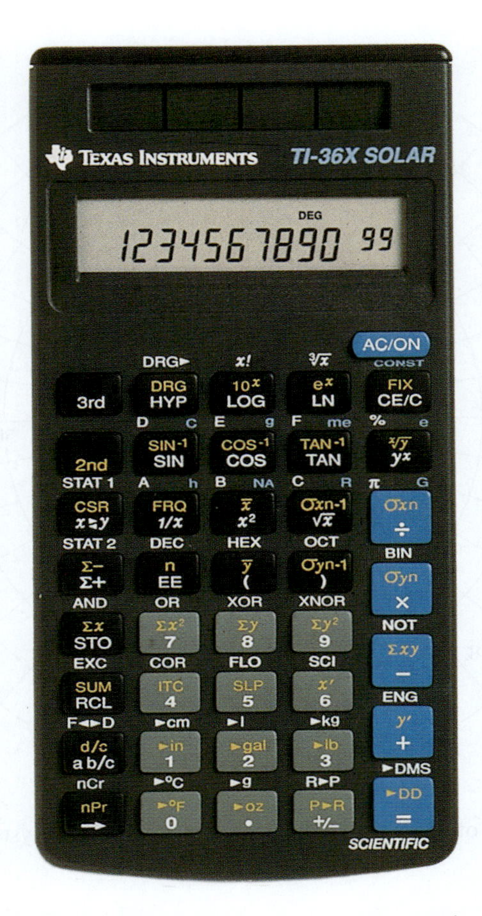

FIGURE 20–24 Sample scientific calculator with *P-R* and *R-P* conversion modes (*Courtesy of Texas Instruments*)

5. Read the number in the calculator display area (92.19). This number represents the vector magnitude for the polar notation. Observe that this number agrees with what we found using the Pythagorean formula.
6. Press the [x⇄y] key again to get the angle information. The readout says 49.3, indicating the angle is 49.3°.
7. Combining the results of steps 5 and 6, you find that the polar notation indicates that $92.19 \angle 49.3°$ is the equivalent of the original $60 + j70$ rectangular notation form.

NOTE: If the imaginary number is a $-j$, then enter the number, followed by the [+/–] key, before pressing the [3rd] key (in step 4).

Verify the following samples of rectangular and equivalent polar form notations.

Rectangular		*Polar*
$3 + j4$	=	$5 \angle 53.1°$
$4 - j3$	=	$5 \angle 36.9°$
$60 + j70$	=	$92.2 \angle 49.3°$ ◆

Now try a couple of conversions on your own, using the calculator technique.

PRACTICE PROBLEMS 3

1. Convert $20 + j30$ to polar form.
2. Convert $30 - j20$ to polar form.

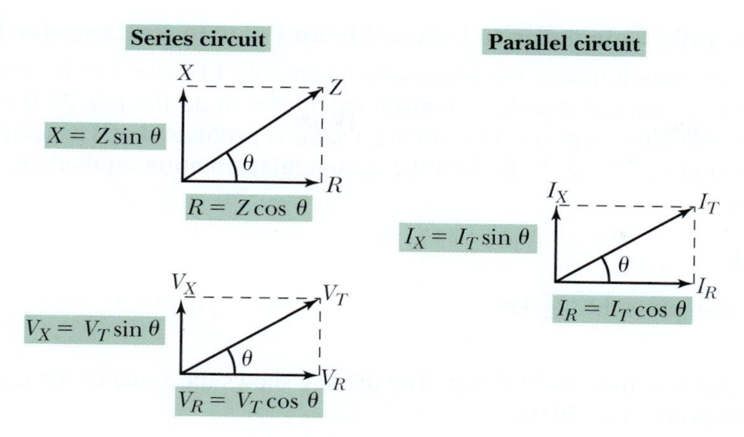

FIGURE 20–25 Converting from polar to rectangular form

Converting from Polar to Rectangular Form

Later in the chapter, you will see the rectangular form is most useful to *add and subtract* vector values and the polar form is useful to *multiply and divide* vector values. For this reason, it is convenient to know how to convert from rectangular to polar form and from polar to rectangular form, as appropriate.

Recall that trig is used to solve for the sides of a right triangle, if the angle and one side are known. Let's put this fact to use.

Since the polar form expresses the hypotenuse of the right triangle plus the angle, we can solve for the other two sides of the triangle by using the appropriate trig functions. Obviously, knowing the sides allows us to express the real and *j* terms appropriately.

Refer to Figure 20–25 and note the horizontal component; that is, the resistive or real term (R in series circuit impedance diagrams), V_R in series circuits, and I_R in parallel circuits is found by:

$$R = Z \cos \theta$$
$$V_R = V_T \cos \theta$$
$$I_R = I_T \cos \theta$$

Recall

$$\cos \theta = \frac{\text{adj}}{\text{hyp}}; \text{ therefore, adj} = \text{hyp} \times \cos \theta$$

In like manner, use trig to solve for the *j* (imaginary) term (i.e., X in series impedance diagrams, V_X in series circuits, and I_X in parallel circuits). Referring to Figure 20–25 again, the reactive term(s) are found by:

$$X = Z \sin \theta$$
$$V_X = V_T \sin \theta$$
$$I_X = I_T \sin \theta$$

Recall

$$\sin \theta = \frac{\text{opp}}{\text{hyp}}; \text{ therefore, opp} = \text{hyp} \times \sin \theta$$

Once the real and *j* terms are known, we can express the answer in rectangular form.

Using a Scientific Calculator to Convert from Polar to Rectangular Form

You can use the trigonometry we have been discussing to solve for these values or, as you might guess, you can use the scientific calculator to do the job. At this point, let's give you the calculator steps used in solving a sample problem. Let's assume that the polar form notation is 50 $\angle 37°$. To find the rectangular notation equivalent, use the following steps:

EXAMPLE

1. Enter 50, then press the [x↔y] key.
2. Enter 37.
3. Press the [2nd] key, then the [P▸R] key. The display shows the value of the real term in the rectangular expression, 39.93.
4. Press the [x↔y] key. The display shows the value of the imaginary (j) component of the rectangular expression, $+j30.09$.
5. Combine the results of steps 3 and 4 to get the answer; that is, 50 $\angle 37$ degrees in polar form $= 39.93 + j30.09$ in rectangular form.

Both the trig-related formulas and the related calculator sequences follow for your convenience.

FORMULA 20–12 $Z \angle \theta° = (Z \cos \theta) + (j\, Z \sin \theta)$

for a series circuit impedance diagram

Z value, [x↔y], angle value (θ), [2nd], [P▸R], [x↔y]

The answer is a combination of the last two readouts; that is, [P▸R] gives the real term and [x↔y] gives the imaginary (j) term.

FORMULA 20–13 $V_T \angle \theta° = V_T \cos \theta + j\, V_T \sin \theta$

for a series circuit V-I vector diagram

V_T value, [x↔y], angle value (θ), [2nd], [P▸R], [x↔y]

The answer is a combination of the last two readouts; that is, [P▸R] gives the real term and [x↔y] gives the imaginary (j) term.

FORMULA 20–14 $I_T \angle \theta° = I_T \cos \theta + j\, I_T \sin \theta$

for a parallel circuit V-I vector diagram

I_T value, [x↔y], angle value (θ), [2nd], [P▸R], [x↔y]

The answer is a combination of the last two readouts; that is, [P▸R] gives the real term and [x↔y] gives the imaginary (j) term.

EXAMPLES Convert 45 $\angle 30°$ to rectangular form (trig method).

$$45 \angle 30° = 45\ (\cos 30°) + j45\ (\sin 30°)$$
$$45 \angle 30° = 45 \times 0.866 + j45 \times 0.5$$
$$45 \angle 30° = 38.97 + j22.5$$

Convert 45 $\angle 30°$ to rectangular form (calculator method).
45, [x↔y], 30, [2nd], [P▸R], (gives R value). Then press [x↔y] again (gives reactive value). Verify the results by drawing a sketch to see if the results look logical.

PRACTICE PROBLEMS 4

1. Convert 66 $\angle 60°$ to rectangular form.
2. Convert 75 $\angle 48°$ to rectangular form.

Now that you are familiar with rectangular and polar notation and their relationships, we will show how these expressions are used mathematically.

20–6 Algebraic Operations

Addition (Rectangular Form)

Basic rule: Add in-phase (resistive) terms and add out-of-phase (reactive) *j* terms.

EXAMPLE Find the sum of V_1 and V_2.

Example 1: $V_1 = 15 + j20$, $V_2 = 20 - j10$

$$V_1 + V_2 = 35 + j10$$

Example 2: $V_1 = 20 + j20$, $V_2 = 30 + j10$

$$V_1 + V_2 = 50 + j30$$

Subtraction (Rectangular Form)

Basic rule: Change the sign of the subtrahend and add.

EXAMPLE Subtract branch current I_2 from branch current I_1, where

$$I_1 = 5 - j6 \text{ A}$$
$$I_2 = 4 + j7 \text{ A}$$

Change the sign of branch current I_2 and add:

$$\begin{array}{r} 5 - j6 \\ \underline{-4 - j7} \\ I_1 - I_2 = 1 - j13 \end{array}$$

EXAMPLE When given total current and one branch current (I_1), find the other branch current (I_2).

$$I_T = I_1 + I_2\text{; therefore, } I_2 = I_T - I_1.$$

where

$$I_T = 13 - j5.5 \text{ mA}$$
$$I_1 = 6 - j4.0 \text{ mA}$$

Change the sign of branch current 1, and add:

$$\begin{array}{r} 13 - j5.5 \\ \underline{-6 + j4.0} \\ I_2 = 7 - j1.5 \end{array}$$

Multiplication (Polar Form)

Basic rule: Multiply the two magnitudes and add the angles algebraically.

EXAMPLE If a current of 5 $\angle 30°$ A flows through a Z of 20 $\angle -15°$ Ω, what is the voltage?

$$V = IZ$$
$$V = (5 \angle 30°) \times (20 \angle -15°)$$
$$V = 100 \angle 15° \text{V}$$

Practical Notes

If addition or subtraction is needed to analyze a problem and the terms are given in polar form, convert them to rectangular form because you cannot use polar form to add or subtract.

On the other hand, even though multiplication and division are possible in rectangular form, polar form is easier for multiplying and dividing. If you know the rectangular form, you will first convert it to the polar form to multiply or divide. For example, $V_T = I_T \times Z_T$.

Division (Polar Form)

Basic rule: Divide one magnitude by the other and subtract the angles algebraically.

◆ **EXAMPLE** Find the current in a circuit if the voltage = 35 ∠60° and the impedance = 140 ∠–20°.

$$I = \frac{V}{Z}$$

$$I = \frac{35\angle 60^\circ}{140\angle -20^\circ}$$

$$I = 0.25\angle 80^\circ \text{ A}$$ ◆

Raising to Powers (Polar Form)

Basic rule: Raise the magnitude to the power indicated and multiply the angle by that power.

◆ **EXAMPLE** If $I = 3\angle 20^\circ$, then $I^2 = 9\angle 40^\circ$. ◆

Taking a Root (Polar Form)

Basic rule: Take the root of the magnitude and divide the angle by the root.

◆ **EXAMPLE** If $I^2 = 100\angle 60^\circ$, then $I = 10\angle 30^\circ$. ◆

PRACTICE PROBLEMS 5

1. Add $20 + j5$ and $35 - j6$
2. Add $34.9 - j10$ and $15.4 + j12$
3. Subtract $20 + j5$ from $35 - j6$
4. Subtract $22.9 - j7$ from $37.8 - j13$
5. Multiply $5\angle 20^\circ \times 25\angle -15^\circ$
6. Multiply $24.5\angle 31^\circ \times 41\angle 23^\circ$
7. Divide $5\angle 20^\circ$ by $25\angle -15^\circ$
8. Divide $33\angle 34^\circ$ by $13\angle 45^\circ$
9. Find I^2 if $I = 12\angle 25^\circ$
10. Find I if $I^2 = 49\angle 66^\circ$
11. Find the result of: $\dfrac{5\angle 50^\circ + (30 + j40)}{2.6\angle -15^\circ}$
12. Find the result of: $\dfrac{10\angle 53.1^\circ + (25 - j25)}{4.8\angle 36^\circ}$

20–7 Application of Rectangular and Polar Analysis

Sample Series *RLC* Circuit Analysis and Applying Vector Analysis

Refer to Figure 20–26 as you study the following discussion on finding various circuit parameters. Note if addition and subtraction are needed, the calculations use rectangular notation; if multiplication or division is involved, the calculations use polar notation.

1. Find Z_T:

Polar form:

Rectangular form: $20 + j10$

Z_T = vector sum of oppositions

$$Z_T = \sqrt{{R_T}^2 + {X_T}^2}$$

$$Z_T = \sqrt{20^2 + 10^2}$$

$$Z_T = \sqrt{500}$$

$Z = 22.36\ \angle 26.56°\ \Omega$ $\qquad$ $Z_T = 22.36\ \Omega$

2. Find I_T (Assume I_T is at 0° because this is a series circuit.):

Polar form: $\qquad$ Rectangular form:

$$I_T = \frac{V_T}{Z_T}$$

$$I_T = \frac{60.00\ \angle 26.56}{22.36\ \angle 26.56}$$

$I_T = 2.68\ \angle 0°\ \text{A}$ $\qquad$ $I_T = 2.68 + j0\ \text{A}$

3. Find individual voltages:

a. $V_{R_1} = I \times R_1 = 2.68\ \angle 0° \times 10\ \angle 0° = 26.8\ \angle 0°\ \text{V}$

b. $V_{R_2} = I \times R_2 = 2.68\ \angle 0° \times 10\ \angle 0° = 26.8\ \angle 0°\ \text{V}$

c. $V_L = I \times X_L = 2.68\ \angle 0° \times 20\ \angle 90° = 53.6\ \angle 90°\ \text{V}$

d. $V_C = I \times X_C = 2.68\ \angle 0° \times 10\ \angle -90° = 26.8\ \angle -90°\ \text{V}$

4. Verify total voltage via sum of individual drops. That is, convert each *V* from polar to rectangular form, then add:

Polar form:	Rectangular form:
$V_{R_1} = 26.8\ \angle 0°$	a. $26.8 + j0$
$V_{R_2} = 26.8\ \angle 0°$	b. $26.8 + j0$
$V_L = 53.6\ \angle 90°$	c. $0.0 + j53.6$
$V_C = 26.8\ \angle -90°$	d. $0.0 - j26.8$

Total voltage = $53.6 + j26.8$ or $60\ \angle 26.56°$

V_T = approximately $60\ \angle 26.56°$ V (polar) or $53.6 + j26.8$ (rectangular)

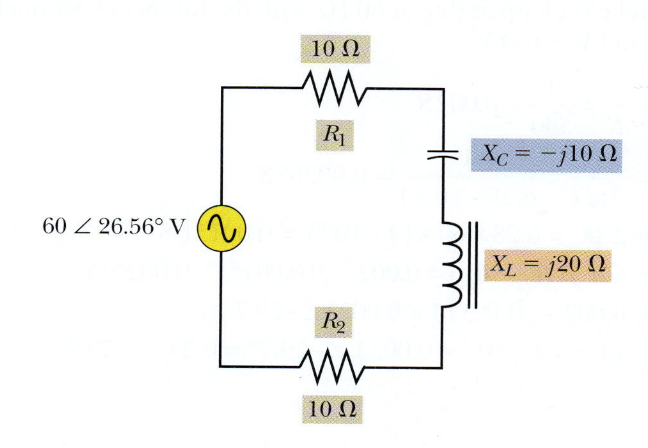

FIGURE 20–26 Sample *RLC* circuit analysis

Sample Parallel *RLC* Circuit Analysis

Remember for parallel circuits in dc, it is easier to use the conductance of branches or to add branch currents, than to combine reciprocal resistances. Recall conductance is the reciprocal of resistance. That is, $G = 1/R$ siemens.

In ac circuits, we have to deal with pure resistance *and* reactance *and* the combination of resistance and reactance (impedance). The comparable term for conductance of a pure reactance is called **susceptance,** where susceptance = $1/\pm X$. The symbol for susceptance is ***B***. The reciprocal of impedance is called **admittance *(Y)*,** which is equal to $1/Z$. Thus:

- For pure resistance, conductance $(G) = 1/R$.
- For pure reactance, susceptance $(B) = 1/\pm X$.
- For combined R and X, admittance $(Y) = 1/Z$.

The phase angle designation for B or Y is opposite that expressed for X or Z, since it is a reciprocal relationship. This means reactance for an inductive branch is $+jX_L$, and inductive susceptance is $-jB_L$. Also, capacitive reactance is $-jX_C$ and capacitive susceptance is jB_C.

FORMULA 20–15 Inductive susceptance: $B_L = \frac{1}{X_L} = \frac{1}{2\pi fL}$

FORMULA 20–16 Capacitive susceptance: $B_C = \frac{1}{X_C} = 2\pi fC$

Notice the inversion of the X_L and X_C formulas. Important admittance relationships are:

$$Y = G \pm jB = G + j(B_C - B_L)$$

FORMULA 20–17 $Y = \sqrt{G^2 + B^2}$

$$\theta = \text{angle whose tangent} = \frac{B}{G}$$

$$\text{NOTE: } Y = \frac{I}{V}$$

$$I = V \times Y$$

$$V = \frac{I}{Y}$$

◆ **EXAMPLE** Refer to Figure 20–27. Use the admittance approach and find the total current of a parallel circuit operating at 60 Hz with the following parameters: $R = 500\ \Omega$; $L = 1$ H; $C = \mu$F; and $V = 100$ V.

$$G = \frac{1}{R} = \frac{1}{500} = 0.002\text{ S}$$

$$B_L = \frac{1}{2\pi fL} = \frac{1}{6.28 \times 60 \times 1} = 0.00265\text{ S}$$

$$B_C = 2\pi fC = 6.28 \times 60 \times (4 \times 10^{-6}) = 0.00151\text{ S}$$

$$Y = G + j(B_C - B_L) = 0.002 + j(0.00151 - 0.00265)$$

$$Y = 0.002 - j0.00114 = 0.0023\angle{-29.7^\circ}\text{ S}$$

$$I_T = VY = 100\angle 0^\circ \times 0.0023\angle{-29.7^\circ} = 0.23\angle{-29.7^\circ}\text{ A}$$ ◆

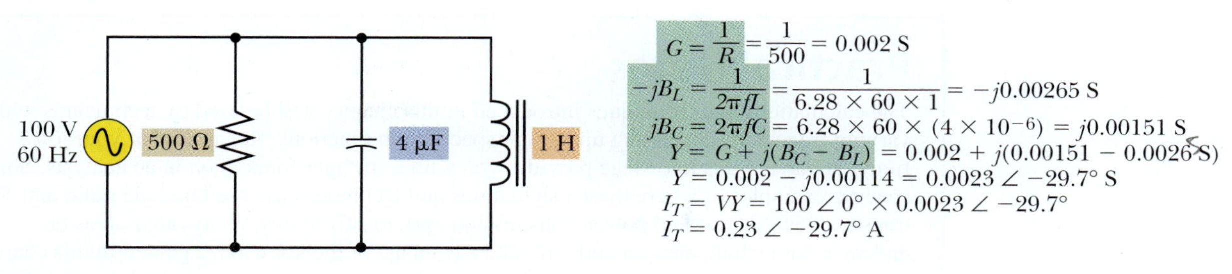

100 V
60 Hz
500 Ω
4 μF
1 H
G = 1/R = 1/500 = 0.002 S
−jB_L = 1/2πfL = 1/(6.28 × 60 × 1) = −j0.00265 S
jB_C = 2πfC = 6.28 × 60 × (4 × 10^−6) = j0.00151 S
Y = G + j(B_C − B_L) = 0.002 + j(0.00151 − 0.00265 S)
Y = 0.002 − j0.00114 = 0.0023 ∠ −29.7° S
I_T = VY = 100 ∠ 0° × 0.0023 ∠ −29.7°
I_T = 0.23 ∠ −29.7° A

FIGURE 20–27 multiSIM

You can calculate the parameters using rectangular forms to add and subtract and polar forms to multiply and divide. When using these notations, it is a good idea to make appropriate conversions so each significant parameter is expressed in both forms. In this way, you use the form most applicable to the calculation you are performing.

Sample Combination Circuit Calculations

Refer to Figure 20–28 and note the following calculations.

1. Find Z_T. Since in a parallel circuit

$$Z_T = \frac{Z_1 Z_2}{Z_1 + Z_2}$$

we need to know both the rectangular and polar forms of each branch impedance. These forms are a good starting point.

$$Z_1 = 10 - j10 \text{ (rectangular) } 14.14\angle{-45^\circ} \text{ (polar)}$$
$$Z_2 = 10 + j20 \text{ (rectangular) } 22.36\angle 63.4^\circ \text{ (polar)}$$
$$Z_T = \frac{Z_1 \times Z_2}{Z_1 + Z_2} = \frac{14.14\angle{-45^\circ} \times 22.36\angle 63.4^\circ}{10 - j10 + 10 + j20} = \frac{316.17\angle 18.4^\circ}{20 + j10}$$
$$Z_T = \frac{316.17\angle 18.4^\circ}{22.36\angle 25.56^\circ} = 14.13\angle{-8.16^\circ}\ \Omega$$

2. Assume 100 V applied voltage at 0° and find the current.

$$\text{Branch 1 current} = \frac{V}{Z_1}$$
$$\text{Branch 1 current} = \frac{100\angle 0^\circ}{14.14\angle -45^\circ} = 7.07\angle 45^\circ\text{ A}$$
$$\text{Branch 2 current} = \frac{V}{Z_2}$$
$$\text{Branch 2 current} = \frac{100\angle 0^\circ}{22.36\angle 63.4^\circ} = 4.47\angle -63.4^\circ\text{ A}$$

3. To find I_T, convert each branch current to rectangular form and add the branch currents.

$I_1 = 7.07\ \angle 45^\circ$ (polar) $\quad 5 + j5$ (rectangular)
$I_2 = 4.47\ \angle{-63.4^\circ}$ (polar) $\quad 2 - j4$ (rectangular)
$I_T = 7.07\ \angle 8.13^\circ$ A (polar) $\quad 7 + j1$ A (rectangular)

4. Individual component voltages are calculated using $I \times R$, $I \times X_L$, and $I \times X_C$, as appropriate. You can practice rectangular and polar notation and appropriate conversion techniques by doing this. If you do this, remember to sketch vector diagram(s), as appropriate, to check the logic of your results.

Practical Notes

The calculations and techniques introduced in this chapter will be used by technicians with varying frequency, depending upon their specific job functions. Regardless of your job function, having this knowlege provides you with a stronger foundation in ac analysis. Some specific examples of where these calculations and techniques are used include radio and TV transmission systems and power transmission systems. Certainly, many other areas of endeavor for technicians can and will take advantage of the knowledge gained in this chapter.

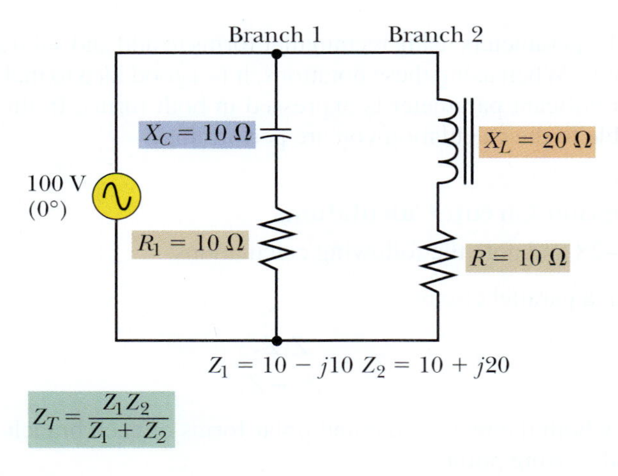

FIGURE 20–28 Sample combination circuit calculations

Summary

- Vectors in opposite directions cancel each other. If one has greater magnitude than the other, there is a net resulting vector equal to the larger value minus the smaller value.
- Power in ac circuits with both resistive and reactive components is described in three ways:
  1. Apparent power (S)—applied voltage times the circuit current, expressed in volt-amperes (VA) and plotted as the hypotenuse of the power triangle
  2. Reactive power (Q) or voltampere-reactive (VAR)—equal to $VI \times \sin\theta$, and plotted on the vertical axis
  3. True power (P) in watts—equal to $VI \cos\theta$, and plotted on the horizontal axis

  Because of the right-triangle relationship:

  $$S = \sqrt{P^2 + Q^2}.$$
- Vector quantities are analyzed in several ways—graphic means, algebraic expressions based on rectangular coordinates, and expressions based on polar coordinates.
- Separating a vector quantity into its components is easily accomplished by simple trigonometric functions. These functions are sine, cosine, and tangent.
- Complex numbers express vector quantities relative to the rectangular notation system. The complex number has a term that defines the horizontal (real or resistive) component and a j term that expresses the reactive component or out-of-phase element. The general term for impedance is $R \pm jX$, for applied voltage is $V_R \pm V_X$, and for total current in a parallel circuit is $I_R \pm jI_X$.
- Vectors expressed algebraically in rectangular coordinates are added, subtracted, multiplied, and divided. However, this form of notation is most easily used to add and subtract. To add, like terms are collected (i.e., resistive terms collected and j terms collected), and added algebraically. To subtract, change the sign of the subtrahend and add algebraically.
- Vectors expressed in polar form *cannot* be added or subtracted algebraically. However, this form is easily used to multiply, divide, and find square roots. To multiply polar expressions, multiply magnitudes and add the angles algebraically. To divide polar expressions, divide the magnitudes and subtract the angles algebraically. To raise a polar expression to a power, raise the magnitude by the given power and multiply the angle by the power. To take a root,

take the root of the magnitude value and divide the angle by the root value.

- Because polar form facilitates multiplication and division and rectangular form expedites addition and subtraction, it is a good idea to convert from one form to the other, as appropriate.
- When two branches are in parallel, with each having resistive and reactive components, the resultant Z is found by the formula $Z = \frac{Z_1 Z_2}{(Z_1 + Z_2)}$
- It is convenient to use the polar form for the numerator (for multiplying) and the rectangular form for the denominator (for adding). Then, convert the denominator to the polar form to allow division into the numerator to get the final answer.
- The ease that current flows through a resistance is called conductance, where $G = 1/R$. A comparable term for reactance is susceptance, where $B = 1/\pm X$. When resistance and reactance are combined in a circuit, the term for ease of current flow is admittance (Y). Admittance is the reciprocal of impedance, where $Y = 1/Z$. The unit of measure for all these terms is siemens (S).
- Some important relationships between admittance, susceptance, and conductance are $Y = G \pm jB$, $Y = \sqrt{G^2 + B^2}$, and θ = angle whose tangent = B/G.

Formulas and Sample Calculator Sequences

FORMULA 20–1
(To find the impedance of a series RLC circuit)

$Z = \sqrt{R^2 + X^2}$ (Series *RLC* circuits)

R value, [x^2], [+], X_{net}, [x^2], [=], [$\sqrt{x}$]

FORMULA 20–2
(To find the total voltage in a series RLC circuit)

$V_T = \sqrt{V_R^2 + V_X^2}$ (Series *RLC* circuits)

V_R value, [x^2], [+], V_X value, [x^2], [=], [$\sqrt{x}$]

FORMULA 20–3
(To find the tangent of the phase angle in a series RLC circuit where reactances and R values are known)

tangent of $\theta = \frac{X \text{ (net reactance)}}{R}$ (Series *RLC* circuits)

X (net reactance) value, [÷], R value, [=]

FORMULA 20–4
(To find the tangent of the phase angle in a series RLC circuit where reactive and resistive voltages are known)

tangent of $\theta = \frac{V_X}{V_R}$ (Series *RLC* circuits)

$V_{X(net)}$ value, [÷], V_R value, [=]

FORMULA 20–5
(To find the total current in a parallel RLC circuit)

$I_T = \sqrt{I_R^2 + I_X^2}$ (Parallel *RLC* circuits)

I_R value, [x^2], [+], $I_{X(net)}$, [x^2], [=], [$\sqrt{x}$]

FORMULA 20–6
(To find circuit power factor)

Power factor ($p.f.$) = cos θ

Angle value, [COS]

FORMULA 20–7
(To find circuit apparent power)

Apparent power (S) = $V \times I$

V value, [×], I value, [=]

FORMULA 20–8
(To find circuit true power when current and resistance are known)

True power (P) = I^2R

I value, [x^2], [×], R value, [=]

FORMULA 20–9
(To find circuit true power when apparent power and power factor are known)

P = Apparent power × power factor

Apparent power value, [×], power factor value, [=]

FORMULA 20–10
(To find circuit power factor when true power and apparent power are known)

$$p.f. = \frac{\text{True power } (P)}{\text{Apparent power } (S)}$$

True power value, [÷], apparent power value, [=]

NOTE: In series circuits $p.f. = \cos\theta = R/Z$. In parallel circuits $p.f. = \cos\theta = I_R/I_T$.

FORMULA 20–11
(To find circuit apparent power when true power and net reactive power are known)

$$S = \sqrt{P^2 + Q^2}$$

True power value, [x^2], [+], net reactive power value, [x^2], [=], [$\sqrt{x}$]

FORMULA 20–12
(To convert an impedance value from polar to rectangular form notation for a series circuit)

$Z\angle\theta° = (Z\cos\theta) + (jZ\sin\theta)$ (For series circuits)

USING THE SCIENTIFIC CALCULATOR CONVERSION TECHNIQUE:

Z value, [x⇄y], angle value (θ), [2nd], [P▸R], [x⇄y]

Answer is combination of last two readouts; that is, [P▸R] gives the real term and [x⇄y] gives the imaginary (j) term.

FORMULA 20–13
(To convert total voltage value from polar to rectangular form notation for a series circuit)

$V_T\angle\theta° = V_T\cos\theta + jV_T\sin\theta$ (For series circuits)

V_T value, [x⇄y], angle value (θ), [2nd], [P▸R], [x⇄y]

Answer is combination of last two readouts; that is, [P▸R] gives the real term and [x⇄y] gives the imaginary (j) term.

FORMULA 20–14
(To convert total current value from polar to rectangular form notation for a parallel circuit)

$I_T\angle\theta° = I_T\cos\theta + jI_T\sin\theta$ (For parallel circuits)

I_T value, [x⇄y], angle value (θ), [2nd], [P▸R], [x⇄y]

Answer is combination of last two readouts; that is, [P▸R] gives the real term and [x⇄y] gives the imaginary (j) term.

FORMULA 20–15
(To find the value of inductive susceptance for a given inductance at a given frequency)

Inductive susceptance: $B_L = \dfrac{1}{X_L} = \dfrac{1}{2\pi fL}$

6.28, [×], frequency value, [×], inductance value, [=], [1/x]

FORMULA 20–16
(To find the value of capacitive susceptance for a given capacitance at a given frequency)

Capacitive susceptance: $B_C = \dfrac{1}{X_C} = 2\pi fC$

6.28, [×], frequency value, [×], capacitance value, [=]

FORMULA 20–17
(To find the value of circuit admittance when conductance and susceptance are known)

$$Y = \sqrt{G^2 + B^2}$$

Value of conductance (G), [x^2], [+], value of susceptance (B), [x^2], [=], [$\sqrt{x}$]

EXCEL AUTOMATED FORMULAS

Helpful problem-solving tools, such as the Excel automated formulas available on the CD, allow automatic calculations of formulas.

Using Excel

RLC Circuit Analysis Formulas

(Excel file reference: FOE20_01.xls)

DON'T FORGET! *It is not necessary to retype formulas once they are entered on the worksheet! Just input new parameters data for each new problem using that formula, as needed.*

- Use the Formula 20–2 spreadsheet sample and the data from Figure 20–9. Find total voltage for the circuit.
- Use the Formula 20–15 spreadsheet sample and the data from Figure 20–27. Solve for the value of inductive susceptance for the circuit.

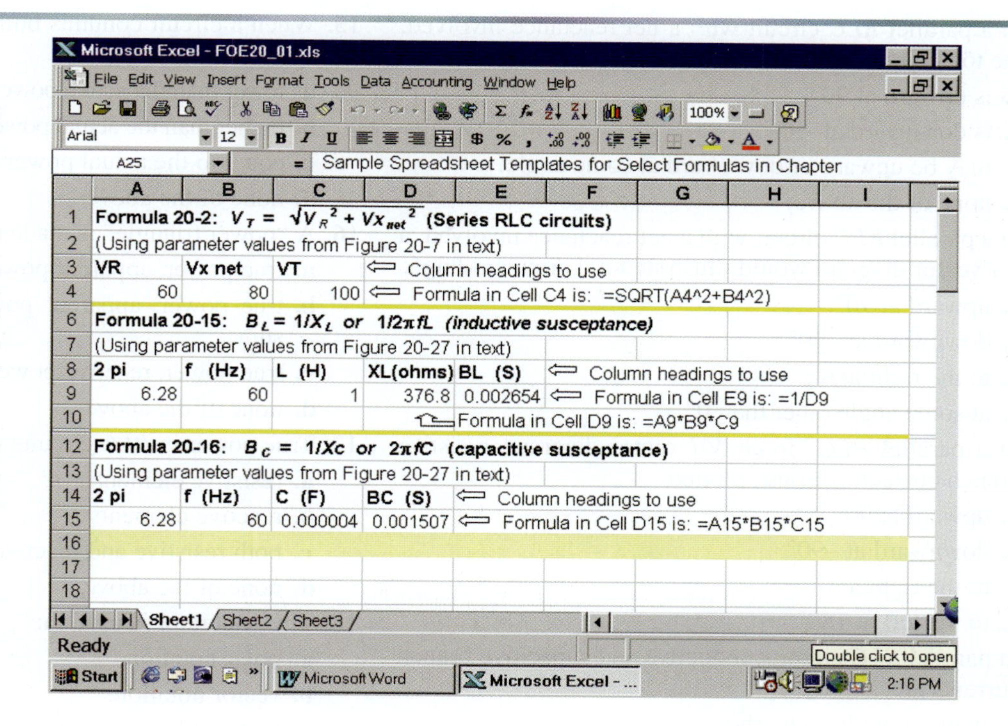

Review Questions

1. In a series *RLC* circuit *V-I* vector diagram, voltage across capacitors is plotted:
   a. upward at 90°.
   b. downward at –90°.
   c. to the right at 0°.
   d. to the left at 180°.
2. In a series *RLC* circuit *V-I* vector diagram, circuit current is plotted:
   a. upward at 90°.
   b. downward at –90°.
   c. to the right at 0°.
   d. to the left at 180°.
3. In a series *RLC* circuit with unequal reactances, applied voltage for a *V-I* vector diagram is plotted:
   a. upward at 90°.
   b. downward at –90°.
   c. to the right at 0°.
   d. none of the above.
4. In a series *RLC* circuit impedance diagram, *R* is plotted:
   a. upward at 90°.
   b. downward at –90°.
   c. to the right at 0°.
   d. none of the above.
5. In a series *RLC* circuit impedance diagram, capacitive reactance is plotted:
   a. upward at 90°.
   b. downward at –90°.
   c. to the right at 0°.
   d. none of the above.
6. In a series *RLC* circuit with unequal reactances, impedance is plotted:
   a. upward at 90°.
   b. downward at –90°.
   c. to the right at 0°.
   d. none of the above.
7. In a parallel *RLC* circuit *V-I* vector diagram, inductor branch current(s) is(are) plotted:
   a. upward at 90°.
   b. downward at –90°.
   c. to the right at 0°.
   d. to the left at 180°.
8. In a parallel *RLC* circuit *V-I* vector diagram, capacitor branch current(s) is(are) plotted:
   a. upward at 90°.
   b. downward at –90°.
   c. to the right at 0°.
   d. to the left at 180°.

9. In a parallel *RLC* circuit with a net reactance involved, the total circuit current plot in a *V-I* vector diagram:

a. is upward at 90°.

b. is downward at –90°.

c. may be upward or downward at some angle.

d. none of the above.

10. In a parallel *RLC* circuit with a net reactance involved, a *V-I* vector diagram would illustrate total circuit voltage:

a. upward at 90°.

b. downward at –90°.

c. to the right at 0°.

d. at some angle other than 0°.

11. In a parallel *RLC* circuit *V-I* vector diagram, resistive branch current(s) is(are) plotted:

a. upward at 90°.

b. downward at –90°.

c. to the right at 0°.

d. to the left at 180°.

12. In parallel *RLC* circuits, inductive and capacitive branch currents are:

a. in phase with each other.

b. 90° out of phase with each other.

c. 180° out of phase with each other.

d. 270° out of phase with each other.

13. For a purely resistive circuit in ac, an instantaneous power graph would show that power is dissipated:

a. only on the positive alternation from the source.

b. only on the negative alternation from the source.

c. during both alternations.

d. none of the above.

14. For ideal reactive devices and a purely reactive circuit, an ac instantaneous power graph would show that average power dissipated over a cycle equals:

a. 2 × Power in.

b. $V \times I$.

c. zero.

d. none of the above.

15. When a circuit contains both resistances and reactances, apparent power is:

a. lower than the actual power dissipated by the circuit.

b. higher than the actual power dissipated by the circuit.

c. equal to the actual power dissipated by the circuit.

d. none of the above.

16. A "power triangle" is made up of:

a. true power, apparent power, and $V \times I$.

b. true power, apparent power, and their mathematical sum.

c. true power, reactive power, and apparent power.

d. none of the above.

17. True power in *RLC* circuits is dissipated by:

a. resistive elements.

b. reactive elements.

c. both resistive and reactive elements.

d. none of the above.

18. The *j* operator is used in:

a. polar notation.

b. vector notation.

c. rectangular notation.

d. none of the above.

19. The advantage of using the polar form of notation is that:

a. it uses the *j* operator to great advantage.

b. it gives the same values for electrical quantities that measuring instruments show.

c. it is easier to write down than the rectangular notation.

d. it is easy to add and subtract with this type of notation.

20. One advantage of rectangular notation over polar notation is that:

a. it is easier to multiply and divide vector quantities in rectangular form.

b. it is easier to add and subtract vector quantities in rectangular form.

c. it provides the angle within its notation.

d. none of the above.

Problems

(Be sure to make appropriate vector diagram sketches to help clarify your thinking.)

1. Refer to Figure 20–29a. Find Z, I, V_C, V_L, V_R, and θ using the Pythagorean theorem, Ohm's law, and appropriate trig functions. Draw impedance and *V-I* vector diagrams, labeling all parts, as appropriate.
2. Refer to Figure 20–29b. Find Z, I_C, I_L, I_R, I_T, and θ using the Pythagorean theorem, Ohm's law, and appropriate trig functions. Draw an appropriate *V-I* vector diagram and label all parts, as appropriate.
3. Calculate true power, apparent power, and reactive power for the circuit of question 1. What is the power factor?
4. Draw a power triangle for the circuit in question 1.
5. A series circuit has a 20-Ω resistance and a 40-Ω inductive reactance. How do you state the circuit Z in rectangular form? If the inductive reactance is changed to an equal value of capacitive reactance, how do you state Z?
6. A series *RLC* circuit is composed of 10-Ω resistance, 20-Ω inductive reactance, and 30-Ω capacitive reactance. How is the circuit Z stated in rectangular form?
7. A parallel circuit with a 20-V source has a resistive branch of 20 Ω and a capacitive reactance branch of 20 Ω. What is the total current stated in polar form? See Figure 20–30.
8. Convert $30 + j40$ to polar form.
9. Convert $14.14 \angle 45°$ to rectangular form.
10. A series *RLC* circuit has a 10-V, 1-kHz source, an R of 8 kΩ, an X_L of 6 kΩ, and an X_C of 4 kΩ. What is the total circuit current? (Express in polar form.) What is the true power?
11. What is the total impedance of a series *RLC* circuit that has a 20-kHz ac source, $C = 100$ pF, $L = 0.159$ H, and $R = 15$ kΩ? (Express in polar form and in rectangular form.)
12. A series *RLC* circuit has 100 kΩ of resistance, a capacitor with 0.0005 μF of capacity, and an inductor with 1.5 H of inductance. What is the value of voltage drop across the inductor if the applied ac voltage is 100 V at a frequency of 10 kHz? (Express in polar form.)
13. If the frequency of the applied voltage increases, what happens to V_C?
14. What is the true power dissipated by a series *RLC* circuit with 100 V applied, if $R = 30$ Ω, $X_L = 40$ Ω, and $X_C = 10$ Ω?

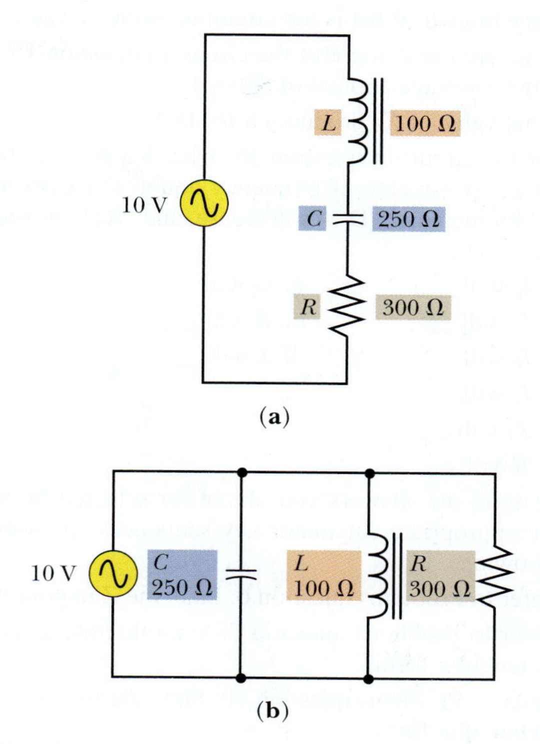

FIGURE 20–29

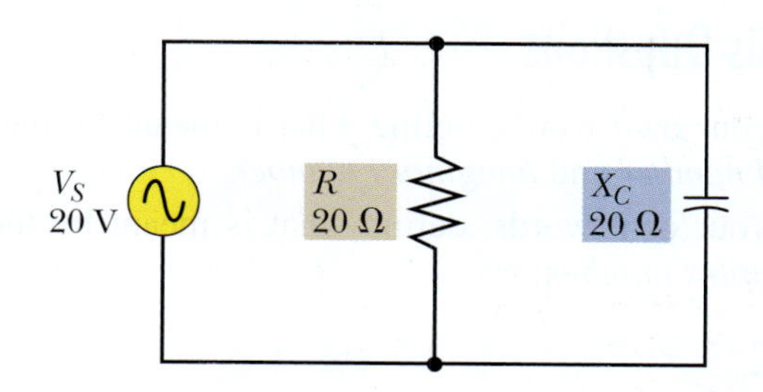

FIGURE 20–30

15. A parallel *RC* circuit has a capacitor with 2 Ω of X_C and a resistor with 4 Ω of resistance. What is the impedance? (Express in polar form.)
16. A parallel *RC* circuit has a power factor of 0.5 and $I_T =$ 5 A, $f = 1$ kHz, and $R = 60$ Ω. What is the value of C?
17. What is the total impedance of a parallel *RLC* circuit comprised of $R = 6$ Ω, $X_L = 12$ Ω, and $X_C = 24$ Ω?

18. A parallel *RLC* circuit has 400-Ω resistance, 500-Ω inductive reactance branch, and 800-Ω capacitive reactance branch. What is the circuit admittance value?
19. What are the Z and θ of the circuit in question 18? (Assume a voltage applied of 800 V.)
20. What value of X_C produces a $\theta = 0°$?
21. For the circuit in question 18, what happens to the following parameters if frequency doubles? (Answer with "I" for increase, "D" for decrease, and "RTS" for remain the same.)
   a. I_R will ___
   b. I_C will ___
   c. I_L will ___
   d. I_T will ___
   e. Z_T will ___
   f. θ will ___
   g. G will ___
   h. B will ___
   i. Y will ___

Referring to the answers you obtain for selected Problems, use the appropriate calculator key sequences and solve for the following.

22. Refer to Problems question 6. State the Z in polar form.
23. Refer to Problems question 7. State the total current in rectangular form.
24. Refer to Problems question 10. State the total current in rectangular form.
25. Refer to Problems question 15. State the impedance in rectangular form.
26. Referring to Figure 20–31 and using rectangular and polar notation techniques, find Z_T, and express your answer in polar form.
27. Express the Z_T of the circuit in Figure 20–31 in rectangular form.
28. Find I_T for the circuit of Figure 20–31 and express your answer in polar form.
29. Express the I_T value for the circuit of Figure 20–31 in rectangular form.

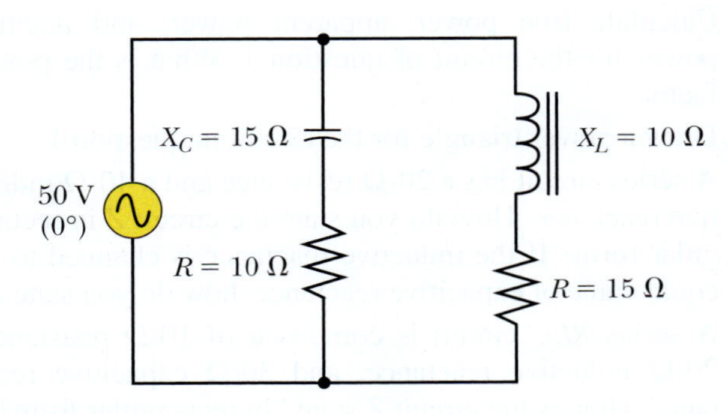

FIGURE 20–31

Analysis Questions

1. In your own words, define what is meant by the terms *real number* and *imaginary number.*
2. In your own words, define what is meant by the term *complex number.*
3. Explain why using measurements of a circuit's voltage with a voltmeter and its current with a current meter does not provide the necessary information to find the actual power dissipated by the circuit.
4. Explain how you can make measurements to determine the actual power dissipated by an ac circuit containing inductors and/or capacitors.

Troubleshooting Challenge

CHALLENGE CIRCUIT 12

(Follow the SIMPLER sequence by referring to inside front cover.)

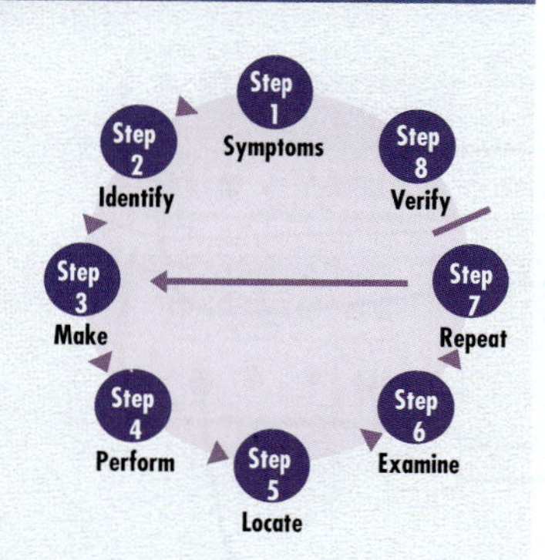

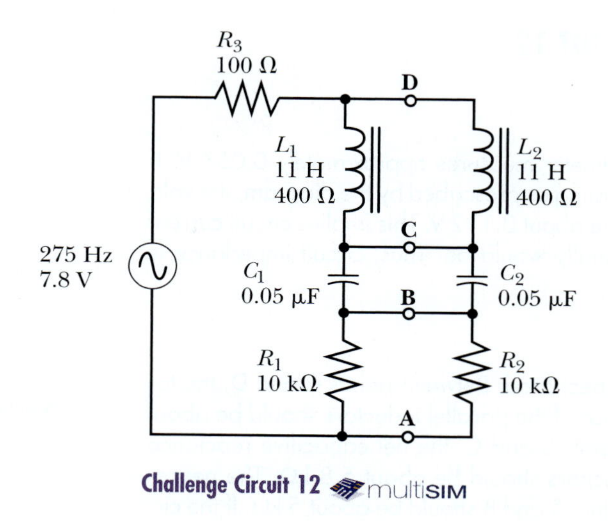

Challenge Circuit 12 multiSIM

STARTING POINT INFORMATION

1. Circuit diagram
2. $V_{R_3} = 0.057$ V

TEST	Results in Appendix C
$V_{A\text{-}B}$	(78)
$V_{B\text{-}C}$	(38)
$V_{C\text{-}D}$	(96)
$R_{A\text{-}B}$	(104)
$R_{B\text{-}C}$	(63)
$R_{C\text{-}D}$	(32)

CHALLENGE CIRCUIT 12

STEP 1

SYMPTOMS The voltmeter measures approximately 0.057 V. If the circuit were operating as prescribed by the diagram, the voltmeter should measure about 0.112 V. This implies circuit current is lower than it normally would be. Thus, circuit impedance is higher than normal.

STEP 2

IDENTIFY initial suspect area: Between points C and D, the total inductive reactance of the parallel inductors should be about 9.5 kΩ. Between points B and C, the net capacitive reactance of the parallel capacitors should be about 5.8 kΩ. The net resistance between points A and B should be about 5 kΩ. If the circuit were operating correctly, the total impedance would be about 6.2 kΩ. However, it's acting like the circuit has about 14 kΩ total impedance. Suspect area includes the whole circuit.

STEP 3

MAKE test decision: The upward shift in Z could be caused by any R, C, or L opening or any one C or L shorting. We'll first look at the resistors. Let's look at $V_{A\text{-}B}$.

STEP 4

PERFORM **1st Test:** Test $V_{A\text{-}B}$. $V_{A\text{-}B}$ is about 3.11 V, again indicating lower-than-normal circuit current. We know the Rs are not shorted or the voltage would read zero volts. Also, since we know the circuit current is about 0.57 mA, then 0.57 mA times the normal equivalent R of approximately 5 kΩ comes close to that value. The resistors appear to be acting normally.

STEP 5

LOCATE new suspect area: The Rs have been eliminated. Thus, the reactive components are the new suspect area.

STEP 6

EXAMINE available data.

STEP 7

REPEAT analysis and testing: Let's look at the inductive branches, isolating as appropriate.

2nd Test: Check $V_{C\text{-}D}$. $V_{C\text{-}D}$ is about 11 V. An open inductor could cause this situation so let's check that out.

3rd Test: Check $R_{C\text{-}D}$. $R_{C\text{-}D}$ is about 400 Ω. This should be about 200 Ω with the parallel circuit situation.

4th Test: Check R_{L_1}. R_{L_1} is about 400 Ω, which is normal.

5th Test: Check R_{L_2}. R_{L_2} is infinite ohms, which is an open coil.

STEP 8

VERIFY **6th Test:** Change L_2 and check the circuit operation. It's normal.

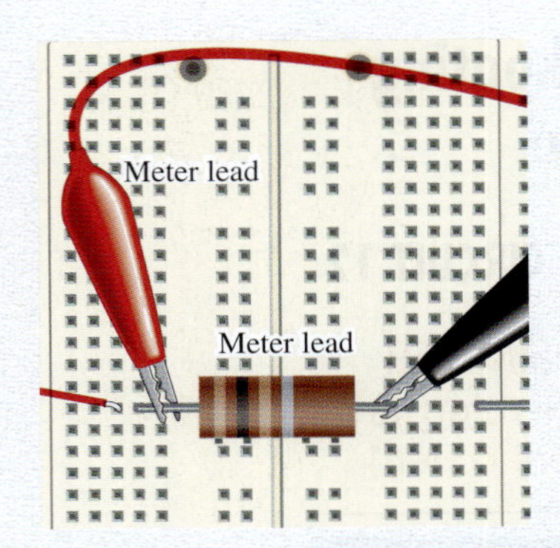

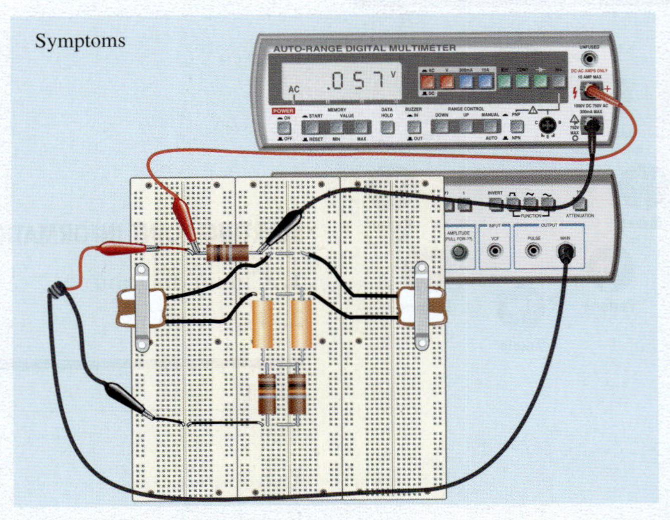

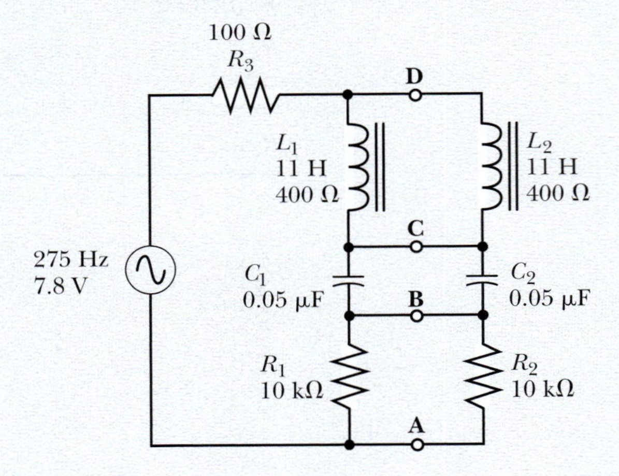

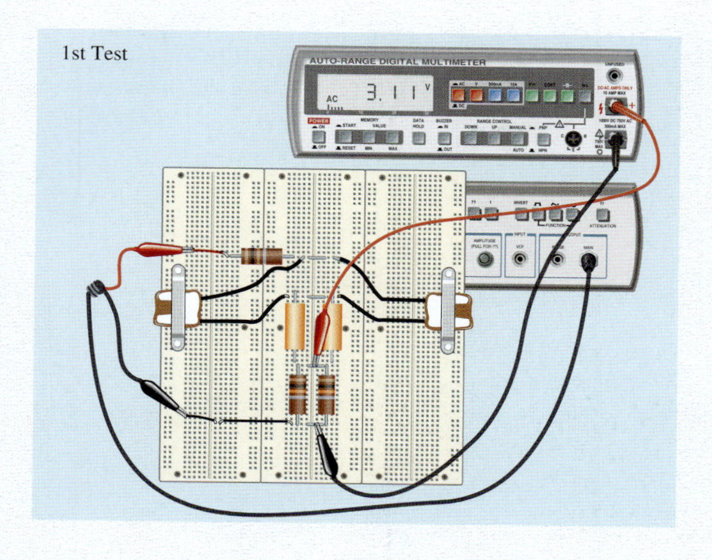
1st Test
AUTO-RANGE DIGITAL MULTIMETER
3.11

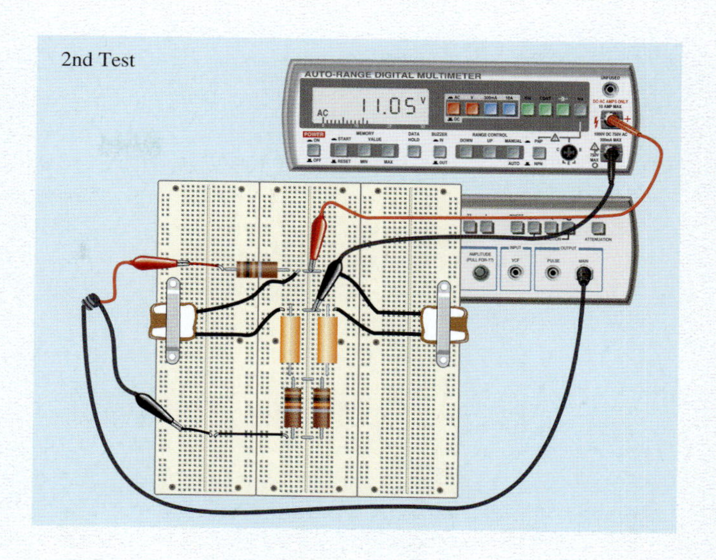
2nd Test
AUTO-RANGE DIGITAL MULTIMETER
11.05

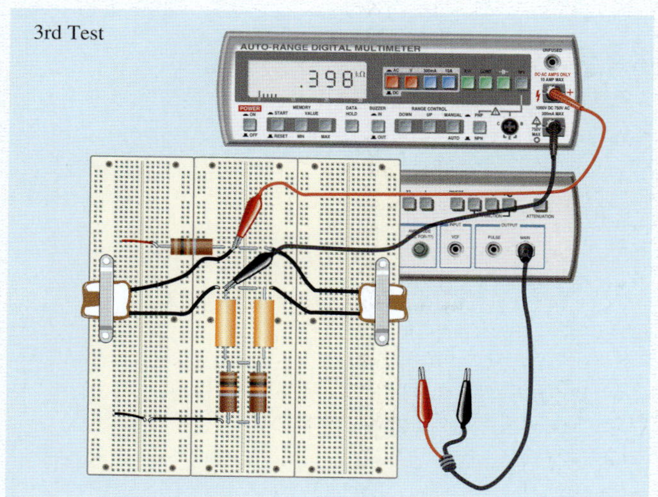
3rd Test
AUTO-RANGE DIGITAL MULTIMETER
.398

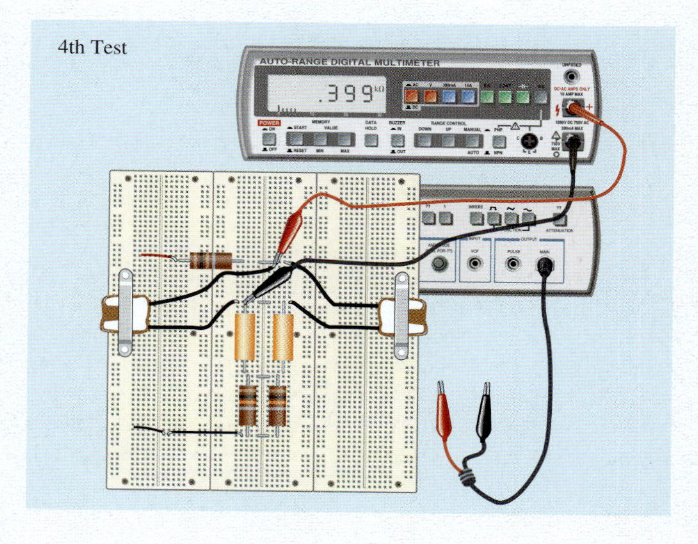
4th Test
AUTO-RANGE DIGITAL MULTIMETER
.399

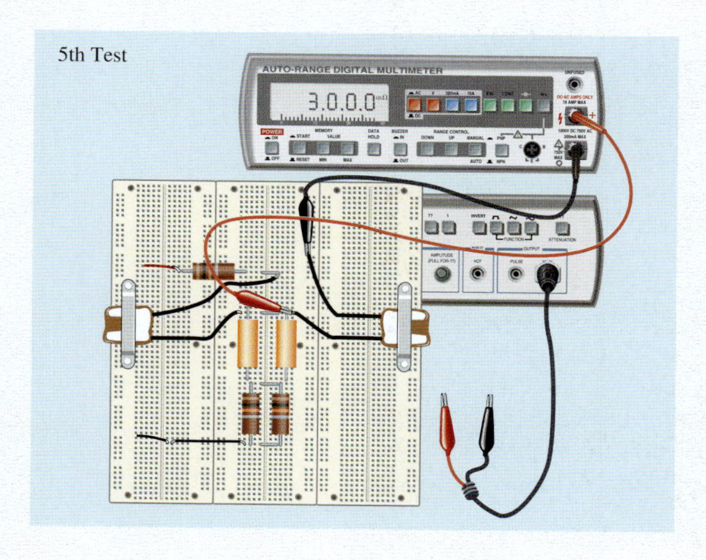
5th Test
AUTO-RANGE DIGITAL MULTIMETER
3.0.0.0

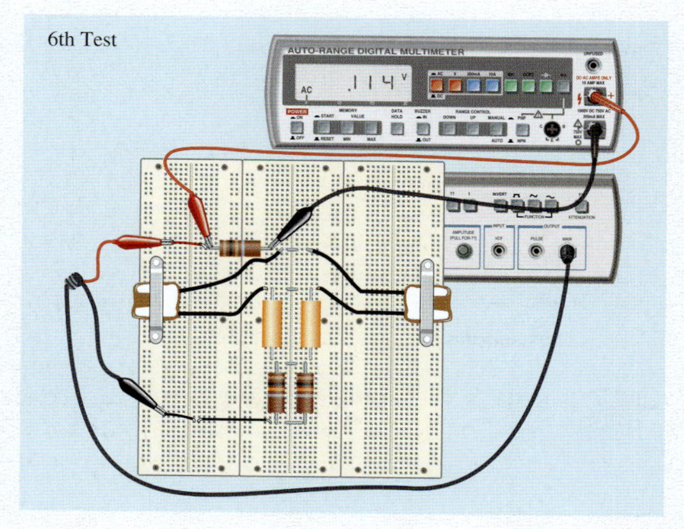
6th Test
AUTO-RANGE DIGITAL MULTIMETER
.114

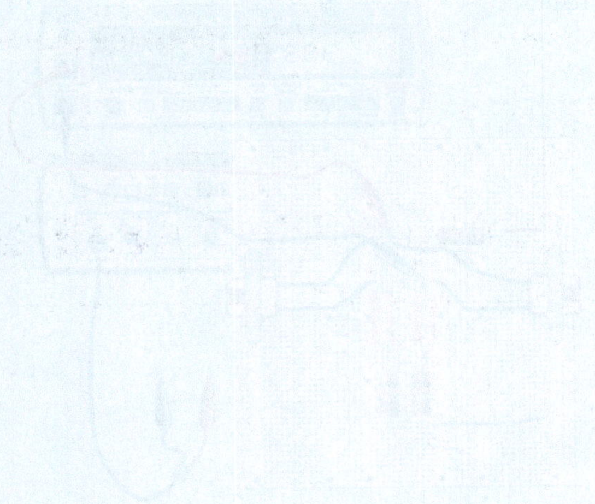

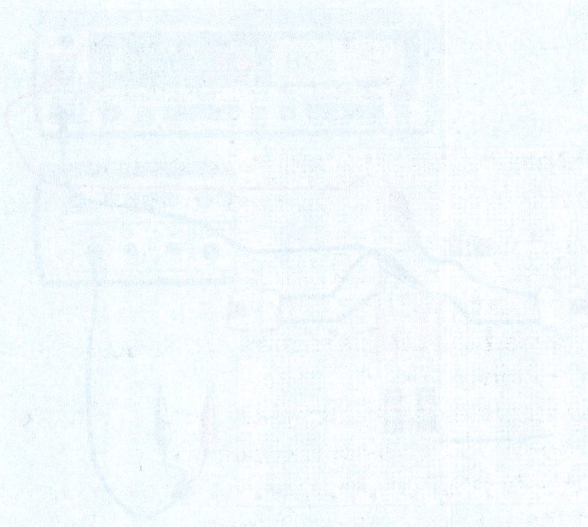

OUTLINE

OBJECTIVES

After studying this chapter, you should be able to:

1. List the key characteristics of series and parallel resonant circuits
2. Calculate the resonant frequency of circuits
3. Calculate ***L*** or ***C*** values needed for **resonance** at a given f_r
4. Calculate the ***Q* factor** for series and parallel resonant circuits
5. Determine **bandwidth** and **bandpass** of resonant circuits
6. Draw circuit diagrams for three types of **filters**
7. Use the computer to solve circuit problems
8. Use the SIMPLER troubleshooting sequence to solve the Troubleshooting Challenge problems

CHAPTER 21

Series and Parallel Resonance

PREVIEW

Series and parallel resonant circuits provide valuable capabilities in numerous electronic circuits and systems. When you select a radio or TV station, you are using the unique capability of a resonant circuit to select a narrow group of frequencies from a wide spectrum of frequencies that impact the receiver's antenna system. The transmitter that is transmitting the selected signals also has a number of "tuned" resonant circuits that assure only the desired frequencies are being transmitted and that spurious signals are not being transmitted. Both the transmitting and receiving antenna systems are further examples of special tuned or resonant circuits.

In addition to the valuable tuning application, resonance allows a related application called *filtering*. Filtering can involve either passing desired signals or voltages, or blocking undesired ones. Tuning and filtering are the most frequent applications for series and parallel resonance.

In this chapter, you will study the key characteristics of both series and parallel resonant circuits. Your knowledge of these circuits will be frequently used throughout your career.

KEY TERMS

Bandpass
Bandpass filter
Bandstop filter
Bandwidth
High-pass filter
L/*C* ratio
Low-pass filter
Q factor
Resonance
Selectivity
Skirts

21–1 X_L, X_C, and Frequency

Previously you learned that as frequency increases, X_L increases proportionately. The formula for inductive reactance expresses this by $X_L = 2\pi fL$. You also learned that as frequency increases, X_C decreases in inverse proportion, expressed in the formula

$$X_C = \frac{1}{2\pi fC}$$

Conversely, as frequency decreases, X_L decreases and X_C increases. In other words, X_L and X_C change in opposite directions as frequency changes.

To review, refer to Figure 21–1 and note the relationships between X_L, X_C, and frequency. Notice X_C is not illustrated at frequencies approaching 0 Hz. This is because X_C is approaching infinite ohms at these frequencies, and therefore is difficult to illustrate on such a graph.

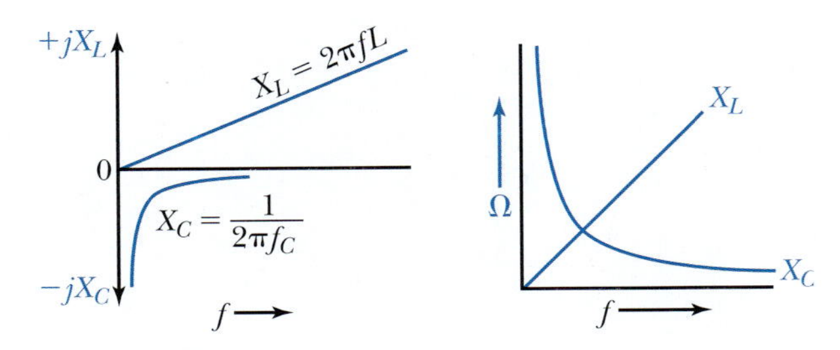

FIGURE 21–1 Relationships between X_L, X_C, and frequency

21–2 Series Resonance Characteristics

The implication of these statements about the relationships of X_L, X_C, and frequency is that for a given set of L and C values, there is a specific frequency where the *absolute* magnitudes of X_L and X_C are equal. Look at Figure 21–2 and notice that for this series *RLC* circuit, there is one frequency where this occurs. In a series circuit containing L and C, the condition when the magnitude of X_L equals the magnitude of X_C is called **resonance** and the frequency where this occurs is the *resonant frequency*. At resonance, some interesting observations can be made about circuit electrical parameters.

Refer to Figure 21–3 as you study the following list of features.

The characteristics of a series *RLC* circuit at resonance are as follows:

1. $X_L = X_C$ (In this case, 100 Ω = 100 Ω.)
2. Z = minimum and equals R in circuit. (In this case, 10 Ω.)
3. I = maximum and equals V/R. (In this case, 50 mV/10 Ω = 5 mA.)
4. V_C and V_L are maximum. ($I \times X_C$ = 500 mV; $I \times X_L$ = 500 mV.)
5. $\theta = 0°$ *(Since circuit is acting purely resistive.)*

Let's examine these statements.

1. When X_L equals X_C, the reactances cancel, leaving 0 Ω of net reactance in the circuit. The remaining opposition to current, under these conditions, is any resistance in the circuit. Generally, most of this resistance is the resistance in the inductor windings (unless we have purposely added an additional discrete resistor component). The coil resistance is shown in Figure 21–3 as r_S, indicating series resistance.

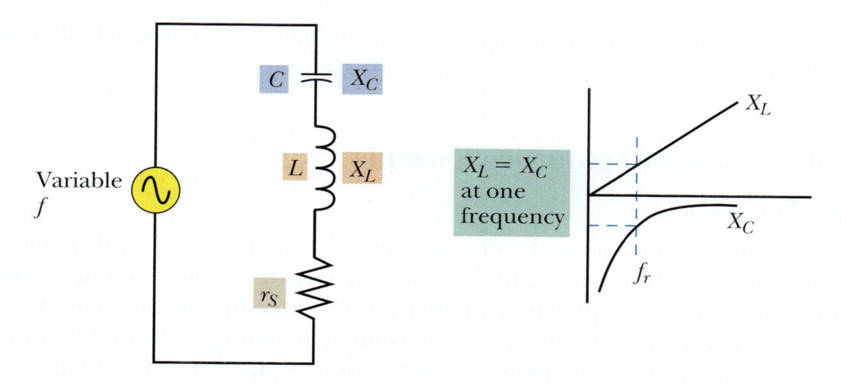

FIGURE 21–2 $X_L = X_C$ at resonant frequency (f_r)

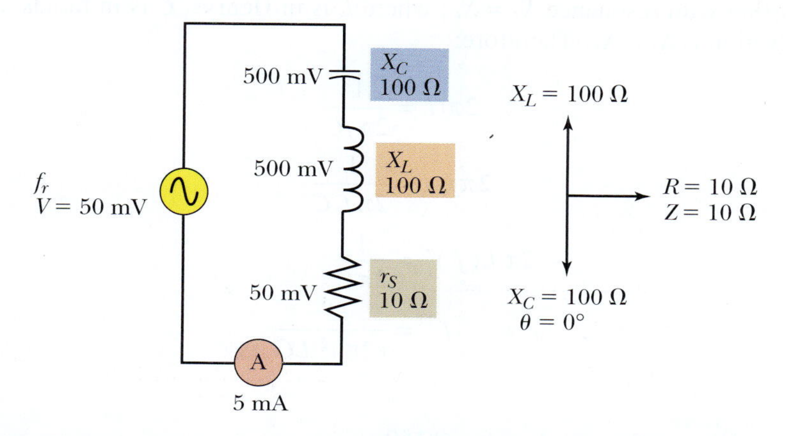

FIGURE 21–3 Sample series circuit parameters at resonance multiSIM

2. Because the reactances cancel each other, the Z is minimum and equals the R in the circuit.
3. When Z is minimum ($Z = R$), I is maximum. That is, $I = V/R$.
4. V_C and V_L are maximum since I is limited only by the R value and is therefore maximum. The voltage across the capacitor equals $I \times X_C$, and the voltage across the inductor equals $I \times X_L$. As noted in our sample circuit, the voltage across the coil and the capacitor is 10 times V applied. There is a *resonant rise* in voltage across the reactive components. We will examine this fact again later in the chapter.
5. Again, because the reactances cancel each other, the circuit acts purely resistive; therefore, phase angle (θ) is zero.

Since you now know some of the series resonance effects, you may wonder how the frequency of resonance is determined.

Let's deviate from examining circuit parameters to look at the basic formulas that allow you to determine the resonant frequency for a circuit with given component values, or to find the appropriate component values for a desired resonant frequency.

IN-PROCESS LEARNING CHECK 1

Fill in the blanks as appropriate.

1. In circuits containing both inductance and capacitance, when the source frequency increases, inductive reactance _____ and capacitive reactance _____.
2. In a series *RLC* (or *LC*) circuit, when X_L equals X_C, the circuit is said to be at _____.
3. In a series *RLC* (or *LC*) circuit, when X_L equals X_C, the circuit impedance is _____; the circuit current is _____; and the phase angle between applied voltage and current is _____ degrees.
4. In a series *RLC* (or *LC*) circuit, when X_L equals X_C, the voltage across the inductor or the capacitor is _____.

21–3 The Resonant Frequency Formula

The Basic Formula

In our previous discussions, a fixed value of L and C were assumed, while f was changed until X_L equaled X_C. With fixed L and C values, there is only one possible resonant frequency. In many practical applications, either the C or L value or both can be varied to achieve resonance at various frequencies. An example is tuning in different radio stations on a receiver. Also, different ***L/C* ratios** for any given frequency can be obtained. You can see that a technician should know how to determine the resonant frequency of any given set of *LC* parameters. Let's look briefly at the derivation and formula for frequency of resonance. Starting with resonance $X_L = X_C$, where L is in Henrys, C is in farads, and f is in Hz, the derivation is $X_L = X_C$. Therefore:

$$2\pi f L = \frac{1}{2\pi f C}$$

$$2\pi f_r L = \frac{1}{2\pi f_r C}$$

$$2\pi L (f_r)^2 = \frac{1}{2\pi C}$$

$$f_r^2 = \frac{1}{(2\pi)^2 LC}$$

FORMULA 21–1 $f_r = \dfrac{1}{2\pi\sqrt{LC}}$ or $\dfrac{0.159}{\sqrt{LC}}$

EXAMPLE What is the resonant frequency of a series *LC* circuit having a 200-μH inductor and a 400-pF capacitor?

$$f_r = \frac{1}{2\pi\sqrt{LC}}$$

$$f_r = \frac{1}{6.28 \times \sqrt{200 \times 10^{-6} \times 400 \times 10^{-12}}}$$

$$f_r = \frac{1}{6.28 \times \sqrt{80{,}000 \times 10^{-18}}}$$

$$f_r = \frac{1}{1{,}776 \times 10^{-9}} = 563.06 \text{ kHz or approximately } 563 \text{ kHz}$$

Now you try one!

PRACTICE PROBLEMS 1

What is the resonant frequency of a series *LC* circuit containing an inductor with an inductance of 250 μH and a capacitor with a capacitance of 500 pF?

Variations of the f_r Formula

As you can see from the basic resonant frequency formula, the *L* and *C* values determine the frequency of resonance. That is, the larger the *LC* product is, the lower is the f_r, and vice versa, Figure 21–4a. Note also in Figure 21–4b how various combinations of *L* and *C* can be resonant at one particular frequency (in this case, 1 MHz).

It is frequently useful to find the *L* value needed with an existing *C* value or the *C* value needed with an existing *L* value to make the circuit resonant at some specified frequency. By rearranging the resonant frequency formula (i.e., squaring both sides to delete the square root portion), we can find *L* or *C*, if the other two variables are known. When this is done, the following useful formulas are derived:

$$L = \frac{1}{(2\pi)^2 f_r^2 C} = \frac{1}{4\pi^2 f_r^2 C} =$$

FORMULA 21–2 $L = \dfrac{0.02533}{f_r^2 C}$

(a)

L (μH)	C (pF)	f_r (MHz)
100	100	1.59
100	200	1.13
400	400	0.398
400	10	2.52

$\left(\text{Using } f_r = \dfrac{1}{2\pi\sqrt{LC}} \text{ OR } \dfrac{0.159}{\sqrt{LC}}\right)$

Different *LC* products cause different f_rs

(b)

L (μH)	C (pF)	f_r (MHz)
500	51	1
400	63.75	1
300	85	1
200	127.5	1

$\left(\text{Using } f_r = \dfrac{1}{2\pi\sqrt{LC}} \text{ OR } \dfrac{0.159}{\sqrt{LC}}\right)$

Different values having same *LC* product yield same f_r

FIGURE 21–4 (a) Values of *L* and *C* determine f_r and (b) different combinations of *L* and *C* can produce same f_r

$$C = \frac{1}{(2\pi)^2 f_r^2} = \frac{1}{4\pi^2 f_r^2 L} =$$

FORMULA 21–3 $C = \dfrac{0.02533}{f_r^2 L}$

NOTE: These formulas are for your information. You need not memorize them!

Other Useful Variations

When dealing in radio frequency circuits, it is sometimes inconvenient to manage Henrys, farads, and hertz, as in the previous formulas. It is often more convenient to deal with microhenrys (μH), picofarads (pF), and megahertz (MHz). The following formulas are based on these convenient units.

$$f^2 = \frac{25{,}330}{LC} \text{ or}$$

FORMULA 21–4 $L = \dfrac{25{,}330}{f^2 C}$

where L is in microhenrys, C is in picofarads, f is in megahertz, or

FORMULA 21–5 $C = \dfrac{25{,}330}{f^2 L}$

NOTE: Again, these formulas are for your information, but you may find occasion to use them.

EXAMPLE A circuit has an L value of 100 μH. What C value will make the circuit resonant frequency equal 1 MHz?

$$\text{Using } C = \frac{0.02533}{f_r^2 L}$$

where L is in Henrys, C in farads, and f in hertz.

$$C = \frac{0.02533}{(1 \times 10^6)^2 \times 100 \times 10^{-6}}$$

$$C = \frac{0.02533}{1 \times 10^{12} \times 100 \times 10^{-6}}$$

$$C = \frac{25{,}330 \times 10^{-12}}{100} = 253.3 \text{ pF}$$

$$\text{Using } C = \frac{25{,}330}{f^2 L}$$

where L is in microhenrys, C is in picofarads, and f is in megahertz.

$$C = \frac{25{,}330}{f^2 L} = \frac{25{,}330}{100} = 253.3 \text{ pF}$$

Incidentally, you can also use the $X_C = X_L$ approach. Since you know the frequency and the L, you can calculate X_L using the $X_L = 2\pi fL$ formula. Then, knowing that X_C is that same value at resonance, you use the X_C formula to solve for C. That is,

$$C = \frac{1}{2\pi f X_C}$$

See Figure 21–5.

Using basic units: L = H, C = F, and f_r = Hz
$L = \frac{0.02533}{f_r^2 C}$ and $C = \frac{0.02533}{f_r^2 L}$
Using convenient units: L = μH, C = pF, and f_r = MHz
$L = \frac{25{,}330}{f_r^2 C}$ and $C = \frac{25{,}330}{f_r^2 L}$
Using the $X_L = X_C$ approach:
If you know X_C, then $L = \frac{X_C}{2\pi f_r}$
If you know X_L, then $C = \frac{1}{2\pi f_r X_L}$

FIGURE 21–5 Some useful variations derived from the f_r and reactance formulas

Now you apply the appropriate formula(s) to solve for an unknown L when C and f_r are known.

PRACTICE PROBLEMS 2

What L value is needed with a 100-pF capacitor to make the circuit resonant frequency equal 5 MHz?

21–4 Some Resonance Curves

Figure 21–6 shows the resonance curves (graphic plots) of current, impedance, and phase angle versus frequency for a series RLC circuit.

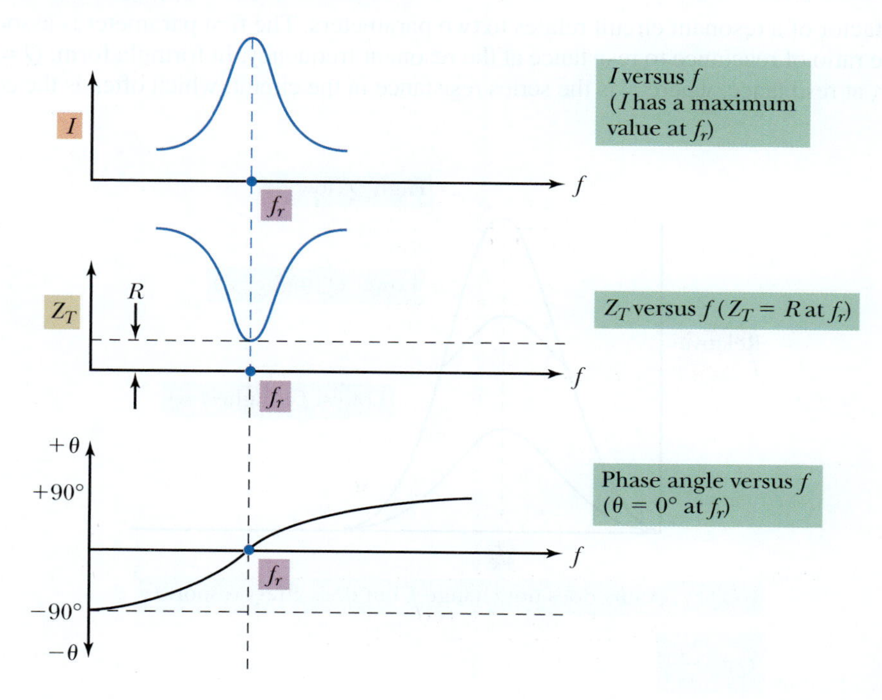

FIGURE 21–6 Plots of I, Z, and θ versus frequency

Plot of Current *(I)* versus Frequency *(f)*

When frequency is considerably below the circuit resonant frequency, the current is low compared with current at resonance. This is because $z = \sqrt{r^2 + (X_C - X_L)^2}$ and the reactances do not completely cancel, leaving a net reactance value; therefore, circuit impedance *(Z)* is much higher than it is at resonance. For series *RLC* circuits below resonance, the net reactance is capacitive reactance. The same pattern is true above resonance, except the net reactance for a series *RLC* circuit is X_L. (NOTE: For series resonant circuits, plotting current rising to maximum value at resonance is the most significant aspect of the response curves.)

Plot of Impedance *(Z)* versus Frequency

As you would expect, impedance is lowest at the resonant frequency, since the reactances cancel each other at resonance. The impedance graph in Figure 21–6 shows that *Z* equals *R* (resistance) at resonance. Thus, for any circuit, when *Z* is low, *I* is higher and when *Z* is high, *I* is lower.

Plot of Phase Angle Versus Frequency

At resonance, the circuit acts resistively and the phase angle is 0°. (This is because the reactances have cancelled.) Below resonance the circuit acts capacitively, since the net reactance is capacitive reactance. Above resonance, the circuit acts inductively since the net reactance is X_L.

21–5 *Q* and Resonant Rise of Voltage

Q and the Response Characteristic

The shape of a given circuit's resonance response relates to an important factor called the ***Q* factor,** or figure of merit, Figure 21–7. The higher the *Q* factor is, the sharper is the response characteristic. This sharp response curve is sometimes described as having steep **skirts,** or as indicating a circuit with high **selectivity** (i.e., one that is very frequency selective). The lower the *Q* factor is, the broader, or flatter the response curve and the less selective the circuit is.

Q Equals X_L/R or X_C/R

The *Q* factor of a resonant circuit relates to two parameters. The first parameter is associated with the ratio of reactance to resistance at the resonant frequency. In formula form, $Q = X_L/r_S$ or X_C/r_S at resonance, where r_S is the series resistance in the circuit, which often is the coil re-

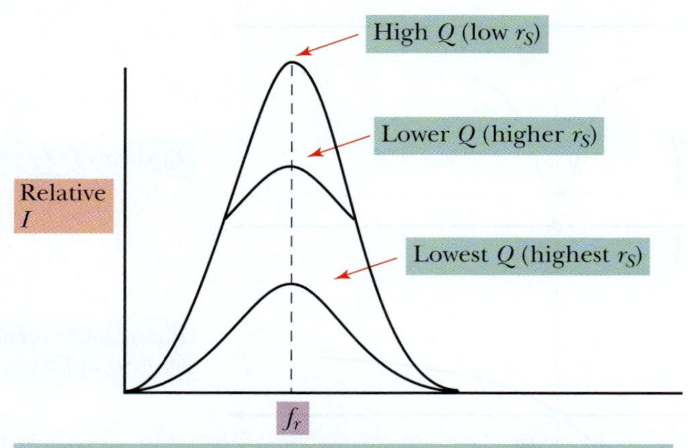

FIGURE 21–7 Response curves versus circuit *Q*

sistance. In other words, the Q of the coil (X_L over R) is frequently the same as the Q of the resonant circuit. Obviously, the lower the series circuit resistance is, the higher the Q is and the sharper the resonant circuit response. The higher the r_s is, the lower the Q is and the flatter the response. Note that since I is the same throughout the circuit, the ratio

$$\frac{I \times X_L}{I \times R}, \text{ or } \frac{V_L}{V_R}$$

is also the same as the Q factor. Likewise, the ratio of reactive power to true power in the circuit also equals the Q factor.

Q Related to L/C Ratio

A second related factor that affects the Q is the ratio of L to C; that is L/C. Since both L and C values affect the frequency of resonance, a *given* resonant frequency can be achieved by various LC values, refer again to Figure 21–4b. However, different combinations of LC values have different reactances at resonance. Therefore, it is best to have a *high L/C* ratio to achieve a high X_L at resonance; hence, a high X_L/R or Q factor, Figure 21–8.

Resonant Rise of Voltage

Refer to Figure 21–9 and note the voltages around the series circuit. The term *magnification factor* describes the fact that voltage across each of the reactive components at resonance is Q times V applied. This is because of the resonant rise of I at resonance, which is due to the reactances cancelling each other at the resonant frequency. Since I is limited only by R at that frequency, and since each reactive voltage drop is I times X with X being very high, the reactive voltage can be very high. In fact, at resonance, the reactive voltages are Q times V applied; thus, voltage has been magnified (Q times) across the reactive components. In other words, the higher the Q is, the greater the magnification is. As stated, the ratio of reactive voltage to applied voltage at resonance equals the Q. If we measure the voltage across one reactive component and call it V_{out} and call V applied V_{in}, then

$$Q = \frac{V_{out}}{V_{in}}$$

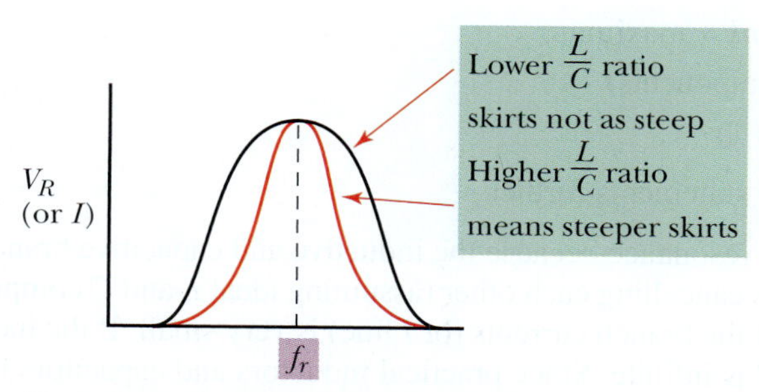

FIGURE 21–8 Q related to L/C ratios

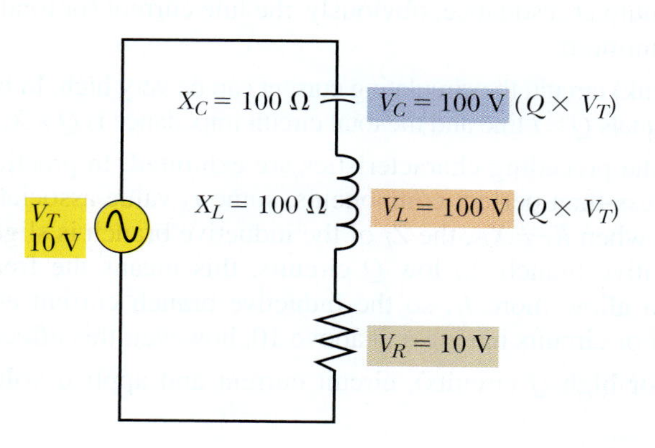

FIGURE 21–9 Magnification factor multiSIM

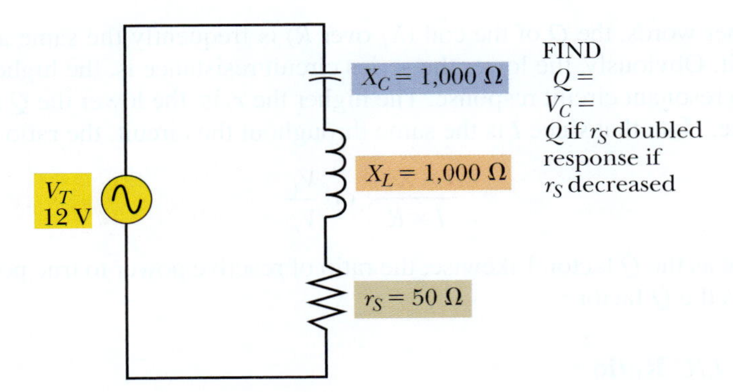

FIGURE 21–10

PRACTICE PROBLEMS 3

1. What is the Q of the circuit in Figure 21–10?
2. What is the V_C value at resonance, if the applied voltage in Figure 21–10 is 12 V?
3. If r_S doubles in this circuit, what is the Q value?
4. If r_S is a smaller value than shown in Figure 21–10, are the response curve skirts more steep or less steep?

21–6 Parallel Resonance Characteristics

A parallel resonant circuit is a frequently used circuit in receivers, transmitters, and frequency measuring devices. As you might expect, many parallel resonant circuit features are opposite to those of a series resonant circuit. Look at Figure 21–11 as you study the following discussion.

Highlighting the basic features of a high-Q parallel resonant circuit:

1. $Z =$ maximum (and is purely resistive)
2. Line current = minimum
3. Current inside *tank* = maximum
4. $X_L = X_C$ (ideal components)
5. $\theta = 0°$ (and *p.f.* = 1)

Let's examine these statements further.

1. Z is maximum at resonance because the inductive and capacitive branch currents are 180° out of phase, thus cancelling each other (assuming ideal L and C components). This means the vector sum of the branch currents (or I line) is very small. If the inductor and capacitor are perfect, the Z is infinite. Since practical inductors and capacitors have resistances and losses, there is an effective R involved, Figure 21–11. (For your information, $Z_T = L/C_{r_S}$.)
2. Since Z is maximum at resonance, obviously, the line current (or total current supplied by the source) is minimum.
3. Within the LC (tank) circuit, the circulating current can be very high. In fact, the current inside the tank circuit equals $Q \times I$ line and the total circuit impedance is $Q \times X_L$.
4. When $X_L = X_C$, the preceding characteristics are exhibited. In practical parallel resonant circuits having less-than-perfect components, some r_S value associated with the coil exists. This means when $X_L = X_C$, the Z_L of the inductive branch is slightly greater than the X_C of the capacitive branch. In low-Q circuits, this means the frequency needs to be slightly lower to allow more I_L, so the inductive branch current equals the capacitive branch current. For circuits having a Q above 10, however, this effect is negligible.
5. At resonance (for high-Q circuits), circuit current and applied voltage are virtually in phase.

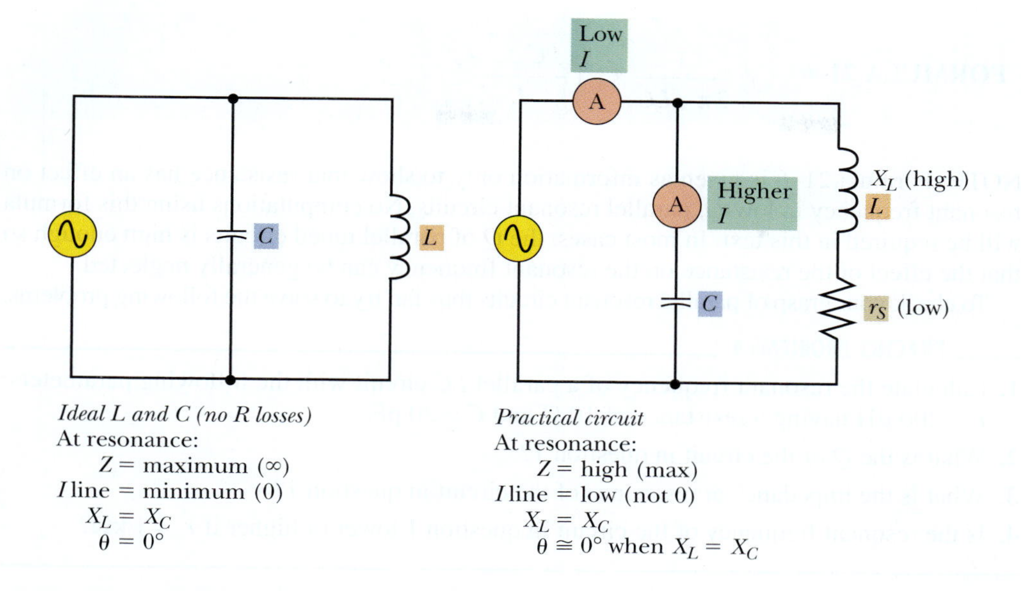

FIGURE 21–11 Parallel resonant circuit characteristics

multiSIM

Practical Notes

For low-Q circuits, the condition of unity power factor occurs at a slightly different frequency than when $X_L = X_C$. The frequency of $p.f. = 1$ is sometimes termed antiresonance, or the antiresonant frequency.

IN-PROCESS LEARNING CHECK 2

1. In a high-Q parallel resonant circuit, Z is _____.
2. In a high-Q parallel circuit using ideal components, X_L _____ equal X_C at resonance.
3. The power factor of an ideal parallel resonant circuit is _____.
4. Phase angle in an ideal parallel resonant circuit is _____.
5. In a parallel resonant circuit, the current circulating in the tank circuit is _____ than the line current.

21–7 Parallel Resonance Formulas

For Q Greater Than 10

When the circuit Q is high (when the r_S is low, or the equivalent R_p is high), the resonant frequency is determined by the LC component values as it is in series resonant circuits. With Qs greater than 10, the resonant frequency is calculated with the same formula used for series resonance. That is,

$$f_r = \frac{1}{2\pi\sqrt{LC}}$$

For Q Less Than 10

When the circuit Q is low (i.e., when r_S is high or effective R_p is low, or when a reflected impedance from a load coupled to the tuned circuit is low causing an effective low value of equivalent R_p), the R, L, and C values enter in the formula for parallel resonance. That is, the resistance has a significant effect on the frequency of resonance so it should be considered. The following formula (only for your information) illustrates that resistance affects resonance in parallel LC circuits.

FORMULA 21–6 $f_r \frac{1}{2\pi\sqrt{LC}} \times \sqrt{1-\frac{C r_S^{2}}{L}}$

NOTE: Formula 21–6 is given as information only to show that resistance has an effect on resonant frequency in low-Q, parallel resonant circuits. No computations using this formula will be required in this text. In most cases, the Q of parallel tuned circuits is high enough so that the effect of the resistance on the resonant frequency can be generally neglected.

To check your grasp of parallel resonant circuits thus far, try to solve the following problems.

PRACTICE PROBLEMS 4

1. Calculate the resonant frequency of a parallel LC circuit with the following parameters: $L = 200\ \mu\text{H}$ having a resistance of $10\ \Omega$ and $C = 20$ pF.
2. What is the Q of the circuit in question 1?
3. What is the impedance at resonance of the circuit in question 1?
4. Is the resonant frequency of the circuit in question 1 lower or higher if $r_S = 1\ \text{k}\Omega$?

21–8 Effect of a Coupled Load on the Tuned Circuit

Whenever a parallel resonant circuit delivers energy to a load (i.e., a load is coupled to the tuned circuit), the Q of the tank circuit is affected, Figure 21–12. If the power dissipated by the coupled load is greater than 10 times the power lost in the tank circuit reactive components, the tank impedance is high compared to the load impedance. Thus, for all practical purposes, the impedance of the parallel combination of the tank and load virtually equals the load resistance value. When a tank is heavily loaded by such a resistive load impedance, the Q of the system equals R over X, or $Q = R/X$, where R is the parallel load resistance and X is the reactance. (NOTE: This is the inverse of the X_L/R formula used for Q in series resonance and high-Q parallel resonant circuits.) For example, if a resistive load of 4,000 Ω is connected across a parallel resonant circuit where X_C (or X_L) = 400 Ω,

$$Q = \frac{R}{X_L} = \frac{4,000}{400} = 10$$

21–9 Q and the Resonant Rise of Impedance

The preceding discussions show us that the current characteristic is the most meaningful response factor in series resonant circuits, where the impedance characteristic is the most significant response factor in parallel resonant circuits. The response characteristics of a parallel tuned circuit are also related to the Q factor. In Figure 21–13 you can see there is a resonant rise of impedance as the resonant frequency is approached. As we previously stated, the impedance at resonance is $Q \times X_L$. You can also see from this illustration that the higher the Q

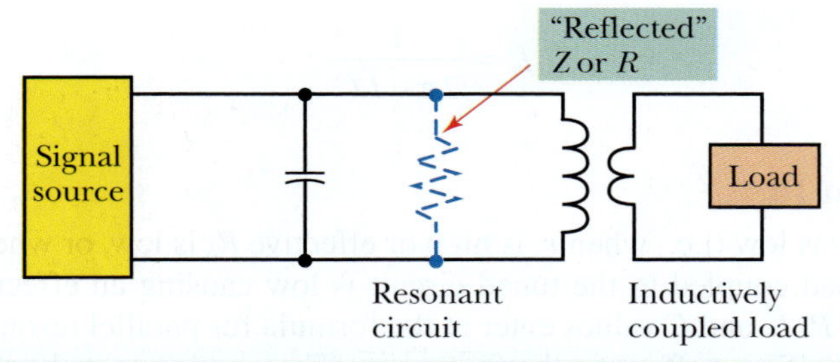
"Reflected" Z or R
Signal source
Load
Resonant circuit
Inductively coupled load
The lower the reflected R value, the lower the Q. If load is "heavy" (i.e., much energy absorbed by load) $Z_T \approx$ Z of load and $Q = \frac{R_p}{X_L}$.

FIGURE 21–12 Effect of coupling a load to a parallel resonant circuit

is, the sharper the response characteristics, and the lower the Q is, the flatter and broader the response.

21–10 Selectivity, Bandwidth, and Bandpass

Selectivity

The sharper the response curve of a resonant circuit, the more selective it is. This means that only frequencies close to resonance appear near the maximum response portion of the resonant circuit response. This indicates that in series resonant circuits, the current value *falls off* rapidly from its maximum value at frequencies slightly below or above the resonant frequency. In parallel resonant circuits, the impedance *falls off* rapidly from the maximum value obtained at resonance. Conversely, resonant circuits that have less selectivity allow a wider band of frequencies close to their maximum response levels. Selectivity, used to describe a resonant circuit, refers to the capability of the circuit to differentiate between frequencies. Refer to Figure 21–14 for examples of various selectivities versus response characteristics.

Bandwidth and Bandpass

You have looked at various response curves and have noticed that none have perfect vertical skirts. Therefore, to define **bandwidth,** some value or level on the response curve is chosen that indicates all frequencies below that level are rejected and frequencies above the chosen value are selected. The accepted level is 70.7% of maximum response level.

For series resonant circuits, we consider those two points on the response curve where current is 70.7% of maximum. For parallel resonant circuits, the points where Z is 70.7% of maximum are used.

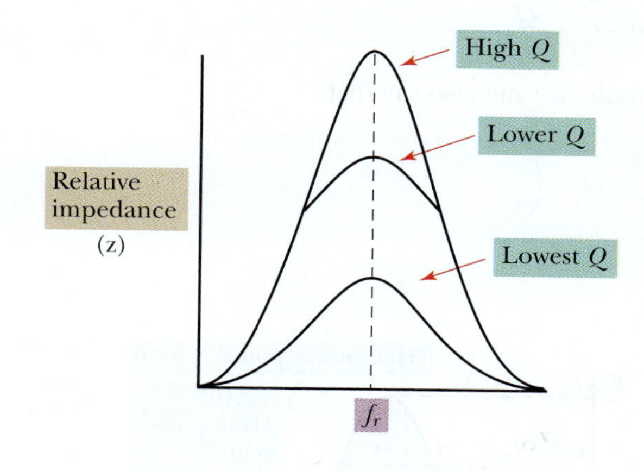

FIGURE 21–13 Resonant rise of Z for a parallel resonant circuit

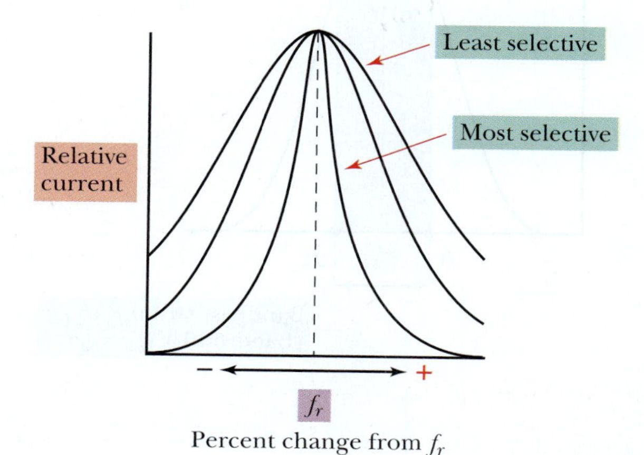

FIGURE 21–14 Examples of various selectivities

Total bandwidth *(BW)* is calculated as:

FORMULA 21–7 $\Delta f = f_2 - f_1$

where: $\Delta f = BW$
f_2 = higher frequency 0.707 point
f_1 = lower frequency 0.707 point on response curve

Also, $f_1 = f_r - \frac{\Delta f}{2}$, and $f_2 = f_r + \frac{\Delta f}{2}$.

Notice that if $P = I^2R$ and at f_1 and f_2 the current level is 0.707 of maximum, then at those points on the response curve, $P = (0.707\ I_{max})^2R = 0.5\ P_{max}$. For this reason, these points (f_1 and f_2) are often called *half-power points,* Figure 21–15.

Practical Notes

The response curves are not perfectly symmetrical above and below resonance. Therefore, the f_1 and f_2 expressions are approximations. They are close to true for very high Q circuits.

Bandwidth Related to *Q*

You have learned the Q of a resonant circuit is the primary factor determining the response curve shape. Thus, BW (Δf) and Q are related. The formula showing this relationship is:

FORMULA 21–8 $\Delta f = \frac{f_r}{Q}$

Rearranging this formula, we can also say that:

FORMULA 21–9 $Q = \frac{f_r}{\Delta f}$

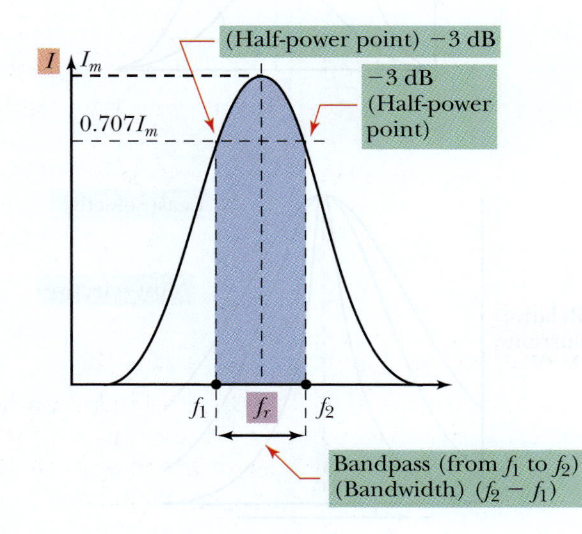

FIGURE 21–15 Bandwidth and bandpass

Parallel Resonant Circuit Damping

Parallel resistance (R_p) can be used to "damp" or lower the Q, and broaden the bandwidth of a parallel resonant circuit.

Recall that series resistance (R_s), typically in the inductor branch, lowers Q. The higher the series resistance present, the lower the Q. In contrast, however, when parallel resistance is present across a parallel resonant circuit (R_p), the lower the parallel resistance, the lower the Q for the circuit. This parallel resistance may be from reflected impedance loading effects (mentioned earlier), or from purposely placing a resistance in parallel with the circuit, for damping purposes.

Bandpass

As has been discussed, the bandwidth is the width of the band of frequencies where a given resonant circuit responds within specified criteria. For example, a given resonant circuit responds over a 20-kHz bandwidth with levels at or above 70.7% of maximum response level.

On the other hand, **bandpass** is the specific frequencies at the upper and lower limits of the bandwidth. For example, if the resonant frequency is 2,000 kHz, and the circuit has a bandwidth of 20 kHz, the bandpass is from 1,990 to 2,010 kHz (assuming a symmetrical response). This indicates the specific band of frequencies passed is from 1,990 to 2,010 kHz, and frequencies outside this limit cause a response that falls below the 70.7% level, or below the half-power points (sometimes expressed as the –3 dB points, as you will learn later). Again, Figure 21–15 illustrates bandwidth and bandpass.

EXAMPLE

1. What is the total bandwidth of a resonant circuit if f_r is 1,000 kHz and f_1 is 990 kHz?
   Since one half-power point is 10 kHz below the resonant frequency, it is assumed that if the response curve is symmetrical, the other half-power point is 10 kHz above the resonant frequency. This means the total $BW = 20$ kHz. That is, $f_1 = 990$ kHz and $f_2 = 1{,}010$ kHz.
2. What is the bandwidth of a resonant circuit if resonant frequency is 5 kHz and Q is 25?

$$\Delta f = \frac{f_r}{Q} = \frac{5{,}000}{25} = 200 \text{ Hz}$$

3. What is the bandpass of the circuit described in question 2?

Bandpass is from 5,000 – 100 Hz to 5,000 + 100 Hz = 4,900 to 5,100 Hz.

PRACTICE PROBLEMS 5

1. In a resonant circuit, f_r is 2,500 kHz and Δf is 12 kHz. What are the f_1 and f_2 values?
2. In a resonant circuit, Q is 50 and bandwidth is 10 kHz. What is the resonant frequency and the bandpass?
3. What is the f_1 value in a circuit whose resonant frequency is 3,200 kHz and whose bandwidth is 2 kHz?

21–11 Measurements Related to Resonant Circuits

Because resonant circuits are so prevalent in electronics, it is worthwhile to know some of their practical measurements. These measurements include measuring resonant frequency; determining Q and magnification factor; determining maximum current (for series resonant circuits) and maximum impedance (for parallel resonant circuits); determining bandwidth and bandpass; measuring the *tuning range* of a tunable resonant circuit (with variable *C* or *L*); and determining the tuning ratio. Let's briefly look at each of these measurements.

1. Measuring resonant frequency
   There are several ways to determine resonant frequency. Of course, if you are trying to measure at what frequency a tunable signal source is operating (what frequency it is tuned to), a digital frequency counter coupled to the output circuit might be used for direct frequency measurement, Figure 21–16. One way to get an approximate resonant frequency measurement of a "stand-alone" *LC* circuit is to use a metering device that reacts to circuit conditions. One type of metering device is called a tunable "dip" meter. This type of device is used with "power-off" conditions to determine the resonant frequency for a given *LC* circuit. When these tunable devices are near (usually inductively coupled) to *LC* circuits, as the tuned circuit in the dip meter is tuned near to the resonant frequency of the *LC* circuit, the *LC* circuit absorbs energy from the dip meter circuit and causes the meter reading to "dip" downward. The frequency at which the dip occurred is then read from the calibrated dial on the dip meter. Other means may include field strength meters or oscilloscopes to monitor frequency outputs of signal sources. It is imperative that the technician be aware of the possibility of "detuning" the circuit under test, and use the proper test probes and/or coupling methods with the scope, or other test equipment to minimize detuning effects.

   Keep in mind the fact that there are various means for checking resonant frequencies of circuits. As you continue in your studies, you will probably have an opportunity to get some practical experience in using some of these devices, and you will learn the proper techniques for making these measurements.

2. Determining *Q* and magnification factor
   In a series resonant circuit, using a high-input impedance DVM, or an oscilloscope allows the measurement of voltage across one reactive component (generally, the capacitor). When the circuit is tuned to resonance, the voltage is maximum. This voltage is then compared with the voltage source to determine the *Q* and the voltage magnification factor. That is,

$$Q = \frac{V_{\text{out}}}{V_{\text{in}}}$$

   See Figure 21–17.

3. Determining maximum current *(I)* for series and maximum impedance *(Z)* for parallel resonant circuits
   In a series resonant circuit, a current meter or a voltmeter reading across a series resistor is used to determine maximum current. Read the current meter directly, or use Ohm's law where

$$I = \frac{V_R}{R}$$

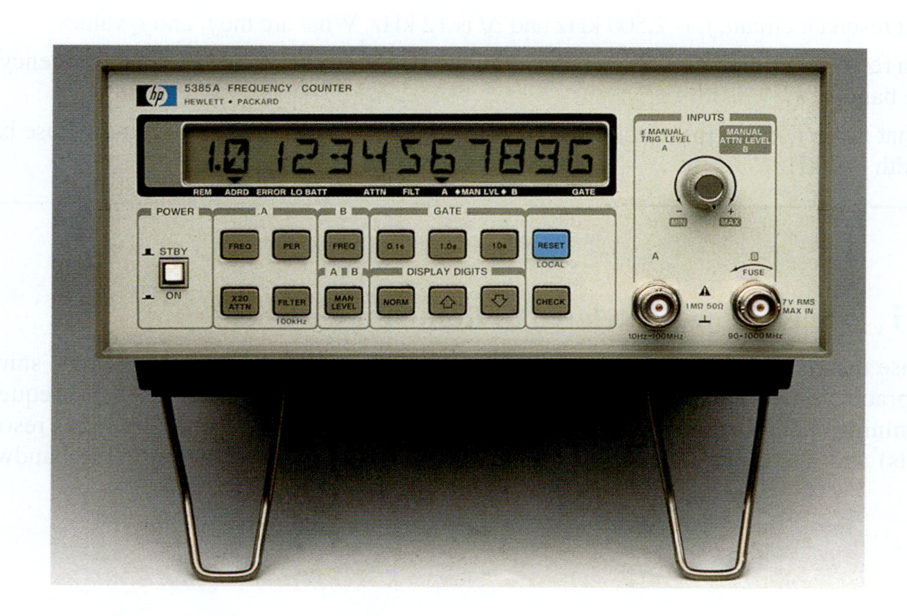

FIGURE 21–16 A device used to determine f_r *(Courtesy of Hewlett-Packard Co.)*

to determine current when the circuit is tuned for maximum current or resonance, Figure 21–18.

In a parallel resonant circuit, use a variable R in series with the source and adjust it so that V_{TANK} across the tuned circuit and V_R across the variable R are equal (at resonance). Remove the R from the circuit and measure its value. Its value is the same as the Z of the tuned circuit at resonance since the V drops were equal. Since we can calculate X_L from the resonant frequency formula and we know that

$$Q = \frac{Z_T}{X_L}$$

we can determine the Q, Figure 21–19.

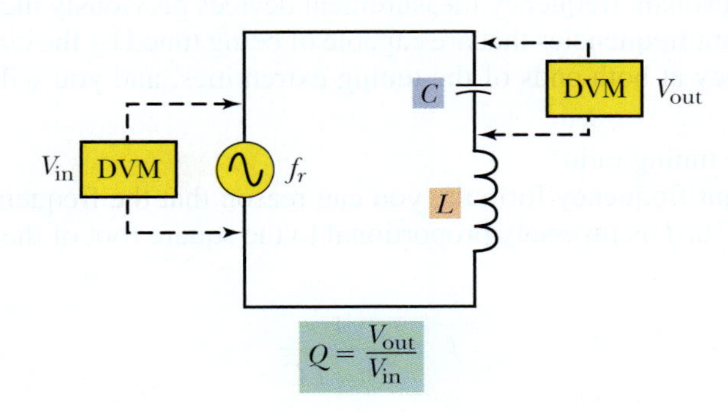

FIGURE 21–17 Determining Q of series resonant circuit

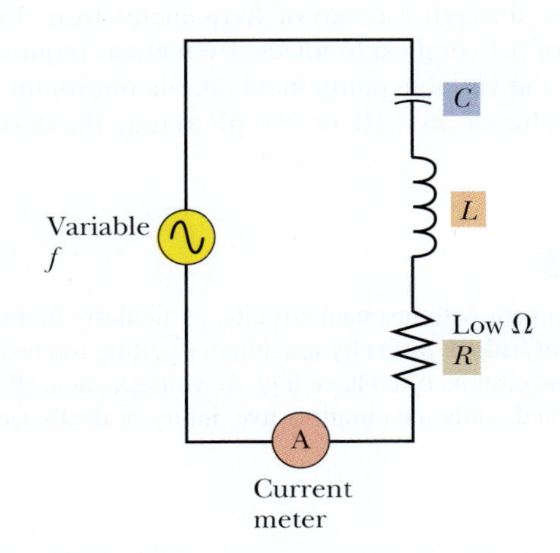

FIGURE 21–18 Determining maximum I for a series resonant circuit

Use a current meter, *or*, insert low R value, measure V_R, and then calculate I. $\left(I = \frac{V_R}{R}\right)$

Adjust R so that $V_R = V_{TANK}$ at resonance (then measure $R \ldots R = Z_{TANK}$)

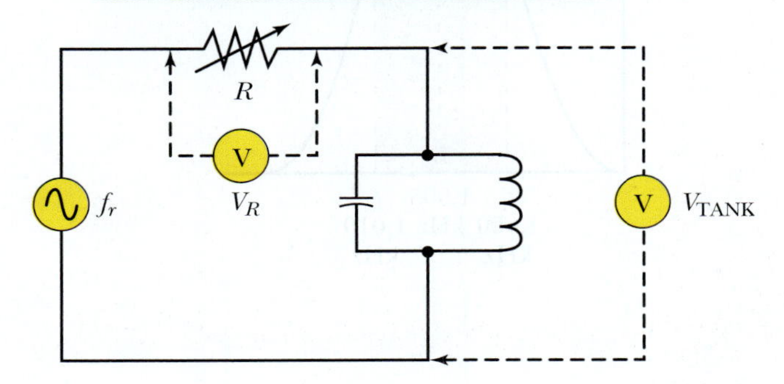

FIGURE 21–19 Determining maximum Z for a parallel resonant circuit

4. Determining bandwidth and bandpass
For a series resonant circuit, tune the circuit through resonance and note the frequencies where current is 70.7% of maximum value. The difference between the lower frequency and the higher frequency is the bandwidth. The actual frequencies at those two points represent the bandpass.
For a parallel resonant circuit, you can use the *Z* determining system and find the frequencies where *Z* is 70.7% of maximum value. Again, the difference between those frequencies is the bandwidth, and the precise lower and upper frequencies are the bandpass, Figure 21–20.
5. Measuring the tuning range
If the tuned circuit has a variable *C* or *L*, or if both are variable, there are limits to the range of resonant frequencies to which the circuit can be tuned.
The various resonant frequency measurement devices previously mentioned can check the actual resonant frequencies that are capable of being tuned by the circuit. Simply measure the frequency at both ends of the tuning extremities, and you will know the circuit tuning range.
6. Determining the tuning ratio
From the resonant frequency formula, you can reason that the frequency change caused by a change in *C* or *L* is inversely proportional to the square root of the *change* in *C* or *L*. That is,

$$f_r = \frac{1}{2\pi\sqrt{LC}}$$

therefore, f_r is proportional to $1/\sqrt{LC}$.
If it is desired to tune through a range of frequencies from 1,000 to 6,000 kHz, this means this tuning ratio of 6:1 (highest to lowest frequency) requires a change of *C* or *L* in the ratio of 36:1. If *C* is the variable component and its minimum *C* value is 10 pF, it has to have a maximum *C* value of 36 × 10, or 360 pF to tune the desired range.

Safety Hints

When working around equipment with resonant circuits, particularly in transmitter circuits, be aware that it is possible to get bad *RF burns* by touching or getting too near a live circuit or component. In addition, these circuits often have high dc voltages, as well as high radio frequency ac energy. If touched while the circuit is live, injury or death can result!

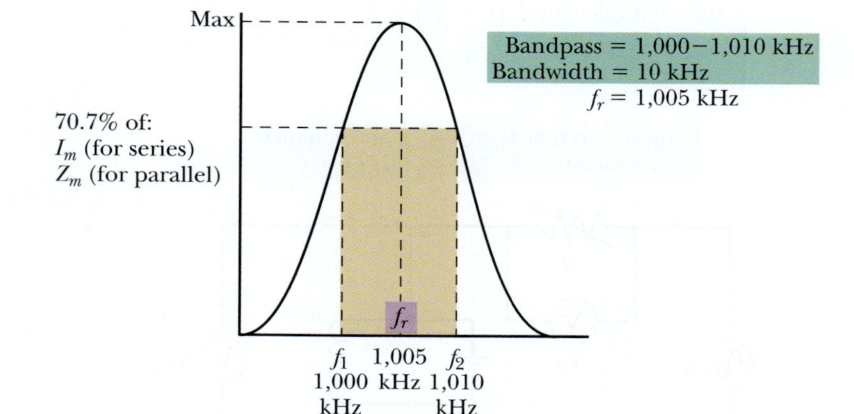

FIGURE 21–20 Determining bandwidth and bandpass

21–12 Special Notations about Single Components

Distributed Capacitance

Distributed capacitance is present in most circuits. It exists anywhere there are two conductors separated by an insulator or dielectric material. For example, there is distributed capacitance between conductor paths on printed circuit boards. Distributed capacitance is also present between wires and chassis in "hard-wired" circuits, and so on.

Self-Resonance of an Inductor

It is important to know that inductors also have distributed capacitance, Figure 21–21a. Because each turn in an operating coil is at a slightly different potential, each pair of turns effectively creates a "parasitic" capacitor. The result of this effective distributed capacitance is that at

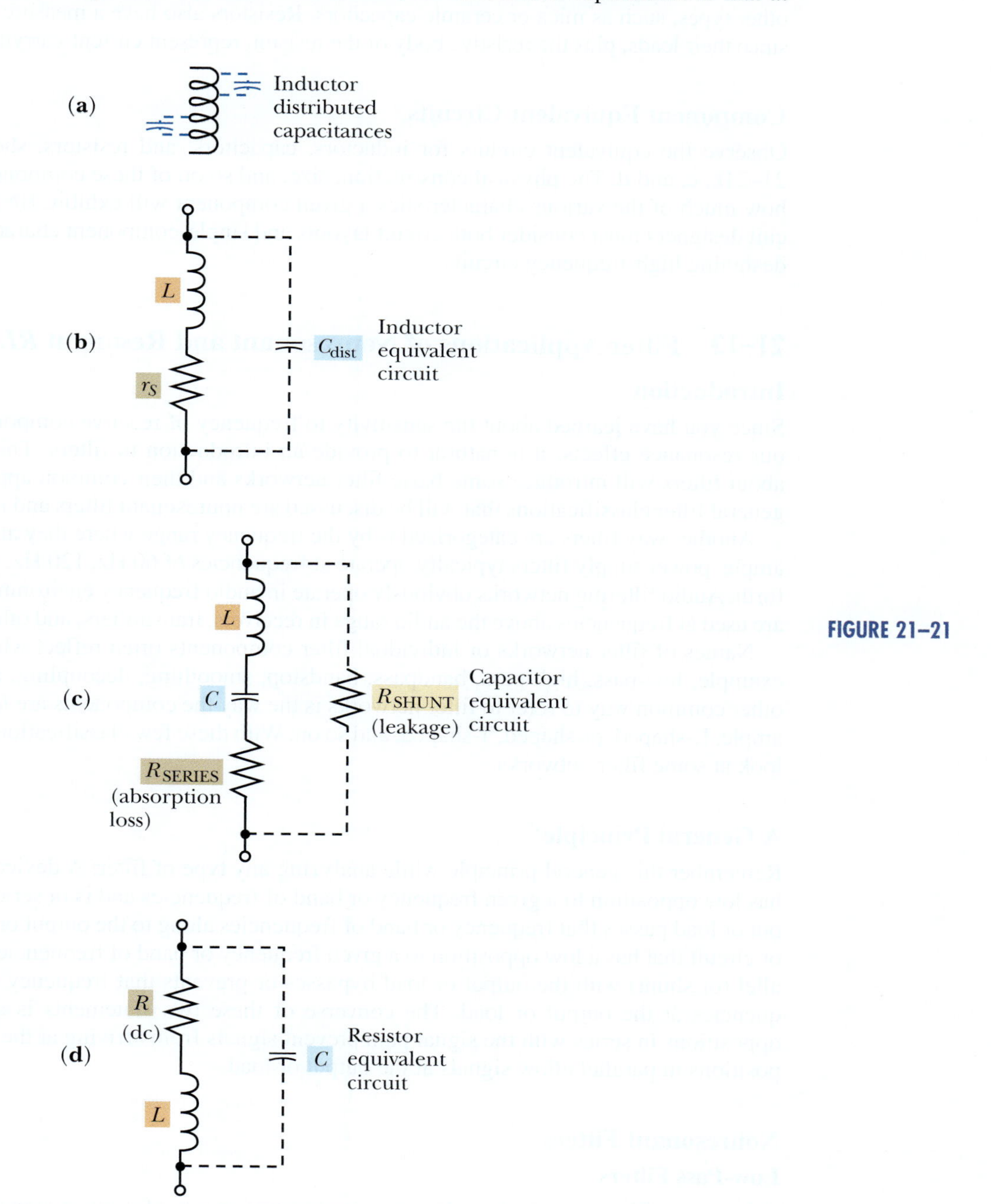

FIGURE 21–21

some frequency, the effective capacitance of the coil will have a reactance equal to the inductive reactance of the inductor. At this frequency, the inductor will show "self-resonance." At frequencies above this self-resonant frequency, the coil will begin to act like a capacitor, rather than an inductor. Typically, this self-resonance effect does not hinder inductor operation except at higher frequencies.

Distributed or Stray Inductance

Anytime there are current-carrying conductors, inductance will be present. The changing flux caused by changing current levels will induce voltage into the conductor itself, or any conductors that are close to, and linked by, the magnetic field. This means that capacitors have some inductance, because their leads and plates are current-carrying conductors. Capacitors whose plates are wrapped in a coil, such as paper and electrolytic types, exhibit more inductance than other types, such as mica or ceramic capacitors. Resistors also have a measure of inductance, since their leads, plus the resistive body of the resistor, represent current-carrying conductors.

Component Equivalent Circuits

Observe the equivalent circuits for inductors, capacitors, and resistors, shown in Figure 21–21b, c, and d. The physical construction, size, and so on of these components determine how much of the various characteristics a given component will exhibit. Be aware that circuit designers must consider both circuit layouts and single-component characteristics when designing high-frequency circuits.

21–13 Filter Applications of Nonresonant and Resonant *RLC* Circuits

Introduction

Since you have learned about the sensitivity to frequency of reactive components and various resonance effects, it is natural to provide an introduction to filters. This brief section about filters will introduce some basic filter networks and their common applications. Two general filter classifications that will be discussed are nonresonant filters and resonant filters.

Another way filters are categorized is by the frequency range where they are used. For example, power supply filters typically operate at frequencies of 60 Hz, 120 Hz, 400 Hz, and so forth. Audio filtering networks obviously operate in audio frequency environments. *RF* filters are used at frequencies above the audio range in receivers, transmitters, and other *RF* systems.

Names of filter networks or individual filter components often reflect what they do; for example, low-pass, high-pass, bandpass, bandstop, smoothing, decoupling, and so on. Another common way to refer to filter networks is the way the components are *laid out;* for example, L-shaped, pi-shaped, T-shaped, and so on. With these few classifications in mind, let's look at some filter networks.

A General Principle

Remember this general principle while analyzing any type of filter: A device or circuit that has low opposition to a given frequency or band of frequencies and is in series with the output or load passes that frequency or band of frequencies along to the output or load. A device or circuit that has a low opposition to a given frequency or band of frequencies and is in parallel (or shunt) with the output or load bypasses or prevents that frequency or band of frequencies at the output or load. The converse of these two statements is also true. High oppositions in series with the signal path prevent signals from arriving at the load. High oppositions in parallel allow signals at the output or load.

Nonresonant Filters

Low-Pass Filters

A **low-pass filter** passes along low-frequency components of a given signal or waveform, while impeding the passage of higher-frequency components. For example, look at the simple *RC* filter network in Figure 21–22. Notice that the output is being taken across the capacitor. The voltage distribution of the signal from the source will depend on the relative values of *R* and X_C of the capacitor. If the X_C is high compared with the *R*, then a greater portion of the signal will be across the capacitor, and the network "output" signal will be greater. If the X_C is low with respect to the *R*, the resistor drops most of the signal and the output is very low. Since the capacitor is frequency-sensitive, its X_C will be high at low frequencies and low at high frequencies.

In Figure 21–23a, the source frequency is dc (or 0 Hz). The X_C is virtually infinite and almost all of the input voltage appears across the capacitor or is present at the output. In Figure 21–23b, the source frequency is such that *R* and X_C are equal; thus, the output signal and the voltage drop across the *R* are equal. (Don't forget, these voltages are out of phase; they

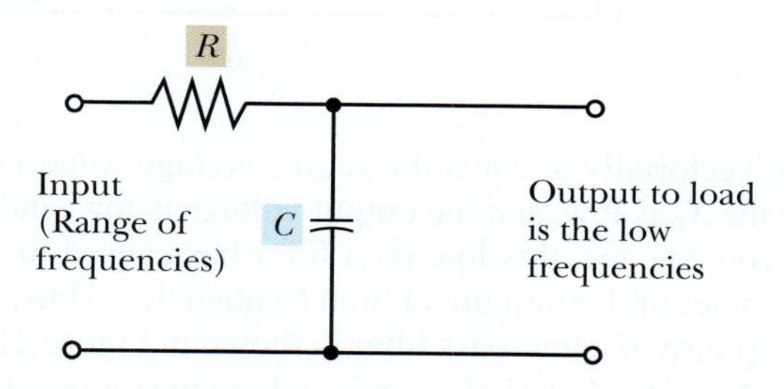

FIGURE 21–22 Example of a low-pass filter

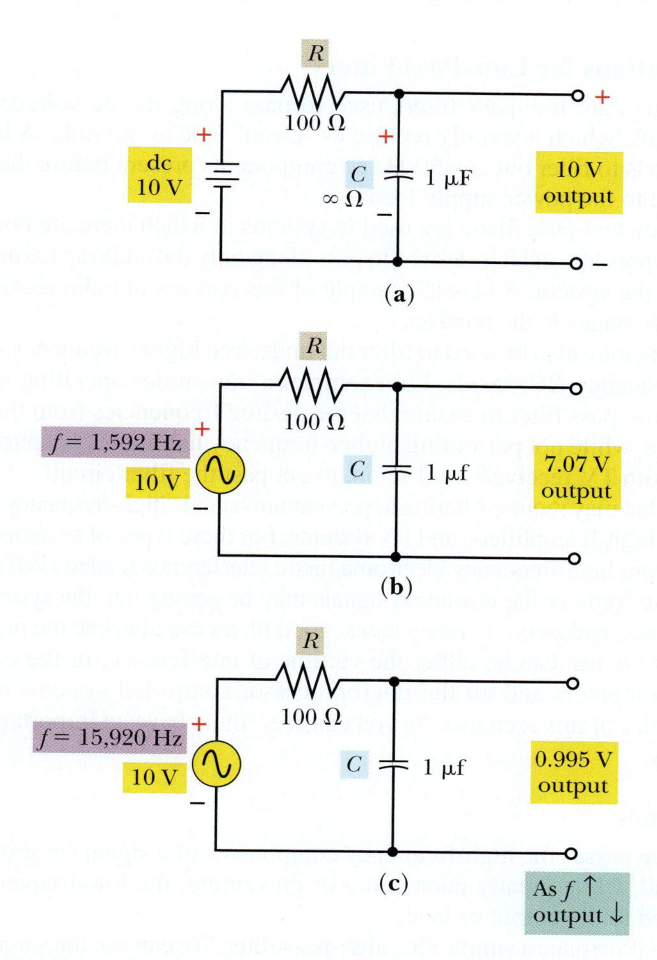

FIGURE 21–23 Example of simple *RC* low-pass filter operation multiSIM

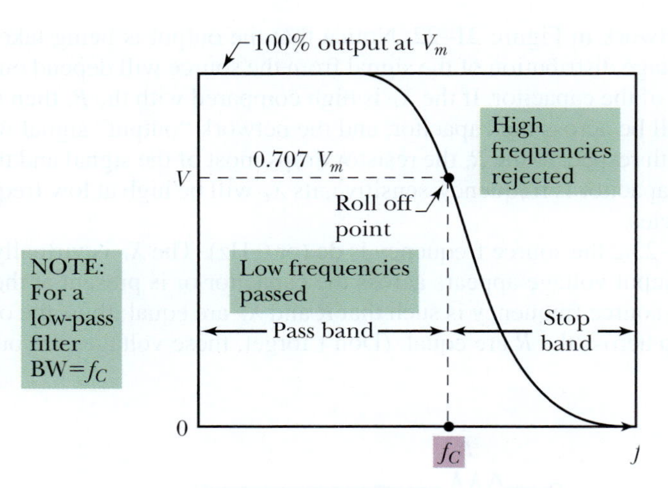

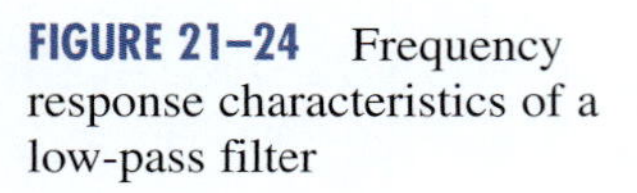
FIGURE 21–24 Frequency response characteristics of a low-pass filter

would have to be added vectorially to equal the source voltage value.) In Figure 21–23c, the frequency is high, thus the X_C is low, and the output voltage is low since the *R* drops most of the signal voltage. As you can see, this low-pass filter has caused the output voltage to be maximum at low frequencies and minimum at high frequencies. Thus, it is called a low-pass filter! The frequency response of a low-pass filter is shown in Figure 21–24. NOTE: f_c stands for "cutoff frequency." That is the 0.707 of V_{in} point where the output starts declining greatly. The steepness of signal fall off depends on the type of filter circuit and its components. Various circuit and component configurations, other than the simple *RC* circuit previously shown, that provide low-pass filtering are shown in Figure 21–25.

Several Applications for Low-Pass Filters

Power-supply filters are low-pass filters used to pass along the dc voltage and current to power-supply loads, which generally require as "clean" a dc as possible. A large part of this type of filter's job is to filter out any 60-Hz ac components present before their getting to the filter's output and to the power supply loads.

Audio frequency low-pass filters are used in systems in which there are both *Rf* and audio-frequency signals present, and it is desired to pass along only the audio spectrum signals to succeeding stages in the system. A classic example of this appears in radio receivers, just before the audio amplifier stages in the receiver.

Low-pass filters may also be used to filter out undesired higher-frequency *RF* signals from desired lower-frequency *RF* signals. For example, a transmitter operating at several megahertz may use a low-pass filter to ensure that the desired frequencies from the transmitter arrive at the antenna, while not permitting higher-frequency harmonics or spurious signals that might interfere with TV receivers, and so on, to get past the filter circuit.

Other devices that may require filtering to prevent unwanted "high-frequency interference" include telephones, high-fi amplifiers, and PA systems. For these types of systems the interference may be coming from high-frequency electromagnetic interference (called EMI) from nearby radio stations, and so forth, or the unwanted signals may be getting into the system via the power lines, telephone lines, and so on. In many cases, good filters can alleviate the problem.

Some systems we use can be either the victims of interference, or the cause of interference, or both. Computers and all the microprocessor-controlled systems used around our homes are examples of this scenario. As you can see, filters have an important role in today's electronic society.

High-Pass Filters

A **high-pass filter** passes the high-frequency components of a signal (or group of signals) to the output or load, while greatly attenuating or preventing, the low-frequency components from being passed to the output or load.

Figure 21–26a illustrates a simple *RC*, high-pass filter. We can use the same concepts to explain this filter as we did in explaining the simple *RC* low-pass filter. That is, the relative val-

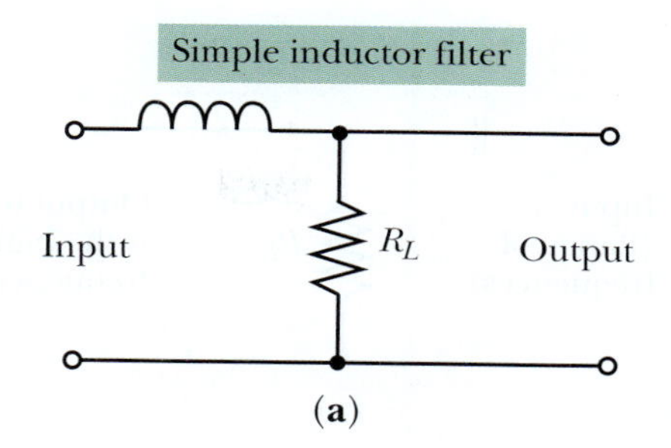

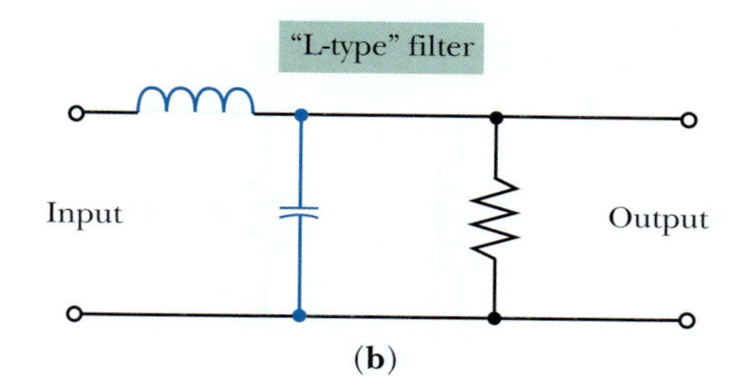

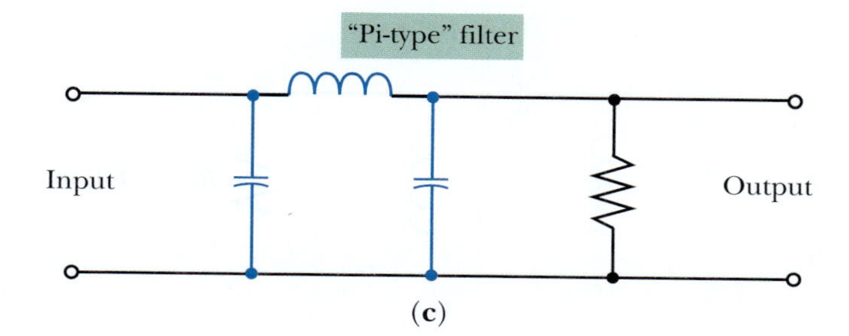

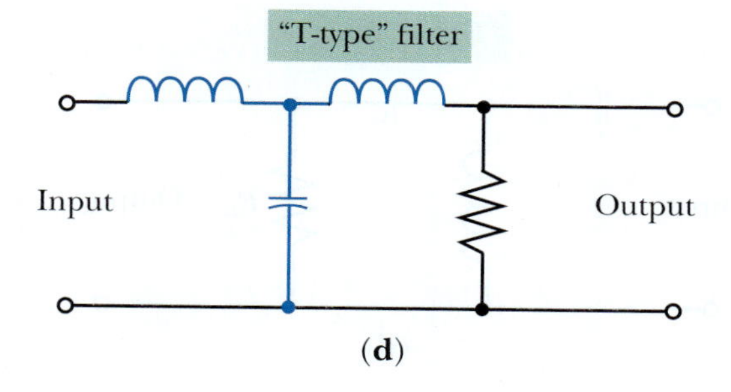

FIGURE 21–25 Examples of circuit configurations for low-pass filters

ues of R and X_C will determine the amount of the input signal that appears at the output. The difference between the two types of filters, is that in this case we take the output across the R, rather than the C. This means that the higher the X_C compared with the R value, the lower the output will be, and vice versa.

Let's assume the same values for R and C in Figure 21–26a as was used in our low-pass filter examples in Figure 21–23. If $R = 100\ \Omega$ and the value of $C = 1\ \mu F$, the following conditions will exist.

1. With a 10-V dc source as input, the output will be zero because the capacitor (once charged) blocks dc. This "ultra-low" frequency (dc) is not passed to the output.
2. At a frequency of 1,592 Hz, R and X_C are equal, and the output will be 7.07 V with a 10-V ac source as the input.
3. At a frequency of 15,920 Hz, $R = 100\ \Omega$ and $X_C =$ about 10 Ω; thus, the output will be about 9 V.

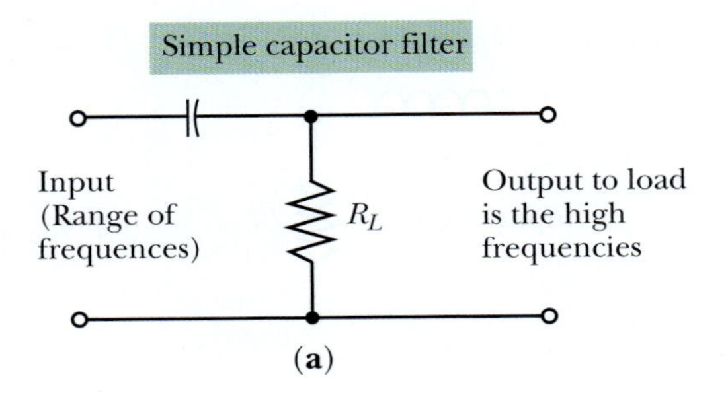

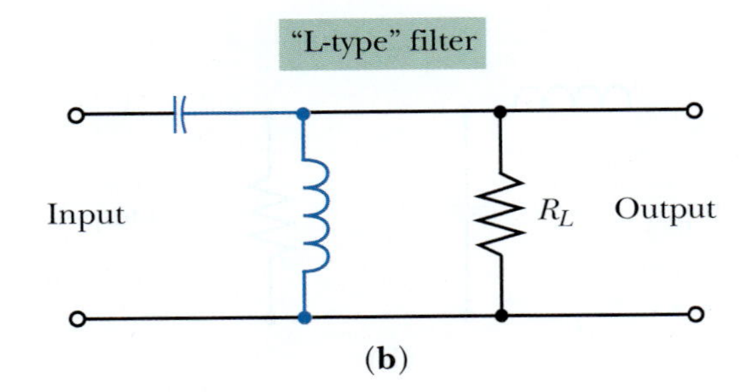

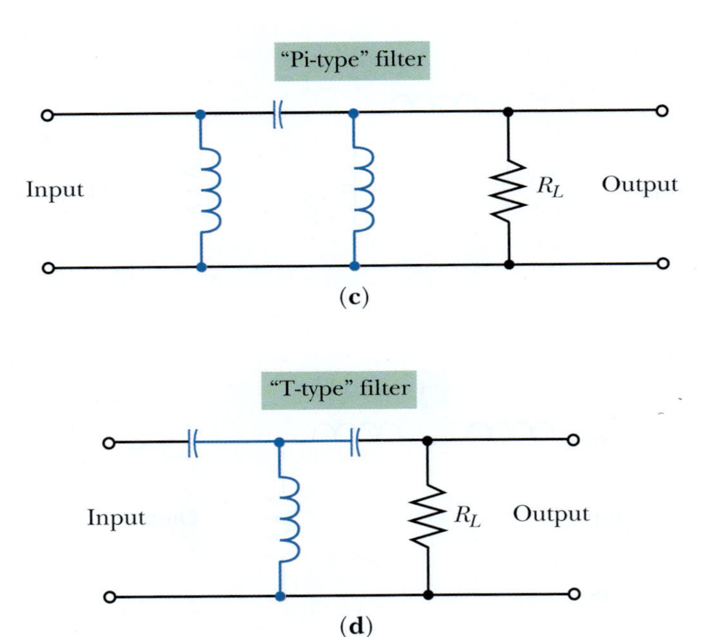

FIGURE 21–26 Examples of different configurations for high-pass filters

Since this filter passes along the higher frequencies to the output, while passing very little signal at low frequencies, it is called a high-pass filter!

Figure 21–26 shows several examples of configurations of *R, L,* and *C* components used as high-pass filters. Typically, the capacitors, which have a high opposition to low frequencies, are in series with the output or load. The inductors, which have a low opposition to low frequencies, are in parallel, or shunt with the output, thus effectively bypassing the low frequencies around the load. The result is that the high-frequency components of the input signal are passed to the load, and the low-frequency components are greatly attenuated or prevented from reaching the load. Thus, the filter is called a high-pass filter.

Figure 21–27 shows the typical response characteristic of a high-pass filter.

Several Applications for High-Pass Filters

Coupling circuits are one common use for networks that will attenuate frequencies below the desired frequency range (including dc) but will pass along the desired frequencies. For exam-

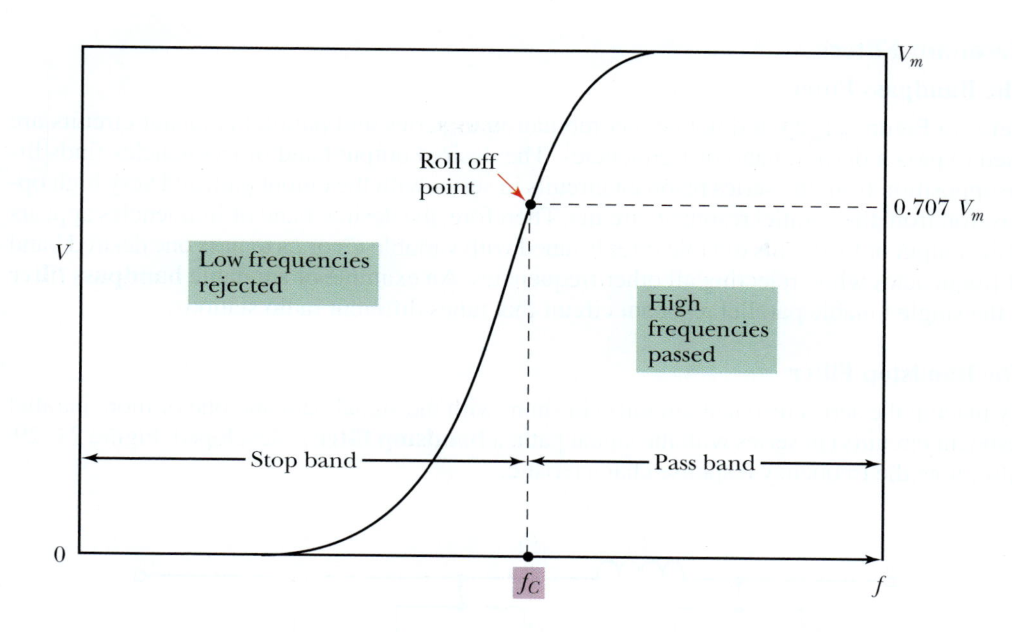

FIGURE 21–27 Typical response characteristic of a high-pass filter

ple, the circuit that couples a signal into the input of an amplifier effectively blocks dc and passes along the signal frequencies to be amplified.

Television receivers may use high-pass filters to prevent unwanted signals below the TV channel frequencies from causing interference. Filtering may also be used in other receiver antenna input circuits to prevent unwanted signals from getting into the system.

Some tone control circuits in radios and audio amplifiers use the concept of bypassing higher audio frequencies to alter the tone of the sound prescribed to the listener.

These are just a few of the many applications for filters in general, and high-pass filters, in particular.

NOTE: Low-pass and high-pass filter networks can be combined to pass a desired band of frequencies. Those frequencies passed along to the output or load are the frequencies that are not stopped by either filter network. This is one application of *bandpass filtering.* Typically, however, this application is often fulfilled by resonant circuits.

Practical Notes

As you know, points falling below 70.7% or the half-power points of a filter response are out of the band of frequencies that the filter will pass. In low-pass filters, Figures 21–24 and 21–25, the high frequencies are rejected. In high-pass filters, Figures 21–26 and 21–27, the lower frequencies are rejected. The frequency where rejection begins is sometimes called the *cutoff frequency* (f_c). For a simple *RC* network, Figure 21–26a, $X_C = R$ at this cutoff frequency. The formula to find the cutoff frequency for this *RC* network is:

FORMULA 21–10 $$f_c = \frac{1}{2\pi RC}$$

To find the cutoff frequency where $X_L = R$ for an *RL* network similar to the one shown in Figure 22–25a, the formula is:

FORMULA 21–11 $$f_c = \frac{1}{2\pi\left(\frac{L}{R}\right)}$$

Resonant Filters

The Bandpass Filter

Refer to Figure 21–28 and notice a combination of series and parallel resonant circuits are used to pass a desired band of frequencies. The desired output band of frequencies finds little opposition from the series resonant circuits in series with the output path and very high opposition from the parallel resonant circuits. Therefore, the desired band of frequencies appears at the output or load. This tunable filter is tuned with variable *C*s or *L*s to pass one desired band of frequencies while rejecting all other frequencies. An example of a tunable **bandpass filter** is the single tunable parallel resonant circuit that tunes different radio stations.

The Bandstop Filter

By placing the series resonant circuit(s) in shunt with the signal path and one or more parallel resonant circuit(s) in series with the signal path, a **bandstop filter** is developed, Figure 21–29. Also, note the frequency response characteristic.

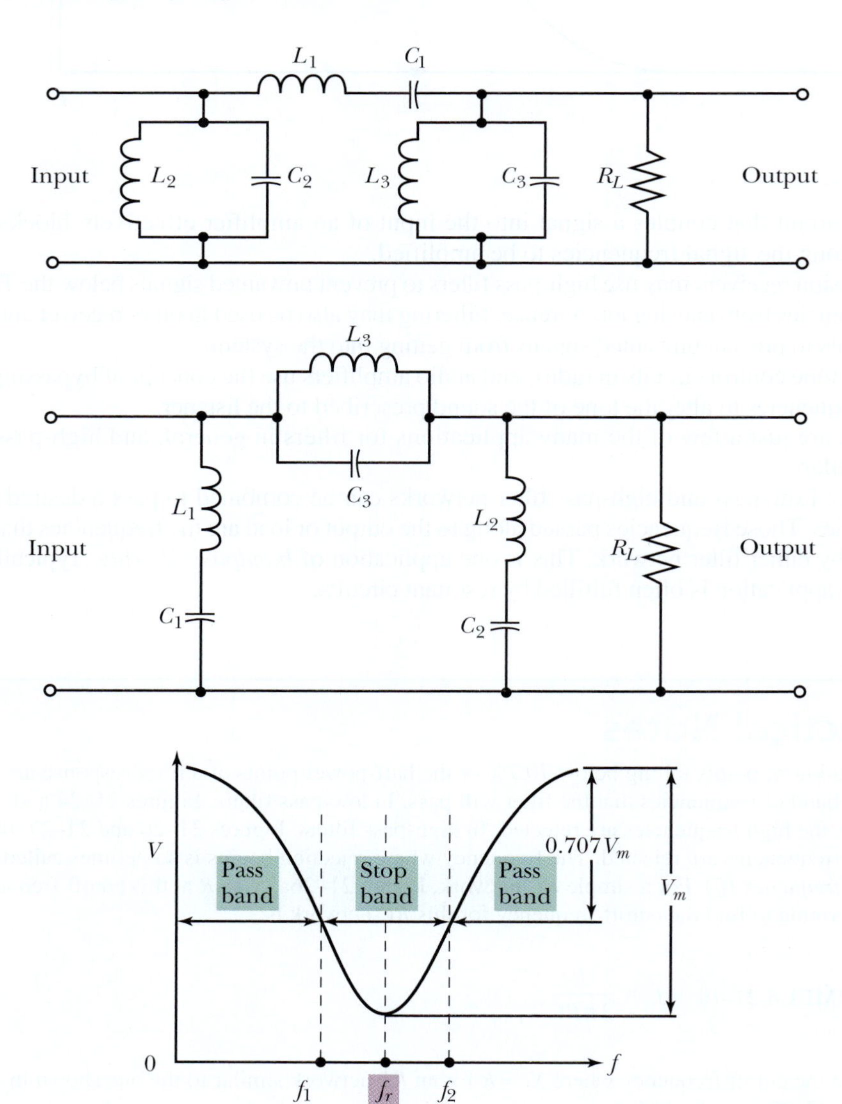

FIGURE 21–28 Resonant-type bandpass filter

FIGURE 21–29 Resonant-type bandstop filter and response characteristic

The previous discussions have only introduced you to filters. We have not attempted to investigate filter design formulas (e.g., for *constant-k* or *m-derived filters*), or the wide spectrum of modern filters, such as YIG (yittrium-iron-garnet) filters, multichannel analog filters, and computer-controlled filters.

Perhaps your interest in studying filters was increased by this simple overview. Filter networks have numerous practical uses, from simple power supply filtering to signal processing in such high-technology areas as speech processing and medical instrumentation. As you continue in your studies, you will have the opportunity to study some of the basic applications of filters, coupling networks, and bypassing networks shown in Figure 21–30. The principles discussed in this chapter should help you in the more detailed studies.

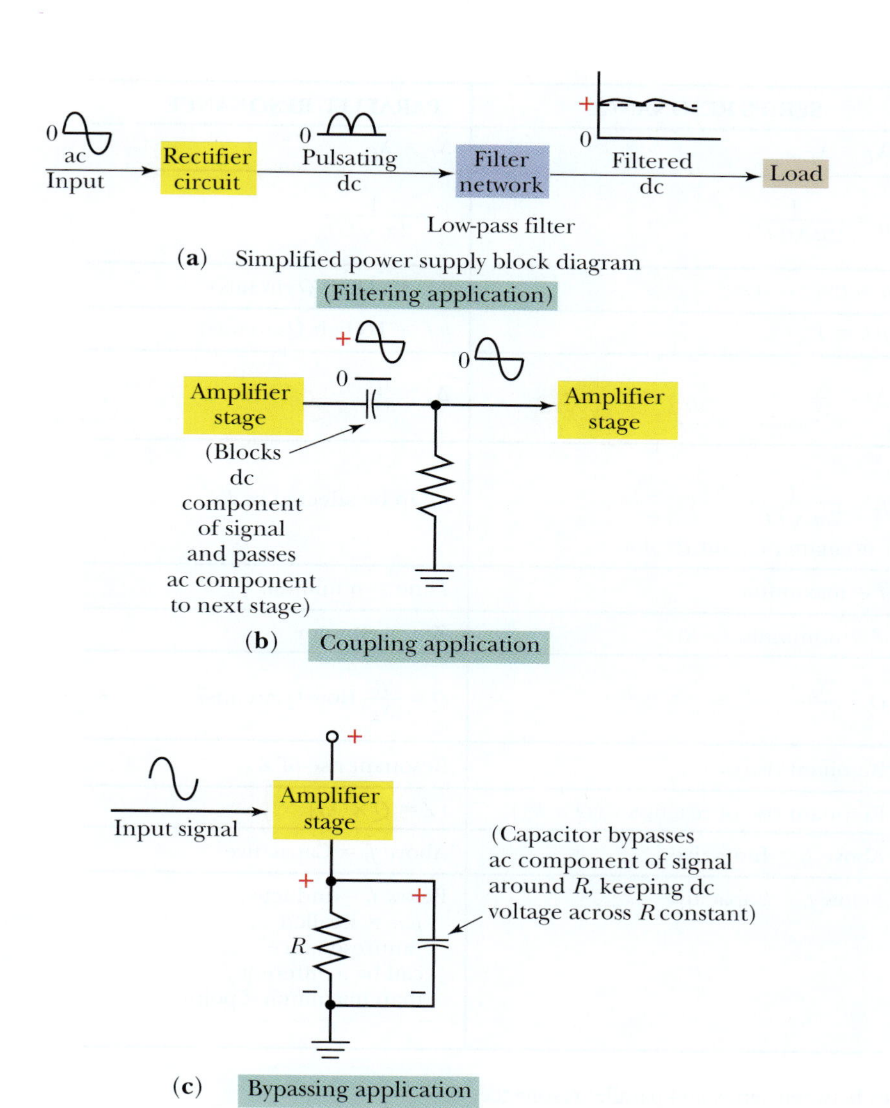

FIGURE 21–30 Sample applications of filters

Summary

- X_L and X_C change in opposite directions for a given frequency change. In other words, as f increases, X_L increases and X_C decreases. Also, as f decreases, X_L decreases and X_C increases.
- For any LC combination, there is one frequency where X_L and X_C are equal: the resonant frequency or the frequency of resonance.
- At resonance, series RLC circuits exhibit the characteristics of: $X_L = X_C$, maximum current, minimum Z (and $Z = R$), θ of 0°, and voltage across each reactance of Q times V applied.
- At resonance, parallel RLC circuits exhibit the characteristics of: $X_L = X_C$, minimum current, maximum Z ($Z = Q \times X_L$), $\theta = 0°$, and current within the tank circuit = Q times I line.
- See Figure 21–31 for a comparison of series and parallel resonant circuits.
- The general formula to find the resonant frequency for series LC circuits and for high-Q parallel LC circuits is: $f_r = \dfrac{1}{2\pi\sqrt{LC}}$

	SERIES RESONANCE	PARALLEL RESONANCE
SIMILARITIES	$X_L = X_C$	$X_L = X_C$
	$f_r = \dfrac{1}{2\pi\sqrt{LC}}$	$f_r = \dfrac{1}{2\pi\sqrt{LC}}$
	$\theta = 0°$	$\theta = 0°$ (High Q circuits)
	$p.f. = 1$	$p.f. = 1$ (High Q circuits)
	$\Delta f = \dfrac{f_r}{Q}$	$\Delta f = \dfrac{f_r}{Q}$
DIFFERENCES	$f_r = \dfrac{1}{2\pi\sqrt{LC}}$ no matter circuit R value	f_r can be affected by R
	I = maximum	I line = minimum
	Z = minimum ($=R$)	Z = maximum
	$Q = \dfrac{X_L}{r_s}$	$Q = \dfrac{R}{X_L}$ (low Q circuits)
	Resonant rise of I	Resonant rise of Z
	Resonant rise of reactive V ($Q \times V_A$)	($Z = Q \times X_L$)
	Above f_r = Inductive	Above f_r = Capacitive
	Below f_r = Capacitive	Below f_r = Inductive $p.f. = 1$ called "antiresonance" . . . can be a different f than maximum Z point.

FIGURE 21–31 Similarities and differences between series and parallel resonance

- The Q of a resonant circuit is determined by the circuit resistance and the L/C ratio. The more the circuit resistive losses are, the lower the Q is. The higher the L/C ratio is, the higher the Q is for a given f_r.
- The higher the Q of a resonant circuit is, the greater the magnification factor is for voltage (series resonant circuit) and impedance (parallel resonant circuit).
- The higher the Q of a resonant circuit is, the sharper is its response curve and the more selective it is to a band of frequencies. High Q means narrow bandwidth and narrow bandpass.
- Bandwidth is often defined by the frequencies where the response curve for the circuit is above the 70.7% of the maximum level. This is sometimes called the half-power or –3 dB points on the response curve.
- The relationship of resonant frequency, Q, and bandwidth (BW) is shown by:

$$BW\,(\Delta f) = \frac{f_r}{Q}, \text{ or } Q = \frac{f_r}{\Delta f}$$

- R, L, and C components can be configured to produce filtering effects. Low-pass filters allow low frequencies to pass to the output, while attenuating high frequencies. High-pass filters pass high frequencies to the output while attenuating the low frequencies.
- Generally, to allow a certain frequency spectrum to pass to the output, a low-impedance circuit or component is placed in series with the signal path, and/or a high-impedance circuit or component is placed in shunt with the signal path.
- To prevent a certain frequency spectrum at the output of the filter, a high-impedance circuit or component is placed in series with the signal path, and/or a low-impedance circuit or component is placed in shunt with the signal path.
- Filters are categorized in various ways. Some classifications are physical layout shape (L, pi, T); usage (bypass, decouple, bandpass, bandstop, smoothing); component types (LC, RC, active, passive); number of sections they have (single-section, two-section); and the frequency spectrum in which they are used (power supply, audio, RF).
- Multiple-section filters generally provide more filtering than single-section filters.

Formulas and Sample Calculator Sequences

FORMULA 21–1
(To find the resonant frequency of a circuit where the values of L *and* C *are known)*

Resonant frequency: $f_r = \dfrac{1}{2\pi\sqrt{LC}}$

Inductance value, [×], capacitance value, [=], [√x], [×], 6.28, [=], [1/x]

FORMULA 21–2
(To find the inductance value needed to achieve a specified resonant frequency when the value of C *is known)*

Inductance value for given f_r and C: $L = \dfrac{0.02533}{f_r^{\,2}C}$

0.02533, [÷], [(], resonant frequency value, [x²], [×], capacitance value, [)], [=]

FORMULA 21–3
(To find the capacitance value needed to achieve a specified resonant frequency when the value of L *is known)*

Capacitance value for given f_r and L: $C = \dfrac{0.02533}{f_r^{\,2}L}$

0.02533, [÷], [(], resonant frequency value, [x²], [×], inductance value, [)], [=]

FORMULA 21–4
(To find the inductance value needed to achieve a specified resonant frequency when the value of C is known)

Inductance value (in μH) for given f_r and C: $L = \frac{25{,}330}{f^2 C}$

25,330, [÷], [(], resonant frequency value in MHz, [x^2], [×], capacitance value in pF, [)], [=]

FORMULA 21–5
(To find the capacitance value needed to achieve a specified resonant requency when the value of L is known)

Capacitance value (in pF) for given f_r and L: $C = \frac{25{,}330}{f^2 L}$

25,330, [÷], [(], resonant frequency value in MHz, [x^2], [×], inductance value in μH, [)], [=]

FORMULA 21–6
(To find resonant frequency for given parallel LC circuit)

$$f_r = \frac{1}{2\pi\sqrt{LC}} \times \sqrt{1 - \frac{C{r_S}^2}{L}}$$

FORMULA 21–7
(To find total bandwidth for a tuned circuit where the half-power point frequencies are known)

Total bandwidth (BW): $\Delta f = f_2 - f_1$

half-power point (higher frequency), [−], half-power point (lower frequency), [=]

FORMULA 21–8
(To find total bandwidth for a tuned circuit where resonant frequency and Q are known)

Bandwidth (BW) related to Q: $\Delta f = \frac{f_r}{Q}$

resonant frequency value, [÷], Q value, [=]

FORMULA 21–9
(To find the circuit Q value when resonant frequency and bandwidth are known)

Q value: $Q = \frac{f_r}{\Delta f}$

resonant frequency value, [÷], bandwidth value, [=]

FORMULA 21–10
(To find the cutoff frequency of an RC filter where the values of R and C are known)

Cutoff frequency $(RC$ filter): $f_c = \frac{1}{2\pi RC}$

6.28, [×], resistor value, [×] capacitance value, [=], [1/x]

FORMULA 21–11
(To find the cutoff frequency of an RL filter where the values of R and L are known)

Cutoff frequency $(RL$ filter): $f_c = \frac{1}{2\pi\left(\frac{L}{R}\right)}$

6.28, [×], [(], inductance value, [÷], resistance value, [)], [=], [1/x]

EXCEL AUTOMATED FORMULAS

Helpful problem-solving tools, such as the Excel automated formulas available on the CD, allow automatic calculations of formulas.

Using Excel

Series and Parallel Resonance Formulas

(Excel file reference: FOE21_01.xls)

***DON'T FORGET!** It is not necessary to retype formulas once they are entered on the worksheet! Just input new parameters data for each new problem using that formula, as needed.*

- Use the Formula 21–1 spreadsheet sample and using the data from Practice Problems 1, find the resonant frequency of the circuit. Check your answer against the answer given in the Appendix for this problem.
- Use the Formula 21–3 spreadsheet sample and determine what capacitance value it will take to cause a circuit to be resonant at 1 MHz if the inductance value is 100 microhenrys. Check your answer against the answer given in the example just under Formula 21–5 in the text.

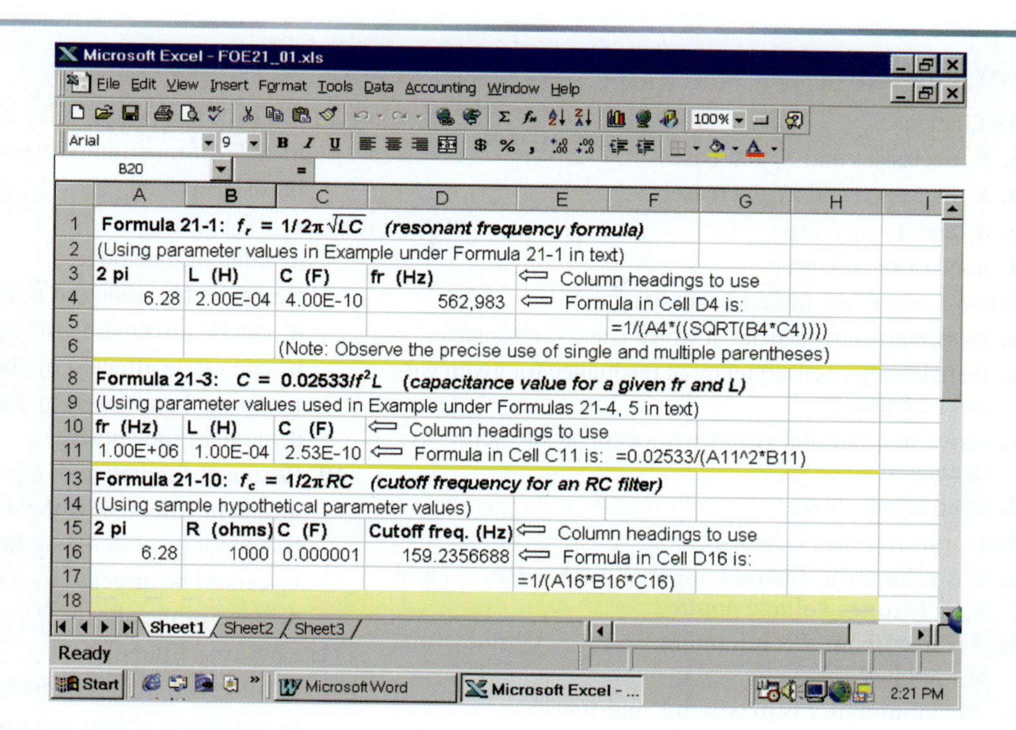

Review Questions

1. As f increases:
   - **a.** X_L decreases and X_C increases.
   - **b.** X_L increases and X_C decreases.
   - **c.** X_L and X_C increase.
   - **d.** none of the above.
2. When L and C are both present in a circuit:
   - **a.** there are two frequencies at which L and C resonate.
   - **b.** there is only one frequency of resonance for given values of L and C.
   - **c.** series and parallel resonance occur at the same frequency in all cases.
   - **d.** none of the above.
3. For series resonant circuits:
   - **a.** Z is minimum, I is maximum, $X_L = X_C$, θ is 0°, and $V_C = Q$ times voltage applied.
   - **b.** Z is maximum, I is minimum, and $V_L = Q$ times voltage applied.
   - **c.** Z is minimum, I is maximum, and $\theta = 45°$.
   - **d.** none of the above.
4. If an LC circuit has a high Q:
   - **a.** the circuit is broadband in nature.
   - **b.** the circuit is not very selective in terms of frequency.
   - **c.** the circuit has a narrow bandpass.
   - **d.** none of the above.
5. If an LC circuit has a low Q:
   - **a.** its bandwidth is narrow.
   - **b.** its bandwidth is broad.
   - **c.** its bandwidth equals its Q.
   - **d.** none of the above.
6. By definition, bandwidth is defined as frequencies wherein the response curve is close to:
   - **a.** 50% of maximum level.
   - **b.** 100% of maximum level.
   - **c.** 70% of maximum level.
   - **d.** none of the above.
7. Bandwidth and Q are:
   - **a.** directly related.
   - **b.** not related.
   - **c.** inversely related.
   - **d.** none of the above.
8. "Half-power points" may be considered:
   - **a.** 50% of maximum response points.
   - **b.** –3 dB points.
   - **c.** 0.5 points.
   - **d.** none of the above.
9. Resonant frequency of a series RLC circuit:
   - **a.** can be affected by the amount of R in the circuit.
   - **b.** cannot be affected by the amount of R in the circuit.
   - **c.** cannot be affected by the amount L in the circuit.
   - **d.** none of the above.
10. Resonant frequency of a parallel RLC circuit:
    - **a.** can be affected by the amount of R in the circuit.
    - **b.** cannot be affected by the amount of R in the circuit.
    - **c.** cannot be affected by the amount of C in the circuit.
    - **d.** none of the above.
11. Low-pass filters:
    - **a.** attenuate low frequencies.
    - **b.** amplify low frequencies.
    - **c.** allow passage of low frequencies.
    - **d.** allow passage of high frequencies.
12. Classification of filter circuits may include:
    - **a.** physical layout and components used.
    - **b.** use, physical layout, components used, and frequency spectrum.
    - **c.** components used, physical layout, use, and number of sections.
    - **d.** number of sections, components used, use, physical layout, and frequency.
13. A bandpass-type filter provides:
    - **a.** high opposition in series with the output.
    - **b.** low opposition in parallel with the output.
    - **c.** low opposition in series with the output.
    - **d.** none of the above.
14. A bandstop-type filter provides:
    - **a.** high opposition in series with the output.
    - **b.** high opposition in parallel with the output.
    - **c.** low opposition in series with the output.
    - **d.** none of the above.

15. Inductors:

a. have inductance, resistance, and capacitance.

b. have inductance and resistance only.

c. must have an external C connected to produce a resonant condition.

d. none of the above.

16. Resistors have the properties of:

a. resistance and capacitance.

b. resistance and inductance.

c. resistance, inductance, and capacitance.

d. resistance only.

17. Several ways of measuring the resonant frequency of a signal source include:

a. a digital frequency counter and a tunable dip meter (with power on).

b. a digital frequency counter, with power on, a dip meter (with power off).

c. an oscilloscope, or a frequency counter, with power on.

d. all of the above.

18. For a simple RC filter, the cutoff frequency is:

a. at a point where R is double X_C.

b. directly related to the value of R and C.

c. inversely related to the value of R and C.

d. only related to the value of R.

19. For a simple RL filter, the cutoff frequency will increase if:

a. the value of L increases and R stays the same.

b. the value of R increases and L stays the same.

c. the value of R and L change by the same factor.

d. none of the above.

20. To create a resonant bandstop filter, you would use:

a. a series resonant circuit in series with the signal path.

b. a series resonant circuit in parallel with the signal path.

c. a parallel resonant circuit in parallel with the signal path.

d. none of the above.

Problems

1. Disregarding the generator or source impedance, a series LC circuit is composed of a 100-pF capacitor and a 1-μH inductor having 10 Ω of resistance.

a. What is the impedance at resonance?

b. What is the resonant frequency?

c. What is the Q?

d. What is the bandwidth?

e. What is the bandpass?

f. Below the resonant frequency, does the circuit act inductively or capacitively?

2. The voltage applied to the circuit in question 1 is 20 V.

a. What is the voltage across the capacitor at resonance?

b. What is the current at resonance?

c. What is the true power at resonance?

d. What is the net reactive power?

3. A parallel LC circuit has a capacitive branch with a 30-pF capacitor and an inductive branch with an inductor having 28 μH and a r_S of 30 Ω.

a. What is the f_r?

b. What is the Q? (Assume a Q greater than 10.)

c. What is the bandwidth?

d. What is the bandpass?

e. At resonance, what is the Z?

f. Below resonance, does the circuit act inductively or capacitively?

4. If the voltage applied to the circuit in question 3 is 20 V, what is the line current? What is the current through the capacitor branch?

5. A series RLC circuit consists of a 100-Ω resistor, a 75-H inductor, and a 300-pF capacitor. What is the resonant frequency?

6. A series RLC circuit has a resonant frequency of 40 kHz and the Q is 20. At what frequency will the phase angle be 45°?

7. A series RLC circuit is resonant at 150 kHz, and has $R = 100\ \Omega$ and $C = 25$ pF. What is the inductance value?

8. A series RLC circuit is resonant at 21 kHz and has $R = 40\ \Omega$ and inductance = 12 H. What is the C value?

9. Given a series RLC circuit that is resonant at 1.5 MHz, and knowing that C has a reactance of 500 Ω, what is the R value for the circuit to have a bandwidth of 10 kHz?

10. A series RLC circuit is composed of a 200-Ω resistor, a 36-mH inductor, and a capacitor. What is the voltage across the R at resonance if V applied is 100 V?

11. What is the resonant frequency of a parallel LC combination having a 200-μH inductor and a 0.0002-μF capacitor?

12. If the frequency of the source in question 11 is adjusted to a new frequency above the resonance frequency, will I_T lead or lag V_A?

13. A 220-μH inductor has a distributed capacitance of 25 pF and an effective resistance of 80 Ω. What impedance will it exhibit at resonant frequency?

Analysis Questions

Perform the required research to determine the answers to the following questions.

1. Is the tuned circuit at the input of a receiver typically a series or parallel LC circuit?

2. What is meant by a "double-tuned" transformer, as used between certain stages of radio or TV receivers?

3. Would the circuits described in questions 1 and 2 be considered resonant or nonresonant filters?

4. Match the frequency response curves (a–f) to the types of filters listed above the curves. Place the letter identifying each response curve in the blank identifying the filter or circuit type that would have that type of response.

_____ Parallel resonant
_____ Low-pass
_____ Bandpass
_____ High-pass
_____ Bandstop
_____ Series resonant

(a) I vs. f (d) Output vs. f

(b) Output
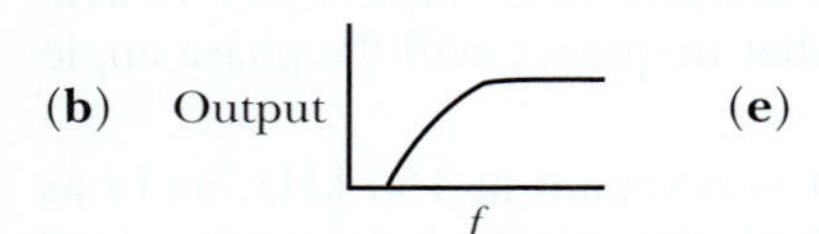
(e) Z
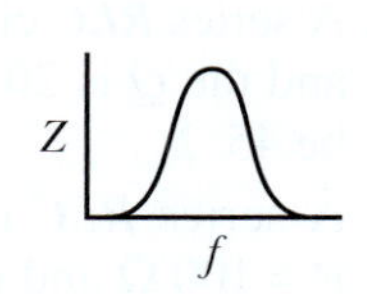

(c) Output (f) Output
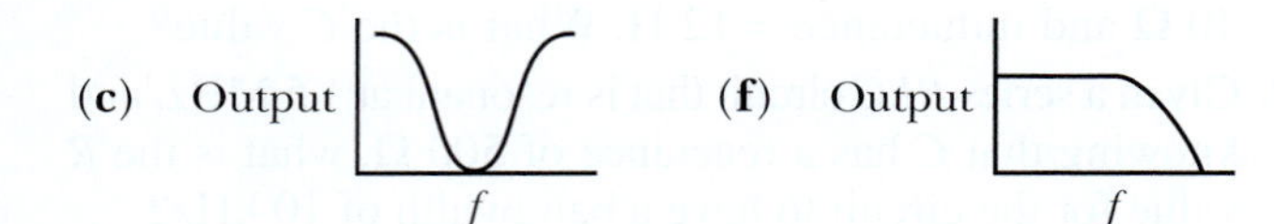

5. In your own words, define the terms *selectivity, bandwidth,* and *bandpass.*

Answer the following questions with "I" for increase, "D" for decrease, and "RTS" for remain the same.

6. For a series resonant RLC circuit, if frequency is slightly increased:

a. X_L will _____
b. X_C will _____
c. R will _____
d. Z will _____
e. I will _____
f. θ will _____
g. V_C will _____
h. V_L will _____
i. V_R will _____
j. V_T will _____
k. BW will _____
l. Q will _____
m. I_C will _____
n. I_L will _____
o. I_T will _____

7. For a series resonant RLC circuit, if frequency is slightly decreased:

a. X_L will _____
b. X_C will _____
c. R will _____
d. Z will _____
e. I will _____
f. θ will _____
g. V_C will _____
h. V_L will _____
i. V_R will _____
j. V_T will _____
k. BW will _____
l. Q will _____
m. I_C will _____
n. I_L will _____
o. I_T will _____

8. For a parallel resonant RLC circuit, if frequency is slightly increased:

a. X_L will _____
b. X_C will _____
c. R will _____
d. Z will _____
e. I will _____
f. θ will _____
g. V_C will _____
h. V_L will _____
i. V_R will _____
j. V_T will _____
k. BW will _____
l. Q will _____
m. I_C will _____
n. I_L will _____
o. I_T will _____

9. For a parallel resonant *RLC* circuit, if frequency is decreased:

   a. X_L will _____
   b. X_C will _____
   c. R will _____
   d. Z will _____
   e. I will _____
   f. θ will _____
   g. V_C will _____
   h. V_L will _____
   i. V_R will _____
   j. V_T will _____
   k. BW will _____
   l. Q will _____
   m. I_C will _____
   n. I_L will _____
   o. I_T will _____

10. Research the terms *m-derived* and *constant-k filters* and write the definition of each.
11. Assuming "perfect" reactive components, state the condition that must exist if a series *RLC* circuit is at resonance.
12. Draw a pi-type *LC* filter that can filter a power-supply output. Is this a low-pass or a high-pass filter?
13. Draw a circuit showing a series resonant and a parallel resonant circuit connected so that the combination passes a band of frequencies close to the resonant frequencies of the circuits.
14. If a parallel *LC* circuit is acting inductively at a frequency of 150 kHz, tuning the circuit to resonance requires which of the following?

    a. Increasing X_C
    b. Decreasing X_L
    c. Decreasing L
    d. Increasing C

MultiSIM Exercise 1 for *Series and Parallel Resonance*

1. Use the MultiSIM program and utilize the circuit shown in Figure 21–9. (NOTE: Connect a ground reference to bottom of source.)
2. Measure and record the voltage drops of the capacitor, inductor, and resistor.

   CAUTION: Wait long enough for each reading to "stabilize" at its maximum level. The final reading does not happen instantaneously!
3. Compare your MultiSIM readings with the data given in Figure 21–9 in the text. Are they reasonably close?
4. What is the Q of this circuit?
5. Double the source frequency and take the readings again. Are the voltage drops across the reactive components still approximately equal? Explain.

MultiSIM Exercise 2 for *Series and Parallel Resonance*

1. Use the MultiSIM program and utilize the circuit similar to the practical circuit shown in Figure 21–11. Let $C = 20$ pF; $L = 200$ μH; $R = 10$ Ω, $V_A = 10$ V rms, and $f = 2.517$ MHz. (NOTE: Connect a ground reference to bottom side of source.)
2. Measure and record the voltages across the inductor and the resistor. Calculate I_L.
3. Calculate the X_L of the inductor; then use the X_L/R formula to calculate Q.
4. Does the ratio of voltage drops in step 2 approximate the Q calculated in step 3?
5. Revise the above circuit by adding a resistor in series with the source (at the top of the diagram between the source and the top of C1). Use the technique shown in Figure 21–19 in the text to find the parallel resonant circuit's Z at resonance. Record your finding from these measurements. $Z =$ __________.
6. Insert a 10-ohm resistor in series with the capacitor in the C branch. NOTE: Connect the 10-Ω R between the bottom of C and the ground reference line at the bottom of the circuit. Use a dual-trace scope simulator and determine the phase difference between the C branch and the L branch. (NOTE: Since the circuit has a ground, DO NOT connect the scope ground terminal.)

A Good Idea

Be sure to use your five senses, plus, easy-to-manipulate "front-panel" controls in gathering and analyzing symptoms, before digging out the test equipment and beginning detailed testing!

Performance Projects Correlation Chart

Suggested performance projects that correlate with topics in this chapter are:

Chapter Topic	Performance Project	Project Number
X_L, X_C, and Frequency	X_L and X_C Relationships to Frequency	54
Series Resonance Characteristics	V, I, R, Z, and θ Relationships when $X_L = X_C$	55
The Resonant Frequency Formula		
Q and the Resonant Rise of Voltage	Q and Voltage in a Series Resonant Circuit	56
Selectivity, Bandwidth, and Bandpass	Bandwidth Related to Q	57
Parallel Resonance Characteristics	V, I, R, Z, and θ Relationships when $X_L = X_C$	58
Parallel Resonance Formulas		
Q and the Resonant Rise of Impedance	Q and Impedance in a Parallel Resonant Circuit	59
Selectivity, Bandwidth, and Bandpass	Bandwidth Related to Q	60
	Story Behind the Numbers: Series Resonance	
	Story Behind the Numbers: Parallel Resonance	

NOTE: It is suggested that after completing the above projects, the student should be required to answer the questions in the "Summary" at the end of this section of projects in the Laboratory Manual.

Troubleshooting Challenge

CHALLENGE CIRCUIT 13

(Follow the SIMPLER sequence by referring to inside front cover.)

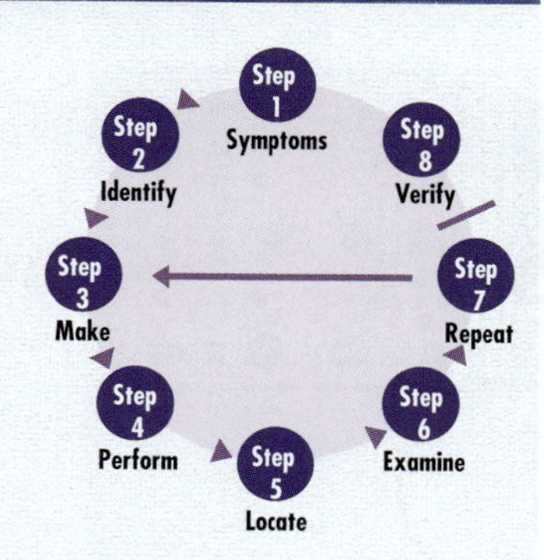

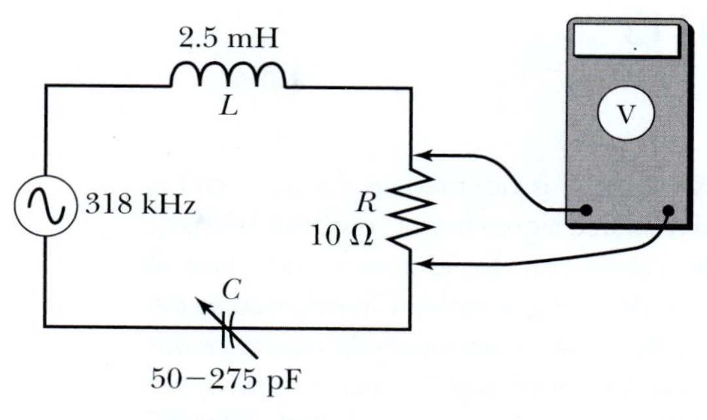

Challenge Circuit 13 multiSIM

STARTING POINT INFORMATION

1. Circuit diagram
2. V_R is lower than it should be when C is adjusted so that the LC resonance is below the 318-kHz source signal frequency.

TEST	Results in Appendix C
V_R (when C set at the position for 318-kHz LC resonance)	(55)
V_R (when C set at a position for LC resonance above 318 kHz)	(85)
V_R (when C set at a position for LC resonance below 318 kHz)	(30)
R_C (C set at a 318 kHz position)	(58)
R_C (C set above 318 kHz position)	(100)
R_C (C set below 318 kHz position)	(72)

NOTE: R_C is resistance measured from stator-to-rotor plates across the capacitor.

CHALLENGE CIRCUIT 13

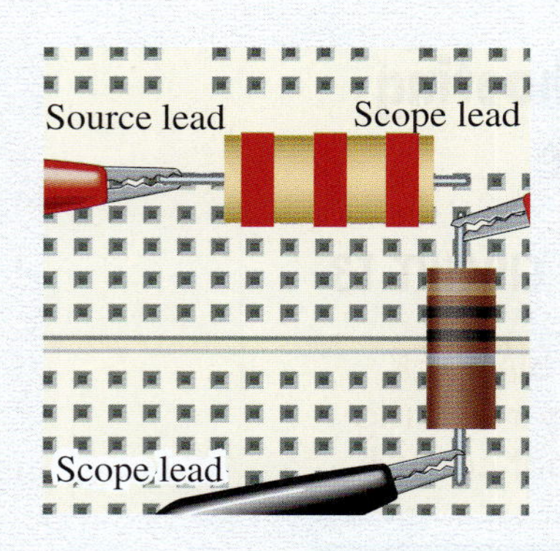

STEP 1

SYMPTOMS When the variable C is adjusted so the LC circuit is resonant at about 318-kHz source signal (i.e., C is about 100 pF), the scope shows maximum deflection due to maximum I_R drop at series resonance (Figure 1). When the variable C is adjusted so the LC circuit resonance is slightly higher in frequency (plates more unmeshed), the V_R (and scope deflection) slightly decrease as would be expected (Figure 2). When the capacitor is adjusted so the LC resonance is slightly lower (plates more meshed), the V_R and scope deflection drop drastically (Figure 3). This indicates some drastic change in the circuit. The off-resonance impedance must have increased at an abnormal rate for this circuit.

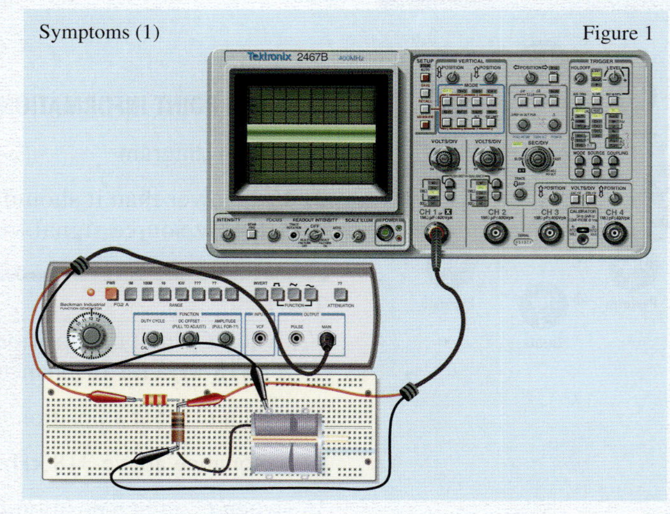

STEP 2

IDENTIFY initial suspect area: Since the only item being varied is the capacitance (because of our tuning efforts), the variable tuning capacitor is the initial suspect area.

STEP 3

MAKE test decision: We can check the tuning capacitor's operation with an ohmmeter to make sure that it isn't shorting at some point in its tuning range. The ohmmeter check is an easy check to find any possible shorts.

STEP 4

PERFORM **1st Test:** Perform a resistance check throughout the tuning range. As the plates approach being more fully meshed, a short between the stator and rotor plates becomes evident. This appears to be our problem. (NOTE: No Figure shown for this test.)

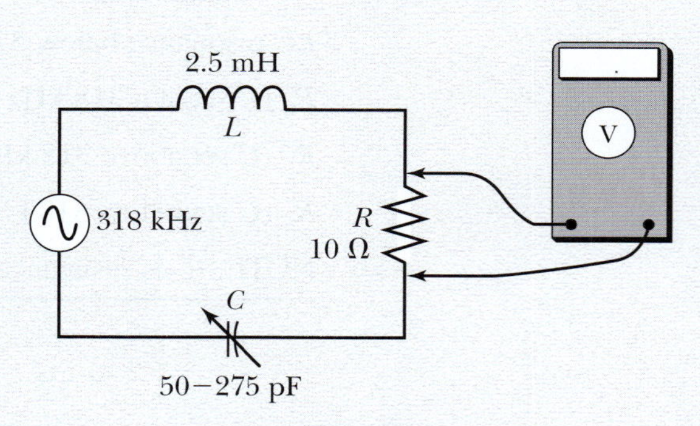

Challenge Circuit 13 multiSIM

STEP 5

LOCATE new suspect area: There is no new area because the variable capacitor has proven to be the likely culprit.

STEP 6

EXAMINE available data.

STEP 7

REPEAT analysis and testing: Let's visually inspect the capacitor to see if we can see some shorted plates.

2nd Test: Inspect the capacitor with strong light. The result is that you see two plates that are rubbing as the plates begin to more fully mesh. The problem is shorting plates on the tuning capacitor. (NOTE: No Figure shown for this test.)

3rd Test: Bend the plates (Figure 4) to eliminate the rubbing. Then, check with the ohmmeter again, while turning the rotor plates through their range.

STEP 8

VERIFY **4th Test:** The result is that everything is normal (Figure 5).

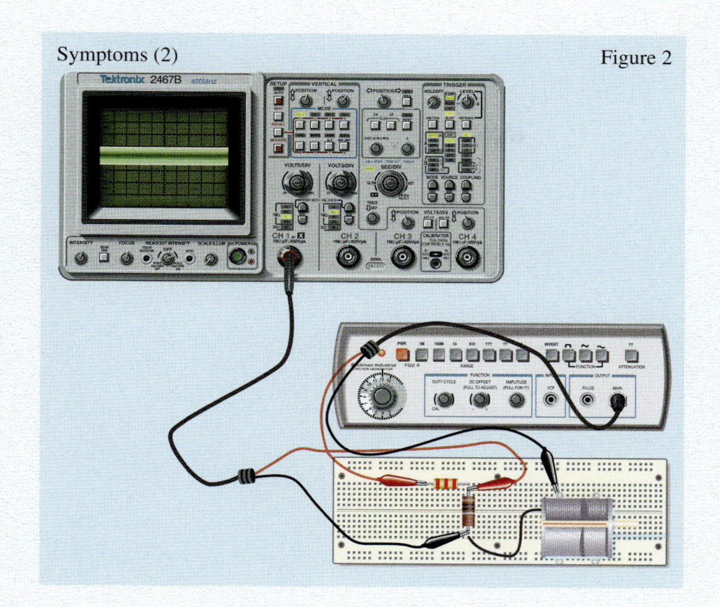

Symptoms (2) Figure 2

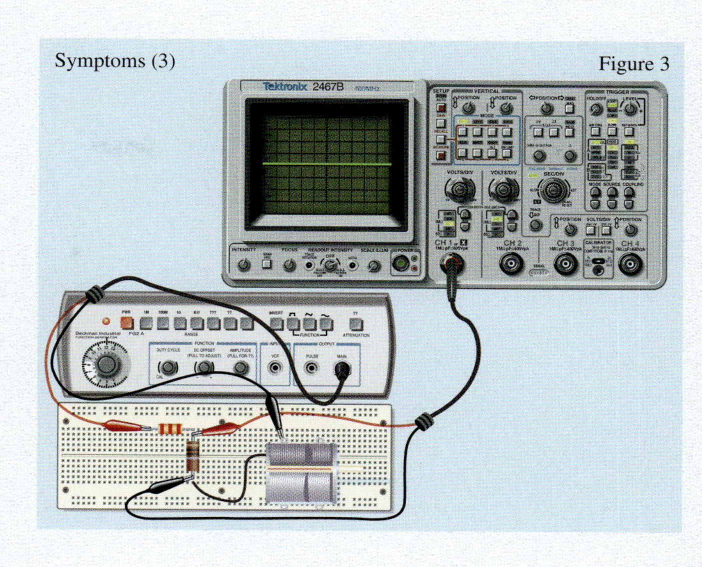

Symptoms (3) Figure 3

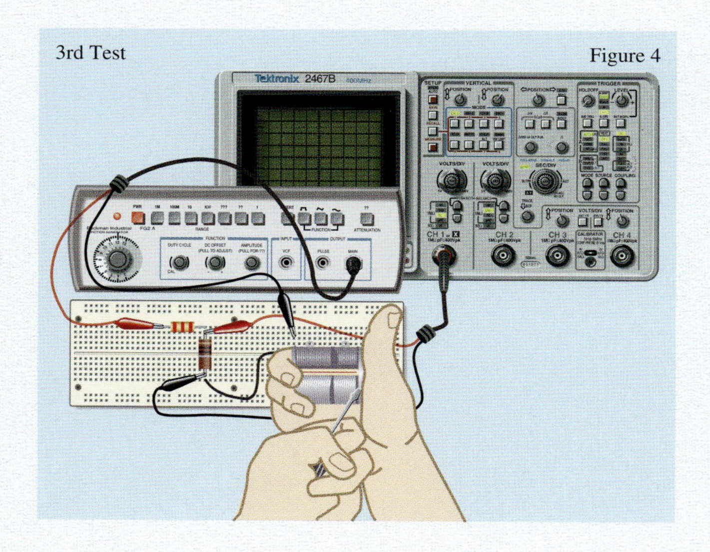

3rd Test Figure 4

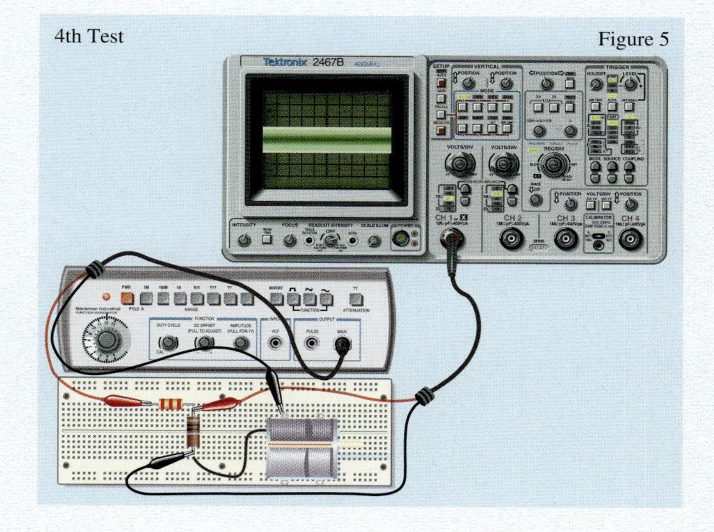

4th Test Figure 5

Troubleshooting Challenge

CHALLENGE CIRCUIT 14

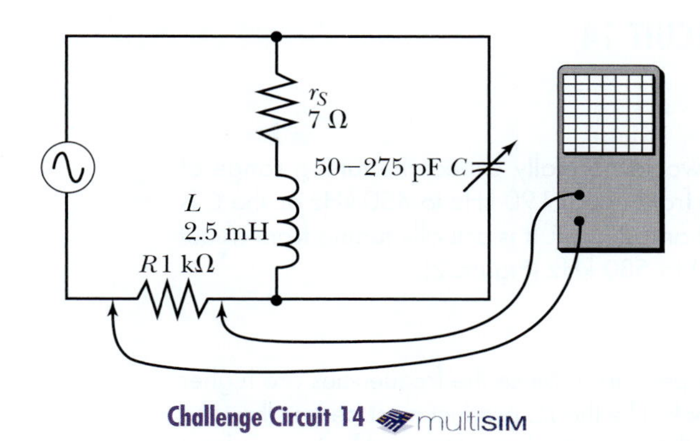

Challenge Circuit 14 multiSIM

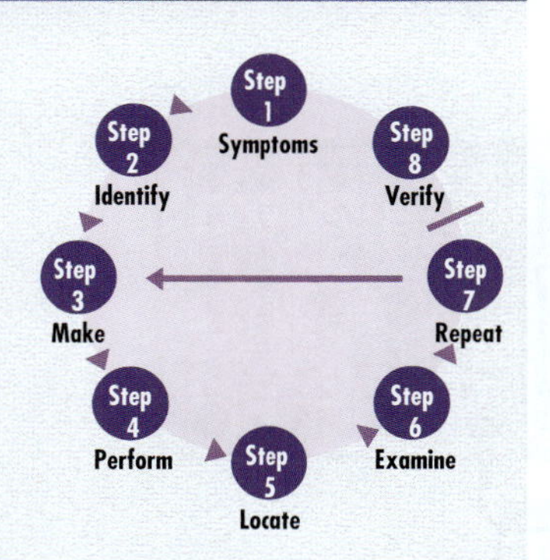

STARTING POINT INFORMATION

1. Circuit diagram
2. Using the scope to monitor when V across the 1-kΩ R is minimum (indicating that the parallel LC impedance is maximum), we find that the range of frequencies tuned from maximum C value to minimum C value is slightly higher in frequency than should be expected from the rated values shown on the diagram.

TEST	Results in Appendix C
$R_{inductor}$	(87)
$R_{capacitor}$	(19)
C of capacitor when "fully meshed"	(60)
C of capacitor when "fully unmeshed"	(107)
L of inductor	(41)

CHALLENGE CIRCUIT 14

STEP [1]

SYMPTOMS One would normally expect the tuning range of the *LC* circuit to be from about 190 kHz to 450 kHz as the *C* is varied between 50 and 275 pF. It is actually tuning from about 220 kHz (Figure 1) to 580 kHz (Figure 2).

STEP [2]

IDENTIFY initial suspect area: Since the frequencies are higher than expected, it looks like the *LC* product must be less than the rated values. Either the capacitor has decreased by losing plates (which is highly unlikely) or the inductor has decreased in value for some reason. We can't be sure at this point. The inductor is the initial suspect. However, we'll keep both the *C* and *L* in the initial suspect area.

STEP [3]

MAKE test decision: What could make the inductor have less inductance than anticipated? Possibly, shorted turns which would make *L* decrease. Since the rated dc resistance of this coil is about 7 Ω, an ohmmeter check (with the inductor isolated from the rest of the circuit) might reveal if enough turns are shorted to make a difference in coil resistance.

STEP [4]

PERFORM **1st Test:** With *L* isolated and source removed, measure its resistance (Figure 3). *R* measures about 5 Ω. Even this little resistance change could be significant, since we're dealing with such a small rated resistance in the first place. A number of turns could be shorted.

STEP [5]

LOCATE new suspect area: The hot zone is still the inductor. However, we need to make another inductor check so we can exclude the capacitor.

STEP [6]

EXAMINE available data.

STEP [7]

REPEAT analysis and testing: Since the coil looks suspicious, let's substitute a known good coil of like ratings and see how the circuit operates.

STEP [8]

VERIFY **2nd Test** and **3rd Test:** Substitute an inductor of the same ratings. Then check the circuit tuning characteristics. The result is that the component substitution worked. The circuit tunes with the expected range of frequencies, 190 kHz (Figure 4) to 450 kHz (Figure 5). The trouble was shorted turns on the inductor.

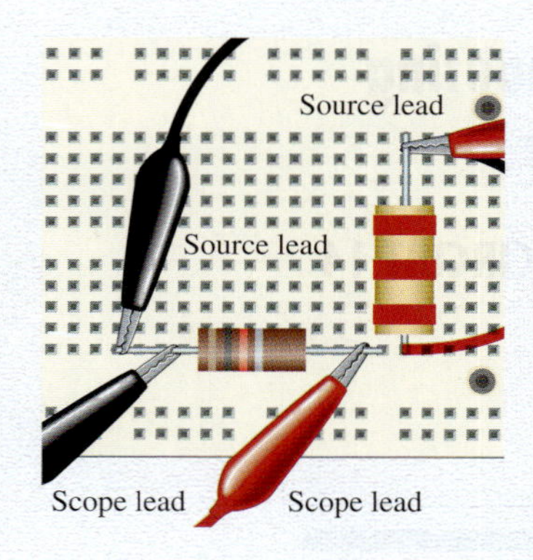

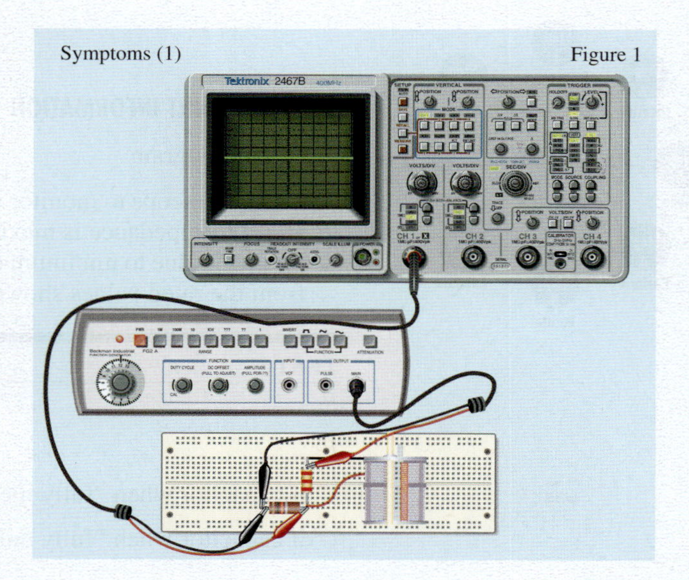

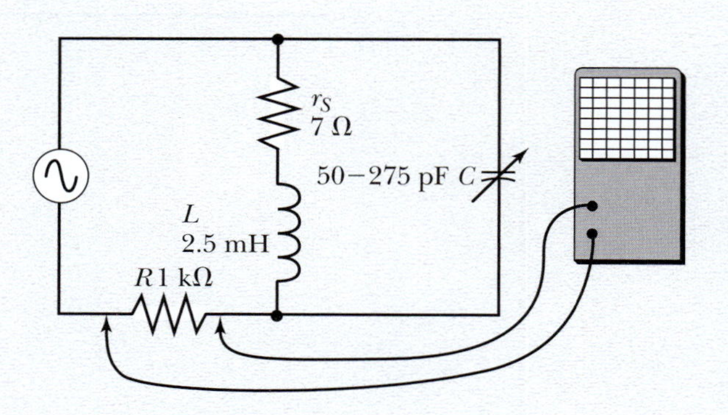

Challenge Circuit 14 multiSIM

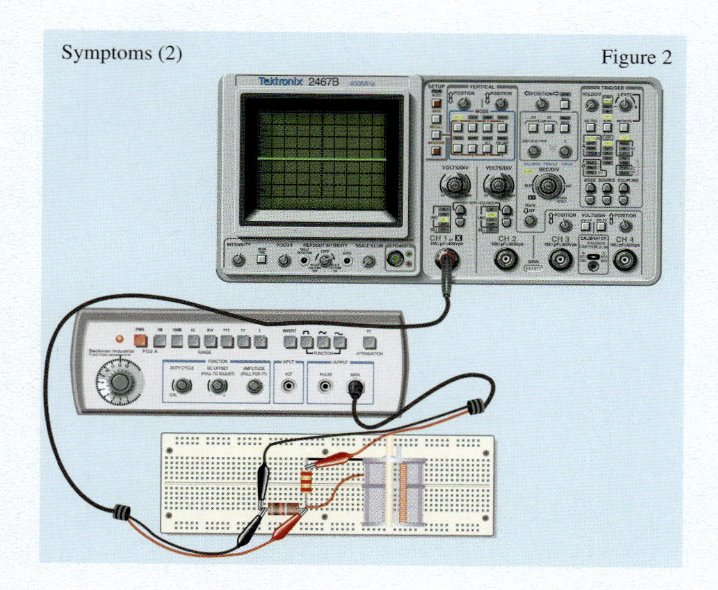

Symptoms (2) Figure 2

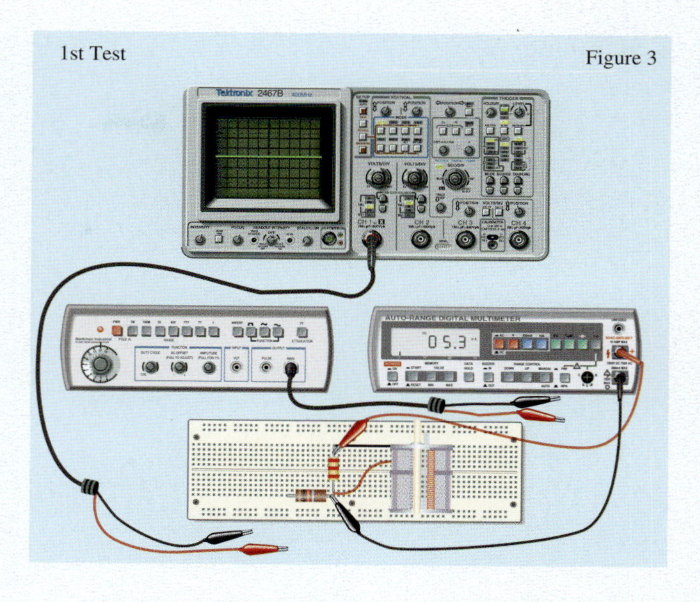

1st Test Figure 3

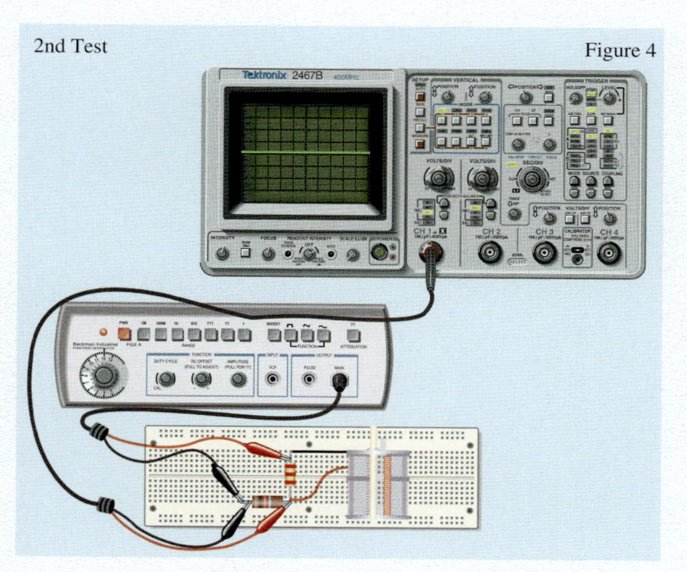

2nd Test Figure 4

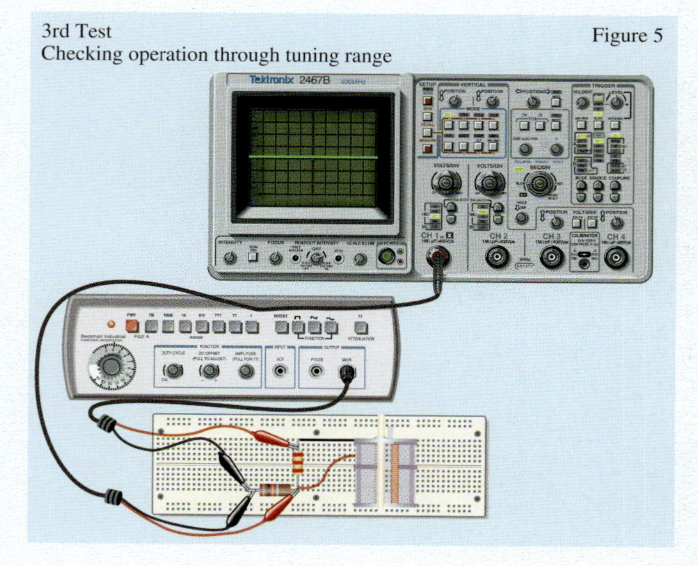

3rd Test Figure 5
Checking operation through tuning range

APPENDIX A

Color Codes

A. RESISTOR COLOR CODES

4-band resistor color code

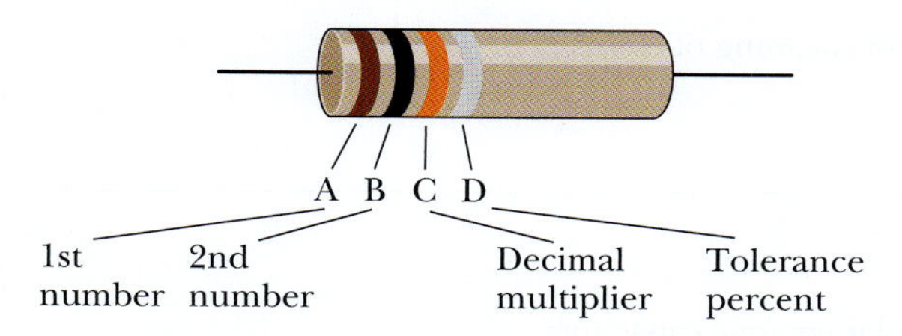

5-band precision resistor color code

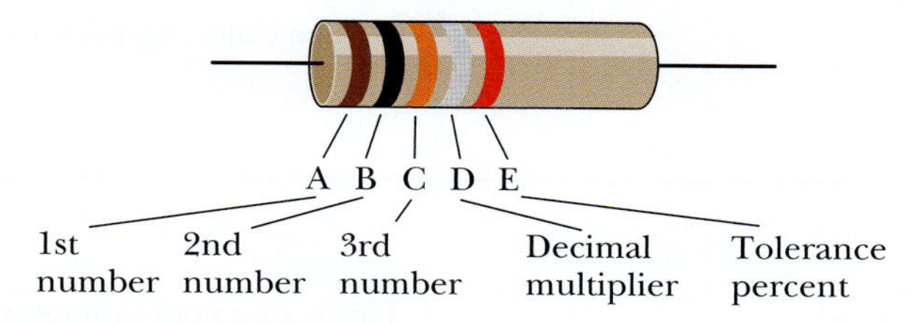

COLOR SIGNIFICANCE CHART

(For resistors and capacitors)

COLOR	NUMBER COLOR REPRESENTS	DECIMAL MULTIPLIER	PRECISION RESISTORS	TOLERANCE PERCENT (±%) GENERAL USE RESISTORS	VOLTAGE RATING	% FAILURE RATE PER 1,000 HRS OPER.
Black	0	1	—	—	—	—
Brown	1	10	1	1*	100*	1%
Red	2	100	2	2*	200*	0.1%
Orange	3	1,000	—	3*	300*	0.01%
Yellow	4	10,000	—	4*	400*	0.001%
Green	5	100,000	0.5	5*	500*	—
Blue	6	1,000,000	0.25	6*	600*	—
Violet	7	10,000,000	0.1	7*	700*	—
Gray	8	100,000,000	—	8*	800*	—
White	9	1,000,000,000	—	9*	900*	—
Gold	—	0.1	—	5	1,000*	—
Silver	—	0.01	—	10	2,000*	—
No Color	—	—	—	20	500*	—

(*Applicable to capacitors only)

B. SAMPLE SURFACE-MOUNT TECHNOLOGY (SMT) "CHIP" RESISTOR CODING

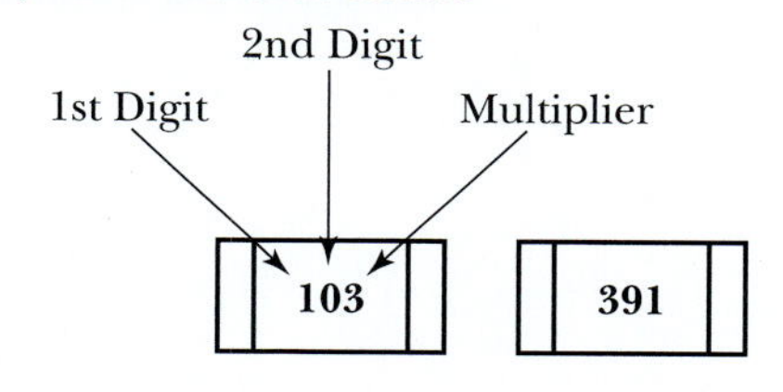

Code: 103 = 1, then 0, then 000 or 10,000 Ω   Code: 391 = 3, then 9, then 0, or 390 Ω

C. CAPACITOR COLOR CODES

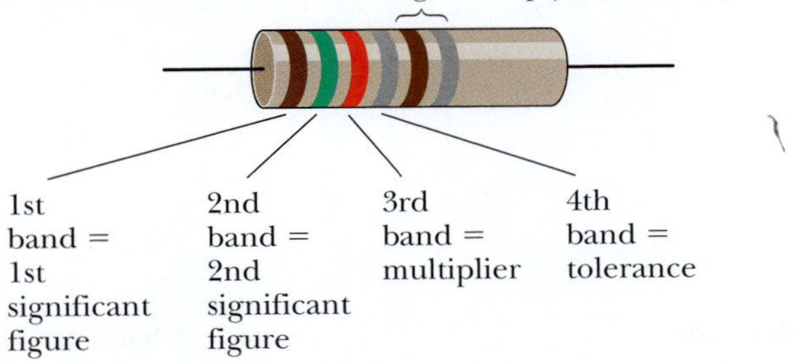

See Color Significance Chart for meaning of colors

Band color coding system for tubular ceramic capacitors

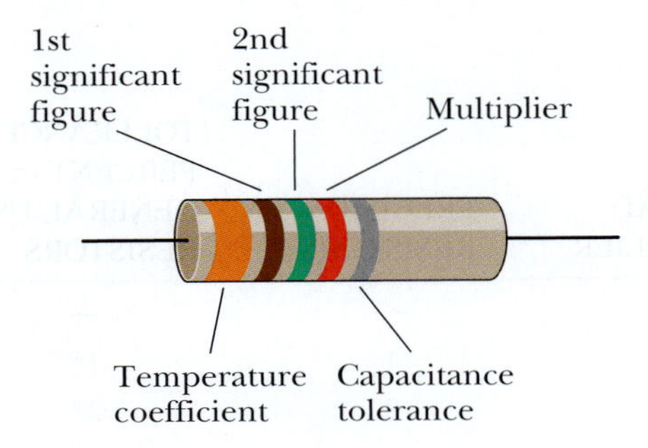

D. SAMPLE CERAMIC DISK CAPACITORS CODED MARKINGS

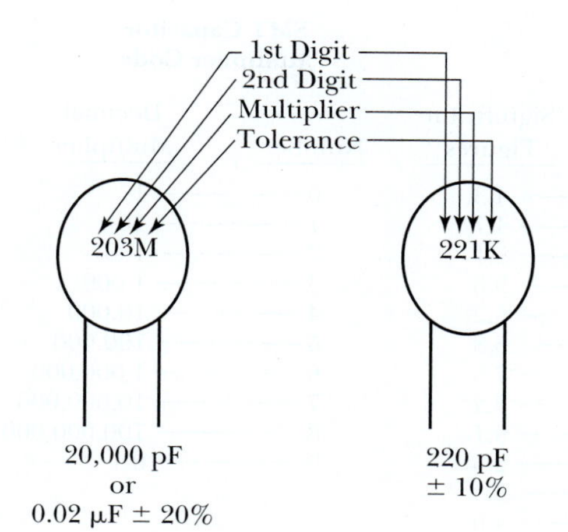

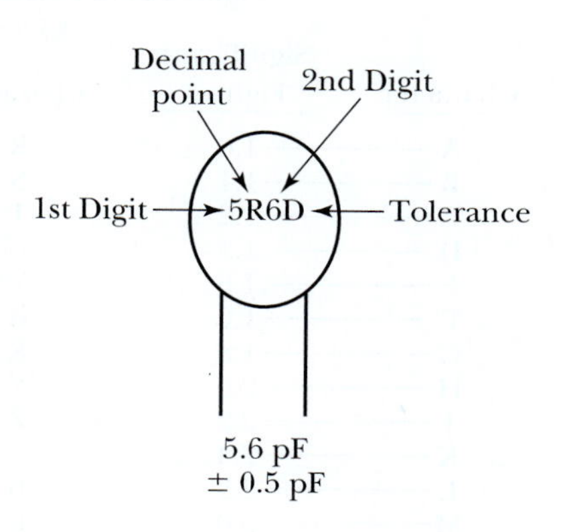

Third-Digit Multiplier

Number	Multiply by
0	Nothing
1	10
2	100
3	1,000
4	10,000

Letter Tolerance Code

Under 10 pF Values:

Letter	Tolerance
B	± 0.1 pF
C	± 0.25 pF
D	± 0.5 pF
F	± 1.0 pF

Over 10 pF Values:

Letter	Tolerance
E	± 25%
F	± 1%
G	± 2%
H	± 2.5%
J	± 5%
K	± 10%
M	± 20%
P	− 0%, + 100%
S	− 20%, + 50%
W	− 0%, + 200%
X	− 20%, + 40%
Z	− 20%, + 80%

E-MOUNT TECHNOLOGY (SMT) "CHIP" CAPACITOR CODING

SMT Capacitor Significant Figures Letter Code

Character	Significant Figures	Character	Significant Figures
A	1.0	R	4.3
B	1.1	S	4.7
C	1.2	T	5.1
D	1.3	U	5.6
E	1.5	V	6.2
F	1.6	W	6.8
G	1.8	X	7.5
H	2.0	Y	8.2
J	2.2	Z	9.1
K	2.4	a	2.5
L	2.7	b	3.5
M	3.0	d	4.0
N	3.3	e	4.5
P	3.6	f	5.0
Q	3.9	m	6.0
		n	7.0
		t	8.0
		y	9.0

SMT Capacitor Multiplier Code

Number	Decimal Multiplier
0	1
1	10
2	100
3	1,000
4	10,000
5	100,000
6	1,000,000
7	10,000,000
8	100,000,000
9	0.1

(a)

Decoding Examples:

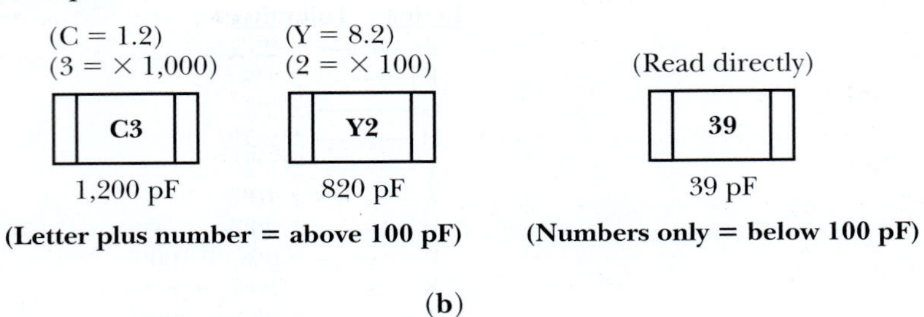

(b)

F. SEMICONDUCTOR DIODE COLOR CODE(S)

(NOTE: Prefix "1N . . . " is understood.)

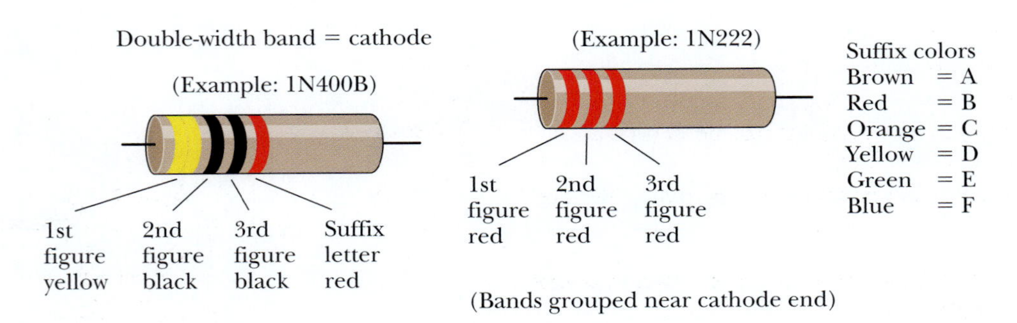

APPENDIX B

Answers

Note: Where current flow directions are shown, solid line arrows are electron flow and dotted line arrows are conventional flow.

CHAPTER 1

In-Process Learning Checks

—1—

1. Matter is anything that has ***weight*** and occupies ***space.***
2. Three physical states of matter are ***solid, liquid,*** and ***gas.***
3. Three chemical states of matter are ***elements, compounds,*** and ***mixtures.***
4. The smallest particle that a compound can be divided into but retain its physical properties is the ***molecule.***
5. The smallest particle that an element can be divided into but retain its physical properties is the ***atom.***
6. The three parts of an atom that interest electronics students are the ***electron, proton,*** and ***neutron.***
7. The atomic particle having a negative charge is the ***electron.***
8. The atomic particle having a positive charge is the ***proton.***
9. The atomic particle having a neutral charge is the ***neutron.***
10. The ***protons*** and ***neutrons*** are found in the atom's nucleus.
11. The particle that orbits the nucleus of the atom is the ***electron.***

—2—

1. An electrical system (circuit) consists of a source, a way to transport the electrical energy, and a ***load.***
2. Static electricity is usually associated with ***nonconductor***-type materials.
3. The basic electrical law is that ***unlike*** charges attract each other and ***like*** charges repel each other.
4. A positive sign or a negative sign often shows electrical ***polarity.***
5. If two quantities are *directly* related, as one increases the other will ***increase.***
6. If two quantities are *inversely* related, as one increases the other will ***decrease.***
7. The unit of charge is the ***coulomb.***
8. The unit of current is the ***ampere.***
9. An ampere is an electron flow of one ***coulomb*** per second.
10. If two points have different electrical charge levels, there is a difference of ***potential*** between them.
11. The volt is the unit of ***electromotive*** force, or ***potential*** difference.

Review Questions

1. Matter is anything that has weight and occupies space, whether it is a solid, liquid, or gas.
3. Three chemical states of matter are elements, compounds, and mixtures. Copper is an element, sugar is a compound, and sand and gold dust are a mixture.
5. A compound is a form of matter that can be chemically divided into simpler substances and that has two or more types of atoms. Examples are water (hydrogen and oxygen) and sugar (carbon, hydrogen, and oxygen).
7. I.D. of charges on particles (electron = negative; proton = positive).
9. Free electrons are electrons in the atomic structure that can be easily moved or removed from their original atom. These are outer-ring or valence electrons in conductor materials.
11. Chemical, mechanical, light, and heat.
13. Unlike charges attract and like charges repel each other.
15. Force of attraction or repulsion will be one-ninth the original force.
17. Unit of measure of electromotive force (emf) is the volt.
19. **a.** Unit of measure for current is the ampere.
    b. 4 amperes of current flow
21. A closed circuit has an unbroken path for current flow. An open circuit has discontinuity or a broken path; therefore, the circuit does not provide a continuous path for current.
23. – sign denotes a negative polarity point.
25. Point B is positive side of source.

Analysis Questions

1. Drawing:

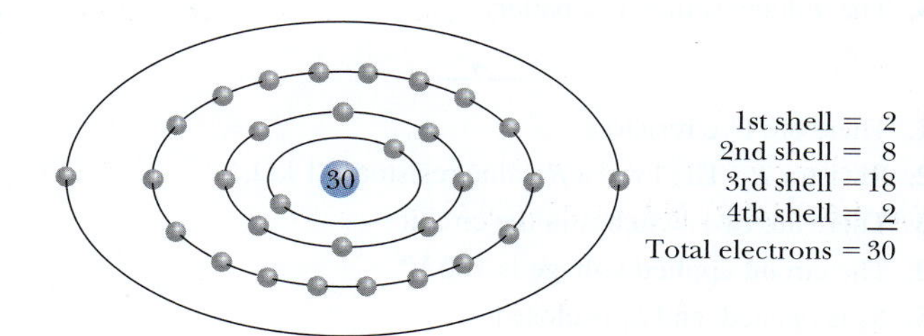

3. Zinc
5. 30
7. Circuit diagram:

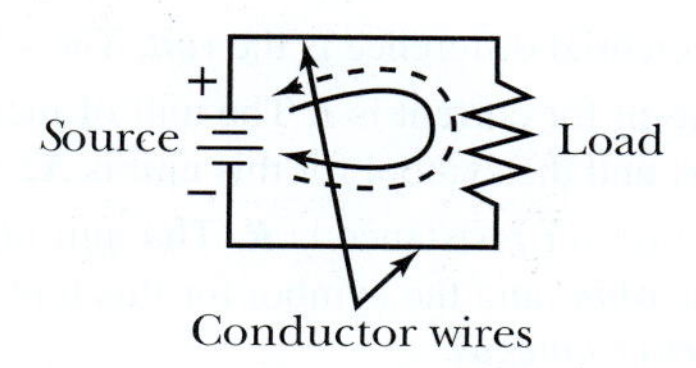

9. Usually an undesired very low resistance path around part or all of a circuit.

CHAPTER 2

Practice Problems

—1—

The circular-mil cross-sectional area = $2.5^2 = 6.25$ CM.

—2—

The resistance of 300 feet of copper wire having a cross-sectional area of 2,048 CM = 1.523 Ω.

—3—

1. 500 feet of #12 wire has a resistance of approximately 0.81 Ω.
2. The wire gauge of wire having a cross-sectional area of 10,380 CM is #10 wire.
3. #10 wire has an approximate resistance of 1.018 Ω per thousand feet of length.

—4—

1. 39 kΩ, 10% tolerance.
2. 18.2 Ω, ± 1% tolerance.

—5—

1. It is a TV receiver system.
2. The signal flows from left to right, as illustrated.
3. The receiver circuitry receives signals from the TV antenna.
4. Two types of outputs are picture information and audio sound information.

—6—

1. There are three kinds of components (resistor, switch, battery).
2. The highest-value resistor is 100 kΩ.
3. An SPST switch is used.
4. The voltage source is a battery.

—7—

1. There are five resistors.
2. Resistor R_3. The lowest R value resistor is 1 kΩ.
3. There are two switches in the circuit.
4. The circuit applied voltage is 100 V.
5. S_2 is opened, and S_1 is closed.

In-Process Learning Checks

—1—

1. Charge is represented by the letter ***Q***. The unit of measure is the ***coulomb,*** and the abbreviation is ***C.***
2. The unit of potential difference is the ***volt.*** The symbol is **V.**
3. The abbreviation for current is ***I.*** The unit of measure for current is the ***ampere,*** and the symbol for this unit is **A.**
4. The abbreviation for resistance is ***R.*** The unit of measure for resistance is the ***ohm,*** and the symbol for this unit is Ω ***(the lowercase Greek letter omega).***
5. Conductance is the ***ease*** with which current flows through a component or circuit. The abbreviation is ***G.*** The unit of measure for conductance is the ***siemens,*** and the symbol for this unit is **S.**
6. 0.0000022 amperes = ***2.2 microamperes (μA).*** Could be expressed as ***2.2 × 10^{-6}A.***
7. One-thousandth is represented by the metric prefix ***milli.*** When expressed as a power of 10, it is ***10^{-3}.***
8. 10,000 ohms might be expressed as ***10 kΩ*** (kilohms). Expressed as a whole number times a power of 10, ***10 × 10^3 ohms.***

—2—

1. The greater the resistivity of a given conductor, the ***higher*** its resistance will be per given length.
2. The circular-mil area of a conductor is equal to the square of its ***diameter*** in ***mils (thousandths of an inch).***
3. The smaller the diameter of a conductor, the ***higher*** its resistance will be per given length.
4. The higher the temperature in which a conductor must operate, the ***higher*** will be its resistance.
5. Wire tables show the AWG wire ***size,*** the cross-sectional area of wires in ***circular mils,*** the resistance per given length of copper wire, and the operating ***temperature*** at which these parameters hold true.

—3—

1. Negative lead connected to ***Point B,*** positive lead connected to ***Point A.***
2. Negative lead connected to ***Point B,*** positive lead connected to ***Point C.***
3. The first thing to do is to remove the source voltage from the circuit.

Review Questions

1. 0.003 V
3. A micro unit is one-thousandth of a milli unit.
5. Schematic diagram of battery, two resistors, and an SPST switch:

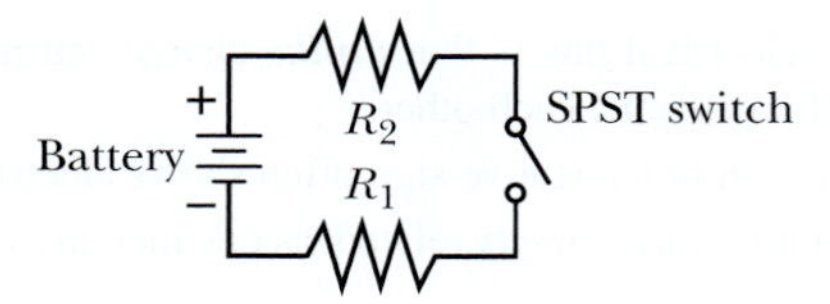

7. Resistor
9. Battery
11. Carbon or composition resistors and wire-wound resistors. Carbon/composition resistors are composed of carbon or graphite mixed with powdered insulating material. Wirewound resistors are constructed with special resistance wire wrapped around a ceramic or insulator core.
13. Value, power rating, tolerance, and physical size
15. 22,200 Ω ± 2%
17. Would likely use a wire-wound resistor
19. Would use the 5th band; expected failure rate = 0.01%
21. Two precautions: Use only one hand; insulate yourself from ground; use only tools with insulated handles.
23. Two dangers: electrical shock, physical injury
25. Causes of shock: Part of body provides current path across points having a potential difference.

Analysis Questions

1. The resistor color code is necessary because many resistors are too small to enable printing their values on them. Also, color code provides a "universal" method, readable by anyone in the world, regardless of their language.
3. Generally, reliability refers to the ability of the component to perform as it is expected to, for the length of time intended, under normal operating conditions. Resistor reliability ratings typically indicate the percentage of failure for 1,000 hours of operation.
5. Block diagram of a simple power distribution system:

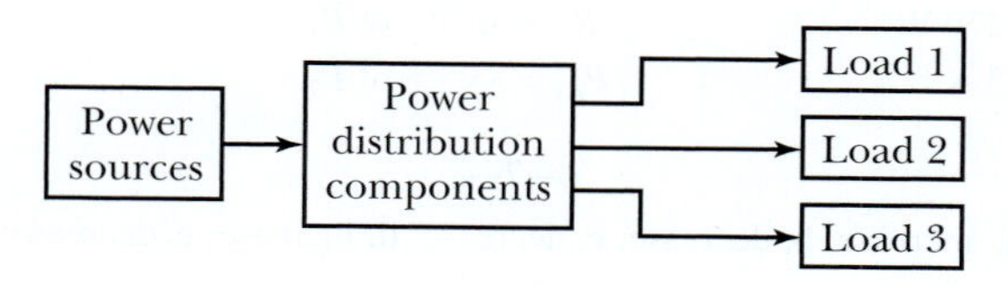

7. **a.** Brown, black, red, silver
   b. Red, violet, orange
   c. Brown, black, green, silver
   d. Brown, black, black, gold
9. Two connection points

CHAPTER 3

Practice Problems

—1—

1. $I = V/R = 150\text{ V}/10\ \Omega = 15\text{ A}$
2. 7.5 A
3. **a.** Remain the same
   b. $300\text{ V}/20\ \Omega = 15\text{ A}$

—2—

1. 25×10^3
2. 1.2335×10^1
3. 1.0×10^{-1}
4. 1.5×10^{-3}
5. 23,500
6. 1,000
7. 10,000
8. 0.015

—3—

1. 1,456,000
2. 0.0333
3. 1.5
4. 1,000,000

—4—

1. P = energy/time = 25 J/2 sec = 12.5 W
2. P = energy/time = 1 J/100 sec = 0.01 W or 10 mW

—5—

1. Energy = Power × Time = 250 W × 8 hr = 2,000 Wh, or 2 kWh

—6—

1. $P = V \times I = 150\text{ V} \times 10\text{ mA} = 1{,}500\text{ mW}$, or 1.5 W
2. **a.** I would decrease to half the original value.
   b. Power diss. $= V \times I = 150 \times 5\text{ mA} = 750\text{ mW}$, or 0.75 W

—7—

1. $V_A = I \times R = 1.5\text{ mA} \times 4.7\text{ k}\Omega = 7.05\text{ V}$
2. $I = V/R = 100\text{ V}/27\text{ k}\Omega = 3.704\text{ mA}$
3. $R = V/I = 100\text{ V}/19.6\text{ mA} = 5.102\text{ k}\Omega$

—8—

1. $P = 40$ W; $I = 2$ A
2. $I = 1$ mA; $R = 100$ kΩ
3. $I = 12$ mA; $P = 1.44$ W (or 1,440 mW)

In-Process Learning Checks

—1—

1. $I =$ ***1.5 A,*** or ***1,500 mA.*** $(I = V/R;\ I = 15\text{ V}/10\ \Omega = 1.5\text{ A})$
2. $V =$ ***30 V.*** $(V = I \times R;\ V = 2\text{ mA} \times 15\text{ k}\Omega = 30\text{ V})$
3. $R =$ ***27 kΩ.*** $(R = V/I;\ R = 135\text{ V}/5\text{ mA} = 27\text{ k}\Omega)$
4. **a.** Current would decrease.
   b. Voltage applied would be higher.
   c. Circuit resistance must have halved.

—2—

1. negative, positive
2. positive, negative
3. conventional
4. one, changes
5. expenditure, work, work, energy
6. joules, seconds
7. kilowatt hour

Review Questions

1. **a.** $I = V/R;\ V = I \times R;\ R = V/I$
   b. $P = V \times I;\ P = I^2R;\ P = V^2/R$
3. **a.** 5.1×10^3
   b. 47×10^3
   c. 0.000001×10^3
   d. 0.039×10^3
5. Energy is the ability to do work.
7. Work is measured in foot poundals, or foot pounds, or horsepower. Energy is measured in joules. Power (electrical) is measured in watts.
9. Power unit = the watt

Problems

1. The new current will be four times the original current.
3. a. Schematic:

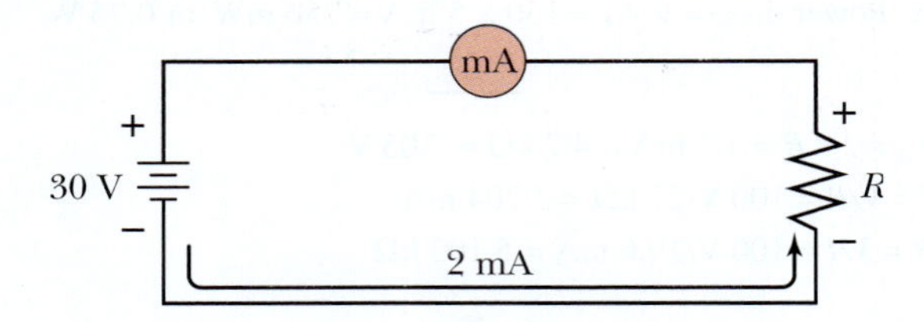

b. $R = V/I = 30\text{ V}/2\text{ mA} = 15\text{ k}\Omega$; $P = V \times I = 30\text{ V} \times 2\text{ mA} = 60\text{ mW}$

c. See diagram.

5. $V = I \times R = 3\text{ mA} \times 12\text{ k}\Omega = 36\text{ V}$
7. $R = V/I = 41\text{ V}/50\text{ mA} = 820\ \Omega$
9. $I = 3.162\text{ A}$
11. $R = V/I = 100\text{ V}/8.5\text{ mA} = 11{,}765\text{ k}\Omega$
13. Power dissipated will increase ($P = V \times I$ and I remained the same while V doubled).
15. 2.8125 W
17. 20.83 mA
19. 33.85 Wh
21. 100 V
23. 31.6 V
25. a. 0.01

b. 10^{12}

c. 8.57×10^4 or 85,700

d. 6.1664×10^3 or 6.17×10^3

Analysis Questions

1. The EE (exponent entry) key allows entering numbers with exponents, such as numbers raised to some power of 10.
3. Current flows only in one direction with the dc source. It alternates direction of flow with ac applied.
5. The direction a + (positive) charge would move.
7. V has doubled.

CHAPTER 4

Practice Problems

—1—

1. Total resistance = 74 kΩ
2. Total resistance = 16 kΩ

—2—

1. $R_1 = 47\ \Omega$; $R_T = 100\ \Omega$
2. New $R_T = 10.5\text{ k}\Omega$
3. $R_2 = 3.9\text{ k}\Omega$

—3—

1. $V_A = 411\text{ V}$; $I_T = 3\text{ mA}$; $V_{R_2} = 81\text{ V}$; $V_{R_1} = 300\text{ V}$
2. $V_A = 342.5\text{ V}$; $I_T = 2.5\text{ mA}$; $V_{R_2} = 67.5\text{ V}$; $V_{R_1} = 250\text{ V}$

—4—

1. $V_X = (R_X/R_T) \times V_T = (2.7\text{ k}\Omega/17.4\text{ k}\Omega) \times 50\text{ V} = 0.155 \times 50 = 7.75\text{ V}$ (V_{R_1})
2. ($V_{R_2} = 13.5$ V, as calculated in the example)
3. $(10\text{ k}\Omega/17.4\text{ k}\Omega) \times 50\text{ V} = 0.575 \times 50 = 28.75\text{ V}$ (V_{R_3})
4. 7.75 + 13.5 + 28.75 = 50 V (Yes, they add up correctly.)

—5—

$V_{R_2} = 30\text{ V}$; $V_A = 75\text{ V}$

—6—

$P_{R_1} = 90\text{ W}$ $V_{R_1} = 30\text{ V}$
$P_{R_2} = 90\text{ W}$ $V_{R_2} = 30\text{ V}$
$P_{R_3} = 90\text{ W}$ $V_{R_3} = 30\text{ V}$
$P_T = 270\text{ W}$ $R_3 = 33.3\%$ of R_T
$I_T = 3\text{ A}$ $P_{R_3} = 33.3\%$ of P_T

—7—

1. a. increase; b. decrease; c. decrease; d. increase; e. decrease
2. Yes

—8—

1. With the 47-kΩ resistor shorted, the voltage drops are:
   10-kΩ resistor drops 14.6 V
   27-kΩ resistor drops 39.4 V
   47-kΩ resistor drops 0 V
   100-kΩ resistor drops 146 V
2. They increase.

—9—

Givens: $R_1 = 20\text{ k}\Omega$; $V_1 = \frac{2}{5}V_A$; $V_2 = \frac{1.5}{5}V_A$;
$V_A = 50\text{ V}$; $R_2 = 15\text{ k}\Omega$; $R_3 = 15\text{ k}\Omega$;
$V_{R_2} = 15\text{ V}$; $V_{R_3} = 15\text{ V}$; $P_T = 50\text{ mW}$; $P_{R_1} = 20\text{ mW}$;
$P_{R_2} = 15\text{ mW}$; $P_{R_3} = 15\text{ mW}$; $I = 1\text{ mA}$

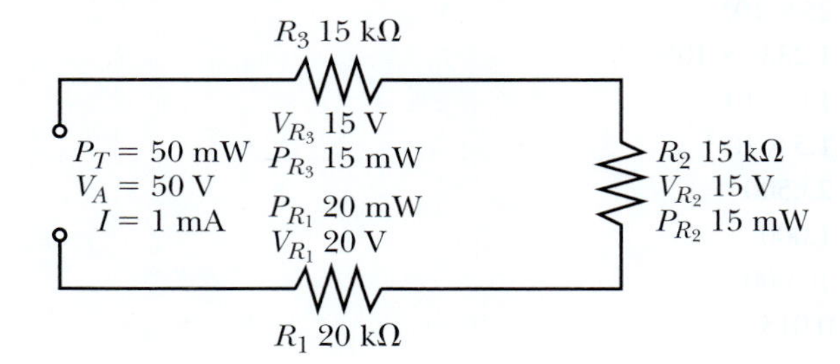

—10—

1. $V_{R_1} = 4.5\text{ V}$
2. $V_{R_2} = 1.5\text{ V}$

—11—

D to C = 94 V; C to B = 54 V; B to A = 40 V; D to B = 148 V

—12—

1. Voltage across $R_2 = 27.49$ V (or approximately 27.5 V).
2. Voltage at top of R_2 is +27.5 V with respect to point C.
3. Voltage across $R_1 = 9.995$ V (or approximately 10 V).
4. Voltage at point C with respect to point A = –37.5 V.

—13—

1. Minimum power rating should be 2 times (60 V × 30 mA), or 2 × 1,800 mW = 3,600 mW (3.6 W).

2. R dropping equals $\frac{110\text{ V}}{30\text{ mA}} = 3.66\text{ k}\Omega$. Minimum power rating should be 2 times (110 V × 30 mA), or 2 × 3,300 mW = 6,600 mW (6.6 W).

In-Process Learning Checks

—1—

1. The primary identifying characteristic of a series circuit is that the ***current*** is the same throughout the circuit.
2. The total resistance in a series circuit must be greater than any one ***resistance*** in the circuit.
3. In a series circuit, the highest value voltage is dropped by the ***largest or highest*** value resistance, and the lowest value voltage is dropped by the ***smallest or lowest*** value resistance.
4. V drop by the other resistor must be ***100 V.***
5. Total resistance ***increases.*** Total current ***decreases.*** Adjacent resistor's voltage drop ***decreases.***
6. Applied voltage is ***130 V.***

Review Questions

1. Current is the same throughout the circuit because there is only one path for current.
3. R_T equals the sum of all the resistor values in series.
5. b (e.g., In a series circuit the applied circuit voltage is equal to the sum of all the individual voltage drops around the circuit.)
7. a (e.g., The power distribution throughout a series circuit is directly related to the resistance distribution around the circuit.)
9. c (e.g., If a short is placed across part of a series circuit, R_T decreases.)

Problems

1. $R_T = 10/5 = 2\ \Omega$; each resistor $= 1\ \Omega$
3. Diagram, labeling, and calculations for a series circuit composed of a 50-, 40-, 30-, and 20-Ω resistor.

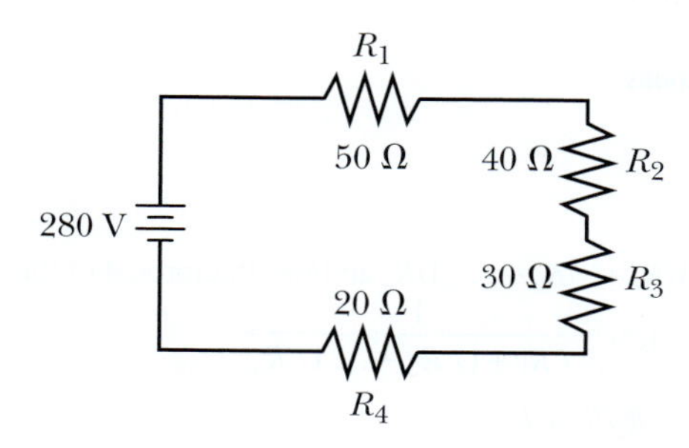

a. $R_T = 50 + 40 + 30 + 20 = 140\ \Omega$
b. If $I = 2$ A, then V applied $= IR = 2\text{ A} \times 140\ \Omega = 280$ V
c. $V_{R_1} = 100$ V; $V_{R_2} = 80$ V; $V_{R_3} = 60$ V; $V_{R_4} = 40$ V
d. $P_T = I_T \times V_T = 2\text{ A} \times 280\text{ V} = 560$ W
e. $P_{R_2} = 2\text{ A} \times 80\text{ V} = 160$ W; $P_{R_4} = 2\text{ A} \times 40\text{ V} = 80$ W
f. 1/7 V_A dropped by R_4
g. If R_3 increased in value:
   (1) total resistance would increase.
   (2) total current would decrease.
   (3) V_{R_1} would decrease; V_{R_2} would decrease; and V_{R_4} would decrease.
   (4) P_T would decrease because I decreased and V stayed the same.
h. If R_2 shorted:
   (1) total resistance would decrease.
   (2) total current would increase.
   (3) V_{R_1} would increase; V_{R_3} would increase; V_{R_4} would increase (because of increased I with Rs staying the same).

5. Diagram of three sources to acquire 60 V if sources are 100 V, 40 V, and 120 V.

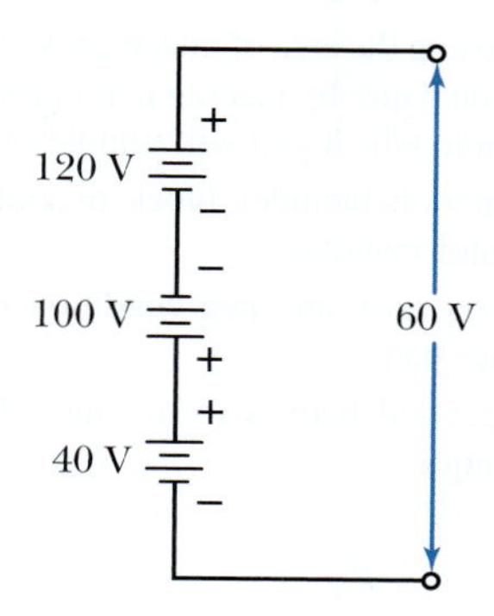

7. a. $V_A = 34$ V (4 V + 20 V + 10 V)
   b. $R_3 = 2.5\text{ k}\Omega$ (to drop 5 V of the 10 V dropped by R_3 and R_4)
   c. $R_2 = 10\text{ k}\Omega$ (to drop 20 V with 2 mA of current)
9. 17.22 V
11. 2.15 V
13. 3.87 V
15. Most power $= R_1$; least power $= R_2$
17. $V_{R_3} = 70$ V
19. a. R_1 is set at 3.81 kΩ
   b. $I_T = 4.2$ mA
   c. R_1 dissipates 16% of the total power
   d. I_T would equal 4 mA
   e. Voltmeter would indicate 50 V
21. a. $P_{R_2} = 423$ mW
   b. $P_T = 819$ mW
   c. $I_T = 3$ mA
23. Circuit with series "dropping resistor" used to drop from a 100-V source to a 50-V source with 50-mA load current:

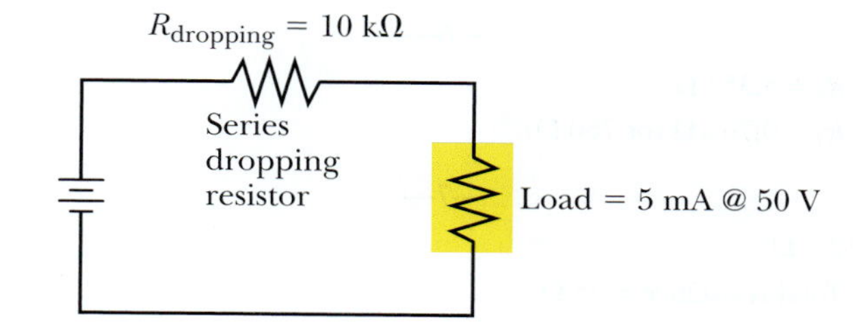

25. Decrease
27. $V_{\text{applied}} = 182$ V

Analysis Questions

1. In the scientific mode, the results always indicate a number between 1 and 10 times some power of 10.

   In the engineering mode, the results show a number times a power of 10 with an exponent that is some multiple or submultiple of 3 (taking advantage of the metric system).

3. The "divide-and-conquer" or "split" rule indicates that where there is a linear, or series, flow it is often good to make the first check at the middle of the questionable area. This enables eliminating half the questionable area with just one test.
5. To identify "initial suspect area": Analyze all symptom information; determine all sections or components in the system that could cause the symptoms; then "bracket" the area of uncertainty, or suspect area in which to make tests and diagnoses.
7. Bracketing: enclosing the area of uncertainty, or "suspect area," either mentally or on paper by placing parenthesis, brackets, etc., to highlight the area in which you will troubleshoot.
9. Useful documentation includes block diagrams, schematic diagrams, and operator manuals.
11. If current drops to zero, an open condition has occurred somewhere in the series path.
13. The method suggested here is to use the "divide-and-conquer" (half-split) technique.

CHAPTER 5

Practice Problems

—1—

$V_{R_1} = V_{R_2} = 2\text{ mA} \times 27\text{ k}\Omega = 54\text{ V}$
$V_A = V_{R_1} = V_{R_2} = 54\text{ V}$

—2—

$I_T = I_1 + I_2 + I_3 + I_4 + I_5$
The lowest R value branch passes the most current.
The highest R value branch passes the least current.

—3—

$V_{R_2} = 50\text{ V}$; $V_{R_1} = 50\text{ V}$; $V_A = 50\text{ V}$; $I_1 = 5\text{ A}$; $I_T = 6\text{ A}$

—4—

1. $R_1 = 15\text{ k}\Omega$; $I_2 = 2.5\text{ mA}$; $I_3 = 2.5\text{ mA}$; $R_2 = 30\text{ k}\Omega$
2. $R_1 = 27\text{ k}\Omega$; $I_2 = 3.78\text{ mA}$; $I_3 = 2.23\text{ mA}$; $R_2 = 33\text{ k}\Omega$; $R_T = 11.75\text{ k}\Omega$.

—5—

1. $V_{R_1} = 30\text{ V}$; $V_T = 30\text{ V}$
2. $R_T = 1.67\text{ k}\Omega$

—6—

1. $R_T = 5.35\text{ k}\Omega$
2. $R_T = 0.76\text{ k}\Omega$ (or 760 Ω)

—7—

1. 2.5 Ω
2. Total resistance = 25 Ω

—8—

1. $R_T = 12\ \Omega$
2. $R_T = 7.67\text{ k}\Omega$

—9—

$R_T = 27.15\ \Omega$

—10—

$R_2 = 100\ \Omega$

—11—

$P_T = 2{,}250\text{ mW}$ or 2.25 W

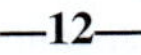

—12—

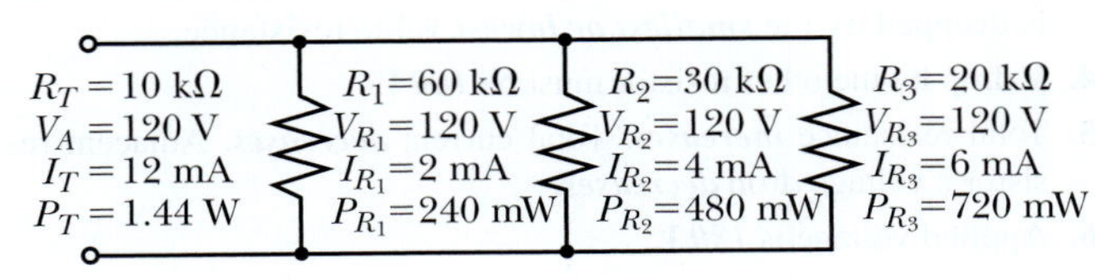

—13—

1. $I_1 = 1.85\text{ mA}$ (through 27-kΩ resistor)
   $I_2 = 1.07\text{ mA}$ (through 47-kΩ resistor)
   $I_3 = 0.50\text{ mA}$ (through 100-kΩ resistor)
2. $R_T = 1.54\text{ k}\Omega$
   $I_1 = 0.616\text{ mA}$ (through 10-kΩ resistor)
   $I_2 = 1.1\text{ mA}$ (through 5.6-kΩ resistor)
   $I_3 = 2.28\text{ mA}$ (through 2.7-kΩ resistor)

—14—

1. $I_1 = 1.65\text{ mA}$
   $I_2 = 0.35\text{ mA}$
2. $I_1 = 16.4\text{ mA}$
   $I_2 = 3.6\text{ mA}$

Review Questions

1. b
3. c
5. a
7. $1/R_T = 1/R_1 + 1/R_2 + \ldots 1/R_n$ and/or "Reciprocal of the reciprocals": $\dfrac{1}{1/R_1 + 1/R_2 + \ldots 1/R_n}$
9. $R_T = R_1 \times R_2/R_1 + R_2$
11. $R_u = R_k \times R_e/R_k - R_e$

Problems

1. $R_T = 17.15\text{ k}\Omega$; product-over-the-sum method
3. $I_2 = 30\text{ mA}$; Kirchhoff's current law

5. a. I_1 will RTS.
   I_2 will D.
   I_3 will RTS.
   I_T will D.
   P_T will D.
   V_{R_1} will RTS.
   V_T will RTS.

   b. I_1 will D.
   I_2 will D.
   I_3 will I.
   I_T will I.

7. $I_1 = 12$ mA
   $I_2 = 6$ mA (since I branch is inverse to R branch)
9. $P_{R_1} = 0.5 \text{ A} \times 50 \text{ V} = 25$ W
11. a. $M_1 = 2.13$ mA
    b. $M_2 = 37$ mA
    c. $M_3 = 39.13$ mA
    d. $M_4 = 10$ mA
    e. $M_5 = 49.13$ mA
    f. $P_{R_1} = 213$ mW
    g. $P_{R_2} = 3.7$ W
    h. $P_{R_3} = 1$ W
    i. $P_T = 4.91$ W
    j. New $P_T = 1.21$ W
13. $R_X = 31.59$ kΩ
15. a. R_4
    b. $P_{R_4} = 71.7$ mW
17. $V_T = 18$ V
19. $I_{R_2} = 5.45$ mA
21. R_1 dissipates the most power.
23. I_T would be 34 mA if R_2 were changed to a 1.5-kΩ resistor value.
25. Power dissipated by R_3 would not change as long as V_T and R_3 were the same values.
27. 120 V
29. $R_e = 5.45$ kΩ

Analysis Questions

1. The purpose of performing "mental approximations" is to provide a rapid check as to whether you might have made a mistake in solving a problem. If your approximation comes out completely different from your other calculations, chances are you have a simple mathematical error that needs correcting.
3. a. 4.864864865 03
   b. 4.864864865 03

5. Circuit diagram of a three-branch parallel circuit.

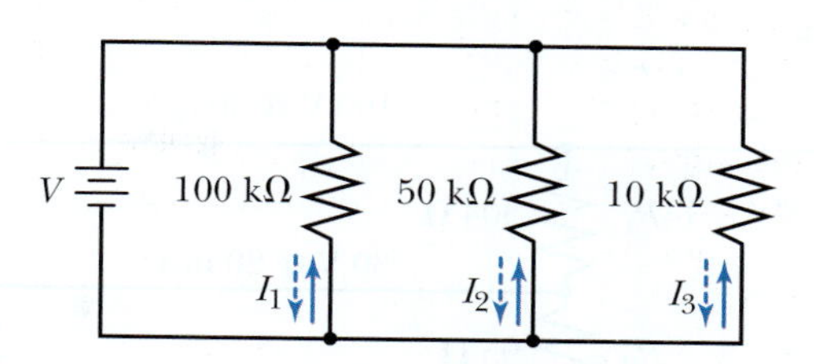

7. A method of isolating parallel components, or an individual branch, when troubleshooting a parallel circuit is to "lift" one end of the branch to be tested from the remainder of the circuit so as to electrically isolate it. (NOTE: This may require unsoldering one lead of the component from the circuit for testing and then resoldering after testing.)

CHAPTER 6

Practice Problems

—1—

1. The only component carrying total current is R_7.
2. The only single components in parallel with each other are R_2 and R_3.
3. The only single components that pass the same current are R_1, R_4, and R_5.

—2—

1. R_T for Figure 6–11 = 12.1 kΩ
2. R_T for Figure 6–12 = 7.63 kΩ

—3—

1. $I_T = 2$ A; $I_1 = 1.33$ A; $I_2 = 0.67$ A; $I_3 = 0.33$ A.
2. $I_T = 6.17$ mA; $I_1 = 2.36$ mA; $I_2 = 1.38$ mA; $I_3 = 2.43$ mA

—4—

1. *Voltage across:* $R_1 = 30$ V; $R_2 = 100$ V; $R_3 = 25$ V; $R_4 = 50$ V; $R_5 = 50$ V; $R_6 = 25$ V.
   Current through: $R_1 = 3$ A; $R_2 = 0.5$ A; $R_3 = 2.5$ A; $R_4 = 0.5$ A; $R_5 = 2$ A; and $R_6 = 2.5$ A.
2. *Voltage across:* $R_1 = 50$ V; $R_2 = 10.7$ V; $R_3 = 8.9$ V; $R_4 = 15.2$ V; $R_5 = 4.45$ V; $R_6 = 0.72$ V; $R_7 = 1.94$ V; $R_8 = 2.0$ V; $R_9 = 2.0$ V.
   Current through: $R_1 = 5$ mA; $R_2 = 2.28$ mA; $R_3 = 2.28$ mA; $R_4 = 2.72$ mA; $R_5 = 2.0$ mA; $R_6 = 0.72$ mA; $R_7 = 0.72$ mA; $R_8 = 0.36$ mA; $R_9 = 0.36$ mA.

—5—

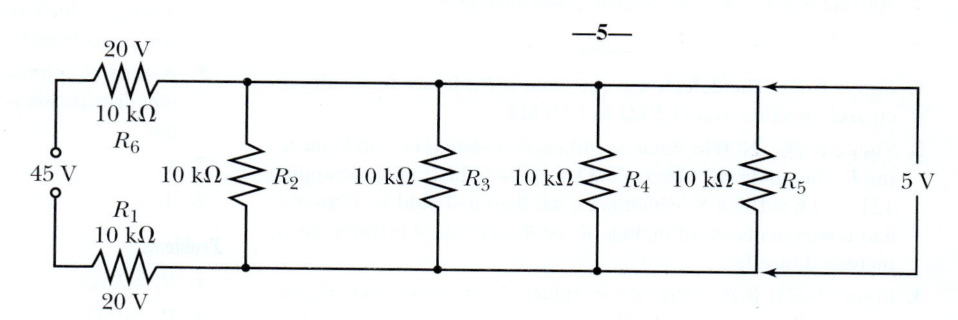

—6—

1. Circuit diagram:

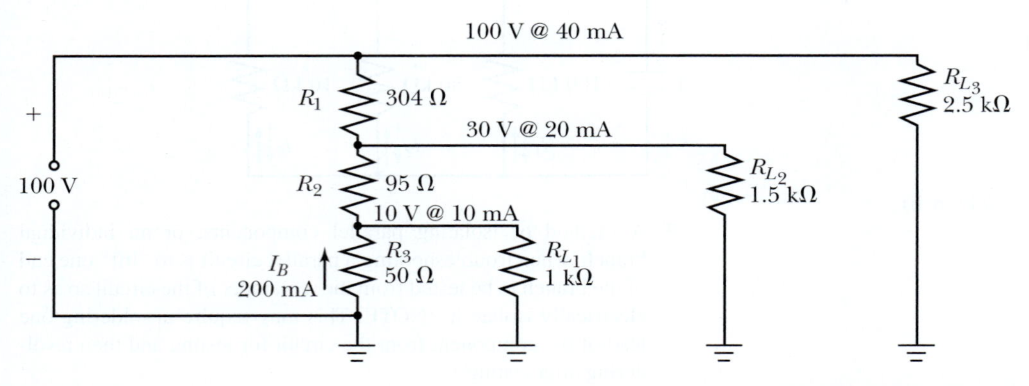

2. Load #1 = 1 kΩ

   Load #2 = 1.5 kΩ

   Load #3 = 2.5 kΩ

   R_1 divider resistor = 304 Ω

   R_2 divider resistor = 95 Ω

   R_3 divider resistor = 50 Ω
3. See labeling of diagram in answer 1.

—7—

1. $V_{R_L} = 150$ V; $P_{R_L} = 300$ mW
2. $R_L = 75$ kΩ
3. $R_2 = 50$ kΩ
4. The change, as specified, means R_2 has decreased.
5. The change, as specified, means R_1 has increased.
6. V_{R_2} equals 166.66 V.
7. If R_1 opens, V_{R_L} drops to 0 V.

   If R_2 shorts, V_{R_L} drops to 0 V.

In-Process Learning Checks

—1—

For Figure 6–13a: $R_T = 10$ Ω

For Figure 6–13b: $R_T = 120$ Ω

—2—

$I_1 = 0.4$ mA; $I_T = 1$ mA; $V_T = 85$ V; $R_T = 85$ kΩ; $P_{R_3} = 36$ mW; and the 100-kΩ resistor (R_3) is dissipating the most power.

—3—

1. Figure 6–25: The defective component is R_5. It has drastically increased in value from 3.3 kΩ to 17.3 kΩ.
2. Figure 6–26: (NOTE: Total circuit current *should be* 2 mA, meaning R_3 should drop 2 mA × 47 kΩ V, or 94 V. Instead it is dropping 121 – 17.6 = 103.4 V, which is higher than it should be.) Possible bad components could include R_1 or R_2 decreased in value, or R_3 increased in value.
3. Figure 6–27: If R_2 increases in value, V_{R_1} increases and V_{R_3} decreases. (This is due to total circuit resistance increasing, which decreases the total current through R_3, resulting in R_3 dropping less voltage. Kirchhoff's voltage law would indicate, therefore, that the voltage across parallel resistors R_1 and R_2 has to increase so the loop voltages equal V applied.)

Answers to Special "Thinking Exercise" (Page 180)

If R_2 opens:

V_T remains the same.

V_{R_1} decreases due to lower total current through R_1.

V_{R_2} increases because V_T stays the same, and V_{R_1} and V_{R_2} decrease due to lower total current through them.

V_{R_3} increases for the same reason V_{R_2} increases.

V_{R_4} decreases due to lower I_T caused by higher R_T.

I_T decreases due to higher R_T with same applied voltage.

I through R_1 decreases due to higher circuit total R.

I through R_2 decreases to zero due to its opening.

I through R_3 increases due to higher V across it.

I through R_4 decreases due to lower I_T through it.

P_T decreases due to lower I_T with same V_T.

P_{R_3} increases due to higher I through same R value.

P_{R_4} decreases due to lower current through same R value.

Review Questions

1. A series-parallel circuit has portions that are connected in series and portions that are connected in parallel. The series portions must be analyzed using series circuit principles, and the parallel portions must be analyzed using parallel circuit principles. The results of the analysis of each type circuit portion must be combined with results from other connected portions to solve for the total circuit quantities.
3. Series portions of series-parallel circuits are identified as those through which the same (common) current passes through the components and by the fact that the components are in "tandem."
5. A general approach for solving for a series-parallel circuit's current distribution is to start at the source end of the circuit and work out.
7. a
9. d

Problems

1. $R_T = 22$ Ω
3. R_T will D.

   R_4 will RTS.

   V_{R_1} will I.

   V_{R_3} will D.

   P_T will I.

5. Minimum voltage is 13 V.
   Maximum voltage is 18 V.
7. $R_4 = 6$ kΩ (so that branches voltage equals 80 V). $I_2 = 10$ mA (since $I_T = 20$ mA and I_1 can be calculated as 10 mA, due to V across the 8-kΩ resistor having to be 80 V. That is, the 100 V applied minus the 20-V drop across the 1-kΩ resistor that has I_T passing through it).
9. Light gets dimmer, due to the additional current that passes through the resistor in series with the source. This drops the voltage across the remaining sections of the circuit.
11.

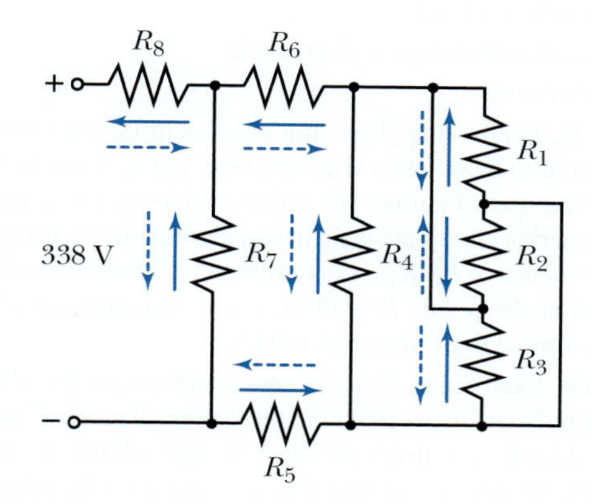

a. $R_T = 169$ kΩ
b. $I_T = 2$ mA
c. $V_{R_1} = 14$ V
d. $V_{R_2} = 14$ V
e. $V_{R_3} = 14$ V
f. $V_{R_4} = 14$ V
g. $V_{R_5} = 62$ V
h. $V_{R_6} = 62$ V
i. $V_{R_7} = 138$ V
j. $V_{R_8} = 200$ V

13. Before connecting voltmeter, $V_{R_1} = 5$ V.
15. Approximately 4 V
17. Yes, connecting the meter changes R_T and I_T. A higher I_T will pass through R_1, causing its voltage to increase from 5 V to 6 V. Less voltage will be dropped across R_2 than was present before the meter was connected.
19. Total current with the meter across one resistor = 60 mA.
21. If the meter were connected from the bottom of R_2 to the top of R_1, the voltages across the two resistors would not be affected.
23. $R_3 = 33.3$ kΩ; 12.5 kΩ; $R_1 = 10$ kΩ.
25. R_1 would have to be decreased in value.
27. Bleeder current would become 4.05 mA.
29. Minimum rating should exceed 400 mW for R_2.
31. P_T being supplied is 15 mA × 200 V = 3 watts.

Analysis Questions

1. If R_2 decreased greatly in value:
   a. voltage across R_3 would I.
   b. voltage across R_1 would I.
   c. current through the load connected across R_4 would I.
3. If the sensor is connected across a balanced bridge circuit, when the bridge is unbalanced by a change in conditions, the sensor's output can then be fed to control elements, which respond to the changed parameter across the sensor, and make appropriate system corrections, based on this feedback.
5. $R_X = 200$ kΩ
7. 0.0555:1 (R_1:R_2) or 18:1 (R_2:R_1)
9. First, set the calculator to the engineering mode. Next, put in one resistor's value, then press the [1/x] key, then the plus ([+]) key. Then, put in the second resistor's value, press the [1/x] key, then the [+] key. Then, input the third resistor's value, press the [1/x] key, then press the equal ([=]) key, then press the [1/x] key, one more time, for the answer.

CHAPTER 7

Practice Problems

—1—

$I = 4$ A
$V_L = 16$ V
$P_T = 80$ W
$P_L = 64$ W
Efficiency = 80%

—2—

1. I through $R_1 = 0$ mA
   V across R_1 with respect to point A = 0 V
   I through $R_2 = 7.5$ mA
   V across R_2 with respect to point A = +75 V
   I through $R_3 = 7.5$ mA
   V across R_3 with respect to point A = –75 V
2. I through $R_1 = 10$ mA
   V across R_1 with respect to point A = +100 V
   I through $R_2 = 12.5$ mA
   V across R_2 with respect to point A = –125 V
   I through $R_3 = 2.5$ mA
   V across R_3 with respect to point A = +25 V

—3—

1. If $R_L = 175$ Ω, $I_L = 0.25$ A and $V_L = 43.75$ V (Figure 7–10)
2. If $I_L = 125$ mA, then $R_L = 375$ Ω (Figure 7–10)
3. $I_L = 1.16$ mA; $V_{R_L} = 0.55$ V
4. In Figure 7–13, $R_{TH} = 4$ kΩ and $V_{TH} = 50$ V.
   If $R_L = 16$ kΩ, then $V_L = 40$ V and $I_L = 2.5$ mA.

—4—

If R_L is 60 Ω, $I_L = 0.588$ A and $V_L = 35.28$ V.

—5—

1. If $R_L = 100$ Ω in Figure 7–18a, $I_L = 0.4$ A, and $V_L = 40$ V.
2. If R_L changes from 25 Ω to 50 Ω, I_L decreases (from 1 A to 0.66 A), and V_L increases (from 25 V to 33.5 V).

In-Process Learning Checks

—1—

1. For maximum power transfer to occur between source and load, the load resistance should ***equal*** the source resistance.
2. The higher the efficiency of power transfer from source to load, the ***greater*** the percentage of total power is dissipated by the load.
3. Maximum power transfer occurs at ***50%*** efficiency.
4. If the load resistance is less than the source resistance, efficiency is ***smaller*** than the efficiency at maximum power transfer.
5. To analyze a circuit having two sources, the superposition theorem indicates that Ohm's law ***can*** be used.

6. What are two key observations needed in using the superposition theorem to analyze a circuit with more than one source? ***Noting the direction of current flow and polarity of voltage drops.***
7. Using the superposition theorem, if the sources are considered "voltage" sources, are these sources considered shorted or opened during the analysis process? ***Shorted.***
8. When using the superposition theorem when determining the final result of your analysis, the calculated parameters are combined, or superimposed, ***algebraically.***

Review Questions

1. Matching the impedance of an antenna system to the impedance of a transmitter output stage, and matching the output impedance of an audio amplifier to the impedance of a speaker.
3. $P_{out}/P_{in} \times 100$
5. To "Thevenize" a circuit:
   (a) Open or remove R_L.
   (b) Determine the open circuit V at points where R_L is to be connected. (This value is V_{TH}.)
   (c) Determine the resistance looking toward the source from the R_L connection points, assuming the source is shorted. (This value is R_{TH}.)
   (d) Draw the Thevenin equivalent circuit with V_{TH} as source, and R_{TH} in series with the source.
   (e) Calculate I_L and V_L for given values of R_L, using a two-resistor series circuit analysis technique.
7. The most important application of the superposition theorem is for analyzing circuits having more than one source.
9. c
11. d

Problems

1. 0 V
3. $V_{TH} = 40.5$ V (voltage at Points A and B with R_L removed)
   $R_{TH} = 14.85$ kΩ (R at Points A and B with source shorted)
   $I_L = 1.63$ mA (V_L/R_L in Thevenin equivalent circuit)
   $V_L = 16.3$ V (from simple series Thevenin equivalent circuit analysis)
5. Norton equivalent circuit is a constant current source with a parallel resistance, called R_N, and the load (R_L) in parallel with R_N and the source. To convert parameters:
   $R_N = R_{TH} = 14.85$ kΩ
   I_N (constant current source) $= V_{TH}/R_{TH} = 40.5$ V/14.85 kΩ $= 2.73$ mA
   The circuit is shown below:

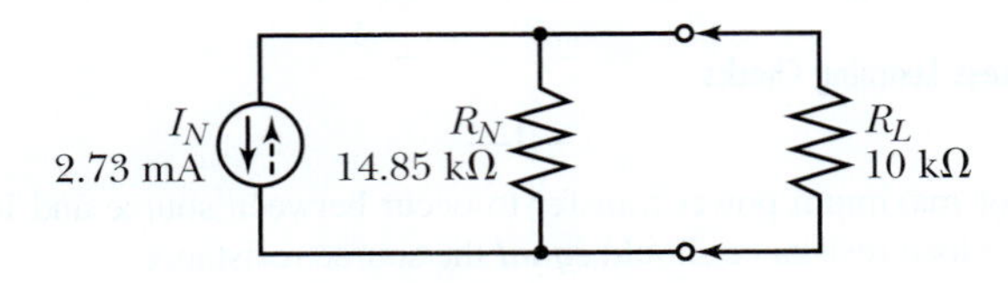

7. The maximum power that can be delivered to the load in Figure 7–22 is 1.25 W.
9. $R_{TH} = 29.49$ Ω; $V_{TH} = 34.45$ V
11. $R_{TH} = 10$ Ω; $V_{TH} = I_N \times R_N = 3$ A $\times$ 10 Ω $= 30$ V
13. Efficiency = 83.2% ($P_{out} = 277$ W; $P_{in} = 333$ W; $P_{out}/P_{in} \times 100 =$ 83.2%)
15. For maximum power transfer, R_L should match the r_{int} of source or should equal 5Ω.

Analysis Questions

1. It is important for a voltage source to have as low an internal resistance (r_{int}) as possible so that source output voltage changes will be minimized as current demands for the connected circuit change. Ideally, a voltage source would have zero internal resistance; thus, it would not have any internal voltage drop for any current demand level.
3. Load resistance should be decreased.
5. Norton's theorem
7. R_{TH} is the fixed-value, Thevenin equivalent-circuit series resistor considered to be in series with whatever the R_L value is, for calculation purposes. Finding this value enables using a simple two-resistor series circuit analysis in calculating circuit parameters for various values of R_L. R_{TH} is determined by "looking back" into the network at the points R_L will be connected, and considering the voltage source to be shorted (or 0 Ω).
9. R_N is the fixed-value, Norton equivalent-circuit parallel resistor considered to be in parallel with the "constant current" source, and R_L. R_N (like R_{TH}) is also found by looking back into the network at the points R_L is to be connected and considering the voltage source as being replaced by its internal R (r_{int}), or 0 Ω in this case. Knowing the R_N value (along with the equivalent circuit I_N value) enables reducing circuit analysis to a simple two-resistor, parallel circuit, current-division approach.
11. **a.** Various answers possible (e.g., some IC circuits)
    b. Superposition theorem
13. Yes, it is possible to measure the circuit R_N. Since this is the same circuit (Figure 7–23), and R_N are determined from the same points, under the same conditions, you would use the same technique as described for measuring R_{TH} in the previous question.
15. Battery resistance would equal 12 V – 11 V/50 A; 1 V/50 A = 0.02 Ω.

CHAPTER 8

Practice Problems

—1—

1. $I_A = 0.55$ A
2. $I_B = 0.25$ A
3. $V_1 = 9.6$ V
4. $V_2 = 8.25$ V
5. $V_3 = 12.1$ V
6. $V_4 = 8.25$ V
7. $V_5 = 11.75$ V
8. $V_{A\text{-}B}$ = close to 0 V

Considered from Points A and B with respect to the opposite ends of resistors R_2 and R_4, respectively: Point A is +8.25 V with respect to the common connection points of R_2 and R_4. Point B is +8.25 V with respect to the common connection points of R_2 and R_4. Thus, Point A and Point B are at the same potential. Considered from Points A and B with respect to the opposite ends of R_3 and R_5, respectively: Point A is

−12.1 V with respect to the common connection points of R_3 and R_5. Point B is −11.75 V with respect to the common connection points of R_3 and R_5. Therefore, Point A is −0.6 V with respect to Point B, or Point B is +0.6 V with respect to Point A. (NOTE: Variance is due to the fact that the math was not carried out to a large number of decimal places.)

—2—

1. $I_A = 0.714$ A
2. $I_B = 0.143$ A
3. $V_{R1} = 7.14$ V; current through $R_1 = 0.714$ A
4. $V_{R2} = 12.85$ V; current through $R_2 = 0.857$ A
5. $V_{R3} = 2.145$ V; current through $R_3 = 0.143$ A

—3—

1. $V_{R_1} = 18.5$ V
2. $V_{R_2} = 41.5$ V
3. $V_{R_3} = 6.5$ V
4. Current through $R_1 = 0.925$ A
5. Current through $R_2 = 4.15$ A
6. Current through $R_3 = 3.25$ A

Solution

$$I_1 + I_2 = I_3$$

Loop with 60-V source:

$$\left(\frac{V_{R1}}{R_1}\right)+\left(\frac{V_{R3}}{R_3}\right)=\frac{V_{R2}}{R_2}$$

$$V_{R1} + V_{R2} = 60$$

$$V_{R1} = 60 - V_{R2}$$

Loop with 48-V source:

$$V_{R_3} + V_{R_2} = 48$$

$$V_{R_3} = 48 - V_{R_2}$$

Substituting these terms into $I_1 + I_2 = I_3$:

$$\left(60-\frac{V_{R2}}{20}\right)+\left(48-\frac{V_{R2}}{2}\right)=\frac{V_{R2}}{10}$$

Multiplying each term by 20 to move and/or remove the denominators:

$$(60 - V_{R_2}) + 10(48 - V_{R_2}) = 2\,V_{R_2}$$

$$60 - V_{R_2} + 480 - 10\,V_{R_2} = 2\,V_{R_2}$$

$$540 = 13\,V_{R_2}$$

$$V_{R_2} = 41.5\text{ V}$$

$$I_{R2} = \frac{V_{R2}}{10} = \frac{41.5\text{ V}}{10\ \Omega} = 4.15\text{ A}$$

$$V_{R_3} = 48 - V_{R_2} = 48 - 41.5 = 6.5\text{ V}$$

$$I_{R3} = \frac{V_{R3}}{2} = \frac{6.5\text{ V}}{2\ \Omega} = 3.25\text{ A}$$

$$V_{R_1} = 60 - 41.5 = 18.5\text{ V}$$

$$I_{R1} = \frac{18.5}{20} = 0.925\text{ A}$$

Verifying:

$$V_{R_1} + V_{R_2} = V_{S_1};\ 18.5\text{ V} + 41.5\text{ V} = 60\text{ V}$$

$$V_{R_2} + V_{R_3} = V_{S_2};\ 41.5\text{ V} + 6.5\text{ V} = 48\text{ V}$$

—4—

1. $R_A = 31.82\ \Omega$; $R_B = 70\ \Omega$; $R_C = 46.67\ \Omega$
2. $R_1 = 22\ \Omega$; $R_2 = 10\ \Omega$; $R_3 = 15\ \Omega$
3. Yes

—5—

$(R_1 = 7\ \Omega;\ R_2 = 3.5\ \Omega;\ R_3 = 2.33\ \Omega)$

$R_T = (R_D + R_2)\|(R_E + R_3) + R_1$

1. $R_T = 13.44\ \Omega$
2. $I_T = V_T/R_T = 84\text{ V}/13.44\ \Omega = 6.25$ A

In-Process Learning Checks

—1—

1. Path
2. Mesh
3. + (positive)
4. +, −

—2—

1. Current
2. Major node
3. Reference node
4. Current

—3—

1. (a) **T** (tee) (b) π (pi)
2. AC (or electrical)
3. Current
4. is not

Review Questions

1. A loop is a complete path in a circuit, from one side of the source, through a series of circuit components, back to the other side of the source.
3. Mesh currents are "assumed" currents (and current directions) used in conjunction with Kirchhoff's voltage law to help in analyzing networks. Mesh currents are different from actual currents in that they are assumed not to branch.
5. c
7. a
9. a
11. b

Problems

1. $I_A = 0.8$ A
3. $V_{R_1} = 8$ V
5. $V_{R_3} = 4$ V
7. $I_{R_2} = 1.2$ A
9. I_1 (current through R_1) = 2.13 A

11. I_3 (current through R_2) = 2.34 A
13. $V_{R_2} = 11$ V
15. $R_A = 45\ \Omega$
17. $R_C = 30\ \Omega$

Analysis Questions

1. Mesh or nodal analysis techniques are used in situations in which basic series and parallel analysis techniques might be cumbersome or very difficult.
3. In industrial settings, you might find delta and wye circuit configurations related to three-phase power, three-phase motors, etc.

CHAPTER 9

Practice Problems

—1—

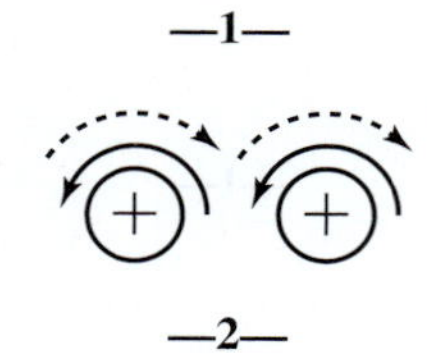

—2—

Figure 9–13a = Move away from each other.
Figure 9–13b = Move toward one another.
Figure 9–13c = Move away from each other.

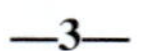

Figure 9–17a = N at right end, S at left end; Figure 9–17b = N at left end, S at right end; and Figure 9–17c = N at left end, S at right end.

In-Process Learning Checks

—1—

1. Small magnets that are suspended and free to move align in a ***North*** and ***South*** direction.
2. Materials that lose their magnetism after the magnetizing force is removed are called ***temporary magnets.*** Materials that retain their magnetism after the magnetizing force is removed are called ***permanent magnets.***
3. A wire carrying dc current ***does*** establish a magnetic field. Once a magnetic field is established, if the current level doesn't change, the field is ***stationary.***
4. A law of magnetism is that like poles ***repel*** each other and unlike poles ***attract*** each other.
5. A maxwell represents ***one*** line of force.
6. A weber represents **10^8** lines of force.
7. Lines of force are ***continuous.***
8. Lines of force related to magnets exit the ***North*** pole of a magnet and enter the ***South*** pole.
9. Nonmagnetic materials do ***not*** stop the flow of magnetic flux lines through themselves.
10. Yes, magnetism ***can*** be induced from one magnetic object to another object, if it is a magnetic material.

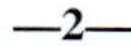

1. For a given coil dimension and core material, what two factors primarily affect the strength of an electromagnet? ***Number of turns*** and ***amount of current.***
2. The left-hand rule for determining the polarity of electromagnets states that when the fingers of your left hand point ***in the same direction*** as the current passing through the coil, the thumb points toward the ***North*** pole of the electromagnet.
3. Adjacent current-carrying conductors that are carrying current in the same direction tend to ***attract each other.***
4. If you grasp a current-carrying conductor so that the thumb of your right hand is in the direction of current through the conductor, the fingers "curled around the conductor" ***will indicate*** the direction of the magnetic field around the conductor.
5. When representing an end view of a current conductor pictorially, it is common to show current coming out of the paper via a ***dot.***

—3—

1. A *B-H* curve is also known as a ***magnetization*** curve.
2. *B* stands for flux ***density.***
3. *H* stands for magnetizing ***intensity.***
4. The point where increasing current through a coil causes no further significant increase in flux density is called ***saturation.***
5. The larger the area inside a "hysteresis loop" the ***larger*** the magnetic losses represented.

Review Questions

1. b
3. b
5. d
7. Permeability
9. ϕ, Weber
11. MMF, Ampere-turns (NI)
13. 10^8
15. The shield simply diverts lines of flux through itself, preventing flux from reaching the items inside the shield enclosure.
17. Drawing of a hysteresis loop, with callouts for saturation, residual magnetism, and coercive force:

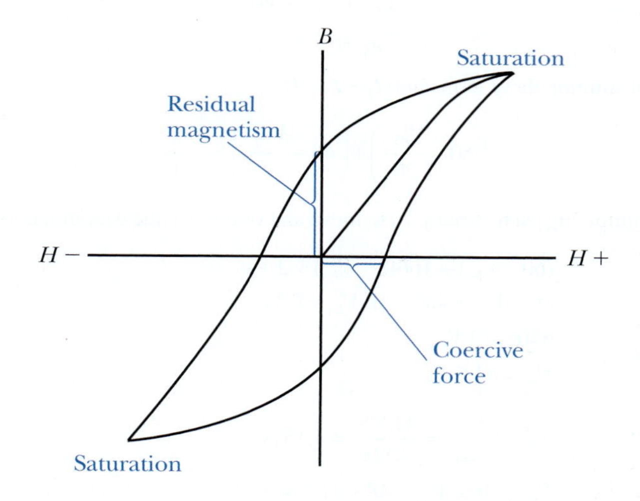

19. b
21. The direction of an induced voltage or current is such that it tends to oppose the change causing it.
23. Motor effect in a generator relates to the induced current causing an opposition to the motion that induced it.
25. Toroidal coil forms are efficient conductors of magnetic flux because of the high permeability materials used in them and because of their geometry not allowing much leakage flux.

Problems

1. Drawing of a typical field pattern for unlike poles facing each other and like poles facing each other:

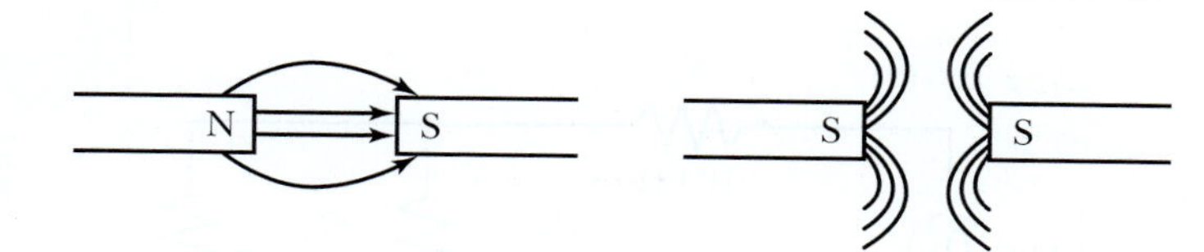

3. Labeling of North and South poles of the electromagnet circuit shown:

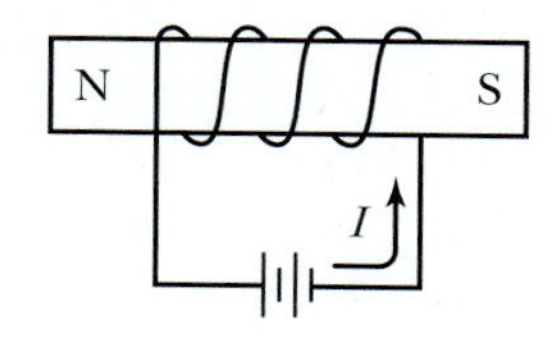

4. **a.** $10^8/10^6 = 100$ lines of flux
   b. Teslas = webers per square meter (Since there are 100 lines in 0.005 m^2, there are $1/0.005 \times 100$ lines in a square meter, or lines = $200 \times 100 = 20{,}000$.)
   Number of webers per square meter: teslas = 20,000/100,000,000 = 0.0002 T
5. 5,000 μV
7. 400,000,000

CHAPTER 10

Review Questions

1. Easy to read; more accurate; less loading effect
3. c
5. c
7. Do not have to break circuit to make the measurement
9. c
11. c

Analysis Questions

1. To minimize changing circuit under test.
3. High, to minimize loading effect.

CHAPTER 11

Practice Problems

—1—

1. The sine of 35° = 0.5735.
2. Cycle 1 ends and cycle 2 begins at Point G.
3. Maximum rate of change for the waveform shown occurs at Points J, N, R, and T.

—2—

1.

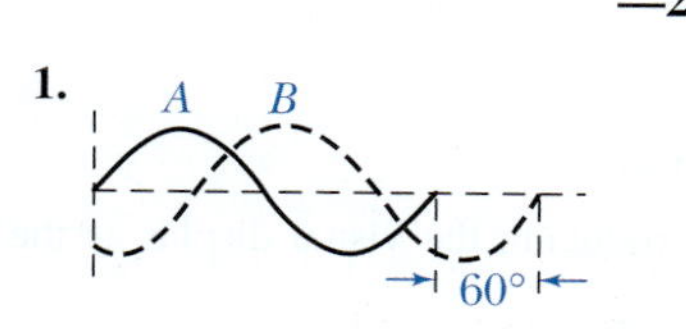

2.

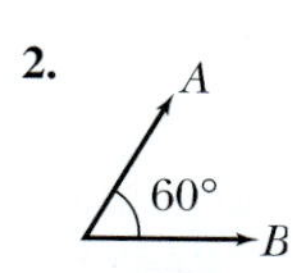

—3—

1. 10 V
2. 9 V
3. 90%
4. No
5. Yes
6. **a.** 120.8 V
   b. 171 V
   c. 120.8 V
   d. 171 V
   e. 108.9 V

—4—

1. **a.** $V_{R_1} = 50$ V
   b. $V_{R_2} = 50$ V
   c. $I_p = 7.07$ mA
   d. $P_p = 999.7$ mW
   e. $P_{R_1} = 250$ mW
2. **a.** $I_1 = 10$ mA
   b. $I_2 = 10$ mA
   c. $I_p = 28.28$ mA
   d. $V_{p\text{-}p} = 282.8$ V
3. **a.** 66.7 V
   b. 33.3 V
   c. 33.3 V
   d. 444.89 mW

—5—

The duty cycle = Pulse width/Time for one period times 100. Duty cycle = (1 μs/15 μs) × 100 = 6.67% duty cycle.

—6—

1. V_{avg} = Baseline value + (duty cycle × p-p amplitude)
   V_{avg} = 0 + (pulse width/period × p-p amplitude)
   $V_{avg} = 0 + [(2 \text{ ms}/30 \text{ ms}) \times 6 \text{ V}] = 0 + 0.066 \times 6 = 0.396$ V
2. –2.604 V

In-Process Learning Checks

—1—

1. The basic difference is that dc (direct current) is in one direction and of one polarity. Alternating current (ac) periodically changes direction and polarity.
2. The reference, or 0° position is horizontally to the right when describing angular motion.
3. The *y*-axis is the vertical axis in the coordinate system.
4. The second quadrant in the coordinate system is that quadrant between 90° and 180°.
5. A vector represents a given quantity's magnitude and direction with respect to location in space.
6. A phasor represents a rotating vector's relative position with respect to time.

—2—

1. 10 kHz
2. 0.0025 seconds, or 2.5 msec
3. 50 μsec
4. Amplitude
5. 45° and 135°; 225° and 315°
6. 1 and 2
7. 0.067 μsec
8. 40-Hz
9. As frequency increases, T ***decreases.***
10. The longer a given signal's period, the ***longer*** the time for each alternation.

Review Questions

1. d
3. b
5. b
7. c
9. b
11. a
13. c
15. c
17. A "periodic wave" is one that repeats itself in time and form.
19. b

Problems

1. t = 1/f = 1/2,000 = 0.0005 seconds, or 0.5 msec
3. 1/120 of a second.
5. 400 Hz
7. The peak-to-peak value would also double.
9. The frequency has decreased to one-third its original frequency.
11. One alternation is half the time of a cycle; therefore, the time for one alternation = 1 ms.
13. Diagram:

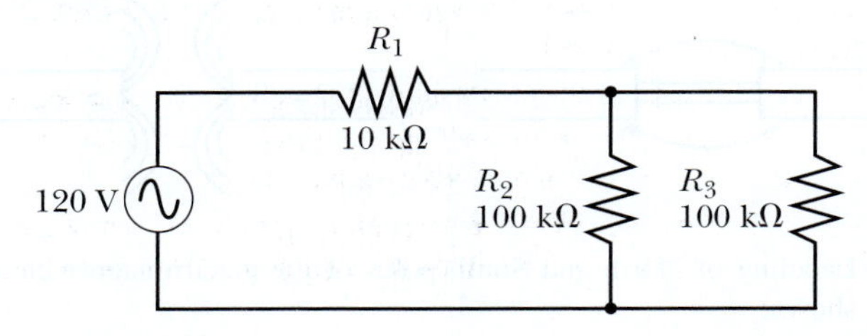

a. $V_{R_1} = 20$ V
b. $I_{R_1} = 2$ mA
c. $P_{R_1} = 40$ mW
d. $I_{R_2} = 1$ mA
e. $V_{R_3} = 100$ V
f. θ between V applied and I total = 0°

15. λ will increase to three times the original.

Analysis Questions

1. Two differences between ac and dc are (1) ac changes polarity, periodically, dc does not; (2) ac continuously varies in amplitude, dc does not; (3) ac can be stepped up or down via transformer action, dc cannot.
3. If rms value = 100 V, then peak value = 141 V and peak-to-peak value = 282 V.
5. 70.7 V
7. Since A = 1/0.002 × 10^{-3} = 500 kHz, B = 250 kHz.
9. **a.** Graphic drawing of rectangular wave:

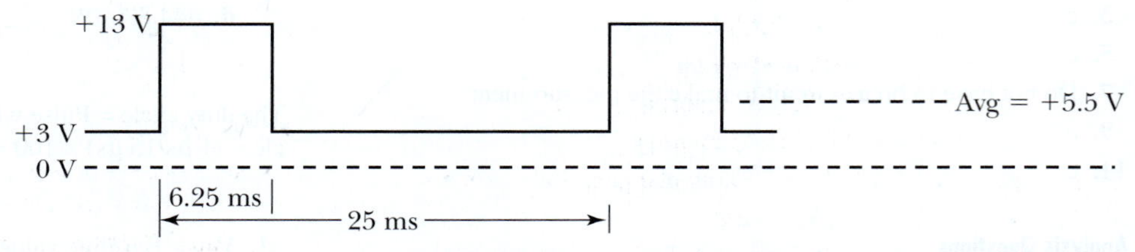

b. The V_{avg} = Baseline value + (Duty cycle × p-to-p amplitude)
V_{avg} = 3 + (Pulse width/period × p-to-p amplitude)
V_{avg} = 3 + [(6.25 ms/25 ms) × 10 V] = 3 + 0.25 × 10 = 5.5 V

CHAPTER 12

Practice Problems

—1—

When the vertical frequency is four times the horizontal frequency, the waveform shows four cycles of the given waveform.

—2—

1. 30 V (p-to-p); 10.6 V (rms)
2. 1.76 V/div (rms)
3. Use the 10 position setting of the vertical volts/div control.

—3—

1. 1.5 V (p-to-p); 0.53 V (rms)
2. 3 V (p-to-p); 1.06 V (rms)

—4—

72 degrees per major division

—5—

1. **a.** Period (T) = 0.25 s
   b. Frequency = 4 Hz
2. **a.** Period (T) = 50 μs
   b. Frequency = 20 kHz

In-Process Learning Checks

—1—

1. The part of the oscilloscope producing the visual display is the ***cathode-ray tube.***

2. The scope control that influences the brightness of the display is the ***intensity*** control.
3. A waveform is moved up or down on the screen by using the ***vertical position*** control.
4. A waveform is moved left or right on the screen by using the ***horizontal position*** control.
5. For a signal fed to the scope's vertical input, the controls that adjust the number of cycles seen on the screen are the horizontal ***time/div.*** and the horizontal ***time variable*** controls.
6. The control(s) that help keep the waveform from moving or jiggling on the display are associated with the ***synchronization*** circuitry.

Review Questions

1. Name is cathode-ray tube and abbreviation is CRT.
3. To protect and add life to coating on CRT screen
5. Vertical input terminal(s)
7. "Ext Trig" terminal (NOTE: Other names are also used.)
9. To retain proper calibration

Problems

1. $4 \times 0.1 = 0.4$ ms/cycle; $f = 1/t = 1/0.4$ ms $= 2{,}500$ Hz
3. 2 divisions
5. T = 0.004 s
7. 0.0004
9. p-to-p = 5 V; rms V = 1.77 V; $I = 1.77$ V/4.7 kΩ = 0.03765 mA

Analysis Questions

1. List of control names should be based on the specific oscilloscope names students can observe on your equipment in your training setting:
   a. Control the trace focus _____.
   b. Control the trace intensity _____.
   c. Control(s) the vertical position _____.
   d. Control(s) the horizontal position _____.
   e. Vertical attenuator/amplifier controls _____.
   f. Horizontal sweep frequency controls _____.
3. Diagram B: vertical gain and horizontal position controls
   Diagram C: vertical gain control
   Diagram D: horizontal sweep frequency controls (horizontal time/div. and time variable)
   Diagram E: vertical gain control
   Diagram F: horizontal sweep frequency controls (horizontal time/div. and time variable)
   Diagram G: vertical gain control

CHAPTER 13

Practice Problems

—1—

Since 1 weber = 10^8 flux lines and cutting 1 Wb/sec induces 1 V, then cutting 2×10^8 flux lines (2 Wb) in 0.5 sec = 4 V.

—2—

$$V_L = L\frac{di}{dt}$$

$$V_L = 5\left(\frac{3}{1}\right) = 15 \text{ V}$$

—3—

10 mH + 6 mH + 100 μH = 16.1 mH total inductance

—4—

1. Three equal value inductors (15 H) in parallel have a total equivalent inductance = 1/3 of one of them. $L_T = 5$ henrys.
2. L_T of 10 mH, 12 mH, 15 mH, and 20 mH inductors in parallel is 3.33 mH.

—5—

a. dc current through coil = 5 A

b. $\text{energy} = \frac{LI^2}{2} = \frac{1(25)}{2} = 12.5 \text{ J}$

—6—

Two seconds = approximately 2.4 TC.
After 2.4 TC, V_R is about 91% of V applied, or 9.1 V.
After 2.4 TC, V_L is about 9% of V applied, or 0.9 V.

—7—

0.92 mA

In-Process Learning Checks

—1—

1. All other factors remaining the same, when an inductor's number of turns increases four times, the inductance value ***increases*** by a factor of ***16 times.***
2. All other factors remaining the same, when an inductor's diameter triples, the inductance value ***increases*** by a factor of ***9 times.*** (NOTE: A is directly proportional to d^2.)
3. All other factors remaining the same, when an inductor's length increases, the inductance value ***decreases.***
4. When the core of an air-core inductor is replaced by material having a permeability of ten, the inductance value ***increases*** by a factor of ***10 times.***

Review Questions

1. Inductance (or self-inductance) is the property of a circuit that opposes a ***change*** in current flow.
3. Number of turns on the coil; length of the coil; cross-sectional area of the coil; relative permeability of the core material
5. The amount of emf induced in a circuit is related to the number of magnetic flux lines cut or linked per unit time.
7. c
9. c
11. b
13. a
15. c
17. Be careful regarding the "high-voltage kick" produced when the test leads break connection with the inductor just tested. Use appropriate personal safety measures.
19. The 63.2% indicates the percentage of change that will occur from the present level toward the final (stable) level or value during the period of one time constant.

Problems

1. Inductance will increase to 400 mH (L varies as the square of the turns).

3. New inductance is $500 \times 100\ \mu H = 0.05$ H.
5. $L_T = L_1 + L_2 + L_3 = 0.1\ H + 0.2\ H + 0.001\ H = 0.301$ H, or 301 mH
7. 1 TC $= L/R = 250 \times 10^{-3}/100 = 0.0025$ sec, or 2.5 ms

   It will take 5 times 2.5 ms, or 12.5 ms for current to complete the change from one level to another.
9. Tapping a coil at midpoint yields an inductance of one-fourth the value of the total inductance of the coil. In the case of the 0.4 H coil, L equals about 0.1 H at the 250-turn tap.
11. V_L after 0.6 µs equals approximately 6.76 V.

Analysis Questions

1. Inductance will decrease.
3. Total inductance will increase.
5. Since the inductance change is proportional to the square of the turns, $L_1/L_2 = (N_1)^2/(N_2)^2$; $4/6 = (1{,}000)^2/(N_2)^2$; $N_2 = 1224.74$ turns.
7. Because the voltage-per-turn is less than the varnish insulation breakdown voltage (The voltage-per-turn in this case is approximately 2 V/turn.)
9. The joules of energy stored by an inductor is directly related to the amount of inductance the inductor has because the strength of the magnetic field built up by a given value of current through the coil is directly related to the inductance value. The energy is stored in the form of the magnetic field.

CHAPTER 14

Practice Problems

—1—

1. 3 kΩ
2. 333.3 Ω

—2—

1. 6,280 Ω
2. 79.6 mH, or approximately 80 mH
3. 238.85 Hz, or approximately 239 Hz

—3—

1. 5 kΩ
2. $X_L = 5{,}024\ \Omega$, so $I \times X_L = 3\ mA \times 5{,}025\ \Omega =$ approximately 15 V
3. 40 mA

In-Process Learning Checks

—1—

1. When current increases through an inductor, the cemf ***hinders*** the current increase.
2. When current decreases through an inductor, the cemf ***hinders*** the current decrease.
3. In a pure inductor, the ***voltage*** leads the ***current*** by 90°.
4. What memory aid helps in recalling the relationship described in question 3? ***"Eli"*** the ice man.
5. The opposition that an inductor shows to ac is termed ***inductive reactance.***
6. The opposition that an inductor shows to ac ***increases*** as inductance increases.
7. The opposition that an inductor shows to ac ***decreases*** as frequency decreases.
8. X_L is ***directly*** related to inductance value.
9. X_L is ***directly*** related to frequency.

Review Questions

1. c
3. c
5. b
7. d
9. b

Problems

1. $L = X_L/2\pi f = 376.8/6.28 \times 100 = 376.8/628 = 0.6$ H; at 450 Hz, $X_L = 1695.6\ \Omega$
3. Total inductance 2 mH × 6 mH/2 mH + 6 mH = 12/8 = 1.5 mH $f = X_L/2\pi L = 39.8$ kHz
5. $Q = X_L/R = 6.28 \times 20\ MHz \times 10\ \mu H/10\ \Omega = 125.6/10 = 12.56$ Q not a dc parameter. Q will increase, because $Q = \frac{X_L}{R}$ and condition is $\frac{3X_L}{2R}$.
7. $V_T = 45$ V

   $V_{L_1} = 15$ V

   $I_T = 0.75$ mA
9. Approximately 6 V
11. L_T will RTS.

    X_{L_T} will I.

    I_T will D.

    V_{L_4} will RTS.

    I_1 will D.
13. $L = 0.318$ H
15. $X_{LT} = 28.89$ kΩ
17. I_T would decrease.
19. Voltage applied would have to decrease to one-third its original value.

Analysis Questions

1. Six times greater
3. Circuit current will increase, because X_L will decrease.
5. $Q = X_L/R$ (as f decreases, X_L decreases while R remains the same); Q will decrease.
7. Less losses (less wasted energy)
9. b

CHAPTER 15

Practice Problems

—1—

5 pounds at an angle of 36.9° from horizontal.

—2—

$$V_S = \sqrt{V_R^2 + V_L^2} = \sqrt{25 + 25} = \sqrt{50} = 7.07\ V$$

—3—

1. $\text{Cos}\ \theta = \frac{\text{adj}}{\text{hyp}} = \frac{5}{10}$ and $\cos^{-1} = 60°$
2. Tangent of 45° = 1.0
3. Sine of 30° = 0.5

—4—

V applied = 150 V

—5—

1. V_L 20 V; $V_A = 36$ V; $\theta = 33.69°$; $V_R = 30$ V

2. $\theta = 33.69°$ (tan θ = opp/adj = 20/30 = 0.6666. Angle whose tangent = 0.6666 is 33.69°.)
3. Tangent function
4. Angle would have been greater (56.3°).

—6—

1. $V_R = 35.35$ V (cos θ = adj/hyp and cos 45° = 0.7071. Therefore, 0.7071 = adj/50 = 35.35.)
2. **a.** $V_R = 32.5$ V
   b. $V_L = 56.29$ V
3. **a.** Angle = 56.3°
   b. $V_T = 43.2$ V

—7—

1. R 15 Ω; 20 Ω

2. $X_L = 20\ \Omega$; $Z = 25\ \Omega$; $\theta = 53.1°$; $R = 15\ \Omega$

3. $Z = \sqrt{15^2 + 20^2}$
   $Z = \sqrt{625}$
   $Z = 25\ \Omega$

4. $\tan\theta = \dfrac{X_L}{R}$
   $\tan\theta = \dfrac{20}{15}$
   $\tan\theta = 1.33$
   Angle = 53.1°
   Yes, answer agrees.

—8—

1. $I_T = \sqrt{I_R^2 + I_L^2}$
   $I_T = \sqrt{5^2 + 3^2}$
   $I_T = \sqrt{34}$
   $I_T = 5.83$ A

2. $Z = \dfrac{V_T}{I_T}$
   $Z = \dfrac{300}{5.83}$
   $Z = 51.45\ \Omega$

3. θ–30.9° (tan θ = opp/adj = 3 A/5 A = 0.6. Angle whose tangent equals 0.6 is 30.9°.)

—9—

$Z = 47 \times 100 / (\sqrt{47^2 + 100^2}) = 4{,}700 / 110.5 = 42.5\ \Omega$

—10—

1. $\boldsymbol{G} = 1/\boldsymbol{R} = 1/60$ ohms $= 0.016$ **S**
2. $\boldsymbol{Bc} = 1X_L = 1/100$ ohms $= 0.01$ **S**
3. $Y = \sqrt{G^2 + Bc^2} = \sqrt{0.016^2 = 0.01^2} = \sqrt{0.000375} = 0.0194$ **S**
4. $\boldsymbol{Z} = 1/\boldsymbol{Y} = 1/0.0194$ **S** $= 51.5$ ohms
5. $\boldsymbol{I_T} = \boldsymbol{VY} = 300$ **V** $\times\ 0.014$ **S** $= 5.82$ **A**
6. $\boldsymbol{V} = \boldsymbol{I/Y} = 5.82$ **A**$/0.0194$ **S** $= 300$ **V**
7. $Y = \boldsymbol{I/V} = 5.82$ **A**$/300$ **V** $= 0.0194$ **S**
8. θ(magnitude) $= \tan^{-1} \boldsymbol{R/X_L} = \tan^{-1} 60/100 = \tan^{-1} 0.6 =$ **30.9°**

In-Process Learning Checks

—1—

1. In a purely resistive ac circuit, the circuit voltage and current are ***in*** phase.
2. In a purely inductive ac circuit, the current through the inductance and the voltage across the inductance are ***90*** degrees out of phase. In this case, the ***voltage*** leads the ***current.***
3. A quantity expressing both magnitude and direction is a ***vector*** quantity.
4. The length of the vector expresses the ***magnitude*** of a quantity.

Review Questions

1. b
3. a
5. a
7. c
9. c
11. c
13. $Z = \sqrt{R^2 + X_L^2}$
15. $Z = R \times X_L / \sqrt{R^2 + X_L^2}$
17. b
19. a

Problems

1. **a.** 36 Ω
   b. Pythagorean result = 36 Ω
   c. Angle between R and Z = 56.3°
3. Diagrams for the specified series RL circuit are shown:

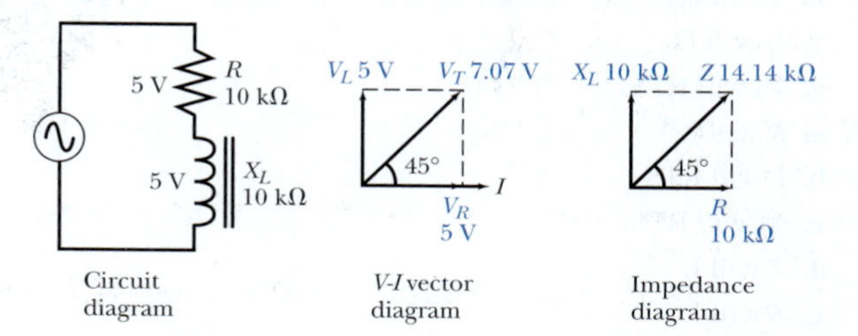

Since $\theta = 45°$, then R must $= X_L$, and V_R must $= V_L$.
$Z = R^2 + X_L^2 = 200 \times 10^6 = 14.14 \times 10^3 = 14.14$ kΩ

5. a. Z = vector sum of 10-kΩ R and approximate 20-kΩ X_L; Z = 22.36 kΩ
   b. $V_R = IR$ = (67 V/22.36 kΩ) × 10 kΩ = approximately 3 mA × 10 kΩ = 30 V
   c. θ = approximately 63.43°
7. I = 5 mA
   Z = 40 kΩ
   V_L = 105.4 V
   X_L = 21.08 kΩ
   $L = X_L/(2\pi f)$ = 56 H
9. Diagrams and calculations for the specified parallel RL circuit are shown:

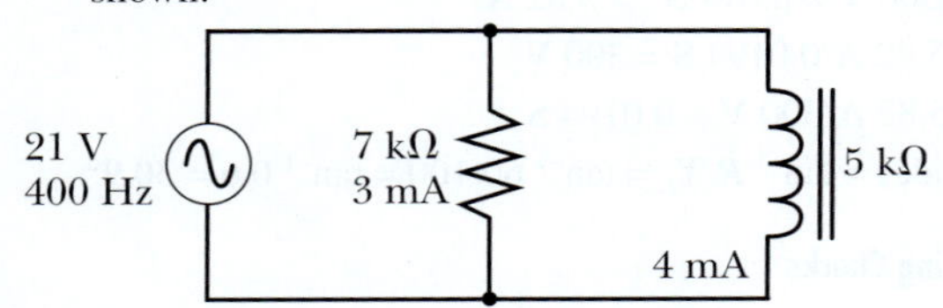

I_R = 3 mA
V_A = 21 V
θ = −53.1°
$I_L \cong$ 4 mA
$I_T \cong$ 5 mA

$X_L = 2\pi fL$ = 5 kΩ (approximately)
I_R = 21 V/7 kΩ = 3 mA
I_L = 21 V/5 kΩ = 4.2 mA (or about 4 mA)
$I_T = I_R^2 + I_L^2$ = 5.16 mA (or about 5 mA)
$Z = V_T/I_T$ = 21 V/5 mA = 4.2 kΩ
cos θ = adj/hyp = 3 mA/5 mA = 0.6; θ = –53.1° (current lags V_T by 53.1°)

Analysis Questions

1. a. R will RTS.
   b. L will RTS.
   c. X_L will I.
   d. Z will I.
   e. θ will D.
   f. I_T will D.
   g. V_T will RTS.
3. a. R will I.
   b. L will RTS.
   c. X_L will RTS.
   d. Z will I.
   e. θ will D.
   f. I_T will D.
   g. V_T will RTS.
5. a. R will I.
   b. L will RTS.
   c. X_L will RTS.
   d. Z will I.
   e. θ will I.
   f. I_T will D.
   g. V_T will RTS.
7. 57, sin
9. 34, cos
11. Not having to know the hypotenuse to calculate

CHAPTER 16

Practice Problems

—1—

$M = k\sqrt{L_1 \times L_2}$
$M = 0.95\sqrt{10 \times 15}$
$M = 0.95\sqrt{150}$
$M = 0.95 \times 12.25 = 11.63$ H

—2—

$L_T = L_1 + L_2 - 2M$
L_T = 10 H + 10 H – (2 × 5 H)
L_T = 20 H – 10 H = 10 H

—3—

1. $L_T = \dfrac{1}{\dfrac{1}{L_1+M}+\dfrac{1}{L_2+M}}$

$L_T = \dfrac{1}{\dfrac{1}{8+4}+\dfrac{1}{8+4}}$

$L_T = \dfrac{1}{\dfrac{1}{12}+\dfrac{1}{12}}$

$L_T = \dfrac{1}{\dfrac{1}{6}} = 1 \times \dfrac{6}{1} = 6$ H

2. $L_T = \dfrac{1}{\dfrac{1}{L_1-M}+\dfrac{1}{L_2-M}}$

$L_T = \dfrac{1}{\dfrac{1}{8-4}+\dfrac{1}{8-4}}$

$L_T = \dfrac{1}{\dfrac{1}{4}+\dfrac{1}{4}}$

$L_T = \dfrac{1}{\dfrac{1}{2}} = 1 \times \dfrac{2}{1} = 2$ H

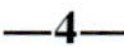

—4—

1. $\dfrac{N_P}{N_S} = \dfrac{V_P}{V_S}$; s = p turns ratio = $\dfrac{100}{300}$ or 1:3
2. Turns ratio equals voltage ratio. Since the transformer steps up the voltage six times, the secondary must have six times the turns of the primary. Therefore, the p-s turns ratio is 1:6.
3. Since the p-s turns ratio is 1:5, the secondary has five times the voltage of the primary, or 5 × 50 V = 250 V.
4. Voltage ratio with p-s turns ratio of 4:1 (primary-to-secondary). This is a step-down transformer, since the primary voltage is four times that of the secondary.

—5—

Since the impedance is transformed in relation to the square of the turns ratio, when the secondary has twice the number of turns of the primary, the impedance at the primary looks like one-fourth the impedance across the secondary, or 2,000/4 = 500 Ω.

—6—

1. The *primary-to-secondary* impedance ratio is related to the square of the turns relationship of primary-to-secondary. Since the primary has five times as many turns as the secondary, the primary impedance is 5^2 times greater than secondary. Therefore, the p-s Z ratio is 25:1.
2. The impedance looking into the primary is $16^2 \times 4\ \Omega = 256 \times 4 = 1{,}024\ \Omega$.

In-Process Learning Checks

—1—

1. Producing voltage via a changing magnetic field is called electromagnetic ***induction.***
2. Inducing voltage in one circuit by varying current in another circuit is called ***mutual inductance.***
3. The fractional amount of the total flux that links two circuits is called the ***coefficient*** of coupling, which is represented by the letter ***k***. When 100% of the flux links the two circuits, the ***coefficient*** of coupling has a value of ***1***.
4. The closer coils are, the ***higher*** the coupling factor produced. Compared with parallel coils, perpendicular coils have a ***lower*** degree of coupling.

Review Questions

1. b
3. b
5. b
7. a
9. c
11. c
13. b
15. c
17. b

Problems

1. $I_p = 200$ mA
3. $N_P/N_S = 20{:}1$
5. $M = 1.2$ H
7. $L_T = 2.5$ H
9. Volts-per-turn = 0.1 volts-per-turn
11. Turns = 2,000 turns
13. Primary-to-secondary turns ratio = 1:4.75
15. Voltage applied to primary = 62.5 V

Analysis Questions

1. A "flyback" transformer (sometimes called a "horizontal transformer") is typically used in the horizontal deflection circuits of televisions. It primarily provides high voltage to the picture tube or CRT. It may also provide low filament voltage for the high voltage rectifier. The scanning circuit in which it operates uses a frequency of 15,750 Hz.
3. Primary winding should have 3,535.5 turns.
5. Number of secondary turns = 200 turns

CHAPTER 17

Practice Problems

—1—

1. $C = \frac{Q}{V} = \frac{100 \times 10^{-6}}{25} = 4 \times 10^{-6} = 4\ \mu\text{F}$
2. $Q = CV = 10 \times 10^{-6} \times 250 = 2{,}500\ \mu\text{C}$
3. $V = \frac{Q}{C} = \frac{50 \times 10^{-6}}{2 \times 10^{-6}} = 25\ \text{V}$

—2—

$E = \frac{1}{2}CV^2 = \frac{10 \times 10^{-6} \times (100)^2}{2} = \frac{100{,}000 \times 10^{-6}}{2}$

$= 50{,}000\ \mu\text{J or } 0.05\ \text{J}$

—3—

1. $C = \frac{2.25\ kA}{10^7_s} \times (n-1) = \frac{2.25 \times 10 \times 1}{10^7 \times 0.1}$

   $C = \frac{22.5}{10^6} \times (3) = 22.5 \times 10^{-6} \times 3 = 67.5\ \mu\text{F}$
2. $C = \frac{8.85\ kA}{10^{12}_s} = \frac{8.85 \times 1 \times 0.2}{10^{12} \times 0.005} = \frac{1.77 \times 10^{-12}}{0.005}$

   $= 354\ \text{pF}$

—4—

1. $C_T = \frac{C_1 \times C_2}{C_1 + C_2} = \frac{(5 \times 10^{-6}) \times (20 \times 10^{-6})}{25 \times 10^{-6}} =$

   $\frac{100 \times 10^{-12}}{25 \times 10^{-6}} = 4 \times 10^{-6}$, or $4\ \mu\text{F}$
2. Total capacitance of four 20-μF capacitors in series is equal to one of the equal-value capacitors divided by 4. Thus, 20/4 = 5 μF, $C_T = C_1/C_N$.
3. V_{C_1} = four-fifths V_T, or 80 V. V_{C_2} = one-fifth V_T, or 20 V.

   $(Q_T = C_T \times V_T;\ 4\ \mu\text{F} \times 100\ \text{V} = 400\ \mu\text{C})$

   $\left(V_{C1} = \frac{Q}{C_1}; = \frac{400\ \mu\text{C}}{5\ \mu\text{F}} = 80\ \text{V}\right)$

   $\left(V_{C2} = \frac{Q}{C_2}; = \frac{400\ \mu\text{C}}{20\ \mu\text{F}} = 20\ \text{V}\right)$

—5—

1. C_T = sum of individual capacitances; therefore = 60 μF.
2. $Q_{C_1} = C_1 \times V = 12\ \mu\text{F} \times 60\ \text{V} = 720\ \mu\text{C}$

   $Q_{C_1} = Q_{C_2} = Q_{C_3} = Q_{C_4} = Q_{C_5}$

   Q_T = sum of all the charges = 5 × 720 μC = 3,600 μC.

—6—

1. One time constant for circuit = $0.001 \times 10^{-6} \times 10 \times 10^3$

   1 TC = 0.01 milliseconds (or 10 microseconds)

   Time allowed for charge = 20 μs, or 2 TC

   After 2 TC, capacitor is charged to 86.5% of V applied.

   In this case, V_C = 86.5% of 100 V, or 86.5 V, after 20 μs.

2. The voltage across the resistor 15 μs after the switch is closed will be equal to the percentage of V applied that V_R drops to after 1.5 TC.

   (15 μs = 1.5 × the 10-μs TC.) According to the TC chart, the resistor voltage is approximately 22% of V applied, or 22 V. Charging current = 2.2 mA.
3. V_C 30 μs after the switch is closed ≅ 95 V.
4. V_R 25 μs after the switch is closed ≅ 8.2 V.

In-Process Learning Checks

—1—

1. A capacitor is an electrical component consisting of two conducting surfaces called ***plates*** that are separated by a nonconductor called the ***dielectric.***
2. A capacitor is a device that stores electrical ***charge*** when voltage is applied.
3. Electrons ***do not*** travel through the capacitor dielectric.
4. During charging action, one capacitor plate collects ***electrons,*** making that plate ***negative.*** At the same time, the other plate is losing ***electrons*** to become ***positive.***
5. Once the capacitor has charged to the voltage applied, it acts like an ***open*** circuit to dc. When the level of dc applied increases, the capacitor ***charges*** to reach the new level. When the level of dc applied decreases, the capacitor ***discharges*** to reach the new level.
6. When voltage is first applied to a capacitor, ***maximum*** charge current occurs.
7. As the capacitor becomes charged, current ***decreases*** through the circuit in series with the capacitor.

Review Questions

1. b
3. a
5. a
7. d
9. b
11. c
13. Observe polarity.
15. Mica
17. Small size; very little lead length provides little stray undesired electrical parameters, such as inductance.
19. b

Problems

1. New capacitance is four times the original capacitance, since C is directly related to the area of the plates facing each other and inversely related to the dielectric thickness.
3. $C = \frac{Q}{V} = \frac{200 \text{ pc}}{50} = 4 \text{ pF}$
5. $C_T\,(\text{pF}) = \dfrac{1}{\frac{1}{C_1}+\frac{1}{C_2}+\frac{1}{C_3}} = \dfrac{1}{\frac{1}{100}+\frac{1}{400}+\frac{1}{1{,}000}}$

   $C_T = \dfrac{1}{0.01+0.0025+0.001} = \dfrac{1}{0.0135} = 74.07 \text{ pF}$

   $C_T = 100 \text{ pF} + 400 \text{ pF} + 1{,}000 \text{ pF} = 1{,}500 \text{ pF}$
7. $Q_{C_1} = C_1 \times V_1 = 10\ \mu\text{F} \times 100 \text{ V} = 1{,}000\ \mu\text{C}$

   $Q_{C_2} = C_2 \times V_2 = 20\ \mu\text{F} \times 100 \text{ V} = 2{,}000\ \mu\text{C}$

   $Q_{C_3} = C_3 \times V_3 = 50\ \mu\text{F} \times 100 \text{ V} = 5{,}000\ \mu\text{C}$

   $C_T = C_1 + C_2 + C_3 = 80\ \mu\text{F}$

   $Q_T = Q_1 + Q_2 + Q_3 = 8{,}000\ \mu\text{C}$
9. $v_R = \text{V applied} \times \varepsilon^{-t/\tan} = 200 \times 2.718^{-1.2} = 60.23 \text{ V}$

   $v_C = V_A - V_R$, or approximately $200 - 60 = 140$ V
11. Voltage across $C_5 = 8$ V
13. 600 V
15. $J = 12{,}800\ \mu\text{J}$
17. $J = C(V^2)/2 = 30\ \mu\text{F} \times (225)^2/2 = 1.518/2 = 0.759375$ J, or 759,375 μJ.

Analysis Questions

1. Students should reflect the ranges and operational procedures for measuring capacitance with a DMM in terms of the equipment or vendor data available to them.
3. Students should reflect the ranges and operational procedures for measuring capacitance with an LCR meter in terms of the equipment or vendor data available to them.

CHAPTER 18

Practice Problems

—1—

1. Yes, the X_C formula verifies the values of X_C shown in Figure 18–3.
2. $X_C = 1/(2\pi fC) = 1/(6.28 \times 1{,}000 \times 10^{-6}) = 159\ \Omega$
3. X_C remains the same if f is doubled and C is halved simultaneously.

—2—

1. $f = 1/(2\pi CX_C) = 1/(6.28 \times 500 \times 10^{-12} \times 1{,}000) = 318$ kHz
2. $C = 1/(2\pi fX_C) = 1/(6.28 \times 2{,}000 \times 3{,}180) = 0.025\ \mu\text{F}$

—3—

$X_{C_1} = 1/(2\pi fC) = 1/(6.28 \times 1{,}000 \times 2 \times 10^{-6}) = 79.5\ \Omega$

$X_{C_2} = 1/(2\pi fC) = 1/(6.28 \times 1{,}000 \times 4 \times 10^{-6}) = 39.75\ \Omega$

$X_{C_T} = X_{C_1} + X_{C_2} + X_{C_3} + X_{C_4} = 79.5 + 39.75 + 200 + 300 = 619.25\ \Omega$

—4—

1. $X_{C_T} = 1/(2\pi fC_T) = 1/(6.28 \times 1{,}000 \times 6 \times 10^{-6}) = 26.5\ \Omega$
2. $X_{C_T} = 1/(2\pi fC_T) = 1/(6.28 \times 500 \times 25 \times 10^{-6}) = 12.7\ \Omega$
3. $X_{C_T} = 1{,}600/4 = 400\ \Omega$

—5—

1. $X_{C_1} = 1{,}000\ \Omega$; $X_{C_2} = 1{,}000\ \Omega$; $X_{C_3} = 2{,}000\ \Omega$; $X_{C_T} = 4{,}000\ \Omega$; $C_2 = 0.795\ \mu\text{F}$; $C_3 = 0.3975\ \mu\text{F}$; $V_2 = 25$ V; $I_T = 25$ mA
2. $C_1 = 0.5\ \mu\text{F}$; $X_{C_1} = 1{,}000\ \Omega$; $X_{C_2} = 1{,}000\ \Omega$; $X_{C_3} = 500\ \Omega$; $I_2 = 2$ mA; $I_T = 8$ mA

In-Process Learning Checks

—1—

1. For a capacitor to develop a potential difference between its plates, there must be a ***charging*** current.
2. For the potential difference between capacitor plates to decrease (once it is charged), there must be a ***discharging*** current.

3. When ac is applied to capacitor plates, the capacitor will alternately ***charge*** and ***discharge.***
4. The value of instantaneous capacitor current directly relates to the value of ***capacitance*** and the rate of change of ***voltage.***
5. The amount of charging or discharging current is maximum when the rate of change of voltage is ***maximum.*** For sine-wave voltage, this occurs when the sine wave is near ***zero*** points.
6. Capacitor current ***leads*** the voltage across the capacitor by ***90*** degrees. The expression that helps you remember this relationship is "Eli the ***ice*** man."

Review Questions

1. c
3. b
5. b
7. b
9. c
11. a
13. c
15. b
17. b
19. b

Problems

1. $X_C = \dfrac{1}{6.28 fC} = \dfrac{1}{7.85 \times 10^{-6}} = 127.38\ \text{k}\Omega$

3. X_C decreases to one-third its original value, since X_C is inverse to frequency $\left(X_C = \dfrac{1}{2\pi fC} \right)$

5. For the circuit shown:

$$C_T = \frac{C_1 \times C_2}{C_1 + C_2} = \frac{0.636 \times 1.59}{0.636 + 1.59} = 0.454\ \mu\text{F}$$

$$X_{C_T} = X_{C_1} + X_{C_2} = 25\ \text{k}\Omega + 10\ \text{k}\Omega = 35\ \text{k}\Omega$$

$V_A = 70$ V (sum of V_{C_1} and V_{C_2})

$$C_1 = \frac{1}{6.28 \times 10\ \text{Hz} \times X_C}; \text{ where } X_C = \frac{V}{I} =$$

$$\frac{50\ \text{V}}{2\ \text{mA}} = 25\ \text{k}\Omega. \text{ Thus, } C_1 =$$

$$\frac{1}{6.28 \times 10 \times 25\ \text{k}\Omega} = \frac{1}{1{,}570 \times 10^3} = 0.636\ \mu\text{F}.$$

$$C_2 = \frac{1}{6.28 \times 10\ \text{Hz} \times X_C}, \text{ where } X_C = \frac{V}{I} = \frac{20\ \text{V}}{2\ \text{mA}} =$$

$$10\ \text{k}\Omega. \text{ Thus, } C_2 = \frac{1}{6.28 \times 10 \times 10\ \text{k}\Omega} =$$

$$\frac{1}{6.28 \times 10^3} = 1.59\ \mu\text{F}$$

$$X_{C_2} = 10\ \text{k}\Omega$$

7. a. C_1 will RTS.
   b. C_2 will RTS.
   c. C_T will RTS.
   d. X_{C_1} will D.
   e. X_{C_2} will D.
   f. X_{C_T} will D.
   g. I will I.
   h. V_{C_1} will RTS.
   i. V_{C_2} will RTS.
   j. V_A will RTS.
9. Total current increases to 12.56 mA, if $C_2 = C_1$ and all other factors remain the same.
11. X_C of $C_1 = 1{,}000\ \Omega$
13. $C_T = 0.063\ \mu\text{F}$
15. Increase

Analysis Questions

1. A capacitor can be said to be an "open circuit" with respect to dc because the insulating dielectric material between the capacitor's plates breaks the dc current path. Only very small "leakage" current can pass through the insulating material.
3. Capacitive voltage divider with the parameters specified as shown.

159 V
2 kHz
C_1 2 μF; $X_{C_1} = 39.75\ \Omega$; $V_{C_1} = 39.75$ V
C_2 1 μF; $X_{C_2} = 79.5\ \Omega$; $V_{C_2} = 79.5$ V
C_3 2 μF; $X_{C_3} = 39.75\ \Omega$; $V_{C_3} = 39.75$ V
$I = 1$A

CHAPTER 19

Practice Problems

—1—

1. $V_T = \sqrt{V_R^2 + V_C^2} = \sqrt{80^2 + 60^2} = \sqrt{10{,}000}$

$V_T = 100$ V

$V_R = 80$ V
θ
$V_C = 60$ V
$V_T = 100$ V

2. $Z = \sqrt{R^2 + X_C^2} = \sqrt{160^2 + 120^2} = \sqrt{40{,}000}$

$Z = 200\ \Omega$

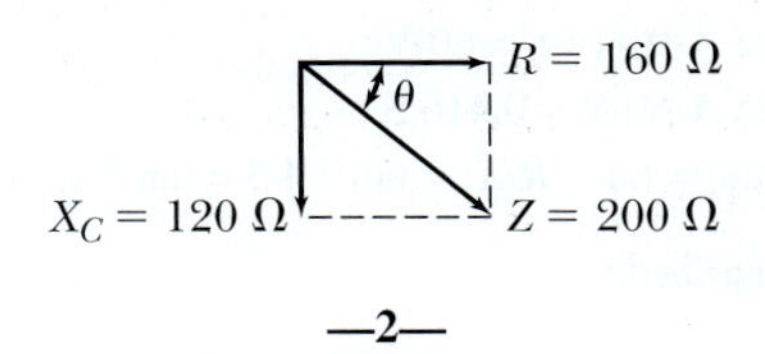

—2—

1. $V_T = 80$ V; $\theta = -48.5°$

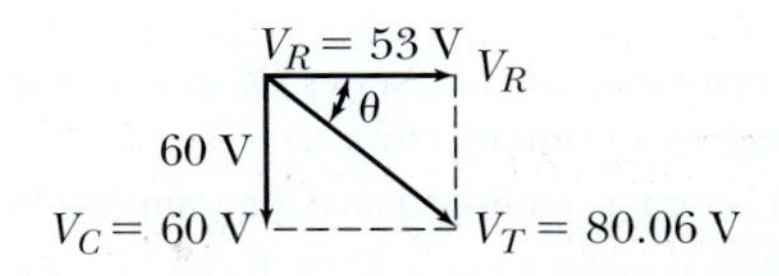

2. $\theta = -48.54°$

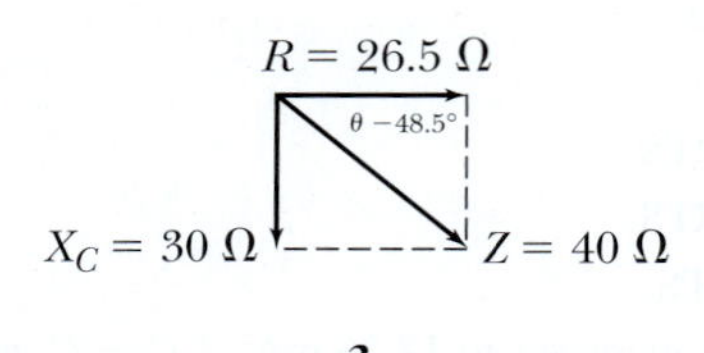

—3—

Angle of output voltage "lead" equals arctan X_C/R. With R set at 500 Ω:

$X_C = 1/(2\pi fC) = 1/(6.28 \times 400 \times 0.5 \times 10^{-6}) = 796\ \Omega$

$R = 500\ \Omega$

$X_C/R = 796/500 = 1.59$

arctan 1.59 = angle of 57.8°

Angle of lead = 57.8°

—4—

Lag angle = 90° minus circuit phase angle

Phase angle = arctan X_C/R

$X_C = 1/(2\pi fC) = 1/(6.28 \times 400 \times 0.5 \times 10^{-6}) = 796\ \Omega$

$R = 750\ \Omega$

$X_C/R = 796/750 = 1.06$

arctan 1.06 = angle of 46.7°

Angle of lag = 90 – 46.7°; angle of lag = 43.3°

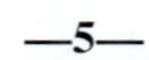

—5—

1. $I_T = 22.36$ mA
2. $\theta = 63.43°$
3. $Z = 7.11$ kΩ

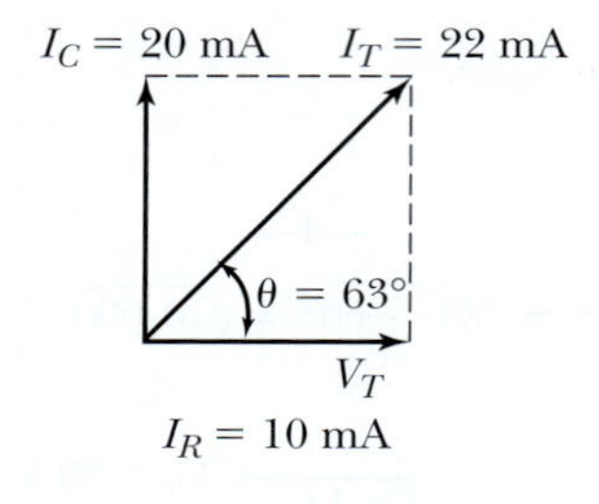

—6—

1. **G** = 1/**R** = 1/4 ohms = 0.25 **S**
2. $\mathbf{B_C}$ = 1/$\mathbf{X_C}$ = 1/3 ohms = 0.33 **S**
3. $\mathbf{Y} = \sqrt{\mathbf{G}^2 + \mathbf{Bc}^2} = \sqrt{0.25^2 + 0.33^2} = \sqrt{0.1714} = 0.1414\ \mathbf{S}$
4. **Z** = 1/**Y** = 1/0.414 **S** = 2.4 ohms.
5. $\mathbf{I_T}$ = **VY** = 60 **V** × 0.414 **S** = 24.84 **A** or approx. 25 **A**
6. **V** = **I**/**Y** = 25 **A**/0.414 **S** = 60 **V**
7. **Y** = **I**/**V** = 25 **A**/60 **V** = 0.416 **S**
8. θ (magnitude) = $\tan^{-1}$ $\mathbf{R/X_c}$ = $\tan^{-1}$ 4/3 = $\tan^{-1}$ 1.333 = **53.1°**

In-Process Learning Checks

—1—

1. In series RC circuits, the ***current*** is the reference vector, shown at 0°.
2. In the V-I vector diagram of a series RC circuit, the circuit voltage is shown ***lagging*** the circuit current.
3. In series RC circuits, an impedance diagram ***can*** be drawn.
4. Using the Pythagorean approach, the formula to find V_T in a series RC circuit is $V_T = \sqrt{V_R^2 + V_C^2}$.
5. Using the Pythagorean approach, the formula to find Z in a series RC circuit is $\mathbf{Z = \sqrt{R^2 + X_C^2}}$.
6. The trig function used to solve for phase angle when you do not know the hypotenuse value is the ***tangent*** function.

—2—

1. In parallel RC circuits, ***voltage*** is the reference vector in V-I vector diagram.
2. In parallel RC circuits, current through the resistor(s) is plotted ***out of phase*** with the circuit current vector.
3. In parallel RC circuits, current through capacitor branch(es) is plotted at ***+90 degrees*** with respect to circuit applied voltage.
4. The Pythagorean formula to find circuit current in parallel RC circuits is $I_T = \sqrt{I_R^2 + I_C^2}$.
5. ***No***
6. ***Yes***
7. ***No***

Review Questions

1. d
3. b
5. b
7. b
9. b
11. b
13. b
15. b
17. b
19. Ohm's law and Pythagorean theorem ($I_T = V_T/Z$; and $I_T = \sqrt{I_R^2 + I_C^2}$)

Problems

1. $Z = \sqrt{R^2 + X_C^2} = \sqrt{(64 \times 10^6) + (225 \times 10^6)} = \sqrt{289 \times 10^6} = 17\ k\Omega$

3. $C = \dfrac{1}{2\ fX_C} = \dfrac{1}{6.28 \times 1{,}060 \times 15\ k\Omega} = \dfrac{1}{99.8 \times 10^6} \times 106 = 0.01\ \mu F$

5. a. R will RTS.
   b. X_C will D.
   c. Z will D.
   d. I_T will I.
   e. θ will D.
   f. V_C will D.
   g. V_R will I.
   h. V_T will RTS.

7. $R = \dfrac{V_R}{I} = \dfrac{150\ \text{V}}{10\ \text{mA}} = 15\ \text{k}\Omega$

9. $f = \dfrac{1}{2\ CX_C} = \dfrac{1}{6.28 \times 0.02 \times 10^{-6} \times 30 \times 10^3} = \dfrac{1}{3.77 \times 10^{-3}} = 265\ Hz$

11. R_T = 20 kΩ + 18 kΩ + 30 kΩ = 68 kΩ

13. $Z = \sqrt{R^2 + X_C^2} = \sqrt{(68 \times 10^3)^2 + (52 \times 10^3)^2} =$
$\sqrt{7{,}328 \times 10^6} = 85.6\ k\Omega$

15. $f = \dfrac{1}{2\pi C X_C} = \dfrac{1}{6.28 \times 0.1 \times 10^{-6} \times 30 \times 10^3} =$
$\dfrac{1}{0.0188} = 53\ Hz$

17. $I_C = \dfrac{320\ V}{16\ k\Omega} = 20\ mA$

19. $\tan \theta = \dfrac{I_C}{I_R} = \dfrac{20}{32} = 0.625$. Angle whose tangent equals 0.625 is 32°. $\theta = 32°$

21. $R = 750\ \Omega$; $X_C = 1{,}000\ \Omega$; $I = 20\ mA$; $V_R = 15\ V$; $V_C = 20\ V$; $\theta = -53.1°$

Analysis Questions

1. Two basic circuit operational differences between series *RC* circuits and series *RL* circuits are as follows:
   - **a.** Circuit current leads circuit voltage in series *RC* circuits, and lags circuit voltage in series *RL* circuits.
   - **b.** As frequency of operation is increased, or *C* is increased in series *RC* circuits, the phase angle between circuit current and circuit voltage decreases. For a series *RL* circuit, as frequency increases, or *L* is increased, the circuit phase angle increases.
3. Yes, there are similar circuit effects when comparing a series *RC* circuit and a parallel *RL* circuit. For example, as frequency of operation is increased, a *series RC circuit* will cause the phase angle to decrease. The same is true for a *parallel RL circuit;* that is, an increase in *f* will cause the phase angle to decrease.

CHAPTER 20

Practice Problems

—1—

$R = 200\ \Omega$
$X_{C_T} = 500\ \Omega$
$X_L = 250\ \Omega$

$Z = \sqrt{R^2 + X^2} = \sqrt{200^2 + 250^2} = \sqrt{102{,}500} = 320.15\ \Omega$

$V_T = I \times Z = 0.25 \times 320.15 = 80.04\ V$
$V_{C_2} = I \times X_{C_2} = 0.25 \times 250 = 62.5\ V$
$V_L = I \times X_L = 0.25 \times 250 = 62.5\ V$
$V_R = I \times R = 0.25 \times 200 = 50\ V$

$\theta = \dfrac{\text{angle whose tangent equals } X(\text{net})}{R}$

Tan $\theta = 250/200 = 1.25$; therefore, $\theta = -51.34°$ (*V* with respect to *I*)

—2—

$I_R = \dfrac{V}{R} = \dfrac{300}{200} = 1.5\ A$
$I_L = \dfrac{V}{X_L} = \dfrac{300}{200} = 1.5\ A$
$I_C = \dfrac{V}{X_C} = \dfrac{300}{300} = 1\ A$
$I_X = I_L - I_C = 1.5 - 1 = 0.5\ A$
$I_T \sqrt{I_R^2 + I_X^2} = \sqrt{1.5^2 + 0.5^2} = \sqrt{2.5} = 1.58\ A$
$Z = \dfrac{V}{I} = \dfrac{300}{1.58} = 189.8\ \Omega$

$\theta =$ angle whose tan equals $I_X/I_R = 0.5/1.5 = 0.333$
Therefore, $\theta = -18.43°$ (*I* with respect to *V*)

—3—

1. $20 + j30$ in rectangular form converts to $36.05\ \angle 56.31°$ in polar form.
2. $30 - j20$ in rectangular form converts to $36.05\ \angle -33.7°$ in polar form.

—4—

1. $66\ \angle 60°$ in polar form converts to $33 + j57.15$ in rectangular form.
2. $75\ \angle 48°$ converts to $50.2 + j55.7$ in rectangular form.

—5—

1. $20 + j5$
   $35 - j6$
   $55 - j1$
2. $34.9 - j10$
   $15.4 + j12$
   $50.3 + j2$
3. $35 - j6$ / $20 + j5$ (change sign) → $35 - j6$ / $-20 - j5$ = $15 - j11$
4. $37.8 - j13$ / $22.9 - j7$ (change sign) → $37.8 - j13$ / $-22.9 + j7$ = $14.9 - j6$
5. $5\ \angle 20° \times 25\ \angle -15° = 125\ \angle 5°$
6. $24.5\ \angle 31° \times 41\ \angle 23° = 1{,}004.5\ \angle 54°$
7. $\dfrac{5\ \angle 20°}{25\ \angle -15°} = 0.2\ \angle 35°$
8. $\dfrac{33\ \angle 34°}{13\ \angle 45°} = 2.54\ \angle -11°$
9. $I = 12\ \angle 25°$, then $I^2 = 144\ \angle 50°$
10. If $I^2 = 49\ \angle 66°$, then $I = 7\ \angle 33°$
11. $\dfrac{5\ \angle 50° + (30 + j40)}{2.6\ \angle -15°} =$
    $\dfrac{(3.2 + j3.8) + (30 + j40)}{2.6\ \angle -15°} = \dfrac{33.2 + j43.8}{2.6 \angle -1.5°} =$
    $\dfrac{54.9 \angle 52.8°}{2.6\ \angle -15°} = 21.1\ \angle 67.8°$
12. $\dfrac{10\ \angle 53.1° + (25 - j25)}{4.8\ \angle 36°} =$
    $\dfrac{31 - j17}{4.8\ \angle 36°} = \dfrac{35\ \angle -28.7°}{4.8\ \angle 36°} = 7.3\ \angle -64.7°$

In-Process Learning Checks

—1—

1. When both capacitive and inductive reactance exist in the same circuit, the net reactance is ***the difference between their reactances.***
2. When plotting reactances for a series circuit containing *L* and *C*, the X_L is plotted ***upward*** from the reference, and the X_C is plotted ***downward*** from the reference.
3. In a series circuit with both capacitive and inductive reactance, the capacitive voltage is plotted ***180 degrees*** out of phase with the inductive voltage.

4. If inductive reactance is greater than capacitive reactance in a series circuit, the formula to find the circuit impedance is $Z = \sqrt{R^2 + (X_L - X_C)^2}$.
5. In a series *RLC* circuit, current is found by $= \frac{V}{Z}$.
6. ***False***
7. If the frequency applied to a series *RLC* circuit decreases while the voltage applied remains the same, the component(s) whose voltage will increase is/are the ***capacitor(s).***

—2—

1. In parallel *RLC* circuits, capacitive and inductive branch currents are ***180 degrees*** out of phase with each other.
2. The formula to find total current in a parallel *RLC* circuit when I_C is greater than I_L is

   $I_T = \sqrt{I_R^2 + (I_C - I_L)^2}$.
3. ***False***
4. If the frequency applied to a parallel *RLC* circuit increases while the applied voltage remains the same, the branch current(s) that will increase is/are the ***capacitive*** branch(es).

—3—

1. Power dissipated by a perfect capacitor over one whole cycle of input is ***zero.***
2. The formula for apparent power is $S = V \times I$, and the unit expressing apparent power is ***volt-amperes.***
3. True power in *RLC* circuits is dissipated by the ***resistances.***
4. The formula for power factor is ***p.f. = cos θ.***
5. If the power factor is one, the circuit is purely ***resistive.***
6. The lower the power factor, the more ***reactive*** the circuit is acting.
7. In the power triangle, true power is on the ***horizontal*** axis, reactive power (VAR) is on the ***vertical*** axis, and apparent power is the ***hypotenuse*** of the right triangle.

Review Questions

1. b
3. d
5. b
7. b
9. c
11. c
13. c
15. b
17. a
19. b

Problems

1. $Z = \sqrt{R^2 + X_1^2} = \sqrt{300^2 + 150^2} = \sqrt{112{,}500} =$ 335.4 Ω

$$I = \frac{V_T}{Z} = \frac{10\text{ V}}{0.3354\text{ k}\Omega} = 29.8\text{ mA}$$

$V_C = I \times X_C = 29.8\text{ mA} \times 250\ \Omega = 7.45\text{ V}$

$V_L = I \times X_L = 29.8\text{ mA} \times 100\ \Omega = 2.98\text{ V}$

$V_R = I \times R = 29.8\text{ mA} \times 300\ \Omega = 8.94\text{ V}$

$$\text{Cos } \theta = \frac{\text{adj}}{\text{hyp}} = \frac{R}{Z} = \frac{300}{335.4} = 0.89$$

θ = angle whose cos = 0.89 = 26.56° (I_T will lead V_T.)

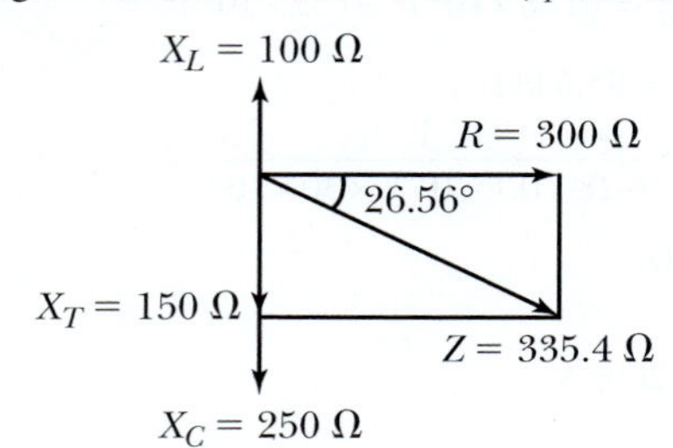

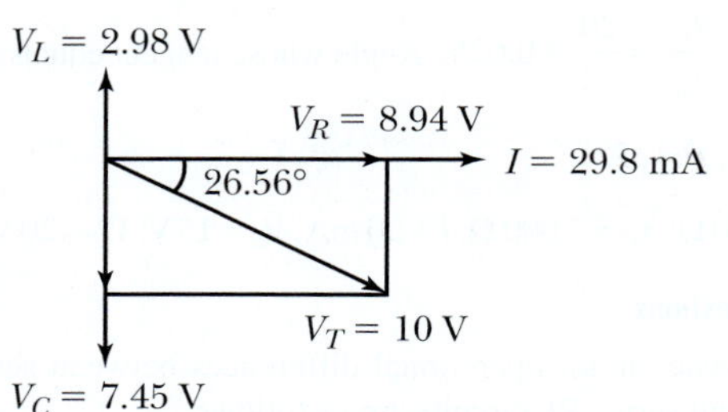

3. True power $= I^2R = 0.0298^2 \times 300 = 0.266$ W

   Apparent power $= V \times I = 10\text{ V} \times 0.0298\text{ mA} = 0.298$ VA

   Reactive power $= V \times I \times \sin\theta = 10\text{ V} \times 0.0298 \times 0.447 =$ 0.133 VAR

$$\text{Power factor} = \frac{\text{true power}}{\text{apparent power}} = \frac{0.266}{0.298} = 0.89$$

   (or p.f. = cos θ = cos 26.56° = 0.89)
5. $Z = 20 + j40$

   If the circuit were an *RC* circuit, it would be stated as

   $Z = 20 - j40$
7. $I_R = \frac{20}{20} = 1\text{ A}$

   $I_C = \frac{20}{20} = 1\text{ A}$

   $I_T = 1.414$ A at an angle of 45°

   $I_T = 1.414\ \angle 45°$
9. Converting 14.14 ∠45° to rectangular form: $10 + j10$. That is, $R = Z \times \cos\theta = 14.14 \times \cos 45° = 10$, since the angle is 45°; *R* and *X* must be equal; thus $Z = R + jX_L$.

   $14.14 = 10 + j10$
11. $Z = 61.43\ \angle{-75.96°}$ kΩ (polar form)

    $Z = 14{,}900 - j59{,}600$ (rectangular form)
13. V_C will decrease.
15. $Z = 1.79\ \angle{-63.4°}\ \Omega$
17. $Z = 5.82\ \angle 14°\ \Omega$
19. $I_R = 800\text{ V}/400\ \Omega = 2\text{ A}$

    $I_C = 800\text{ V}/800\ \Omega = 1\text{ A}$

    $I_L = 800\text{ V}/500\ \Omega = 1.6\text{ A}$

    $I_{XT} = 1.6\text{ A} - 1\text{ A} = 0.6\text{ A}$

    $I_T = \sqrt{2^2 + 0.6^2} = 2.09\text{ A}$

    $Z = V_T/I_T = 800\text{ V}/2.09\text{ A} = 383.8\ \Omega$

    $\theta = 16.69°$
21. I_R will RTS.

    I_C will I.

    I_L will D.

    I_T will I.

Z_T will D.

θ will I.

23. $1.414 \angle 45°$ A $= 1 + j1$ A in rectangular form.

25. $1.79 \angle -63.4°\ \Omega = 0.8 - j1.6\ \Omega$ in rectangular form.

27. $Z_T = 12.5 - j2.5\ \Omega$ (rectangular)

29. $I_T = 3.84 + j0.8$ A (rectangular)

Analysis Questions

1. A "real number" in math can be any rational or irrational number. In complex numbers related to the coordinate system, the real term in the complex number is plotted on the horizontal (or x) axis of the coordinate system. Imaginary numbers are complex numbers; they involve the square roots of negative numbers. The imaginary unit, or j operator, represents the square root of -1. For our purposes, the j operator is the vertical element used in ac analysis.
3. The reason simply measuring voltage with a voltmeter and the current with a current meter does not provide adequate information to find the actual power dissipated by the circuit is that in ac circuits containing reactances, currents and voltages are not generally in phase with one another. Thus, taking the product of values that are not in phase will not yield the actual power being dissipated.

CHAPTER 21

Practice Problems

—1—

$$f_r = \frac{1}{(2\ \sqrt{LC})} =$$

$$\frac{1}{(6.28 \times \sqrt{250 \times 10^{-6} \times 500 \times 10^{-12}})} = 450.4 \text{ kHz}$$

—2—

$$L = \frac{25{,}330}{f^2 C} = 10.13\ \mu\text{H}$$

—3—

1. $Q = \frac{X_L}{R} = \frac{1{,}000}{50} = 20$
2. $V_C = I \times X_C$

   $I = \frac{V_T}{Z} = \frac{12}{50} = 0.24$ A

   $V_C = 0.24 \text{ A} \times 1{,}000\ \Omega = 240$ V
3. If r_s doubles, $Q = \frac{1{,}000}{100} = 10.$
4. If r_s is made smaller, the skirts are steeper.

—4—

1. $f_r = \frac{1}{(2\pi\sqrt{LC})} = 2.517$ MHz
2. $Q = \frac{X_L}{R} = 316.13$
3. $Z = Q \times X_L =$ approximately 999 kΩ
4. If r_s is 1 kΩ, the f_r is lower.

—5—

1. $f_1 = 2{,}494$ kHz and $f_2 = 2{,}506$ kHz
2. $f_r = 500$ kHz and bandpass = from 495 kHz to 505 kHz
3. $f_1 = 3{,}200 \text{ kHz} - \frac{2 \text{ kHz}}{2} = 3{,}200 \text{ kHz} - 1 \text{ kHz} = 3{,}199$ kHz

In-Process Learning Checks

—1—

1. In circuits containing both inductance and capacitance, when the source frequency increases, inductive reactance ***increases*** and capacitive reactance ***decreases.***
2. In a series RLC (or LC) circuit, when X_L equals X_C, the circuit is said to be at ***resonance.***
3. In a series RLC (or LC) circuit, when X_L equals X_C, the circuit impedance is ***minimum;*** the circuit current is ***maximum;*** and the phase angle between applied voltage and current is ***zero*** degrees.
4. In a series RLC (or LC) circuit, when X_L equals X_C, the voltage across the inductor or the capacitor is ***maximum.***

—2—

1. In a high-Q parallel resonant circuit, Z is ***maximum.***
2. In a high-Q parallel circuit using ideal components, X_L ***does*** equal X_C at resonance.
3. The power factor of an ideal parallel resonant circuit is ***1.***
4. Phase angle in an ideal parallel resonant circuit is ***0 degrees.***
5. In a parallel resonant circuit, the current circulating in the tank circuit is ***higher*** than the line current.

Review Questions

1. b
3. a
5. b
7. c
9. b
11. c
13. c
15. a
17. b, c
19. b

Problems

1. **a.** At resonance $Z = R$; therefore, $Z = 10\ \Omega$
   b. $f_r = 0.159 / \sqrt{LC} = 15.9$ MHz
   c. $Q = X_L/R$
   $X_L = 2\pi fL = 99.8\ \Omega$
   $Q = 99.8/10 = 9.98$
   d. $B.W. = 1.59$ MHz
   e. Bandpass = 15.105 – 16.695 MHz
   f. Below f_r acts capacitively
3. **a.** $f_r \cong 5.5$ MHz
   b. $Q \cong 32.155$
   c. $BW = 0.17$ MHz
   d. Bandpass $\cong$ 5.415 MHz to 5.585 MHz
   e. At resonance, $Z = 31{,}018.6\ \Omega$
   f. Below resonance the circuit acts inductively.
5. Resonant frequency is 1,060 Hz.

7. $L = 45$ mH
9. R value must be 3.33 Ω.
11. Resonant frequency is 796 kHz.
13. Exhibited impedance equals 104 kΩ.

Analysis Questions

1. The tuned circuit in a receiver input stage is typically a parallel LC circuit.
3. The circuits described would be considered "resonant filters."
5. *Selectivity* is the ability to select desired signal frequencies, while rejecting undesired signal frequencies. Referring to resonant circuits, selectivity indicates the sharpness of the response curve. A narrow and more pointed response curve results in a higher degree of frequency selectiveness, or selectivity.

   Bandwidth is the frequency limits (total difference in frequency) between upper and lower frequencies over which a specified response level is shown by a frequency-sensitive circuit, such as a resonant circuit. The cutoff level is defined at 70.7% of peak response, sometimes called the half-power points on the response curve.

   Bandpass is the band of frequencies that passes through a filter or circuit with minimal attenuation or degradation, often designated by specific frequencies. When specified, they are designated indicating that the lower frequency is at the lower frequency "half-power point" of the response curve and the upper frequency, which is at the upper "half-power point" on the response curve. Bandpass is often expressed as *the number of hertz between these two frequency response limits.*
7. a. X_L will D.
   b. X_C will I.
   c. R will RTS.
   d. Z will I.
   e. I will D.
   f. θ will I.
   g. V_C will I.
   h. V_L will D.
   i. V_R will D.
   j. V_T will RTS.
   k. BW will RTS.
   l. Q will RTS (close to the same).
   m. I_C will D.
   n. I_L will D.
   o. I_T will D.
9. a. X_L will D.
   b. X_C will I.
   c. R will RTS.
   d. Z will D.
   e. I will I.
   f. θ will I.
   g. V_C will RTS.
   h. V_L will RTS.
   i. V_R will RTS.
   j. V_T will RTS.
   k. BW will RTS.
   l. Q will RTS.
   m. I_C will D.
   n. I_L will I.
   o. I_T will I.
11. The condition is such that $X_L = X_C$.
13. Circuit diagram of a bandpass filter using a series resonant and a parallel resonant circuit:

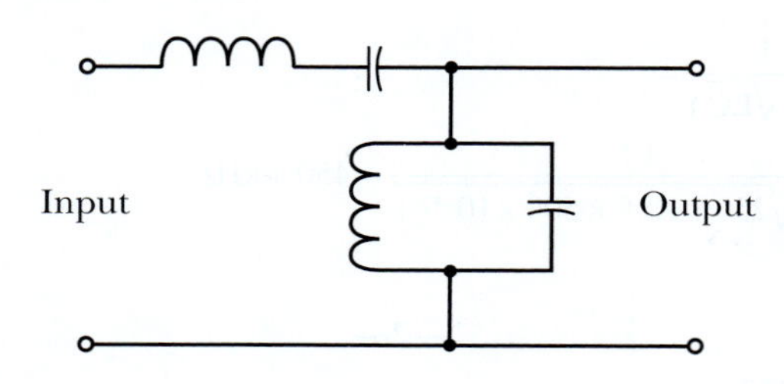

APPENDIX C

Troubleshooting Challenge Test Results

Find the number listed next to the test you chose and record the result.

1. —
2. signal normal
3. 2 kΩ
4. 2 V
5. 1.5 V
6. high
7. —
8. 0 V
9. —
10. —
11. 0 V
12. 7 V
13. —
14. 0 Ω
15. ≅ 0 V
16. 1 kΩ
17. 400 Ω
18. —
19. infinite Ω
20. —
21. (h.w. rect. waveform)
22. 10 V
23. no signal
24. circuit operates normally
25. 10 kΩ
26. slightly high
27. —
28. 0 V
29. 0 Ω
30. greatly below max value
31. 5 V
32. 400 Ω
33. infinite Ω
34. 6.4 V
35. noticeably high
36. 4 kΩ
37. 10 kΩ
38. 3.3 V
39. 14 V
40. —
41. ≅ 1.5 mH
42. 3.0 V
43. 7 V
44. —
45. high
46. 0 Ω
47. —
48. 2.3 V
49. 14 V
50. —
51. —
52. 10 V
53. (Inv. sine wave) (14 VAC)
54. 10 kΩ
55. max value
56. 10 V
57. signal normal
58. infinite Ω
59. —
60. 275 pF
61. 14 V
62. 1 kΩ
63. infinite Ω
64. —
65. ≅ 6 VDC
66. 0 V
67. 7.5 V
68. 10 kΩ
69. low
70. signal normal
71. no change in operation
72. 0 Ω
73. —
74. (sine wave) (120 V)
75. 10 V
76. slightly low
77. —
78. 3.1 V
79. 7.5 V
80. 7 V
81. —
82. 0 V
83. 10 V
84. —
85. slightly below max value
86. 1.5 V
87. 5 Ω
88. 14 V
89. 2.3 V
90. infinite Ω
91. —
92. normal
93. 10 kΩ
94. ≅ 5 V
95. 10 V
96. ≅ 11 V
97. —
98. low
99. —
100. infinite Ω
101. —
102. 0 V
103. 0 V
104. ≅ 5 kΩ
105. —
106. no signal
107. 50 pF
108. —
109. 12 kΩ
110. —
111. no change in operation
112. —
113. signal normal
114. 297 Ω
115. —
116. ≅ 5 V
117. within normal range
118. 150 mA
119. 17 mA
120. 100 Ω
121. distorted
122. clipped sine wave
123. slightly lower
124. slightly high
125. positive saturation
126. normal
127. normal
128. low
129. high
130. high
131. low
132. high
133. high
134. low
135. high
136. high
137. low
138. high
139. high
140. low
141. high
142. low
143. low
144. normal
145. low
146. low
147. normal
148. low
149. low
150. normal
151. low
152. low
153. normal
154. normal
155. low
156. normal
157. normal
158. low
159. normal
160. normal
161. high

162. normal
163. normal
164. normal
165. low
166. normal
167. normal
168. normal
169. normal
170. normal
171. low
172. normal
173. normal
174. normal
175. normal
176. normal
177. low
178. normal
179. normal
180. normal
181. normal
182. high
183. low
184. normal
185. high
186. normal
187. normal
188. normal
189. low
190. normal
191. normal
192. normal
193. high
194. high
195. normal
196. high
197. low
198. normal
199. low
200. low
201. low
202. high
203. low
204. low
205. low
206. high
207. low
208. low
209. high
210. low
211. low
212. high
213. low
214. low
215. high
216. low
217. normal
218. high
219. low
220. low
221. high
222. low
223. normal
224. high
225. low
226. normal
227. high
228. low
229. normal
230. normal
231. normal
232. normal
233. normal
234. high
235. normal
236. high
237. normal
238. low
239. low
240. low
241. normal
242. low
243. normal
244. normal
245. low
246. low
247. normal
248. abnormal
249. low
250. low
251. normal
252. abnormal
253. normal
254. normal
255. normal
256. normal
257. low
258. abnormal
259. low
260. low
261. normal
262. abnormal
263. normal
264. normal
265. normal
266. normal
267. low
268. normal
269. low
270. high
271. abnormal
272. low
273. low
274. normal
275. abnormal
276. high
277. normal
278. high
279. normal
280. normal
281. high
282. high
283. normal
284. normal
285. normal
286. normal
287. normal
288. normal
289. normal
290. none

APPENDIX D

Using the Excel Formulas Templates Files

General Information

On the CD there are useful and powerful "pre-programmed" spreadsheet (Excel) formula templates. For the convenience of both instructors and students, every formula in each text has been translated into Excel spreadsheet formula format and is available for use. (A total of 272 Excel formulas are pre-programmed and ready to go!) Near the end of this appendix, a shortcut method to find any given formula in the long list of data is provided to aid in using these files.

The files containing these helpful tools on the CD are titled:

Meade Text Excel Formulas Templates—Chapters 1–21

Meade Text Excel Formulas Templates—Chapters 22–31

(The advantage of using the "pre-programmed" Excel formula files is that it is much faster than inputting math operators and numbers on a calculator!)

These files are useful to both students and instructors for several reasons, including the following:

1. They provide pre-programmed Excel formulas for every formula/equation in both texts mentioned previously.
2. Students learn to use the computer as a tool in solving electronic circuit problems, whether to help with homework or in-class problems, or in their jobs beyond graduation.
3. The CD file worksheet can be applied as is to any problem in either text, simply by inserting the parameters from the problem(s) into columns D through G, as appropriate. The answers automatically appear in column H.
4. Any given formula or set of formulas can be copied from the CD file(s) worksheets to a separate or a blank worksheet to solve problems on the worksheet to which the formulas are copied.
5. They can be used as teaching tools to help students learn how to format Excel formulas. Excel formulas shown in column C as text are used in column H as functional spreadsheet formulas that produce the desired answers.
6. They provide easy cross-referencing to the textbook formulas by chapter, formula number, and purpose of the formula or equation in the CD worksheet files, as shown on the sample excerpt that follows.
7. Hard-copy versions can be printed out from the CD Excel files so students and instructors can have a handy, useable reference for all the text formulas and Excel formulas contained in the CD files.

Sample Excerpt from CD Files' Worksheets

Observe:

Column A = Chapter number and formula number information

Column B = Purpose of formula, and the formula (in style of text in book)

Column C = Excel-equivalent formulas in textual format (so Excel formula can be read or copied by students)

Column D–G = Cells into which parameter values needed for each formula are entered, as appropriate

Column H = The same Excel formula as illustrated in Column C is written in Column H, which produces answer information for that equation and the parameters used. (Only the answers show in the H column.) On the worksheet, you can place the "cell pointer" on a specific "answer" cell in H, and see the formula used in that cell in the f_x Formula Bar (near the top of the worksheet).

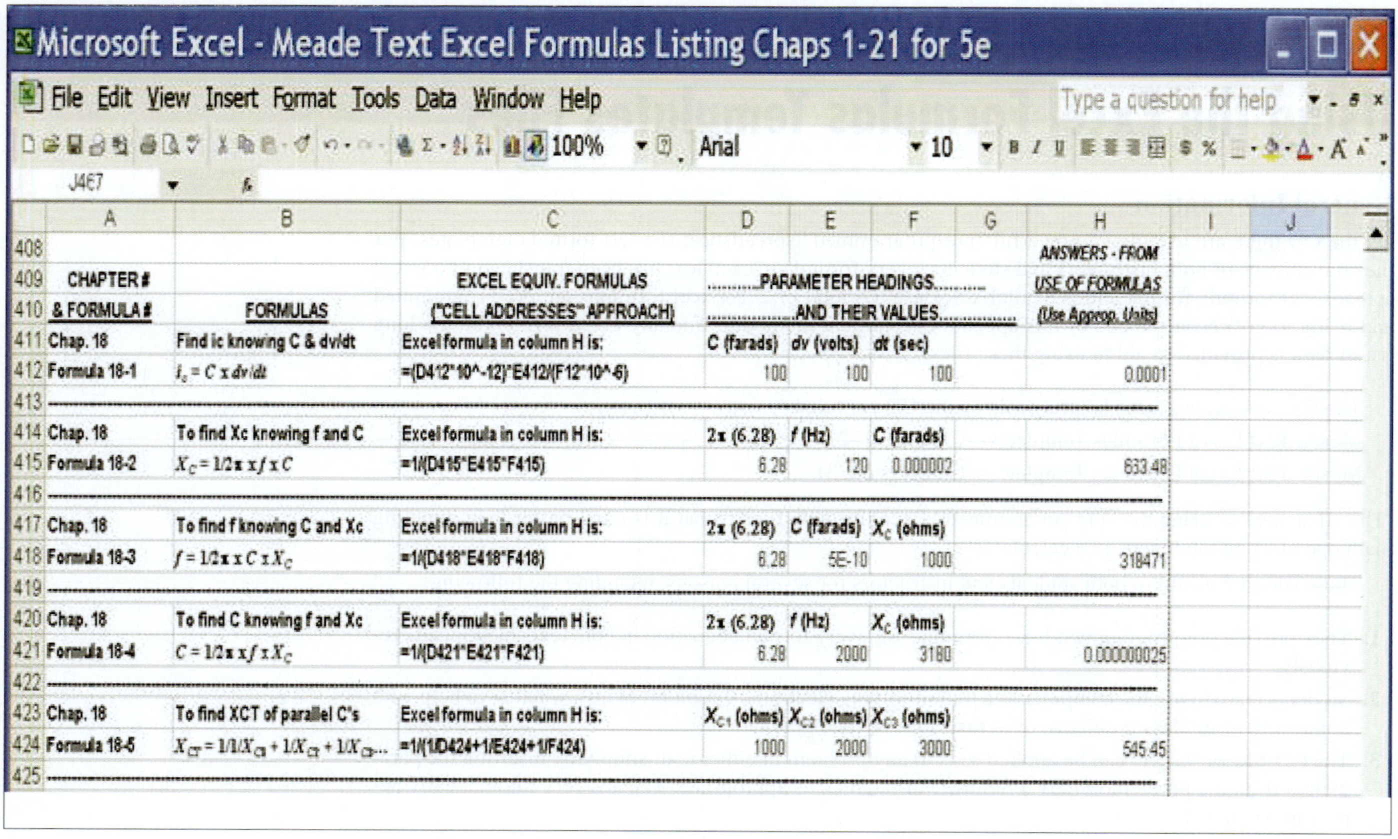

Microsoft Excel - Meade Text Excel Formulas Listing Chaps 1-21 for 5e

File Edit View Insert Format Tools Data Window Help — Type a question for help

J467

	A	B	C	D	E	F	G	H
408								ANSWERS - FROM
409	CHAPTER #		EXCEL EQUIV. FORMULAS	..........PARAMETER HEADINGS..........				USE OF FORMULAS
410	& FORMULA #	FORMULAS	("CELL ADDRESSES" APPROACH)	..................AND THEIR VALUES................				(Use Approp. Units)
411	Chap. 18	Find ic knowing C & dv/dt	Excel formula in column H is:	C (farads)	dv (volts)	dt (sec)		
412	Formula 18-1	$i_c = C \times dv/dt$	=(D412*10^-12)*E412/(F12*10^-6)	100	100	100		0.0001
413	----------							
414	Chap. 18	To find Xc knowing f and C	Excel formula in column H is:	2π (6.28)	f (Hz)	C (farads)		
415	Formula 18-2	$X_C = 1/2\pi \times f \times C$	=1/(D415*E415*F415)	6.28	120	0.000002		633.48
416	----------							
417	Chap. 18	To find f knowing C and Xc	Excel formula in column H is:	2π (6.28)	C (farads)	X_C (ohms)		
418	Formula 18-3	$f = 1/2\pi \times C \times X_C$	=1/(D418*E418*F418)	6.28	5E-10	1000		318471
419	----------							
420	Chap. 18	To find C knowing f and Xc	Excel formula in column H is:	2π (6.28)	f (Hz)	X_C (ohms)		
421	Formula 18-4	$C = 1/2\pi \times f \times X_C$	=1/(D421*E421*F421)	6.28	2000	3180		0.000000025
422	----------							
423	Chap. 18	To find XCT of parallel C's	Excel formula in column H is:	X_{C1} (ohms)	X_{C2} (ohms)	X_{C3} (ohms)		
424	Formula 18-5	$X_{CT} = 1/1/X_{C1} + 1/X_{C2} + 1/X_{C3}...$	=1/(1/D424+1/E424+1/F424)	1000	2000	3000		545.45
425	----------							

How to Use the Pre-Programmed Features

Refer to the sample graphic as you read the following:

1. The information in both Columns A and B is useful for finding the formula needed to help solve the problem being addressed.

   NOTE: The CD file worksheets provide every formula in the text from which to choose.

2. Column C shows how the text formula in Column B can be written for Excel. This information can help you learn how to create Excel formulas, and provide information regarding how the "pre-programmed" formula is actually written in Column H, which produces the answer information you see in that location.

   CAUTION: Observe that both the Excel formula, shown in text form in Column C, and the functional Excel formula in Column H, illustrate that you must use proper "cell addresses" in Excel formulas for the Excel program to produce the answer information being sought. In our sample graphic, note that the Excel formula to solve for capacitive reactance (derived from text Formula 18–2) refers to the values shown in cell addresses D415, E415, and F415, for the Excel formula (in Column H, cell address H415) to produce the answer based on the values shown in D415, E415, and F415. In similar fashion, the Excel formula to solve for total capacitive reactance of parallel capacitors (derived from text Formula 18–5) uses parameter values shown in cell addresses D424, E424, and F424 to produce the right answer in cell address H424.

3. Two ways to use the pre-programmed formulas in Column H are as follows:

   a. Simply plug the values for your problem into the correct parameters' cell addresses in Columns D, E, F, and G, replacing the sample values data in the CD file worksheet. The answer to your problem will automatically appear in Column H. (IMPORTANT: Students must assign appropriate value units to the answers shown in Column H (e.g., amperes, milliamperes, volts, etc.) based on the value units given in Columns D–G).

   b. Copy data from the CD file worksheet to your own separate worksheet, as follows:

      1) Highlight the data cells you want to copy from the CD file worksheet to your separate worksheet by moving the cell pointer (while holding down the Shift key). NOTE: The data cells of interest will be Columns D through H in the two rows containing

         a) parameter identifier headings used in the formula of interest

         b) parameter values under those headings and the formula cell in row H

2) Press Control (Ctrl key) and the letter *C* key to initiate the copy command.
3) Bring up a blank worksheet, or the worksheet you want to transfer the data to.
4) Identify the place on your worksheet where you want the copied data from the CD file to be inserted or pasted on your worksheet.
5) Press the Enter key, or point at the paste button at the top of the screen and click to paste the copied information onto your worksheet at the selected location.

   NOTE: Because the formulas on the CD worksheet are written with "the relative cell address approach," the copied formula (from CD file worksheet, cell H) will automatically change to addresses appropriate to the location into which you placed the formula on your worksheet.

c. Replace parameter values data that were copied and transferred from the CD file worksheet to your worksheet with the appropriate parameter values for your problem. NOTE: Be sure the desired parameter values are placed in the correct cell addresses under the parameter identifier headings, so the Excel formula references, in Column H, will work properly.

4. Another use of the CD files worksheets is to help facilitate student practice in inputting data and formulas in the Excel spreadsheet program. Instructors may have students practice inputting specified data and formulas on practice worksheets, then cross-reference their work to the CD files Excel data to see if they are learning the Excel worksheet mathematical operators, cell addressing, and data entry properly.

A Handy Shortcut

As mentioned earlier, there are many formulas listed in the CD Excel Formulas Templates files—148 from Chapters 1–21 of both texts, and 124 from the last 10 chapters of the FOE-C&D text, making a total of 272 formulas. This means that there are a number of pages and screens full of information to scan, or cursor, through to get to the formula needed for your specific problem.

On the following page is a shortcut for finding the correct formula.

1. With the Excel Formulas Templates file pulled up on your computer screen at the first page of the file—place (and click) the cell pointer onto a blank cell address near the top of the page (i.e., somewhere above "Formula 1-1" for the Chaps 1–21 file, or above "Formula 22-1" for the Chaps 22–31 file).
2. Use the mouse pointer; point and click on the *Edit* command at the top of the screen.
3. On the *Edit* submenu, point and click on the "*Find*" command.
4. In the "*Find and Replace*" dialogue box (under the "*Find*" tab): Type the formula number you want to find in the blank space next to the "*Find what:*" statement.
   (For example, "Formula 19-2" or "Formula 16-3", or the formula number you need to find.)
5. Press "*Enter*" and you should automatically be taken to Column A, at the called-for location in the worksheet's file without the necessity of scrolling through numerous screens and pages to get there!

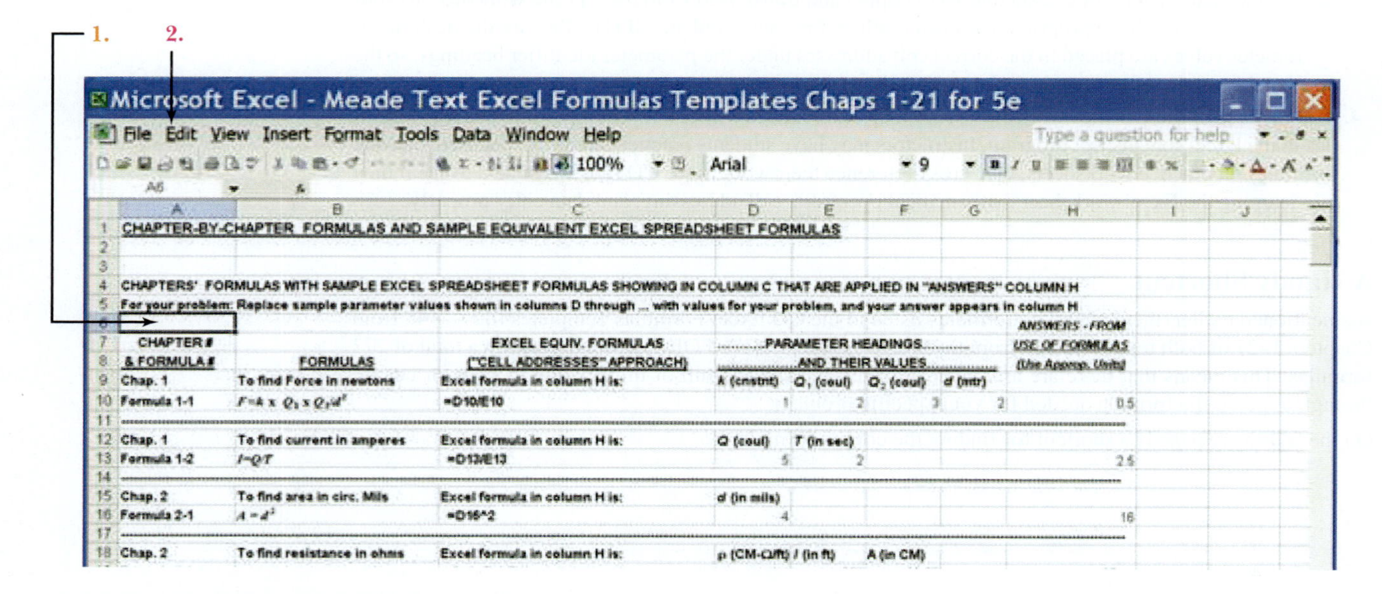

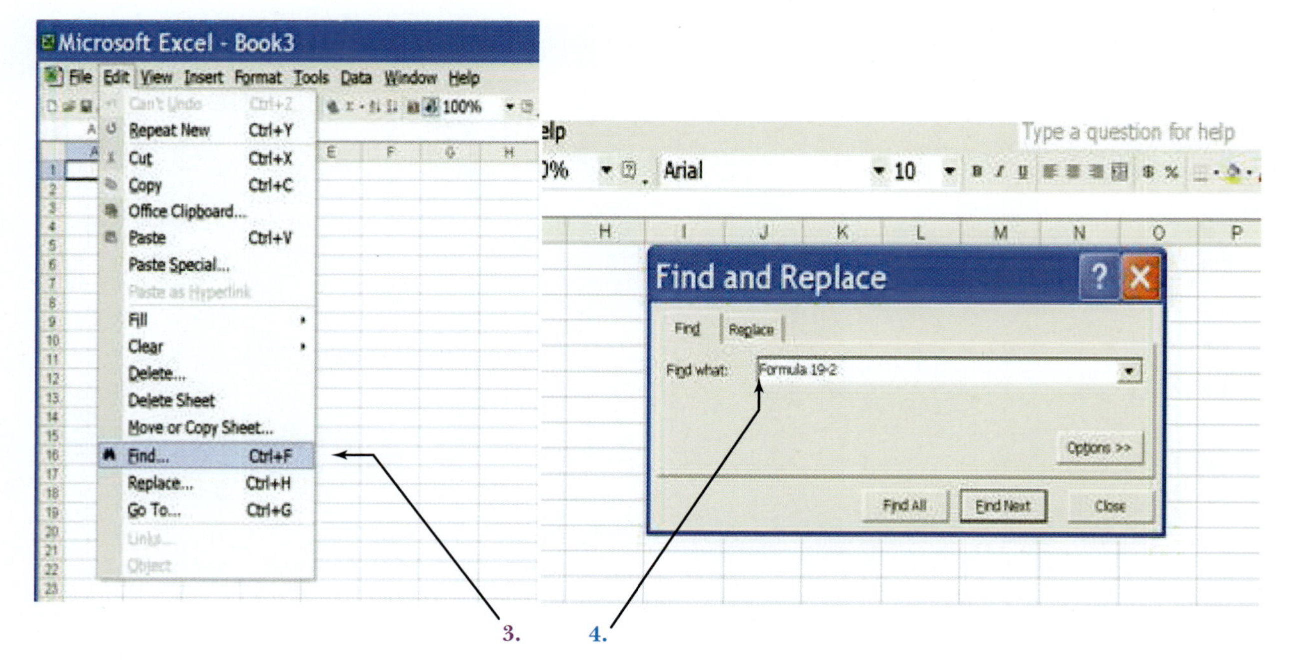

APPENDIX E

Copper-Wire Table

Wire Size A.W.G. (B&S)	Diam. in Mils[1]	Circular Mil Area	Turns per Linear Inch (25.4 mm)[2]			Cont.-duty current[3] single wire in open air	Cont.-duty current[3] wires or cables in conduits or bundles	Feet per Pound (0.45 kg) Bare	Ohms per 1000 ft. 25°C	Current-Carrying Capacity[4] at 700 C.M. per Amp.	Diam. in mm.	Nearest British S.W.G. No.
			Enamel	S.C.E.	D.C.C.							
1	289.3	83,690	—	—	—	—	—	3.947	0.1264	119.6	7.348	1
2	257.6	66,370	—	—	—	—	—	4.977	0.1593	94.8	6.544	3
3	229.4	52,640	—	—	—	—	—	6.276	0.2009	75.2	5.827	4
4	204.3	41,740	—	—	—	—	—	7.914	0.2533	59.6	5.189	5
5	181.9	33,100	—	—	—	—	—	9.980	0.3195	47.3	4.621	7
6	162.0	26,250	—	—	—	—	—	12.58	0.4028	37.5	4.115	8
7	144.3	20,820	—	—	—	—	—	15.87	0.5080	29.7	3.665	9
8	128.5	16,510	7.6	—	7.1	73	46	20.01	0.6405	23.6	3.264	10
9	114.4	13,090	8.6	—	7.8	—	—	25.23	0.8077	18.7	2.906	11
10	101.9	10,380	9.6	9.1	8.9	55	33	31.82	1.018	14.8	2.588	12
11	90.7	8,234	10.7	—	9.8	—	—	40.12	1.284	11.8	2.305	13
12	80.8	6,530	12.0	11.3	10.9	41	23	50.59	1.619	9.33	2.053	14
13	72.0	5,178	13.5	—	12.8	—	—	63.80	2.042	7.40	1.828	15
14	64.1	4,107	15.0	14.0	13.8	32	17	80.44	2.575	5.87	1.628	16
15	57.1	3,257	16.8	—	14.7	—	—	101.4	3.247	4.65	1.450	17
16	50.8	2,583	18.9	17.3	16.4	22	13	127.9	4.094	3.69	1.291	18
17	45.3	2,048	21.2	—	18.1	—	—	161.3	5.163	2.93	1.150	18
18	40.3	1,624	23.6	21.2	19.8	16	10	203.4	6.510	2.32	1.024	19
19	35.9	1,288	26.4	—	21.8	—	—	256.5	8.210	1.84	0.912	20
20	32.0	1,022	29.4	25.8	23.8	11	7.5	323.4	10.35	1.46	0.812	21
21	28.5	810	33.1	—	26.0	—	—	407.8	13.05	1.16	0.723	22
22	25.3	642	37.0	31.3	30.0	—	5	514.2	16.46	0.918	0.644	23
23	22.6	510	41.3	—	37.6	—	—	648.4	20.76	0.728	0.573	24
24	20.1	404	46.3	37.6	35.6	—	—	817.7	26.17	0.577	0.511	25
25	17.9	320	51.7	—	38.6	—	—	1,031	33.00	0.458	0.455	26
26	15.9	254	58.0	46.1	41.8	—	—	1,300	41.62	0.363	0.405	27
27	14.2	202	64.9	—	45.0	—	—	1,639	52.48	0.288	0.361	29
28	12.6	160	72.7	54.6	48.5	—	—	2,067	66.17	0.228	0.321	30
29	11.3	127	81.6	—	51.8	—	—	2,607	83.44	0.181	0.286	31
30	10.0	101	90.5	64.1	55.5	—	—	3,287	105.2	0.144	0.255	33
31	8.9	80	101	—	59.2	—	—	4,145	132.7	0.114	0.227	34
32	8.0	63	113	74.1	61.6	—	—	5,227	167.3	0.090	0.202	36
33	7.1	50	127	—	66.3	—	—	6,591	211.0	0.072	0.180	37
34	6.3	40	143	86.2	70.0	—	—	8,310	266.0	0.057	0.160	38
35	5.6	32	158	—	73.5	—	—	10,480	335	0.045	0.143	38–39
36	5.0	25	175	103.1	77.0	—	—	13,210	423	0.036	0.127	39–40
37	4.5	20	198	—	80.3	—	—	16,660	533	0.028	0.113	41
38	4.0	16	224	116.3	83.6	—	—	21,010	673	0.022	0.101	42
39	3.5	12	248	—	86.6	—	—	26,500	848	0.018	0.090	43
40	3.1	10	282	131.6	89.7	—	—	33,410	1,070	0.014	0.080	44

[1]A mil is 0.001 inch. A circular mil is a square mil $\times \frac{\pi}{4}$. The circular mil (c.m.) area of a wire is the square of the mil diameter.

[2]Figures given are approximate only; insulation thickness varies with manufacturer.

[3]Max. wire temp. of 212°F (100°C) and max. ambient temp. of 135°F (57°C).

[4]700 circular mils per ampere is a satisfactory design figure for small transformers, but values from 500 to 1,000 c.m. are commonly used.

APPENDIX F

Useful Conversion Factors

To Convert	Into	Multiply By
ampere-turns/cm	amp-turns/inch	2.54
ampere-turns/inch	amp-turns/cm	0.3937
ampere-turns/inch	amp-turns/meter	39.37
ampere-turns/meter	amp-turns/inch	0.0254
Centigrade	Fahrenheit	$(°C \times 9/5) + 32$
centimeters	feet	3.281×10^{-2}
centimeters	inches	0.3937
centimeters/sec	feet/sec	0.03281
circular mils	sq mils	0.7854
circumference	radians	6.283
circular mils	sq inches	7.854×10^{-7}
coulombs	faradays	1.036×10^{5}
degrees	radians	0.01745
dynes	joules/cm	10^{-7}
dynes	joules/meter (newtons)	10^{-5}
ergs	foot-pounds	7.367×10^{-6}
ergs	joules	10^{-7}
farads	microfarads	10^{6}
faradays	ampere-hours	26.8
faradays	coulombs	9.649×10^{4}
feet	centimeters	30.48
foot-pounds	ergs	1.356×10^{7}
gausses	lines/sq inch	6.452
gausses	webers/sq meter	10^{-4}
gilberts	ampere-turns	0.7958
grams	dynes	980.7
grams	pounds	2.205×10^{-3}
henrys	millihenrys	1,000
horsepower	foot-lbs/sec	550
horsepower	kilowatts	0.7457
horsepower	watts	745.7
inches	centimeters	2.54
inches	millimeters	25.4
joules	ergs	10^{7}
joules	watt-hrs	2.778×10^{-4}
kilolines	maxwells	1,000
kilowatts	foot-lbs/sec	737.6
kilowatts	horsepower	1.341
kilowatts	watts	1,000
kilowatt-hrs	joules	3.6×10^{6}
lines/sq inch	webers/sq meter	$1{,}550 \times 10^{-5}$
maxwells	kilolines	0.001
maxwells	webers	10^{-8}
microfarad	farads	10^{-6}
microns	meters	1×10^{-6}
millihenrys	henrys	0.001
mils	inches	0.001
nepers	decibels	8.686
ohms	megohms	10^{-6}
quadrants	degrees	90
quadrants	radians	1.571
radians	degrees	57.3
square inches	sq cms	6.452
temperature	temperature (°C)	$5/9(°F) - 32$
watts	horsepower	1.341×10^{-3}
watts (Abs.)	joules/sec	1
webers	maxwells	10^{8}
webers/sq meter	webers/sq inch	6.452×10^{-4}
yards	meters	0.9144

APPENDIX G

Schematic Symbols

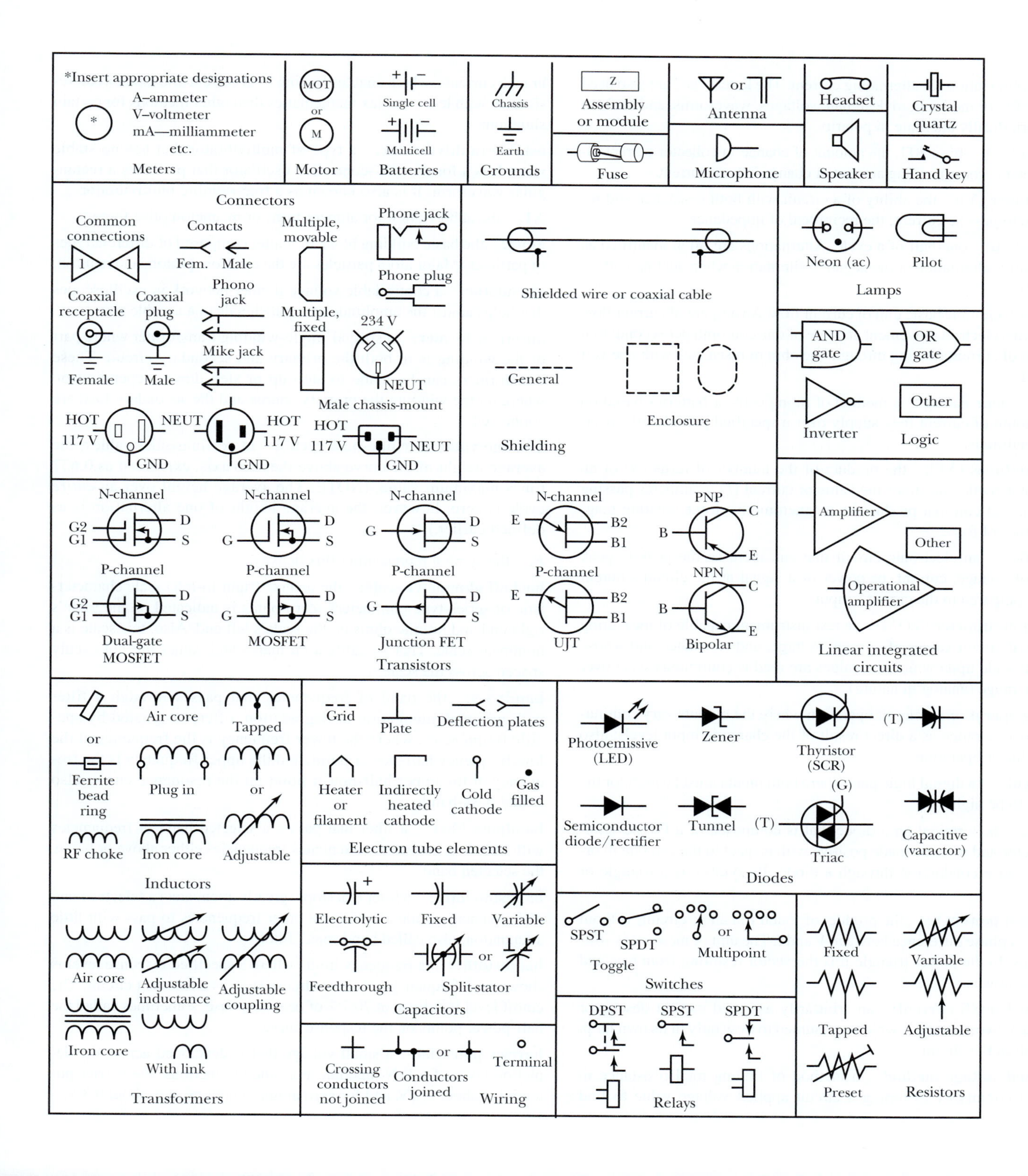

APPENDIX H

Glossary

ac: abbreviation for alternating current. The letters "ac" are also used as a prefix, or modifier, to designate voltages, waveforms, and so forth that periodically alternate in polarity.

ac beta (β_{ac}): for a BJT, the amount of change in collector current divided by the corresponding amount of change in base current.

admittance (Y): the ability of a circuit with both resistance and reactance to pass ac current; the reciprocal of impedance.

alternation: one-half of a cycle. Alternations are often identified as the positive alternation or the negative alternation when dealing with ac quantities.

ampere (A): the basic unit of current flow. An ampere of current flow represents electron movement at a rate of one coulomb per second; that amount of current flowing through one ohm of resistance with one volt applied.

ampere-hour rating: a method of rating cells or batteries based on the amount of current they supply over a specified period with specified conditions.

ampere-turns (AT): the product of the number of turns (N) of an electromagnetic coil times the value of current (I) in amperes passing through that coil that produces a magnetomotive force creating magnetic lines of force.

amplifier: an electronic circuit that can increase the peak-to-peak value of voltage, current, or power of a signal at the circuit's output when compared to the circuit's input.

analog multimeter (VOM): a test instrument capable of measuring electrical circuit values of current, voltage, and resistance and whose readout scale upon which the values are read is continuous scale over a given range (analog in nature).

analog operation: circuit operation where the outputs vary continuously over a range as a direct result of the changing input levels; also called linear operation.

AND gate: a digital logic gate where both inputs must be high for the output to be high.

anode: the terminal on a diode that is composed of a P-type semiconductor and must be made positive with respect to the cathode in order to cause conduction through a diode. Its symbol is a triangle or arrow.

apparent power (S): the product of V and I in an ac circuit, without regard to phase difference between V and I. The unit of measure is volt-amperes. In the power triangle it is the vector resulting from true and reactive power.

assumed mesh current: an arbitrarily assigned current descriptor used in network analysis which is assumed to flow only in its own loop, or mesh of the circuit.

assumed voltage method: a method of finding total resistance in parallel circuits by assuming a circuit applied voltage value to find branch currents, which enables finding total current and total circuit resistance with less complex mathematics than other methods for certain situations.

astable multivibrator: a type of multivibrator that has no stable state. It is a form of RC-controlled oscillator that produces a rectangular waveform. It is also known as a free-running multivibrator.

AT: the abbreviation for ampere-turns or magnetomotive force.

atom: the basic building block of matter composed of different types of particles. Major atom particles are the electron, proton, and neutron.

attenuator: a controllable voltage divider network in oscilloscopes that helps adjust the amplitude of input signals to a useable size.

autotransformer: a special single-winding transformer where part of the winding is in both the primary and secondary circuit. These transformers can be made to step up or step down, depending on where on the winding the primary source and the secondary load are connected.

average value (for one-half cycle): in ac sinusoidal values, the average height of the curve above the zero axis, expressed as 0.637 times maximum value. (NOTE: The average height over an *entire* cycle is zero; however, the average height of one alternation is as defined earlier.)

***B*:** the symbol representing flux density.

backoff ohmmeter scale: the reverse (right-to-left) scale characteristic of series-type ohmmeters. Zero ohms is indicated on the scale's right end and infinite ohms on the scale's left end. Also, this scale is a nonlinear scale. That is, calibration marks and values are not evenly spaced across the scale.

bandpass: the band of frequencies that passes through a filter with minimal attenuation or degradation; often designated by specific frequencies, where the lower frequency is the frequency at the low-frequency half-power point, and the upper frequency is the frequency at the upper half-power point on the resonance curve; also called pass-band.

bandpass filter: a filter that passes a selected band of frequencies with little loss but greatly attenuates frequencies either above or below the selected band.

bandstop filter: a filter that stops (greatly attenuates) a selected band of frequencies while allowing all other frequencies to pass with little attenuation; also called band-reject filter.

bandwidth: the frequency limits where a specified response level is shown by a frequency-sensitive circuit, such as a resonant circuit. The cutoff level is defined at 70.7% of peak response, sometimes called the half-power points on the response curve.

barrier potential: a small voltage that is developed across the depletion layer of a P-N junction. For silicon junctions, the barrier potential is about 0.7 V, and for germanium junctions, it is about 0.3 V.

base: the center of the NPN or PNP sandwich in a BJT that separates the emitter and the collector transistor regions. The emitter-base junction is forward biased and shows low impedance to current flow. The collector-base junction is reverse-biased and displays high resistance to current flow.

base current (I_B): the level of current that flows into or out of the base terminal on a BJT. This current is normally due to the forward biasing of the emitter-base P-N junction.

base-emitter voltage (V_{BE}): the level of voltage applied between the base and emitter terminals on a BJT. Forward base current flow is enabled when the polarity of V_{BE} is such that it forward biases the base-emitter junction.

battery: a dc voltage source containing a combination of cells, connected to produce higher voltage or current than a single cell produces alone.

bidirectional: the ability of a device to conduct current in two directions. Diacs and triacs are bidirectional, whereas SCRs and rectifier diodes are unidirectional (conduction current in only one direction).

bilateral resistance: any resistance having equal resistance in either direction; that is, R is the same for current passing either way through the component.

bipolar junction transistor (BJT): a three-terminal, bipolar semiconductor containing both majority and minority current carriers. The device has alternate layers of P- and N-type semiconductor materials (which form two P-N junction diodes). This transistor requires the flow of both types of charge carriers (electrons and holes) to complete the path for current flow through it. Examples are the NPN and the PNP transistor.

bleeder: the resistive component assuring that the minimum current drawn from a power-supply system is sufficient to achieve reasonable voltage regulation over the range of output currents demanded from the supply; serves to discharge the filter capacitors when the power supply is turned off, which adds safety for people working on electronic equipment fed by the power supply. This resistive component is also called bleeder resistor.

bleeder current: the fixed current through a resistive system that helps keep voltage constant under varying load conditions.

bleeder resistor: the resistor connected in parallel with power supply to draw a bleeder current.

block diagrams: graphic illustrations of systems or subsystems by means of blocks (or boxes) that contain information about the function of each block. These blocks are connected by lines that show direction of flow (signal, fluid, electrical energy, etc.) and how the blocks interact. Block diagrams can illustrate any kind of system: electrical, electronic, hydraulic, mechanical, and so forth. "Flow" generally moves from left to right through the diagram but is not a requirement.

block-level troubleshooting approach: is troubleshooting a system by referring to the block diagram for that system and testing inputs and outputs of each block in order to draw troubleshooting conclusions to the "block" level as opposed to troubleshooting down to the discrete components within that block.

bridge circuit: a special series-parallel circuit usually composed of two, two-resistor branches. Depending on the relative values of these resistors, the bridge output (between the center points of each branch) may be balanced or unbalanced.

bridge rectifier: a full-wave rectifier circuit that requires only four diodes arranged in series-parallel such that two of them conduct on one ac input half-cycle and the other two conduct on the opposite ac input half-cycle.

bridging or shunting: connecting a component, device, or circuit in parallel with an existing component, device, or circuit.

capacitance: the ability to store electrical energy in the form of an electrostatic field.

capacitive reactance (X_C): the opposition a capacitor gives to ac or pulsating dc current. The symbol used to represent capacitive reactance is X_C.

capacitor: a device consisting of two or more conductors, separated by nonconductor(s). When charged, a capacitor stores electrical energy in the form of an electrostatic field and blocks dc current.

cathode: the part of a diode that is composed of an N-type semiconductor and must be made negative with respect to the anode in order to cause conduction through the diode. Its symbol is a bar with a perpendicular connecting lead. (This bar symbol is also perpendicular to the anode arrow symbol.)

cell: (relates to voltage sources) a single (stand-alone) element or unit within a combination of similar units that convert chemical energy into electrical energy. The chemical action within a cell produces dc voltage at its output terminals or contacts.

cemf (counter-emf or back-emf): the voltage induced in an inductance that opposes a current change and is present due to current changes causing changing flux linkages with the conductor(s) making up the inductor.

charge (electrical): for our purposes, charge can be thought of as the electrical energy present where there is an accumulation of excess electrons, or a deficiency of electrons. For example, a capacitor stores electrical energy in the form of "charge" when one of its plates has more electrons than the other.

charging a capacitor: moving electrons to one plate of a capacitor and moving electrons off the other plate; resultant charge movement is called charging current.

chip: a semiconductor substrate where active devices (transistors) and passive devices (resistors and capacitors) form a complete circuit.

"chip" capacitor: a surface-mount device (SMD), sometimes called a "surface-mount" capacitor. Characterized by its small size, almost zero lead-length connection method, and its use in printed circuit board surface-mount applications.

"chip" resistor: a surface-mount device (SMD), sometimes called "surface-mount" resistor designed to be used in surface-mount technology applications. Typically, very small and rectangular in shape, with metallic end electrodes as the points of connection to the component.

circuit: a combination of elements or components that are connected to provide paths for current flow to perform some useful function.

clamp circuit: a circuit that offsets the zero baseline of an input waveform to a different positive or negative value.

Clapp oscillator: an LC resonant-circuit oscillator that requires a single inductor and three capacitors. One capacitor is in the inductor leg of the tank circuit, and two capacitors are connected in series and center-tapped to ground in the other leg of the tank circuit.

Class A, B, and C: in reference to amplifier operation, the classifications that denote the operating range of the circuit. Class A operates halfway between cutoff and saturation. Class B operates at cutoff. Class C operates at below cutoff bias.

clipper circuit: a type of circuit that clips off the portion of a waveform that extends beyond a prescribed voltage level. Clipper circuits can be designed to clip positive or negative levels, or both.

coefficient of coupling (k): the amount or degree of coupling between two circuits. In inductively coupled circuits, it is expressed as the fractional amount of the total flux in one circuit that links the other circuit. If all the flux links, the coefficient of coupling is 1. The symbol for coefficient of coupling is k. When k is multiplied by 100, the percentage of total flux linking the circuits is expressed.

collector: the transistor electrode that receives current carriers that have transversed from the emitter through the base region; the electrode that removes carriers from the base-collector junction.

collector-base voltage (V_{CB}): the amount of voltage present between the collector and the base of a BJT. For proper operation of the BJT, this voltage must have a polarity that reverse biases the P-N junction between the base and collector regions of the transistor.

collector characteristic curves: a family of curves that shows how BJT collector current increases with an increasing amount of emitter-collector voltage at several different base-current levels. These curves can be used for constructing load lines for BJT amplifier circuits.

collector current (I_C): the value of current measured at the collector terminal of a BJT.

collector-emitter voltage (V_{CE}): the amount of voltage present between the emitter and collector terminals of a BJT. For proper operation of the BJT, this voltage must have a polarity such that the flow of majority charge carriers is from emitter to collector.

Colpitts oscillator: an *LC* resonant-circuit oscillator that uses a single inductor in parallel with two series-connected capacitors that are grounded at their common point.

common-base: relating to transistor amplifiers, the circuit configuration where the base is common to both input and output circuits.

common-base amplifier (CB): a BJT amplifier configuration where the input signal is applied to the emitter and the output is taken from the collector.

common-collector: the transistor amplifier circuit configuration where the collector is common for input and output circuits.

common-collector amplifier (CC): a BJT amplifier configuration where the input signal is applied to the base and the output is taken from the emitter. Also known as an emitter-follower circuit.

common-drain amplifier (CD): an FET amplifier configuration where the input signal is applied to the gate and the output is taken from the source. Also known as a source-follower circuit.

common-emitter: in reference to transistor amplifiers, the circuit configuration where the emitter is common to both input and output circuits.

common-emitter amplifier (CE): a BJT amplifier configuration where the input signal is applied to the base and the output is taken from the collector.

common-gate amplifier (CG): an FET amplifier configuration where the input signal is applied to the source terminal and the output is taken from the drain.

common-mode gain: comparison (ratio) of the output voltage of a differential amplifier to the common-mode input voltage. Ideally, the output should be zero when the same input is fed to both differential amplifier inputs.

common-mode input: an input voltage to a differential amplifier that is common to both its inputs.

common-mode rejection ratio (CMRR): a measure expressing the amount of rejection a differential amplifier shows to common-mode inputs; the ratio between the differential-mode gain and the common-mode gain equals the CMRR.

common-source amplifier (CS): an FET amplifier configuration where the input signal is applied to the gate and the output is taken from the drain.

complex numbers: term that expresses a combination of a *real* and an *imaginary* term that must be added vectorially or with phasors. The real term is either a positive or a negative number displayed on the horizontal axis of the rectangular coordinate system. The imaginary term, *j operator,* expresses a value on the vertical or j-axis. The value of the real term is expressed first. For example, $5 + j10$ means plus 5 units (to the right) on the horizontal axis and up 10 units on the vertical axis.

component-level troubleshooting approach: is troubleshooting by testing and analysis down to a discrete component within a circuit, block, or subsystem.

compound: a form of matter that can be chemically divided into simpler substances and that has two or more types of atoms.

conductance: the ease with which current flows through a component or circuit; the opposite of resistance.

conduction band: is used to describe energy levels in the atoms of semiconductor or insulator materials, and is considered a range of electron energy (an energy band) higher than the valence band of an atom in which electrons are free to move under the influence of an applied electric field or external energy; and thus enables electric current flow throught the material.

conductor: a material that has many free electrons due to its atoms' outer rings having less than four electrons, which is less than the eight needed for chemical and/or electrical stability.

constant current source: a current source whose output current is constant even with varying load resistances. Its output voltage varies to enable constant output current, as appropriate.

constant voltage source: a voltage source where output voltage (voltage applied to circuitry connected to source output terminals) remains virtually constant even under varying current demand (load) situations.

conventional current: is a descriptor used for defining current when it is assumed that the movement of charges (or current flow) through a circuit is from the more positive point in a circuit toward the less positive (or negative) point in a circuit.

coordinate system: in ac circuit analysis, a system to indicate phase angles within four, 90° quadrants. The vertical component is the "Y" axis. The horizontal component is the "X" axis.

coordinates: the vertical and horizontal axes that describe the position and magnitude of a vector.

copper losses: the I^2R losses in the copper wire of the transformer windings. Since current passes through the wire and the wire has resistance to current flow, an I^2R loss is created in the form of heat, rather than this energy being a useful part of energy transfer from primary source to secondary load.

core: soft iron piece to help strengthen magnetic field; helps produce a uniform air gap between the permanent magnet's pole pieces and itself.

core losses: energy dissipated in the form of heat by the transformer core that subtracts from the energy transferred from primary to secondary; basically comprised of eddy-current losses (due to spurious induced currents in the core material creating an I^2R loss) and hysteresis losses (due to energy used in constantly changing the polarity of the magnetic domains in the core material).

cosine: a trigonometric function of an angle of a right triangle; the ratio of the adjacent side to the hypotenuse.

cosine function: a trigonometric function used in ac circuit analysis; relationship of a specific angle of a right triangle to the side adjacent to that angle and the hypotenuse where cos θ= adjacent/ hypotenuse.

coulomb: the basic unit of charge; the amount of electrical charge represented by $6.25 = 10^{18}$ electrons.

covalent bonding: the pairing of adjacent atom electrons to create a more stable atomic structure.

CRT (cathode-ray tube): the oscilloscope component where electrons strike a luminescent screen material causing light to be emitted. Waveforms are displayed on an oscilloscope's CRT.

crystal oscillator: an oscillator circuit that uses a special element (usually quartz-crystal material and holder) for the primary frequency-determining element in the circuit.

current (*I*): the progressive movement of charges through a conductor. Current is measured in amperes.

current divider circuit: any parallel circuit divides total current through its branches in inverse relationship to its branch *R* values.

current leading voltage: sometimes used to describe circuits where the circuit current leads the circuit voltage, or the circuit voltage lags the circuit current.

current ratio: the relationship of the current values of secondary and primary, or vice versa. In an ideal transformer, the current ratio is the *inverse* of the voltage and turns ratios; $I_P/I_S = N_S/N_P = V_S/V_P$.

cycle: a complete series of values of a periodic quantity that recur over time.

damping (electrical): the process of preventing an instrument's pointer from oscillating above and below the measured value. Electrical damping is due to the magnetic field produced by induced current opposing the motion that induced it.

dc beta (β_{dc}): for a BJT, the ratio of collector current (I_C) to the corresponding amount of base current (I_B).

delta (Δ): when used in mathematical equations, a term meaning "a small change in." When used in conjunction with an electrical network configuration, it describes the general triangular appearance of the schematic diagram layout, which depicts the electrical connections between components. Also, sometimes, "delta networks" are laid out in such a fashion that they are called pi (π) networks.

delta network: an interconnection of components in a closed-ring configuration that schematically can be represented as a triangular shape, similar to the Greek letter delta. A set of three branches connected so as to provide one closed mesh. Delta networks often appear in three-phase power systems, but also appear in other electrical and electronic circuits.

delta-wye conversion: the mathematical process of converting network component values, used in a delta network, to appropriate values to be used in a "wye" network, and yet present the same impedance and load conditions to a source feeding such a network.

depletion mode: a mode of operation for FET devices where the gate bias tends to reduce the availability of charge carriers in the channel.

depletion region: the region near the junction of a P-N junction where the current carriers are depleted due to the combining effect of holes and electrons.

diac: a type of bidirectional thyristor that is switched on by exceeding the peak forward blocking voltage or peak reverse-breakdown voltage. Diacs do not require gate terminals.

dielectric: nonconductive material separating capacitor plates.

dielectric constant: the relative permittivity of a given material compared with the permittivity of vacuum or air; the relative ability of a given material to store electrostatic energy (per given volume) to that of vacuum or air.

dielectric strength: the highest voltage a given dielectric material tolerates before electrically breaking down or rupturing; rated in volts per thousandths of an inch, V per mil.

differential amplifier: an amplifier circuit configuration having two inputs and one output; produces an output that is related to the difference (or differential) between the two inputs.

differential-mode gain: the gain of a differential amplifier when operated in the differential mode; the ratio of its output voltage to the differential mode (difference between the two input voltages) input voltage.

differentiator circuit: an op-amp configuration that produces an output voltage proportional to the rate of change of the input signal.

diffusion current: can be considered as a temporary current flowing through a semiconductor device's P-N junction without an externally applied source of energy.

digital multimeter (DMM): a test instrument that is capable of measuring electrical values of current, voltage, and resistance and whose readout is an actual discrete number, or numeric value representing the value of the electrical parameter being measured. In some cases DMMs can measure other circuit parameters as well, such as frequency, capacitance values, inductance values, etc.

digital operation: circuit operation where the output varies in discrete steps or in an on-off, high-low, 1-0 category of operation.

discharging a capacitor: balancing (or neutralizing) the charges on capacitor plates by allowing excess electrons on the negative charged plate to move through a circuit to the positive charged plate. When opposite capacitor plates have the same charge, the voltage between them is zero and the capacitor is fully discharged.

doping: introducing specific amounts of P- or N-type elements (called impurities) into a semiconductor crystal (lattice) structure to aid current conduction through the material.

doping material: a material that is added to a pure semiconductor material that enables the existence of charge carriers.

double-ended input: a type of op-amp configuration that uses two inputs.

drain: the portion of an FET that receives charge carriers from the channel.

dropping resistor: a resistor connected in series with a given load that drops the circuit applied voltage to the level required by the load when passing the rated load current through it. (Recall a *voltage drop* is the difference in voltage between two points caused by a loss of pressure [or loss of emf] as current flows through a component offering opposition to current flow.)

duty cycle: is a comparison of the pulse width with the time for one period for a given nonsinusoidal waveform; that is duty cycle equals the pulse width divided by the time for one period of the waveform being analyzed.

effective value (rms): the value of ac voltage or current that produces the same heating effect in a resistor as that produced by a dc voltage or current of equal value. In ac, this value is 0.707 times maximum value.

efficiency: percentage, or ratio, of useful output (such as power) delivered compared with amount of input required to deliver that output. For example, in electrical circuits, the amount of power input to a device, component stage, or system compared with the amount of power delivered to the load. Generic formula is Efficiency (%) = $(P_{out}/P_{in}) \times 100$.

electrical charge: an excess or deficiency of electrons compared with the normal neutral condition of having an equal number of electrons and protons in each atom of the material.

electrode: generally, one of the electrical terminals that is a conducting path for electrons in or out of a cell, battery, or other electrical or electronic device.

electrolyte: a liquid, paste, or substance through which electrical conduction occurs due to chemical processes of the electrolyte and other materials immersed or exposed to it.

electromagnetic frequency spectrum: is a listing or a diagram showing the various frequency ranges for different types of electromagnetic radiation. For example the diagram may show the wavelength ranges for radio waves, microwaves, infrared, visible light, ultraviolet light, and x-rays.

electromagnetism: the property of a conductor to produce a magnetic field when current is passing through it.

electromotive force: is an electrical force that can cause movement of electrons due to a difference of potential between two points.

electron: the negatively charged particle in an atom orbiting the atom's nucleus.

electrostatic (electric) field: the region surrounding electrically charged bodies; area where an electrically charged particle experiences force.

element: a form of matter that cannot be chemically divided into simpler substances and that has only one type of atom.

emitter: the portion of a BJT that emits charge carriers into the device.

emitter current (I_E): the amount of current found at the emitter terminal of a BJT.

energy: the ability to do work, or that which is expended in doing work. The basic unit of energy is the erg, where 980 ergs = 1 gram cm of work. A common unit in electricity/electronics is the joule, which equals 10^7 ergs.

enhancement mode: a mode of operation for FET devices where the gate bias tends to enhance the availability of charge carriers in the channel.

ϵ: the Greek letter epsilon, used to signify the mathematical value of 2.71828. This number, raised to various exponents that relate to time constants, can be used to determine electrical circuit parameter values in circuits where exponential changes occur.

equation: a mathematical statement of equality between two quantities or sets of quantities. Example, $A = B + C$.

equivalent resistance: the total resistance of two or more resistances in parallel.

Faraday's law: the law states that the amount of induced voltage depends on the rate of cutting magnetic flux lines. In other words, a greater number of flux lines cut per unit time produces greater induced voltage.

feedback: with signal amplifying or signal generating circuits, that small portion of the output signal that is coupled back to the input circuit of the stage.

fiber-optic cable: a cable made up of thin, glass, optical fibers capable of carrying signals created with laser light.

field-effect transistor (FET): a unipolar transistor that depends on a voltage field to control the flow of current through a channel.

filter: in power supplies, the component or combination of components that reduce the ac component of the pulsating dc output of the rectifier prior to the output fed to a load.

flux: relating to magnetism; the magnetic lines of force.

flux density (B): the magnetic flux per unit cross-sectional area; unit of measure is the tesla (T) representing 1 Wb per square meter.

focus control: a scope control that adjusts the narrowness of the electron stream striking the screen to create a clear and sharp presentation.

force: in electrical charges, the amount of attraction or repulsion between charges. In electromotive force, the potential difference in volts.

forward bias: the external voltage applied to a semiconductor junction that causes forward conduction; for example, positive to P-type material and negative to N-type material.

frame/movable coil: the parts in a moving-coil meter movement that move in response to the interaction between the fixed magnetic field and the field of the current-carrying, movable coil.

free electrons: the electrons in the outer ring of conductor materials that are easily moved from their original atom. Sometimes termed valence electrons.

free-running multivibrator: a type of multivibrator that runs continuously. It is a form of an *RC*-controlled oscillator circuit that produces a rectangular waveform. It is technically known as an astable multivibrator.

frequency (f): the number of cycles occurring in a unit of time. Generally, frequency is the number of cycles per second, called hertz (Hz), which means cycles per second. In formulas, frequency is expressed as f, where $f = 1/T$.

full-scale current: the value of current that causes full-scale, or full-range deflection of the pointer or indicator device in a current meter movement; the maximum current that should pass through the moving-coil in an analog-type current meter movement.

full-wave rectifier: a type of rectifier that produces a dc output for both ac half-cycles of input.

gain: the change in the peak-to-peak value of voltage, current, or power of a signal associated with an amplifier circuit, frequently expressed as the ratio of amplifier output signal level compared to the amplifier's input signal level.

gain controls: oscilloscope controls that adjust the size of the image displayed.

gate: (1) the terminal on a thyristor, such as an SCR or triac, that is used for the voltage that turns on the device. (2) The terminal on an FET where the voltage is applied that will control the amount of current flowing through the channel.

gate-controlled switch (GCS): a type of SCR that can be gated off as well as gated on.

ground reference: a common reference point for electrical/electronic circuits; a common line or conducting surface in electrical circuits used as the point where electrical measurements are made; a common line or conductor where many components make connection to one side of a power source. Typically, ground reference is the chassis, or a common printed circuit "bus" (or line) called chassis ground. In power line circuits, may also be "earth ground."

H: the symbol representing magnetic field intensity.

half-wave rectifier: a type of rectifier that produces a dc output for only one of the half-cycles in each cycle of input.

Hartley oscillator: an *LC* resonant-circuit oscillator that uses a single capacitor in parallel with two series-connected inductors that are grounded at their common point.

heat sink: a mounting surface for semiconductor devices that is intended to carry away excessive heat from the semiconductor, by dissipating it into the surrounding air.

henry: the basic unit of inductance, abbreviated "H." One henry of inductance is that amount of inductance having one volt induced cemf when the rate of current change is one ampere per second.

Hertz (Hz): the unit of frequency expressing a frequency of one cycle per second.

high-pass filter: a filter that passes frequencies *above* a specified cutoff frequency with little attenuation. Frequencies below the cutoff are greatly attenuated.

holding current: the minimum amount of forward current that a thyristor can carry and still remain switched on.

horizontal (X) amplifier: the scope amplifier circuitry that amplifies signals fed to the horizontal deflection circuitry.

hybrid circuits: generally, circuits composed of components created by more than one technique, such as thin-film, discrete transistors, and so on.

hysteresis: the difference in voltage between the amount of turn-on voltage and turn-off voltage for an amplifier that is used as a switch. A Schmitt-trigger amplifier, for example, might switch on when the input exceeds +2 V, but not switch off until the input drops below +1.25 V. The amount of hysteresis in this case is 0.75 V.

I_m or I_{fs}: full-scale current value of an analog-type meter movement.

imaginary numbers: complex numbers; the imaginary part of a complex number that is a real number multiplied by the square root of minus one.

impedance: the total opposition an ac circuit displays to current at a given frequency; results from both resistance and reactance. Symbol representing impedance is *Z* and unit of measure is the ohm.

impedance ratio: the relationship of impedances for the transformer windings; directly related to the *square* of the turns ratio; $Z_P/Z_S = N_P^2/N_S^2$. Reflected impedance is the impedance reflected back to the primary from the secondary. This value depends on the square of the turns ratio and value of the load impedance connected across the secondary.

inductance (self-inductance): the property of a circuit to oppose a change in current flow.

induction: the ability to induce voltage in a conductor; ability to induce magnetic properties from one object to another by magnetically linking the objects.

inductive reactance: the opposition an inductance gives to ac or pulsating current.

in-line: typically means in series with the source.

instantaneous value: the value of a sinusoidal quantity at a specific moment, expressed with lowercase letters, such as *e* or *v* for instantaneous voltage and *i* for instantaneous current.

insulator: a material in which the atoms' outermost ring electrons are tightly bound and not free to move from atom to atom, as in a conductor. Typically, these materials have close to the eight outer shell electrons required for chemical/electrical stability.

integrated circuit (IC): a combination of electronic circuit elements. In relation to semiconductor ICs, this term generally indicates that all of these components are fabricated on a single substrate or piece of semiconductor material.

integrator circuit: an op-amp configuration that produces an output voltage proportional to the "area under the curve" for the input waveform.

intensity control: a scope control that adjusts the number of electrons striking the screen per unit time; thus controlling the brightness of the screen display.

inverse parallel: a way of interconnecting two diodes or triacs such that they are in parallel with the anode of one connected to the cathode of the other. Also known as a back-to-back connection.

inverter: a digital logic gate where the output is the opposite of the input. If the input is low, the output is high. If the input is high, the output is low.

inverting amplifiers: amplifier configurations (especially in reference to op-amps) that produce an output waveform having the opposite polarity of the input waveform.

ion: atoms that have gained or lost electrons and are no longer electrically balanced, or neutral atoms. Atoms that lose electrons become positive ions. Atoms that gain electrons become negative ions.

I_S: current through the shunt resistance of an analog meter.

isolation transformer: a 1:1 turns ratio transformer providing electrical isolation between circuits connected to its primary and secondary windings.

joule: the unit of energy. One joule of energy is the amount of energy moving one coulomb of charge between two points with a potential difference of one volt of emf; the work performed by a force of one newton acting through a distance of one meter equals one joule of energy. Also, 3.6×10^6 joules = 1 kilowatt hour.

junction field-effect transistor (JFET): FETs that use the voltage field around a reverse-biased P-N junction to control the flow of charge carriers through the channel region.

Kirchhoff's current law: the value of current entering a point must equal the value of current leaving that same point in the circuit.

Kirchhoff's voltage law: the arithmetic sum of voltage drops around a closed loop equals V applied; the algebraic sum of voltages around the entire closed loop, including the source, equals zero.

L, L_T: L is the abbreviation representing inductance or inductor. L_T is the abbreviation for total inductance. L_1, L_2, and so on represent specific individual inductances.

L/C ratio: the ratio of inductance to capacitance in an *LC* resonant circuit. Generally, the higher this ratio is, the higher the *Q* of the circuit and the more selective its response.

large-signal amplifiers: amplifiers designed to operate effectively with relatively large ac input signals (e.g., power levels greater than 1 watt).

laser diode: a light-emitting diode that produces light of a single frequency (coherent light) when it is forward conducting.

leading in phase: in ac circuits the electrical quantity that first reaches its maximum positive point is specified as leading the electrical quantity that reaches that point later.

left-hand rule: related to magnetism; when the thumb and fingers of the left hand determine the direction of magnetic fields of current-carrying conductors to find the polarity of an electromagnet's poles.

Lenz's law: the law states that the direction of an induced voltage or current opposes the change causing it.

light-activated SCR (LASCR): an SCR that can be gated on by directing light energy onto its gate-cathode junction.

light-emitting diode (LED): a junction diode that emits light energy when it is forward biased and conducting.

limiter circuit: sometimes called a clipper circuit, performs the function of limiting or preventing some portion of the circuit input from reaching the output of the circuit.

linear amplifier: an amplifier that produces an output waveform that has the same general shape as the input waveform.

linear network: a circuit whose electrical behavior does not change with different voltage or current values.

lines: in magnetic circuits, lines represent flux.

lines of force (flux): imaginary lines representing direction and location of the magnetic influence of a magnetic field relative to a magnet.

load: the amount of current or power drain required from the source by a component, device, or circuit.

load current: is the current required by the components or circuits that are connected to the circuit power source, or its output voltage divider.

load line: a straight line drawn between the points of cutoff voltage and saturation current on a family of collector characteristic curves.

loaded voltage divider: a network of resistors designed to create various voltage levels of output from one source voltage; where parallel loads, which demand current, are connected.

loading effect: changing the electrical parameters of an existing circuit when a load component, device, or circuit is connected. With voltmeters, the effect of the meter circuit's resistance when in parallel with the portion of the tested circuit; thus, causing a change in the circuit's operation and some change in the electrical quantity values throughout the circuit.

long time-constant circuit: when the circuit conditions are such that the t/RC ratio = 0.1 (or less). This means the RC time required to charge or discharge the capacitor is long compared with the time allowed during each alternation or cycle.

loop: in electrical networks, a complete electrical circuit or current path.

low-pass filter: a filter that permits all frequencies *below* a specified cutoff frequency to pass with little attenuation. Frequencies above the cutoff frequency are attenuated greatly.

***L/R* time constant:** the time required for an exponentially changing quantity to change by 63.2% of the total value of change to be achieved. For a circuit containing inductance, it is the time (in seconds) required for the current to acquire a level that is 63.2% of its final value. One time constant (in seconds) in a series RL or inductor circuit equals L/R, where L is in henrys and R is in ohms.

magnetic field: the field of influence or area where magnetic effects are observed surrounding magnets; a region of space where magnetic effects are observed.

magnetic field intensity (*H*): the amount of magnetomotive force per unit length. In the MKSA system, it is the ampere-turns per meter (AT/meter).

magnetic polarity: the relative direction of a magnet's flux with respect to its ends or poles; defined as the North-seeking pole and the South-seeking pole of a magnet.

magnetism: the property causing forces of attraction or repulsion in certain materials; property causing voltage to be induced in nearby conductors when there is relative motion between the magnetic object's field and the conductor.

magnetomotive force (mmf or F): the force causing magnetic lines of force through a medium, thus establishing a magnetic field; developed by current through a coil, where the amount of mmf relates to the amount of current and the number of turns on the coil ($F = AT$).

magnitude: the comparative size or amount of one quantity with respect to another quantity of the same type.

majority carriers: the electrons or holes in extrinsic semiconductor materials that enable current flow. In N-type materials, the majority carriers are electrons. In P-type materials, the majority carriers are holes.

matter: anything that has weight and occupies space, whether it is a solid, liquid, or gas.

maximum power transfer theorem: a theorem that states when the resistance (or impedance) of the load is properly matched to the internal resistance of the power source, maximum power transfer can take place between the source and the load.

Maxwell (Mx): one line of force, or one flux line.

mesh: sometimes called loops. In electrical networks, a set of branches that form a complete electrical path. If any branch is omitted, the remainder of the circuit does not form a complete path.

metal-oxide semiconductor FET (MOSFET): a unipolar silicon transistor that uses a very thin layer of metal-oxide insulation between the gate and channel.

meter shunt (ammeter): the parallel resistance used with analog-type ammeter movements to allow them to measure currents higher than their full-scale current rating; a percentage of the total current being measured is shunted, or bypassed around the meter by the meter shunt.

microprocessor: typically, a VLSI chip that acts as the central processing unit (CPU) for a computer system.

minority carriers: the electrons in P-type material and the holes in N-type material; called minority since there are fewer electrons in P-type material than holes and fewer holes in N-type material than electrons.

mixture: a combination of substances where the individual elements in the mixture possess the same properties as when each element in the mixture was alone.

mode: related to semiconductor ICs; the type of operation for which the IC is designed. The linear mode is commonly used for amplifiers. The digital mode is often used for gates.

molecule: the smallest particle of a compound that resembles the compound substance itself.

monolithic: one piece as in a monolithic IC formed on one semiconductor substrate (or one stone).

monostable multivibrator: a type of multivibrator that is stable in only one state. An input trigger pulse sets the circuit to its unstable state where it remains for a period of time determined by an RC circuit. This circuit is also known as a one-shot multivibrator.

MOS: abbreviation for metal-oxide semiconductor; relates to the fabrication method.

moving-coil (d'Arsonval) movement: a type of analog meter movement. A coil is suspended within a permanent magnetic field and caused to move when current passes through it. Movement is due to the interaction of its magnetic field with the permanent magnet's field.

μ_r: the symbol for relative permeability.

multimeter: an instrument designed to measure several types of electrical quantities, for example, current, voltage, and resistance; commonly found in the forms of the analog VOM and the digital multimeter (DMM).

multiple-source circuit: a circuit containing more than one voltage or current source.

multiplier resistor(s) (voltmeter): the current-limiting series resistance(s) used with analog-type current meter movements, enabling them to be used as voltmeters that measure voltages much greater than the basic movement's voltage drop at full-scale current. The value of the multiplier resistance is such that its resistance (plus the meter's resistance) limits current to the full-scale value at the desired voltage-range level.

multivibrator: a circuit that produces a waveform that represents a condition of being fully switched on or fully switched off. The three basic types of multivibrators are the monostable (one-shot), the bistable, and the astable (free-running) multivibrators.

mutual inductance (M or L_M): the property that enables a current change in one conductor or coil to induce a voltage in another conductor or coil, and vice versa. This induced voltage results from magnetic flux of one conductor or coil linking the other conductor or coil. The symbol for mutual inductance is M or L_M and the unit of measure is the henry.

NAND gate: an AND gate with inverted output. That is, if both inputs are high, the output is low. For all other input conditions, the output is high.

network: a combination of electrically connected components.

neutron: a particle in the nucleus of the atom that displays no charge. Its mass is approximately equal to the proton.

node: a current junction or branching point in a circuit.

noninverting amplifiers: amplifiers that produce an output that has the same polarity or phase as their input.

nonlinear amplifier: an amplifier that produces an output waveform that is different from its input waveform.

NOR gate: an OR gate with inverted output. That is, if either or both inputs are high, the output is low. Only when both inputs are low is output high.

Norton's theorem: a theorem regarding circuit networks stating that the network can be replaced (and/or analyzed) by using an equivalent circuit consisting of a single "constant-current" source (called I_N) and a single shunt resistance (called R_N).

***NPN* transistor:** a transistor made of P material sandwiched between two outside N material areas. Then, going from end-to-end, the transistor is an NPN type. The symbol for the NPN transistor shows the emitter arrow going away from the base.

N-type material: a semiconductor material that has an excess of electrons compared with holes.

ohm (Ω): the basic unit of resistance; the amount of electrical resistance limiting the current to one ampere with 1 V applied.

Ohm's law: a mathematical statement describing the relationships among current, voltage, and resistance in electrical circuits. Common equations are $V = I \times R$, $I = V/R$, and $R = V/I$.

ohms-per-volt rating: a voltmeter rating used with analog meters indicating how many ohms of resistance are needed to limit the current through the meter to its full-scale value when one volt is applied to the meter circuit.

one-shot multivibrator: a popular name for a monostable multivibrator. (*See* monostable multivibrator.)

op-amp: abbreviation for operational amplifier; a versatile amplifier circuit with very high gain, high input impedance, and low output impedance.

open circuit: any break in the current path that is undesired, such as a broken wire or component, or designed, such as open switch contacts in a lighting circuit.

operational amplifier (op-amp): a versatile differential amplifier circuit with very high gain, high input impedance, and low output impedance.

optoelectronic device: a group of electronic devices that generate or respond to light.

opto-isolator: a device often consisting of an LED used as a source of light, and a photodiode used as a detector of light, configured so as to provide isolation between circuits or processes.

OR gate: a digital logic gate where if either input is high, or if both inputs are high, the output is high.

oscillator: a circuit that generates and sustains an output signal without an input signal being supplied by another circuit or source.

oscilloscope: a device that visually displays signal waveforms so time and amplitude parameters are easily determined.

parallel branch: a single current path within a circuit having two or more current paths where each of the paths are connected to the same voltage points; a single current path within a parallel circuit.

parallel circuit: a circuit with two or more paths for current flow where all components are connected between the same voltage points.

peak-inverse-voltage (PIV): the maximum voltage across a diode in the reverse (nonconducting) direction. The maximum voltage that a diode tolerates in this direction is the peak-inverse-voltage rating.

peak-to-peak value (p-p): the difference between the positive peak value and the negative peak value of a periodic waveform; computed for the sine wave as 2.828 times effective value.

peak value (V_{pk} or maximum value) (V_{max}): the maximum positive or negative value that a sinusoidal quantity reaches, sometimes referred to as peak value. It is calculated as 1.414 times effective value.

pentavalent atoms: an atom normally having five electrons in the valence shell. Such materials are often used as a donor impurity in the manufacture of N-type semiconductor materials. Examples are phosphorus, arsenic, and antimony.

period (T): the time required for one complete cycle. In formulas, it is expressed as T, where $T = 1/f$.

permeability (mu or μ): the ability of a material to pass, conduct, or concentrate magnetic flux.

phase angle (θ): the difference in time or angular degrees between ac electrical quantities (or periodic functions) when the two quantities reach a certain point in their periodic function. This difference is expressed by time difference (the fraction of a period involved), or by angular difference (expressed as degrees or radians). (NOTE: A radian = 57.3°. $2\pi \times$ a radian = 360°.) An example is the difference in phase between circuit voltage and circuit current in ac circuits that contain reactances, such as inductive reactance.

phase-shift oscillator: an RC sine-wave oscillator that uses a sequence of three RC circuits to achieve a total phase shift of 180° at the frequency of oscillation.

phasor: a quantity that expresses position relative to time.

photodiode cell: a device where resistance between terminals is changed by light striking its light-sensitive element; a light-controlled variable resistance. Typically, the more light striking it, the lower its terminal resistance is.

photo-reactive device: (sometimes called a photo-detector). A device sensitive to light energy wherein the device's electrical properties (such as current through the device) is affected by light falling on certain portions of the device.

piezoelectric effect: an effect enabled by certain kinds of materials whereby applying a voltage to the material causes it to change size and shape, and bending the material causes it to generate a voltage.

pinch-off voltage (V_P): the source-drain voltage level (of an FET) where a further increase in source-drain voltage causes virtually no increase in drain current.

P-N junction: the location in a semiconductor where the P- and N-type materials join; where the transition from one type material to the other type occurs.

P-N junction diode: a semiconductor diode that consists of a single P-N junction.

PNP transistor: a transistor where the middle section is N-type material and the two outside sections are composed of P-type material, thus called a PNP transistor. The symbol for the PNP transistor shows the emitter arrow going toward the base.

pointer: the element in an analog-type meter that provides the indicating function of how much movement has taken place by the movable coil.

polar form notation: expressing circuit parameters with respect to polar coordinates; expressing a point with respect to distance and angle (direction) from a fixed reference point on the polar axis. For example, $25 \angle 45°$.

polarity: in an electrical circuit, a means of designating differences in the electrical charge condition of two points, or a means of designating direction of current flow. In a magnetic system, a means of designating the magnetic differences between two points or locations.

pole pieces: special meter pieces used to help linearize the magnetic field in a moving-coil movement.

position control(s) (vertical and horizontal): scope controls that move the display vertically or horizontally; thus enabling centering the display on the screen.

positive feedback: a portion of the output of an amplifier system that is fed back to the input with the same phase or polarity as the original input signal.

potential: in electrical circuits, the potential, or ability, to move electrons. In electromotive force, the difference in charge levels at two points. This potential difference is measured in volts.

power: the rate of doing work. In mechanical terms, a horsepower equals work done at a rate of 550 foot-pounds per second. In terms of electrical power, it is electrical work at a rate of one joule per second.

power dissipation: frequently considered the amount of heat energy created/dissipated by a component or device when current passes through the component or device.

power factor (p.f.): the ratio of true power to apparent power. It is equal to the cosine of the phase angle (θ).

powers of 10: working with small or large numbers can be simplified by converting the unit to a more manageable number times a power of ten. An example is 0.000006 ampere expressed as 6×10^{-6} amperes. Some of the frequently used powers of 10 are 10^{-12} for pico (p) units, 10^{-6} for micro (μ) units, 10^{-3} for milli units, 10^{3} for kilo units, and 10^{6} for mega units.

power supply: a circuit that converts ac power into dc power required for the operation of most kinds of electronic systems.

product-over-the-sum method: a method of finding total resistance in parallel circuits by using two resistances at a time in the formula $R_T = R_1 \times R_2/R_1 + R_2$.

proton: the positively charged particle in an atom located in the nucleus of the atom. Its weight is approximately 1,836 times that of the electron, and approximately the same weight as the neutron.

P-type material: a semiconductor material that has an excess number of holes compared with electrons.

pulsating dc: a varying voltage that is one polarity, as opposed to reversing polarity as ac does.

push-pull amplifier: an amplifier circuit that uses two Class B amplifiers, one to amplify one alternation of the ac input waveform and a second to amplify the opposite alternation. The circuit has the advantage of Class B efficiency and the low distortion of a Class A amplifier.

Pythagorean theorem: a theorem expressing the mathematical relationship of the sides of a right triangle, where the square of the hypotenuse is equal to the sum of the squares of the other two sides. That is, $c^2 = a^2 + b^2$. Equation used in electronics is $c = \sqrt{a^2 + b^2}$.

Q (figure of merit): a number indicating the ratio of the amount of stored energy in the inductor magnetic field to the amount of energy dissipated. Found by using the ratio of the inductance's inductive reactance (X_L) to the resistance of the inductor (R). That is, $Q = X_L/R$.

Q factor: the relative quality or figure of merit of a given component or circuit; often related to the ratio of X_L/R of the inductor in the resonant circuit. Other circuit losses can also enter into the effective Q of the circuit, such as the loading that a load couples to the resonant circuit causes and the effect of R introduced in the circuit. The less the losses are, the higher the Q is. The sharpness of the resonant circuit response curve is directly related to the Q of the circuit.

Q point: the operating point of an amplifier when there is no input signal.

quadrants: segments of the coordinate system representing 90° angles, starting at a 0° reference plane.

range select (switch): the switch used with meter movements that connects or disconnects various shunt resistances values (for current or ohmmeter measurements), or multiplier resistances for the voltage measurement mode.

RC time constant: the time, in seconds, represented by the product of R in ohms and C in farads; time required for a capacitor to charge or discharge 63.2% of the change in voltage level applied. Five RC time constants are needed for a capacitor to completely charge or discharge to a changed voltage level.

reactance: the opposition a capacitor or inductor gives to alternating current or pulsating direct current.

reactive power (Q): the vertical component of the ac power triangle that equals $VI \sin \theta$; expressed as VAR (voltampere-reactive).

real numbers: in math, any rational or irrational number.

rectangular form notation: expressing a set of parameters with the complex number system. For example, $R \pm jX$ indicates a combination of resistance and reactance. Thus, $5 + j10$ means 5 ohms of resistance in series with 10 ohms of inductive reactance.

rectangular waveforms: a "square" waveform that has only two voltage levels, often one level that is close to the main supply voltage and one that is close to zero volts.

rectifier circuit: a circuit that converts an ac waveform to pulsating dc.

rectifier diode: a junction diode that is designed for the higher current and higher reverse-voltage conditions encountered in power-supply rectifier circuits.

rectify (rectification): the process of changing ac to pulsating dc.

regulator: a section of some power supplies that keeps the dc output voltage at a desired level in spite of changes in the input voltage level and/or changes in the output loading.

relative permeability (μ_r): comparing a material's permeability with air or a vacuum; relative permeability is not a constant, since it varies for a given material depending on the degree of magnetic field intensity present; relative permeability

$$\frac{\text{flux density with given core}}{\text{flux density with vacuum core}}$$

reluctance ($\Re$): the opposition of a given path to the flow of magnetic lines of force through it.

residual magnetism: the magnetism remaining in the core material of an electromagnet after the magnetizing force is removed.

resistance (R): in an electrical circuit, the opposition to electron movement or current flow. Its value is affected by many factors, including the dimensions and material of the conductors, the physical makeup and types of the components in the circuit, and temperature.

resistor: an electrical component that provides resistance in ohms and that controls or limits current or distributes or divides voltage.

resistor color code: a system to display the ohmic value and tolerance of a resistor by means of colored stripes or bands around the body of the resistor.

resonance: a special circuit condition existing when opposite types of reactances in a circuit are equal and have cancelling effects (i.e., $X_L = X_C$).

reverse bias: the external voltage applied to a semiconductor junction that causes the semiconductor not to conduct; for example, negative to P-type material and positive to N-type material.

reverse-breakdown voltage: a rating for semiconductors that specifies the maximum amount of reverse voltage that a P-N junction can withstand without breaking down and conducting in the reverse direction.

reverse Polish notation (RPN): an operational system, based on a technique derived from the work of a Polish scientist, used in some calculators to simplify and minimize the number of input steps required to solve certain complex mathematical problems. Since the notation method used is somewhat reversed from the scientist's original technique, the system, or mode, is sometimes called reverse Polish notation (RPN).

ripple: the ac component in a pulsating dc signal, such as the output of a power supply.

R_M: meter movement resistance value.

R_{Mult}: resistance value of the resistor used that enables a basic current meter to measure voltage by its current-limiting effect.

saturation: in magnetism, the condition when increasing the magnetizing force does not increase the flux density in a magnetic material.

sawtooth wave: a nonsinusoidal sawtooth-shaped wave or waveform consisting of a fundamental frequency signal and a large number of harmonics (multiples) of that fundamental signal frequency, producing the resultant sawtooth-shaped waveform.

scalar: a quantity expressing only quantity, such as a "real number."

scale: the face of an analog-type meter containing calibration marks that interpret the pointer's position in appropriate units of measure.

schematic diagrams: graphic illustrations of circuits or subcircuits and/or systems or subsystems. Schematics are more detailed than block diagrams and give specific information regarding how components are connected. Schematics show the types, values, and ratings of component parts, and in many cases show the electrical values of voltage, current, and so forth at various points throughout the circuit.

Schmitt trigger: a nonlinear amplifier circuit that converts an ac input waveform into a rectangular waveform. The output waveform goes to its high level when the input exceeds a certain voltage level, and the output goes to its low level when the input falls below a certain voltage level.

SCR: abbreviation for silicon-controlled rectifier; often used to control the time during an ac cycle the load receives power from the power source. A four-layer, P-N-P-N device where conduction is initiated by a voltage to the gate. Conduction doesn't cease until anode voltage is reversed, removed, or greatly reduced although the initiating signal has been removed.

selectivity: referring to resonant circuits, selectivity indicates the sharpness of the response curve. A narrow and more pointed response curve results in a higher degree of frequency selectiveness or selectivity.

self-bias: a method of biasing an amplifier such that the bias voltage is generated by the main flow of current through the device, itself. The method is usually used where the biasing voltage has a polarity opposite the main supply voltage (as with depletion-mode FETs).

semiconductor: a material with four valence (outermost ring) electrons. It is neither a good conductor, nor a good insulator.

semiconductor diode: a semiconductor P-N junction that passes current easily in one direction while not passing current easily in the other direction.

sensitivity rating: a means of rating analog-type meters, or meter movements to indicate how much current is required to cause full-scale deflection; how much voltage is dropped by the meter movement at full-scale current value. Also, voltmeters are rated by the ohms-per-volt required to limit current to full-scale value. (This is inversely related to the current sensitivity rating. That is, the reciprocal of the full-scale current value yields the ohms-per-volt rating.)

series circuit: a circuit in which components are connected in tandem or in a string providing only one path for current flow.

series-dropping resistor: (sometimes called a voltage-dropping resistor) is a resistor used in series with a load, with the purpose of dropping a source voltage that is higher than the load needs down to the value needed by the load, and limiting the current through the load portion of the circuit to the desired load current value.

series-parallel circuit: a circuit comprised of a combination of both series-connected and parallel-connected components and/or subcircuits.

series-type ohmmeter circuit: the ohmmeter circuit where the meter, its zero-adjust resistance, and its internal voltage source are in series. When the test leads are connected across an unknown resistance to be measured, that resistance is placed in series with the meter circuit.

short circuit: generally, an undesired very low resistance path across two points in a circuit. It may be across one component, several components, or the entire circuit. There can be designed shorts in circuits, such as purposeful "jumpers" or closed switch contacts.

short time-constant circuit: describes a circuit where the frequency is such that the capacitor has at least ten times one *RC* time constant to charge or discharge. A short time-constant circuit is sometimes described as one where: $t/RC = 10$ (or greater).

siemens: the unit used to measure or quantify conductance; the reciprocal of the ohm ($G = 1/R$).

silicon-controlled rectifier (SCR): four-layer (P-N-P-N) thyristor device that is gated on by forward biasing the gate-cathode junction and is turned off only by eliminating or reversing the polarity of the cathode-anode voltage.

silicon-controlled switch (SCS): an SCR that has two gate terminals that can be used for switching on the device.

SIMPLER Troubleshooting Sequence: is a sequence of troubleshooting steps that enable a logical approach to troubleshooting and consists of a series of words where the first letter in each word in the sequence spells the word "SIMPLER." The sequence of words used in this acrostic suggests the actions to be performed at each step of troubleshooting. For example, the S relates to **symptoms** of the circuit or system problem and suggest the first troubleshooting step is to analyze the known symptoms of the problem. The I relates to **identifying** the questionable area of the circuit or system—and so on.

sine: a trigonometric function of an angle of a right triangle; the ratio of the opposite side to the hypotenuse.

sine function: a trigonometric function used in ac circuit analysis; expresses the relationship of a specified angle of a right triangle to the side opposite that angle and the hypotenuse where $\sin \theta =$ opposite/hypotenuse.

sine wave: a wave or waveform expressed in terms of the sine function relative to time.

single-ended input: a type of op-amp configuration that uses a single input.

sinusoidal quantity: a quantity that varies as a function of the sine or cosine of an angle.

skirts: term typically used to describe the portions of a resonance response curve that descend down from the peak response point. The steepness of the skirts reflects the selectivity of the circuit.

small-signal amplifiers: amplifiers designed to operate effectively with small ac input signals; that is, power levels less than 1 watt, and peak-to-peak ac current values less than 0.1 times the amplifier's input bias current.

source: in electrical circuits, the source is the device supplying the circuit applied voltage, or the electromotive force causing current to flow through the circuit. Sources range from simple flashlight cells to complex electronic circuits. The portion of an FET that emits charge carriers into the channel.

source/gate/drain: the electrodes in a field-effect transistor device; gate controls current flow between the source and the drain.

springs: the elements in a moving-coil meter movement that conduct current to the movable coil and provide a mechanical force against which the moving coil must operate.

square wave: a nonsinusoidal square-shaped wave or waveform consisting of a fundamental frequency signal and a large number of harmonics (multiples) of that fundamental signal frequency, producing the resultant square-shaped waveform.

SSI, MSI, LSI, and VLSI: SSI is small-scale integration (less than 10 gates on a single chip). MSI is medium-scale integration (less than 100 gates on a single chip). LSI is large-scale integration (more than 100 gates on a single chip). VLSI is very-large-scale integration (over 1,000 gates on a single chip).

steady-state condition: the condition where circuit values and conditions are stable or constant; opposite of transient conditions, such as switch-on or switch-off conditions, when values are changing from one state to another.

stops: pegs that limit the left and right movement of the analog-type meter's pointer and related moving elements.

subcircuit: a part or portion of a larger circuit.

summing amplifier: an op-amp configuration that mathematically adds (or sums) the voltage levels found at two or more inputs.

superconductivity: the flow of electrical current through an electrical conductor whose resistance has decreased to almost $0\ \Omega$ due to the conductor material and the super-cold temperature at which it is being operated. The discovery was made in 1911 that this "superconductivity" phenomenon appears in certain materials that are operated below a critical specific cold temperature level. (Usually close to absolute zero.)

superposition theorem: superimposing one set of values on another to create a set of values different from either set of the superimposed values alone; the algebraic sum of the superimposed sets of values.

surface-mount technology: the technology in which discrete electrical components or devices (such as resistors and capacitors) are manufactured in a way that allows miniaturization of the component elements and minimizing of lead lengths. These components are designed to be mounted and soldered directly to the "surface" of printed circuit boards.

susceptance (B): the relative ease that ac current passes through a pure reactance; the reciprocal of reactance; the imaginary component of admittance.

sweep circuits: circuits in an oscilloscope that cause the horizontal trace and the rapid retrace of the electron beam across the CRT.

sweep frequency control(s): scope controls that set the number of times the horizontal trace occurs per unit time. Frequently, this control is calibrated in time/div. to measure the time per division traced on the display. (NOTE: The formula $f = 1/T$ is then used to determine frequency.)

switches: devices for making or breaking continuity in a circuit.

switching diode: a junction diode that is designed to allow a very short recovery time when the applied voltage changes polarity.

synchronization: the process of bringing two entities into time coherence. In a scope, the circuitry and process that cause the beginning of the horizontal trace to be synchronized with the vertical input signal.

synchronization control(s): scope controls that control the timing of the horizontal sweep in synchronization with the signal(s) being tested.

tangent function: a trigonometric function used in ac circuit analysis; expresses the relationship of a specified angle of a right triangle to the side opposite and side adjacent to that angle where $\tan \theta =$ opposite/adjacent.

TC: the abbreviation often used to represent time constant.

tesla: a unit of flux density representing one weber per square meter in the MKSA system of units.

tetravalent atoms: an atom normally having four electrons in the outer, or valence, shell. Such materials are often used as the base material for semiconductors. Examples include silicon and germanium.

Thevenin's theorem: a theorem used to simplify analysis of circuit networks because the network can be replaced by a simplified equivalent circuit consisting of a single voltage source (called V_{TH}) and a single series resistance (called R_{TH}).

thin-film process: a process where thin layers of conductive or nonconductive materials are used to form integrated circuits.

thyristor: a semiconductor device that is composed of four layers, P-N-P-N. The conduction is either all or nothing; it is either switched fully on or turned fully off.

transconductance (g_m): for FETs, the ratio of a change in drain current to the corresponding change in gate voltage. Unit of measurement is siemens.

transformers: devices that transfer energy from one circuit to another through electromagnetic (mutual) induction.

triac: a special semiconductor device, sometimes called a 5-layer N-P-N-P-N device, that controls power over an entire cycle, rather than half the cycle, like the SCR device; equivalent of two SCRs in parallel connected in opposite directions but having a common gate.

trivalent materials: materials that have three valence electrons; used in doping semiconductor material to produce P-type materials where holes are the majority current carriers. Examples include boron, aluminum, gallium, and indium.

true (or real) power (P): the power expended in the ac circuit resistive parts that equals $VI \cos \theta$, or apparent power $\times \cos \theta$. The unit of measure is watts.

tunnel diode: a special P-N junction diode that has been doped to exhibit high-speed switching capability and a special negative-resistance over part of its operating range; sometimes used in oscillator and amplifier circuits.

turns ratio: the number of turns on the separate windings of a transformer; expressed as the number of turns on the secondary versus the number of turns on the primary (N_S/N_P); or frequently stated as the ratio of primary-to-secondary turns.

valence electrons: the electrons in the outer shell of an atom.

varactor diode: a silicon, voltage-controlled semiconductor capacitor where capacitance is varied by changing the junction reverse bias; used in various applications where voltage-controlled, variable capacitance can be advantageous.

vector: a quantity that illustrates both magnitude and direction. A vector's direction represents position relative to space.

vertical (Y) amplifier: the scope amplifier circuitry that amplifies signals fed to the vertical input jack(s).

vertical volts/div. control(s): scope controls that adjust the amplification or attenuation of the input signal fed to the vertical deflection system. Generally, this control has positions that provide calibrated vertical deflections in ranges of millivolts per division or volts per division; thus enabling measurement of waveform voltages.

volt (V): the basic unit of potential difference or electromotive force (emf); the amount of emf causing a current flow of one ampere through a resistance of one ohm.

voltage: the amount of emf available to move a current.

voltage divider action: the dividing of a specific source voltage into various levels of voltages by means of series resistors. The distribution of voltages at various points in the circuit is dependent on what proportion each resistor is of the total resistance in the circuit.

voltage follower: an amplifier configuration where the output voltage is slightly less than the input voltage level, and the output has the same phase or polarity as the input.

voltage rate of change (dV/dt): the amount of voltage level change per unit time. For example, a rate of change of one volt per second, 10 V per μ second, and so forth.

voltage ratio: in an ideal transformer with 100% coupling ($k = 1$), the voltage ratio is the same as the turns ratio, since the voltage induced in each turn of the secondary and the primary windings is the same for a given rate of flux change; $V_S/V_P = N_S/N_P$.

voltampere-reactive (VAR): the unit of reactive power in contrast to real power in watts; 1 VAR is equal to one reactive voltampere.

watt: the basic unit of electrical power. The basic formula for electrical power is P (in watts) equals V (in volts) times I (in amperes).

watthour: is a unit of energy consumption and is a measure of the electrical energy used over a given time. In the case of the watthour, or the electrical energy used is computed as power in watts times the number of hours over which that power was consumed.

wattsecond: is a unit of energy consumption and is a measure of the electrical energy used over a given time. In the case of the wattsecond, the electrical energy used is computed as power in watts times the number of seconds over which that power was consumed.

weber: the unit of magnetic flux indicating 10^8 maxwells (flux lines).

Wheatstone bridge circuit: a special application of the bridge circuit commonly used to measure unknown resistances. It utilizes a sensitive current or voltage meter to determine when the bridge is balanced.

Wien-bridge oscillator: a type of RC sine-wave oscillator that uses a lead-lag network to maintain a zero-degree phase shift of the feedback signal at the frequency of oscillation.

work: the expenditure of energy, where work = force × distance. The basic unit of work is the foot poundal. NOTE: At sea level, 32.16 foot-poundals = one foot-pound.

wye network: also sometimes called a "T" network. A wye network consists of three branches whose connections all join at one end of each branch, and are separate electrical points at the other end of each branch. This electrical connection system can be schematically represented in the shape of a Y or a T, hence the names Wye and Tee are used in describing this electrical network.

wye-delta conversion: the mathematical process of converting network component values, used in a wye network, to appropriate values to be used in a delta network, and yet present the same impedance and load conditions to a source feeding such a network.

zener diode: a junction diode that is designed to allow reverse breakdown at a certain voltage level, thereby acting as a voltage regulator device.

zero-adjust control: the adjustable series resistor in an analog-type ohmmeter used to regulate current through the meter movement so that with the test probes across zero ohms (or touching each other), there is full-scale current through the meter, and the pointer is located above the zero-ohms mark on the scale.

INDEX

Entries in **Bold** indicate figure entry.

–G–

–R–